Enhanced Guide to Oracle8i

Joline Morrison
Mike Morrison
University of Wisconsin-Eau Claire

**COURSE
TECHNOLOGY**
———✳———™
THOMSON LEARNING

Australia • Canada • Mexico • Singapore • Spain • United Kingdom • United States

COURSE TECHNOLOGY
THOMSON LEARNING
™

Enhanced Guide to Oracle8i is published by Course Technology.

Senior Vice President, Publisher:
Kristen Duerr

Managing Editor:
Jennifer Locke

Development Editor:
Marilyn R. Freedman

Production Editor:
Kristen Guevara

Associate Product Manager:
Matthew Van Kirk

Marketing Manager:
Toby Shelton

Editorial Assistant:
Janet Aras

Text Designer:
GEX Publishing Services

Cover Designer:
Betsy Young

Cover Art:
Rakefet Kenaan

Disclaimer
Course Technology reserves the right revise this publication and make changes from time to time in its cont without notice.

ISBN 0-619-03549-8

BRIEF
Contents

TABLE OF
Contents

Preface

Enhanced Guide to Oracle8i provides a comprehensive guide for developing database applications using the Oracle8i relational database and the Developer 6/6i application development utilities. This book also addresses all of the concepts covered on the Oracle Certified Application Developer Track exams.

The Intended Audience

This book is intended to support individuals in database courses in which the instructor has chosen to illustrate database development concepts using an Oracle8i relational database. No prior knowledge of programming concepts or relational database concepts is necessary. The book is also intended to support individuals who are preparing for the Oracle Certified Application Developer Track Exams.

Overview and Organization of This Book

The text, examples, tutorials, review questions, and problem-solving cases in *Enhanced Guide to Oracle8i* will help you achieve the following objectives:

- Become familiar with relational database concepts and terms, and understand the architecture of a client-server database
- Understand the Oracle SQL commands used to create and manage database tables and to insert, update, and delete database records
- Understand how to use SQL commands to retrieve data from a relational database
- Learn how to create PL/SQL programs and named program units that are stored in the database and in the workstation file system
- Learn how to use the Developer 6/6i utilities (Form Builder, Report Builder, Graphics Builder, and Project Builder) to create an integrated database application
- Develop project applications for databases that contain large data sets
- Become familiar with the object-oriented features in Form Builder and create reusable form objects

Chapter 1 introduces client/server and relational database concepts. This chapter also introduces three case study databases that are used throughout the book. These sample databases contain a variety of realistic data values and relationships to enhance your learning.

Chapters 2 through 5 provide hands-on instructions for performing a variety of SQL command-line operations using SQL*Plus, Oracle's command-line SQL utility. **Chapter 2** covers creating database tables; covers inserting, updating, and deleting records; and addresses using the Oracle large object (LOB) data types that store binary data values, such as images and sounds. **Chapter 3** provides a comprehensive overview of retrieving data. New sections have been added on outer joins, nested queries, and self-joins. **Chapter 4** provides an in-depth introduction to PL/SQL, Oracle's procedural programming language. **Chapter 5** provides hands-on instructions for creating named PL/SQL programs. It also provides instructions for creating and saving PL/SQL stored program units, libraries, packages, and triggers using both SQL*Plus commands and the Procedure Builder utility. This edition adds in-depth coverage on using Oracle's built-in packages and on writing Dynamic SQL programs.

Chapters 6 through 12 provide instructions for using the Oracle Developer application set, which includes Form Builder, Report Builder, Graphics Builder, and Project Builder. Chapters 6, 7, and 8 describe how to use Form Builder to create form applications. **Chapter 6** addresses the basic steps for creating data block forms, and **Chapter 7** describes how to create custom forms that allow developers to create more flexible user applications. **Chapter 8** explores advanced form concepts, including storing and retrieving image and sound data in form applications, and creating form applications for databases that contain many records. **Chapter 9** provides an overview of how to use the Report Builder utility to create reports. **Chapter 10** addresses using the Graphics Builder utility to create charts based on database data. **Chapter 11** provides extensive coverage on how to create an integrated database application, and discusses technical project management issues in the Oracle Developer environment. This chapter provides coverage of Project Builder, which is the Oracle utility for managing database development projects. **Chapter 12** provides material on how to use Form Builder's object-oriented features to create reusable form objects and flexible code to improve developer productivity.

Oracle Certified Application Developer

The Oracle Certified Professional (OCP) Program was developed by Oracle Corporation to recognize standards of competence in specific areas based on depth of knowledge and hands-on skill. An Oracle certification is acknowledged by the information technology industry as a valuable measure of knowledge and ability.

To become an Oracle Certified Application Developer (Release 2), you must pass the following five separate exams:

- Exam 1: Introduction to Oracle: SQL and PL/SQL (Exam #1Z0-001)
- Exam 2: Develop PL/SQL Program Units (Exam #1Z0-101)
- Exam 3: Developer/2000: Build Forms I (Exam #1Z0-121)
- Exam 4: Developer/2000: Build Forms II (Exam #1Z0-122)
- Exam 5: Developer/2000: Build Reports (Exam #1Z0-123)

This text contains coverage of all of the objectives associated with these five exams, and may be used as a learning aid in preparing for certification.

You can download a document at *www.oracle.com/education/certification/* that provides specific exam details.

The Approach

Enhanced Guide to Oracle8i emphasizes sound database design and development techniques and GUI design skills in the Windows environment. It addresses real-world considerations for creating and managing realistic database development projects.

The Oracle client/server database software provides a rich environment for illustrating multi-user and client/server database concepts, such as managing concurrent users and sharing database resources, and allows users to develop database applications in a production environment using the Developer utilities. *Enhanced Guide to Oracle8i* distinguishes itself from other Oracle books because it is designed specifically for users and instructors in educational environments. It is written in a clear and easy-to-read style. Each chapter is divided into separate lessons that break down the concepts into manageable sections. The exercises at the end of each lesson emphasize the issues that must be considered when developing applications using the Oracle software, and are easy to follow, comprehensive, and have been extensively tested.

The Course Technology Kit for Oracle Developer Software, Second Edition, available when purchased as a bundle with this book, provides the Personal Oracle Database and the Developer utilities, so users can install all of the software needed to complete the tutorials and case problems on their own computers. The installation instructions, available at *www.course.com*, are clear and concise, and have been extensively tested.

Features

- **Chapter Objectives** Each chapter begins with a list of the important concepts that you will master in each chapter. This list provides a quick reference to the contents of the chapter as well as a useful study aid.

- **Step-By-Step Methodology** Each chapter introduces new concepts, explains why the concepts are important, and provides tutorial exercises to illustrate the concepts. When you are learning a new concept, detailed instructions lead you through each step, and the numerous illustrations include labels that direct your attention to how your screen should look. As you proceed through the tutorials, less detailed instructions are provided for familiar tasks, and more detailed instructions are provided for new concepts.

- **Case Approach** Three running cases address database-related problems that you could reasonably expect to encounter in business, followed by a demonstration of an application that could be used to solve the problem. The sample databases referenced in the three ongoing cases represent realistic client/server applications with several database tables, and require supporting multiple users simultaneously at different physical locations. The Clearwater Traders database represents a standard sales order and inventory system. The Northwoods University database illustrates a student registration system. The Software Experts database illustrates a project management system.

- **Data Disks** Data Disks provide files for creating the sample databases used in the chapters, as well as files needed to complete the tutorials and Problem-Solving Cases.

- **HELP?** These paragraphs anticipate problems your may encounter and help you to resolve these problems on your own. This feature facilitates independent learning and frees the instructor to focus on substantive issues rather than on common procedural errors.

- **Tips** These notes provide additional information about a procedure—for example, an alternative method of performing the procedure.

- **CAUTION** These paragraphs anticipate common errors and provide strategies to help you avoid these errors before they occur.

- **Summaries** Following each lesson is a Summary that recaps the programming concepts and commands covered in the lesson.

- **Review Questions and Problem-Solving Cases** Each lesson concludes with meaningful, conceptual Review Questions that test your understanding of the concepts you learned. Problem-Solving Cases provide you with additional practice of the skills and concepts you learned in the lesson. These cases increase in difficulty and are designed to allow you to explore the language and programming environment independently.

The Oracle Server and Client Software

This book was written and tested using the following software:

- **Database Server** Oracle8i Enterprise Edition Server, Version 8.15, installed on a Windows 2000 Server database server; Personal Oracle8i, Version 8.1.7, installed on a Windows 2000 Professional workstation; and Personal Oracle8i, Version 8.1.6, installed on a Windows 98 workstation.

- **Client Utilities** Oracle Developer 6/6i, Version 6.0.8.8.0, installed on Windows 98 and Windows 2000 Professional client workstations.

Teaching Tools

The following teaching tools are available to instructors when this book is used in a classroom setting. All of the teaching tools are available on CD-ROMs.

- **Instructor's Manual** The authors wrote the Instructor's Manual, and it was quality assurance tested. It is available in printed form and through the Course Technology Faculty Online Companion on the World Wide Web at *www.course.com*. (Call your customer service representative for the specific URL and your password.) The Instructor's Manual contains the following items:

  - Complete instructions for installing and configuring the server and client software for a client/server installation, provided you obtain the software using the Oracle Academic Initiative

  - Complete instructions for installing and configuring the server and client software for a Personal Oracle Software installation, using the software that may be included with the book

  - Answers to all of the review questions

  - Solutions files for all of the problem-solving cases

  - Teaching notes to help introduce and clarify the material presented in the chapters.

  - Quick Quizzes, which are quick (five-minute) quizzes that you can give your students to assess their understanding of lesson concepts

- **ExamView®** This textbook is accompanied by ExamView, a powerful testing software package that allows instructors to create and administer printed, computer (LAN-based), and Internet exams. ExamView includes hundreds of questions that correspond to the topics covered in this text, enabling students to generate detailed study guides that include page references for further review. The computer-based and Internet testing components allow students to take exams at their computers, and also save the instructor time by grading each exam automatically.

- **PowerPoint Presentations** This book comes with Microsoft PowerPoint slideshows for each chapter. These are included as a teaching aid for classroom presentation, to make available to users for chapter review, or to be printed for classroom distribution. Instructors can add their own slides or edit the existing slides to customize the presentations for their own use.

ACKNOWLEDGMENTS

We would like to thank all of the people who helped to make this book a reality. Thanks to Marilyn Freedman, who has been a pleasure to work with and works as hard as we do. Thanks also to Kristen Duerr, Senior Vice President and Publisher; Kristen Guevera, Production Editor; and John Bosco and his Quality Assurance testing team. A special thanks to Managing Editor Jennifer Locke, who always tries to makes things easier for us.

We are grateful to the many reviewers who provided helpful and insightful comments during the development of this book, including Tina Shaner, Evergreen Valley College; Carol Rosenow, Creighton University; Paula Worthington, Northern Virginia Community College; and Richard Scudder, University of Denver. Thanks to Lauren Morrison, who designed and developed the art used in the software applications—your persistence and perfectionism is amazing. And a sincere thanks to all of the past and current students at the University of Wisconsin-Eau Claire who continue to teach us how to teach Oracle.

Joline Morrison

Mike Morrison

Read This Before You Begin

TO THE USER

Data Disks

To complete the tutorials and cases in this book, you will need Data Disks that contain source files that have been created for this book. Some of the files are very large and take a long time to load and execute if they are stored on a floppy disk, so we recommend that you store the Data Disk files on a hard drive. If you cannot do this, then you can save your work on floppy disks. The following paragraphs describe how to set up your Data Disks both for a hard disk and a floppy disk installation. If you are installing the Data Disk files on a hard disk, your instructor will tell you how to access the Data Disk source files. If you are using floppy disks, your instructor will provide you with Data Disks or ask you to make your own.

Hard Disk Installation

If you are going to copy the Data Disk source files from a file server, your instructor will tell you the drive letter or folder path where the files are located. Or, you can download the source files directly from the Course Technology Web site by connecting to *www.course.com*, and searching for this book title.

To create your Data Disks, start Windows Explorer, navigate to the folder that contains all of the source files, select all of the subfolders, and copy them to the folder on your hard drive where you want to store your Data Disk files. There is a folder for each of the chapters in the textbook. All of the folders contain files you will need to complete the chapters and exercises in the book, and some of the folders contain subfolders.

Floppy Disk Installation

If you are asked to create your own Data Disks on floppy disks, you will need 12 blank, formatted high-density disks. You will need to copy a set of folders from a file server or stand-alone computer onto your disks. Your instructor will tell you which computer, drive letter, and folders contain the files you need. The following table shows which folders go on each of your disks, so that you will have enough disk space to store the Data Disk files for all the tutorials and cases.

When you begin each chapter, make sure you are using the correct Data Disk.

Data Disk	Write this on the disk label	Create or copy these folders and the files they contain on the disk
1	Enhanced Guide to Oracle8i Chapter2, Chapter3, Chapter4, Chapter5, Chapter6, Chapter10	Chapter2 Chapter3 Chapter4 Chapter5 Chapter6 Chapter10
2	Enhanced Guide to Oracle8i Chapter7 Disk 1	Chapter7
3	Enhanced Guide to Oracle8i Chapter7 Disk 2	Chapter7
4	Enhanced Guide to Oracle8i Chapter8 Disk 1	Chapter8
5	Enhanced Guide to Oracle8i Chapter8 Disk 2	Chapter8
6	Enhanced Guide to Oracle8i Chapter9 Disk 1	Chapter9
7	Enhanced Guide to Oracle8i Chapter9 Disk 2	Chapter9
8	Enhanced Guide to Oracle8i Chapter11 Disk 1	Chapter11\CWPROJECT
9	Enhanced Guide to Oracle8i Chapter11 Disk 2	Chapter11\CWPROJECT Chapter11\NWPROJECT
10	Enhanced Guide to Oracle8i Chapter11 Disk 3	Chapter11\SWEPROJECT
11	Enhanced Guide to Oracle8i Chapter12 Disk 1	Chapter12
12	Enhanced Guide to Oracle8i Chapter12 Disk 2	Chapter12

Solution Disks

As you complete the tutorials and cases, you will be instructed to save the files you create on Solution Disks. Some of the files are very large and take a long time to load and execute if they are stored on a floppy disk, so we recommend that you store the Solution Disk files on a hard drive. If you cannot do this, then you can save your work on floppy disks. The following paragraphs describe how to set up your Solution Disks both for a hard disk and a floppy disk installation.

Hard Disk Installation

To create your Solution Disks, start Windows Explorer, navigate to the folder where you will store your solutions, and make the following folders:

Chapter2\TutorialSolutions
Chapter2\CaseSolutions
Chapter3\TutorialSolutions
Chapter3\CaseSolutions
Chapter4\TutorialSolutions
Chapter4\CaseSolutions
Chapter5\TutorialSolutions
Chapter5\CaseSolutions
Chapter6\TutorialSolutions
Chapter6\CaseSolutions
Chapter7\TutorialSolutions
Chapter7\CaseSolutions
Chapter8\TutorialSolutions
Chapter8\CaseSolutions
Chapter9\TutorialSolutions
Chapter9\CaseSolutions
Chapter10\TutorialSolutions
Chapter10\CaseSolutions
Chapter11\TutorialSolutions\CWPROJECT_DONE
Chapter11\CaseSolutions\NWPROJECT_DONE
Chapter11\CaseSolutions\SWEPROJECT_DONE
Chapter12\TutorialSolutions
Chapter12\CaseSolutions

Floppy Disk Installation

To store your solution files on floppy disks, you will need 16 blank, formatted high-density disks. The following table shows which folders go on each of your disks, so that you will have enough disk space to store the solution files for all the tutorials and cases. (The solution files for Chapter 11 must be stored on a hard disk, because they involve creating a database application that integrates multiple files, and the files occupy too much disk space to fit on a single floppy disk.)

Solution Disk	Write this on the disk label	Create these folders on the disk
1	Enhanced Guide to Oracle8i Solutions Chapter2, Chapter3, Chapter4, Chapter5	Chapter2\TutorialSolutions Chapter2\CaseSolutions Chapter3\TutorialSolutions Chapter3\CaseSolutions Chapter4\TutorialSolutions Chapter4\CaseSolutions Chapter5\TutorialSolutions Chapter5\CaseSolutions
2	Enhanced Guide to Oracle8i Solutions Chapter6 Disk 1	Chapter6\TutorialSolutions Chapter6\CaseSolutions
3	Enhanced Guide to Oracle8i Solutions Chapter6 Disk 2	Chapter6\CaseSolutions
4	Enhanced Guide to Oracle8i Solutions Chapter7 Disk 1	Chapter7\TutorialSolutions Chapter7\CaseSolutions
5	Enhanced Guide to Oracle8i Solutions Chapter7 Disk 2	Chapter7\CaseSolutions
6	Enhanced Guide to Oracle8i Solutions Chapter8 Disk 1	Chapter8\TutorialSolutions Chapter8\CaseSolutions
7	Enhanced Guide to Oracle8i Solutions Chapter8 Disk 2	Chapter8\CaseSolutions
8	Enhanced Guide to Oracle8i Solutions Chapter9 Disk 1	Chapter9\TutorialSolutions
9	Enhanced Guide to Oracle8i Solutions Chapter9 Disk 2	Chapter9\TutorialSolutions Chapter9\CaseSolutions
10	Enhanced Guide to Oracle8i Solutions Chapter9 Disk 3	Chapter9\CaseSolutions
11	Enhanced Guide to Oracle8i Solutions Chapter10 Disk 1	Chapter10\TutorialSolutions
12	Enhanced Guide to Oracle8i Solutions Chapter10 Disk 2	Chapter10\TutorialSolutions Chapter10\CaseSolutions
13	Enhanced Guide to Oracle8i Solutions Chapter10 Disk 3	Chapter10\CaseSolutions
14	Enhanced Guide to Oracle8i Solutions Chapter12 Disk 1	Chapter12\TutorialSolutions
15	Enhanced Guide to Oracle8i Solutions Chapter12 Disk 2	Chapter12\TutorialSolutions Chapter12\CaseSolutions
16	Enhanced Guide to Oracle8i Solutions Chapter12 Disk 3	Chapter12\CaseSolutions

Using Your Own Computer

You can use your own computer to complete the tutorials and cases. To use your own computer, you must have the following:

- **The Windows 98 or Windows 2000 Professional operating system.**

- **Configuration to support TCP/IP networking operations.** (If you can connect to the Internet, then your computer is configured to support TCP/IP networking operations.)

- **Hardware Requirements.** If your computer is running Windows 98, you must have at least 64 MB of main memory, and approximately 600 MB of free hard disk space. If you computer is running Windows 2000 Professional, you must have at least 128 MB of main memory, and approximately 800 MB of free hard disk space.

- **The Course Technology Kit for Oracle Developer Software, Second Edition.** This includes three CDs: the Oracle Developer 6/6i CD, which can be installed on Windows 98 and 2000 computers; the Oracle8i Personal Edition release 8.1.6 CD, which is used to install the Personal Oracle database on Windows 98 computers; and the Oracle8i Personal Edition release 8.1.7 CD, which is used to install the Personal Oracle database on Windows 2000 computers. Detailed installation and configuration instructions are provided at *www.course.com/cdkit* on the Web page for this title.

- **Data Disks.** You can get the Data Disk files from your instructor, or by connecting to the Course Technology Web site at *www.course.com*, and then searching for this book title. You will not be able to complete the tutorial and case problems in this book using your own computer unless you have the Data Disk files. See the Data Disk section for instructions on setting up your Data Disk files.

TO THE INSTRUCTOR

To complete the chapters in this book, your users must have the Data Disk files. These files are included in the Instructor's Resource Kit. They may also be obtained electronically at *www.course.com* by searching for this title. Follow the instructions in the Readme file to copy the Data Disk files to your server or standalone computer. You can view the Readme file using a text editor such as WordPad or Notepad.

Once the files are copied, you can make Data Disks for the users yourself, or tell users where to find the files so they can make their own Data Disks. Make sure the files get copied correctly onto the Data Disks by following the instructions in the Data Disks section, which will ensure that users have all of the required files to complete the tutorials and cases in this book.

Course Technology Data Files

You are granted a license to copy the Data Disk files to any computer or computer network used by individuals who have purchased this book.

Visit Our World Wide Web Site

Additional materials designed especially for you might be available for your course on the World Wide Web. Go to *www.course.com* and search for this title in the online catalog.

1

CLIENT/SERVER DATABASES AND THE ORACLE8I RELATIONAL DATABASE

Objectives

♦ Develop an understanding of the purpose of database systems

♦ Become familiar with the structure of a relational database and review relational database concepts and terms

♦ Explore the differences between personal databases and client/server databases

♦ Learn about the Clearwater Traders sales order database, the Northwoods University student registration database, and the Software Experts project management database

lient/server databases like Oracle are used to create and maintain data in an organized way. The Oracle8i database environment includes a client/server database management system (DBMS) and utilities for developing and managing database applications. In this book, you will learn about some of these utilities and practice using them to build database applications. When you have finished with this chapter, you will have the background you need to begin your exploration of Oracle.

DATABASE SYSTEMS

When organizations first began converting from manual to computerized processing systems, each application had its own set of data files that were used only for that application. For example, in a bank's computer system, the checking account processing system would have its own data files, the auto loan system would have its own files, and the savings account system would have its own files. This file-based approach to data processing is shown in Figure 1-1.

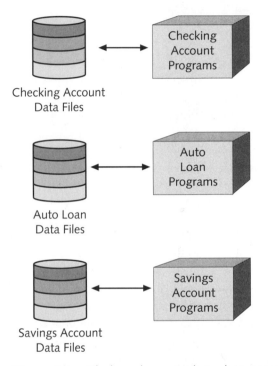

Figure 1-1   File-based approach to data processing

This approach had several problems. One problem was that each set of data files required specialized programs for performing routine data-handling operations, such as adding new data, updating data, and deleting data. Another problem was that the same data was likely to be in multiple files. For example, the information about a customer (such as the customer's name, address, and phone number) is needed in files used for checking, auto, and savings systems (see Figure 1-1). This redundancy takes up extra storage space and causes a problem when information needs to be updated. When a customer changes his or her address, the information might not get updated in every file.

To address this problem, database systems have been developed that view data as an organizational resource. In a database system, all data is stored in a central repository, which eliminates redundant data. All applications, like the ones to support the checking, auto, and savings systems, interface with a single set of data in the database. All of the routine data-handling operations are performed by a single application, called the **database management system (DBMS)**, which provides a central set of common functions for inserting, updating, retrieving, and deleting data. The database approach to system architecture is illustrated in Figure 1–2.

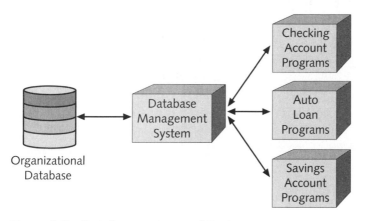

Organizational
Database

Figure 1-2   Database system architecture

OVERVIEW OF RELATIONAL DATABASES

Early databases used a **hierarchical** structure, in which all related data had a parent-to-child relationship: A "parent" data item, such as a customer, could have multiple "child" data items, such as a checking account, an auto loan, and a savings account. You created relationships between related data using **pointers**, which are links to the physical locations where data is written on a disk. Figure 1-3 illustrates a hierarchical database structure, in which the customer data represents the parent records, and the checking account and auto loan data represents the child records. Each customer's checking account and auto loan data is linked using a pointer. A **pointer** is a number that references the physical location of the associated record on a hard disk. For example, in Maria Chavez's customer data record, there is a pointer that indicates the physical location of her checking account record and another pointer that indicates the physical location of her auto loan record. This approach solved the problems of redundant applications and inconsistent data in file-based systems, but it introduced a new problem: The data was physically dependent on its location on the storage media, and it was difficult to migrate a database to a new medium when the disk became full or new hardware was purchased. Another problem with this approach is that all data was accessed within application programs written in languages like COBOL. Any change to the structure of the database required rewriting these programs, which contained commands used to navigate through the database that were based on the structure of the table and the table pointers. These, as well as other problems, led to the development of relational databases, such as Oracle.

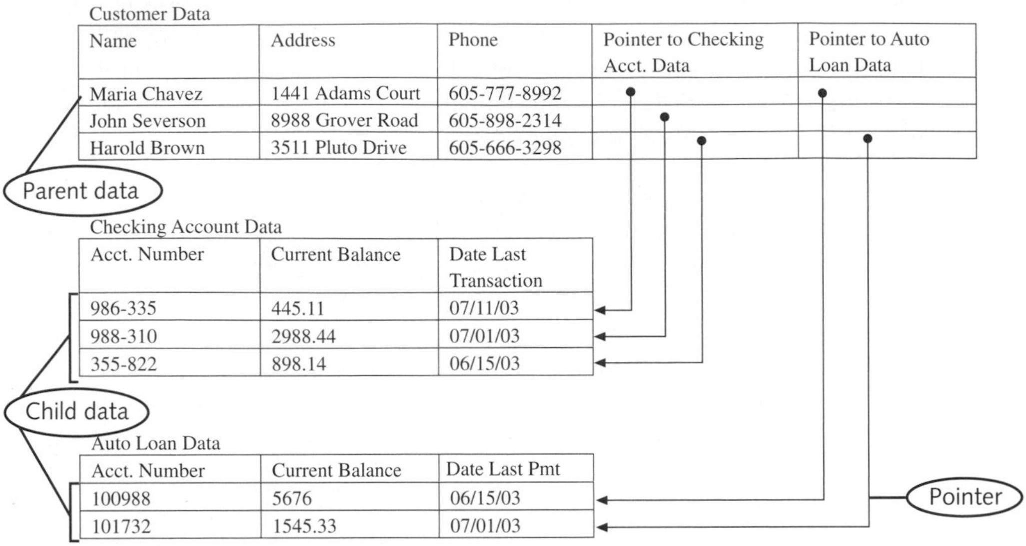

Customer Data

Name	Address	Phone	Pointer to Checking Acct. Data	Pointer to Auto Loan Data
Maria Chavez	1441 Adams Court	605-777-8992	●	●
John Severson	8988 Grover Road	605-898-2314	●	
Harold Brown	3511 Pluto Drive	605-666-3298	●	●

Parent data

Checking Account Data

Acct. Number	Current Balance	Date Last Transaction
986-335	445.11	07/11/03
988-310	2988.44	07/01/03
355-822	898.14	06/15/03

Child data

Auto Loan Data

Acct. Number	Current Balance	Date Last Pmt
100988	5676	06/15/03
101732	1545.33	07/01/03

Pointer

Figure 1-3   Hierarchical database example

A relational database organizes data in **tables**, or matrixes, with columns and rows. **Columns** represent different data categories, and **rows** contain the actual data values. Columns are also called **fields**, and rows are also called **records**. Figure 1-4 shows an example of a relational database that contains two tables.

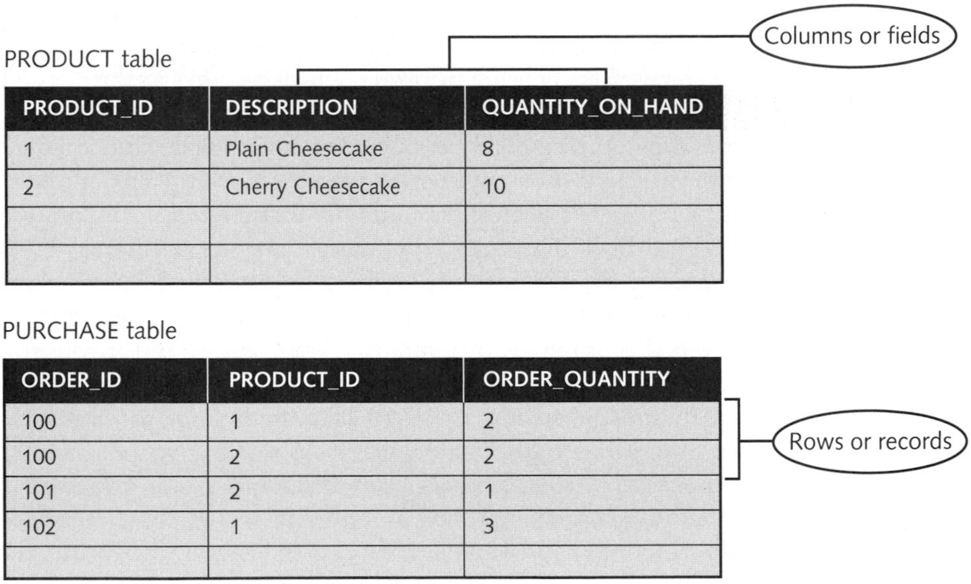

Columns or fields

PRODUCT table

PRODUCT_ID	DESCRIPTION	QUANTITY_ON_HAND
1	Plain Cheesecake	8
2	Cherry Cheesecake	10

PURCHASE table

ORDER_ID	PRODUCT_ID	ORDER_QUANTITY
100	1	2
100	2	2
101	2	1
102	1	3

Rows or records

Figure 1-4   A relational database with two tables

Individual records in relational database tables are identified using primary keys. A **primary key** is a field whose value must be unique for each record. Every record must have a primary key, and the primary key cannot be NULL. **NULL** means a value is indeterminate or undefined. In the PRODUCT table shown in Figure 1-4, the PRODUCT_ID field is a good choice for the primary key because a unique value can be assigned for each product. The DESCRIPTION field might be another choice for the primary key, but there are two drawbacks to using this field. First, two different products might have the same description. Second, the field is a text field and is prone to typographical, spelling, and punctuation data entry errors, as well as variations due to being typed in all uppercase, lowercase, or mixed-case letters. Being prone to data entry errors or upper- and lowercase variations is a problem, because primary key values are used to create links, or relationships, with other database tables, and the values must exactly match in the different tables. The best thing to use for a primary key, then, is a number. Using numbers for primary keys also enables database application developers to use **sequences**, which are sequential lists of numbers automatically generated by the database that guarantee that each primary key value will be unique.

When you are looking at a relational database table, you might not know which key is the primary key, but you can identify fields called **candidate keys** that could be used as the primary key. If the candidate keys are all text fields, you should create a new, numeric field and add it to the table as the primary key. This kind of key is called a **surrogate key**. Once you have selected the field that will be used as the primary key, you must specify this field as the primary key when you create the table. Other users can use special commands to identify the primary keys in existing database tables.

 In this textbook, table and field names appear in all capital letters: for example, PRODUCT, PRODUCT_ID, and QUANTITY_ON_HAND.

Relationships among database tables are created by matching key values. Suppose ORDER_ID 100 is for two plain cheesecakes and two cherry cheesecakes. Figure 1-5 shows how PRODUCT_ID 1 and PRODUCT_ID 2 in the PRODUCT table relate to ORDER_ID 100 in the PURCHASE table. A field in a table that is a primary key in another table and thus creates a relationship between the two tables is called a **foreign key**. PRODUCT_ID is a foreign key in the PURCHASE table. Note that, unlike in a hierarchical database, the PRODUCT_ID foreign key fields in the PRODUCT and PURCHASE tables do not refer to physical storage locations on a hard disk for related records. The DBMS is free to store this information anywhere, so long as the DBMS can find the information when necessary.

The foreign key value must exist in the table where it is a primary key. For example, suppose you have a new record for ORDER_ID 103 that specifies that the customer ordered one unit of PRODUCT_ID 3. There is no record for PRODUCT_ID 3 in the PRODUCT table, so the purchase record does not make sense. Foreign key values must match the value in the primary key table *exactly*. That is why it is not a good idea to use text fields and risk typographical, punctuation, spelling, and case variation errors in the primary key.

PRODUCT table

PRODUCT_ID	DESCRIPTION	QUANTITY_ON_HAND
1	Plain Cheesecake	8
2	Cherry Cheesecake	10

PURCHASE table

ORDER_ID	PRODUCT_ID	ORDER_QUANTITY
100	1	2
100	2	2
101	2	1
102	1	3
		Foreign key

Figure 1-5   Creating a relationship using a foreign key

Sometimes you have to combine multiple fields to create a unique primary key. Figure 1-6 shows an example of this situation. For ORDER_ID 100, the customer purchased two units of PRODUCT_ID 1 and two units of PRODUCT_ID 2, so the ORDER_ID value is not unique for each record. However, the combination of the ORDER_ID and PRODUCT_ID values is unique. The combination of fields to create a unique primary key is called a **composite primary key**, or a **composite key**. Note that PRODUCT_ID, which is a foreign key, can also be part of a composite key.

PURCHASE table

ORDER_ID	PRODUCT_ID	ORDER_QUANTITY
100	1	2
100	2	2
101	2	1
102	1	3
		Composite primary key

Figure 1-6   Example of a composite primary key

Sometimes a database table does not have a field data value that would make a good primary key. Figure 1-7 shows a CUSTOMER table with this problem. The LAST_NAME or FIRST_NAME fields, or even the combination of the two fields, are not good candidate keys because many people have the same name. Multiple people can share the same address and phone number as well. Another reason phone numbers are not good candidate keys is that people often get new phone numbers. If the phone number is updated in the table where it is the primary key, it must also be updated in every table where it is a foreign key. If the phone number is not updated, relationships are lost. To address this problem, a good database development practice is to create a surrogate key as the record's primary key identifier. In Figure 1-4, PRODUCT_ID and ORDER_ID are examples of surrogate keys. For the table in Figure 1-7, you would probably call the surrogate key CUSTOMER_ID, and you might start the CUSTOMER_ID numbers at 1 or 100. The CUSTOMER_ID numbers will not change, and every customer gets a unique number. Surrogate keys are always numerical fields, because their values are usually generated automatically by the database. In Oracle, you can use sequences to automatically generate unique surrogate key values.

CUSTOMER table

LAST_NAME	FIRST_NAME	ADDRESS	PHONE
Brown	John	101 Main Street	7155554321
Brown	John	3567 State Street	7155558901
Carlson	Mike	233 Water Street	7155557890
Carlson	Martha	233 Water Street	7155557890
Davis	Carol	1414 South Street	7155555566

Figure 1-7   Table lacking suitable candidate keys

CLIENT/SERVER DATABASE SYSTEMS

The first databases were housed on large centralized mainframe computers that users accessed from **terminals**, which are devices that do not perform any processing, but only send keyboard input and display video output from a central computer. As distributed computing and microcomputers became popular during the 1980s, two new kinds of databases emerged: personal databases and client/server databases.

Personal database systems, such as Microsoft Access, are aimed toward single-user database applications that are usually stored on a single user's desktop computer, or **client workstation**. When a personal DBMS is used for a multiuser application, the database application files are stored on a file server and transmitted to the individual users across a network, as shown in Figure 1-8. **Server**, broadly defined, is a computer that is willing to accept requests from other computers and to share some or all of its resources. **Resources** could include printers connected to a server, files stored on hard disks attached to a server, programs running in a server's main memory, and more. A **network** is an infrastructure of telecommunications hardware and software that enables computers to transmit messages to each other.

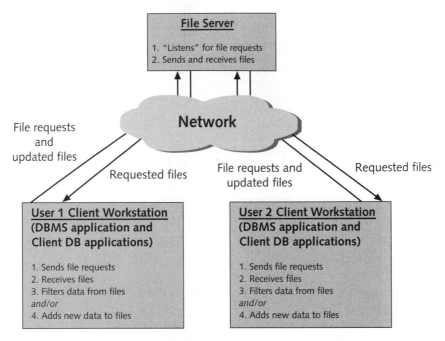

Figure 1-8   Personal database used for multiuser application

With a personal DBMS, each client workstation must load the entire DBMS into main memory, along with any applications used to view, insert, update, or print data. A client request for a small amount of data from a large database might require the server to transmit the entire database (which might be hundreds of megabytes in size) to the client's workstation. Newer personal databases use indexed files that enable the server to send only part of the database, but in either case, these database management systems put a heavy demand on client workstations and on the network. The network must be fast enough to handle the traffic generated when transferring database files to the client workstation and sending them back to the server for database additions and updates.

In contrast, **client/server databases**, like Oracle, split the DBMS and applications accessing the DBMS into a server DBMS process running on the server and one or more client processes running on the client, as shown in Figure 1-9. The **client process** sends data requests across the network. When the server receives a request, the server **DBMS process** retrieves the data from the database, performs the requested functions on the data (sorting, filtering, and so on), and sends *only* the final query result (not the entire database) back via the network to the client. As a result, multiuser client/server databases generate less network traffic than non–client/server databases and are less likely to bog down due to an overloaded network.

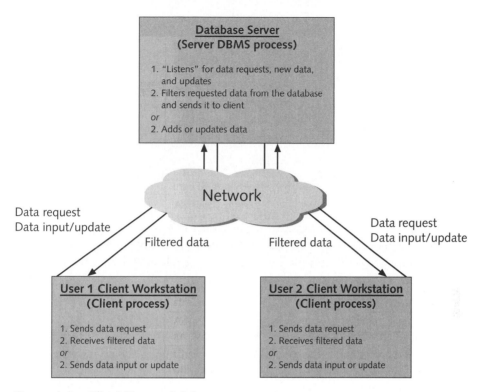

Figure 1-9   Client/Server database

Another important difference between client/server and personal databases is how they handle client failures. In a personal database system, when a client workstation fails due to a software malfunction or power failure, the database is likely to become damaged due to interrupted updates, insertions, or deletions. Records in use at the time of the failure are **locked** by the failed client, which means they are unavailable to other users. The database might be reparable, but all users must log off during the repair process, which could take several hours. Updates, deletions, and insertions taking place at the time of the failure often cannot be reconstructed. If repair is not possible, the person responsible for installing, administering, and maintaining the database, called the **database administrator** or **DBA**, can restore the database to its state at the time of the last regular backup, but transactions that occurred since the backup are lost.

Most client/server DBMS servers have extra features to minimize the chance of failure, and when they do fail, they have powerful recovery mechanisms that often operate automatically. In addition, a client/server database is not affected when a client workstation fails. The failed client's in-progress transactions are lost, but the failure of a single client does not affect other users. In case of a server failure, a central synchronized **transaction log** contains a record of all current database changes. The transaction log enables in-progress transactions from all clients to be either fully completed or rolled back. **Rolling back** a database transaction means that the transaction is undone, and the database is made to look like the transaction never took place. Using the transaction log, a DBA can notify users with rolled back transactions to resubmit them.

Client/server systems also differ from personal database systems in how they handle transaction processing. **Transaction processing** refers to grouping related database changes into units of work that must either all succeed or all fail. For example, assume a customer writes a check to deposit money from a checking account into a money market account. The bank must ensure that the checking account is debited for the amount and the money market account is credited for the same amount. If any part of the transaction fails, then neither account balance should change. Microsoft Access provides procedures to group related changes, keep a record of these changes in main memory on the client workstation, and roll back the changes if the grouped transactions fail. However, if the client making the changes fails in the middle of a group of transactions, then the transaction log in the *client's* main memory is lost. There is no centrally located, file-based transaction log on which to base a rollback, and the partial changes cannot be reversed. Depending on the order of the transactions, a failed client could result in a depleted checking account and unchanged money market account, or an enlarged money market account and unchanged checking account.

The e-commerce boom is fueling the demand for data-intensive Web sites that merge Web and database technologies. Customers need to be able to get information about a vendor's products and services, submit inquiries, select items for purchase, and submit payment information. Vendors need to be able to track customer inquiries and preferences and process customer orders. This demand for data-intensive Web sites is driving the integration of Web site and database technologies. Client/server databases are appropriate for Web-based systems because they can handle multiple simultaneous users, process transactions, and have fault-tolerance and backup capabilities.

Figure 1-10 illustrates a Web-based client/server database architecture. A **data-based Web page**, which is a Web page that contains data that is retrieved from a database, is requested when a user clicks a hyperlink or a "Submit" button on a Web page form. The Web server translates the request into a query that is forwarded to the client/server database for processing. The database retrieves the data and returns it to the Web server, which then sends the output back to the client's browser as a Web page.

In addition to their improved fault-tolerance and error-handling capabilities, client/server databases are desirable for Web-based installations because they have a built-in security system to prevent unauthorized access that personal database systems do not. In addition, client/server database vendors like Oracle provide features and tools for creating and administering Web-based applications and for converting existing applications for deployment over the Internet and display in a Web browser.

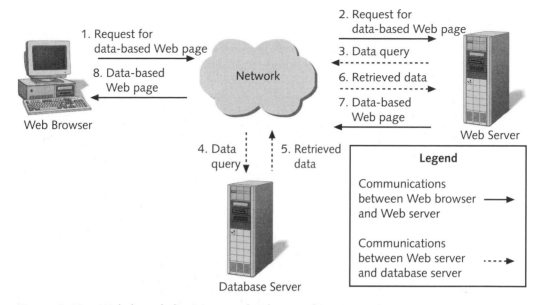

Figure 1-10   Web-based client/server database architecture

In summary, client/server databases are preferred for database applications that retrieve and manipulate small amounts of data from databases containing large numbers of records, because they minimize network traffic and improve response times. Client/server systems are essential for mission-critical applications because of their failure handling, recovery mechanisms, and transaction management control. They also have a rich set of database management and administration tools for handling large numbers of users. They also are ideal for Web-based database applications that require increased security and fault tolerance. Some general guidelines are to use a client/server database if your database will have more than 10 simultaneous users, if your database is mission critical, or if you need a rich set of administration tools.

THE ORACLE8I ENVIRONMENT

Oracle8i runs in a client/server environment, so the environment must be considered in terms of programs that run on the database server, called **server-side** programs, and programs that run on the client workstation, which are called **client-side** programs. On the server side, the Oracle8i database reliably and efficiently manages large volumes of data in a multiuser environment. Large numbers of users must be able to concurrently access, insert, update, and delete records in a timely manner. It prevents unauthorized access and provides fast recovery in case of system failure. Oracle8i provides DBMSs for a variety of configurations and operating systems. Another Oracle8i server–side product is the Oracle Application Server, which is used for creating a World Wide Web site that allows users to access Oracle databases and create dynamic Web pages that serve as a database interface.

The Oracle server products are installed and configured by a DBA. In this book, you will not be performing database administration tasks. Rather, you will be learning to use the Oracle database as an application developer. Therefore, you will be using client-side products that connect to server-side products that have already been installed and configured.

On the client side, Oracle8i provides development tools for designing and creating custom applications, as well as specific applications for supporting tasks such as accounting, finance, production, and sales. In this book, you will use client-side applications to learn how to interact with an Oracle database, create and manage database tables, and create database applications. Specifically, you will use the following Oracle client products:

- *SQL*Plus,* for creating and testing command-line SQL queries and executing PL/SQL procedural programs. PL/SQL is Oracle's sequential programming language that is used to create programs that process database data.

- *Developer 6/6i,* for developing database applications

Developer consists of the following tools:

- *Procedure Builder,* for creating procedural programs, functions, packages, libraries, and triggers that perform actions on database data and objects

- *Form Builder,* for creating graphical forms and menus for user applications

- *Report Builder,* for creating reports for displaying and printing data

- *Graphics Builder,* for creating graphics charts based on database data

THE CASE STUDY DATABASES

In this book, you will encounter tutorial exercises and end-of-chapter cases that illustrate the Oracle utilities using databases developed for Clearwater Traders, Northwoods University, and Software Experts, which are three fictional organizations. The focus of this textbook is on database development rather than on database design, so the text does not go into depth on the rationale behind the design of these database tables.

The described database systems require a client/server database application, because they will have many simultaneous users accessing the system from different locations. Each database has data examples to illustrate the tasks that are addressed in this textbook, including the development of data entry and maintenance forms, output reports, and chart displays.

The Clearwater Traders Sales Order Database

Clearwater Traders markets a line of clothing and sporting goods via mail-order catalogs and a site on the World Wide Web. Clearwater Traders accepts customer orders via phone, mail, and fax. The company recently experienced substantial growth and, as a result, has decided to offer 24-hour customer order service. The existing microcomputer-based database system cannot handle the current transaction volume generated by sales representatives processing incoming orders. Management is also concerned that it does not have the failure-handling and recovery capabilities needed for an ordering system that cannot tolerate failures or downtime.

When a customer orders an item, the sales representative must check to see if the ordered item is in stock. If the item is in stock, the representative must update the available quantity on hand to reflect that the item has been sold. If the item is not in stock, the representative needs to advise the customer when the item will be available. When new inventory shipments are received, a receiving clerk must update the inventory to show the new quantities on hand. The system must produce invoices that can be included with customer shipments and must print reports showing inventory levels, possibly using charts or graphs. Marketing managers would like to be able to track each order's source (particular catalog number or Web site) to help plan future promotions.

The following data items have been identified:

1. Customer name, address, and daytime and evening phone numbers

2. Order date, payment method (check or credit card), order source (catalog description or Web site), and associated item numbers, sizes, colors, and quantities ordered

3. Item descriptions and photo images, as well as item categories (women's clothing, outdoor gear, etc.), prices, and quantities on hand. Many clothing items are available in multiple sizes and colors. Sometimes the same item has different prices, depending on the item size.

4. Information about incoming product shipments

Figure 1-11 shows sample data for Clearwater Traders. Five customer records are shown. Customer 107 is Paula Harris, who lives at 1156 Water Street, Apt. #3, Osseo, WI, and her ZIP code is 54705. Her daytime phone number is 715-555-8943, and her evening phone number is 715-555-9035. CUST_ID has been designated as the table's primary key.

CUSTOMER

CUST_ID	LAST	FIRST	MI	ADD	CITY	STATE	ZIP	D_PHONE	E_PHONE
107	Harris	Paula	E	1156 Water Street, Apt. #3	Osseo	WI	54705	7155558943	7155559035
232	Edwards	Mitch	M	4204 Garner Street	Washburn	WI	54891	7155558243	7155556975
133	Garcia	Maria	H	2211 Pine Drive	Radisson	WI	54867	7155558332	7155558332
154	Miller	Lee		699 Pluto St. NW	Silver Lake	WI	53821	7155554978	7155559002
179	Chang	Alissa	R	987 Durham Rd.	Sister Bay	WI	54234	7155557651	7155550087

CUST_ORDER

ORDER_ID	ORDER_DATE	METH_PMT	CUST_ID	ORDER_SOURCE_ID
1057	5/29/2003	CC	107	2
1058	5/29/2003	CC	232	6
1059	5/31/2003	CHECK	133	2
1060	5/31/2003	CC	154	3
1061	6/01/2003	CC	179	6
1062	6/01/2003	CC	179	3

ORDER_SOURCE

ORDER_SOURCE_ID	SOURCE_DESC
1	Winter 2002
2	Spring 2003
3	Summer 2003
4	Outdoor 2003
5	Children's 2003
6	Web Site

ITEM

ITEM_ID	ITEM_DESC	CATEGORY_ID	ITEM_IMAGE
894	Women's Hiking Shorts	1	shorts.jpg
897	Women's Fleece Pullover	1	fleece.jpg
995	Children's Beachcomber Sandals	2	sandals.jpg
559	Men's Expedition Parka	3	parka.jpg
786	3-Season Tent	4	tents.jpg

CATEGORY

CATEGORY_ID	CATEGORY_DESC
1	Women's Clothing
2	Children's Clothing
3	Men's Clothing
4	Outdoor Gear

INVENTORY

INV_ID	ITEM_ID	ITEM_SIZE	COLOR	PRICE	QOH
11668	786		Sky Blue	259.99	16
11669	786		Light Grey	259.99	12
11775	894	S	Khaki	29.95	150
11776	894	M	Khaki	29	147
11777	894	L	Khaki	29.95	0
11778	894	S	Navy	29.95	139
11779	894	M	Navy	29.95	137
11780	894	L	Navy	29.95	115
11795	897	S	Eggplant	59.95	135
11796	897	M	Eggplant	59.95	168
11797	897	L	Eggplant	59.95	187
11798	897	S	Royal	59	0
11799	897	M	Royal	59.95	124
11800	897	L	Royal	59.95	112
11820	995	10	Turquoise	15.99	121
11821	995	11	Turquoise	15.99	111
11822	995	12	Turquoise	15.99	113
11823	995	1	Turquoise	15.99	121
11824	995	10	Bright Pink	15.99	148
11825	995	11	Bright Pink	15.99	137
11826	995	12	Bright Pink	15.99	134
11827	995	1	Bright Pink	15.99	123
11845	559	S	Spruce	199.95	114
11846	559	M	Spruce	199.95	17
11847	559	L	Spruce	209.95	0
11848	559	XL	Spruce	209.95	12

Figure 1-11 Clearwater Traders database

SHIPMENT

SHIPMENT_ID	DATE_EXPECTED
211	09/15/2003
212	11/15/2003
213	06/25/2003
214	06/25/2003
215	08/15/2003

SHIPMENT_LINE

SHIPMENT_ID	INV_ID	SHIP_QUANTITY	DATE_RECEIVED
211	11668	25	09/10/2003
211	11669	25	09/10/2003
212	11669	25	
213	11777	200	
213	11778	200	
213	11779	200	
214	11798	100	08/15/2003
214	11799	100	08/25/2003
215	11845	50	08/15/2003
215	11846	100	08/15/2003
215	11847	100	08/15/2003

ORDER_LINE

ORDER_ID	INV_ID	ORDER_QUANTITY
1057	11668	1
1057	11800	2
1058	11824	1
1059	11846	1
1059	11848	1
1060	11798	2
1061	11779	1
1061	11780	1
1062	11799	1
1062	11669	3

COLOR

COLOR
Sky Blue
Light Grey
Khaki
Navy
Royal
Eggplant
Blue
Red
Spruce
Turquoise
Bright Pink

Figure 1-11   Clearwater Traders database (continued)

The CUST_ORDER table shows six customer orders. The first is 1057, dated 5/29/2003, method of payment CC (credit card), and ordered by customer 107, Paula Harris. ORDER_ID is the table's primary key, and CUST_ID is a foreign key that creates a relationship to the CUSTOMER table. ORDER_SOURCE_ID is a foreign key that creates a relationship to the ORDER_SOURCE table, which has ORDER_SOURCE_ID as the primary key, and SOURCE_DESC, which describes the order source as a specific catalog or the company Web site. In the first record in the CUST_ORDER table, the ORDER_SOURCE_ID foreign key indicates that the Spring 2003 catalog was the source for the order.

The ITEM table contains five items: Women's Hiking Shorts, Women's Fleece Pullover, Children's Beachcomber Sandals, Men's Expedition Parka, and 3-Season Tent. ITEM_ID is the primary key for this table. The CATEGORY_ID field is a foreign key that creates a relationship with the CATEGORY table. Item 894, Women's Hiking Shorts, is in the Women's Clothing category. The ITEM_IMAGE field contains the name of the JPEG image file that displays a photograph of each item. This image will ultimately be stored in the Oracle database.

JPEG (Joint Photographic Experts Group) is a standardized graphic file format. JPEG files typically have .jpg or .jpeg extensions.

The INVENTORY table contains inventory numbers for specific item sizes and colors. It also shows the current price and quantity on hand (QOH) for each item. Items that are not available in different sizes contain no values in the ITEM_SIZE column, which indicates their values are NULL, or undefined. Notice that some items have different prices for different sizes. INV_ID is the primary key of this table, and ITEM_ID is a foreign key that creates a relationship with the ITEM table.

In some database systems, the word NULL appears when a data value is undefined. In Oracle the data value NULL appears as blank.

The SHIPMENT table contains a schedule of expected shipments and the date each shipment is expected. The primary key of the SHIPMENT table is SHIPMENT_ID. A shipment can include multiple inventory items, so the SHIPMENT_LINE table records the corresponding shipment ID, the inventory ID, the quantity of each item, and the date each item was received. The first record shows that the first line of SHIPMENT_ID 211 is for 25 Sky Blue 3-Season Tents. The second line of SHIPMENT_ID 211 is for 25 Light Grey 3-Season Tents. Both items were received on 09/10/2003. Notice that the primary key of this table cannot be SHIPMENT_ID, because some shipments have multiple lines. The primary key cannot be INV_ID, since the same inventory item might be in several shipments. The primary key must be a composite key comprised of SHIPMENT_ID and INV_ID, because each shipment might consist of multiple inventory items, but each inventory item is listed only once per shipment. Therefore, each SHIPMENT_ID/INV_ID combination is unique for each record. SHIPMENT_ID and INV_ID are also foreign keys in this table, because they reference records in the SHIPMENT and INVENTORY tables.

The ORDER_LINE table represents the individual inventory items in a customer order. The first line of order 1057 specifies one Sky Blue 3-Season Tent, and the second line of this order specifies two large Women's Fleece Pullovers in Royal. This information is used to create the printed customer order invoice and to calculate sales revenues. Note that the primary key of this table is not ORDER_ID, because more than one record might have the same ORDER_ID. The primary key is a composite key made up of the combination of ORDER_ID and INV_ID. An order might have several different inventory items, but it will never have the same inventory item listed more than once. Along with being part of the primary key, ORDER_ID and INV_ID are also foreign keys, because they create relationships to the CUST_ORDER and INVENTORY tables.

The COLOR table is a lookup table, which is also sometimes called a pick list. A **lookup table** is a list of legal values for a field in another table. Notice the variety of colors shown in the INVENTORY table (Sky Blue, Light Grey, Khaki, Navy, Royal, etc.). If users are allowed to type these colors each time an inventory item is added to the table, typing errors might occur. For example, a query looking for sales of items with the Light Grey color will not find instances if Light Grey is spelled as Light Gray, or is specified with a different combination of upper- and lowercase letters, such as Light grey or LIGHT GREY. Typically, a user entering a new inventory item will select a color from a pick list that displays values from the COLOR table, so that the color is not typed directly, thus reducing errors. Small lists that are unlikely to change over time might be coded directly into an application, but lists with many items that might be added to over time are usually stored in a separate lookup table.

The Northwoods University Student Registration Database

Northwoods University has decided to replace its aging mainframe-based student registration system with a more modern client/server database system. School officials want to provide students with the capability to retrieve course availability information, register for courses, and print transcripts using personal computers located in the student computer labs. In addition, faculty members must be able to retrieve student course lists, drop and add students, and record course grades. Faculty members must also be able to view records for the students they advise. Security is a prime concern, so student and course records must be protected by password access.

Students will log onto the Northwoods system using their student ID and PIN (personal identification number). They will be given the option of viewing current course listings or viewing information on courses they have completed. They can check what courses are available during the current term by viewing course information, such as course names; call IDs (such as MIS 101); section numbers, days, times, locations; and the availability of open seats in the course. They can also view information about the courses they have taken in the past and print a transcript report showing past course grades and grade point averages. Faculty members will log onto the system by entering their faculty IDs and PINs. Then they can select from a list of the courses they are teaching in the current term and retrieve a list of students enrolled in the selected course. A faculty member can also retrieve a list of his or her student advisees, select one, and then retrieve that student's past and current course enrollment information.

 In an actual database installation, PIN or password values will probably be stored as encrypted values so that no one can view them.

The data items for the Northwoods database include:

1. Student name, address, phone number, class (freshman, sophomore, etc.), date of birth, PIN number, and advisor ID

2. Course call number, name, credits, maximum enrollment, instructor, and term offered

3. Instructor name, office location, phone number, and rank

4. Student enrollment and grade information

Figure 1-12 shows sample data for the Northwoods database. Six student records are shown. Student 100 is Sarah Miller, who lives at 144 Windridge Blvd., Eau Claire, WI, and her ZIP code is 54703. Her phone number is 715-555-9876, she is a senior, her date of birth is 07/14/82, and her faculty advisor is Kim Cox. Note that the S_PIN (student PIN) field stores her student personal identification number to control data access. S_ID (student ID) is the table's primary key. F_ID (faculty ID) is a foreign key that refers to the F_ID field in the FACULTY table.

The FACULTY table describes five faculty members. The first record shows faculty member Kim Cox, whose office is located at BUS 424, and her phone number is 715- 555-1234. She has the rank of associate professor, and her PIN is 1181. This F_PIN (faculty PIN) will be used as a password to determine if a faculty member can update specific student or course records. F_ID is the primary key, and LOC_ID is a foreign key that references LOC_ID in the LOCATION table. The F_IMAGE field specifies the name of a JPEG image file that displays a photograph of each faculty member that will be stored in the Oracle database.

The LOCATION table identifies building codes, room numbers, and room capacities. LOC_ID is the primary key. After a user makes a selection from the LOCATION table, the LOC_ID is inserted into the COURSE_SECTION and FACULTY tables. The user never needs to see the LOC_ID field in the LOCATION table.

The TERM table provides a textual description of each term along with an ID number used to link the semester to different course offerings and a STATUS field that shows whether enrollment is open or closed. The first record shows a TERM_ID of 1 for the Fall 2002 term, with enrollment status as CLOSED. TERM_ID is the primary key.

The COURSE table shows five courses. The first, COURSE_ID 1, has the CALL_ID MIS 101 and is named "Intro. to Info. Systems". It provides three credits. COURSE_ID is the primary key, and there are no foreign keys.

The COURSE_SECTION table shows the course offerings for specific terms and includes fields for course ID, section number, ID of the instructor teaching the section, and course day, time, location, and maximum allowable enrollment. C_SEC_ID is the primary key, and COURSE_ID, TERM_ID, F_ID, and LOC_ID are all foreign key fields. The F_ID field is an example of a foreign key field that can have a NULL value. The first record shows that C_SEC_ID 1000 is section 1 of MIS 101. It is offered in the Fall 2003 term and is taught by John Blanchard. The section meets on Mondays, Wednesdays, and Fridays at 10:00 A.M. in room CR 101. It has a maximum enrollment of 140 students.

STUDENT

S_ID	S_LAST	S_FIRST	S_MI	S_ADD	S_CITY	S_STATE	S_ZIP	S_PHONE	S_CLASS	S_DOB	S_PIN	F_ID
100	Miller	Sarah	M	144 Windridge Blvd.	Eau Claire	WI	54703	7155559876	SR	07/14/82	8891	1
101	Umato	Brian	D	454 St. John's Place	Eau Claire	WI	54702	7155552345	SR	08/19/82	1230	1
102	Black	Daniel		8921 Circle Drive	Bloomer	WI	54715	7155553907	JR	10/10/79	1613	1
103	Mobley	Amanda	J	1716 Summit St.	Eau Claire	WI	54703	7155556902	SO	09/24/81	1841	2
104	Sanchez	Ruben	R	1780 Samantha Court	Eau Claire	WI	54701	7155558899	SO	11/20/81	4420	4
105	Connoly	Michael	S	1818 Silver Street	Elk Mound	WI	54712	7155554944	FR	12/4/83	9188	3

FACULTY

F_ID	F_LAST	F_FIRST	F_MI	LOC_ID	F_PHONE	F_RANK	F_PIN	F_IMAGE
1	Cox	Kim	J	53	7155551234	ASSO	I181	cox.jpg
2	Blanchard	John	R	54	7155559087	FULL	1075	blanchard.jpg
3	Williams	Jerry	F	56	7155555412	ASST	8531	williams.jpg
4	Sheng	Laura	M	55	7155556409	INST	1690	sheng.jpg
5	Brown	Phillip	E	57	7155556082	ASSO	9899	brown.jpg

LOCATION

LOC_ID	BLDG_CODE	ROOM	CAPACITY
45	CR	101	150
46	CR	202	40
47	CR	103	35
48	CR	105	35
49	BUS	105	42
50	BUS	404	35
51	BUS	421	35
52	BUS	211	55
53	BUS	424	1
54	BUS	402	1
55	BUS	433	1
56	LIB	217	2
57	LIB	222	1

TERM

TERM_ID	TERM_DESC	STATUS
1	Fall 2002	CLOSED
2	Spring 2003	CLOSED
3	Summer 2003	CLOSED
4	Fall 2003	CLOSED
5	Spring 2004	CLOSED
6	Summer 2004	OPEN

Figure 1-12   Northwoods University database

COURSE

COURSE_ID	CALL_ID	COURSE_NAME	CREDITS
1	MIS 101	Intro. to Info. Systems	3
2	MIS 301	Systems Analysis	3
3	MIS 441	Database Management	3
4	CS 155	Programming in C++	3
5	MIS 451	Web-Based Systems	3

COURSE_SECTION

C_SEC_ID	COURSE_ID	TERM_ID	SEC_NUM	F_ID	DAY	TIME	LOC_ID	MAX_ENRL
1000	1	4	1	2	MWF	10:00 AM	45	140
1001	1	4	2	3	TTH	9:30 AM	51	35
1002	1	4	3	3	MWF	8:00 AM	46	35
1003	2	4	1	4	TTh	11:00 AM	50	35
1004	2	5	2	4	TTh	2:00 PM	50	35
1005	3	5	1	1	MWF	9:00 AM	49	30
1006	3	5	2	1	MWF	10:00 AM	49	30
1007	4	5	1	5	TTh	8:00 AM	47	35
1008	5	5	1	2	MWF	2:00 PM	49	35
1009	5	5	2	2	MWF	3:00 PM	49	35
1010	1	6	1	1	M-F	8:00 AM	45	50
1011	2	6	1	2	M-F	8:00 AM	50	35
1012	3	6	1	3	M-F	9:00 AM	49	35

ENROLLMENT

S_ID	C_SEC_ID	GRADE
100	1000	A
100	1003	A
100	1005	B
100	1008	B
101	1000	C
101	1004	B
101	1005	A
101	1008	B
102	1000	C
102	1011	
102	1012	
103	1010	
103	1011	
104	1000	B
104	1004	C
104	1008	C
104	1010	
104	1012	
105	1010	
105	1011	

Figure 1-12   Northwoods University database (continued)

The ENROLLMENT table shows the students currently enrolled in each course section and their associated grade, if one has been assigned. The primary key for this table is a composite key comprised of S_ID and C_SEC_ID.

The Software Experts Project Management Database

Software Experts is a consulting firm that specializes in creating custom software applications for clients. When a client approaches Software Experts with a project, the managers at Software Experts determine the staffing needs in terms of the number of consultants needed and the required skill sets of the consultants. They then locate available consultants with the necessary skills and assign them to the project. One of the consultants is assigned to be the project manager. When the project is completed, the project manager evaluates all of the other consultants who worked on the project, and all of the project consultants evaluate the project manager. A project can be subdivided into multiple subprojects.

The data items for the Software Experts database include:

1. Consultant information, including name, address, city, state, ZIP code, phone number, and e-mail address

2. Descriptions of employee skills

3. Client information, including the client name, client contact name, and contact phone number

4. Project information, including the project name, the client associated with the project, associated subprojects, the project manager, and required skill sets

5. Dates that a consultant started and finished working on a specific project and total hours that the consultant spent working on the project

6. Consultant evaluation information, including the date the evaluation was completed, who performed the evaluation, and the evaluation score and comments

Figure 1-13 shows sample data for the Software Experts database.

The CONSULTANT table lists information about each consultant, including first and last name; middle initial; address, city, state, and ZIP code; phone number; and e-mail address. The primary key of this table is C_ID. The SKILL table lists project skills, such as Visual Basic Programming and COBOL Programming. The SKILL_ID field is a surrogate key that is the table's primary key, and SKILL_DESCRIPTION describes the associated skill.

The CONSULTANT_SKILL table shows the skills associated with each consultant, as well as whether the consultant has been certified in a particular skill. For example, the first record of the CONSULTANT_SKILL table shows that consultant Mark Myers is a certified Visual Basic programmer. Since a particular consultant can have multiple skills, and a specific skill can be associated with multiple consultants, this table has a composite key comprised of the combination of C_ID and SKILL_ID. C_ID is also a foreign key that references the CONSULTANT table, and SKILL_ID is a foreign key that references the SKILL table.

The CLIENT table shows information about current clients. CLIENT_ID is a surrogate key that is the primary key of this table. The other fields provide the first and last name of the client contact, as well as the contact phone number.

CONSULTANT

C_ID	C_LAST	C_FIRST	C_MI	C_ADD	C_CITY	C_STATE	C_ZIP	C_PHONE	C_EMAIL
100	Myers	Mark	F	1383 Alexander Ave.	Eau Claire	WI	54703	7155559652	mmyers@swexpert.com
101	Hernandez	Sheila	R	3227 Brian Street	Eau Claire	WI	54702	7155550282	shernandez@earthware.com
102	Zhang	Brian		2227 Calumet Place	Altoona	WI	54720	7155558383	zhang@swexpert.com
103	Carlson	Sarah	J	1334 Water Street	Eau Claire	WI	54703	7155558008	carlsons@swexpert.com
104	Courtlandt	Paul	R	1911 Pine Drive	Eau Claire	WI	54701	7155555225	courtlpr@yamail.com
105	Park	Janet	S	2333 157th St.	Chippewa Falls	WI	54712	7155554944	jpark@swexpert.com

SKILL

SKILL_ID	SKILL_DESCRIPTION
1	Visual Basic Programming
2	COBOL Programming
3	Java Programming
4	Project Management
5	Web Applications Programming
6	Oracle Developer Programming
7	Oracle Database Administration
8	Windows NT Network Administration
9	Windows 2000 Network Administration

CONSULTANT_SKILL

C_ID	SKILL_ID	CERTIFICATION
100	1	Y
100	3	N
100	6	Y
101	4	N
101	5	N
102	7	Y
103	1	Y
103	6	Y
103	8	Y
103	9	Y
104	8	N
104	9	Y
105	2	N
105	3	N
105	4	Y

CLIENT

CLIENT_ID	CLIENT_NAME	CONTACT_LAST	CONTACT_FIRST	CONTACT_PHONE
1	Crisco Systems	Martin	Andrew	5215557220
2	Supreme Data Corporation	Martinez	Michelle	5205559821
3	Lucid Technologies	Brown	Jack	7155552311
4	Morningstar Bank	Wright	Linda	9215553320
5	Maverick Petroleum	Miller	Tom	4085559822
6	Birchwood Mall	Brenner	Nicole	7155550828

Figure 1-13   Software Experts database

PROJECT

P_ID	PROJECT_NAME	CLIENT_ID	MGR_ID	PARENT_P_ID
1	Hardware Support Intranet	2	105	
2	Hardware Support Interface	2	103	1
3	Hardware Support Database	2	102	1
4	Teller Support System	4	105	
5	Internet Advertising	6	105	
6	Network Design	6	104	5
7	Exploration Database	5	102	

PROJECT_SKILL

P_ID	SKILL_ID
1	8
1	9
2	3
3	6
3	7
4	2
4	7
5	5
5	9
6	9
7	6
7	7

PROJECT_CONSULTANT

P_ID	C_ID	ROLL_ON_DATE	ROLL_OFF_DATE	TOTAL_HOURS
1	101	06/15/2002	12/15/2002	175
1	104	01/05/2002	12/15/2002	245
1	103	01/05/2002	06/05/2002	50
1	105	01/05/2002	12/15/2002	45
2	105	07/17/2002	09/17/2002	25
2	100	07/17/2002	09/17/2002	0
3	103	09/15/2002	03/15/2003	125
3	104	10/15/2002	12/15/2002	50
4	105	06/05/2002	06/05/2003	25
4	104	06/15/2002	12/15/2002	125
4	102	07/15/2002	12/15/2002	30
5	105	09/19/2002	03/19/2003	15
5	103	09/19/2002	03/19/2003	15
6	103	09/19/2002	03/19/2003	5
6	104	09/19/2002	03/19/2003	10
7	102	05/20/2002	12/20/2002	125
7	100	05/25/2002	12/20/2002	100

EVALUATION

E_ID	E_DATE	P_ID	EVALUATOR_ID	EVALUATEE_ID	SCORE	COMMENTS
100	01/07/2003	1	105	101	90	
101	01/07/2003	1	105	104	85	
102	01/08/2003	1	105	103	90	
103	12/20/2002	1	103	105	100	
104	12/29/2002	1	104	105	75	
105	01/15/2003	1	101	105	90	

Figure 1-13   Software Experts database (continued)

The PROJECT table has the P_ID field as the primary surrogate key and lists project names in the PROJECT_NAME field. The project client is represented as a foreign key link to the CLIENT table, and the consultant who is acting as the project manager is represented as a foreign key link to the CONSULTANT table in the MGR_ID field. The field name for the project manager in the PROJECT table is MGR_ID, and the field name in the CONSULTANT_TABLE is C_ID. However, the values in the MGR_ID field in the PROJECT table are associated with values in the C_ID field in the CONSULTANT table, and a foreign key relationship is specified when the tables are created. The field in the PROJECT table could be named C_ID, but doing so might cause confusion, because a project can have multiple consultants associated with it, but only one consultant who is acting as project manager. When a field that represents the same values as another field in the same database is given a different name, the new name is called a database **synonym**. You should avoid using database synonyms unless they add clarity to the database structure, as they do in the Software Experts database.

The PARENT_P_ID field represents the parent project ID if the current project is a subproject of another project. For example, the second record in the PROJECT table shows that the Hardware Support Interface project for Supreme Data Corporation is managed by consultant Sarah Carlson and is a subproject of the Hardware Support Intranet project. This is another example of a synonym, which occurs when a foreign key field does not have the same name as the field in which it is a primary key. The field name is P_ID when it is a primary key and PARENT_P_ID when it is a foreign key.

The PROJECT_SKILL table shows the skills that are required for each project. A project might require multiple skills, and the same skill might be required for many different projects. Therefore, the link between projects and skills is created using this table, which has a composite primary key comprised of the project ID (P_ID) and skill ID (SKILL_ID). For example, this table shows that project 3 (Hardware Support Database) requires skill ID 6 (Oracle Developer Programming), and skill ID 7 (Oracle Database Administration).

The PROJECT_CONSULTANT table describes the consultants who are working on a particular project. The project ID (P_ID) and consultant ID (C_ID) fields form a composite primary key for this table, since each project can have several consultants, and a specific consultant can work on multiple projects. However, the project/consultant combination is always unique. This table also shows the date the consultant started work on the project (ROLL_ON_DATE), stopped work on the project (ROLL_OFF_DATE), and the total number of hours that the consultant has worked on the project. For example, the first record shows that consultant 101 (Sheila Hernandez) started work on project 1 on 06/15/2002, ended work on 12/15/2002, and worked a total of 175 hours on the project.

The final table is the EVALUATION table, which contains consultant evaluation information for projects. The evaluation ID (E_ID) field is the surrogate primary key of this table, and the E_DATE field shows the date that the evaluation was completed. P_ID is a foreign key field that references the PROJECT table and shows the project with which the evaluation is associated. EVALUATOR_ID and EVALUATEE_ID are foreign key fields that use synonyms and reference the C_ID column in the CONSULTANT table. The EVALUATOR_ID field specifies the C_ID who is performing the evaluation, and the EVALUATEE_ID field specifies the C_ID who is being evaluated. The SCORE field specifies the evaluation score, and the COMMENT field provides text comments associated with the evaluation. The first record of this table shows that evaluation ID 100 was completed on 01/07/2003, and the evaluator was consultant 105 (Janet Park). This evaluation is associated with project 1 (Hardware Support Intranet), and the evaluatee is consultant 101 (Sheila Hernandez), who receives a score of 90. Currently, the comment field is NULL.

CHAPTER SUMMARY

- Database systems view data as an organizational resource in which data is stored in a central repository, and applications interface with the database using the database management system (DBMS).

- A relational database views data in tables, or matrixes, with columns and rows. Columns represent different data categories, and rows contain the actual data values.

- A primary key is a field that uniquely identifies a specific record in a database table. Primary key values must be unique within a table and cannot be NULL.

- A composite key is a primary key composed of the combination of two or more fields.

- Foreign keys link related records. A foreign key field in a table must exist as a primary key in another table.

- A surrogate key can be created if no single existing data field uniquely identifies each record in a table or if existing primary key candidates are unsuitable because they are text fields or their values might change. A surrogate key is a numeric field that provides no real information about the record and is usually generated automatically by the database.

- Personal database management systems (DBMSs) are best suited for single-user database applications that are usually stored on a single user's desktop computer, while a client/server DBMS runs on a network server, and the user applications run on the client workstations.

- Nonclient/server databases download some or all of the data a user needs to the user's client workstation, manipulate the data, and then upload it to the server. This process can be slow and cause network congestion.

- Client/server databases send data requests to the server and the results of data requests back to the client workstation. This process minimizes network traffic and congestion.

- Client/server databases have better failure recovery mechanisms than non-client/server databases.

- Client/server databases maintain a file-based transaction log that is not lost in the event of a client workstation failure.

- Client/server databases automatically handle competing user transactions.

- Client/server databases have utilities for managing systems with many users.

- The Oracle database development environment has programs that run on the database server and on the client workstation. Server-side programs include the database and the Oracle Application Server.

- Client-side programs include utilities for creating database tables and queries; developing data forms, reports, and graphics based on database table data; and performing database administration tasks.

❏ The Clearwater Traders sales order database includes the following data tables:

- CUSTOMER
- CUST_ORDER
- ORDER_SOURCE
- ITEM
- CATEGORY
- INVENTORY
- SHIPMENT
- SHIPMENT_LINE
- ORDER_LINE
- COLOR

❏ The Northwoods University student registration database includes the following database tables:

- STUDENT
- FACULTY
- LOCATION
- TERM
- COURSE
- COURSE_SECTION
- ENROLLMENT

❏ The Software Experts project management database includes the following database tables:

- CONSULTANT
- SKILL
- CONSULTANT_SKILL
- CLIENT
- PROJECT
- PROJECT_SKILL
- PROJECT_CONSULTANT
- EVALUATION

REVIEW QUESTIONS

1. What is the main difference between the way that client/server and non–client/server databases handle and process data?

2. Why can client/server databases handle client failures better than non–client/server systems?

3. Give two examples of database applications that might be appropriate for a client/server system. Give two examples of database applications that might be more appropriate for a non–client/server system. (*Hint*: Non-client/server systems are appropriate for situations in which a client/server system is not required.)

4. In a relational database table, what is a field, and what is a record?

5. What is a primary key?

6. What is a foreign key?

7. What is a surrogate key?

8. What is a composite key?

9. What does it mean when a data value is NULL?

10. What is the difference between client-side and server-side programs? Which ones will you be using in this book?

PROBLEM-SOLVING CASES

1. Answer the following questions using the Clearwater Traders database tables shown in Figure 1-11. No computer work is necessary; you will learn how to do these queries using SQL commands with the Oracle database in another chapter.

    a. Identify every primary and foreign key field in every table in the Clearwater Traders database.

    b. Find the name of every customer who placed orders using the Web site as an order source.

    c. Find the name of every item that currently is out of stock (QOH = 0).

    d. Find the customer name, item description, size, color, quantity, extended total (quantity × price), and total order amount for every item in order 1057.

    e. Find the name and address of every customer who ordered items in the Women's Clothing category.

    f. Find the total amount of every order generated from the Web site source.

    g. Find the total amount of every order generated from the Spring 2003 catalog.

2. Answer the following questions using the Northwoods University database tables shown in Figure 1-12. No computer work is necessary; you will learn how to do these queries using the Oracle database in another chapter.

   a. Identify every primary and foreign key field in every table in the Northwoods University database.

   b. Find the name of every course offered during the Summer 2004 term.

   c. List the name of every student that faculty member Kim Cox advises.

   d. List the number of students enrolled in all courses offered during the Summer 2004 term.

   e. List the course call ID, section number, term description, and grade for every course that student Sarah Miller has taken.

   f. Calculate the total credits earned to date by student Brian Umato.

   g. Calculate the total number of students taught by John Blanchard during the Spring 2004 term.

   h. Calculate the total number of student credits generated during the Spring 2004 term (student credits equal number of students enrolled in a course times course credits).

3. Answer the following questions using the Software Experts database tables shown in Figure 1-13. No computer work is necessary; you will learn how to do these queries using the Oracle database in another chapter.

   a. Identify every primary and foreign key field in every table in the Software Experts database.

   b. List the descriptions of all skills required by the Internet Advertising project.

   c. List the names of all consultants who are certified Visual Basic programmers.

   d. Find the total number of hours that consultants have spent on projects for Supreme Data Corporation.

   e. List the names of all projects that consultant Janet Park has either managed or worked on.

   f. List the names of all of the subprojects associated with the Hardware Support Intranet project.

   g. Find the average value of all of the evaluation scores given by all consultants to the project manager on the Hardware Support Intranet project.

2

CREATING AND MODIFYING DATABASE OBJECTS

Objectives

♦ Develop an understanding of Oracle user accounts and system privileges

♦ Identify data types used in Oracle database tables

♦ Learn how integrity and value constraints are defined in Oracle database tables

♦ Create database tables using SQL*Plus

♦ View information about database tables using Oracle Data Dictionary views

♦ Modify and delete database tables using SQL*Plus

U sers interact with relational databases using high-level **query languages** that have commands containing standard English words, such as CREATE, ALTER, INSERT, UPDATE, and DELETE. The standard query language for relational databases is **structured query language (SQL)**. The basic SQL language consists of about 30 commands that enable users to create database structures and manipulate and view data. The American National Standards Institute (ANSI) has published standards for the SQL language, with the most recent version being SQL-92. Almost all relational database applications support the SQL-92 standard, but most vendors have added extensions to make their implementations more powerful or easier to use. These extensions are fairly similar across different platforms, and once you have become proficient with the SQL language on one DBMS platform, it is fairly easy to move to other platforms.

SQL commands fall into two basic categories:

■ **Data definition language (DDL)** commands, which are used to create new database objects (such as user accounts and tables) and modify or delete existing objects. DDL commands immediately change the database objects and do not need to be explicitly saved.

■ **Data manipulation language (DML)** commands, which are used to insert, update, delete, and view database data. DML commands that change the data values in the database need to be explicitly saved.

In this chapter, you will be working with DDL commands to create new database tables. DDL commands are usually created and executed only by database administrators, and their operations are sometimes accomplished using utilities that automatically generate the underlying SQL commands. However, it is important to have a basic understanding of the structure, syntax, and use of SQL DDL commands. In this chapter, you will use DDL commands to create, modify, and drop Oracle database objects using a text editor, and you will execute the commands in SQL*Plus, the Oracle command-line SQL utility.

USER ACCOUNTS AND SYSTEM PRIVILEGES

Before you begin creating database tables, it is useful to understand how a client/server database is structured. When you create a new table using a personal database, you start the database application and create a new database, which is saved in a file in the file system of your workstation. You are usually the only person who uses this database file. When a database administrator creates a new Oracle database, the files that make up the database are stored on the database server and will be shared by many different users. To keep each user's tables and data separate and secure, each user has a user account that is identified by a unique username and password. Each user account owns tables and other data objects within its area of the database, which is called its **user schema**. The data objects within a user schema are called **schema objects**.

An Oracle database has two levels of security. The first level controls access to the Oracle database itself. To connect to an Oracle database, you must enter a username and a password. The password can be authenticated by the database, where it is stored in an encrypted format, or by the network operating system. The second security level controls what privileges a user has once he or she connects to the database. These privileges fall into two categories: system privileges and object privileges. **System privileges** are granted to an individual user and control what operations the user may perform within the database. Examples of system privileges include connecting to the database, creating new tables, creating new users, and performing a system start-up or shutdown. **Object privileges** are granted to a user on an individual database object, such as a table, and control how the user can access and manipulate the object. Usually, Oracle users want to share their schema objects with other users, so they grant object privileges on their schema objects to other users. Examples of object privileges on a database table include being able to insert new data into the table, view the table data, or delete records from the table. In this section, you will learn about system privileges.

A convention used in this book is that in SQL commands, all command words, also called reserved words, appear in all uppercase letters. All user-supplied variable names appear in lowercase letters, although Oracle displays them in uppercase letters on the screen display. Use these conventions when entering your own commands. Although SQL*Plus commands are not case sensitive, these conventions make queries, as well as the examples in this book, easier to interpret. When a table or field or other database object is referenced outside of a command, its name will appear in all uppercase letters. When the general syntax of a command is given, user-supplied words will be written in italics. In Oracle, each individual SQL command is terminated with a semicolon(;).

Creating User Accounts and Granting System Privileges

You cannot create or administer Oracle user accounts unless you are a database administrator. The commands in this section are provided for informational purposes only.

You can create a new Oracle user account using the following general SQL command:

```
CREATE USER username IDENTIFIED BY password;
```

Oracle usernames and passwords must follow the **Oracle naming standard** that Oracle has established for naming all database objects. This standard states that object names must be from 1 to 30 characters long; can contain characters, numbers, and the special symbols $, _, and #; and must begin with a character. Names cannot contain blank spaces. The Oracle naming standard will be referenced throughout this book.

For Oracle 8i databases, usernames and passwords are not case sensitive. You use the following command to create a user named *lhoward*, identified by the password *asdf*:

```
CREATE USER lhoward IDENTIFIED BY asdf;
```

When you create a new user account, the user account does not have any system privileges. You must explicitly grant system privileges to new user accounts for them to be able to perform such actions as connecting to the database or creating database tables. Some common system privileges are summarized in Table 2-1.

Privilege	Level	Purpose
CREATE SESSION	User	Connecting to the Oracle database
CREATE TABLE	User	Creating new tables in the current user's schema
DROP TABLE	User	Deleting tables in the current user's schema
UNLIMITED TABLESPACE	User	Allowing the user to create schema objects using as much space as is available in the database
CREATE USER	DBA	Creating new database users
GRANT ANY PRIVILEGE	DBA	Granting system privileges to users
CREATE ANY TABLE	DBA	Creating new tables in any user's schema
DROP ANY TABLE	DBA	Deleting tables in any user's schema

Table 2-1   Oracle system privileges

In Table 2-1, the Level column indicates the type of user who is usually granted the associated privilege. Some privileges are granted to all users, such as being able to connect to the database and create database tables. Other privileges are usually reserved for database administrators, such as creating new users and granting privileges. Also, note the difference between the privileges that are granted with and without the ANY option. When a DBA has a privilege that is granted with the ANY option, such as DROP ANY TABLE, the DBA can perform the action, such as deleting a table, on any table in any user schema. When the privilege is granted without the ANY option, the action can only be performed on tables in the current user's schema. The ANY option gives a user a lot of privileges for database objects belonging to other users, so the ANY option is usually reserved for database administrators.

The general syntax for the command to grant system privileges to a user is:

```
GRANT privilege1, privilege2, … TO username;
```

You would use the following command to grant user LHOWARD the privileges to connect to the database and create database tables:

```
GRANT CREATE SESSION, CREATE TABLE
TO lhoward;
```

Using Roles to Manage System Privileges

Often, groups of similar database users require the same system privileges. For example, database administrators need to be able to create new users, grant privileges to users, and create and modify other database objects. Rather than having to explicitly grant each individual privilege to each DBA, Oracle provides **roles**, which are database objects that can be created and then granted system privileges. A role can then be assigned to a specific user, and the user assumes all of the privileges that have been granted to that role.

2

The following commands show how to create a role named ORACLE_STUDENT, and assign the privilege for creating tables to this role:

```
CREATE ROLE oracle_student;
GRANT CREATE TABLE
TO oracle_student;
```

The following command assigns the ORACLE_STUDENT role to user LHOWARD:

```
GRANT oracle_student TO lhoward;
```

Combining related privileges into a role simplifies the process of assigning the same privileges to many different users. All Oracle databases automatically contain a role named DBA, which has been granted all of the privileges needed by database administrators. When an account for a new database administrator is created, it can be granted the DBA role, and the user automatically receives the privileges he or she needs to perform administrative tasks.

Revoking System Privileges

You revoke system privileges using the REVOKE command, which has the following general syntax:

```
REVOKE privilege1, privilege2, … FROM username;
```

For example, the following command revokes the CREATE TABLE privilege from user LHOWARD:

```
REVOKE CREATE TABLE FROM lhoward;
```

Similarly, you can use the REVOKE command to revoke a role that has been previously assigned to a user. The following command is used to revoke the ORACLE_STUDENT role from user LHOWARD:

```
REVOKE oracle_student FROM lhoward;
```

Administering System Privileges

In order to grant privileges to a user, or revoke privileges from a user, the user who is granting or revoking the privileges must have the privilege and must have been granted that privilege **WITH ADMIN OPTION**. This means that the user has the authority to grant the privilege to or revoke the privilege from other users. The following code shows how a user with admin option on the CREATE TABLE privilege would grant this privilege WITH ADMIN OPTION to user LHOWARD:

```
GRANT CREATE TABLE
TO lhoward
WITH ADMIN OPTION;
```

DATABASE TABLE DEFINITIONS AND DATA TYPES

Tables are the primary data objects in a relational database. When you create an Oracle database table, you specify the table name, the name of each data field, and the data type and size of each data field. You also define **constraints**, which are restrictions on the data values that can be stored in a field. Examples of constraints include whether the field is a primary key, whether it is a foreign key, whether NULL values are allowed, and whether data is restricted to certain values, such as an M or F for a gender field.

Table and field names must follow the Oracle naming standard described earlier. Within an individual user schema, every table name must be unique. Table 2-2 shows some invalid SQL table and field names and their descriptions.

Invalid SQL Table and Field Names	Description
STUDENT TABLE	Spaces not permitted
STUDENT-TABLE	Hyphens not permitted
#CUST	Must begin with a letter
US_SOCIAL_SECURITY_NUMBERS_COLUMN	Cannot exceed 30 characters

Table 2-2   Invalid SQL table and field names

The SQL command used to create a new table is CREATE TABLE. The general syntax of the CREATE TABLE command follows. The constraint declarations are optional.

```
CREATE TABLE tablename
(fieldname1 data_type_specification,
 fieldname2 data_type_specification, ...);
```

The CREATE command is an example of a SQL DDL command. (Recall that DDL commands are used to create and modify database objects. The other commonly used DDL commands in SQL are DROP and ALTER, which you will learn later.) With a CREATE command, the database table is created as soon as the query executes and does not need to be explicitly saved. In the CREATE TABLE command, you must specify the table name and the column declarations, which include the column name and the data type specification.

Data Types

When you create a table, you must assign to each column a **data type,** which specifies what kind of data will be stored in the field. Data types are used for two reasons. First, assigning a data type provides a means for error checking. For example, you cannot store the character data "Chicago" in a field assigned a DATE data type. Second, data types cause storage space to be used more efficiently by optimizing the way specific types of data are stored. The main Oracle data types include characters, numbers, dates, and large objects.

Character Data Types

Alphanumeric fields containing text and numbers not used in calculations, such as telephone numbers and postal codes, are stored in character data fields. The main Oracle 8i character data types, VARCHAR2, CHAR, and NCHAR, each store data in a different way.

VARCHAR2 Character Data Type   The **VARCHAR2** data type stores variable-length character data up to a maximum of 4,000 characters in Oracle 8i, expanded from 2,000 characters in previous Oracle versions. When you declare a field using VARCHAR2, you must specify the field size. If inserted data values are smaller than the specified size, only the inserted values are stored, and trailing blank spaces are not added to the end of the entry to make it fill the specified column length. If an entered data value is wider than the specified size, an error occurs. Because no trailing blank spaces are added, this data type is recommended for most character data fields. For example, the S_LAST field in the Northwoods University database STUDENT table is declared as:

```
s_last VARCHAR2(30)
```

This statement means that the field in which a student's last name is entered is a variable-length character field with a maximum of 30 characters. Examples of data stored in this field include the last names Miller and Umato.

CHAR Character Data Type   The **CHAR** data type holds fixed-length character data up to a maximum size of 2,000 characters. (In previous Oracle versions, the maximum size of a CHAR data field was 255 characters.) You should use the CHAR data type only for fixed-length character fields that have a restricted set of values. An example of an appropriate CHAR field definition is the S_CLASS field in the STUDENT table in the Northwoods University database. The declaration of the S_CLASS field is:

```
s_class CHAR(2)
```

The data that will be stored in the S_CLASS field are SR, JR, SO, or FR (senior, junior, sophomore, or freshman). If no field size is specified, the default size for a CHAR field is one character.

Using the CHAR data type on fields that might not fill the column width forces the DBMS to add trailing blank spaces, which causes inconsistent query results in other Oracle applications. For example, if you declare the S_LAST field in the STUDENT table using the CHAR data type as **s_last CHAR(30)**, and insert a data value of Umato, then the actual value that will be stored is Umato, plus 25 blank spaces to the right of the last character.

The CHAR data type uses data storage space more efficiently than VARCHAR2 and can be processed faster. However, if there is any chance that all column spaces will not be filled, use the VARCHAR2 data type.

NCHAR Character Data Type Typically, character data values are stored using American Standard Code for Information Interchange (ASCII) coding, which represents each character as an 8-digit (one-byte) binary value and can represent a total of 256 different characters. To represent data in other alphabets, such as the Japanese Kanji character set, which has thousands of different characters, each character must be encoded using a 16-digit binary value. The **NCHAR** data type supports 16-digit (two-byte) binary character codes, and it is similar to the CHAR data type in all other ways.

NUMBER Data Types

The **NUMBER** data type stores negative, positive, fixed, and floating-point numbers between 10^{-130} and 10^{126} with precision up to 38 decimal places. **Precision** is the total number of digits both to the left and to the right of the decimal point.

The precision length includes only the digits and does not include formatting characters, such as the dollar sign ($) or the decimal point.

You use the NUMBER data type for all numerical data, and you must use it for fields that will be used in mathematical operations. When you declare a NUMBER field, you can specify the precision and the scale. **Scale** is the number of digits to the right of the decimal point. There are three specific NUMBER data types: integer, fixed-point, and floating-point.

Integers An **integer** is a whole number with no digits to the right of the decimal point. When you declare an integer, you specify the precision, which is the total number of digits. You omit the scale specification, since there are no digits to the right of the decimal point. For example, the S_ID field in the STUDENT table stores only integers (student identification numbers), so it is declared as:

```
s_id NUMBER(5)
```

Examples of values in this field are 100, 101, 102, and so on. NUMBER(5) specifies that the S_ID field has a maximum length of five digits and has no digits to the right of the decimal point. When you are defining any field, it is important to allow room for growth. Northwoods University begins its S_ID numbers at 100, but there will be more than 999 students enrolled, so the data type is defined to store five digits, or up to 99,999 students.

Fixed-Point Numbers A **fixed-point number** contains a specific number of decimal places. An example is the PRICE field in the INVENTORY table in the Clearwater Traders database. The PRICE field is declared as follows:

```
price NUMBER(5,2)
```

Examples of data values in the PRICE field are 259.99 and 59.99. NUMBER(5,2) specifies that all values have exactly two digits to the right of the decimal point and that there are a total of five digits. Therefore, no merchandise prices can exceed $999.99. Note that the decimal point and dollar sign are *not* included in the precision value.

Floating-Point Numbers A **floating-point number** is a number with a variable number of decimal places. The decimal point can appear anywhere, from before the first digit to after the last digit, or can be omitted. Floating-point values are defined in Oracle by *not* specifying either the precision or scale in the field declaration. You declare a floating point variable using the general declaration `variable_name NUMBER`. Although no floating-point values exist in any of the case study databases, a potential floating-point field in the Northwoods database could be student grade point average, which might be declared as follows:

    s_gpa NUMBER

A student's GPA might include one or more decimal places, such as 2.7045, 3.25, or 4.0.

DATE Data Type

The **DATE** data type stores dates from December 31, 4712 B.C. to December 31, 4712 A.D. The DATE data type stores the century, year, month, day, hour, minute, and second. The default date format is DD-MON-YY, which indicates the day of the month, a hyphen, the month (abbreviated using three capital letters), another hyphen, and the last two digits of the year. The default time format is HH:MI:SS A.M., which indicates the hours, minutes, and seconds using a 12-hour clock. If no time is specified in a date data value, the default time value is specified as 12:00:00 A.M. If no date is specified when a time is entered in a DATE data value, the default date is the first day of the current month.

The S_DOB field (student date of birth) in the STUDENT table is declared as:

    s_dob DATE

This field stores a value similar to 07-OCT-82 12:00:00 A.M. DATE fields are stored in a standard internal format in the database, so no length specification is required.

DATE is a reserved word, so it cannot be used as a field name.

Large Object (LOB) Data Types

Sometimes you might need to store binary data, such as digitized sounds or images, or references to binary files from a word processor or spreadsheet. In these cases, you can use one of the large object (LOB) data types available in Oracle 8i. The four LOB data types are summarized in Table 2-3.

Large Object (LOB) Data Type	Description
BLOB	Binary LOB, storing up to 4 GB of binary data in the database
CLOB	Character LOB, storing up to 4 GB of character data in the database
BFILE	Binary File, storing a reference to a binary file located outside the database in a file maintained by the operating system
NCLOB	Character LOB that supports 2-byte character codes, stored in the database

Table 2-3   Large object (LOB) data types

When you declare a LOB data field, you do not need to specify the object size. For example, suppose you want to store the image of each item in the ITEM_IMAGE field of the Clearwater Traders ITEM database table. You could store the actual binary image data using a BLOB data type, which would be declared as:

```
item_image BLOB
```

Alternately, you could store a reference to the location of an external image file using the BFILE data type, using the following declaration:

```
item_image BFILE
```

You will store and retrieve binary objects using the BLOB data type in the case study databases.

Constraints

Constraints are rules that restrict the data values that can be entered in a column. There are two types of constraints: **integrity constraints**, which are used to define primary and foreign keys; and **value constraints**, which define specific data values or data ranges that must be inserted into columns and whether values must be unique or not NULL. There are two levels of constraints: table constraints and column constraints. A **table constraint** restricts the data value with respect to all other values in the table. Examples of table constraints are primary key constraints, which must have unique values for each record, and unique constraints, which specify that a field value must be unique and cannot be repeated in any other table record. A **column constraint** limits the value that can be placed in a specific column, irrespective of values that exist in other table records. Examples of column constraints are value constraints, which specify that a certain value or set of values must be used, and not NULL constraints, which specify that a value cannot be NULL.

Constraint definitions can be placed at the end of the CREATE TABLE command after all of the columns are declared. They can also be placed immediately following the data column declaration for the column associated with the constraint.

Integrity Constraints

An integrity constraint allows you to define primary key fields and specify foreign keys and their corresponding table and column references. The general syntax for defining a primary key constraint within a column declaration is:

CONSTRAINT *constraint_name* PRIMARY KEY

The syntax for defining a primary key constraint outside of a column declaration is:

CONSTRAINT *constraint_name* PRIMARY KEY (fieldname)

Note that in the first syntax the column name does not need to be specified, because the constraint is defined within the column declaration. In the second syntax, the column name must be specified, because the constraint is defined independently of the column declaration.

The ***constraint_name*** is the internal name that Oracle uses to identify the constraint. Every constraint name must be unique for each user schema. In other words, a specific user cannot have two constraints with the same name in the database tables that she or he creates. Constraint names must adhere to the Oracle naming standard. A convention for assigning unique constraint names is to use the following general format:

tablename_fieldname_ID

The ID is a two-character abbreviation that specifies the type of constraint. Commonly used constraint ID abbreviations are shown in Table 2-4.

Constraint Type	Constraint ID Abbreviation
Primary Key	pk
Foreign Key	fk
Check Condition	cc
Not Null	nn
Unique	uk

Table 2-4   Common constraint ID abbreviations

Sometimes constraint names become longer than 30 characters using the naming convention. When this happens, you must truncate either the table name or the field name. Be sure to do this in a way that allows you to still identify the table and field name that the constraint is associated with.

An example of a primary key field is LOC_ID in the Northwoods LOCATION table. You would use the following code to create the LOCATION table, and specify LOC_ID as the primary key. This example uses the syntax in which the constraint definition is independent of the column declarations:

```
CREATE TABLE location
(loc_id NUMBER(6),
bldg_code VARCHAR2(10),
room VARCHAR2(10),
capacity NUMBER(3),
CONSTRAINT location_loc_id_pk PRIMARY KEY (loc_id));
```

The following code would be used to define LOC_ID as the primary key of the table, using the syntax in which the constraint definition is created within the column definition. Note that in this approach, the field name is not repeated after the constraint definition.

```
CREATE TABLE location
(loc_id NUMBER(6)
CONSTRAINT location_loc_id_pk PRIMARY KEY,
bldg_code VARCHAR2(10),
room VARCHAR2(10),
capacity NUMBER(3));
```

The constraint name is optional in a constraint definition. If you omit the constraint name, Oracle will assign a system-generated name to the constraint. It is wise, to name the constraint yourself, using the constraint naming convention, so you can easily interpret the meaning of each constraint. Recall that each constraint name must be unique for your entire user schema. For example, you cannot have two constraints named `location_loc_id_pk`. By using the naming convention, you ensure that each constraint name is unique.

Like primary key constraints, foreign key constraints can be defined independently of a column, or they can be defined within the declaration of the column that has the constraint. The general syntax for specifying a foreign key constraint independently of a column is:

```
CONSTRAINT constraint_variable_name
FOREIGN KEY (constraint_fieldname)
REFERENCES table_where_field_is_a_primary_key
(name_of_field_in_table_where_it_is_the_primary_key)
```

The general syntax for specifying a foreign key constraint within a column declaration is:

```
CONSTRAINT constraint_variable_name
REFERENCES table_where_field_is_a_primary_key
(name_of_field_in_table_where_it_is_the_primary_key)
```

Note that in the second syntax, the keyword FOREIGN KEY is omitted, along with the fieldname on which the constraint is placed. This is because the foreign key constraint is being declared within the column declaration, and the keyword REFERENCES implicitly defines the constraint as a foreign key constraint.

An example of a foreign key is the LOC_ID field in the FACULTY table. The command to create the LOC_ID foreign key constraint in the FACULTY table independently of the column declaration is:

```
CONSTRAINT faculty_loc_id_fk FOREIGN KEY (loc_id)
REFERENCES location (loc_id)
```

The command to create the LOC_ID foreign key constraint in the FACULTY table within the LOC_ID column declaration is:

```
loc_id NUMBER(6) CONSTRAINT faculty_loc_id_fk
REFERENCES location (loc_id),
```

 Before you can create a foreign key reference, the table in which the field is a primary key must already exist, and the field must be defined as a primary key. In this example, before creating the FACULTY table, you must create the LOCATION table and define LOC_ID as the primary key.

Recall that a composite primary key is a primary key comprised of two or more data fields. A composite primary key constraint is defined by listing all of the fields that comprise the composite primary key, with each field name separated by a comma. The constraint name should consist of the table name, the names of the fields that comprise the composite key, and the identifier "pk." A composite key can only be defined after the columns that comprise the key are declared.

In the ENROLLMENT table, the primary key consists of both the S_ID and the C_SEC_ID fields. S_ID and C_SEC_ID are also foreign keys in this table, and their associated values must also exist in the STUDENT and COURSE_SECTION tables. The command to create the ENROLLMENT table would include the following field and table constraint definitions:

```
CREATE TABLE enrollment
(s_id NUMBER(5),
c_sec_id NUMBER(8),
CONSTRAINT enrollment_s_id_c_sec_id_pk
PRIMARY KEY (s_id, c_sec_id),
CONSTRAINT enrollment_s_id_fk FOREIGN KEY (s_id)
REFERENCES student (s_id),
CONSTRAINT enrollment_c_sec_id_fk FOREIGN KEY (c_sec_id)
REFERENCES course_section (c_sec_id));
```

Value Constraints

Value constraints are column-level constraints that restrict what data values can be entered into a given field, enable you to specify whether a field can or cannot be NULL, allow you to specify a default value for a field, and allow you to specify that a field value must be unique for all records. To restrict the data values that can be inserted into a given field, you use a value constraint called a **check condition**. For example, you can create a check condition to specify that numeric data must fall within a specific range (like specifying that entries must be greater than zero but less than 1,000) or that character data must be from a set of specific values (like specifying that entries must be SR, JR, SO, or FR). Check conditions should be used only when the number of allowable values is limited and not likely to change. You must specify check conditions in table definitions prudently, because once the table is populated with data, it is difficult or impossible to modify the constraint—all records must satisfy the constraint. In general, you should avoid using value constraints in database table definitions. A better approach is to enforce data value restrictions when you create individual applications that access the database.

An example of an appropriate use of a check condition is for a gender field in which the values are restricted to M or F. An inappropriate use would be for the COLOR field in the INVENTORY table in the Clearwater Traders database—there are many possible values, and the values change constantly. A better approach is to create a lookup table, such as the COLOR table used in the Clearwater Traders database.

Each expression in a check condition must be able to be evaluated as true or false, and you can combine expressions using the logical operators AND and OR. When two expressions are joined by the AND operator, both expressions must be true for the expression to be true. When two expressions are joined by the OR operator, only one expression needs to be true for the expression to be true. A check condition can validate specific values, such as whether a gender field is equal to M or F, or validate ranges of allowable values. An example of a range check condition is used in the CREDITS field when creating the COURSE table in the Northwoods University database to specify that course credits must be greater than 0 and less than 12. The constraint definition is:

```
CONSTRAINT course_credits_cc CHECK((credits > 0) AND
(credits < 12))
```

Note that both of the expressions (credits > 0) and (credits < 12) can be evaluated as either true or false. The AND condition specifies that both conditions must be true for the check condition to be true.

Another check condition in the database checks to ensure that the value entered in the S_CLASS field of the STUDENT table is FR, SO, JR, or SR:

```
CONSTRAINT student_s_class_cc CHECK ((s_class = 'FR') OR
(s_class = 'SO')
OR (s_class = 'JR') OR (s_class 'SR'))
```

Again, each of the expressions in parentheses can be evaluated as true or false. The OR condition specifies that if any one condition is true, then the check condition is satisfied.

The **NOT NULL** value constraint specifies whether a field must have a value entered for every record or whether it can be NULL (indeterminate or unknown). When a field is specified as a primary key, it automatically has a NOT NULL constraint. From a business standpoint, some fields, such as a customer's name, should not be NULL. In general, NOT

NULL constraints should be enforced in user applications rather than in the database, because these restrictions make the database very inflexible.

An exception to this rule is for foreign key fields. Although foreign keys can be NULL (meaning there is no link between a given record and another table), sometimes foreign keys should not be NULL. For example, in the Northwoods University database, it doesn't make sense for a COURSE_SECTION record not to have an associated value for a term ID. Foreign key fields that are not allowed to be NULL must have an explicit NOT NULL constraint. The following example creates the COURSE_SECTION table, and includes examples of NOT NULL constraints:

```
CREATE TABLE COURSE_SECTION
(c_sec_id NUMBER(6),
c_id NUMBER(6) CONSTRAINT course_section_c_id_nn NOT NULL,
term_id NUMBER(6) CONSTRAINT course_section_term_id_nn NOT NULL,
sec_num NUMBER(2),
f_id NUMBER(5),
day VARCHAR2(10),
time DATE,
loc_id NUMBER(6),
max_enrl NUMBER(4),
CONSTRAINT course_section_c_sec_id_pk PRIMARY KEY (c_sec_id),
CONSTRAINT course_section_c_id_fk FOREIGN KEY (c_id)
REFERENCES course(c_id),
CONSTRAINT course_section_loc_id_fk FOREIGN KEY (loc_id)
REFERENCES location(loc_id),
CONSTRAINT course_section_term_id_fk FOREIGN KEY (term_id)
REFERENCES term(term_id),
CONSTRAINT course_section_f_id_fk FOREIGN KEY (f_id)
REFERENCES faculty(f_id));
```

The **DEFAULT** value constraint specifies that a particular field will have a default value that is inserted automatically for every table record. For example, if most Northwoods University students live in Wisconsin, the default S_STATE field could be declared with a default value of WI, using the following field declaration:

```
s_state CHAR(2) DEFAULT 'WI'
```

The default will only be used if a null value is inserted into S_STATE. It will be over-ridden if the user specifies a value.

The **UNIQUE** value constraint is a table constraint that specifies that a specific value can exist in only one table record. All primary keys automatically have a unique constraint. An example of a field that might require a unique constraint is the consultant e-mail (C_EMAIL) field in the CONSULTANT table in the Software Experts database. You would use the following command to create this constraint:

```
CONSTRAINT consultant_c_email_uk UNIQUE (c_email)
```

CREATING DATABASE TABLES WITH SQL*PLUS

Now you will create database tables using SQL*Plus, Oracle's command-line SQL utility. Starting an Oracle application is a two-step process. First, you start the application on your client workstation, and then you log onto the Oracle database. Your instructor will advise you of the correct logon procedures for your lab. Now you will start SQL*Plus and create some database tables.

If you are using the Personal Oracle database, you may need to first start the database by following the steps in the installation instructions for your particular configuration.

To start SQL*Plus:

1. Click **Start** on the Windows taskbar, point to **Programs**, point to **Oracle for Windows 95**, and then click **SQL*Plus 8.0.** After a few moments, Oracle SQL*Plus starts, opens the Log On dialog box, and requests your user name and password. You must also enter your host string (or connect string), as shown in Figure 2-1.

In most cases, Windows NT and Windows 2000 users will see "Windows NT" instead of "Windows 95" in the Start menu program names.

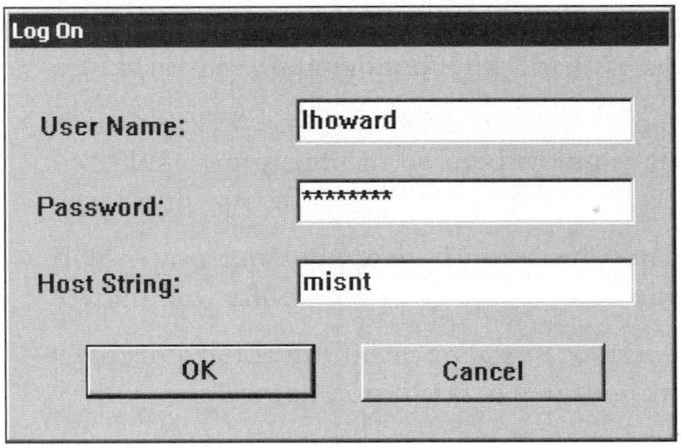

Log On	
User Name:	lhoward
Password:	********
Host String:	misnt
OK	Cancel

Figure 2-1   SQL *Plus Log On dialog box

Another way to start SQL*Plus is to start Windows Explorer, navigate to your Developer *Oracle_Home* folder, and then double-click the plus80w.exe file.

2. Type your user name, press **Tab**, type your password, press **Tab**, type your host string, and then click **OK**. The SQL*Plus program window appears, as shown in Figure 2-2.

```
Oracle SQL*Plus                                                    _ □ X
File  Edit  Search  Options  Help

SQL*Plus: Release 8.0.6.0.0 - Production on Fri Nov 10 13:45:14 2000

(c) Copyright 1999 Oracle Corporation.  All rights reserved.


Connected to:
Oracle8i Enterprise Edition Release 8.1.5.0.0 - Production
With the Partitioning and Java options
PL/SQL Release 8.1.5.0.0 - Production

SQL>
```

Figure 2-2   SQL *Plus program window

If necessary, click the Maximize button on the Oracle SQL*Plus title bar to maximize the program window.

Your version of SQL*Plus might have a different version number and show different application options, but the steps in this book will work the same.

Creating a Database Table Using SQL*Plus

Suppose you need to create the Northwoods University STUDENT table. To do this, you must create the FACULTY table first, because the F_ID field in the STUDENT table is a foreign key reference to the F_ID field in the FACULTY table. However, note that the FACULTY table also has a foreign key reference to LOC_ID in the LOCATION table. The LOCATION table does not have any foreign key references, so it can be created before the FACULTY and STUDENT tables. You will create the LOCATION table next.

To create the LOCATION table:

1. To create the LOCATION table, type the command shown in Figure 2-3. Press the **Enter** key after typing each line to go to the next line. Do not type the line numbers. SQL*Plus adds line numbers to your command after you press the Enter key. Each line of the command defines a different column of the database. You can use the line numbers to reference specific command lines using the SQL*Plus online editing facility, which you will learn about later.

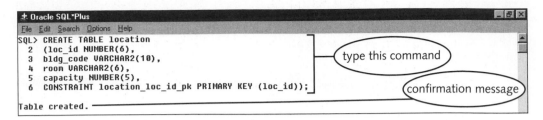

Figure 2-3   SQL command to create LOCATION table

2. If necessary, type **;** after the last line to end the command. The semicolon is the signal to the SQL*Plus compiler that it has reached the end of the SQL command.

3. Press the **Enter** key to execute the query. You should see the confirmation message "Table created," as shown in Figure 2-3.

SQL is not case sensitive, and spaces between characters for formatting are ignored, as are line breaks. If you did not receive the "Table created" confirmation message (or even if you did), proceed to the next section to learn how to edit and debug SQL commands.

Editing and Debugging SQL Commands Using an Alternate Text Editor

Many SQL commands are long and complex, and it is easy to make typing errors. You can enter your commands in SQL*Plus by typing the commands in an alternate environment, such as Notepad or any other Windows text editor, and then copying and pasting them into SQL*Plus. While SQL*Plus is running, start Notepad or any other text editor as a separate application, type the command, select and copy the command text, then switch to SQL*Plus and paste the selected command text, and execute the command. If the command fails to execute correctly, simply switch back to the text editor, edit the command, copy and paste the edited text back into SQL*Plus, and reexecute the command.

When you are creating database tables, it is a good idea to save the text of all of your CREATE TABLE queries in a single Notepad text file so you have a record of the original code. Then you can easily re-create the tables if changes are required later. To save a text file in Notepad, click File on the menu bar, and then click Save and specify the drive, folder, and filename for the file. You can save more than one CREATE TABLE command in a text file, if you want. Just make sure that they are in the proper order so that foreign key references are made after their parent tables are created.

2

Next, you will start Notepad, copy the text that you typed to create the LOCATION table, edit it and rerun it if necessary, and save the Notepad file so you have a record of your query text. You will continue to use this file to record and edit your SQL commands for the rest of this chapter.

To copy the query into Notepad and save the file:

1. In SQL*Plus, drag the mouse pointer over the text of the command to create the LOCATION table so the command text is highlighted, as shown in Figure 2-4. Do not highlight the command line numbers, the SQL prompt, confirmation, or error message.

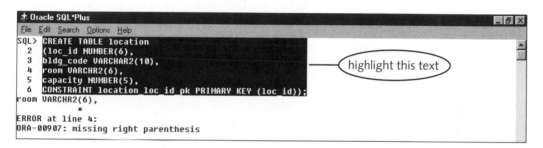

Figure 2-4   Highlighting the query text in SQL *Plus

2. Click **Edit** on the menu bar, and then click **Copy**.

3. Start Notepad (or an alternate text editor) on your computer.

4. In Notepad, click **Edit** on the menu bar, and then click **Paste**. The SQL command text appears in Notepad. If necessary, correct the command text, copy it, switch to SQL*Plus, paste the corrected text at the SQL prompt, and press **Enter** to run the command again. If an error message still appears, switch back to Notepad, debug the command, and run it again until the LOCATION table is successfully created.

5. Switch to Notepad, click **File**, and then click **Save**. Navigate to your Chapter2\TutorialSolutions folder, type **2Queries.sql** in the File name text box, open the Save as type list, select All files (*.*), and then click Save.

Text files that contain SQL commands normally have an .sql extension. If you do not open the Save as type list and select All files (*.*), Notepad automatically appends a .txt extension onto the filename.

6. Minimize the Notepad window.

Using Oracle Help Resources

Suppose that when you tried to create the LOCATION table, you got the error message shown in Figure 2-5. The error message shows the line; the position on the line where the error occurred, indicated by the position of the asterisk above the erroneous line; the error code number; and a brief description of the error.

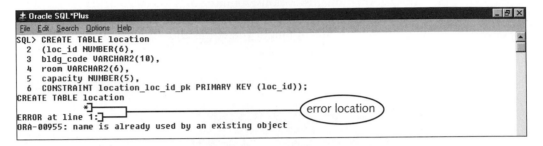

Figure 2-5 Query error message

The error message indicates that the error occurred on Line 1, and the asterisk indicates that it was generated when the table name (LOCATION) was specified. There are two ways to get further information: you can use the online Help system, or you can connect to the Oracle Technology Network Web site and look up the error message. First, you will look up the error message using online Help.

To identify the error message using online Help:

1. Start the Oracle8i Error Message Help application using the instructions provided by your instructor. The Oracle Messages and Codes window appears. Oracle does not have a central online Help application for all of its products, but it has specific online Help applications for each product. The Oracle8i Error Message online Help application provides explanations for DBMS generated (ORA-) error messages, while other Help applications provide other types of information.

 You start the Oracle8i Error Message online Help application by running a file named ora.hlp, which you can download from the Course Technology Web site, *www.course.com*

2. Click **Index**, type **ORA-00955** in the first text box, and then click **Display**. The explanation of the error message that appears in the Oracle Messages and Codes window indicates that you attempted to create a database object (in this case, the LOCATION table) using a name that is already used by an existing object.

3. Click the **Close** button on the title bar to close the Oracle Messages and Codes window.

Another way to get more information about errors encountered in Oracle applications is using the Oracle Technology Network (OTN). This is a Web-based resource that is provided free of charge by Oracle Corporation. Next, you will use the Oracle Technology Network Web site to get more information about the error message.

To identify the error message:

1. Start your Web browser, and type **otn.oracle.com** in the Address box. The Oracle Technology Network Web page appears. Search the Oracle Technology Network Web site using the search text **Oracle8i error messages**.

2. Click the **Oracle8i Error Messages** link. This Web page provides explanations for DBMS-generated (ORA-) error messages.

 The first time you use OTN, you may need to become a member by clicking the Membership link and then following the instructions for specifying a username and password.

3. Click the **Index** link, and then follow the links to find the error explanation for ORA-00955, which is as follows:

ORA-00955 name is already used by an existing object

> **Cause:** An attempt was made to create a database object (such as a table, view, cluster, index, or synonym) that already exists. A user's database objects must have distinct names.

> **Action:** Enter a unique name for the database object or modify or drop the existing object so it can be reused.

Note that the error explanation, whether you obtained it using online Help or via the OTN, indicates that the error occurred because the user has tried to create a database object (in this case, the LOCATION table) that has the same name as an existing object.

Some errors are harder to find than others. For example, note the error shown in Figure 2-6. The error message indicates the problem is caused by a "missing right parenthesis," but examination of the command indicates that each left (opening) parenthesis in the command has an associated right (closing) parenthesis. Further examination indicates that the error is caused by CONSTRAINT being misspelled as CONSTRANT in Line 6. To debug SQL queries, always *start* looking for an error at the line referred to in the error message, and keep in mind that the error won't necessarily be on that line. If the referenced line is correct, examine the error message, and if necessary, look up the error message using Online Help or the OTN Web site to identify what type of problem might cause the error. Then, examine all the query lines to look for typographical errors, omitted or misplaced commas or parentheses, misspelled words, or repeated constraint variable names.

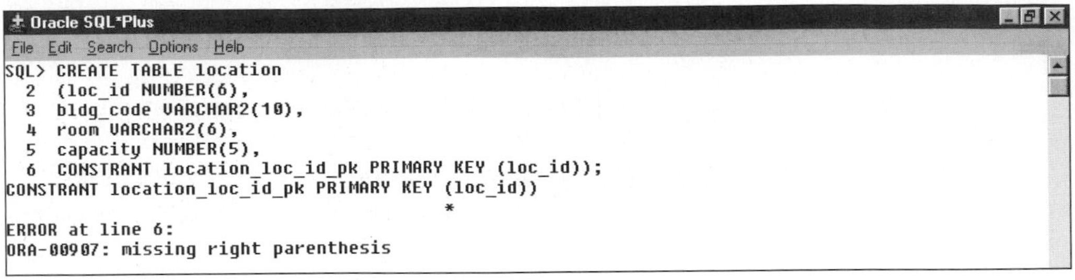

```
± Oracle SQL*Plus                                                          _ ⊡ ✕
File  Edit  Search  Options  Help
SQL> CREATE TABLE location
  2  (loc_id NUMBER(6),
  3  bldg_code VARCHAR2(10),
  4  room VARCHAR2(6),
  5  capacity NUMBER(5),
  6  CONSTRANT location_loc_id_pk PRIMARY KEY (loc_id));
CONSTRANT location_loc_id_pk PRIMARY KEY (loc_id))
                                 *
ERROR at line 6:
ORA-00907: missing right parenthesis
```

Figure 2-6   Misleading error message

When you have an error that you cannot locate, a "last resort" debugging technique is to create the table multiple times and add one additional column each time, until you find the column causing the error. First paste your nonworking query in a Notepad file, and modify it so that it creates the table with only the first column declaration. Copy the modified query, and paste it into SQL*Plus. If the table is created successfully, you now know that the error was not in the first column declaration. Delete the table using the DROP TABLE command, which has the following general syntax: DROP TABLE tablename;. Then modify the query in Notepad to create the table using only the first and second column declarations. If this works, you know that the problem was not in either the first or second column declaration. Drop the table again, and modify the query so that the table is created using only the first, second, and third column declarations. Continue this process of adding one more column declaration to the CREATE statement until you locate the column declaration that is causing the error.

Exiting SQL*Plus

There are three ways to exit SQL*Plus. You can type *exit* at the SQL prompt; click File on the menu bar, and then click Exit; or click the Close button on the program window title bar. When you connect to a database on a server, your database connection disconnects automatically when you exit SQL*Plus.

To exit SQL*Plus:

1. In SQL*Plus, type **exit** at the SQL prompt; click **File** on the menu bar, and then click **Exit**; or click the **Close** button on the program window title bar.

2. Your database connection disconnects, and SQL*Plus closes.

Creating a Table with a Foreign Key Constraint

Recall that a foreign key creates a link between two database tables. In the FACULTY table in the Northwoods University database, LOC_ID is a foreign key that references a record in the LOCATION table. Now you will create the FACULTY table.

To create the FACULTY table:

1. Switch to Notepad, where your 2Queries.sql file should be open. Type the command shown in Figure 2-7, highlight the text, click **Edit**, and then click **Copy**.

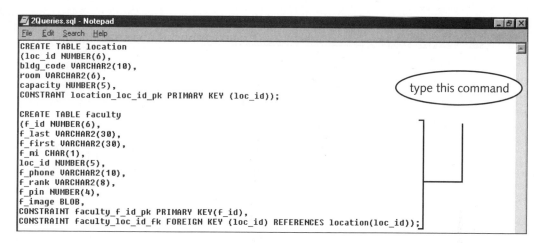

Figure 2-7   Command to create FACULTY table

2. If necessary, start SQL*Plus, log onto the database, paste the copied command at the SQL prompt, and then press **Enter**. If necessary, debug the command until you successfully create the FACULTY table.

For the rest of the tutorial, you should type all SQL commands in the 2Queries.sql file in Notepad, paste them into SQL*Plus, and debug them if necessary.

VIEWING INFORMATION ABOUT TABLES

After you create database objects like tables, you often need to review information like table names, field and data type values, and constraint definitions. To get information about the field names and data types of an individual table, you can use the DESCRIBE command, which has the following general syntax:

    DESCRIBE *tablename*;

Now you will use the DESCRIBE command to view information about the fields in the LOCATION and FACULTY tables that you created earlier.

To view information about the table fields:

1. In SQL*Plus, type the command **DESCRIBE location;**, and then press **Enter**. The field names and data types should appear as shown in Figure 2-8.

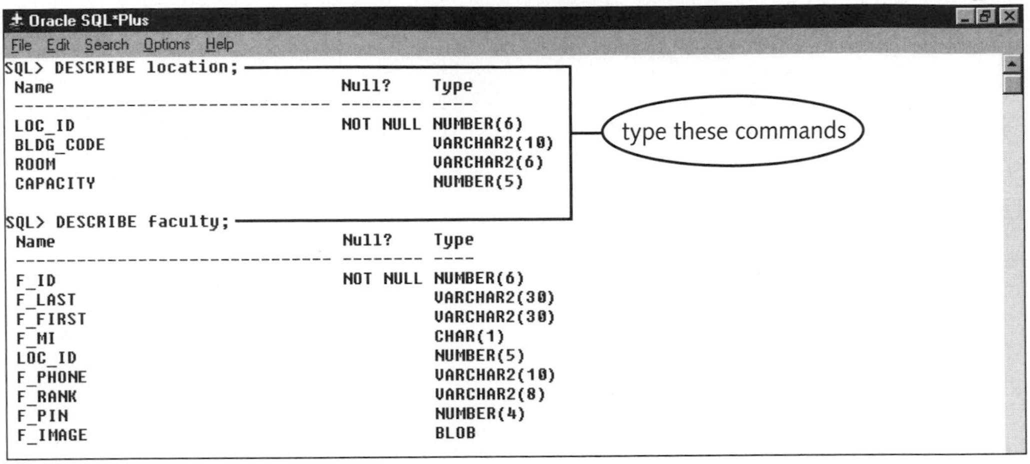

Figure 2-8   Using the DESCRIBE command

2. Next, type the command **DESCRIBE faculty;**, and then press **Enter**. The field names and data types should appear as shown in Figure 2-8.

Note that the output lists the field names and data types. It also shows the fields that have NOT NULL constraints: LOC_ID in the LOCATION table and F_ID in the FACULTY table. These fields cannot be NULL because they are the primary key fields of the tables. However, note that the foreign key constraint (LOC_ID in the FACULTY table) is not listed. To get information about constraints other than the NOT NULL constraint, you have to use the Oracle Data Dictionary views.

USING ORACLE DATA DICTIONARY VIEWS

The Oracle Data Dictionary consists of tables that contain information about the structure of the database. The data dictionary is created in a user schema named SYS, which is created when the database is created. The Oracle DBMS updates the data dictionary tables automatically as users create, update, and delete database objects. As a general rule, users and database administrators do not directly view, update, or delete information in the data dictionary tables. Rather, they interact with the data dictionary using views. A view is based on an actual database table but presents the table data in a different format based on the needs of the users. A view can hide some table fields that the user isn't interested in or doesn't have required object privileges to view. (Recall that object privileges have to be explicitly granted to allow a user to manipulate database objects that the user didn't create.) For an individual user, the data dictionary views present information only about the user's database objects or objects that the user has specific object privileges to manipulate. For a database administrator, the data dictionary views can present information about all of the objects in the database.

The data dictionary views are divided into three general categories based on the user's object privileges:

- **USER**, which shows objects that the user created.

- **ALL**, which shows both objects that the user created and objects that the user has privileges to manipulate through object privileges granted by the user who created the object. For example, another user might create a table, and then give you the privilege to update the table.

- **DBA**, which allows users with DBA privileges to view information about all database objects.

Data dictionary views are named using these categories as prefixes to indicate the level of information that the view contains. The general command to retrieve information from a data dictionary view is:

```
SELECT view_fieldname1, view_fieldname2, …
FROM category_prefix_object;
```

The category prefix corresponds to the three view categories and can have the values USER, DBA, or ALL. The object corresponds to the type of object being examined, like TABLE or CONSTRAINT. For example, you would use the following command to retrieve the names of all of the database tables that a user has created from the TABLE_NAME field in the USER_TABLES view:

```
SELECT table_name
FROM user_tables;
```

Similarly, the following command uses the ALL prefix and retrieves the names of all database tables that a user has either created or has been given object privileges to manipulate:

```
SELECT table_name
FROM all_tables;
```

Now you will use data dictionary view commands to view information about your tables.

To view information about your tables:

1. Type the command **SELECT table_name FROM user_tables;**, and then press **Enter**. The output should appear similar to the output shown in Figure 2-9. Note that your output may be different if you have created additional tables or if other users of your database have created tables and given you object privileges to manipulate these tables.

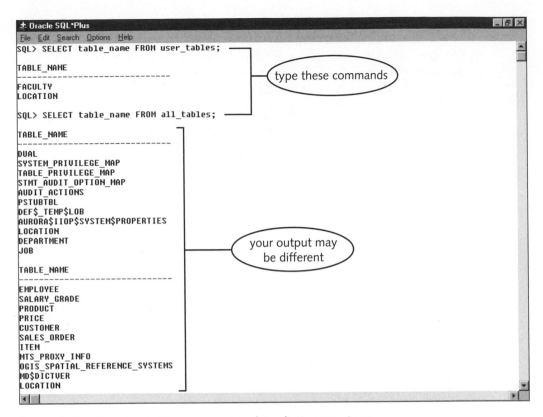

Figure 2-9   Viewing table names using data dictionary views

2. Next, type the command **SELECT table_name FROM all_tables;**, and then press **Enter**. Again, the output should appear similar to the output shown in Figure 2-9. Like the output of the SELECT table_name FROM user_tables; command, the output of this command may be different, and some output appears off the screen.

Unless your user account has the DBA role, you cannot retrieve information from any views using the DBA prefix.

You can retrieve information about a variety of database objects using the data dictionary views. Table 2-5 shows database object type names that can be used in the FROM clause of the command to query a data dictionary view. For example, a possible query on the first object type would be **SELECT * FROM user_objects**.

Object Name	Object Type
OBJECTS	All database objects
TABLES	Database tables
INDEXES	Table indexes created to improve query retrieval performance
VIEWS	Database views
SEQUENCES	Sequences used to automatically generate surrogate key values
USERS	Database users
CONSTRAINTS	Table constraints
CONS_COLUMNS	Table columns that have constraints
IND_COLUMNS	Table columns that have indexes
TAB_COLUMNS	All table columns

Table 2-5   Database objects with data dictionary views

To determine the column names that exist in a database view, you can use the DESCRIBE command along with the view name. Figure 2-10 shows a description of the USER_CONSTRAINTS data dictionary view, showing the names of the fields that can be queried.

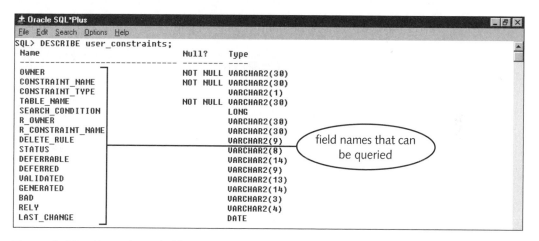

Figure 2-10   Describing field names in USER_CONSTRAINTS view

Now you will list of all of the constraints that have been created on tables in your user account by selecting the CONSTRAINT_NAME and CONSTRAINT_TYPE fields from the USER_CONSTRAINT view. From this view, you can determine what constraints exist for each table.

To list your table constraints:

1. In SQL*Plus, type the command
   SELECT constraint_name, table_name FROM user_constraints;.
   The output should appear similar to the output shown in Figure 2-11. Your
   output may be different, however, if you have created additional database
   tables with other constraints.

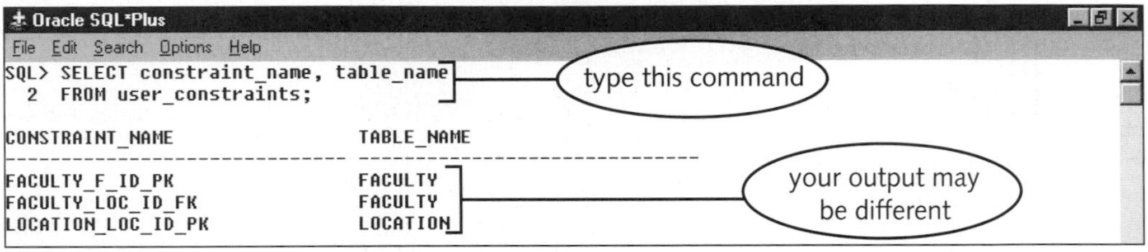

Figure 2-11   Viewing table constraint information

The output shows that you have three constraints: two in the FACULTY table and
one in the LOCATION table. The constraint names that you entered when you
created the tables make it easy to determine the field name that the constraint is associated with, and the constraint identifiers (pk and fk) make it easy to determine the
type of each constraint.

MODIFYING TABLES USING SQL*PLUS

After using the DESCRIBE command or selecting the USER_CONSTRAINTS
information, you might find that you need to make changes to the column lengths or
data types in your database tables. You should plan your tables carefully to avoid having to make changes to the overall table structure, but inevitably, you will need to
make changes. There are some specifications of an Oracle database table, called
unrestricted actions, that you can always modify; and others, called **restricted
actions**, that you can modify only in certain situations. Table 2-6 shows the unrestricted and restricted actions for a table.

Unrestricted Actions	Restricted Actions
Renaming tables	Dropping a table from a user schema; allowed only if the table does not contain any fields that are referenced as foreign keys in other tables.
Increasing column widths	Changing an existing column's data type; allowed only if existing data in the column being modified is compatible with the new specification. For example, you could change a VARCHAR2 column to a CHAR column, but you could not change a VARCHAR2 column to a NUMBER column.
Dropping columns	Decreasing the width of an existing column; allowed only if existing column values are NULL.
Dropping constraints	Adding a primary key constraint to an existing column; allowed only if current field values are unique (no duplicate entries) and not NULL.
	Adding a UNIQUE constraint to a column; allowed only if current field values are all unique.
	Adding a CHECK CONDITION constraint; allowed, but the constraint is only applied to new values that are inserted.
	Adding a foreign key constraint; allowed only if current field values are NULL or exist in the referenced tables.
	Changing a column's default value; allowed, but the change is only applied to new values that are inserted.

Table 2-6   Unrestricted and restricted table changes

Dropping and Renaming Existing Tables

When you drop a table from the database, you delete the table structure and all the data it contains. To delete a table, use the DROP TABLE *tablename*; command, where *tablename* is the name of the table that you want to drop. This command will not be successful if the table being dropped contains fields referenced as foreign keys in other tables. In this case, you can either drop all of the tables that contain the foreign key references first, or you can drop the foreign key constraints in all of the associated tables using the CASCADE CONSTRAINTS option. When the DROP TABLE command is issued with the CASCADE CONSTRAINTS option, the system first drops all of the constraints associated with the table, and then drops the table. This allows tables to be dropped in any order, regardless of foreign key constraints.

In the following set of steps, you will first try to drop the LOCATION table without the CASCADE CONSTRAINTS option. Then you will drop the table using the CASCADE CONSTRAINTS option, which drops the table constraint, which is the foreign key reference that references the LOC_ID field.

To drop the LOCATION table along with all of the constraints that reference it:

1. Type **DROP TABLE location;** and press **Enter;.** The error message "ORA-02449: unique/primary keys in table referenced by foreign keys" appears, indicating that one or more fields in the LOCATION table are referenced as foreign keys in other tables.

2. Type **DROP TABLE location CASCADE CONSTRAINTS;** and press **Enter;.** The "Table dropped" message confirms that the table was successfully dropped.

To rename a table, use the **RENAME** *old_tablename* **TO** *new_tablename;* command, and replace *old_tablename* and *new_tablename* with the appropriate names.

Adding Fields to Existing Tables

The basic syntax of the statement to add a new field to a table is:

```
ALTER TABLE tablename
ADD(fieldname data declaration
constraint definitions);
```

Suppose that you need to add a column to the FACULTY table that specifies each faculty member's employment start date. You will add the START_DATE column to the FACULTY table next.

To add the START_DATE column to the FACULTY table:

1. **At the SQL prompt**, type the following command and press **Enter**. The "Table altered" confirmation message appears and confirms that the table was successfully altered.

```
ALTER TABLE faculty
ADD (start_date DATE);
```

Modifying Existing Fields

The general syntax of the command to modify an existing field's data declaration is:

```
ALTER tablename
MODIFY(fieldname new_data_declaration);
```

Suppose that you decide to modify the FACULTY table's F_RANK field to a data type of VARCHAR2 with a length of six.

To change the data type and size of the F_RANK field:

1. Type **ALTER TABLE faculty MODIFY (f_rank VARCHAR2 (6));**, and then press **Enter** to change the data type and size of the F_RANK field in the FACULTY table. The "Table altered" message confirms that the table was successfully altered.

Dropping a Column

Sometimes you need to drop an existing column because the column is no longer needed, or you might want to change a column's name. Renaming a column is not allowed, but an alternative is to drop the existing column, and then add a new column that has the desired name. The general command to drop an existing column is `ALTER TABLE tablename DROP COLUMN columnname;`.

Suppose you decide that you need to change the name of the F_RANK column in the FACULTY table from F_RANK to FACULTY_RANK. To make this change, you need to drop the existing F_RANK column, and then add a new column named FACULTY_RANK.

To change the F_RANK column name to FACULTY_RANK in the FACULTY table:

1. At the SQL prompt, type
   `ALTER TABLE faculty DROP COLUMN f_rank;`, and then press Enter to drop the F_RANK column. The message "Table altered" appears.

2. Type `ALTER TABLE faculty ADD (faculty_rank VARCHAR2(8));`, and then press Enter to add the new column to the table. The message "Table altered" appears again.

3. Exit SQL*Plus.

4. If Notepad is still open from earlier chapter exercises, switch to Notepad, save your file, and then exit Notepad.

SUMMARY

- SQL is a high-level language that is used to query relational databases.

- Database objects are created, dropped, and modified using data description language (DDL) commands.

- Each user account owns table and data objects in its own area of the database. This specific area owned by the user is called the user schema, and the objects within it are referred to as schema objects.

- Security within an Oracle database is enforced using system privileges, which control the operations that a user can perform on the database, and object privileges, which control the operations that a user can perform on schema objects.

- Oracle database objects, like user names and table names, can be 1 to 30 characters in length and consist of only alphanumeric characters and the special characters $, _, and #. Object names must begin with a letter and cannot contain spaces.

- Roles are database objects that can be granted one or more system privileges. A role can then be granted to a user and the user assumes the privileges granted to the role.

❏ When you create a table, you must assign a data type to each column to specify the kind of data that will be stored in the field. Data types provide error checking by making sure the correct data values are entered, and data types also cause storage space to be used more efficiently by optimizing how specific types of data are stored.

❏ Character data can be stored using the CHAR, VARCHAR2, and NCHAR data types.

❏ Numerical data can be stored using the NUMBER data type. The precision of a NUMBER data type declaration specifies the total number of digits both to the left and to the right of the decimal point, and the scale specifies the number of digits to the right of the decimal point.

❏ The DATE data type stores date and time data.

❏ The large object (LOB) data type stores large amounts of binary or character data, or references to external binary files.

❏ In an Oracle database, integrity constraints are enforced through primary key specifications and foreign key references.

❏ In an Oracle database, value and check constraints on fields specify which values can be entered into those fields and whether the values can be NULL.

❏ In an Oracle database, you can specify whether a field has a default value that will be inserted automatically for every record.

❏ To create a new database table in SQL*Plus, use the CREATE TABLE command followed by the new table name, an opening parenthesis, and then a list of each of the field names, their data type declarations, and constraint declarations, followed by a closing parenthesis and a semicolon.

❏ The CREATE, ALTER, and DROP commands are examples of data definition language (DDL) statements that change the database structure and do not need to be explicitly saved.

❏ SQL is not case sensitive, and spaces and line breaks for formatting are ignored.

❏ You can edit SQL*Plus commands using command-line editing or an alternate text editor. The edit buffer is a memory area in SQL*Plus that stores the text of the last executed query.

❏ You can change a column's data type, size, or default value only if there is no data in the column. You can add a primary key constraint to a column only if all of the existing column values are unique. You can add a foreign key constraint to a column only if the existing column values are NULL or if they exist in the referenced table.

❏ A constraint name is the internal name used to identify integrity and value constraints. Every constraint in a user schema must have a unique name.

❏ The DESCRIBE command can be used to display a table's field names, data types, and NOT NULL constraints.

2

❏ The Oracle data dictionary consists of tables that contain information about the database structure. Data dictionary views provide a way for users to access information about the data dictionary.

❏ You can drop and rename tables, add new columns to tables, increase column widths, or drop existing constraints with no restrictions.

❏ You can only modify a column data type or constraint if the data that has been inserted in the column is compatible with the updated data type or constraint.

❏ You cannot change a column name, but you can drop the column and then add a new column with the desired name.

REVIEW QUESTIONS

1. Which of the following declarations are legal Oracle table and field names?
   a. SALES_REP_NAME
   b. INVENTORY_TABLE
   c. CLEARWATER_TRADERS_INVENTORY_TABLE
   d. SalesRepID
   e. _ITEM_PRICE
   f. SALESREP TABLE

2. List which data type you would recommend for the following data items:
   a. Social Security number
   b. gender (M or F)
   c. telephone number
   d. an image of a city map that is stored in binary format in the database
   e. sales representative last name
   f. sales representative date of birth
   g. a reference to a file containing a spreadsheet

3. Write the declarations for the following fields. Include the data type and length. Do not include value constraints.
   a. CUST_AGE, containing integer values ranging from 1 to 99
   b. KILOS_PURCHASED, containing numeric values ranging from 0 to 100, and having a variable number of decimal places
   c. PRODUCT_PRICE, containing numeric values ranging from 0 to 999.99, and rounded to two decimal places
   d. APPT_START_TIME, containing times ranging from 8:00 A.M. to 5:00 P.M.
   e. FACULTY_IMAGE, containing a binary image that is stored directly in the database

4. What happens when you omit the constraint name in a data declaration?

5. List an order in which the tables in the Clearwater Traders database can be created so that all fields that are referenced as foreign keys are first created as primary keys.

6. List an order in which the tables in the Software Experts database can be created so that all fields that are referenced as foreign keys are first created as primary keys.

7. Using the constraint naming convention, write valid names for the following constraints in the Northwoods University database. Refer to the database tables in Figure 1-12 in Chapter 1 to find the appropriate column names as needed.

   a. primary key in the COURSE_SECTION table

   b. S_ID foreign key reference in the ENROLLMENT table

   c. check condition requiring the STATUS column in the TERM table to have a value of either OPEN or CLOSED

   d. constraint requiring the TERM_DESC column in the TERM table to not be NULL

8. Write the constraint declarations for the following variables in the Software Experts database. Refer to Figure 1-13 in Chapter 1 to determine field and table names as needed. Include the correct constraint name as well as the constraint specification.

   a. primary key constraint for the CONSULTANT table

   b. primary key constraint for the CONSULTANT_SKILL table

   c. Foreign key constraints for the PROJECT_CONSULTANT table

   d. Constraint to restrict the values in the CERTIFICATION field in the CONSULTANT_SKILL table to the values 'Y' or 'N'

PROBLEM-SOLVING CASES

Use Notepad to write the SQL commands to create the specified tables. Choose appropriate data types and field widths for each field based on the guidelines in this chapter. Be sure to create appropriate constraint definitions for all primary and foreign keys, and create the tables in the correct order for the foreign key references. Test your commands in SQL*Plus to make sure they are correct, and do not enter any data into the tables. Save all solutions in the Chapter2\TutorialSolutions folder.

1. Using Figure 1-12 in Chapter 1 as a reference, write the SQL command to create the STUDENT, FACULTY, and LOCATION tables in the Northwoods University database. If you already created any of these tables in the Chapter 2 tutorial material, you will need to drop the existing tables before you can create the new ones. Add a check condition constraint to the STUDENT table to restrict values in the S_CLASS field to 'FR', 'SO', 'JR', or 'SR'. Save your command in a file named 2Case1.sql in your Chapter2\CaseSolutions folder.

You will not be able to complete Case 2 until you have completed Case 1.

2. Using Figure 1-12 in Chapter 1 as a reference, write the SQL commands to create the TERM, COURSE, COURSE_SECTION, and ENROLLMENT tables in the Northwoods University database. Create a check condition on the STATUS field in the TERM table to restrict allowable values to either 'OPEN' or 'CLOSED'. Create a check condition constraint on the ENROLLMENT table to restrict values for the GRADE field to 'A', 'B', 'C', 'D', or 'F'. Save your commands in a file named 2Case2.sql in the Chapter2\CaseSolutions folder.

3. Using Figure 1-11 in Chapter 1, write the SQL commands to create the CUSTOMER, CUST_ORDER, ORDER_SOURCE, ITEM, CATEGORY, INVENTORY, and COLOR tables in the Clearwater Traders database. Declare the ITEM_IMAGE field in the ITEM table as a BLOB data type. Save your commands in a file named 2Case3.sql in your Chapter2\CaseSolutions folder.

You will not be able to complete Case 4 until you have completed Case 3.

4. Using Figure 1-11 in Chapter 1 as a reference, create the SHIPMENT, SHIPMENT_LINE, and ORDER_LINE tables in the Clearwater Traders database. Save your commands in a file named 2Case4.sql in your Chapter2\CaseSolutions folder.

5. Using Figure 1-13 in Chapter 1 as a reference, create the CONSULTANT, SKILL, and CONSULTANT_SKILL tables in the Software Experts database. Save your commands in a file named 2Case5.sql in your Chapter2\CaseSolutions folder.

You will not be able to complete Case 6 until you have completed Case 5.

6. Using Figure 1-13 in Chapter 1 as a reference, create the CLIENT, PROJECT, PROJECT_SKILL, PROJECT_CONSULTANT, and EVALUATION tables in the Software Experts database. Use the CLOB data type for the COMMENTS field in the EVALUATION table. Add a check condition constraint to the EVALUATION table that restricts the values for the SCORE field to values greater than or equal to 0, and less than or equal to 100. Save your commands in a file named 2Case6.sql in your Chapter2\CaseSolutions folder. (*Hint:* To create the foreign key constraint for the PARENT_P_ID field in the PROJECT table, you will need to create the PROJECT table first, and then add the constraint by modifying the table.)

3

USING ORACLE TO ADD, VIEW, AND UPDATE DATA

◀ LESSON A ▶

Objectives

- ♦ Automate SQL commands using scripts
- ♦ Insert data into database tables
- ♦ Commit data to the database and create database transactions
- ♦ Update and delete database records
- ♦ Create sequences
- ♦ Grant and revoke object privileges

The next step in developing a database is to insert, modify, and delete data records, and learn how to retrieve data. Ultimately, database users will perform these operations using forms and reports that automate the data entry, modification, and summarization processes. To make these forms and reports operational, database developers often must translate user input into SQL queries that will be submitted to the database. As a result, database developers must be very proficient with SQL command-line operations. Therefore, it is important that you know how to add, view, and update data records using SQL*Plus. You will also learn how to create and use three new kinds of database objects: sequences, indexes, and views. These objects make it easier to manipulate database data and improve the speed at which data is retrieved for users. You will also learn how to grant the object privileges needed to allow other users to manipulate data in your database tables using SQL*Plus.

Using Scripts to Automate SQL Commands

A **script** is a text file that contains a sequence of SQL commands that can be executed in SQL*Plus. Usually, scripts contain commands for creating, modifying, or deleting database tables, or inserting or updating data records. You created a script in Chapter 2 when you typed a series of SQL commands in your Notepad text file. Usually, SQL scripts have an .sql file extension.

To run a script, you type the START command at the SQL prompt, followed by a space, and then the full path and filename of the script text file. For example, to start a script saved in the text file A:\Chapter3\myscript.sql, you would enter the command: **START a:\chapter3\myscript.sql**. The pathname and text filename and extension can be any legal Windows filename, but they *cannot* contain any blank spaces.

You do not have to type the file extension if it is .sql, since this is the default extension for script files. Also, you can type @ instead of START to run a script. Therefore, an alternate way to run the script myscript.sql is to type **@a:\chapter3\myscript**.

Before you can complete the exercises in this chapter, you need to run a script named emptynorthwoods.sql that will delete all of the Northwoods University database tables that you already have created. Then, the script will re-create the tables. Now, you will examine the script file in Notepad, and then run it.

To examine and run the emptynorthwoods.sql script file:

1. Start Notepad, and open the emptynorthwoods.sql file that is stored in the Chapter3 folder on your Data Disk. The file contains a DROP TABLE command to drop each of the tables in the Northwoods University database. It also contains a CREATE TABLE command to re-create each table. These commands are similar to the ones you learned in Chapter 2.

2. Start SQL*Plus and log onto the database.

3. Run the emptynorthwoods.sql script by typing **START a:\Chapter3\emptynorthwoods.sql** at the SQL prompt and pressing **Enter**.

You could also type START a:\chapter3\emptynorthwoods (omitting the .sql file extension), or @a:\chapter3\emptynorthwoods (using @ instead of START).

This book assumes that your Data Disk is stored on the A: drive. If your Data Disk files are stored on an alternate drive, type the appropriate drive letter and pathname when you are instructed to open or save a file on your Data Disk.

Don't worry if you receive an error message in the DROP TABLE command, as shown in Figure 3-1. This error message indicates that the script is trying to drop a table that does not exist. The script is written so it drops all existing Northwoods University database table definitions before creating the new tables. If a script tries to create a table that already exists, SQL*Plus will generate an error and will not re-create the table.

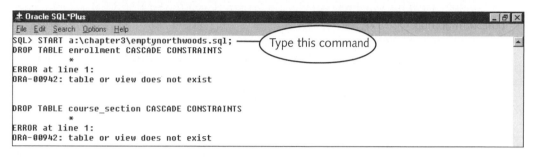

Figure 3-1   Error message when attempting to drop nonexistent table

INSERTING DATA INTO TABLES

After successfully running the script to create the tables, you are ready to begin adding data to the tables. In a business setting, programs called **forms** are used to automate the data entry process. Program developers who create forms often use the SQL INSERT statement in the form program code to insert data into tables.

The INSERT statement can be used two ways: to insert all fields into a record at once, or to insert a few selected fields. The basic syntax of the INSERT statement for inserting all record fields is:

```
INSERT INTO tablename
VALUES (column1_value, column2_value, …);
```

When you are inserting a value for all record fields, the VALUES clause of the INSERT statement must contain a value for each column in the table, or the word NULL instead of a data value if the data value is currently unknown or undefined. Column values must be listed in the same order in which the columns were defined in the CREATE TABLE command and must be the same data type as the associated column. You can use the DESCRIBE command to determine the correct field order in a table.

The basic syntax of the INSERT statement for inserting selected table fields is:

```
INSERT INTO tablename (column1_name, column2_name, …)
VALUES (column1_value, column2_value, …);
```

In this version of the INSERT command, you can list the column names in any order. You must list the data values, however, in the same order as their associated columns, and the data values must be the same data type as the associated column. A NULL value will automatically be inserted for columns omitted from the INSERT INTO clause.

Before you can insert a new data record, you must ensure that all of the foreign key records referenced by the new record have been added. Looking at the Northwoods database (see Figure 1-12), you see that in the first STUDENT record, Sarah Miller's F_ID value is 1. This value refers to F_ID 1 (Kim Cox) in the FACULTY table. Therefore, the FACULTY record must be added before you can add the first STUDENT record, or you will get a foreign key reference error. Now, look at Kim Cox's FACULTY record, and note that it has a foreign key value of LOC_ID 53. Similarly, this LOCATION record must be added before you can add the FACULTY record. Thankfully, the LOC_ID 53 record in the LOCATION table has no foreign key values to reference. Therefore, you can insert LOC_ID 53 in the LOCATION table, and then insert the associated FACULTY record, and finally, add the STUDENT record. Now you will insert LOC_ID 53 into the LOCATION table.

To insert a record into the Northwoods University database:

1. In Notepad, create a new file, and then type the SQL command shown in Figure 3-2. Save the Notepad file as **3AQueries.sql** in your Chapter3\TutorialSolutions folder on your Data Disk. Do not close the file.

2. Copy the query text and paste it into SQL*Plus, and then press **Enter**. The message "1 row created" should appear, indicating that the record has been added to the LOCATION table.

```
Oracle SQL*Plus                                                    _ 8 X
File  Edit  Search  Options  Help
SQL> INSERT INTO location VALUES          ( Type this command )
  2  (53, 'BUS', '424', 1);

1 row created.

SQL>
```

Figure 3-2   Query to insert a record into the LOCATION table

For the remainder of the tutorial, assume that you will enter all query commands into your Notepad text file, and then copy and paste them into SQL*Plus.

Note that the LOCATION table has four columns—LOC_ID, BLDG_CODE, ROOM, and CAPACITY—and that their associated data types are NUMBER, VARCHAR2, VARCHAR2, and NUMBER, respectively. The NUMBER fields are entered as digits, while the VARCHAR2 fields are enclosed within single quotation marks ('). Data stored in CHAR and VARCHAR2 fields must be enclosed in single quotation marks, and the text within the single quotation marks is case sensitive.

You enter the field values in the INSERT statement in the same order as the fields in the table, and you must include a value for every column in the table. When you insert a record into the LOCATION table, the DBMS expects a NUMBER data type, then a VARCHAR2, another VARCHAR2, and then another NUMBER. An error occurs if you try to insert the values in the wrong order or if you omit a column value. If you cannot remember the order of the columns or their data types, use the DESCRIBE command to verify the table's structure.

The next step is to insert Kim Cox's record into the FACULTY table. To practice the command for inserting selected fields instead of inserting all of the table fields, you will enter values only for the F_ID, F_LAST, F_FIRST, F_MI, LOC_ID, and F_IMAGE columns, and you will omit the values for F_PHONE, F_RANK, and F_PIN. This method of inserting data is useful when you are inserting data into only a few selected fields in a record.

Recall that the F_IMAGE field is a large object (LOB) data field of type BLOB, so the binary data for the image is stored in the database. To make data retrievals for both LOB and non-LOB data more efficient, Oracle stores LOB data in a separate physical location from other types of data in a record. Therefore, whenever you insert a record that contains a LOB field, you must initially insert a **locator** for the LOB field, which is a structure that contains information about the LOB type and points to the alternate memory location. (You will learn how to load the BLOB data into the database using Form Builder in Chapter 8.) A locator for a BLOB data field is created using the syntax **EMPTY_BLOB( )**. Now you will insert the record for faculty member Kim Cox that includes the BLOB locator.

To enter the record for faculty member Kim Cox:

1. Insert the first record into the FACULTY table by typing the SQL command shown in Figure 3-3. The message "1 row created" indicates that the record has been added to the table.

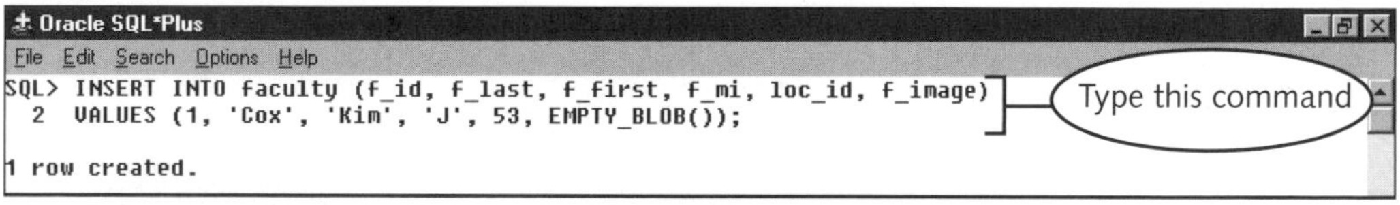

Figure 3-3 Inserting specified column values

Next you will add the FACULTY table record for John Blanchard. First, however, you must insert the associated LOCATION record.

To insert the record for John Blanchard:

1. Type the commands shown in Figure 3-4 to insert the records into the LOCATION and FACULTY tables. The "1 row created" messages for each INSERT command indicate that the records have been added to the tables.

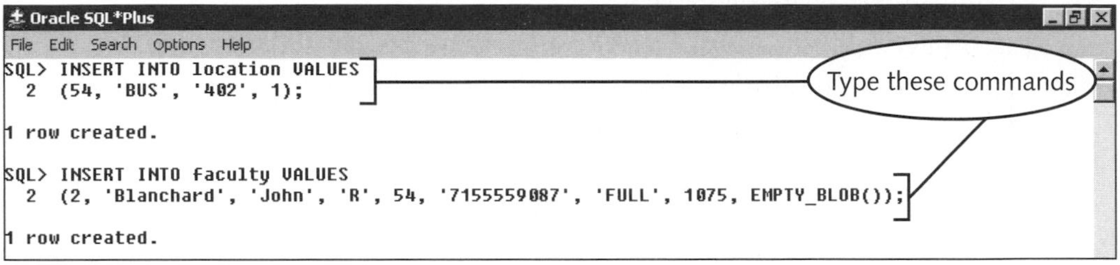

Figure 3-4 Adding another record to the LOCATION and FACULTY tables

FORMAT MASKS

Data values in Oracle database tables are stored in a binary internal format, and you can display them using a variety of output formats. For example, a DATE data field might contain a value that represents 07/22/2003 9:29:00 P.M. The default display format for DATE data fields is DD-MON-YY, so by default, this value would be displayed as 22-JUL-03. (The time portion of the date is stored in the database, but does not appear in the default display format.) Alternately, you could specify an alphanumeric character string called a **format mask** to specify that the value should appear in a different display format, such as July 22, 2003 9:29 P.M., or 21:29:00 P.M. (which uses 24–hour clock notation). Similarly, a NUMBER value of 1257.33 could appear as $1,257.33, or as 1257, depending on the specified format mask. Format masks affect only the way the data is displayed or printed—the full data value is stored in the database. Table 3-1 lists how you can use some common numerical data format masks to display the numeric data value stored as 059783. Within a format mask, the digit "9" is a placeholder that represents how number data values appear within the formatting characters.

Format Mask	Description	Displayed Value
99999	Number of 9s determines display width	59783
099999	Displays leading zeros	059783
$99999	Prefaces the value with dollar sign	$59783
99999MI	Displays "-" before negative values	-59783
99999PR	Displays negative values in angle brackets	<59783>
99,999	Displays a comma in the indicated position	59,783
99999.99	Displays a decimal point in the indicated position	59783.00

Table 3-1 Common numerical format masks

Table 3-2 shows some common date format masks using the example date of 5:45:35 P.M., Friday, January 15, 2003.

3

Format Mask	Description	Displayed Value
YYYY	Displays all four digits of year	2003
YYY or YY or Y	Displays last three digits, last two digits, or last digit of year	003, 03, 3
RR	Last two digits of the year, but modified to store dates from different centuries using two digits. If the current year's last two digits are 0 to 49, then years numbered 0 to 49 are assumed to belong to the current century, and years numbered 50 to 99 are assumed to belong to the previous century. If the current year's last two digits are from 50 to 99, then years numbered 0 to 49 are assumed to belong to the next century, and years numbered 50 to 99 are assumed to belong to the current century.	03
MM	Displays month as digits (01–12)	01
MONTH	Displays name of month, spelled out, uppercase (for months with fewer than nine characters in their name, trailing blank spaces are added to pad the name to nine characters)	JANUARY
Month	Displays name of month, spelled out, mixed case (trailing blank spaces added to pad name to nine characters)	January
DD	Displays day of month (01–31)	15
DDD	Displays day of year (01–366)	15
DAY	Displays day of week, spelled out, uppercase	FRIDAY
Day	Displays day of week, spelled out, mixed case	Friday
DY	Displays name of day as a three-letter abbreviation	FRI
AM, PM, A.M., P.M.	Meridian indicator (without or with periods)	PM P.M.
HH	Displays hour of day using 12-hour clock	05
HH24	Displays hour of day using 24-hour clock	17
MI	Displays minutes (0–59)	45
SS	Displays seconds (0–59)	35

Table 3-2 Common date format masks

Front slashes, hyphens, and colons can be used as separators between different date elements. For example, the format mask MM/DD/YY would appear as 01/15/03, and the mask HH:MI:SS would appear as 05:45:35. You can also include additional characters such as commas, periods, and blank spaces. For example, the format mask DAY, MONTH DD, YYYY would appear as FRIDAY, JANUARY 15, 2003.

You can create format masks that embed characters within character data fields to create formatting for character fields, such as Social Security numbers and telephone numbers. To create format masks for character fields, you must precede the field's format mask with the characters "FM," and enclose formatting characters within double quotation marks. Table 3-3 lists some commonly used character data format masks.

Field Type	Format Mask	Data Value	Displayed Value
Social Security	FM999"-"99"-"9999	123456789	123-45-6789
Phone Number	FM"("999") "999"-"9999	1234567890	(123) 456-7890

Table 3-3   Common character format masks

Inserting Values into DATE Columns

You must use format masks when you insert values into DATE columns, because you specify date values as character strings. You then convert the date character string to an internal DATE format in the INSERT command using the TO_DATE function. The general syntax for the TO_DATE function is:

```
TO_DATE('character_string_representing_date',
'date_format_mask')
```

Recall that date format masks, summarized in Table 3-2, can be combined using embedded characters, such as -, /, and :. For example, the character string '08/24/2003' is converted to a DATE data type using the command **TO_DATE('08/24/2003','MM/DD/YYYY')**. To enter the same date using the character string '24-AUG-2003', you would use the command **TO_DATE('24-AUG-2003', 'DD-MON-YYYY')**. Next, you will add records to the STUDENT table, which contains the S_DOB field, which has a DATE data type.

To add the STUDENT records:

    1. Type the command shown in Figure 3-5 to insert the first record into the STUDENT table. The message "1 row created" indicates that one record has been added to the table.

Figure 3-5   Adding a STUDENT record

    2. Type the command shown in Figure 3-6 to add the second record to the STUDENT table. The error message shown in Figure 3-6 appears.

Oracle SQL*Plus
Type this command
File Edit Search Options Help
SQL> INSERT INTO student VALUES
2 (101, 'Umato', 'Brian', 'D', '454 St. John's Street', 'Eau Claire',
3 'WI', '54702', '7155552345', 'SR', TO_DATE('08/19/1982', 'MM/DD/YYYY'), 1230, 1);
ERROR:
ORA-01756: quoted string not properly terminated

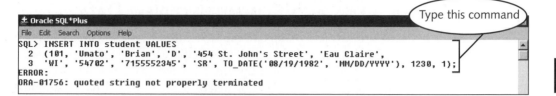

Figure 3-6 Character string termination error

When you entered the command in Figure 3-6, the result was an error message. What happened? Notice that the S_ADD value '454 St. John's Street' has a single quotation mark within the text. When the DBMS reached this single quotation mark, it assumed that this was the end of the S_ADD field, and then it expected a comma. When it found the letter *s* instead, it generated an error. To add text strings with embedded single quotation marks, you need to enter the single quotation mark twice. Next, you will add the correct record.

To enter text that contains a single quotation mark:

1. In Notepad, change the S_ADD text to **'454 St. John''s Street'**, and then copy, paste, and execute the command again. Note that '' is two single quotation marks, not a double quotation mark ("). The message "1 row created" indicates that the record has been added to the table.
2. Insert the record for S_ID 102, as shown in Figure 3-7. Daniel Black does not have a middle initial, so NULL is inserted for the S_MI field because a value must be included in the INSERT statement for every column in the table.

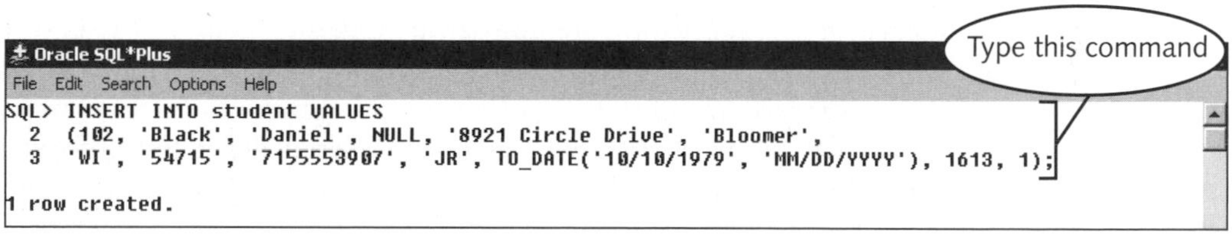

Oracle SQL*Plus
Type this command
File Edit Search Options Help
SQL> INSERT INTO student VALUES
2 (102, 'Black', 'Daniel', NULL, '8921 Circle Drive', 'Bloomer',
3 'WI', '54715', '7155553907', 'JR', TO_DATE('10/10/1979', 'MM/DD/YYYY'), 1613, 1);
1 row created.

Figure 3-7 Adding a STUDENT record with a NULL column value

Remember that the DATE data type stores time values as well as date values. Note that in the COURSE_SECTION table, the TIME field stores values for the times when course sections are scheduled. To convert the 10:00 A.M. start time to a DATE for C_SEC_ID 1000, you would use the command **`TO_DATE('10:00 AM', 'HH:MI AM')`**.

CREATING TRANSACTIONS AND COMMITTING NEW DATA

When you create a new table or update the structure of an existing table, the change is immediately recorded in the database and is immediately visible to other users. However, this is not the case when you insert, update, or delete data records. Recall that commands for operations that add, update, or delete data are called data manipulation language (DML) commands. An Oracle database allows users to execute a series of DML commands as a **transaction**, which is a logical unit of work. When the user has entered all of the commands in a transaction, she or he can either **commit** (save) all of the changes, or roll back (discard) all of the changes. As each DML command is executed, the Oracle DBMS updates the data in the database, and it also records information that enables it to undo the changes made by the command if the user chooses to roll back the transaction that includes the command. This "undo" information is saved until the user commits or rolls back the transaction.

The purpose of transaction processing is to enable every user to see a consistent view of the database. To achieve this, a user cannot view or update data values that are involved in another user's uncommitted transactions, because these pending transactions might be rolled back. Transaction processing is implemented by locking data records that are involved in uncommitted update or delete operations, which prohibits other users from viewing or modifying them. When the transaction is committed or rolled back, the locks on the data records are released, and the changed data values are visible to other users.

An example of a transaction is the entry of a new customer order at Clearwater Traders. The associated CUST_ORDER record must be inserted to record the order ID, date, payment method, customer ID, and order source. Then, one or more ORDER_LINE records must be inserted to record the inventory ID and quantity ordered for each item ordered. If all parts of the transaction are not completed, the database will contain inconsistent data. For example, suppose you successfully insert the CUST_ORDER record, but before you can insert the ORDER_LINE record, your workstation malfunctions. The CUST_ORDER record exists, but has no associated order line information, and the database contains inconsistent data.

SQL*Plus automatically commits all DML changes when you exit the program. However, it is a good idea to explicitly commit your changes often so that the records will be available to other users, and your changes will be saved if you do not exit normally because of a power failure or workstation malfunction. Now you will commit the DML changes that you have made during your current SQL*Plus session using the COMMIT command.

To commit your changes:

1. At the SQL prompt, type **COMMIT;**. The message "Commit complete" indicates that your changes have been made permanent, and are now visible to other users.

The rollback process enables users to return the database to its original state by discarding all of the changes made by DML commands executed since the last commit was executed. An example of a rolled back transaction is shown in Figure 3-8. A new record is inserted into the Clearwater Traders CUST_ORDER table, and then the user issues the ROLL-BACK command. The final SELECT statement shows that the record was not inserted into the CUST_ORDER table and that the INSERT command was effectively rolled back.

```
Oracle SQL*Plus                                                           _ 8 X
File  Edit  Search  Options  Help
SQL> INSERT INTO cust_order VALUES                         ⎫ Command to insert
  2  (1063, TO_DATE('06/02/2003', 'MM/DD/YYYY'), 'CC', 154, 6); ⎭    Order 1063

1 row created.

SQL> ROLLBACK;─────────  Transaction
                          rolled back
Rollback complete.

SQL> SELECT * FROM cust_order;

  ORDERID ORDERDATE METHPMT      CUSTID ORDERSOURCEID
--------- --------- -------   --------- -------------
     1057 29-MAY-03 CC           107          2
     1058 29-MAY-03 CC           232          6
     1059 31-MAY-03 CHECK        133          2   ⎫ Order 1063 is
     1060 31-MAY-03 CC           154          3   ⎬   not listed
     1061 01-JUN-01 CC           179          6   ⎭
     1062 01-JUN-01 CC           179          3

6 rows selected.
```

Figure 3-8   Rolling back a transaction

 You will learn more about the SELECT command later in the chapter, but for now, note that to view the current data in a table, you type SELECT *column1*, *column2*, … FROM *tablename* or SELECT * FROM *tablename*. The * is a wildcard character that specifies to display the values of all table columns.

You can use rollbacks with **savepoints** that mark the beginning of individual sections of a transaction. By using savepoints, you can roll back part of a transaction. Figure 3-9 shows how to create savepoints, and then roll back a SQL*Plus session to an intermediate savepoint. A savepoint named ORDER_SAVE is created. Then, a record is inserted for CUST_ORDER 1063. Next, the ORDER_LINE_SAVE savepoint is created, and an ORDER_LINE record for CUST_ORDER 1063 is inserted. Finally, a rollback command to the ORDER_LINE_SAVE savepoint is issued.

Figure 3-10 shows the effect of rolling back to the second savepoint, ORDER_LINE_SAVE. The record inserted into the ORDER_LINE table (which was inserted after the ORDER_LINE_SAVE savepoint was created) was rolled back, while the record for Order 1063 in the CUST_ORDER table (which was inserted by the first command in the transaction) still exists. If the command ROLLBACK TO order_save; is issued, then the record for Order 1063 would be rolled back also.

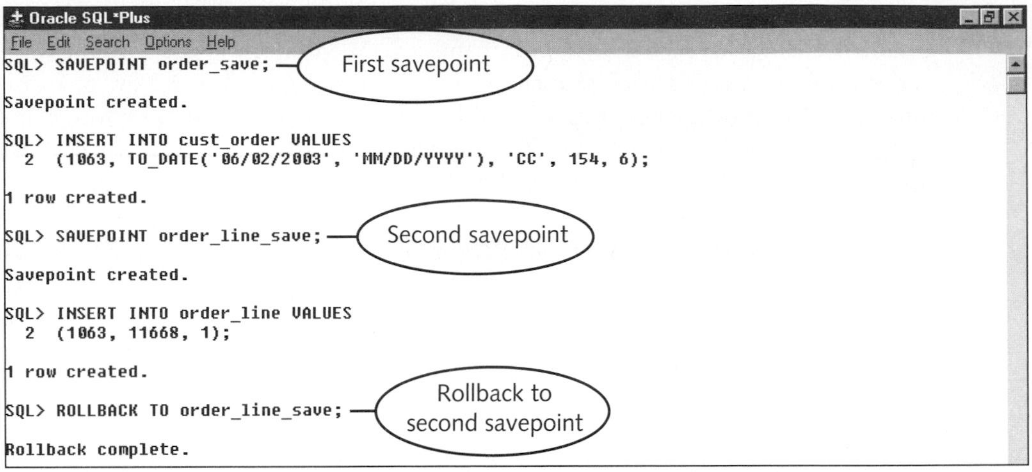

Figure 3-9   Using the ROLLBACK command with savepoints

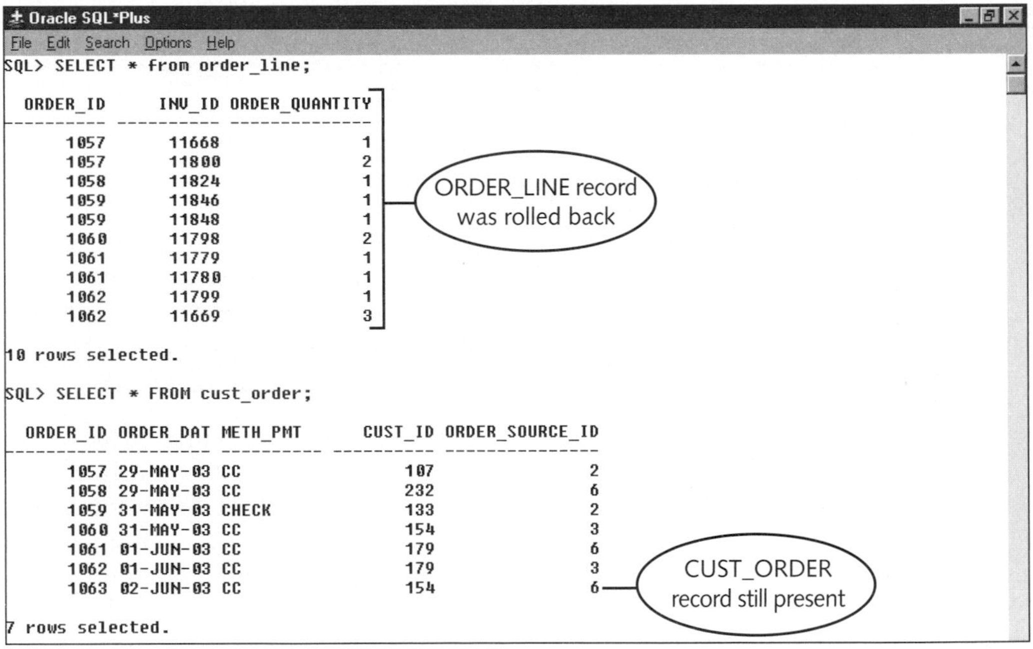

Figure 3-10   Result of rolling back to ORDER_LINE_SAVE savepoint

3

UPDATING EXISTING TABLE RECORDS

An important data maintenance operation is updating existing data records. Student addresses and telephone numbers change often, and every year students (hopefully) move up to the next class. In the UPDATE command, you specify the name of the table to be updated and list the name of the column (or columns) to be updated, along with the new data value (or values). You also specify a search condition to identify the row to be updated. The general syntax of the UPDATE statement is:

```
UPDATE tablename
SET column1 = new_value1, column2 = new_value2, …
WHERE search_condition;
```

You can update multiple columns using a single UPDATE command, but you can update records in only one table at a time. Search conditions are specified in the WHERE clause to make the command update specific records.

The general syntax for a search condition is **WHERE fieldname comparison_operator search_expression**. Every search condition must be able to be evaluated as either TRUE or FALSE. Search expressions are constants, like the number 10 or the character string 'SR'. Search conditions can be used to compare numerical, date, and character values. Search conditions for character values are case sensitive, so character expressions must be enclosed in single quotation marks. Table 3-4 lists common comparison operators used in SQL expressions.

Operator	Description	Example
=	Equal to	s_class = 'SR'
>	Greater than	capacity > 50
<	Less than	capacity < 100
>=	Greater than or equal to	max_enrl >= 35
<=	Less than or equal to	max_enrl <= 30
< > != ^=	Not equal to	status <> 'CLOSED' status != 'CLOSED' status ^= 'CLOSED'
LIKE	Uses pattern matching in text strings	term_desc LIKE 'Summer%'. Is usually used with the wildcard character (%), which indicates that part of the string can contain any characters. Is case sensitive

Table 3-4   Common search condition comparison operators

Operator	Description	Example
BETWEEN	Range comparison to find values that fall between specific inputs, inclusive of the inputs	`capacity BETWEEN 100 AND 150 (equivalent to capacity >= 100 AND capacity <= 150)`
IN	Determines if a value is a member of a specific search set	`s_class IN ('FR', 'SO')`
NOT IN	Determines if a value is not a member of a specific search set	`s_class NOT IN ('FR', 'SO')`

Table 3-4   Common search condition comparison operators (continued)

Oracle pads CHAR values with blank spaces if an entered data value does not fill all of the declared variable's size. For example, suppose you want to declare S_CLASS as a CHAR field with size 2 and then enter a data value of 'U' (for unclassified). To search for this value in the WHERE clause, you would have to type S_CLASS = 'U' (with a blank space after the U).

Don't forget that values enclosed within single quotation marks are case sensitive. If you type S_CLASS = 'sr', you will not retrieve rows in which the S_CLASS value is 'SR'.

Now you will update student Daniel Black's S_CLASS value from 'JR' to 'SR'. You will use his S_ID value (102) as the search condition.

To use the UPDATE command:

1. Type the command shown in Figure 3-11 to update student Daniel Black's S_CLASS value. The "1 row updated" message indicates that the row was updated.

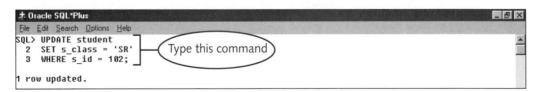

Figure 3-11   Updating a single field in a record

2. You can also update multiple fields in a record using a single UPDATE command. Type the query shown in Figure 3-12 to change Daniel Black's S_CLASS back to JR and update his F_ID value to 2.

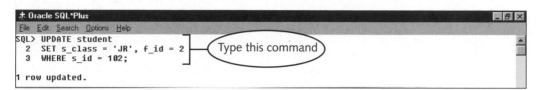

Figure 3-12   Updating multiple fields in a record

You can update multiple records in a table using a single UPDATE command by specifying a search condition that matches multiple records.

You can also combine multiple search conditions using the AND and OR operators. When you use the **AND** operator to connect two search conditions, both conditions must be true for the row to satisfy the search conditions. If no rows exist that match *both* conditions, then no matching records are found. For example, the following search condition would find all records in which the BLDG_CODE is 'CR', and the capacity is greater than 50:

```
WHERE bldg_code = 'CR' AND capacity > 50
```

When you use the **OR** operator to connect two search conditions, only one of the conditions must be true for the row to satisfy the search conditions. For example, the following search condition would find all records for the course sections that meet either on Tuesday and Thursday, or on Monday, Wednesday, and Friday:

```
WHERE day = 'TTh' OR day = 'MWF'
```

You can also use the **NOT** operator to indicate the logical opposite of a search condition that is enclosed in parentheses. For example, the following search condition would find all rows in the STUDENT table in which the S_CLASS field has any value other than 'FR':

```
WHERE NOT (s_class = 'FR')
```

If you omit the search condition in the UPDATE command, then all table rows are updated. For example, the following command would update the value of S_CLASS to 'SR' for all records in the STUDENT table:

```
UPDATE student
SET s_class = 'SR';
```

DELETING RECORDS AND TRUNCATING TABLES

Another table maintenance operation is deleting records. Deleting records is a dangerous business, because if the values are accidentally deleted, and the deletion is committed, the deleted records are difficult or impossible to recover. Therefore, the privilege to delete records from tables should be granted to users with a good deal of caution and exercised with caution when granted.

The general syntax for the DELETE command is:

```
DELETE FROM tablename
WHERE search_condition;
```

Always include a WHERE clause when deleting a record from a table to ensure that the correct record is deleted. If the WHERE clause is omitted, *all* table records are deleted. Now, you will delete student ID 102 (Daniel Black) from the STUDENT table.

To delete a selected record from a table:

1. Type the command shown in Figure 3-13 to delete Daniel Black's record from the STUDENT table. The message "1 row deleted" appears.

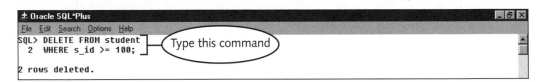

Figure 3-13   Deleting a single record

2. Type the command shown in Figure 3-14 to delete records in the STUDENT table, where S_ID is greater than or equal to 100. The message "2 rows deleted" appears, indicating you have deleted the records for students Sarah Miller and Brian Umato.

Figure 3-14   Deleting multiple records

You can delete multiple records from a table at one time if your search condition matches several rows. You can delete every row in a table by omitting the WHERE clause.

You cannot delete a record if it contains a value that is referenced as a foreign key by another record. For example, you cannot delete the record for LOC_ID 53 (Kim Cox's office) unless you delete the record in which the foreign key value exists (F_ID 1, faculty member Kim Cox) first. Now you will attempt to delete a record that contains a field that is a foreign key reference to see the error message that appears.

To try to delete a record with a foreign key:

    1. Type the command shown in Figure 3-15 to delete LOC_ID 53 (Kim Cox's office) from the LOCATION table. Note the error message that appears.

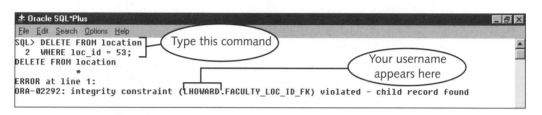

Figure 3-15   Deleting a record with a foreign key reference

Recall that if you omit the search condition in the DELETE command, *all* table records are deleted. Now, you will delete all of the records you inserted into the FACULTY table by omitting the search condition.

To delete all of the FACULTY records:

    1. Type the command shown in Figure 3-16 to delete all of the records in the FACULTY table. The message "2 rows deleted" should appear as shown.

Figure 3-16   Deleting all of the records from the FACULTY table

Recall that the DELETE command is a DML statement. After a DELETE command is issued, you must commit the transaction before the deletion is made permanent. Whenever a record is deleted, a process in the Oracle DBMS records rollback information, which is used to restore the record if the user chooses to roll back the transaction. For a DELETE command, rollback information includes all of the deleted data values. As a result, deleting large volumes of records can take a long time, because the DBMS must make a copy of each deleted value.

As an alternative, you can **truncate** the table, which removes all of the table data without saving any rollback information. The space in the database that was used to store the table data is reclaimed by the database and can be used to store other data. When you truncate a table, the table structure and constraints remain intact. As a result, truncation is a high-speed delete operation that does not save a backup copy of the data.

The general syntax for the TRUNCATE command is:

    TRUNCATE TABLE *tablename*;

The TRUNCATE command is implicitly committed as soon as it executes, so you do not have to type COMMIT. As a result, however, it cannot be rolled back. You cannot truncate a table if it has fields that are defined as foreign key constraints, even if no records exist that reference any fields in the table. For example, the LOC_ID field in the LOCATION table is a foreign key in both the FACULTY table and the COURSE_SECTION table. Even though no records currently exist in the FACULTY or COURSE_SECTION tables, you still cannot truncate the LOCATION table. Now you will attempt to truncate the LOCATION table to see what happens.

To truncate the LOCATION table:

1. Type the command shown in Figure 3-17 to truncate the LOCATION table. An error message indicating that the table contains foreign key references appears.

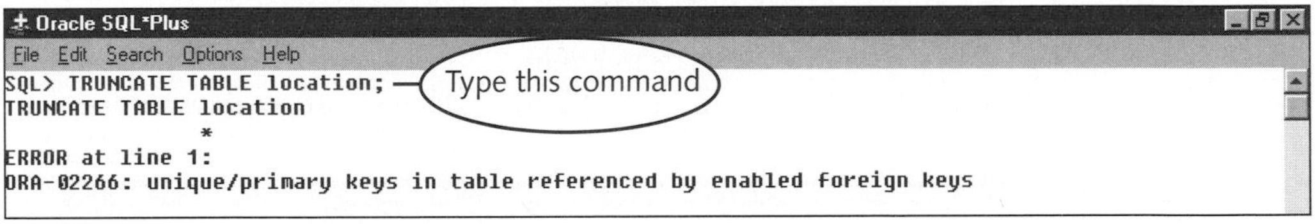

Figure 3-17   Truncating a table with foreign key references

Because many tables contain foreign key references, the usefulness of the TRUNCATE command is limited. To get around this limitation, you can **disable** the table's foreign key constraints before you truncate it. A constraint can be in one of two states: enabled or disabled. When a constraint is **enabled**, the DBMS checks to make sure that new or updated data values obey the rules of the constraint. When a constraint is disabled, the constraint still exists, but the DBMS does not perform this checking.

The general syntax for disabling a constraint is:

    ALTER *tablename* DISABLE CONSTRAINT *constraint_name*;

In the command, *tablename* is the name of table in which the constraint was created, and *constraint_name* is the name that you gave to the constraint in the CREATE TABLE command. The general syntax for enabling a constraint is:

```
ALTER tablename ENABLE CONSTRAINT constraint_name;
```

Now you will disable the foreign key constraints in the FACULTY and COURSE_SECTION tables and then truncate the LOCATION table.

To disable the foreign key constraints and then truncate the LOCATION table:

1. Type the commands shown in Figure 3-18 to disable the foreign key constraints on the LOC_ID fields in the FACULTY and COURSE_SECTION tables.

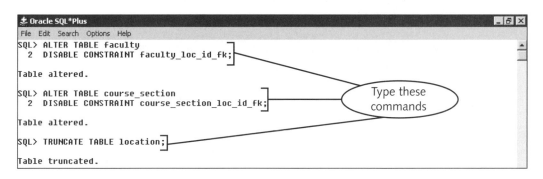

Figure 3-18   Disabling foreign key constraints

2. Type the command **TRUNCATE TABLE location;** at the SQL prompt to truncate the LOCATION table. The message "Table truncated" appears, indicating that the truncation was successful.

SEQUENCES

Sequences are database objects that automatically generate sequential lists of numbers. Sequences are useful for creating unique surrogate key values for primary key fields when no field exists that is suitable to use as the primary key. For example, you might assign a sequence for a customer ID field so that each record is unique. Fields using sequences must have a NUMBER data type. Sequences can be created and manipulated using SQL*Plus.

Creating New Sequences

The general command for creating a new sequence is as follows. Optional parameters are enclosed in square brackets, and parameters in which only one of the two values shown would be used are separated by a bar (|).

```
CREATE SEQUENCE sequence_name
[INCREMENT BY number]
[START WITH start_value]
[MAXVALUE maximum_value] | [NOMAXVALUE]
[MINVALUE minimum_value] | [NOMINVALUE]
[CYCLE] | [NOCYCLE]
[CACHE number_of_sequence_values_to_cache] | [NOCACHE]
[ORDER] | [NOORDER];
```

Every sequence must have a unique name and must follow the Oracle naming standard presented in Chapter 2. The CREATE SEQUENCE command is the only command required to create a sequence; the rest of the commands are optional. Table 3-5 describes these optional commands and shows their default value if the parameter specification is omitted. The CREATE SEQUENCE command is a DDL command. Recall that DDL commands do not need to be explicitly saved, because an implicit COMMIT operation is performed when you execute the DDL command. Therefore, you do not need to commit the transaction to save the sequence.

Parameter	Description	Default Value
INCREMENT BY	Specifies the value by which the sequence is incremented	INCREMENT BY 1
START WITH	Specifies the sequence start value. You must specify a start value if you already have data in the fields where the sequence will be used to generate primary key values. For example, in the ITEM table, the highest existing ITEM_ID is 995, so you would start the sequence with 996.	START WITH 1
MAXVALUE	Specifies the maximum value to which you can increment the sequence	NOMAXVALUE, which allows the sequence to be incremented to 1 * 1027
MINVALUE	Specifies the minimum value of the sequence for a decrementing sequence	NOMINVALUE
CYCLE	Specifies that, when the sequence reaches its MAXVALUE, it cycles back and starts again at the MINVALUE. For example, if you specify a sequence with a maximum value of 10 and a minimum value of 5, the sequence will increment up to 10 and then start again at 5	NOCYCLE, which specifies that the sequence will continue to generate values until it reaches its MAXVALUE and will not cycle

Table 3-5   Optional sequence parameters

Parameter	Description	Default Value
CACHE	Specifies that, whenever you request a sequence value, the server automatically generates several sequence values and stores them in a server memory area called a cache, to improve system performance. The default number of sequence values stored in the cache is 20. To specify a different number of sequence values to be cached, use the CACHE command along with the number of sequence values you want to generate and store.	20 sequence numbers are cached
NOCACHE	Directs the server not to cache any sequence values	
ORDER	Ensures that the sequence numbers are granted in the exact chronological order in which they are requested. For example, the first user who requests a number from the sequence would be granted sequence number 1, the second user who requests a number from the sequence would be granted sequence number 2, and so forth. This is useful for tracking the order in which the sequence numbers were requested	NOORDER, which specifies that the sequence values are not necessarily granted in chronological order; therefore, although users are guaranteed to get unique sequence numbers, the order of the values might not correspond with the order in which specific requests occur

Table 3-5   Optional sequence parameters (continued)

Next, you will create a sequence named LOC_ID_SEQUENCE for the LOC_ID field in the LOCATION table. You will specify that the sequence will start with 45, since that is the first LOC_ID value listed in the sample data. You will accept the defaults for the rest of the parameters: the sequence will increment by 1, the sequence will not CYCLE, you will not have the server cache sequence values, and the sequence will not have a maximum value.

To create the LOC_ID_SEQUENCE:

1. Type the SQL command shown in Figure 3-19. The confirmation message "Sequence created" indicates that the LOC_ID_SEQUENCE sequence was created.

Figure 3-19   Creating a sequence

Viewing Sequence Information Using the Data Dictionary Views

Sometimes you might need to review the names and properties of your sequences after you create them. You can do this by querying the USER_SEQUENCES data dictionary view. (Recall that the data dictionary views contain information about the database structure.) Now you will review the properties of your sequences to confirm that the LOC_ID_SEQUENCE was successfully created.

To review the properties of your sequences:

1. Type the command shown in Figure 3-20. This view has a field named SEQUENCE_NAME, which displays the names of all of the sequences in your user schema. The query output lists your current sequence names and properties, including the minimum and maximum allowable values that were established when you created the sequence and the value by which the sequence increments. The CYCLE and ORDER parameter values of "N" indicate that cycling and ordering are not used. The cache size is shown as 20, because you accepted the default cache value of 20 sequence values.

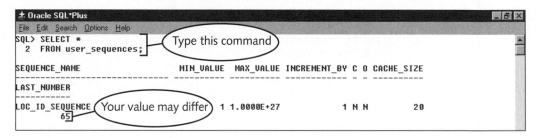

Figure 3-20   Current dequence properties

 Your query output might wrap to two lines. You will learn how to customize the SQL*Plus environment to format query output later in this chapter.

Pseudocolumns

To use sequences to insert data values, you must understand pseudocolumns. A **pseudo-column** is an object that acts like a column but is actually a command that returns a specific value. You can use pseudocolumns within SQL commands in place of column names to perform functions such as retrieving the current system date, the name of the current database user, or the next value in a sequence. Table 3-6 summarizes the names of commonly used pseudocolumns and their output.

Pseudocolumn Name	Output
CURRVAL	Most recent sequence value retrieved during the current user session
LEVEL	Current data level in table with hierarchical relationships
NEXTVAL	Next available sequence value
SYSDATE	Current system date from database server
USER	Username of current database user

Table 3-6   Examples of Oracle pseudocolumns

Many of the values retrieved by pseudocolumns are independent of a specific database table, so you can use a pseudocolumn name in a SELECT command along with any table name that you have privileges to select data from. For example, you could retrieve the current system date by selecting the SYSDATE pseudocolumn from the LOCATION table in the Northwoods University database, using the following command: `SELECT SYSDATE FROM location;`. There is no column named SYSDATE in the LOCATION table, but Oracle returns the current system date because it recognizes SYSDATE as a pseudocolumn.

In a SELECT command, pseudocolumn values are often retrieved from the DUAL system database table, using the following syntax: `SELECT pseudocolumn FROM DUAL;`. **DUAL** is a simple table that belongs to the SYSTEM database user. It has only one row and one column that contains the value 'X'. All database users can use DUAL in SELECT commands, but they cannot modify or delete it. Whenever you perform a SELECT command using a pseudocolumn on this small, simple table, a value is returned for each table row. It is more efficient to retrieve pseudocolumns from DUAL, which only has one record, than from other tables that usually contain many records.

Using Sequences in the INSERT Command

To retrieve the next value in a sequence, you use the NEXTVAL pseudocolumn.

The following syntax accesses the next sequence value using the NEXTVAL pseudocolumn and inserts it into a new record:

```
INSERT INTO tablename
VALUES(sequence_name.NEXTVAL, column1_value,
column2_value,…);
```

This command assumes that the primary key associated with the sequence is the first table field.

Next, you will insert the first record into the LOCATION table using the LOC_ID_SEQUENCE. Recall that you truncated the LOCATION table in the previous set of steps, so there are currently no records in the LOCATION table.

To insert a new LOCATION record using a sequence:

1. Type the SQL command shown in Figure 3-21. The confirmation message "1 row created" indicates that you inserted one record.

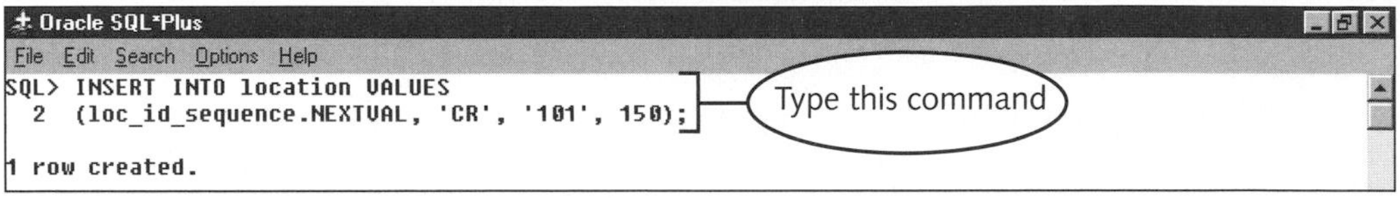

Figure 3-21 Inserting a new record using a sequence

2. To see what value was inserted for LOC_ID, type the following command: **SELECT * FROM location;**. This command allows you to view all of the records in the LOCATION table. The new record has LOC_ID 45, which was the first value in the sequence, along with the other values you specified in the INSERT command.

3. To insert the next LOCATION record using the sequence and then view the inserted record to confirm that the sequence increments correctly, type the SQL commands shown in Figure 3-22. The records for LOC_ID 45 and LOC_ID 46 appear, confirming that the sequence is incrementing correctly.

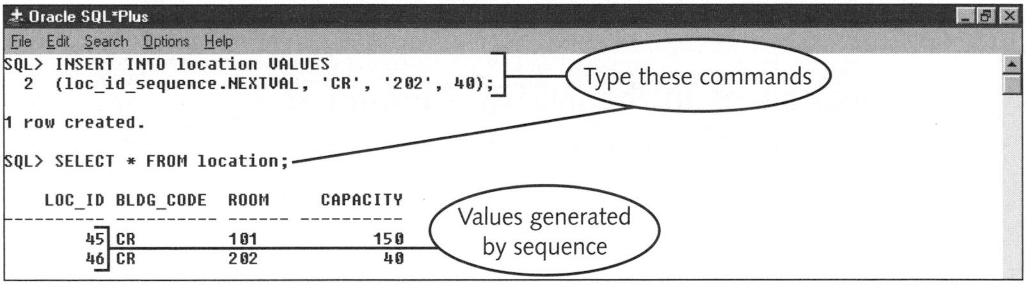

Figure 3-22 Inserting another record and viewing the sequence-generated values

Accessing Sequence Values

Sometimes you need to access the next value of a sequence, but you don't want to insert a new record. For example, suppose you want to create a new customer order, display the ORDER_ID on the order form, and have the user enter more information before you actually insert the new record. To do this, you use the NEXTVAL pseudocolumn and create a SELECT command using the DUAL system database table as follows: **SELECT *sequence_name*.NEXTVAL FROM DUAL;**.

To view the current value of the sequence, you use the CURRVAL pseudocolumn, and enter the command **SELECT *sequence_name*.CURRVAL FROM DUAL;**.

Now you will use these commands to increment the LOC_ID_SEQUENCE to its next value, and then view the current value.

To use the SELECT command to access sequence values:

1. Type the first SQL command shown in Figure 3-23. Your query result should show 47 as the NEXTVAL in the LOC_ID_SEQUENCE. Now that it has been accessed, 47 is the current value of the sequence, and the next value will be 48.

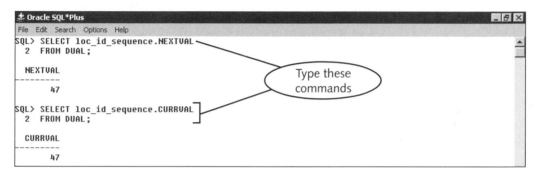

Figure 3-23 Viewing the next and curent sequence values

2. Now use CURRVAL to display the current value. Type the second SQL command shown in Figure 3-23 to confirm that 47 is now the current value of the LOC_ID_SEQUENCE.

Once you move on to the next sequence value, the previous sequence value cannot be accessed using the sequence commands. This restriction prevents you (or anyone else) from accidentally using the same sequence value as the primary key for two different records. The CURRVAL command can be used only in the same database user session and immediately after using the NEXTVAL command. This restriction prevents two database user sessions from assigning the same sequence value to two different records.

 A user session is created when a user starts a client application like SQL*Plus and logs onto the database and is terminated when the user exits the application.

Next, you will confirm the CURRVAL database session restriction by exiting SQL*Plus, starting it again, and then trying to use the CURRVAL command.

To examine the CURRVAL database session restriction:

1. Type **SELECT loc_id_sequence.NEXTVAL FROM DUAL;** at the SQL prompt. The value 48 should appear as the next value of **loc_id_sequence**.

2. Exit SQL*Plus.

3. Start SQL*Plus again, and log onto the database.

4. Type **SELECT loc_id_sequence.CURRVAL FROM DUAL;** at the SQL prompt. The error message "Error at line 2: ORA-08002: sequence LOC_ID_SEQUENCE.CURRVAL is not yet defined in this session" appears. This indicates that no CURRVAL exists for the sequence because you exited SQL*Plus and started a new SQL*Plus session but have not yet retrieved a sequence value in this session. A CURRVAL can be selected from a sequence only after selecting a NEXTVAL.

Dropping Sequences

To drop a sequence from the database, you use the DROP SEQUENCE command. For example, to drop the LOC_ID_SEQUENCE, you would use the command **DROP SEQUENCE loc_id_sequence;**. Do not drop the LOC_ID_SEQUENCE right now, because you will need to use it for other exercises in the tutorial.

OBJECT PRIVILEGES

When you create database objects, such as tables or sequences, other users cannot access or modify your objects unless you give them explicit privileges to do so. Table 3-7 lists some common object privileges for tables and sequences, and their descriptions.

Object Type(s)	Privilege	Description
Table, Sequence	ALTER	Allows the user to change the object's structure using the ALTER command
Table, Sequence	DROP	Allows the user to drop the object
Table, Sequence	SELECT	Allows the user to view table records or select sequence values
Table	INSERT	Allows the user to insert new records into the table
Table	UPDATE	Allows the user to update table records
Table	DELETE	Allows the user to delete table records
Any object	ALL	Allows the user to perform all possible operations on the object

Table 3-7   Common object privileges

Granting Object Privileges to Other Users

Object privileges are granted to other users using the SQL GRANT command. The general syntax for the SQL GRANT command is:

```
GRANT privilege1, privilege2, …
ON object_name
TO user1, user2, …;
```

Note that you can grant privileges for only one object at a time, but you can grant privileges to many users at once. If you want to grant privileges to every database user, you can use the word PUBLIC in the TO clause. Now, you will grant object privileges to other database users.

To grant privileges:

1. Type the first command shown in Figure 3-24 to grant SELECT and ALTER privileges on your STUDENT table to two other students in your class by substituting the students' Oracle usernames for MORRISJP and MORRISCM in the query. When your query executes successfully, the confirmation message "Grant succeeded" appears.

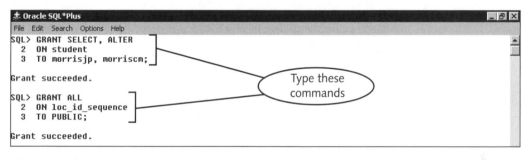

Figure 3-24   Granting object privileges

2. Type the second command shown in Figure 3-24 to grant all privileges on your sequence named LOC_ID_SEQUENCE to all database users.

Revoking Table Privileges

To cancel a user's object privileges, you use the REVOKE command. The general syntax for the REVOKE command is:

```
REVOKE privilege1, privilege2, …
ON object_name
FROM user1, user2, …;
```

Now, you will revoke the privileges on the STUDENT table that you granted in the previous set of steps.

To revoke privileges:

1. Type the command shown in Figure 3-25 to revoke the SELECT and ALTER privileges that you granted to your two classmates on the STUDENT table. Remember that you will have to change the user names from MORRISJP and MORRISCM to the names of the users to whom you granted these privileges in Figure 3-24.

```
± Oracle SQL*Plus                                                        _ |&| X
File  Edit  Search  Options  Help
SQL> REVOKE SELECT, ALTER
  2   ON student                    ( Type this command )
  3   FROM morrisjp, morriscm;

Revoke succeeded.
```

Figure 3-25   Revoking the SELECT and ALTER privileges

2. Exit SQL*Plus.

3. Exit Notepad, and save the changes you made to 3AQueries.sql.

SUMMARY

❑ Database developers need to understand the SQL commands for selecting, inserting, updating, and deleting data, because they use these commands extensively when developing Oracle database applications.

❑ A script is a text file with an .sql extension that contains a series of SQL commands for creating or modifying database objects or manipulating data records.

❑ The INSERT command is used to insert new data records into a table. The INSERT command can be used to insert all of the data fields into a record, or to insert only specific fields.

❑ Before you can add a new data record, you must ensure that all records containing foreign key values that it references have already been added.

❑ Character fields in Oracle are case sensitive, and character values must be enclosed within single quotation marks when they are inserted.

❑ To add text strings with embedded single quotation marks, type the single quotation mark twice.

❑ DATE fields are inserted as characters and then converted to dates using the TO_DATE function. DATE fields also store time values.

❑ To insert a LOB data field, you initially insert a locator, which contains information about the LOB type and points to the memory location in the database where the LOB will eventually be stored.

□ A format mask is a text string that specifies how a data value appears to the user.

□ A transaction is one or more related SQL commands that constitute a logical unit of work. All parts of the transaction must be completed, or the database will contain inconsistent data.

□ It is a good idea to commit your changes often (just like saving work in other applications) so that new or modified records will be available to other users and so that your changes will be saved if you do not exit normally because of a power failure or workstation malfunction.

□ A rollback enables you to return the database to its original state by undoing the effects of all the commands since the last commit.

□ A savepoint marks the beginnings of individual sections of a transaction. You can roll back transactions to specific savepoints.

□ You can update records in only one table at a time using a single UPDATE command. You can update multiple records in a table using a single UPDATE command by specifying a search condition that matches multiple records.

□ Search conditions must be able to be evaluated as either TRUE or FALSE. Search conditions can be combined using the AND and OR operators.

□ Data records should be deleted with caution, because records deleted in error are difficult to recover.

□ You cannot delete a record if one of its fields is a foreign key reference to another record.

□ Table constraints can be enabled or disabled. When a constraint is enabled, the DBMS ensures that new or updated data values obey the constraint rules. When a constraint is disabled, it still exists, but the DBMS does not enforce the constraint rules.

□ You can delete a lot of records quickly by truncating a table, which removes all of the table data and does not store any rollback information. You cannot truncate a table that contains enabled foreign key constraints.

□ A sequence is a database object that automatically generates sequential numbers that are used for surrogate key values. Sequence values can be retrieved using either the INSERT command or the SELECT command.

□ When you create a database object, other database users cannot manipulate the object unless you give them explicit privileges to do so. Object privileges are granted using the GRANT command and revoked using the REVOKE command.

REVIEW QUESTIONS

1. What is the purpose of creating a script?

2. In an INSERT command when you are inserting a data value for every table field, how do you determine the order to list the data values in the VALUES clause?

3. Write the INSERT command to insert C_SEC_ID 1000 (see Figure 1-12) into the COURSE_SECTION table of the Northwoods University database. What other records must be inserted prior to running this command?

4. Write the TO_DATE function that converts the value 09/16/04 to a date using the format DD-MON-YYYY.

5. Write the TO_DATE function to convert the time 10:45:07 P.M. so it appears as 22:45:07.

6. Write the format mask to display the data value –2897.23 as <$2,897.23>.

7. List the SQL commands that enable transaction processing.

8. The purpose of transaction processing is to ensure that every user has a consistent view of the database. Why is this important?

9. What is the difference between deleting all of the records in a table and truncating the table?

10. Why can the CURRVAL command only be used immediately after using the NEXTVAL command?

11. When do you have to assign object privileges to other database users?

12. What does the keyword PUBLIC signify in relation to object privileges?

PROBLEM-SOLVING CASES

For all cases, use Notepad or another text editor to write a script using the specified file-name. Place the commands in the order listed, and save the script files in your Chapter3\CaseSolutions folder.

Cases 1, 2, and 3 involve executing DML commands using the Northwoods University database. Before you start Case 1, 2, or 3, run the emptynorthwoods.sql script in the Chapter3 folder on your Data Disk to re-create the sample database tables with no data in them.

1. Create a script named 3ACase1.sql that contains commands to perform the following actions:

   a. Insert C_SEC_ID 1000 into the COURSE_SECTION table of the Northwoods University database. Use the data values shown in Figure 1-12.

   b. Insert all records in other tables that have values that are referenced as foreign keys in this command. Be sure to insert the records in the correct order so the script successfully inserts the records when it is run.

   c. Grant the SELECT privilege on all tables modified by the script to all database users. (You will have to write a separate GRANT command for each table.)

2. Create a script named 3ACase2.sql that contains commands to perform the following actions:

   a. Create a sequence named COURSE_ID_SEQUENCE that automatically generates primary key values for the COURSE table in the Northwoods University database.

   b. Insert the first two records in the COURSE table, using the values shown in Figure 1-12. Use the form of the INSERT command that only inserts selected fields, and only insert values for the COURSE_ID and CALL_ID fields.

   c. Remove the COURSE_ID_SEQUENCE from the database.

3. Create a script named 3ACase3.sql that contains commands to perform the following actions:

   a. Create a savepoint named INSERT_STUDENTS.

   b. Insert the first four records into the STUDENT table of the Northwoods University database, using the data values shown in Figure 1-12.

   c. Insert all records in other tables that have values referenced as foreign keys in the records you insert into the STUDENT table. Be sure to insert the records in the correct order so the script runs correctly.

   d. Create a savepoint named UPDATE_CLASS, and then write commands to update the S_CLASS field in each record as follows:

Current S_CLASS value	Updated S_CLASS value
FR	SO
SO	JR
JR	SR

   Do not update any record more than once.

   e. Roll back to the UPDATE_CLASS savepoint, and then roll back to the INSERT_STUDENT savepoint.

Cases 4, 5, 6, and 7 involve executing DML commands using the Clearwater Traders database. Before you start any of these cases, run the emptyclearwater.sql script in the Chapter3 folder on your Data Disk to re-create the sample database tables with no data in them.

4. Create a script named 3ACase4.sql to perform the following actions:

   a. Create a sequence named ORDER_ID_SEQUENCE that generates primary key values for the ORDER_ID field in the CUST_ORDER table in the Clearwater Traders database. Specify that the sequence should start with the value 1057.

   b. Grant the SELECT privilege to all database users for the ORDER_ID_SEQUENCE.

   c. Insert the first record into the CUST_ORDER table, using the data values shown in Figure 1-11 and retrieving the ORDER_ID value from the sequence.

d. Insert all records in other tables that have values that are referenced as foreign keys in the first record in the CUST_ORDER table, using the data values shown in Figure 1-11. Be sure to insert the records in the correct order so the script successfully executes.

f. Remove the ORDER_ID_SEQUENCE.

5. Write a script named 3ACase5.sql to perform the following actions:

a. Insert the first three records into the ITEM table of the Clearwater Traders database, using the data shown in Figure 1-11. Insert the values for the ITEM_IMAGE field as BLOB locators.

b. Insert all records in other tables that have values referenced as foreign keys in the records inserted into the ITEM table, using the data values shown in Figure 1-11. Be sure to insert the records in the correct order so the script executes correctly.

c. Update the ITEM_DESC value for ITEM_ID 995 to 'Kid's Beachcomber Sandals'.

6. Create a script named 3ACase6.sql to perform the following actions:

a. Create a savepoint named INSERT_INVENTORY.

b. Insert the first record into the INVENTORY table, using the data values shown in Figure 1-11.

c. Insert all records in other tables that have values that are referenced as foreign keys, using the data values shown in Figure 1-11. Be sure to insert the records in the correct order so the script successfully executes.

d. Create a second savepoint named UPDATE_COLOR, and then write the command to update the color of INV_ID 11668 to 'Bright Pink'.

e. Roll back to the UPDATE_COLOR savepoint, and then roll back to the INSERT_INVENTORY savepoint.

7. Create a script named 3ACase7.sql to perform the following actions:

a. Insert the first two records into the ORDER_LINE table of the Clearwater Traders database, using the data values shown in Figure 1-11.

b. Insert all records in other tables that have values referenced as foreign keys in the records inserted into the ORDER_LINE table, using the data values shown in Figure 1-11. Be sure to insert the records in the correct order so the script successfully executes.

c. Delete all of the records that you have just inserted.

Cases 8, 9, and 10 involve executing DML commands using the Software Experts database. Before you start Case 8, 9, or 10, run the emptysoftwareexp.sql script in the Chapter3 folder on your Data Disk to re-create the sample database tables with no data in them.

8. Create a script named 3ACase8.sql to perform the following actions:

   a. Create a sequence named SKILL_ID_SEQUENCE to generate primary key values for the SKILL_ID field in the SKILL table.

   b. Insert the first three records into the SKILL table of the Software Experts database, using the data values shown in Figure 1-13. Use the SKILL_ID_SEQUENCE to generate the primary key values.

   c. Update the SKILL_DESCRIPTION field of SKILL_ID 3 to 'J++ Programming'.

   d. Grant the SELECT, INSERT, and UPDATE privileges on the SKILL table to all database users.

   e. Remove the SKILL_ID_SEQUENCE from the database.

9. Create a script named 3ACase9.sql to perform the following actions:

   a. Insert the first record into the CONSULTANT_SKILL table of the Software Experts database, using the data shown in Figure 1-13. Use the form of the INSERT command that only inserts selected field values, and only insert values for the C_ID and SKILL_ID fields.

   b. Insert all records in other tables that have values referenced as foreign keys in the record you inserted into the CONSULTANT_SKILL table, using the data values shown in Figure 1-13. Be sure to insert the records in the correct order so the script successfully executes.

   c. Delete all of the records that you just inserted.

10. Write a script named 3ACase10.sql to perform the following actions:

   a. Create a sequence named P_ID_SEQUENCE to automatically generate surrogate key values for the PROJECT table.

   b. Insert the first two records into the PROJECT table of the Software Experts database. Use the data values shown in Figure 1-13, and use the P_ID_SEQUENCE to generate the primary key values.

   c. Insert all records in other tables that have values referenced as foreign keys in the records inserted into the PROJECT table, using the data values shown in Figure 1-13. Be sure to insert the records in the correct order so the script successfully executes.

   d. Write a single UPDATE command to update the CLIENT_ID field in both PROJECT records to CLIENT_ID 1.

   e. Remove the P_ID_SEQUENCE from the database.

◄ LESSON B ►

Objectives

♦ Write SQL queries to retrieve data records from a single database table using exact and inexact searches
♦ Sort and group query output
♦ Use mathematical calculations, number functions, and date arithmetic in queries
♦ Create queries with input variables
♦ Format query output in SQL*Plus

Now that you have learned how to insert, update, and delete data records using SQL*Plus, the next step is to learn how to retrieve data. SQL enables you to view data in relational database tables, sort the output, and perform calculations on data, such as calculating a person's age from data stored in a date of birth column or calculating a total for an invoice. You will sort query outputs in a variety of ways and perform group functions on retrieved records, such as finding the sum or average of a set of values. You will also learn how to create queries that enable users to enter input parameters that can be used for searching or sorting records, and how to format query output in SQL*Plus.

RETRIEVING DATA FROM A SINGLE TABLE

The basic syntax for a SQL query to retrieve data from a single table is:

```
SELECT column1, column2, …
FROM ownername.tablename
WHERE search_condition;
```

3

The SELECT clause lists the columns that you want to display in your query. The FROM clause lists the name of each table involved in the query. If you are querying a table that you did not create, then you must preface it with the creator's user name, and the creator must have given you the privilege to SELECT data from that table. Therefore, to query another user's database tables, you must qualify the table name with the user schema name using the syntax *ownername.tablename*. For example, if you want to query user LHOWARD's LOCATION table, you would write the table name as LHOWARD.LOCATION.

The WHERE clause is an optional clause used to identify which rows to display by applying a search condition. For example, if you want to display only the records associated with the CR building in the LOCATION table, you would include the search condition WHERE bldg_code = 'CR'.

You will perform SELECT queries in this section using fully populated Northwoods University, Clearwater Traders, and Software Experts database tables. You will create these tables next by running the northwoods.sql, clearwater.sql, and softwareexp.sql scripts on your Data Disk.

To run the script file:

1. If necessary, start SQL*Plus and log onto the database. Run the northwoods.sql script by typing the following command at the SQL prompt: **START a:\chapter3\northwoods.sql**.

Recall that you might receive an error message if the script tries to drop a table that has not yet been created (see Figure 3-1). You can ignore this error message.

2. Run the clearwater.sql script by typing the following command at the SQL prompt: **START a:\chapter3\clearwater.sql**.

3. Run the softwareexp.sql script by typing the following command at the SQL prompt: **START a:\chapter3\softwareexp.sql**.

To retrieve every row in a table, the data values do not need to satisfy a search condition, so you omit the WHERE clause. Next, you will retrieve the first name, middle initial, and last name from every row in the STUDENT table in the Northwoods University database.

To retrieve every row from the STUDENT table:

1. Start Notepad and create a new file named **3BQueries.sql** in your Chapter3\TutorialSolutions folder. Then, type the command shown in Figure 3-26 to retrieve the student first name, middle initial, and last name from every row in the STUDENT table, and execute the command in SQL*Plus. Your output should look like Figure 3-26. Note that the column names appear as column headings. Only the columns included in the SELECT statement appear, and the columns appear in the same order they are listed in the SQL command.

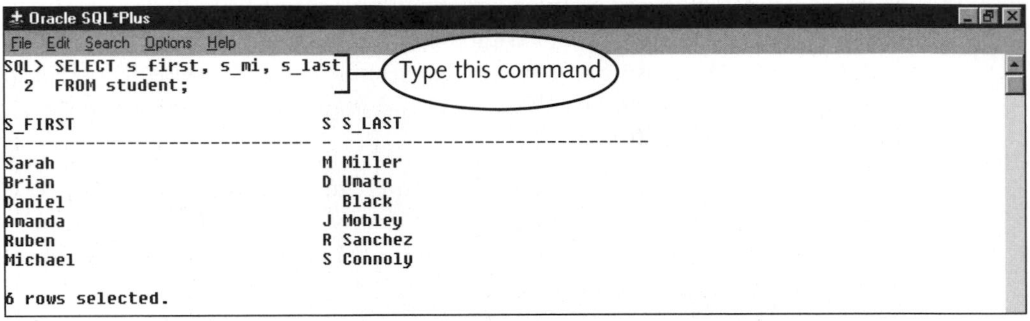

```
± Oracle SQL*Plus                                                    _ 8 X
File  Edit  Search  Options  Help
SQL> SELECT s_first, s_mi, s_last ┐   Type this command
  2  FROM student;

S_FIRST                              S S_LAST
------------------------------------ - ------------------------------
Sarah                                M Miller
Brian                                D Umato
Daniel                                 Black
Amanda                               J Mobley
Ruben                                R Sanchez
Michael                              S Connoly

6 rows selected.
```

Figure 3-26   Retrieving selected values from all rows in the STUDENT table

You will enter all of the tutorial commands in your 3BQueries.sql file, and then execute them in SQL*Plus.

If you want to retrieve all of the columns in a table, you can use an asterisk (*) as a wild-card character in the SELECT statement instead of typing every column name.

To retrieve every row and column from a table:

1. Type the command shown in Figure 3-27 to select all rows and columns from the LOCATION table.

```
Oracle SQL*Plus                                                                    _ 8 X
File  Edit  Search  Options  Help
SQL> SELECT *
  2  FROM location;                              ( Type this command )

     LOC_ID BLDG_CODE  ROOM     CAPACITY
---------- ---------- ------  ----------
         45 CR          101          150
         46 CR          202           40
         47 CR          103           35
         48 CR          105           35
         49 BUS         105           42
         50 BUS         404           35
         51 BUS         421           35
         52 BUS         211           55
         53 BUS         424            1
         54 BUS         402            1
         55 BUS         433            1

     LOC_ID BLDG_CODE  ROOM     CAPACITY
---------- ---------- ------  ----------
         56 LIB         217            2
         57 LIB         222            1

13 rows selected.
```

Figure 3-27   Selecting all rows and columns in the LOCATION table

> The SQL*Plus environment is configured to show query output one screen at a time. If the query output is wider than the current SQL*Plus line width, then the output will wrap to the next line. If the query output is longer than the current SQL*Plus page size, then the column headings will be redisplayed, and the remaining output will appear under the repeated column headings. You will learn to format output in SQL*Plus later in this lesson.

Sometimes a query will retrieve duplicate rows. For example, suppose you want to see the different ranks that faculty members at Northwoods University have. Some of the faculty members have the same rank, so the query SELECT f_rank FROM faculty; will retrieve duplicate values. You can suppress duplicate column values in query output using the **DISTINCT** qualifier, which examines the query output before it appears and displays each distinct value only once. The DISTINCT qualifier has the following general syntax in the SELECT command: SELECT DISTINCT column_name;. Now you will execute the query to retrieve all of the faculty ranks and display the duplicate values. Then, you will execute it again using the DISTINCT qualifier to suppress the duplicates.

To retrieve and suppress duplicate rows:

1. Type the first command shown in Figure 3-28. The ASSO value is listed twice because there are two records in the table with the ASSO value in the F_RANK field. To suppress duplicate values, use the DISTINCT qualifier immediately after the SELECT command. The DISTINCT qualifier tells SQL to display each value only once.

Oracle SQL*Plus
File Edit Search Options Help
SQL> SELECT f_rank
2 FROM faculty;
F_RANK

ASSO
FULL
ASST
INST
ASSO
SQL> SELECT DISTINCT f_rank
2 FROM faculty;
F_RANK

ASSO
ASST
FULL
INST
Type these commands
Output with duplicates
Output with duplicates suppressed

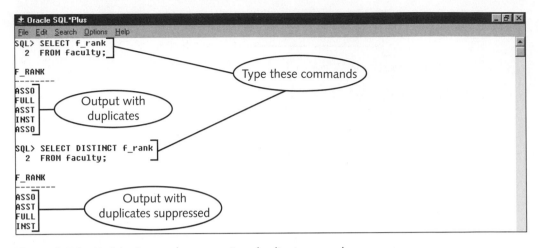

Figure 3-28 Retrieving and suppressing duplicate records

2. Type the second command shown in Figure 3-28. The output now appears with the duplicate rows suppressed.

Using Search Conditions in SELECT Commands

So far, the SELECT commands that you have written have retrieved all of the records in the table. You can use search conditions to retrieve selected rows that match specific criteria. Search conditions in the SELECT command work just the same as they do in the UPDATE and DELETE commands. Recall that search conditions use comparison operators, and each search condition must be able to be evaluated as TRUE or FALSE. Now, you will use a search condition to retrieve specific records in the FACULTY table.

To use a search condition in a SQL command:

1. Type the command shown in Figure 3-29. The query output lists the first name, middle initial, last name, and rank of all faculty members with a rank of ASSO.

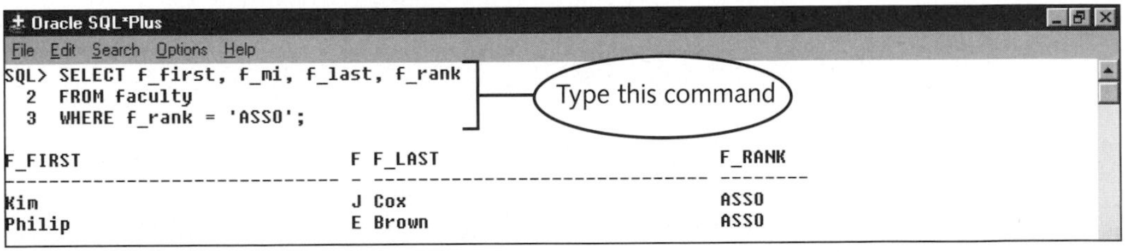

Oracle SQL*Plus
File Edit Search Options Help
SQL> SELECT f_first, f_mi, f_last, f_rank
2 FROM faculty
3 WHERE f_rank = 'ASSO';
F_FIRST F F_LAST F_RANK
Kim J Cox ASSO
Philip E Brown ASSO
Type this command

Figure 3-29 Query with search condition

In the following sections, you will learn how to create search conditions to retrieve records in a variety of ways.

Combining Search Conditions Using AND and OR

You can combine multiple search conditions using the AND and OR operators. Recall that when you use the AND operator to connect two search conditions, both conditions must be true for the row to appear in the query output. If no data rows exist that match both conditions, then no data values are retrieved. When you use the OR operator to connect two search conditions, only one of the conditions must be true for the row to appear in the query output. You will use the AND operator in the next query.

To use the AND operator in a search condition:

1. Type the command shown in Figure 3-30 to find the room numbers of all rooms in the BUS building that have a capacity greater than or equal to 40. The output displays values that match both conditions in the query.

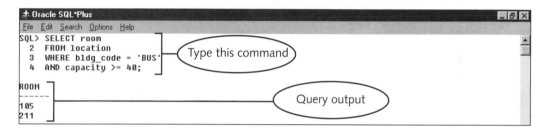

Figure 3-30   Query with multiple search conditions using AND operator

2. Type the command shown in Figure 3-31 to list the first and last names of all students who are freshmen or who were born before January 1, 1981. The output retrieves the records for student Daniel Black, who was the only student born before January 1, 1981, and for student Michael Connoly, who is the only freshman. When the OR operator is used to connect two conditions, a row is returned if *either* the first *or* second condition is true. If neither condition is true, no data values are returned. Note that whenever you use a data field with the DATE data type in a search condition, you must convert the date from a character string to a date data type, using the TO_DATE function.

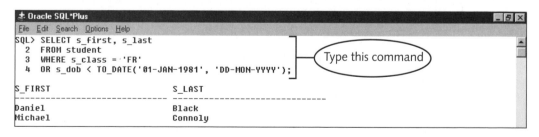

Figure 3-31   Query with multiple search conditions using OR operator

You can combine the AND and OR operators in a single query. This is a very powerful operation, but it can be tricky to use. SQL evaluates AND conditions first, and then SQL evaluates *the result* of the AND condition query against the OR condition.

To use the AND and OR operators in a single condition:

1. Type the command shown in Figure 3-32 to find the location ID, building code, room number, and capacity of every room in either the BUS or CR building whose capacity is greater than 35.

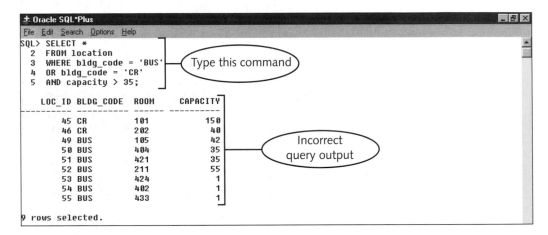

Figure 3-32   Incorrect query output when combining the AND and OR operators

The query output shown in Figure 3-32 is puzzling. Why were rooms with capacity of less than 35 returned? SQL first evaluates the AND condition (bldg_code = 'CR' AND capacity > 35), and retrieves rows that contain LOC_ID values of 45 and 46. Then, SQL evaluates the first half of the OR condition (bldg_code = 'BUS') and returns all rows where the bldg_code = 'BUS' (LOC_ID values of 49, 50, 51, 52, 53, 54, and 55). Finally, SQL combines these results with the result of the AND condition (LOC_ID values 45 and 46). Therefore, it returns rows 45, 46, and 49 through 55, which is not the result you wanted.

The best way to overcome AND/OR ordering problems is to put the operation that should be performed first in parentheses. SQL always evaluates operations in parentheses first, regardless of whether they contain an AND or OR operator.

To control the order in which AND and OR operators are evaluated in a query:

1. Type the command shown in Figure 3-33, which includes parentheses around the search conditions joined by the OR operation. Now the query output is correct. SQL performs the OR operation first (by returning all LOC_ID values in either CR or BUS), and then it evaluates the AND condition by returning only those locations that have a capacity of greater than 35.

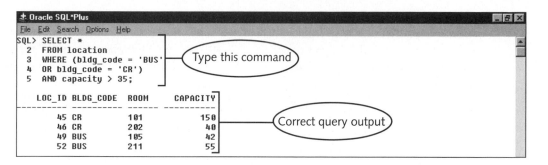

Figure 3-33   Using parentheses to force ordering of AND/OR operators

Using the NOT Operator

You also can use the **NOT** operator in a condition to retrieve rows that do not match the search condition, using the general syntax WHERE NOT (*search_condition*). Now, you will use the NOT operator to retrieve the names of all students at Northwoods University who are not seniors.

To use the NOT operator in a search condition:

1. Type the command shown in Figure 3-34 to list the first and last name of every student who is not a senior. The query output lists students who are not seniors.

Figure 3-34   Using the NOT operator in a search condition

Another way to write the search condition in the query shown in Figure 3-34 is WHERE s_class <> 'SR'.

Searching for NULL and NOT NULL Values

Sometimes you need to create a query to return records in which the value of a particular field is NULL. For example, you might want to retrieve enrollment records for courses in which a grade has not yet been assigned or find shipments in the SHIPMENT table in the Clearwater Traders database that have not been received yet, so the DATE_RECEIVED value is NULL. To do this, you use the following general syntax for the search condition: **WHERE *fieldname* IS NULL**. Similarly, to return records in which the value of a particular field is not NULL, you use the syntax **WHERE *fieldname* IS NOT NULL**. Next, you will create a query to find all records in the ENROLLMENT table in which the grade field has not been assigned and is currently NULL.

To create queries using the IS NULL and IS NOT NULL search conditions:

1. Type the first command shown in Figure 3-35. Note that the returned records show all enrollment records where the grade value has not yet been entered.

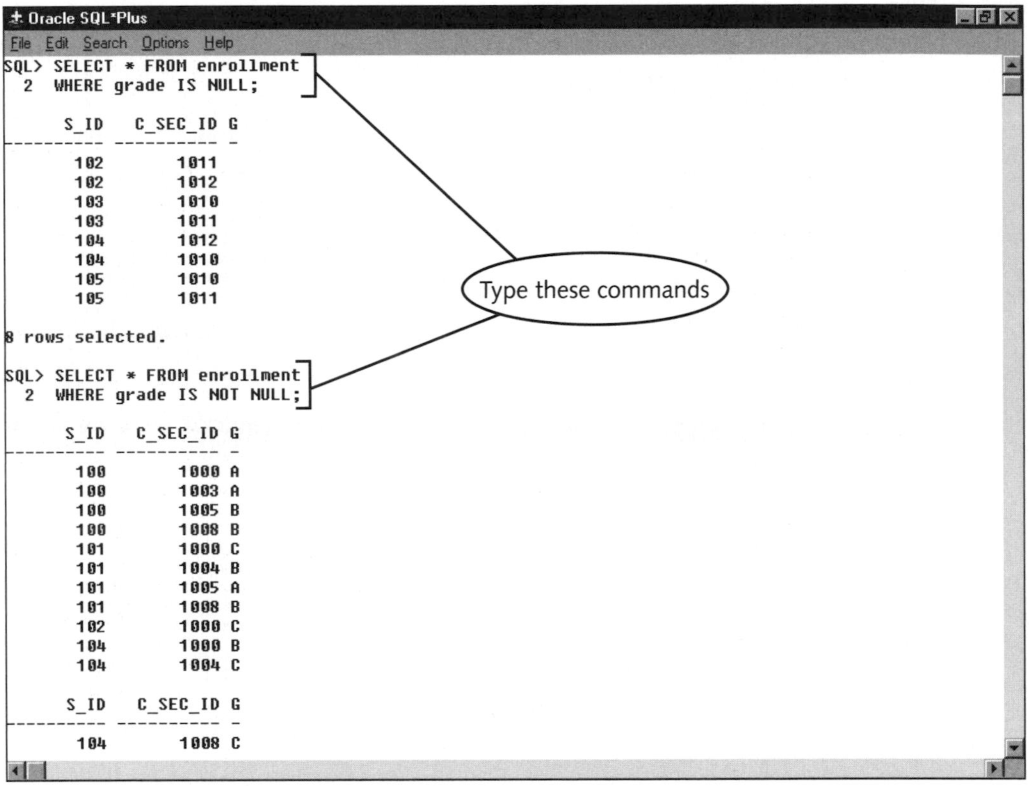

```
± Oracle SQL*Plus                                                    _ 8 X
File  Edit  Search  Options  Help
SQL> SELECT * FROM enrollment
  2  WHERE grade IS NULL;

      S_ID    C_SEC_ID G
---------- ----------- -
       102        1011
       102        1012
       103        1010
       103        1011
       104        1012
       104        1010
       105        1010
       105        1011

8 rows selected.

SQL> SELECT * FROM enrollment
  2  WHERE grade IS NOT NULL;

      S_ID    C_SEC_ID G
---------- ----------- -
       100        1000 A
       100        1003 A
       100        1005 B
       100        1008 B
       101        1000 C
       101        1004 B
       101        1005 A
       101        1008 B
       102        1000 C
       104        1000 B
       104        1004 C

      S_ID    C_SEC_ID G
---------- ----------- -
       104        1008 C
```

Type these commands

Figure 3-35 Queries using IS NULL and IS NOT NULL search conditions

2. Type the second command shown in Figure 3-35, which uses the IS NOT NULL search condition to retrieve the enrollment records where a grade value has been assigned. Note that the returned records show all enrollment records where the grade value has been entered.

Using the IN and NOT IN Search Operators

You can use the IN and NOT IN search operators to retrieve data values that are members of a set of search values. For example, you could retrieve all enrollment records where the GRADE field value is a member of the set ('A', 'B'). Similarly, you could retrieve all enrollment records, where the GRADE field is not a member of this set.

To retrieve records using the IN and NOT IN search operators:

1. Type the first command in Figure 3-36 to retrieve every enrollment record, where the GRADE value is either A or B.

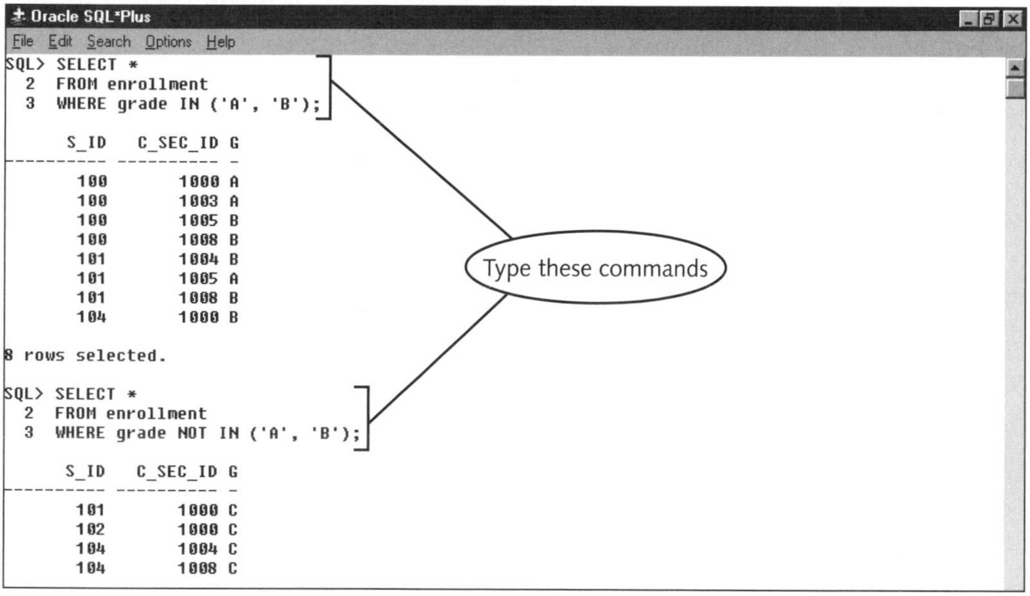

Figure 3-36   Queries using IN and NOT IN search operators

2. Type the second command in Figure 3-36 to retrieve every enrollment record where the grade value is not either A or B.

Performing Inexact Searches

Sometimes, you need to perform searches by matching part of a character string. For example, you might want to retrieve records for students whose last name begins with the letter *M*, or find all courses with MIS in the call ID field. To do this, you use the LIKE operator. The general syntax of a search condition that uses the LIKE operator is WHERE *fieldname* LIKE *character_string*. The character_string to be matched must always be in single quotation marks and must contain either the percent sign (%) or underscore (_) wildcard characters.

The percent sign (%) wildcard character represents multiple characters, and the underscore (_) represents a single character. If the wildcard character (%) is placed on the left edge of the character string to be matched, then SQL searches for an exact match on the right-most characters and allows an inexact match for the characters represented by the wildcard character. For example, the search condition **WHERE term_desc LIKE '%2003'** retrieves all term records where the last four characters in the TERM_DESC field are '2003'. The characters that make up the left side of the character string up to the substring '2003' are ignored. Similarly, the search condition **WHERE term_desc LIKE 'Fall%'** retrieves all term records where the first four characters are 'Fall', regardless of the value of the rest of the string. The search condition **WHERE course_name LIKE '%Systems%'** retrieves every record in the COURSE table where the course name field has the character string 'Systems' in any part of the string: the beginning, the middle, or the end.

 In the search condition LIKE '%Systems%', the % wildcard character does not require that characters have to be present in front of 'Systems' or after 'Systems'.

The single-character wildcard character (_) is often used in conjunction with the multiple-character wildcard character (%). For example, the search condition **WHERE S_CLASS LIKE '_R%'** would retrieve all values for S_CLASS where the first character can be any value, the second character must be the letter *R*, and the rest of the characters can be any values. Now you will create some queries with inexact search conditions using the LIKE operator.

To create queries using the LIKE operator:

1. Type the first command in Figure 3-37 to retrieve all records from the TERM table where the last four characters of the TERM_DESC field are '2003'.

2. Type the second command shown in Figure 3-37 to retrieve all records from the TERM table where the first four letters are "Fall". Note that the characters in single quotes are always case sensitive.

3. Type the third command shown in Figure 3-37 to search for the substring 'Systems' anywhere within the COURSE_NAME field in the COURSE table. Note that all records with the string 'Systems' anywhere in the course name are retrieved.

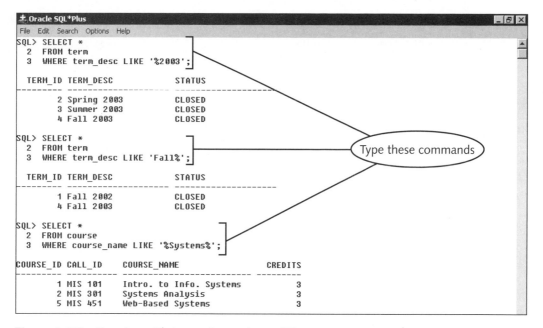

Figure 3-37   Queries with inexact search conditions

Sorting Query Output

The query output you have retrieved so far has not displayed the data values in any particular order. You can sort query output by using the ORDER BY clause and specifying the **sort key**, which is the column SQL will use as a basis for ordering the data. The syntax for the ORDER BY clause is **ORDER BY *columnname***. By default, records are sorted in numerical ascending order if the sort key is a NUMBER column and in alphabetical ascending order if the sort key is a CHAR or VARCHAR2 column. To sort the records in descending order, you need to insert the DESC command at the end of the ORDER BY clause.

To use the ORDER BY clause to sort data:

1. Type the first command shown in Figure 3-38 to list the building code, room number, and capacity for every room with a capacity that is greater than or equal to 40, sorted in ascending order by capacity.

2. Type the second command shown in Figure 3-38 to repeat the same query, but add the DESC command at the end of the ORDER BY clause to sort the records in descending order.

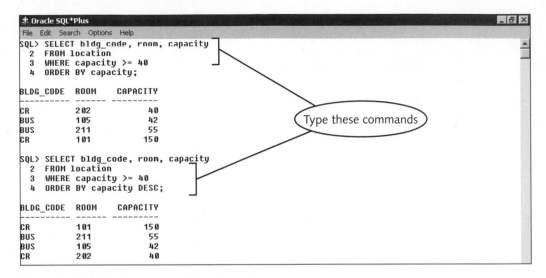

Figure 3-38   Sorting query output

You can specify multiple sort keys to sort query output on the basis of multiple columns. You specify which column gets sorted first, second, and so forth by the order that the sort keys are listed in the ORDER BY clause. The next query lists all building codes, rooms, and capacities, sorted first by building code and then by room number.

To sort data on the basis of multiple columns:

1. Type the command shown in Figure 3-39. The query output lists every row in the LOCATION table, first sorted alphabetically by building code, and then sorted within building codes by ascending room numbers.

```
Oracle SQL*Plus
File  Edit  Search  Options  Help
SQL> SELECT bldg_code, room, capacity       Type this command
  2  FROM location
  3  ORDER BY bldg_code, room;

BLDG_CODE  ROOM   CAPACITY
---------- ------ ---------
BUS        105          42
BUS        211          55
BUS        402           1
BUS        404          35
BUS        421          35
BUS        424           1
BUS        433           1       Output sorted on
CR         101         150       multiple keys
CR         103          35
CR         105          35
CR         202          40
LIB        217           2
LIB        222           1

13 rows selected.
```

Figure 3-39   Query with multiple sort keys

Using Calculations in a Query

You can perform basic mathematical calculations on retrieved data. Table 3-8 lists the operations and their associated SQL operators.

Operator	Description
+	Addition
-	Subtraction
*	Multiplication
/	Division

Table 3-8   Operators used in SQL query calculations

For example, suppose that you want to display the inventory ID, size, color, price, quantity on hand, and inventory value, which is calculated as price times quantity on hand, for each record in the Clearwater Traders INVENTORY table that is associated with ITEM_ID 897. You can use the multiplication operator to calculate the inventory value for each retrieved record.

To calculate values in a SQL query:

1. Type the command shown in Figure 3-40. Note that the query output shows the calculated value in a column whose heading is the calculation formula.

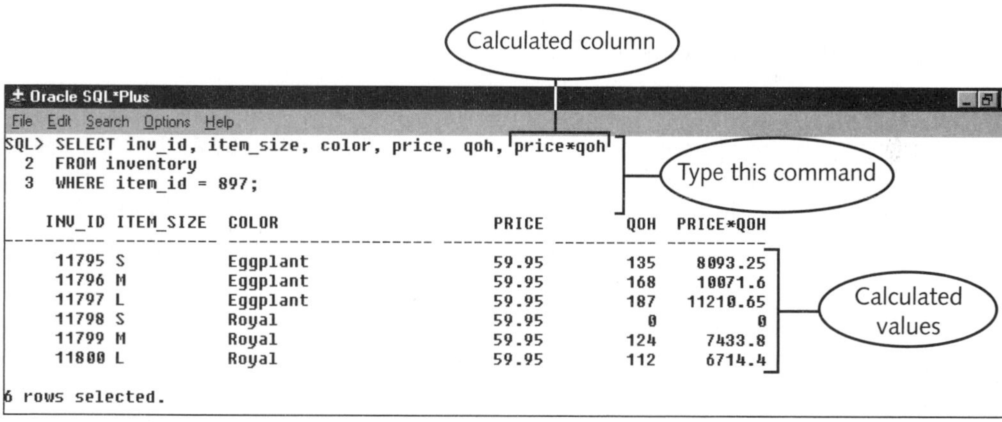

Figure 3-40   Query with calculated values

In a database, you usually do not store a person's age, because ages change from year to year. Rather, you store the person's date of birth, and calculate his or her age based on the current date. In Oracle SQL, you can calculate the difference between two dates by subtracting one date from another. The result shows the number of days between the two dates. To calculate someone's age based on his or her date of birth, you use the SYS-DATE pseudocolumn, which retrieves the current system date and time from the database server. Recall that to form a query where only the SYSDATE value is retrieved, you use the following command: **SELECT SYSDATE FROM DUAL**. If you retrieve the SYSDATE value along with other database fields from another table, you can omit DUAL from the FROM clause.

Now you will create a query that retrieves student information and calculates the student's age based on the student date of birth (S_DOB) data field and the current date.

To use the SYSDATE function to calculate student ages:

1. Type the command shown in Figure 3-41 to list the student ID, last name, and age for each student. The student age is calculated by subtracting the student's date of birth from the current date.

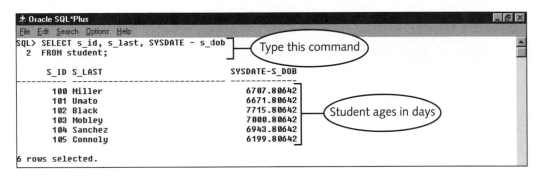

Figure 3-41   Query using the SYSDATE function

Your query output will look different from that in Figure 3-41 because the output depends on the current system date. Note that the query output lists the calculated ages in days rather than in years. To express the age in years, you must divide this value by the number of days in a year, which is approximately 365.25 (including leap years). You can do this calculation in SQL by combining multiple arithmetic operations in a single query.

In mathematics and in programming languages, expressions that contain more than one operator must be evaluated in a specific order. SQL evaluates division and multiplication operations first, and addition and subtraction operations last. SQL always evaluates expressions enclosed within parentheses first. Therefore, to calculate the difference between the current date and the student's date of birth, and then divide the result by 365.25, you use the following expression: (SYSDATE - S_DOB)/365.25.

To evaluate multiple arithmetic operations in a specific order:

1. Type the command shown in Figure 3-42, with the order of the arithmetic operations specified by parentheses. The query output now lists the students' ages in years.

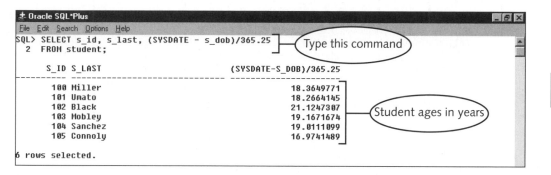

Figure 3-42   Query specifying multiple arithmetic operations in a specific order

Your output will be different from the output in Figure 3-42, depending on the current system date

Oracle Single Row Number Functions

In Figure 3-42, each student's age is shown in years with a fraction that represents the time since the student's last birthday. Usually, ages are displayed as whole numbers, without a fractional component. To remove the fraction, you need to use a SQL number function. Oracle SQL has several number functions that you can use to manipulate retrieved data. These functions are called **single-row functions**, because they return a single result for each row of data retrieved. Table 3-9 summarizes some of the commonly used SQL single-row number functions.

Function	Description	Example Query	Result
ABS(*number*)	Returns the absolute value of a number	`SELECT ABS(capacity) FROM location WHERE loc_id = 45;`	ABS(150) = 150
CEIL(*number*)	Returns the value of a number, rounded up to the next highest integer	`SELECT CEIL(price) FROM inventory WHERE inv_id = 11668;`	CEIL(259.99) = 260
FLOOR(*number*)	Returns the value of a number, rounded down to the next integer	`SELECT FLOOR(price) FROM inventory WHERE inv_id = 11668;`	FLOOR(259.99) = 259

Table 3-9   Oracle SQL single-row number functions

Function	Description	Example Query	Result
MOD(*number, divisor*)	Returns the remainder (modulus) for a number and its divisor	`SELECT MOD(qoh, 10) FROM inventory WHERE inv_id = 11668;`	MOD(16,10) = 6
POWER(*number, power*)	Returns the value representing a number raised to the specified power	`SELECT POWER(qoh, 2) FROM inventory WHERE inv_id = 11669;`	POWER(12, 2) = 144
ROUND(*number, precision*)	Returns a number, rounded to the specified precision	`SELECT ROUND (price, 0) FROM inventory WHERE inv_id = 11668;`	ROUND(259.99,0) = 260
SIGN(*number*)	Identifies if a number is positive or negative by returning 1 if the value is positive, -1 if the value is negative, or 0 if the value is 0	`SELECT SIGN(qoh) FROM inventory WHERE inv_id = 11668;`	SIGN(16) = 1
SQRT(*n*)	Returns the square root of *n*	`SELECT SQRT(qoh) FROM inventory WHERE inv_id = 11668;`	SQRT(16) = 4
TRUNC(*n, precision*)	Returns n truncated to the specified precision, where all digits beyond the specified precision are removed. The default precision value is 0	`SELECT TRUNC (price, 1) FROM inventory WHERE inv_id = 11668;`	TRUNC(259.99,1) = 259.9

Table 3-9   Oracle SQL single-row number functions (continued)

To use a SQL single-row number function, list the function name followed by the required parameter (or parameters) in parentheses. The next query demonstrates how to use a number function with a calculated database value. You will use the TRUNC function to truncate the fraction portion of the calculated student ages.

To list the students' ages in years without fractions:

1. Type the command shown in Figure 3-43. The query output shows the students' ages in years without fractional values. Your values might be different due to different system dates.

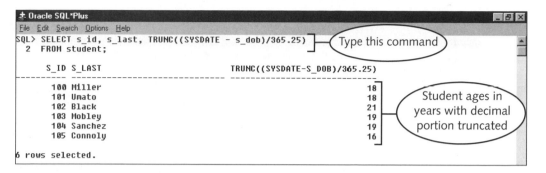

Figure 3-43   Query using the TRUNC number function

Oracle Single Row Character Functions

Oracle SQL also provides a variety of single-row character functions that are used to format character output. These character functions are summarized in Table 3–10.

Function	Description	Example Query	String Used in Function	Function Result
CONCAT (*string1, string2*)	Concatenates (joins) two strings	SELECT CONCAT (f_last, f_rank) FROM faculty WHERE f_id = 1;	'Cox' and 'ASSO'	'CoxASSO'
INITCAP(*string*)	Returns the string, with the initial letter only in upper-case	SELECT INITCAP (bldg_code) FROM location WHERE loc_id = 45;	'CR'	'Cr'
LENGTH(*string*)	Returns an integer representing the string length	SELECT LENGTH (meth_pmt) FROM cust_order WHERE order_id = 1057;	'CC'	2
LPAD(*string, number_of_ characters_to_ add, padding_ character*), RPAD(*string, total_length_of_ return_value, padding_ character*)	Returns the value of the string, with sufficient padding characters added to the left/right edge so return value equals total length specified	SELECT LPAD (meth_pmt, 5, '*'), RPAD (meth_pmt, 5, '*') FROM cust_order WHERE order_id = 1057;	'CC'	***CC CC***

Table 3-10   Oracle SQL single-row character functions

Function	Description	Example Query	String Used in Function	Function Result
LTRIM(*string, search_string*), RTRIM(*string, search_string*)	Returns the string with all occurrences of the search string characters trimmed on the left/right side. The order of the search string characters does not matter	SELECT LTRIM (call_id, 'MIS ') FROM course WHERE course_id = 1;	'MIS 101'	'101'
REPLACE(*string, search_string, replacement_string*)	Returns the string with every occurrence of the search string replaced with the replacement string	SELECT REPLACE (term_desc, '200', '199') FROM term WHERE term_id = 1;	'Fall 2002'	'Fall 1992'
SUBSTR(*string, start_position, length*)	Returns a substring of the specified string, starting at the start position, and of the specified length	SELECT SUBSTR (term_desc, 1, 4) FROM term WHERE term_id = 1;	'Fall 2002'	'Fall'
UPPER(*string*), LOWER(*string*)	Returns the string, with all characters converted to upper- or lower case	SELECT UPPER (term_desc) FROM term WHERE term_id = 1;	'Fall 2002'	'FALL 2002'

Table 3-10   Oracle SQL single-row character functions (continued)

Now, you will create some queries that use the Oracle SQL character functions. First, you will use the CONCAT function to display the Northwoods University location building codes and rooms as a single text string. Then, you will use the INITCAP function to display the values in the STATUS field in the TERM table in mixed-case letters, with the first letter capitalized.

To create queries that use the Oracle SQL character functions:

1. Type the first command shown in Figure 3-44. The building codes and rooms appear as a single text string.

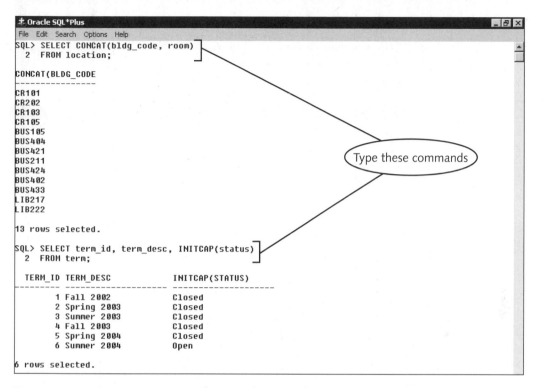

Figure 3-44   Queries using Oracle SQL character functions

2. Type the second command shown in Figure 3-44. The STATUS values are shown in mixed-case letters, with the first letter capitalized. (Recall that the STATUS values are stored in the database in all capital letters.)

Date Arithmetic and Oracle Single-Row Date Functions

Often when creating database reports, it is useful to be able to perform arithmetic calculations on dates in order to retrieve records corresponding to a date that is within the next week, next month, or next year. To specify a date that is a specific number of days after a known date, you add the number of days, as an integer, to the known date. Similarly, to specify a date that is a specific number of days before a known date, you subtract the number of days, as an integer, from the known date. For example, suppose today is August 10, 2003, and you would like to determine all shipments that are expected to arrive at Clearwater Traders during the next 60 days. To calculate this date, you add the number 60 to the known date.

To write a query that uses date arithmetic to specify a date after a known date:

1. Type the command shown in Figure 3-45. Note that the query output specifies shipments that are expected to arrive in the period from 8/11/2003 through 10/10/2003 (60 days later).

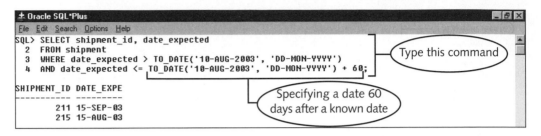

Figure 3-45   Query specifying a date after a known date

To write a query that specifies a date that occurs a specific number of days before a known date, you subtract the number of days from the date. Suppose today is August 31, 2003, and you would like to view all shipments that were received within the past 60 days. You would write the search condition as:

```
WHERE date_expected > TO_DATE('10-AUG-2003', 'DD-MON-YYYY')
AND date_expected <= TO_DATE('10-AUG-2003', 'DD-MON-YYYY') - 60;
```

Oracle SQL provides a variety of single-row date functions to support additional date manipulation capabilities. These date functions are summarized in Table 3-11.

Function	Description	Example Query	Date(s) Used in Function	Function Result
ADD_MONTHS (date, number_of_ months_to_add)	Returns a date that is the specified number of months after the input date	SELECT ADD_ MONTHS(date_ expected, 2) FROM shipment WHERE shipment_id = 211;	9/15/2003	11/15/2003
LAST_DAY(date)	Returns the date that is the last day of the month specified in the input date	SELECT LAST_ DAY(date_ expected) FROM shipment WHERE shipment_id = 211;	9/15/2003	9/30/2003
MONTHS_ BETWEEN date1, date2)	Returns the number of months, including decimal fractions, between two dates. If date1 is afterdate2, a positive number is returned, and if date1 is before date2, a negative number is returned	SELECT MONTHS_ BETWEEN (date_ expected, TO_ DATE('10-AUG-2003', 'DD-MON-YYYY')) FROM shipment WHERE shipment_id = 211;	9/15/2003	1.1612903

Table 3-11   Oracle SQL single row date functions

ORACLE SQL GROUP FUNCTIONS

An Oracle SQL **group function** performs an operation on a group of queried rows and returns a single result, such as a column sum. Table 3-12 describes commonly used Oracle SQL group functions.

Function	Description	Example Query	Result
AVG (*fieldname*)	Returns the average value of a numeric column's returned values	`SELECT AVG(capacity)` `FROM location;`	33.230769

COUNT (*fieldname*) COUNT(*)	Returns an integer representing a count of the number of returned rows. COUNT(*field_name*) counts only rows where the specified field is not NULL, while COUNT(*) counts all rows	`SELECT COUNT(grade)`
		`FROM enrollment;`
		
		`SELECT COUNT(*)`
		`FROM enrollment;`
			Rows for which a GRADE value exists = 12;
			
			All rows = 20

MAX(*fieldname*)	Returns the maximum value of a numeric column's returned values	`SELECT MAX(max_enrl)` `FROM course_section;`	140
MIN(*fieldname*)	Returns the minimum value of a numeric column's returned values	`SELECT MIN(max_enrl)` `FROM course_section;`	30
SUM(*fieldname*)	Sums a numeric column's returned values	`SELECT SUM(capacity)` `FROM location;`	432

Table 3-12   Oracle SQL group functions

To use a group function in a SQL query, list the function name followed by the column name on which to perform the calculation in parentheses. Next, you will enter a query that uses group functions to sum the total maximum enrollment for each course section in the Summer 2004 term and calculate the average, maximum, and minimum current enrollments.

To use group functions in a query:

1. Type the command shown in Figure 3-46 to sum the total maximum enrollment, and calculate the average, maximum, and minimum current enrollment for each course section for the Summer 2004 term (TERM_ID = 6).

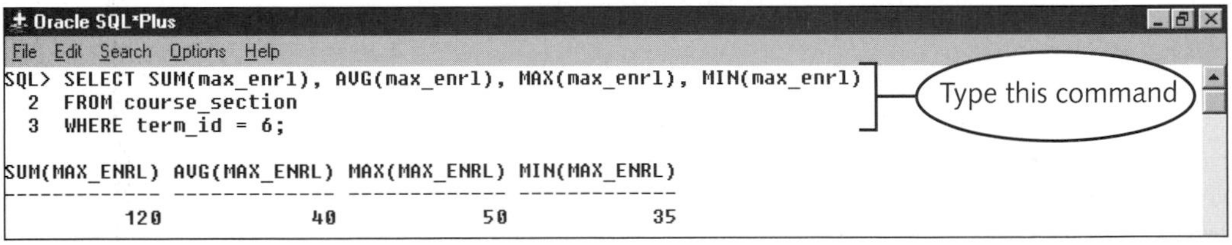

Figure 3-46   Query using group functions

The COUNT group function returns an integer that represents the number of records that are returned by a given query. The COUNT(*) version of this function calculates the total number of rows in a table that satisfy a given search condition. The COUNT(*fieldname*) version calculates the number of rows in a table that satisfy a given search condition and also contain a non-null value for the given column. Next, you will use both versions of the COUNT function to count the total number of students in a course section and to count the number of students who have received a grade (GRADE is not NULL) in the course section.

To use the COUNT group function:

1. Type the first command shown in Figure 3-47 to count the total number of students enrolled in C_SEC_ID 1010.

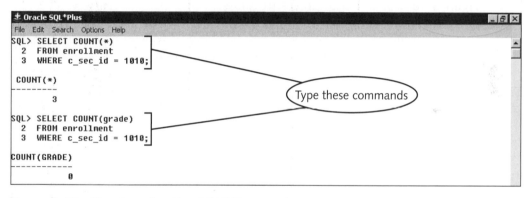

Figure 3-47   Queries using the COUNT group function

2. Using the COUNT(*fieldname*) group function, type the second query shown in Figure 3-46 to count the total number of students who have received a value for GRADE in C_SEC_ID 1010. Notice the difference between output for the first and second queries. There are three students enrolled in the course, but no student has received a value for GRADE.

Using the GROUP BY Clause with Group Functions

If a query retrieves multiple records and the records in one of the retrieved columns have duplicate values, you can group the output by the column with duplicate values and apply group functions to the grouped data. For example, you might want to list the names of the different building codes at Northwoods University and list the sum of the capacity of each building. To do this, you use the GROUP BY clause, which has the following syntax: GROUP BY *group_fieldname*;. Now, you will create a query that uses a group function with the GROUP BY clause.

To use the GROUP BY clause to group rows:

1. Type the command shown in Figure 3-48 to list the building code and the total capacity of each building in the LOCATION table.

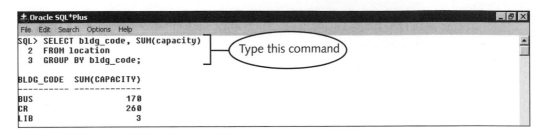

Figure 3-48   Query using GROUP BY clause

It is important to note that if you create a query in which one or more of the columns in the SELECT clause involve a group function and one or more columns do not, then the columns that are *not* in the group function must always be placed in a GROUP BY clause. If they are not placed in a GROUP BY clause, an error will occur. This error occurs because SQL cannot display single-value results and group function results in the same query output unless the single-value results are grouped using the GROUP BY clause. Next, you will discover what happens when you repeat the query to return building codes and the sums of their capacities but omit the GROUP BY clause.

To repeat the query without the GROUP BY clause:

1. Type the command shown in Figure 3-49 to list the building code name and the total capacity of each building in the LOCATION table, omitting the GROUP BY clause.

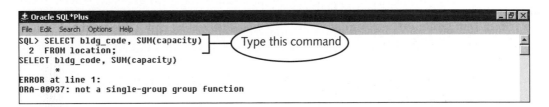

Figure 3-49   Query with GROUP BY clause omitted

The error message "ORA-00937: not a single-group group function" indicates that a SELECT clause cannot contain both a group function and an individual column expression unless the individual column expression is in a GROUP BY clause. To solve this problem, you would include BLDG_CODE in the GROUP BY clause, as in the query shown in Figure 3-48.

When you create a query with one or more group functions, you can group the output by only one column, and the only columns you can list in the SELECT clause are the columns by which the values are grouped. Suppose you want to display room numbers along with building codes in the previous query. You will be trying to return a single value for each record, along with a group value for all records.

To display room numbers in the group function query:

1. Type the first command shown in Figure 3-50, adding ROOM to the SELECT clause and grouping only by building code. The error message "ORA-00979: not a GROUP BY expression" appears. This indicates that the ROOM field must also be in the GROUP BY clause, or else the query will not execute. All SELECT clause columns must either be within a group function or be listed in the GROUP BY clause.

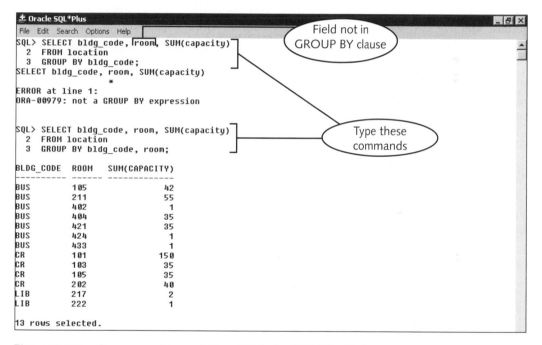

Figure 3-50   Query requiring addtional field in GROUP BY clause

2. Type the second command shown in Figure 3-50, this time including ROOM in the GROUP BY clause.

Adding ROOM to the GROUP BY clause enables the query to execute, but the output is grouped by both the building code and the room. Because there are no duplicate building code/room records, the SUM function operates on each record individually, and the sum of the capacity for each room is the same as each room's actual capacity.

Filtering Grouped Data Using the HAVING Clause

You can filter the results of queries that retrieve values calculated by group functions using the HAVING clause. This clause effectively places a search condition on the results of a group function using the following format: **HAVING** *group_function operator search_constraint*. For example, suppose you want to retrieve the total capacity of each building at Northwoods University, but you are not interested in the data for buildings that have a capacity of less than 100. You would use the HAVING clause **HAVING SUM(capacity) >= 100**. You will now write a query to retrieve this data.

To filter grouped data using the HAVING clause:

1. Type the command shown in Figure 3-51. Compare this result with the query result in Figure 3-48, and note that the result for the LIB building code has been filtered because the sum of the capacity does not meet the search criteria in the HAVING clause.

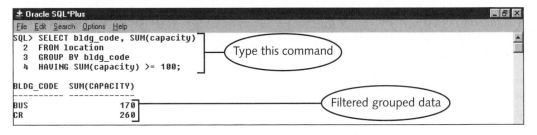

Figure 3-51   Group function query using HAVING clause

CREATING ALTERNATE COLUMN HEADINGS AND COLUMN ALIASES

In SQL*Plus query output, column headings for selected columns are the names of the column fields as defined in the database table. When you create new output columns using arithmetic functions and single-value and group functions, the output columns appear as the formula or function used to create them. For example, in Figure 3-51, note that the heading for the column that shows the sum of the capacity for each building appears as SUM(CAPACITY). Similarly, when you calculated the student ages (see Figure 3-43), the column heading appeared as the full formula for the calculation. You can display an alternate column heading in the SELECT clause using the following command syntax: **SELECT** *column1 "heading1_text", column2 "heading2_text", ...*. Now you will create a query that specifies an alternate column heading.

To specify an alternate column heading:

1. Type the command shown in Figure 3-52. The output appears with the alternate column heading. Note that the alternate column heading can contain blank spaces.

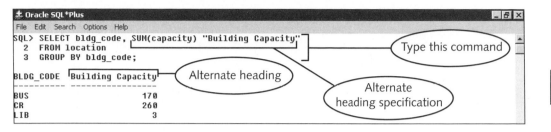

Figure 3-52   Query with alternate column heading

A column heading only changes the SQL*Plus output display—you cannot use it to reference the column in the GROUP BY or ORDER BY clause in the query. For example, in the query in Figure 3-52, if you wanted to order the query output by the building capacity value, you would need to specify the ORDER BY clause as **ORDER BY SUM(capacity);**. You could not order the query output by specifying the heading "Building Capacity" in an ORDER BY clause.

If you want to reference a calculated column in a GROUP BY or ORDER BY clause, you can reference the column using the calculation formula that you used in the SELECT clause. This is tedious to do for columns that involve complex calculations. To simplify referencing columns that specify complex arithmetic operations or group functions, you can create an alternate column name called an **alias** that can be referenced in other parts of the query. The general syntax for creating an alias is: **SELECT *fieldname* AS *alias_name*.** The alias name must follow the Oracle naming standard. You will now repeat the query to sum the capacity of all of the buildings but modify it to create an alias for the summed capacity column. Then, you will use the alias to specify the sort order of the output.

To create an alias:

    1. Type the command shown in Figure 3-53. The alias appears as the new column heading.

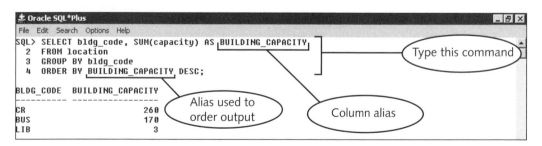

Figure 3-53   Query with column alias

CREATING DYNAMIC SQL QUERIES

So far, all of the queries that you have created are static—the values that are inserted, updated, deleted, and selected are defined when the query is written. Queries can be made more flexible if they let you change a search condition parameter or specify different display fields while the query is running. To do this, you can write dynamic SQL commands that prompt users for inputs and accept user inputs at runtime by using substitution values and by defining runtime variables.

Substitution Values

One way to create an interactive SQL command is to replace a search expression or data value with a **substitution value**, which has the following syntax: **&*prompt_variable*.** The ampersand (&) signals the SQL compiler to prompt the user for a value that is then substituted into the query. The prompt variable is the name of the database field for which the user is prompted to specify a value. Now you will create a query that prompts the user to enter a value for building code. Then, the query displays verification information showing how the input value is substituted into the query and retrieves the total capacity for the building.

To create an interactive SQL command using a substitution value:

1. Type the following command at the SQL prompt, and then press **Enter** to execute the command in SQL*Plus:

```
SELECT bldg_code, SUM(capacity)
FROM location
WHERE bldg_code = &Building_Code
GROUP BY bldg_code;
```

The prompt "Enter value for building_code" appears, as shown in Figure 3-54. Note that the prompt variable value appears in all lowercase characters.

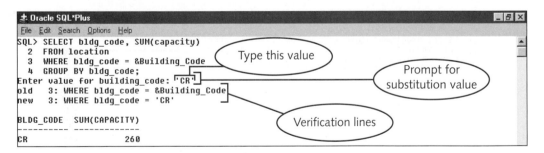

Figure 3-54 Query with substitution value

2. Type **'CR'** for the substitution value, and then press **Enter**. In queries that use substitution values, users must enter values for character data columns, such as BLDG_CODE, in single quotation marks. The query output for the specified search condition appears. Note that verification lines ("old 3:" and "new 3:") indicate how the user input ('CR') was substituted into Line 3 of the query.

If you specify multiple substitution values in a single command, the prompts appear in the order the substitution values are listed in the command.

You can also use substitution values in DML commands for inserting, updating, and deleting records. Next, you will use multiple substitution values in a DML command to insert a new record in the COURSE table.

To use multiple substitution values in an INSERT command:

1. Type the following command at the SQL prompt to insert a new record in the COURSE table, using substitution values instead of the column names in the VALUES clause:

```
INSERT INTO course VALUES
(&course_id, &call_id, &course_name, &credits);
```

2. Type the user inputs at the appropriate prompts to specify the data values for the record, as shown in Figure 3-55. Be sure to enclose the values for the CALL_ID and COURSE_NAME columns in single quotation marks. The message "1 row created" indicates that the row was successfully inserted.

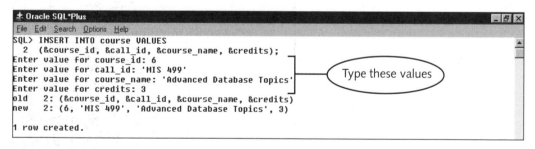

Figure 3-55   Using multiple substitution values in an INSERT command

Runtime Variables

Another approach for creating dynamic SQL queries is to define runtime variables in the SQL*Plus environment, and then use these variables in SQL queries. The command to define a SQL runtime variable is **DEFINE *variable_name* = *variable_value***. You can then substitute the variable name, preceded by an ampersand, in a SQL query. You can then change the values of these variables to customize the output of the query, depending on your needs. For example, suppose you are searching for rooms at

Northwoods University that have a particular maximum capacity. You could define the maximum capacity variable value as 50, run the query, and examine the results. You could then change the variable value to 100 and run the query again to get different results. Now you will define a runtime variable named MAX_CAPACITY that specifies a specific room capacity and then write a query using the defined variable.

To define and use a runtime variable in a query:

1. Type the first command shown in Figure 3-56 to define the runtime variable and assign its value to be 35.

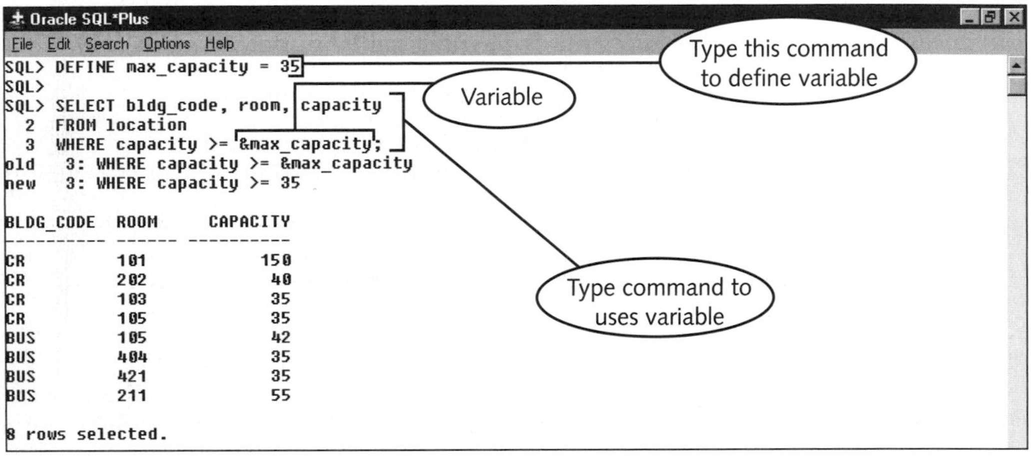

Figure 3-56   Defining and using a runtime variable

2. Type the SELECT command shown in Figure 3-56 that uses the runtime variable within the search condition. Note that the variable value is inserted into the search condition in place of the variable, and the correct output appears.

Once defined, runtime variables remain available during an entire SQL*Plus session, unless the runtime variable is undefined. **Undefined** means the variable is no longer available, and the memory space used to save its value is released. The command to undefine a runtime variable is **UNDEFINE _variable_name_**. To change the value of a runtime variable, you redefine the variable by defining it again using the DEFINE command with a new value. Figure 3-57 shows examples of commands that redefine the MAX_CAPACITY value to 45, execute the query again with the changed value, and then undefine the variable.

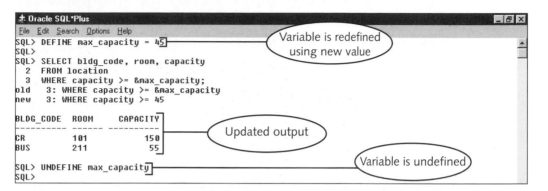

Figure 3-57   Redefining and undefining a runtime variable

An alternate way to assign values to runtime variables is to use the ACCEPT command, which prompts the user to enter a value that is then assigned to the runtime variable. The ACCEPT command must be used in a script that also contains the query in which the runtime variable is used. The advantage of using runtime variables in conjunction with the ACCEPT command over using substitution values is that you can customize the prompt that the user sees, and make it more understandable and informative.

The syntax of the ACCEPT command is **ACCEPT** *variable_name* **PROMPT** *'prompt_text'*. When the script runs, the prompt text appears, and the value that the user enters is then assigned to the variable name in the command. Now you will create a script named bldg_capacity.sql that prompts a user to enter the maximum capacity needed and the desired building code. These values will be assigned to runtime variables that are used in the query.

To use the ACCEPT command to get user input for runtime variables:

1. In Notepad, close the 3BQueries.sql file where you are saving your queries, and create a new file.

2. Type the ACCEPT commands shown in Figure 3-58 to get user input for the maximum capacity and desired building code and assign the input values to the runtime variables.

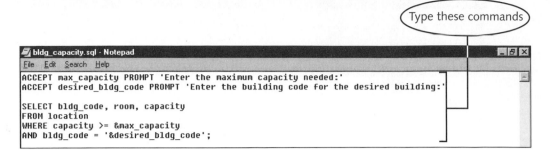

Figure 3-58   Script using ACCEPT commands to assign values to runtime variables

3. Type the SELECT command shown in Figure 3-58 to retrieve location information based on the input values for capacity and building code, and then save the file as **bldg_capacity.sql** in your Chapter3\TutorialSolutions folder.

 In the Notepad Save As dialog box, remember to open the Save as type list, and select All Files. Otherwise, Notepad will automatically place a .txt extension on your file.

4. In SQL*Plus, type **start a:\chapter3\tutorialsolutions\ bldg_code.sql** to run the script. When prompted, type **45** for the maximum capacity needed, and **CR** for the desired building code, as shown in Figure 3-59. Your query output should look like Figure 3-59. Note that when you use runtime variables and the ACCEPT command, you do not need to place input text strings in single quotes.

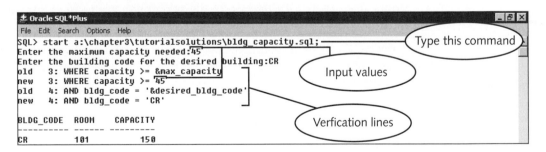

Figure 3-59   Output for script using ACCEPT commands

The verification lines that show how the input values are substituted into the query text appear in the output. To suppress these verification lines, type the following command at the SQL command prompt: **SET VERIFY OFF**. When you want to display verification lines again during your SQL*Plus session, type **SET VERIFY ON** at the SQL prompt. Now you will suppress verification lines, and observe the results.

To suppress verification lines:

1. In SQL*Plus, type **SET VERIFY OFF** at the SQL prompt.

2. Type `start a:\chapter3\tutorialsolutions\bldg_capacity.sql` to run the script. When prompted, type **45** for the maximum capacity needed, and **CR** for the desired building code. The verification lines should no longer appear.

3. Type **SET VERIFY ON** to enable verification lines to appear again during your SQL*Plus session.

FORMATTING OUTPUT IN SQL*PLUS

So far, you have accepted the default output formats in SQL*Plus—output column headings are the same as the database column names, and the column widths are the same as the database column widths. You have also accepted the default screen widths and lengths. Now you will learn how to modify the format of output data.

Creating SQL*Plus Reports

Consider the output for a query that retrieves all of the columns from the CUSTOMER table of the Clearwater Traders database, which is shown in Figure 3-60.

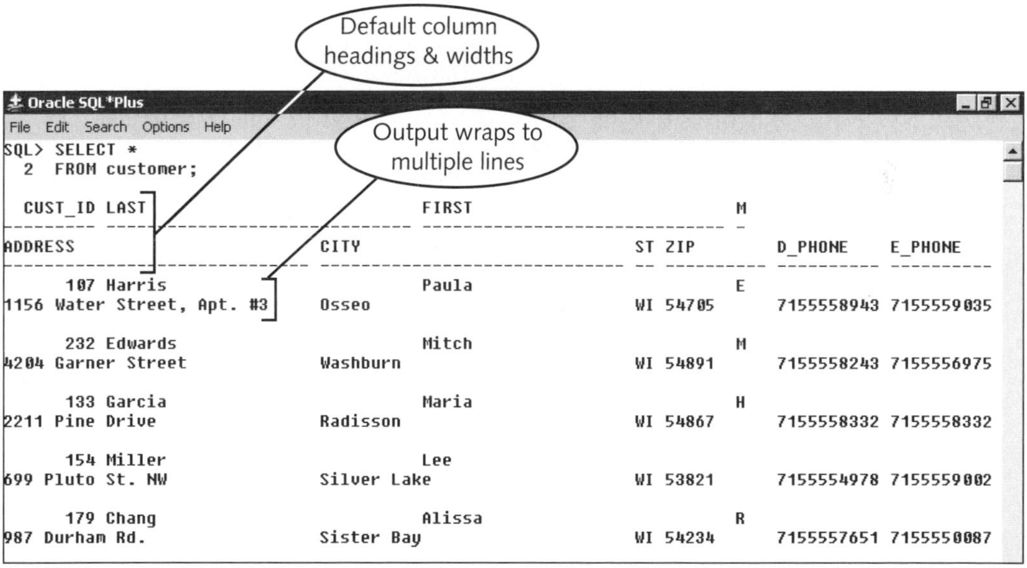

Figure 3-60   Default SQL*Plus output format

The default SQL*Plus output format displays the data using the column names and widths that were specified when the table was created. The default column widths are wider than necessary for most of the data values, so the output wraps to multiple lines. You can customize SQL*Plus output to create a formatted report by adding commands in a script to specify a report title, customized column headings and widths, and customized page lengths and widths. SQL*Plus report formatting commands are summarized in Table 3-13.

Command	Description	Example	Output
TTITLE *'text_of_title'*	Title displayed at the top of each report page. A title can span multiple lines	TTITLE 'Customer Report'	Customer Report
BTITLE *'text_of_title '*	Title displayed at the bottom of each report page	BTITLE 'Confidential'	Confidential
COLUMN *column_name* HEADING *'text_ of_heading'* FORMAT *width_ format_ specification*	Custom heading and width specification for associated column. Format specifications for NUMBER columns are expressed as numerical format masks; format specifications for character columns can specify the column width using the format A *maximum_width*	COLUMN price HEADING 'Current Price' FORMAT $9,999.99 COLUMN color HEADING "Color" FORMAT A10	Current Price $259.99 Color 'Sky Blue ' (10 characters wide)
SKIP *number_of_lines*	Skips the specified number of lines to create white space on report	SKIP 2	
LEFT, RIGHT, CENTER	Specifies whether titles are left, right, or center justified	TTITLE CENTER 'Customer Report'	Customer Report (centered on page)
SET LINESIZE *number*	Specifies the number of characters that are displayed per line.	SET LINESIZE 100	
SET PAGESIZE *number*	Specifies the number of lines that appear per page. Titles and column headings reappear at the top of each page	SET PAGESIZE 40	
Hyphen (-)	Line continuation character, specifies that a single command appears on multiple script lines	TTITLE — 'Customer Report'	Customer Report

Table 3-13   SQL*Plus report formatting commands

The TTITLE and BTITLE commands specify a top and bottom title for every query executed during the current SQL*Plus session. The report titles can span multiple lines by skipping lines using the SKIP command. The command defining the title must be specified on a single command line using the line continuation (-) character. After the report executes, you must disable the titles, or they will appear for all future queries. To disable titles, you use the commands **TTITLE OFF** and **BTITLE OFF**.

Now you will create a report to format the information shown in the CUSTOMER table. The design for the report is shown in Figure 3-61.

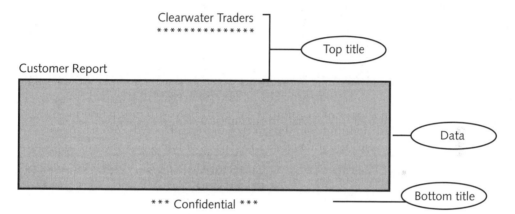

Figure 3-61   Design of Clearwater Traders customer report

The top title will span multiple lines. The first two lines, "Clearwater Traders" and the border underlining the text, will be centered. Two lines below, the subheading, "Customer Report", will appear. The bottom title, "* * * Confidential * * *", will be centered. Now you will create the script that will generate this formatted report.

To create the script to generate the formatted report:

1. Switch to Notepad, create a new file, and type the commands shown in Figure 3-62 to create the formatted report. Note that the TTITLE command spans multiple lines and uses the line continuation character (-). Save the file as **customer_report.sql** in your Chapter3\TutorialSolutons folder.

customer_report.sql - Notepad
File Edit Search Help
TTITLE CENTER 'Clearwater Traders' SKIP 1 -
CENTER '******************' SKIP 2 -
LEFT 'Customer Report'
BTITLE CENTER '* * * Confidential * * *'
SET LINESIZE 100
SET PAGESIZE 25
COLUMN cust_id HEADING 'ID' FORMAT 999
COLUMN last HEADING 'Last Name' FORMAT A10
COLUMN first HEADING 'First Name' FORMAT A10
COLUMN mi HEADING 'MI' FORMAT A2
COLUMN address HEADING 'Address' FORMAT A18
COLUMN city HEADING 'City' FORMAT A12
COLUMN state HEADING 'St'
COLUMN zip HEADING 'Zip' FORMAT A5
COLUMN d_phone HEADING 'Day Phone' FORMAT A12
COLUMN e_phone HEADING 'Eve. Phone' FORMAT A12
SELECT *
FROM customer;
TTITLE spans multiple lines
Page width and length
Column definitions

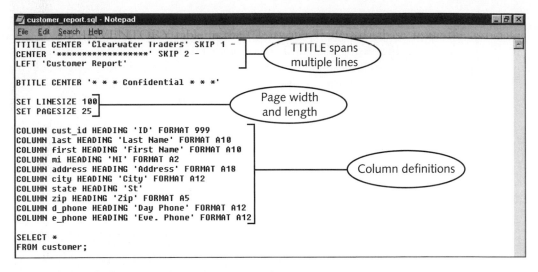

Figure 3-62 Script to create customer report

2. Switch to SQL*Plus, and type **`start a:\chapter3\tutorialsolutions\customer_report.sql`** to run the script. The report appears, as shown in Figure 3-63.

3. At the SQL prompt, type **TTITLE OFF** and then press **Enter** to disable the top report title.

4. At the SQL prompt, type **BTITLE OFF** and then press **Enter** to disable the bottom report title.

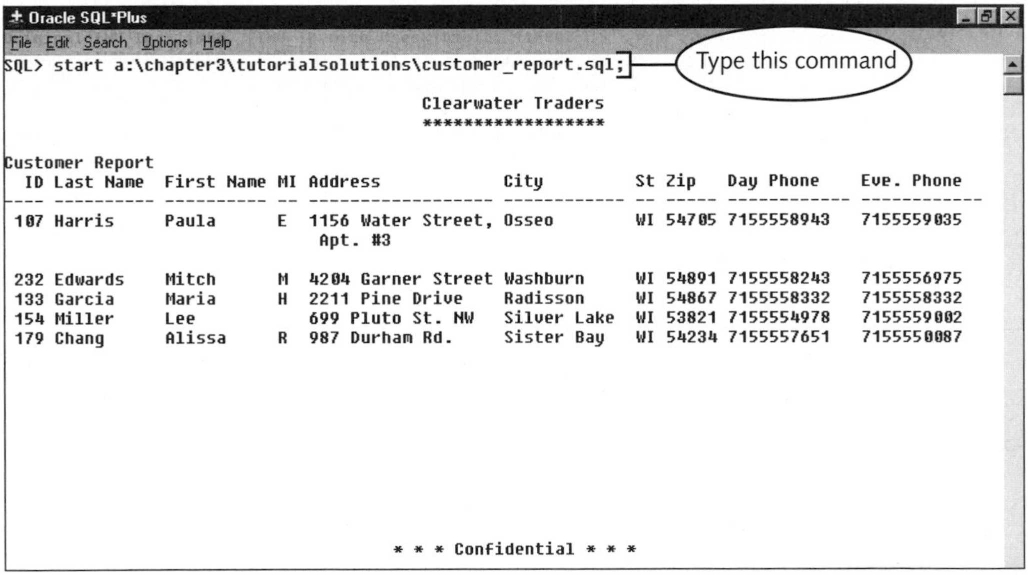

Oracle SQL*Plus
File Edit Search Options Help
SQL> start a:\chapter3\tutorialsolutions\customer_report.sql;
Type this command
Clearwater Traders
****
Customer Report
ID Last Name First Name MI Address City St Zip Day Phone Eve. Phone
107 Harris Paula E 1156 Water Street, Apt. #3 Osseo WI 54705 7155558943 7155559035
232 Edwards Mitch M 4204 Garner Street Washburn WI 54891 7155558243 7155556975
133 Garcia Maria H 2211 Pine Drive Radisson WI 54867 7155558332 7155558332
154 Miller Lee 699 Pluto St. NW Silver Lake WI 53821 7155554978 7155559002
179 Chang Alissa R 987 Durham Rd. Sister Bay WI 54234 7155557651 7155550087
* * * Confidential * * *

Figure 3-63 Formatted customer report

Formatting Data Using the TO_CHAR Function

When you retrieve data in SQL*Plus from NUMBER or DATE data fields, the data values appear using the SQL*Plus default date and number formats. Sometimes, you might want to display number fields using a currency format mask that shows a currency symbol, or you might want to display DATE fields using an alternate format mask. To use an alternate format mask, you can convert the field to a character string and then apply the desired format mask using the TO_CHAR function. The TO_CHAR function has the following format: **TO_CHAR(*data_field*, '*format_mask*')**.

One instance in which it is necessary to convert output fields to characters and apply a specific format mask is when you store time values using the DATE data type. Recall that DATE data fields store time as well as date information. However, the default DATE output format mask is DD-MON-YY, which does not have a time component. Therefore, you need to format time output using an alternate format mask. Now you will retrieve the TIME field from the COURSE_SECTION table to examine the default output for a DATE field in which only a time component was inserted.

To examine the TIME field:

1. In your 3BQueries.sql file, type the command shown in Figure 3-64 to retrieve the TIME field from the COURSE_SECTION table.

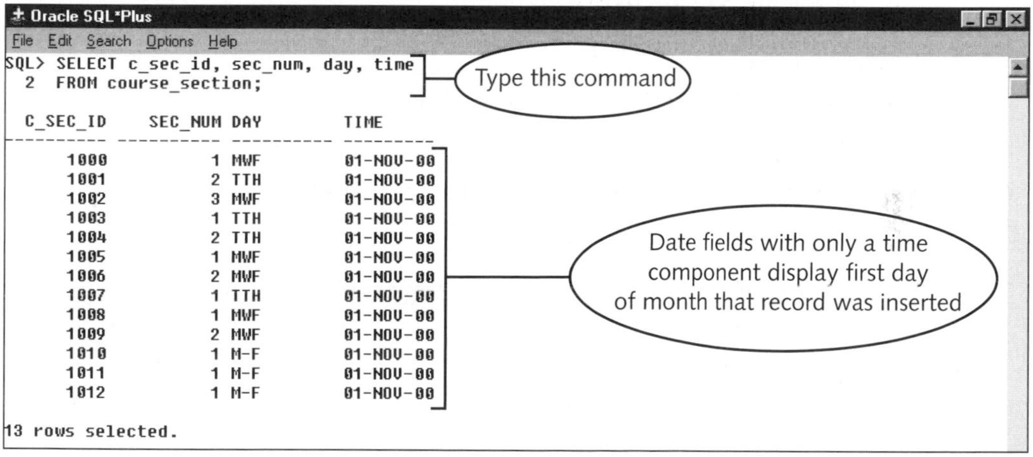

Figure 3-64   Default date values for TIME field

When the records were inserted for the TIME field in the COURSE_SECTION table, the time values were specified. However, no date component was specified, so the system inserted the default date value, which is the first day of the month in which the time values were inserted in the database. Now you will repeat this query, but convert the date field to a character string with a time format mask.

To display time data in a SQL query:

1. Type the command shown in Figure 3-65. Note that the course section times now appear.

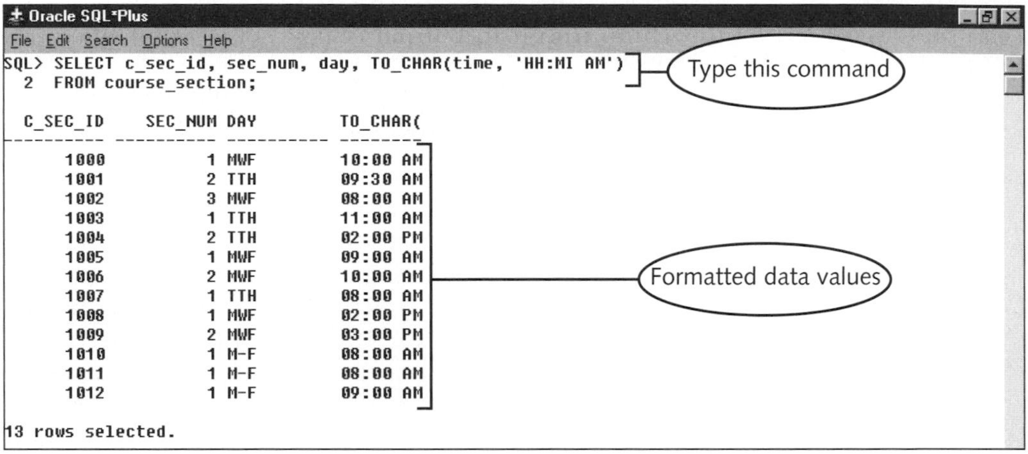

Figure 3-65 Displaying formatted time values

You can use the TO_CHAR function to apply format masks to NUMBER fields in a similar way. For example, the PRICE field in the INVENTORY table in the Clearwater Traders database represents currency values. Now you will create a query that formats output values using a currency format mask.

To create a query that formats number values as currency:

1. Type the command shown in Figure 3-66. The PRICE values appear using the currency format mask.

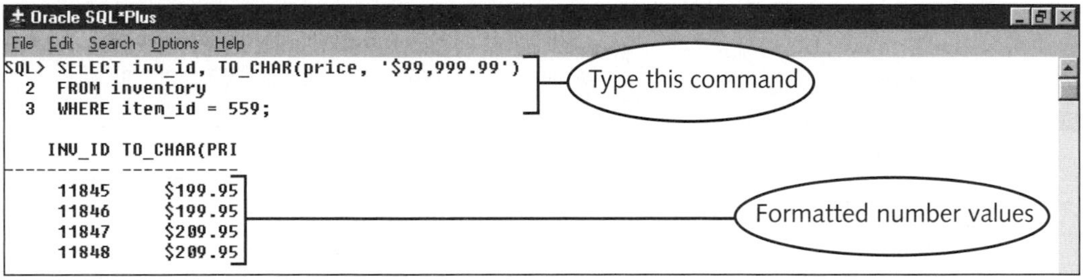

Figure 3-66 Displaying formatted number values

2. Close SQL*Plus and all other open applications.

3

SUMMARY

- In a single-table SELECT query, the SELECT clause lists the columns that you want to display in your query, the FROM clause lists the name of the table involved in the query, and the WHERE clause specifies a search condition.

- If you are querying a table that you did not create, then you must preface the table name with the creator's username, and the creator must have given you the privilege to retrieve data from that table.

- To retrieve every row in a table, the data values do not need to satisfy a search condition, so you omit the WHERE clause.

- If you want to retrieve all of the columns in a table, you can use an asterisk (*) as a wildcard character in the SELECT clause instead of typing every column name.

- The SELECT DISTINCT command suppresses duplicate records in query output.

- To retrieve specific records, you specify a search condition in the WHERE clause. The search condition must be able to be evaluated as TRUE or FALSE. Multiple search conditions can be combined using the AND and OR operators.

- To return records where the value of a particular field is NULL, you use the search condition `WHERE fieldname IS NULL`. To return records where the value of a particular field is not NULL, you use the search condition `WHERE fieldname IS NOT NULL`.

- You can perform inexact searches that retrieve values that match partial text strings using the LIKE operator and the % and _ wildcard characters.

- You can sort query outputs using the ORDER BY clause.

- SQL has operations for adding, subtracting, multiplying, and dividing retrieved data values.

- The SYSDATE function returns the current system (server) date and time.

- SQL has single-value number, character, and date functions that allow you to manipulate retrieved values on each row of data retrieved in a query in a variety of ways.

- SQL has group functions for performing operations on a group of retrieved records. Functions exist for summing, averaging, finding the maximum or minimum value of, or counting the number of records returned.

- An alias is an alternate output column name that describes the column data and hides the underlying complexity of the query.

- Dynamic SQL queries allow you to modify query parameters while a query is executing and prompt the user for input values. You can create dynamic SQL commands using substitution values, the ACCEPT command, or runtime variables.

❏ You can format SQL*Plus output by adding report headings, changing column widths and headings, and changing the line size and page size.

❏ To format values stored in DATE and NUMBER fields in a SQL query, you can convert the field to a character field using the TO_CHAR function and then format the field using a format mask.

REVIEW QUESTIONS

1. How do you retrieve all records in a table? How do you retrieve all fields in a table without writing the name of each field in the SELECT clause?

2. What is the purpose of the DISTINCT qualifier?

3. In what order are AND and OR operations evaluated? How can you force a different order?

4. In what order are arithmetic operations evaluated in SQL?

5. What is the purpose of the SYSDATE function?

6. Predict the output of the following queries that use single-row number functions. Refer to the data values from the Clearwater Traders database (Figure 1-11), the Northwoods University database (Figure 1-12), and the Software Experts database (Figure 1-13) as necessary to determine retrieved data values.

   a. `SELECT ABS(total_hours) FROM project_consultant WHERE p_id = 1 AND c_id = 101;`

   b. `SELECT POWER(credits, 2) FROM course WHERE course_id = 3;`

   c. `SELECT ROUND(price, 1) FROM inventory WHERE inv_id = 11668;`

   d. `SELECT TRUNC(price) FROM inventory WHERE inv_id = 11668;`

7. Predict the outputs of the following queries that use single-row character functions. Refer to the data values in the Clearwater Traders database (Figure 1-11), the Northwoods University database (Figure 1-12), and the Software Experts database (Figure 1-13) as necessary to determine retrieved data values.

   a. `SELECT DISTINCT INITCAP(f_rank) FROM faculty;`

   b. `SELECT LPAD(price, 8, '-') FROM inventory WHERE inv_id = 11668;`

   c. `SELECT RTRIM(d_phone, '715') FROM customer;`

   d. `SELECT CONCAT(c_first, c_last) FROM consultant WHERE c_id = 102;`

8. Predict the output of the following queries that use date functions. Refer to the data values from the Clearwater Traders database (Figure 1-11), the Northwoods University database (Figure 1-12), and the Software Experts database (Figure 1-13) as necessary to determine retrieved data values.

a. `SELECT s_last, s_first FROM student`
   `WHERE s_dob > TO_DATE('01/01/1977', 'MM/DD/YYYY')`
   `AND s_dob <= ADD_MONTHS(TO_DATE('01/01/1977',`
   `'MM/DD/YYYY'), 11);`

b. `SELECT order_id, order_date FROM cust_order`
   `WHERE order_date >= TO_DATE('06/01/2003', 'MM/DD/YYYY')`
   `AND order_date <= TO_DATE('06/30/2003', 'MM/DD/YYYY');`

c. `SELECT roll_off_date - roll_on_date FROM project_consultant`
   `WHERE p_id = 1 AND c_id = 101;`

9. Predict the output of the following queries that use group functions. Refer to the data values from the Clearwater Traders database (Figure 1-11), the Northwoods University database (Figure 1-12), and the Software Experts database (Figure 1-13) as necessary to determine retrieved data values.

a. `SELECT AVG(capacity) FROM location WHERE bldg_code = 'CR';`

b. `SELECT COUNT(*) FROM project_consultant WHERE p_id = 1;`

c. `SELECT COUNT(date_received) FROM shipment_line;`

d. `SELECT MAX(max_enrl) FROM course_section WHERE term_id = 6;`

e. `SELECT MIN(capacity) FROM location WHERE bldg_code = 'LIB';`

f. `SELECT SUM(capacity) FROM location WHERE bldg_code = 'CR'`
   `OR bldg_code = 'LIB';`

10. When is it necessary to use the GROUP BY clause in a SELECT query that contains a group function?

11. When creating dynamic SQL queries, what is the advantage of using runtime variables and the ACCEPT command over using substitution values?

PROBLEM-SOLVING CASES

For all cases, use Notepad or another text editor to write a script using the specified filename. Place the commands in the order listed, and save the script files in your Chapter3\CaseSolutions folder.

1. Create a script named 3BCase1.sql that contains the commands for the following queries that retrieve data from the Northwoods University database (see Figure 1-12).

a. Retrieve the S_CLASS field for all rows in the STUDENT table and suppress duplicate output. Sort the values in alphabetical order.

b. Retrieve the S_FIRST, S_LAST, and S_MI fields for every student who does not have a value for S_MI.

c. Retrieve the S_FIRST, S_LAST, and S_DOB for all students whose S_CLASS value is either 'JR' or 'SR'. Use the IN operator in the search condition. Format the S_DOB field as 'MM/DD/YYYY', using the TO_CHAR function to convert the date field to a character and apply the format mask.

d. Retrieve the S_FIRST, S_LAST, and S_DOB of all students who were born during 1982.

e. Retrieve the C_SEC_ID, SEC_NUM, DAY, and TIME for all courses in the COURSE_SECTION table that are held on Monday. (*Hint*: Use the LIKE search operator.) Format the TIME field so the time values appear as times and not as dates.

2. Create a script named 3BCase2.sql that contains the commands for the following queries that retrieve data from the Northwoods University database (see Figure 1-12).

a. Retrieve all of the fields from the ENROLLMENT table in which a grade either has not yet been assigned or in which the grade is not 'C'. Order the output by grade value, listing the As first, Bs second, and NULL values last. (*Hint*: Use the NOT operator in the second search condition.)

b. Calculate the difference between the current SYSDATE and student Michael Connoly's date of birth. Calculate the difference so it appears in years, and truncate the fraction so the age is shown as a whole number. Display the output using the column heading "Age". Use 'Michael' and 'Connoly' as the search condition in the query's WHERE clause.

c. Retrieve the C_SEC_ID, SEC_NUM, DAY, and TIME values for all course sections. Convert the TIME field so the values appear as times, and remove the leading zeros from values like '08:00 A.M.' using the LTRIM function.

d. Retrieve the number of Bs assigned in course section 1008.

e. Calculate the average maximum enrollment for all sections of COURSE_ID 1 during TERM_ID 4.

3. Create a script named 3BCase3.sql that contains the commands for the script whose output is shown in Figure 3-67. The script should use runtime variables and the ACCEPT command to prompt the user to input a value for term ID and course ID, and the associated query should display the sum of the maximum enrollment for all course sections that were offered during the specified term for the specified course.

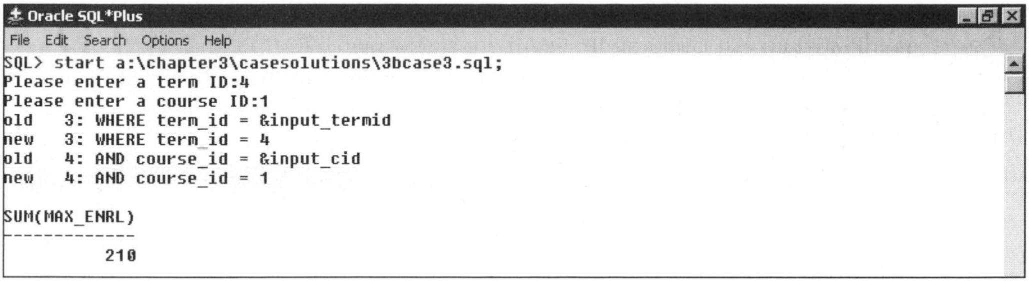

```
Oracle SQL*Plus                                                    _ 8 X
File  Edit  Search  Options  Help
SQL> start a:\chapter3\casesolutions\3bcase3.sql;
Please enter a term ID:4
Please enter a course ID:1
old    3: WHERE term_id = &input_termid
new    3: WHERE term_id = 4
old    4: AND course_id = &input_cid
new    4: AND course_id = 1

SUM(MAX_ENRL)
-------------
          210
```

Figure 3-67

4. Create a script named 3BCase4.sql that contains the commands for the following queries that retrieve data from the Clearwater Traders database (see Figure 1-11).

   a. Retrieve all fields from the CUST_ORDER table associated with orders placed on May 29, 2003, where the method of payment was 'CC'.

   b. Retrieve the COLOR field from every row in the INVENTORY table, sorted in alphabetical order, with duplicate output suppressed. Display the retrieved values in all uppercase characters. (*Hint*: In the ORDER BY clause, you will have to include the character function along with the field name.)

   c. Retrieve the SHIPMENT_ID, INV_ID, and DATE_RECEIVED values for all shipments received during August 2003.

   d. Calculate the number of days between the current SYSDATE and the DATE_EXPECTED value for SHIPMENT_ID 211. Structure the query so the result appears as a positive value. Round the output to the nearest whole day, and specify the output column using a column alias named DIFFERENCE_IN_DAYS.

   e. Retrieve the ITEM_ID and ITEM_DESC for all items where the text string 'Women's' appears anywhere in the ITEM_DESC field.

5. Create a script named 3BCase5.sql that contains the commands for the following queries that retrieve data from the Clearwater Traders database (see Figure 1-11).

   a. Sum the quantity on hand for every inventory item associated with ITEM_ID 995 in the INVENTORY table.

   b. Display the number of records in the INVENTORY table for which no size value is specified.

   c. Display the item ID and the sum of the inventory value (price times quantity on hand) for each associated item in the INVENTORY table. Display the inventory values using the column heading "Inv.Value", and use the TO_CHAR function to format the number values using a currency format mask with two decimal places.

   d. Display the order ID and total number of items ordered in each order. Only display the results for orders where three or more items were ordered. (*Hint*: You will need to use the GROUP BY function and the HAVING clause.)

6. Create a script named 3BCase6.sql that contains the commands to create the Clearwater Traders Inventory Report shown in Figure 3-68. (*Hint:* You will need to assign an alias to the column that calculates the inventory value, and then use the alias when you specify the column heading and format.)

```
± Oracle SQL*Plus                                                        _ 8 X
File  Edit  Search  Options  Help
SQL> start a:\chapter3\casesolutions\3bcase6.sql;

Clearwater Traders
Inventory Report
================================================================
Inv. ID  Item ID Size Color             Price  QOH      Value
-------  ------- ---- ------------      -----  ---  ----------
  11668    786        Sky Blue        $259.99   16   $4,159.84
  11669    786        Light Grey      $259.99   12   $3,119.88
  11775    894 S      Khaki            $29.95  150   $4,492.50
  11776    894 M      Khaki            $29.95  147   $4,402.65
  11777    894 L      Khaki            $29.95    0       $.00
  11778    894 S      Navy             $29.95  139   $4,163.05
  11779    894 M      Navy             $29.95  137   $4,103.15
  11780    894 L      Navy             $29.95  115   $3,444.25
  11795    897 S      Eggplant         $59.95  135   $8,093.25
  11796    897 M      Eggplant         $59.95  168  $10,071.60
  11797    897 L      Eggplant         $59.95  187  $11,210.65
  11798    897 S      Royal            $59.95    0       $.00
  11799    897 M      Royal            $59.95  124   $7,433.80
  11800    897 L      Royal            $59.95  112   $6,714.40
  11820    995 10     Turquoise        $15.99  121   $1,934.79
  11821    995 11     Turquoise        $15.99  111   $1,774.89
  11822    995 12     Turquoise        $15.99  113   $1,806.87
  11823    995 1      Turquoise        $15.99  121   $1,934.79
  11824    995 10     Bright Pink      $15.99  148   $2,366.52
  11825    995 11     Bright Pink      $15.99  137   $2,190.63
  11826    995 12     Bright Pink      $15.99  134   $2,142.66
  11827    995 1      Bright Pink      $15.99  123   $1,966.77
  11845    559 S      Spruce          $199.95  114  $22,794.30
  11846    559 M      Spruce          $199.95   17   $3,399.15
  11847    559 L      Spruce          $209.95    0       $.00
  11848    559 XL     Spruce          $209.95   12   $2,519.40

26 rows selected.
```

Figure 3-68

7. Create a script named 3BCase7.sql that contains the commands for the following queries that retrieve data from the Software Experts database (see Figure 1-13).

   a. Retrieve the name of each city where a consultant lives. Suppress duplicate output, and display the values in alphabetical order and in all uppercase characters.

   b. Retrieve the project ID and project name of all projects that have parent projects.

   c. Retrieve the consultant ID, skill ID, and certification status for every consultant who is proficient with skill ID 1, skill ID 2, or skill ID 3.

   d. Retrieve the project ID and project name for all projects that have the text string "Internet" or "Intranet" anywhere in the project name.

3

8. Create a script named 3BCase8.sql that contains the commands for the following queries that retrieve data from the Software Experts database (see Figure 1-13).

a. Display the average evaluation score for consultant ID (EVALUATEE_ID) 105. Round the retrieved value to one decimal place.

b. Retrieve the project ID, consultant ID, and total number of months that each consultant has spent on each project. Assign an alias named TOTAL_MONTHS to the calculated field. (*Hint*: Calculate the number of days as ROLL_OFF_DATE minus ROLL_ON_DATE, and convert the value to months by assuming that an average month has 30.4 days.) Truncate the number of months to two decimal places.

c. Count the number of consultants who are certified in skill ID 1.

d. Display each project ID and the total number of hours that all consultants have spent on that project. Only display data for projects that have accumulated over 100 hours. (*Hint*: Use the GROUP BY clause and the HAVING clause.)

e. Display the maximum, minimum, and average number of hours that consultants have spent on Project 1.

9. Create a script named 3BCase9.sql that contains the command for the query of the CONSULTANT_PROJECT table in the Software Experts database (see Figure 1-13) that displays the output shown in Figure 3-69. The script should use substitution values to prompt the user for a value for project ID, and the associated query should display the consultant ID, roll on date, roll off date, months spent on the project, and total hours. Format the date values as shown using the TO_CHAR function. Use the MONTHS_BETWEEN function to calculate the months spent on the project, and round the number of months to one decimal place.

```
± Oracle SQL*Plus                                                    _ 8 X
File  Edit  Search  Options  Help
SQL> start a:\chapter3\casesolutions\3bcase9.sql;
Enter value for project_id: 1
old    3: WHERE p_id = &PROJECT_ID
new    3: WHERE p_id = 1

Clearwater Traders
Inventory Report
===============================================================

    C_ID ROLL_ON     ROLL_OFF    MONTHS_ON TOTAL_HOURS
    ---- ----------  ----------  --------- -----------
     101 06/15/2002  12/15/2002          6         175
     103 01/05/2002  06/05/2002          5          50
     104 01/05/2002  12/15/2002       11.3         245
     105 01/05/2002  12/15/2002       11.3          45
```

Figure 3-69

10. Create a script named 3BCase10.sql that contains the commands to create the Software Experts Consultant Evaluation Report shown in Figure 3-70. Retrieve the results using four separate queries. The first query will retrieve the project ID, evaluation date, and score values, and the other three queries will retrieve the average, highest, and lowest score. After the top title appears once, suppress it from being displayed again by using the TTITLE OFF command in the script.

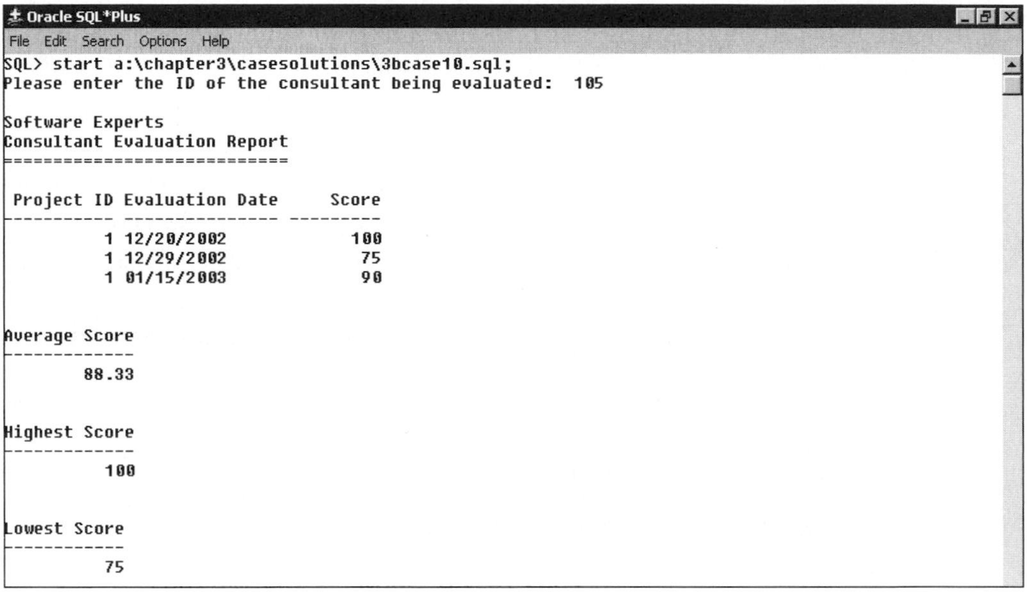

```
Oracle SQL*Plus                                                    _ 8 X
File  Edit  Search  Options  Help
SQL> start a:\chapter3\casesolutions\3bcase10.sql;
Please enter the ID of the consultant being evaluated:  105

Software Experts
Consultant Evaluation Report
=============================

 Project ID Evaluation Date      Score
----------- ---------------- ---------
          1 12/20/2002             100
          1 12/29/2002              75
          1 01/15/2003              90


Average Score
-------------
        88.33


Highest Score
-------------
          100


Lowest Score
------------
           75
```

Figure 3-70

◄ LESSON C ►

Objectives

♦ Write SQL queries to join multiple database tables
♦ Create subqueries and nested queries
♦ Combine query results using set operators
♦ Select records for update
♦ Create database views and indexes
♦ Understand public and private synonyms

In this lesson, you will create queries that combine data values from multiple tables using a variety of join operations. You will learn how to create nested SQL queries. You will also learn techniques for creating queries that use set operators and that use the SELECT FOR UPDATE function to lock records while the user is viewing them. In addition, you will be introduced to database views, indexes, and synonyms.

JOINING MULTIPLE TABLES

All of the queries you have created so far have retrieved data from a single table. However, one of the strengths of SQL is its ability to **join**, or combine, data from multiple database tables using foreign key references. First, you will refresh the values in your database tables and customize the page width and length in your SQL*Plus environment.

To refresh your database tables and customize the SQL*Plus environment:

1. If necessary, start SQL*Plus and log onto the database. Run the northwoods.sql script by typing the following command at the SQL prompt: **START a:\chapter3\northwoods.sql**.

2. Run the clearwater.sql script by typing the following command at the SQL prompt: **START a:\chapter3\clearwater.sql**.

3. Run the softwareexp.sql script by typing the following command at the SQL prompt: **START a:\chapter3\softwareexp.sql**.

4. Type **SET LINESIZE 90** at the SQL prompt to increase the width of the SQL*Plus display to 90 characters.

5. Type **SET PAGESIZE 35** at the SQL prompt to increase the length of the SQL*Plus display to 35 lines.

The general syntax of a SELECT statement that joins two tables is:

```
SELECT column1, column2, …
FROM table1, table2
WHERE table1.join_column = table2.join_column;
```

The SELECT clause contains the names of the columns to display in the query output. If you display a column that exists in more than one of the tables in the FROM clause, you must write the table name, followed by a period, and then write the column name. Otherwise, the DBMS will issue an error. Listing the table name before a column name is known as **qualifying** the column name. Because the column exists in more than one table, you can qualify the column name using the name of any table that is listed in the FROM clause that contains the field. For example, suppose you write a query that lists the ITEM_ID field in the SELECT clause, and the FROM clause includes both the ITEM and INVENTORY tables. Because ITEM_ID is in both tables, you would qualify the ITEM_ID field in the SELECT clause as **ITEM.ITEM_ID**.

The FROM clause contains the name of each table involved in the join operation. The WHERE clause contains the **join condition**, which specifies the table and column names on which to join the tables. The join condition contains the foreign key reference in one table and the primary key in the other table. Additional search conditions can be listed in the WHERE clause using the AND and OR operators. In the following paragraphs, you will learn how to write queries that join multiple tables, and you will learn about the different ways that tables can be joined.

Inner (Equality) Joins

The simplest type of join is where two tables are joined based on values in one table being equal to values in another table. This type of join is called an **inner join**, **equality join**, or **equijoin**. For example, suppose you want to retrieve student last and first names, along with each student's advisor ID and last name. Each student's advisor ID is specified in the F_ID column in the STUDENT table, and each F_ID value in the STUDENT table corresponds to a specific F_ID value in the FACULTY table. This query requires you to retrieve data from both the STUDENT and FACULTY tables, and to join the tables on the F_ID column, which is the primary key in the FACULTY table and a foreign key in the STUDENT table. Now you will execute the command to retrieve these values.

To retrieve records from two tables by joining them through a foreign key reference using an inner join:

1. Start Notepad and create a new file name **3CQueries.sql** in your Chapter3\TutorialSolutions folder. Type the command shown in Figure 3-71 to retrieve the student ID, last name, first name, advisor ID, and advisor last name, and then execute the command in SQL*Plus. Note that you must qualify the F_ID field in the SELECT clause, because the F_ID field exists in both the STUDENT and FACULTY tables. You could qualify F_ID using either the STUDENT table or the FACULTY table.

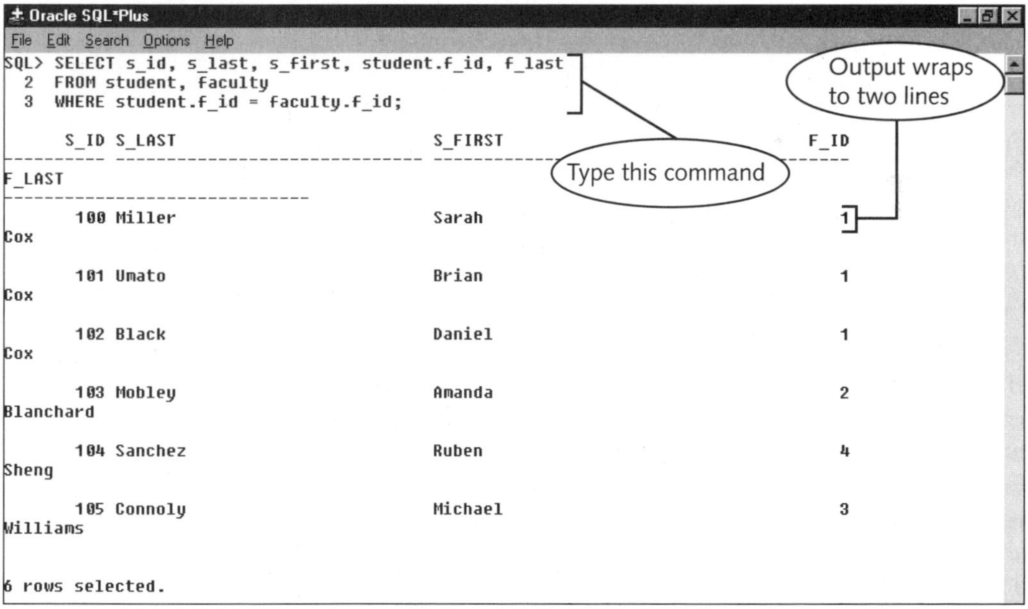

Figure 3-71   Inner join of two tables

You will enter all of the tutorial commands in your 3CQueries.sql file, and then execute them in SQL*Plus.

2. To add additional search conditions in the WHERE clause using the AND operator, type the query shown in Figure 3-72 to list the location ID, build-ing code, and room number for faculty member Laura Sheng. Note that in the SELECT clause, LOC_ID needs to be qualified with a table name, because the LOC_ID field exists in both the FACULTY and LOCATION tables. You could preface LOC_ID with either LOCATION or FACULTY, since the value is the same in both tables for the joined records.

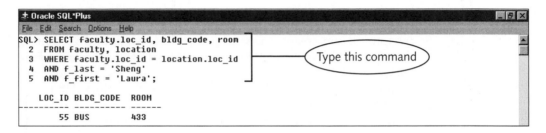

Figure 3-72   Inner join query with additional search conditions

Although you joined two tables in the previous two examples, you can join any number of tables in a SELECT command. When you join tables, each table involved in the query must be listed in the FROM clause, and each table in the FROM clause must be listed in the WHERE clause in a join condition. Suppose that you want to create a query to dis-play the call ID and grade for each of Sarah Miller's courses. This query requires you to join four tables: STUDENT (to search for S_FIRST and S_LAST), ENROLLMENT (to display GRADE), COURSE (to display CALL_ID), and COURSE_SECTION. You join the STUDENT and ENROLLMENT tables using the S_ID field, because S_ID is the primary key in the STUDENT table and a foreign key in the ENROLLMENT table. You need to include the COURSE_SECTION table in order to join the ENROLLMENT table to the COURSE table, using the COURSE_ID foreign key link between COURSE and COURSE_SECTION and the C_SEC_ID foreign key link between COURSE_SECTION and ENROLLMENT. For complex queries like this, it is often helpful to draw a diagram like the one shown in Figure 3-73, which shows the columns and associated tables that you need to display and search, as well as the required joining columns and their links.

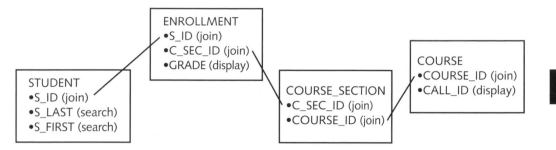

Figure 3-73   Join query design diagram

The process for deriving the SQL query based on this join query design diagram is described in Table 3-14.

Step	Process	Result
1	Create the SELECT clause by listing the display fields	`SELECT grade, call_id`
2	Create the FROM clause by listing the table names	`FROM student, enrollment,` `course_section, course`

3	Create a join condition for every link between the tables	`WHERE student.s_id =`
		`enrollment.s_id`
		`AND enrollment.c_sec_id =`
		`course_section.c_sec_id`
		`AND course_section.course_id =`
		`course.course_id`

4	Create additional search conditions for remaining search fields	`AND s_last = 'Miller'` `AND s_first = 'Sarah'`

Table 3-14   Deriving SQL query from join query design diagram

Note that in Figure 3-73, there are four tables and three links between the tables. Because there are three links, there must be three join conditions. You must always have one fewer join condition than the total number of tables joined in the query. In this query, you are joining four tables, so you have three join conditions.

Now you will type the command derived in Table 3-14 to join the four database tables.

To join four tables in a single query:

    1. Type the command shown in Figure 3-74.

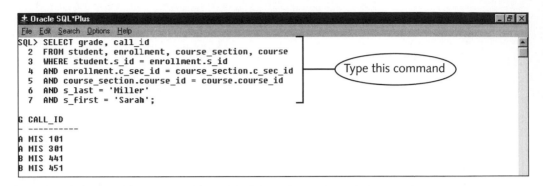

Figure 3-74   Query joining four tables

If you accidentally omit a join condition in a multiple-table query, the result is a **Cartesian product**, whereby every row in one table is joined with every row in the other table. For example, suppose you repeat the query to show each student record along with each student's advisor, but you omit the join condition. Every row in the STUDENT table (six rows) is joined with every row in the FACULTY table (five rows). The result is 6 times 5 rows, or 30 rows. You will create this query next.

To create a Cartesian product by omitting a join condition:

    1. Type the command shown in Figure 3-75. Each row in the STUDENT table is first joined with the first row in the FACULTY table (F_LAST 'Cox'). Next, each row in the STUDENT table is joined with the second row in the FACULTY table (F_LAST 'Blanchard'). This continues until each STUDENT row is joined with each FACULTY row, for a total of 30 rows returned. (Some rows scroll off-screen.) When a multiple-table query returns more records than you expect, look for missing join conditions.

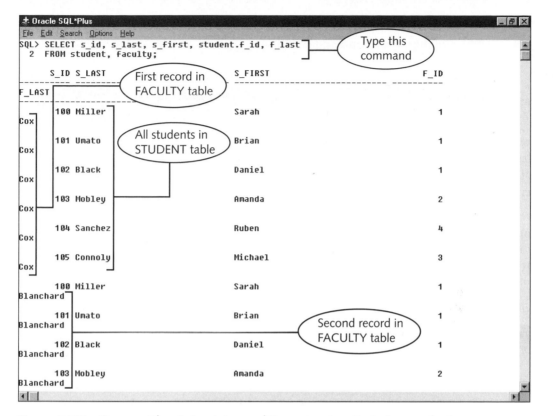

Figure 3-75   Query with missing join condition returning Cartesian product

Outer Joins

A limitation of an inner join is that it might omit records that you want to retrieve if no records exist in one of the tables involved in the join condition. For example, suppose you want to retrieve the item descriptions, sizes, color, quantity on hand, dates shipments were received, and shipment quantities from the Clearwater Traders database. This query requires joining records in the ITEM, INVENTORY, and SHIPMENT_LINE tables. Now you will type the command to retrieve this data using an inner join.

To retrieve inventory and shipment information using an inner join:

1. Type the command shown in Figure 3-76. Note that 11 records were retrieved.

```
Oracle SQL*Plus
File  Edit  Search  Options  Help
SQL> SELECT inventory.inv_id, item_size, color, qoh, shipment_id, date_received, ship_quantity
  2  FROM inventory, shipment_line
  3  WHERE inventory.inv_id = shipment_line.inv_id;

   INV_ID ITEM_SIZE  COLOR                            QOH SHIPMENT_ID DATE_RECE SHIP_QUANTITY
---------- ---------- -------------------- ---------- ----------- --------- -------------
    11668             Sky Blue                         16         211 10-SEP-03            25
    11669             Light Grey                       12         211 10-SEP-03            25
    11669             Light Grey                       12         212                      25
    11777 L           Khaki                             0         213                     200
    11778 S           Navy                            139         213                     200
    11779 M           Navy                            137         213                     200
    11798 S           Royal                             0         214 15-AUG-03           100
    11799 M           Royal                           124         214 25-AUG-03           100
    11845 S           Spruce                          114         215 15-AUG-03            50
    11846 M           Spruce                           17         215 15-AUG-03           100
    11847 L           Spruce                            0         215 15-AUG-03           100

11 rows selected.
```

Type this command

Figure 3-76 Retrieving incomplete information using an inner join

The query output seems incomplete. Information is shown for only 11 items, while the INVENTORY table (see Figure 1-11) contains more than eleven records. Too few records were retrieved because information was retrieved only for inventory ID values that have records in the SHIPMENT_LINE table. To retrieve information for all inventory items, regardless of whether they have associated shipment lines or not, you must use an outer join.

An **outer join** returns all rows in the first table, which is called the **inner table**, along with their matching rows in the second table, which is called the **outer table**. What makes an outer join different is that it also returns values from the inner table that do not have a match in the outer table. To create an outer join in Oracle, you use the syntax **outer_table.join_column(+)** in the join condition. The plus sign (+), which is called the **outer join marker**, indicates that a NULL value is inserted for the fields in the outer table that do not have matching rows in the inner table. Now you will modify this query so all inventory values are retrieved, even values that do not have associated records in the SHIPMENT_LINE table. In this query, the INVENTORY table is the inner table, and the SHIPMENT_LINE table is the outer table.

To retrieve inventory and shipment information using an outer join:

1. Type the command shown in Figure 3-77. The output now shows values for 27 rows, which represents every record in the INVENTORY table. Note that the inventory items that do not have associated shipment lines contain NULL values in the SHIPMENT_ID, DATE_RECEIVED, and SHIP_QUANTITY columns.

Some records do not display
SHIPMENT_LINE information

Outer join marker

Type this
command

3

```
SQL> SELECT inventory.inv_id, item_size, color, qoh, shipment_id, date_received, ship_quantity
  2  FROM inventory, shipment_line
  3  WHERE inventory.inv_id = shipment_line.inv_id (+);

  INV_ID ITEM_SIZE COLOR                QOH SHIPMENT_ID DATE_RECE SHIP QUANTITY
  ------ --------- -----              ----- ----------- --------- ------------
   11668           Sky Blue              16         211 10-SEP-03           25
   11669           Light Grey            12         211 10-SEP-03           25
   11669           Light Grey            12         212                     25
   11775 S         Khaki                150
   11776 M         Khaki                147
   11777 L         Khaki                  0         213                    200
   11778 S         Navy                 139         213                    200
   11779 M         Navy                 137         213                    200
   11780 L         Navy                 115
   11795 S         Eggplant             135
   11796 M         Eggplant             168
   11797 L         Eggplant             187
   11798 S         Royal                  0         214 15-AUG-03          100
   11799 M         Royal                124         214 25-AUG-03          100
   11800 L         Royal                112
   11820 10        Turquoise            121
   11821 11        Turquoise            111
   11822 12        Turquoise            113
   11823 1         Turquoise            121
   11824 10        Bright Pink          148
   11825 11        Bright Pink          137
   11826 12        Bright Pink          134
   11827 1         Bright Pink          123
   11845 S         Spruce               114         215 15-AUG-03           50
   11846 M         Spruce                17         215 15-AUG-03          100
   11847 L         Spruce                 0         215 15-AUG-03          100
   11848 XL        Spruce                12

27 rows selected.
```

Figure 3-77   Retrieving inventory and shipment information using outer join

In Oracle, NULL (undefined) values are represented by blank spaces rather
than the word NULL or an alternate symbol.

Self Joins

In the PROJECT table in the Software Experts database, projects can have subprojects.
The PARENT_P_ID field is a foreign key in each PROJECT record that represents the
project ID of the current project's parent project. To create a query that lists the names
of all projects and the names of their associated subprojects, you must join the PROJECT
table to itself. When a table contains a foreign key reference to its own primary key, it has
a **hierarchical relationship**. When you create a query based on this relationship by joining a table to itself, you create a **self-join**.

To create a self-join, you must create a table alias and structure the query as if you are
joining the table to a copy of itself. A **table alias** is an alternate name assigned to a table

in the FROM clause of a query. When you create a table alias, you must then use the alias, rather than the table name, to qualify field names in the SELECT clause and in join conditions.

The syntax for creating a table alias in the FROM clause is **FROM _table1 alias1_**. To clarify the process for joining the PROJECT table to itself, you will create two table aliases, as shown in Figure 3-78. One alias will be named PARENT_PROJECT, and the second alias will be named SUB_PROJECT. Because the PARENT_P_ID field represents the project ID of the parent project in the subproject record, the join condition will be as follows: WHERE sub_project.parent_p_id = parent_project.p_id.

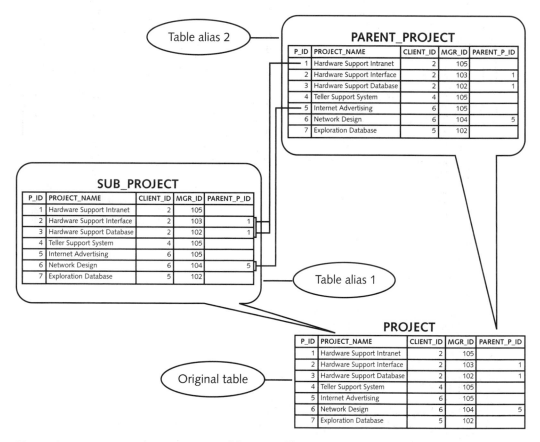

Figure 3-78 Joining the PROJECT table to itself using table aliases

Now you will type the query to list project and subproject names by creating a self-join on the PROJECT table.

To create a self-join on the PROJECT:

1. Type the command shown in Figure 3-79. The SELECT command specifies alternate column headings for the output columns to clarify the output. Note that the output only shows projects that have subprojects.

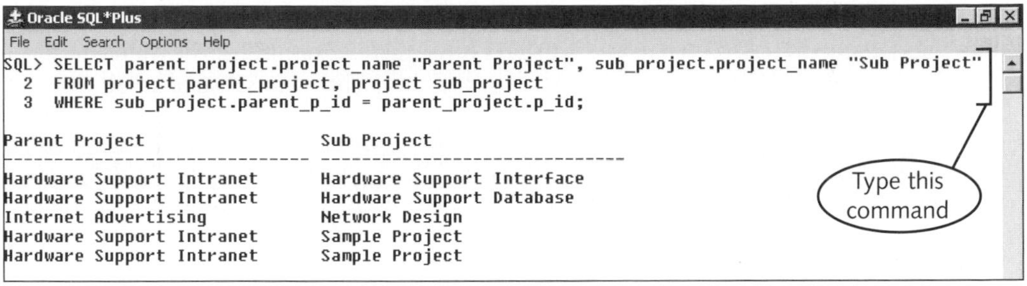

Figure 3-79   Query to join PROJECT table to itself

2. To list all projects, even projects that do not have parent projects, you need to perform an outer join. In this case, the SUB_PROJECT table alias is the outer table, so you will modify the query by placing the outer join marker (+) beside the SUB_PROJECT table in the join condition. Type the command shown in Figure 3-80 to make the self-join query an outer join. The output now displays all projects, and projects that do not have subprojects display NULL values in the SUB_PROJECT column.

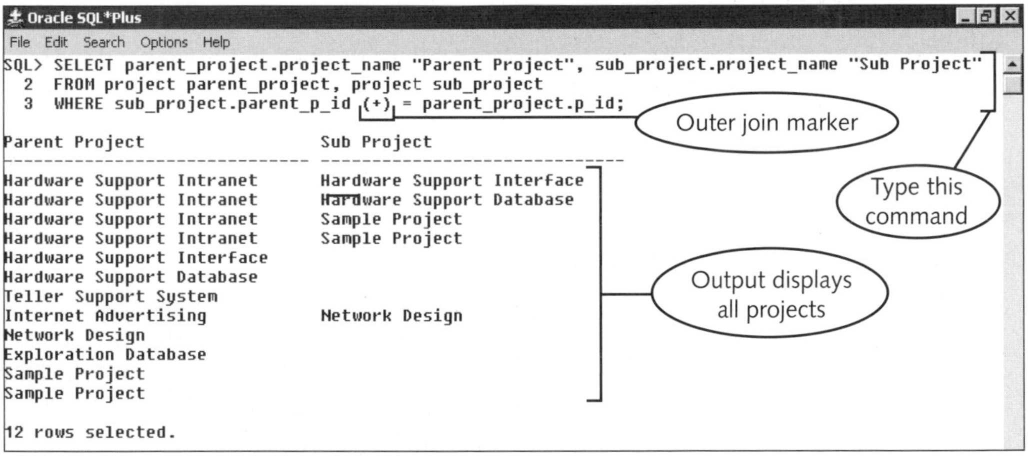

Figure 3-80   Combining self-join with outer join to display all projects

Inequality Joins

Tables can also be joined using an inequality statement, where each record in one table is joined with every record in the second table where the join condition satisfies an inequality condition. For example, Figure 3-81 shows an example of an inequality join, where every record in the FACULTY table is joined with every record in the LOCATION table where the LOC_ID value in the FACULTY table is less than the LOC_ID in the LOCATION table. In the query output, the first record in the FACULTY table (Cox, LOC_ID 53) appears with LOC_ID values 54, 55, 56, and 57 from the LOCATION table, because these values satisfy the inequality condition that the LOC_ID in the FACULTY table must be less than the LOC_ID in the LOCATION table. Examining the values returned for the other FACULTY records indicates that the inequality was satisfied in all cases.

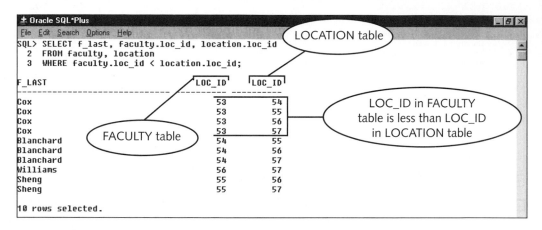

Figure 3-81   Inequality join

The usefulness of query output from an inequality join is open to debate. Inequality joins are only useful for database tables where primary key values (and associated foreign key values) represent actual data. In a database where key values are all represented by surrogate keys that have no meaning associated with the underlying data, inequality joins are not appropriate.

CREATING NESTED QUERIES

A **nested query** is created when a second query, called a subquery, is nested within a main query. The **main query** is the first query that is listed in the SELECT command. The **subquery** is used to specify values that are used in a search condition in the main query. The purpose of creating nested queries is to limit the output of a main query by using one or more subqueries to create intermediate results that are used within the main query's search conditions. Subqueries can return one or more values and can also be nested queries.

Creating Subqueries That Return a Single Value

The general syntax for a creating a nested query with a subquery that returns a single value is shown in Figure 3-82.

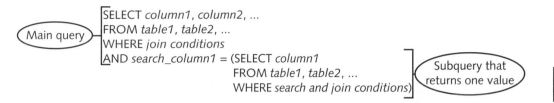

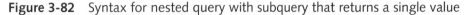

Figure 3-82   Syntax for nested query with subquery that returns a single value

In this syntax, the subquery result specifies the search condition value used in the main query. If the subquery retrieves no rows or multiple rows, an error message appears. You can use subqueries in any DML command's WHERE clause.

For example, suppose you want to retrieve the last and first names of all students who have the same S_CLASS value as student Amanda Mobley. One approach is to first write a query to retrieve Amanda's S_CLASS value, and then write a second query that retrieves the names of all students who have that value for S_CLASS. Alternately, you can combine these steps into a single nested query in which the subquery retrieves Amanda's S_CLASS value, which serves as the search condition for the main query, which retrieves the names of all students with that S_CLASS value. You will now create a nested query.

To create a nested query with a subquery that returns a single value:

1. Type the command shown in Figure 3-83 that retrieves the names of all students who have the same S_CLASS value as student Amanda Mobley. Note that the subquery must retrieve the field in its SELECT clause that corresponds to the field it is searching for in the main query's WHERE clause.

Figure 3-83   Nested query with subquery

When this query is executed in SQL*Plus, the subquery is evaluated first, the result is then substituted in the main query, and the result appears.

When you are debugging nested queries with subqueries, run the subquery separately to ensure it is retrieving the correct results before you combine it with the main query.

Creating Subqueries that Return Multiple Values

The nested query in Figure 3-83 uses an equality condition because the subquery returns a single value, which is Amanda Mobley's S_CLASS value. However, what if the subquery might return multiple results, each of which satisfies the search condition? For example, suppose you want to retrieve the names of all students who have ever been enrolled in the same course section as Amanda Mobley. The subquery will retrieve all C_SEC_ID values in the ENROLLMENT table for Amanda, which will probably include multiple values. The main query will then retrieve the names of all students who have enrolled in any of these same course sections. To structure this query, you must use the IN operator. (You learned earlier that the IN operator enables a search condition to be satisfied using a set of values.) The general syntax for a nested query with a subquery that might return multiple values is shown in Figure 3-84.

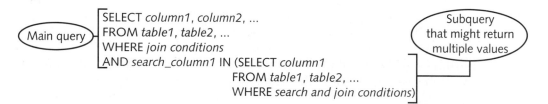

Figure 3-84   Syntax for nested query with subquery that might return multiple values

 You can also use the NOT IN operator to specify for the main query to retrieve all records except those that satisfy the search condition.

Now you will write the command to create a nested query that uses the IN operator, with a subquery that might return multiple values, to retrieve the names of all students who have ever been enrolled in a course with Amanda Mobley. You will use the DISTINCT qualifier in the main query's SELECT clause to suppress duplicate names.

To write a nested query with a subquery that uses the IN operator:

1. Type the command shown in Figure 3-85 to retrieve the names of all students who have enrolled in the same course sections as Amanda Mobley.

Figure 3-85   Nested query using IN operator in a subquery

Using Multiple Subqueries

You can use multiple subqueries within a nested query by joining the search conditions associated with the subqueries using the AND and OR operators. For example, suppose you want to retrieve the names of all students who have the same S_CLASS value as Amanda Mobley and have also been enrolled in a course section with her. The query will join the two search conditions using the AND operator, and each search condition will be specified using a separate subquery.

To write a query that uses multiple subqueries:

1. Type the command shown in Figure 3-86 to retrieve the names of all students who have the same S_CLASS value as Amanda Mobley and have also been enrolled in a course section with her. The output shows that only one student other than Amanda herself, Ruben Sanchez, satisfies both these conditions.

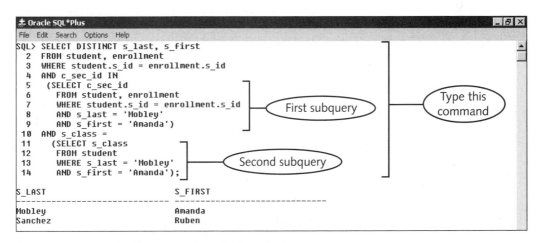

Figure 3-86   Nested query with multiple subqueries

Creating Nested Subqueries

When a subquery contains a subquery, it is called a **nested subquery**. Now you will use a nested subquery to create a query to retrieve the names of students who have taken courses with Amanda Mobley in the CR building. The innermost subquery will retrieve the course section ID values for all course sections located in the CR building. Its main query will retrieve the course sections from this subset that have been taken by Amanda Mobley. The outermost main query will then retrieve the names of all students who enrolled in the course sections that satisfy both of these conditions.

To create a query with a nested subquery:

1. Type the command shown in Figure 3-87. Note that the output excludes student Daniel Black, because the course section that he took with Amanda Mobley (C_SEC_ID 1011) was located in the BUS building.

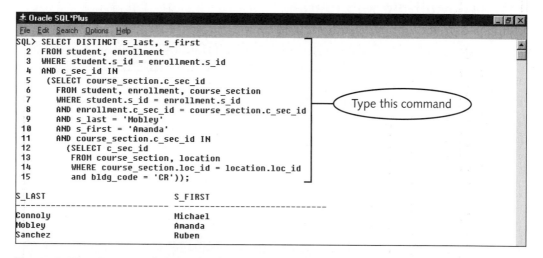

Figure 3-87   Query with nested subquery

Sometimes, you can retrieve the same records either by joining multiple tables and using additional search conditions, or by creating a query that contains subqueries. In the previous example, you could have retrieved the same records by retrieving only the course sections that Amanda Mobley enrolled in that were located in the CR building, and then using this C_SEC_ID set as the search condition in the nested query. You will retrieve the same records by joining multiple tables next.

To retrieve the same records by joining multiple tables in the nested query:

1. Type the command shown in Figure 3-88, which joins the LOCATION and COURSE_SECTION tables in the nested query and eliminates the need for the nested subquery.

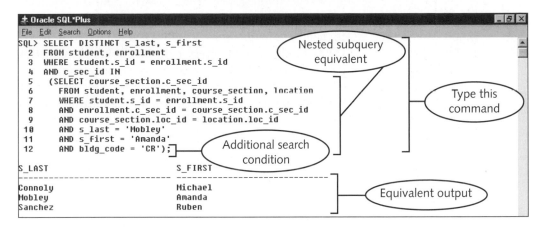

Figure 3-88   Query with additional search condition in nested query

Subqueries can be nested to multiple levels. However, as a general rule, nested queries or queries with nested subqueries execute slower than queries that join multiple tables, so you should probably not use nested queries unless you cannot retrieve the desired result using a non-nested query.

Creating Subqueries That Use the EXISTS Operator

Subqueries are sometimes used with the EXISTS search operator. In this type of query, the subquery must be related to the main query using a table alias. The EXISTS operator has the following syntax in a search condition: **WHERE EXISTS (*subquery*);** If the subquery returns one or more rows, then the search condition evaluates as TRUE, and the related row in the main query is retrieved. If the subquery returns no rows, then the search condition evaluates as FALSE, and the related row in the main query is not retrieved.

For example, suppose you want to retrieve the consultant last name and skill descriptions for every consultant who has more than two skills. Figure 3-89 shows how you can use a simple query to easily retrieve the consultant last name for every consultant satisfying this criteria.

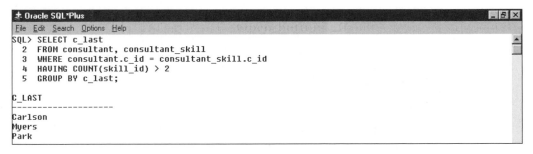

Figure 3-89   Retrieving names of consultants who have two or more skills

You cannot use this query to display the skill descriptions for each consultant, however. When you add the SKILL table to the query, the COUNT search condition will never be true, because each skill appears only once for every consultant/skill combination. To construct this query, you must create a subquery that retrieves the consultant ID for each consultant that satisfies the condition of having two or more skills. You then need to use the EXISTS operator to instruct the database to retrieve the consultant name and skill description of every consultant ID for which this condition is true. You must use a table alias to associate the consultant ID value in the subquery to the consultant ID value in the main query. You create a table alias for the CONSULTANT_SKILL table in the main query, and then join the table alias with the CONSULTANT_SKILL table in the subquery.

To create a query using the EXISTS operator within a subquery:

1. Type the command shown in Figure 3-90 to retrieve the names and associated skill descriptions of all consultants who have more than two skills.

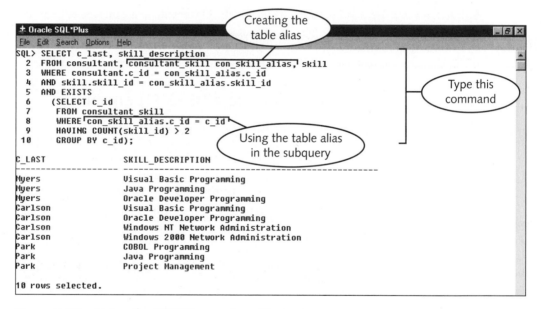

Figure 3-90   Using the EXISTS operator with a subquery

If you do not use the table alias to associate the main query with the subquery, the subquery always evaluates as TRUE, because consultants Carlson, Myers, and Park always satisfy the criteria (see Figure 3-89). As a result, the output shown in Figure 3-91 appears, which retrieves a value for every record in the CONSULTANT_SKILL table. The criteria for returning only values for consultants with more than two skills is ignored. Therefore, when you use the EXISTS operator in a subquery, you must always use a table alias to correlate the parent query to the subquery.

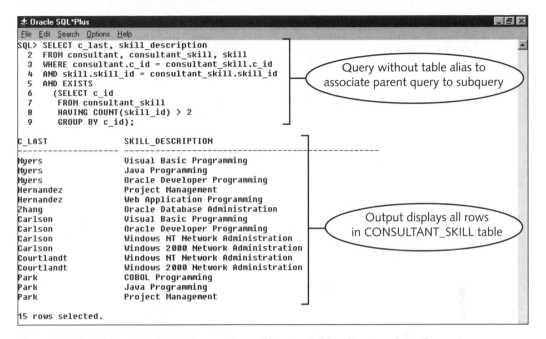

Figure 3-91   Using the EXISTS operator without a table alias to relate the main query and subquery

USING SET OPERATORS TO COMBINE QUERY RESULTS

Sometimes, you need to combine the results of two separate queries in ways that are not specified by foreign key relationships. You can use common set operators to combine the results of two separate queries into a single result. Table 3-15 lists the Oracle SQL set operators and their purpose.

Set Operator	Purpose
UNION	Returns all rows from both queries, but displays duplicate rows only once
UNION ALL	Returns all rows from both queries, and displays all duplicate rows
INTERSECT	Returns only rows returned by both queries
MINUS	Returns the rows returned by the first query minus the matching rows returned by the second query

Table 3-15   Oracle SQL set operators

UNION and UNION ALL

A **UNION** query combines the output of two unrelated queries into a single output result. Suppose you need to create a phone directory of every student and faculty member at Northwoods University. To do this, you would need to create a query to list the last name, first name, and phone number of every student and faculty member in the Northwoods University database. The query shown in Figure 3-92 attempts to retrieve this output using a simple SELECT command. No join clause was included in the command, because the relationship between the STUDENT and FACULTY records is not relevant for this query. A list of all student and faculty names and phone numbers has nothing to do with student–advisor relationships. Nevertheless, a single SELECT command cannot return data from two unrelated queries as a single output. Instead, you must use a UNION.

```
± Oracle SQL*Plus                                                          _ 5 X
File  Edit  Search  Options  Help
SQL> SELECT s_last, s_first, s_phone, f_last, f_first, f_phone
  2  FROM student, faculty;

S_LAST                         S_FIRST                      S_PHONE
------------------------------ ---------------------------- ----------
F_LAST                         F_FIRST                      F_PHONE
------------------------------ ---------------------------- ----------
Miller                         Sarah                        7155559876
Cox                            Kim                          7155551234

Umato                          Brian                        7155552345
Cox                            Kim                          7155551234

Black                          Daniel                       7155553907
Cox                            Kim                          7155551234

Mobley                         Amanda                       7155556902
Cox                            Kim                          7155551234

Sanchez                        Ruben                        7155558899
Cox                            Kim                          7155551234

Connoly                        Michael                      7155554944
Cox                            Kim                          7155551234

Miller                         Sarah                        7155559876
Blanchard                      John                         7155559087

Umato                          Brian                        7155552345
Blanchard                      John                         7155559087

Black                          Daniel                       7155553907
Blanchard                      John                         7155559087

Mobley                         Amanda                       7155556902
Blanchard                      John                         7155559087
```

Figure 3-92   Output that is a Cartesian product of the STUDENT and FACULTY tables

A UNION requires that both queries have the same number of display columns in the SELECT statement, and that each column in the first query has the same data type as the corresponding column in the second query. The general syntax of the UNION query is *query1* **UNION** *query2*;. You will create a UNION query next.

To create a UNION query:

1. Type the command shown in Figure 3-93. The output shows the student and faculty names and phone numbers in a single list. Note that by default the column titles are taken from the column names in the first query's SELECT statement. Also note that output results appear in order, based on the first column of the first SELECT statement.

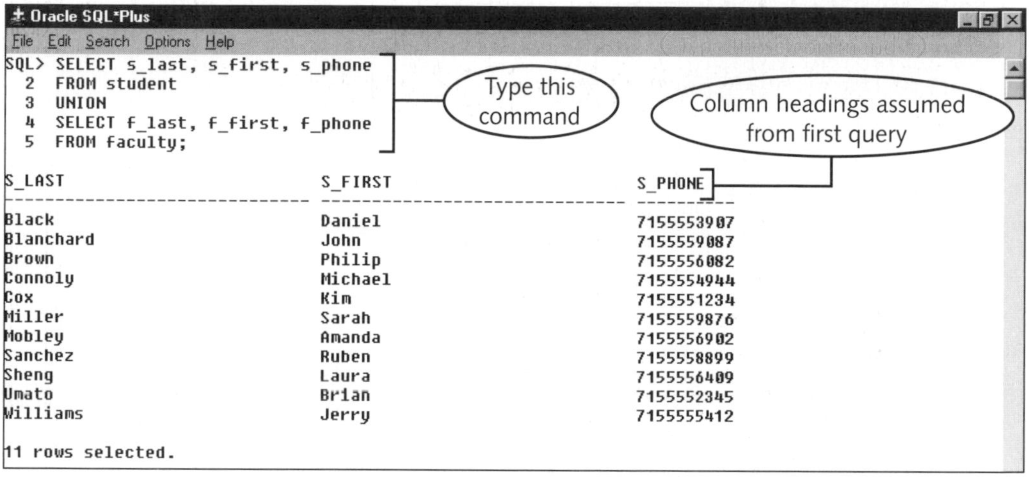

Figure 3-93 UNION query

If one record in the first query exactly matches a record in the second query, the UNION output suppresses duplicates and shows the duplicate record only once. To display duplicate records, you must use the UNION ALL operator. For example, suppose you need to create a query that displays the name of every consultant who has worked on project ID 5 or who has expertise with skill ID 1 (Visual Basic Programming). You must use a UNION to create this query, because the two data sets are not related by foreign key relationships. Some consultants might both have worked on Project 5 and have Visual Basic Programming expertise. If you create a UNION of these two queries, the duplicate records will be suppressed. If you use the UNION ALL operator, then duplicate records will appear. Now you will create the query with the UNION operator and then with the UNION ALL operator, and examine the differences in the results.

To examine UNION and UNION ALL query results:

1. Type the first command shown in Figure 3-94, using the UNION operator. Note that duplicate records do not appear.

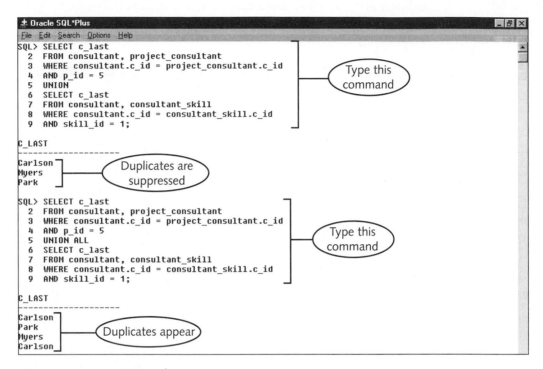

Figure 3-94   UNION and UNION ALL queries

2. Type the second command shown in Figure 3-94, using the UNION ALL operator. This time the duplicate records in both queries appear, showing that consultant Carlson satisfies the search requirements in both queries.

INTERSECT

Some queries require an output that finds the intersection, or matching rows, in two unrelated queries. Like a UNION, an INTERSECT query requires that both queries have the same number of display columns in the SELECT statement and that each column in the first query has the same data type as the corresponding column in the second query. In an INTERSECT query, duplicate values are automatically suppressed.

Suppose you want to create a query that retrieves the first and last names of faculty members who have offices in the BUS building and have taught a class in BUS. This query cannot be created using join conditions, because the two relationships are independent: The building where faculty members have their offices is not related to the building where they teach courses. First, you will type the individual queries and examine the results. Then, you will combine the queries using the INTERSECT set operator to find the names that are returned by both queries.

To create a query using the INTERSECT operator:

    1. First, type the commands shown in Figure 3-95. Faculty members Cox, Blanchard, and Sheng each have an office in BUS, while faculty members Sheng, Cox, Blanchard, and Williams have taught courses in BUS.

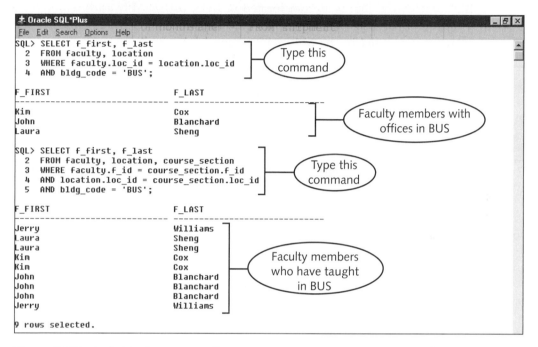

Figure 3-95   Individual query results

    2. Next, type the command shown in Figure 3-96, which uses the INTERSECT operator to find the names on both lists. Note that the INTERSECT operator suppresses duplicate records in the query output.

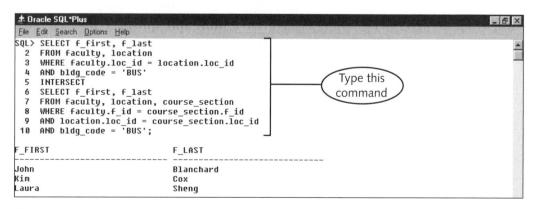

Figure 3-96   Query result using INTERSECT operator

MINUS

The MINUS operator allows you to find the difference between two unrelated query result lists. Duplicate output is also suppressed in queries that use the MINUS operator. Suppose you need to find the names of all faculty members who have taught courses in BUS, but whose offices are not located in BUS. To find this information, you will create a query that returns the names of all faculty members who have taught in BUS, then subtracts the names of the faculty members whose offices are in BUS.

To create a query using the MINUS operator:

1. Type the command shown in Figure 3-97. The query output shows that the only faculty member whose name is not on both lists is Jerry Williams. Note that duplicate output is suppressed—Jerry Williams's name appears twice in the second query in Figure 3-95 but only once in the MINUS query.

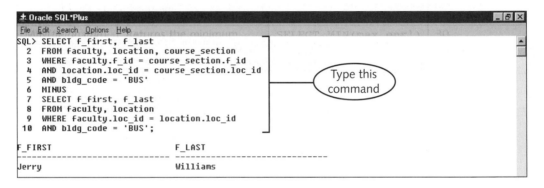

Figure 3-97 Query using MINUS operator

SELECTING RECORDS FOR UPDATE

When you enter a SELECT command to query database records, no locks are placed on the selected records, and other database users can view and update these records at the same time you are viewing them. This is necessary for databases with many concurrent users. Otherwise, many or all of the database records would be locked most of the time, and work would grind to a halt. However, records that have been changed with an INSERT or UPDATE statement are locked until the user holding the lock releases it by issuing a COMMIT or ROLLBACK command.

Sometimes you might want to view a record and then update it in the same transaction. For example, when a Clearwater Traders customer wants to order an item, the sales representative must first determine if the item is in stock before placing the order. Suppose that a particular item is in stock and the sales representative verbally confirms this fact to the customer. Before the sales representative can place the order and commit the update, however, another sales representative sells the entire inventory on hand. To avoid this situation, you must be able to view the quantity on hand and then update it in a single transaction. You can do this in Oracle using the SELECT FOR UPDATE command. The SELECT FOR UPDATE command places a **shared lock** on the record, which means that other users can view the data but cannot update or delete the data.

The general syntax of the SELECT FOR UPDATE command is:

```
SELECT column1, column2,…
FROM table1, table2,…
WHERE search and join conditions
FOR UPDATE OF column1, column2,…
NOWAIT;
```

The column names listed in the FOR UPDATE OF clause do not restrict which columns can be updated in the record, because the entire record is locked. However, listing these column names helps document which fields are to be updated. Normally, when you perform an UPDATE, DELETE, or SELECT FOR UPDATE query, if the record is currently locked by another user, you must wait until the other user commits his or her transaction and releases the record before your transaction can be processed. The NOWAIT command instructs Oracle to not wait if the selected record is currently locked by another user. Instead, a message appears telling you the resource is busy. You can use the NOWAIT command in any UPDATE or DELETE command that requires a record to be unlocked before the command can be processed.

Next, you will select and lock an inventory item by selecting item 11668 from the Clearwater Traders database and using the SELECT FOR UPDATE command.

To select and lock an inventory item in a single transaction:

1. Type the command shown in Figure 3-98 to select and lock the record for INV_ID 11668.

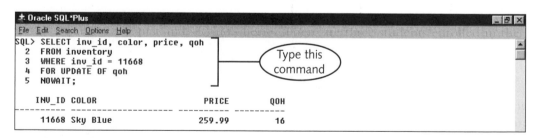

Figure 3-98   Selecting a record FOR UPDATE

After entering this query, your SQL*Plus session has the record for INV_ID 11668 locked with a shared lock. Other users can view the record, but they cannot update or delete the record until you release the record with a COMMIT or ROLLBACK command. To confirm that the record is locked with a shared lock, you will start a second SQL*Plus session and attempt to view and update the record.

To attempt to view and update the record in a second SQL*Plus session:

1. Click **Start** on the Windows taskbar, point to Programs, Oracle for Windows 95 or Windows NT, and click **SQL*Plus 8.0** to start a new SQL*Plus session. Log onto SQL*Plus in the usual way.

 Records are locked by database sessions, not by individual users. You can lock a record from yourself if you log onto SQL*Plus or other Oracle applications multiple times. Each time you log onto Oracle, you create a new session. If you lock a record in one session, you must release the lock by committing or rolling back the transaction to make the record available to other sessions.

2. In the second SQL*Plus session, type the first command shown in Figure 3-99 to view the record, and then press **Enter**. The output appears as expected.

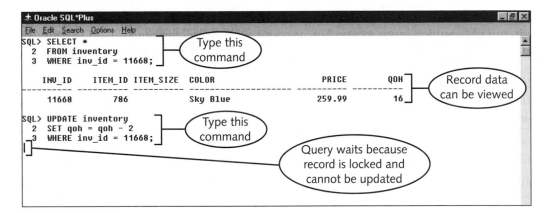

Figure 3-99   SELECT and UPDATE commands in second SQL*Plus session

3. Type the second command shown in Figure 3-99 to update the record. Your SQL*Plus query "waits" because the command cannot be processed, because the current record is locked by your first SQL*Plus session.

4. Click the **Oracle SQL*Plus** button on the taskbar to return to your first SQL*Plus session. Type **COMMIT;** to unlock the record. The "Commit complete" message appears.

5. Click the second **Oracle SQL*Plus** button on the taskbar to return to your second SQL*Plus session when you tried to update the record. The message "1 row updated" now appears. As soon as you committed and unlocked the record in the first SQL*Plus session, the second session was able to update the record.

6. Exit the second SQL*Plus session.

When a record is locked by a SELECT FOR UPDATE command, a second user cannot acquire the lock on the record using a second SELECT FOR UPDATE command. An error message will appear informing the second user that the resource is busy. Sometimes you might inadvertently lock records from yourself if you are multitasking between different Oracle applications. If a query "waits" and will not execute while you are working, determine if you have an uncommitted INSERT, UPDATE, or SELECT FOR UPDATE command in any other database session, and then issue a commit or rollback.

DATABASE VIEWS

In the last chapter, you learned that a database view is based on an actual database table, but presents the table data in a different format based on the needs of the users. A view does not store data, but presents it in a format different from the one in which it is stored in the underlying tables. To an application, a view looks and acts like a table and can be queried like a table. Sometimes applications are not able to display data from multiple tables, but you can create a view to join these tables into what looks like a single table. For example, suppose you are creating a report to list the item description, size, color, price, and quantity ordered for all items associated with ORDER_ID 1057. This report involves joining four tables (CUST_ORDER, ORDER_LINE, INV_ID, and ITEM_ID). However, the application you are using to generate the report will only display data from a single table. You can create a view, then attach the view to the application, and it will work just like a table.

A view does not physically exist in the database. It is derived from other database tables. When the data in its source tables is updated, the view reflects the updates as well. If the structure of a view's source table is altered, or if a view's source table is dropped, then the view becomes invalid and can no longer be used until it is dropped and re-created or replaced.

Views are useful because you do not have to reenter frequently used, complex query commands. They also can be used by the DBA to enforce database security and enable certain users to view only selected table fields or records. There are two types of views: **simple views**, which are based on a single database table and can be used to insert, update, and delete data from that table; and **complex views**, which can be based on multiple tables.

Creating and Manipulating Simple Views

The general command to create a new view is **CREATE VIEW** *view_name* **AS** *SQL_query*;. The view name must follow the Oracle naming standard. The view name cannot already exist in the user's database schema. If there is a possibility that the view already exists, then the view must first be dropped and then re-created, or the command **CREATE OR REPLACE VIEW** *view_name* **AS** *SQL_query*;, must be used to create the view. This command either creates a new view or replaces an existing view with the new specification if a view with the specified name already exists. When you replace an existing view, you only replace the view column definitions. All of the existing object privileges that were granted on the view remain intact.

To create a simple view, the query specification is limited to retrieving fields from a single table. The query to specify a simple view can contain only data values; it cannot contain single-row functions and arithmetic functions. The query also cannot contain the ORDER BY, DISTINCT, or GROUP BY clauses, or group functions.

Simple views are usually used to enforce security measures within a single table. A specific user might be given privileges to view and edit data in a view based on a table, but she or he may not be given privileges to manipulate all of the data in the table. For example, suppose that a data entry person is needed to insert, update, and view data in the FACULTY table, but for security reasons, the DBA does not want this person to have access to the F_PIN column. The DBA would create a view based on the FACULTY table that contains all table fields except F_PIN. In the following example, you will create a simple view named FACULTY_VIEW based on fields in the FACULTY table. This view contains all of the FACULTY columns except F_PIN and F_IMAGE.

To create the simple view FACULTY_VIEW:

1. Type the command shown in Figure 3-100 to create the FACULTY_VIEW. The "View created" confirmation message indicates that the view was created and is similar to the confirmation message generated when a table is created.

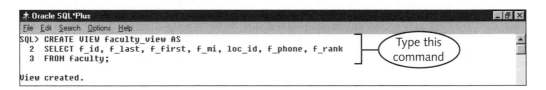

Figure 3-100   Creating a simple view

After creating a simple view, you can use the view to insert data into the underlying database table, as long as the insert operation does not violate any underlying table constraints. For example, suppose the FACULTY table's F_PIN field has a NOT NULL constraint. You could not use the FACULTY_VIEW to insert data for the F_PIN field because the F_PIN field does not exist in the view. In addition, if you did not create the view, then you must also have sufficient object privileges to be able to insert records into the view. Now you will insert a record using the FACULTY_VIEW.

To insert a record using the FACULTY_VIEW:

1. Type the command shown in Figure 3-101 to insert a new record into the FACULTY_VIEW. The message "1 row created" confirms that the record was successfully inserted. Although you inserted the record using the view, the record was actually inserted into the FACULTY table.

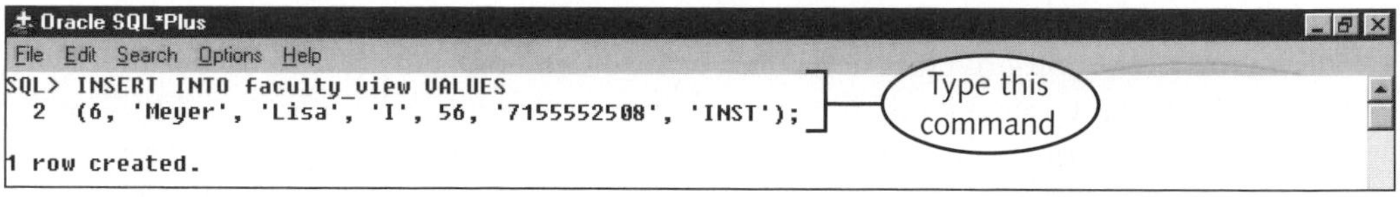

Figure 3-101 Inserting a record using a view

You can also perform update and delete operations on the view just as if it were a database table, again providing you do not violate any constraints that exist for the underlying database table and you have sufficient object privileges for updating or deleting records in the view.

You can query the view using a SELECT statement, just as with a database table, and use the view in complex queries that involve join operations and subqueries. To explore this functionality of the FACULTY_VIEW, you will now create a query that lists the names of each faculty member, along with the building code and room number of the faculty member's office.

To query the FACULTY_VIEW:

1. Type the command shown in Figure 3-102 to retrieve the faculty information by joining the FACULTY_VIEW with the LOCATION table.

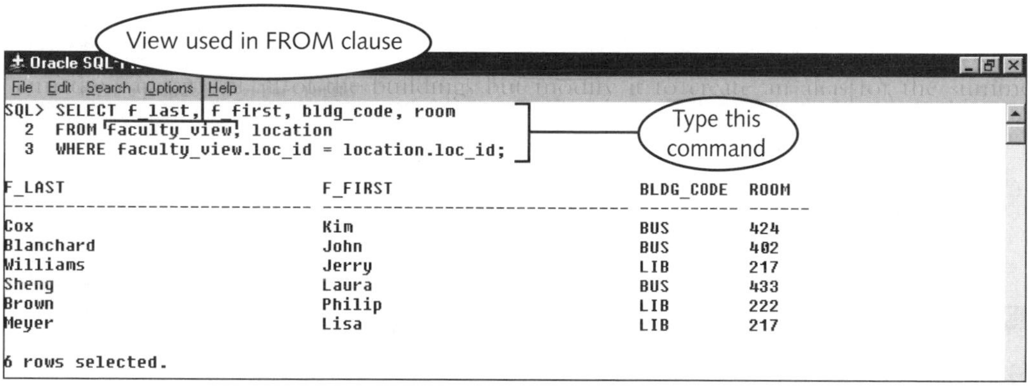

Figure 3-102 Query joining a view with a database table

Creating and Manipulating Complex Views

Complex views can be based on data from multiple tables and can contain columns specified by arithmetic operations, single-row functions, and group functions. When you use a calculated column in a view specification, the column must always be assigned a column alias, which determines how the column will be referenced within the view. The query used to specify a complex view can include the DISTINCT, ORDER BY, and GROUP BY clauses.

Complex views, unlike simple views, cannot be used to directly insert, update, or delete data in the underlying database tables. Complex views provide an excellent way to easily view data while hiding the underlying complexity of multiple-table join queries and complex arithmetic and single-row and group function operations. Now you will create and then retrieve values from a complex view that contains inventory information from the Clearwater Traders database. The view will list the item description, item size, color, price, quantity on hand, and current value of each inventory item, which is calculated as price times quantity on hand. This calculated column will be assigned the column alias VALUE. The view will be ordered based on the item description field.

To create and query a complex view:

1. Type the command shown in Figure 3-103 to create the complex view. The message "View created" confirms that the view was successfully created. Now you will query the view to display its values.

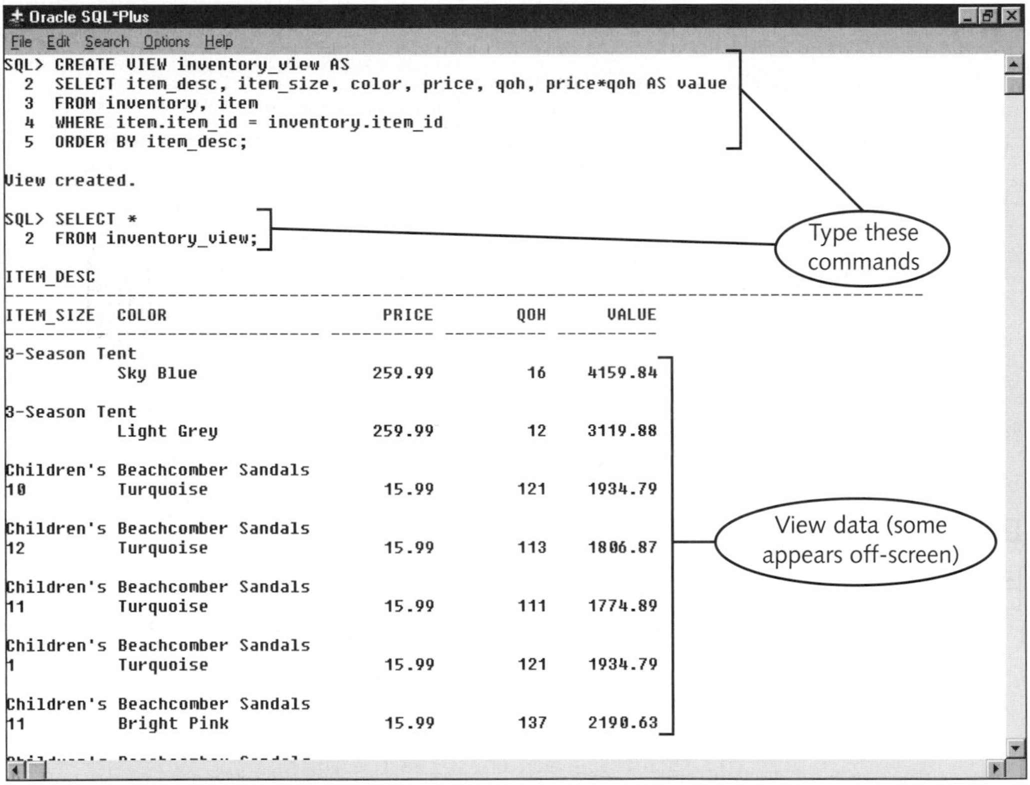

Figure 3-103   Creating a complex view and displaying its contents

2. Type the second command shown in Figure 3-103 to display all of the fields of the INVENTORY_VIEW. Note that the query to display the view is much simpler than the query would be to retrieve the data from the underlying database tables.

Dropping Views

You use the DROP VIEW command to delete a view from your user schema. This command has the syntax **DROP VIEW *view_name*;**. Remember that a view is based on a query that executes to display the requested data from the underlying database table. When you drop a view, you do not delete the data values that appear in the view—you only drop the view definition. Now you will drop the FACULTY_VIEW that you created earlier.

To drop the FACULTY_VIEW:

1. Type **DROP VIEW faculty_view;** to drop FACULTY_VIEW. The "View dropped" confirmation message indicates that the view was dropped.

Using Data Dictionary Views to Retrieve Information About Views

Recall that the Oracle data dictionary views provide information about the overall structure of the database, as well as information about objects in a specific user's database schema. There are two data dictionary views that can be used to retrieve information about database table views. The ALL_VIEWS data dictionary view contains information about views created by all database users, and the USER_VIEWS data dictionary view contains information about only a specific user's views. You can query the ALL_VIEWS table using a specific user name as a search condition, and you can query the USER_VIEWS view to see information about your views only. In the ALL_VIEWS view, you can only see information about views for which you have been granted object privileges. Now you will write a query to display the names of another user's views and your own views.

To display the names of views:

1. Type the first command shown in Figure 3-104 to query the ALL_VIEWS view and show the names of the views that are in the SYSTEM schema. You could substitute any other user name for SYSTEM to see the names of the views created by another database user. Your query output might be different, depending on the database views that you have privileges to view in your database's SYSTEM schema.

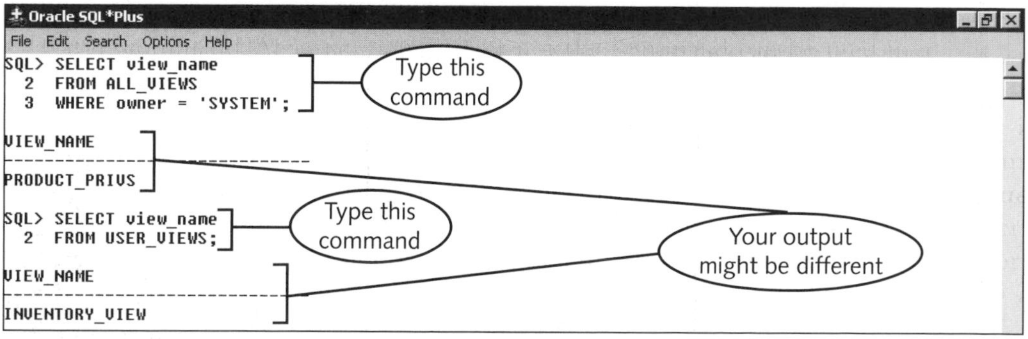

Figure 3-104 Displaying view information from the data dictionary views

2. Type the second command shown in Figure 3-104 using the USER_VIEWS system table to display the names of the views you have created. Your query output might be different, depending on the views that you have created and dropped.

INDEXES

An index in a book contains topics or keywords, along with page references for the locations of materials associated with each topic or keyword. Although you could search through every page in a book to find a particular topic, consulting the index speeds up the process considerably. Similarly, a database **index** is a table that contains data values, along with a corresponding column that specifies the physical locations of the records that contain the associated data values.

Consider the queries shown in Figure 3-105. The first query shows all of the records in the PROJECT_CONSULTANT table, and the second query shows the records in the PROJECT_CONSULTANT table where the TOTAL_HOURS value is greater than 100. The PROJECT_CONSULTANT table doesn't have an index on the TOTAL_HOURS field, so to process the second query, the DBMS must read every record in the table and then retrieve the ones where the TOTAL_HOURS value is greater than 100. In actual practice, the PROJECT_CONSULTANT table might contain hundreds or even thousands of records, so this process might take a long time.

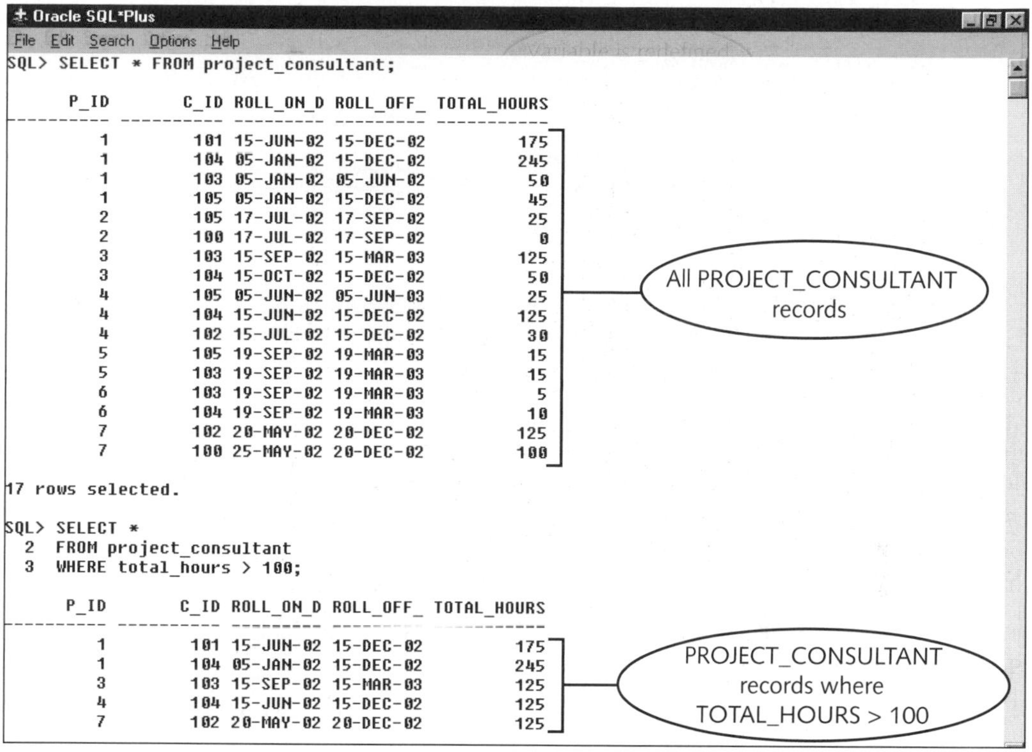

Figure 3-105   Viewing the PROJECT_CONSULTANT records

If the PROJECT_CONSULTANT table had an index on the TOTAL_HOURS field, the index would have the following columns: ROWID, which specifies the internal location of the row; and TOTAL_HOURS, which contains the sorted data values. Every table has a ROWID column, and you can retrieve its values by querying the ROWID pseudocolumn in the SELECT clause. Figure 3-106 shows the values that would be stored in the index on the TOTAL_HOURS field for the PROJECT_CONSULTANT table.

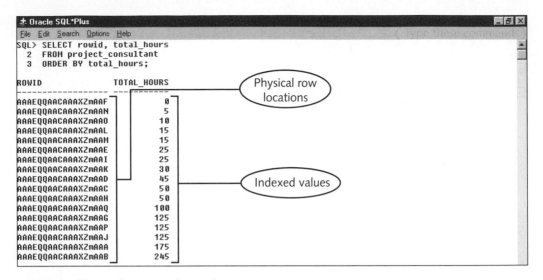

Figure 3-106 Values stored in index

The ROWID column shows the encoded values that correspond to the physical location where each record is stored on the database server. After the index has been created, when you execute the second query in Figure 3-105 to retrieve the PROJECT_CONSULTANT information where TOTAL_HOURS is greater than 100, the DBMS can quickly search the index, find the five records that satisfy the search condition, and based on the associated ROWID value, retrieve and display the output.

Index values are actually stored in a data structure called a B-tree, which speeds up the search process. A B-tree is an inverted tree that uses a "divide and conquer" search strategy. It has search nodes that contain values that direct the search process and leaf nodes that contain the actual data values. A discussion of the mechanics of a B-tree is beyond the scope of this book, but you can find information on B-trees in most computer science books that describe data structures.

When you create a table, Oracle automatically creates an index on the primary key of the table. To improve query performance, you should create indexes on fields that are often specified in query search conditions and on foreign key fields that often have to be matched in queries that join multiple tables. You should create indexes after a table is created and populated with data, not before. Inserting, updating, and deleting records in an indexed table can be a slow process because the DBMS must modify the index every time the record values change. A table can have an unlimited number of indexes, but keep in mind that indexes add considerable overhead on insert, update, and delete operations. In general, indexes are only needed for tables that contain many records. When developing a database application, it is a good practice to estimate the maximum number of records that each table might contain, and then create and load enough sample data records to test and then identify where indexes are needed.

Creating an Index

To create an index, you use the following command: **CREATE INDEX** *index_name*
ON *tablename* **(***index_column***);**. The index name should use the following naming convention: **index_tablename_fieldname**. If these values make the object name violate the Oracle naming standard and exceed 30 characters, then abbreviate the table and field names. It is important to give indexes descriptive names, because when you retrieve the names of your indexes from the data dictionary views, the names must clearly indicate the tables and fields on which the index is based. Now you will create an index named PROJECT_CONSULTANT_TOTAL_HOURS. This index will be on the PROJECT_CONSULTANT table and will index values in the TOTAL_HOURS field.

To create the PROJECT_CONSULTANT_TOTAL_HOURS index:

1. Type the command shown in Figure 3-107 to create the index.

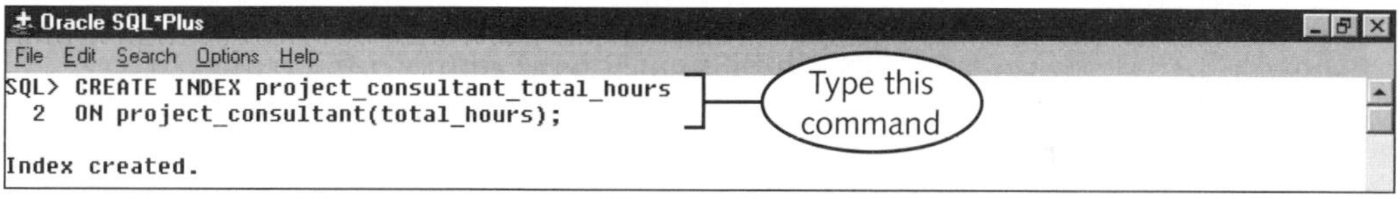

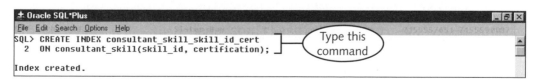

Figure 3-107   Creating an index

You can also create **composite indexes**, which can contain multiple columns (up to 16) that the DBMS can use for identifying the row location. An example of a table that might be appropriate for a composite index is CONSULTANT_SKILL, where searches are often performed to determine if a consultant has a specific skill ID and if the consultant is certified in that skill. You would create the index based on the most selective column first, which is SKILL_ID. This is called the **primary search column**. Then, you would specify the next search column, which is the CERTIFICATION column. This is called the **secondary search column**. The syntax to create a composite index is **CREATE INDEX** *index_name* **ON** *tablename***(***index_column1*, *index_column2*, **…);**. Now you will create this composite index.

To create the composite index:

1. Type the command shown in Figure 3-108 to create the index. The message "Index created" confirms that the index was successfully created.

```
± Oracle SQL*Plus                                                    _ 8 X
File  Edit  Search  Options  Help
SQL> CREATE INDEX consultant_skill_skill_id_cert          Type this
  2  ON consultant_skill(skill_id, certification);        command
Index created.
```

Figure 3-108   Creating a composite index

Viewing Index Information Using the Data Dictionary Views

You can retrieve information about your indexes using the USER_INDEXES data dictionary view. Now you will retrieve the names of the indexes in your user schema by querying the USER_INDEXES data dictionary view.

To retrieve index information:

1. Type the command shown in Figure 3-109 to retrieve the names of all of the indexes in your user schema.

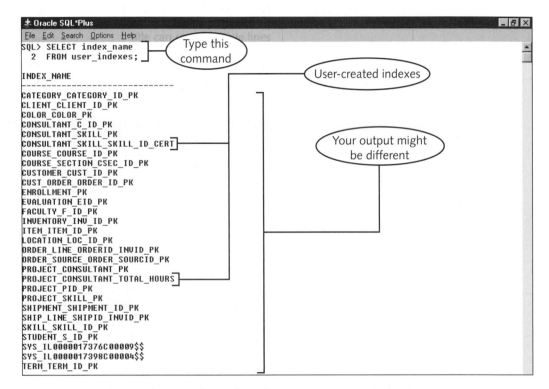

Figure 3-109   Viewing user index information

Note that the output lists the indexes that you created, along with the indexes on the primary keys of all of the tables that were created in your user schema when you ran the scripts to create the case study databases. The indexes on the primary keys of each table were created automatically when the CREATE TABLE command was executed. Your list might be different than the one shown in Figure 3-109, depending on the tables and indexes that you have created or dropped.

Dropping an Index

You might need to drop an index that you have created if applications no longer use queries that are aided by the index or if the index does not improve query performance enough to justify the overhead it creates on insert, update, and delete operations. To drop an index, you use the command **DROP INDEX *index_name*;**. Now you will drop the PROJECT_CONSULTANT_TOTAL_HOURS index that you created earlier.

To drop an index:

1. At the SQL prompt, type **DROP INDEX project_consultant_total_hours;** and then press **Enter**. The message "Index dropped" confirms that the index was successfully dropped.

SYNONYMS

Recall that database objects, like tables, views, and sequences, are owned by the users who created them. Owners can grant object privileges to other users to enable other users to use their objects. When you reference an object in a SQL command that is owned by another user, you must preface the object name with the owner's username. Now you will retrieve all of the fields from a table named DEPT that is owned by database user SCOTT.

 The SCOTT username and the tables used in this query should automatically be created when the Oracle8i database is installed. If they do not exist, ask your instructor or technical support person for help.

To query a table owned by another database user:

1. Type the command shown in Figure 3-110. Note that you must preface the table name with the username.

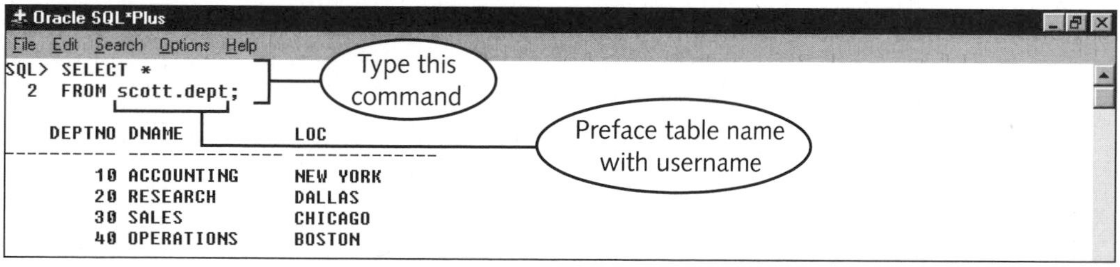

```
± Oracle SQL*Plus                                          _ |8|X
File  Edit  Search  Options  Help
SQL> SELECT *
  2  FROM scott.dept;

   DEPTNO DNAME          LOC
--------- -------------- --------------
       10 ACCOUNTING     NEW YORK
       20 RESEARCH       DALLAS
       30 SALES          CHICAGO
       40 OPERATIONS     BOSTON
```

Type this command

Preface table name with username

Figure 3-110   Querying a table owned by another user

To make it so database users do not have to preface the name of a table, view, sequence, or other database object with a username, you can create synonyms. A **synonym** is an alternate name for an object that references the owner name, as well as the object name. There are two types of synonyms: public synonyms and private synonyms.

Public Synonyms

A **public synonym** can be created only by a DBA or by a user who has the privilege to create public synonyms. After a public synonym is created, all database users can use the synonym to reference the database object, as long as they have sufficient privileges to access the object. The general command to create a public synonym is **`CREATE PUBLIC SYNONYM`** ***`synonym_name`*** **`FOR`** ***`owner_name.tablename`***. Figure 3-111 shows the command that a DBA executes to create a public synonym named PUBLIC_DEPT on the DEPT table.

Oracle SQL*Plus
File Edit Search Options Help
SQL> CREATE PUBLIC SYNONYM public_dept
2 FOR scott.dept;
Synonym created.
This command must be typed by a DBA

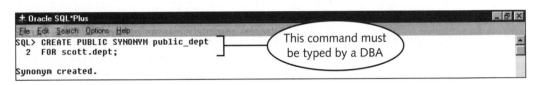

Figure 3-111 Public synonym created by DBA

After this public synonym has been created, any user who has privileges to view records in the SCOTT.DEPT table can now reference the table in commands using the public synonym rather than having to preface the table with the username. Figure 3-112 shows an example of a query that uses the public synonym to retrieve records from the SCOTT.DEPT table.

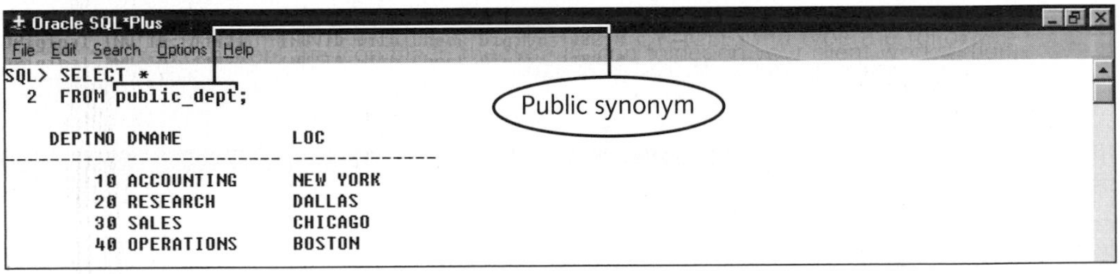

Oracle SQL*Plus
File Edit Search Options Help
SQL> SELECT *
2 FROM public_dept;
Public synonym
DEPTNO DNAME LOC
10 ACCOUNTING NEW YORK
20 RESEARCH DALLAS
30 SALES CHICAGO
40 OPERATIONS BOSTON

Figure 3-112 Using the public synonym in a query

Private Synonyms

Any user who has the privilege to use a database object can create and use a **private synonym** in place of a *username.object_name* specification for the object. When a user creates a private synonym, only that user can use the private synonym—other users must create their own private synonyms if they want to use a synonym in place of the *username.object_name* specification. The command to create a private synonym is **`CREATE SYNONYM`** ***`synonym_name`*** **`FOR`** ***`owner_name.object_name`***.

Now you will create a private synonym for the SCOTT.DEPT table and then use the private synonym in a query.

To create and use a private synonym:

1. Type the first command shown in Figure 3-113 to create the private synonym. The message "Synonym created" confirms that the synonym was successfully created.

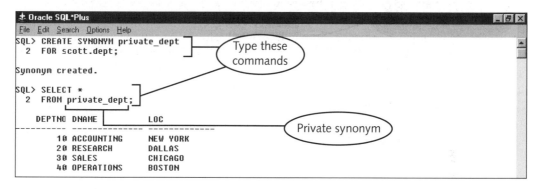

Figure 3-113   Creating and using a private synonym

2. Type the second command shown in Figure 3-113 to query the SCOTT.DEPT table using the private synonym. The table contents appear as shown.

3. Close SQL*Plus and all other open applications.

SUMMARY

❑ To join multiple tables in a SELECT command, you create an equality join, which is also called an inner join or an equijoin. You must include every involved table in the FROM clause and include a join condition that specifies every link between those tables in the WHERE clause.

❑ If you display a column in a multiple-table query that exists in more than one of the tables, you must qualify the column name by writing the table name, followed by a period, and then the column name.

❑ If you accidentally omit a join condition in a multiple-table query, it results in a Cartesian product, whereby every row in one table is joined with every row in the other table.

❑ An outer join is used to include all records in a table being joined with a second table, even if no corresponding values exist in the second, or outer, table.

❑ To join a table to itself, you must create a table alias and structure the query as if you are joining the table to a copy of itself.

❏ An inequality join is created by using an inequality operator in the join condition. Inequality joins are usually not appropriate in tables that are linked using surrogate key values.

❏ Nested queries contain subqueries that determine an intermediate result that is used as a search condition. Subqueries can be nested to multiple levels.

❏ A UNION query combines the results of two queries that are not related through foreign key relationships, and suppresses duplicate rows. A UNION ALL query is the same as a UNION query, except that it displays duplicate rows. Both types of UNIONS require that both queries have the same number of display columns in the SELECT clause, and each column in the first query must have the same data type as the corresponding column in the second query.

❏ An INTERSECT query returns the intersection, or matching rows, in two unrelated queries.

❏ A MINUS query allows you to find the difference between two unrelated query result outputs.

❏ The SELECT FOR UPDATE command enables you to view and lock a record with a single command. You must COMMIT or use the ROLLBACK command to unlock the record.

❏ A database view is a logical table based on a query. Views can be used to enforce security measures within a table or to simplify the process of displaying data that must be retrieved using a complex query.

❏ A simple view is based on a single table and can be used to insert, update, and delete values in the underlying database table.

❏ A complex view is based on complex queries that involve multiple tables, calculated columns, and other complex query specifications. You cannot insert, update, or delete values in a complex view.

❏ An index is used to speed up data retrieval for fields often used in search and join operations. An index contains the ROWID for each record, which specifies the physical location of the record in that database, along with data values of the indexed field.

❏ Oracle automatically creates indexes for primary key fields when you create a table.

❏ A synonym is an alternate name for an object that references the object's owner's name, as well as the object name.

❏ A public synonym can only be created by DBAs and users with special privileges, and it can be used by any database user who has privileges to use the database object associated with the synonym.

❏ A private synonym can be created by any user, but it can only be used by the user who created it.

REVIEW QUESTIONS

1. If you are creating a query that joins six tables, how many join conditions are required?

2. If the result of a query joining multiple tables is many more records than you expected, what probably happened?

3. When do you need to qualify a field in the SELECT clause of a query?

4. What is the difference between an inner join and an outer join?

5. List two situations in which you must use a table alias to retrieve required information in a query.

6. When you can retrieve the same information using either a query that joins multiple tables or a query that has one or more subqueries, which approach should you use, and why?

7. When do you need to use a UNION query?

8. What happens when a record is retrieved using the SELECT FOR UPDATE command?

9. What is the difference between a simple view and a complex view?

10. How can a view become invalid, and how do you fix it?

11. On what types of fields should you create indexes? Why should you limit the number of indexes that you create on a table?

12. What is a synonym? What is the difference between a public synonym and a private synonym?

PROBLEM-SOLVING CASES

For all cases, use Notepad or another text editor to write a script using the specified filename. Always use the search condition text exactly as it is specified in the question. Place the commands in the order listed, and save the script files in your Chapter3\ CaseSolutions folder.

1. Create a query design diagram like the one shown in Figure 3-73 for the following queries for the Northwoods University database (see Figure 1-12). Then, derive the associated SQL queries, and save the text of the SQL queries in a file named 3CCase1.sql.

   a. Display CALL_ID, COURSE_NAME, SEC_NUM, DAY, and TIME for every course in the Spring 2004 term. In the SQL query, format the TIME field so the values appear using a time format mask.

   b. Display F_LAST, F_FIRST, F_PHONE, DAY, and TIME for every faculty member who teaches in BUS Room 105 during the Spring 2004 term. In the SQL query, format the TIME field so the values appear using a time format mask.

2. Create a query design diagram like the one shown in Figure 3-73 and derive the associated SQL query text for the following queries for the Clearwater Traders database (see Figure 1-11). Then, derive the associated SQL queries, and save the text of the queries in a file named 3CCase2.sql.

   a. Display the ITEM_DESC, ITEM_SIZE, and COLOR of every item ever ordered by customer Alissa Chang.

   b. Display the ITEM_DESC, ITEM_SIZE, COLOR, SHIP_QUANTITY, and DATE_EXPECTED for all unreceived shipments (DATE_RECEIVED in the SHIPMENT_LINE table is NULL) of every item where QOH = 0 in the INVENTORY table.

3. Create a query design diagram like the one shown in Figure 3-73, and derive the associated SQL query text for the following queries for the Software Experts database (see Figure 1-13). Save the text of the SQL queries in a file named 3CCase3.sql.

   a. Display the first and last names of all consultants who have worked on any projects for Morningstar Bank.

   b. Display the name of each project that consultant Mark Myers has ever worked on, the name of the project's client, and the number of hours that Mark contributed to each project.

4. Create a script named 3CCase4.sql that contains the commands for the following queries that retrieve data from the Clearwater Traders database (see Figure 1-11).

   a. Display each source description from the ORDER_SOURCE table, along with a count of the total number of orders that have been derived from that source. (*Hint*: You will need to use an outer join.)

   b. List the first and last name, order ID, source description, and order date of every customer who has ever placed an order using the same order sources as customer Alissa Chang. Use a query that contains a subquery.

   c. List the inventory ID, item description, item size, color, and shipment quantity for every item that was received on the same date as inventory item 11798 in shipment ID 214. Use a query that contains a subquery, and do not include the information for inventory item 11798 in the output.

5. Create a script named 3CCase5.sql that contains the commands for the following queries that retrieve data from the Northwoods University database (see Figure 1-12).

   a. List the building code and room of every record in the LOCATION table, and also list the term description, call ID, section number, day, and time of every course ever taught at that location. Format the TIME field so the values appear as times, not dates. (*Hint*: You will need to use multiple outer joins.)

   b. List the building code and room number of all locations that have the same capacity as CR 103. Use a query that contains a subquery.

   c. List the first and last names of all faculty members who have ever taught any of the same courses as faculty member Kim Cox or who have ever taught a class in the same rooms as Kim. Use a query with multiple subqueries.

6. Create a script named 3CCase6.sql that contains the commands for the following queries that retrieve data from the Software Experts database (see Figure 1-13).

   a. List all of the skill descriptions in the SKILL table, as well as the ID of every consultant who is certified in that skill. Use an outer join so that if no consultants are certified for a particular skill, the query retrieves NULL values for the consultant ID value. (*Hint*: Create a view named CERTIFICATION_VIEW that contains IDs for consultant/skill combinations that are certified, and then perform the outer join using this view.)

   b. List the first and last name of every consultant who has ever worked on a project with consultant Mark Myers. Use a query with a subquery.

   c. Display the name of each project that consultant Mark Myers has ever worked on, and the first and last name of the consultant who was the manager of the project. Concatenate the manager fields so the values appear like "Sarah Carlson", and assign the column heading "Project Manager" to the manager name output column. (*Hint*: Nest the CONCAT function.)

7. Create a script named 3CCase7.sql that contains the commands for the following queries that require set operations.

   a. Write a query that uses the UNION operator to list the building code and room of every location at Northwoods University (see Figure 1-12) that is either currently in use as a faculty office, or in use as a classroom during the Summer 2004 term.

   b. Write a query that uses the INTERSECT operator to retrieve the skill descriptions that are the same for Software Experts (see Figure 1-13) consultants Sarah Carlson and Mark Myers.

   c. Write a query that uses the MINUS operator to retrieve the skill descriptions that represent the difference between the skills possessed by Software Experts (see Figure 1-13) consultant Sarah Carlson and the skills possessed by consultant Mark Myers.

8. Create a script named 3CCase8.sql that contains the SQL commands for the following queries that use the SELECT FOR UPDATE command and then the command that releases the locks on the selected records.

   a. Retrieve all of the fields in the ENROLLMENT table in the Northwoods University database (see Figure 1-12) that are associated with course section ID 1000, for update of the GRADE field.

   b. Retrieve the project ID, roll-on date, roll-off date, and total hours fields from the PROJECT_CONSULTANT table in the Software Experts database (see Figure 1-13) for records that are associated with consultant Janet Park, for update of the TOTAL_HOURS field.

9. Create a script named 3CCase9.sql that contains the commands to create the following database views in the Northwoods University database (see Figure 1-12).

   a. Create a simple view named STUDENT_VIEW that contains the S_ID, S_LAST, S_FIRST, S_MI, and S_CLASS fields in the STUDENT table. Then, using the view, write the command to insert the record for student ID 106, Carla J. Johnson, who is a freshman.

   b. Create a complex view named TERM_VIEW that contains the CALL_ID, SEC_NUM, DAY, TIME, BLDG_CODE, and ROOM for all courses offered during the Summer 2004 term. Then, write a command to view all of the fields in all of the view records, and a command to drop the view.

   c. Write the command to retrieve the names of all of your views using the data dictionary views.

10. Create a script named 3CCase10.sql that contains the commands to create the following indexes in the Northwoods University database (see Figure 1-12).

   a. Create an index on the STATUS field of the TERM table. Specify the index name using the naming convention specified in the chapter text.

   b. Create a composite index on the DAY and TIME fields in the COURSE_SECTION table. Use DAY as the primary search value and TIME as the secondary search value. Specify the index name using the naming convention specified in the chapter text.

   c. Write the command to retrieve the names of all of your indexes using the data dictionary views.

   d. Write the commands to drop the indexes that you have created.

11. Create a script named 3CCase11.sql that contains the commands to create the following synonyms in the Northwoods University database (see Figure 1-12).

   a. Write the command to create a public synonym named PUBLIC_STUDENT for the STUDENT table in your database schema. (You will not be able to successfully execute this command in SQL*Plus unless your user account has the privileges required to create public synonyms.)

   b. Create a private synonym named PRIVATE_FACULTY for the FACULTY table in your database schema.

   c. Write the command to select the F_ID, F_LAST, and F_FIRST fields from all of the records in your FACULTY table using your private synonym.

INTRODUCTION TO PL/SQL

◀ LESSON A ▶

Objectives

♦ Learn about PL/SQL variables and data types
♦ View the structure of PL/SQL program blocks and their basic operations
♦ Manipulate variables and perform number and character string operations
♦ Become familiar with PL/SQL program control structures
♦ Learn techniques for debugging PL/SQL programs

You have learned how to create Oracle database tables and write commands to insert, update, delete, and view records. However, most database users don't use SQL commands to interact with a database. They use applications that automatically perform these functions. Oracle provides several different utilities for developing user applications and for writing programs that help DBAs manage the database. To effectively use these utilities, database developers need to understand PL/SQL, which is the procedural programming language that Oracle utilities use to manipulate database data.

A **procedural programming language** is a programming language that uses detailed, sequential instructions to process data. A PL/SQL program combines SQL commands (such as SELECT and UPDATE) with procedural commands for tasks, such as manipulating variable values, evaluating IF/THEN logic structures, and creating loop structures that repeat instructions multiple times until an exit condition is reached. Although other procedural programming languages can contain SQL commands and interact with an Oracle database, PL/SQL was expressly designed for this purpose. As a result, it interfaces directly with the Oracle database and is available within Oracle development environments. Also, PL/SQL programs can be stored directly in the Oracle database, making them available to all database users, which simplifies application management. This chapter presents an introduction to the PL/SQL programming language. It assumes you have already used a programming or scripting language.

Basics of PL/SQL Programming

To introduce the components and syntax of the PL/SQL programming language, the following sections discuss PL/SQL variables and data types, and the structure of PL/SQL program blocks. They also address how to create comment lines to document programs, how to use arithmetic operators and assignment statements, and how to enable interactive output from PL/SQL programs in the SQL*Plus environment.

PL/SQL Variables and Data Types

PL/SQL variable names follow the Oracle naming convention for database objects. Although variables names can be reserved words, such as NUMBER, VALUES, BEGIN, or other words used in SQL or PL/SQL commands, it is not a good practice to use these names, because they can cause unpredictable results. Variable names can also use the same names as database table names and column names, but again, to avoid unpredictable results, you should not use these names. Variable names should be as descriptive as possible, such as current_s_id rather than X. Variable names are normally expressed in lowercase letters, with words within phrases separated by underscores. PL/SQL capitalization styles are summarized in Table 4-1.

Item Type	Capitalization	Example
Reserved words	Uppercase	`BEGIN, DECLARE`
Built-in functions	Uppercase	`COUNT, TO_DATE`
Predefined data types	Uppercase	`VARCHAR2, NUMBER`
SQL commands	Uppercase	`SELECT, INSERT`
Database objects	Lowercase	`student, f_id`
Variable names	Lowercase	`current_s_id, current_f_last`

Table 4-1   PL/SQL capitalization styles

PL/SQL is a **strongly typed** language, which means that all variables and their associated data types must be declared prior to use. Strong typing also means that you can perform assignments and comparisons only between variables with the same data type. When you declare a variable the system sets up a memory location for storing the value that is assigned to the variable, and associates the variable name with that memory location. The general syntax for declaring a variable in a PL/SQL program is ***variable_name data_type_declaration;***. PL/SQL has the following types of variables: scalar, composite, reference, and large binary object (LOB). The following sections discuss the different variable types.

Scalar Variables

Scalar variables reference a single value, such as a number, date, or character string. Scalar variables can have data types that correspond to Oracle database data types, and these data types are usually used to specify variables that are used to reference data values that are retrieved from the database. PL/SQL scalar database data types, descriptions, and sample declarations are summarized in Table 4-2.

Database Data Type	Description	Sample Declaration
VARCHAR2	Variable-length character strings	`current_s_last VARCHAR2(30);`
CHAR	Fixed-length character strings	`student_gender CHAR(1);`
DATE	Dates	`todays_date DATE;`
LONG	Text, up to 32,760 bytes	`evaluation_summary LONG;`
NUMBER	Floating-point, fixed-point, or integer numbers	`current_price NUMBER(5,2);`

Table 4-2   Scalar database data types

The PL/SQL **VARCHAR2** data type is used to reference variable-length character data up to a maximum length of 32,767 characters. It is different from its database counterpart, which can only hold a maximum of 4,000 characters when it is used as an Oracle8 database field. When you declare a VARCHAR2 variable in PL/SQL, you must also specify the maximum number of characters that the variable can reference.

The PL/SQL **CHAR** data type is used for fixed-length character strings, to a maximum length of 32,767 characters. When you declare a CHAR variable, specifying the maximum field width is optional. If no maximum width is specified for a CHAR variable, then the default width is 1. When the variable is assigned a data value, if the data value has fewer characters than the maximum width specification, the rest of the width is padded with blank spaces so the value completely fills the maximum width. If the assigned value is larger than the maximum field width, an error occurs.

The **DATE** data type in PL/SQL is the same as its Oracle database counterpart and stores both date and time values.

The **LONG** data type is different than its database counterpart. Recall that a LONG field in a database can store up to two gigabytes of data. In PL/SQL, a LONG variable can reference a maximum of 32,760 characters, so it is approximately equivalent to a VARCHAR2 field.

The PL/SQL **NUMBER** data type is identical to the Oracle database NUMBER data type, with the general syntax **NUMBER(_precision, scale_)**. The precision specifies the total length of the number, including decimal places; the scale specifies the number of digits to the right of the decimal point. When you declare a NUMBER variable, you specify only the precision for integer values and both the precision and scale for fixed-point values. For floating-point values, you omit both the precision and scale specifications.

PL/SQL has other scalar data types that do not correspond to database data types and are used for general program operations. These data types are summarized in Table 4-3.

Nondatabase Data Type	Description	Sample Declaration
Integer number subtypes (BINARY_INTEGER, INTEGER, INT, SMALLINT)	Integers	`counter BINARY_INTEGER;`
Decimal number subtypes (DEC, DECIMAL, DOUBLE PRECISION, NUMERIC, REAL)	Numeric values with varying precision and scale	`student_gpa REAL;`
BOOLEAN	True/False values	`order_flag BOOLEAN;`

Table 4-3   General scalar data types

The **integer number subtypes** reference integer values. The differences among the subtypes are the sizes of the values that each subtype can reference and how the values are represented internally. The **BINARY_INTEGER** data type is usually used for loop counters. BINARY_INTEGER data values are stored internally in binary format, which takes slightly less storage space than the NUMBER data type, and calculations can be performed on BINARY_INTEGER data values more quickly than on integer NUMBER values.

The **decimal number subtypes** can be used to reference numeric values of varying precision and scale. They are included in PL/SQL to provide ANSI compatibility and compatibility with other database systems like IBM DB/2.

The **BOOLEAN** data type is used to reference values that are TRUE, FALSE, or NULL. When a BOOLEAN variable is declared, it has a value of NULL until it is assigned a value of TRUE or FALSE.

Composite Variables

A **data structure** is a data item that is composed of multiple individual data elements. A **composite** variable references a data structure that contains multiple scalar variables, such as a record or a table. In PL/SQL, these data types include **RECORD**, which is used to specify a structure that contains multiple scalar values, and is similar to a table record; **TABLE**, which specifies a tabular structure with multiple columns and rows; and **VARRAY**, which specifies a variable-sized array, which is a tabular structure that can expand or contract based on the data values it contains. You will learn how to declare and use the RECORD and TABLE composite variables later in the chapter.

Reference Variables

Reference variables directly reference a specific database field or record and assume the data type of the associated field or record. The different reference data types are summarized in Table 4-4.

Reference Data Type	Description	Sample Declaration
%TYPE	Assumes the data type of a database field	`cust_address customer.c_add%TYPE;`
%ROWTYPE	Assumes the data type of a database row	`cust_order_row cust_order%ROWTYPE;`

Table 4-4   PL/SQL reference data types

The **%TYPE** reference data type is used to declare a variable that will reference a single database field. The general format for a %TYPE data declaration is *variable_name tablename.fieldname%TYPE;*. For example, the declaration for a variable named current_f_last that will reference values retrieved from the F_LAST field in the FACULTY table would be written as **current_f_last FACULTY.F_LAST%TYPE;**. The current_f_last variable would assume a data type of VARCHAR2(30), since this is the data type that was used when the F_LAST field was declared in the CREATE TABLE command for the FACULTY table.

The **%ROWTYPE** reference data type is used to create composite variables that reference an entire row of data. The general format for a %ROWTYPE data declaration is *row_variable_name tablename%ROWTYPE;*. The following code will declare a variable that references an entire row of data that was retrieved from the FACULTY table: **faculty_row FACULTY%ROWTYPE;**. This variable would reference all nine fields in the FACULTY table, and each field would have the same data type as the associated database field.

LOB Data Types

LOB data types are used to declare variables that reference binary data objects, such as images or sounds. The **large object (LOB)** data types can store either binary or character data up to four gigabytes in size. LOB values in PL/SQL programs must be manipulated using a special set of programs, called the DBMS_LOB package, which you will learn about in a later chapter.

PL/SQL Program Blocks

The structure of a PL/SQL program block is:

```
DECLARE
 variable declarations
BEGIN
 program statements
EXCEPTION
 error-handling statements
END;
```

PL/SQL program variables are declared in the program's **declaration** section using the data declarations syntax shown earlier. The beginning of the declaration section is marked with the reserved word DECLARE. You can declare multiple variables in the declaration section, with each variable declaration separated by a semicolon (;). The **body** of a PL/SQL block consists of program statements, which can be assignment statements, conditional statements, looping statements, and so on, that lie between the BEGIN and EXCEPTION statements. The **exception** section contains program statements for error handling. PL/SQL programs end with the END; statement.

In a PL/SQL program block, the DECLARE and EXCEPTION sections are optional. If there are no variables to declare, you can omit the DECLARE section and start the program with the BEGIN command. If there are no error handling statements, then you can omit the EXCEPTION section.

PL/SQL Commands and Comment Statements

PL/SQL commands include assignment statements that assign values to variables, commands to perform arithmetic operations, conditional (IF/THEN) structures and looping structures to control program flow, SQL commands that retrieve database data, and other types of commands to handle advanced processing operations. Every PL/SQL program command ends with a semicolon. PL/SQL commands can span many lines in a text editor, but when the code is interpreted, the interpreter considers everything up to an ending semicolon to be part of the same line of code. Forgetting to include a semicolon at the end of a code line will cause an interpreter syntax error. PL/SQL commands are not case sensitive, except for character strings, which must be enclosed in single quotation marks.

 PL/SQL is an interpreted language, which means that when it is executed, the program is translated into machine language one line at a time and executed one line at a time.

Comment statements are text comments within a computer program that do not contain program commands, but are used to explain or document a program step or series of steps. Comment statements must be marked, or **delimited**, so that the interpreter does not try to interpret them as program commands. PL/SQL program comment statements can be delimited in two ways. A block of comments that spans several lines can begin with the symbols /* and end with the symbols */. For example, the following code would be interpreted as a comment block:

```
/* Script: orcl_cold_backup
Purpose: To perform a complete cold backup on the ORCL database instance
Revisions: 9/8/2003 JM Script */
```

If a comment statement appears on a single line, you can delimit it by typing two hyphens at the beginning of the line, as shown in the following example:

```
DECLARE
--variable to hold current value of S_ID
current_S_ID NUMBER;
```

PL/SQL Arithmetic Operators

The arithmetic operators used in PL/SQL are similar to those used in most programming languages. Table 4-5 describes the PL/SQL arithmetic operators in the order in which they are evaluated. As with most programming languages, you can force operators to be evaluated in a specific order by placing the operation in parentheses.

4

Operator	Meaning	Example	Result
**	Exponentiation	2**3	8
*	Multiplication	2 * 3	6
/	Division	9/2	4.5
+	Addition	3 + 2	5
−	Subtraction	3 − 2	1
−	Negation	−5	negative 5

Table 4-5   PL/SQL arithmetic operators

PL/SQL Assignment Statements

In programming, an **assignment statement** assigns a value to a variable. In PL/SQL, the assignment operator is :=. The name of the variable that is being assigned the new value is placed on the left side of the assignment operator, and the value is placed on the right side of the operator. The value can be a **literal**, which is an actual data value, such as 'John', or another variable that has been assigned the desired value previously. Examples of both types of assignments are shown in Table 4-6.

Description	Example
Variable assigned a literal value	`current_s_first_name := 'John';`
Variable assigned to another variable	`current_s_first_name := s_first_name;` (assume s_first_name has previously been assigned a value)

Table 4-6   PL/SQL assignment statements

You can use an assignment statement within a variable declaration to assign an initial value to a new variable. For example, to declare a variable named current_student_ID with data type NUMBER and assign to it an initial value of 100, you would write the combined variable declaration/assignment statement as follows:

```
DECLARE
  current_student_ID NUMBER :=100;
```

Displaying PL/SQL Program Output in SQL*Plus

You can create and execute PL/SQL program blocks within a variety of Oracle development environments. To display program output in the SQL*Plus environment, you must use the **DBMS_OUTPUT.PUT_LINE** procedure, which has the syntax:

```
DBMS_OUTPUT.PUT_LINE('text to be displayed');
```

The text to be displayed can contain literal character strings, like 'My name is Jennifer', or variable values, like current_s_first. An example command to display the current value stored in a variable named current_s_first is:

```
DBMS_OUTPUT.PUT_LINE(current_s_first);
```

The PUT_LINE procedure can display a maximum of 255 characters of text data. If you try to display more than 255 characters, an error occurs.

 DBMS_OUTPUT is an Oracle built-in package, which is a set of related programs that users can access. PUT_LINE is the name of the DBMS_OUTPUT procedure that displays output in SQL*Plus.

Whenever you start a new SQL*Plus session and plan to display PL/SQL program output, you must first set up a **buffer**, or memory area, on the database server that stores the output values, which are then displayed to the user. If you don't set up this buffer, DBMS_OUTPUT commands will not display any output. The command to set up the buffer has the syntax:

```
SET SERVEROUTPUT ON SIZE buffer_size
```

The *buffer_size* value specifies the buffer_size in bytes. For example, the following command sets the output buffer size to 4,000 bytes:

```
SET SERVEROUTPUT ON SIZE 4000;
```

If you do not specify the buffer size, then the default buffer size is 2,000 bytes. If you write a PL/SQL program that displays more than 2,000 bytes, an error message appears stating that a buffer overflow error has occurred. Therefore, if you are displaying more than just a few lines of output, it is a good idea to make the output buffer larger than the default size.

WRITING AND EXECUTING A PL/SQL PROGRAM

Now that you have learned the basics of the PL/SQL programming language, you are ready to write a simple PL/SQL program that will be executed in SQL*Plus. First, you will start SQL*Plus and set up the DBMS_OUTPUT buffer.

To start SQL*Plus and set up the output buffer:

1. Start SQL*Plus and log onto the database.

2. Type **SET SERVEROUTPUT ON SIZE 4000;** at the SQL prompt to set up the output buffer.

Now you will write a PL/SQL program that will declare a DATE variable named todays_date. The program will then assign the current system date to todays_date and display the variable value as output. It is a good programming practice to place the DECLARE, BEGIN, and END commands flush with the left edge of the text editor window, and then indent the commands within each section. The blank spaces do not affect the program's functionality, but they make it easier to read and understand. You can indent the lines by pressing the Tab key, or by pressing the spacebar to add blank spaces.

To write and execute a PL/SQL program:

1. Start Notepad or an alternate text editor, and type the following program code. Save your Notepad text file as **4APrograms.sql** in your Chapter4\TutorialSolutions folder. Indent the program lines as shown to format the different program blocks.

```
--PL/SQL program to display the current date
DECLARE
   todays_date DATE;
BEGIN
   todays_date := SYSDATE;
   DBMS_OUTPUT.PUT_LINE('Today"s date is ');
   DBMS_OUTPUT.PUT_LINE(todays_date);
END;
```

2. Copy the text, and then paste it into SQL*Plus. Press **Enter** after the last line of the program, type /, and then press **Enter** again to execute the program. The output is shown in Figure 4-1. If your program does not run correctly, examine your code carefully to make sure you typed it correctly. If you think you typed it correctly, read the section on Debugging PL/SQL Programs later in this lesson to learn about debugging techniques.

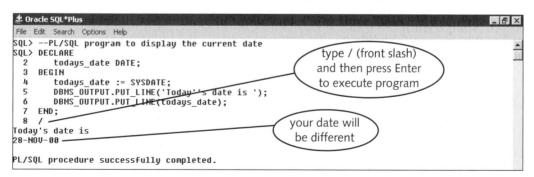

Figure 4-1   PL/SQL program output

DATA TYPE CONVERSION FUNCTIONS

Sometimes you need to convert a value from one data type to another data type. For example, if the value '2' is input as a character variable, it must be converted to a number before you can assign it to a NUMBER variable and use it to perform arithmetic calculations. PL/SQL performs implicit data conversions, which means that the interpreter automatically converts one data type to another when it makes sense to do so. For example, in the statement DBMS_OUTPUT.PUT_LINE(todays_date) in your PL/SQL program in the preceding section, the DATE variable todays_date was automatically converted to a text string for output. However, it is risky to rely on implicit conversions, because their output is unpredictable. Therefore, you should always perform explicit data conversions, which are data conversions performed using specific conversion functions that are built into PL/SQL. A **function** is a built-in program that receives one or more inputs, called **parameters**, and returns a single output value. The main PL/SQL data conversion functions are listed in Table 4-7.

Function	Description	Example
TO_DATE	Converts a character string to a date using a specific format mask	`TO_DATE('07/14/2003', 'MM/DD/YYYY');`
TO_NUMBER	Converts a character string to a number	`TO_NUMBER('2');`
TO_CHAR	Converts either a number to a character string, or a date to a character string using a specific format mask	`TO_CHAR(2);` `TO_CHAR(SYSDATE, 'MM/DD/YYYY');`

Table 4-7   PL/SQL data conversion functions

To convert a date to a character string, you use the TO_CHAR function, which has the following general syntax: *character string := TO_CHAR(date_value, 'date_format_mask');*. For example, to convert the current SYSDATE to a character string with the format MM/DD/YYYY, and store the result in a variable named date_string, you would use the command `date_string := TO_CHAR (SYSDATE, 'MM/DD/YYYY');`. Recall that to convert a character string to a date, you use the TO_DATE function, which has the following general syntax: `date_value := TO_DATE (date_string, 'date_format_mask');`. Common date format masks are listed in Chapter 3.

CHARACTER FUNCTIONS

Manipulating **character strings**, which are character data values that consist of one or more characters, is an important topic in any programming language. For example, you might want to create a single character string by **concatenating**, or joining, two separate character strings. Or, you might want to take a single character string consisting of two data items separated by commas or spaces and **parse**, or separate, it into two individual character strings. SQL*Plus has a variety of built-in string-handling functions to concatenate and parse.

Concatenating Character Strings

Suppose you are working on a PL/SQL program that uses the following variable names and values:

Variable Name	Value
s_first_name	Sarah
s_last_name	Miller

You would like to combine these two values into one variable named s_full_name that has the value "Sarah Miller." To concatenate two character strings in PL/SQL, you use the double bar (||) operator. The command to concatenate these values is **s_full_name := s_first_name || s_last_name;**.

This command looks pretty simple, but there is a catch. This command puts the two character strings together with no spaces between them, so the value of s_full_name is now "SarahMiller," which is not what you wanted. You need to insert a blank space between the first and last name. To insert a blank space, you use the same command but concatenate a character string consisting of a blank space between the two variable names, as in the following command: **s_full_name := s_first_name || ' ' || s_last_name;**.

Literals are actual data values coded directly into a program, such as 3.14159 for the value of pi, or 'Customer Name:' for a column heading. You must always enclose a **string literal**, which is a literal value that is a character string, in single quotation marks. A string literal can contain any combination of valid PL/SQL characters (such as 'Student Name:'). If a single quotation mark is used within a string literal, you must type two single quotation marks (for example, 'Sarah''s Computer'). Suppose you saved the following values for a building code, room, and capacity in three variables:

Variable Name	Data Type	Value
building_code	VARCHAR2	CR
room_num	VARCHAR2	101
room_capacity	NUMBER	150

Now suppose you need to display the following text string: "CR Room 101 has 150 seats." Note that the ROOM_CAPACITY value is a NUMBER data type and not a CHARACTER data type. Before you can use the number in a concatenation operation, you must use the TO_CHAR function to convert it to a character. The required code to create this character string, which is saved in a variable named room_message, is:

```
room_message := building_code ||' Room '|| room_num || ' has
'|| TO_CHAR (room_capacity) ||' seats.';
```

Always remember to add spaces before and after the variable values so that the strings don't run together.

Now you will modify the PL/SQL program you wrote earlier so that it displays the lead-in message 'Today is' and the date as a single concatenated string. You will need to convert the todays_date variable to a character data type.

 In the following exercises, you will be modifying the code that you have already written. To keep a record of all of your PL/SQL programs, make a copy of the existing program, and then make modifications to the copy. Save all program code in the 4APrograms.sql file in your Chapter4\Tutorial Solutions folder on your Data Disk.

To modify the PL/SQL program to display the date as a single string:

1. In Notepad, copy your current code, and then paste the copied code below the existing code. Then, change the copied code so it looks like the code shown in Figure 4-2. Remember not to type the line numbers; they will automatically be added when you paste the code into SQL*Plus.

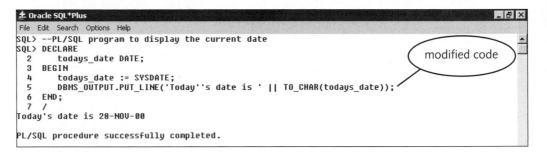

Figure 4-2   Program output displayed on a single line

2. Copy the modified code, and paste it into SQL*Plus. Press **Enter** after the last line of the program, type /, and then press **Enter** again to execute the program. The output should appear as shown in Figure 4-2. If your program does not execute correctly, examine your code carefully to make sure that you did not make any typing mistakes. If you think that you typed it correctly, go to the section on Debugging PL/SQL Programs later in the lesson, and follow the instructions for locating syntax errors.

Removing Blank Leading and Trailing Spaces from Strings

Sometimes when you store retrieved data values in variables or convert numeric or date values to characters, the data values contain blank spaces that pad out the value to its maximum column width. Suppose you have a variable named cust_address, which is a CHAR data type of size 20. Its current value is '2103 First St   ' (13 characters followed by seven spaces). To remove all spaces from the right side of the variable named cust_address, use the **RTRIM** function as follows: **cust_address := RTRIM (cust_address);**. Similarly, you can use the **LTRIM** function to remove blank leading spaces from character strings.

Finding the Length of Character Strings

When you manipulate character strings, you often need to find the number of characters in a string to perform concatenation and parsing operations, or to avoid errors by trying to insert a character string that is too wide into the database. To do this, you use the PL/SQL **LENGTH** function, which returns an integer that is the length of a character string. Suppose you have a variable named building_code, and its current value is 'CR'. To find the length of the string, you first declare an integer variable named code_length, and then enter the command **code_length := LENGTH(building_code);**. Because the value of building_code is 'CR', the LENGTH function would return the number 2 to the code_length variable. If the variable contains additional padded spaces to the right, these also will be counted in the length. For example, if building_code is retrieved from a CHAR type database field of size six, the value of building_code will be 'CR   ' (CR followed by four blank spaces), and the LENGTH function will return the number 6.

Character String Case Functions

Sometimes, you need to modify the case of character strings. For example, in the TERM table in the Northwoods University database, the STATUS values are stored in all uppercase letters, as 'OPEN' and 'CLOSED'. You might want to display the STATUS value in program output using mixed-case letters with the initial letter of each word capitalized, as 'Open' and 'Closed', or in all lowercase letters, as 'open' and 'closed'. PL/SQL has functions to display character strings using lowercase, uppercase, and mixed-case letters. The **UPPER** function converts lowercase or mixed-case characters to all uppercase characters. For example, if a variable named s_full_name stores the value 'Sarah Miller', the following code will convert the s_full_name value to all uppercase characters: **s_full_name := UPPER(s_full_name);**. This function changes the variable value to 'SARAH MILLER'.

Similarly, the **LOWER** function converts uppercase or mixed-case characters to all lowercase characters. Using the previous example's data values, the command **s_full_name := LOWER(s_full_name);** would change the variable value to 'sarah miller'.

The **INITCAP** function returns a string in mixed case, with the first letter of each word in uppercase letters and the remaining characters in lowercase letters.

Next, you will add the INITCAP and LENGTH functions to your PL/SQL program so that it displays the current day of the week and the number of characters in the name of the current day. You will declare a VARCHAR2 variable named current_day and assign to it the current day of the week. You will declare another VARCHAR2 variable named current_day_length and assign to it the length of the current day's character string. The program will then convert the day of the week to mixed-case characters with the first letter capitalized, trim the blank trailing spaces from the day using the RTRIM function, and display the day and the length of the name of the day.

To modify your PL/SQL program to use the INITCAP, RTRIM, and LENGTH functions:

1. In Notepad, make a copy of your program, and then change the copied code so it looks like the code shown in Figure 4-3. Remember not to type the line numbers, because they will automatically be added when you paste the code into SQL*Plus.

2. Copy the text, and then paste it into SQL*Plus and execute the program. The output is also shown in Figure 4-3. If your program does not execute correctly, go to the section on Debugging PL/SQL Programs later in the lesson.

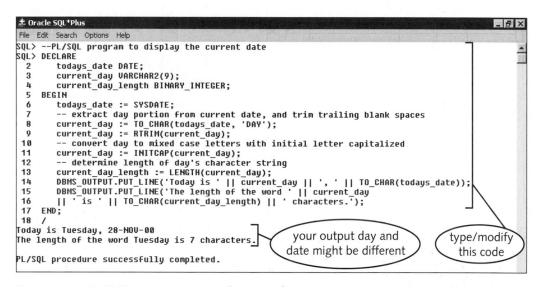

Figure 4-3   PL/SQL program using character functions

The INSTR and SUBSTR String Functions

Sometimes, you need to search through an existing character string to determine if it contains a specific substring. For example, the CALL_ID column in the COURSE table in the Northwoods University database contains department abbreviations and course numbers combined in a single character string, such as 'MIS 101' and 'CS 163'. You might want to search through the CALL_ID values to determine which course call ID values contain the substring 'MIS'. To do this, you use the **INSTR** function, which searches one string and looks for a matching substring. If it finds a matching substring, the function returns the starting position of the substring within the original string as an integer. If a matching substring is not found, the function returns the value 0. The general syntax for the INSTR function is:

```
return_value := INSTR(string_being_searched, search_string);
```

The following code example uses the INSTR function to return the starting position of the single blank space in the s_full_name variable that currently contains the value 'Sarah Miller' and returns the value to an integer number variable named blank_position:

```
blank_position := INSTR(s_full_name, ' ');
```

If the s_full_name string variable contains the value 'Sarah Miller', blank_position will be set to 6, since the blank space is the sixth character in the string.

Sometimes during string handling operations, you need to extract a specific number of characters from a character string. For example, in the CALL_ID values in the COURSE table, you might want to extract the department abbreviations (like MIS or CS) from the rest of the string. The **SUBSTR** function extracts a specific number of characters from a character string, starting at a given point. The general syntax for the SUBSTR function is:

```
extracted_string := SUBSTR(string_being_searched,
starting_point, number_of_characters_to_extract);
```

To extract the string 'Sarah' from the s_full_name variable and set it equal to a character variable named s_first_name, you would use the following code:

```
s_first_name := SUBSTR(s_full_name, 1, 5);
```

The 1 represents the starting place (the first character in the string), and the 5 represents the number of characters to extract (five, one for every letter in 'Sarah').

You can use the INSTR function within the SUBSTR function to parse out part of a string. For example, you could use the INSTR and SUBSTR functions together to return the student's first name (up to the blank space) in a single command. You would use the following command:

```
s_first_name :=
SUBSTR(s_full_name, 1, (INSTR(s_full_name,' ') -1));
```

In this command, the last parameter (the number of characters to extract) is specified by the INSTR function, which returns the position of the blank space. You must subtract 1 from the result of the INSTR function, since you only want to return the characters up to, but not including, the blank space.

Now, you will modify your PL/SQL program so that it will use the INSTR function to display the position in the name of the day of the week where the substring 'day' starts. It will also use the SUBSTR function to extract the characters in the day of the week that occur before the substring 'day'. The SUBSTR function will start on the first character of the string. The length of the string to be extracted will be the length of the string minus 3 (which is the number of characters in 'day'). You will declare a variable named position_of_day to reference the position where the substring 'day' starts, and a variable named string_before_day to reference the substring of the name of the day of the week.

To modify your PL/SQL program to use the INSTR and SUBSTR functions:

1. In Notepad, make a copy of your program, and change your code so it looks like the code shown in Figure 4-4.

2. Execute the program in SQL*Plus. The output is also shown in Figure 4-4. If your program does not execute correctly, use the debugging strategies that will be described next.

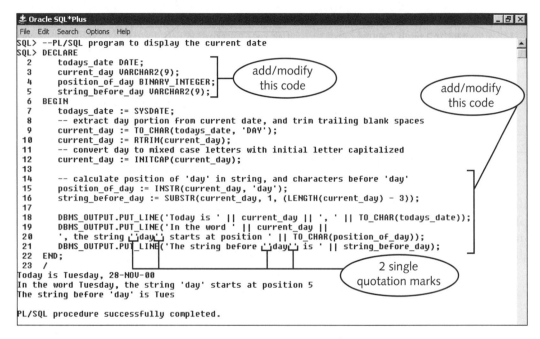

Figure 4-4   PL/SQL program using INSTR and SUBSTR functions

DEBUGGING PL/SQL PROGRAMS

Programming errors usually fall into one of two categories: syntax and logic. A **syntax error** occurs when code does not follow the guidelines of the programming language and results in a compiler or interpreter error message. A **logic error** does not stop a program from running, but results in an incorrect result. For example, in the previous exercise, you used the following program line to find the string before 'day' in the current day of the week:

```
string_before_day :=
SUBSTR(current_day, 1, (LENGTH(current_day) -3));
```

Suppose that for the last argument in the SUBSTR function, which is the number of characters to extract from the string, you used the expression `(LENGTH(current_day - 2));`. Now, instead of returning the string representing the current day minus 'day', the SUBSTR function would return the current day minus 'ay', and the output would look like Figure 4-5. The program ran successfully, but gave an incorrect output.

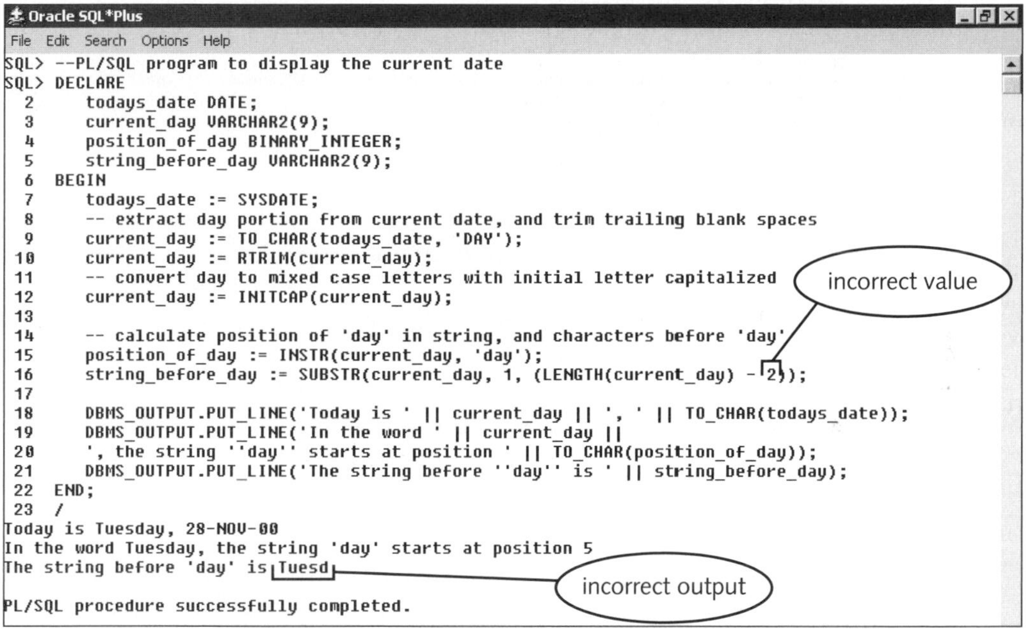

Figure 4-5   Logic error example

The following sections present examples of common syntax and logic errors, and tips for debugging them.

Common Syntax Errors

Syntax errors, also called compile errors, might involve misspelling a reserved word; omitting a required character in a command, such as a parenthesis, single quote, or semicolon; or using a built-in function improperly. In SQL*Plus, the line number and character location of these errors are flagged by the interpreter, and an error code and message appear. Figure 4-6 shows an example of a syntax error where the semicolon was omitted from the current_day variable declaration.

The ORA-06550 error message indicates that a compile error has occurred and reports the line number and location. The PLS-00103 error message is a specific PL/SQL error message, indicating that the interpreter was expecting the semicolon end-of-line marker. You can find explanations and solution strategies for PLS error codes in the Oracle8 Error Message online Help application, which you used in Chapter 2 to interpret ORA-error codes.

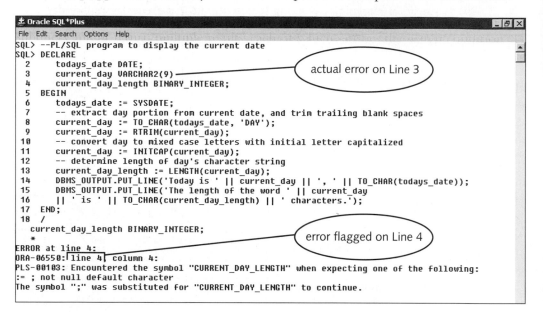

```
Oracle SQL*Plus                                                              _ 8 X
File  Edit  Search  Options  Help
SQL> --PL/SQL program to display the current date
SQL> DECLARE
  2      todays_date DATE;
  3      current_day VARCHAR2(9)                        actual error on Line 3
  4      current_day_length BINARY_INTEGER;
  5  BEGIN
  6      todays_date := SYSDATE;
  7      -- extract day portion from current date, and trim trailing blank spaces
  8      current_day := TO_CHAR(todays_date, 'DAY');
  9      current_day := RTRIM(current_day);
 10      -- convert day to mixed case letters with initial letter capitalized
 11      current_day := INITCAP(current_day);
 12      -- determine length of day's character string
 13      current_day_length := LENGTH(current_day);
 14      DBMS_OUTPUT.PUT_LINE('Today is ' || current_day || ', ' || TO_CHAR(todays_date));
 15      DBMS_OUTPUT.PUT_LINE('The length of the word ' || current_day
 16      || ' is ' || TO_CHAR(current_day_length) || ' characters.');
 17  END;
 18  /
    current_day_length BINARY_INTEGER;
    *                                                  error flagged on Line 4
ERROR at line 4:
ORA-06550: line 4 column 4:
PLS-00103: Encountered the symbol "CURRENT_DAY_LENGTH" when expecting one of the following:
:= ; not null default character
The symbol ";" was substituted for "CURRENT_DAY_LENGTH" to continue.
```

Figure 4-6   Syntax error example

Recall that the Oracle8 Error Message online Help application is started by running a file named ora.hlp, which can be downloaded from the Course Technology Web site at www.course.com. You can also find information about error codes by connecting to the Oracle Technology Network Web site at *otn.oracle.com*, and then searching for Oracle8i error messages.

Sometimes the location flagged for a syntax error is not the line where the error actually occurs. For example, Figure 4-6 shows an error where a semicolon is omitted after the current_day variable declaration. The error is flagged on Line 4, but the error actually occurs on Line 3. When you receive an error message and the flagged line appears correct, look for the error on lines preceding the flagged line.

Another common error message is illustrated in Figure 4-7: 'Encountered the symbol ";" when expecting one of the following: …'. This error message indicates that the interpreter was expecting another character when it encountered the end-of-statement semicolon. In this case, the second closing parenthesis at the end of the DBMS_OUTPUT.PUT_LINE command on Line 14 was omitted. You must carefully examine the line where the error was flagged to determine which character was omitted.

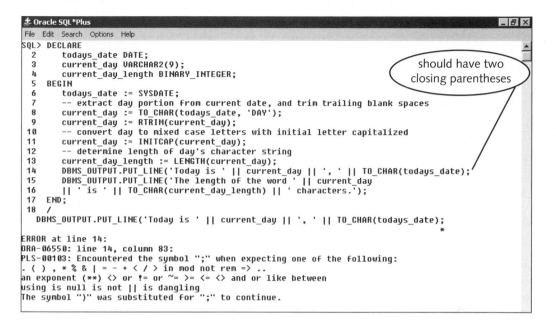

Figure 4-7   Missing character syntax error

4

Sometimes errors are very difficult to find visually, and the interpreter does not specify the error line location. For example, Figure 4-8 displays the error message "quoted string not properly terminated," but does not specify the line or column location.

```
± Oracle SQL*Plus                                                    _ 8 X
File  Edit  Search  Options  Help
SQL> --PL/SQL program to display the current date
SQL> DECLARE
  2       todays_date DATE;
  3       current_day VARCHAR2(9);
  4       current_day_length BINARY_INTEGER;
  5    BEGIN
  6       todays_date := SYSDATE;
  7       -- extract day portion from current date, and trim trailing blank spaces
  8       current_day := TO_CHAR(todays_date, 'DAY');
  9       current_day := RTRIM(current_day);
 10       -- convert day to mixed case letters with initial letter capitalized
 11       current_day := INITCAP(current_day);
 12       -- determine length of day's character string
 13       current_day_length := LENGTH(current_day);
 14       DBMS_OUTPUT.PUT_LINE('Today is ' || current_day || ', ' || TO_CHAR(todays_date));
 15       DBMS_OUTPUT.PUT_LINE('The length of the word ' || current_day
 16       ||  is ' || TO_CHAR(current_day_length) || ' characters.');
 17    END;
 18    /
ERROR:
ORA-01756: quoted string not properly terminated
```

Figure 4-8   Syntax error with location not specified

If you cannot find a compile error visually, or if the location is not specified, you need to systematically determine which program line is generating the error. A useful debugging technique for isolating program errors is to **comment out** program lines, which means to turn program code statements into comment statements so the interpreter does not attempt to interpret the commands. This way, you can determine which line is causing the error. In the program in Figure 4-8, character strings are being concatenated in the DBMS_OUTPUT commands, so there is a good chance that the error is occurring on one of these command lines. Figure 4-9 shows the program with the second DBMS_OUTPUT command commented out. No error message is generated, so this isolates the error—it is occurring somewhere in the second DBMS_OUTPUT command, on Line 15 or Line 16.

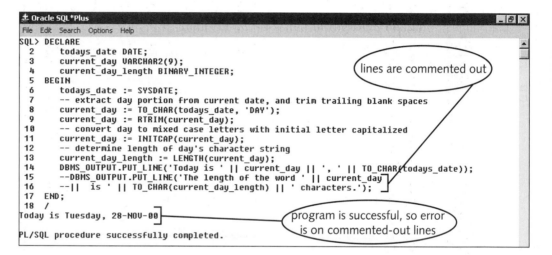

Figure 4-9 Isolating error by commenting out the second DBMS-OUTPUT command

To determine if the error is occurring on Line 15 or on Line 16, you can break the DBMS_OUTPUT command into two lines to determine if the error is occurring in the first or second line. Figure 4-10 shows how the command on Line 15 is ended properly by deleting the concatenation (||) operator and adding the end-of-function parenthesis and a semicolon. Line 16 remains commented out. The program again executes successfully, so the error must be on Line 16.

Careful examination of Line 16 locates the error: the initial single quote before the word *is* was omitted at the beginning of the line.

Oracle SQL*Plus
File Edit Search Options Help
SQL> DECLARE
2 todays_date DATE;
3 current_day VARCHAR2(9);
4 current_day_length BINARY_INTEGER;
5 BEGIN
6 todays_date := SYSDATE;
7 -- extract day portion from current date, and trim trailing blank spaces
8 current_day := TO_CHAR(todays_date, 'DAY');
9 current_day := RTRIM(current_day);
10 -- convert day to mixed case letters with initial letter capitalized
11 current_day := INITCAP(current_day);
12 -- determine length of day's character string
13 current_day_length := LENGTH(current_day);
14 DBMS_OUTPUT.PUT_LINE('Today is ' || current_day || ', ' || TO_CHAR(todays_date));
15 DBMS_OUTPUT.PUT_LINE('The length of the word ' || current_day);
16 --|| is ' || TO_CHAR(current_day_length) || ' characters.');
17 END;
18 /
Today is Tuesday, 28-NOV-00
The length of the word Tuesday
PL/SQL procedure successfully completed.
line is terminated
line is commented out
program is successful, so error is on Line 16

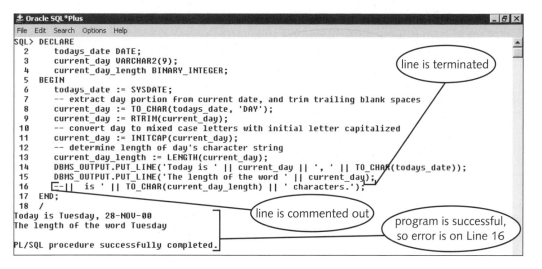

Figure 4-10 Further isolating the error

Figure 4-11 shows an example of an error that is generated when a built-in function is used improperly. The TO_CHAR data conversion function is applied to the current_day variable, but current_day is already a character data field, so it cannot be converted, and the resulting error message is displayed.

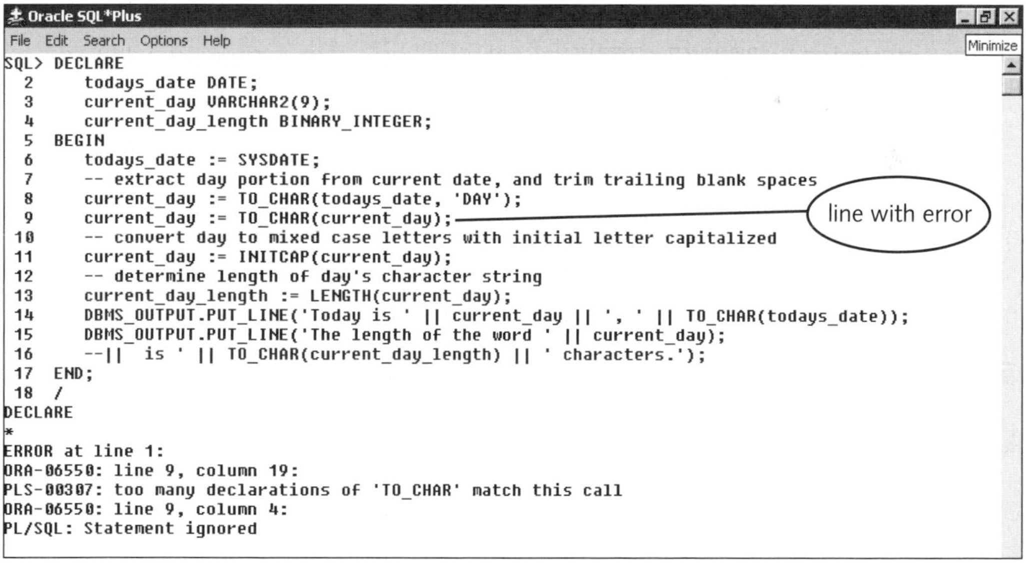

Oracle SQL*Plus
File Edit Search Options Help
Minimize
SQL> DECLARE
2 todays_date DATE;
3 current_day VARCHAR2(9);
4 current_day_length BINARY_INTEGER;
5 BEGIN
6 todays_date := SYSDATE;
7 -- extract day portion from current date, and trim trailing blank spaces
8 current_day := TO_CHAR(todays_date, 'DAY');
9 current_day := TO_CHAR(current_day);
10 -- convert day to mixed case letters with initial letter capitalized
11 current_day := INITCAP(current_day);
12 -- determine length of day's character string
13 current_day_length := LENGTH(current_day);
14 DBMS_OUTPUT.PUT_LINE('Today is ' || current_day || ', ' || TO_CHAR(todays_date));
15 DBMS_OUTPUT.PUT_LINE('The length of the word ' || current_day);
16 --|| is ' || TO_CHAR(current_day_length) || ' characters.');
17 END;
18 /
DECLARE
*
ERROR at line 1:
ORA-06550: line 9, column 19:
PLS-00307: too many declarations of 'TO_CHAR' match this call
ORA-06550: line 9, column 4:
PL/SQL: Statement ignored
line with error

Figure 4-11 Error resulting from improper use of TO_CHAR function

Sometimes one syntax error can generate many more errors, called **cascading errors**. For example, Figure 4-12 shows a program in which DECLARE is misspelled. As a result, none of the variable declarations are executed, and several syntax errors are generated. A good debugging strategy is to locate and fix the first error and then rerun the program. Don't try to fix all of the syntax errors before rerunning the program, because one error might generate several others, and fixing that one error may correct the others as well.

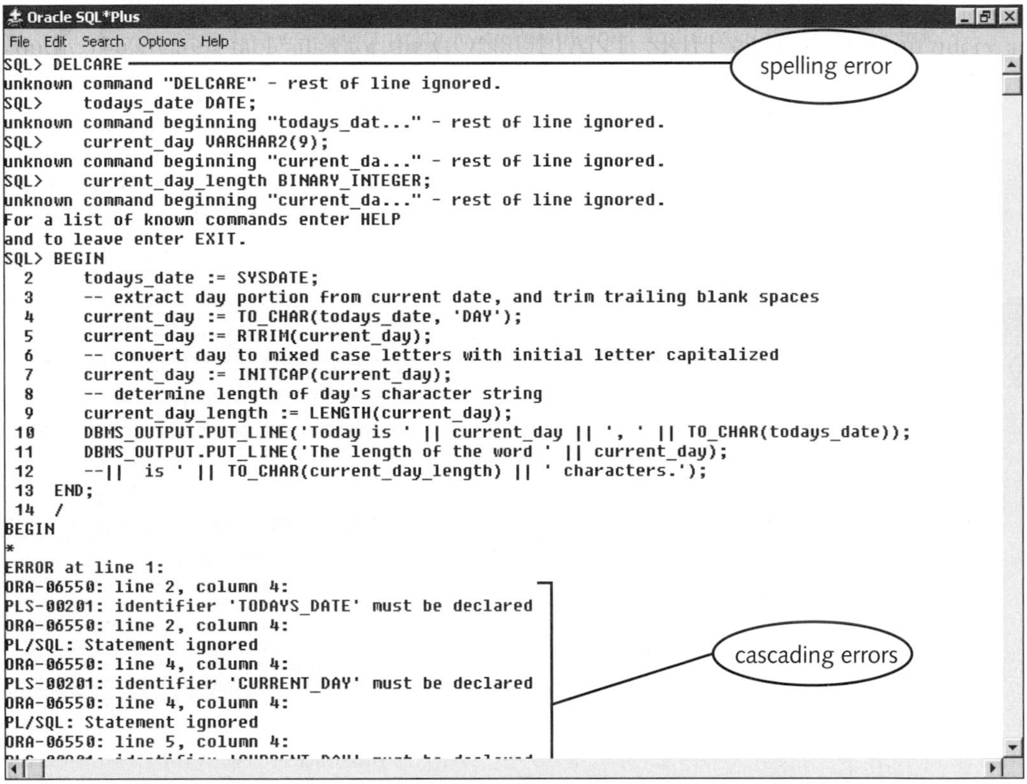

Figure 4-12   Syntax error causing cascading errors

In summary, keep the following points in mind when resolving syntax errors:

- Error locations in error messages might not correspond to the line of the actual error.

- Try to isolate the error location systematically by commenting out and/or modifying suspect lines.
- One error might result in several other cascading errors.

Logic Errors

Logic errors that result in incorrect program output can be caused by many things: not using the proper order of operations in arithmetic functions, passing incorrect parameter values to built-in functions, creating loops that do not terminate properly, using data values that are out of range or not of the right data type, and so forth. In complex commercial software applications, logic errors are located by legions of professional testers during long beta-testing processes. For the programs you will write while you are learning PL/SQL and developing Oracle database applications, you will usually be able to locate logic errors by determining what the output *should* be and comparing it to your actual program output.

The best way to locate logic errors is to view variable values during program execution using a **debugger**, which is a program that enables software developers to pause program execution and examine current variable values. The SQL*Plus environment does not provide a PL/SQL debugger. However, you can track variable values using DBMS_OUTPUT statements. For example, Figure 4-13 shows a program that is supposed to display the string that is before 'day' in the name of the current day of the week. However, the program is currently displaying the string 'Tu' instead of 'Tues'.

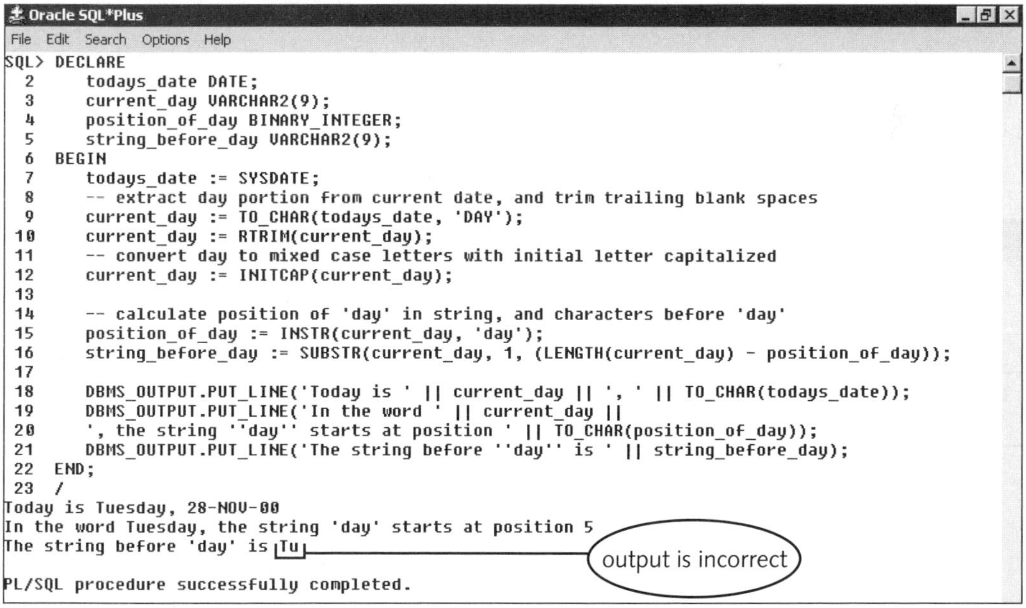

```
SQL> DECLARE
  2     todays_date DATE;
  3     current_day VARCHAR2(9);
  4     position_of_day BINARY_INTEGER;
  5     string_before_day VARCHAR2(9);
  6  BEGIN
  7     todays_date := SYSDATE;
  8     -- extract day portion from current date, and trim trailing blank spaces
  9     current_day := TO_CHAR(todays_date, 'DAY');
 10     current_day := RTRIM(current_day);
 11     -- convert day to mixed case letters with initial letter capitalized
 12     current_day := INITCAP(current_day);
 13
 14     -- calculate position of 'day' in string, and characters before 'day'
 15     position_of_day := INSTR(current_day, 'day');
 16     string_before_day := SUBSTR(current_day, 1, (LENGTH(current_day) - position_of_day));
 17
 18     DBMS_OUTPUT.PUT_LINE('Today is ' || current_day || ', ' || TO_CHAR(todays_date));
 19     DBMS_OUTPUT.PUT_LINE('In the word ' || current_day ||
 20     ', the string ''day'' starts at position ' || TO_CHAR(position_of_day));
 21     DBMS_OUTPUT.PUT_LINE('The string before ''day'' is ' || string_before_day);
 22  END;
 23  /
Today is Tuesday, 28-NOV-00
In the word Tuesday, the string 'day' starts at position 5
The string before 'day' is Tu

PL/SQL procedure successfully completed.
```


Figure 4-13 Program with logic error

The variable value with the error is string_before_day. Therefore, the error is most likely occurring in the assignment statement on Line 16, where string_before_day is assigned the value returned by the SUBSTR function. To debug the program, you can place a DBMS_OUTPUT statement before the line that uses the SUBSTR function to display the values of the variables that are being passed to the SUBSTR function. When you put debugging statements in your code, always include strings that label the variables, so you do not become confused about which variable values you are viewing. Figure 4-14 shows the DBMS_OUTPUT debugging statements and their resulting output.

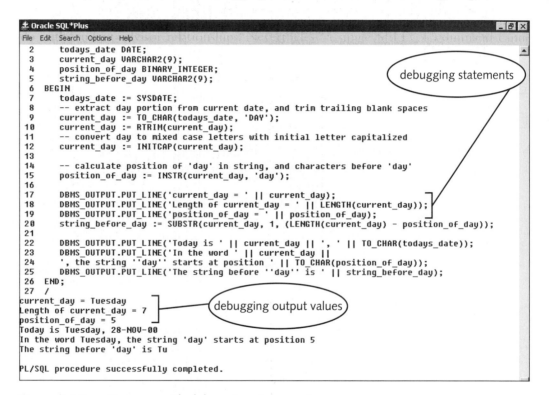

Figure 4-14   Program with debugging statements

The part of program Line 20 that contains the SUBSTR function is:

```
SUBSTR(current_day, 1,
(LENGTH(current_day) — position_of_day));
```

The output shows that the value of current_day is 'Tuesday', the value of LENGTH (current_day) is 7, and the value of position_of_day is 5. Therefore, the SUBSTR function receives the following parameters: ('Tuesday', 1, (7 − 5)). As a result, the function returns 7 - 5, or two characters ('Tu'), starting with the first character of the string. However, the function should return four characters ('Tues'), starting with the first character of the string. This output reveals the logic error: the third parameter in the SUBSTR function should be the length of current_day minus 3, not the length of current_day minus the position_of_day variable value, which is 5.

Here is a checklist to use when trying to debug a logic error:

1. Identify the output variable(s) that are involved in the error.

2. Identify the inputs and calculations that contribute to the invalid output.

3. Use DBMS_OUTPUT statements to find the values of the inputs that are contributing to the invalid output.

4. If you still can't locate the problem, take a break and look at it again later.

5. If you still can't locate the problem, ask a fellow student for help.

6. If you still can't locate the problem, ask your instructor for help.

NULL VALUES IN PL/SQL ASSIGNMENT STATEMENTS

Recall that when a variable is declared in the PL/SQL DECLARE block, the variable name is associated with a memory area on the database server that will eventually store the variable's value. In some programming languages, variables are automatically assigned default values. For example, in Visual Basic, variables with numerical data types are given a default value of zero, and character strings are given a default value of an empty string. In PL/SQL programs, until a variable is explicitly assigned a value, its value is NULL. Numerical operations on NULL values result in NULL values: For example, if you add a number to a NULL value, the resulting sum is also NULL. Now you will write a program to examine how PL/SQL handles assignment statements that involve variables with unassigned (NULL) values.

To write a program with an assignment statement involving a variable that has a NULL value:

1. In Notepad, type the program code shown in Figure 4-15 that calculates the sum of two numbers. Note that the value for the second number is not initialized in the DECLARE block or anywhere in the program. Execute the program in SQL*Plus.

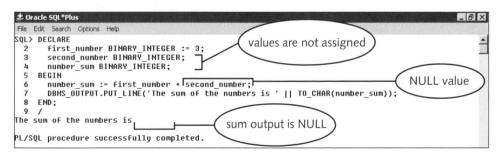

Figure 4-15   Result of assignment statement that involves an unassigned (NULL) variable value

The output does not display a value for the sum. What happened? No value was ever assigned for the second number, and when you add a value to a NULL value, the result is a NULL value. The result of all arithmetic operations on NULL values will always be another NULL value.

PL/SQL SELECTION STRUCTURES

The programs you have written so far use sequential processing, in which statements are processed one after another. However, most programs require selection structures in the form of IF/THEN/ELSE statements that make the processing order change, depending on the values of certain variables.

The IF/THEN Structure

The PL/SQL **IF/THEN** selection structure has the following general syntax:

```
IF condition THEN
    program statements that execute when condition is TRUE;
END IF;
```

Every IF must have a corresponding END IF. The *condition* has to be able to be evaluated as either TRUE or FALSE and can be either a comparison or a Boolean variable. Table 4-8 shows the PL/SQL comparison operators and examples of their usage.

Operator	Description	Example
=	Equal	count = 5
<>	Not equal	count <> 5
!=	Not equal	count != 5
>	Greater than	count > 5
<	Less than	count < 5
>=	Greater than or equal to	count >= 5
<=	Less than or equal to	count <= 5

Table 4-8   PL/SQL comparison operators

Remember that the operator that compares two values is =, while the assignment operator that assigns a value to a variable is :=.

If the condition in the IF/THEN structure evaluates as TRUE, one or more program statements execute. If the condition evaluates as FALSE or NULL, the program skips the statements. It is good programming practice to format IF/THEN structures by indenting the program statements that execute if the condition is TRUE so that the structure is easier to read and understand. You can indent these program statements by pressing the Tab key or by pressing the spacebar to add blank spaces.

Next, you will add an IF/THEN structure to your PL/SQL program so that it displays the current day of the week if today happens to be Friday. For the comparison to work properly, you will need to apply the RTRIM function to the current_day variable to remove blank spaces that are added during the TO_CHAR conversion. Note that the comparison in the condition is case sensitive.

To add an IF/THEN structure to your PL/SQL program:

1. In Notepad, copy the PL/SQL program you wrote to output the current day, and then modify the copied code as shown in Figure 4-16. Do not include the line numbers.

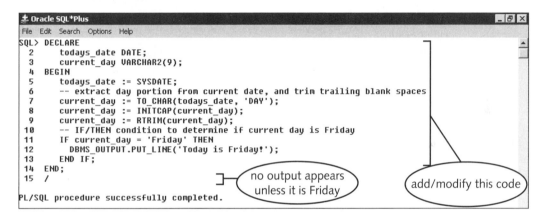

Figure 4-16   PL/SQL with IF/THEN structure

2. Copy the code into SQL*Plus, and execute the program. If it happens to be Friday (which it is not in the example), the condition evaluates to TRUE, and no output is generated, as shown.

3. In Notepad, copy your program code, modify the copied code as shown in Figure 4-17 so the condition will evaluate to TRUE if it is not Friday, and execute the modified program. Unless you were one of the lucky ones whose condition evaluated to TRUE in the previous step, your output should now look like the output in Figure 4-17.

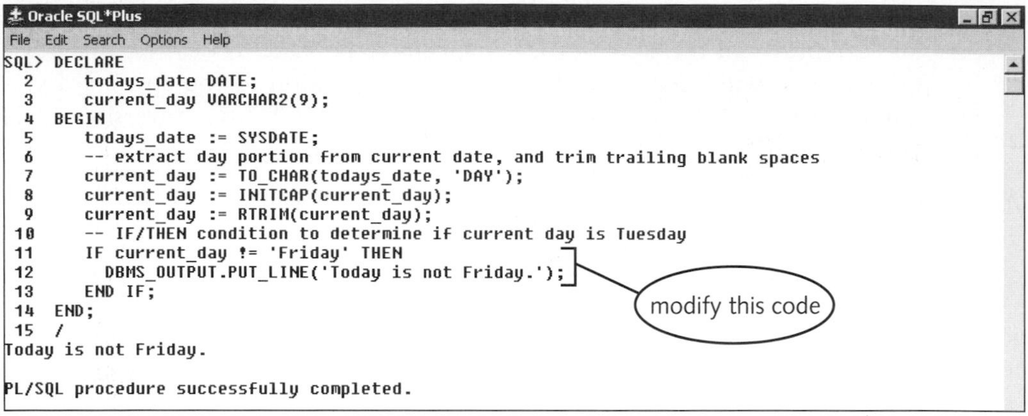

```
± Oracle SQL*Plus                                                          _ 5 X
File  Edit  Search  Options  Help
SQL> DECLARE
  2      todays_date DATE;
  3      current_day VARCHAR2(9);
  4  BEGIN
  5      todays_date := SYSDATE;
  6      -- extract day portion from current date, and trim trailing blank spaces
  7      current_day := TO_CHAR(todays_date, 'DAY');
  8      current_day := INITCAP(current_day);
  9      current_day := RTRIM(current_day);
 10      -- IF/THEN condition to determine if current day is Tuesday
 11      IF current_day != 'Friday' THEN
 12        DBMS_OUTPUT.PUT_LINE('Today is not Friday.');
 13      END IF;                                              modify this code
 14  END;
 15  /
Today is not Friday.

PL/SQL procedure successfully completed.
```

Figure 4-17   PL/SQL program with alternate IF/THEN comparison

The IF/THEN/ELSE Structure

The previous example suggests the need for a selection structure that executes alternate program statements when a condition evaluates to FALSE. This is usually called an IF/THEN/ELSE structure. In PL/SQL, the IF/THEN/ELSE structure has the following general syntax:

```
IF condition THEN
   program statements that execute when condition is TRUE;
ELSE
   alternate program statements that execute when
   condition is FALSE;
END IF;
```

Next, you will modify the program with the IF/THEN structure so it uses an IF/THEN/ELSE structure that displays one output if the current day is Friday and a different output if the current day is not Friday.

To modify your PL/SQL program to use an IF/THEN/ELSE structure:

1. In Notepad, copy your program code, and modify the copied code as shown in Figure 4-18 to use the IF/THEN/ELSE structure. (Do not include the line numbers.)

2. Copy the code into SQL*Plus, and execute the program. Depending on the current day, the appropriate output appears.

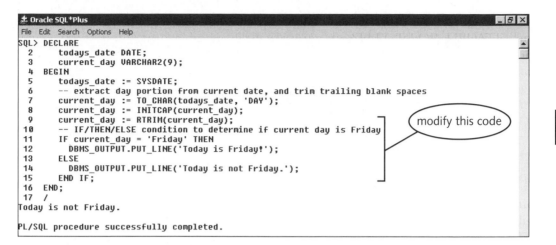

Figure 4-18   PL/SQL IF/THEN/ELSE structure

 In an IF/THEN/ELSE structure, a semicolon is only used to terminate the program statements that are executed if the condition is evaluated as TRUE or FALSE, and after the END IF;.

Nesting IF/THEN/ELSE Statements

IF/THEN/ELSE structures can be **nested**, which means that one or more additional IF/THEN/ELSE statements are included within the program statements, below either the IF or ELSE. It is especially important to properly indent the program lines following the THEN and ELSE commands in nested IF/THEN/ELSE structures—correct formatting enables you to better understand the program logic and spot syntax errors, such as missing END IF commands. Next, you will modify your PL/SQL program that uses the IF/THEN/ELSE structure so that if the current day is not Friday, the program also tests to see if the current day is Saturday and displays the appropriate output.

To modify your PL/SQL program using a nested IF/THEN/ELSE structure:

1. In Notepad, copy your PL/SQL code and modify it as shown in Figure 4-19 to include the nested IF/THEN/ELSE structure.

2. Copy the code into SQL*Plus, and execute the program. Depending on the current day, the appropriate output appears.

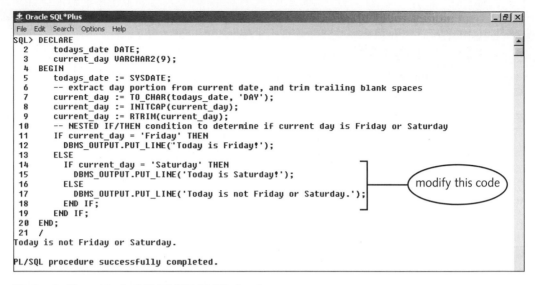

Figure 4-19 Nested IF/THEN/ELSE structure

The IF/ELSIF Structure

The **IF/ELSIF** structure allows you to test for many different conditions and is similar to the CASE or SELECT CASE structure used in other programming languages. The general syntax for this structure is as follows:

```
IF condition1 THEN
   program statements that execute when condition1 is TRUE;
ELSIF condition2 THEN
   program statements that execute when condition2 is TRUE;
ELSIF condition3 THEN
   program statements that execute when condition3 is TRUE;
...
ELSE
   program statements that execute when none of the
   conditions are TRUE;
END IF;
```

If the first condition is true, then its program statement(s) execute, and the IF/THEN structure is exited. Otherwise, the first ELSIF condition is evaluated. If it is true, then its associated program statement(s) execute, and the IF/THEN structure is exited. If the first ELSIF condition is not true, then the next ELSIF condition is evaluated, and so forth. If all ELSIF conditions are false, then the ELSE program statement(s) execute. Next, you will modify your PL/SQL program with the nested IF/THEN/ELSE structure so that it uses the ELSIF to test for each day of the week and display the appropriate output. Note that the program statements following all of the ELSIF commands and the program statement following the ELSE command are all indented to make the structure easier to read and interpret.

To modify your PL/SQL program to use the IF/ELSIF structure:

1. In Notepad, copy your PL/SQL code, and modify it as shown in Figure 4-20 to include the IF/ELSIF structure.

2. Copy the code into SQL*Plus, and execute the program. Depending on the current day, the appropriate output appears.

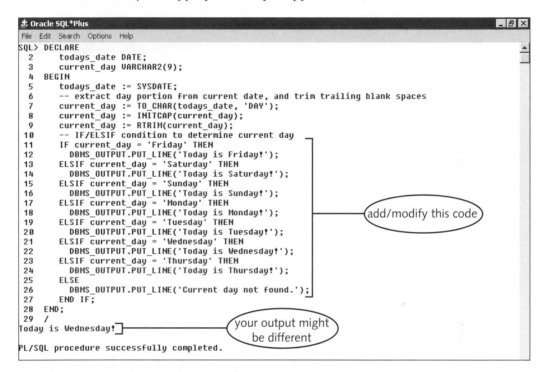

Figure 4-20   IF/ELSIF structure

In the ELSIF command, there is no E in ELSE and no space between ELS and IF. This syntax comes from the Ada programming language.

Evaluating NULL Conditions in IF/THEN/ELSE Structures

Recall that the first set of program statements in an IF/THEN/ELSE structure execute if the condition evaluates to TRUE, and the second set of statements (following the ELSE command) execute if the condition evaluates to FALSE. If a condition evaluates to NULL, then the structure behaves the same as if the condition evaluated to FALSE. A condition can evaluate to NULL under two circumstances:

1. If the condition is specified using a Boolean variable that has not been initialized.

2. If any variables used in the condition currently have a value of NULL.

Now you will create a PL/SQL program that examines the behavior of an IF/THEN/ELSE structure, first using a Boolean variable that has been initialized to TRUE and then using a Boolean variable that has not been initialized and is currently NULL.

To write a PL/SQL program that uses an initialized and an uninitialized Boolean variable:

1. In Notepad, type the program shown in Figure 4-21, and then execute the program in SQL*Plus. Note that the value of the Boolean variable (condition_flag) is initialized as TRUE in the DECLARE section of the program, and the output of the IF/THEN/ELSE structure confirms this.

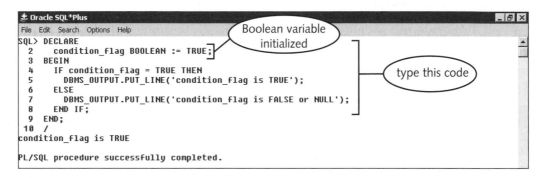

Figure 4-21   IF/THEN/ELSE structure using initialized Boolean variable

2. In Notepad, copy the program, modify it as shown in Figure 4-22, and then execute the program in SQL*Plus. Note that the value of the Boolean variable (condition_flag) is not initialized in the DECLARE section of the program, so the value is currently NULL. The IF/THEN/ELSE structure executes the second set of program statements, indicating the condition is FALSE or NULL.

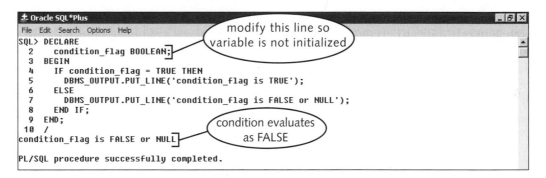

Figure 4-22   Condition using noninitialized (NULL) Boolean variable

A similar output will result if you use a comparison in the condition statement and any of the variables in the comparison have NULL values. Now you will write a PL/SQL program that uses comparison variables with NULL values to examine the result.

To write a PL/SQL program that uses variables with NULL values in a comparison:

1. In Notepad, type the program shown in Figure 4-23, and then execute the program in SQL*Plus. Note that the second variable (second_number) is not initialized, and is therefore currently NULL. As a result, the condition in the IF/THEN/ELSE structure is evaluated as FALSE.

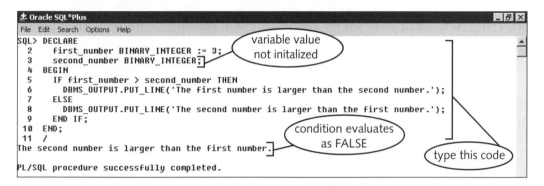

Figure 4-23   Condition using noninitialized (NULL) variable value

2. Close SQL*Plus.

3. Save your file in Notepad, and then close Notepad.

SUMMARY

- ☐ PL/SQL is the procedural language used to write procedures and functions for Oracle applications and is used in most Oracle database application development tools.

- ☐ PL/SQL variables must follow the Oracle naming standard.

- ☐ PL/SQL is a strongly typed language, which means all variables and their associated data types must be declared prior to use, and assignments and comparisons can be performed only between variables with the same data type.

- ☐ PL/SQL has scalar data types, which can be used by variables that reference a single value; composite data types, which can be used by variables that reference multiple scalar values; reference data types, which can be used by variables that reference retrieved data values; and LOB data types, which can be used by variables that reference large binary objects.

- ☐ Reference data types assume the same data type as a referenced database column or row.

- ☐ A PL/SQL program block consists of the declaration section, where variables are defined or described; the body, where program statements are placed; and the exception section, where error handling statements are placed. The declaration and exception sections are optional.

❏ In PL/SQL, a block of comment statements can be delimited by beginning the block with the characters /* and ending the block with the characters */. A single comment line is delimited by beginning it with two hyphens (--).

❏ In PL/SQL, values are assigned to variables using the := assignment operator. The variable name is placed on the left side of the operator, and the value is placed on the right side of the operator.

❏ The DBMS_OUTPUT.PUT_LINE procedure displays output from PL/SQL programs in the SQL*Plus environment.

❏ Before using the DBMS_OUTPUT.PUT_LINE function, you must type SET SERVER OUTPUT ON at the SQL prompt to set up a memory buffer on the server to store output statements.

❏ PL/SQL performs implicit data conversions, but the output is unpredictable, so you should always use the explicit data conversion functions.

❏ To concatenate two character strings, use the double bar (||) operator.

❏ A syntax error is generated when the program code does not follow the guidelines of the programming language. A logic error does not stop a program from compiling, but it results in incorrect output.

❏ To correct syntax errors in PL/SQL programs in the SQL*Plus environment, you must systematically identify the line that is generating the error. You can do this by commenting out any program line that you suspect might be causing the problem.

❏ To correct logic errors in PL/SQL programs in the SQL*Plus environment, you must identify the program line that is generating the incorrect output, and then examine the values that are being used by that line to create the incorrect output. You can do this by temporarily displaying intermediate output values.

❏ In PL/SQL programs, a variable's value is NULL until it is explicitly assigned a value. Numerical operations involving NULL values (such as adding a value to a NULL variable value) result in NULL values.

❏ The IF/THEN structure evaluates a given condition. Every IF statement must have a corresponding END IF command.

❏ The IF/THEN/ELSE structure is similar to the IF/THEN structure, except that when the condition evaluates to FALSE, one or more alternate program statements execute.

❏ The IF/ELSIF structure allows you to test for many different conditions in a single IF structure.

❏ If a condition in an IF/THEN/ELSE structure is evaluated as NULL, then the structure treats the condition just as if it was evaluated as FALSE.

REVIEW QUESTIONS

1. What is a procedural programming language? List two advantages of using PL/SQL over other procedural programming languages for developing Oracle applications.

2. Write the PL/SQL declaration statement for the following variables:

Variable Name	Data Description
mystery_text	Up to 4,000 characters of variable-length text
course_credit	An integer with maximum value of 99
order_status	TRUE or FALSE
order_price	A fixed-point number with two decimal places, with a maximum value of 999.99
date_received	Date
student_class	Text field that always contains two characters

3. What is the difference between the = and the := operators?

4. When should you issue the command SET SERVEROUTPUT ON?

5. What are referential data types used for?

6. What are the two approaches for delimiting comment blocks in a PL/SQL program?

7. What should you do before asking your instructor for help with a logic programming error?

8. Why is it important to always initialize the values of numerical variables that are used in arithmetic calculations?

9. What is the difference between a syntax error and a logic error? What is the difference in the approach you use to debug each type of error?

10. What is a cascading error?

PROBLEM-SOLVING CASES

Use Notepad or an alternate text editor to create the following PL/SQL programs. Save all solution files in your Chapter4\CaseSolutions folder.

1. Declare a numeric variable named counter and a variable-length character string variable named my_string. Assign the value 100 to counter, and the text 'Hello world' to my_string, and display the values on two separate lines using the DBMS_OUTPUT procedure. Save the file as 4ACase1.sql.

2. Declare the following variables and assign to them the given values:

Variable Name	Data Type	Value
inventory_ID	Numeric	11668
inventory_color	Character	Sienna
inventory_price	Numeric	259.99
inventory_QOH	Numeric	16

Write program statements to display the output shown in Figure 4-24, using the DBMS_OUTPUT function and the variable values. Format the output so it appears exactly as shown in Figure 4-24. Save the file as 4ACase2.sql. Do not insert, or hard-code, the actual data values in the output procedure.

Figure 4-24

3. Declare the following variables using appropriate data types, and assign them the given values in the DECLARE block.

Variable Name	Data Type	Value
student_last_name	Character	Miller
student_first_name	Character	Sarah
student_MI	Character	M
student_address	Character	144 Windridge Blvd.
student_city	Character	Eau Claire
student_state	Character	WI
student_zip	Character	54703

Write program statements to display the output exactly as shown in Figure 4-25, using the DBMS_OUTPUT.PUT_LINE procedure and variable values. Save the file as 4ACase3.sql. Do not insert, or hard-code, the actual data values in the output procedure.

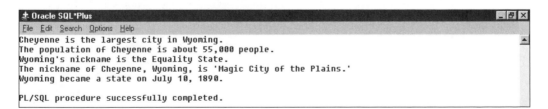

Figure 4-25

4. Declare character variables named faculty_last_name, faculty_first_name, and faculty_phone. Assign the value 'COX' to faculty_last_name, 'KIM' to faculty_first_name, and '7155551234' to faculty_phone. Write program commands so that the program displays the output exactly as follows:

```
Kim Cox's phone number is (715) 555-1234.
```

   Declare and use additional variables as needed. Save your file as 4ACase4.sql. Do not insert, or hard-code, the actual data values in the output function.

5. Declare the following variable names and assign the associated values, using the data types specified.

Variable Name	Value	Data Type
city	Cheyenne	Character
state	Wyoming	Character
state_nickname	Equality State	Character
city_nickname	Magic City of the Plains	Character
city_population	55,000	Numeric
statehood_date	07/10/1890	Date

   Write a PL/SQL statement that will concatenate the variables with the necessary literal values to give the output exactly as shown in Figure 4-26. Declare and use additional variables as needed. Save the file as 4ACase5.sql. Do not insert, or hard-code, the actual data values in the output procedure.

Oracle SQL*Plus
File Edit Search Options Help
```
Cheyenne is the largest city in Wyoming.
The population of Cheyenne is about 55,000 people.
Wyoming's nickname is the Equality State.
The nickname of Cheyenne, Wyoming, is 'Magic City of the Plains.'
Wyoming became a state on July 10, 1890.

PL/SQL procedure successfully completed.
```

Figure 4-26

6. Write a PL/SQL program that calculates the diameter and radius of a circle. The formula to calculate the diameter of a circle is $2\pi r$, and the formula to calculate

the area of a circle is πr^2. The program should declare the following variables with the specified data types and initial values:

pi   NUMBER   3.1415926
current_radius   NUMBER   3

Display the output exactly as follows:

For a circle with radius 3, the diameter is 18.85 and the area is 28.27.

Declare and use additional variables as needed. Save the file as 4ACase6.sql. (*Hint*: Use the Oracle SQL ROUND function to round the output values.)

7. Declare a date variable named current_date, and assign SYSDATE to it. Depending on the day of the month, your program should display the following output:

Day	**Output**
1–10	It is Day <day number> of <month name>. It is early in the month.
11–20	It is Day <day number> of <month name>. It is the middle of the month.
21–31	It is Day <day number> of <month name>. It is nearly the end of the month.

For example, if it is currently November 30, your program should display "It is Day 30 of November. It is nearly the end of the month." Declare and use additional variables as needed. Save the file as 4ACase7.sql.

8. According to the Chinese zodiac, birth years are associated with the following animals:

Birth Year	Animal
1924, 1936, 1948, 1960, 1972, 1984, 1996	Rat
1925, 1937, 1949, 1961, 1973, 1985, 1997	Cow
1926, 1938, 1950, 1962, 1974, 1986, 1998	Tiger
1927, 1939, 1951, 1963, 1975, 1987, 1999	Rabbit
1928, 1940, 1952, 1964, 1976, 1988, 2000	Dragon
1929, 1941, 1953, 1965, 1977, 1989, 2001	Snake
1930, 1942, 1954, 1966, 1978, 1990, 2002	Horse
1931, 1943, 1955, 1967, 1979, 1991, 2003	Sheep
1932, 1944, 1956, 1968, 1980, 1992, 2004	Monkey
1933, 1945, 1957, 1969, 1981, 1993, 2005	Chicken
1934, 1946, 1958, 1970, 1982, 1994, 2006	Dog
1935, 1947, 1959, 1971, 1983, 1995, 2007	Pig

Declare a date variable named birth_date, and assign to it your birth date. Use an IF/ELSIF structure to test every year and determine the animal associated with your birth year. Then display your birth year and the associated animal name. For example, the program would display the following output for someone born in 1984:

```
I was born in 1984, which is the year of the Rat.
```

Declare and use additional variables as needed. Save the file as 4ACase8.sql.

9. PL/SQL has a function named NEW_TIME that converts an input date and time from one time zone to another. The general syntax of the function is
time_in_desired_time_zone := NEW_TIME(*time_in_current_time_zone*, *current_time_zone_abbreviation*, *desired_time_zone_abbreviation*). Some of the time zone abbreviations follow:

CST   Central Standard Time
EST   Eastern Standard Time
GMT   Greenwich Mean Time
HST   Alaska/Hawaii Standard Time
MST   Mountain Standard Time
PST   Pacific Standard Time
YST   Yukon Standard Time

For example, to convert the current time from Eastern Standard Time to Central Standard Time and assign the result to a variable named tulsa_time, you would use the following function: `tulsa_time := NEW_TIME(SYSDATE, 'EST', 'CST');`.

Write a program that declares a character variable named time_zone and assign it to the abbreviation corresponding to your time zone. (If your time zone is not listed, use 'EST'.) Then, write the code to display the output shown in Figure 4-27.

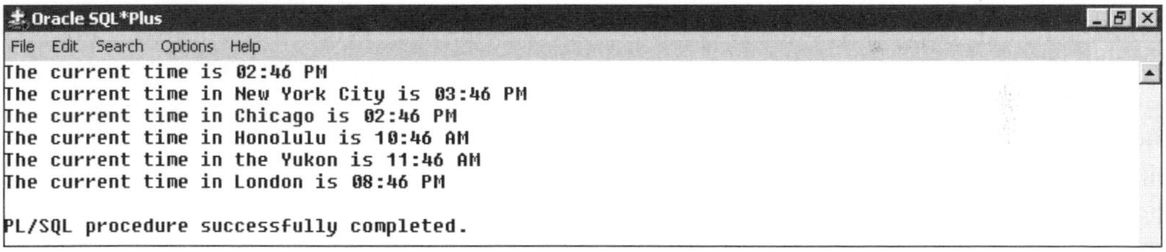

```
Oracle SQL*Plus
File  Edit  Search  Options  Help
The current time is 02:46 PM
The current time in New York City is 03:46 PM
The current time in Chicago is 02:46 PM
The current time in Honolulu is 10:46 AM
The current time in the Yukon is 11:46 AM
The current time in London is 08:46 PM

PL/SQL procedure successfully completed.
```

Figure 4-27

Declare and use other variables as needed. Save the file as 4ACase9.sql.

10. Develop a PL/SQL program to calculate the yearly bonus that a company gives to its employees. The bonus is determined by considering three factors: (1) how many years the employee has worked at the company, (2) employee job type, and (3) whether or not the employee has been recommended for extra merit.

If an employee has worked from 1–5 years, then he or she receives a $100 bonus; 6–10 years merits a $200 bonus; 11–15 years merits a $300 bonus, and over 16 years merits a $500 bonus. For job type, employees in Job Type "A" receive an additional $100 bonus, and employees in Job Type "B" receive an additional $200 bonus. Finally, if the employee has been commended for Extra Merit, he or she receives an additional 5% raise based on his or her current salary.

Write the code to calculate the employee bonus. Create a DATE variable named employee_start_date, and calculate the employee's years of service based on the difference between the start date and the current year. Truncate the value to the nearest whole number. Create a CHAR variable named job_type that specifies if the employee is employed in Job Type "A" or Job Type "B." Create a BOOLEAN variable named extra_merit; if the value of this variable is TRUE, then the employee should receive the extra merit bonus compensation. Finally, create a NUMBER variable named current_salary to represent the employee's current salary. Create three test programs, with each copy using the following test values:

Program Filename	employee_start_date	job_type	extra_merit	current_salary	Bonus Amount
4ACase10a.sql	05/01/1993	A	FALSE	55,000	$300
4ACase10b.sql	01/05/1982	B	TRUE	75,800	$4,490
4ACase10c.sql	07/15/2000	A	TRUE	37,800	$1,990

Declare and use additional variables as needed. Display the calculated bonus for each program as follows:

4ACase10a.sql output:

The bonus for this employee is $300

4ACase10b.sql output:

The bonus for this employee is $4,490

4ACase10c.sql output:

The bonus for this employee is $1,990

◀ LESSON B ▶

Objectives

♦ Use SQL commands in PL/SQL programs
♦ Create loops in PL/SQL programs
♦ Create PL/SQL tables and tables of records
♦ Use cursors to retrieve database data into PL/SQL programs
♦ Use the exception section to handle errors in PL/SQL programs

USING SQL COMMANDS IN PL/SQL PROGRAMS

Recall that SQL data definition language (DDL) commands, such as CREATE, ALTER, and DROP, change the structure of database objects. Data manipulation language (DML) commands, such as SELECT and INSERT, query or manipulate data in database tables. Transaction control commands, such as COMMIT and ROLLBACK, organize DML commands into logical transactions and commit them to the database or roll them back. DDL commands cannot be used in PL/SQL programs, but both DML and transaction control commands can be used in PL/SQL programs. Table 4-9 summarizes the SQL command categories and shows which can and cannot be used in PL/SQL programs.

Category	Purpose	Examples	Can Be Used in PL/SQL Programs
DDL	Changes the structure of database objects	CREATE, ALTER, DROP	No
DML	Manipulates data in tables	SELECT, INSERT, UPDATE, DELETE	Yes
Transaction Control	Organizes DML commands into logical transactions	COMMIT, ROLLBACK, SAVEPOINT	Yes

Table 4-9   Categories of SQL commands

Now you will write a PL/SQL program that uses DML and transaction control commands to insert the first three records into the TERM table of the Northwoods University database. First, you will start SQL*Plus and run the emptynorthwoods.sql script to delete your current Northwoods University database tables and re-create the Northwoods University database tables without any data in them. Then, you will check to make sure the TERM table is empty.

To run the script to create the database tables and check the TERM table:

1. If necessary, start SQL*Plus and type **SET SERVEROUTPUT ON SIZE 4000** to enable interactive output and set up the output buffer.

2. Type **START a:\chapter4\emptynorthwoods.sql** at the SQL prompt to run the emptynorthwoods.sql script from the Chapter4 folder on your Data Disk. This script drops your existing Northwoods University tables and creates the empty Northwoods University database tables.

3. Type the following command at the SQL prompt: **SELECT * FROM term;**. The message "no rows selected" appears, indicating that the table is empty.

Now you will write a PL/SQL program to insert the first three records into the TERM table. You will declare variables corresponding to each of the table fields, assign the variables to the appropriate values, insert the records, and then commit the records.

To write the program to insert the TERM records:

1. Start Notepad, and type the PL/SQL program commands shown in Figure 4-28. Do not type the line numbers. Save the Notepad text file as **4BPrograms.sql** in your Chapter4\TutorialSolutions folder.

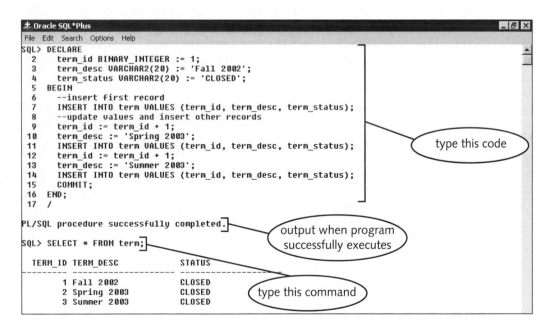

Figure 4-28   PL/SQL program using INSERT command

2. Copy your code into SQL*Plus, and run the program. The message "PL/SQL procedure successfully completed" appears when your program runs correctly. If a syntax error appears, debug and rerun your program until it successfully executes.

3. Type the SELECT command shown in Figure 4-28 at the SQL prompt to view the records in the TERM table. The output shows that the records were inserted correctly.

LOOPS

Loops are program structures that perform one or more commands, called **looping statements**, multiple times, until a specific exit condition is reached. This exit condition is often specified by incrementing a variable that is used as a counter within the looping statements and then testing to see if the counter has reached a desired value. A loop can be a **pretest loop**, in which the condition is tested before the looping program statements execute. Alternatively, a loop can be a **posttest loop**, in which the condition is tested after the looping program statements execute, and the looping program statements always execute at least once. You use a pretest loop if there is a case in which the looping program statements might never execute. You use a posttest loop when you are sure that the looping program statements always need to execute at least once. PL/SQL has five looping structures: LOOP...EXIT, LOOP...EXIT WHEN, WHILE...LOOP, numeric FOR loops, and cursor FOR loops. The first four loop structures will be discussed in the following paragraphs. Cursor FOR loops will be discussed in the section on cursors later in this lesson.

To illustrate the different types of loops, you will create a database table named COUNT_TABLE that has one numerical data field named COUNTER. You will create PL/SQL programs that use the different loop structures to automatically insert the numbers 1, 2, 3, 4, and 5 into the COUNTER field in COUNT_TABLE, as shown in Table 4-10.

COUNTER
1
2
3
4
5

Table 4-10 COUNT_TABLE

You will now create COUNT_TABLE in SQL*Plus.

To create COUNT_TABLE:

1. In SQL*Plus, type the following command at the SQL prompt:

```
CREATE TABLE count_table
(counter NUMBER(2));
```

The message "Table created." appears if the table is successfully created.

The LOOP...EXIT Loop

The basic syntax of a **LOOP...EXIT** loop is:

```
LOOP
   program statements
   IF condition THEN
       EXIT;
   END IF;
   more program statements
END LOOP
```

When the IF/THEN condition is true, the EXIT program statement directs the program to exit the loop and resume execution of program statements after the END LOOP statement. The looping program statements, which are the program statements that are repeated if the condition is TRUE, can be placed before or after the IF/END IF exit condition evaluation section. As a result, the loop can be either a pretest loop or a posttest loop.

How do you decide whether to use a pretest loop or a posttest loop? If the looping program statements might never need to be executed, use a pretest loop. If the looping program statements will always need to be executed at least once before the loop exit condition is evaluated, use a posttest loop.

Now, you will write a program to insert records 1 through 5 into COUNT_TABLE using a LOOP...EXIT loop. The program statements in the loop will execute five times, once for each record. Because the looping program statements will always execute at least once, you will use a posttest loop. The test condition will evaluate the value of a variable named loop_count. The program will exit the loop when loop_count is equal to 5, indicating that 5 records have been inserted.

To insert the COUNT_TABLE records using a posttest LOOP...EXIT loop:

1. In Notepad, type the code shown in Figure 4-29. It is a good programming practice to format the loop as shown by indenting the program lines between the LOOP and END LOOP commands.

2. Paste your code into SQL*Plus, run it, and debug it if necessary.

3. Type the SELECT command shown in Figure 4-29 to view the records in COUNT_TABLE and confirm that the records were inserted correctly.

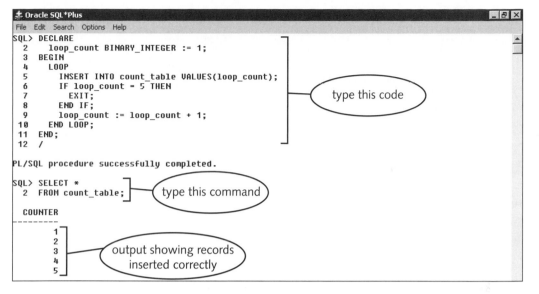

Figure 4-29   PL/SQL program with LOOP...EXIT loop

The LOOP...EXIT WHEN Loop

The **LOOP...EXIT WHEN** loop is a posttest loop, and the looping program statements always execute at least once. The general syntax of the LOOP...EXIT WHEN loop is:

```
LOOP
    program statements
    EXIT WHEN condition;
END LOOP;
```

This loop executes the program statements, and then tests for the exit condition, using the EXIT WHEN command. Now, you will delete the records you inserted into COUNT_TABLE, and then write a PL/SQL program to insert the records using a LOOP...EXIT WHEN loop.

To delete the records and write the program using a LOOP...EXIT WHEN loop:

1. Delete the existing COUNT_TABLE records by typing the following command at the SQL prompt in SQL*Plus: **DELETE FROM count_table;**. The message "5 rows deleted" appears.

2. In Notepad, type the code shown in Figure 4-30. Note that the program lines between the LOOP and END LOOP commands are indented.

3. Paste your code into SQL*Plus, run it, and debug it if necessary.

4. Type the SELECT command shown in Figure 4-30 to view the records in COUNT_TABLE and confirm that the records were inserted correctly.

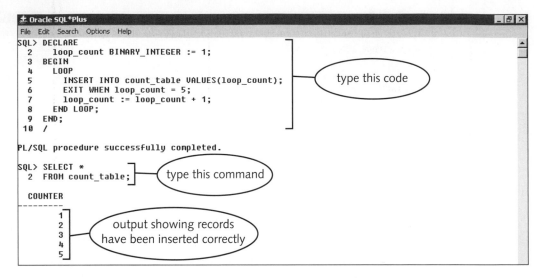

Figure 4-30   PS/SQL program with LOOP...EXIT WHEN loop

The WHILE...LOOP

The **WHILE...LOOP** is always a pretest loop in which the exit condition is evaluated before any program statements execute. The general syntax of the WHILE...LOOP is:

```
WHILE condition
LOOP
    program statements
END LOOP;
```

You will again delete the records you inserted into COUNT_TABLE and then write a PL/SQL program to insert the records using the WHILE...LOOP structure. Because this is a pretest loop, you must write the condition so the loop is exited when the loop_count value is one more than the number of records you want to insert, or in this case 6.

To delete the records and write the program using the pretest WHILE...LOOP:

1. In SQL*Plus, delete the existing COUNT_TABLE records by typing the following command at the SQL prompt: **DELETE FROM count_table;**. The message "5 rows deleted" appears.

2. Type the code shown in Figure 4-31. Again, note that the program lines between the LOOP and END LOOP commands are indented.

3. Paste your code into SQL*Plus, run it, and debug it if necessary.

4. Type the SELECT command shown in Figure 4-31 to confirm that the records were inserted correctly.

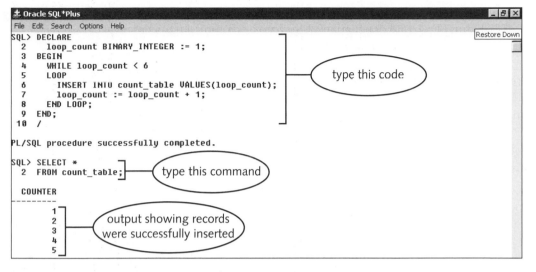

Figure 4-31   PL/SQL program with WHILE...LOOP

The Numeric FOR Loop

In all of the loops you have used so far, you have had to declare a counter variable and explicitly increment it within your program statements. With the **numeric FOR** loop, you do not need to declare the counter variable. The counter is defined by start and end numbers in the FOR statement, and it automatically increments each time the loop repeats. The general syntax of the numeric FOR loop is:

```
FOR counter_variable IN start_value .. end_value
LOOP
    program statements
END LOOP;
```

The start and end values must be integers, and they are always incremented by one. Now, you will delete the COUNT_TABLE records and reinsert the records using a numeric FOR loop.

To insert the records using a numeric FOR loop:

1. In SQL*Plus, delete the existing COUNT_TABLE records by typing the following command at the SQL prompt: **DELETE FROM count_table;**. The message "5 rows deleted" appears.

2. Type the code shown in Figure 4-32. Because this program does not require any variables to be declared, the DECLARE section has been omitted. Again, indent the statements between the LOOP and END LOOP commands for clarity.

3. Paste your code into SQL*Plus, run it, and debug it if necessary.

4. Type the SELECT command shown in Figure 4-32 to confirm that the records were inserted correctly.

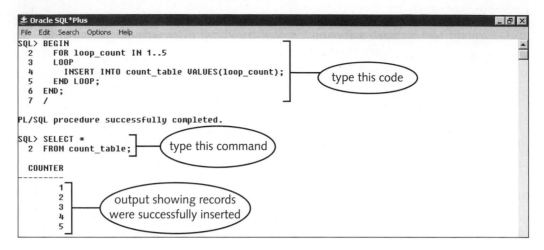

Figure 4-32   PL/SQL program wih numeric FOR loop

CURSORS

When Oracle processes a SQL command, it allocates a memory location in the database server's memory known as the **context area**. This memory location contains information about the command, such as the number of rows processed by the statement, a parsed (machine language) representation of the statement, and in the case of a query that returns data, the **active set**, which is the set of data rows returned by the query. In Chapter 1, you learned that a pointer is a link to a physical location where data is stored. A **cursor** is a pointer that references the memory location that contains the SQL command's context area. Figure 4-33 illustrates a cursor that points to a context area for a query that returns all of the rows from the COURSE table in the Northwoods University database.

Server Memory

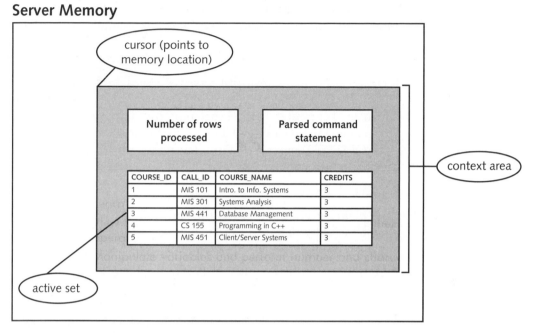

Figure 4-33   Cursor to retrieve all rows in COURSE table

There are two kinds of cursors: implicit and explicit. The following sections will discuss both cursor types.

Implicit Cursors

An **implicit** cursor is created automatically within the Oracle environment and does not need to be declared in the DECLARE section of a program. Oracle creates an implicit cursor every time you issue an INSERT, UPDATE, or DELETE command. You can use an implicit cursor when you want to assign the output of a simple SELECT query to a PL/SQL variable and you are sure that the query will return one—and only one—record. If the query returns more than one record, or does not return any records, an error message is generated.

The general syntax for creating an implicit cursor is:

```
SELECT data_field(s)
INTO declared_variable_name(s)
FROM tablename(s)
WHERE join_conditions_if_needed
AND search_condition_that_will_return_a_single_record;
```

The variables receiving the data from the query in the INTO clause must be the same data types as the returned field(s). To declare these variables, it is useful to use the %TYPE referential data type, which is used to declare a variable that has the same data type as a table field.

Recall that to declare a variable using the %TYPE data type, you use the following syntax: *variable_name tablename.fieldname*%TYPE. For example, to declare a variable named current_f_last that has the same data type as the F_LAST field in the FACULTY table, you would use the command: **current_f_last faculty.f_last%TYPE;**.

Now, you will create an implicit cursor that retrieves a specific faculty member's last and first names from the FACULTY table in the Northwoods University database into declared variables named current_f_last and current_f_first. These variables will be declared using the %TYPE referential data type. Then, the program displays the retrieved values using the DBMS_OUTPUT procedure. First, you will run a script to refresh your case study database tables.

To refresh your database tables and create a program with an implicit cursor:

1. Run the **clearwater.sql**, **northwoods.sql**, and **softwareexp.sql** scripts in the Chapter4 folder of your Data Disk.

2. Type the code shown in Figure 4-34 in Notepad. Note that a blank space enclosed in single quotes is inserted before the faculty member's first name, and between the faculty member's first and last names, using single quotes.

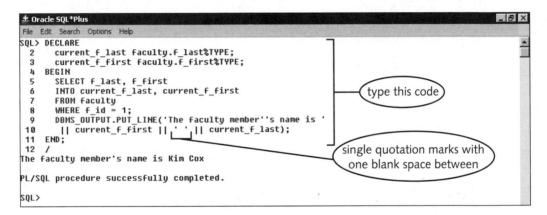

Figure 4-34   PL/SQL program with implicit cursor

3. Paste your code into SQL*Plus, run it, and debug it if necessary. The output displays the retrieved data values.

Recall that an implicit cursor can only be used when the SQL query retrieves one and only one record. In Figure 4-35, the search condition was deleted in the previous implicit cursor query, so the query returned all five records in the FACULTY table. The figure shows the error message that appears when an implicit cursor retrieves multiple records.

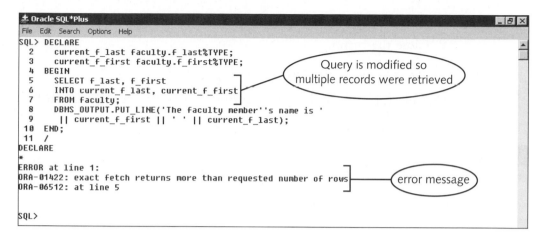

Figure 4-35   Error resulting from implicit cursor that attempts to retrieve multiple records

Figure 4-36 shows the result of an implicit cursor that returns no records. The search condition was modified so the query searched for F_ID 6, but the database does not contain a record where F_ID = 6. An error message stating "no data found" appears.

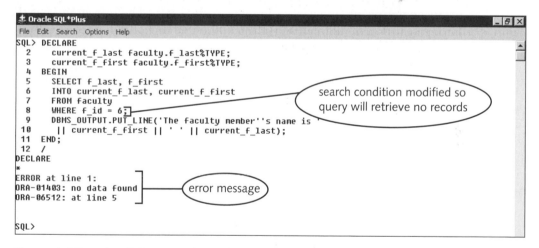

Figure 4-36   Implicit cursor that attempts to return no records

Explicit Cursors

Explicit cursors must be used with SELECT statements that might retrieve a variable number of records or might return no records. Explicit cursors are declared in the DECLARE section and processed in the program body. The steps for using an explicit cursor are:

1. Declare the cursor.

2. Open the cursor.

3. Fetch the cursor results into PL/SQL program variables.

4. Close the cursor.

Declaring an Explicit Cursor

Declaring an explicit cursor names the cursor, defines the query associated with the cursor, and establishes the server memory area that is used by the cursor. The general syntax for declaring an explicit cursor is:

```
CURSOR cursor_name IS select_statement;
```

The *cursor_name* can be any valid PL/SQL variable name. The *select_statement* is the query that will retrieve the desired data values. You can use any valid SQL SELECT statement, including queries that involve set operators, and queries with subqueries or nested subqueries. The query search condition can contain PL/SQL variables, as long as the variables are declared before the cursor is declared, and assigned values before the cursor is processed.

Opening an Explicit Cursor

Opening the cursor causes the SQL compiler to parse the SQL query, or examine its individual components and check them for syntax errors. The parsed query is then stored in the context area in an internal machine-language format, and the active set values are determined. The general syntax of the command to open an explicit cursor is OPEN cursor_name;. The OPEN command causes the cursor to identify the data rows that satisfy the SELECT query. However, the data values are not actually retrieved yet.

Fetching the Cursor Results

The FETCH command retrieves the query data from the database into the active set, one row at a time, and associates each field value with a program variable. Because a query might return several rows, the FETCH command is executed within a loop. The general syntax of the FETCH command is FETCH cursor_name INTO variable_name(s);. The *variable_names(s)* parameter is either a single variable or a list of variables that will receive data from the field or fields currently being processed. Usually, you declare this variable using either the %TYPE or %ROWTYPE reference data types.

Closing the Cursor

A cursor should be closed after processing completes so that its memory area and resources can be made available to the system for other tasks. The general syntax for the cursor CLOSE command is: CLOSE cursor_name;. If you forget to close a cursor, it will automatically be closed when the program in which the cursor is declared ends.

Processing Explicit Cursors

You process explicit cursors using a loop that terminates after all rows are processed. Two different looping structures may be used: the LOOP…EXIT WHEN loop or the cursor

FOR loop. When you process an explicit cursor, a pointer called the **active set pointer** indicates which record will be fetched next. When the FETCH command executes past the last record of the query, the active set pointer points to an empty record. When you use the LOOP...EXIT WHEN structure to process a cursor, you use the exit condition **WHEN *cursor_name*%NOTFOUND** to determine if the last cursor record has been fetched and if the active set pointer is now pointing to an empty record.

Now you will write a program to create an explicit cursor and display its output using a LOOP...EXIT WHEN processing structure. This cursor will retrieve and display the ROOM value for every record in the LOCATION table where BLDG_CODE is 'LIB'. The program will store retrieved values in a variable named current_room that is declared using the %TYPE data type. The search condition string 'LIB' will be assigned and referenced as a variable value.

To process an explicit cursor using a LOOP...EXIT WHEN structure:

1. Type the code shown in Figure 4-37.

2. Run the program, and debug it if necessary. The formatted output should appear as shown in Figure 4-37.

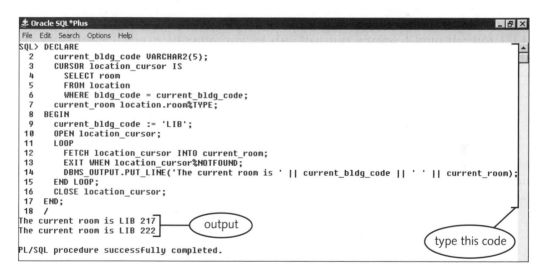

Figure 4-37   Processing an explicit cursor using a LOOP...EXIT WHEN structure

This cursor retrieved only one field, ROOM. When a cursor retrieves multiple data fields, it is convenient to fetch the output into a variable that is declared using the %ROWTYPE reference data type. When processing a cursor, a %ROWTYPE variable assumes the same data type as a row retrieved by the cursor. To declare a cursor %ROWTYPE variable, you use the command ***row_variable_name cursor_name*%ROWTYPE;**.

Now you will modify the program you just wrote so the query retrieves both the ROOM and CAPACITY fields from the LOCATION table and fetches the value into a %ROWTYPE variable named location_row. In the program output, the individual fields within the row are referenced using the syntax **_row_variable_name.table_fieldname_**. For example, to reference the ROOM field in the row variable, you would use the variable name location_row.room.

To modify the explicit cursor so it retrieves multiple field values:

1. In Notepad, make a copy of the explicit cursor program code, and then modify it as shown in Figure 4-38. When you execute the modified program, the output should appear as shown.

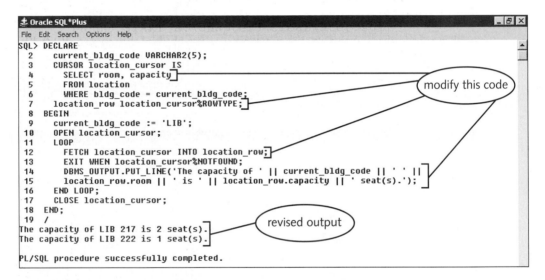

Figure 4-38   Using the % ROWTYPE data type in cursor processing

An easier way to process explicit cursors is to use the cursor FOR loop. With this loop, you do not need to explicitly open, fetch the rows, or close the cursor. By using the cursor FOR loop, Oracle implicitly performs these operations. However, you cannot reference the cursor values outside of the cursor FOR loop. The general syntax of a cursor FOR loop is:

```
FOR cursor_variable(s) IN cursor_name LOOP
    additional processing statements
END LOOP;
```

Now you will write a program to use a cursor FOR loop to process the location_cursor. You do not need to explicitly open, fetch into, or close the cursor, so the processing requires less code.

To process a cursor using a cursor FOR loop:

1. In Notepad, copy your explicit cursor program, and modify it as shown in Figure 4-39.

2. Run the program, and debug it if necessary. The LOCATION records appear as before.

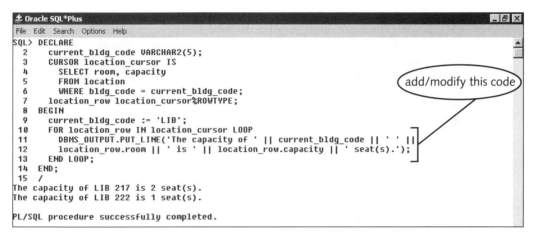

Figure 4-39   Processing an explicit cursor using a cursor FOR loop

Other Explicit Cursor Attributes

Explicit cursors have **attributes** that describe the cursor's current state, such as whether the cursor is currently open, whether records are found after a fetch operation, and how many records are fetched. You append these attributes to the cursor name. Explicit cursor attributes are described in Table 4-11.

Attribute	Description
%NOTFOUND	Evaluates as TRUE when a cursor has no rows left to fetch and FALSE when a cursor has remaining rows to fetch
%FOUND	Evaluates as TRUE when a cursor has rows remaining to fetch and FALSE when a cursor has no rows left to fetch
%ROWCOUNT	Returns the number of rows that a cursor has fetched so far
%ISOPEN	Returns TRUE if the cursor is open and FALSE if the cursor is closed

Table 4-11   Explicit cursor attributes

You can use the %ISOPEN attribute before or after the cursor is opened. You can use the other attributes only while the cursor is open. You used the %NOTFOUND attribute in the LOOP...EXIT WHEN processing structure to exit the loop when all cursor rows were fetched and processed. The %NOTFOUND and %FOUND attributes are also useful for detecting when a cursor query returns no rows. Figure 4-40 shows how to use the %NOTFOUND attribute within an IF/THEN structure to notify the user when a cursor returns no rows. Note that the test to see if the cursor was found is made after the cursor was opened, but before a FETCH operation is performed. This operation can only be performed using a LOOP...EXIT WHEN cursor processing structure.

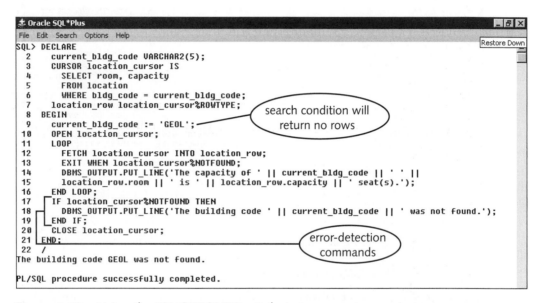

Figure 4-40   Using the %NOTFOUND attribute to generate a custom error message

Sometimes a cursor is used to determine whether or not a SELECT query will retrieve any records. Figure 4-41 shows how the %FOUND cursor attribute can be used in an IF/THEN structure to determine whether or not a cursor returns any records.

The %ROWCOUNT attribute returns the total number of records fetched by the cursor. Figure 4-42 shows how you use this attribute to display the number of records retrieved by the location_cursor.

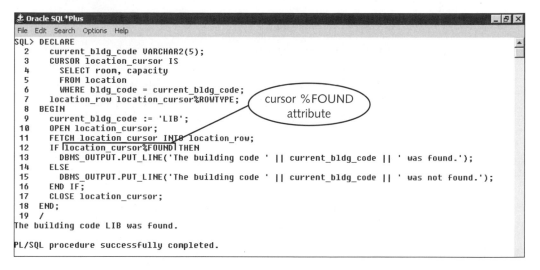

Figure 4-41   Using the %FOUND attribute to signal whether or not a cursor returns records

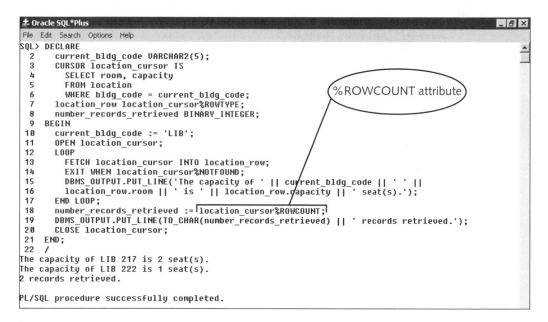

Figure 4-42   Using the %ROWCOUNT attribute to display the number of records fetched

SELECT FOR UPDATE Cursors

A normal SELECT command does not place any locks on records, and other users are free to view and update the retrieved data values. As a result, when a cursor is opened, the active set is defined based on the current values in the database records. If other users change these values through INSERT, UPDATE, and DELETE commands, the cursor values will not be updated unless the cursor is closed and then reopened.

Often when records are retrieved by a cursor, additional processing takes place that involves updating the records. In this case, the cursor needs to be created using the SELECT FOR UPDATE command, which places shared locks on the retrieved rows. These locks allow other users to view the data, but they do not allow other users to modify the data values until the transaction associated with the cursor has been committed or rolled back. The syntax for the SELECT FOR UPDATE command for a query used in a cursor is:

```
SELECT cursor_fieldnames
FROM tablenames
WHERE search_condition(s)
AND join_conditions
FOR UPDATE; | FOR UPDATE OF names_of_fields_to_be_updated;
additional processing statements
COMMIT; | ROLLBACK;
```

You must either COMMIT or ROLLBACK the transaction when cursor processing completes to release the shared locks and make the records associated with the cursor available for other users to modify. Listing the field names in the FOR UPDATE OF version of the last clause is optional; it is used to document the field names being updated. You can update any fields in the locked record, regardless of whether they are listed in the FOR UPDATE OF clause. If you do not want to list the field names to be updated, use the shortened (FOR UDPATE;) version of the command.

When you perform an update operation during cursor processing, you can use the **WHERE CURRENT OF *cursor_name*** clause to reference the table row currently being processed in the UPDATE command search condition. For example, to add one seat to the capacity of the location row that is currently being processed by the location_cursor, you would use the following command:

```
UPDATE location
SET capacity = capacity + 1
WHERE CURRENT OF location_cursor;
```

Now you will create a cursor that uses a SELECT FOR UPDATE query with the WHERE CURRENT OF clause. The cursor will retrieve each location in the LIB building and update its capacity by one seat.

To create a SELECT FOR UPDATE cursor:

1. Type the code shown in Figure 4-43, then run the program, and debug it if necessary. The CAPACITY values are updated in the database, but the updated values cannot be viewed with the current cursor, because its active set is associated with the data values as they existed when the cursor was first opened.

2. Type the SELECT command shown in Figure 4-43 to confirm that the LOCATION records were updated as shown.

4

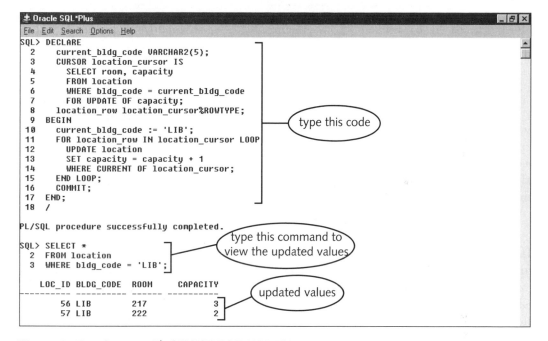

```
± Oracle SQL*Plus                                                    _ 8 X
File  Edit  Search  Options  Help
SQL> DECLARE
  2     current_bldg_code VARCHAR2(5);
  3     CURSOR location_cursor IS
  4       SELECT room, capacity
  5       FROM location
  6       WHERE bldg_code = current_bldg_code
  7       FOR UPDATE OF capacity;
  8     location_row location_cursor%ROWTYPE;
  9  BEGIN
 10     current_bldg_code := 'LIB';
 11     FOR location_row IN location_cursor LOOP
 12       UPDATE location
 13       SET capacity = capacity + 1
 14       WHERE CURRENT OF location_cursor;
 15     END LOOP;
 16     COMMIT;
 17  END;
 18  /

PL/SQL procedure successfully completed.

SQL> SELECT *
  2  FROM location
  3  WHERE bldg_code = 'LIB';

   LOC_ID BLDG_CODE ROOM    CAPACITY
---------- --------- ------  ----------
       56 LIB       217              3
       57 LIB       222              2
```

type this code

type this command to view the updated values

updated values

Figure 4-43   Cursor with SELECT FOR UPDATE query

PL/SQL Tables and Tables of Records

A **PL/SQL table** is a data structure used in a PL/SQL program that contains multiple data items that are all of the same data type. Each table item is comprised of two elements: the **key**, which is a BINARY_INTEGER value that uniquely identifies the item within the table, and the **value**, which is the actual data value. Table 4-12 shows an example of a PL/SQL table named MY_PETS that consists of names of pets:

Key	Value
1	Shadow
2	Dusty
3	Sassy
4	Bonnie
5	Clyde
6	Nadia

Table 4-12   MY_PETS PL/SQL table

The first key value does not have to be 1, and the keys and associated values do not have to be inserted sequentially.

You can reference an item in a PL/SQL table using the following syntax: ***tablename(item_key)***. For example, to assign the second value in the PL/SQL table shown in Table 4-12, to a variable named current_dog, you would use the assignment statement: **current_dog := MY_PETS(2);**.

A PL/SQL table is similar to an array, which is a composite data type used in many programming languages.

The elements in a PL/SQL table are not stored in any particular order, and the key values do not need to be sequential. The restrictions on a table key are that it must be an integer, and it must be unique within the table. PL/SQL table values are stored in main memory on the database server. A PL/SQL table can grow as large as needed, based on available server memory space. The memory space used by the table is allocated as new values are inserted into the table. Because a PL/SQL table is stored in main memory, not on disk, PL/SQL table values are available only to the PL/SQL program in which the table is created, and are no longer available when the program terminates.

In a PL/SQL program, a PL/SQL table provides a way to store and manipulate data that needs to be structured in a list or tabular format but is not stored in the database and, as a result, cannot be processed using cursors. For example, you might want to create a PL/SQL program that calculates the grade point averages of all students and then stores the values in a PL/SQL table, with the key value being the student ID and the data value being the calculated grade point average. You could not store or process this tabular data that associates a specific student ID value with a specific grade point average using any other PL/SQL data type.

To create a PL/SQL table, you must first define a user-defined data subtype. A **user-defined data subtype** is a special data type that a user can define based on existing PL/SQL data types. To declare a user-defined subtype for a table, you use the following type declaration command in the DECLARE block:

```
TYPE PL/SQL_table_data_type_name IS TABLE OF item_data_type
INDEX BY BINARY_INTEGER;
```

The *PL/SQL_table_data_type_name* parameter is any legal variable name. The *item_data_type* parameter specifies the data type of the values that will be stored in the table.

After you create the user-defined subtype, you must declare a table variable that uses this user-defined subtype using the following syntax:

```
table_variable_name PL/SQL_table_data_type_name;
```

The table variable name parameter can be any legal variable name. The *PL/SQL_table_data_type_name* parameter is the user-defined subtype that was defined for the TABLE data type.

For example, the following code creates a user-defined data subtype named PET_NAMES_TABLE and then creates a table named MY_PETS:

```
DECLARE
    TYPE pet_names_table IS TABLE OF VARCHAR2(30);
    my_pets PET_NAMES_TABLE;
```

To insert a value into a PL/SQL table, you use the following assignment statement:

```
table_variable_name(item_index_value) := item_value;
```

For example, to insert the first value in the table in Table 4-12, you would use the assignment statement **my_pets(1) := 'Shadow';**.

Now you will write a PL/SQL program that creates the MY_PETS PL/SQL table shown in Table 4-12 and inserts the data values into the table.

To create a PL/SQL table:

1. Type the code shown in Figure 4-44, then run the program, and debug it if necessary.

```
± Oracle SQL*Plus                                                        _ 8 X
File  Edit  Search  Options  Help
SQL> DECLARE
  2      --declare the user-defined subtype for the table
  3      TYPE pet_names_table IS TABLE OF VARCHAR2(30)
  4        INDEX BY BINARY_INTEGER;
  5      --declare the table based on the user-defined subtype
  6      my_pets PET_NAMES_TABLE;
  7   BEGIN
  8      --populate the table
  9      my_pets(1) := 'Shadow';
 10      my_pets(2) := 'Dusty';
 11      my_pets(3) := 'Sassy';
 12      my_pets(4) := 'Bonnie';
 13      my_pets(5) := 'Clyde';
 14      my_pets(6) := 'Nadia';
 15      DBMS_OUTPUT.PUT_LINE('The first value in the MY_PETS table is ' || my_pets(1));
 16   END;
 17   /
The first value in the MY_PETS table is Shadow          output

                                                                type these commands
PL/SQL procedure successfully completed.
```

Figure 4-44   Creating a PL/SQL table

Dynamically Populating a PL/SQL Table Using Database Values

You can populate PL/SQL tables using database data values retrieved with a cursor. To see how this works, you will create a table that contains the item descriptions from the ITEM table in the Clearwater Traders database. The table will use the ITEM_ID value as the table key. When you declare the table's user-defined data subtype, you will use the %TYPE reference data type for the ITEM_DESC field, which automatically assigns the correct data type to the table value.

To create a PL/SQL table and populate it with item descriptions from the Clearwater Traders database ITEM table:

1. Type the code shown in Figure 4-45, then run the program, and debug it if necessary.

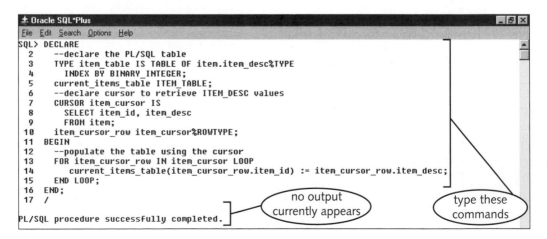

```
± Oracle SQL*Plus                                                        _ 8 X
File  Edit  Search  Options  Help
SQL> DECLARE
  2      --declare the PL/SQL table
  3      TYPE item_table IS TABLE OF item.item_desc%TYPE
  4        INDEX BY BINARY_INTEGER;
  5      current_items_table ITEM_TABLE;
  6      --declare cursor to retrieve ITEM_DESC values
  7      CURSOR item_cursor IS
  8        SELECT item_id, item_desc
  9        FROM item;
 10      item_cursor_row item_cursor%ROWTYPE;
 11   BEGIN
 12      --populate the table using the cursor
 13      FOR item_cursor_row IN item_cursor LOOP
 14        current_items_table(item_cursor_row.item_id) := item_cursor_row.item_desc;
 15      END LOOP;
 16   END;
 17   /
                                           no output           type these
                                        currently appears       commands
PL/SQL procedure successfully completed.
```

Figure 4-45   Creating a PL/SQL table using database values

Currently, the program does not display any output showing the table values. Before you can add output statements to this program, you need to learn about PL/SQL table attributes, which are discussed in the next section.

PL/SQL Table Attributes

A PL/SQL table has several attributes that you can use to retrieve information about tables and access individual table items. PL/SQL table attributes are described in Table 4-13.

4

Attribute	Description	Examples	Results
COUNT	Returns the number of rows in the table	my_pets.COUNT	6
DELETE (row_key) DELETE (first_ key_to_be_ deleted,last_ key_to_be_ deleted)	Deletes all table rows, a specified table row, or a range of rows	my_pets.DELETE my_pets.DELETE(1) my_pets.DELETE (1,3)	Deletes all table rows Deletes the row associated with index value 1 Deletes the rows associated with index values 1 through 3
EXISTS (row_key)	Used in conditional statements to return the Boolean value TRUE if the specified row exists and FALSE if the specified row does not exist	IF my_pets.EXISTS(1)	Returns TRUE if the item associated with key value 1 exists
FIRST	Returns the value of the key of the first item in the table	my_pets.FIRST	1
LAST	Returns the value of the key of the last item in the table	my_pets.LAST	6
NEXT (row_key)	Returns the value of the key of the next row after the specified row	my_pets.NEXT (3)	If current row key = 3, returns 4
PRIOR (row_key)	Returns the value of the key of the row immediately before the specified row	my_pets.PRIOR (3)	If current row key = 3, returns 2

Table 4-13   PL/SQL table attributes

Now you will modify the Clearwater Traders ITEM table program that created the table of item description values. You will use the COUNT attribute to display the total number of records in the table, and the FIRST and NEXT attributes to display the table values. You will create a variable named current_table_key that references the table key. You will then use this variable within a loop to step through and display all of the table values.

To use PL/SQL table attributes to display table values:

1. In Notepad, copy and paste the code for the program that creates a PL/SQL table using database values, and then modify the code as shown in Figure 4-46. Run the program, and debug it if necessary. The output should show the number of rows in the table and the value for each row.

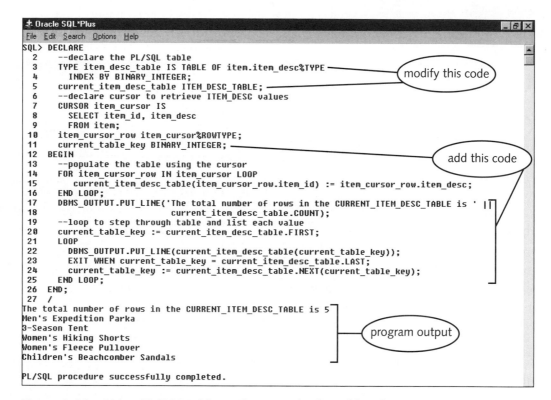

Figure 4-46   Using PL/SQL table attributes to display table values

Creating a PL/SQL Table of Records

So far, the PL/SQL tables that you have created consist of a key value associated with a single data value. You can also create a **table of records**, which is a PL/SQL table that stores multiple data values that are referenced by a unique key. PL/SQL tables of records are useful for storing lookup information from small database tables that are used many times within a program. By storing the information in a PL/SQL table, the program does not have to repeatedly query the database, which improves processing performance.

The user-defined data subtype for a table of records is declared using the %ROWTYPE reference data type, as shown in the following command:

```
TYPE PL/SQL_table_data_type_name
IS TABLE OF database_table_name%ROWTYPE
INDEX BY BINARY_INTEGER;
```

For example, to declare a user-defined subtype named ITEM_TABLE and an associated PL/SQL table of records named CURRENT_ITEM_TABLE based on the ITEM table in the Clearwater Traders database, you would use the following commands:

```
TYPE item_table IS TABLE OF item%ROWTYPE
   INDEX BY BINARY_INTEGER;
current_item_table ITEM_TABLE;
```

To assign a value to a specific field in a table of records, you use the following syntax:

```
PL/SQL_tablename(key_value).database_fieldname := field_value;
```

For example, to assign the value 'Women's Hiking Shorts' to the ITEM_DESC field in the first record in a table of records named CURRENT_ITEM_TABLE that is based on the ITEM database table, you would use the following command:

```
CURRENT_ITEM_TABLE(1).item_desc := 'Women''s Hiking Shorts';
```

Now you will create a table of records based on the ITEM database table. You will use an explicit cursor to retrieve all of the records from the ITEM table, and then write commands to insert the retrieved values into the table of records. Then, you will display the contents of the table of records using a loop that displays each item ID value, five blank spaces, and the associated item description value.

To create the table of records:

1. In Notepad, type the code shown in Figure 4-47. Run the program, and debug it if necessary. The output should show the ITEM_ID and ITEM_DESC values for each table row.

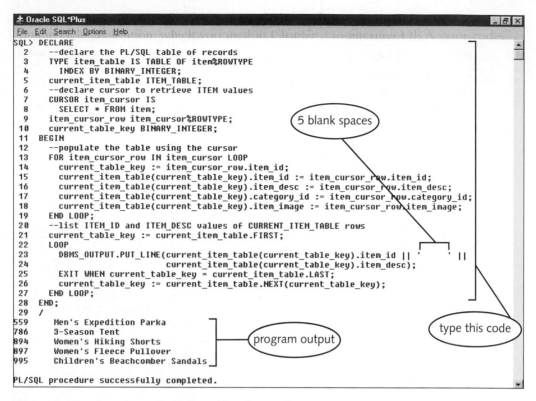

Figure 4-47   Creating a PL/SQL table of records

HANDLING ERRORS IN PL/SQL PROGRAMS

Programmers prefer to concentrate on the positive aspects of their programs. Creating programs that generate invoices, process checks, track time usage, and so forth is difficult and time-consuming. As a result, programmers tend to breathe a sigh of relief once a program works and overlook or ignore the consequences of a user pressing the wrong key or a critical network link going down. The reality, however, is that users enter incorrect data and press wrong keys, networks fail, computers crash, and anything else that can go wrong (eventually) will go wrong. Programmers can't do much for the user if the user's computer fails, but they can and should do everything possible to prevent incorrect data from being entered and wrong keystrokes from damaging the system. They should also do everything possible to inform the user of errors and how to correct them when they occur. PL/SQL supports **exception handling**, whereby all code for displaying error messages and giving users options for fixing errors is placed in the EXCEPTION section of the code. The EXCEPTION section is PL/SQL's primary method for processing errors and informing users of corrective actions to take.

Program errors can be classified as compile errors or runtime errors. Compile errors usually involve spelling or syntax problems. Compile errors are reported by the PL/SQL interpreter and have to be corrected by the programmer before the program will execute. Figure 4-48 shows an example of a compile error, where the comparison operator (=) was used instead of the assignment operator (:=). The error code ORA-06550 indicates that the error is a compile error and shows the error's line and column location. The error code PLS-00103 provides specific details about the type of error that occurred.

 Error codes with the ORA- prefix are generated by the Oracle DBMS and usually involve constraint errors, such as trying to insert a record with a NULL value in a field with a NOT NULL constraint. Error codes with the PLS- prefix are generated by the PL/SQL compiler and involve PL/SQL syntax errors.

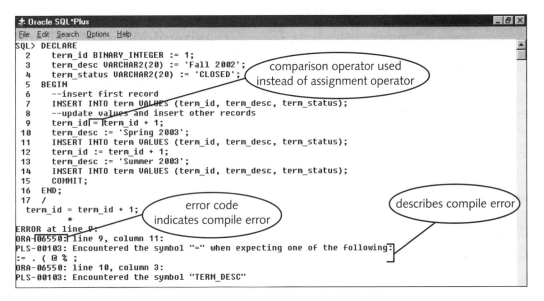

Figure 4-48   Example of a compile error

Runtime errors are reported by the PL/SQL runtime engine, and they must be handled in the exception-handling section of the program. Runtime errors usually involve problems with data values, such as trying to retrieve no rows or several rows using an implicit cursor, trying to divide by zero, or trying to insert a value into a database table that violates the table's constraints. Figure 4-49 shows a runtime error generated when an attempt to insert a record into the TERM table was unsuccessful, because the user tried to insert a record with a primary key value (TERM_ID 1) that already existed in the table. This violated the TERM table constraint that ensures that the primary key for each record must be unique. The error code ORA-00001 gives details about the nature of the error. The error code ORA-6512 indicates that it is a runtime error rather than a compile error and shows the line number that generated the error.

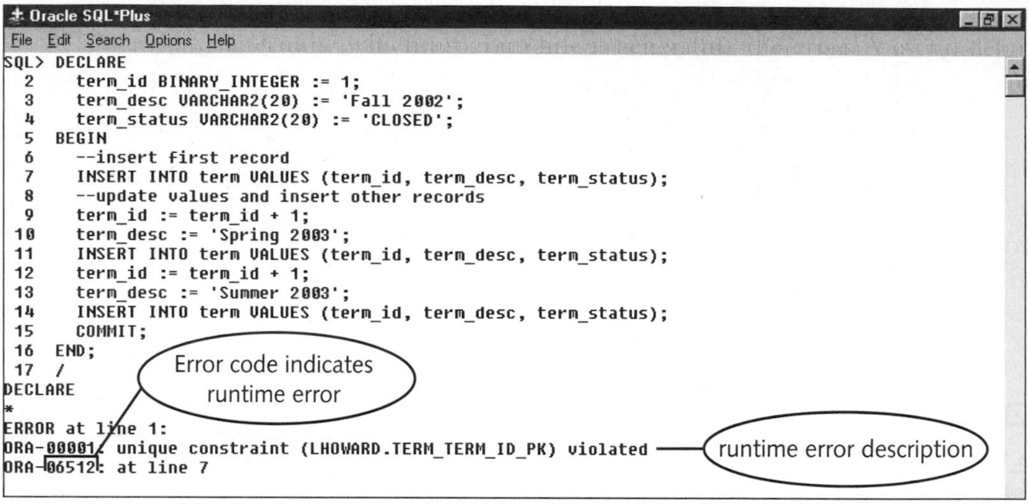

```
± Oracle SQL*Plus                                                        _ □ ×
File  Edit  Search  Options  Help
SQL> DECLARE
  2      term_id BINARY_INTEGER := 1;
  3      term_desc VARCHAR2(20) := 'Fall 2002';
  4      term_status VARCHAR2(20) := 'CLOSED';
  5  BEGIN
  6      --insert first record
  7      INSERT INTO term VALUES (term_id, term_desc, term_status);
  8      --update values and insert other records
  9      term_id := term_id + 1;
 10      term_desc := 'Spring 2003';
 11      INSERT INTO term VALUES (term_id, term_desc, term_status);
 12      term_id := term_id + 1;
 13      term_desc := 'Summer 2003';
 14      INSERT INTO term VALUES (term_id, term_desc, term_status);
 15      COMMIT;
 16  END;
 17  /                          Error code indicates
DECLARE                           runtime error
*
ERROR at line 1:
ORA-00001 unique constraint (LHOWARD.TERM_TERM_ID_PK) violated ——  runtime error description
ORA-06512: at line 7
```

Figure 4-49   Example of a runtime error

When a runtime error occurs, an exception, or unwanted event, is **raised,** or occurs. Program control immediately transfers to the EXCEPTION section of the PL/SQL program, where special sections of code called exception handlers must exist to deal with all error situations. An **exception handler** is a specific series of commands that provide operation instructions when a specific exception is raised. An exception handler might correct the error without notifying the user of the problem, or it might inform the user of the error without taking corrective action. An exception handler could also correct the error and inform the user of the error, or it could inform the user of the error and allow the user to decide what action to take.

There are three kinds of exceptions: predefined, undefined, and user-defined. The following sections describe these exceptions and how they are handled.

Predefined Exceptions

Predefined exceptions correspond to the most common errors that are seen in many programs, have been given a specific exception name, and contain a built-in exception handler. Table 4-14 presents a list of the most common predefined exceptions.

Oracle Error Code	Exception Name	Description
ORA-00001	DUP_VAL_ON_INDEX	Unique constraint on primary key violated
ORA-01001	INVALID_CURSOR	Illegal cursor operation
ORA-01403	NO_DATA_FOUND	Query returns no records
ORA-01422	TOO_MANY_ROWS	Query returns more rows than anticipated
ORA-01476	ZERO_DIVIDE	Division by zero
ORA-01722	INVALID_NUMBER	Invalid number conversion (like trying to convert '2B' to a number)
ORA-06502	VALUE_ERROR	Error in truncation, arithmetic, or conversion operation

Table 4-14   Common Oracle predefined exceptions

With predefined exceptions, Oracle automatically displays an error message informing the user of the nature of the problem. For example, in Figure 4-49, the ORA-00001 (DUP_VAL_ON_INDEX) exception was raised, and Oracle displayed the message "unique constraint (LHOWARD.TERM_TERM_ID_PK) violated." The expression in parentheses is the name of the constraint that was violated, which consists of the user schema name (LHOWARD), and the constraint name (_TERM_TERM_ID_PK).

You can create exception handlers to display custom error messages for predefined exceptions that are easier to understand and provide instructions to the user for correcting the error. The general syntax of exception handlers that trap predefined errors and replace the system error messages with custom error messages is:

```
EXCEPTION
  WHEN exception1_name THEN
   exception1 handling statements;
  WHEN exception2_name THEN
   exception2 handling statements;
  ...
  WHEN OTHERS THEN
   other handling statements;
END;
```

The *exception_name* parameter refers to the predefined exception name, as shown in Table 4-14. The *exception handling statements* are the code lines that inform the user of the error, make suggestions for corrective actions, and so forth. The WHEN OTHERS THEN statement is a catch-all exception handler that allows you to present a general message to describe errors not handled by a specific error-handling statement. After an exception handler finishes processing, program execution terminates. The exception-handling statements are indented to make the exception section easier to read and understand.

Now you will write a program that displays custom messages for predefined exceptions. In this program, an implicit cursor tries to select an item description from the ITEM_DESC field in the ITEM table using an invalid search condition. When an implicit cursor query returns no records, the predefined exception NO_DATA_FOUND is raised. First, you will write the program without a custom exception handler to view the error message that is generated by the PL/SQL runtime engine. Then, you will add a custom exception handler that will inform the user of the nature of the error and suggest a corrective action.

To write the program to display a custom message for a predefined exception:

1. Type the code shown in Figure 4-50. Run the program. The Oracle error code ORA-06512 should appear, indicating that there is a runtime error. The specific Oracle error code ORA-01403 and associated predefined error message "no data found" also appear. (If another error appears, debug the program until your output matches the output shown in Figure 4-50.)

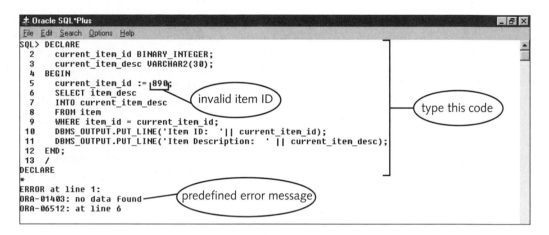

Figure 4-50   PL/SQL program with predefined exception

2. Copy and modify your program code by adding the exception-handling section shown in Figure 4–51. Run the program, and debug it if necessary. The more informative and instructive error messages specified by the exception handler now appear.

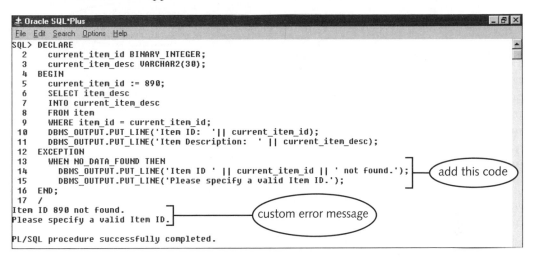

Figure 4-51   PL/SQL program that handles a predefined exception

During program development, it is helpful to use the WHEN OTHERS exception handler to display the associated Oracle error code number and error message for unanticipated errors. To do this, you use the **SQLERRM** function. This function returns a character string that contains the Oracle error code and the text of the error code's error message for the most recent Oracle error generated. To use this function, you must first declare a VARCHAR2 character variable to which the text of the error message will be assigned. The maximum length of the character variable will be 512, because the maximum length of an Oracle error message is 512 characters.

You will now modify your program so that the implicit cursor query returns all of the rows from the ITEM table. Because an implicit cursor query cannot handle more than one row, the query will generate an unhandled exception. You will add the WHEN OTHERS error handler so that it displays the associated Oracle error code and message when an unhandled error is generated.

To add a WHEN OTHERS error handler to the program to display other error codes and messages:

1. Modify the cursor SELECT command and add the WHEN OTHERS error handler, as shown in Figure 4-52, and then run the program. The error code and message appear as shown in the figure.

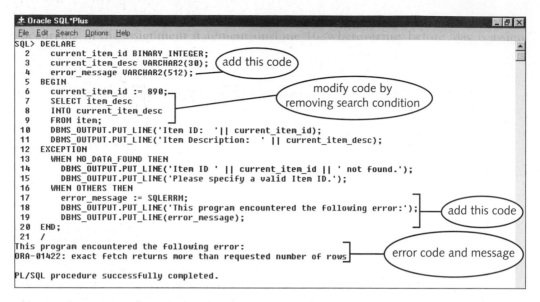

Figure 4-52 Using the WHEN OTHERS exception handler

Undefined Exceptions

Recall that defined exceptions are the common errors that have been given explicit names, which are listed in Table 4-14. **Undefined exceptions** are the less common errors that have not been given an explicit exception name. Figure 4-53 shows an example of an undefined exception. Here, a program tries to insert a record into the FACULTY table where the data value for the LOC_ID field is 60. Recall that LOC_ID is a foreign key field in the FACULTY table, and values that are inserted as foreign keys must already exist in the parent table. If you examine the current values in the LOCATION table in the Northwoods University database (see Figure 1-12), you will see that there is no value for LOC_ID 60. As a result, the foreign key constraint is violated, as indicated by the Oracle error code and associated error message.

To handle an undefined exception, you must explicitly declare the exception in the DECLARE section of the program and associate it with a specific Oracle error code. Then, you can create an error handler as you did for predefined exceptions. The general syntax for declaring an exception is:

```
DECLARE
 e_exception_name EXCEPTION;
PRAGMA EXCEPTION_INIT(e_exception_name, Oracle_error_code);
```

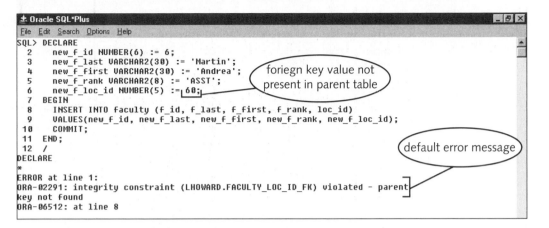

Figure 4-53   PL/SQL program with undefined exception

The *e_exception_name* parameter can be any legal variable name. Usually, user–declared exception names are prefixed with e_ to keep them from being confused with other variables. The PRAGMA EXCEPTION_INIT command tells the interpreter to associate the given exception name with a specific Oracle error code. The *Oracle_error_code* parameter is the numeric error code that Oracle assigns to runtime errors. Recall that you can find a complete listing of Oracle error codes and messages in the Oracle8 Error Message online Help application, which you can start by running a file named ora.hlp, or from the Oracle Technology Network Web site. Usually, you are familiar with the error codes that your programs generate, because you see them so often during development! The error code number must be preceded by a hyphen (-), and you can omit leading zeroes.

Now, you will write a program that declares an exception named e_foreign_key_error that is associated with Oracle error code ORA-02291. When the user tries to insert a foreign key value where the value does not exist in its parent table, an exception handler will display a message.

To write a program to handle an undefined exception:

1. Type the code shown in Figure 4-54. Run the program and debug it if necessary. The exception handler error message appears instead of the Oracle error codes and messages.

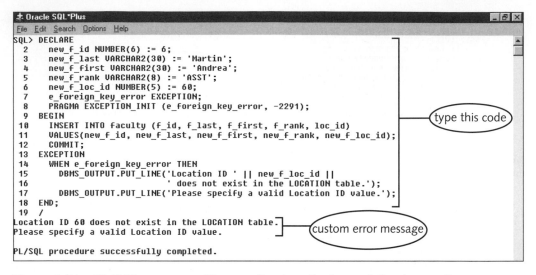

Figure 4-54   PL/SQL program with exception handler for undefined exception

User-Defined Exceptions

User-defined exceptions are used to handle exceptions that will not cause an Oracle runtime error but that require exception handling to enforce business rules or to ensure the integrity of the database. Suppose that a database program is used to delete records from the ENROLLMENT table. Northwoods University has a business rule that states that the only records that can be deleted from the ENROLLMENT table are records in which the GRADE value is NULL. Records for which grades have been assigned cannot be deleted. However, suppose that a user tries to delete an ENROLLMENT record where GRADE is not NULL. The program should raise an exception to advise the user that the grade field is not NULL, and then not delete the record.

The general syntax for declaring, raising, and handling a user-defined exception is:

```
DECLARE
  e_exception_name EXCEPTION;
  other variable declarations;
BEGIN
  other program statements
IF undesirable condition THEN
   RAISE e_exception_name;
END IF;
  other program statements;
EXCEPTION
  WHEN e_exception_name THEN
    error-handling statements;
END;
```

Now you will write a program that includes a user-defined exception handler to avoid deleting an ENROLLMENT record with an assigned grade.

To write a program with a user-defined exception handler:

1. Type the code shown in Figure 4-55. Run the program, and debug it if necessary. The user-defined exception message appears. This exception-handling message could also be placed directly in the ELSE portion of the IF/THEN statement. However, it is a good programming practice to place all error-handling statements in the EXCEPTION section, because this keeps all of the error-handling program statements in the same place and makes the program easier to understand and maintain.

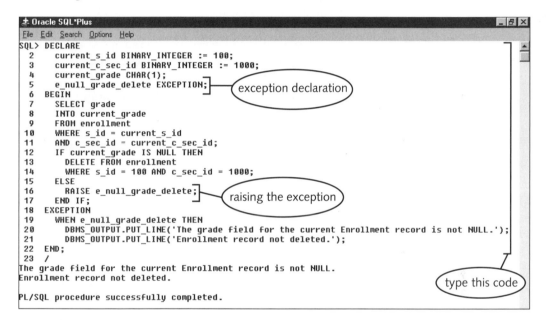

Figure 4-55   Creating and raising a user-defined exception

NESTED PL/SQL PROGRAM BLOCKS

Recall that with exception handling, as soon as an exception is raised, program control jumps to the EXCEPTION section, the exception is handled, and then the program terminates. Sometimes, you need a program that has an exception handler that notifies the user when an exception is raised but continues execution. To do this, you create nested PL/SQL program blocks. With **nested PL/SQL program blocks**, one program block, called an **outer block**, contains another program block, called an **inner block**. An inner block must always be within the body (after the BEGIN command) of the outer block. Figure 4-56 illustrates nested program blocks. A PL/SQL program can contain multiple nested program blocks.

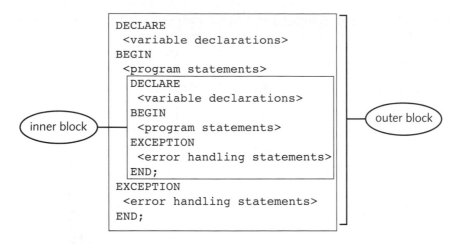

Figure 4-56   Nested PL/SQL program block

Recall that when a variable is declared in the DECLARE section, a memory location associated with the variable name is established in main memory. When the program block's END statement is reached, the variable's memory location is returned to the system and made available to other program blocks. With this in mind, you must consider variable scope and persistence within a nested PL/SQL program block. Variable **scope** refers to the visibility of a variable to different program blocks. Variable **persistence** refers to when memory is set aside for a variable and when this memory is returned to the system. When a variable is declared in the outer block of a nested program block, memory is allocated to store the value associated with this variable. This outer block variable persists through all of the inner (nested) blocks. Its memory is only returned to the system when the END statement for the outer block is reached. In most cases, a variable declared in the outer block is visible to all nested blocks, no matter how deep the nesting goes. This means an outer block variable's scope extends through its nested blocks.

Now you will create a PL/SQL program with nested blocks. In the outer block, a variable named current_bldg_code will be declared, and initialized with the value 'BUS'. Then, this variable will be used as a search condition for an explicit cursor in the inner block.

To create a PL/SQL program with nested blocks:

1. Type the code shown in Figure 4-57. Run the program, and debug it if necessary. The output should appear as shown, indicating that the value of the variable declared in the outer block is visible to the program code in the inner block.

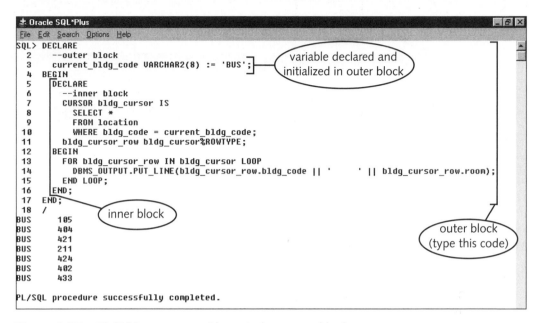

Figure 4-57 PL/SQL program with nested program blocks

In rare cases, a variable declared in an outer block will not be visible in a nested block. This happens when a variable in an inner block is given the same name as a variable in an outer block. In this situation, the inner block's variable is allocated its own memory location. In program statements within the inner block, referencing this variable will return the inner block's variable value rather than the outer block's variable value. The outer block variable "persists" through inner block program execution, but its visibility, or scope, does not extend to the inner block in this situation.

Now you will modify your program so that a second variable named current_bldg_code is declared in the inner block and initialized to the value 'CR', which is a different value than was used for the variable in the outer block.

To see the effect of using variables with the same names in nested blocks:

1. In Notepad, copy your code for the program with nested blocks, and modify it as shown in Figure 4-58. Run the program, and debug it if necessary.

From the output shown in Figure 4-58, you can see that the program statements in the inner block use the value for current_bldg_code that was assigned in the inner block. To avoid potential confusion and scope errors, you should avoid naming inner block variables with the same name used for variables in an outer block.

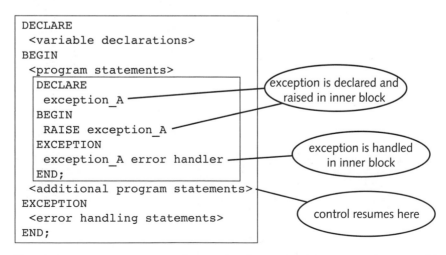

```
add this code

± Oracle SQL*Plus                                                              _ 8 X
File Edit Search Options Help
SQL> DECLARE
  2    --outer block
  3    current_bldg_code VARCHAR2(8) := 'BUS';       same variable name is
  4  BEGIN                                           declared in both blocks
  5    DECLARE
  6      --inner block
  7      current_bldg_code VARCHAR2(8) := 'CR';
  8      CURSOR bldg_cursor IS
  9        SELECT *
 10        FROM location
 11        WHERE bldg_code = current_bldg_code;
 12      bldg_cursor_row bldg_cursor%ROWTYPE;
 13    BEGIN
 14      FOR bldg_cursor_row IN bldg_cursor LOOP
 15        DBMS_OUTPUT.PUT_LINE(bldg_cursor_row.bldg_code || '        ' || bldg_cursor_row.room);
 16      END LOOP;
 17    END;
 18  END;
 19  /
CR      101           output reflects value
CR      202           assigned in inner block
CR      103
CR      105

PL/SQL procedure successfully completed.
```

Figure 4-58   Using variables with the same names in nested blocks

EXCEPTION HANDLING IN NESTED PROGRAM BLOCKS

Recall that one of the main reasons for creating nested blocks is to facilitate exception handling. When an exception is raised in a nonnested PL/SQL program block, program control immediately shifts to the EXCEPTION section, and then the program terminates. Figure 4-59 illustrates that with nested program blocks, if an exception is raised in an inner block, program control resumes in the outer block.

```
DECLARE
 <variable declarations>
BEGIN
 <program statements>
  DECLARE
   exception_A                    exception is declared and
  BEGIN                           raised in inner block
   RAISE exception_A
  EXCEPTION
   exception_A error handler      exception is handled
  END;                            in inner block
 <additional program statements>
EXCEPTION
 <error handling statements>      control resumes here
END;
```

Figure 4-59   Program execution path when exception is raised in inner block

Sometimes you need to write a program in which an exception will be raised in some situations, but program execution should still continue. For example, suppose that the capacity of some of the classrooms at Northwoods University changes, and you want to create a program to determine which course sections need different classrooms to accommodate the course section's maximum enrollment. An exception will be raised when a classroom cannot accommodate the maximum enrollment in a course section, and an error message will appear. Then, program execution will continue and examine the maximum enrollment and capacity of the next course section. Before creating and running this program, you will update the CAPACITY field for LOC_ID 49 to ensure that an exception will be raised.

To update the CAPACITY field:

1. In SQL*Plus, type the following commands at the SQL prompt:

```
UPDATE location
SET capacity = 30
WHERE loc_id = 49;
COMMIT;
```

Now you will write the program to use exception handling within nested blocks so program execution resumes in the outer block. In the outer block, you will declare a cursor to retrieve the course section ID, location ID, maximum enrollment, and capacity of the location for each course section. You will then open the cursor in the outer block using a cursor FOR loop. In the DECLARE section of the inner block, you will declare a user-defined exception named e_capacity_error. In the body of the inner block, you will compare the course section maximum enrollment to the location capacity. If the maximum enrollment is greater than the location capacity, an exception will be raised, and a user-defined exception handler in the EXCEPTION section of the inner block will display a message showing the course section ID of the course section with the capacity error. After the exception is raised, execution will resume in the outer block by examining the next cursor row. If the maximum enrollment is equal to or less than the location capacity, no exception will be raised in the inner block, and execution will resume in the outer block by examining the next cursor row.

To write a program that uses exception handling within a nested block:

1. Type the code shown in Figure 4-60, run the program, and debug it if necessary. The output appears as shown, confirming that the exception is raised for some cursor rows, while execution continues for rows without a capacity error.

Suppose that an exception is raised in an inner block, but no exception handler exists for this exception in the exception handler for the inner block. However, an exception handler for this particular exception exists in the EXCEPTION section of the outer block. The outer block's exception handler will handle the exception, but the program will immediately terminate, as illustrated in Figure 4-61.

```
± Oracle SQL*Plus                                                                    _ |8| X
File Edit Search Options Help
SQL> DECLARE
  2    CURSOR c_sec_cursor IS
  3      SELECT c_sec_id, location.loc_id, max_enrl, capacity
  4      FROM course_section, location
  5      WHERE location.loc_id = course_section.loc_id;
  6    c_sec_row c_sec_cursor%ROWTYPE;
  7  BEGIN
  8    FOR c_sec_row IN c_sec_cursor LOOP
  9      --inner block
 10      DECLARE
 11        e_capacity_error EXCEPTION;
 12      BEGIN
 13        IF c_sec_row.max_enrl > c_sec_row.capacity THEN
 14          RAISE e_capacity_error;
 15        END IF;
 16      --exception handler in inner block
 17      EXCEPTION
 18        WHEN e_capacity_error THEN
 19          DBMS_OUTPUT.PUT_LINE('Capacity error in C_SEC_ID ' || c_sec_row.c_sec_id);
 20      END;
 21    END LOOP;
 22  END;
 23  /
Capacity error in C_SEC_ID 1008
Capacity error in C_SEC_ID 1009
Capacity error in C_SEC_ID 1012

PL/SQL procedure successfully completed.
```

error messages displayed by
exceptions raised in inner block

type these commands

Figure 4-60   Program with exceptions in which execution resumes in outer block

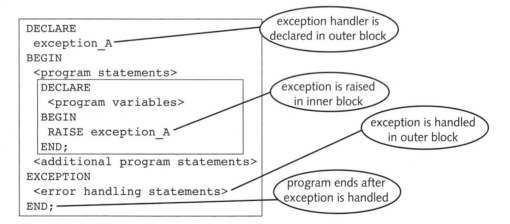

exception handler is
declared in outer block

exception is raised
in inner block

exception is handled
in outer block

program ends after
exception is handled

Figure 4-61   Execution path in which exceptions raised in inner block are handled in
outer block

To confirm this pattern, you will modify your program that has nested blocks. You will move
the user-defined exception declaration and exception handler to the outer block. You will also
define a variable named current_c_sec_id in the outer block that will be assigned the value of
the current course section ID in the cursor FOR loop. You need to define this new variable
because the exception handler will be outside of the cursor FOR loop, so you cannot display
the name of the course section IDs that have capacity errors using the cursor row variables.

To write a program that handles an exception raised in an inner block with an exception handler that is in the outer block:

1. In Notepad, copy the code for your program that uses nested blocks, and then modify the program as shown in Figure 4-62. Note that the exception handler is now in the outer block.

2. Run the program. Note that the output shows only the first course section ID with a capacity error (C_SEC_ID 1008), as compared with the output shown in Figure 4-60, where course sections 1008, 1009, and 1012 are flagged as having capacity errors. This occurs because the program terminated immediately after the first exception was raised.

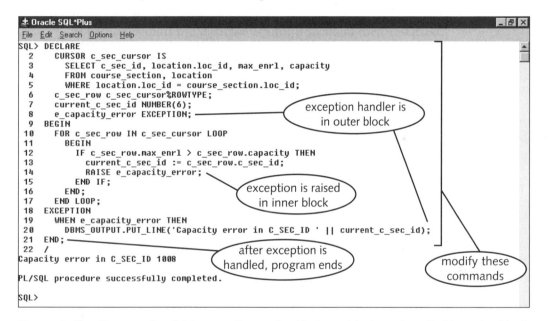

Figure 4-62   Program in which exceptions raised in inner block are handled in outer block

3. Exit SQL*Plus.

4. Save your program files in Notepad, and then close Notepad.

SUMMARY

- ❐ SQL commands can be classified as data definition language (DDL) commands, which change the database structure; data manipulation language (DML) commands, which query or manipulate the data in database tables; or transaction control commands, which organize commands into logical transactions and commit them to the database or roll them back.

- ❐ DML and transaction control commands can be used in PL/SQL programs, but DDL commands cannot.

- ❐ A loop repeats an action multiple times until an ending condition is reached.

- ❐ PL/SQL has five different loops: LOOP…EXIT, LOOP…EXIT WHEN, WHILE…LOOP, numeric FOR loops, and cursor FOR loops.

- ❐ The LOOP…EXIT loop can be either a pretest or posttest loop.

- ❐ You do not need to declare a counter for a numeric FOR loop. The counter is defined in a series of numbers in the FOR statement and automatically increments each time the loop repeats.

- ❐ A cursor is a pointer to a memory area, called the context area, that Oracle uses to process a query. The context area contains the number of rows processed, the parsed query, and the active set of records that have been fetched.

- ❐ Implicit cursors do not need to be formally declared. They are used with SELECT statements that return one and only one record.

- ❐ Explicit cursors must be formally declared. They are used with SELECT statements that might retrieve a variable number of records or that might return no records.

- ❐ To use an explicit cursor, you must declare the cursor, open it, fetch the records, and then close the cursor.

- ❐ Usually, an explicit cursor fetches records into a variable that is declared using either the %TYPE or %ROWTYPE reference variable data type.

- ❐ A cursor should be closed after processing completes to release its memory area and resources.

- ❐ When you process an explicit cursor using a cursor FOR loop, you do not need to explicitly open the cursor, fetch the rows, or close the cursor.

- ❐ Explicit cursors have attributes that can be used to describe the cursor's state, whether records are found after a fetch operation, and how many records are fetched.

- ❐ You can create an explicit cursor using a SELECT FOR UPDATE command, which places shared locks on the rows associated with the cursor.

❑ A PL/SQL table is a data structure that is similar to an array. A PL/SQL table contains a key that uniquely identifies each table item, along with an associated data value. A PL/SQL table of records is a PL/SQL table that can store multiple data items in each row. It can be used to store the contents of small database tables used for lookup operations.

❑ PL/SQL uses exception handling, whereby all code for displaying error messages and giving users options for fixing errors is placed in the EXCEPTION section of the program block.

❑ When a runtime error occurs, an exception (or unwanted event) is raised.

❑ Runtime errors are reported by the PL/SQL runtime engine and must be handled in the exception-handling section of the program.

❑ When an exception is raised, program control immediately transfers to the program's EXCEPTION block.

❑ Predefined exceptions correspond to common errors seen in many programs and are handled using predefined exception names in error handlers.

❑ Undefined exceptions are less-common errors that must be given an explicit exception name before they can be handled in the EXCEPTION section.

❑ User-defined exceptions are used to handle exceptions that will not cause an Oracle runtime error but require exception handling to enforce business rules or ensure the integrity of the database.

❑ PL/SQL program blocks can be nested, and the body of one program block can contain another program block. Variables declared in higher level blocks are visible and persistent within inner blocks.

❑ When an exception is raised and handled within a nested block, program execution resumes in the outer block. When an exception is raised in an inner block and is then handled in an outer block, the program exits after the exception is handled.

REVIEW QUESTIONS

1. Specify whether each of the following is a DDL, DML, or transaction control SQL command. Also state whether or not the command can be used in a PL/SQL program.

   a. CREATE SEQUENCE

   b. UPDATE

   c. ROLLBACK TO SAVEPOINT

   d. GRANT PRIVILEGE

   e. SELECT FOR UPDATE

2. Of the four basic (noncursor) PL/SQL loop types (LOOP…EXIT, LOOP…EXIT WHEN, WHILE…LOOP, numeric FOR LOOP), which are pretest loops and which are posttest loops?

3. What is the difference between the data that can be retrieved by an implicit cursor and the data that can be retrieved by an explicit cursor?

4. When should you use a SELECT FOR UPDATE cursor?

5. What is the advantage of using a cursor FOR loop for cursor processing?

6. When should you create a PL/SQL table of records?

7. What is the difference between a compile error and a runtime error?

8. What is an exception? What happens to program execution when an exception is raised?

9. Do exception handlers handle compile errors or runtime errors?

10. What is the difference between a predefined exception and an undefined exception?

11. When should you create a user-defined exception?

PROBLEM-SOLVING CASES

Use Notepad or an alternate text editor to create the following PL/SQL programs. Save all solution files in your Chapter4\CaseSolutions folder. Cases refer to the sample databases for Clearwater Traders (see Figure 1-11), Northwoods University (see Figure 1-12), and Software Experts (see Figure 1-13).

1. Write a PL/SQL program that uses an implicit cursor to display the date expected, quantity expected, item description, and color of SHIPMENT_ID 212 in the Clearwater Traders database. Format the program output so that it appears as shown in Figure 4-63. Use variable values rather than hard-coded values wherever possible in the DBMS_OUTPUT function. Save your file as 4BCase1.sql.

```
Oracle SQL*Plus
File  Edit  Search  Options  Help
Shipment 212 is expected to
arrive on 11/15/2003
and will contain 25
3-Season Tents, Color Light Grey

PL/SQL procedure successfully completed.
```

Figure 4-63

2. Write a PL/SQL program that uses an implicit cursor to display the call ID, course name, term description, faculty first and last name, day, time, and building code and room for C_SEC_ID 1011 in the Northwoods University database. Include exception handlers for the cases where no data is returned or where multiple records are returned (ORA-01422). Format the program output so that it appears as shown in Figure 4-64. Use variable values rather than hard-coded values wherever possible in the DBMS_OUTPUT function. Save your file as 4BCase2.sql.

```
± Oracle SQL*Plus                                              _ 8 X
File  Edit  Search  Options  Help
MIS 301   Systems Analysis
M-F 08:00 AM
BUS 404
Instructor:   John Blanchard

PL/SQL procedure successfully completed.
```

Figure 4-64

3. Write a PL/SQL program that uses an implicit cursor to display the project name, client name, and project manager first and last name for P_ID 5 in the Software Experts database. Include the WHEN OTHERS exception handler. Format the program output as shown in Figure 4-65. Use variable values rather than hard-coded values wherever possible in the DBMS_OUTPUT function. Save your file as 4BCase3.sql.

```
± Oracle SQL*Plus                                              _ 8 X
File  Edit  Search  Options  Help
Project: Internet Advertising
Client:  Birchwood Mall
Manager: Janet Park

PL/SQL procedure successfully completed.
```

Figure 4-65

4. Write a PL/SQL program that uses an explicit cursor to display the item description, color, size, price, quantity on hand, and value (PRICE * QOH) for item ID 559 in the Clearwater Traders database. (You will need to create a column alias for the calculated column.) Also, calculate the sum of the values for all items, and display the output as shown in Figure 4-66. Save the file as 4BCase4.sql.

```
± Oracle SQL*Plus                                              _ 8 X
File  Edit  Search  Options  Help
Men's Expedition Parka  S   Spruce   $199.95  114  $22,794.30
Men's Expedition Parka  M   Spruce   $199.95   17  $3,399.15
Men's Expedition Parka  L   Spruce   $209.95    0  $.00
Men's Expedition Parka  XL  Spruce   $209.95   12  $2,519.40
Total Value:   $28,712.85

PL/SQL procedure successfully completed.
```

Figure 4-66

5. Write a PL/SQL program that displays the call ID and section number for each course section offered during the Summer 2004 term at Northwoods University. Under the call ID and course name, list the first and last name of each student enrolled in the course. Format the output as shown in Figure 4-67. Save the file as 4BCase5.sql. (*Hint:* You will need to create two explicit cursors and use nested loops to process the cursors.)

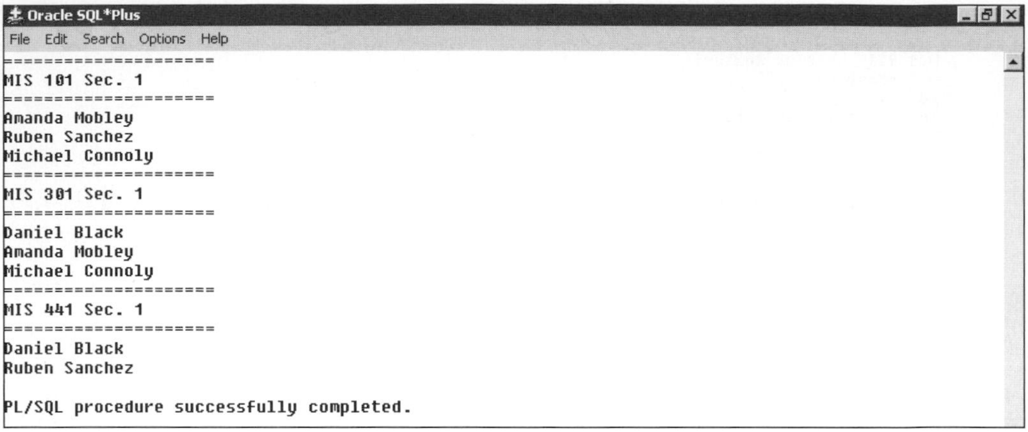

Figure 4-67

6. Write a PL/SQL program that lists the project name and client name for each project in the Software Experts database, along with the first and last name of each consultant who is working on the project, each consultant's total hours, and the sum of the total hours for all project consultants. Format the output as shown in Figure 4-68. For consultants whose project hours are zero, display the alternate output as shown in Figure 4-68 by specifying a user-defined exception. (*Hint*: Use nested blocks, and place the user-defined exception in the inner block.) Save the file as 4BCase6.sql.

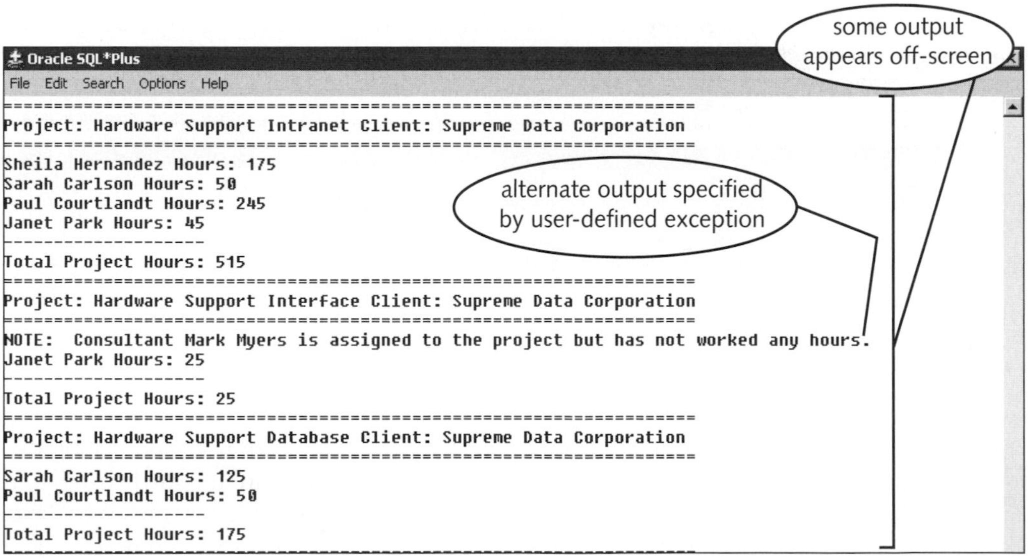

Figure 4-68

7. Write a PL/SQL program that displays the full name and address of Clearwater Traders customer Alissa Chang and then summarizes the details of all of her orders, as shown in Figure 4-69. Use an implicit cursor to retrieve and display the customer information, and use two nested explicit cursors to display the order information. Save the file as 4BCase7.sql.

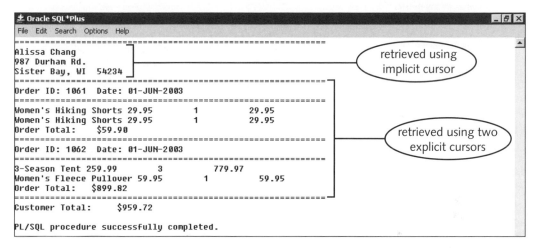

Figure 4-69

8. Write a PL/SQL program that creates a PL/SQL table, based on the SKILL table in the Software Experts database, that contains the skill ID values as the table keys and the associated skill descriptions as the table values. Then, create a loop that steps through the PL/SQL table sequentially. Within the loop, process an explicit cursor that retrieves the name and certification status of each consultant who has the associated skill, using the current PL/SQL table key as the cursor search condition for the SKILL_ID. Format the output as shown in Figure 4-70. Save the file as 4BCase8.sql.

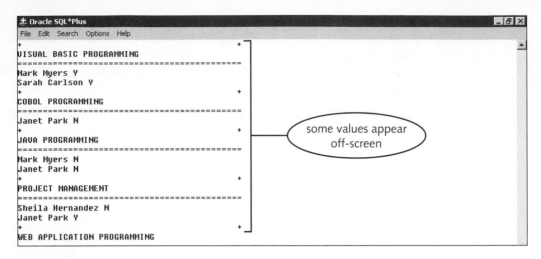

Figure 4-70

9. Write a PL/SQL program that calculates the cumulative grade point average for every student in the STUDENT table of the Northwoods University database. Grade point average is calculated as follows:

SUM(Course Credits * Course Grade Points)
SUM(Course Credits)

Course grade points are awarded as follows:

Grade	Grade Points
A	4
B	3
C	2
D	1
F	0

Create an explicit cursor that retrieves the student data for all students who have earned a grade in at least one course. (*Hint:* Do an outer join using the STUDENT and ENROLLMENT tables.) Create a second explicit cursor that is processed within the loop for the first cursor. The second cursor retrieves the course information for these students and calculates each student's grade point average. Format the output as shown in Figure 4-71. Save the file as 4BCase9.sql.

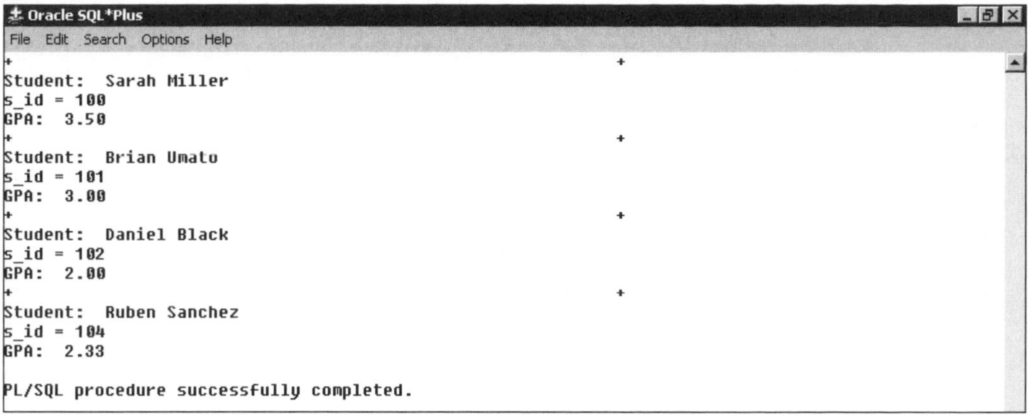

```
±Oracle SQL*Plus                                                    _ 8 X
File Edit Search Options Help
+                                              +
Student:  Sarah Miller
s_id = 100
GPA:  3.50
+                                                     +
Student:  Brian Umato
s_id = 101
GPA:  3.00
+                                                     +
Student:  Daniel Black
s_id = 102
GPA:  2.00
+                                                     +
Student:  Ruben Sanchez
s_id = 104
GPA:  2.33

PL/SQL procedure successfully completed.
```

Figure 4-71

10. Computer programs often use check digits as an error-checking technique for input data. A check digit is an extra reference number that is appended to the end of a numerical data value, such as a product identification code. This extra digit has a mathematical relationship to the data value. The check digit is input with the data value and then recomputed by the computer program. The result is compared with the input, and an exception is raised if the input is incorrect. The most common check digit system is the Modulus 11 system. Suppose that Clearwater Traders uses a check digit system in which each digit of a product's INV_ID is multiplied by the corresponding digit of the product's ITEM_ID. The result is summed and then divided by a fixed number—in this case, the number 11. The resulting remainder, called the modulus, is subtracted from 11, with the result being the check digit. Here is an example of a check digit computation for an INV_ID in the Clearwater Traders database:

Sample INV_ID	1	1	6	6	8	
Corresponding ITEM_ID	0	0	7	8	6	
Multiply each associated digit in INV_ID by corresponding ITEMID digit	0	0	42	48	48	
Sum the results	$0 + 0 + 42 + 48 + 48 = 138$					
Divide the sum by 11, and take the modulus	$138/11 = 12$ with remainder of 6					
Subtract modulus from 11 to get the check digit	$11 - 6 = 5$					
Append check digit to original INV_ID to get new INV_ID	1	1	6	6	8	5

Write a PL/SQL program that generates a check digit for each INV_ID value in the INVENTORY table in the Clearwater Traders database, using the associated ITEM_ID and a modulus value of 11. Use an explicit cursor to retrieve each INV_ID and corresponding ITEM_ID, compute the check digit, and then output the value of the new INV_ID with the check digit appended to it. The remainder of a division operation can be returned using the PL/SQL MOD function, which has the following general syntax:

```
MOD(number_being_divided, divisor)
```

For example, MOD(9,2) would return 1, since 9/2 = 4 with a remainder of 1.

(*Hint*: Change the numeric values to strings, and parse them using the SUBSTR function. Assume that all ITEM_ID values have three digits, and all INV_ID values have five digits.)

(*Hint*: In the case where the remainder is 1 and the check digit is 10, change the check digit to 0. In the case where the remainder is 0 and the check digit is 11, change the check digit to 1.)

Format the output as shown in Figure 4-72. Save your file as 4BCase10.sql.

```
± Oracle SQL*Plus                                               _ 8 X
File  Edit  Search  Options  Help
Original INU_ID: 11668    INU_ID with Check Digit: 116685
Original INU_ID: 11669    INU_ID with Check Digit: 116690
Original INU_ID: 11775    INU_ID with Check Digit: 117754
Original INU_ID: 11776    INU_ID with Check Digit: 117761
Original INU_ID: 11777    INU_ID with Check Digit: 117777
Original INU_ID: 11778    INU_ID with Check Digit: 117783
Original INU_ID: 11779    INU_ID with Check Digit: 117790
Original INU_ID: 11780    INU_ID with Check Digit: 117804
Original INU_ID: 11795    INU_ID with Check Digit: 117954
Original INU_ID: 11796    INU_ID with Check Digit: 117968
Original INU_ID: 11797    INU_ID with Check Digit: 117971
Original INU_ID: 11798    INU_ID with Check Digit: 117985
Original INU_ID: 11799    INU_ID with Check Digit: 117999
Original INU_ID: 11800    INU_ID with Check Digit: 118002
Original INU_ID: 11820    INU_ID with Check Digit: 118209
Original INU_ID: 11821    INU_ID with Check Digit: 118214
Original INU_ID: 11822    INU_ID with Check Digit: 118220
Original INU_ID: 11823    INU_ID with Check Digit: 118235
Original INU_ID: 11824    INU_ID with Check Digit: 118241
Original INU_ID: 11825    INU_ID with Check Digit: 118256
Original INU_ID: 11826    INU_ID with Check Digit: 118261
Original INU_ID: 11827    INU_ID with Check Digit: 118277
Original INU_ID: 11845    INU_ID with Check Digit: 118455
Original INU_ID: 11846    INU_ID with Check Digit: 118467
Original INU_ID: 11847    INU_ID with Check Digit: 118479
Original INU_ID: 11848    INU_ID with Check Digit: 118481

PL/SQL procedure successfully completed.
```

Figure 4-72

5

ADVANCED PL/SQL PROGRAMMING

◀ LESSON A ▶

Objectives

- ◆ Understand named PL/SQL program units
- ◆ Create server-side stored procedures and functions in SQL*Plus
- ◆ Create client-side procedures and functions in Procedure Builder
- ◆ Use the Procedure Builder debugging environment
- ◆ Create libraries of program units

In the preceding chapter, you learned to create PL/SQL programs that performed a variety of processing tasks. These were **anonymous** PL/SQL programs: They were submitted to the PL/SQL interpreter and run, but were not established as named database objects that could be available to other users or called by other procedures. In this chapter, you will learn to create **named** PL/SQL programs. Named programs are database objects that can be referenced by other programs and can be used by other database users. Named programs can also receive input values and pass output values to other programs.

In addition to learning how to create named programs, you will also become familiar with Procedure Builder, which is an Oracle utility that can be used to create named PL/SQL programs that can be stored and executed on the user's computer or stored in the database and executed on the database server. You will also learn about libraries and packages, which are collections of related programs. In addition, you will learn to create database triggers, which are PL/SQL programs that execute when a specific database event occurs, such as inserting or updating a record in a particular table. You will also become familiar with using built-in Oracle packages and libraries that are supplied with the Oracle database. Finally, you will learn how to use Dynamic SQL, which allows you to execute DDL statements within SQL programs.

It is a good practice to break complex programs into smaller program units because it is easier to conceptualize, design, and debug a small procedure than a large program. A **program unit** is a self-contained group of program statements that can be used within a larger program. When all of the smaller program units work correctly, you can link them into the large program. Program units can also be reused, saving valuable programming time. In PL/SQL, a **named program unit** is a database object consisting of PL/SQL code that can be used for processing tasks.

NAMED PROGRAM UNITS

In a client/server database, such as Oracle, the database usually runs on a central database server. (If you are using Personal Oracle, the database actually runs on your workstation but acts like a database running on a server.) The workstation where the user runs applications (such as SQL*Plus and Procedure Builder) that connect to the database on the server is called the **client workstation**, or simply the client. **Server-side program units** are stored in the database as database objects and execute on the database server. **Client-side program units** are stored in the file system of the client workstation and execute on the client workstation.

The advantage of using server-side program units is that because they are stored in a central location that is accessible to all database users, they are always available whenever a database connection is made. Users can easily access and run server-side program units without having to locate files on remote workstations. The disadvantage of using server-side program units is that it forces all processing to be done on the database server. If the database server is very busy, this can result in applications with very slow response times. If you are creating a program unit that only you or a few co-located users will use, and system performance is a consideration, then you should create a client-side program unit.

Table 5-1 describes the different types of Oracle named program units and shows where they are stored and where they execute in the client/server architecture.

Program Unit Type	Description	Where Stored	Where Executed
Procedure	Can accept multiple input parameters, and returns multiple output values	Operating system file or database	Client-side or server-side
Function	Can accept multiple input parameters, and returns a single output value	Operating system file or database	Client-side or server-side
Library	Contains code for multiple related procedures or functions	Operating system file or database server	Client-side

Table 5-1   Types of Oracle named program units

Program Unit Type	Description	Where Stored	Where Executed
Package	Contains code for multiple related procedures, functions, and variables and can be made available to other database users	Operating system file or database server	Client-side or server-side
Trigger	Contains code that executes when a specific database action occurs, such as inserting, updating, or deleting records	Database server	Server-side

Table 5-1   Types of Oracle named program units (continued)

The first kind of named program units you will learn about are procedures and functions. (You will learn about libraries, packages, and triggers later in the chapter.)

PROCEDURES AND FUNCTIONS

A **procedure** can receive multiple input parameters, can return multiple output values, or can return no output values. A **function** can receive multiple input parameters and always returns a single output value. A function works well when you want to compute a single value, because you can assign the return value of the function to a variable within the calling program. A procedure works well when you want to manipulate the values of several variables.

As you can see from Table 5-1, procedures and functions accept inputs and return outputs. Inputs and outputs are handled using parameters, which are variables that pass information from one program to another.

Program Unit Parameters

Recall that a PL/SQL program block has three sections: declaration, body, and exception. Instead of a declaration section, procedures have a section called a **header**, in which the program unit's name and parameters are specified and the program unit variables are declared. The general syntax of a program unit header is:

```
unit_type unit_name
      (parameter1 mode datatype,
       parameter2 mode datatype,
          ...) IS
declaration statements
```

The *unit_type* can be PROCEDURE or FUNCTION. The *unit_name* must conform with the Oracle naming standard. Note that in the header, the parameter declarations are indented so that they align vertically. The parameter *mode* describes how the parameter value can be changed in the program unit. The mode of a parameter can be **IN**, **OUT**, or **IN OUT**. Table 5-2 describes the different modes.

Mode	Description
IN	Parameter is passed to the program unit as a read-only value that cannot be changed within the program unit.
OUT	Parameter is a write-only value that can only appear on the left side of an assignment statement in the program unit.
IN OUT	Combination of IN and OUT; the parameter is passed to the program unit, and its value can be changed within the program unit.

Table 5-2   Parameter modes

If a parameter is declared with no mode specification, it is by default an IN parameter. When you specify the parameter data type, you do not include the precision or scale value for a numerical data type, or the maximum width for a character data type.

Calling Program Units and Passing Parameter Values

The syntax to call a procedure differs slightly from that used to call a function. To call a procedure from the SQL command line, you use the EXECUTE command, then list the procedure name, followed by the parameter values, which are delimited by commas and enclosed in parentheses. The general syntax for the EXECUTE command is:

```
EXECUTE procedure_name
(parameter1_value, parameter2_value, ...);
```

When you are calling a procedure, you must distinguish between formal parameters and actual parameters. **Formal parameters** are the parameters that are declared in the header of the procedure. **Actual parameters** are the values placed in the procedure parameter list when the procedure is called. Actual parameters can be constants or variables that have assigned values. When variables are used as actual parameters, the variable names do not have to be the same as the formal parameter names.

Let's look at an example. Suppose you have a procedure named CALC_GPA that calculates the grade point average of a student for a given term. The procedure header would look like this:

```
PROCEDURE calc_gpa (student_id IN NUMBER,
                    current_term_id IN NUMBER,
                    calculated_gpa OUT NUMBER) IS
```

Note that the procedure has three formal parameters: STUDENT_ID, CURRENT_TERM_ID, and CALCULATED_GPA.

You would call the procedure using the following program statement:

```
EXECUTE calc_gpa(current_s_id, 4, current_gpa);
```

This statement passes three actual parameters to the procedure. The student ID parameter is passed as a variable value (CURRENT_S_ID), and the term ID value is passed as a constant, the value 4. The calculated GPA parameter will be returned as a variable value that is referenced as CURRENT_GPA. Notice that the actual parameter variable names in the command calling the procedure are different from the formal parameter names in the procedure declaration.

Figure 5-1 shows the relationship between the formal parameters and the actual parameters. It is important to remember that the variables or constants that are passed for each parameter must be in the order in which the parameters are declared in the procedure declaration, because that is how the values are associated. The first variable value in the procedure calling statement will be assigned to the first parameter in the procedure declaration; the second variable will be assigned to the second parameter, and so forth.

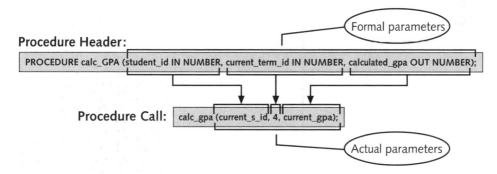

Figure 5-1   Relationship between formal and actual parameters

Calling a function requires assigning its return value to a variable. The general command for calling a function is ***variable_name := function_name(parameter1, parameter2, ...);***. As with procedures, the variables or constants that are passed for each parameter must be in the same order in which the parameters are declared in the function declaration, because that is how the values are associated. The first variable value in the function assignment statement will be assigned to the first parameter in the function declaration; the second variable will be assigned to the second parameter, and so forth.

CREATING STORED PROGRAM UNITS IN SQL*PLUS

A **stored program unit**, which is also called a **stored procedure** regardless of whether it is a procedure or function, is a server-side PL/SQL program unit that is stored in the database and executed on the database server. Creating a stored program unit makes the program unit available at all times to all database users who have the privilege to use it, so you do not have to load it from an operating system file.

To create stored procedures in an Oracle database, you must have the CREATE PROCEDURE and CREATE FUNCTION system privileges.

In the following sections, you will learn how to create and debug stored procedures and functions in SQL*Plus.

Creating Stored Procedures in SQL*Plus

You can create a stored procedure in SQL*Plus using the CREATE PROCEDURE command, which has the following syntax:

```
CREATE OR REPLACE PROCEDURE procedure_name
    (parameter1 mode datatype,
    parameter2 mode datatype,
    ...) IS | AS
    declarations
BEGIN
    program statements
EXCEPTION
    exception handlers
END;
```

When you create a procedure in SQL*Plus, the parameter declarations in the procedure header can be followed by the keyword IS or AS—either will work. When you create a procedure in Oracle applications other than SQL*Plus, the parameter declarations must be followed by the keyword IS. In all environments, the OR REPLACE clause in the CREATE PROCEDURE command is optional. Once a procedure is created, however, an error is generated if you attempt to create another procedure with the same name. Therefore, it is a good practice to always include the OR REPLACE clause, which automatically replaces an existing procedure if one exists. In addition, if other users have been granted privileges to use the procedure, the users retain these privileges when you replace the existing code with new code.

The procedure program statements are enclosed between the BEGIN and the END commands. If an exception is raised in a procedure and handled in the procedure's EXCEPTION section, the procedure immediately exits, and program execution resumes in the calling program with the program statement following the command that called the procedure.

Now you will create a stored procedure in SQL*Plus named UPDATE_INVENTORY that will update the quantity on hand of an inventory item in the Clearwater Traders database INVENTORY table. The procedure will accept two input parameters: the inventory ID and the amount by which the quantity on hand (QOH) field is to be updated.

To create a stored procedure in SQL*Plus:

1. Start SQL*Plus, log onto the database, and type the following commands at the SQL prompt to run the **clearwater.sql**, **northwoods.sql**, and **software-exp.sql** scripts from the Chapter5 folder on your Data Disk to refresh your database tables.

   ```
   @a:\chapter5\clearwater.sql;
   @a:\chapter5\northwoods.sql;
   @a:\chapter5\softwareexp.sql;
   ```

2. At the SQL prompt, type **SET SERVEROUTPUT ON SIZE 4000;** to initialize the DBMS_OUTPUT buffer.

3. If necessary, start Notepad, and type the commands as shown in Figure 5-2. Save the file as **5APrograms.sql** in your Chapter5\TutorialSolutions folder.

 As you perform the tutorial exercises in this lesson, save all of your code in the 5Aprograms.sql file.

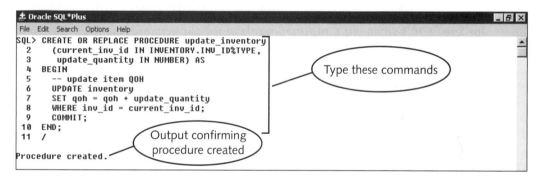

Figure 5-2   Creating a stored procedure in SQL*Plus

4. Paste the code into SQL*Plus, and then run the program by typing /, and then pressing **Enter**. If your program executes successfully, the message "Procedure created" appears, as shown in Figure 5-2. If your program has compile errors, proceed to the section titled "Debugging Named Program Units in SQL*Plus."

Now the stored procedure is created and exists as a database object in your user schema. To test your stored procedure, you will use the EXECUTE command to call the procedure and pass the parameter values for the inventory ID to be updated, which will be 11668, and the quantity by which the inventory will be updated, which will be –3. First, you will query the database to determine the current quantity on hand for INV_ID 11668. Then you will execute the procedure. Finally, you will query the database again to confirm that the stored procedure updated the inventory QOH correctly.

To test the stored procedure:

1. Type the first command shown in Figure 5-3 to display the current QOH of INV_ID 11668.

2. Type the second command to run the procedure and pass the required parameter values.

3. Type the third command to display the updated QOH value, confirming that the stored procedure successfully updated the inventory.

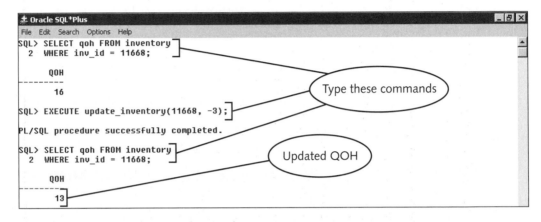

Figure 5-3   Testing the stored procedure

Debugging Named Program Units in SQL*Plus

Debugging named program units in SQL*Plus is similar to debugging any program: You must isolate the program lines causing the errors and then fix them. Figure 5-4 shows a named program unit with compile errors. Finding and fixing compile errors are a little more difficult with named programs in the SQL*Plus environment, because when a named program unit has compile errors, the SQL*Plus interpreter displays only the error warning message shown in Figure 5-4.

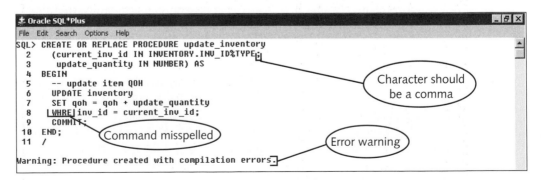

Figure 5-4   Named program unit with compile errors

As you can see, the interpreter does not automatically display compile error messages and locations like it does for anonymous program units, which makes debugging more of a challenge. Thankfully, all compile errors are written to a system table that you can access using the USER_ERRORS data dictionary view. To display a summary listing of compile errors generated by the last program unit that was compiled, you can use the SHOW ERRORS command. Figure 5-5 shows the listing provided by this command.

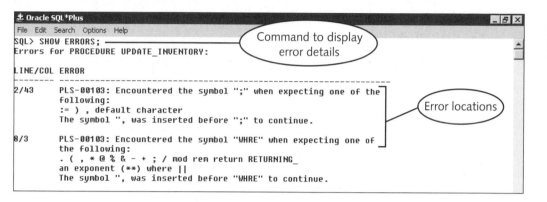

Figure 5-5   Using the SHOW ERRORS command to view compile error details

The LINE/COL column displays the line number of the error and the character column (position) where the error was encountered. The ERROR column summarizes the error code number and associated error message. The first error message indicates that an error occurred on Line 2, and the message indicates that a semicolon (;) was used instead of a comma (,) in the parameter declarations. The second error message indicates that the WHERE clause on Line 8 was formed improperly.

The advice provided in Chapter 4 regarding debugging compile errors in anonymous program blocks also applies to debugging named program blocks. Remember that error locations might not correspond to the line of the actual error, so you might need to isolate the error location systematically by commenting out and modifying suspect lines.

Recall that when you are trying to locate a logic error, you need to identify the output variables that have the error, identify the inputs and calculations that contribute to the invalid output, and then find the values of the inputs that are contributing to the invalid output during program execution. Later in this lesson, you will learn how to debug named program units using the debugger in Procedure Builder, which is a utility that provides a development environment for creating PL/SQL named programs.

To debug procedures in the SQL*Plus environment, you can place DBMS_OUTPUT.PUT_LINE statements in the procedure code to display variable values during execution. Then, when you call the procedure from the calling program, the debugging statements appear while the procedure executes. When you use this error-finding approach, you should always add text within the debugging statement to indicate which variable value is being displayed. Figure 5-6 shows the UPDATE_INVENTORY procedure with added debugging statements. The statements include text that provides the names of the variables being displayed (such as "current_inv_id value = "), along with the associated values.

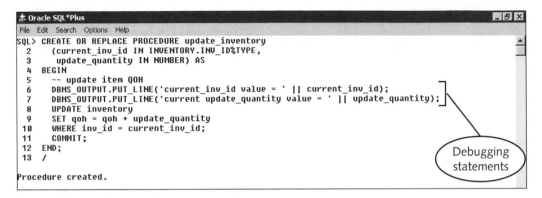

Figure 5-6   Procedure with debugging statements

Creating Functions in SQL*Plus

A function is similar to a procedure, except that it returns a single value that is assigned to a variable in the calling program. The general syntax for creating a function header is:

```
CREATE OR REPLACE FUNCTION function_name
    (parameter1  mode datatype,
    parameter2 mode datatype,
    ...)
RETURN function_return_value_datatype IS
    variable declarations
```

A function can receive multiple input parameters. Note that the function parameters are indented so they are vertically aligned in the function header. The RETURN command at the end of the header (before the variable declarations) specifies the data type of the value that the function returns. You call a function using the syntax *variable_name := function_name(parameter1, parameter2, ...);*, and the *variable_name* must have the same data type as the function return value data type.

The general syntax of the body of a function is:

```
BEGIN
    program statements
    RETURN return_value;
```

The RETURN command specifies the actual value that the function will return. Usually, the *return_value* is a variable that has been assigned to a value as a result of the function computations.

The exception section for a function has the syntax:

```
EXCEPTION
    exception handlers
    RETURN EXCEPTION_NOTICE;
END;
```

The RETURN EXCEPTION_NOTICE command instructs the function to display the exception notice in the program that calls the function.

Now you will create a function named AGE that receives a parameter that is a person's date of birth. The function calculates the person's age by subtracting the date of birth from the current system date and returns the value as a NUMBER data type. To format the age in years, the function divides the result of the difference between the birth date and the system date, which is in days, by 365.25 and then truncates the decimal portion.

To create a function that returns a person's age:

1. Type the commands shown in Figure 5-7, and then execute them in SQL*Plus to create the AGE function. Debug the program if necessary until you have successfully created the function.

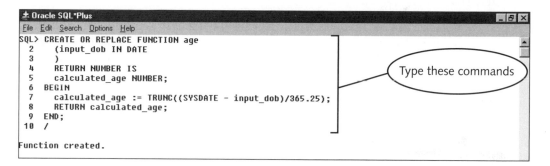

Figure 5-7   Creating a user-defined function

Recall that to call a function, you assign its return value to a variable. Now you will write an anonymous program block that retrieves student Daniel Black's date of birth using an implicit cursor and stores the retrieved value in a variable named CURRENT_S_DOB. The program then calls the AGE function and passes the CURRENT_S_DOB value to the function as a parameter. The value returned by the function is assigned to a program variable named CURRENT_AGE. Because the function was declared using the data type NUMBER, the CURRENT_AGE variable will be declared using the NUMBER data type also.

To write the program to call the function:

1. Type the commands shown in Figure 5-8 to call the AGE function and display student Daniel Black's current age. Note that your output will be different, depending on your current system date.

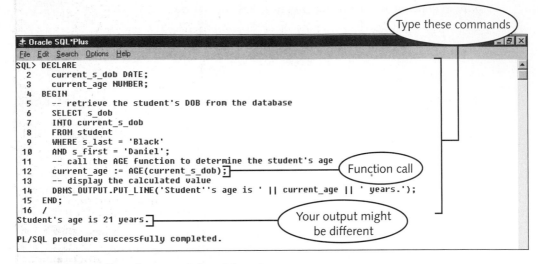

Figure 5-8 Calling the user-defined function

Function Purity Levels

When you write functions that are stored in the database, you can use these functions in PL/SQL programs. You can also sometimes use these functions directly in SQL commands, depending on the purity level. A function's **purity level** determines whether or not it can be used directly within a SQL command.

Recall that in Chapter 3, you learned to use several single-row functions, such as ROUND, TO_CHAR, and LENGTH, in SQL commands. These functions are called **inline functions**, because they can be used directly within SQL commands. To be used as an inline function, a user-defined function must follow these basic rules:

- The function can only use IN mode parameters, since to be used in a SQL command, it must return only a single value to the calling program.

- The data types of the function input variables and the function return value must be the PL/SQL data types that correspond to the Oracle database data types (VARCHAR2, CHAR, NUMBER, DATE, and so forth). You cannot use the PL/SQL data types that have no corresponding database data type, such as BOOLEAN and BINARY_INTEGER.

- The function must be stored in the database as a database object.

Function purity levels are summarized in Table 5-3.

Purity Level	Abbreviation	Description
Writes No Database State	WNDS	Function does not perform any DML commands.
Reads No Database State	RNDS	Function does not perform any SELECT commands.
Writes No Package State	WNPS	Function does not change values of any package variables.
Reads No Package State	RNPS	Function does not read any package variables.

Table 5-3   Oracle function purity levels

A package is a code library that provides a way to manage several related programs. You will learn how to create and use packages later in this chapter.

The function purity levels place the following restrictions on whether or not a function can be called within a SQL command:

- All inline functions must meet the WNDS purity level.

- Inline functions stored on the database and executed from a SQL query in a program running on the user's workstation (like an anonymous PL/SQL program) must meet the RNPS and WNDS purity levels. In contrast, an inline function called from a SQL query in a stored procedure does not have to meet these two purity levels.

- Functions called from the SELECT, VALUES, or SET clauses of a SQL query can write package variables, so they do not need to meet the WNPS purity level. Functions called from any other clause of a SQL query must meet the WNPS purity level.

- A function is only as pure as the purity of the subprograms it calls.

You can write functions that you do not intend to use as inline functions. In this case, you don't have to consider the purity levels. If you plan to create user-defined inline functions, however, you must consider function purity levels.

The AGE function that you created in the previous section can be used as an inline function, because it meets the criteria: it only contains IN parameters, it returns a database data type, and it satisfies the RNPS and WNDS function purity levels. Now you will use the AGE function as an inline function within a SQL command.

To use the AGE function in a SQL command:

1. Type the SQL command shown in Figure 5-9 that uses the AGE function as an inline function in a SQL command. The output displays the student ages as shown.

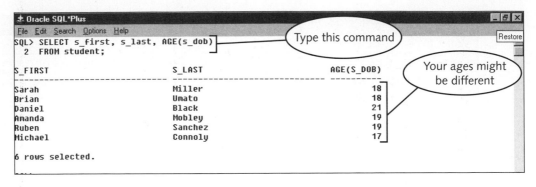

Figure 5-9   Using the AGE function as an inline function

STORED PROGRAM UNIT OBJECT PRIVILEGES

When you create a stored program unit, it exists as a database object in your user schema. Like all database objects, you own it, and other users cannot execute it unless you explicitly grant them the EXECUTE privilege on the procedure or function. The general command to grant this privilege is **GRANT EXECUTE ON** *unit_name* **TO** *username***;**. Now you will grant the EXECUTE privilege on your AGE function to all database users.

To grant the execute privilege on the AGE function:

1. Type the following command at the SQL prompt: **GRANT EXECUTE ON age TO PUBLIC;**. The confirmation message "Grant succeeded" confirms that the privilege was granted.

To execute your procedure or function, another user would need to preface the function name with your username. For example, Figure 5-10 shows the command that another user would use to execute user LHOWARD's AGE function.

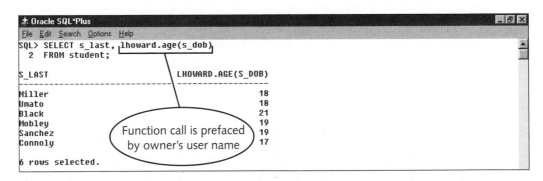

Figure 5-10   Executing a function owned by another user

USING PROCEDURE BUILDER TO CREATE NAMED PROGRAM UNITS

You have learned how to use SQL*Plus to create procedures and functions that are stored as database objects. In this section, you will learn how to use Procedure Builder, which is a graphical environment for creating, compiling, testing, and debugging procedures and other PL/SQL named program units. Procedure Builder provides enhanced support for managing and debugging stored program units. Before you start working with Procedure Builder, you will refresh your Clearwater Traders database tables in SQL*Plus. Then you will start Procedure Builder and become familiar with its environment.

Procedure Builder is part of the Oracle Developer utilities, so you must have Developer 6/6i installed on your computer.

To refresh your database tables, and then start Procedure Builder:

1. If necessary, start SQL*Plus, log onto the database, and refresh your Clearwater Traders database tables by running the clearwater.sql script in the Chapter5 folder on your Data Disk.

2. Click **Start** on the Windows taskbar, point to **Programs**, point to **Oracle Forms & Reports 6i**, and then click **Procedure Builder**. The Procedure Builder windows open, as shown in Figure 5-11. If necessary, maximize the outer window, and resize the inner windows to configure your environment as shown.

Another way to start Procedure Builder is to start Windows Explorer, and then double-click the file oracle_home\BIN\DE60.EXE.

Procedure Builder has two main windows: the **Object Navigator**, which provides access to different Procedure Builder objects, and the **PL/SQL Interpreter**, which has two panes. The upper pane of the PL/SQL Interpreter, called the **source pane**, displays the source code for existing PL/SQL programs and provides a debugging environment. The lower pane, called the **PL/SQL command prompt**, provides a SQL command-line environment that you can use like SQL*Plus.

The Object Navigator provides a hierarchical view of the different types of objects that you can create and modify in Procedure Builder. To view the different objects that you can access and modify, you click the **plus** beside the object type, and a list of current objects of that type appears. If there is no ➕ in front of an object type, then there are no objects of that type currently available to edit or execute. To close a list of objects, you click ➖ beside the object type. When you first start Procedure Builder, the only available object type is Built-in Packages.

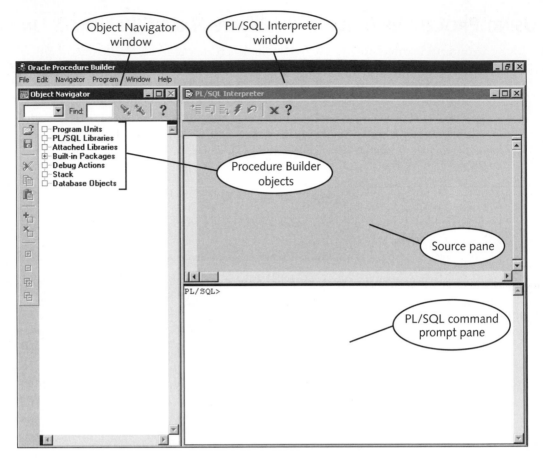

Figure 5-11   Procedure Builder environment

 Built-in packages are special PL/SQL program libraries that are stored in the database when the database is created They contain the code for built-in procedures and functions, such as TO_DATE and ROUND. You will learn more about Oracle built-in packages later in this chapter.

To become familiar with opening and closing objects in the Object Navigator, you will view the different built-in packages that are available.

To open an object in the Object Navigator:

1. Click ➕ beside the Built-in Packages node. A list of different packages appears. Notice that ➕ changes to ➖ to indicate that the object is open.

2. Click ➖ beside the Built-in Packages node to close the object. The list of Built-in Packages disappears, and only the different object type categories are displayed.

Creating a Client-Side Procedure in Procedure Builder

When you created stored procedures in SQL*Plus, you created server-side procedures that are stored in the database as database objects and execute on the database server. In contrast, you can use Procedure Builder to create client-side procedures, which are stored in the file system of the client and executed on the client. (You cannot create client-side procedures in SQL*Plus.)

Suppose you need to modify the Clearwater Traders INVENTORY table so that it stores the current value of each inventory item, which is equal to PRICE * QOH. To make these modifications, you will add a column named INV_VALUE to the INVENTORY table. Then, you will write a client-side procedure to calculate the value for each inventory item and insert the value into the associated record.

When you first start Procedure Builder, you are not automatically prompted to connect to the database, so you have to explicitly make a connection. First, you will connect to the database. Then, you will modify the INVENTORY database table by typing the ALTER TABLE command at the Procedure Builder PL/SQL command prompt.

To connect to the database and modify the INVENTORY table using Procedure Builder:

1. Click **File** on the Procedure Builder menu bar, and then click **Connect**. Log onto the database in the usual way.

2. Type the following code at the PL/SQL prompt to modify the INVEN-TORY table, and then press **Enter**. The PL/SQL prompt will appear again, indicating that the command executed successfully.

```
ALTER TABLE inventory
ADD inv_value NUMBER (11,2);
```

Now you will create a procedure named UPDATE_INV_VALUE that calculates the value for each inventory item and updates the INV_VALUE field.

To create the procedure:

1. Click **Program Units** in the Object Navigator window, and then click the **Create button** on the Object Navigator toolbar. The New Program Unit dialog box opens.

2. Make sure that the **Procedure** option button is selected, and then type **update_inv_value** in the Name text box to define the procedure name. Click **OK** to create the procedure. The Program Unit Editor window opens, as shown in Figure 5-12. Maximize the window.

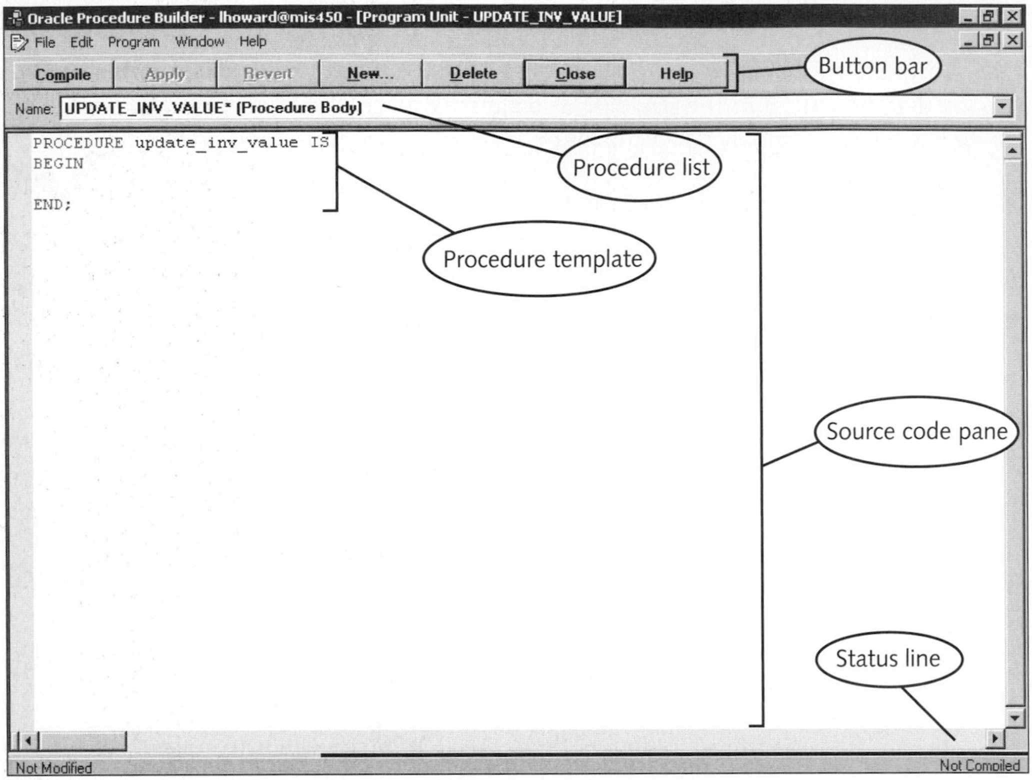

Figure 5-12   Program Unit Editor

The Program Unit Editor is an environment for writing, compiling, and editing PL/SQL programs. The **procedure list** shows the name of the current program unit and allows you to access other program units. The **source code pane** is where you type PL/SQL program statements. The **status line**, which is also called the status bar, displays the program unit's current modification status (Modified or Not Modified) and compile status (Not Compiled, Successfully Compiled, or Compiled with Errors). The Program Unit Editor button bar has the following buttons:

- **Compile**, which compiles the program statements in the source code pane. The compiler detects syntax errors and reports runtime errors.

- **Apply**, which saves changes made since the Program Unit Editor was opened or since the last time the Apply button was clicked

- **Revert**, which reverts the source code to its status as of the last time the Apply or Revert button was clicked

- **New**, which creates a new program unit

- **Delete**, which deletes the current program unit

- **Close**, which closes the Program Unit Editor
- **Help**, which accesses the Procedure Builder online Help system

When you create a new procedure, a **procedure template** is automatically inserted into the source code pane to define the procedure header and body. Text in the source code pane uses color highlighting to define different command elements. Reserved words (such as BEGIN and UPDATE) appear in blue. Command operators for arithmetic, assignment, and comparison are red. User-defined variables are black, literal values (such as the number 11) are blue-green, and comment statements are light green.

Now you will enter the procedure code for updating the Clearwater Traders INVENTORY table so that it stores the current value of each inventory item. You will need to create an explicit cursor that retrieves each record in the INVENTORY table. The procedure will then calculate the inventory value (PRICE * QOH) for each record, and update the record.

To enter the procedure code for modifying the Clearwater Traders INVENTORY table so that it stores the current value of each inventory item:

1. Click the mouse pointer before the B in BEGIN, and then press **Enter**.

2. Type the code shown in Figure 5-13.

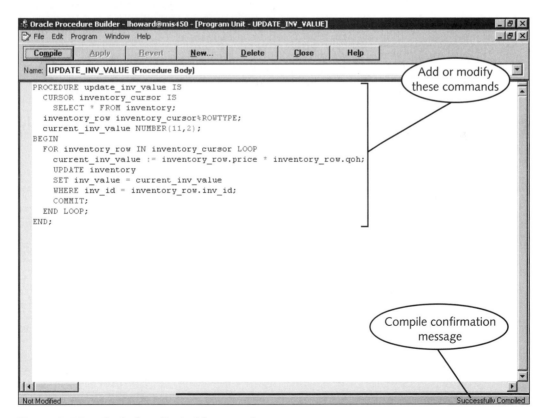

Figure 5-13   Code for client-side procedure

3. Click **Compile** on the Program Unit Editor button bar. If your procedure successfully compiles, then the message "Successfully Compiled" appears on the right side of the status line, as shown in Figure 5-13. If your procedure did not compile successfully, the message "Compiled with Errors" appears. The next section discusses how to interpret and correct compile errors in the Program Unit Editor environment.

4. When your procedure compiles successfully, click **Close** on the button bar to close the Program Unit Editor. The Procedure Builder environment should have the tiled window configuration shown in Figure 5-14, with the procedure source code displayed in the source code pane.

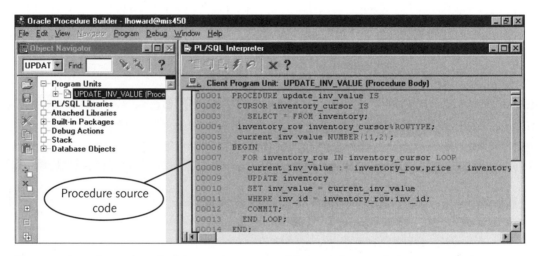

Figure 5-14   Procedure Builder environment with procedure code in source code pane

 If your window configuration does not look like Figure 5-14, click the Restore button  on the menu bar to display the tiled window configuration again. If you accidentally close the PL/SQL Interpreter window, click Program on the menu bar, click PL/SQL Interpreter to redisplay the window, and then resize the windows as necessary.

Interpreting and Correcting Compile Errors in Procedure Builder

When a compile error occurs, the Program Unit Editor's **compilation messages pane** displays the line number of the error and the error message, as well as correction suggestions. For example, in Figure 5-15, a compile error occurs because the compiler expected the equal sign (=) instead of the assignment operator (:=) in the SQL SET command. This error can be corrected by changing := to = and then clicking Compile again to recompile the procedure. Notice that you can select an error in the compilation messages pane. Double-clicking the selected error will place the insertion point in the source code pane in the line of code that is producing the error.

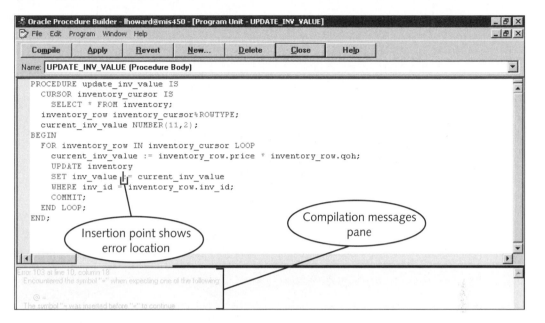

Figure 5-15   Procedure with compile error

If there are multiple errors, you can select a different error message by clicking the alternate message in the compilation messages pane. The insertion point then moves to the location of the alternate error message. Figure 5-16 shows a procedure with multiple compile errors. Along with the previous error, the user also forgot the semicolon after the COMMIT command. When the user clicks the second error message in the compilation messages pane, the insertion point moves to the location described by the second error message.

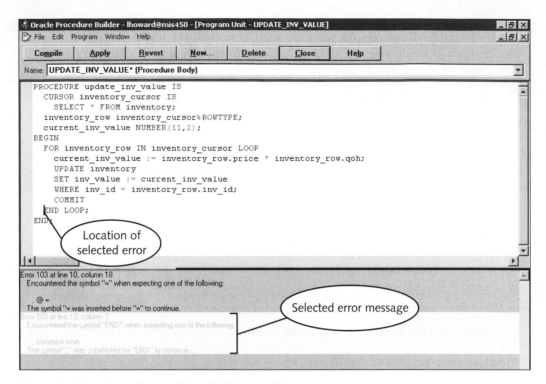

Figure 5-16   Procedure with multiple compile errors

When a procedure is compiled, the PL/SQL interpreter verifies that all database objects referenced in the PL/SQL code exist in the user's database schema or can be accessed by the user through privileges granted to the user by other users. Figure 5-17 shows an example of a compile error message that appears when a database object referenced in a procedure is not found or its name is entered incorrectly. In this example, the INVENTORY table is mistyped as INVENTRY. Because the interpreter cannot verify that this table exists, it is flagged as an undeclared variable.

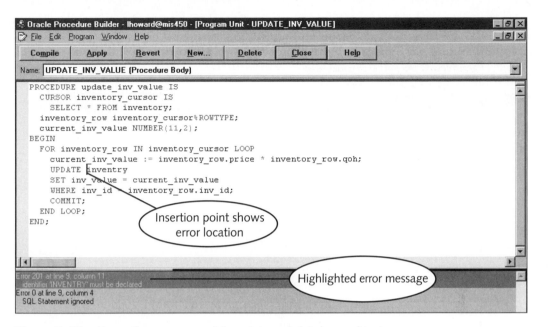

Figure 5-17   Compile error caused by mistyped database object name

Figure 5-18 shows the same error message, but for a different reason. The database table name is spelled correctly, but the user forgot to connect to the database when he or she started Procedure Builder. To correct this error, click File on the top menu bar, click Connect, connect to the database, and then recompile the program unit.

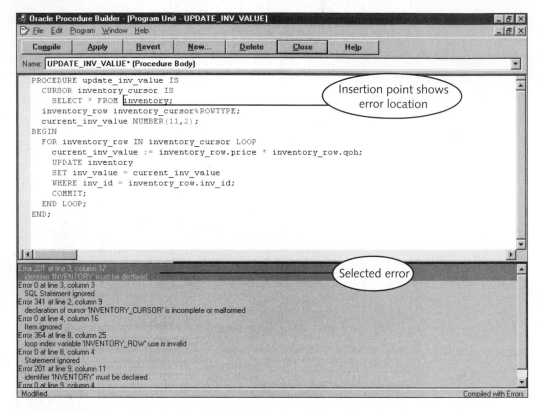

Figure 5-18   Compile errors generated when user forgets to connect to database

There are many more possible compile errors, but these are some of the more common ones. The best way to locate errors is to examine the code carefully and isolate the line that is causing the error by commenting out program lines. It is also a good idea to copy SQL commands into SQL*Plus and test them by themselves to see if they are generating compile errors.

Saving a Client-Side Procedure in Procedure Builder

Recall that a client-side procedure is stored in the file system of the client workstation. There are two ways to store a client-side procedure in Procedure Builder: as an uncompiled file, which saves the source code as text, or as a library file, which saves both the source code and the compiled executable code. To save a procedure as an uncompiled source code file, you export the PL/SQL commands to a text file with a .pls extension, which stands for PL/SQL source. You would save a program unit as an uncompiled source code file while the program unit is under development and before it can be compiled. First, you will save your procedure as an uncompiled source code file in your Chapter5\TutorialSolutions folder. You can only save a procedure as a source code file when the procedure is displayed in the Program Unit Editor, so you will reload the procedure into the Program Unit Editor.

To save the procedure source code as an uncompiled .pls file:

1. Display the procedure in the Program Unit Editor by double-clicking the **Program Unit icon** beside UPDATE_INV_VALUE in the Object Navigator window. If necessary, maximize the Program Unit Editor window.

 You can also open the Program Unit Editor by clicking Program on the menu bar, and then clicking Program Unit Editor. Alternately, you can right-click in the upper pane of the PL/SQL Interpreter window, and then click Edit.

2. Click **Edit** on the menu bar, and then click **Select All**.

3. Click **File** on the menu bar, click **Export Text**, save the selected text in a file named **update_inv_value.pls** in your Chapter5\TutorialSolutions folder, and then close the Program Unit Editor.

You will remove the procedure from the Procedure Builder environment. Then, you will reload it using the exported source code file.

To remove the procedure and then reload it:

1. Remove the UPDATE_INV_VALUE program unit from Procedure Builder by selecting the procedure in the Object Navigator window, clicking the **Delete button** on the Object Navigator toolbar, and then clicking **Yes** to confirm removing the unit. This unloads the procedure from Procedure Builder but does not delete it from your file system.

2. Click **File** on the menu bar, click **Load**, click **Browse**, navigate to your saved file in your Chapter5\TutorialSolutions folder, click **Open**, and then click **Load**. Click **+** beside Program Units to confirm that the procedure is again listed as a program unit in the Procedure Builder environment.

To save the procedure in a compiled format, you must save it within a PL/SQL **client-side library**, which is a file with a .pll extension that can contain one or more related procedures or functions. Saving a client-side library in Procedure Builder is a three-step process:

1. Create the library in Procedure Builder;

2. Save the library file in the file system on the client workstation;

3. Add the procedure to the library.

First, you will create the library.

To create a library:

1. Select the **PL/SQL Libraries** node in the Object Navigator window, and then click the **Create button**. A new library appears under the PL/SQL Libraries node in the Object Navigator window.

   You might need to click Window on the menu bar, and then click Object Navigator to redisplay the Object Navigator window.

5

The next step is to save the library. When you create new objects in Oracle applications, they are often given default names. To change the library name, you must save the library either in the database or in the file system of your computer. You will save the library in the file system in your Chapter5\TutorialSolutions folder.

To save the library:

1. Make sure that the new library is selected in the Object Navigator window, and then click the **Save button** 🔲 on the Object Navigator toolbar. The Save Library dialog box opens.

2. Type **update_inventory** as the library name, and make sure that the File System option button is selected.

3. Click **Browse**, and navigate to your Chapter5\TutorialSolutions folder. Type **update_inventory.pll** as the library filename, click **Open**, and then click **OK** to save the library. The new library name is appears in the Object Navigator window.

The final step is to add the program unit to the library. To add a program unit to a library, the program unit must currently be loaded in the Procedure Builder environment. Then, you simply drag the program unit from the Program Units category to the library's program unit subcategory. Now, you will add the UPDATE_INV_VALUE program unit to the UPDATE_INVENTORY library.

To add the program unit to the library:

1. Make sure that UPDATE_INV_VALUE program unit appears in the top-level Program Units node in the Object Navigator window. If it does not appear, click File, click Load, and then load the procedure from your Chapter5\TutorialSolutions folder.

2. Click ➕ beside the UPDATE_INVENTORY library if necessary, so its objects are displayed.

3. Select the UPDATE_INV_VALUE program unit, and then drag it to the UPDATE_INVENTORY library. The pointer changes to the move program unit pointer 🐾 when it moves over the Program Units node.

4. When the tip of the pointer arrow is on the Program Units category under the UPDATE_INVENTORY library, release the mouse button and drop the program unit. It now appears under the library, as shown in Figure 5-19.

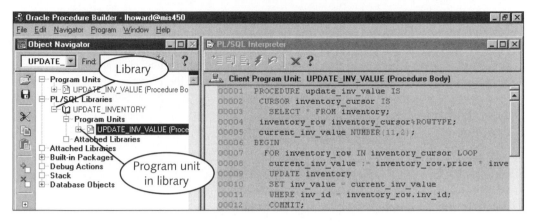

Figure 5-19   Adding a program unit to a library

5. Click the **UPDATE_INVENTORY** library to select it, and then click the **Save button** 🔲 to save the changes to the library.

Finally, you will learn how to unload and reload the library file. First, you will unload the current program unit and library.

To unload the program unit and library:

1. Select the **UPDATE_INV_VALUE** program unit under the top-level Program Units node (not under the UPDATE_INVENTORY library), click the **Delete button** 🔳 to unload the procedure, and then click **Yes** to confirm unloading the procedure.

2. Select the **UPDATE_INVENTORY** library, and then click 🔳 to unload the library from the Procedure Builder environment. There should currently be no program units or libraries loaded in Procedure Builder.

Now you will reload the UPDATE_INVENTORY library. Then, you will be able to edit the UPDATE_INV_VALUE procedure.

To reload the library:

1. If necessary, select the **PL/SQL Libraries** node in the Object Navigator window, and then click the **Open button** 🔲 on the Object Navigator toolbar. The Open Library dialog box opens.

2. Make sure the File System option button is selected, click **Browse**, navigate to your Chapter5\TutorialSolutions folder, select **update_inventory.pll**, click **Open**, and then click **OK**. The UPDATE_INVENTORY library appears in the Object Navigator window.

If you try to load a library .pll file before you have connected to the database, or if the library references database objects are not accessible to the current database connection account, an error message will appear asking if you want to load the library source code only. If this happens, click No to confirm not loading the source code. Then, connect to the database using a valid user account, and reload the library.

 If an error message appears even when you are connected to the database, you need to recompile the library program units and resave the library. Click Yes to confirm loading the source code, move all program units to the top-level Program Unit node, and recompile all program units in the Program Unit Editor to regenerate the compiled files. Delete the existing program unit files in the library, and then move the recompiled program units to the library. Finally, save the library.

3. Click ➕ beside the Program Units node under UPDATE_INVENTORY. The UPDATE_INV_VALUE program unit appears.

4. Click **UPDATE_INV_VALUE** to display the program unit source code in the PL/SQL Interpreter window.

5. Double-click the **Program Unit icon** 🖹 to display the source code in the Program Unit Editor. The procedure is available for editing and debugging in the Procedure Builder environment.

6. Click **Close** on the Program Unit Editor button bar to close the Program Unit Editor.

Executing a Procedure in Procedure Builder

Before you can execute a program unit in Procedure Builder, you must load the program unit into the Procedure Builder environment as a top-level Program Unit object. If you are using a program unit that is stored in a library, you must explicitly move the program unit from the library to the Procedure Builder top-level Program Units node. Now you will load the UPDATE_INV_VALUE procedure into Procedure Builder.

To load a library program unit into Procedure Builder:

1. In the Object Navigator, select **UPDATE_INV_VALUE** under the UPDATE_INVENTORY library, drag it to the top-level Program Units node, and then drop it by releasing the mouse button when the tip of the pointer arrow is on the Program Units label. UPDATE_INV_VALUE appears as a top-level program unit in the Procedure Builder environment.

When you create libraries and load and unload program units, text commands appear at the PL/SQL command prompt. These commands are the text equivalent of the operations you perform using menu selections and mouse operations. However, there is no menu selection that allows you to execute a procedure in Procedure Builder; to run the procedure, you must type the procedure name at the PL/SQL command prompt. To execute a procedure in Procedure Builder, you type the following command at the PL/SQL prompt: *procedure_name(parameter1_value, parameter2_value, ...);*. Now you will clear the command prompt window of all current commands, and run the procedure. Then, you will type a SQL command at the PL/SQL prompt to verify that the records were inserted correctly.

To clear the command prompt window, run the procedure, and verify the output:

1. Right-click anywhere in the command prompt pane, and then click **Clear** to clear the window of all current commands. The PL/SQL prompt appears.

2. Place the insertion point in the command prompt pane, type **update_inv_value;** at the PL/SQL prompt, as shown in Figure 5-20, and then press **Enter** to run the procedure. If the procedure runs successfully and no runtime errors appear, then the PL/SQL prompt reappears. If the procedure did not run successfully, runtime error messages appear. If your procedure did not run correctly, refer to the next section, which discusses how to use the PL/SQL Interpreter to detect runtime errors. If your procedure ran successfully, you should now view the INVENTORY records to make sure that they were updated correctly.

This procedure updates every record in the INVENTORY table, so it might take a few minutes to run, especially if you are connecting to a remote database link.

3. Type the SELECT command shown in Figure 5-20 to view the updated records. Use the scrollbar in the PL/SQL command prompt pane to view all of the retrieved records. The command should display a total of 26 records.

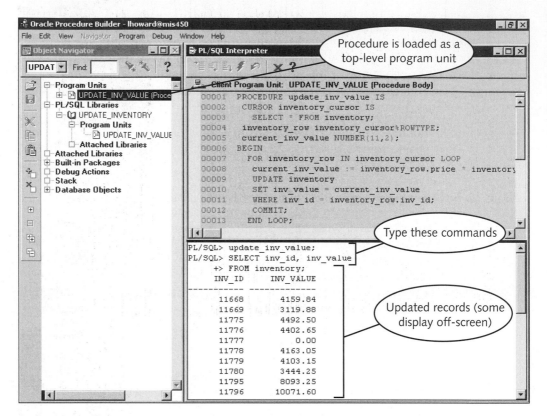

Figure 5-20   Running the procedure and viewing the updated records

Using the PL/SQL Interpreter to Find Runtime Errors

Now you will learn how to use the PL/SQL Interpreter debugging environment. When a program unit is loaded in the PL/SQL Interpreter, you can set a **breakpoint**, which is a place in a program where execution pauses in a debugging environment. You can then step through the procedure to observe the execution path and examine variable values. You can set breakpoints only on program lines that contain executable program statements and on SQL statements. You cannot set breakpoints on comment lines or variable declarations. Now you will set a breakpoint in the UPDATE_INV_VALUE procedure, and step through the program statements to observe variable values during execution.

As you use the PL/SQL Interpreter debugging environment, you might need to periodically maximize and adjust the widths of the Object Navigator and PL/SQL Interpreter windows.

To set a breakpoint and single-step through the program:

1. Place the pointer on the number 00007 in the PL/SQL Interpreter window, or in the program line directly below the BEGIN command, if your line numbers are different. Double-click to set a breakpoint on that line. The program line number changes to B (01) to indicate that Breakpoint 01 occurs on Line 7, as shown in Figure 5-21.

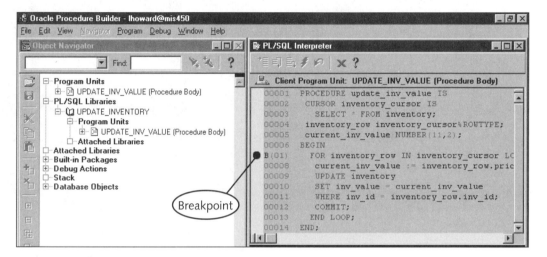

Figure 5-21 Setting a breakpoint on Line 7

You can also set a breakpoint by right-clicking on the program line, and clicking Break. Alternately, you can double-click the mouse pointer in the space on the left side of the line number to set a breakpoint.

2. Type **update_inv_value;** at the PL/SQL prompt in the command prompt window, and then press **Enter** to run the procedure. Program execution pauses at the breakpoint on Line 7. Figure 5-22 shows the position of the execution arrow on Line 7 of the procedure. The execution arrow shows the line of the procedure that will execute next. As you step through the program, the execution arrow stops on the PL/SQL statements and skips comment lines and SQL statements.

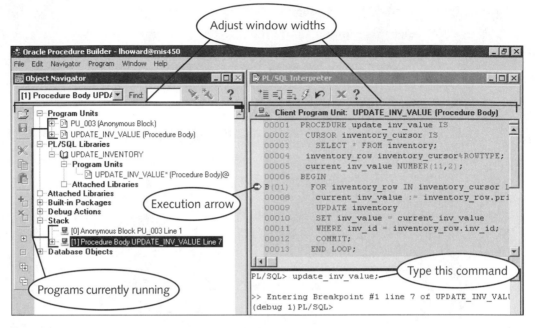

Figure 5-22   PL/SQL Interpreter window during program execution

Notice the objects that appear in the Object Navigator window. The Program Units node now displays objects representing every program and subprogram that is currently running: Anonymous Block PU_OO3 and Procedure Body UPDATE_INV_VALUE. (Your Anonymous Block number might be different.) The **Anonymous Block** is an unnamed program unit that is automatically created whenever a program unit runs in Procedure Builder. The **Stack** node shows the names of all programs and subprograms that are currently running, along with the line number that is currently being executed. Note that under the Stack node, the second object, Procedure Body UPDATE_INV_VALUE, is highlighted. This indicates that UPDATE_INV_VALUE is the procedure that is currently running, and that it is paused at Line 7. Also, note that this procedure has a ➕ beside its name, which indicates that it contains variable values that can be examined during program execution. You will examine variable values during program execution next.

To examine the current values of the procedure variables:

1. Click ➕ beside Procedure Body UPDATE_INV_VALUE under the Stack node. The procedure variables appear, as shown in Figure 5-23. The value for the variable CURRENT_INV_VALUE (NUMBER) has not been assigned yet, so no value is shown. Notice that there are two INVENTORY_ROW variables with data type RECORD. The first INVENTORY_ROW variable lists the items in the record. The second INVENTORY_ROW variable lists the items along with the most recently fetched values.

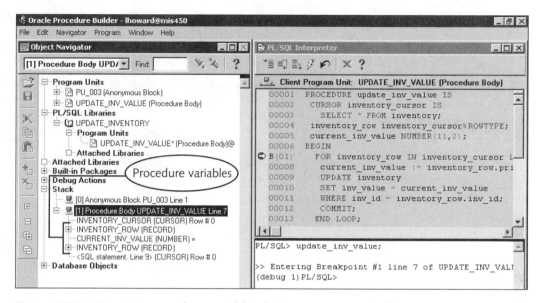

Figure 5-23 Viewing procedure variables during program execution

2. Click ➕ beside both INVENTORY_ROW variable items. The individual items that are fetched by the cursor appear under both items, as shown in Figure 5-24. Because no rows have been fetched yet, no values appear.

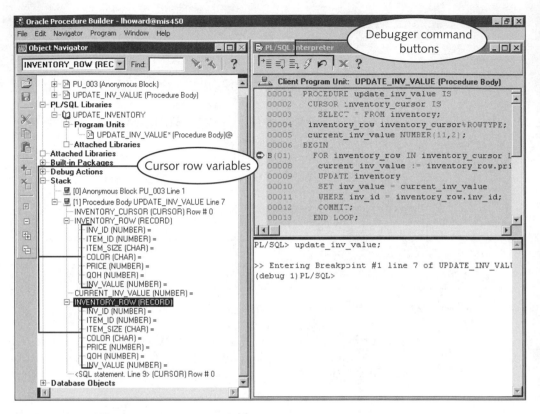

Figure 5-24 Viewing cursor row variables

Notice the debugger command buttons that are active on the PL/SQL Interpreter window toolbar. The **Step Into button** ⌐ allows you to step through the program one line at a time. The **Step Over button** ⌐ allows you to bypass a procedure that is called by the current program unit. The **Step Out button** ⌐ executes all program lines to the end of the current program unit. The **Go button** ⌐ allows you to pass over the current breakpoint and run the program until the next breakpoint or until the program terminates. The **Reset button** ⌐ terminates execution. Now you will single-step through the procedure and observe how the variable values change.

To step through the procedure:

1. Click the **Step Into button** ⌐. The execution arrow jumps to Line 3 to process the cursor and fetch the first record.

2. Click ⌐ again. The execution arrow jumps to Line 8 to process the command after the first cursor record has been fetched. Now you will examine the cursor row values.

3. Click ➕ beside Procedure Body UPDATE_INV_VALUE, and then click ➕ beside both INVENTORY_ROW (RECORD) nodes. The values fetched by the INVENTORY_ROW cursor appear in the second INVENTORY_ROW node, as shown in Figure 5-25. These data values correspond with the first data row in the INVENTORY table.

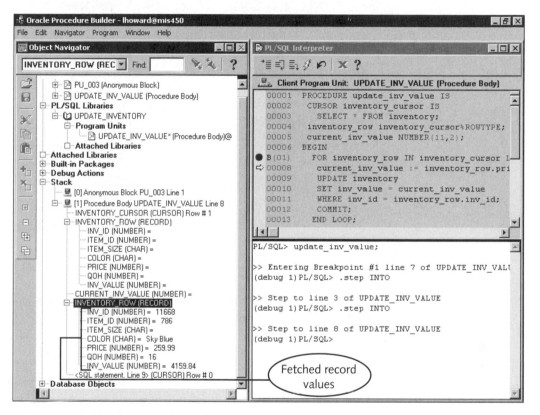

Figure 5-25   Viewing variable values

4. Click 📋 to execute Line 8 that calculates and assigns the value of the CURRENT_INV_VALUE variable. After the line is executed, the CURRENT_INV_VALUE (NUMBER) variable updates to 4159.84 in the Object Navigator window.

5. Click 📋 to execute the UPDATE command.

6. Click 📋 to execute the COMMIT command. The execution arrow moves back to Line 7 to fetch the next record.

7. Click the **Go button** 🔧 to finish running the program without stopping on the breakpoint again. The PL/SQL prompt should appear when execution is completed.

8. Double-click Line 7 again to remove the breakpoint.

Now you will unload the UPDATE_INVENTORY library and UPDATE_INV_VALUE program unit from the Procedure Builder environment.

To unload the library and program unit:

1. Select the **UPDATE_INVENTORY** library, click the **Delete button** [✗], and then click **Yes** to confirm saving the changes to the library. The library unloads.

2. Select the **UPDATE_INV_VALUE** program unit, click [✗], and then click **Yes** to confirm removing the program unit.

Running a program in a debugging environment is useful for locating program lines causing runtime errors or for identifying the cause of logic errors. Use the following strategy to debug programs with the PL/SQL Interpreter window debugger:

1. Single-step through the program to identify the program line that is generating the error.

2. Set a breakpoint on the line that is generating the error.

3. Run the program again, and examine the variable values that are used in the command that is generating the error to determine the cause of the error.

Now you will open a program unit source code file named UPDATE_INVENTORY_ERR that is stored in the Chapter5 folder on your Data Disk. This program unit currently generates a runtime error. You will determine which line is generating the error and then determine the cause of the error. First, you will open the file, run the procedure, and observe the runtime error message.

To open the program unit and run the procedure:

1. Click **File** on the menu bar, click **Load**, click **Browse**, navigate to the Chapter5 folder on your Data Disk, select **update_inventory_err.pls**, click **Open**, and then click **Load**.

2. Click [+] beside the top-level Program Units node. The program unit appears in the Object Navigator. Select the **UPDATE_INV_VALUE** program unit to load the source code into the PL/SQL Interpreter.

3. Right-click in the command prompt window, and then click **Clear** to clear the command prompt window.

4. Type **update_inv_value;** at the PL/SQL prompt, and then press **Enter** to run the procedure. A runtime error message appears, as shown in Figure 5-26.

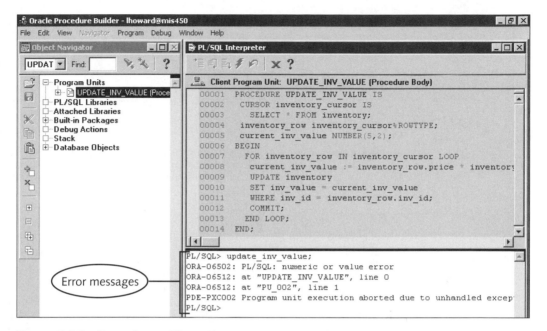

Figure 5-26 Procedure with runtime error

The error message states that the error is a numeric or value error, but the error location is given as Line 0. To find the specific program line causing the error, you will set a breakpoint at the beginning of the procedure, and step through the program commands until you find the line that is generating the error.

To set the breakpoint and find the location of the runtime error:

1. Place the mouse pointer on Line 7, and then double-click to set the breakpoint.

2. Type **update_inv_value;** at the PL/SQL prompt, and then press **Enter** to run the program. Execution should pause with the execution arrow on the breakpoint on Line 7.

3. Click the **Step Into button** ⬚. The execution arrow jumps to Line 3 to fetch the first record.

4. Click ⬚ again. The execution arrow jumps to Line 8, which is the program statement that calculates the inventory value.

5. Click ⬚ again. The error message shown in Figure 5-27 appears. This indicates that the runtime error is being generated by the commands on Line 8.

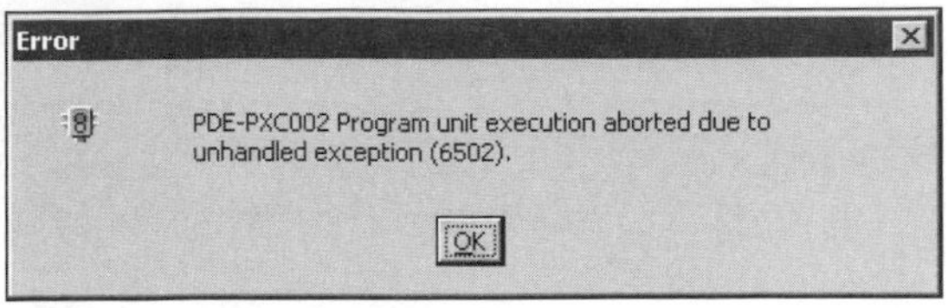

> **Error**   [×]
>
> 🛑   PDE-PXC002 Program unit execution aborted due to
>   unhandled exception (6502).
>
>   [OK]

Figure 5-27   Runtime error message

6. Click **OK** to acknowledge the error message and stop program execution.

The next step is to run the program again with a breakpoint on Line 8. Then, before you execute Line 8, you will examine the variable values that are being used in the command on this line to see if you can determine the cause of the error.

To examine the variable values used in the error-generating command:

1. Double-click Line 7(or in the program line directly below the BEGIN command, if your line numbering is different) to remove the current breakpoint, and then double-click Line 8 to set a new breakpoint on Line 8.

2. Type **update_inv_value;** at the PL/SQL prompt, and then press **Enter.** Execution should pause with the execution arrow on the breakpoint on Line 8.

3. Click ➕ beside Procedure Body UPDATE_INV_VALUE under the Stack node, and then click ➕ beside the second INVENTORY_ROW (RECORD) node to display the current record values. Your display should look like Figure 5-28.

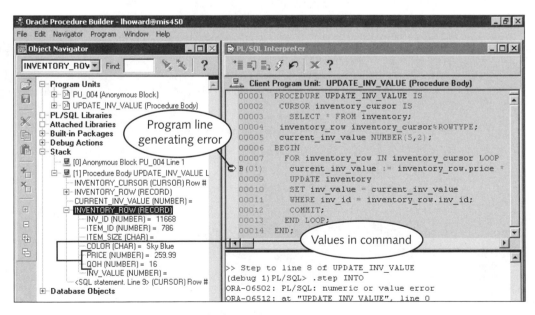

Figure 5-28   Variable values just before runtime error

The command that generates the error on Line 8 calculates the product of the cursor record's PRICE value and QOH value, and assigns the calculated value to the CURRENT_INV_VALUE variable. The Stack node indicates that the current value of PRICE is 259.99, and the current value of QOH is 16, so the product of these two values is 259.99 times 16, or 4159.84. Upon closer inspection, you will see that when the CURRENT_INV_VALUE variable was declared in the procedure, it was specified as a NUMBER with precision 5 and scale 2. Recall that the precision of a number specifies its total number of digits including decimal places, so the highest value that can currently be stored in the CURRENT_INV_VALUE variable is 999.99. CURRENT_INV_VALUE must have a precision of at least 6 for the value 4159.84 to be successfully assigned, so that is the cause of the runtime error.

Now you will correct the error by changing the CURRENT_INV_VALUE variable declaration to NUMBER(10,2). You use a larger value than is required for the first record to be sure that the field is wide enough for all values. You will then recompile and rerun the procedure to confirm that it runs correctly.

To correct the program unit:

1. Click the **Step Into button** . The runtime error message appears as expected, so click **OK**.

2. Remove the breakpoint on Line 8, and then double-click the **Program Unit icon** beside the UPDATE_INV_VALUE procedure to display the source code in the Program Unit Editor.

3. Change the fifth program line (the declaration for the current_inv_value variable) as follows:

```
current_inv_value NUMBER(10,2);
```

4. Compile the modified program unit, and then close the Program Unit Editor. The modified program code should appear in the PL/SQL Interpreter window.

5. Type **update_inv_value;** at the PL/SQL prompt to execute the program. The program should now execute with no errors.

Now you will save the corrected program unit. You will save the source code in a file named update_inventory_corr.pls in your Chapter5\TutorialSolutions folder.

To save the corrected procedure:

1. Double-click the **Program Unit icon** beside the UPDATE_INV_VALUE procedure to display the source code in the Program Unit Editor, and then maximize the Program Unit Editor window.

2. Click **File** on the menu bar, click **Export Text**, and then save the file as **update_inventory_corr.pls** in your Chapter5\TutorialSolutions folder.

   3. Close Procedure Builder, SQL*Plus, Notepad, and all other open Oracle applications.

You can use a similar debugging process for locating logic errors, which occur when a program runs successfully but does not generate correct output. You should determine which program line is generating the incorrect output, place a breakpoint on that line, and then examine the input variables that are used in the command.

SUMMARY

- Anonymous PL/SQL blocks are run by the PL/SQL interpreter but are not available to other users, cannot be called by other program units, and cannot receive input parameters. Named PL/SQL program units are database objects that can be referenced by other users and other program units, and can receive input parameters.

- Named program units can be server-side program units, which are stored in the database as database objects and execute on the database server, or client-side program units, which are stored in the file system of the client and execute on the client.

- Oracle named program units include procedures, functions, libraries, packages, and triggers.

- Procedures can manipulate multiple data values, and functions return a single data value.

- Procedures receive and deliver output values using parameters. By default, parameters are passed as IN parameters. Even if IN is omitted from the procedure declaration IN, parameters can be read but not modified. A parameter declared using the OUT mode can be written to, but cannot be read within the procedure. A parameter declared as both IN and OUT can be read and modified.

- Formal parameters are the parameters that are declared in the header of the procedure, and actual parameters are the values placed in the procedure parameter list when the procedure is called.

- Stored program units are PL/SQL program units that are stored in the database and executed on the database server. They can be created both in SQL*Plus and in Oracle Procedure Builder.

- Inline functions have the highest function purity level and can be used within SQL commands. These functions can use only IN parameters and must use only PL/SQL data types that correspond with Oracle database data types. They cannot perform any DML or DDL operations, or read or write to any packages.

- To allow another user to execute one of your stored program units, you must explicitly grant the EXECUTE privilege to the user for the unit.

- Oracle Procedure Builder provides a graphical interactive environment for creating, compiling, testing, and debugging named PL/SQL program units.

❐ In Procedure Builder, the Object Navigator provides access to different objects. The PL/SQL Interpreter enables you to single-step through programs to observe execution and examine variable values at runtime, and to directly issue SQL commands.

❐ The Procedure Builder Program Unit Editor is an environment for writing, compiling, and editing PL/SQL programs.

❐ A client-side procedure can be saved as an uncompiled source code text file with a .pls extension or as a library file with a .pll extension, which stores both the source code text and the compiled code.

❐ To debug programs using the PL/SQL Interpreter, you should set a breakpoint to pause execution on a program line before the error occurs. Then, you should single-step through the program and examine the variable values in the Object Navigator to determine the cause of the error.

5

REVIEW QUESTIONS

1. List two differences between anonymous PL/SQL blocks and named PL/SQL blocks.

2. When should you use a procedure, and when should you use a function?

3. What is a parameter?

4. What is the difference between formal and actual parameters, and what is the relationship between formal and actual parameters?

5. When should you create a server-side PL/SQL program unit, and when should you create a client-side program unit?

6. When do you need to consider function purity levels?

7. How can you run a procedure in Oracle Procedure Builder? How can you run a procedure that expects an input parameter?

8. What is the difference between a .pls and a .pll file? When would you have to save a client-side procedure as a .pls file rather than as a .pll file?

9. What is a breakpoint? On what types of program lines can you place breakpoints when you are debugging a program?

10. During program execution, what does the Object Navigator Stack node show?

PROBLEM-SOLVING CASES

Cases refer to the Clearwater Traders (see Figure 1-11), Northwoods University (see Figure 1-12), and Software Experts (see Figure 1-13) databases. Save all solution files in your Chapter5\CaseSolutions folder.

1. Use SQL*Plus to create a server-side procedure named UPDATE_ENROLLMENT that receives a specific Northwoods University student ID and course section ID as input parameters named CURRENT_S_ID and CURRENT_C_SEC_ID. This

procedure then inserts the S_ID and C_SEC_ID values into the ENROLLMENT table in the Northwoods University database. Specify that the grade value is NULL. Save the procedure in a file named 5ACase1.sql.

2. Use SQL*Plus to create a server-side inline function named ADD_DAYS that returns a date value representing a specific number of days after a given date. The function should receive input variables of a date and an integer representing the number of days to add to the date. Save the function code in a file named 5ACase2.sql. Below the code for the function, write a SQL query that uses the ADD_DAYS function to retrieve the date that shipment ID 213 in the SHIPMENT table in the Clearwater Traders database is expected, along with dates that are 30, 45, and 60 days after the expected date.

3. Use SQL*Plus to create a server-side procedure named DISPLAY_SCHEDULE that receives a Northwoods University student ID and a term ID as input variables, and then displays the student's name and current schedule, as shown in Figure 5-29. Save the procedure in a file named 5ACase3.sql.

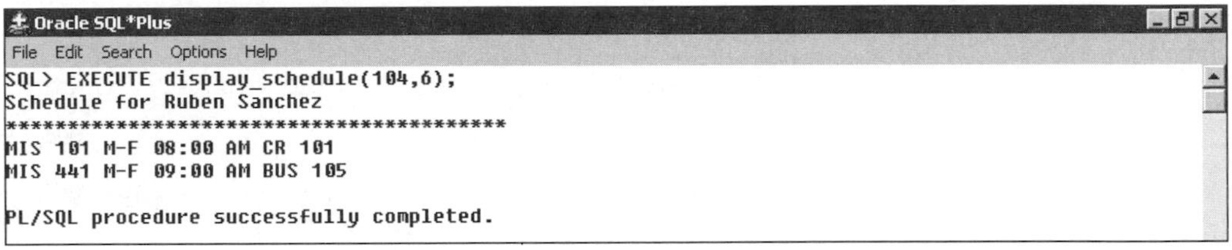

```
Oracle SQL*Plus                                                    _ 8 X
File  Edit  Search  Options  Help
SQL> EXECUTE display_schedule(104,6);
Schedule for Ruben Sanchez
*****************************************
MIS 101 M-F 08:00 AM CR 101
MIS 441 M-F 09:00 AM BUS 105

PL/SQL procedure successfully completed.
```

Figure 5-29

4. Use Procedure Builder to create a client-side procedure that inserts a new record in the SHIPMENT table in the Clearwater Traders database and then inserts a corresponding record in the SHIPMENT_LINE table.

a. In SQL*Plus, create a new sequence named SHIPMENT_ID_SEQUENCE that starts at 300. Save the command in a file named 5ACaseQueries.sql.

b. In Procedure Builder, create a new procedure named UPDATE_SHIPMENT, and specify that the procedure receives input parameters corresponding to the inventory ID and shipment quantity expected.

c. In the UPDATE_SHIPMENT procedure, add the command to insert a new record in the SHIPMENT table that retrieves the next value in the SHIPMENT_ID_SEQUENCE for SHIPMENT_ID and uses the current system date for the DATE_EXPECTED value.

d. In the UPDATE_SHIPMENT procedure, add the command to insert a new record in the SHIPMENT_LINE table that uses the current sequence value for SHIPMENT_ID and the inventory ID and shipment quantity values that were input as parameters. Specify that the DATE_RECEIVED value is NULL.

e. Test the procedure to confirm that it works correctly.

f. Save the source code in a file named 5ACase4.pls.

g. Save the compiled procedure in a library named 5ACASE4, and save the library file as 5ACase4.pll.

5. Use Procedure Builder to create a client side procedure that inserts new records in the PROJECT and PROJECT_CONSULTANT tables in the Software Experts database.

a. In SQL*Plus, create a new sequence named PROJECT_ID_SEQUENCE that starts with 10. Save the command in a file named 5ACaseQueries.sql.

b. In Procedure Builder, create a new procedure named CREATE_NEW_PROJECT that receives input parameters for the project name, client ID, manager ID, and expected duration (in months) that the manager will spend on the project. The procedure then inserts a record into the PROJECT table and a corresponding record into the PROJECT_CONSULTANT table for the project manager. The roll-on date should be the current system date, and the roll-off date should be calculated using the expected duration input parameter using the ADD_MONTHS function. The parent project ID will be initialized as NULL, and the total consultant hours will be initialized as 0.

c. Save the procedure source code in a file named 5ACase5.pls.

d. Save the procedure in a client-side library named 5ACASE5, and save the library file as 5ACase5.pll.

6. A client-side procedure named PROCESS_SHIPMENT has been created to process incoming shipments at Clearwater Traders. The procedure receives input parameters of the shipment ID, inventory ID, and shipment quantity. The procedure is supposed to update the corresponding SHIPMENT_LINE record by placing the current system date in the DATE_RECEIVED field, and then update the INVENTORY table by adding the shipment quantity to the inventory quantity on hand (QOH) field. The code for the procedure is currently stored in a file named 5ACase6ERR.pls in the Chapter5 folder on your Data Disk. The code currently contains compile, runtime, and logic errors.

Open 5ACase6ERR.pls in Procedure Builder, and debug it. Use the PL/SQL Interpreter debugging environment if necessary. When you find and correct errors, add comment statements to the code indicating what you changed. Verify that the program works correctly for updating shipment ID 212 for inventory ID 11669. After the procedure runs, the updated QOH value for INV_ID 11669 should be 37. Save the corrected code as 5ACase6CORR.pls.

7. A client-side procedure named UPDATE_GRADE has been created to enable faculty members at Northwoods University to update student grades in the course sections that they teach. The procedure receives input parameters of the faculty ID, faculty PIN, student ID, course section ID, and letter grade. The procedure first verifies the faculty PIN in the FACULTY table. If the PIN is correct, then the procedure verifies that the F_ID represents the instructor for the course section in the

COURSE_SECTION table. If the F_ID is verified, then the procedure updates the grade in the ENROLLMENT table. If either the PIN or F_ID is not correct, then the grade is not updated. The code for the procedure is currently stored in a file named 5ACase7ERR.pls in the Chapter5 folder on your Data Disk. The code currently contains compile and logic errors.

Open 5ACase7ERR.pls in Procedure Builder, and debug it. Use the PL/SQL Interpreter debugging environment if necessary. When you find and correct errors, add comment statements to the code indicating what you changed. Verify that the program works correctly. Save the corrected code as 5ACase7CORR.pls.

8. In this case, you will modify the ORDER_SOURCE table in the Clearwater Traders database so that it has a field to represent the total revenue for all orders that have been placed using each source. Then, you will create a client-side procedure that calculates the total order revenue for each source and then updates the ORDER_SOURCE table with the calculated value. After the procedure runs, the ORDER_SOURCE table should have the values shown in Figure 5-30.

   a. In SQL*Plus, add a column to the ORDER_SOURCE table named TOTAL_REVENUE, which will store NUMBER values with a precision of 10 and a scale of two. Save the command in a file named 5ACaseQueries.sql.

   b. In Procedure Builder, create a procedure named UPDATE_SOURCE_REVENUE that calculates the total revenue (order quantity times price) for each order source, and then updates the ORDER_SOURCE table using the calculated value. If an order source has no revenue, then insert a value of 0.

   c. Save the procedure source code in a file named 5ACase8.pls.

   d. Save the procedure in a library named 5ACASE8. Save the file as 5ACase8.pll.

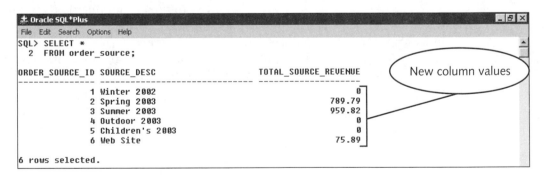

Figure 5-30

9. In this case, you will modify the PROJECT table in the Software Experts database so that it has a field to store the total consultant hours spent on each project. Then, you will create a client-side procedure that calculates the total number of hours that all consultants have spent on each project and then updates the field in

the PROJECT table with the calculated value. After the procedure runs, the PROJECT table should have the values shown in Figure 5-31.

a. In SQL*Plus, add a column to the PROJECT table named TOTAL_PROJECT_HOURS, which will hold NUMBER values to a maximum width of 6, with no decimal places. Save the command in a file named 5ACaseQueries.sql.

b. In Procedure Builder, create a procedure named UPDATE_PROJECT_HOURS that calculates the total number of consultant hours that have been spent on each project and then updates the TOTAL_PROJECT_HOURS field for the project in the PROJECT table. If a project has no consultant hours, insert a value of 0.

c. Save the procedure source code in a file named 5ACase9.pls.

d. Save the procedure in a library named 5ACASE9. Save the library in a file named 5ACase9.pll.

```
Oracle SQL*Plus
File  Edit  Search  Options  Help
SQL> SELECT *
  2  FROM project;

     P_ID PROJECT_NAME                    CLIENT_ID     MGR_ID PARENT_P_ID TOTAL_PROJECT_HOURS
--------- ------------------------------ --------- ---------- ----------- -------------------
        1 Hardware Support Intranet              2        105                             515
        2 Hardware Support Interface             2        103           1                  25
        3 Hardware Support Database              2        102           1                 175
        4 Teller Support System                  4        105                             180
        5 Internet Advertising                   6        105                              30
        6 Network Design                         6        104           5                  15
        7 Exploration Database                   5        102                             225

7 rows selected.
```

New column values

Figure 5-31

10. In this case, you will modify the STUDENT table in the Northwoods University database so that it has an additional field to store the student's grade point average. Then, you will create a library that contains a procedure and a function to insert the appropriate records into this field. After the procedure runs, the student GPA values should be stored as shown in Figure 5-32.

a. In SQL*Plus, add a column to the STUDENT named S_GPA, which will store floating-point numbers. Save the command in a file named 5ACaseQueries.sql.

b. In Procedure Builder, create a function named CALC_GPA that receives a specific S_ID value as an input parameter and returns the student's current grade point average. If the student has not earned a grade in any classes, the function should return 0. Save the function source code in a file named 5ACase10Function.pls. (To create a function in Procedure Builder, create a new program unit, and select the Function option button. Use the same function header format that you used to create functions in SQL*Plus.)

Recall that grade point average is calculated as follows:

$$\frac{\text{SUM(Course Credits * Course Grade Points)}}{\text{SUM(Course Credits)}}$$

Course grade points are awarded as follows:

Grade	Grade Points
A	4
B	3
C	2
D	1
F	0

c. In Procedure Builder, create a procedure named UPDATE_STUDENT_GPA that updates the S_GPA field for each student in the STUDENT table using the CALC_GPA function. Save the procedure source code in a file named 5ACase10Procedure.pls.

d. Create a library named 5ACASE10 that contains UPDATE_STUDENT_GPA and CALC_GPA. Save the library in a file named 5ACase10.pll.

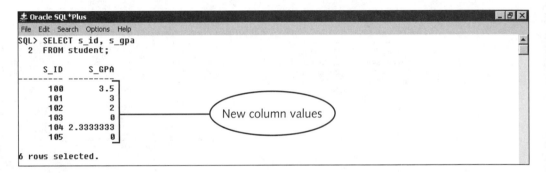

Figure 5-32

◀ LESSON B ▶

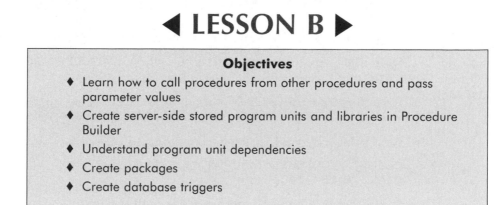

Objectives

♦ Learn how to call procedures from other procedures and pass parameter values

♦ Create server-side stored program units and libraries in Procedure Builder

♦ Understand program unit dependencies

♦ Create packages

♦ Create database triggers

CALLING PROCEDURES AND PASSING PARAMETERS

PL/SQL database applications can become very large and involve many lines of code. To make applications more manageable, you often decompose applications into logical units of work and then write individual program units for each logical unit. To integrate all of the individual program units, you must be able to call procedures from other procedures and pass parameter values to the called procedures. For example, suppose when a customer places an order at Clearwater Traders, a record must be inserted into the CUST_ORDER table that contains general information about the order, and associated records must be inserted into the ORDER_LINE table that contain information about each individual item ordered. It is logical to break this into two procedures: one to insert the order information, and another to insert the order line information.

Suppose that customer Paula Harris places an order on the Clearwater Traders Web site for one Women's Fleece Pullover, color eggplant, size small. You will write a procedure named CREATE_NEW_ORDER that receives the customer ID, order source ID, method of payment, inventory ID, and order quantity, and then inserts the order information into the CUST_ORDER table. Then, the procedure will call a second procedure named CREATE_NEW_ORDER_LINE and passes to it the values for the order ID, inventory ID, and quantity ordered. The second procedure will then insert a new record into the ORDER_LINE table. You must always create the called procedure before you create the calling procedure. If you try to compile a procedure that calls a procedure that does not yet exist, a compile error will be generated. Therefore, you will create the procedure that inserts the new order line first.

To begin, you will run the scripts to refresh your database tables. Then, you will create the sequence to generate the ORDER_ID surrogate key, and then create the calling procedure. The sequence will start with the value 1100, which is higher than the current highest ORDER_ID value (1062). Finally, you will create the procedure to insert the new order line. The new order line procedure receives the inventory ID and order quantity as input parameters. It will use the next value of the ORDER_ID_SEQUENCE for the ORDER_ID value.

To refresh your database tables, create the sequence, and create the called procedure:

1. Start SQL*Plus, log onto the database, and type **SET SERVEROUTPUT ON SIZE 4000** to initialize the SQL*Plus output buffer.

2. Refresh your database tables by running the northwoods.sql, clearwater.sql, and softwareexp.sql scripts from the Chapter5 folder on your Data Disk.

3. In SQL*Plus, type the following command to create the sequence:

   ```
   CREATE SEQUENCE order_id_sequence
   START WITH 1100;
   ```

4. Start Procedure Builder, and connect to the database.

5. Click **File** on the menu bar, point to **New**, and then click **Program Unit**. The New Program Unit dialog box opens.

6. Type **create_new_order_line** for the program unit name, make sure that the Procedure option button is selected, and then click **OK**. The procedure template appears in the PL/SQL Editor.

7. Type the code shown in Figure 5-33. Compile the code, debug it if necessary, and then close the Program Unit Editor window.

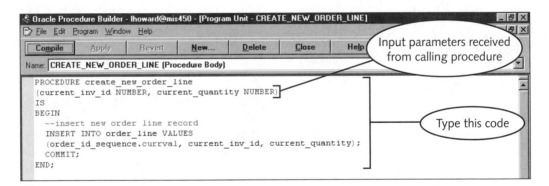

Figure 5-33   Procedure to create a new order line

Next, you will create the CREATE_NEW_ORDER procedure. This procedure will receive the customer ID, payment method, order source ID, inventory ID, and order quantity as input parameters. It will insert a new CUST_ORDER record based on these parameters, using the next value in the ORDER_ID_SEQUENCE for the order ID and the system date for the ORDER_DATE. It will then call the procedure to insert the ORDER_LINE record.

To write the CREATE_NEW_ORDER procedure:

1. In Procedure Builder, select the top-level Program Units node in the Object Navigator, and then click the **Create button** to create a new program unit. Create a new procedure named **CREATE_NEW_ORDER**.
2. Type the code shown in Figure 5-34. Compile the procedure, debug it if necessary, and then close the Program Unit Editor.

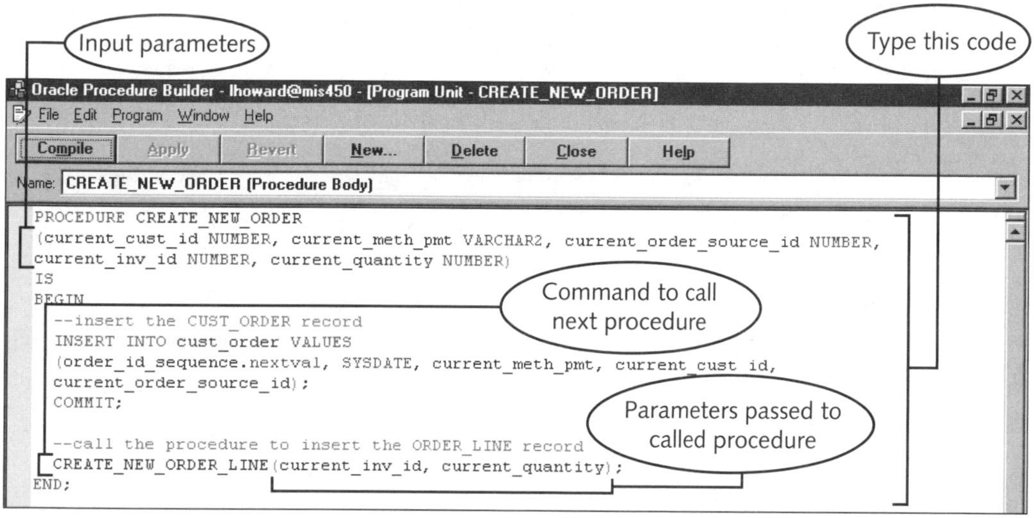

Figure 5-34 Procedure to create new order and call procedure to insert new order line

Now you will run the procedures. You will call the CREATE_NEW_ORDER procedure at the PL/SQL command prompt and pass to it the required parameter values to insert the new order and order line. Then, you will examine the values that are inserted into the CUST_ORDER and ORDER_LINE tables.

To run the procedures:

1. Maximize the PL/SQL Interpreter window, and resize the windowpanes so your screen looks like Figure 5-35.
2. Type the commands shown in Figure 5-35 at the PL/SQL prompt to call the CREATE_NEW_ORDER procedure and pass the required input parameter values, and then press **Enter**. The PL/SQL prompt appears again if the program successfully executes. If runtime error messages appear, debug the procedures in the Program Unit Editor window, and then recompile them.
3. Type the SQL commands shown in Figure 5-35 to confirm that the new records were inserted as shown.

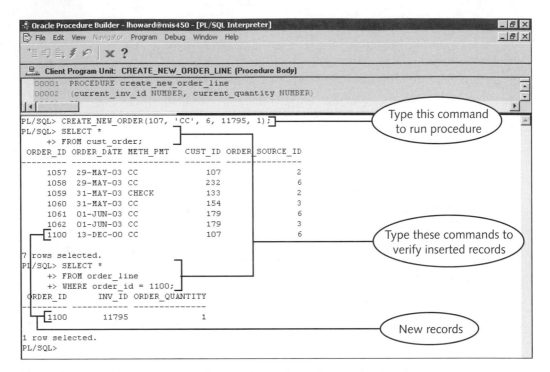

Figure 5-35 Calling the procedure and examining inserted values

 Your ORDER_ID values might be different than 1100, as shown in Figure 5-35. If this happens, type the second query using the ORDER_ID that appears in the first query.

4. Click the **Windows restore button** 🗗 on the PL/SQL Interpreter window to restore the Procedure Builder environment to the tiled window configuration.

Now, you will create a library named NEW_ORDERS to store the procedures you just created.

To create the NEW_ORDERS library:

1. In the Object Navigator, select the **PL/SQL Libraries** node, and then click the **Create button** 🗄. A new library appears.

2. Click **CREATE_NEW_ORDER** under the top-level Program Units node, press **Shift**, and then click **CREATE_NEW_ORDER_LINE**. Both program units are selected. Drag them to the Program Units node under the new library.

3. Click the **Save button** 🔲 on the toolbar, type **new_orders** for the library name, make sure the File System option button is selected, click **Browse**, and save the library as **new_orders.pll** in your Chapter5\TutorialSolutions folder.

CREATING STORED PROGRAM UNITS AND SERVER-SIDE LIBRARIES USING PROCEDURE BUILDER

Recall that stored program units are PL/SQL programs that are stored in the database and run on the database server. You created stored program units earlier when you created procedures and functions in SQL*Plus using the CREATE PROCEDURE and CREATE FUNCTION commands. All of the program units and libraries you have created so far using Procedure Builder have been client-side programs units that are stored in the client file system and run on the client workstation.

You can create a server-side stored program unit in Procedure Builder by creating and compiling the program unit in the Procedure Builder environment and then storing it in the database. To store the procedure in the database, you drag it to the Database Objects node in the Object Navigator window. You can use this node to access information about your database objects (tables, views, and user-defined data types) in the Procedure Builder Object Navigator pane, as well as to store and access server-side program units. First, you will view your database objects and stored program units.

To view your database objects and stored program units:

1. In the Object Navigator window, click ✚ beside the Stored Program Units node. A list of all database users appears. Scroll down in the window until you see your database username.

2. Click ✚ beside your username. A list of the available database object types (Stored Program Units, PL/SQL Libraries, Tables, Views, Types) appears.

3. Click ✚ beside Stored Program Units. The stored program units you created earlier in SQL*Plus (the AGE and AGE_IN_YEARS functions, and the UPDATE_INVENTORY procedure) should appear.

Your stored program units will be different if you did not create these stored program units in SQL*Plus or if you have created other stored program units.

4. Click ✚ beside the PL/SQL Libraries node. No libraries should be listed, because all of the libraries you have created so far are client-side libraries and are stored in the file system of your client workstation rather than in the database.

You will not be able to create a server-side library unless your database administrator has run the Developer 6/6i initialization scripts. If you are using the Personal Oracle database, and you have a problem creating a server side library, run the configuration script by clicking Start on the taskbar, pointing to Programs, pointing to Oracle Forms 6i Admin, and then clicking Build. The SQL*Plus window opens. Type "manager" for the SYSTEM password, press Enter, press Enter again to accept the LOCAL connection, and then wait a few moments while the script runs. SQL*Plus will automatically exit when the script completes.

5. Click ➕ beside the Tables node. A list of all of your database tables appears.

6. Click ➕ beside the Views node. A list of all of your database views appears.

Now, you will save the CREATE_NEW_ORDER and CREATE_NEW_ORDER_LINE procedures (which are currently stored in a .pll file on your client workstation) to server-side stored program units stored in the database. Then, they will be accessible to other database users and execute on the database server.

To save the procedures as server-side stored program units:

1. Click **CREATE_NEW_ORDER** under the top-level Program Units category, press **Shift**, and then click **CREATE_NEW_ORDER_LINE**. Both program units are selected.

2. Drag the program units to the Stored Program Units subcategory under your username. After a moment, the procedures appear as stored program units, as shown in Figure 5-36.

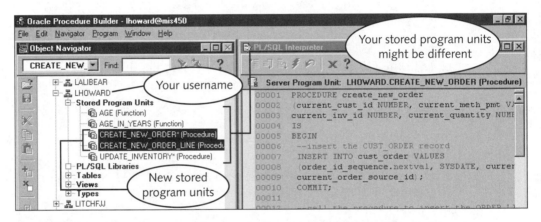

Figure 5-36   Creating server-side stored program units in Procedure Builder

 When you are dragging the objects down to your username, the Object Navigator window will automatically scroll down if you place the pointer with the selected objects just above the horizontal scrollbar.

You can also create **server-side libraries**, which are PL/SQL libraries (collections of program units) stored on the server. Server-side libraries are different from server-side stored program units because the programs in server-side libraries always run on the client workstation, while server-side stored program units always run on the database server. In client/server systems, this is a desirable trait because by using both libraries and stored program units, you can distribute processing across several machines without overloading the server. Some programs, however, have to be processed on the server, because the client workstation might not have the required applications or hardware to run the program. Now you will save the NEW_ORDERS PL/SQL library as a server-side library.

To save the NEW_ORDERS library as a server-side library:

1. In the Object Navigator window, select the **NEW_ORDERS** library under the top-level PL/SQL Libraries node.

2. Click **File** on the menu bar, and then click **Save As**. The Save Library dialog box appears.

 You cannot save server-side libraries by dragging and dropping the library under the Database Objects node.

3. Confirm that new_orders appears as the library name. If it does not, click **Browse**, navigate to the Chapter5 folder on your Data Disk, select **new_orders.pll**, and then click **Open**.

4. Select the **Database option button**, and then click **OK**. After a few moments, the library is saved as a server-side library.

You will not be able to create a server-side library unless your database account has been granted the CREATE ANY LIBRARY system privilege.

To verify that the library is saved correctly, you will need to disconnect from the database and then reconnect. This is because Procedure Builder only displays a snapshot of database objects that existed when you connected to the database. To refresh your database objects, you must make a new connection. Now you will disconnect from the database, reconnect, and then view your server-side library.

To view your server-side library:

1. Click **File** on the menu bar, and then click **Disconnect**. Your database connection terminates.

2. Click **File** on the menu bar, click **Connect**, and then connect in the usual way.

3. Click ➕ beside the Database Objects node, click ➕ beside your username, and then click ➕ beside the PL/SQL Libraries node under your username. The library should appear as a database object, as shown in Figure 5-37.

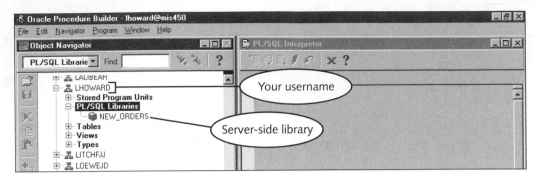

Figure 5-37   Viewing a server-side library

PROGRAM UNIT DEPENDENCIES

When a procedure or function is compiled, the database objects that it references (such as tables, views, or sequences) are verified to make sure that they exist and that the user has sufficient object privileges to use the objects as specified in the program code. If the objects do not exist or the user has insufficient privileges, a compile error occurs. The program unit has **object dependency** on these referenced database objects. When a program unit calls another program unit, the calling program has **procedure dependency** on the called program. The calling program also indirectly has object dependency on the database objects referenced in the called program unit. Figure 5-38 shows the dependencies of the CREATE_NEW_ORDER procedure that you just created.

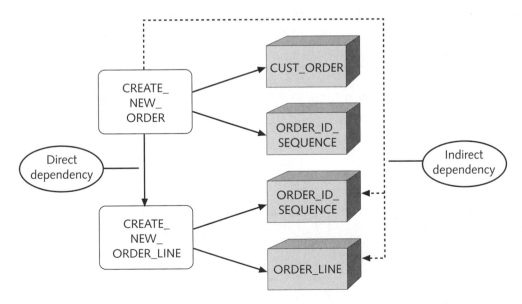

Figure 5-38   Program unit dependencies

A program unit is **directly dependent** on an item if it references it directly within its procedure code. A program unit is **indirectly dependent** on an item if the item is referenced by an item that the procedure references. Note that the CREATE_NEW_ORDER program unit is directly dependent on the ORDER_ID_SEQUENCE and the CUST_ORDER table. It is also directly dependent on the CREATE_NEW_ORDER _LINE procedure. The CREATE_NEW_ORDER_LINE program unit is directly dependent on the ORDER_LINE table and the ORDER_ID_SEQUENCE. The CREATE_NEW_ORDER program unit is indirectly dependent on the ORDER_LINE table and the ORDER_ID_SEQUENCE. If a referenced database object is deleted from the database or changed in such a way that the program unit can no longer use it, the program unit is **invalidated**, which means that the program unit must be modified and recompiled before it will execute correctly.

As you develop PL/SQL applications and work with existing applications, you will find that applications tend to be very large in terms of the number of lines of source code they contain. To make large applications more manageable, it is desirable to break them into modules that represent logical work divisions. An example is the application you created that has one procedure to insert the customer order record and another to insert each order line associated with the customer order. The challenge with this approach is tracking and managing the program unit dependencies.

Data dictionary views named DBA_DEPENDENCIES, USER_DEPENDENCIES, and ALL_DEPENDENCIES exist to help manage program unit dependencies. The **DBA_DEPENDENCIES** view shows dependency information for program units for all users; the **USER_DEPENDENCIES** view shows program unit dependency information for only the current user; and, the **ALL_DEPENDENCIES** view shows program unit dependency information for the current user, as well as dependency information for program units that the current user has privileges to execute. Table 5-4 describes the columns in the DBA_DEPENDENCIES view.

Column Name	Description
OWNER	Username of the program unit owner
NAME	Program unit name in all uppercase characters
TYPE	Program unit type (procedure, function, library, package, or package body)
REFERENCED_OWNER	Username of the owner of the referenced object
REFERENCED_NAME	Name of the referenced object
REFERENCED_TYPE	Object type (table, sequence, view, procedure, package)
REFERENCED_LINK_NAME	The name of the database link (connect string) used to access the referenced object. This column has a value only when the referenced object is on a different Oracle database.

Table 5-4   DBA_DEPENDENCIES view columns

The ALL_DEPENDENCIES view has the same columns as the DBA_DEPENDENCIES view. The USER_DEPENDENCIES view has the same columns with the exception of the OWNER column, because all of the objects in the USER_DEPENDENCIES view belong to the current user.

You can use these views only to retrieve information about program units stored in the database. Now you will use the USER_DEPENDENCIES view to retrieve the dependency information about the CREATE_NEW_ORDER stored procedure. To retrieve this information, you will write a script named dependencies.sql that creates a formatted report based on the view query. You will create a formatted report because the output wraps to multiple lines, and it is hard to read the output of the view query if the columns are not formatted.

To create a formatted report to query the USER_DEPENDENCIES view:

1. Start Notepad, and create a new file named **dependencies.sql**. Save the file in your Chapter5\TutorialSolutions folder.

2. Type the commands shown in Figure 5-39 to define the report columns and query the data dictionary view, and then save the file.

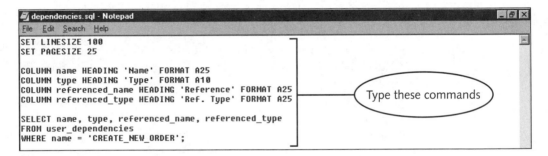

Figure 5-39   Script to format output for USER_DEPENDENCIES query

3. Switch to SQL*Plus, type the command shown in Figure 5-40 at the SQL prompt, and then press **Enter** to run the script. The output should appear as shown in Figure 5-40.

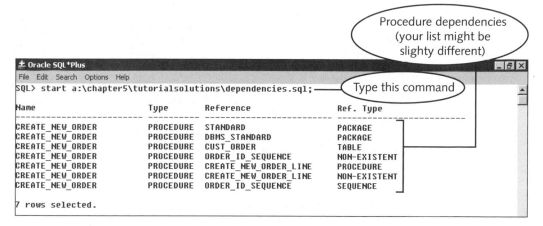

Figure 5-40   Output showing procedure dependencies

The output shows that the CREATE_NEW_ORDER procedure has multiple dependencies. It is dependent on the STANDARD and DBMS_STANDARD packages, which are built-in packages used to define and process cursors. It is dependent on the CUST_ORDER table and the ORDER_ID_SEQUENCE, as well as the CREATE_NEW_ORDER_LINE procedure. Note that these are the same dependencies as shown in Figure 5-38. Also note that the USER_DEPENDENCIES view only shows direct dependencies. To determine indirect dependencies, you would need to write a query to determine the dependencies of the CREATE_NEW_ORDER_LINE procedure.

PACKAGES

Sometimes you need to make multiple PL/SQL procedures and functions available to other database users and other programs. An easy way to do this is to combine the procedures and functions into a **package**, which is a code library that provides a way to manage large numbers of related programs and variables. With a package, you only have to assign object privileges for other users for the entire package, rather than having to grant privileges for each individual program unit. A package has two components: the package specification and the package body.

The Package Specification

All of the variables that you have used so far are **private variables**. Private variables are visible only in the program in which they are declared. As soon as the program terminates, the memory used to store the private variables is made available for other programs to use, and the variable values cannot be accessed again. You can use a package to declare **public variables**, which are visible to many different PL/SQL programs. Public variable values remain in memory even after the programs that declare and reference them terminate. You declare a public variable using the same syntax as you use for declaring a private variable. However, while private variables are declared in the DECLARE section of an individual program, public variables must be declared in the DECLARE section of a package. You use public variables in programs in the same way that you use private variables.

The **package specification**, also called the package header, can be used to declare public variables, cursors, procedures, and functions that are in the package and are to be made public. This means that they can be referenced by program units outside of the package. The general syntax for a package specification is:

```
PACKAGE package_name
IS
    public variable declarations;
    public cursor declarations;
    public procedure and function declarations;
END;
```

You can declare the elements in the package specification (variables, cursors, procedures, and functions) in any order. A package does not have to contain all four types of elements; for example, a package can just consist of variable declarations, or it can just consist of procedure or function declarations.

You declare public variables in a package using the same syntax as variable declarations in any PL/SQL block. You declare public cursors in a package using the following syntax:

```
CURSOR cursor_name
IS cursor_SELECT_statement;
```

To declare a procedure in a package, you must specify the procedure name, followed by the parameters and variable types, using the following syntax:

```
PROCEDURE procedure_name (parameter1 parameter1_data_type,
parameter2 parameter2_data_type, ...);
```

To declare a function in a package, you must specify the function name, parameters, and return variable type, as follows:

```
FUNCTION function_name (parameter1 parameter1_data_type,
parameter2 parameter2_data_type, ...)
RETURN return_datatype;
```

The declarations for procedures and functions in a package are called **forward declarations**, because the program units are only declared in the package specification. The actual program unit code is specified in another program unit.

You can create a package specification using either SQL*Plus or Procedure Builder. When you create a package specification using SQL*Plus, the package specification is automatically stored as a database object. When you create a package specification in

Procedure Builder, it can be tested on the client workstation and then stored either in the database or on the client workstation in a .pll file. Now you will create a package specification for a package named ORDER_PACKAGE that can be used to insert new orders and order lines in the Clearwater Traders database.

Earlier in this lesson, you created a procedure named CREATE_NEW_ORDER that inserts a new order record in the CUST_ORDER table and then calls the CREATE_NEW_ORDER_LINE procedure, which inserts a new record in the ORDER_LINE table. In the package, the values for the inventory ID and order quantity will be declared as public variables, and will be initialized by the program that uses the package. This way, they will not have to be passed as parameters. (Recall that public variables can be referenced by any procedure.) This program will call the CREATE_NEW_ORDER procedure and pass to it the values for customer ID, payment method, and order source ID. The calling program will then call the CREATE_NEW_ORDER_LINE procedure. Now you will create the ORDER_PACKAGE specification using SQL*Plus.

To create the package specification using SQL*Plus:

1. Switch to Notepad, and create a new file named **5BPrograms.sql**. Type the code shown in Figure 5-41 to create the package specification.

2. Copy and execute the code in SQL*Plus, and debug the code if necessary. The message "Package created" confirms that the package specification was successfully created.

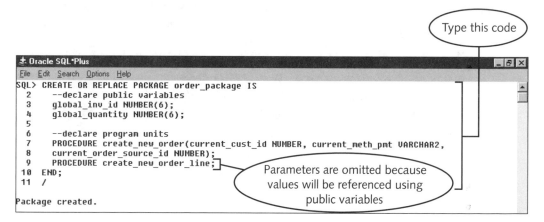

Figure 5-41   Package specification

The Package Body

The **package body** contains the code for the program units declared in the package specification. You must always create the package specification before the package body, or an error will occur. The package body is optional, because sometimes a package contains only variable declarations and no program units. The general syntax for the package body is:

```
PACKAGE BODY package_name
IS
     private variable declarations
     private cursor specifications
     program unit blocks
END;
```

The *package_name* in the package body must be the same as the package name in the package specification. Variables declared at the beginning of the body are private to the package, which means that they are visible to all modules in the package body, but are not visible to modules outside of the package body. Variables declared within the individual module blocks are only visible to the individual module blocks in which they are declared. Each program unit block has its own variable declarations section and BEGIN and END statements. Each program unit declared in the package body must have a matching program unit forward declaration in the package specification, with an identical parameter list. Otherwise, an error will occur.

Like the package specification, you can create the package body in either SQL*Plus or Procedure Builder. Now you will create the package body in SQL*Plus.

To create the package body in SQL*Plus:

1. Switch to Notepad, and type the code shown in Figure 5-42 to create the package body.

2. Copy and execute the code in SQL*Plus, and debug the code if necessary. The message "Package body created." confirms that the package body was successfully created.

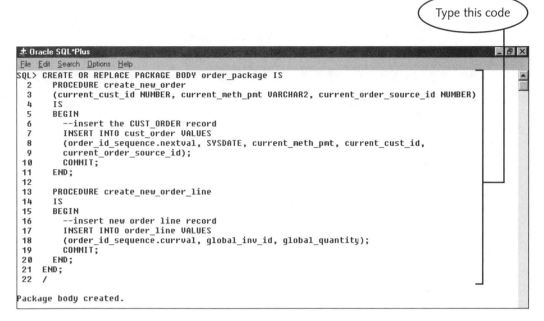

Type this code

```
± Oracle SQL*Plus                                                                    _ 6 X
File  Edit  Search  Options  Help
SQL> CREATE OR REPLACE PACKAGE BODY order_package IS
  2     PROCEDURE create_new_order
  3     (current_cust_id NUMBER, current_meth_pmt VARCHAR2, current_order_source_id NUMBER)
  4     IS
  5     BEGIN
  6       --insert the CUST_ORDER record
  7       INSERT INTO cust_order VALUES
  8       (order_id_sequence.nextval, SYSDATE, current_meth_pmt, current_cust_id,
  9       current_order_source_id);
 10       COMMIT;
 11     END;
 12
 13     PROCEDURE create_new_order_line
 14     IS
 15     BEGIN
 16       --insert new order line record
 17       INSERT INTO order_line VALUES
 18       (order_id_sequence.currval, global_inv_id, global_quantity);
 19       COMMIT;
 20     END;
 21  END;
 22  /

Package body created.
```

Figure 5-42   Package body

Referencing Package Items

The first time an item in a package is called, the package is loaded into memory, and space is allocated to store its variable values. From then on, the package remains in memory and is available to all users who have the privilege to execute the package. To grant other users the privilege to execute a package, you use the general command **GRANT EXECUTE ON *package_name* TO *username*;**.

To reference an item that is in a package, you must preface the item with the package name. For example, to assign the value 11668 to the GLOBAL_INV_ID variable in the ORDER_PACKAGE, you would use the command: **ORDER_PACKAGE.GLOBAL _INV_ID := 11668;**. Similarly, to call a procedure that is within a package, you preface the procedure call with the package name. For example, you have already used the DBMS_OUTPUT built-in package to display output in SQL*Plus: the package name is DBMS_OUTPUT, and the procedure name is PUT_LINE, so you call the procedure using the command **DBMS_OUTPUT.PUT_LINE('*message text*');**.

Now you will write an anonymous PL/SQL program to test ORDER_PACKAGE. The program will initialize the values of GLOBAL_INV_ID to 11668 and GLOBAL_QUANTITY to 2. Then, the program will call the CREATE_NEW_ORDER and CREATE_NEW_ORDER_LINE procedures from within the package. After you run the program, you will query the database to confirm that the package worked correctly.

To create the program to reference the package items:

1. In Notepad, type the code shown in Figure 5-43 to initialize the package variables and call the package procedures.

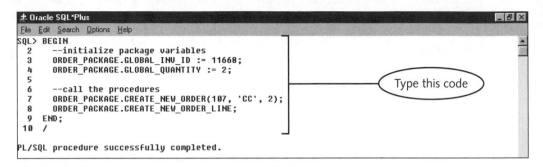

Figure 5-43   Program to reference package items

2. Execute the program in SQL*Plus, and debug it if necessary.

3. Type the commands shown in Figure 5-44 to confirm that the new order and order line records were inserted correctly.

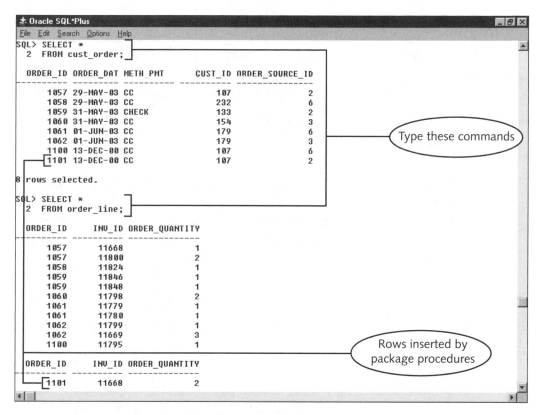

Figure 5-44   Records inserted by the package procedures

Overloading Program Units in Packages

In a package, program units can be **overloaded**, which means that multiple program units with the same name but different input parameters exist. Overloading allows a user to issue the same command to perform similar tasks but pass different parameter values depending on the circumstances. For example, it might be useful to be able to record a student's enrollment in a course section by specifying either the student's ID or by specifying the student's first and last names. Now you will create a package named ENROLLMENT_PACKAGE that has two procedures named ADD_ENROLLMENT. One of the procedures will accept input parameters of the student ID and course section ID, and the other procedure will accept parameters of the student's first name, last name, and the course section ID. You will create the package using Procedure Builder. First, you will create the package specification to declare the procedure names and input parameters.

To create the package specification:

1. In Procedure Builder, select the top-level **Program Units** node in the Object Navigator window, and then click the **Create button** to create a new program unit. The Program Unit dialog box opens.

2. Type **enrollment_package** for the package specification name, select the **Package Spec option button**, and then click **OK**. The Package Specification template appears in the Program Unit Editor.

3. Type the package specification shown in Figure 5-45. Compile the package specification, and debug it if necessary. Do not close the PL/SQL Editor.

4. Click **File** on the menu bar, click **Export Text**, and save the package specification source code as **enrollment_package_spec.pls** in your Chapter5\TutorialSolutions folder. Then close the Program Unit Editor.

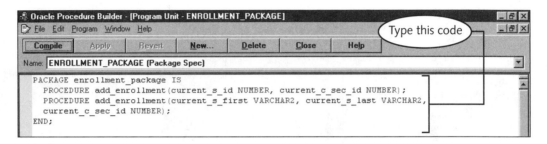

Figure 5-45   Package specification with overloaded program units

Next, you will create the package body. The first ADD_ENROLLMENT procedure will directly insert a record into the ENROLLMENT table using the student ID and course section ID values that it receives as input parameters. The second ADD_ENROLLMENT procedure will retrieve the student ID based on the student's first and last names, and then insert the ENROLLMENT record.

To create the package body:

1. In Procedure Builder, select the top-level **Program Units** node in the Object Navigator window, and then click the **Create button** to create a new program unit. The Program Unit dialog box opens.

2. Type **enrollment_package** for the package body name, select the **Package Body option button**, and then click **OK**. The package body template appears in the Program Unit Editor.

3. Type the code shown in Figure 5-46 to create the package body procedures. Compile the code, and debug it if necessary. Do not close the PL/SQL Editor.

4. Click **File** on the menu bar, click **Export Text**, and save the package body source code as **enrollment_package_body.pls** in your Chapter5\TutorialSolutions folder, and then close the Program Unit Editor.

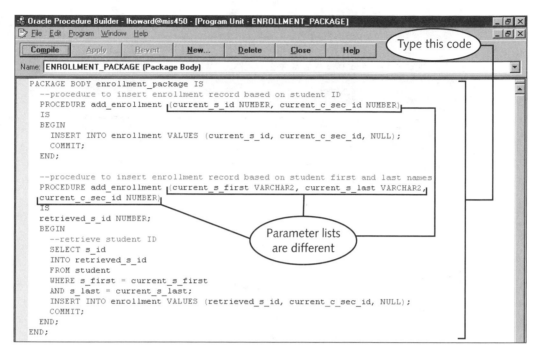

Figure 5-46   Package body with overloaded procedures

Help

Remember that the procedure heading in the package body has to exactly match the procedure declaration in the package specification. If you have trouble compiling your package body, make sure that your procedure headings exactly match the procedure declarations in the package specification in Figure 5-45.

Now you will test the overloaded procedures in the ENROLLMENT_PACKAGE. You will use the first procedure in the package to insert a record to enroll student _ID 100 in course section 1012. Then, you will use the second procedure in the package to enroll student Daniel Black in course section 1001. Finally, you will view the contents of the ENROLLMENT table to verify that the records were successfully inserted.

To test the overloaded procedures:

1. Maximize the PL/SQL Interpreter window, and resize the panes as shown in Figure 5–47.

2. In the command prompt pane, type the first command to insert the record using the procedure that accepts the student ID and course section ID as input parameters.

3. Type the second command to insert the record using the procedure that accepts the student's first and last name, and course section ID as input parameters.

4. Type the third command to query the ENROLLMENT table and verify that the records were successfully inserted.

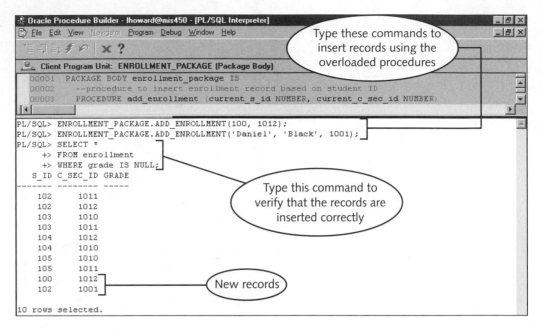

Figure 5-47   Testing the overloaded procedures

5. Restore the Procedure Builder environment to the tiled window configuration.

Normally, program units are overloaded by specifying that they have a different number of parameters. Sometimes, however, it is necessary to overload two program units that have the same number of parameters. In this case, the following restrictions apply:

- You can overload two procedures that use the same number of parameters if the parameters differ in data type. The following overloaded procedure declarations are legal:

```
PROCEDURE legal_overload (current_value1 NUMBER);
PROCEDURE legal_overload (current_value1 VARCHAR2);
```

- You cannot overload two procedures if they have the same number of parameters and the parameters have the same data type and differ only in name. The following overloaded procedure declarations are invalid:

```
PROCEDURE illegal_overload (current_value1 NUMBER);
PROCEDURE illegal_overload (current_value2 NUMBER);
```

- You cannot overload two procedures that use the same number of parameters if the parameter data types come from the same data type **family**, such as numbers or characters. The following overloaded procedure declarations are invalid:

```
PROCEDURE illegal_overload (current_value1 CHAR);
PROCEDURE illegal_overload (current_value1 VARCHAR2);
```

- You cannot overload two functions if their only differences are their return types. The following overloaded function declarations are invalid:

```
FUNCTION illegal_overload RETURN NUMBER;
FUNCTION illegal_overload RETURN DATE;
```

Saving Packages as Database Objects in Procedure Builder

When you created the package specification and package body in SQL*Plus, the package was saved as a database object on the database server. Currently, ENROLLMENT_PACKAGE, which you created in Procedure Builder is saved only as source code on your client workstation. Recall that packages are usually saved as database objects so they can be accessible to other users. Now you will save the ENROLLMENT_PACKAGE specification and body as database objects.

To save the package as a database object:

1. If necessary, click ➕ beside the Database Objects node to display the database users. Scroll down in the Object Navigator window, click ➕ beside your username, and then click ➕ beside the Stored Program Units node. Note that the package specification and body for the ORDER_PACKAGE that you created earlier in SQL*Plus are listed as stored program units.

2. Scroll up to the top-level Program Units node, click **ENROLLMENT_PACKAGE (Package Spec)**, press and hold the **Shift** key, and then click **ENROLLMENT_PACKAGE (Package Body)** to select both items.

3. Drag the items to the Stored Program Units node under your username. After a few moments, the items appear as stored program units in the database.

4. Close the Database Objects node, and unload the package components from the top-level Program Units node in Procedure Builder.

DATABASE TRIGGERS

Database triggers are program units that execute in response to the database events of inserting, updating, or deleting a record. Triggers are useful for maintaining integrity constraints. For example, when you delete an order from the CUST_ORDER table in the Clearwater Traders database, you might create a trigger to automatically delete all associated records in the ORDER_LINE table. Triggers are also useful for creating auditing information, such as recording the username of every user who modifies the GRADE field in the ENROLLMENT table in the Northwoods University database.

Database triggers are similar to all PL/SQL program units in the sense that they have declaration, body, and exception sections. One difference between triggers and other program units is that triggers cannot accept input parameters. Triggers and program units also differ in the way they execute. You must explicitly execute program units by typing the program unit name at the command prompt or in a command in a calling procedure. In contrast, a trigger executes only when its triggering event occurs. When a trigger executes, it is said to have **fired**.

Types of Triggers

A trigger is categorized by the type of SQL statement that causes it to fire, the timing of when it fires, and the level at which it fires. Table 5-5 summarizes the different types of triggers and shows the values used in the command to define the trigger.

Type of Trigger	Values	Description
Statement	INSERT, UPDATE, DELETE	Defines statement that causes trigger to fire
Timing	BEFORE, AFTER	Defines whether trigger fires before or after statement is executed
Level	ROW, STATEMENT	Defines whether trigger fires once for each triggering statement, or once for each row affected by triggering statement

Table 5-5   Types of triggers

SQL INSERT, DELETE, and UPDATE commands can cause triggers to fire. A trigger can fire either before or after its associated SQL command. A BEFORE trigger is often used to create an audit trail. For example, you might create a BEFORE trigger whenever the GRADE field is updated in the ENROLLMENT table in the Northwoods University database, to record the grade value before it is updated, along with the date the grade is updated. You might create an AFTER trigger to update the student's grade point average after the grade is changed.

Statement-level triggers fire once, either before or after the SQL triggering statement executes. For example, you would use a statement-level trigger to record an audit trail showing each time the ENROLLMENT table is changed, regardless of how many rows are affected. Row-level triggers fire once for each row affected by the triggering statement. In the audit trail example, you would use a row-level trigger to record exactly which rows in the ENROLLMENT table are changed, and how the values were changed.

Creating New Triggers

To create a trigger, you must specify the trigger level (ROW or STATEMENT), the trigger timing (BEFORE or AFTER), the SQL command type (INSERT, UPDATE, or DELETE) and table that that the trigger is associated with, and the trigger body, which is the PL/SQL code that executes when the trigger fires.

The general syntax for the SQL command for creating a trigger is:

```
CREATE OR REPLACE TRIGGER trigger_name
    [BEFORE|AFTER] [INSERT|UPDATE|DELETE] ON
    table_name
    [FOR EACH ROW] [WHEN(condition)]
BEGIN
    trigger body
END;
```

The *trigger_name* must follow the Oracle naming standard. A trigger can have the same name as existing PL/SQL program units in a database schema. This is because trigger names are stored in a namespace that is different from the namespace that other Oracle database objects use. A **namespace** is a place in the database that stores object identifiers for a user's database objects. For statement-level triggers, you omit the FOR EACH ROW clause. The BEFORE and AFTER commands specify the trigger timing. The INSERT, UPDATE, and DELETE commands specify the statement type that causes the trigger to fire. To create a trigger that fires for multiple statement types, you join the statement types using the OR operator. For example, to specify a trigger that fires for both UPDATE and DELETE statements, you specify **UPDATE OR DELETE ON tablename** in the trigger definition.

You can reference the value of a field in the current record both before and after the triggering statement executes. To reference a value in the trigger body before the triggering statement executes, you use the syntax: **:OLD.fieldname**. For example, to reference the value of GRADE before it is updated, you would use the statement **:OLD.grade**. To reference a field value after the triggering statement executes, you use the syntax **:NEW.fieldname**.

The **[WHEN (condition)]** clause is optional, and you use it only for row-level triggers. It specifies that the trigger will fire only for rows that satisfy a specific search condition. For example, in the trigger to record when a grade value is changed, you would only want the trigger to fire when the current value of the GRADE field is changed, but not when the grade is initially inserted. In this example, you would specify the following WHEN clause: **WHEN OLD.grade IS NOT NULL**. Note that in the WHEN clause, you omit the colon before the OLD or NEW qualifier.

The trigger body is a PL/SQL code block that contains sections for the declarations, body, and exceptions. It cannot contain any transaction control statements, such as COMMIT, ROLLBACK, and SAVEPOINT. You can only reference the NEW and OLD field values in a row-level trigger.

Now you will create statement-level and row-level triggers attached to the ENROLL-MENT table in the Northwoods University database. Whenever one or more ENROLLMENT records change, the statement-level trigger will insert a single audit trail record into a table named ENRL_AUDIT. The record will store the date the change was made and the name of the user who made the change. The row-level trigger will insert an audit trail record into a table named ENRL_ROW_AUDIT for each individual record that changes. The ENRL_ROW_AUDIT record will store the student ID, course section ID, and the old grade value of the record, along with the date the change was made and the name of the user who made the change.

When you create a trigger that is used to create an audit trail, you must first create one or more tables to store the audit trail values. Possible audit trail information can include when an action occurs, who performs the action, and the nature of the action. First, you will create the ENRL_AUDIT and ENRL_ROW_AUDIT tables in your database schema by running a script file named trigger_tables.sql that contains the CREATE TABLE commands for these tables. This script also contains commands to create sequences named ENRL_AUDIT_ID_SEQUENCE and ENRL_AUDIT_ROW_ID_SEQUENCE, which automatically generate values for the primary keys for the new trigger tables.

To create the ENRL_AUDIT and ENRL_ROW_AUDIT tables and sequences:

1. In SQL*Plus, run the **trigger_tables.sql** script in the Chapter5 folder on your Data Disk. The "Table created" and "Sequence created" confirmation messages should appear twice, indicating that both tables and sequences were successfully created.

Next, you will create the statement-level trigger that fires whenever a change is made to the ENROLLMENT table. The trigger timing will be AFTER. The command type will be INSERT, UPDATE, or DELETE, and the table is ENROLLMENT. When the trigger fires, it will insert a record into the ENRL_AUDIT table that retrieves the next value in the table's sequence and inserts the date the change was made, along with the name of the user who made the change. To insert the name of the current database user, you use the system USER variable in the INSERT command. After you create the trigger, you will change the grade for S_ID 100 and C_SEC_ID 1000 from 'A' to 'B', and query the ENRL_AUDIT table to confirm that the trigger fired correctly and inserted the audit trail record.

To create and test the statement-level trigger:

1. Type the commands shown in Figure 5-48, execute the commands in SQL*Plus, and debug them if necessary. The message "Trigger created" confirms that the trigger was successfully created.

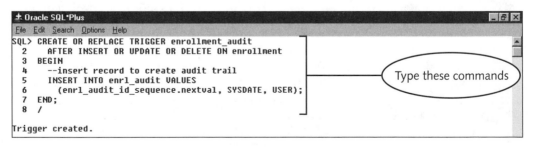

Figure 5-48 Creating a statement-level trigger in SQL*Plus

To create a trigger, you must have the CREATE TRIGGER database privilege, and you must be the owner of the table associated with the trigger.

2. Type the following command at the SQL prompt to update the GRADE value in the ENROLLMENT table for S_ID 100 and C_SEC_ID 1000, and then press **Enter**. The message "1 row updated" should appear.

```
UPDATE enrollment
SET grade = 'B'
WHERE s_id = 100 AND c_sec_id = 1000;
```

3. Type **SELECT * FROM enrl_audit;** at the SQL prompt to verify that the trigger inserted the audit trail record in the ENRL_AUDIT table, and then press **Enter**. The output should show a record of the update, including the ENRL_AUDIT_ID value generated by the sequence, the current date, and your username.

Note that this trigger fired once when only one row was updated. It would also fire only once if multiple rows were updated, because it is a statement-level trigger.

You can create triggers in Procedure Builder as well as in SQL*Plus. Next, you will use Procedure Builder to create a row-level trigger named ENROLLMENT_ROW_AUDIT that records information concerning how an individual row is changed in the ENROLLMENT table. To do this, you open the Tables node under the Database Objects node in the Object Navigator window, select the table that the trigger will be attached to, and then create the trigger. The trigger timing will be BEFORE, because you want to record the current (old) values of S_ID, C_SEC_ID, and GRADE. The command type will be UPDATE, and the table is ENROLLMENT. The WHEN clause will specify that the trigger will only fire when the grade value is changed. When the trigger fires, it will insert a record into the ENRL_ROW_AUDIT table, and record the current (old) values of S_ID, C_SEC_ID, and GRADE, as well as the date the change was made and who made the change.

To create the row-level trigger in Procedure Builder:

1. In Procedure Builder, open the following nodes in the Object Navigator: Database Objects, *your_username*, Tables, ENROLLMENT, and Triggers. The ENROLLMENT_AUDIT trigger that you just created in SQL*Plus should appear.

2. Be sure that the Triggers node is selected, and then create a new trigger by clicking the **Create button** on the Object Navigator toolbar. The Database Trigger dialog box opens.

3. Click **New** on the lower-left area of the dialog box. Replace the default trigger name with **ENROLLMENT_ROW_AUDIT**, as shown in Figure 5-49.

4. Be sure the **Before** option button is selected to specify the trigger timing, check **UPDATE** to specify the trigger statement, and select GRADE in the

5

Of Columns list to specify the update field that will fire the trigger. Select the **Row** option button to specify that this is a row-level trigger. The trigger's type and action specifications should look like Figure 5-49.

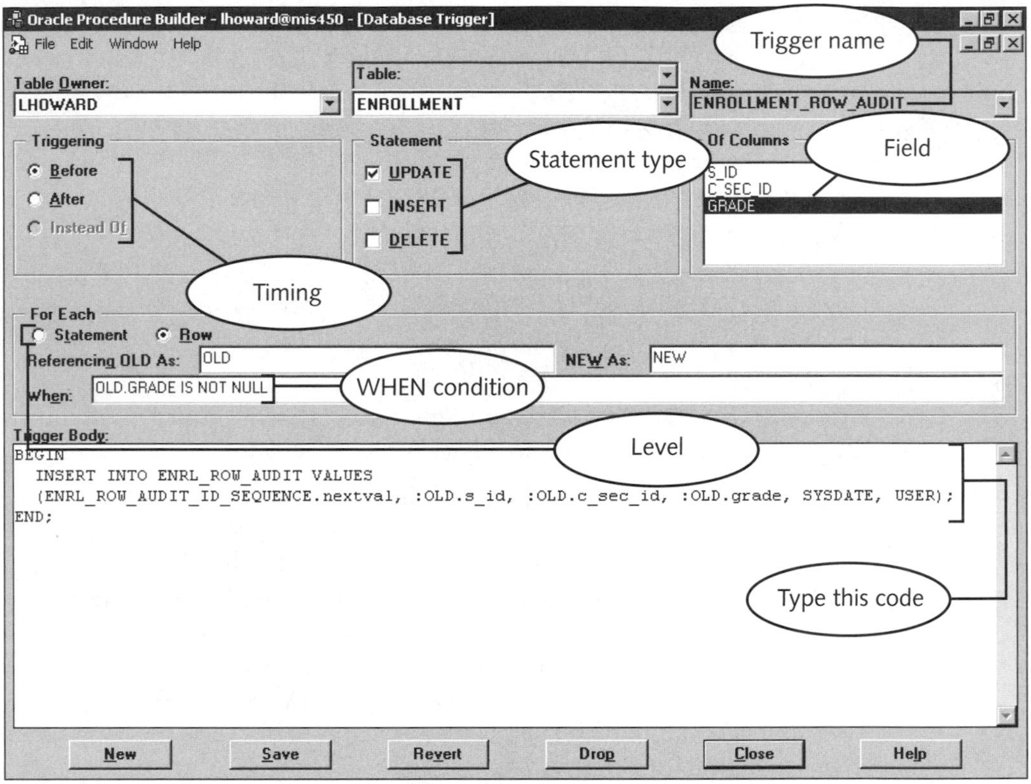

Figure 5-49   Procedure Builder row-level trigger specifications

The final steps are to specify the WHEN condition and to enter the code that executes when the trigger fires. The Database Trigger dialog box allows you to enter alternate aliases for OLD (table values before the trigger fires) and NEW (field values after the trigger fires), because you might have database tables named OLD or NEW. Because you don't have tables named OLD or NEW, you will specify the default alias names of OLD and NEW, and enter the trigger code.

To enter the trigger aliases and trigger body code:

1. Type **OLD** in the Referencing OLD As: text box, and type **NEW** in the NEW As: text box, as shown in Figure 5-49.

2. Type the WHEN clause as shown in Figure 5-49. Note that in the WHEN clause, the OLD reference is not prefaced with a colon.

3. Type the trigger body code shown in Figure 5-49. Note that in the trigger body code, the OLD references are prefaced with a colon.

4. Click **Save** to create the trigger.

5. Click **Close** to close the Database Trigger dialog box. The new trigger appears in the Object Navigator window under the ENROLLMENT table.

This trigger could also have been created in SQL*Plus, using the code shown in Figure 5-50. The advantage of creating triggers in Procedure Builder is that a lot of the trigger properties specified in the trigger code are generated automatically using option buttons and check boxes. The disadvantage of creating triggers in Procedure Builder is that if there is an error somewhere in the code, it is harder to locate and debug because the Procedure Builder Database Trigger dialog box provides very terse error messages.

```
± Oracle SQL*Plus                                                    _ 8 X
File  Edit  Search  Options  Help
SQL> CREATE OR REPLACE TRIGGER enrollment_row_audit
  2     BEFORE UPDATE ON enrollment
  3     FOR EACH ROW WHEN (OLD.grade IS NOT NULL)
  4  BEGIN
  5     --insert record to create audit trail
  6     INSERT INTO enrl_row_audit VALUES
  7     (enrl_row_audit_id_sequence.nextval, :OLD.s_id, :OLD.c_sec_id, :OLD.grade, SYSDATE, USER);
  8  END;
  9  /

Trigger created.
```

Figure 5-50   SQL*Plus trigger code for row-level trigger

Do not type the code shown in Figure 5-50, because you have already created this trigger in Procedure Builder. This code is provided for informational purposes only.

Now you will change the grade for S_ID 100 and C_SEC_ID 1000 from 'B' back to 'A' and query the ENRL_ROW_AUDIT table to confirm that the trigger fires correctly and inserts the audit trail record providing row-level information.

To test the row-level trigger:

1. In Procedure Builder, maximize the PL/SQL Interpreter window.

2. Type the following command at the PL/SQL prompt, and then press **Enter**:

```
UPDATE enrollment
SET grade = 'A'
WHERE s_id = 100 AND c_sec_id = 1000;
```

3. Type **SELECT * FROM enrl_row_audit;** at the PL/SQL prompt to verify that the trigger inserted the audit trail record in the
ENRL_ROW_AUDIT table, and then press **Enter**. The output should show
a record of the update, including the ENRL_ROW_AUDIT_ID value generated by the sequence, the student ID and course section ID, the former grade
value, the current date, and your username.

INSTEAD-OF Triggers

In Chapter 3, you learned about views, which are logical tables based on queries that
are used to enforce database security and simplify making complex queries. A limitation
of views is that you cannot perform DML operations (INSERT, UPDATE, or DELETE)
on views created by joining multiple tables. To circumvent this limitation, you can create **INSTEAD-OF triggers**, which fire when a user issues a DML command associated with a view. When the user tries to insert, update, or delete a record in the view,
the trigger specifies to insert, update, or delete the record in a table underlying the view,
which effectively performs the operation on the view.

To learn about INSTEAD-OF triggers, you will work with a view named
ENROLLMENT_VIEW that specifies enrollment information for all students at
Northwoods University. The view contains the student ID, first and last name, term
description, course section ID, course call ID, section number, day, and time associated with
enrollment records for which the GRADE field is NULL. Now you will run a script to
create this view.

To run the script to create the view:

1. In SQL*Plus, type **start a:\chapter5\enrollment_view.sql** at the SQL prompt.
The message "View created" confirms that the view was successfully created.

Now you will create an INSTEAD-OF trigger named ENROLLMENT_DELETE that
fires when the user deletes a record in the ENROLLMENT_VIEW. Because you cannot directly delete records from this view because it was created by joining multiple
tables, the trigger deletes the associated record from the ENROLLMENT table instead,
which effectively deletes the record from the view. INSTEAD-OF triggers can only be
created in SQL*Plus and are attached to a specific database view. There is no BEFORE
or AFTER specification, and INSTEAD-OF triggers are always row-level triggers,
because they correspond to DML commands associated with specific table rows. The
trigger body specifies one or more SQL commands that modify the tables underlying
the view. The table fields to be modified must be contained in the view and are referenced using the OLD qualifier.

To create the INSTEAD-OF trigger:

1. Type the commands shown in Figure 5-51, execute them in SQL*Plus, and
debug them if necessary. The "Trigger created" confirmation message appears.

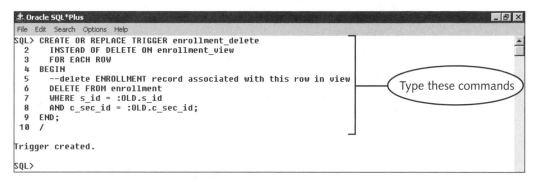

Figure 5-51   INSTEAD-OF trigger specification

Now you will test the INSTEAD-OF trigger. You will view the records in the ENROLLMENT_VIEW and then delete the record for student Ruben Sanchez's (S_ID 104) enrollment in course section ID 1010 by deleting the value from the view. Then, you will confirm that the record was deleted from the ENROLLMENT table.

To test the trigger:

1. In SQL*Plus, type the following command to view student Ruben Sanchez's current enrollment. The output confirms that Ruben is enrolled in course section ID values 1010 and 1012.

   ```
   SELECT *
   FROM enrollment_view
   WHERE s_id = 104;
   ```

2. Delete the enrollment in course section 1010 from the view using the following command:

   ```
   DELETE FROM enrollment_view
   WHERE s_id = 104
   AND c_sec_id = 1010;
   ```

3. Confirm that the record was deleted from the ENROLLMENT table by typing **SELECT * FROM enrollment WHERE s_id = 104 AND grade IS NULL;**. The output should show that Ruben is now only enrolled in course section ID 1012.

Disabling and Dropping Triggers

Triggers can be dropped from the database using the general command **DROP TRIGGER *trigger_name*;**. Triggers can also be disabled, which means they still exist in the user's database schema, but they no longer fire when their triggering event occurs. When you first create a trigger, it is automatically enabled. The SQL command to disable or enable a trigger is:

```
ALTER TRIGGER trigger_name [ENABLE | DISABLE];
```

Now you will disable the ENROLLMENT_DELETE trigger and then enable it again. Then, you will drop the trigger.

To disable, re-enable, and drop a trigger:

1. Type the ALTER TRIGGER commands shown in Figure 5-52 to disable and then reenable the ENROLLMENT_DELETE trigger.

2. TYPE the DROP TRIGGER command shown in Figure 5-52 to drop the trigger.

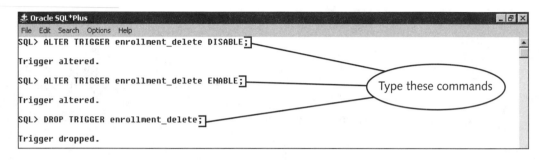

Figure 5-52   Disabling, enabling, and dropping a trigger

Viewing Information About Triggers

You can view information about triggers using the USER_TRIGGERS data dictionary view, which contains columns describing the trigger name, type, triggering events, owner of the database table or other objects referenced in the trigger, trigger status, and so forth. Now you will create a query that lists the names, types, and events of your database triggers.

To query the USER_TRIGGERS data dictionary view:

1. Type the command shown in Figure 5-53 to display information about your database triggers. Your output might be different if you have created or dropped different triggers.

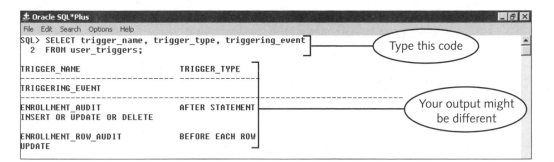

Figure 5-53   Displaying information about database triggers

2. Close SQL*Plus, Procedure Builder, and all other open Oracle applications.

SUMMARY

- ❐ To make applications more manageable, you should decompose them into logical units of work, write individual program units for each unit, and call program units from other program units.

- ❐ A server-side library is a collection of program units that is stored on the server, but runs on the client workstation.

- ❐ A program unit has object dependency on the objects that it references and procedure dependency on the program units that it calls. A program unit directly depends on the objects and procedures that it references and indirectly depends on the objects and procedures referenced by programs that it calls.

- ❐ If a program unit depends on an object or program unit that is subsequently changed, the program unit is invalidated and must be recompiled.

- ❐ A package is a code library that provides a way to manage large numbers of related programs and variables and to create public variables and program units.

- ❐ A package specification is used to declare variables and program units that are public and available to any user who has privileges to execute the package. A package body contains code for program units declared in the package specification.

- ❐ To access items in a package, you preface the variable or procedure name with the package name.

- ❐ Overloaded program units within packages have the same name but different numbers and types of parameters. The correct procedure executes based on the parameters that are passed.

- ❐ Database triggers are procedures that execute in response to the database events of inserting, updating, or deleting a record.

□ An INSTEAD-OF trigger can be used to effectively insert, update, or delete records in a view that is based on a query that joins multiple tables.

REVIEW QUESTIONS

1. Where is a server-side library stored, and where is it executed?

2. What is the difference between a direct dependency and an indirect dependency?

3. How does a program unit become invalidated?

4. What is the difference between a package and a library?

5. Where are private package variables declared? Where are public package variables declared?

6. Is it possible to create a package with a package specification but with no package body?

7. Where are packages stored and executed?

8. What is the purpose of creating an overloaded program unit?

9. What is the difference between a statement-level and a row-level trigger?

10. When should you create an INSTEAD-OF trigger?

PROBLEM-SOLVING CASES

Cases refer to the Clearwater Traders (see Figure 1-11), Northwoods University (see Figure 1-12), and Software Experts (see Figure 1-13) databases. Save all solution files in your Chapter5\CaseSolutions folder.

1. In this case, you will create program units that process items that are received at Clearwater Traders. The first program unit will update the SHIPMENT_LINE table and then call a second program unit that updates the quantity on hand (QOH) of the item in the INVENTORY table.

   a. In Procedure Builder, create a procedure named UPDATE_INV_QOH that receives input parameters of an inventory ID and an update quantity, and then uses these values to update the QOH field in the INVENTORY table. Save the procedure source code in a file named 5BCase1a.pls.

   b. In Procedure Builder, create a second procedure named UPDATE_SHIPMENT_LINE that receives input parameters of an existing shipment ID and inventory ID, and the shipment quantity. The procedure should update the associated SHIPMENT_LINE record based on the input values, using the system date as the date received. It should then call the UPDATE_INV_QOH procedure to update the associated inventory quantity on hand. Save the procedure source code in a file named 5BCase1b.pls.

   c. Save both procedures as stored program units in your database schema.

2. In this case, you will create program units that initialize a new project in the Software Experts database. The first program unit will insert a new record into the PROJECT table and then call a second program unit that inserts the associated record for the project manager in the PROJECT_CONSULTANT table.

a. In Procedure Builder, create a procedure named ADD_PROJ_MGR that receives input parameters of a project ID, a consultant ID, and a project duration (in months), and then uses these values to insert a new record into the PROJECT_CONSULTANT table. Use the current system date for the roll-on date, and use the ADD_MONTHS function to calculate the roll-off date. Initialize the project hours to zero. Save the procedure source code in a file named 5BCase2a.pls.

b. In Procedure Builder, create a second procedure named ADD_PROJECT that receives input parameters of the project name, client name, project manager's first and last name, and project ID of the project's parent project, if one exists. The procedure should insert a record in the PROJECT table based on the input values. The P_ID value should be retrieved from a sequence named PROJECT_ID_SEQUENCE. (You will need to create this sequence if you have not created it in a previous case.) Then, the procedure should call the ADD_PROJ_MGR procedure to insert the information about the project manager. Save the procedure source code in a file named 5BCase2b.pls.

c. Save both procedures as stored program units in your database schema.

3. In this case, you will create a procedure to insert new records into the ENROLLMENT table of the Northwoods University database that first verifies that the maximum course section enrollment has not been exceeded.

a. In Procedure Builder, create a function named MAX_ENRL_EXCEEDED that returns a BOOLEAN value. The function should receive an input value of a course section ID. If the current enrollment of the course is greater than or equal to the maximum allowable enrollment specified in the COURSE_SECTION table, then the function should return the value TRUE. If the current enrollment is less than the maximum allowable enrollment, then the function should return FALSE. (*Hint*: Calculate the current enrollment for the course section by counting the number of associated enrollment records in the ENROLLMENT table.) Save the function source code as 5BCase3a.pls.

b. In Procedure Builder, create a procedure named NEW_ENROLLMENT that receives input values of a student ID and course section ID. The procedure should call the MAX_ENRL_EXCEEDED function to determine if the maximum enrollment has been exceeded. If the maximum enrollment has not been exceeded, then the procedure should insert the new record into the ENROLLMENT table, with a NULL value for the GRADE field. If the maximum enrollment has been exceeded, then the procedure should not insert the record. Save the procedure source code as 5BCase3b.pls.

c. Save the function and the procedure as stored program units in your database schema.

5

4. In this case, you will create a package named SHIPMENT_PKG that contains procedures associated with the Clearwater Traders database for creating new shipments and shipment lines when orders are placed with suppliers and processing incoming shipments when orders are received.

   a. In Procedure Builder, create a package specification that declares a variable named CURR_SHIPMENT_ID as a public variable in the package and declares the procedures and associated input variables shown in Table 5-6. Save the package specification source code as 5BCase4a.pls.

Procedure Name	Input Variables	Procedure Commands
ADD_SHIPMENT	CURR_DATE_EXPECTED	Insert new SHIPMENT record using CURR_SHIPMENT_ID and input variables.
ADD_SHIP_LINE	CURR_INV_ID, CURR_QUANTITY	Insert new SHIPMENT_LINE record using CURR_SHIPMENT_ID and input variables.
UPDATE_SHIP_LINE	CURR_INV_ID, CURR_QUANTITY	Update associated SHIPMENT_LINE record using CURR_SHIPMENT_ID and input variables, and system date for DATE_RECEIVED.
UPDATE_INV_QOH	CURR_INV_ID, CURR_QUANTITY	Update associated INVENTORY record using input variables.

Table 5-6

   b. In Procedure Builder, create a package body with the specified procedures. Save the package body source code as 5BCase4b.pls.

   c. Store the package specification and body as stored program units in the database.

   d. In SQL*Plus, create an anonymous PL/SQL program that uses the package's ADD_SHIPMENT and ADD_SHIP_LINE procedures to add shipment ID 300 for 100 units of inventory ID 11846 and 50 units of inventory ID 11847. Set the date expected as one month from the current system date, using the ADD_MONTHS function. Save the code in a file named 5BCase4.sql.

   e. In SQL*Plus, create an anonymous PL/SQL program that uses the package's UPDATE_SHIP_LINE and UPDATE_INV_QOH procedures to record that 100 units of inventory ID 11846 and 25 units of inventory ID 11847 were received for shipment ID 300. Add the code to the file named 5BCase4.sql that you created in the previous step.

5. In this case, you will use SQL*Plus to create a package named CUST_ORDER_PKG associated with the Clearwater Traders database that contains an overloaded procedure that allows users to create a new customer order based on three different input parameter sets: the customer ID, the customer first and last names, or the customer daytime telephone number. Save all source code for the package in a file named 5BCase5.sql.

a. Create a package specification that declares three procedures named CREATE_NEW_ORDER. The first procedure has input parameters of order ID, customer ID, payment method, and order source ID; the second procedure has input parameters of order ID, customer first and last name, payment method, and order source ID; and the third procedure has input parameters of order ID, customer daytime phone number, payment method, and order source ID.

b. Create the package body for the three overloaded procedures. Use the payment method and order source ID input parameters for the CUST_ORDER field values, and derive the customer ID based on the other customer values when necessary. Use the current system date for the order date value for all procedures.

c. Create an anonymous PL/SQL block that uses the overloaded procedures to insert the following records:

❐ Order ID 2000 for customer ID 154, payment method 'CC', order source ID 6

❐ Order ID 2001 for customer Maria Garcia, payment method 'CHECK', order source ID 5

❐ Order ID 2002 for the customer with daytime phone number 715-555-8943, payment method 'CC', order source ID 6

6. In this case, you will create a package named GPA_PKG that contains an overloaded procedure that allows users to calculate a Northwoods University student's cumulative grade point average for all terms or a student's grade point average for a specific term.

a. In Procedure Builder, create a package specification that declares two procedures named CALC_GPA. The first procedure has an input parameter of a student ID and an output parameter of the student GPA. The second procedure has input parameters of a student ID and a term ID, and an output parameter of the student GPA. Save the source code in a file named 5BCase6a.pls.

b. In Procedure Builder, create the package body for the two overloaded procedures. (The algorithm for calculating grade point average is presented in Case 10 in Lesson A of this chapter.) If the student has not earned a grade in any courses in a given term, specify the grade point average as zero. Save the source code in a file named 5BCase6b.pls.

c. Save the package specification and body as stored program units in the database.

d. In SQL*Plus, create an anonymous PL/SQL block that uses the overloaded procedures to calculate and display the grade point average for each student for each term and the student's cumulative grade point average. Display the output as shown in Figure 5-54. If a student did not earn a grade in any courses in a given term, do not display any output for the student for that term. If a student has not earned a grade in any courses, display 0.00 for the student's cumulative grade point average. Save the program in a file named 5BCase6.sql.

```
± Oracle SQL*Plus                                                    _ 8 X
File Edit Search Options Help
+                                    +
Student ID: 100
=========================================
Term: Fall 2003    GPA:  4.00
Term: Spring 2004    GPA:  3.00
Cum. GPA:  3.50
+                                    +
Student ID: 101
=========================================
Term: Fall 2003    GPA:  2.00
Term: Spring 2004    GPA:  3.33
Cum. GPA:  3.00
+                                    +
Student ID: 102
=========================================
Term: Fall 2003    GPA:  2.00
Cum. GPA:  2.00
+                                    +
Student ID: 103
=========================================
Cum. GPA:    .00
+                                    +
Student ID: 104
=========================================
Term: Fall 2003    GPA:  3.00
Term: Spring 2004    GPA:  2.00
Cum. GPA:  2.33
+                                    +
Student ID: 105
=========================================
Cum. GPA:    .00
```

Figure 5-54

7. In this case, you will use SQL*Plus to create a series of triggers that automatically update the quantity on hand field in the INVENTORY table in the Clearwater Traders database by the correct amount whenever an ORDER_LINE record is inserted, updated, or deleted. Save all source code in a file named 5BCase7.sql.

   a. Create a trigger named QOH_INSERT that is associated with the ORDER_LINE table and subtracts the order quantity of the specified inventory item from the QOH field in the INVENTORY table when a new item is sold to a customer.

   b. Create a trigger named QOH_UPDATE that is associated with the ORDER_LINE table and correctly adjusts the QOH field in the INVENTORY table whenever the ORDER_QUANTITY field is updated for the associated inventory item.

   c. Create a trigger named QOH_DELETE that is associated with the ORDER_LINE table and correctly adjusts the QOH field in the INVENTORY table when a record is deleted from the ORDER_LINE table.

8. In SQL*Plus, create a trigger named COLOR_UPDATE that fires whenever a record in the COLOR table in the Clearwater Traders database is updated. The trigger should automatically update all INVENTORY records that have the old COLOR value with the new COLOR value. For example, if the color "Bright Pink" is updated to "Tropical Pink," then inventory records that currently have

"Bright Pink" for their COLOR value should be updated to "Tropical Pink." Save the source code for the trigger in a file named 5BCase8.sql.

9. In this case, you will use SQL*Plus to create triggers that maintain an audit trail to record when the PROJECT_CONSULTANT table in the Software Experts database is modified. Save all code for creating the tables and triggers in a file named 5BCase9.sql.

   a. Create the following two database tables:

      ❑ PROJ_CONS_CHANGE, with a field named CHANGE_DATE to record the date that the table was changed by an insert, update, or delete command, and a field named CHANGE_USER to record the username of the person making the change;

      ❑ PROJ_CONS_ROW_UPDATE, with P_ID and C_ID fields to record the project ID and consultant ID of the row that is updated, a CHANGE_DATE field to record the date the change is made, a CHANGE_USER field to record the username of the person making the change, and fields named OLD_ROLL_ON, OLD_ROLL_OFF, OLD_HOURS, NEW_ROLL_ON, NEW_ROLL_OFF, and NEW_HOURS to record both the old and new values for roll-on date, roll-off date, and total hours.

   b. Create a statement-level trigger named PROJ_CONS_CHANGE to insert a record into the PROJ_CONS_CHANGE table whenever a record is changed.

   c. Create a row-level trigger named PROJ_CONS_ROW_UPDATE that inserts a record into the PROJ_CONS_ROW_UPDATE table to record the old and new roll-on and roll-off values, and old and new hour values whenever a row is updated.

10. In this case, you will use SQL*Plus to create a view that joins consultant and skill information in the Software Experts database. Then, you will create an INSTEAD-OF trigger to enable users to modify the underlying database tables using the view. Save the code to create the view and trigger in a file named 5BCase10.sql.

    a. Create a view named CONSULTANT_SKILL_VIEW that lists the consultant ID, consultant first and last name, associated skill IDs, skill descriptions, and whether the consultant is certified for each skill.

    b. Create a trigger named CERTIFICATION_UPDATE that enables users to update a consultant's skill certification using the view by specifying the consultant first and last name and the skill description. (*Hint*: You will need to use nested queries in the search conditions of the trigger's UPDATE command.)

◄ LESSON C ►

Objectives

♦ Become familiar with the different types of Oracle built-in packages

♦ Understand how to use the DBMS_JOB built-in package

♦ Use the DBMS_PIPE and DBMS_DDL built-in packages

♦ Understand Dynamic SQL

♦ Use Dynamic SQL to create programs with query parameters that are defined when the program is executed

BUILT-IN PACKAGES

In addition to user-defined packages like the ones you created in the previous lesson, Oracle provides many built-in packages that are accessible to both client and server programs and provide support for basic database functions. These packages are in the SYS database schema, and the DBA must sometimes initialize them by running scripts that are stored in the RDBMS\ADMIN folder on the database server. The packages fall into five basic categories: transaction processing, application development, database administration, application administration, and internal support.

This overview of built-in packages is a survey for certification exam preparation purposes, and detailed information is given only for specific packages that will be covered on the exam. To find out more about other packages, open the Database Objects node in Procedure Builder, open the SYS user node, and then open the package specification.

Transaction Processing Packages

The transaction processing packages contain procedures related to locking and releasing database objects to support transaction processing, supporting the COMMIT and ROLLBACK commands, and providing a way to specify query properties in PL/SQL programs at runtime. Important packages in this category are summarized in Table 5-7.

Package Name	Description
DBMS_ALERT	Dynamically sends messages to other database sessions
DBMS_LOCK	Creates user-defined locks on tables and records
DBMS_SQL	Implements Dynamic SQL
DBMS_TRANSACTION	Provides procedures for transaction management, such as creating read-only transactions, generating a rollback, and creating a savepoint

Table 5-7　Built-in packages to support transaction processing

A database session is created when a user connects to an Oracle database in a client application. If you are connected to the database using both SQL*Plus and Procedure Builder, then you have created two separate database sessions.

Application Development Packages

The application development built-in packages aid developers in creating and debugging PL/SQL applications. Important packages in this category are summarized in Table 5-8. You have used the PUT_LINE procedure in the DBMS_OUTPUT package many times to create output in PL/SQL programs.

Package Name	Description
DBMS_DESCRIBE	Returns information about the parameters of any stored program unit
DBMS_JOB	Schedules PL/SQL programs to run at specific times
DBMS_OUTPUT	Provides text output in PL/SQL programs
DBMS_PIPE	Sends messages to other database sessions asynchronously
DBMS_SESSION	Dynamically changes properties of a database session
UTL_FILE	Enables PL/SQL output to be written to a binary file

Table 5-8   Built-in packages to support application development

Using the DBMS_JOB Package

The DBMS_JOB package enables PL/SQL program units to be scheduled to run at specific times. This is useful when you need to run certain programs at predicted intervals. For example, you might need to run a PL/SQL procedure that creates and prints a report containing information about incoming shipments at Clearwater Traders on the first day of every month. The package creates a **job queue**, which is a list of program units to be run and the details for when and how often they are to be run. The package's SUBMIT procedure is used to submit a job to the job queue, using the following syntax: **DBMS_JOB.SUBMIT (*job_number, call_to_stored_program_unit, next_run_date, interval_to_run_job_again, no_parse_value*);**. This command has the following parameters:

- *job_number* is a unique job number that is automatically assigned to the job by the procedure.

- *call_to_stored_program_unit* is a string representing the code used to call the program unit, including parameter values.

- *next_run_date* is a date that specifies the next date and time when the job is scheduled to run.

- *interval_to_run_job_again* is a text string that specifies a time interval when the job will be run again. For example, if the job is to run for the first time on the current system date and the next time one day after the current system date, the interval parameter would be specified as 'SYSDATE + 1'.

- *No_parse_value* is a Boolean parameter that specifies whether or not the program unit is to be parsed and validated the next time it executes.

The DBMS_JOB package's RUN procedure can be used to run a previously submitted job immediately. This procedure has the syntax **DBMS_JOB.RUN(*job_number*);**. The *job_number* parameter corresponds to the job number that is assigned to the job when it is submitted to the job queue using the SUBMIT procedure. The DBMS_JOB package's REMOVE procedure can be used to remove a job from the job queue. The REMOVE procedure has the syntax **DBMS_JOB.REMOVE(*job_number*);**.

Using the DBMS_PIPE Package

A **pipe** is a program that directs information from one destination (such as the screen display) to a different destination (such as a file or database). The DBMS_PIPE package implements **database pipes**, which are pipes that are implemented entirely in the Oracle database and are independent of the operating system of the database server or client workstation. These pipes can be used to transmit data values to different database sessions. A DBA might use a pipe to send a message to database users to notify them about system actions, such as shutting down the database. A pipe might also be created by a process to automatically notify users or DBAs when an error occurs.

A specific pipe has a single **writer**, which is the session that creates the message, and multiple **readers**, which are sessions that receive the message. Pipes are **asynchronous**, which means that they operate independently of transactions. Once a message is sent using a pipe, it cannot be taken back using a rollback process.

To send a message using a pipe, you use the PACK_MESSAGE and SEND_MESSAGE procedures. The PACK_MESSAGE procedure places the outgoing message in a buffer using the syntax: **DBMS_PIPE.PACK_MESSAGE(*message*);**. The message can be any data type, and multiple items of different data types can be sent in a single message. Each item is packed into the message using a separate PACK_MESSAGE procedure call. The pipe is automatically created the first time you call the PACK_MESSAGE procedure.

The SEND_MESSAGE function sends the contents of the pipe using the following syntax: *return_value* := **DBMS_PIPE.SEND_MESSAGE('*pipe_name*',*timeout_interval*, *maximum_pipe_size*);**. The parameters for this function are:

- *return_value*, which is an integer variable that is assigned the following values depending on the pipe status after the function is called:

Value	Meaning
0	Pipe message successfully sent
1	The call timed out, possibly because the pipe was too full to be sent or a lock could not be obtained
3	The call was interrupted due to an internal error

- *pipe_name*, which identifies the pipe and can be any name that conforms to the Oracle naming standard.

- *timeout_interval*, which is an optional parameter that specifies the time interval, in seconds, that the procedure should try to send the pipe before quitting and returning an error message. Because the default value is 1,000 days, it is a good idea to specify a smaller value when developing new pipes. Otherwise, the system will "hang," and you will have to shut down SQL*Plus and restart it if you make a mistake in your code.

- *maximum_pipe_size*, which specifies the amount of buffer space, in bytes, that is allocated to the pipe, with a default value of 8,192 bytes. Most messages need a larger value, because some of the pipe buffer space is occupied with values specifying the pipe properties. If the pipe buffer size is too small, the pipe will not be successfully received and the receiver's SQL*Plus session will lock up and have to be restarted. Once a pipe is sent, it remains in the pipe buffer until the database is shut down and then restarted.

Now you will use the DBMS_PIPE package to create a pipe that sends three items: a text message, the username of the person who sent the pipe, and the date the pipe was sent.

To create a pipe:

1. Start Notepad, create a new file named **5CPrograms.sql**, and type the commands shown in Figure 5-55.

2. Start SQL*Plus.

3. Copy the program you created in Step 1, and execute it. The outgoing pipe status message displays that the pipe status is 0, which indicates that the pipe was successfully sent.

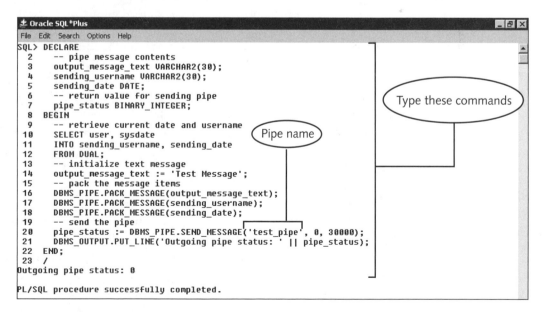

Figure 5-55 Sending a pipe

To use the DBMS_PIPE package, you must have the EXECUTE_ANY_PROCEDURE system privilege.

To receive a pipe, you use the RECEIVE_MESSAGE function, which has the following syntax: **return_value := DBMS_PIPE.RECEIVE_MESSAGE('pipe_name', timeout_interval);**. The *return_value* is an integer variable assigned the following values based on the outcome of receiving the message:

Value	Meaning
0	Message successfully received
1	No message received and the function timed out
2	The message in the pipe was too large for the buffer
3	An internal error occurred

The *pipe_name* parameter is the name of the pipe that was specified in the SEND_MESSAGE function when the sender sent the pipe. The *timeout_interval* is an optional parameter that specifies the time interval, in seconds, that the procedure should continue to try to receive the pipe. The default value is 1,000 days, so it is a good practice to use a shorter value when developing new programs to receive pipes, or else the program will appear to lock up in SQL*Plus if there is an error in your code.

Once you have received the pipe, you need to unpack it, using the UNPACK_MESSAGE procedure. This procedure has the syntax: **DBMS_PIPE.UNPACK_MESSAGE (output_variable_name);**. The *output_variable_name* parameter specifies the name of the variable that is used to reference each item in the pipe. The receiver must unpack the pipe message for each item that was sent. For example, if a pipe message was packed using three PACK_MESSAGE statements, then the message must be unpacked using three separate calls to the UNPACK_MESSAGE procedure. The message items must be unpacked in same order they were packed, using output variables of the correct data type. For example, if a message was packed with three message items that had the DATE, NUMBER, and VARCHAR2 data types, respectively, it must be unpacked three times, using variables that have the DATE, NUMBER, and VARCHAR2 data types also. If you attempt to unpack an item into a variable of the wrong data type, an error will be generated.

Now you will exit SQL*Plus and then start SQL*Plus again to create a new database session. Then, you will write and execute a program to receive the pipe you just created.

To receive the pipe:

1. Exit SQL*Plus, then start SQL*Plus, log on as usual, and type **SET SERVEROUTPUT ON SIZE 4000** at the SQL prompt.

2. In Notepad, type the commands shown in Figure 5-56 to receive the pipe. Execute the program in SQL*Plus, and debug it if necessary. Your output should show your username as the sender and the current date for the date sent.

If an error message appears when you try to receive the pipe, it might be because you executed the commands to send the pipe multiple times. If a pipe already exists, the PACK_MESSAGE procedure adds to the contents of the existing pipe, so the contents are different than the receiving program expects. If this happens, you can delete the pipe by stopping and then re-starting the database. Alternately, you can create a new pipe using a pipe name that is different from the one shown in Figure 5-55, and receive the pipe using a pipe name that is different from the one shown in Figure 5-56.

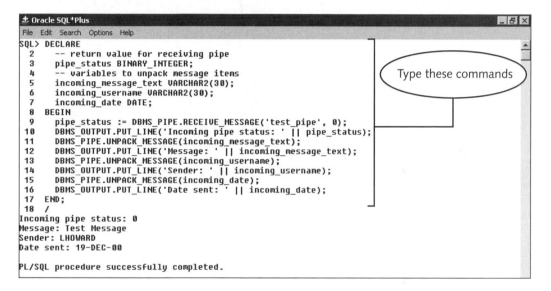

Figure 5-56   Receiving a pipe

Database and Application Administration Packages

Database and application administration packages support database administration tasks, such as managing memory on the database server, managing how disk space is allocated and used, and recompiling and managing stored program units and packages. Table 5-9 summarizes primary built-in packages in this category.

Package Name	Description
DBMS_APPLICATION_INFO	Registers information about programs being run by individual user sessions
DBMS_DDL	Provides procedures for compiling program units and analyzing database objects
DBMS_SHARED_POOL	Used to manage the shared pool, which is a server memory area that contains values that can be accessed by all users
DBMS_SPACE	Provides information for managing how data values are physically stored on the database server disks
DBMS_UTILITY	Provides procedures for compiling all program units and analyzing all objects in a specific database schema

Table 5-9   Built-in packages to support database and application administration

The DBMS_DDL Package

The DBMS_DDL package has a procedure named ALTER_COMPILE, which is used to compile PL/SQL program units. You would use this to recompile a program unit that is invalidated due to changes in tables or procedures on which it has dependencies. The procedure makes it easy to quickly recompile several program units. To call the procedure, you use the syntax: **DBMS_DDL.ALTER_COMPILE(*program_unit_type, owner_name, program_unit_name*);**. This procedure has the following parameters:

- *program_unit_type*, which specifies the type of program unit. Valid values are 'PROCEDURE', 'FUNCTION', 'PACKAGE', and 'PACKAGE BODY'. The value must be enclosed in single quotation marks and must be in all capital letters.

- *owner_name*, which specifies the username of the user who owns the program unit. The username is enclosed in single quotation marks and must be in all capital letters.

- *program_unit_name*, which specifies the name of the program unit, enclosed in single quotation marks and in all capital letters.

Now you will use the DBMS_DDL package to recompile the CREATE_NEW_ORDER procedure, the AGE function, and the ENROLLMENT_PACKAGE, which are all stored program units that you created earlier in the chapter.

To recompile the program units:

1. Type the commands shown in Figure 5-57, execute the program in SQL*Plus, and debug it if necessary. The confirmation message "PL/SQL procedure successfully completed" indicates that the program units were successfully recompiled.

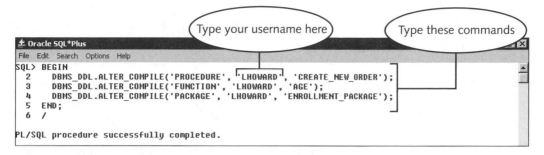

Figure 5-57   Using the DBMS_DDL.ALTER_COMPILE procedure

Oracle Internal Support Packages

The internal support packages provide the underlying functionality of the Oracle database. This code is placed in packages to enable users to view the package specifications, which are useful for understanding how to use the items, while preventing users from viewing or modifying the underlying package body code. Table 5-10 summarizes primary built-in packages in this category.

To call many procedures in commonly used packages like STANDARD and DBMS_STANDARD, you do not need to preface the procedure name with the package name. For example, the COMMIT command is actually a procedure in the DBMS_STANDARD package.

Package Name	Description
STANDARD	Defines all built-in functions and procedures, database data types, and PL/SQL data type extensions
DBMS_SNAPSHOT	Used to manage snapshots of databases, which capture the database state at a specific point in time
DBMS_REFRESH	Used to create groups of snapshots, which can be refreshed simultaneously
DBMS_STANDARD	Contains common processing functions of the PL/SQL language

Table 5-10   Internal support packages

DYNAMIC SQL

All of the SQL commands you have used so far in PL/SQL programs have been **static**, which means that the structure of the commands is established and the database objects are validated when the program containing the SQL command is compiled. When you create PL/SQL programs using **Dynamic SQL**, the SQL commands are created as text strings and then compiled and validated at runtime. The advantage of this approach is that you can dynamically structure SQL queries based on user inputs, and you can include DDL commands like CREATE, ALTER, and DROP in PL/SQL programs. This is useful when you want to create programs that contain SQL queries that use search conditions that are based on dynamic conditions, such as the current system date or time. Also, it allows you to create programs that create or alter the structure of database tables. For example, you might want to write a program that creates a temporary database table, uses the table to generate a report, and then drops the table.

All Dynamic SQL processing is performed using a cursor that defines the server memory area where the processing takes place. The program units for creating and manipulating this cursor are in a built-in package named DBMS_SQL. Table 5-11 summarizes these program units.

Program Unit Name	Description	Type	Input Variables
OPEN_CURSOR	Opens the cursor that defines the processing area	Function, returns *cursor_ID*	None
PARSE	Sends statement to the server, where syntax is verified	Procedure	*cursor_ID*, *SQL_statements*, *language_flag*, which is set to 'DBMS_SQL.V7'
BIND_VARIABLE	Associates program variables with input or output values	Procedure	*cursor_ID*, *variable_name*, *variable_value*, *maximum_character_ column_size*
DEFINE_COLUMN	Specifies the type and length of output variables	Procedure	*cursor_ID*, *column_position*, *column_name*, *maximum_character_ column_size*
EXECUTE	Executes the Dynamic SQL statement	Function, returns the number of rows processed	*cursor_ID*
FETCH_ROWS	Fetches rows for a SELECT operation	Function, returns the number of rows fetched	*cursor_ID*
VARIABLE_VALUE	Retrieves values of output variables	Procedure	*cursor_ID*, *variable_name*, *variable_value*
COLUMN_VALUE	Associates fetched values with program variables	Procedure	*cursor_ID*, *column_position*, *output_variable*
CLOSE_CURSOR	Closes the cursor when processing is complete	Procedure	*cursor_ID*

Table 5-11 DBMS_SQL program units used for Dynamic SQL processing

Dynamic SQL programs are structured differently and use different procedures depending on whether they are processing DML commands, DDL commands, SQL SELECT commands, or anonymous SQL blocks. The following sections explore how to create stored procedures that use Dynamic SQL to process these different command types.

Dynamic SQL Programs That Contain DML Commands

It is useful to be able to create a program that can insert, update, or delete records based on user inputs or based on current system variables like the system date. For example, in the Software Experts database, you might want to create a lookup table that contains values that represent every date during the next year, to allow the user to select future project roll-on and roll-off dates.

Processing Dynamic SQL programs that involve DML commands involves the following steps:

1. Open the cursor using the OPEN_CURSOR function. This function has the syntax *cursor_ID* := **DBMS_SQL.OPEN_CURSOR;**. The *cursor_id* is a variable that has been declared using the NUMBER data type. The function returns an integer that references the cursor in subsequent processing steps.

2. Define the SQL command as a text string, using placeholders for dynamic values. The SQL command is assigned to a program variable that is defined using the VARCHAR2 data type that must be wide enough to accommodate all of the SQL command text. This command uses the syntax *SQL_command_string_variable* := '**SQL_command_text';**. Note that the ending semicolon is placed outside of the text string single closing quotation mark. To embed a single quotation mark within a SQL command that is created as a text string, type the single quotation mark twice.

   A **placeholder** is a variable that is prefaced with a colon and is not formally declared in the procedure. Placeholders must follow the Oracle naming standard. A placeholder is used within a SQL command in place of a value and is then dynamically associated with the desired value at runtime. For example, you would dynamically specify the field name and the table name using placeholders in the simple SELECT command: **SELECT :*field_placeholder_name* FROM :*table_placeholder_name*;**. The *field_placeholder_name* is a placeholder for the field name that will be used in the SELECT clause, and the *table_placeholder_name* is a placeholder for the table name that will be used in the FROM clause. These placeholders will be bound to actual data values later.

3. Parse the SQL command using the PARSE procedure. The PARSE procedure has the syntax **DBMS_SQL.PARSE(***cursor_ID, SQL_command_string_variable, language_flag***);**. The *SQL_command_string_variable* is the variable that you defined in Step 2. The *language_flag* parameter specifies the version of the DBMS_SQL package being used. Possible values are DBMS_SQL.V6, which is used with Oracle Version 6 databases, and DBMS_SQL.V7, which is used with Oracle Version 7 and higher databases.

4. Bind input variables to placeholders using the BIND_VARIABLE procedure. This procedure is used to bind each placeholder to an actual data value, using either a variable or a constant. The general syntax is **DBMS_SQL.BIND _VARIABLE (*cursor_ID*, ':*placeholder*', *placeholder_value*, *maximum_character_column_size*);**. For example, to bind the value 'STUDENT' to the placeholder *:table_pholder*, you would use DBMS_SQL. BIND_VARIABLE(cursor_ID, ':table_pholder', 'STUDENT', 30);. Note that when you are binding a placeholder to a character value, you must specify the maximum width of the character column, and you must place the value in single quotation marks. When you are binding any other data type to a placeholder, you do not need to specify maximum column size.

5. Execute the SQL command using the EXECUTE function. This function has the syntax ***number_of_rows_processed* := DBMS _SQL.EXECUTE (*cursor_ID*);**. The *number_of_rows_processed* return value is a variable that is declared using the NUMBER or INTEGER data type.

6. Close the cursor using the CLOSE_CURSOR procedure. This procedure has the syntax **DBMS_SQL.CLOSE_CURSOR(*cursor_ID*);**.

Now you will create a procedure named UPDATE_PRICES that uses Dynamic SQL to update the prices in the INVENTORY table of the Clearwater Traders database for a specific item by a specific percentage. The item ID and percentage increase will be passed to the procedure as input parameters and will be associated with the query at runtime using placeholders.

To create and execute the Dynamic SQL procedure that uses a DML command:

1. Type the commands shown in Figure 5-58 to create the procedure, execute it in SQL*Plus, and debug it if necessary. The message "Procedure created" confirms that the procedure was successfully created.

2. Type the following command at the SQL prompt to call the procedure and pass to it the values for the percent price change and item ID: **EXECUTE UPDATE_PRICES(.1, 786);**. The message "PL/SQL procedure successfully completed" appears, indicating that the procedure was successfully executed.

3. Type the following command at the SQL prompt to confirm that the INVENTORY price values were successfully updated for ITEM_ID 786: **SELECT price FROM inventory WHERE item_id = 786;**. The price should appear as 285.99, which is 10% higher than the previous price, confirming that the Dynamic SQL procedure worked correctly. The price will appear twice, because ITEM_ID 786 appears twice in the INVENTORY table.

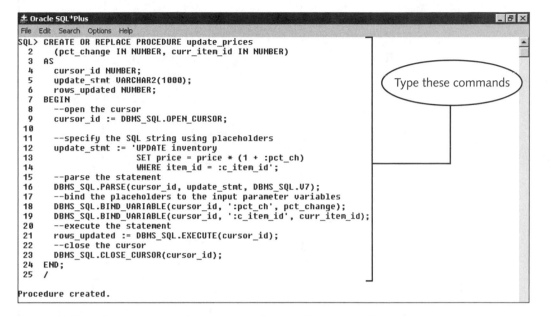

Figure 5-58   Processing a DML command using Dynamic SQL

Processing Dynamic SQL Programs That Contain DDL Commands

Recall that you cannot use DDL commands, such as CREATE, ALTER, or DROP, within PL/SQL programs. Dynamic SQL programs allow you to perform these operations, which are useful for such tasks as creating temporary tables, and modifying the structure of or deleting existing tables based on user inputs.

Creating Dynamic SQL programs that involve DDL commands that create, drop, or alter database objects is similar to creating Dynamic SQL programs that use DML commands, with the following differences: You cannot use placeholders in a DDL command, so you cannot dynamically bind parameter values at runtime. In addition, DDL statements are executed in the PARSE procedure, so the call to the EXECUTE procedure is not needed. The steps for processing a Dynamic SQL program that uses DDL commands are:

1. Open the cursor.

2. Define the SQL command as a text string.

3. Parse the SQL command.

4. Close the cursor.

Now you will create a Dynamic SQL procedure named CREATE_TEMP_TABLE that creates a table for which the name of the table is specified as an input parameter. The table has one field named TABLE_ID, which has a NUMBER data type of size 6.

To create a Dynamic SQL procedure that creates a table:

1. Type the commands shown in Figure 5-59 to create the procedure. Execute the commands in SQL*Plus, and debug them if necessary. The message "Procedure created" confirms that the procedure was successfully created.

Type these commands

```
Oracle SQL*Plus                                                          _ | 6 | X
File  Edit  Search  Options  Help
SQL> CREATE OR REPLACE PROCEDURE create_temp_table (table_name UARCHAR2)
  2  AS
  3     cursor_id NUMBER;
  4     ddl_stmt UARCHAR2(500);
  5  BEGIN
  6     --open the cursor
  7     cursor_id := DBMS_SQL.OPEN_CURSOR;
  8     --specify the SQL string to create the table
  9     ddl_stmt := 'CREATE TABLE ' || table_name || '(table_id NUMBER(6))';
 10     --parse and execute the statement
 11     DBMS_SQL.PARSE(cursor_id, ddl_stmt, DBMS_SQL.U7);
 12     --close the cursor
 13     DBMS_SQL.CLOSE_CURSOR(cursor_id);
 14  END;
 15  /

Procedure created.
```

Figure 5-59 Processing a DDL command using Dynamic SQL

To be able to create Dynamic SQL programs, your database account must have been granted the EXECUTE privilege on the DBMS_SYS_SQL package, which is in the SYS database schema. If an insufficient privileges error appears when you try to execute this procedure, and you are working on a remote database, contact your instructor or technical support person. If this error appears and you are working on a Personal Oracle database, start SQL*Plus, connect using the username INTERNAL and the password ORACLE, and type the following command to grant the privilege:

GRANT execute on DBMS_SYS_SQL to PUBLIC;.

2. Type the following command at the SQL prompt to call the procedure and pass to it the value 'my_table' for the table name: **EXECUTE CREATE_TEMP_TABLE ('my_table');**. The message "PL/SQL procedure successfully completed" appears, indicating that the procedure was successfully executed.

If you are using Personal Oracle, and you receive an error stating you have insufficient privileges to create the table, log onto SQL*Plus as username INTERNAL, password ORACLE, and use this account for all Dynamic SQL procedures that require creating a new table.

3. Type the following command at the SQL prompt to confirm that MY_TABLE was successfully created: **DESCRIBE my_table;**. The table field TABLE_ID should appear, with data type NUMBER(6), as specified in the procedure.

Processing Dynamic SQL Programs That Contain SELECT Commands

Creating a Dynamic SQL program that contains a SELECT command is useful when you are integrating a different user interface environment with an Oracle database. You might have the user specify the text of the SQL query in a text editor, and then pass the query text to the Dynamic SQL program. The steps for creating a Dynamic SQL procedure that contains a SELECT command are:

1. Open the cursor.

2. Define the SQL command as a text string.

3. Parse the SQL command.

4. Bind input variables to placeholders.

5. Define output variables using the DEFINE_COLUMN procedure. This procedure specifies the variables within the procedure that will be used to reference the retrieved columns. The procedure syntax is **DBMS_SQL.DEFINE_ COLUMN (*cursor_id, column_position, variable_name, maximum_ character_column_size*);**. Each column is defined by its position: The first column is in position 1, the second column is in position 2, and so forth. If the SELECT command retrieves three columns, then you need to execute the DEFINE_COLUMN procedure three times, for positions 1, 2, and 3. The *variable_name* parameter is a variable in the procedure that has been declared using the same data type and size as the retrieved column. The *maximum_character_column_size* parameter is only used for CHAR and VARCHAR2 columns and defines the maximum column width.

6. Execute the query.

7. Fetch the rows using the FETCH_RECORD function. This function fetches each record into a buffer that temporarily stores the values and has the syntax ***number_of_rows_left_to_fetch* := DBMS_SQL.FETCH _RECORD (*cursor_id*);**. The FETCH_RECORD function must be processed using a loop, because the query might retrieve multiple records. If the FETCH_RECORD function returns the value 0, then the SELECT query has no more rows to return. Therefore, the comparison **IF DBMS_SQL. FETCH_RECORD (*cursor_id*) = 0** is used to test for the loop exit condition.

8. Associate the fetched rows with the output columns using the COLUMN_VALUE procedure. This procedure associates the fetched values with the appropriate output columns within the FETCH_RECORD loop. It has the syntax **DBMS_SQL.COLUMN_VALUE (*cursor_id, column _position, variable_name*);**. The parameters for the COLUMN_VALUE procedure are the same as those used in the DEFINE_COLUMN procedure, except that you do not define the maximum column width for character columns. When all records have been fetched and all column values have been assigned, you exit the processing loop.

9. Close the cursor.

Now you will create a Dynamic SQL procedure that contains a SELECT command. The procedure receives the value of a Software Experts consultant ID as an input parameter. The procedure's SELECT statement retrieves the names of all projects the consultant is associated with and the total hours spent on the project, using the input consultant ID as the search condition. It then writes the project names and consultant hours to the screen using the DBMS_OUTPUT.PUT_LINE procedure. After you have created the procedure, you will execute the procedure to retrieve project information for consultant 102.

To create a Dynamic SQL procedure that contains a SELECT command:

1. Type the commands shown in Figure 5-60 to create the procedure. Debug the procedure if necessary. The message "Procedure created" confirms when the procedure is successfully created.

Type these commands

```
Oracle SQL*Plus
File  Edit  Search  Options  Help
SQL> CREATE OR REPLACE PROCEDURE retrieve_consultant_details(curr_c_id IN NUMBER)
  2  AS
  3    cursor_id NUMBER;
  4    select_stmt VARCHAR2(500);
  5    cursor_return_value INTEGER;
  6    curr_proj_name VARCHAR2(30);
  7    curr_hours NUMBER(6);
  8  BEGIN
  9    --open the cursor
 10    cursor_id := DBMS_SQL.OPEN_CURSOR;
 11    --specify the SQL SELECT command
 12    select_stmt := 'SELECT project_name, total_hours FROM project, project_consultant
 13                    WHERE project.p_id = project_consultant.p_id AND c_id = :in_c_id';
 14    --parse the statement
 15    DBMS_SQL.PARSE(cursor_id, select_stmt, DBMS_SQL.V7);
 16    --bind the placeholder to the input parameter variable
 17    DBMS_SQL.BIND_VARIABLE(cursor_id, ':in_c_id', curr_c_id);
 18    --define the output columns
 19    DBMS_SQL.DEFINE_COLUMN(cursor_id, 1, curr_proj_name, 30);
 20    DBMS_SQL.DEFINE_COLUMN(cursor_id, 2, curr_hours);
 21    --execute the statement
 22    cursor_return_value := DBMS_SQL.EXECUTE(cursor_id);
 23    --fetch rows and associate fetched values with output columns
 24    LOOP
 25      IF DBMS_SQL.FETCH_ROWS(cursor_id) = 0 THEN
 26        EXIT;
 27      END IF;
 28      DBMS_SQL.COLUMN_VALUE(cursor_id, 1, curr_proj_name);
 29      DBMS_SQL.COLUMN_VALUE(cursor_id, 2, curr_hours);
 30      DBMS_OUTPUT.PUT_LINE(curr_proj_name || ' ' || curr_hours);
 31    END LOOP;
 32    DBMS_SQL.CLOSE_CURSOR(cursor_id);
 33  END;
 34  /

Procedure created.
```

Figure 5-60   Processing a SELECT command using Dynamic SQL

2. Type the command shown in Figure 5-61 to execute the procedure, and pass consultant ID 102 to the procedure as an input parameter. The output should appear as shown.

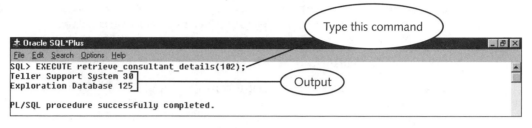

Figure 5-61   Executing the Dynamic SQL procedure and viewing the SELECT command output

Using Dynamic SQL to Create an Anonymous PL/SQL Block

A final way to use Dynamic SQL is to create a procedure that creates an anonymous PL/SQL block as a text string and defines its parameters when the procedure executes. This provides a way to use Dynamic SQL in PL/SQL programs when you do not want to create a named program unit or when you do not have the necessary privileges to create named program units. This approach also provides an easier way to process SELECT commands in Dynamic SQL programs when only one record is retrieved. You can retrieve query values using an implicit cursor and associate them with procedure variables.

To create a procedure that processes an anonymous PL/SQL block using Dynamic SQL, you use the following steps:

1. Open the cursor using the OPEN_CURSOR function.

2. Define the PL/SQL block as a text string, using placeholders for dynamic values. The entire PL/SQL block, including the declarations, body, and exception section, is assigned to a text string variable. Output variables are defined in the INTO clause of an implicit cursor in the SQL query using placeholders. For example, to select a specific student first name and last name into placeholder variables, you would use the query
`SELECT s_first, s_last INTO :sf_place,`
`:sl_place FROM student WHERE s_id = :s_id_place;`. You cannot use the double-hyphen (--) method of defining comment statements, because this will cause the rest of the block to be treated as a comment statement. You must use the /* ... */ method of creating comments instead. You must include the final semicolon after the END statement within the text string.

3. Parse the SQL command using the PARSE procedure.

4. Bind input and output variables to placeholders using the BIND_VARIABLE procedure.

5. Execute the SQL command using the EXECUTE function.

6. Retrieve the values of output variables using the VARIABLE_VALUE procedure. This procedure has the syntax **DBMS_SQL.VARIABLE_VALUE**

(cursor_ID, ':placeholder_name' , output_variable_name);.
The *output_variable_name* parameter is a declared variable that has the same
data type as the associated output placeholder variable.

7. Close the cursor using the CLOSE_CURSOR procedure.

Now you will create a procedure that receives an input variable of a consultant ID. The
procedure defines and processes an anonymous PL/SQL block that involves a SQL
query with an implicit query that retrieves the consultant's first and last name and the
total hours that the consultant has spent on all projects.

To create a Dynamic SQL procedure that processes an anonymous PL/SQL block:

1. Type the commands shown in Figure 5-62 to create the procedure. Debug
   the procedure if necessary. The message "Procedure created" indicates that the
   procedure is successfully created.

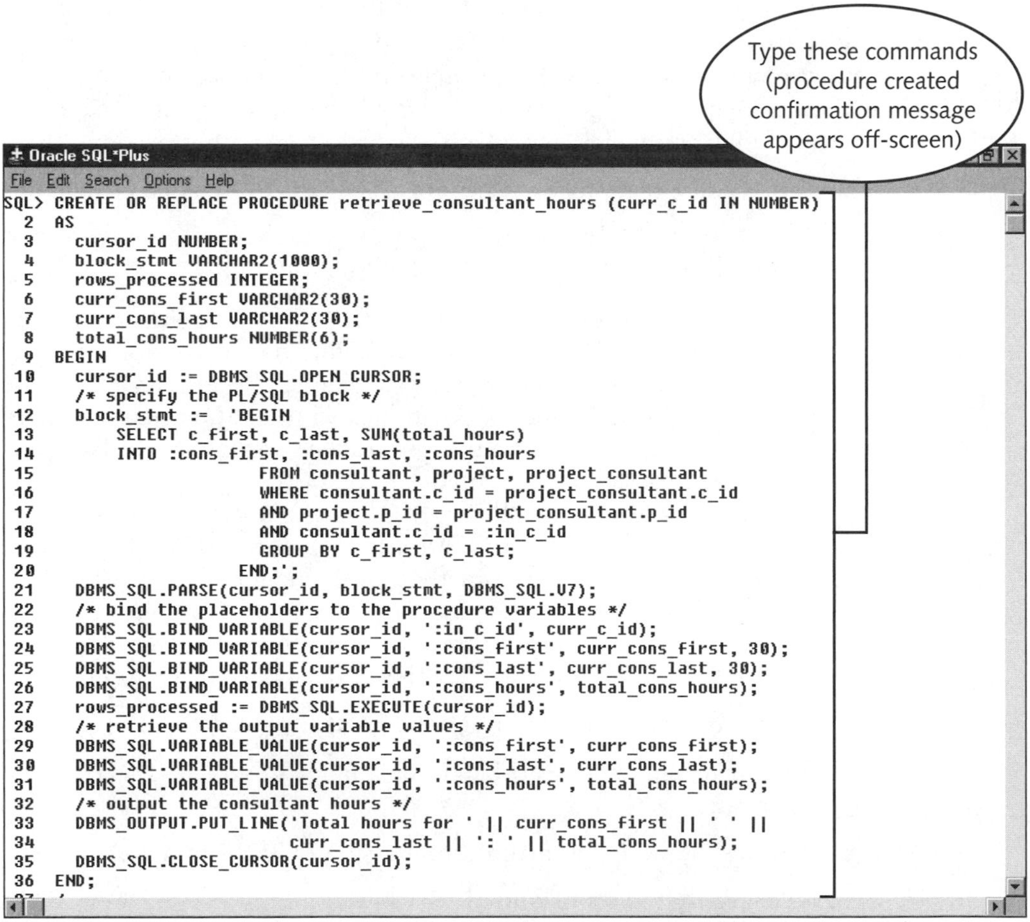

Type these commands
(procedure created
confirmation message
appears off-screen)

```
± Oracle SQL*Plus                                                              [_] [□] [X]
File  Edit  Search  Options  Help
SQL> CREATE OR REPLACE PROCEDURE retrieve_consultant_hours (curr_c_id IN NUMBER)
  2  AS
  3    cursor_id NUMBER;
  4    block_stmt VARCHAR2(1000);
  5    rows_processed INTEGER;
  6    curr_cons_first VARCHAR2(30);
  7    curr_cons_last VARCHAR2(30);
  8    total_cons_hours NUMBER(6);
  9  BEGIN
 10    cursor_id := DBMS_SQL.OPEN_CURSOR;
 11    /* specify the PL/SQL block */
 12    block_stmt :=   'BEGIN
 13        SELECT c_first, c_last, SUM(total_hours)
 14        INTO :cons_first, :cons_last, :cons_hours
 15                  FROM consultant, project, project_consultant
 16                  WHERE consultant.c_id = project_consultant.c_id
 17                  AND project.p_id = project_consultant.p_id
 18                  AND consultant.c_id = :in_c_id
 19                  GROUP BY c_first, c_last;
 20                END;';
 21    DBMS_SQL.PARSE(cursor_id, block_stmt, DBMS_SQL.V7);
 22    /* bind the placeholders to the procedure variables */
 23    DBMS_SQL.BIND_VARIABLE(cursor_id, ':in_c_id', curr_c_id);
 24    DBMS_SQL.BIND_VARIABLE(cursor_id, ':cons_first', curr_cons_first, 30);
 25    DBMS_SQL.BIND_VARIABLE(cursor_id, ':cons_last', curr_cons_last, 30);
 26    DBMS_SQL.BIND_VARIABLE(cursor_id, ':cons_hours', total_cons_hours);
 27    rows_processed := DBMS_SQL.EXECUTE(cursor_id);
 28    /* retrieve the output variable values */
 29    DBMS_SQL.VARIABLE_VALUE(cursor_id, ':cons_first', curr_cons_first);
 30    DBMS_SQL.VARIABLE_VALUE(cursor_id, ':cons_last', curr_cons_last);
 31    DBMS_SQL.VARIABLE_VALUE(cursor_id, ':cons_hours', total_cons_hours);
 32    /* output the consultant hours */
 33    DBMS_OUTPUT.PUT_LINE('Total hours for ' || curr_cons_first || ' ' ||
 34                    curr_cons_last || ': ' || total_cons_hours);
 35    DBMS_SQL.CLOSE_CURSOR(cursor_id);
 36  END;
```

Figure 5-62   Processing an anonymous PL/SQL block using Dynamic SQL

2. Type the following command at the SQL prompt to test the procedure: **`EXECUTE retrieve_consultant_hours(102);`**. The output "Total hours for Brian Zhang: 155" should appear, indicating that the procedure executed correctly.

3. Close SQL*Plus, Notepad and all other open Oracle applications.

SUMMARY

◻ Oracle provides many built-in packages in the SYS database schema that are accessible to both client and server programs, and provide support for basic database functions.

◻ The main transaction processing packages are DBMS_ALERT, DBMS_LOCK, DBMS_SQL, and DBMS_TRANSACTION. They provide procedures related to locking and releasing database objects to support transaction processing, supporting the COMMIT and ROLLBACK commands, and specifying query properties at runtime.

◻ The application development packages aid developers in creating and debugging PL/SQL applications. The main packages in this category are DBMS_DESCRIBE, DBMS_JOB, DBMS_OUTPUT, DBMS_PIPE, DBMS_SESSION, and UTL_FILE.

◻ The DBMS_JOB package enables PL/SQL program units to be scheduled to run at specific times by creating a job queue, which is a list of program units to be run, and the details for when and how often they are to be run.

◻ The DBMS_PIPE package implements database pipes, which are used to transfer information from one database session to another through the database and independently of transactions.

◻ A pipe message is specified using the PACK_MESSAGE procedure. The message can be any data type, and you can send multiple items of different data types using multiple calls to this procedure.

◻ A pipe message is sent using the SEND_MESSAGE procedure. Once a message is sent, it remains in a buffer on the database server until the database is shut down and then restarted.

◻ You can receive a pipe message any time the pipe is available on the database server. You receive a message using the RECEIVE_MESSAGE procedure and unpack it using the UNPACK_MESSAGE procedure.

◻ The database and application administration packages support database administration tasks, such as managing memory on the database server, managing how disk space is allocated and used, and recompiling and managing stored program units and packages. The main packages in this category are DBMS_APPLICATION_INTO, DBMS_DDL, DBMS_SHARED_POOL, DBMS_SPACE, and DBMS_UTILITY.

❑ The DBMS_DDL package has a procedure named ALTER_COMPILE that you can use to compile PL/SQL program units. You use this procedure to create a program unit that quickly and easily recompiles server program units at once.

❑ The internal support packages provide the underlying functionality of the Oracle database. Because these packages are frequently used, users often do not need to preface the procedure name with the package name. Primary packages in this category are STANDARD, DBMS_REFRESH, DBMS_SNAPSHOT, and DBMS_STANDARD.

❑ Dynamic SQL enables users to create procedures that define SQL commands in which the query parameters are specified at runtime rather than at compile time. This allows you to dynamically structure queries based on user inputs and use DDL commands in PL/SQL programs.

❑ Procedures for manipulating cursors within Dynamic SQL programs are included in the DBMS_SQL built-in package.

❑ You must structure Dynamic SQL programs differently depending on whether they are processing DML commands, DDL commands, SQL SELECT commands, or anonymous SQL blocks.

REVIEW QUESTIONS

1. What user schema owns the Oracle database built-in packages?
2. What is the purpose of the DBMS_JOB built-in package?
3. What is a database pipe?
4. By default, how long is a message associated with a database pipe available for retrieval?
5. What is the ALTER_COMPILE procedure used for?
6. What is the purpose of the STANDARD built-in package?
7. What is the difference between a static SQL command and a dynamic SQL command?
8. What built-in package contains the procedures and functions used to process Dynamic SQL commands?
9. What is a placeholder?
10. Describe how placeholders are used in Dynamic SQL commands.

PROBLEM-SOLVING CASES

Cases refer to the Clearwater Traders (see Figure 1-11), Northwoods Universtiy (see Figure 1-12), and Software Experts (see Figure 1-13) databases. Save all solution files in your Chapter5\CaseSolutions folder.

1. In this case, you will create a database pipe that sends the minimum, maximum, and average age of each student in the Northwoods University database. Then, you will create an anonymous PL/SQL program that receives the pipe. Save the source code for both programs in a file named 5CCase1.sql.

   a. Create an anonymous program unit that first calculates the minimum, maximum, and average age of the students in the Northwoods University database, based on the difference between the student date of birth and the current system date. Truncate the values so the decimal portion of the age is not included. Then, add code to create a pipe named STUDENT_AGE_PIPE that packs the minimum, maximum, and average age, and then sends the pipe.

   b. Create a second anonymous PL/SQL program that retrieves the STUDENT_AGE_PIPE, and displays the contents using the DBMS_OUTPUT.PUT_LINE procedure. The program output should be as follows:

   ```
   Maximum student age: 21
   Minimum student age: 17
   Average student age: 18
   ```

2. In this case, you will create a database pipe that sends the total number of order sources in the Clearwater Traders database and the description of each order source. Then, you will create an anonymous PL/SQL program that receives the pipe. Save the source code for both programs in a file named 5CCase2.sql.

   a. Create an anonymous PL/SQL program that contains code to create a pipe named ORDER_SOURCE_PIPE that packs and sends a pipe that contains the total number of order sources and the description of each order source. Use an implicit cursor to count the total number of order sources and an explicit cursor to retrieve each order source description from the ORDER_SOURCE table. Pack the total number of order sources and each individual order source as a separate pipe message, and then send the message.

   b. Create a second anonymous PL/SQL program that retrieves the ORDER_SOURCE_PIPE and displays the contents using the DBMS_OUTPUT.PUT_LINE procedure. When you unpack the pipe, unpack and display the number of order sources first. Then, use a FOR loop to unpack the source descriptions, using the total number of order sources to terminate the FOR loop. The program output should be as follows:

   ```
   Number of order sources: 6
   Winter 2002
   Spring 2003
   Summer 2003
   Outdoor 2003
   Children's 2003
   Web Site
   ```

3. In SQL*Plus, create a procedure named UPDATE_CONSULTANT_SKILL that uses Dynamic SQL to update the CONSULTANT_SKILL table in the Software Experts database. The procedure will receive input parameters of the consultant

ID, skill ID, and new certification status. These values should be replaced by place-holders in the SQL UPDATE command and bound to the input parameters at runtime. Save the procedure source code in a file named 5CCase3.sql. Below the source code for the procedure, include an EXECUTE command to update the certification status of consultant ID 105 for skill ID 3 to 'Y'.

4. In SQL*Plus, create a procedure named LIST_PROJECT_SKILLS that uses Dynamic SQL to retrieve all of the skills required for a specific project in the Software Experts database. The procedure will receive an input parameter of a project ID and will retrieve the skill descriptions associated with the project. If the project is a parent project that has subprojects, the query should retrieve the subproject skills as well. Save the procedure source code in a file named 5CCase4.sql. Below the source code, include an EXECUTE command to list the project skills associated with project ID 4. The output should appear as follows:

```
Windows NT Network Administration
Windows 2000 Network Administration
Java Programming
Oracle Developer Programming
Oracle Database Administration
```

5. In SQL*Plus, create a procedure named REFRESH_EVALUTION_TABLE that uses Dynamic SQL to drop the EVALUATION table in the Software Experts database, recreate the table, and insert the first record into the table. When you recreate the table, omit the COMMENTS field. Save the source code in a file named 5CCase5.sql. (*Hint*: You can find the SQL commands to create the EVALUATION table in the softwareexp.sql file in the Chapter5 folder on your Data Disk. To place embedded single quotation marks within a text string, type the single quotation mark twice.)

6. In this case, you will use SQL*Plus to create a procedure that dynamically adds a new column to a table. Then, you will use the procedure to modify the INVENTORY table in the Clearwater Traders database so it includes a column to represent the value of each inventory item. Finally, you will create a second procedure that uses Dynamic SQL to automatically refresh the inventory values.

   a. Create a procedure named ADD_TABLE_COLUMN that uses Dynamic SQL to add a new column to a table. The procedure should accept input parameters of the table name, the column name, and a text string representing the new column's data type. Then, execute a command that uses this procedure to alter the INVENTORY table so it includes a new column named INV_VALUE, which will contain the value of each inventory item, which is calculated as QOH * PRICE. Save the ADD_TABLE_COLUMN procedure and the EXECUTE command that calls the procedure in a file named 5CCase6.sql.

   b. Create a procedure named UPDATE_INV_VALUE that receives an input parameter of an inventory ID and then uses Dynamic SQL to update each record in the INVENTORY table so the INV_VALUE column contains a value representing the current QOH * PRICE. Use a placeholder for the inventory ID search condition value in the UPDATE command, and bind the value to the placeholder at runtime. Save the source code for the procedure in the 5CCase6.sql file. Include the EXECUTE command that calls the procedure to update the value of inventory ID 11668.

7. In SQL*Plus, create a procedure named RETRIEVE_SOURCE_SALES that receives an input parameter of a Clearwater Traders order source ID and then uses Dynamic SQL to retrieve each order ID associated with that order source, along with the order date and the order total amount, calculated as order quantity times price. Use a placeholder in the SQL command and dynamically bind the placeholder to the order source ID parameter value at runtime. Use the DBMS_OUTPUT.PUT_LINE procedure to display the output on the screen. Save the procedure source code in a file named 5CCase7.sql. Include the EXECUTE command to retrieve the data for order source 2. The output should look like this:

```
1057 29-MAY-03      $379.89
1059 31-MAY-03      $409.90
```

(You may have additional output lines if you have inserted additional records in your database tables.)

8. In SQL*Plus, create a procedure named RETRIEVE_FACULTY_COURSES that receives input parameters of a Northwoods University faculty ID and a term ID and then uses Dynamic SQL to retrieve each course section ID and the associated call ID, section number, day, time, building code, and room for the specified faculty ID and term. Use placeholders in the SQL command for the faculty ID and term ID search conditions and dynamically bind the placeholders to the parameter values at runtime. Use the DBMS_OUTPUT.PUT_LINE procedure to display the output on the screen. Save the procedure source code in a file named 5CCase8.sql. Include the EXECUTE command to retrieve the data for faculty ID 1 for term ID 5. The output should look like this:

```
+                                            +
Section ID: 1005 MIS 441
Section 1 MWF 09:00 AM BUS 105
+                                            +
Section ID: 1006 MIS 441
Section 2 MWF 10:00 AM BUS 105
```

9. Create a procedure named RETRIEVE_STUDENT_CREDITS that receives an input parameter of a Northwoods University student ID and then uses Dynamic SQL to create an anonymous block. This block will use an implicit cursor to retrieve the student's first and last name and the total number of credits that have been earned by that student (if the GRADE field in the ENROLLMENT table is not NULL). Use placeholders in the SQL command for the student ID search condition value and for the values retrieved into the implicit cursor and dynamically bind the placeholders at runtime. Use the DBMS_OUTPUT.PUT_LINE procedure to display the output on the screen. Save the procedure source code in

a file named 5CCase9.sql. Include the EXECUTE command to retrieve the data for student ID 100. The output should look like this:

```
Student: Sarah Miller
Total Credits: 12
```

10. Create a procedure named CONSULTATION_CERTIFICATIONS that receives an input parameter of a Software Expert skill ID and uses Dynamic SQL to create an anonymous PL/SQL block with an implicit cursor to retrieve the skill description and the number of consultants who are certified with that skill. Bind all input and output variables at runtime. Then, add code to the procedure that uses a Dynamic SQL SELECT command to retrieve the first and last names of all of the consultants who are certified for the specified skill. (To embed a single quotation mark within a PL/SQL text string, type the single quotation mark twice.) Save the procedure source code in a file named 5CCase10.sql. Include the EXECUTE command to retrieve the data for skill ID 1. The procedure output should look like this:

```
Visual Basic Programming Certifications: 2
Consultants:
Mark Myers
Sarah Carlson
```

6

CREATING ORACLE DATA BLOCK FORMS

◀ LESSON A ▶

Objectives

- ◆ View, insert, update, and delete data records using a data block form
- ◆ Create a single-table data block form that displays one record at a time
- ◆ Understand form components
- ◆ Use the Object Navigator within Form Builder to change the names of form components
- ◆ Create a single-table data block form that displays multiple records

Y ou have learned to use SQL commands to insert, update, delete, and view database data. It is not practical, however, to expect users to regularly interact with a database by creating SQL queries. Instead, users use applications called forms to interact with a database. A form looks like a paper form and provides a graphical interface that allow users to easily insert new database records and to modify, delete, or view existing records. Programmers use an Oracle application named Form Builder to create Windows-based forms. Form Builder is part of the Oracle application development utility called Developer 6/6i. In this chapter, you will learn how to create and use data block forms which are forms that are explicitly associated with specific database tables.

USING A DATA BLOCK FORM

Before you learn how to create a data block form, you will run an existing form to become familiar with its appearance and functionality. This form is associated with the CUSTOMER table in the Clearwater Traders database (see Figure 1-11). First, you will use a form to insert a new record into the CUSTOMER table. Using a form to add records is the equivalent of adding records in SQL using the INSERT command.

Before you open the form, you will run the SQL scripts to refresh your database tables.

To refresh your database tables:

1. Start SQL*Plus and log onto the database.

2. Run the **clearwater.sql**, **northwoods.sql**, and **softwareexp.sql** scripts from the Chapter6 folder on your Data Disk, and then close SQL*Plus.

If you are storing your Data Disk files on floppy disks, you will need to use two floppy disks for this chapter. When your first disk becomes full, save your solutions files on the second disk.

Next, you will run the form application file that is stored on your Data Disk.

To run the form application:

1. Start Windows Explorer, navigate to the Chapter6 folder on your Data Disk, and then double-click the **CustomerDemo.fmx** file. The Developer Forms Runtime Logon window opens. Log onto the database, and then click **Connect**. The Oracle Forms Runtime application opens and displays the Clearwater Traders Customers form, as shown in Figure 6-1.

When you start an FMX file by double-clicking it in Windows Explorer, the filename and pathname cannot contain any spaces.

If the Open With dialog box opens, it means that files with an .fmx extension have not yet been associated with the Forms Runtime application on your computer. Type Forms Builder Executable in the Description box, click Other, navigate to your *Developer_Home*\BIN folder, click ifrun60.exe, click Open, and then click OK.

The outer window is the Forms Runtime window, which is where all form applications run. The inner window displays the Clearwater Traders customer form, which you can use to insert a new record and to modify, delete, or view existing records in the CUS-TOMER table. The Forms Runtime window has pull-down menus and a toolbar that you can use to insert, view, modify, and delete records.

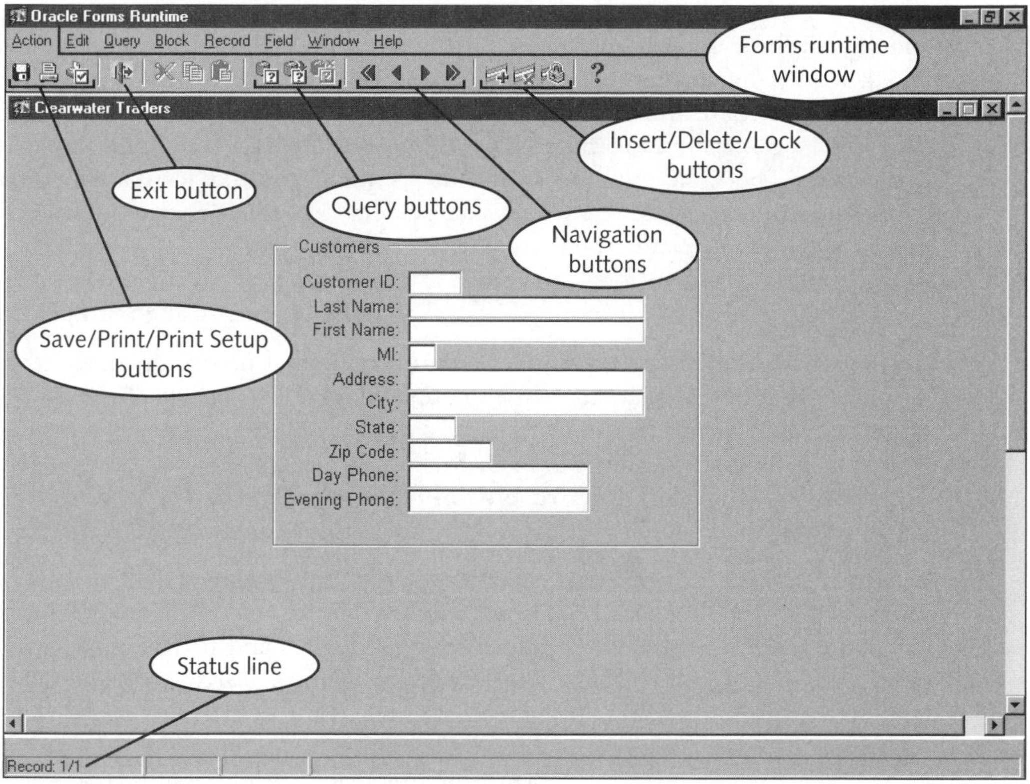

Figure 6-1   Clearwater Traders Customers form

The toolbar has buttons for calling application functions and manipulating data. The first button group includes the **Save button** 🖫, used to save a new record or to save changes made to an existing record; the **Print button** 🖺, to print the current form; and the **Print Setup button** 🖼, to reconfigure the printer setup. The **Exit button** ⬆ exits the form.

The query button group is used for querying data. When a form is running, it can be in one of two modes: Normal or Enter Query. When a form is in **Normal** mode, you can view data records, sequentially step through the records, and change data values. When a form is in **Enter Query** mode, you can enter a search parameter in the form fields, and then retrieve the associated records. To move into Enter Query mode, you click the **Enter Query button** 🔲 on the toolbar. This clears the form fields and allows you to enter a search condition. To return to Normal mode, you must either execute the query or cancel the query. To execute the query, you click the **Execute Query button** 🔲. This retrieves the records associated with the search condition and returns the form to Normal mode. To cancel the query, you click the **Cancel Query button** 🔲. This also returns the form to Normal mode without retrieving any data.

The navigation button group is for navigating among different records and different blocks. A **block** is a group of related form items, such as text fields and buttons. A data block form contains one or more data blocks. Each **data block** is a group of related form items that are associated with a specific database table. A data block contains items that correspond to one or more of the table's data fields. These items display the field's data values. The most common kind of block items are **text items**, which display text data values in text boxes. The Clearwater Traders Customers form contains a single data block, with text items to represent all of the fields in the CUSTOMER table. Since this form displays data from only one table, it is a single-block form. Later, you will learn how to create a form that displays data from multiple tables using multiple data blocks.

In the navigation button group, the **Previous Block button** ◀◀ moves the insertion point to the previous data block in a multiple block form, and the **Next Block button** ▶▶ moves the insertion point to the next data block. When the results of a query appear, you can sequentially step forward and backward through the records to view them one at a time. The **Previous Record button** ◀ moves back to the previous record in the table, and the **Next Record button** ▶ moves to the next record in the table.

The final button group is used to insert, delete, and lock records. The **Insert Record button** ⬚ clears the form fields and creates a blank record into which the user may enter new data. The user must enter all required fields and then save the record or else delete the new record by clicking the **Remove Record button** ⬚. The ⬚ button can also be used to remove an existing record. The **Lock Record button** ⬚ locks the current record so that other users cannot update or delete it. The bottom of the window has a status line, which provides information about how many records have been retrieved and shows the current form mode (Normal or Enter Query).

Using a Form to Insert, Update, and Delete Records

You can use a form to perform the DML operations of inserting, updating, and deleting records. Now you will use the form to insert a new record into the CUSTOMER table.

To insert a new record using the form:

1. If necessary, place the insertion point in the Customer ID text item, and then type **1000**.

2. Press **Tab** to navigate from field to field and enter the data values shown in Figure 6-2.

3. Click the **Save button** ⬚ on the toolbar. The confirmation message "FRM-40400: Transaction complete: 1 records applied and saved" appears on the status line in the lower-left corner of the screen. (If your mouse pointer is still on ⬚, then this message will be overwritten with the "Save" ToolTip.)

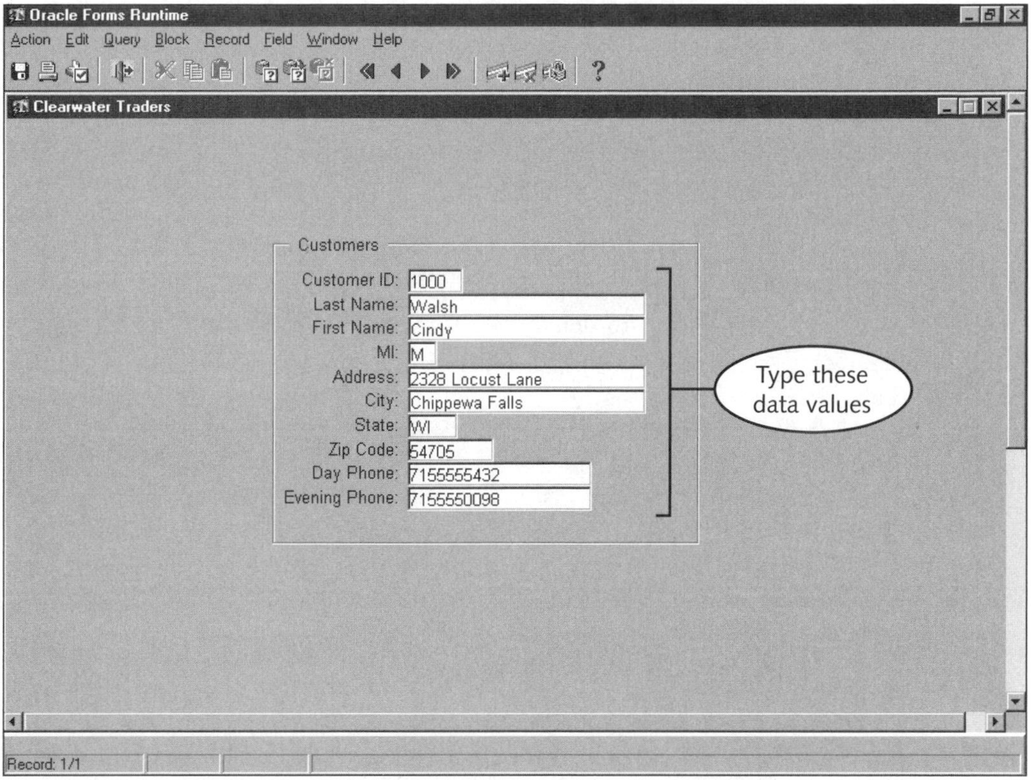

Figure 6-2   Inserting a new record

Next, you will modify an existing record in the CUSTOMER table. Using a form to modify records is the equivalent of modifying records using the SQL UPDATE command. You will retrieve the record for customer Paula Harris and then change her evening telephone number.

To update an existing record:

1. Click the **Enter Query button** ![icon] on the toolbar. The form fields are cleared, and the insertion point appears in the Customer ID field. Notice the message "Enter Query" appears in the status line. Also, notice that the mode indicator in the status line at the bottom of the form indicates that the form is in Enter Query mode.

2. Type **107** in the Customer ID field, which is customer Paula Harris's customer ID.

3. Click the **Execute Query button** ![icon] to execute the query. The complete data record for customer Paula Harris appears in the form. Notice that the messages in the status line no longer appear.

Pressing F8 is equivalent to clicking the Execute Query button 🔲 on the toolbar, and pressing Ctrl + Q is equivalent to clicking the Cancel Query button 🔲.

4. Place the insertion point in the Evening Phone field, and then change the phone number to **7155558975**.

5. Click the **Save button** 🔲. The confirmation message "FRM-40400: Transaction complete: 1 records applied and saved" appears on the status line.

You can also use a form to delete data records. You will do this next.

To delete a record from the CUSTOMER table:

1. Click the **Enter Query button** 🔲, do not enter a search condition, click the **Execute Query button** 🔲, and click the **Next Record button** ▶ on the toolbar until you see customer Cindy Walsh's record.

2. Click the **Remove Record button** 🔲 on the toolbar to delete Cindy's record. Alissa Chang's record now appears in the form because her record was just before Cindy's record in the database.

Using a Form to View Table Records

You can also use a form to view table records. One approach is to retrieve all of the records in the table and step through them sequentially. To view the first record in a table, place the form in Enter Query mode, but do not enter a search condition. Instead, click the Execute Query button 🔲. Since no search condition was specified, the first record in the table appears. Then, you can use the record navigation buttons to step through the records sequentially.

To display the first CUSTOMER record and step through the records sequentially:

1. Click the **Enter Query button** 🔲, do not enter a search condition, and then click the **Execute Query button** 🔲 on the toolbar. Paula Harris's data record, which is the first CUSTOMER record, appears on the form.

2. Click the **Next Record button** ▶ on the toolbar. The record displaying data for the next customer in the database, customer Mitch Edwards, appears.

3. Click the **Previous Record button** ◀ on the toolbar. Paula Harris's record appears again because it is the record directly before Mitch Edwards's record.

4. Click ▶ again. Continue clicking ▶ until you scroll through all of the CUSTOMER records.

Sequentially viewing all table records works well for the small sample databases you are using in this book, but it won't work when you are looking for a specific record in a database that contains thousands of records. When you are working with large databases, you can use another approach. You can enter search conditions in a form to retrieve specific records, just as when you use the WHERE clause in a SQL command. In Form Builder, as in SQL*Plus, searches involving text strings are always case sensitive.

To use a query to find a specific record:

1. Click the **Enter Query button** 🔲 on the toolbar. The message in the message area prompts you to enter a query.

2. Place the insertion point in the Zip Code field, type **53821**, and then click the **Execute Query button** 🔲 to execute the query. Lee Miller's customer record appears. Queries can retrieve multiple records, as you will see next.

3. Click 🔲 again, and note that the fields are cleared. Type **WI** in the State field, and then click 🔲 to execute the query.

4. Click the **Next Record button** ▶ to scroll through the records, and confirm that all five CUSTOMER records have been retrieved.

So far you have done exact searches. You can also do **restricted searches**, which retrieve records that contain data values that fall within a range of values or contain partial text strings. Table 6-1 summarizes the form restricted search operators.

Search Operator	Function	Example
_ (underscore)	Wildcard replacing a single character	715_____ (715 followed by seven underscores)
%	Wildcard replacing any number of characters	715%
> and <	Greater than or less than	>200
>= or <=	Greater than or equal to, or less than or equal to	>=200
<> or !=	Not equal to	<>WI

Table 6-1   Form restricted search operators

You can use the restricted search operators with fields of any data type. For example, you could perform a restricted search looking for all dates in the year 2003 using the following search condition: %2003. Now you will use the % wildcard character to perform a restricted search for all customers whose customer ID is greater than 200. Then, you will perform a restricted search for all customers who have the string "Apt" in their address.

To perform restricted searches:

1. Click the **Enter Query button** 🔲, type **>200** in the Customer ID text item as the search condition, and then click the **Execute Query button** 🔲. The record for customer Mitch Edwards, customer ID 232, appears.

2. Click 🔲 again, type **%Apt%** in the Customer Address text item as the search condition, and then click 🔲. The query retrieves the record for Paula Harris, who is the only customer with the string "Apt" in her address.

Viewing and Interpreting Forms Runtime Errors

When a user makes an error while entering form data, an error code and message appear in the status line. Now, you will learn how to get more information about user errors. You will try to enter a record that uses the incorrect data type in the Customer ID text item and a record that uses a primary key value that already exists in the database, and view information about the errors.

To view error information in Forms Runtime:

1. Click the **Insert Record button** ; make sure the insertion point is in the Customer ID: field; type **aaa**, which is the wrong data type for the Customer ID field; and then press **Tab**. The error message "FRM-50016: Legal characters are 0 − 9 - + E" appears in the status line, indicating that the entered value must be a number. This error is trapped within the form, and the data will not be sent to the database until the error is corrected.

2. Change the Customer ID value to **107**, press **Tab**, type **James** for the Customer Last Name, press **Tab** again, and type **Henry** for the Customer First Name. Note that 107 already is the CUST_ID value for customer Paula Harris, so it is a duplicate primary key. Click the **Save button** to save the record. The error message "FRM-40508: ORACLE error: unable to INSERT record" appears in the status line. An ORA error indicates that the data values were submitted to the database, and an error was returned by the database.

3. To get information about the cause of the error, click **Help** on the menu bar, and then click **Display Error**. The Database Error dialog box opens and indicates that the error is caused by a unique primary key violation. Click **OK** to close the Display Error dialog box.

4. Click the **Remove Record button** to remove the record with the erroneous values from the form.

Closing a Form and Committing Changes

If you have any unsaved changes when you close a form, you are prompted to commit or roll back your changes. Now you will update a record, close the form, and then commit your changes.

To update a record and then close the form and commit the changes:

1. Click the **Enter Query button** , type **107** in the Customer ID text item, and then click the **Execute Query button** . The data for customer Paula Harris appears.

2. Change Paula's Zip Code value to **54703**.

3. Click the **Exit button** on the Forms Runtime toolbar. A message box asking if you want to save your changes appears. You can click Yes to commit your changes, No to roll back your changes, or Cancel to return to the form.

You can also close the Forms Runtime window to exit the form, or click Action on the menu bar, and then click Exit.

 4. Click **Yes** to commit your changes, and if necessary, click **OK** to save your changes and close Forms Runtime.

CREATING THE CUSTOMER FORM

The customer form is a data block form. Recall that a data block is a block that is related to a database table. When you create a data block form, the system automatically generates the text items and labels for data fields in that table and then provides the code for inserting, modifying, deleting, and viewing data records. Now, you will start Form Builder and begin to create the CUSTOMER form.

To start Form Builder:

 1. Click **Start** on the Windows taskbar, point to **Programs**, point to **Oracle Forms 6i**, and then click **Form Builder**. The Welcome to the Form Builder dialog box appears. If you cannot find the Form Builder program, ask your instructor or technical support person for help.

 Another way to start Form Builder is to start Windows Explorer, navigate to your *Developer_Home*\BIN folder, and then double-click ifbld60.exe (Windows NT users click the ORANT\BIN\ifbld60.exe file).

This dialog box gives you options for creating new forms and opening existing forms. It also allows you to go directly to the Quick Tour and Cue Cards learning features. You will first create a new data block form using the **Data Block Wizard**, which displays a series of pages that automatically guide you through the form-building process.

 If the Object Navigator window appears instead of the Welcome to the Form Builder dialog box, click Tools on the menu bar, then click Data Block Wizard, and continue with the next set of steps.

To use the Data Block Wizard to create a new form:

 1. Be sure that the Use the Data Block Wizard option button is selected, and then click **OK**. The Data Block Wizard Welcome page appears. This page describes how you can use the Data Block Wizard to create a data block based on a table, view, or set of stored procedures.

 2. Click **Next**. The Data Block Wizard Type page appears.

3. The **Type page** allows you to choose whether you will create the data block by using a table or view, or by using a stored procedure. Since this data block will be associated with the CUSTOMER table, confirm that the Table or View option button is selected, and then click **Next**. The Table page appears.

The Data Block Wizard **Table page** allows you to select the database table that you will use to create the data block. It also allows you to select the specific table fields that you will include in the block and whether you want to enforce integrity constraints in the form application. Integrity constraints (such as unique primary keys and foreign key references) are always enforced by the database, but you can enforce them in the form application as well, so data values that violate integrity constraints are flagged before they are submitted to the database.

Before you can select a database table, you must connect to the database using your usual user name, password, and connect string. Now you will connect to the database, select the CUSTOMER table, and include all of its fields in the data block.

To connect to the database and select the CUSTOMER table and all of its fields:

1. Click **Browse**. The Connect dialog box opens. Type your username, password, and connect string as usual, and then click **Connect**. The Tables dialog box opens, as shown in Figure 6-3.

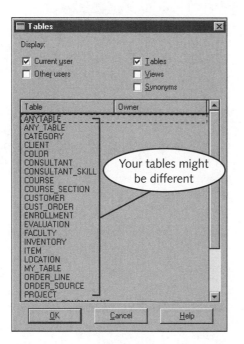

Figure 6-3   Tables dialog box

Tip   If the Current user check box is checked, then every table that you own is listed in the Table list. If the Other users check box is checked, every table in the database is listed. You would check the Other users check box only when you want to use another user's database table in your form. If you use a table that was created by another user, that user must first grant you the required privileges for inserting, updating, deleting, or viewing data. You also are given the option of displaying different kinds of database objects: Tables, Views, or Synonyms. In this lesson, you will use only tables.

2. Make sure that the Current user and Tables check boxes are checked and that the other check boxes are cleared. Note that the list displays every database table you own.

3. Click **CUSTOMER**, and then click **OK** to select the CUSTOMER table and return to the Data Block Wizard Table page, as shown in Figure 6-4. CUSTOMER appears as the table, and the customer fields appear in the Available Columns list. The icon beside each field indicates whether the field has a number or character data type.

To include a specific column in the data block, you click the column name to select it and then click the **Select button** ⏵ . To remove a column from the data block, you select it in the Database Items list and click the **Deselect button** ⏴ . To select all of the table columns, click the **Select All button** ⏵⏵ . To remove all of the selected columns, click the **Deselect All button** ⏴⏴ . To select several adjacent columns, click the first column, press and hold the Shift key, then click the last column. To select several nonadjacent columns, click the first column, press and hold the Ctrl key, and then click the other columns. Now, you will select all CUSTOMER column fields to be included in the data block.

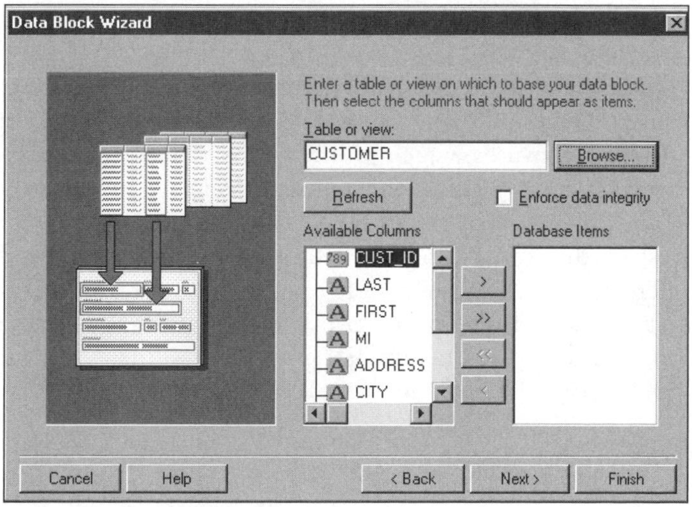

Figure 6-4   Data Block Wizard Table page

To select all of the column fields:

1. Click the **Select All button** ⟫ . The selected columns appear in the Database Items list.

If you check the Enforce data integrity check box on the Table page, the form will flag violations to the integrity constraints that you established when you created the table (unique primary keys, foreign key references, and so forth). This means that when an error occurs, you will see FRM-error codes generated by Form Builder, rather than ORA-error codes generated by the DBMS. The integrity constraints will be enforced by the DBMS even if you do not check the Integrity Constraints check box, and users will receive ORA-error codes directly from the DBMS if they violate a table integrity constraint. In this application, you will not enforce data integrity in the form, so you will not check the Enforce data integrity check box. Now, you will finish creating the block.

To finish creating the block:

1. Click **Next**. The Data Block Wizard Finish page appears.

The **Finish page** presents the options of creating the data block, then starting the Layout Wizard, or just creating the data block. The form layout specifies how the form looks to the user. The basic components of a layout include the specific data block fields that appear, the field labels, the number of records that appear at one time, and the form title. The Layout Wizard uses a series of pages to help you specify the form layout. Now you will start the Layout Wizard.

To start the Layout Wizard:

1. Be sure that the Create the data block, then call the Layout Wizard option button is selected, and then click **Finish**. The Layout Wizard Welcome page appears. Click **Next**. The Layout Wizard Canvas page appears.

A **canvas** is the area on a form where you place graphical user interface (GUI) objects, such as command buttons and text items. The Layout Wizard **Canvas page** lets you specify which canvas the block will appear on and the canvas properties. Since you have not created any canvases yet, you will accept the default (New Canvas) selection. You can select from the following canvas types: **content canvases**, which fill the entire window; **stacked canvases**, which are stacked on top of a content canvas and are used to change the appearance of the canvas by making items disappear or reappear as needed; **toolbar canvases**, which have horizontal or vertical toolbars that can be displayed or hidden programmatically; or **tab canvases**, whereby different related canvases appear on a tab page, and each canvas can be accessed by clicking a tab that is labeled with a description of the particular canvas. Since you want the CUSTOMER data to appear on a single canvas that fills the entire window, you will create a content canvas that will display all of the CUSTOMER fields.

To create a content canvas:

1. Confirm that (New Canvas) is selected as the canvas and that Content is selected as the canvas type, and then click **Next**. The **Data Block page**, which allows you to select the display fields, appears.

2. Click the Select All button ⟩⟩ to select all block fields for display, and then click **Next**. The Items page appears.

The **Items page** allows you to specify the column prompts, widths, and heights. A **column prompt** is a label that describes the data value that appears in the associated text items, such as Customer ID or Evening Phone. By default, the prompts are the same as the database column names, with blank spaces inserted in place of underscores. The column widths and heights are specified using **points**, which correspond to font sizes, as the default measurement unit. The default column widths correspond to the maximum column data widths specified in the database tables and the default form font. Now you will modify the column prompts so that they are more descriptive than the table column names.

To modify the column prompts:

1. Click in the first Prompt row, and change the text from Cust Id to **Customer ID:**.

2. Modify the rest of the prompts, as shown in Figure 6-5, and accept the default width values, even if they are different than the ones shown. Then, click **Next**. The Style page appears.

Since the form user of the form is being prompted to enter a value, prompts should end with a colon (:).

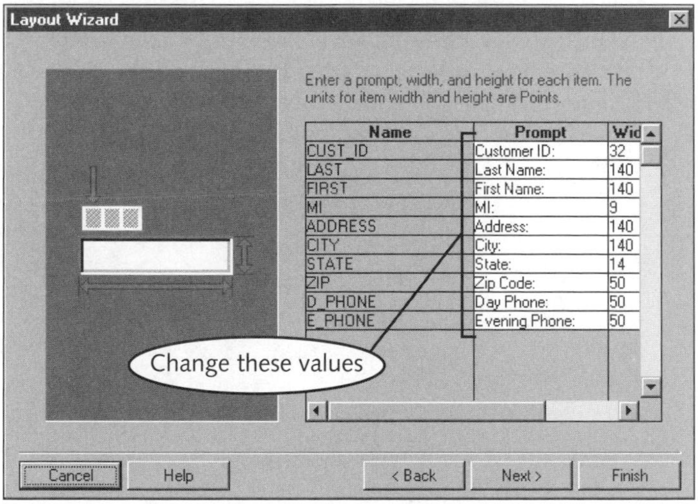

Layout Wizard

Enter a prompt, width, and height for each item. The units for item width and height are Points.

Name	Prompt	Wid
CUST_ID	Customer ID:	32
LAST	Last Name:	140
FIRST	First Name:	140
MI	MI:	9
ADDRESS	Address:	140
CITY	City:	140
STATE	State:	14
ZIP	Zip Code:	50
D_PHONE	Day Phone:	50
E_PHONE	Evening Phone:	50

Change these values

Cancel   Help   < Back   Next >   Finish

Figure 6-5 Modifying the column prompts

The **Style page** allows you to specify the layout style and properties, which determine how the data values appear. In a **form–style layout**, only one record appears at a time. In a **tabular–style layout**, multiple records appear on the form. On a tabular layout, if more records exist than can be displayed at one time, the user can use a scroll bar to view records. Since you will display only one customer record at a time, you will specify a form-style layout. Then, you will enter the title that appears in the frame around the record, specify how many records appear on the form at one time, and enter the distance between successive records.

To specify the layout style and frame title:

1. Be sure that the Form option button is selected, and then click **Next**. The Rows page appears.

2. The Rows page allows you to specify the frame title and the number of rows that will appear on the form. Type **Customers** in the Frame title box. Since you specified a form layout style, only one record appears at a time. Be sure that Records Displayed is 1, and Distance Between Records is 0, so the fields are stacked one directly after another. You do not need a scroll bar, so you will leave the scroll bar check box cleared. Click **Next**, and then click **Finish** to finish the layout.

3. Maximize the Layout Editor window, so the form layout appears as shown in Figure 6-6.

If the Layout Editor window does not open, click Tools on the menu bar, and then click Layout Editor.

Sometimes inner windows (such as the Layout Editor) do not open in a maximized state. If an inner window does not have scrollbars or appears clipped, use the outer window's vertical scroll bar to scroll to the top of the inner window, and then maximize the inner window.

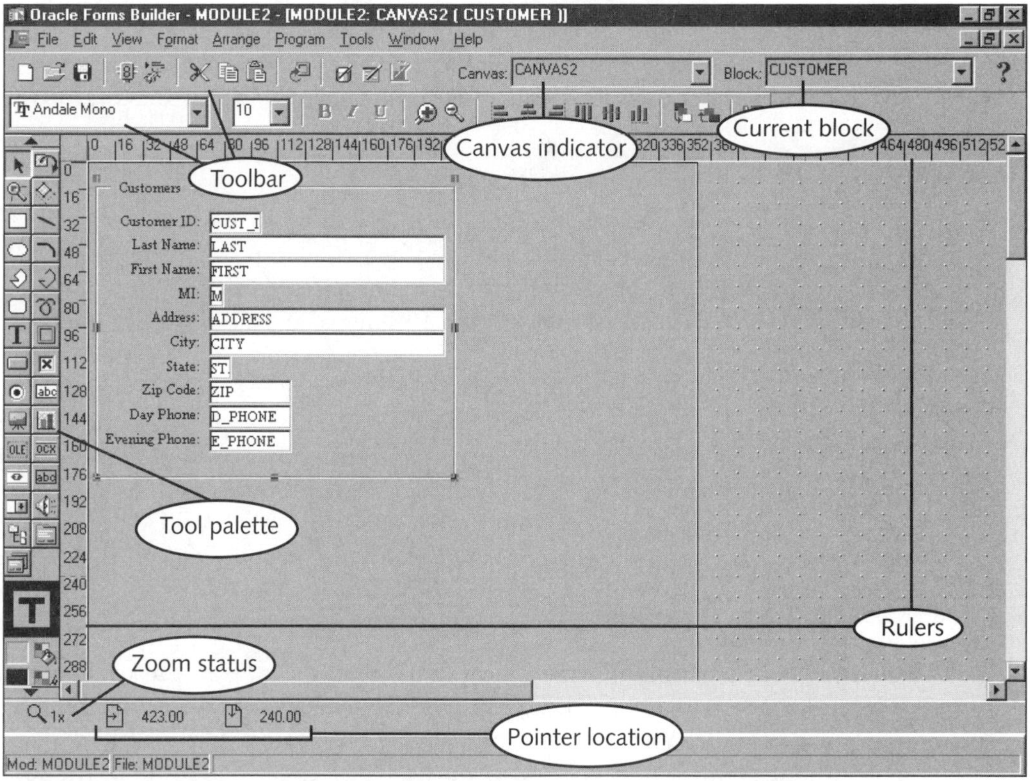

Figure 6-6 Form layout in Layout Editor

If you don't see one of the Layout Editor window elements in Figure 6-6 on your screen display, maximize the window, click View on the menu bar, and then click the name of the element that you need to display.

The **Layout Editor** provides a graphical display of the form canvas that you can use to draw and position form items, and to add boilerplate objects such as labels, titles, and graphic images. **Boilerplate objects** are objects that do not contribute directly to the form's functionality but are created to enhance its appearance. The Canvas indicator at the top of the Layout Editor window shows the name of the current canvas. When you create a new canvas using the Layout Wizard, the canvas receives a default name. In Figure 6-6, this default name is CANVAS2, but your default name might be different. The current block is the CUSTOMER block.

The Layout Editor toolbar allows you to save and edit the form and modify the form's text properties. The tool palette provides tools for creating boilerplate objects and other objects. Rulers appear along the top and left edges of the canvas, and the zoom status and pointer location indicators are on the bottom-left edge of the window. Now you will save the form.

To save the form:

1. Click the **Save button** 🖬 on the toolbar. Navigate to your Chapter6\TutorialSolutions folder, and save the file as **6ACustomer.fmb**.

Running the Form

Form Builder design files are saved as files with .fmb (form module binary) extensions. Before you can run a form to test and see if it works, you must compile the .fmb file to create an executable form file, which has an .fmx (form module executable) extension. As you develop a new form using Form Builder, you will periodically "run" the form to test it. When you run a form in Form Builder, two things must happen: First, the form design (.fmb) file must be compiled to generate a new executable (.fmx) file that contains your latest design specifications. Second, the .fmx file needs to be executed. Form Builder .fmx files are not directly executable from your workstation's operating system, but must be run using another application. There are three options for running Form Builder .fmx files: in the Forms Runtime application; as a World Wide Web application in a Web browser; and in the Form Builder Debugger environment. In this chapter, you will run forms in the Forms Runtime application. (You will run forms in the Form Builder debugger in Chapter 7.)

When you click the Run Form Client/Server button 🗐 on the toolbar in either the Object Navigator or Layout Editor, Form Builder automatically compiles your form, and then it starts Forms Runtime and displays the current form in the Forms Runtime environment. Before you run your form, you need to change some of the setup options to instruct the system to automatically save your current Form Builder design (.fmb) file and regenerate your .fmx file each time you run the form.

To change the setup options:

1. Click **Tools** on the menu bar, and then click **Preferences**. The Preferences dialog box opens. On the General tab, the Save Before Building check box instructs the system to save your Form Builder design (.fmb) file automatically each time you run a form. The Build Before Running check box instructs the system to automatically generate your .fmx file again before running it. It is advisable to check both of these boxes so that your design file will be saved automatically and generated each time you run a form.

2. Make sure that the **Save Before Building** and **Build Before Running** check boxes are checked, and then click **OK**. Do not check or clear any other check boxes or change any other options.

Now you will run the form. When you run a form, Form Builder creates a new .fmx file and sometimes also creates a text file named *form_filename*.err in the same folder as the .fmb file. The *form_filename*.err file records compilation messages and errors. If there are compilation errors, the error messages automatically appear on the screen, so you probably will not need to use this file. However, you can open it in text editor to review its contents.

To run the form:

1. Click the **Run Form Client/Server button** on the Layout Editor toolbar. After a few moments, your form should appear in the Oracle Forms Runtime window.

 You can run a form by clicking on the Object Navigator toolbar, or by pressing Ctrl + R.

2. Click the **Exit button** to close the form in Forms Runtime.

When users run your form, they will not run it in the Form Builder environment, but by opening the Forms Runtime application on their workstation and then opening the form .fmx file. You will do this next.

To open the form .fmx file:

1. Click **Start** on the taskbar, point to **Programs**, point to **Oracle Forms6i**, and then click **Forms Runtime**. The Forms Runtime Options window opens. This window allows the user to select the form .fmx file, specify his or her database username, password, and connect string, and specify runtime options.

2. Click **Browse**, navigate to your Chapter6\TutorialSolutions folder, select **6ACustomer.fmx**, and then click **Open**.

 If the 6ACustomer.fmx file is not in your Chapter6\TutorialSolutions folder, confirm that you saved the 6ACustomer.fmb file in your Chapter6\TutorialSolutions folder, and then run the form again in Form Builder to generate the .fmx file.

3. Type your username in the Userid text box, your password in the Password text box, and your connect string in the Database text box, and then click **OK**. The form appears in the Forms Runtime window.

4. Close the Forms Runtime window.

Users can also run forms directly from the operating system by double-clicking the form's .fmx file in Windows Explorer or by clicking on a shortcut that has been created for the form. For the form to run successfully by double-clicking the .fmx file, the .fmx file type must be registered on the user's workstation. Registering a file type involves specifying a file extension and then associating this extension with a particular application. When a user double-clicks an .fmx file, the operating system starts Forms Runtime and then loads the .fmx file that was double-clicked.

The .fmx file type must be registered with the Forms Runtime application, which is an executable file named ifrun60.exe and is located in your *Developer_home*\BIN\ folder on the user's workstation. This file registration is usually created when Form Builder is installed.

If you want to take a break, you can save your form file, and then close Form Builder. You can reopen your file later and continue the lesson.

To save the file and close Form Builder:

1. Click the **Save button** 🖫 on the toolbar to save the file.

2. Click **File** on the menu bar, and then click **Exit** to close Form Builder.

You also can click the Close button on the Form Builder window to close Form Builder.

FORM COMPONENTS AND THE OBJECT NAVIGATOR

Now that you have created a form, you need to become familiar with its components by viewing the components using a Form Builder window called the **Object Navigator**, which provides a hierarchical display of all of the form components. Now you will view the form components in the Object Navigator.

To view the form components:

1. If necessary, start Form Builder, select the **Open an existing form** option button, and then click **OK**. Navigate to your Chapter6\TutorialSolutions folder, select **6ACustomer.fmb**, and then click **Open**. The form opens in the Object Navigator window, as shown in Figure 6-7. If necessary, maximize the window.

2. Click **File** on the menu bar, and then click **Connect**. Connect to the database as usual.

3. If you did not restart Form Builder, click **Window** on the menu bar in the Layout Editor, and then click **Object Navigator**. The Object Navigator windows opens, as shown in Figure 6-7. If necessary, maximize the Object Navigator window.

If your Object Navigator window shows different objects than the ones shown in Figure 6-7, click MODULE2 (or the module name directly under the Forms node), and click the Collapse All button 🖭 on the Object Navigator toolbar. Then, click ➕ beside MODULE2 so that only the top-level form components (which are also called objects) appear.

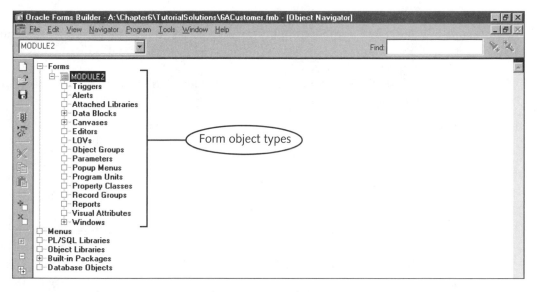

Figure 6-7   Object Navigator window in Ownership View

The Object Navigator window shows the top-level form object, which is called the form module. Below the form module are the individual form components. Each component is represented by an object node. If an object node has **+** to the left of its name, it means that objects of that type are present in the form. Object nodes that appear with a **+** symbol beside their name are currently **collapsed**, which means the node components are hidden. Object nodes that appear with a **–** symbol beside their name are currently **expanded**, which means the node components are displayed. An empty node ☐ appears beside a node that does not contain lower-level objects.

A Form Builder application, such as the customer form you created, is called a form module, or just a form. A form can contain all of the form object types shown in Figure 6-7. The customer form specifically contains data blocks, canvases, and windows, as indicated by the **+** to the left of the node names. These particular node types are the basic building blocks of every form—a form must have a window, canvas, block, and at least one block item.

A **window** is the familiar rectangular area on a computer screen that has a title bar at the top. Windows usually have horizontal and vertical scroll bars, and they usually can be resized, maximized, and minimized. In Form Builder forms, you can specify window properties, such as title, size, and position on the screen. Recall that a canvas is the area in a window where you place graphical user interface (GUI) objects, such as buttons and text fields. A form window can have multiple canvases. A block is a structure that contains a group of GUI objects. A canvas might display one or more blocks. A data block is a block that is related to a database table. When you create a data block, the system automatically generates the text items and labels for data fields in that table and provides the code for inserting, modifying, deleting, and viewing data records.

A form can contain one or more windows. Simple applications usually have only one window, and more complex applications might have several windows. A window can have multiple canvases. A canvas can have multiple blocks, and individual items within a block can appear on different canvases. It is useful to think of the form as a painting, with the window as the painting's frame, the canvas as the canvas of the painting, and a block as a particular area of the painting.

You can expand each object node in the Object Navigator window by clicking ➕ at the left of the object's name. Now you will expand the form objects.

To expand the form objects:

1. Click ➕ beside the Data Blocks node, click ➕ beside the Canvases node, and click ➕ beside the Windows node.

Figure 6-8 shows the expanded view of the form objects in the Object Navigator window. The form has one block (CUSTOMER), one canvas (CANVAS2), and one window (WINDOW1). CANVAS2 and WINDOW1 are the default names that were given to the canvas and window when you made the form using the Layout Wizard, and your default names might be different.

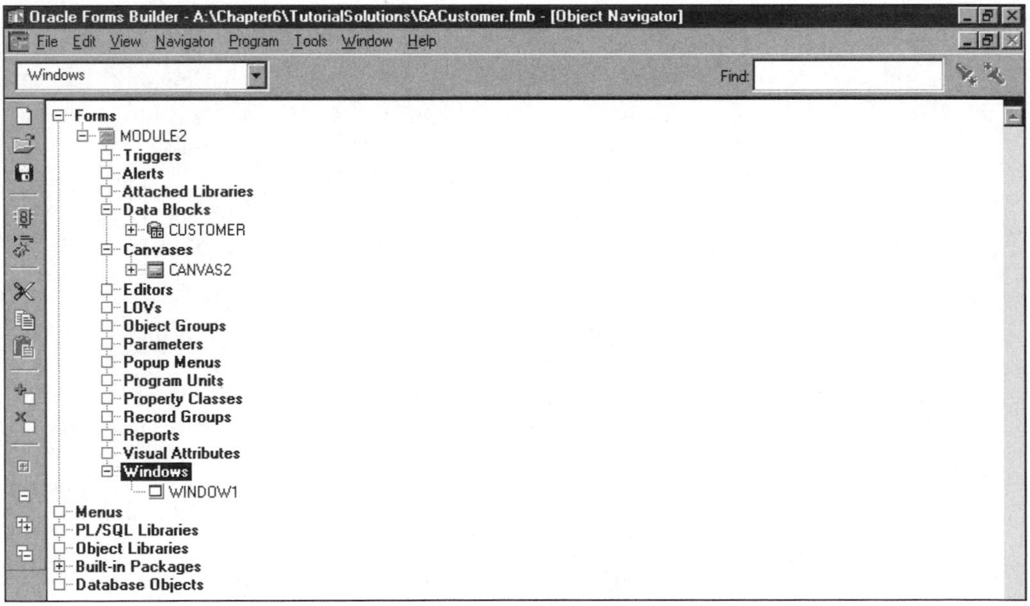

Figure 6-8   Expanding the form objects

Each object in the Object Navigator displays an icon that indicates its object type. Double-clicking the icon allows you to edit the object's properties. For example, the **Canvas icon** 🖼 appears beside a canvas, and double-clicking 🖼 opens the Layout Editor. The ➕

symbols to the left of the CUSTOMER block and CANVAS2 indicate that these objects contain more objects that you can view by expanding the object node.

Now you will expand the block objects further.

To expand the block objects:

1. Click ✚ beside the CUSTOMER node under Data Blocks, and click ✚ beside the Items node under CUSTOMER. The form block objects appear as shown in Figure 6-9.

Figure 6-9 shows that the CUSTOMER block contains items. Blocks also can contain triggers and relations, although the empty nodes beside Triggers and Relations in the CUSTOMER block indicate that the block does not contain any triggers or relations.

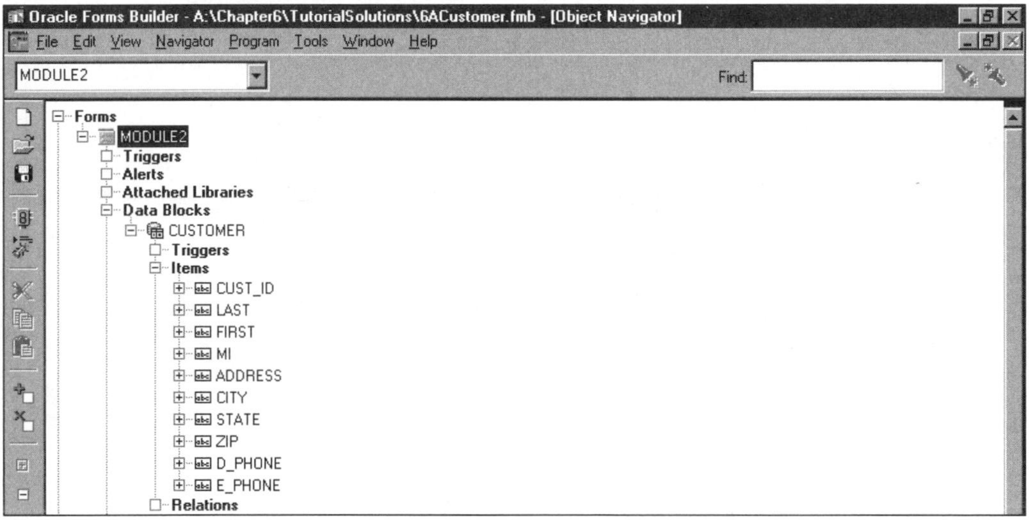

6

Figure 6-9 Form block objects

Windows applications respond to user actions, such as clicking a button, or to a system action, such as loading a form. These actions are called **events**. In Form Builder, events start programs called **triggers** that are written in PL/SQL. Triggers that respond to form events are not the same as the database triggers you created in Chapter 5. Database triggers respond to an event in the database, such as inserting, updating, or deleting a record. Form event triggers respond to user events, such as clicking a button. You will create triggers when you create custom forms in the next chapter.

A **relation** is a form object that is created when you specify a relationship between two data blocks with a foreign key relationship. To create a relationship, you must specify that the value of a primary key field in one block is equal to the corresponding foreign key field in the second block. Since the customer form involves only one database table, it has no relations.

Items are the GUI objects that a user sees and interacts with on the canvas. All of the items in the customer form are text items, which display text values stored in database fields.

Now you will expand the form canvas and examine its objects.

To expand the form canvas:

1. Click ⊞ beside CANVAS2 (or the default name of your canvas). The canvas contains a **Graphics** node, which represents boilerplate objects, such as frames, lines, and graphic images.

2. Click ⊞ beside the Graphics node. The FRAME4 (your default frame name might be different) object appears, which represents the frame around the data block items. There are no other boilerplate objects in the form.

So far you have been using the Object Navigator in **Ownership View**, which presents the form as the top-level object, and then lists all form object nodes in the next level. The Object Navigator also has a **Visual View** that shows how form objects "contain" other objects: a form contains windows, a window contains canvases, and a canvas contains data blocks. Figure 6-10 shows the form components in Ownership View, and Figure 6-11 shows the form components in Visual View.

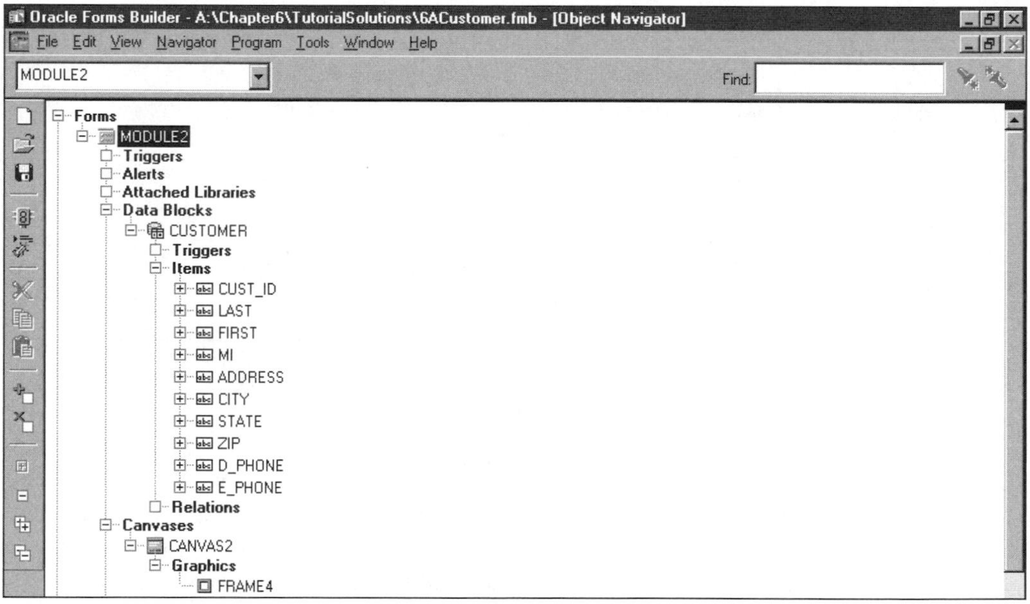

Figure 6-10 Customer form in Ownership View

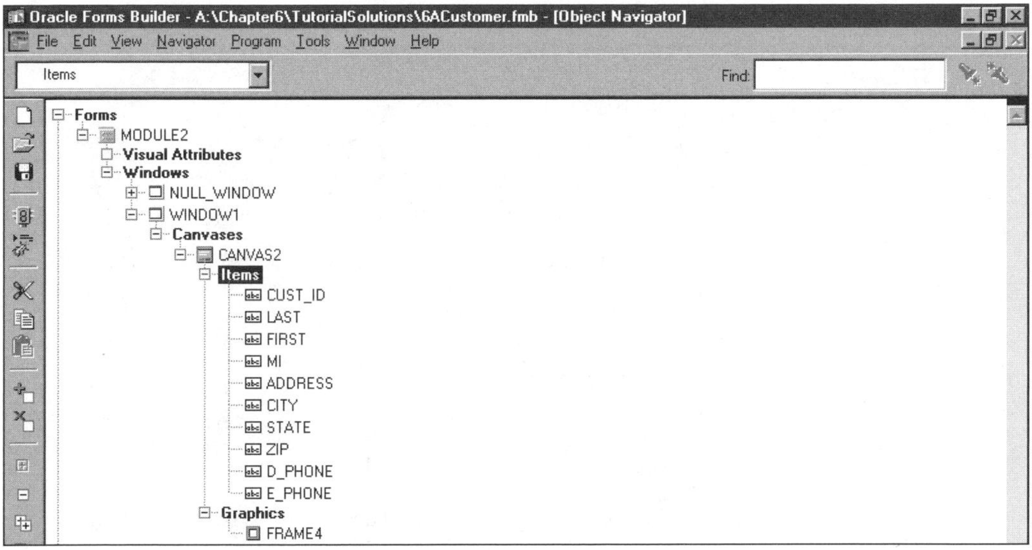

Figure 6-11   Customer form in Visual View

Note that in Ownership View, data blocks, canvases, and windows all appear as first-level objects, and the hierarchical relationships among windows, canvases, and blocks are not shown. (Ownership View does provide one hierarchical relationship: it shows which items are in a specific data block.) Conversely, Visual View shows that WINDOW1 contains CANVAS2, and that the form items and graphics are on CANVAS2. Note that data blocks are not shown in Visual View. Ownership View is useful for quickly accessing specific objects without having to open all of the objects' parents, while Visual View is useful for viewing and understanding form object relationships.

Now you will examine the form objects in Visual View.

To examine the form objects in Visual View:

1. Select **MODULE2** (or the top-level object under Forms if yours is named differently), and then click the **Collapse All button** 🔳 to close all form objects.

2. Click **View** on the menu bar, and then click **Visual View**. The top-level object—the form—still appears.

3. Click ➕ beside MODULE2 to display the next-level object. The Windows node appears, since it is the highest-level object within a form. The ➕ to the left of Windows indicates that the form contains Windows objects.

4. Click ➕ beside the Windows node. The windows within the 6ACUSTOMER form appear.

In Visual view, the form's Windows node contains all of the other form objects. Two Windows objects appear: the **NULL_Window**, which is created automatically when the form is created but does not appear on the form, and **WINDOW1** (your window name might be different), which is automatically created by the Layout Wizard and contains the rest of the form objects. Now you will expand WINDOW1 to examine the rest of the form objects.

To expand WINDOW1:

1. Click ⊞ beside WINDOW1. The Canvases node appears.

2. Click ⊞ beside the Canvases node. The default canvas (CANVAS2) appears. (Your canvas name might be different.)

3. Click ⊞ beside CANVAS2. The Object Navigator shows that the canvas contains items and graphics.

4. Click ⊞ beside the Items node. The form text items appear.

5. Click ⊞ beside the Graphics node. The form frame appears.

The canvas contains the final level of the Visual View, which are the form items and graphics. Note that data blocks are not shown in Visual View.

CHANGING OBJECT NAMES IN THE OBJECT NAVIGATOR

After you create a form using the Data Block and Layout Wizards, it is a good idea to change the default object names to more descriptive names. When you have multiple forms open in the Object Navigator and when you start creating forms with multiple windows, canvases, and frames, it is hard to visually distinguish between different objects in the Object Navigator unless they have descriptive names. Now you will change the form module name to CUSTOMER_FORM, the window name to CUSTOMER_WINDOW, the canvas name to CUSTOMER_CANVAS, and the frame name to CUSTOMER_FRAME.

To change the object names:

1. Click **MODULE2** (or whatever your default form module name is) to select it. Click it again so its background color changes to blue, type **CUSTOMER_FORM**, and then press **Enter** to save the change.

2. Click **WINDOW1** (or whatever your default window name is) to select the window, click it again so its background color changes to blue, and type **CUSTOMER_WINDOW**. Press **Enter** to save the change.

3. Click **CANVAS2** (or whatever your default canvas name is) to select the canvas, click it again so its background color changes to blue, type **CUSTOMER_CANVAS**, and then press **Enter**.

4. Click **FRAME3** (or whatever your default frame name is), and change its name to **CUSTOMER_FRAME**.

5. Click **View** on the menu bar, and then click **Ownership View** to redisplay the form objects in Ownership view.

6. Save the form.

MODIFYING A FORM USING THE DATA BLOCK WIZARD AND LAYOUT WIZARD

A powerful characteristic of the Data Block Wizard and Layout Wizard is that they are **reentrant**, which means that they can be used to modify the properties of an existing block or layout. To use a wizard to modify an existing block or layout, the wizard must be in reentrant mode. To start a wizard in reentrant mode, you must first select the data block or layout frame that you wish to modify in either the Layout Editor or the Object Navigator and then start the Data Block Wizard or Layout Wizard. (If you do not want to start the Data Block Wizard or Layout Wizard in reentrant mode, you must be sure that no data block or layout frame is currently selected when you start the wizard.)

You can visually tell when a wizard is in reentrant mode, because the different wizard pages appear as tabs on the top of the wizard pages. Now you will modify the CUSTOMER data block so that it enforces database integrity constraints directly in the form. You will do this by selecting the block and then starting the Data Block Wizard in reentrant mode.

To modify a block using the Data Block Wizard in reentrant mode:

1. In the Object Navigator, click **CUSTOMER** under the Data Blocks node to select the data block.

2. Click **Tools** on the menu bar, and then click **Data Block Wizard** to open the Data Block Wizard in reentrant mode. The Data Block Wizard appears with the Type and Table tabs displayed at the top of the window, indicating that it is in reentrant mode.

 You could also start the Data Block Wizard in reentrant mode by selecting one of the block items in the Layout Editor and then starting the Data Block Wizard.

3. Click the **Table** tab, and then click the **Enforce data integrity** check box so it is checked. Log onto the database if necessary. Click **Finish** to save the change and close the Data Block Wizard.

 Clicking Apply in the Data Block Wizard saves the change, but it does not close the Data Block Wizard. Clicking Finish saves the change and also closes the Data Block Wizard. Clicking Cancel closes the Data Block Wizard without saving any changes.

Next, you will modify the form layout so that the prompt for the Customer ID field appears as "ID" instead of "Customer ID." To do this, you will open the Layout Wizard in reentrant mode by first selecting the layout frame, and then starting the Layout Wizard.

 If you delete the frame around a layout, you cannot revise the layout using the Layout Wizard in reentrant mode.

To modify the form layout using the Layout Wizard in reentrant mode:

1. In the Object Navigator window, click **Tools** on the menu bar, and then click **Layout Editor** to open the Layout Editor window.

2. Click the **Customers** frame that is around the form fields to select it. Selection handles appear on its left and right edges.

3. Click **Tools** on the menu bar, and then click **Layout Wizard**. The Layout Wizard opens with the Data Block, Items, Style, and Rows tabs displayed at the top of the page, indicating that it is in reentrant mode.

 Another way to start the Layout Wizard in the Layout Editor is to click the Layout Wizard button ⧄ on the toolbar.

4. Click the **Items** tab to move to the Items page, and then change the prompt for the CUST_ID field to **ID:**.

5. Click **Finish**. The modified prompt appears in the Layout Editor.

You can also modify form layout properties manually in the Layout Editor window. Currently, some of the customers' middle initials will not appear, because the MI text item is not wide enough. Next, you will make the MI text item wider, using the Layout Editor.

To make the MI text item wider:

1. In the Layout Editor, click the **MI** text item in the layout. Selection handles appear on its left and right edges.

2. Click the middle selection handle on the right edge, and drag it to the right so that the MI text item is slightly wider.

3. Run the form, click the **Enter Query button** 🔳, do not enter a search condition, and then click the **Execute Query button** 🔳 to step through all of the records sequentially to verify that the MI value for each customer appears correctly. (Recall that Lee Miller does not have a middle initial, so no value should appear for this record.)

4. Close Forms Runtime.

5. Click **File** on the menu bar, and then click **Close** to close the Customer form. If you are asked if you want to save your changes, click **Yes**.

CREATING A FORM TO DISPLAY MULTIPLE RECORDS

The customer form that you created displays a single record at a time using a form-style layout. Now you will create a tabular data block form that displays multiple records on the same form. In this form, you will display all of the records in the INVENTORY table. You will create a new form module, and then create a new data block.

To create a new form module and data block:

1. In the Object Navigator, click the **Forms** node, and then click the **Create button** to create a new form module.

 Another way to create a new form module is to click File on the top menu bar, point to New, and then click Form.

2. Make sure that the new form module is selected in the Object Navigator. Click **Tools** on the menu bar, and then click **Data Block Wizard**. The Welcome to the Data Block Wizard page appears. Click **Next**.

3. Make sure that the Table or View option button is selected, and then click **Next**.

4. Click **Browse** on the Type page, and connect to the database if necessary. Be sure that the Current user and Tables check boxes are checked, select **INVENTORY** on the table list, and then click **OK**.

5. Select all of the columns in the INVENTORY table to be included in the data block. Do not check the Enforce data integrity box. Click **Next**.

6. Make sure that the Create the Data Block, then call the Layout Wizard option button is selected, and then click **Finish**.

Now that the data block has been created, you must use the Layout Wizard to specify the layout properties. This form layout will be different from the one you made before, because you will specify a tabular style layout and display five records on the form at one time. You will also include scroll bars on the layout so the user can scroll up and down through the data records. You will now create the INVENTORY block layout.

To create the layout:

1. On the Layout Wizard Welcome page, click **Next**.

2. Accept **(New Canvas)** as the layout canvas and **Content** as the canvas type, and then click **Next**.

3. Select all of the INVENTORY block items to appear in the layout, and then click **Next**.

4. Modify the item prompts as follows, and then click **Next**:

Item	Prompt
INV_ID	**Inventory ID**
ITEM_ID	**Item ID**
ITEM_SIZE	**Size**
COLOR	**Color**
PRICE	**Price**
QOH	**QOH**

 Since the data items will appear in columns, you will not put a colon after the column headings.

5. On the Style page, select the **Tabular** option button to specify that multiple records appear on the form, and then click **Next**.

6. On the Rows page, type **Inventory** for the Frame Title and **5** for Records Displayed. Leave Distance Between Records as **0**. Check the **Display Scrollbar** check box.

7. Click **Next**, and then click **Finish**. The completed Inventory form layout appears in the Layout Editor, as shown in Figure 6-12. (Your layout may look slightly different if your default system font is different.)

 If the Object Navigator window opens, click Tools on the menu bar, and then click Layout Editor to view the form in the Layout Editor.

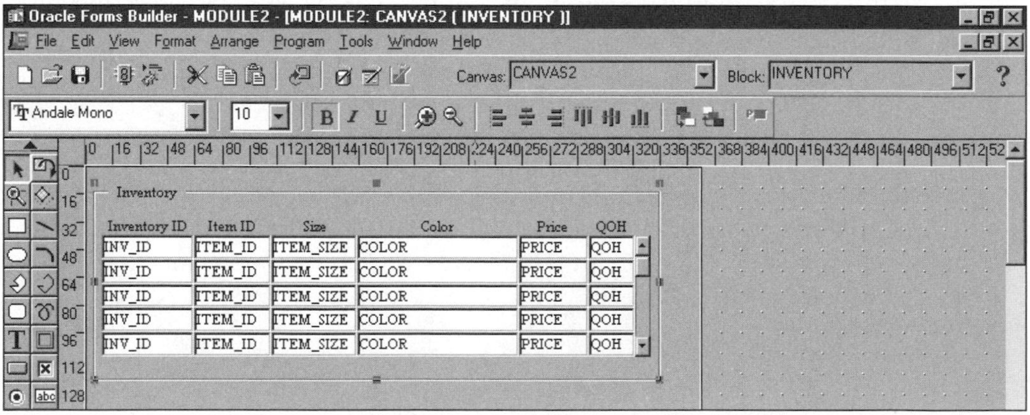

Figure 6-12   Tabular layout displaying five records

The form layout in Figure 6-12 shows that five records from the INVENTORY table will appear on the form at one time. The scroll bar at the right edge of the records will allow the user to scroll up and down through the records if more than five records are retrieved into the form. Because you specified the distance between records as 0 in the layout properties, the records are stacked directly on top of one another.

Now you will save and run the form.

To save and run the form:

1. Click the **Save button** 🖫, and save the file as **6AInventory.fmb** in your Chapter6\TutorialSolutions folder.

2. Click the **Run Form Client/Server button** 🔢 to compile the form and display it in the Forms Runtime window. Maximize the Forms Runtime window.

You can perform insert, update, delete, and query operations in a multiple-record form just as you did in a single-record form. First, you will view all of the INVENTORY records and then add a new record.

To view the INVENTORY records and add a new record:

1. Click the **Enter Query button** 🔳, do not enter a search condition, and then click the **Execute Query button** 🔳. All of the INVENTORY records are retrieved into the form.

2. Scroll down through the records using the scroll bar, and then click Inventory ID **11848** to select it. Now you will insert a new record below Inventory ID 11848.

3. Click the **Insert Record button** 🔳. A new blank record appears below Inventory ID 11848.

4. Type the information for the new INVENTORY row, as shown in Figure 6-13, to add Inventory ID 11849 (50 Spruce-colored size XXL Mountain Parkas), and then click the **Save button** 🖫 to save the new record.

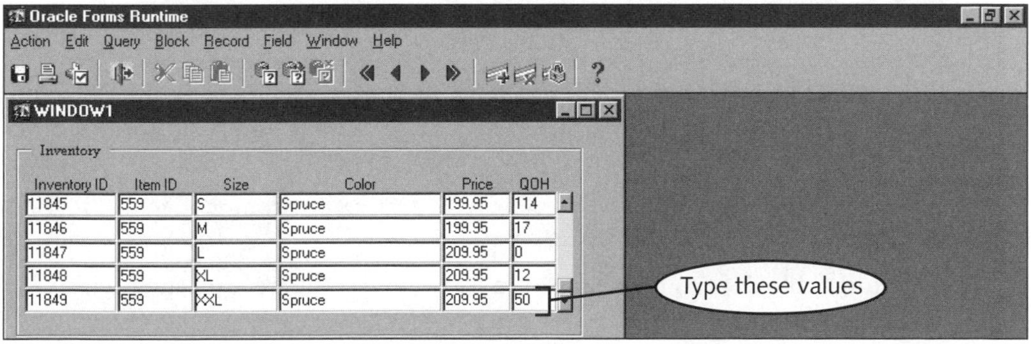

Figure 6-13   Adding a new record on a tabular form

You can also search for specific records on a tabular form. Next, you will retrieve all inventory items corresponding to ITEM_ID 559. Then, you will delete the record you just created.

To search for specific records and delete a record:

1. Click the **Enter Query button** 🔲 to start Enter Query mode. The insertion point appears in the first Inventory ID field.

2. Press **Tab**, type **559** in the Item ID field for the query search condition, and then click the **Execute Query button** 🔲 to execute the query. The five records corresponding to Item ID 559 appear.

3. Click Inventory ID **11849** to select it, and then click the **Remove Record button** 🔲 to delete the record. The record is deleted.

4. Click the **Save button** 🔲 to save your changes.

5. Close Forms Runtime and Form Builder.

SUMMARY

- ❐ A form is a database application that looks like a paper form and provides a graphical interface that allows users to add new database records and to modify, delete, or view existing records.

- ❐ A form is created in Form Builder and runs in the Forms Runtime environment.

- ❐ When a form is running, it can be in either Normal or Enter Query mode. In Normal mode, you can view data records and sequentially step through the records. In Enter Query mode, you can enter search parameters in the form fields, and then retrieve the associated records.

- ❐ In a form, you can perform exact searches to retrieve records that contain specific data values. You can perform restricted searches to retrieve records that contain data values that fall within a range of values or contain partial text strings.

- ❐ Using a form to add records is the equivalent of adding records in SQL using the INSERT command. Using a form to modify records is the equivalent of updating records in SQL using the UPDATE command. Using a form to delete records is the equivalent of deleting records in SQL using the DELETE command.

- ❐ FRM errors are raised by the Forms Runtime environment, and ORA errors are raised by the Oracle database.

- ❐ A data block is a group of related form items that are associated with a single database table.

- ❐ When you create a data block form using the Data Block Wizard and Layout Wizard, the system automatically generates the text items and prompts for data fields in that table and supplies the code for inserting, modifying, deleting, and viewing data records.

❐ If you enforce data integrity constraints in a form, the form will flag violations to the integrity constraints that you established when you created the table (unique primary keys, foreign key references, and so forth), and you will see Form Builder FRM error messages.

❐ If you do not enforce integrity constraints in a form, the DBMS will still enforce the integrity constraints, and users will receive ORA error messages directly from the DBMS if they violate a table integrity constraint.

❐ The form layout specifies how the form appears to the user.

❐ A canvas is the area on a form where you place graphical user interface (GUI) objects, such as buttons and text items.

❐ In a form-style layout, only one record appears at a time, while in a tabular-style layout, multiple records appear on the form at one time.

❐ Boilerplate objects are customizable text, graphics, and other objects that do not contribute to the functionality of a form, but enhance its appearance and ease of use.

❐ The Layout Editor provides a graphical display of the form canvas that you can use to draw and position form items, and to add boilerplate objects, such as labels, titles, and graphic images.

❐ While running a form, you can perform approximate searches by using the percent sign (%) to indicate that there can be any number of wildcard characters either before or after a search string.

❐ The Form Builder Object Navigator enables you to access all form objects either by object type in Ownership View, or hierarchically by opening objects that contain other objects in Visual View.

❐ Form modules contain windows, windows contain canvases, and canvases contain items such as text items and buttons.

❐ Form Builder design files are saved with an .fmb extension, and executable Forms Runtime files are saved with an .fmx extension. Whenever you make a change to an .fmb file, you must also create an updated .fmx file. You can configure the Form Builder environment to automatically generate an updated .fmx file every time a form is run.

❐ You can use the Data Block Wizard and Layout Wizard to modify existing data blocks and form layouts by selecting the data block or layout frame and then starting the wizard in reentrant mode.

REVIEW QUESTIONS

1. Describe the purpose of each of the following buttons on the Forms Runtime toolbar:

Button	Purpose

2. When a form is running, it can be in one of two modes: Normal or Enter Query. Which mode must it be in for you to:

a. enter a search condition?

b. insert a new record?

c. step through table records sequentially?

3. Describe two ways to change from Enter Query mode back to Normal mode.

4. What is the maximum number of database tables that a data block can be associated with?

5. List how the following form objects appear from top to bottom level in Visual View in the Object Navigator: Canvas, Item, Form, Window, Data Block.

6. How do you rename objects in the Object Navigator?

7. What is the difference between the form and tabular layout styles in a data block form?

8. Why is it not necessary to check the Enforce Data Integrity check box when you create a new data block?

9. What is a boilerplate object?

10. What is the difference between the files named customer.fmb and customer.fmx?

PROBLEM-SOLVING CASES

The case problems reference the Clearwater Traders (see Figure 1-11), Northwoods University (see Figure 1-12), and Software Experts (see Figure 1-13) sample databases. Save all case solution files in your Chapter6\CaseSolutions folder.

1. Create a form for inserting, updating, and viewing records in the LOCATION table in the Northwoods University database. Name the form LOCATION_FORM, the window LOCATION_WINDOW, and the canvas LOCATION_CANVAS. Use a form-style layout, create descriptive item prompts for the form text items, and use the frame label "Building Locations." Do not enforce integrity constraints in the form. Save the Form Builder file in a file named 6ACase1.fmb. Use the form to execute the following operations. The changes made as a result of these operations will be reflected in your database tables, and are not saved in the form file.

   a. Add a new LOCATION record with LOC_ID = 65, BLDG_CODE = BUS, Room = 100, and Capacity = 150.

   b. Modify the capacity of LOC_ID 52 to 75.

   c. Use an exact search to retrieve the records where BLDG_CODE = BUS.

2. Create a form for inserting, updating, and viewing records in the ITEM table in the Clearwater Traders database. Rename the form module ITEM_FORM, the window ITEM_WINDOW, and the canvas ITEM_CANVAS. Use a form-style layout, create descriptive item prompts for the form fields, and use the frame label "Clearwater Traders Items." Enforce integrity constraints in the form. Save the Form Builder file as 6ACase2.fmb. Use the form to perform the following operations. The changes made as a result of these operations will be reflected in your database tables, and are not saved in the form file.

   a. Update the ITEM_DESC of ITEM_ID 786 to "4-Season Tent."

   b. Add the following record: ITEM_ID = 800, ITEM_DESC = Fleece Vest, Category = 1 (Women's Clothing).

   c. Use an exact search to retrieve the items in the category Women's Clothing.

3. Create a form for inserting, updating, and viewing records in the FACULTY table in the Northwoods University database. Name the form FACULTY_FORM, the window FACULTY_WINDOW, and the canvas FACULTY_CANVAS. Use a form-style layout, create descriptive item prompts for the form text items, and use the frame label "Faculty." Do not enforce integrity constraints in the form. Save the Form Builder file as 6ACase3.fmb. Use the form to execute the following operations. The changes made as a result of these operations will be reflected in your database tables, and are not saved in the form file.

   a. Insert the following record: F_ID = 10, F_LAST = Wilson, F_FIRST = Ephraim, F_MI = V, LOC_ID = 57, F_PHONE = 7155556023, F_RANK = ASST, F_PIN = 4433.

   b. Update the phone number of faculty member Laura Sheng to 7155559878.

   c. Retrieve the records where F_RANK = ASSO. Use 'ASSO' as the exact search condition.

4. Create a form for inserting, updating, and viewing records in the COURSE table in the Northwoods University database. Name the form COURSE_FORM, the

window COURSE_WINDOW, and the canvas COURSE_CANVAS. Use a form-style layout, create descriptive item prompts for the form text items, and use the frame label "Northwoods University Courses." Do not enforce integrity constraints in the form. Save the Form Builder file as 6ACase4.fmb. After you have created and tested the form, change the frame label to "Courses" using the appropriate wizard in reentrant mode.

5. Create a form for inserting, updating, and viewing records in the PROJECT table in the Software Experts database. Name the form PROJECT_FORM, the window PROJECT_WINDOW, and the canvas PROJECT_CANVAS. Use a form-style layout, create descriptive item prompts for the form text items, and use the frame label "Projects." Do not enforce integrity constraints in the form. Save the Form Builder file as 6ACase5.fmb. After you have created and tested the form, modify the form to enforce integrity constraints within the form using the appropriate wizard in reentrant mode.

6. Create a form for inserting, updating, and viewing records in the TERM table in the Northwoods University database. Name the form TERM_FORM, the window TERM_WINDOW, and the canvas TERM_CANVAS. Use a tabular-style layout that displays five records on the canvas and uses a scrollbar. Create descriptive item prompts for the form text items, and use the frame label "Terms." Save the Form Builder file as 6ACase6.fmb. Use the form to execute the following operations. The changes made as a result of these operations will be reflected in your database tables, and are not saved in the form file.

   a. Insert the following new record: TERM_ID = 7, TERM_DESC = "Fall 2004," STATUS = "OPEN."

   b. Modify the status of term ID 6 to "CLOSED."

   c. Use a restricted search to retrieve the term records with the text string "Summer" anywhere in the term name.

7. Create a form for inserting, updating, and viewing records in the CUST_ORDER table in the Clearwater Traders database. Name the form CUST_ORDER_FORM, the window CUST_ORDER_WINDOW, and the canvas CUST_ORDER_CANVAS. Use a tabular-style layout that displays five records on the canvas and uses a scrollbar. Create descriptive item prompts for the form text items, and use the frame label "Customer Orders." Do not enforce integrity constraints in the form. Save the Form Builder file as 6ACase7.fmb. Use the form to execute the following operations. The changes made as a result of these operations will be reflected in your database tables, and are not saved in the form file.

   a. Insert the following new record: ORDER_ID = 1070, ORDER_DATE = 15-JUN-2003, METH_PMT = "CHECK," CUST_ID = 179, ORDER_SOURCE_ID = 3.

   b. Modify the order source ID of order 1060 to 2.

   c. Use an exact search to retrieve all orders for customer ID 179.

8. Create a form for inserting, updating, and viewing records in the SKILL table in the Software Experts database. Name the form SKILL_FORM, the window SKILL_WINDOW, and the canvas SKILL_CANVAS. Use a tabular-style layout that displays five records on the canvas and uses a scrollbar. Create descriptive item prompts for the form text items, and use the frame label "Consultant Skills." Enforce integrity constraints in the form. Save the Form Builder file as 6ACase8.fmb. Use the form to execute the following operations. The changes made as a result of these operations will be reflected in your database tables, and are not saved in the form file.

   a. Insert the following new record: SKILL_ID = 20, SKILL_DESCRIPTION = "C++ Programming."

   b. Modify the description of skill 5 to "Web Application Development."

   c. Use a restricted search to retrieve all skills with "Oracle" anywhere in the skill description.

9. Create a form for inserting, updating, and viewing records in the SHIPMENT_LINE table in the Clearwater Traders database. Name the form SHIPMENT_LINE_FORM, the window SHIPMENT_LINE_WINDOW, and the canvas SHIPMENT_LINE_CANVAS. Use a tabular-style layout that displays five records on the canvas and uses a scrollbar. Create descriptive item prompts for the form text items, and use the frame label "Shipment Lines." Do not enforce integrity constraints in the form. Save the Form Builder file as 6ACase9.fmb. After you have created and tested the form, modify the form using the appropriate wizard in reentrant mode so it displays 10 records at one time.

10. Create a form for inserting, updating, and viewing records in the PROJECT_CONSULTANT table in the Software Experts database. Name the form PROJECT_CONSULTANT_FORM, the window PROJECT_CONSUL-TANT_WINDOW, and the canvas PROJECT_CONSULTANT_CANVAS. Use a tabular-style layout that displays five records on the canvas and uses a scrollbar. Create descriptive item prompts for the form text items, and use the frame label "Project Consultants." Do not enforce integrity constraints in the form. Save the Form Builder file as 6ACase10.fmb. After you have created and tested the form, modify the frame label to "Consultant Project Assignments" using the appropriate wizard in reentrant mode.

6

◀ LESSON B ▶

Objectives

- ♦ Create a data block form that is based on a database view
- ♦ Learn how to modify form properties to improve form appearance and function
- ♦ Create a master-detail form that contains multiple data blocks
- ♦ Format form text items using format masks
- ♦ Learn how to delete data blocks

CREATING A FORM BASED ON A DATABASE VIEW

Recall that a database view looks and acts like a database table but is derived from other database tables. It is useful to create views that retrieve data values for records that are related by foreign keys and then display the actual data values in the form, rather than display the foreign key value. For example, in the inventory form you created in the previous lesson, it would probably be more informative to show the item description for each inventory ID rather than just the item ID. Keep in mind that when you create a data block using a view, you can only examine the data. You cannot modify the data by inserting new records or updating or deleting existing records.

Now you are going to create a view that retrieves all of the INVENTORY records, along with their associated item descriptions. Then, you will create a tabular form to display the information. First, you will create the view using SQL*Plus.

To create the view in SQL*Plus:

1. Start SQL*Plus, log onto the database, and type the following command to create the view. The message "View created" appears.

   ```
   CREATE VIEW inventory_view AS
   SELECT inv_id, item_desc, item_size, color, price, qoh
   FROM inventory, item
   WHERE inventory.item_id = item.item_id;
   ```

2. Exit SQL*Plus.

Now you will create a new form that uses the INVENTORY_VIEW as its data source and displays five records at one time. You will select the view from your list of database objects when you create the data block.

To create a new form based on a view:

1. Start Form Builder. When the Welcome to the Form Builder page appears, make sure the Use the Data Block Wizard option button is selected, and then click **OK**. The Data Block Wizard Welcome page appears. Click **Next**.

2. Make sure the Table or View option button is selected as the data block type, and then click **Next**.

3. When the Table page appears, click **Browse**, and then log onto the database. Make sure that the Current User check box is checked. Clear the Tables check box, and then check the **Views** check box so that only your database views appear. Select **INVENTORY_VIEW**, and then click **OK**.

4. Select all of the view columns to be included in the data block, do not check the Enforce data integrity check box, and then click **Next**. Make sure the Create the data block, then call the Layout Wizard option button is selected, and then click **Finish**.

The form layout will display five records and include scroll bars. Now you will create the layout using the Layout Wizard.

To create the form layout:

1. Click **Next** on the Layout Wizard Welcome page.

2. Accept the **(New Canvas)** and **Content** defaults, and then click **Next**.

3. Select all of the block items for display in the layout, and then click **Next**.

4. Modify the item prompts and field widths as follows, and then click **Next**.

Name	Prompt	Width
INV_ID	Inv. ID	54
ITEM_DESC	Description	177
ITEM_SIZE	Size	50
COLOR	Color	95
PRICE	Price	41
QOH	QOH	27

5. Select the **Tabular** layout style option button, and then click **Next**.

6. Type **Inventory** for the Frame Title, change Records Displayed to **5**, leave Distance Between Records as 0, and check the **Display Scrollbar** check box. Click **Next**, and then click **Finish**.

Now you will save the form and change the form's window, canvas, and frame names in the Object Navigator.

To save the form and change the form item names:

1. In the Object Navigator, click the **Save button** 🖫, and save the form as **6BInventoryView.fmb** in your Chapter6\TutorialSolutions folder.

2. Change the form module name to **INVENTORY_VIEW_FORM**.

3. Change the form window name to **INVENTORY_VIEW_WINDOW**.

4. Change the form canvas name to **INVENTORY_VIEW_CANVAS**.

Now you will run the form and retrieve all of the view records to test the form.

To test the form:

1. Click the **Run Form Client/Server button** 🔳 to run the form. The Forms Runtime window opens. Maximize the Forms Runtime window, and resize the form to make it wider by dragging its right edge and bottom edge so it completely fills the Forms Runtime window, as shown in Figure 6-14.

2. Click the **Enter Query button** 🔳, and then click the **Execute Query button** 🔳 to retrieve all of the view records into the form. The form looks like Figure 6-14. (Your form might look slightly different, depending on your display settings.)

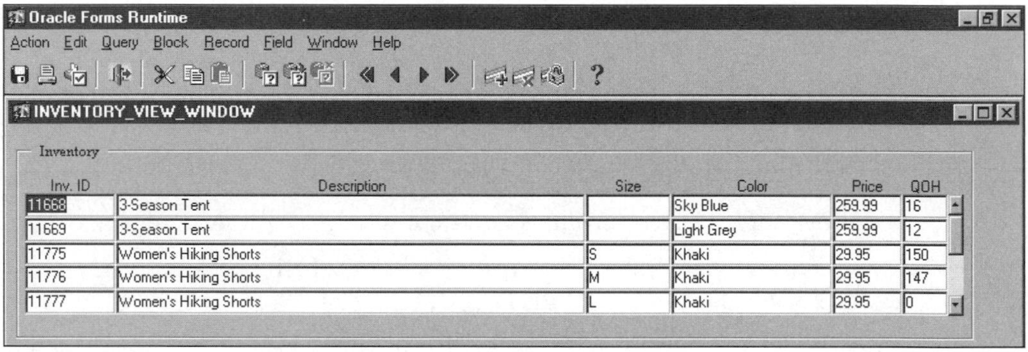

Figure 6-14   Form based on a view

3. Scroll down to view all of the records.

4. Close the Forms Runtime window.

5. Click **File** on the menu bar, and then click **Close** to close the form in Form Builder.

Recall from Chapter 3 that you cannot directly insert, update, or delete view records for a view that you create by joining multiple tables. This limitation is also true for data block forms based on views: If you create a data block form based on a view that joins multiple tables, you can only view the data. You cannot use the form to insert, update, or delete records.

MODIFYING FORM PROPERTIES

Form applications should be attractive, easy to use, and configurable. Often, you need to modify some of the properties of data block forms you create using the Data Block and Layout Wizard to meet these standards. For example, the data block form based on the COURSE table in the Northwoods University database shown in Figure 6-15 requires some modifications. The form would be more attractive if the text items were stacked on top of one another in a single column, the frame was centered on the form, and the window was sized correctly for the Forms Runtime window. The form window title should be changed from the Object Navigator window name (COURSE_WINDOW) to a more descriptive name that uses mixed-case letters.

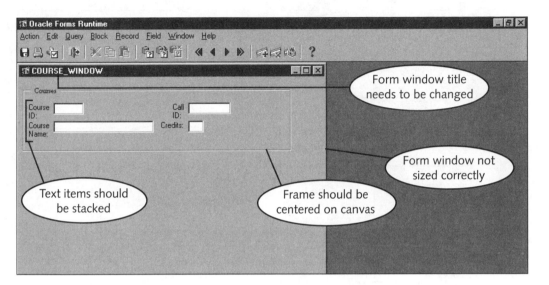

Figure 6-15 Data block form requiring modifications

To change these window properties, you will need to modify the properties in the window's Property Palette. Every form object has a **Property Palette** that allows you to configure the object's properties. Now you will open the form and examine the window Property Palette.

To open the form and examine the window Property Palette:

1. In Form Builder, open the **Course.fmb** file from the Chapter6 folder on your Data Disk, and save the file as **6BCourse.fmb** in your Chapter6\ TutorialSolutions folder.

2. In the Object Navigator, right-click the **Window icon** ☐ beside COURSE_WINDOW in the Object Navigator, and then click **Property Palette**. The Property Palette for the window opens, as shown in Figure 6-16.

 You can also open the Property Palette for most objects by double-clicking the object icon in the Object Navigator window, or by selecting the object, clicking Tools on the menu bar, and then clicking Property Palette.

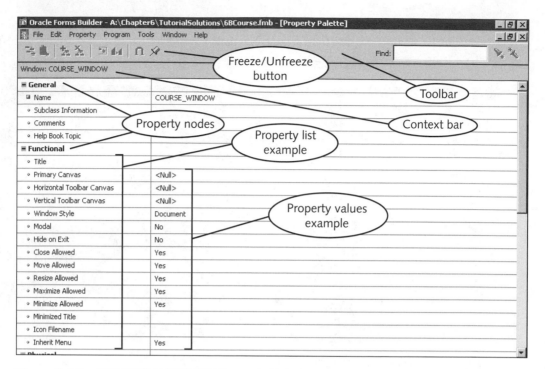

Figure 6-16 COURSE_WINDOW Property Palette

The Property Palette window has a toolbar that provides functions for working with the Property Palette; a **context bar**, which describes the object associated with the Property Palette; and a **property list**, which lists object properties and their associated values. Since a property is not currently selected, the only button that is activated on the Property Palette toolbar is the Freeze/Unfreeze button. Normally, whenever you open the Property Palette window, it displays the property list for the object that is currently selected in the form. When you click, the current Property Palette window remains open, and you can open additional Property Palette windows and work in multiple Property Palette windows at the same time.

Properties within the property list are grouped according to related functions. For example, properties related to the window's functions are grouped together, and properties related to the window's appearance are grouped together. Each group of related properties is called a **property node**. To expand a property node to view the individual properties, click [+] beside the property node. To collapse a property node to hide the individual properties, click [−] beside the property node. By default, all property nodes are expanded when the Property Palette opens.

Different form object types have different properties. For example, a text item has different properties than a window or a canvas. To specify a property value you first select the property in the Property Palette. To select a property, you click the space next to the property name. Then, you can specify the property value by typing alphanumeric characters, selecting a value from a list, or specifying a value using a dialog box that opens by clicking the **More button** [...] that appears on the right edge of the property space when the property is selected.

Modifying Form Properties Using the Property Palette

You should always configure form windows using the following checklist:

1. Change the form window title to a descriptive title.

2. Make the form window large enough to fill the Forms Runtime window.

3. Always allow the user to be able to minimize the form window.

4. Do not allow the user to maximize the form window unless you intend for the window to run maximized, since form objects might appear off-center in a maximized window.

5. Do not allow the user to resize the form window, since resizing might cause some canvas objects to be clipped off.

6. Include horizontal and vertical scroll bars if canvas objects extend beyond the window boundaries.

Now you will change the form window title, disable the window's Maximize button and resize property, make the form window larger, and add horizontal and vertical scroll bars to the window so that you can scroll to view the form areas that are off-screen.

To modify the window properties:

1. In the COURSE_WINDOW Property Palette, scan the properties under the Functional property node. Click in the space next to the Title property to select the Title property, and type **Northwoods University** for the new window title.

When you select or change a property value, the node marker next to the property name changes to a green color.

2. Click the space next to the Resize Allowed property to select the property. A drop-down list appears. Open the list, and select **No** to disable the Resize property.

3. Scroll down in the Property Palette window if necessary, click the space next to the Maximize Allowed property, open the list, and select **No**.

4. Scroll down in the Property Palette window to the Physical property node, and change the Width property to **575** and the Height property to **375**.

These are approximate values for window width and height for a display setting of 600 × 800. You might need to adjust these values for your display so the form window correctly fills the Forms Runtime window.

5. Click the space next to Show Horizontal Scrollbar, and select **Yes**.

6. Click the space next to Show Vertical Scrollbar, and select **Yes**.

7. Click the **Window Close button** ☒ on the top-right corner of the inner Property Palette window to close the Property Palette. Your changes are saved automatically.

 If the message "Save changes to COURSE_FORM" appears when you try to close the Property Palette, you closed the outer Form Builder window instead of the inner Property Palette window. Click Cancel, and then click ☒ on the inner Property Palette window.

8. Run the form, and maximize the Forms Runtime window. The modified form window properties should look like Figure 6-17. Close the Forms Runtime window.

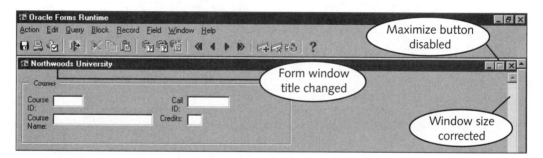

Figure 6-17   Form with modified window properties

 If the right edge of your form window is not visible without scrolling, open the COURSE_WINDOW Property Palette again, and make the Width property smaller, so your display looks like Figure 6-17 when you run the form.

Recall that the canvas is like the canvas in a painting, and the window is like the painting's frame. Currently, the canvas in the course form is smaller than the window. You cannot place form items off of the canvas, even though there is extra room in the window to move objects to the right and bottom edge. Since you changed the window width and height, you will also need to change the canvas width and height so it fills the larger window. Now you will open the course form canvas Property Palette and change the canvas width and height.

To change the canvas width and height:

1. In the Object Navigator, double-click the **Canvas icon** beside COURSE_CANVAS to display the canvas in the Layout Editor. Notice the position of the right edge of the canvas, as shown in Figure 6-18. You cannot place form objects beyond this edge, or an error will occur.

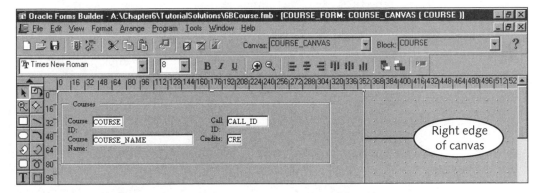

Figure 6-18   Current canvas edge position

2. To open the canvas Property Palette, right-click anywhere on the canvas except on the frame or on a text item or label, and then click **Property Palette**. The canvas Property Palette opens.

> **Tip** You can also open a canvas Property Palette in the Object Navigator by right-clicking the Canvas icon , and then clicking Property Palette, or by selecting the canvas in the Object Navigator, clicking Tools on the menu bar, and then clicking Property Palette.

3. Change the canvas Width to **575** and the Height to **375**, and then close the canvas Property Palette. The right edge of the canvas should no longer appear in the Layout Editor.

Modifying Item Prompts and Frame Properties

Recall that a frame encloses the text items that are in a specific data block. When you create a form layout, you specify the frame title and the distance between the records. Usually, the text items are stacked directly on top of each other in a single column. In the course form, the frame defaults to a wider size and contains two columns of text items. This happened because some of the text item prompts are wider than the Form Builder default value, so the prompts wrap to two text lines. When prompts wrap to two lines, you have to manually modify the frame size and item prompts if you want the text items to stack on top of each other in a single column. First, you will modify the prompts so they are on a single line of text. To do this, you select the prompt text, remove the hard return that wraps the text to the second line, and replace it with a space.

To modify the item prompts:

1. In the Layout Editor, click the **Course ID:** item prompt, and then click again to begin editing the prompt text. The insertion point appears in the prompt text.

2. Use the arrow keys to move the insertion point so it is just in front of the *I* in ID, and then press the **Backspace** key to delete the hard return that is wrapping the text to two lines.

3. Press the spacebar to place a blank space between Course and ID: in the prompt. The prompt appears on a single line, as shown in Figure 6-19. Click anywhere on the canvas to end editing, and then click anywhere again to deselect the prompt. Do not worry if the text is on top of the frame—you will modify the frame size later.

Figure 6-19   Modifying the prompt text

4. Modify the item prompts for the Course Name: and Call ID: prompts so the text appears on a single line also.

5. Save the form.

Currently, the prompts appear in the font that was selected when Form Builder started. It is a good practice to always use a sans serif (block) font, such as Arial, on screen displays to improve legibility. Now, you will modify the font for all of the item prompts. You can modify all of the prompts in one step by selecting all of the prompts at the same time and then changing their font type and size.

To change the prompt font:

1. In the Layout Editor, click the **Course ID:** prompt, press and hold the **Shift** key, and then click the **Course Name:**, **Call ID:**, and **Credits:** prompts. Selection handles appear around all of the prompts, indicating the prompts are selected as an object group.

You can also select form objects as an object group by clicking the mouse pointer on the canvas at a point below the left-bottom corner of the objects and then dragging the mouse pointer to draw a rectangle around the objects to select them.

2. Open the font list on the Layout Editor toolbar, and select **Arial**. If necessary, change the font size to **8**. The prompts appear in the new font.

Now you need to change the frame properties. First, you will examine the frame properties in the frame Property Palette.

To examine the frame Property Palette:

1. In the Layout Editor, select the **Courses frame**, right-click, and then click **Property Palette**. The frame Property Palette opens, as shown in Figure 6-20.

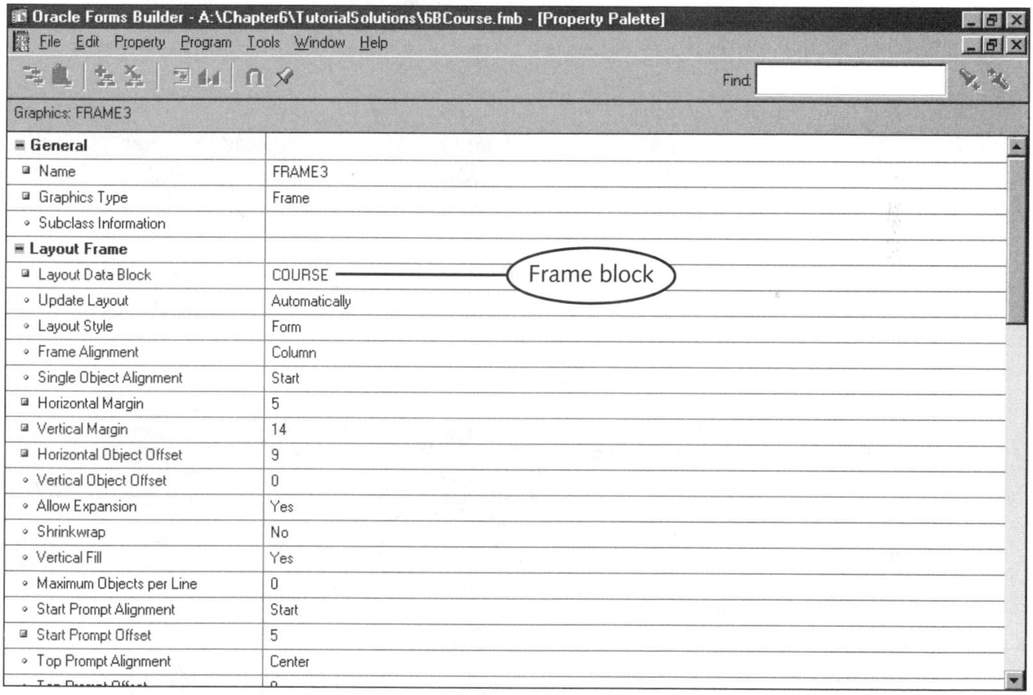

Figure 6-20   Frame Property Palette

The Layout Data Block property value indicates the name of the data block that is associated with the frame. The Update Layout property indicates how the frame behaves. Currently, the Update Layout property value is **Automatically**, which means that when you resize or move the frame, the items within the frame are automatically moved and repositioned. The items are always positioned in the order they appear in the Data Block list in the Object Navigator. If the frame is not long enough to accommodate all of the items in a single vertical column, then the items appear in multiple columns. Other possible values of the Update Layout property are **Manually**, which enables you to resize the frame and then position the items manually, and **Locked**, which behaves the same as Manually, except that the item positions do not update when you change the layout properties using the Layout Wizard in reentrant mode. When you are developing simple data block forms, you usually leave the Update Layout property as the default (Automatically) value, so you do not have to manually position and align the form items. Now you will resize the frame and observe how the frame items are automatically repositioned. Then, you will reposition the frame in the center of the canvas.

To resize and reposition the form frame:

1. Close the frame Property Palette.

2. Click anywhere on the frame line or frame label to select the frame, and then drag the lower right selection handle down and to the left edge of the screen display to make the frame narrower and longer, as shown in Figure 6-21.

If the message "FRM-10855: Not enough horizontal space in frame *frame_name* for object *object_name*." appears, you made the frame too narrow to enclose one or more of the text items. Click OK, and then make the frame wider so the frame fully encloses all text items.

3. After you have resized the frame, click anywhere on the frame line or label to select the frame, and then drag the frame to reposition it in the center of the canvas, as shown. Note that the items in the frame are automatically aligned in a single column, and stacked on top of one another.

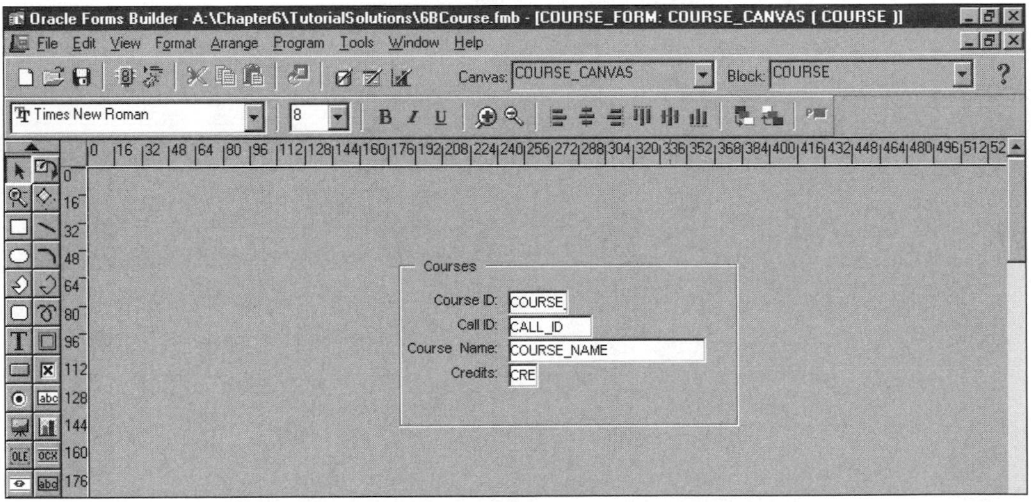

![Screenshot of Oracle Forms Builder showing the COURSE_CANVAS with a Courses frame containing Course ID, Call ID, Course Name, and Credits fields]

Figure 6-21 Resizing and repositioning the frame

Another important frame property is the frame **bevel**, which controls the frame appearance. Figure 6-22 shows the different frame bevel options. You will continue using the default, which is the inset bevel.

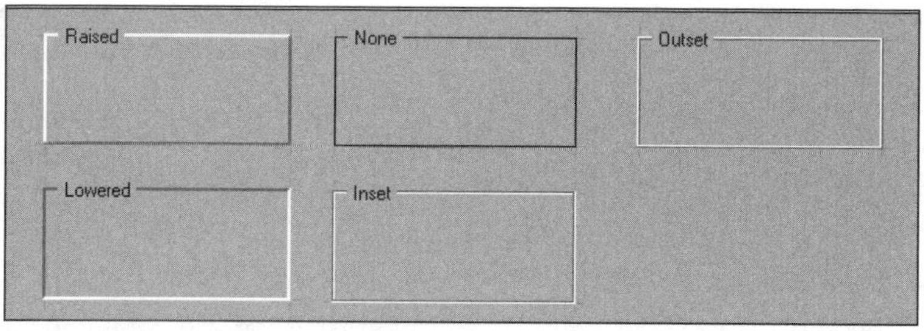

Figure 6-22   Frame bevel options

Many frame properties, such as the frame caption, the number of records in the frame, and the space between the records, were specified when you created the form layout using the Layout Wizard. You can modify these properties by using the Layout Wizard in reentrant mode or by modifying the properties directly in the frame Property Palette. Now you will change the frame caption font using the Property Palette.

> **Caution** Be careful if you modify the frame properties using the Layout Wizard in reentrant mode, because this restores all frame properties to their default values. Any manual formatting you have done, such as resizing the frame, modifying the prompts directly, or modifying frame properties in the Property Palette, will be lost.

To change the frame caption font:

1. Open the frame Property Palette, scroll down to the Title Font property node, and click in the space beside the Frame Title Font Name property. The **More button** [...] appears. Click [...]. The Fonts dialog box opens. Select **Arial**, and then click **OK**.

2. Change the Frame Title Font Size property to **8**.

3. Close the Property Palette, and then save the form.

Text Item Properties

You can control the appearance of the data value that appears in a text item using the text item's Property Palette. Table 6-2 summarizes some of the important text item appearance properties.

Property Node	Property Name	Description
Physical	Visible	Determines whether the item appears when the form is running
Color	Foreground Color, Background Color, Fill Pattern	Specifies the text color, background color, and fill pattern
Font	Font Name, Font Size, Font Weight, Font Style, Font Spacing	Specifies the appearance of the text item's font
Prompt	Prompt	Specifies the text item's prompt (label)
Prompt	Prompt Display Style, Prompt Justification, Prompt Alignment	Specifies the appearance of the prompt text

Table 6-2   Text item appearance properties

Now you will modify the font of all of the form text items. You will modify this property for all of the items in one step by selecting all of the items and formatting them by using the Property Palette in intersection mode. A Property Palette can be opened in **intersection mode** by selecting multiple form objects and then opening the Property Palette window. You can tell that a Property Palette is in intersection mode because the Intersection/Union button ⊓ on the toolbar appears as an inverted U. An intersection Property Palette can be used to modify a property of a group of objects to the same value. When all group objects currently have the same value for a property, the common property value appears. When the objects have different values, the property value appears as "****" in the intersection Property Palette, and the property node appears as a question mark (?)

To modify the text item fonts using the Property Palette in intersection mode:

1. In the Layout Editor, select the **COURSE_ID** text item (not the item prompt), press and hold the **Shift** key, and then select the **CALL_ID**, **COURSE_NAME** and **CREDITS** text items. All of the text items should display selection handles.

2. Right-click, and then click **Property Palette**. The intersection Property Palette opens, as shown in Figure 6-23. Note that the Intersection/Union button ⊓ appears as an inverted *U*, and the Name property appears as "****", indicating that multiple items with different names are selected.

3. Scroll down the Property Palette to the Font property node, and change the Font Name to **Arial** and the Font Size to **8**. Make sure that the new selections appear on the Property Palette.

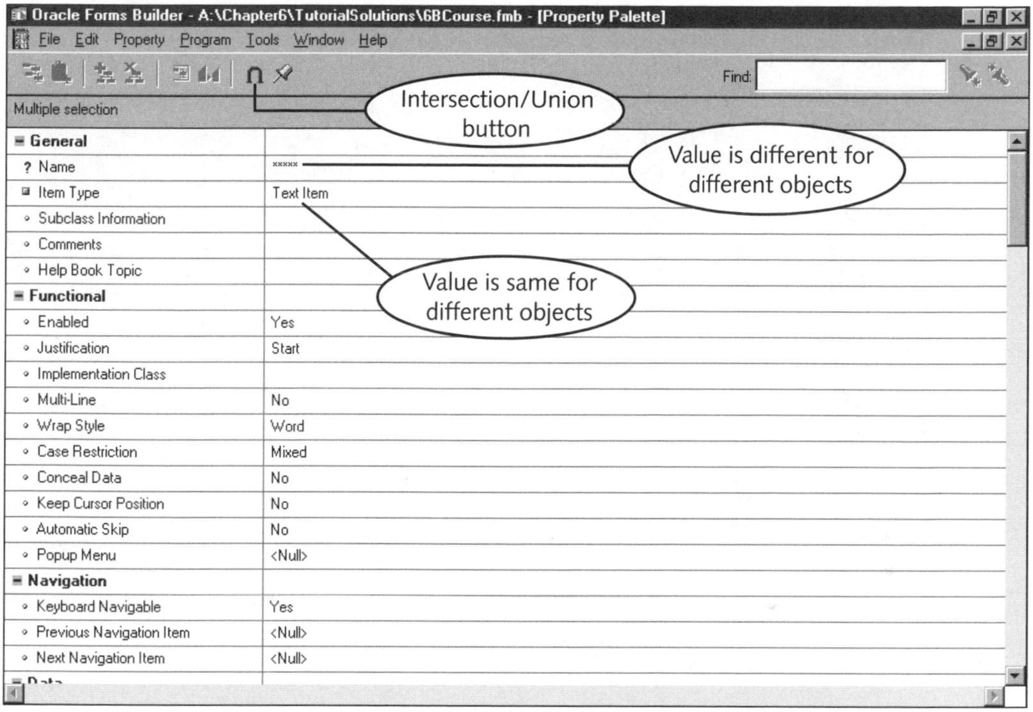

Oracle Forms Builder - A:\Chapter6\TutorialSolutions\6BCourse.fmb - [Property Palette]

File Edit Property Program Tools Window Help

Find:

Multiple selection

Intersection/Union button

≡ **General**	
? Name	xxxxx
▣ Item Type	Text Item
○ Subclass Information	
○ Comments	
○ Help Book Topic	
≡ **Functional**	
○ Enabled	Yes
○ Justification	Start
○ Implementation Class	
○ Multi-Line	No
○ Wrap Style	Word
○ Case Restriction	Mixed
○ Conceal Data	No
○ Keep Cursor Position	No
○ Automatic Skip	No
○ Popup Menu	<Null>
≡ **Navigation**	
○ Keyboard Navigable	Yes
○ Previous Navigation Item	<Null>
○ Next Navigation Item	<Null>
≡ **Data**	

Value is different for different objects

Value is same for different objects

6

Figure 6-23   Intersection Property Palette

Help

Sometimes when you change a text item font to a different style or size, the change does not appear on the form layout. If this happens, change the font to an alternate font style or size, redisplay the layout, and then change the font back to the desired style and size.

4. Close the Property Palette, and save the form.

A text item has several other properties that control how data values can be entered and displayed. These properties are summarized in Table 6-3.

In data block forms, many of these properties are determined by the properties of the corresponding database field and are usually not changed by the form developer. Now, you will modify the format mask of the CREDITS text item so the output appears with one decimal place. Then, you will run the form and confirm that the data value appears correctly.

To modify the format mask of the CREDITS text item:

1. Select the **CREDITS** text item, right-click, and then click **Property Palette**.

2. Scroll down to the Data property node and change the Format Mask property to **9.9** to indicate that the data value will be numeric and formatted to one decimal place. Close the Property Palette.

Property Node	Property Name	Description
Functional	Multi-Line	Specifies if the text item allows multiple-line input
Functional	Case Restriction	Converts entered value to uppercase or lowercase letters
Functional	Enabled	Specifies if the user can navigate to the item using the Tab key or mouse pointer
Functional	Conceal Data	Specifies that characters entered in the field are hidden and replaced by "*"
Data	Data Type	Specifies the text item data type
Data	Initial Value	Specifies a default value that is inserted into the text item each time the user creates a new record. This can be a value or a variable.
Data	Maximum Length	Specifies the maximum number of characters that the user can enter
Data	Fixed Length	Specifies that the value must be of a specified length
Data	Required	When set to Yes, specifies that the user must enter a value in the text item in order to save a new record
Data	Format Mask	Specifies the format mask of the item value

Table 6-3   Text item data value properties

3. Save the form, and then run the form. Click the **Enter Query button** 🔳, do not enter a search condition, and then click the **Execute Query button** 🔳 to retrieve all of the data records. The formatted form, with the CREDITS field formatted to one decimal place, appears as shown in Figure 6-24.

4. Close Forms Runtime.

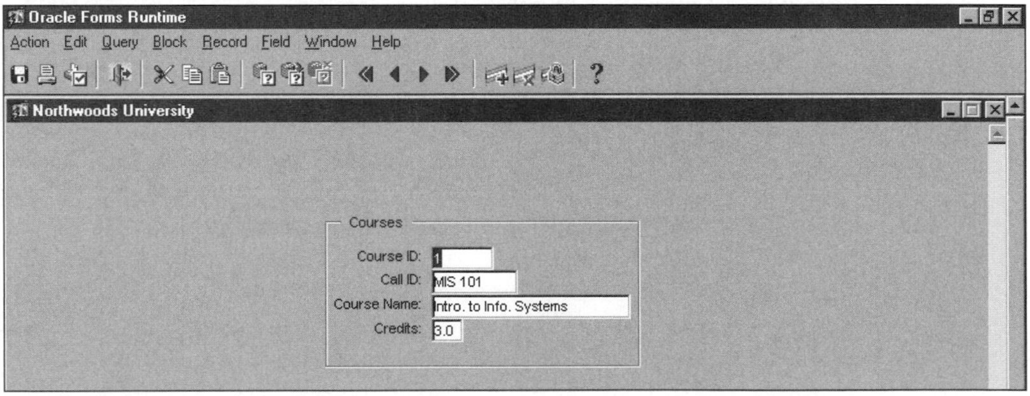

Figure 6-24   Formatted form

Text items also have properties that specify how the item interacts with the database. Table 6-4 summarizes these properties, which appear under the Database property node.

Property Name	Description
Primary Key	Specifies that the item corresponds to a database table's primary key
Query Only	Specifies that the item can appear in the form, but cannot be used for an insert or update operation
Query Allowed	Determines whether or not the user can perform queries using this block
Query Length	Specifies the maximum length allowed in the text item for a restricted query search condition. A value of 0 means there is no length limit.
Insert Allowed	Determines whether the user can manipulate the item value when inserting a new record
Update Allowed	Determines whether the user can change the item value when performing an update

Table 6-4   Text item database properties

These properties are automatically set to default values that allow the user to insert, update, delete, and view database records. The only time a developer changes them is when the form needs to restrict user operations.

The Help text item properties make it easy to add information that makes the form easier to use as the user navigates among form text items. You can add two different types of information: hints and ToolTips. A **hint** appears in the message line on the bottom-left edge of the screen display when the insertion point is in the text item. A **ToolTip** appears beside the mouse pointer when the user moves the mouse pointer over the text item. Table 6-5 summarizes the text item hint and ToolTip properties in the Help property node.

Property Name	Description
Hint	Specifies the text of the hint help information
Display Hint Automatically	Specifies whether the hint appears automatically when the user places the insertion point in the text item, or whether the hint only appears when the insertion point is in the text item and the user presses the F1 function key or selects the Help command on the menu bar
ToolTip	Stores the text of the ToolTip help information

Table 6-5   Text item Help properties

Now you will add hints and a ToolTip to the course form. The hints will describe the contents of the COURSE_ID and CALL_ID text items, and the ToolTip will describe the allowable range of values for the CREDITS item. Then, you will run the form and view the hints and ToolTip.

To add and view the text item hints and ToolTip:

1. Open the Property Palette of the specified text item, and change its properties to the following values:

Text Item	Property	New Value
COURSE_ID	Hint	**Unique identifier for the record**
COURSE_ID	Display Hint Automatically	**Yes**
CALL_ID	Hint	**Course department and number, such as MIS 101**
CALL_ID	Display Hint Automatically	**Yes**
CREDITS	ToolTip	**Number between 1 and 9**

2. Save the form, and then run the form. When the form appears in the Forms Runtime window, the insertion point is in the COURSE_ID field, and the hint automatically appears in the message bar.

3. Type **10** for the COURSE_ID value, and then press **Tab** to move the insertion point to the CALL_ID text item. The hint for the CALL_ID text item appears.

4. Place the mouse pointer on the CREDITS text item. The text item's ToolTip appears.

5. Close Forms Runtime, and click **No** when you are asked if you want to save your changes.

6. Click **File** on the menu bar, and then click **Close** to close the form in Form Builder.

CREATING A MULTIPLE-TABLE FORM

You can create data block forms that display data from multiple database tables that have master-detail relationships. In a **master-detail relationship**, one database record (the master record) can have multiple related (detail) records through foreign key relationships. For example, a master-detail relationship exists between the CUSTOMER and CUST_ORDER tables in the Clearwater Traders database, in which one CUSTOMER record can have multiple associated CUST_ORDER records. For instance, customer Alissa Chang (CUST_ID 179) has two different CUST_ORDER records (ORDER_ID 1061 and ORDER_ID 1062). The CUSTOMER record is the master side of the relationship, since a master (CUSTOMER) can have many detail (CUST_ORDER) records, but a detail record can have only one associated master record. Now you will create a form that allows users to select a master customer record and then display and edit detail customer order records.

When you create a form that contains a master-detail relationship, you always create the master block, which will be the list of customers, first. For the master block, you will use a file on your Data Disk named Customer.fmb that displays a single customer record on a form. You will modify this file by creating the detail block, which is the list of orders for a specific customer. You specify the master-detail relationship when you create the detail block. First, you will open the Customer.fmb file, save it using a different filename, and change the form module name.

To open Customer.fmb, save it using a different filename, and change the form module name:

1. In the Object Navigator window, click the **Open button** 🖼, and then open **Customer.fmb** from the Chapter6 folder on your Data Disk. Save the file as **6BCustomerOrder.fmb** in your Chapter6\TutorialSolutions folder.

 Your system might not show the .fmb file extensions in the Open File dialog box. If you try to open an .fmx instead of an .fmb file in Form Builder, the error message "FRM-10043: Cannot open file" will appear.

2. Select **CUSTOMER_FORM** under the Forms node in the Object Navigator, and change the form module name to **CUSTOMER_ORDER_FORM**.

3. Save the form.

Creating the Detail Data Block

When the user selects a customer in the CUSTOMER block, the form will show the selected customer's order information. Next, you will create a detail block called CUST_ORDER that is associated with the Clearwater Traders CUST_ORDER table to show the customer's related order information. To create a new block on an already existing form using the Data Block Wizard, you must select the form in the Object Navigator, and then start the Data Block Wizard. This block will be the detail block, so you will specify the master-detail relationship when you create it. The master-detail relationship information will be stored in a relation object within the master block.

To create the CUST_ORDER data block and specify the master-detail relationship:

1. Make sure the CUSTOMER_ORDER_FORM module is selected in the Object Navigator, click **Tools** on the menu bar, click **Data Block Wizard**, and then click **Next**.

2. Make sure the Table or View option button is selected, and then click **Next**.

3. Click **Browse**, make sure the Tables check box is checked, select **CUST_ORDER** as the database table, and then click **OK**.

4. Select all of the block fields to be included in the data block, and then click **Next**. The Wizard Master-Detail page appears, as shown in Figure 6-25.

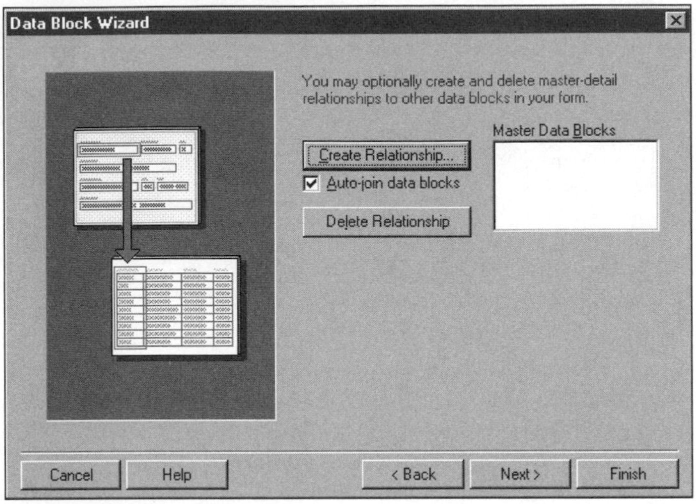

Figure 6-25   Data Block Wizard Master-Detail page

The **Master-Detail page** allows you to specify master-detail relationships among data blocks. This page appears when you create a new data block in a form that already contains at least one other data block. First, you will select the name of the master block, which will appear in the Master Data Blocks list. If the **Auto-join data blocks** check box is checked, then the system will automatically create the join condition between the text item in the selected master block that has a foreign key relationship with a text item in the current (CUST_ORDER) block. In this form, this is the CUST_ID text item. (Recall that CUST_ID is the primary key in the CUSTOMER table, and a foreign key in the CUST_ORDER table.)

If the Auto-join data blocks box is cleared, then you will need to manually select the text item in the master block and the text item in the detail block that will join the two blocks. First, you will create the relationship using the Auto-join data blocks feature. You will click the Create Relationship button and be presented with a list of the current form blocks. You will select the CUSTOMER block to be the master block. The join condition will automatically be created based on the foreign key relationship between the CUST_ID field in the CUSTOMER table and the CUST_ID field in the CUST_ORDER table.

To create the relationship using the Auto-join data blocks feature:

1. Make sure the Auto-join data blocks check box is checked, and then click **Create Relationship**. The CUSTOMER_ORDER_FORM: Data Blocks dialog box opens, which shows all of the form blocks.

2. Select **CUSTOMER** (which is the only choice), and then click **OK**. CUSTOMER appears in the Master Data Blocks list, and the join condition for the two blocks appears in the Join Condition box, as shown in Figure 6-26.

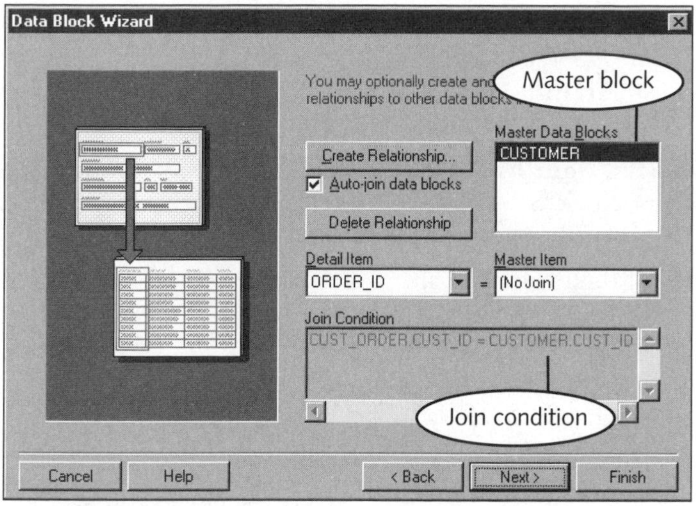

Figure 6-26 Master-Detail page with master block and join condition specified

The Join Condition box shows the join condition, using the general syntax *detail_block.join_item = master_block.join_item* to specify the block and item names on which you are joining the two blocks. For this form, the join condition is CUST_ORDER.CUST_ID = CUSTOMER.CUST_ID.

Now, you will create a master-detail relationship manually. You need to create the join condition manually if you have multiple blocks on the form, and the automatic join feature cannot create the join condition, because the Master Block could be one of many blocks. First, you will delete the relationship you just created and clear the Auto-join data blocks check box. Then, you will create the join condition manually.

To delete the relationship and create the join condition manually:

1. On the Data Block Wizard Master-Detail page, click **Delete Relationship**. The master data block and join condition no longer appear.

2. Clear the **Auto-join data blocks** check box.

3. Click **Create Relationship**. The Relation Type dialog box opens. You can create the relation based on a join condition or on a REF item. (Recall from Chapter 2 that an REF item is a data type that contains pointers, or physical addresses, of related data objects.) Since your tables do not contain any REFs, you will base the relation on a join condition.

4. Confirm that the Based on a join condition option button is selected, and then click **OK**. The CUSTOMER_ORDER_FORM: Data Blocks dialog box opens, which shows the CUSTOMER block, which is the only other block in the form.

5. Select **CUSTOMER**, and then click **OK**. Notice that no join condition appears in the Join Condition box.

6. Open the **Detail Item** list box. This lists all of the text items in the detail (CUST_ORDER) block. Select **CUST_ID**.

7. Open the **Master Item list box**. This lists all of the text items in the master (CUSTOMER) block. Select **CUST_ID**. (You might need to scroll up to display CUST_ID.) The join condition appears in the Join Condition box.

8. Click **Next**, and then click **Finish**.

Sometimes when you click Create Relationship when the Auto-join data blocks check box is checked, the message "FRM-10757: No master blocks are available" appears. If it does, click OK, clear the Auto-join data blocks check box, and then create the join condition manually.

Next, you will create the layout for the CUST_ORDER block. You will use a tabular layout, display five order records, and use a scroll bar.

To create the CUST_ORDER layout:

1. Click **Next** on the Welcome to the Layout Wizard dialog box. Since you want the CUST_ORDER block fields to appear on the same canvas as the CUSTOMER block fields, accept **CUSTOMER_CANVAS** as the layout canvas, and then click **Next**.

2. Select all block fields to appear in the layout, and then click **Next**.

3. Accept the default values for the item prompts. Change the field widths as follows, and then click **Next**.

Prompt	Width
Order ID	40
Order Date	50
Meth Pmt	40
Cust ID	30
Order Source ID	30

4. Select the **Tabular option button**, and then click **Next**.

5. Type **Customer Orders** for the frame title, change Records Displayed to **5**, and check the **Display Scrollbar** check box. Click **Next**, and then click **Finish**. The Customer Orders frame and text items appear on the canvas. Move the frame so the records appear on the right side of the Customer frame text items, as shown in Figure 6-27.

6. Open the Object Navigator window, click ➕ beside the CUSTOMER block, and then click ➕ beside the Relations node. Note that a relation object named CUSTOMER_CUST_ORDER has been created, which specifies the join condition created on the Master–Detail page.

7. Save the form.

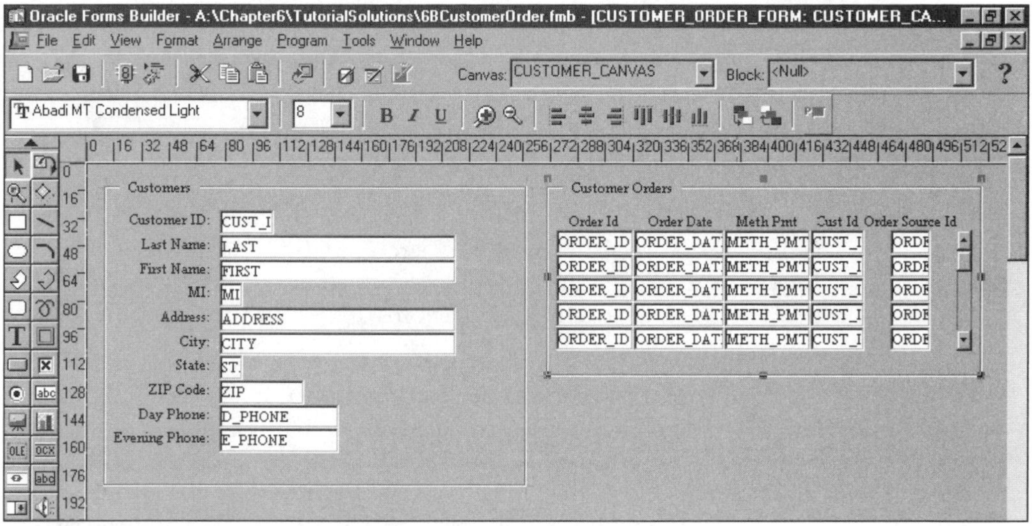

Figure 6-27   Repositioning the detail block frame

Running the Master-Detail Form

Now you will run the form to see how a master-detail form works. First, you will run the form, and retrieve all of the customer records along with their associated customer orders.

To run the form and retrieve the customer and order records:

1. Run the form, and maximize the Forms Runtime window.

2. Click the **Enter Query button** 🔳, do not enter a search condition, and then click the **Execute Query button** 🔳. Paula Harris's data values appear in the Customers frame text items, and her associated customer order record appears in the Customer Orders frame.

3. Click the **Next Record button** ▶. The record for customer Mitch Edwards appears in the Customers frame, and his associated order record appears in the Customer Orders frame.

4. Continue to click ▶ to scroll through the rest of the customer records, noting how the customer order information changes to reflect the current customer.

You can add, update, delete, and query database records in a form that is based on multiple tables, just as you did in the single-table form. The only difference is that when a form displays multiple data blocks, you can only insert, update, delete, or query records in the block that is currently selected and displays the insertion point. For example, to enter a new CUST_ORDER record for the current customer, you must place the form insertion point in the CUST_ORDER block. (Currently, the insertion point is in the CUSTOMER block.) There are two ways to move the form insertion point: click the

Next Block button ⏩ on the toolbar to move from the Customer block to the CUST_ORDER block, or click the mouse pointer on any field in the CUST_ORDER block. Now you will retrieve all of the customer orders for Alissa Chang, place the insertion point in the CUST_ORDER block, and insert a new order record.

To retrieve Alissa Chang's records and enter a new customer order record:

1. Click the **Enter Query button** 🔲, press the **Tab** key to move to the Last Name field, type **Chang** as the search condition, and then click the **Execute Query button** 🔲. Alissa Chang's data appears in the Customers frame, and her associated customer order records appear in the Customer Orders frame.

2. Click the **Next Block button** ⏩ to move to the CUST_ORDER block. The insertion point moves to the first record in the CUST_ORDER block.

3. Click the **Insert Record button** 🔳 to insert a new record. A blank record appears below Order ID 1061.

4. Enter the following new record: Order ID **1063**, Order Date **10-JUN-2003**, Pmt. Method **CHECK**, Customer ID **179**, Order Source ID **6**.

5. Click the **Save button** 🔲 to save the record. The message "FRM-40400: Transaction complete: 1 records applied and saved" appears on the message line to confirm that the new record is saved.

6 Close the Forms Runtime.

Adding Another Detail Data Block to the Form

A block can be the master block in multiple master-detail relationships. For example, an inventory item can be in many customer orders, and an inventory item can also have many associated shipments. The inventory block is the master block in the inventory-customer order relationship, and it is also the master block in the inventory-shipment relationship.

A block can also be the detail block in one master-detail relationship, and the master block in a second master-detail relationship. For example, a customer can have many orders, and an order can have many order lines. In this scenario, the order is the detail block in the customer-order relationship, and it is the master block in the order-order line relationship. Currently in your form, CUSTOMER is the master block, and CUST_ORDER is the detail block. Next, you will create a third block that displays data from the ORDER_LINE table to show the order line details that correspond to a particular customer order. CUST_ORDER will be the master block in this new relationship, and ORDER_LINE will be the detail block.

To create the second master-detail data block relationship:

1. In the Object Navigator, select **CUSTOMER_ORDER_FORM**, click **Tools** on the menu bar, and then click **Data Block Wizard** to create a new data block.

2. Create a new data block using all of the columns in the ORDER_LINE table.

3. For the master-detail relationship, select CUST_ORDER as the master block, and ORDER_ID as the join field. Your master-detail specification page should look like Figure 6-28.

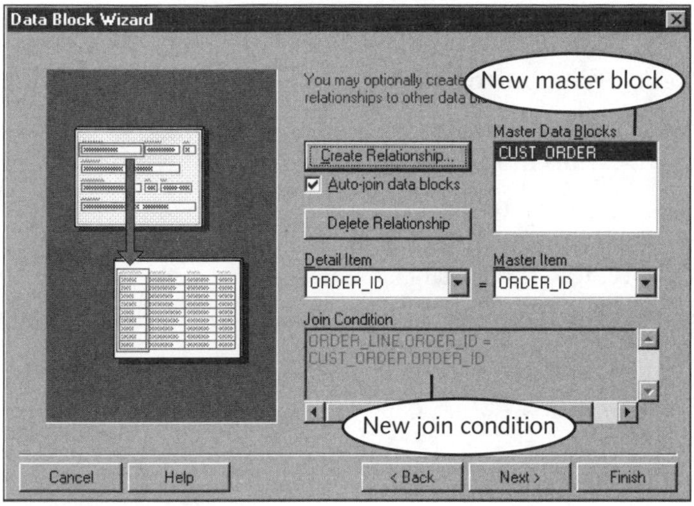

Figure 6-28   Specifying the second master-detail relationship

4. Accept **CUSTOMER_CANVAS** as the form canvas, select all of the ORDER_LINE data block fields for the layout, and accept the default prompt values.

5. Use a tabular-style layout, type **Order Line Detail** for the frame title, show **5** records, and display a scroll bar.

6. Move the Order Line Detail frame so it is directly below the Customer Orders frame.

7. Save the form.

When you select a particular CUSTOMER record, the associated customer order records appear in the CUST_ORDER block, and the associated ORDER_LINE records appear for the record that is currently selected in the CUST_ORDER block. The insertion point position indicates the record that is currently selected in a block. Now you will run the form and view Alissa Chang's orders and their associated order lines.

To run the form and view the customer orders and order lines:

1. Run the form, and maximize the Forms Runtime window.

2. Click the **Enter Query button** 🖼, press **Tab**, type **Chang** as the search condition in the Last Name field, and then click the **Execute Query button** 🖼.

Alissa Chang's CUSTOMER record appears in the CUSTOMER block, and her associated CUST_ORDER records appear in the CUST_ORDER block. The ORDER_LINE block displays the ORDER_LINE records associated with CUST_ORDER 1061, because it is the first record in the block, which makes it the current block record. Now you will display the ORDER_LINE records for CUST_ORDER 1062 by moving to the CUST_ORDER block and selecting Order ID 1062.

To view the ORDER_LINE records for Order 1062:

1. Click the **Next Block button** ⏩ to move to the CUST_ORDER block.

2. Click the **Next Record button** ▶ to select Order ID 1062. Notice that the ORDER_LINE records update to show information for Order ID 1062.

3. Click ▶ until Order ID 1063 appears. This is the new order that you entered before, and no ORDER_LINE records have been inserted, so none appear in the ORDER_LINE block.

You can insert, update, and delete data records just as before. Next, you will add an order line to Order ID 1063.

To add an order line to the order:

1. Click the mouse pointer in the Inv Id text item next to Order ID 1063 in the Order Line Detail frame, and enter the following data in the form: Inventory ID = **11846**, Quantity = **1**.

2. Click the **Save button** 💾. Note the confirmation message at the bottom left of the screen.

3. Close Forms Runtime.

WORKING WITH FORMAT MASKS TO FORMAT TEXT ITEMS

Recall that you can change a text item's format mask property in its Property Palette to format how the text item's data values appear. Figure 6-29 shows that some of the data fields in the customer order form need to be formatted. The daytime phone number for customer Paula Harris appears as 7155558943, but it would be more understandable if the digits were separated as (715) 555-8943. In addition, you will change the date format from the default DD-MON-YYYY format to MM/DD/YYYY.

Now you will change the format masks for the customer daytime and evening phone numbers, and the order date. Recall that you can embed characters in character string data fields by preceding the format mask with the characters FM, and placing embedded characters in quotation marks. For the phone number, you will embed an opening parenthesis ("("), list the first three numbers, embed a closing parenthesis and a blank space (") "), list the next three numbers, embed a hyphen ("-"), then list the last four numbers. Now you will enter the format mask for the phone number.

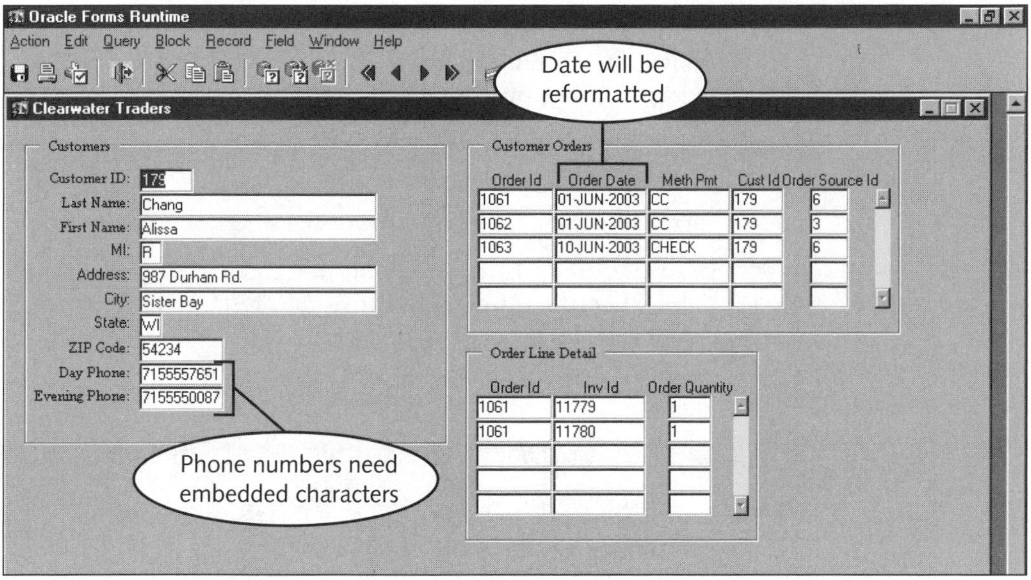

Figure 6-29   Form fields requiring formatting

To enter the phone number format mask:

1. If necessary, click **Tools** on the menu bar, and then click **Layout Editor** to open the form in the Layout Editor.

> You can also open a form in the Layout Editor by double-clicking the form's Canvas icon 🖼 in the Object Navigator.

2. Select the **D_PHONE** text item, press **Shift**, select the **E_PHONE** text item, right-click, and then click **Property Palette**. Note that the name property value appears as *****, which indicates that this is an intersection Property Palette.

3. Scroll down to the Data property node, and type the following format mask for the telephone number: **FM"("999") "999"-"9999**. (There are three embedded character strings in the format mask: an opening parenthesis, a closing parenthesis followed by a space, and a hyphen.) Close the Property Palette.

Now you will format the ORDER_DATE text item. Recall that you can use front slashes, hyphens, and colons as separators between different date and time elements in DATE data fields. For this format mask, you will embed front slashes between the month and the day fields. Then, you will run the form and view the formatted text items.

To enter the date format mask and view the formatted text items:

1. In the Layout Editor, right-click any one of the **ORDER_DATE** fields. The fields are selected as a group. Right-click, and then click **Property Palette**.

2. Type the following format mask: **MM/DD/YYYY**, and then close the Property Palette.

3. Save the form, and then run it, and maximize the Forms Runtime window.

4. Click the **Enter Query button** 🔍, do not enter a search condition, and then click the **Execute Query button** 🔍 to retrieve all of the database records. The first record appears as shown in Figure 6-30.

5. Close Forms Runtime.

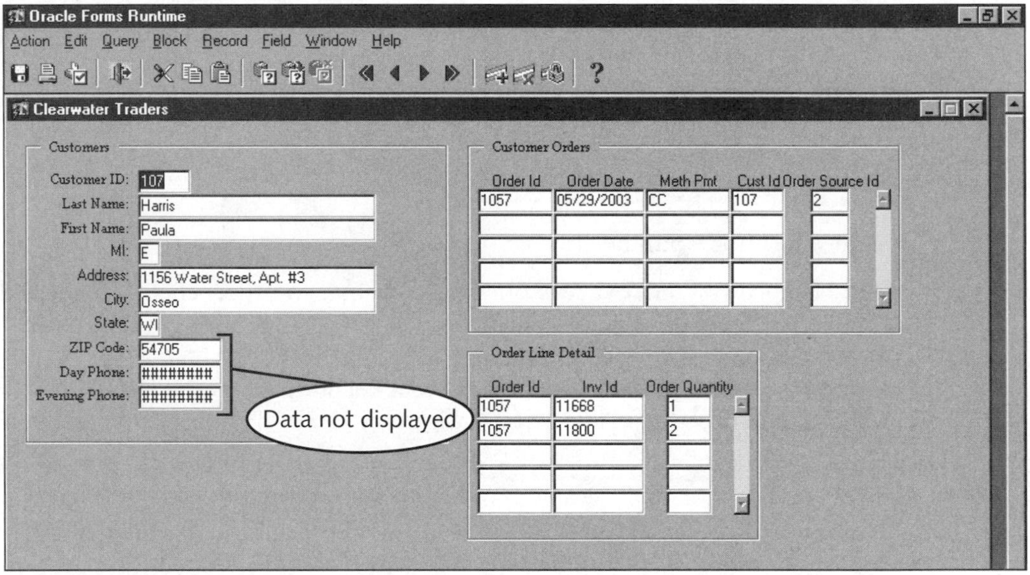

Figure 6-30 Data not displayed due to text item width error

Note that the date format mask for the ORDER_DATE text item appears correctly, but the phone number text items appear as pound signs (#). The pound signs appear because the form text items are not wide enough for the extra formatting characters specified in the format masks. Currently, the D_PHONE and E_PHONE text items are just wide enough to display the phone numbers' digits but are not wide enough to display any extra formatting characters. When the embedded formatting characters are included, the fields are not wide enough. You will need to change two properties in the phone number text item Property Palettes: the Data Width, which determines how many data characters can be displayed, and the Visible Width, which determines how wide the displayed field is on the form. If a text item's Data Width property value is not wide enough for a format mask, the field value will appear as pound signs (####). If a field's Visible Width is not wide enough, the field value will be clipped. Now you will change these properties using an intersection Property Palette.

To change the phone number field width properties:

1. In the Layout Editor, select **D_PHONE**, press and hold the **Shift** key, select **E_PHONE**, right-click, and then click **Property Palette**.

2. Scroll down to the Data property node, and find the Maximum Length property, which currently has a value of 10. You are adding four additional formatting characters, so the maximum length must be at least 14. It is a good idea to always add extra characters so you are sure all of your data appears, so you will change the width to 16.

3. Change the Maximum Length value to **16**.

4. Scroll down to the Physical property node, and change the Width to **70**. Close the Property Palette. In the Layout Editor, note that the D_PHONE and E_PHONE fields appear wider.

You can also change the Width property of a text item by selecting the item in the Layout Editor, and then resizing it by dragging the selection handles to the right or left.

5. Save the form, run the form, and maximize the Forms Runtime window. Click the **Enter Query button** 🔲, do not enter a search condition, and then click the **Execute Query button** 🔲 to retrieve all of the customer records. Your formatted text items should look like Figure 6-31.

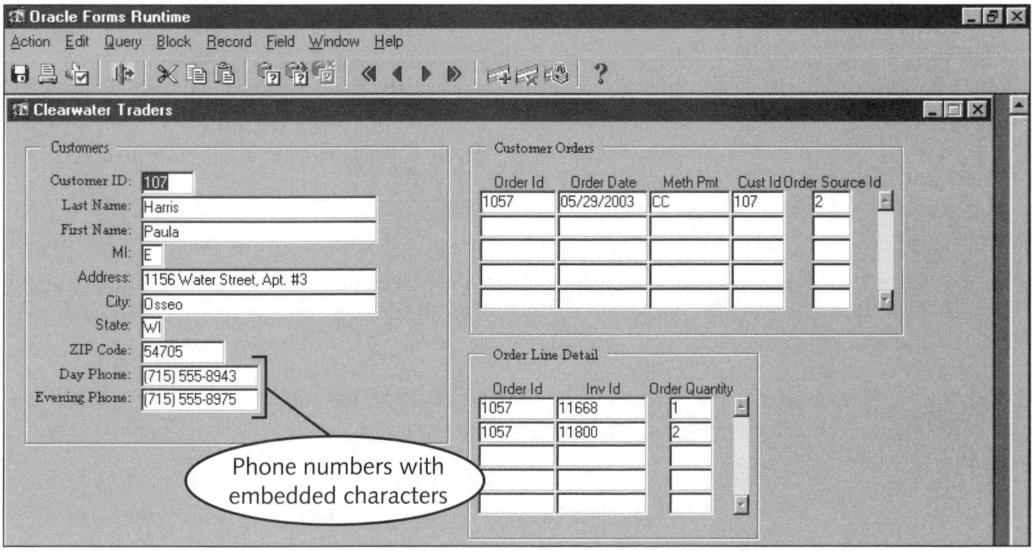

Figure 6-31   Formatted phone number text items

6. Close the Forms Runtime window.

DELETING DATA BLOCKS

Sometimes you need to delete a form data block. The only way to delete a data block is to select the block in the Object Navigator, and delete it from the form. You can delete text items within a block in the Layout Editor, but you cannot delete the block itself or the block scroll bars in the Layout Editor. To illustrate this, you will first delete the text items of the ORDER_LINE block in the Layout Editor.

To delete the text items in the Layout Editor:

1. Click the **Save button** 🖫 to confirm that the CUSTOMER_ORDER_FORM is saved.

2. In the Layout Editor, click the **ORDER_ID** text item in the Order Line Detail frame, press and hold the **Shift** key, and then click the **INV_ID** and **ORDER_QUANTITY** text items, the block **scroll bar**, and the **Order Line Detail frame**. Selection handles appear around all of the block items.

3. Press **Delete**. The block text items and frame no longer appear. Note that the scroll bar still appears, because you cannot delete a scroll bar in the Layout Editor

4. Click **Window** on the menu bar, and then click **Object Navigator** to open the Object Navigator window for the form. Note that the ORDER_LINE block still exists, but all of its objects have been deleted.

 To hide a block's scroll bar, open the block's Property Palette, and change the Show Scroll Bar property value to No.

Now you will use the revert feature in Form Builder to restore the ORDER_LINE text items. The **revert** feature restores the current form to its state the last time it was saved. Then, you will delete the ORDER_LINE block in the Object Navigator.

To revert to the last saved copy of the form, and then delete the block in the Object Navigator:

1. Click **File** on the menu bar, click **Revert**, and then click **Revert** in the dialog box to confirm restoring to the last saved copy. The restored form appears in the Object Navigator.

2. Display the form in the Layout Editor to confirm that the Order Line Detail frame text items have been restored.

3. In the Object Navigator, click ➕ beside the Data Blocks node, select the **ORDER_LINE** data block, and then click the **Delete button** 🗙 on the

Object Navigator toolbar to delete the block. Click **Yes** to confirm removing the block. The block is removed from the Object Navigator window.

 You can also delete form objects in the Object Navigator by selecting the object and then pressing the Delete key.

4. Open the Layout Editor to confirm that the block items no longer appear on the canvas. Note that the Order Line Detail frame still appears—the frame is a boilerplate object on the canvas and must be deleted independently of the block.

5. Click **File** on the menu bar, and then click **Revert** to again restore the file to the last saved state.

6. Close Form Builder and all other open Oracle applications.

When you delete individual form objects in the Layout Editor, you have the option of immediately undoing the operation by clicking Edit on the menu bar and then clicking Undo. This will only work as long as the Layout Editor window is still open. If you open another window, the Undo information is no longer saved. You cannot undo delete operations that you perform in the Object Navigator.

6

SUMMARY

❐ When you create a data block based on a view, you can view data, but you cannot directly modify the database by inserting new records or updating or deleting existing records.

❐ Every form object has a Property Palette that allows you to modify the object's properties. Different objects have different property lists, which you can modify by typing new values, selecting from predefined lists, or specifying values using dialog boxes.

❐ When a Property Palette window's Freeze/Unfreeze button 🗷 is pressed, the window can be opened with other Property Palette windows at the same time.

❐ You should not allow users to maximize or resize a form window, because the form objects might appear off-center or clipped off. Users should always be able to minimize a form window.

❐ You should place form items within the boundaries of the form canvas or an error will appear.

❐ You should display form text using a sans serif (block) font for increased legibility.

❐ You can configure frames three ways: to automatically position the items enclosed in the frame; to allow the developer to manually position the items; or to lock the items and not have the custom formatting removed when the layout is updated by opening the Layout Wizard in reentrant mode and applying changes.

❑ You can open a Property Palette in intersection mode and use it to modify the properties of multiple objects to the same value.

❑ A text item can have a hint, which is informative text that appears in the message line when you place the insertion point in the text item, or a ToolTip, which is informative text that appears directly beside the text item when you place the mouse pointer over the item.

❑ In a master-detail relationship, one database record can have multiple related records through foreign key relationships. A master record can have many detail records, but a detail record always has only one associated master record.

❑ A block can be the master block in multiple master-detail relationships.

❑ A block can be the detail block in one master-detail relationship and the master block in a second master-detail relationship with a different block.

❑ When you create a master-detail relationship, you always create the master block first and then specify the master-detail relationship when you create the detail block. The relationship information is stored in a relation object within the master block.

❑ When you create a master-detail relationship, checking the Auto-join data blocks check box instructs the system to automatically create the join condition between the text items in the selected master block that has a foreign key constraint with a text item in the current block. You also can create the join condition manually.

❑ You can add, update, delete, and query database records in master-detail blocks just as in single-table data blocks, except that you need to first place the insertion point in the data block before performing the insert, update, or query.

❑ You can specify format masks for a form text item by entering the desired format mask in the item Property Palette.

❑ After you have specified a format mask for a text item, you must always make sure that the text item is wide enough to display the data, as well as extra formatting characters. If it is not wide enough, you need to adjust the item's Data Width and Visible Width properties.

❑ You can only delete a data block in the Object Navigator window. You can delete individual items within a data block in the Layout Editor. When you delete an item in the Object Navigator, Form Builder does not save information for undoing the delete operation.

REVIEW QUESTIONS

1. When should you create a data block form using a database view?
2. What is the limitation of a form created from a view?
3. Why do you need to change the canvas size to match the window size?

4. What is the difference between a frame with the Update Layout property set as Automatically, and one with the Update Layout property set as Manually?

5. How do you open a Property Palette in intersection mode? When is this useful?

6. For the following data fields, indicate the required values for the text item format mask and width to display the desired formatted data:

Variable Name	Data Type	Data Value	Desired Format
CUST_SSN	VARCHAR2(9)	888993452	888-99-3452
APPT_TIME	DATE	12-DEC-2003 10:00 AM	10:00 AM
ACCT_BALANCE	NUMBER(6,2)	-1,221.56	<$1,221.56>

7. What is the difference between a hint and a ToolTip?

8. Where is the information about a master-detail relationship between two blocks stored?

9. How do you delete a data block in a form?

10. What is the advantage of deleting form items in the Layout Editor rather than in the Object Navigator?

6

PROBLEM-SOLVING CASES

The case problems reference the Clearwater Traders (see Figure 1-11), Northwoods University (see Figure 1-12), and Software Experts (see Figure 1-13) sample databases. For all cases, save the solution files in your Chapter6\CaseSolutions folder. For all forms, rename the window and canvas using descriptive object names, and use the guidelines in the lesson to format the window and canvas properties. Use a sans serif font for all form text item data values and prompts. Specify descriptive item prompts, and if necessary, format the item prompts so they appear on a single line.

1. In this case, you will create a form based on a view that displays information about Clearwater Traders customer orders.

   a. In the SQL*Plus, create a view named CUST_ORDER_VIEW that retrieves the order ID, order date, payment method, and associated customer, first and last name, and order source description for every record in the CUST_ORDER table in the Clearwater Traders database. Save the command to create the view in a file named 6BCase1.sql.

   b. Create a form named CUST_ORDERS_FORM based on the CUST_ORDER_VIEW. Title the frame "Customer Orders," use a tabular layout, and display five records on the form at a time with a scrollbar. Resize the layout text items if necessary so all items appear on your screen without scrolling horizontally. Format the order date so the date for Order ID 1057 appears as "May 19, 2003." Save the form in a file named 6BCase1.fmb.

2. In this case, you will create a form based on a view that displays information about course sections at Northwoods University.

   a. In the SQL*Plus, create a view named COURSE_SECTION_VIEW that retrieves the section ID, course name, term description, section number, faculty first and last name, day, time, and building code and room, and maximum enrollment for every record in the COURSE_SECTION table in the Northwoods University database. Save the command to create the view in a file named 6BCase2.sql.

   b. Create a form named COURSE_SECTIONS_FORM based on the COURSE_SECTION_VIEW. Title the frame "Course Sections," and use a form-style layout that displays one record on the form at a time. Add a format mask so the TIME value for Section ID 1000 appears as "10:00 AM." Save the form in a file named 6BCase2.fmb.

3. Create a master-detail form named FACULTY_STUDENT_FORM that displays records from the FACULTY and STUDENT tables in the Northwoods University database. Format the form as shown in Figure 6-32. Save the form in a file named 6BCase3.fmb.

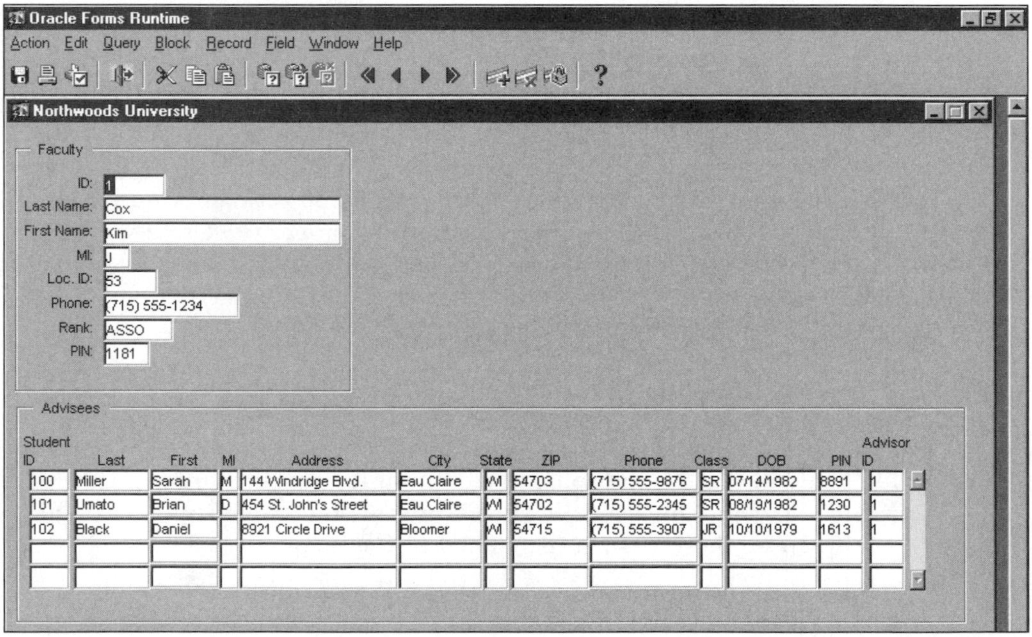

Figure 6-32

4. Create a master-detail form named SHIPMENT_FORM that displays records from the SHIPMENT and SHIPMENT_LINE tables in the Clearwater Traders database. Save the form in a file named 6BCase4.fmb. Format all DATE data fields using the MM/DD/YYYY format mask.

a. Create a master block that displays all of the fields from the SHIPMENT table. Display one record at a time. Use "Shipments" as the frame title.

b. Create a detail block that displays all of the fields in the SHIPMENT_LINE table. Use a tabular layout, and display five records at one time. Use "Shipment Lines" as the frame title.

5. Create a master-detail form named PROJECT_CONSULTANTS_FORM that displays records from the PROJECT, CONSULTANT, and PROJECT_CONSULTANT tables in the Software Experts database. Save the form in a file named 6BCase5.fmb.

a. Create a master block that displays all of the fields from the PROJECT table. Display one project record at a time. Use "Projects" as the frame title.

b. Create a detail block that displays all of the fields in the PROJECT_CONSULTANT table. Use a tabular layout, and display five records at one time. Use "Project Consultants" as the frame title. Format the roll-on and roll-off dates so the roll-on date for project ID 1 and consultant ID 101 appear as 06/15/2002.

c. Create another detail block that displays all of the fields in the CONSULTANT table for the consultant who is currently selected in the PROJECT_CONSULTANT block. Use a form layout to display one record at a time. Use "Consultant Detail" as the frame title. Format the consultant phone number so the number for consultant Sheila Hernandez is displayed as (715) 555-0282.

6. Create a master-detail form named FACULTY_COURSE_SECTION_FORM that displays records from the FACULTY and COURSE_SECTION tables in the Northwoods University database. Save the form in a file named 6BCase6.fmb.

a. Create a master block that displays all of the fields from the FACULTY table except the F_IMAGE field. Display one faculty record at a time. Use "Faculty" as the frame title. Format the phone number field with parentheses and hyphens so that it is easier to read; for example, Kim Cox's phone number should appear as (715) 555-1234.

b. Create a related detail block that displays all of the related fields from the COURSE_SECTION table for the current faculty member. Show five records at one time, and use a scroll bar to display additional records. Format the COURSE_SECTION TIME field so it appears as "10:00 AM."

7. Create a master-detail form named CONSULTANT_FORM that displays records from the CONSULTANT, SKILL, CONSULTANT_SKILL, and PROJECT tables in the Software Experts database. Format the form so all information appears on a single screen without scrolling. Save the form in a file named 6BCase7.fmb.

a. Create a master block that displays all of the fields from the CONSULTANT table. Display one consultant record at a time. Use "Consultants" as the frame title. Format the phone number field with parentheses and hyphens so that it is easier to read; for example, Mark Myers' phone number should appear as (715) 555-9652.

6

b. Create a related detail block that displays all of the related fields from the CONSULTANT_SKILL table for the current consultant. Show five records at one time, and use a scroll bar to display additional records. Use "Consultant Skills" for the frame title.

c. Create a related detail block that displays the related skill description for the skill ID that is currently selected in the CONSULTANT_SKILL block. Use "Skill Description" for the frame title.

d. Create a related detail block that displays all of the project information from the PROJECT_CONSULTANT table for the current consultant. Use "Projects" for the frame title. Format the date fields using the MM/DD/YYYY format mask.

8. Create a master-detail form named STUDENT_ENROLLMENT_FORM in which the detail block is created based on a database view. Save the form in a file named 6BCase8.fmb.

a. Create a master block that displays all of the fields from the STUDENT table in the Northwoods University database. Display one record on the form at a time. Format the phone number field so Sarah Miller's phone number appears as 715-555-9876. Change the frame label to "Students."

b. In SQL*Plus, create a view named COURSE_ENRL_VIEW that displays the course section ID, student ID, call ID, course name, course credits, and grade for all course section records in which a grade has been recorded. Save the command to create the view in a file named 6BCase8.sql.

c. Create a detail block that displays five COURSE_ENRL_VIEW records for the current student at one time, and use a scroll bar to display additional records. Change the frame label to "Student Courses."

9. Create a multiple-table form named COURSE_SECTIONS_FORM that shows courses and course sections in the Northwoods University database. Enrollment information for each course section will be displayed using a view. Save the form in a file named 6BCase9.fmb.

a. Create a master block that shows all records in the COURSE table. Show five records at one time, and use a scroll bar to display additional records. Change the frame title to "Courses."

b. Create a detail block that shows all records in the COURSE_SECTION table for the selected course. Show three records at one time, and use a scroll bar to display additional records. Change the frame title to "Course Sections." Format the TIME text item so the value for course section ID 1000 is appears as "10:00 AM."

c. In SQL*Plus, create a view named STUDENT_ENRL_VIEW that displays the student ID, last name, first name, section ID, and grade for all students. Save the command used to create the view in a file named 6BCase9.sql.

d. Create a detail block that displays all fields from STUDENT_ENRL_VIEW for a selected course section. Show three records at one time, and use a scroll bar to display additional records. Change the frame label to "Enrolled Students."

10. Create a multiple-table form named LOCATIONS_FORM that shows locations, and the faculty member or course section associated with a specific location, using data from the Northwoods University database. Format the form as shown in Figure 6-33. Course section information will be displayed using a view. Save the form in a file named 6BCase10.fmb.

a. Create a master block that shows all records in the LOCATION table and that is formatted as shown.

b. Create a detail block that shows the record in the FACULTY table that is associated with the selected location. Do not include the F_IMAGE field in the block. Format the block as shown, and format the phone number so Kim Cox's phone number appears as 715-555-1234.

c. In SQL*Plus, create a view named COURSE_LOCATIONS_VIEW that displays the course section ID, call ID, term description, section number, location ID, day, and time for all course sections. Save the command used to create the view in a file named 6BCase10.sql.

d. Create a detail block that displays all fields from COURSE_SECTION_VIEW for a selected location. Format the block as shown.

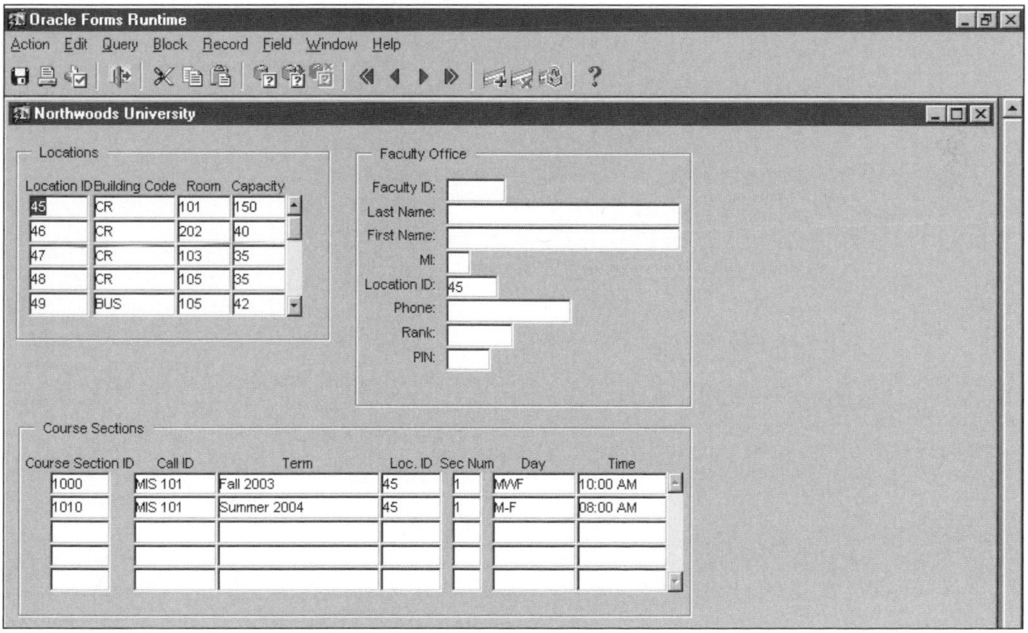

Figure 6-33

◀ LESSON C ▶

Objectives

- ◆ Use sequences to automatically generate primary key values in a form
- ◆ Create single-table and multiple-table lists of values (LOVs) to provide lists for foreign key values
- ◆ Create editors to aid entering and editing blocks of text
- ◆ Become familiar with different form item types that you can use to enter and modify data values

So far, you have used forms to insert, update, delete, and view data. All of these actions occur as a result of programs that exist in the Forms Runtime environment that you did not need to write. In this lesson, you will learn how to write PL/SQL programs called triggers that make your forms more functional and easier to use. You will use triggers to automatically retrieve sequence values into forms for primary key values. You will also use triggers to display a list of values, which is a pick list from which the user can select a value for a form text item. You will learn how to create editors for entering large blocks of text, and you will explore how to use alternate controls, such as radio buttons and check boxes, to display and manipulate data.

USING SEQUENCES IN FORMS TO GENERATE PRIMARY KEY VALUES

With the forms you have created so far, whenever you insert a new record into a database by entering the values in the form, you have to type in the value for the record's primary key. This could be a source of errors if two users use the same primary key value for different records, and it is inconvenient for the user to always have to query the database or look up the next available primary key value using a printed report.

Recall that in Chapter 3, you learned how to create sequences to automatically generate surrogate key values for primary key fields. Now you are going to learn how to create a form for inserting and viewing records that automatically retrieves the next value in a sequence and displays it in the form text item that is associated with the database table's primary key. Specifically, you will create a form for the INVENTORY table in the Clearwater Traders database that retrieves a value from a sequence and displays the value in the Inventory ID text item. (The Inventory ID text item corresponds to the INV_ID database field, which is the primary key of the INVENTORY table.) First, you will run the script that refreshes the database tables. Then, you will create the sequence to generate the INV_ID values that the form will retrieve. Since the highest current INV_ID value is 11848, you will set the sequence start value at 11900.

To run the scripts and create the sequence:

1. Start SQL*Plus and log onto the database.

2. Run the **clearwater.sql**, **northwoods.sql**, and **softwareexp.sql** scripts in the Chapter6 folder on your Data Disk.

3. Type the following code to create the INV_ID_SEQUENCE. Do not close SQL*Plus, because you will use it later in this lesson.

```
CREATE SEQUENCE inv_id_sequence
START WITH 11900;
```

Now you will start Form Builder and open a form named Inventory.fmb that is in the Chapter6 folder on your Data Disk. This is a data block form associated with the INVENTORY table in the Clearwater Traders database that displays one inventory record at a time. First, you will start Form Builder, open the form, and save it with a different filename.

To open the form and save it as a different filename:

1. Start Form Builder, select the **Open an existing form** option button, click **OK**, navigate to the Chapter6 folder on your Data Disk, and open **Inventory.fmb**.

2. Save the file as **6Cinventory.fmb** in your Chapter6\TutorialSolutions folder.

3. Click **File** on the menu bar, click **Connect**, and connect to the database.

To automatically insert the next INV_ID_SEQUENCE value into the Inventory ID form field, you will create a form trigger. The next section describes form triggers.

Creating Form Triggers

A **form trigger** is a program unit in a form that runs when a user performs an action, such as clicking a button. Form triggers also can be associated with specific actions performed by the system, such as loading the form or exiting the form.

 Form triggers are different from the database triggers that you learned about in an earlier chapter. Database triggers run as a result of an action that occurs with a database table, such as inserting a new record. In the Form Builder environment, form triggers are simply called triggers, and they run as a result of actions that occur within a form. Keep the distinction between database triggers and form triggers in mind.

Triggers are associated with specific form objects, such as text items, command buttons, data blocks, or the form itself. For example, when you click the Save button 💾 on the Forms Runtime toolbar, a trigger containing a SQL INSERT command runs. The trigger's form object was the Save button, and the action was clicking the button. To write a trigger, you must specify the object that the trigger is associated with, the action, or event, that starts the trigger, and the code that executes when the trigger fires. Every form object has specific **events** that can be used to fire triggers. For example, buttons have a WHEN-BUTTON-PRESSED event that fires a trigger when the user clicks the button. Forms have a PRE-FORM event that fires a trigger just before the form first opens.

Now you will create a trigger that automatically inserts the next sequence value into the Inventory ID text item on the inventory form whenever the user creates a new block record. You will associate this trigger with the INVENTORY block. The trigger is associated with a block event called WHEN-CREATE-RECORD. This block event occurs whenever a user creates a new blank record in a data block.

To create the trigger:

1. Open the Object Navigator window in Ownership View, click ➕ beside the Data Blocks node, click ➕ beside INVENTORY, and then select the **Triggers node** under the INVENTORY data block.

2. Click the **Create button** 📄 to create a new trigger. The INVENTORY: Triggers dialog box opens.

The Triggers dialog box shows all of the block events that can have associated triggers. An object can have several associated events, so the Find text box allows you to enter a restricted search string, which helps you to quickly find the event that you want to associate with the trigger. You want to attach the trigger code to the WHEN-CREATE-RECORD event, so you will perform a restricted search using the string "WHEN" to retrieve all events that start with the text string "WHEN".

To find the trigger event:

1. Click the mouse pointer just before the % in the Find text box, type **when**, and then press **Enter**. The list box automatically scrolls down to the beginning of the events that start with the word *when*.

In the Triggers dialog box, searches for event names are not case sensitive.

2. Select the **WHEN-CREATE-RECORD** event, and then click **OK**. The PL/SQL Editor opens, as shown in Figure 6-34. If necessary, maximize the window.

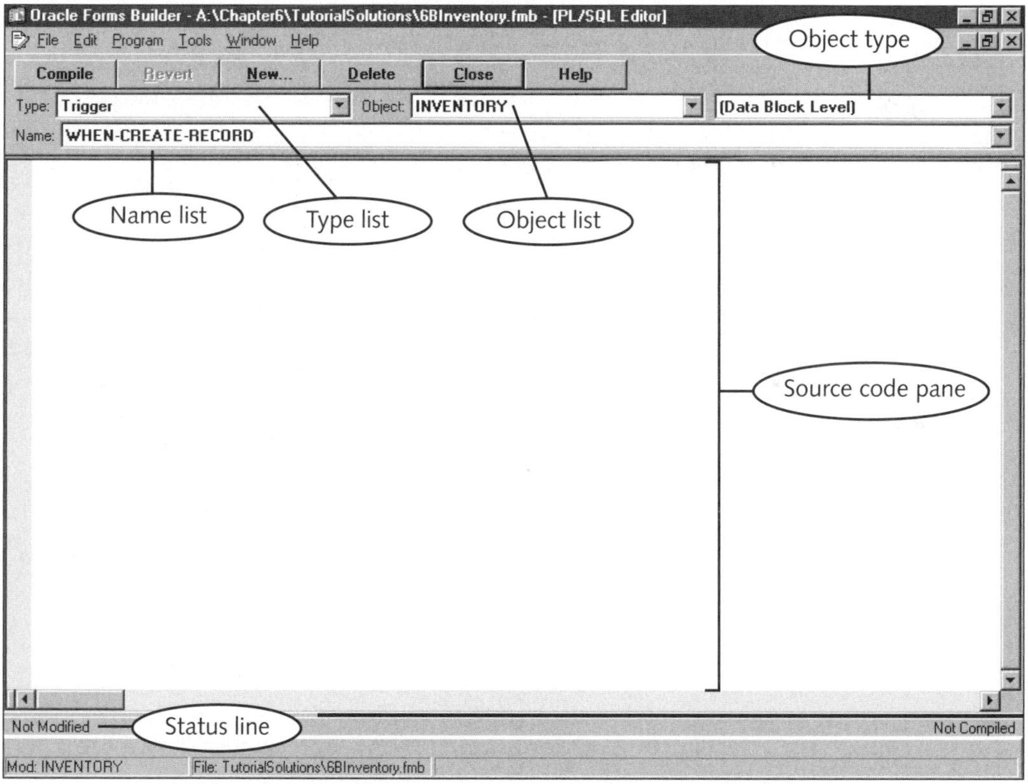

6

Oracle Forms Builder - A:\Chapter6\TutorialSolutions\6BInventory.fmb - [PL/SQL Editor]

File Edit Program Tools Window Help

Object type

| Compile | Revert | New... | Delete | Close | Help |

Type: Trigger   Object: INVENTORY   (Data Block Level)

Name: WHEN-CREATE-RECORD

Name list   Type list   Object list

Source code pane

Not Modified ─── Status line   Not Compiled

Mod: INVENTORY   File: TutorialSolutions\6BInventory.fmb

Figure 6-34   Form builder PL/SQL Editor

The PL/SQL Editor is similar to the editor you used in Procedure Builder in Chapter 5, except that it has been customized to run in the Form Builder environment. The **type list** shows the procedure type, which is a trigger. (Procedures can also be program units, which are subprograms that are called by triggers.) The **object list** displays the name of the form object that the trigger is attached to, and the list beside the Object list shows the object type. The **name list** shows the event associated with the trigger. The **source code pane** is where you type the PL/SQL program statements. The **status line** shows the trigger's current modification status (Modified or Not Modified) and compile status (Not Compiled, Successfully Compiled, or Compiled with Errors).

The PL/SQL Editor button bar has the following buttons:

- **Compile**, which compiles the program statements in the Source code pane. The compiler detects syntax errors and references to database tables, fields, and form objects that do not exist or are referenced incorrectly.

- **Revert**, which reverts the source code to its status prior to the last compile or since the last time the Revert button was clicked

- **New**, which creates a new procedure

- **Delete**, which deletes the current procedure

- **Close**, which closes the PL/SQL Editor

- **Help**, which allows the user to access the Help System while working in the PL/SQL Editor

Now you will type the code for the WHEN-CREATE-RECORD trigger. This trigger needs to retrieve the next value in the INV_ID_SEQUENCE and insert it into the INV_ID item on the form. To do this, you will use an implicit cursor, which can be used to retrieve a single query value and assign the result to a variable. You will select the next value of the sequence and display it in the form inventory ID text item. Form data items can be referenced in a PL/SQL program using the following syntax: *:block_name.item_name*. The *block_name* is the name of the block that contains the item. The *item_name* is the name of the text item as it appears in the Object Navigator. The block name is always preceded with a colon (:). In this form, the block name is INVENTORY, and the text item name is INV_ID. Therefore, this item will be referenced as `:inventory.inv_id`. Now you will enter the PL/SQL code for the WHEN-CREATE-RECORD trigger.

To type the WHEN-CREATE-RECORD trigger code:

1. Click anywhere in the Source code pane in the PL/SQL Editor, and then type the code shown in Figure 6-35. Note that PL/SQL and SQL commands appear in blue text, and user-entered variables appear in black text. This makes the code easier to understand and debug.

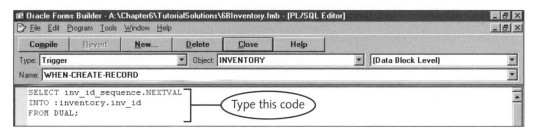

Figure 6-35   Trigger code to enter next sequence value into form text item

Like other PL/SQL programs, triggers must be **compiled**, or converted to executable code. When you compile a trigger, the compiler checks the code for syntax errors and for references to database or form objects that are not referenced correctly or do not exist. Form Builder automatically compiles all form triggers when you generate a form to create the executable (.fmx) file, but it is better to compile each trigger just after you enter the source code, so you can find and correct errors immediately. When the compiler detects an error, it displays the line number of the statement causing the error, with an error description. Next, you will compile the WHEN-CREATE-RECORD trigger.

To compile the trigger:

1. Click **Compile** on the PL/SQL Editor button bar. If the trigger compiles successfully, the "Successfully Compiled" message appears on the status bar.

2. If your trigger compiles successfully, click **Close** on the PL/SQL button bar to close the PL/SQL Editor. If it does not compile successfully, you will need to debug it. Debugging form triggers is similar to debugging other PL/SQL programs, except that there are some different kinds of errors that you might encounter. The next section discusses common form trigger errors.

3. Save the form.

Debugging Form Triggers

Figure 6-36 shows an example of a compile error that appears when compiling the WHEN-CREATE-RECORD trigger. The compilation messages pane displays the line numbers and error messages. The selected error is the error message that is shaded in the compilation messages pane. The insertion point shows the location for the selected error in the source code pane.

The error shown in Figure 6-36 is a common compile error. The error message "bad bind variable 'inventry.inv_id'" indicates that the indicated form item does not exist. The probable cause is that the block name or text item name was entered incorrectly. In this case, the block name is INVENTORY, not INVENTRY. To correct the error, you would correct the block name and then recompile the trigger.

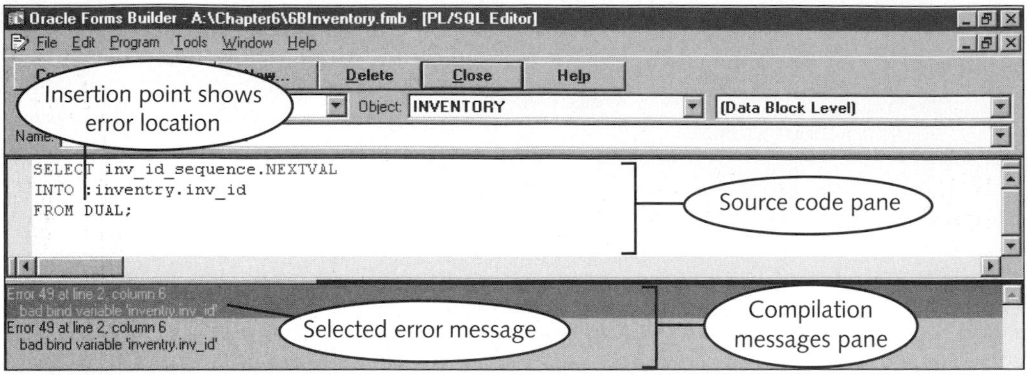

Figure showing Oracle Forms Builder PL/SQL Editor with compile error and labeled callouts: "Insertion point shows error location", "Source code pane", "Selected error message", "Compilation messages pane".

```
SELECT inv_id_sequence.NEXTVAL
INTO :inventry.inv_id
FROM DUAL;
```

Error 49 at line 2, column 6
bad bind variable 'inventry.inv_id'

Figure 6-36 Trigger compile error generated by referring to a nonexistent form object

Another common compile error message, shown in Figure 6-37, is "identifier 'DUAL' (or some other table name used in the trigger) must be declared."

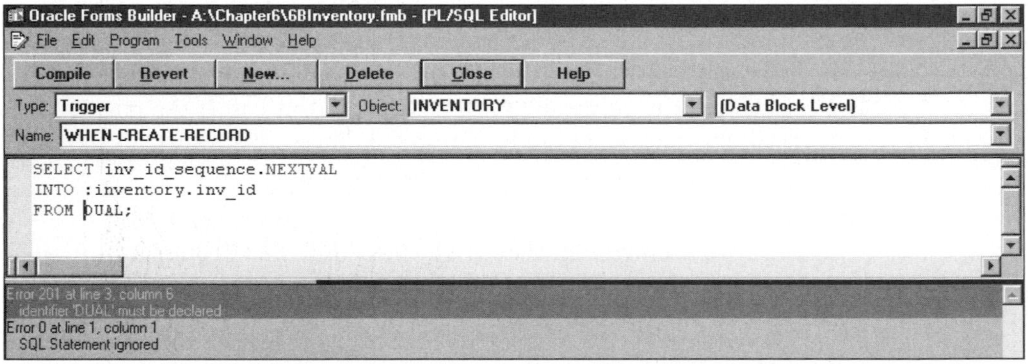

Figure showing Oracle Forms Builder PL/SQL Editor for Trigger WHEN-CREATE-RECORD.

```
SELECT inv_id_sequence.NEXTVAL
INTO :inventory.inv_id
FROM DUAL;
```

Error 201 at line 3, column 6
identifier 'DUAL' must be declared
Error 0 at line 1, column 1
SQL Statement ignored

Figure 6-37 Trigger compile error generated when user is not connected to the database

This error (or one naming a database table other than DUAL) occurs when you are not connected to the database at the time you compile the trigger. This happens when you start Form Builder, open an existing form, revise a trigger, and then try to compile it. Since you are not explicitly prompted to connect to the database, you might forget to connect before compiling. The solution is to click File on the menu bar, click connect, connect to the database in the usual way, and then recompile the trigger.

 The error message shown in Figure 6-37 will appear if you are not connected to the database when you compile a trigger with a SELECT command for *any* database table, not just DUAL.

Figure 6-38 shows another common error message. The error message "identifier 'INVID_SEQUENCE.NEXTVAL' must be declared" indicates that the compiler cannot identify the given variable name. The variable INVID_SEQUENCE is a database object, and the compiler cannot find it in any of the user's database objects, which include tables, fields, and sequences. If the user checked the names of his or her sequences using SQL*Plus, he or she would find that the name of the sequence is INV_ID_SEQUENCE, not INVID_SEQUENCE. Entering the sequence name correctly and then recompiling the trigger corrects this error.

6

```
Oracle Forms Builder - A:\Chapter6\TutorialSolutions\6BInventory.fmb - [PL/SQL Editor]
 File  Edit  Program  Tools  Window  Help

   Compile      Revert      New...      Delete      Close      Help

Type: Trigger                       ▼   Object: INVENTORY        ▼   (Data Block Level)      ▼
Name: WHEN-CREATE-RECORD                                                                    ▼

   SELECT invid_sequence.NEXTVAL
   INTO  :inventory.inv_id
   FROM DUAL;

Error 201 at line 1, column 8
   identifier 'INVID_SEQUENCE.NEXTVAL' must be declared
Error 0 at line 1, column 1
   SQL Statement ignored
```

Figure 6-38   Trigger compile error generated by referencing a nonexistent database object

Errors often occur when a SQL command within the PL/SQL trigger code is not correct. If you suspect that this is the case, copy the SQL command into a text editor, and remove any PL/SQL commands in the SQL statement. Paste the edited SQL query into SQL*Plus, run it, and see if the query output is correct. Figure 6-39 shows the SELECT command from Figure 6-38 running in SQL*Plus with its INTO command commented out. The SQL*Plus error message immediately identifies that the error occurred because the sequence name does not exist.

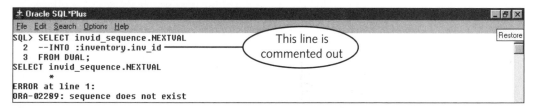

Figure 6-39   Testing a trigger SQL command in SQL * Plus

There are many more possible compile errors, but these are some of the most common ones. As always, the best way to locate errors is to isolate the source code line that is generating the error, make sure that the form objects and database objects are specified correctly, and make sure that the SQL command or PL/SQL syntax is correct. Test suspected SQL commands in SQL*Plus.

Testing the Trigger

After you successfully compile the trigger, the next step is to test it to be sure that it works correctly. It is a good idea to run the form each time you create a new trigger or add new code to an existing trigger, so if there are errors, you know exactly what code caused them. Now you will run the form to test the trigger.

To test the trigger:

1. If necessary, correct any syntax errors in your trigger, recompile the code until it compiles successfully, and then save the form.

2. If necessary, click **Close** on the PL/SQL Editor button bar to close the PL/SQL Editor. The WHEN-RECORD-CREATED trigger appears under the INVENTORY block.

3. Click the **Run Form Client/Server button** 📳 to run the form, and maximize the Forms Runtime window. The form appears with the next sequence number (11900) inserted in the Inventory ID field. Now you will enter the rest of the data fields and save the new record to confirm that the form inserts new records correctly.

4. Type the new INVENTORY item data fields as follows:

Item ID:	**559**
Size:	**XXL**
Color:	**Navy**
Price:	**209.95**
QOH:	**50**

5. Click the **Save button** 🔒. The confirmation message appears, confirming that the record was successfully inserted.

6. Click the **Insert Record button** 📇 to insert another record. A new blank record appears, with the next sequence value (11901) inserted into the Inventory ID field.

7. Click the **Remove Record button** 📇 to remove the blank record.

8. Close Forms Runtime.

9. Close the form.

CREATING A LIST OF VALUES (LOV)

Currently, when you insert a new record into the INVENTORY table using the INVENTORY form, you must type data values for foreign keys, such as ITEM_ID and COLOR. To make data entry easier and avoid errors, it is a good practice to guide the user to select from a list of allowable values for data fields that contain foreign key values or restricted data values. (For example, allowable values for ITEM_ID are 894, 897,

995, 559, and 786.) To allow users to select from a list of allowable values, you will create a list of values. A **list of values (LOV)** (pronounced Ell Oh Vee) displays a list of data
values when the user places the insertion point in the text item associated with the LOV
and presses F9. The LOV display also opens when the user clicks a button called the **LOV
command button** that is beside the data field that has restricted values. The user can
select one of the records from the **LOV display**, and the selected record values will be
inserted into one or more form text items. An example of an LOV display is shown on
the inventory form in Figure 6-40. When the user clicks the LOV command button, the
LOV display shows a list of records returned by a query. The user can do exact or restricted
searches using the Find text box. When the user selects a record and clicks OK, one or
more values from the selected record are returned to specified form text items.

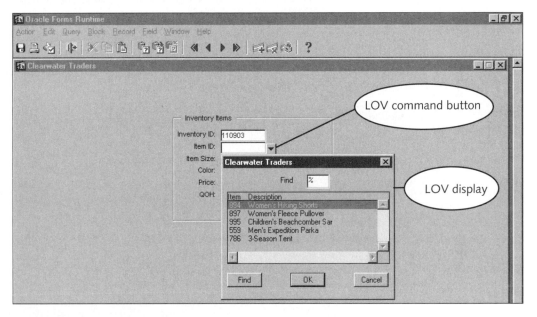

Figure 6-40   LOV display on form

The values in the LOV display are derived from a **record group**, which is a form object
that is a data structure that represents data in a tabular format. However, unlike database
tables, record groups are separate objects that belong to the form module in which they
are defined and only exist as long as the form is running. A record group object can have
a maximum of 255 columns of the following data types: CHAR, LONG, NUMBER,
or DATE. Record group column names must follow the Oracle naming standard. A
record group is similar to a PL/SQL table of records, which you learned about in
Chapter 4.

You can create a new LOV using the LOV Wizard, which automates the process. You
can also create an LOV manually by creating an LOV object and a new record group in
the Object Navigator, and then defining their properties.

Creating an LOV Using the LOV Wizard

Creating an LOV using the LOV Wizard is a five-step process:

1. **Specify the LOV display values.** When you create a new LOV, you enter a SQL query that returns the data values that will appear in the LOV display. This query creates a new record group. You can also specify for the LOV to use an existing record group.

2. **Format the LOV display.** This involves changing the column titles and widths, specifying which display fields are returned to form data fields, modifying the LOV display title, and changing the position of the LOV display on the screen.

3. **Attach the LOV to a text item.** An LOV is always associated with a specific form text item. When the form insertion point is in this text item and the user presses F9 or clicks the LOV command button, the LOV display opens. Usually, the LOV command button is beside this text item.

4. **Change the new LOV and record group default names to descriptive names in the Object Navigator.** When you create an LOV using the LOV Wizard, the LOV and record group objects are automatically created and given default names. It is a good practice to change these to descriptive names, especially in forms that contain multiple LOVs.

5. **Create an optional command button to activate the LOV display.** To display an LOV, users can place the form insertion point in the text item associated with the LOV and press the F9 key, or they can click Edit on the menu bar and then click Display List. You can also create an LOV command button that looks like a Windows list button. The LOV command button has a trigger that places the form insertion point in the text item associated with the LOV and then issues a command to open the LOV display.

The **LOV Wizard** automates the first three steps for creating an LOV. Now you will use the LOV Wizard to create the LOV shown in Figure 6-40 to allow users to select from a list of Item ID and Descriptions in the inventory form. You will use the LOV Wizard to specify the LOV display values, format the display, and attach the LOV to a text item.

To use the LOV Wizard:

1. In Form Builder, open **Inventory.fmb** from the Chapter6 folder on your Data Disk, and save the file as **6CInventory_LOV.fmb** in your Chapter6\TutorialSolutions folder.

2. Click **Tools** on the menu bar, and then click **LOV Wizard**. The LOV Source page appears, which allows you to create a new record group or use an existing record group. Since there are no existing record groups in the form, this option is not available. Click **Next**. The SQL query page opens. The SQL query page allows you to type a SQL query in the SQL Query

statement box, build a SQL query button using Query Builder, or import a query from a text file.

3. Type the following command to retrieve the item ID and item description fields for display. *Do not* type a semicolon(;) at the end of the command or an error will occur. Click **Next**. The Column Selection page appears.

```
SELECT item_id, item_desc FROM item
```

 If an error message appears, it means you did not enter the SQL command correctly. Here are some debugging tips:

- Immediately copy and paste the command text into a text editor by clicking and dragging over the command text to select it, pressing Ctrl + C, switching to the text editor, and pressing Ctrl + V.
- Errors usually are caused by writing the command incorrectly. If you cannot spot the error visually, copy the command from the text editor, paste it into SQL*Plus, and run and debug it.
- Do not type a semicolon (;) at the end of the command.

4. The Column Selection page specifies which record group columns will appear in the LOV display. For this LOV, all of the retrieved record group columns will appear in the LOV display, so click the **Select All button** ⟫ .

5. Click **Next**. The Column Display page appears. Click the horizontal scroll bar to view all of the display column properties.

The Column Display page enables you to format the LOV display. The Title column shows the column title that will appear for each column in the LOV display, and the Width column shows the width of each column in the LOV display. By default, the LOV display fields have the same titles and column widths as the corresponding database fields, which might result in column titles that are hard to understand and column widths that are wider than required for the LOV display.

The Return Value column specifies the form text item (preceded by its block name and a period) where the user's LOV selection will be transferred for the current column. For example, if the user selects Women's Hiking Shorts in the LOV display and clicks OK, the Item ID for Women's Hiking Shorts (894) will be transferred to the ITEM_ID text item on the form. On the column display page, you will select the form block and text item name (INVENTORY.ITEM_ID) as the return value for the ITEM_ID column. In this LOV, the selected ITEM_DESC value will not be returned to the form, because the item description does not appear in the INVENTORY table or on the form. Now you will specify the LOV display column titles and specify the return value.

To specify the LOV Wizard Display Column page:

1. Modify the column titles and widths, as follows:

Title	Width
Item ID:	45
Description:	100

2. Scroll horizontally to the right, click the mouse pointer in the Return value space next to Item ID, and then click **Look up return item**. The Items and Parameters dialog box opens, which is a list of all the items in the INVENTORY block. Each item is preceded by the block name and a period (INVENTORY.). These are the form text items to which the selected item ID values can be returned.

3. Click **INVENTORY.ITEM_ID**. This is the text item in which the selected item ID will appear after the user makes a selection from the LOV display.

4. Click **OK**. INVENTORY.ITEM_ID appears as the Return value for the Item ID text item on the Column Display page. Click **Next**. The LOV Display page appears.

 The Return value in the LOV Wizard Column Display page cannot exceed 20 characters, or it will be truncated, and an error will occur when you click Next. You must always make sure that the combination of the block name and the item name for return values does not exceed 20 characters.

5. The LOV Display page specifies the title, size, and position of the LOV display. Type **Clearwater Traders Items** in the Title box, and accept the default values for Width and Height. Be sure the Yes, Let Forms position my LOV automatically option button is selected.

6. Click **Next**. The Advanced Options page appears.

The Advanced Options page specifies how many records the LOV retrieves. For an LOV that might return hundreds or thousands of records, it is desirable to limit the number of records to improve system response time. The default value for number of records to retrieve is 20.

 The maximum number of records that an LOV can display is 32,767.

The first check box on the Advanced Options page (Refresh record group data before displaying LOV) allows you to specify whether the record values are refreshed each time the LOV display opens. If this check box is cleared, the LOV retrieves the data values only once from the database when the user first opens the LOV display, and the data values are stored. When the user opens the LOV display subsequent times, he or she is presented with the stored data values. This would be a desirable option for an LOV that retrieves many data values, and the values don't change very often. The second check box (Let the user filter records before displaying them) allows the user to type a search condition in the Find text box on the LOV display before any records appear. This will reduce the number of records that the LOV retrieves and improve system response time for an LOV that retrieves many records. You will learn more about the LOV advanced options in Chapter 8.

Since your LOV display will retrieve only the five records in the ITEM table, you will accept the LOV defaults (retrieve 20 records, refresh the data each time the LOV is displayed, and do not let the user filter the data). Now you will complete the LOV Advanced Options page and finish creating the LOV.

To complete the LOV Advanced Options page and finish the LOV:

1. Accept the default selections on the Advanced Options page, and then click **Next**. The Items page appears. Recall that an LOV is attached to a form text item, and when the form insertion point is in this text item and the user presses F9 or clicks the LOV command button, the LOV display opens. The field the LOV is attached to is usually one of the LOV return values.

2. Select **INVENTORY.ITEM_ID** in the Return Items list if necessary, and then click the **Select button** to place it in the Assigned Items list.

3. Click **Next**, and then click **Finish** to finish creating the LOV.

4. If necessary, open the Object Navigator window to display the new LOV and record group.

5. Save the form.

> LOVs only appear in the Object Navigator Ownership View. They do not appear in the Object Navigator Visual View.

After you create the LOV using the LOV Wizard, two new objects appear in the Object Navigator Ownership View window under the LOVs and Record Groups nodes. The LOV object specifies the LOV display properties, and the record group object specifies the data that the LOV displays. It is a good practice to replace the default names with descriptive names, especially in forms that contain multiple LOVS. Descriptive names allow you to distinguish among different LOVs and record groups if you want to modify their properties. Now you will give descriptive names to the LOV and record group. Usually, you name the LOV and record group the same name, so you can easily see the association between each LOV and record group.

To change the names of the LOV and record group:

1. In the Object Navigator, change the name of the LOV to **INVENTORY_LOV**, and the name of the record group to **INVENTORY_LOV**.

Creating the LOV Command Button

You can create an optional LOV command button, which is a list button that the user can click to open the LOV display. Alternately, the user can place the insertion point in the text item associated with the LOV, and then press F9 or click Edit on the menu bar, and then click Display List. The LOV command button will be a small square button positioned on the right edge of the ITEM_ID text item that displays a Windows list arrow. Buttons that display pictures, such as this one, are called **iconic buttons**, because they display bitmapped images called icons.

An icon file is a file with an .ico extension that contains a special kind of bitmap image file that is sized so it can appear as an icon. Icon images are usually 16 pixels wide by 16 pixels long, or 32 pixels wide by 32 pixels long. To create a new icon image or convert a regular bitmap image to an .ico image, you must have a special graphics application for creating .ico files. This book does not address creating or editing icons.

To create the LOV command button, you will open the form in the Layout Editor and draw a command button on the canvas using the Button tool 🔲 on the Layout Editor tool palette. Then you will open the button's Property Palette, resize the button, change the button's Iconic property to True, and change its Icon Name property to the name of the icon file that you want to display on the button. To specify an icon file, you must enter the complete path (including drive letter and folder path) to the icon file. In the filename specification, you include only the filename, and not the .ico extension.

You can store the icon file in the default icon folder on your client workstation and all of the client workstations where your application will be used and omit the path information. To do this, you will need to open the Registry editor and create a new string value in the HKEY_LOCAL_MACHINE/Software/ORACLE folder named UI_ICON. Set the associated value to the folder path, including drive letter, where the icons will be stored.

To create the LOV command button:

1. Click **Tools** on the menu bar, and then click **Layout Editor** to open the Layout Editor window.

2. Click the **Button tool** 🔲 on the tool palette, and draw a small square button positioned just to the right of the ITEM_ID data field, as shown in Figure 6-41.

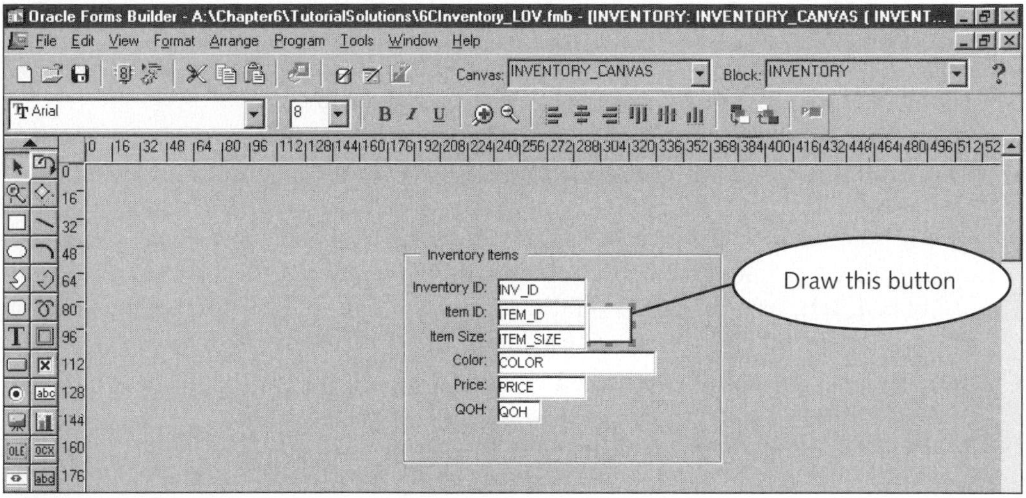

Figure 6-41   Drawing the LOV command button

Sometimes when you try to draw a small button, Form Builder will automatically transform it to a default larger-sized button. If this happens, resize the button to match Figure 6-41.

3. Select the button, right-click, and then click **Property Palette**. The new command button's Property Palette opens. Change the button's properties as follows:

Name	**ITEM_LOV_BUTTON**
Label	(Delete the default label—it will have an icon instead)
Iconic	**Yes**
Icon Filename	**a:\chapter6\down3**
Width	**12**
Height	**14**

If your Chapter6 Data Disk folder is in a different location than a:\, type the correct drive letter and path specification.

You could also double-click the button to open the button's Property Palette.

4. Close the ITEM_LOV_BUTTON Property Palette. The icon should appear on your button.

If the icon does not appear, verify that you have changed the button's Iconic property to Yes, and entered the correct drive letter and path specification for your down3.ico file in the Icon Filename property. Verify that the down3.ico file is in the Chapter6 folder on your Data Disk.

To complete the LOV command button, you must create the trigger that opens the LOV display when the user clicks the LOV command button. This trigger will be associated with the button's WHEN-BUTTON-PRESSED event. The trigger program code will first place the form insertion point in the text item that the LOV is attached to (ITEM_ID), using the GO_ITEM command. The general format of this command is `GO_ITEM ('block_name.item_name');`. In this command, the block and item name specification is placed in single quotation marks. The block name is optional in the item name specification, but sometimes you have a form with multiple text items that have the same name but are in different blocks, so it is a good practice to include the block name. In this form, you want to place the insertion point in the ITEM_ID text item, which is in the ITEM block, so the command will be `GO_ITEM('item.item_id');`. The trigger will then issue the **LIST_VALUES** command, which instructs the system to display the LOV that is attached to the text item that contains the form insertion point.

To create the LOV command button trigger:

1. Right-click the **LOV command button** you just created, point to **SmartTriggers**, and then click **WHEN-BUTTON-PRESSED** to create a new trigger. The PL/SQL Editor opens.

The Smart Triggers feature allows you to quickly make triggers for the events most often associated with an object.

2. Type the following code in the Source code pane:

```
GO_ITEM ('inventory.item_id');
LIST_VALUES;
```

3. Compile the trigger, correct any errors, and then close the PL/SQL Editor.

4. Save the form.

Testing the LOV

Next, you will test the form to verify that the LOV display appears and that the LOV returns the selected record's item ID to the ITEM_ID form text item.

To test the LOV:

1. Run the form, and maximize the Forms Runtime window.

2. Type **11902** for the inventory ID value.

3. Click the **LOV command button** to display the LOV. The LOV display should appear, as shown in Figure 6-42. Note that the message "List of Values" appears on the form status line to indicate that the current text item has an associated LOV. Also note that you will need to reformat the column widths of the LOV display later. First, you will confirm that the selected data value is returned to the form text item correctly.

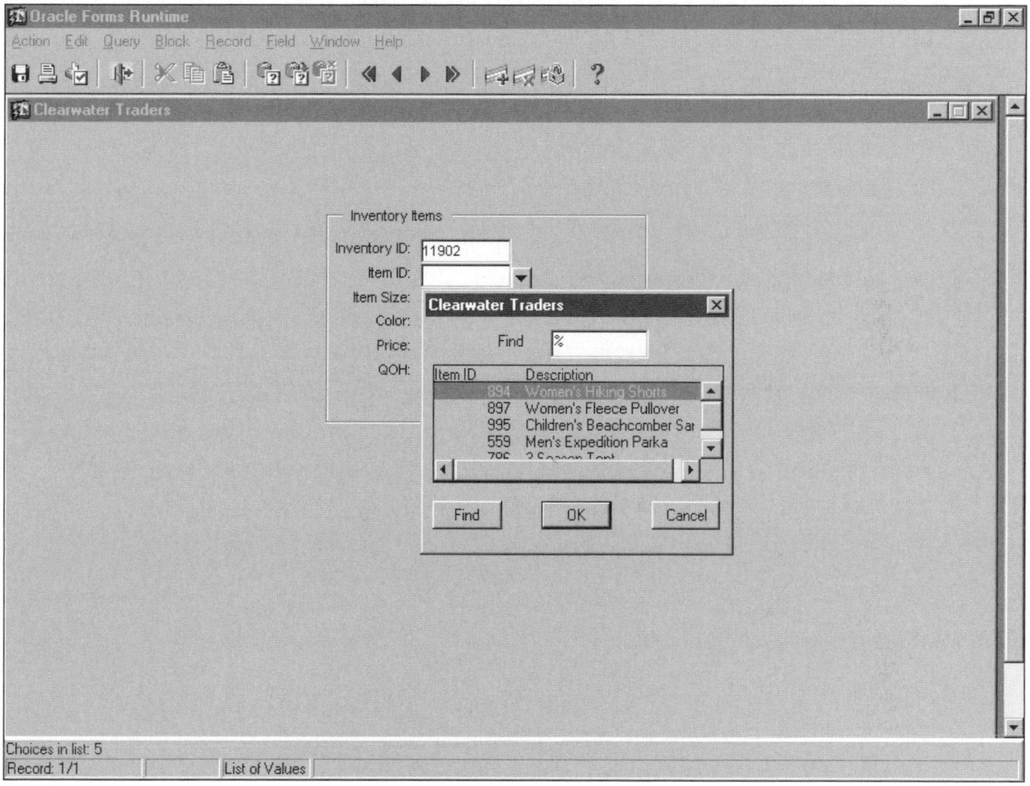

Figure 6-42   Viewing the LOV display

4. Make sure Item ID 894 (Women's Hiking Shorts) is selected, and then click **OK**. Item ID 894 appears in the Item ID text item on the form.

5. Close Forms Runtime, and click **No** to discard your changes.

If you click Cancel on the LOV display, no value is returned for Item ID, and the message "FRM-40202: Field must be entered" appears on the status line. This is because you are inserting a new record, and the Item ID field is a required (NULL not allowed) field in the INVENTORY table. To display the LOV again, you must click the Remove Record button 🖳 to remove the current record, and then insert a new blank record.

Reformatting the LOV Display

You need to adjust a few final formatting details so that the LOV display is more attractive and functional. Currently, the item ID text item is too wide for the data values, and the description field is clipped off, because the overall LOV display is not wide enough to display both fields. Also, some of the records do not appear, because the LOV display is not quite long enough. Whenever possible, you should always format an LOV display so all of the records and columns are visible without scrolling. Now, you will reformat the LOV display to shorten the width of the Item ID field, widen the Description field, and make the LOV display longer. To do this, you will select the LOV in the Object Navigator, and then open the LOV Wizard in reentrant mode. Then you will modify the Width and Height of the LOV display.

To modify the format of the LOV display:

1. Open the Object Navigator window, and select the LOV named **INVENTORY_LOV**.

2. Click **Tools** on the menu bar, and then click **LOV Wizard**. The LOV Wizard opens in reentrant mode, with tabs across the top of the page to allow you to access each of the Wizard pages.

3. Click the **Column Display tab**, and change the width of the Item ID column to **15**. Click **Apply** to save the change, but leave the LOV Wizard open.

4. Click the **LOV Display tab**, change the LOV display Width to **200**, and the Height to **150**, and then click **Finish** to save your changes and close the LOV Wizard.

5. Save the form, run the form, type **11903** for the Inventory ID, and then click the **LOV command button** to display the LOV. The formatted display should look like Figure 6-40. Click **Cancel** to close the LOV display.

6. Close Forms Runtime, and click **No** to discard your changes.

7. Close the 6CInventory_LOV.fmb form.

Creating an LOV Manually

The LOV Wizard automates the process of creating the LOV, formatting the LOV display, and attaching the LOV to a text item, but you can create LOVs manually also. Now you will manually create an LOV. This LOV will be used on a data block form associated with the CUST_ORDER form and will be used to display fields in the CUSTOMER table and allow the user to select a value for the customer ID. You will format the LOV display using the LOV Property Palette and attach the LOV to the text item using the text item Property Palette. First, you will open the form and save it using a different name. Then, you will create the LOV and its associated record group in the Object Navigator and associate the record group with the LOV. (Recall that the Record

Group determines the records that appear in the LOV display.) Finally, you will specify the title for the LOV display.

To open the form and create an LOV and Record Group in the Object Navigator:

1. Open **Cust_Order.fmb** from the Chapter6 folder on your Data Disk, and save the form as **6CCust_Order.fmb** in your Chapter6\TutorialSolutions folder.

2. In the Object Navigator, select the **LOVs node**, and then click the **Create button** 🔳. The CUST_ORDER: New List-of-Values (LOV) dialog box opens. Select the **Build a new LOV manually** option button, and then click **OK**. A new LOV appears under the LOVs node. Change the LOV name to **CUST_ID_LOV**.

3. In the Object Navigator, select the **Record Groups node**, and then click 🔳. The CUST_ORDER: New Record Group dialog box opens. Type the following command in the Query Text box:

   ```
   SELECT cust_id, last, first, mi, address, city, state
   FROM customer;
   ```

4. Click **OK**. A new record group appears under the Record Groups node. Change the record group name to **CUST_ID_LOV**.

5. Right-click **CUST_ID_LOV** under the LOVs node, and then click **Property Palette** to open the LOV Property Palette.

6. Select the **Record Group property**, open the list, and select **CUST_ID_LOV** to associate the record group with the LOV.

7. Type **Customers** as the Title property value to specify the LOV display title. Do not close the Property Palette.

The next step is to specify the record group columns that will appear in the LOV. You need to manually map each record group column to an LOV column by typing the name of each record group column in the LOV's Column Names list.

To specify the record group columns in the LOV:

1. In the LOV Property Palette, click **Column Mapping Properties**, and then click **More**. The LOV Column Mapping dialog box opens. You use this dialog box to list the columns in the LOV display, specify the column titles and widths, and specify the text item to which the column value is returned.

2. Place the insertion point in the first item in the Column Names list, delete the current value and type **CUST_ID**, because this is the first column in the LOV record group. Press **Tab** to move to the next Column Name.

3. Type **LAST** for the next column name, and then press **Tab**. Type all of the record group column names, as shown in Figure 6-43. You will need to scroll down in the Column Names list to enter the STATE column.

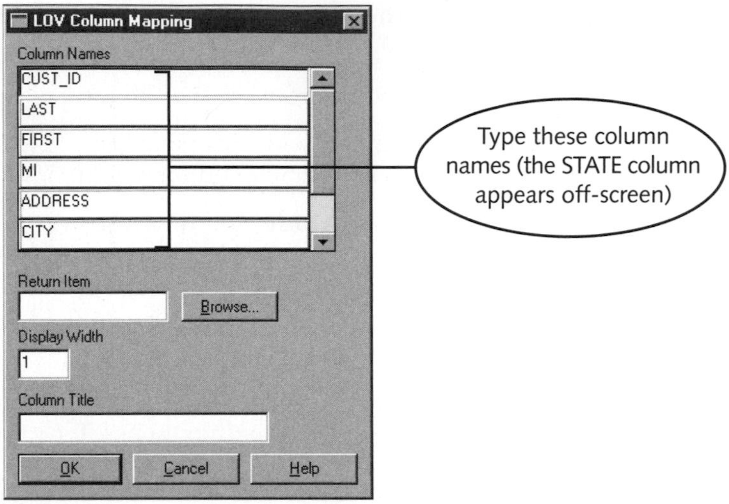

Figure 6-43   Entering the LOV record group columns

The Return Item, Display Width, and Column Title text boxes on the LOV Column Mapping dialog box are associated with the column name that is currently selected. Now, you will specify the return item, display width, and column title for the CUST_ID column in the LOV display.

To specify the CUST_ID return item, display width, and column title:

1. Select **CUST_ID** in the Column Names list.

2. Click **Browse** to select the return item. The Items and Parameters dialog box opens, which shows a list of the form text items. Select **CUST_ORDER.CUST_ID**, which is the form text item to which the selected CUST_ID value from the LOV will be returned, and then click **OK**. The return item is appears in the Return Item text box.

3. Place the insertion point in the Display Width text box, and change the value to **15**.

4. Press **Tab** to place the insertion point in the Column Title text box, and type **Customer ID** for the column title.

Now you will specify the display widths and column titles for the rest of the display columns. To specify the display width and column title for each column, you will select the column in the Column Names list, and then enter the column Display Width and Column Title.

To specify the display widths and column titles for the rest of the display columns:

1. Click **LAST** in the Column Names list, click the Display Width text box, change the value to **50**, press **Tab**, and change the Column Title to **Last Name**.

2. Click **FIRST** in the Column Names list, click the Display Width text box, change the value to **50**, press **Tab**, and change the Column Title to **First Name**.

3. Change the display width and column title for the rest of the record group column names as follows:

Column Name	Display Width	Column Title
MI	10	MI
ADDRESS	50	Address
CITY	25	City
STATE	10	State

4. Click **OK** to save your column mapping property changes, close the Property Palette, and save the form.

Next, you need to associate the LOV with the CUST_ID text item. You will open the Property Palette of the CUST_ID text item and specify that the text item's LOV property is CUST_ID_LOV.

To associate the LOV with the CUST_ID text item:

1. If necessary, click **+** beside the Data Blocks node, click **+** beside CUST_ORDER, and click **+** beside the Items node. Right-click **CUST_ID**, and then click **Property Palette** to open the text item Property Palette.

2. Scroll down to the List of Values (LOV) property node, click **List of Values**, and select **CUST_ID_LOV** from the list.

3. Close the Property Palette, and save the form.

Now, you will run the form and view the LOV display. Instead of creating an LOV command button, you will place the insertion point in the CUST_ID text item manually, and then press F9 to open the LOV display.

To view the LOV display:

1. Run the form, and maximize the Forms Runtime window.

2. Type **2000** for the Order ID, and then press **Tab** three times to move the form insertion point to the CUST_ID text item.

3. Press **F9** to open the LOV display. The display appears on the form as shown in Figure 6-44. The LOV display needs to be wider to display all of the column values, and the customer ID column needs to be wider to display the column heading. (Your LOV display might appear wider or narrower, depending on your default system font.)

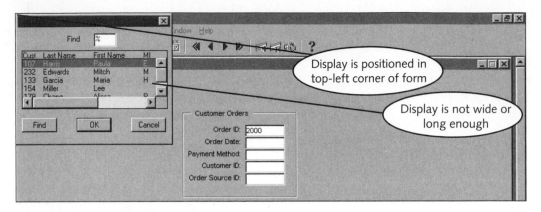

Figure 6-44   LOV display of LOV created manually

4. Scroll to the right edge of the LOV display. The address and state columns need to be wider also.

5. Click **Cancel** to close the LOV display, close the Forms Runtime window, and click **No** to discard your changes.

The LOV display needs further formatting: the customer ID, address, and state columns are not wide enough, the overall LOV display is not wide or long enough to show all of the columns, and the LOV display appears in the top left corner of the form instead of beside its associated text item. First, you will modify the LOV column widths. Sizing LOV display columns is a trial and error process—you make your best guess at the correct widths, and then view the LOV display and adjust the widths as necessary.

To modify the column widths:

1. Open the LOV Property Palette, click **Column Mapping Properties**, click **More**, and adjust the column widths as follows:

Column	Width
CUST_ID	25
ADDRESS	75
STATE	25

2. Click **OK** to save your changes.

Next, you will modify the LOV display size and position. Recall that you should always make the LOV display wide and long enough so the user can see all of the columns and records without scrolling. You also should place the LOV display directly below the text item to which it returns a value, to minimize the distance that the user has to move the mouse pointer. Sizing the LOV display is also a trial and error process, but determining the LOV position is a little easier. To determine the LOV display position, you will place the mouse pointer on the canvas just below the lower left edge of the Customer ID text item and note the coordinate position. Then, you will enter this position in the LOV Property Palette.

To modify the LOV display size and position:

1. In the LOV Property Palette, change the Width to **350**, and the Height to **175**, and then close the Property Palette.

2. Click **Tools** on the menu bar, and then click **Layout Editor** to view the form in the Layout Editor.

3. Place the mouse pointer on the upper-left corner of the CUST_ID text item, as shown in Figure 6-45, and note the pointer position in the status bar at the bottom of the screen. In the figure, the X position is 268.5, and the Y position is 106.5.

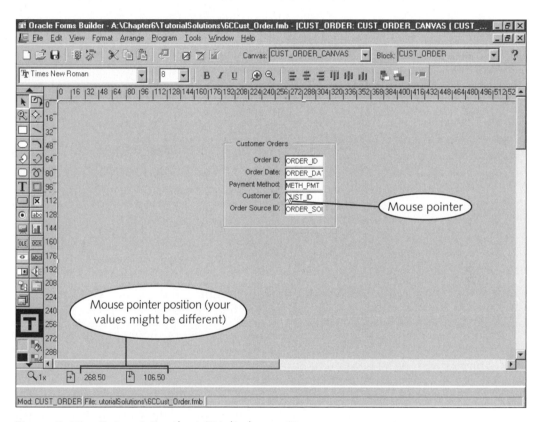

Figure 6-45   Determining the LOV display position

4. Open the Object Navigator, and then open the Property Palette for the LOV named CUST_ID_LOV.

5. Change the X Position and Y Position properties to the values you noted in Step 3, rounded down to the nearest whole number. For example, for the form in Figure 6-45, you would type **268** for the X Position and **106** for the Y Position.

6. Close the Property Palette, and then save the form.

7. Run the form, and then maximize the Forms Runtime window.

8. Type **2000** for the Order ID, press **Tab** three times to place the insertion point in the CUST_ID text item, and then press **F9** to open the LOV display. The LOV display size and position is still not perfect, but it is better.

9. Click **Cancel** to close the LOV display, close Forms Runtime, and click **No** to discard your changes.

Creating an LOV with Static Values

So far, the LOVs you have created have displayed values retrieved from the database. You can create an LOV that displays values from a **static list**, which is a list that you specify when you create the form. You might use a static list to store values that do not change and are not foreign key values. Examples would be two-character state abbreviations (like WI or FL) or values for faculty rank. Now you will create an LOV for the Payment Method field in the CUST_ORDER form that is derived from a static list. The Payment Method will be selected from two values: CHECK or CC (credit card). You could create the LOV manually, or you could use the LOV Wizard to create the LOV and base the LOV on a manually created record group that contains the specified list of static values. You will use the LOV Wizard, and base the LOV on a static record group. First, you will manually create the static record group and specify that its values are restricted to CHECK or CC.

To manually create the record group that contains the static values:

1. In the Object Navigator, select **Record Groups**, and then click the **Create button** . The CUST_ORDER: New Record Group dialog box opens.

2. Select the **Static Values** option button, and then click **OK**. The CUST_ORDER: Column Specification dialog box opens.

This dialog box lets you specify the column names of the record group and the associated column values. This record group will contain a single column named PAYMENT_METHOD, with two values, CHECK and CC.

To specify the record group column names and values:

1. Place the insertion point in the first item in the Column Names list, delete the current value, and type **PAYMENT_METHOD** for the column name.

2. Confirm that the Data Type value is Character and the Length is 30. These default values are compatible with the column values of CHECK or CC, so you will accept them.

3. Place the insertion point in the first item in the Column Values list, and type **CHECK**.

4. Place the insertion point in the second item in the Column Values list, and type **CC**.

5. Click **OK** to save the column specification.

6. Change the name of the new record group to **PAYMENT_METHOD_LOV**.

Now, you will create an LOV that is based on this record group. You will use the LOV Wizard to create the LOV and associate the LOV with the Payment Method text item.

To create the LOV using the LOV Wizard:

1. In the Object Navigator, select **CUST_ORDER** under the Forms node, click **Tools** on the menu bar, and then click **LOV Wizard**. The LOV Wizard Source page appears.

6

 If you open the LOV Wizard when an existing LOV is selected in the Object Navigator, the wizard will open in reentrant mode.

2. Select the **Existing Record Group** option button, select the **PAYMENT_METHOD_LOV** record group from the list, and then click **Next**. The Column Selection page opens.

3. Select the **PAYMENT_METHOD** column for the LOV display column, and then click **Next**. The Column Display page appears.

4. Scroll to the right edge of the Column Display list, place the insertion point in the Return value space for the Payment_Method column, click **Look up return value**, select **CUST_ORDER.METH_PMT** from the list of form text items, and then click **OK**. This returns the selected value to the METH_PMT text item in the form. The text item name appears in the Return value column. Click **Next**. The Column Display page appears.

5. Type **Payment Methods** for the Title property, accept the rest of the default values, and click **Next**. The Advanced Options page appears. Accept the default values, and then click **Next**. The Items page appears.

6. Select **CUST_ORDER.METH_PMT** to associate this text item with the LOV, click **Next**, and then click **Finish**. The new LOV appears under the LOVs node in the Object Navigator window.

7. Change the LOV name to **PAYMENT_METHOD_LOV**, and then save the form.

Now you will run the form and confirm that the LOV based on static values works correctly.

To test the LOV:

1. Run the form, and maximize the Forms Runtime window.

2. Type **2000** for the Order ID, and then press **Tab** two times to move to the Payment Method text item.

3. Press **F9** to open the LOV display. The static-value LOV appears, as shown in Figure 6-46. Select **CC**, and then click **OK**. The selected value (CC) appears in the Payment Method text item.

4. Close Forms Runtime, and click **No** to discard your changes.

5. Close the CUST_ORDER form.

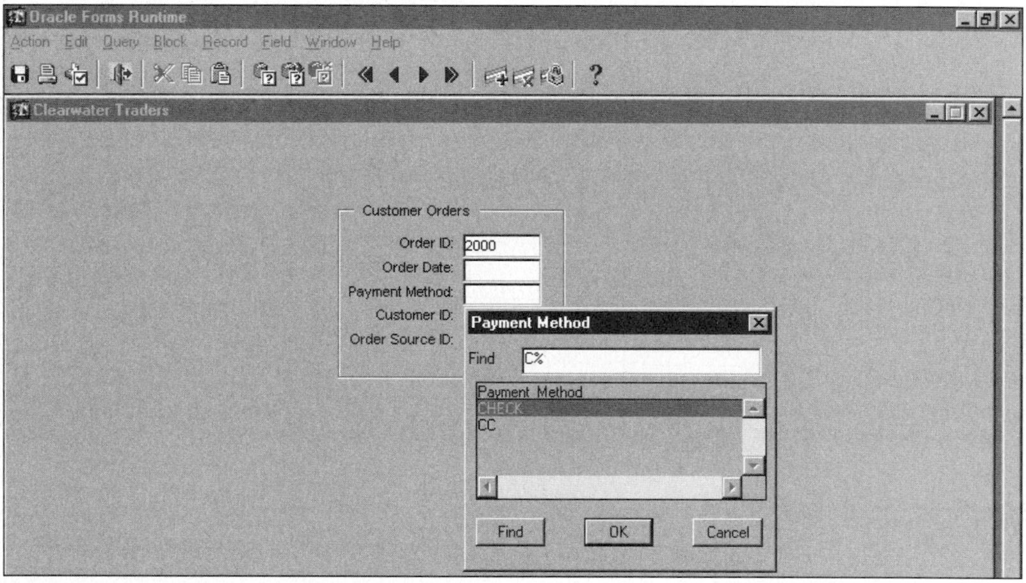

Figure 6-46   Static value LOV display

CREATING TEXT ITEM EDITORS

In an Oracle form, an **editor** is a dialog box that the user can open and use to edit character, number, or date values that appear in text items. Every text item has a **default editor**. The default editor is a text editing dialog box that provides Find and Replace functions, as well as a larger user workspace. To open the default forms editor, the user places the insertion point in the text item, and presses Ctrl + e. Now you will open a form named Evaluation.fmb, which is associated with the EVALUATION table in the Software Experts database and use the default editor to enter a value for the COMMENTS text item. Recall that the COMMENTS database column has the LONG data type, which stores up to 2 GB of variable length character data. You will use the editor to enter data into the COMMENTS text item for evaluation ID 100. Currently, the database column does not contain any data.

To open the form and use the default editor:

1. In Form Builder, open **Evaluation.fmb** from the Chapter6 folder on your Data Disk, and save the file as **6CEvaluation.fmb** in your Chapter6\TutorialSolutions folder.

2. Run the form, and maximize the Forms Runtime window.

3. Click the **Enter Query button** , type **100** for the search condition in the Evaluation ID text item, and then click the **Execute Query button** to retrieve the record. The values for Evaluation ID 100 appear.

4. Place the insertion point in the Comments field, and then press **Ctrl + e**. The default editor opens, as shown in Figure 6-47.

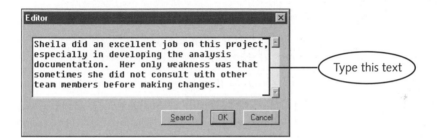

Figure 6-47   Default editor

 In Forms, all Ctrl + KEY commands are case sensitive. You must press Ctrl + e (lower-case). If the Caps Lock key is on, the editor will not appear.

5. Type the text shown in Figure 6-47 for the evaluation comment, and then click **OK** to close the editor. The text appears in the form text item.

6. Click the **Save button** to save the changes, and then close Forms Runtime.

You can create a **custom editor** that provides the same editing functions as the default editor but has a customized window size, title, and appearance. To do this, you create an editor object in the Object Navigator, configure its properties, and then change the text item's Editor property to specify the name of the custom editor. When the user places the insertion point in a text item that has a custom editor and then presses Ctrl + e, the custom editor opens in place of the default editor. Now you will create a new editor and configure the editor window in its Property Palette. You will specify the title that will appear in the title bar of the editor window and specify the editor window size and position.

To create and configure a custom editor:

1. In the Object Navigator, select the **Editors node**, and then click the **Create button** [icon] to create a new editor. A new editor object appears.

2. Change the editor name to **COMMENTS_EDITOR**.

3. Double-click the **Editor icon** [icon] beside the COMMENTS_EDITOR object to open the editor's Property Palette.

4. Change the Title property to **Evaluation Comments**.

5. Change the X Position value to **175** and the Y Position value to **150** to place the editor just below the COMMENTS text item.

6. Change the Width to **300** and the Height to **200**.

7. Close the Property Palette, and then save the form.

The final task is to associate the custom editor with the text item. To do this, you change the Editor property of the text item to the name of the editor you just created.

To associate the custom editor with the text item:

1. Click [+] beside the Data Blocks node, click [+] beside EVALUATION, and then click [+] beside the Items node. Select the **COMMENTS** text item in the Object Navigator, right-click, and then click **Property Palette** to open the text item Property Palette.

2. Scroll down to the Editor property node, select the Editor property, open the list, and select **COMMENTS_EDITOR**.

3. Close the Property Palette, and then save the form

Now you will run the form and test the custom editor.

To test the custom editor:

1. Run the form, and maximize the Forms Runtime window.

2. Click the **Enter Query button** [icon], type **100** for the search condition in the Evaluation ID text item, and then click the **Execute Query button** [icon] to retrieve the record. The values for Evaluation ID 100 appear.

3. Place the insertion point in the Comments field, and then press **Ctrl + e**. The custom editor window appears, as shown in Figure 6-48. The custom editor has a customized window title, displays more lines of text, and is in a different position than the default editor.

4. Change the comment text, as shown in Figure 6-48, click **OK** to close the editor, and then click the **Save button** [icon] to save the change.

5. Close Forms Runtime, and then close the form in Form Builder.

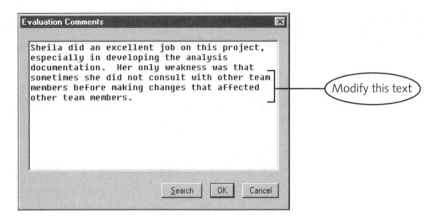

Figure 6-48   Custom editor

To provide a more powerful editing environment than is provided by the default Forms editor, you can specify SYSTEM_EDITOR for the Editor property in a text item's Property Palette. If you specify SYSTEM_EDITOR, when the user places the insertion point in the text item and presses Ctrl + e, the default system editor, which is Notepad or WordPad for most Windows-based systems, opens.

ALTERNATE FORM INPUT ITEMS

Up to this point, you have entered and viewed form data values as text items. Forms Builder supports a variety of other item types that you can use to simplify entering and modifying database data values. Table 6-6 summarizes these item types.

Item Type	Description	Usage Examples
Radio Button	Displays groups of two or more mutually exclusive choices that do not change	Gender (M or F); Faculty Rank (ASST, ASSO, FULL, INST); Grade (A, B, C, D or F)
Check Box	Represents a data value that can have only one of two values	Certification (Y or N); Term Status (OPEN or CLOSED)
Poplist	Allows users to select a value from a static drop-down list. Displays 10 items at a time, and displays a scrollbar if list contains more than 10 items.	State abbreviations

Table 6-6   Form item types for entering and modifying data

Item Type	Description	Usage Examples
T-List	Displays a single value at a time from a list populated by static values. User can sequentially scroll through list items using the up and down arrow on the list box. Should only be used with lists that have a limited number of choices (approximately 10 or less).	Payment Methods; Building Codes
Combo Box	Allows users to select from a poplist or to enter an alternate choice	Payment Methods; Building Codes

Table 6-6   Form item types for entering and modifying data (continued)

The following sections describe how to create forms that use these alternate input item types.

Creating Form Radio Buttons

When a data field has a limited number of possible values, usually five choices or less, you can use radio buttons to enter and display data values. **Radio buttons**, also called option buttons, limit the user to only one of two or more related choices. In Form Builder, individual radio buttons exist within a radio group. Only one button in a **radio group** can be selected at a time. Each radio button has an associated data value, and the radio group has the data value of the currently selected radio button.

When you use the Layout Wizard to create a form layout, you can select the type of form item that is used to represent each data field. So far, you have used the default item type, which is a text item, that displays data values as text. Now you will create a form that displays a data field using a radio group. You will make a form for inserting, updating, viewing, and deleting records in the STUDENT table of the Northwoods University database. In this table, the S_CLASS field currently is limited to four values: FR, SO, JR, and SR. You will display the S_CLASS field using a radio group. First, you will create a new form and data block associated with the STUDENT table. Then, you will create the form layout and specify that the S_CLASS field will be displayed in a radio group.

To create the new form, data block, and layout:

1. Create a new form, and save the form as **6CStudent.fmb** in your Chapter6\TutorialSolutions folder.

2. Use the Data Block Wizard to create a new data block associated with the STUDENT table that includes all STUDENT data fields. Do not check the Enforce data integrity check box.

3. Use the Layout Wizard to create a form layout. Accept the **(New Canvas)** Canvas and **Content** type default options, and then click **Next**.

4. On the Layout Wizard Data Block page, select all data block items for display.

5. Click the mouse pointer on the **S_CLASS** item in the Displayed Items list to select it by itself. The Item Type list box is activated, as shown in Figure 6-49. Open the list to view the list choices. The default item type is Text Item.

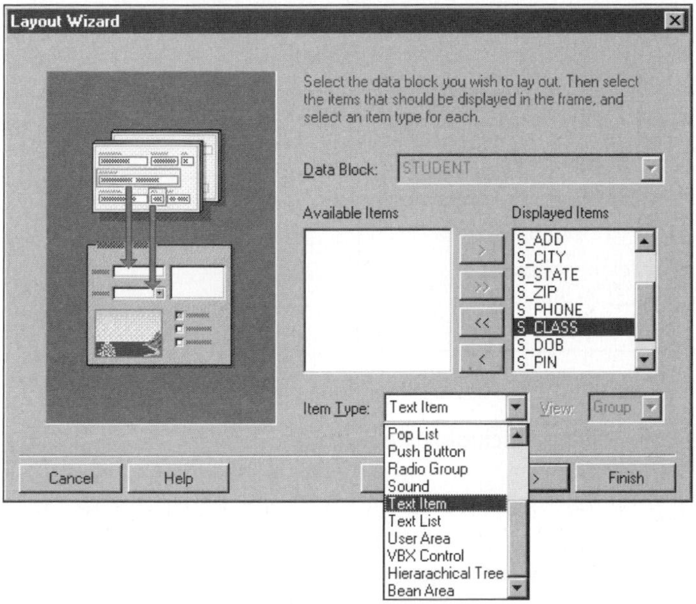

Figure 6-49 Selecting a layout item type

6. Select **Radio Group** as the item type, and then click **Next**.

7. Accept the default item prompts and widths, and then click **Next**.

8. Make sure that the Form option button is selected for the layout style, and then click **Next**.

9. Type **Students** for the frame title, leave Records Displayed as 1, click **Next**, and then click **Finish**.

The layout appears in the Layout Editor. However, no radio buttons appear, and nothing appears for the S_CLASS data field. When you specify to use a radio group for a data block field, the Layout Wizard creates the radio group in the Object Navigator, as you will see next.

To view the new radio group in the Object Navigator:

1. Open the Object Navigator, click ➕ beside the Data Blocks node. If necessary, click ➕ beside the STUDENT object, and, if necessary, click ➕ beside the Items node. The data blocks items appear, as shown in Figure 6-50.

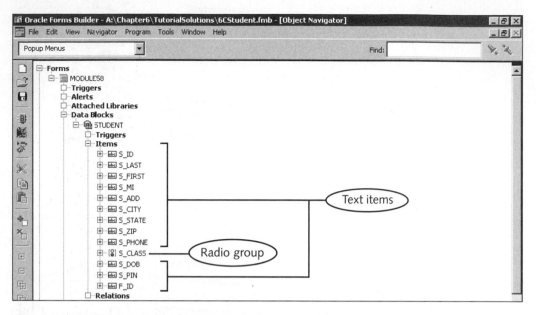

Figure 6-50   STUDENT block items, including new radio group

The text item icon ![ab] beside the object names indicates that all of the fields in the STUDENT block except S_CLASS are represented as text items. The radio button group icon ![radio] beside S_CLASS indicates that it is represented as a radio group. Currently, there are no radio buttons in the radio group, and the radio group does not appear on the form layout. You need to manually create the individual radio buttons to represent the four data values that can be used in the S_CLASS data field. To do this, you will draw the radio buttons on the canvas using the Radio Button tool ![tool] on the Layout Editor tool palette. First, you will format the form window and canvas, and save the form.

To format the window and canvas:

1. In the Object Navigator, format the form window so it cannot be maximized or resized, and resize the window so it fills the Forms Runtime window. Change the window name to **STUDENT_WINDOW** and the window title to **Northwoods University**.

2. Resize the canvas so it is the same size as the window. Change the canvas name to **STUDENT_CANVAS**. Change the font of all of the form items to **8-point Arial**.

3. Change the form name to **STUDENT_FORM**.

4. Save the form.

Now you will format the text items on the canvas and make the data block frame larger to accommodate the radio buttons. You will also change the frame's Update Layout property. Currently, the frame automatically positions the block items within the frame. You will change the Update Layout property to Locked, so when you update the form using the Layout Wizard later, you will not lose your custom formatting. Then, you will create the first radio button.

To resize and format the frame, and make the radio button:

1. Click **Tools** on the menu bar, and then click **Layout Editor** to open the Layout Editor. Select the frame, and make it narrower so all of the text items appear in a single column and are stacked on top of one another. If necessary, format the text item labels so the prompts all appear on one line, as shown in Figure 6-51.

2. Select the frame, open its Property Palette, select the **Update Layout** property, and select **Locked** from the property list. Close the Property Palette.

 If you do not change the Update Layout frame property to Locked, you will lose all custom formatting if you open the Layout Wizard in reentrant mode.

3. If necessary, select the **Students frame**, and drag the lower-right selection handle to the right to make the frame wider.

4. Click the **Radio Button tool** ⊙ on the tool palette, and draw a rectangle on the canvas to correspond with the first radio button, as shown in Figure 6-51. After you release the mouse button, the Radio Groups dialog box opens, asking which radio group to place the radio button in. This dialog box allows you to place a new radio button in an existing radio group or to create a new radio group. You will place the new radio button in the S_CLASS radio group created by the Layout Wizard when you made the form layout.

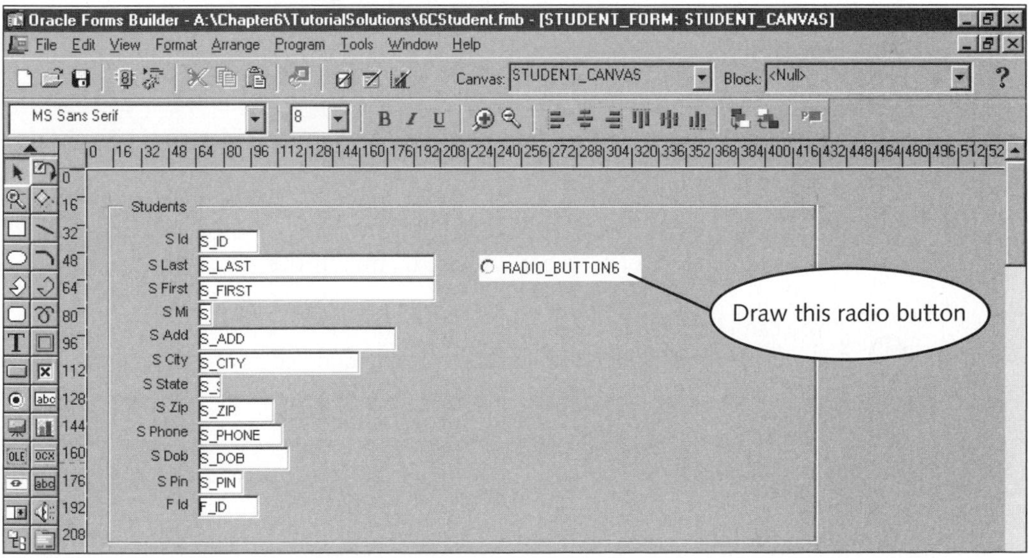

Figure 6-51   Drawing the first radio button

5. Make sure the S_CLASS radio group is selected, and then click **OK**. The radio button and its corresponding label appear, as shown in Figure 6-51.

Now you will change some of the radio button's properties. You will specify values for the button's Name property, which is how it is referenced within the form, and its Label property, which is the description that appears beside the button on the canvas. You also must specify the button's Radio Button Value, which corresponds to the actual data value that the radio button represents. Recall that when a radio button is selected, its data value becomes the data value for the radio group.

To change the radio button properties:

1. Double-click the new radio button to open its Property Palette, and then change the following properties. When you are finished, close the Property Palette, and maximize the Layout Editor if necessary.

Property	Value
Name	FR_RADIO_BUTTON
Label	Freshman
Radio Button Value	FR

The Value property is the actual database data value that is associated with the radio button, while the Label property is the label that appears beside the radio button on the form.

Now you need to change the background color of the radio button so that it is the same color as the canvas. To do this, you will use the **Fill Color tool**. This tool allows you to change the fill color of any form item. The tool appears at the bottom of the tool palette. Depending on your screen display setting, it might not currently be visible. If it is not visible, you will need to scroll down on the tool palette to find it.

To change the radio button fill color:

1. If the **Fill Color tool** is not visible, click the down arrow on the tool palette to scroll down.

2. In the Layout Editor, select the radio button, and then click. The Fill Color palette opens. Select a color to match the form background.

You cannot change the Fill Color property of a radio button to No Fill. Instead, you need to match its fill color to the background color of the canvas.

Now that you have formatted the Freshman radio button, you can easily create the rest of the buttons in the radio group by copying this button, pasting the copy onto the canvas, and changing the individual properties of each button to correspond to each of the S_CLASS data values. First, you will copy the radio button and paste it three times, so you have a total of four radio buttons.

To copy and paste the radio button:

1. Select the **Freshman radio button**, and click the **Copy button** 🗎 on the Layout Editor toolbar to make a copy of the radio button.

You also can press Ctrl + C to copy and Ctrl + V to paste, or click Copy or Paste on the Edit menu on the menu bar.

2. Click the **Paste button** 🗎 on the Layout Editor toolbar, make sure S_CLASS is the selected radio group, and then click **OK** to paste the copied radio button to the canvas. The new radio button is pasted on top of the first radio button.

3. Select the newly pasted radio button, and drag it down so it is positioned below the first radio button.

4. Click 🗎, select the **S_CLASS** radio group and click **OK**, and then drag the new button down so it is below the current radio buttons. Repeat this process one more time to create one more radio button. Your pasted radio buttons should look something like Figure 6-52.

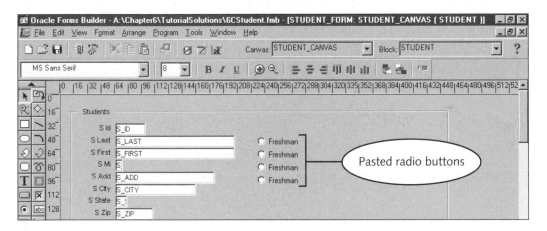

Figure 6-52   Pasting the new radio buttons

You will need to reposition the radio buttons so that they are aligned and spaced evenly. To do this, you will select all of the radio buttons into an object group and then align the left edges and stack them vertically.

To reposition the radio buttons:

1. Select the radio buttons as an object group by selecting the first radio button, pressing and holding the **Shift** key, and then selecting the remaining radio buttons. Selection handles appear around the object group.

2. With the buttons still selected as an object group, click **Arrange** on the menu bar, click **Align Objects**, and then click the **Horizontally Align Left** and **Vertically Stack** option buttons. Click **OK**. The four radio buttons should now appear with their left edges aligned and stacked on top of each other.

3. Click anywhere on the canvas to deselect the object group.

4. Change the properties of the second, third, and fourth buttons in the radio group as follows:

Property	Button 2	Button 3	Button 4
Name	**SO_RADIO_ BUTTON**	**JR_RADIO_ BUTTON**	**SR_RADIO_ BUTTON**
Label Radio	**Sophomore**	**Junior**	**Senior**
Button Value	**SO**	**JR**	**SR**

5. Save the form.

Next, you will specify the radio group's initial value. The **initial value** of a radio group is the Radio Button Value property of the radio button that is selected when the form first appears, or when you click ![icon] on the Forms Runtime toolbar to create a new blank record. If you do not specify the initial value of a radio group, an error will occur when you run the form.

To specify the initial value, you must change the Initial Value property on the radio group's Property Palette so it has the same Radio Button Value as the radio button that is selected when the form first opens, which is usually the first button in the group. (Recall that the Radio Button Value is the database value that corresponds to the radio button.) For example, in the S_CLASS radio group, the Radio Button Value for the first radio button is "FR." If you want the first radio button to be selected when the form first appears, you specify the Initial Value property of the S_CLASS radio group to be "FR" also.

To change the radio group's initial value:

1. Open the Object Navigator window, right-click the **S_CLASS** radio group, and then click **Property Palette**.

2. Scroll down to the Data property node, and type **FR** for the Initial Value property. Close the Property Palette.

3. Save the form.

Next, you will format the radio group to enhance its appearance. You will create a frame around the radio group and reposition the radio buttons symmetrically on the form.

To create a frame around the radio group and reposition the radio buttons:

1. Open the Layout Editor, click the **Frame tool** ![icon] on the tool palette, and then draw a frame around the radio buttons, as shown in Figure 6-53.

2. If the background color of the new frame is not the same as the form's background color, click anywhere on the line of the new frame to select it, click the **Fill Color tool** on the tool palette, and change the frame's fill color to **No Fill**.

3. With the frame still selected, click **Format** on the menu bar, select **Bevel**, and then click **Inset**.

4. Right-click the frame to open its Property Palette, and change the frame's Name property to **S_CLASS_FRAME**, and change the Frame Title property to **Class**. Close the Property Palette.

5. Reposition the radio buttons, and adjust the frame size and form objects so your form looks like Figure 6-53. You might need to make your radio buttons narrower so they do not cover the right edge of the frame.

> To resize all of the buttons in one operation, select the buttons as an object group, and drag the group selection handle to the desired size.

6. Save the form.

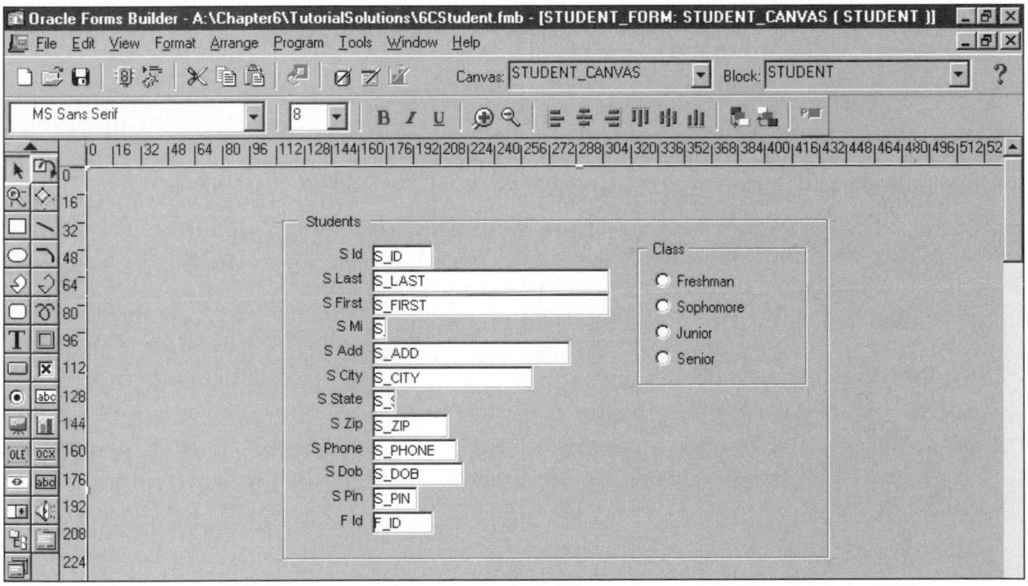

Figure 6-53   Formatted radio group

Now you will test the form to confirm that the radio buttons display the student data correctly. First, you will run the form and step through all of the table records. Then, you will insert a new record and use the radio buttons to specify the student class value.

To test the form:

1. Run the form, and maximize the Forms Runtime window.

2. Click the **Enter Query button** 🔲, do not enter a search condition, and then click the **Execute Query button** 🔲. The first STUDENT record (Sarah Miller) appears, with the Senior radio button selected. Click the **Next Record button** ▶.

3. Click ▶ until you have stepped through all of the STUDENT records and confirmed that the radio group displays the correct data value.

4. Click the **Insert Record button** 🔲 to insert a new blank record. Note that the Freshman radio button is selected, because it is the radio group's initial value.

5. Enter the following data values, and then click the **Save button** 🔲 to save the new record. The save confirmation message appears, confirming that the record was inserted.

Student ID:	**200**
S_Last:	**Andersen**
S_First:	**Ashley**
Class:	**Sophomore**

Next, you will use the radio group to create a query. You will enter a query with a search condition that returns all items for which the S_CLASS value is "JR."

To create a query using the radio group:

1. Click the **Enter Query button** 🔲, and then click the **Junior** radio button.

2. Click the **Execute Query button** 🔲. The record for the only student who is a junior (Daniel Black) appears.

What happens if a database table field is displayed as a radio group and the current field value has no corresponding radio button? To find out, you will update student Brian Umato's S_CLASS value to 'GR' (for graduate), which is not one of the radio button values. You will need to update the value in SQL*Plus, because in the form, you cannot update the S_CLASS value to anything other than the values represented by the radio buttons. Then, you will retrieve all of the student records using the form.

To see what happens when a record value has no corresponding radio button:

1. Switch to SQL*Plus, and type the following command at the SQL prompt:

```
UPDATE student
SET s_class = 'GR'
WHERE s_id = 101;
```

2. When the message "1 row updated." appears, type **COMMIT;** to commit the change.

If you update a record in SQL*Plus, your changes will not be visible in your form database session until you commit the change.

3. Switch to the student form in the Forms Runtime window, click the **Enter Query button** 🔲, do not enter a search condition, and then click the **Execute Query button** 🔲 to retrieve all of the student records. The record for the first student (S_ID 100, Sarah Miller) appears.

4. Click the **Next Record button** ▶. The record for student ID 102 (Daniel Black) appears. Note that student ID 101 (Brian Umato) appears to have been skipped.

5. Continue to click ▶ to confirm that Brian Umato's record was not retrieved.

6. Close Forms Runtime.

As you have seen, if there is no radio button value that maps to a record's data value, the record will not appear in the form.

Creating Form List Items

You should use a form list item when the user can select from a limited number of choices that do not change very often. Use lists instead of radio buttons when there are more than five choices or when there is a limited amount of space on the form to display radio buttons. All of the form list types can only be populated using static values—they cannot retrieve database values. As a result, do not use lists for form items that contain foreign key references. Lists are useful for enabling the user to select values from a predefined list when the values are not linked to a database table. Form Builder supports three types of lists, as illustrated in Figure 6-54.

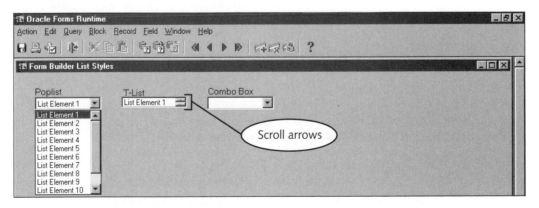

Figure 6-54   Form Builder list types

A **Poplist** is a drop-down list that the user opens when needed. When the user opens the list, the drop-down box displays up to 10 items at once. If the list contains more than 10 items, the Poplist displays a vertical scrollbar to allow the user to scroll and view all of the list items. When the user selects a value, the value appears in the list selection box, and the list closes. A **T-List** always displays the current selection. It has up and down arrows on its right edge, and the user can use these arrows to sequentially scroll through the list items and select a different value. T-Lists should not be used for lists with more than five to ten selections, since it becomes tedious to scroll through every value. A **Combo box** works like a Poplist, except the user has the option of entering a value if the desired value does not appear in the list.

Now, you will modify the STUDENT form and change the S_STATE text item to a Poplist item that displays choices for the state value. You will save the form file using a different filename and then change the item type of the item associated with the S_STATE field, using the Layout Wizard in reentrant mode. Then, you will populate the list with static values.

To modify the form to use a list item:

1. In Form Builder, save the student form as **6CStudent_List.fmb** in your Chapter6\TutorialSolutions folder.

2. Open the Layout Editor, select the **Students frame**, click **Tools** on the menu bar, and then click **Layout Wizard** to open the wizard in reentrant mode. The Layout Wizard window appears with tabs across the top, indicating it is in reentrant mode.

3. Be sure the Data Block tab is selected, select **S_STATE** in the Displayed Items list, and then select **Pop List** from the Item Type list. Click **Finish** to close the Layout Wizard, and save your changes. The state field now appears as a very narrow list item, with a drop-down arrow on its right edge.

 If your form layout appears with the radio buttons stacked below the advisor text item, it is because you did not change the frame Update Layout property to Locked. Click File on the menu bar, click Revert to revert back to your last saved copy, change the frame Update Layout property to Locked, and then repeat steps 1 through 3.

4. Select the list item, and drag its right edge to the right so the text in the list is visible, as shown in Figure 6-55.

5. Save the form.

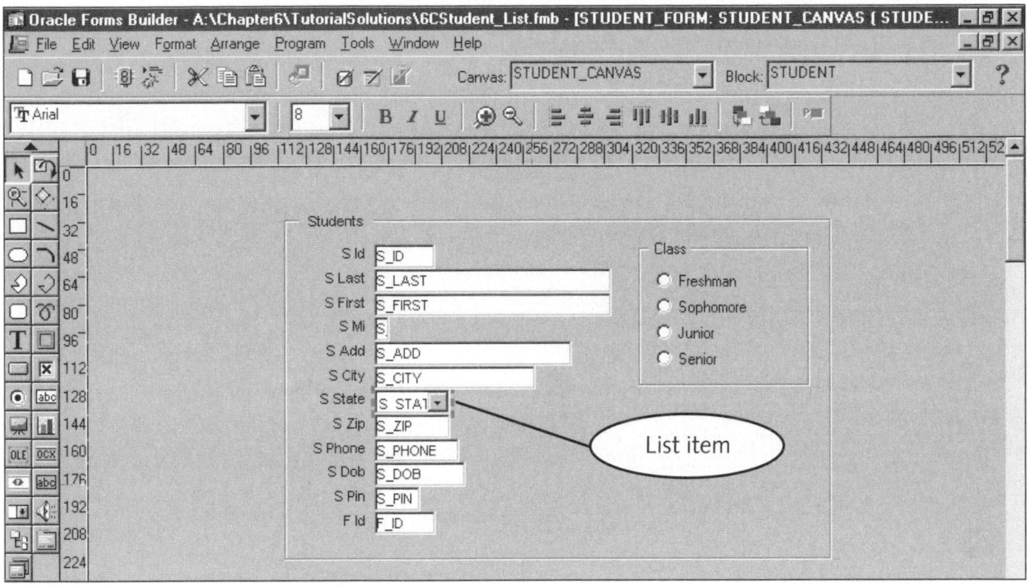

Figure 6-55   Form layout with list item

Now, you will configure the state list item. Since Northwoods University is a regional university in the Midwestern United States, you will only include the state abbreviations for Wisconsin (WI), Minnesota (MN), and Illinois (IL) as the state list choices. You will enter the items so they appear in alphabetical order.

To configure the S_STATE list item:

1. Double-click the list to open its Property Palette. Note that the Item Type property value is List Item, and the List Style property value is Poplist.

2. Select the **Elements in List** property, and then click **More**. The List Elements dialog box opens. The List Elements list shows the items that will appear in the list. The List Item Value text box displays the value that will be returned to the form when the list item is selected. These items always have the same value: The element that displays in the list is the same as the data value that is returned to the form. Now you will specify the list elements and their associated values.

3. Delete the current text in the first item in the List Elements list, and then type **IL**.

4. Place the insertion point in the List Item Value text box, and type **IL**.

5. Place the insertion point in the second item in the List Elements list, and type **MN**.

6. Place the insertion point in the List Item Value text box, and type **MN**.

7. Place the insertion point in the third item in the List Elements list, and type **WI**.

8. Place the insertion point in the List Item Value text box, and type **WI**. Your completed List Elements specification should look like Figure 6-56. When you select a different list element (like "IL"), its associated List Item Value ("IL") should appear in the List Item Value text box.

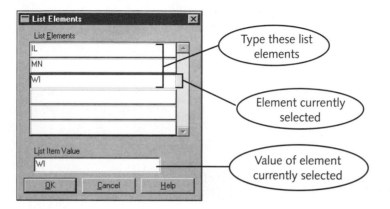

Figure 6-56   Specifying the list elements

9. Click **OK** to close the List Elements dialog box and save your changes, and then close the Property Palette.

10. Save the form.

Now you will run the form and test the list. You will retrieve all of the STUDENT records and then change the state data value for student Sarah Miller to "MN".

To test the list:

1. Run the form, and maximize the Forms Runtime window.

2. Click the **Enter Query button** , do not enter a search condition, and then click the **Execute Query button** to retrieve all of the student records. The first student record (Sarah Miller) appears.

3. Open the state list. The three state choices (IL, MN, WI) appear.

4. Select **MN**. MN appears in the list text box.

5. Click the **Save button** to save the record. The confirmation message appears, confirming that the change was successfully saved.

6. Close Forms Runtime.

You can easily modify an existing list to be a different list style by changing the List Style property in the list Property Palette. Now you will change the S_STATE list item to a Combo box and then run the form and enter ND for the S_STATE value, which is a value that is not in the list.

To change the list to a Combo box and test the list:

1. Open the Property Palette for the state list, and change the List Style property to **Combo Box**. Close the Property Palette.   •

2. Save the form, run the form, and maximize the Forms Runtime Window.

3. Click the **Enter Query button** 🔲, do not enter a search condition, and then click the **Execute Query button** 🔲 to retrieve all of the STUDENT records. The first STUDENT record appears.

4. Open the **State list**. The current list items appear. Do not select a value. The current value in the state text box, which is MN, appears highlighted.

5 Type **ND** in the State list box to overwrite the current value. Click the **Save button** 🔲. The confirmation message appears, indicating the change was successful.

6. Close Forms Runtime.

7. Close the form.

Creating a Check Box Item

Windows applications use check boxes to represent data values that can have only one of two values. The check box caption is usually interpreted as True or False: If the check box is checked, the data value represented by the caption is true. If the check box is cleared, the data value represented by the caption is false. For example, a check box might represent the STATUS field in the TERM table of the Northwoods University database. This field can have two values: CLOSED or OPEN. The check box might be labeled as "Closed?" If the check box is checked, then the term status value is CLOSED. If the check box is cleared, then the user would understand that the term status value is not closed, but OPEN. Similarly, a check box might represent the CERTIFICATION field in the CONSULTANT_SKILL table in the Software Experts database. The CERTIFICATION field can have the value Y or N. The check box would be labeled as "Certified?" If the box is checked, the certification value is Y, and if the box is cleared, the certification value is N.

Now you will modify a data block form corresponding to the TERM table in the Northwoods University database. Currently, the STATUS field is represented by a text item. You will change the STATUS field's item type to a check box, and then change the check box label to CLOSED?. When new terms are added to the table, their STATUS value is usually OPEN, so the default value for the check box will be Unchecked.

6

To modify the data block form to use a check box:

1. Open the **Term.fmb** file from the Chapter6 folder on your Data Disk, and save the file as **6CTerm.fmb** in your Chapter6\TutorialSolutions folder.

2. Open the Layout Editor. The records appear in a tabular layout style, with five records displayed at one time. You will change the text item to a check box item using the item Property Palette, rather than using the Layout Wizard in reentrant mode.

3. Double-click anywhere on the STATUS text items to open the item Property Palette. In the Property Palette, confirm that the item name is STATUS, and that the Item Type value is Text Item.

4. Click the **Item Type property**, open the list, and select **Check Box** to change the item type to a check box. The Property Palette properties change to display properties for a Check Box item.

5. Change the Label property value to **CLOSED?**.

6. Change the Value when Checked property value to **CLOSED**. This represents the data value of the STATUS item when the check box is checked.

7. Change the Value when Unchecked property to **OPEN**. This represents the data value of the STATUS item when the check box is cleared.

8. Click the **Check Box Mapping of Other Values property**, open the list, and select **Unchecked**. This specifies that the check box is cleared when the form first opens, or when the user adds a new blank record to the form.

9. Close the Property Palette. The STATUS items now appear as check boxes on the canvas.

Now you will run the form and confirm that the check boxes display the correct values for the STATUS field in the term table.

To run the form and test the check boxes:

1. Save the form, run the form, and maximize the Forms Runtime window.

2. Click the **Enter Query button** 🔲, do not enter a search condition, and then click the **Execute Query button** 🔲 to retrieve all of the TERM records. Scroll down to display the record for Term ID 6, as shown in Figure 6-57. Note that the check boxes are checked for all terms where the STATUS value is CLOSED, and the check box is cleared for Term ID 6, where the STATUS value is OPEN.

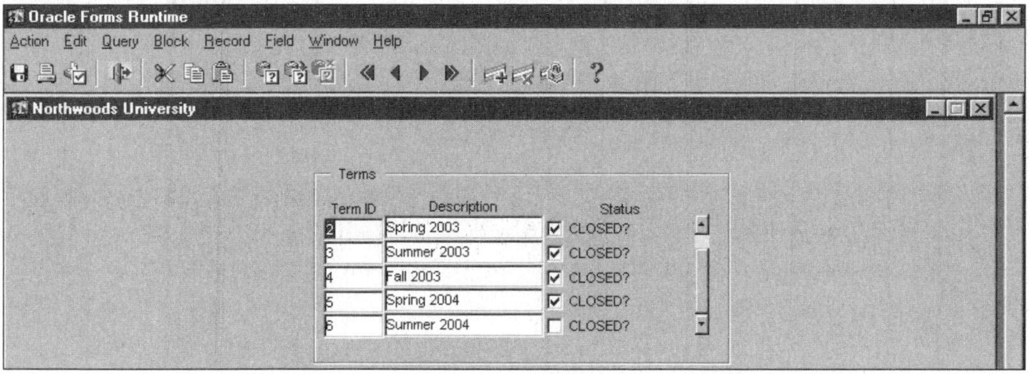

Figure 6-57   Data values represented by check boxes

3. Clear the check box by the record for Term ID 5 (Spring 2004), and then click the **Save button**. The confirmation message appears, indicating that the change was successfully made.

> If you are unable to update the record, open the Property Palette for the STATUS check box, and confirm that the Value when Unchecked property is set to **OPEN**. The property must be specified in all uppercase letters and cannot have any blank spaces at the end of the value.

4. Select the record for Term ID 6, and then click the **Insert Record button**. A new blank record appears under the record for Term ID 6.

5. In the blank record, type **7** for Term ID, **Fall 2004** for Description, and leave the Status check box cleared to indicate that the status of the new term is OPEN. Click 🖫 to save the record. The confirmation message appears, indicating that the record was saved.

6. Close Forms Runtime.

7. Close the TERM form in Form Builder.

8. Close Form Builder, and close all other open Oracle applications.

Check boxes are similar to radio buttons in the way they handle records that do not have a value that corresponds to checked and unchecked values. If a database table field is displayed using a check box and the field value for a particular record does not correspond with either the check box's checked or unchecked value, then the record does not appear in the form and is not retrieved by form queries.

SUMMARY

- ❑ Triggers are program units that run when a user performs an action, such as clicking a button, or when the system performs an action, such as loading a form.

- ❑ Triggers are associated with specific form objects, such as text items, buttons, data blocks, or the form itself.

- ❑ To write a trigger, you must specify the object that the trigger is associated with, the action, or event, that starts the trigger; and the PL/SQL code that executes when the trigger fires.

- ❑ A list of values (LOV) displays a list of data values that can be used to populate foreign key fields or fields with restricted data values.

- ❑ To create an LOV, you create a record group using a SQL command. The LOV display shows records returned by the SQL command. You can open the LOV display by placing the form insertion point in the text item associated with the LOV and pressing F9, or by clicking the LOV command button.

- ❑ You can use the LOV Wizard to create the LOV display and record group, specify the display properties, and associate the LOV with a form text item.

- ❑ When you create an iconic button, you must specify the complete path to the .ico file for the icon image that appears on the button. If you do not specify the complete path to the file, then the file containing the button icon must be in the default icon folder on your workstation.

- ❑ You should always format an LOV display so all of the records and columns are visible without scrolling. Formatting the LOV display width and the column widths is a trial-and-error process.

- ❑ An editor is a dialog box that provides search and replace functions and a larger workspace that the user can open and use to edit character, number, or date values that appear as text items.

- ❑ You can use radio buttons, also called option buttons, to enter and modify data values when the field has a limited number of related choices.

- ❑ In a form, individual radio buttons exist within a radio group. Each individual radio button has an associated data value, and the radio group has the data value of the currently selected radio button.

- ❑ You should display data using a form list item when the user can select from a limited number of choices that do not change very often.

❑ Form Builder supports three list types: Poplists, which are standard drop-down lists; T-Lists, which always display the selected value and allow the user to scroll sequentially through the list items using up and down arrows; and Combo boxes, which are like Poplists, except they allow the user to enter a value if the desired value is not in the list.

❑ You can only populate form lists using static values. You cannot use lists to display values retrieved from the database.

❑ You can use check boxes to represent data values that can have only one of two values. The check box caption is usually interpreted as True or False: If the check box is checked, the data value represented by the caption is true. If the check box is cleared, the data value represented by the caption is false.

6

REVIEW QUESTIONS

1. Why is it a good idea to use a sequence to automatically generate primary key values in a form?

2. What three things must be specified when you create a trigger?

3. When you compile a trigger, you receive the following error message: "'DUAL' must name a table to which the user has access." What is the cause of the error?

4. List three ways that a user can open an LOV display.

5. List the five steps for creating an LOV.

6. What is the difference between the default Forms editor, a custom editor, and the SYSTEM_EDITOR?

7. What is a radio group? What does the Initial Value property of a radio group specify?

8. What does the Radio Button Value property for a radio button specify?

9. What is the relationship between the form insertion point and the LIST_VALUES command?

10. When should you use radio buttons to display data values, and when should you use a list?

11. When should you use an LOV rather than a list to display selection values for a text item?

PROBLEM-SOLVING CASES

The case problems reference the Clearwater Traders (see Figure 1-11), Northwoods University (see Figure 1-12), and Software Experts (see Figure 1-13) sample databases. For all cases, save the solution files in your Chapter6\CaseSolutions folder. For all forms, rename the window and canvas using descriptive names, and format the window and canvas properties using the guidelines described in the chapter. Change the font of all form items to a sans serif font. Use descriptive item prompts, and if necessary, format item prompts so they appear on a single line.

1. Create a form named ORDER_SOURCE_FORM that displays all of the fields in the ORDER_SOURCE table in the Clearwater Traders database. Save the file as 6CCase1.fmb.

   a. Create a new data block and a tabular-style layout. Display five records on the form, and display a scrollbar. Change the frame title to "Order Sources"

   b. In SQL*Plus, create a sequence named ORDER_SOURCE_ID_SEQUENCE that starts at 10 and has no maximum value.

   c. In the form, create a form trigger to automatically insert the next sequence value when the user creates a new record.

2. Create a form named COURSE_FORM that displays all of the fields in the COURSE table in the Northwoods University database. Save the file as 6CCase2.fmb.

   a. Create a new data block and form-style layout. Change the frame title to "Courses".

   b. In SQL*Plus, create a sequence named COURSE_ID_SEQUENCE that starts at 10 and has no maximum value.

   c. In the form, create a form trigger to automatically insert the next sequence value when the user creates a new record.

   d. Create a custom editor named COURSE_NAME_EDITOR that you can use to enter and edit values in the COURSE_NAME text item. Change the editor window title to "Course Description," and position the editor so its top-left corner is approximately at the same position as the top-left corner of the COURSE_NAME text item.

3. Create a form named SKILL_FORM that displays all of the fields in the SKILL table in the Software Experts database. Save the file as 6CCase3.fmb.

   a. Create a new data block and form-style layout. Change the frame title to "Skills."

   b. In SQL*Plus, create a sequence named SKILL_ID_SEQUENCE that starts at 10 and has no maximum value.

c. In the form, create a form trigger to automatically insert the next sequence value when the user creates a new record.

d. Specify that the system editor opens when the user opens an editor for the SKILL_DESCRIPTION text item.

4. Create a form named LOCATION_FORM that displays all of the fields in the LOCATION table in the Northwoods University database. Save the file as 6CCase4.fmb.

a. Create a new data block and form-style layout. Change the frame title to "Locations."

b. In SQL*Plus, create a sequence named LOC_ID_SEQUENCE that starts at 60 and has no maximum value. (If you have already created this sequence in a case from a previous chapter, use the existing sequence.)

c. In the form, create a form trigger to automatically insert the next sequence value when the user creates a new record.

d. Create a Combo box that provides values for the BLDG_CODE text item. Specify the current list choices as CR, BUS, and LIB. Place the list elements in alphabetical order.

5. Create a form named ITEM_FORM that displays all of the fields except ITEM_IMAGE in the ITEM table in the Clearwater Traders database. Save the file as 6CCase5.fmb.

a. Create a new data block and tabular-style layout that displays five records at one time and has a scrollbar. Change the frame title to "Items."

b. Create an LOV to display the allowable values for the CATEGORY_ID foreign key field. In the LOV display, show the CATEGORY_ID and CATEGORY_DESC columns. Format the LOV using the instructions provided in the lesson. Allow users to open the LOV display by placing the insertion point in the CATEGORY_ID text item and pressing F9.

6. Create a form named FACULTY_FORM that displays all of the fields except F_IMAGE in the FACULTY table in the Northwoods University database. Format the form as shown in Figure 6-58, and save the file as 6CCase6.fmb.

a. Create a new data block and form-style layout. Change the frame title to "Faculty."

b. Display the F_RANK data values using radio buttons with the following labels and associated data values. Configure the radio group so the "Assistant" radio button is selected when the form first appears or when a new blank record is created.

Label	Value
Assistant	ASST
Associate	ASSO
Full	FULL
Instructor	INST

6

c. Create an LOV named LOC_ID_LOV to display the allowable values for the LOC_ID foreign key field, and return the selection to the text item. In the LOV display, show the LOC_ID, BLDG_CODE, and ROOM columns. Create a command button to open the LOV display. Format the LOV display using the instructions provided in the lesson.

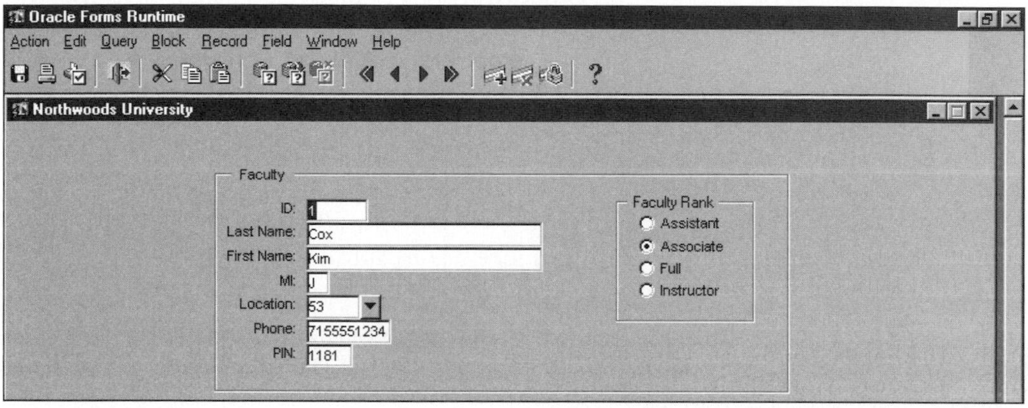

Figure 6-58

7. Create a form named PROJECT_FORM that displays all of the fields in the PROJECT table in the Software Experts database. Save the file as 6CCase7.fmb.

a. Create a new data block and form-style layout. Change the frame title to "Projects."

b. In SQL*Plus, create a sequence named P_ID_SEQUENCE that starts at 10 and has no maximum value. (If you have already created this sequence in a case from a previous chapter, use the existing sequence.)

c. In the form, create a trigger to automatically insert the next sequence value when the user creates a new record.

d. Create an LOV named CLIENT_ID_LOV to display the allowable values for the CLIENT_ID text item. The LOV should display the client IDs and client names, and it should return the selected CLIENT_ID value to the form text item. Format the LOV display using the guidelines in the lesson. Create a command button to display the LOV.

e. Create an LOV named MGR_ID_LOV to display the allowable values for the MGR_ID text item. The LOV should display the consultant ID and last and first name values, and it should return the selected C_ID value to the form text item. Format the LOV display using the guidelines in the lesson. Create a command button to open the LOV display.

8. Create a form named CONSULTANT_SKILL_FORM that displays all of the fields in the CONSULTANT_SKILL table in the Software Experts database. Save the file as 6CCase8.fmb.

   a. Create a new data block and tabular-style layout that displays five records at a time and has a scrollbar. Change the frame title to "Consultant Skills."

   b. Display the CERTIFICATION field using a check box with an appropriate label. Configure the check box so the box is cleared when the user creates a new record.

   c. Create an LOV named C_ID_LOV to display the allowable values for the C_ID text item. The LOV should display the consultant ID and last and first name values, and it should return the selected C_ID value to the form text item. Format the LOV display using the guidelines in the lesson. Allow the user to open the LOV display by placing the form insertion point in the C_ID text item, and then pressing F9.

   d. Create an LOV named SKILL_ID_LOV to display the allowable values for the SKILL_ID text item. The LOV should display the skill ID and description, and it should return the selected SKILL_ID value to the form text item. Format the LOV display using the guidelines in the lesson. Allow the user to open the LOV display by placing the form insertion point in the SKILL_ID text item, and then pressing F9.

9. Create a form named ENROLLMENT_FORM that displays all of the fields in the ENROLLMENT table in the Northwoods University database. Save the file as 6CCase9.fmb.

   a. Create a new data block with a tabular-style layout that displays five records at a time and has a scrollbar. Change the frame title to "Enrollments."

   b. Display the GRADE field using radio buttons corresponding to each letter grade (A, B, C, D, or F). Format the radio buttons as shown in Figure 6-59. Specify that the "A" radio button is selected when the user creates a new record.

   c. Create an LOV named S_ID_LOV to display the allowable values for the S_ID text item. The LOV should display the student ID and student first and last name values, and it should return the selected S_ID value to the form text item. Format the LOV display using the guidelines provided in the lesson. Allow users to open the LOV display by placing the form insertion point in the S_ID text, and then pressing F9.

   d. Create an LOV named C_SEC_ID_LOV to display the allowable values for the C_SEC_ID text item. The LOV should display the course section ID, call ID, and section number, day, and time. Format the LOV display using the guidelines in the lesson. Format the TIME field using a time format mask by using the TO_CHAR function in the LOV SQL query. Allow users to open the LOV display by placing the form insertion point in the C_SEC_ID text item, and then pressing F9.

   e. Create hints and a tooltip for the S_ID and C_SEC_ID text items to tell the user how to open the LOV display. Display the hints automatically.

6

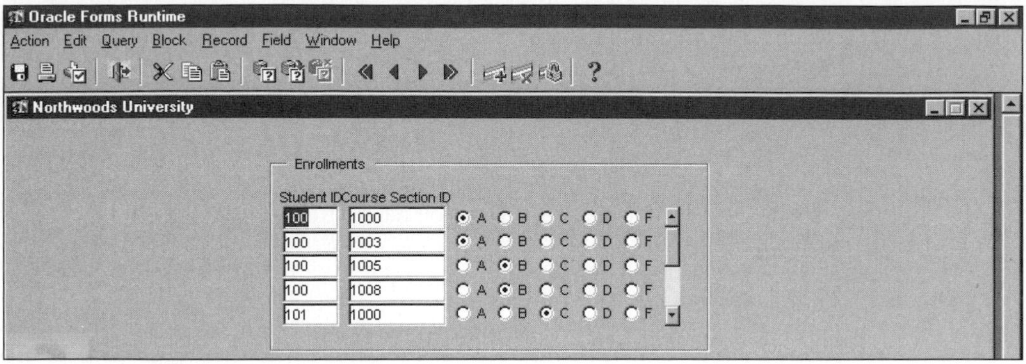

Figure 6-59

10. Create a multiple-table form named COURSE_SECTION_FORM that is based on the COURSE_SECTION and TERM tables in the Northwoods University database. Save the file as 6CCase10.fmb.

a. Create a new data block based on the COURSE_SECTION table that has a form-style layout. Change the frame title to "Course Sections." Format the TIME field so the values appear as times.

b. Create an LOV named COURSE_ID_LOV to display the allowable values for the COURSE_ID text item. The LOV should display the course ID and call ID, and it should return the selected COURSE_ID value to the form text item. Format the LOV display using the guidelines provided in the lesson. Create an LOV command button to open the LOV display.

c. Create an LOV named TERM_ID_LOV to display the allowable values for the TERM_ID text item. The LOV should display the term ID and term description, and it should return the selected TERM_ID value to the form text item. Format the LOV display using the guidelines provided in the lesson. Allow users to open the LOV display by placing the form insertion point in the TERM_ID text item, and then pressing F9.

d. Create an LOV named F_ID_LOV to display the allowable values for the F_ID text item. The LOV should display the faculty ID and faculty first and last name, and it should return the selected F_ID value to the form text item. Format the LOV display using the guidelines provided in the lesson. Create an LOV command button to open the LOV display.

e. Create an LOV named LOC_ID_LOV to display the allowable values for the LOC_ID text item. The LOV should display the location ID, building code, room, and room capacity, and it should return the selected LOC_ID value to the form text item. Format the LOV display using the guidelines in the lesson. Create an LOV command button to open the LOV display.

f. Create a detail data block based on the TERM table that displays the term ID, description, and status using a form–style layout. Change the frame label to "Term." Display the STATUS field using a check box that specifies whether the term enrollment is open or closed. (Hint: change the Item Type property of the text item on the Property Palette.)

6

◀ LESSON A ▶

Objectives

♦ Create a custom form
♦ Write PL/SQL triggers to process database records
♦ Create program units that are called by a trigger
♦ Use the Forms Debugger
♦ Understand navigational triggers
♦ Learn how to control user navigation in a form

The data block forms that you created in the previous chapter were based on a single database table or were based on multiple tables related by foreign key values. The appearance of the forms reflected the structure of the database tables. However, sometimes you need to create forms to insert, update, delete, and display data from many related tables in a seamless fashion that supports the organization's processes instead of reflecting the structure of the database tables. In this chapter, you will learn how to create custom forms that are not associated with specific database tables. These forms use PL/SQL triggers to process data using SQL commands. While developing these forms, you will learn how to use the Form Builder Debugger.

CUSTOM FORMS

A **custom form** is a form that, unlike a data block form, is not associated with a specific database table. Rather, the developer designs the form to display the desired data and perform the required operations, regardless of which database tables the data values are stored in. All form processing is performed using triggers that process SQL commands that retrieve, insert, update, and delete data values as needed. To create a custom form, you must identify the business processes and related database operations involved, design the interface, and then create the form.

Identifying the Business Processes and Database Operations

To create a custom database form, first you must identify the processes you wish to support and the corresponding database tables that are involved. The best way to start is to describe the process. For example, the first custom form you will develop supports the merchandise-receiving function at Clearwater Traders. When Clearwater Traders purchases new merchandise from its suppliers, it keeps track of information about anticipated shipments in the SHIPMENT and SHIPMENT_LINE tables. This information includes the shipment ID, the date the shipment is expected, the inventory ID of the purchased item, the shipment quantity, and the date the shipment is expected. When a new shipment arrives, the receiving clerk retrieves the shipment line record associated with the received merchandise by identifying the shipment ID and the inventory ID. The clerk confirms the shipment quantity, or changes it if necessary, and updates the DATE_RECEIVED field to the current date. The system must also update the QOH field in the INVENTORY table with the shipment quantity to reflect the updated quantity on hand for the item.

Designing the Interface

The next step is to visualize how the user interface will look. Figure 7-1 shows a design sketch of the interface design. The user will select a record from the SHIPMENT_LINE table for a shipment that has not yet been received (DATE_RECEIVED is NULL) using an LOV associated with the Shipment ID text item. (Users can open the LOV display by placing the insertion point in the Shipment ID text item and then pressing F9.) After the user selects a shipment line record, the form displays the selected shipment ID and inventory ID, as well as the inventory description, size, and color information. The form also contains text items to display the quantity that was received and the date on which the shipment was received. The date received text item will automatically display the current system date, which the user can change if the date the shipment was received is not the current date. The form will have an Update button to update the shipment and inventory tables, a Cancel button to cancel the current operation and clear the form text items, and an Exit button to exit the form.

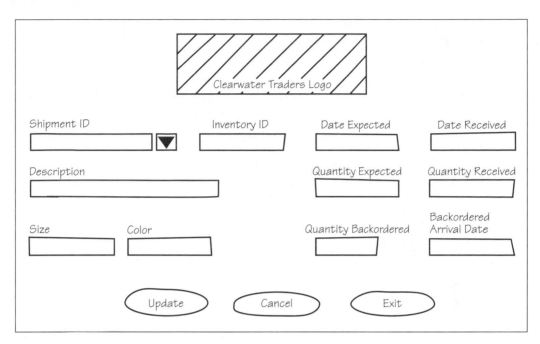

Figure 7-1   Design of the receiving form

Before you create the form, you will refresh your database tables by running the scripts that re-create the tables and insert all of the sample data records.

To refresh the database tables:

1. Start SQL*Plus, log onto the database, and then run the **clearwater.sql**, **northwoods.sql**, and **softwareexp.sql** scripts from the Chapter7 folder of your Data Disk.

Creating a Custom Form

To create a custom form, you have to create the form manually, rather than use the Data Block and Layout Wizards, because the form displays fields from multiple tables. When you create a form manually, you create the form window and canvas in the Object Navigator. Then, you create the form items manually by "painting" them on the canvas, using tools on the Layout Editor tool palette. Finally, you write the code that controls the form functions. Now you will start Form Builder and select the option for building a form manually.

To create a form manually:

1. Start Form Builder, select the **Build a new form manually option button**, and then click **OK**. The Object Navigator window opens and displays a new form module. Change the form module name to **RECEIVING_FORM**.

If Form Builder is already running, open the Object Navigator window, select the Forms node, and then click the Create button to create a new form module.

2. Click **File** on the menu bar, click **Connect**, and connect to the database in the usual way.

3. Save the form as **7AReceiving.fmb** in your Chapter7\TutorialSolutions folder.

If you are storing your Data Disk files on floppy disks, you will need to use two floppy disks for this chapter. When your first disk becomes full, save your solutions files on the second disk.

Next, you will rename the default form window, create a new canvas, and modify the form window properties. You will perform these operations in the Object Navigator Visual View, because Visual View displays the hierarchical relationships among form objects.

To modify the form objects in Visual View:

1. Click **View** on the menu bar, and then click **Visual View**. The form appears, along with the Visual Attributes node and the Windows node.

Visual attributes are attributes that determine the appearance of the form or the appearance of form objects, such as fonts, colors, and line bevels. You will learn how to create a Visual Attribute object that specifies form appearance properties in Chapter 12.

2. Click ➕ beside the Windows node. Two windows appear: the NULL_WINDOW and WINDOW1. Whenever you create a new form manually, Form Builder creates these two windows automatically. The NULL_WINDOW is a default window object that cannot be deleted or modified. WINDOW1 is the window in which the form will appear, so you will modify its properties.

3. Change the name of WINDOW1 to **RECEIVING_WINDOW**.

4. Click ➕ beside RECEIVING_WINDOW, select the **Canvases node**, and then click the **Create button** to create a new canvas. Change the canvas name to **RECEIVING_CANVAS**.

5. Select the **RECEIVING_WINDOW**, open its Property Palette, and change the window title to **Clearwater Traders**. Change the Resize Allowed and Maximize Allowed properties to **No**.

6. Change the window width to **580**, and change the height to **375** (or whatever values are necessary to completely display the window in the Forms Runtime window on your display without scrolling), and then close the RECEIVING_WINDOW Property Palette.

7. Select the **RECEIVING_CANVAS**, open its Property Palette, and change its width and height properties so that they are the same as the window's width and height. Close the RECEIVING_CANVAS Property Palette.

8. Save the form.

Next, you need to create a control block. A **control block** is a data block that is not associated with a particular database table. Instead, it contains text items, command buttons, radio buttons, and other form items that you manually draw on the form canvas and then control through triggers that you write using PL/SQL. To create a control block, you create a new data block in the Object Navigator and specify to create the data block manually, rather than use the Data Block Wizard. You will create the control block in Ownership View, because the Data Blocks node does not appear in Visual View.

To create the control block:

1. Click **View** on the menu bar, and then click **Ownership View**.

2. If necessary, click ⊞ beside RECEIVING_FORM to display the form objects.

3. Select the **Data Blocks node**, and then click the **Create button** ⬚ to create a new data block. The New Data Block dialog box opens. Select the **Build a new data block manually option button**, and then click **OK**. The new data block appears in the Data Blocks list in the Object Navigator window. Since you did not create the data block using the Data Block Wizard and you have not associated the block with a database table, it is a control block rather than a data block.

4. Change the block name to **RECEIVING_BLOCK**, and then save the form.

Creating the Form Text Items

Next, you will open the Layout Editor and create text items that will appear on the form shown in Figure 7-1. The text item is the box that will contain the actual data retrieved from the database, and the prompt is the label that indicates the type of data that appears in the corresponding text item (such as "Last Name"). First, you will create the Shipment ID text item.

To create the Shipment ID text item:

1. Click **Tools** on the menu bar, and then click **Layout Editor**. The Layout Editor displays a blank canvas and shows the current canvas and block in the Canvas and Block lists at the top of the Layout Editor window. Be sure that RECEIVING_CANVAS appears as the current canvas and RECEIVING_BLOCK appears as the current block.

If you are not working in RECEIVING_BLOCK, or if RECEIVING_BLOCK is a data block rather than a control block, your form will not work correctly. To move to a different block or canvas in the Layout Editor window, open the Block or Canvas list, and select the desired block or canvas.

2. Select the **Text Item tool** [abc] on the tool palette, and then draw a rectangular box for the SHIPMENT_ID text item, as shown in Figure 7-2. The default name of your text item will be different, because this is a system-generated name that you will change. The prompt will not appear yet.

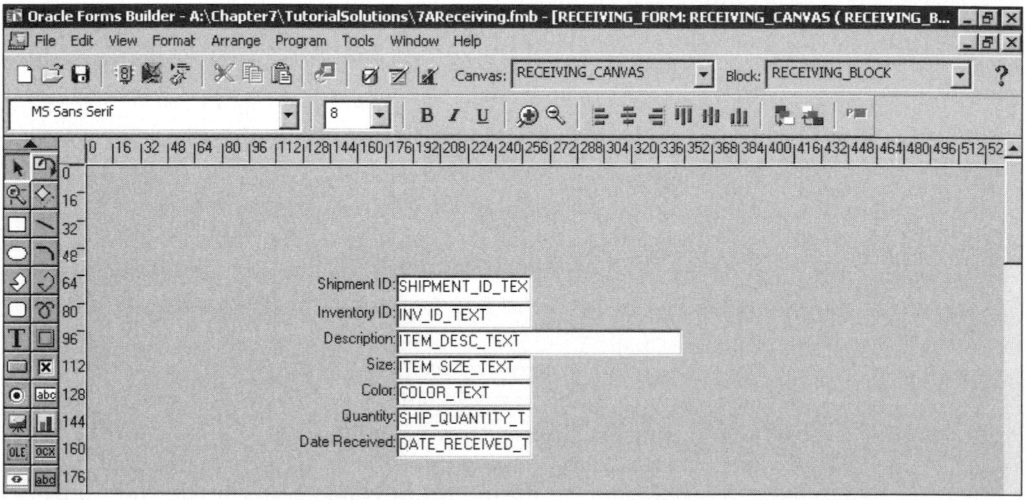

Figure 7-2   Form text items

Next, you will change the properties of the text item and configure the text item prompt (label). You will change the text item's Name property, which you use to reference the text item in triggers. You will also change the text item's Data Type and Maximum Length property values. Text items that display database records must have the same data type as the corresponding database data field. The maximum length must be large enough to accommodate the maximum length of the data, plus any characters included in a format mask. When the SHIPMENT_LINE table was created, SHIPMENT_ID was specified as a NUMBER field of size 10, so you will change the text item's data type to NUMBER and its maximum length to 10.

To specify the text item properties:

1. Right-click the new text item, and then click **Property Palette**. Confirm that the Item Type property value is Text Item, and then change the Name property to **SHIPMENT_ID_TEXT**.

2. Scroll down to the Data property node, click in the space next to the Data Type property to open the value list, and select **Number**.

3. Change the Maximum Length property to **10**.

4. Scroll down to the Prompt property node in the Property Palette, and change the Prompt property to **Shipment ID:**.

5. Close the Property Palette. The new prompt and text item name appear. Save the form.

Now you will create the rest of the form text items. After you create each item, you will change the item name and data type to correspond with the item's associated database field.

To create the rest of the form text items:

1. Create text items with the following properties, and position the text items as shown in Figure 7-2.

Name	Data Type	Maximum Length	Prompt
INV_ID_TEXT	Number	10	Inventory ID:
ITEM_DESC_TEXT	Char	100	Description:
ITEM_SIZE_TEXT	Char	10	Size:
COLOR_TEXT	Char	20	Color:
SHIP_QUANTITY_TEXT	Number	4	Quantity:
DATE_RECEIVED_TEXT	Date	20	Date Received:

To position the text items, select all of the text items as an object group, click Arrange on the menu bar, click Align Objects, then select Align Left for the horizontal alignment and Stack for the vertical alignment.

2. Save the form.

Creating the LOV

Next, you will create the LOV to retrieve data for existing shipments that have not yet been received. To determine if an existing shipment has not been received, test to determine if the DATE_RECEIVED field in the SHIPMENT_LINE table is NULL. The LOV will display data from the SHIPMENT_LINE, INVENTORY, and ITEM tables, and return the selected values to the form. You will create the LOV using the LOV Wizard.

To create the LOV:

1. Click **Tools** on the menu bar, and then click **LOV Wizard**. The LOV Source page appears. Accept the default values, and then click **Next**.

2. Type the following SQL command to retrieve the LOV display fields, and then click **Next**. The Column Selection page appears.

```
SELECT shipment_ID, inventory.inv_id, item_desc,
item_size, color, ship_quantity
FROM shipment_line, inventory, item
WHERE shipment_line.inv_id = inventory.inv_id
AND inventory.item_id = item.item_id
AND date_received IS NULL
```

3. Select all of the query fields for the LOV display, and then click **Next**. The Column Display page appears.

4. Specify the following column display properties:

Column	Title	Width	Return Value
SHIPMENT_ID	**Shipment ID**	45	**RECEIVING_BLOCK. SHIPMENT_ID_TEXT**
INV_ID	**Inv. ID**	40	**RECEIVING_BLOCK. INV_ID_TEXT**
ITEM_DESC	**Description**	100	**RECEIVING_BLOCK. ITEM_DESC_TEXT**
ITEM_SIZE	**Size**	40	**RECEIVING_BLOCK. ITEM_SIZE_TEXT**
COLOR	**Color**	70	**RECEIVING_BLOCK. COLOR_TEXT**
SHIP_QUANTITY	**Quantity**	30	**RECEIVING_BLOCK. SHIP_QUANTITY_TEXT**

5. Click **Next**. The LOV Display page appears. Type **Expected Shipments** for the LOV title, change the Width to **425**, change the Height to **150**, and then click **Next**. The Advanced Options page appears.

6. Accept the default options, and then click **Next**. The Items page appears. This page enables you to specify the form text item to which the LOV is assigned. (Recall that to display the LOV, you must place the form insertion point in the text item the LOV is assigned to, then press F9.) You will assign the LOV to SHIPMENT_ID_TEXT.

7. Assign the LOV to **RECEIVING_BLOCK.SHIPMENT_ID_TEXT**, click **Next**, and then click **Finish** to finish the LOV.

8. Open the Object Navigator window, and change the name of the new LOV item and record group to **SHIPMENT_ID_LOV**.

9. Save the form.

Now you will test the form. You will run the form and display the LOV to confirm that the columns appear correctly and that the values of the selected shipment appear correctly in the form text items.

To test the LOV:

1. Run the form, maximize the Forms Runtime window, make sure that the insertion point is in the Shipment ID text item, and then press **F9**. The LOV display opens.

The first two digits of the Shipment ID appear in the Find window because the first two digits are the same value (21) for every record returned in the LOV display.

2. Select the record for Shipment **213** and Inventory ID **11777**, and then click **OK**. The selected values appear in the form text items, as shown in Figure 7-3.

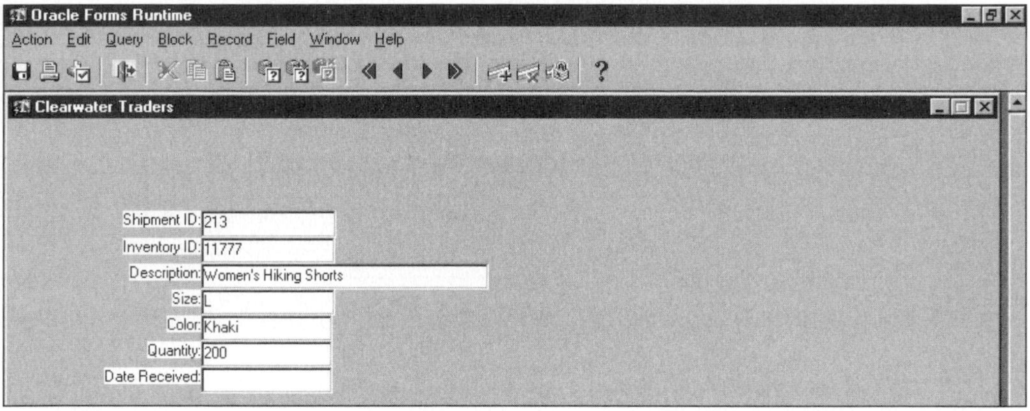

Figure 7-3   Viewing the selected values

7

If the returned values do not display correctly or if only some of the values appear, it might be because you did not specify the LOV return values correctly. Close the Forms Runtime window, open the Object Navigator in Form Builder, select the SHIPMENT_ID_LOV object under LOVs node, open the LOV Wizard in reentrant mode, and then examine the return values on the Column Display page to confirm that the column names are listed correctly and that each column corresponds with the correct return value.

LOV values also might not appear in the form text items if the form text item Data Type and Maximum Length properties are not exactly the same as the database field properties. Open the Property Palette for each text item, and make sure that the data type and maximum width for each form text item are the same as for the text item's corresponding database field.

3. Close Forms Runtime.

Displaying System Date and Time Values in Form Text Items

Recall that the Date Received text item will automatically display the current date, and the user can enter a different date if necessary. You can display system date and time values in form text items using the system variables summarized in Table 7-1. In the Windows environment, the operating system values are retrieved from your local workstation.

System Variable	Return Value
$$DATE$$	Current date from operating system
$$TIME$$	Current time from operating system
$$DATETIME$$	Current date and time from operating system
$$DBDATE$$	Current date from database server
$$DBTIME$$	Current time from the database server
$$DBDATETIME$$	Current date and time from database server

Table 7-1   System date and time variables that can be displayed in form items

Now you will change the Initial Value property of the DATE_RECEIVED_TEXT text item to $$DATE$$ so it displays the current operating system date.

To display the current operating system date in the text item:

1. Open the Property Palette for the DATE_RECEIVED_TEXT text item, scroll down to the Data property node, and change the Initial Value property to **$$DATE$$**.

2. Close the Property Palette, and save the form.

3. Run the form. The current system date appears in the Date Received text item.

4. Close Forms Runtime.

Creating the Form Command Buttons

The next step is to create the Update, Cancel, and Exit buttons. First, you will draw the buttons on the canvas. Whenever you create a button group in a graphical user interface, the buttons should all be the same size and should be wide enough to accommodate the longest button label, which is the text that appears on the button. You will draw the first button and change its name and label properties. Then you will copy the button and paste it on the canvas twice to create the other two buttons.

To create the Update, Cancel, and Exit buttons:

1. In the Layout Editor, use the **Button tool** 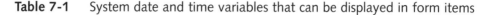 to draw a command button that will be used as the Update button, as shown in Figure 7-4.

2. Open the new button's Property Palette; change its Name property to **UPDATE_BUTTON**, Label to **Update**, Width to **60**, and Height to **16**; and then close the Property Palette.

3. In the Layout Editor, select the button, if necessary. Then click the **Copy button** on the toolbar to copy the button, and click the **Paste button** two times to paste the button twice. The new buttons are pasted directly on top of the first button.

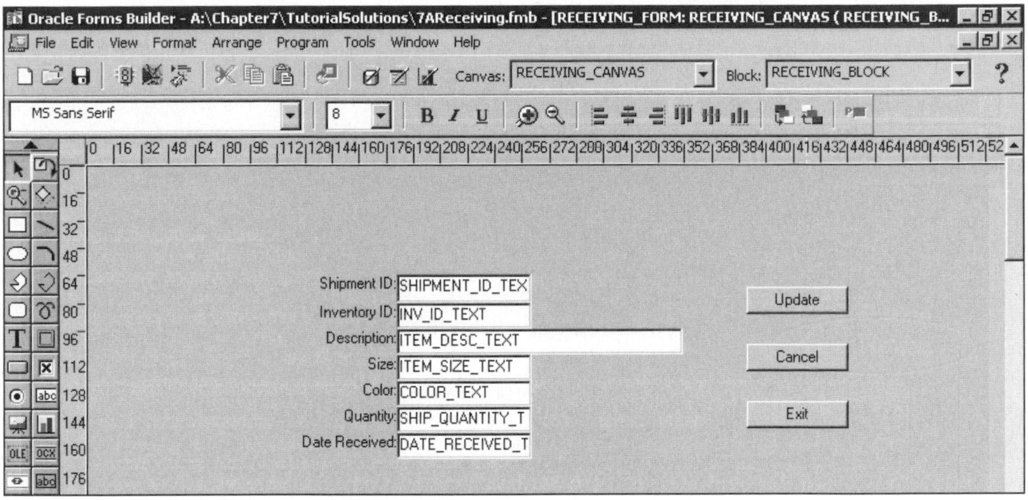

Figure 7-4 Form command buttons

4. Select the top button, and drag it below the Update button. Select the next button, and drag it below the second button.

5. Select all three buttons as an object group, click **Arrange** on the menu bar, select **Align Objects**, click the Horizontally **Align Left** and Vertically **Distribute** option buttons, and then click **OK**. Position the button group as shown in Figure 7-4.

6. Select the second button, open its Property Palette, change its Name property to **CANCEL_BUTTON,** change its Label property to **Cancel**, and then close the Property Palette.

7. Select the third button, open its Property Palette, change its Name property to **EXIT_BUTTON,** change its Label property to **Exit**, and then close the Property Palette.

8. Save the form.

For data block forms like the ones you created in Chapter 6, the Forms Runtime environment automatically provides triggers for retrieving, inserting, updating, and deleting data. For the custom forms you will create in this chapter, you need to create these triggers manually. First, you will create the trigger for the Update button. Recall that the user retrieves the shipment information, updates the quantity received and date received if necessary, and then clicks the Update button. The UPDATE_BUTTON trigger must update the SHIP_QUANTITY and DATE_RECEIVED fields in the SHIPMENT_LINE table and then add the shipment quantity to the current QOH value in the INVENTORY table.

To update the database fields with the trigger, you use the same SQL commands that you would use in SQL*Plus, and replace the search and update values with references to the form text item values. Recall that to reference form text item values, you use the syntax

:block_name.item_name. For example, :RECEIVING_BLOCK.SHIPMENT_ID_TEXT references the current value in the SHIPMENT_ID_TEXT text item. Now you will create a trigger for the Update button that contains the SQL code to perform these actions.

To create the Update button trigger:

1. In the Layout Editor, select the **Update button**, right-click, point to **SmartTriggers**, and select **WHEN–BUTTON–PRESSED**. The PL/SQL Editor window opens.

2. Type the code shown in Figure 7-5 in the PL/SQL Editor Source code pane, compile the trigger, and then close the PL/SQL Editor.

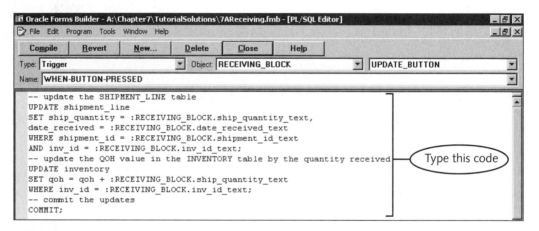

Figure 7-5 Update button trigger code

 If your trigger does not compile correctly, refer to the Debugging Form Triggers section in Chapter 6. If you still cannot compile the trigger, refer to the Using the Forms Debugger section later in this lesson.

3. Save the form.

Now you will test the trigger. You will retrieve the record for shipment 212 and inventory ID 11669, and update the database to indicate that all 25 units were received today.

To test the trigger:

1. Run the form, place the insertion point in the Shipment ID text item, press **F9** to open the LOV display, and select Shipment ID **212** and Inventory ID **11669**. The shipment details appear in the form.

2. Confirm that Quantity Received is 25 and Date Received is the current date, and then click **Update**. The message "FRM-40401: No changes to save" appears in the status line.

3. Close Forms Runtime.

Although the confirmation message ("FRM-40401: No changes to save") seems counter-intuitive, it is the confirmation message issued by the Form Builder when a COMMIT command successfully executes in a PL/SQL trigger in a control block. Before you can be totally confident that this operation succeeded, you should double-check the values in the SHIPMENT_LINE and INVENTORY tables. You will confirm that the SHIPMENT_LINE table has been updated so the DATE_RECEIVED field is the current date, the QUANTITY_RECEIVED field is 25, and the INVENTORY QOH field has been incremented by 25 units. The original QOH for item 11669 in the INVENTORY table was 12. You just received 25 additional items, so the new QOH should be 37. Next, you will use SQL*Plus to double-check the values in the SHIPMENT_LINE and .INVENTORY tables.

To confirm that the database records have been updated:

7

1. Switch to SQL*Plus, and type the following command to confirm that the SHIP_QUANTITY and DATE_RECEIVED fields were updated correctly:

```
SELECT ship_quantity, date_received FROM shipment_line
WHERE shipment_id = 212
AND inv_id = 11669;
```

The output should show that the date received is the current date and the shipment quantity is 25.

2. Type the following command to confirm that the QOH for inventory ID 11669 was updated correctly:

```
SELECT qoh
FROM inventory
WHERE inv_id = 11669;
```

The output should show an updated QOH value of 37.

Clearing Form Fields Using a Program Unit

Before you closed the Forms Runtime window, the values for shipment ID 212 and inventory ID 11669 still appeared in the form text items. If the user had clicked the Update button again, the SHIPMENT_LINE and INVENTORY records would be updated again. This second update would increase the inventory QOH for the item by another 25 units, which would be incorrect. Therefore, after a record is processed, the form text items need to be cleared. There are two ways to clear the form text items: use the **CLEAR_FORM** built-in procedure, which clears all of the form text items; or create a program unit to set the value of the text items to a blank text string. The second approach can be used to clear selected text items, while retaining the current values in others. Since you would like to retain the current date value in the Date Received text item, you will create a program unit to clear the values in selected text items.

It is a good coding practice to divide programs into smaller program units that are easier to understand and debug. Recall that a program unit is a self-contained program that can be used within a larger program. A program unit can be a procedure, which is a series of statements that change variable values. A program unit can also be a function that returns a single value. You can pass variables to and from program units. Program units are useful for creating code modules that will be executed by multiple triggers. For example, in the RECEIVING_FORM, the Cancel button will clear the form text items also, so it can use the same program unit that you will build next.

Now you will create a program unit in Form Builder named CLEAR_RECEIVING_FORM. This program unit will clear the current values in all of the form text items except the Date Received text items by assigning a blank text string as the text item value.

To create the program unit:

1. In the Object Navigator, select the **Program Units** node, and then click the **Create button** 🔓 to create a new program unit. The New Program Unit dialog box opens.

2. In the Name text box, type **CLEAR_RECEIVING_FORM**, make sure the Procedure option button is selected, and then click **OK**. The PL/SQL Editor opens, with the heading template for the program unit displayed.

3. Type the code in Figure 7-6 to define the program unit, then compile the program unit and close the PL/SQL Editor.

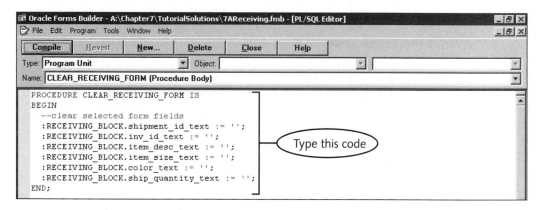

Figure 7-6 Program unit code to clear form text items

To call a program unit, you list the name of the program unit, along with the parameter values, enclosed in parentheses, in the code of the calling program. Now you will modify the Update button trigger to call the program unit.

To modify the trigger to call the program unit:

1. In the Layout Editor, right-click the **Update** button, and select **PL/SQL Editor**. The PL/SQL Editor opens, with the WHEN-BUTTON-PRESSED trigger code displayed.

When an object only has one associated trigger, right-clicking the object and selecting PL/SQL Editor opens the trigger. If an object has multiple triggers, this operation displays a list of the object's triggers.

2. Modify the trigger by adding the following command as the very last line of the trigger code. Compile the trigger, and then close the PL/SQL Editor and save the form.

    `CLEAR_RECEIVING_FORM;`

Now you will test the program unit. You will run the form, retrieve the record for Shipment ID 213 and Inventory ID 11777, and confirm that all 200 units were received today. Then, you will check to make sure that the form text items are cleared when you click the Update button.

To test the program unit:

1. Run the form, open the Shipment ID LOV display, select Shipment ID **213** and Inventory ID **11777**, and then click **OK**.

2. Click **Update**. The confirmation message appears, and all of the form text items except Date Received are cleared.

3. Close Forms Runtime.

Creating the Triggers for the Cancel and Exit Buttons

Now you will create and test the triggers for the Cancel and Exit buttons. The Cancel button will clear the form text items using the CLEAR_RECEIVING_FORM program unit. The Exit button exits the form using the EXIT_FORM procedure, which is a built-in procedure that exits Forms Runtime.

To create and test the Cancel and Exit button triggers:

1. Create and compile a WHEN-BUTTON-PRESSED trigger for the Cancel button with the following code: **CLEAR_RECEIVING_FORM;**.

2. Create and compile a WHEN-BUTTON-PRESSED trigger for the Exit button with the following code: **EXIT_FORM;**.

3. Save and run the form.

4. Open the LOV display, select the record for Shipment ID **213** and Inventory ID **11778**, and click **OK**.

5. Click **Cancel**. The form text items are cleared.

6. Click **Exit**. The Forms Runtime window closes.

To complete the form so it looks like the design sketch in Figure 7-1, you will draw a frame around the form and modify its properties. You will change the frame title to "Merchandise Receiving" and change the frame bevel to Inset.

To draw the frame:

1. Select the **Frame tool** on the tool palette, and draw a frame around the form text items, as shown in Figure 7-7.

> **?**
> **Help**
>
> If your frame background color is different than the form color, select the frame, select the Fill Color tool 🖋 on the tool palette, and click No Fill.

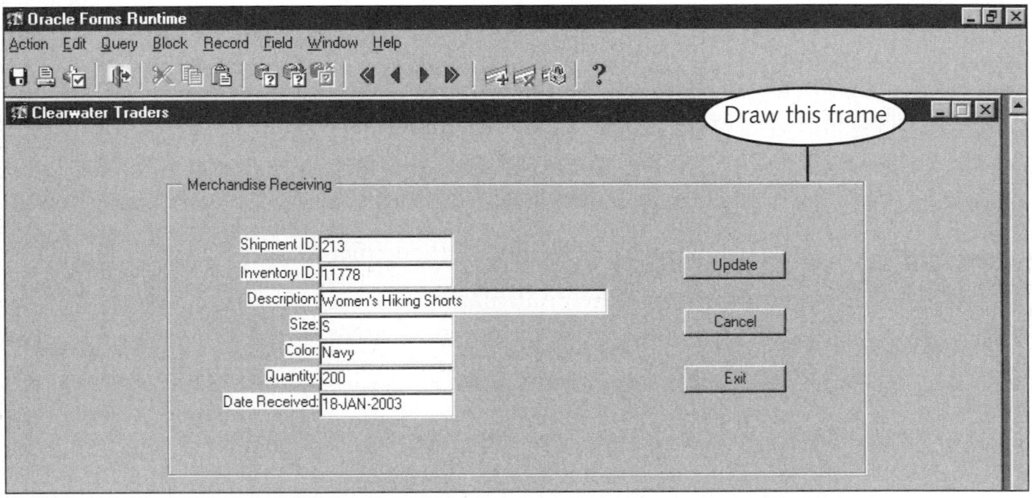

Figure 7-7   Completed receiving form

2. Open the frame Property Palette, scroll down to the Physical property node, and change the Bevel to **Inset**.

3. In the Property Palette, scroll down to the Frame Title property node, and change the Frame Title property to **Merchandise Receiving**, and then close the Property Palette.

4. Click **Edit** on the menu bar, click **Select All**, and then move the form objects so they are centered on the canvas.

5. Save and run the form. The completed form should look like Figure 7-7.

6. Click **Exit** to close the form, and then close the form in Form Builder.

USING THE FORMS DEBUGGER TO FIND RUNTIME ERRORS

In the previous chapter, you learned how to find and correct some of the more common compile errors that are seen in Form Builder triggers. As you start writing more complex triggers and program units to process custom forms, you might need to use the

Forms Debugger to find runtime errors. Recall that the strategy for finding runtime errors is to identify the program line that is causing the error and examine the variable values used within the command to determine the cause of the error. The Forms Debugger, like the debugger in Procedure Builder, allows you to step through triggers and other PL/SQL programs one line at a time to examine variable values during program execution. To use the Forms Debugger, you run the form by clicking the **Run Form Debug button** , which is on the toolbar in both the Object Navigator and Layout Editor. Then, you specify breakpoints that temporarily halt execution so you can step through program lines to examine program flow and variable values.

Now you will open a form named Receiving_ERR.fmb that is stored in the Chapter7 folder on your Data Disk. The code in this form's Update button trigger generates a runtime error. Before working with the Receiving_ERR.fmb file, you will refresh your Clearwater Traders database tables by running the clearwater.sql script in SQL*Plus. Then, you will open the form, save the form using a different name, run it the usual way, and observe the error message.

To run the script and then open, save, and run the form:

1. In SQL*Plus, run the **clearwater.sql** script.

2. In Form Builder, open **Receiving_ERR.fmb** from the Chapter7 folder on your Data Disk, and save the file as **7AReceiving_CORR.fmb** in your Chapter7\TutorialSolutions folder.

3. Run the form. The receiving form appears. Maximize the Forms Runtime window.

 A Compile dialog box showing the percentage that the compilation process is complete might appear. This dialog box always appears the first time you compile a new form that has multiple triggers.

4. Make sure that the insertion point is in the Shipment ID text item, and then press **F9** to open the LOV display. Select the record associated with Shipment ID **212** and Inventory ID **11669**, and then click **OK**.

5. Accept the Quantity Received value and the current system date for the Date Received value, and then click **Update**. The error message "FRM-40735: WHEN-BUTTON-PRESSED trigger raised unhandled exception ORA-01722" appears in the status line.

6. Click **Help** on the menu bar, and then click **Display Error**. The message "FRM-42100: No errors encountered recently" appears in the status line. As you can see, the Display Error facility does not provide error information about errors encountered in custom form triggers.

7. Click **Exit** to close the form.

The first step toward finding and fixing a runtime error like this is to investigate the nature of the error by looking up the error code explanation using other sources. Recall that error codes prefaced with FRM- are Form Builder error codes, and error codes prefaced with ORA- are generated by the database. First, you will look up the description of the FRM- error code.

To find the description of the FRM- error message:

1. In Form Builder, click **Help** on the menu bar, and then click **Form Builder Help Topics**. Click the **Index** tab, type **FRM-40735** in the search text box, click **Display**, make sure that **FRM-40735** is selected in the Topics Found list, and then click **Display** again. The Reference window appears, which describes the meaning of the error and possible actions you can take to correct the error.

The FRM-40735 error message indicates that the generated error needed to be handled in the EXCEPTION section of the trigger. However, you still need to learn what caused the error so you can write the code to handle it. Next, you will look up the description of the error code that appeared on the form. To do this, you will look in the database error code Help system that you used in the earlier chapters on SQL*Plus and PL/SQL.

To look up the ORA- error description:

1. In Form Builder, close the Reference window.

2. Start the Oracle8i Error Message Help application.

 Recall that you start the Oracle8i Error Message Help application by running a file named ora.hlp, which you can find at the Course Technology Web site at www.course.com. You can also look up Oracle error message codes by logging onto the Oracle Technology Network at otn.oracle.com.

3. Click **Index**, type **ORA-01722** in the search text box, and then click **Display** to view the error explanation. The ORA-01722 message explanation indicates that the code in the trigger attempted to convert a character string to a number but failed because the character string was not a valid numeric literal. This suggests that the trigger attempted to store a character string in a numeric data field, or it tried to use a character field in an arithmetic calculation.

4. Close the Oracle Messages and Codes window.

Now you have an idea that you are looking for an error that involves a character data type that is used incorrectly, but you don't know which command in the Update button trigger is causing the error. Next, you will use the Forms Debugger to locate the line on which the error is occurring. First, you will start the Forms Debugger and examine the environment.

To start the Forms Debugger:

1. Click the **Run Form Debug** button on the Layout Editor toolbar.

> You can also start the Debugger by clicking on the Object Navigator toolbar or by clicking Program on the menu bar, pointing to Run Form, and then clicking Debug.

2. Maximize the Forms Runtime window and the Forms Debugger window so your display looks like Figure 7-8.

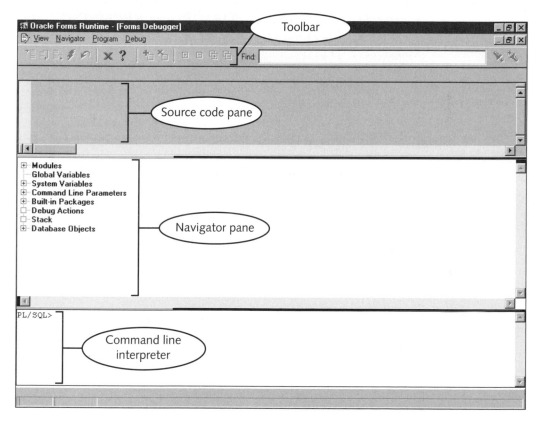

Figure 7-8   Forms Debugger window

The Forms Debugger window has a toolbar with buttons that perform debugger commands. None of the buttons are currently enabled, because no source code has been loaded for debugging. The window also has a source code pane to show the code you are debugging, a navigator pane similar to the Object Navigator that allows you to navigate among form and database objects, and a command-line interpreter for entering command-line debugger and SQL*Plus commands.

The navigator pane lists the system objects that you might want to access while you are debugging. An object node with ⊞ to the left of its name indicates that objects of this type might exist in the form you are currently running. To access the trigger code for the Update button, you will first open the Modules node, which lists the forms and form objects in your current application. Then you will navigate to the RECEIVING_BLOCK, since this is the block that contains the Update button. Then you will select the UPDATE_BUTTON item and display its WHEN-BUTTON-PRESSED trigger, since it contains the code that you want to debug.

To access the Update button code:

1. Click ⊞ beside the Modules node in the navigator pane, and then open the following objects: **RECEIVING_FORM**, **Blocks**, **RECEIVING_BLOCK**, **Items**, **UPDATE_BUTTON**, **Triggers**, **WHEN-BUTTON-PRESSED**. The source code for the UPDATE_BUTTON trigger appears in the Source code pane.

Next, you need to reconfigure the Debugger window. You will enlarge the source code and navigator panes so that you can see most of the trigger code in the source code pane. Since you will control the Debugger by clicking buttons rather than typing commands, you can close the command-line interpreter pane.

To reconfigure the Forms Debugger window:

1. Move the pointer over the line between the navigator and command-line interpreter panes so the pointer changes to the Precision pointer ╋. Click and drag down so the command-line interpreter pane is no longer visible.

2. Move the pointer over the line between the source code and navigator panes so the pointer changes to ╋ again. Click and drag down so each pane occupies about one-half of the screen display.

Setting Breakpoints

The first step in using the Forms Debugger is to set breakpoints to halt program execution. Now you will set a breakpoint on the first executable line of the UPDATE_BUTTON trigger so you can step through the trigger code and identify the program line that is causing the error. As with the Procedure Builder debugger, you can only set breakpoints on executable SQL or PL/SQL statements—you cannot place a breakpoint on variable declaration statements, comment statements, or other nonexecutable commands.

To set a breakpoint:

1. Double-click anywhere on Program line 00003 in the source code pane. A breakpoint icon appears beside the program line number, and the program line number changes to B (01), as shown in Figure 7-9, to indicate that the first breakpoint occurs on this line.

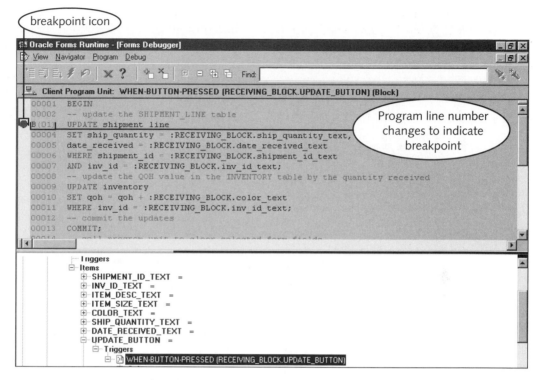

Figure 7-9   Creating a breakpoint

2. Click the **Close button** ☒ on the Debugger toolbar to close the Debugger window. The Debugger window closes, and the form appears in the Forms Runtime window.

Monitoring Program Execution and Variable Values

Now you will select a shipment line record and click the Update button. As the trigger runs, you will step through the source code, one line at a time, and examine the values of the form variables.

To step through the source code and monitor program execution:

1. Make sure that the insertion point is in the Shipment ID text item, press **F9** to open the LOV display, select Shipment ID **212** and Inventory ID **11669**, and then click **OK**. The shipment information appears in the form text items.

If Shipment 212 does not appear in your list of shipments, run the clearwater.sql script in SQL*Plus to refresh your database tables.

2. Accept the current Quantity Received and Date Received values, and then click **Update**. The Debugger window opens, as shown in Figure 7-10.

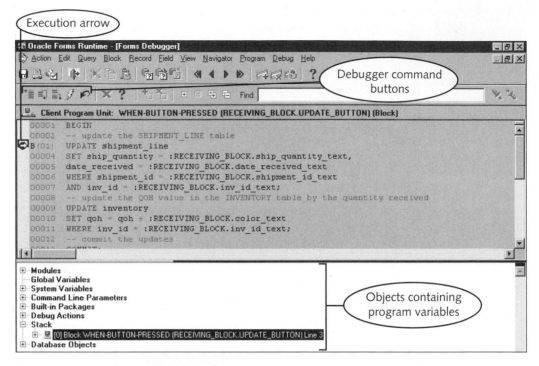

Figure 7-10   Debugger window during program execution

Notice that the Debugger command buttons that are now enabled on the toolbar are the same as the ones used in the Procedure Builder debugger. The **Step Into button** allows you to step through the program one line at a time. The **Step Over button** allows you to bypass a call to a program unit. The **Step Out button** executes all program lines to the end of the current trigger. The **Go button** allows you to pass over the current breakpoint and run the program until the next breakpoint occurs or the program ends. The **Reset button** terminates execution and returns the form to the state it was in before you clicked the Update button and started the current trigger.

Also notice the position of the execution arrow on Line 3. The execution arrow shows the program line that the Forms Debugger will execute next. As you step through the program, the execution arrow stops on SQL and PL/SQL statements, and skips comment lines. Finally, notice the objects that appear in the Navigator pane. You can open these objects during program execution to examine the current values of program variables. The three most important object types are stack, modules, and global variables. The **stack** object contains stack variables, which are variables that are initialized in the DECLARE statements of individual triggers and procedures. Stack variables sometimes are called local variables, because they can only be referenced within the procedure or trigger in which they are declared. **Global variables** are variables that are visible to all forms that are running. (You will learn how to use global variables in forms in

Chapter 11.) No stack or global variables are used in the Update button trigger, so there will be no values listed in these nodes.

Another important variable type is **modules**, which allows you to select a specific block and then examine the block variables, which are the current values of block items in the form. Since you have already selected a shipment, many of the form block items have values. You will now view these values.

To view the form module block variables:

1. In the Navigator pane, click ➕ beside the Modules node. Open the following objects: **RECEIVING_FORM**, **Blocks**, **RECEIVING_BLOCK**, **Items**. The current form block text item values appear.

Now you will step through the trigger code and watch how the form variable values change. You will also identify the program line that is causing the ORA-01722 error and determine the cause of the error by examining the values of the form variables.

To step through the module block item values:

1. Click the **Step Into button** ⬚ on the Debugger toolbar. The program executes the first UPDATE command (for the SHIPMENT_LINE table), and the execution arrow stops on the second UPDATE command (for the INVENTORY table). Note that the Navigator pane displays the values for the block text items that are used in the UPDATE command.

If necessary, maximize the Debugger window every time you click ⬚.

2. Click ⬚ again. The form appears in the Forms Runtime window again, along with the error message. This indicates that the error occurred on the program line that was just executed, which was the line containing the second UPDATE command.

When you are stepping through a program using the Forms Debugger, you can identify the program line that contains the runtime error because when the program line with the error executes, execution immediately halts, and an error message appears. After you identify the program line that is causing the error, you can run the program using the Debugger again and set a breakpoint on the program line causing the error. Before you execute the program line that causes the error a second time, you should examine the variable values used in the program line carefully to identify the cause of the error.

Now you will exit the form, run it again using the Debugger, and set the breakpoint on the second UPDATE command. Then, just before you execute this line, you will carefully examine the variable values used in the command.

To run the program again and examine the variable values:

1. Click **Exit** to exit the form.

2. In Form Builder, click the **Run Form Debug button** again. Maximize the Forms Runtime and Forms Debugger windows, and resize the panes within the Forms Debugger window.

3. Open the module objects, and load the WHEN-BUTTON-PRESSED trigger for the UPDATE_BUTTON in the Debugger Source code pane.

4. Create a breakpoint on program line 00009 (the second UPDATE command).

5. Click the **Close button** on the Debugger toolbar to close the Forms Debugger window and display the form.

6. Open the LOV display, and select the first record. Accept the current form values, and then click **Update** to execute the trigger to the breakpoint.

7. In the Navigator pane, click ➕ beside the Modules node. Open the following objects: **RECEIVING_FORM, Blocks, RECEIVING_BLOCK, Items**. The current form block text item values appear.

At this point, you can reconstruct the UPDATE command that is submitted to the database by substituting all of the text item values that are used in the command. Figure 7-11 highlights the variable values used in the UPDATE command causing the error.

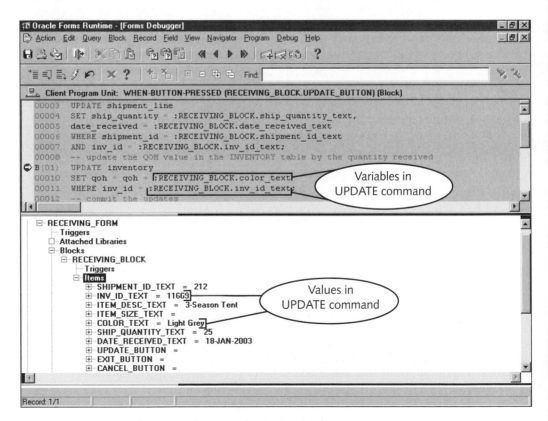

Figure 7-11   Variables and values used in command causing error

If you reconstruct the command by substituting the values for the variables, you note that the command is as follows:

```
UPDATE inventory
SET qoh = qoh + 'Light Grey'
WHERE inv_id = 11669;
```

Obviously, trying to add a text string ('Light Grey') to the QOH value is causing an error. This program line should add the SHIP_QUANTITY_TEXT value to the existing inventory QOH, not the COLOR_TEXT value. You cannot correct the error directly in the Debugger—you must exit the form and then correct the code in Form Builder. Now you will correct the error and confirm that the form runs correctly.

To correct the error:

1. Click the **Step Into button** ⁎≣ to execute the program line and display the error message, and then click **Exit** to exit the form.

2. In Form Builder, open the trigger for the Update button, and change the program line `SET qoh = qoh + :RECEIVING_BLOCK.color_text` to `SET qoh = qoh + :RECEIVING_BLOCK.ship_quantity_text` in the second UPDATE command.

3. Compile the trigger, correct any syntax errors, and then close the PL/SQL Editor.

4. Save the form, run the form, open the LOV display, select the first record in the LOV display, and click **OK**.

5. Accept the retrieved values, and click **Update**. The "No changes to save" confirmation message confirms that the trigger worked correctly. If an error message appears instead of the confirmation message, repeat the debugging steps to locate the error and determine its cause. Fix your code, and then test it again.

6. Click **Exit** to exit the form.

In summary, use the following process for debugging runtime errors in form triggers:

1. Look up explanations for FRM- and ORA-error messages to determine the nature of the error.

2. Run the form in the Forms Debugger, set a breakpoint on the first line of the trigger that generates the error, and step through the trigger to identify the program line causing the error. Execution immediately halts when the line causing the error executes.

3. Run the form in the Forms Debugger again, and set a breakpoint on the line causing the error. Before the line executes, examine the variable values used in the command to determine the cause of the error.

FORM TRIGGERS

Form triggers are a very important part of custom forms. You have created triggers in both data block and custom forms. This section explores trigger categories, timing, scope, and execution hierarchy, and describes form navigational triggers in detail.

Recall that a trigger is a block of PL/SQL code attached to a form data block, to a block item, or to a form itself. The trigger is activated, or fires, in response to an event, such as clicking a button. The trigger name defines the event that activates it. For example, a button's WHEN-BUTTON-PRESSED trigger activates the trigger's code.

Trigger Categories

Table 7-2 summarizes the different categories of form triggers.

Trigger Category	Fires in response to	Example
Block processing	Events in form data blocks, such as inserting new records or deleting existing records	ON-DELETE
Interface event	User actions, such as clicking a button or a mouse	WHEN-BUTTON-PRESSED
Master-detail processing	Master-detail relationship processing, such as updating a detail block when a new master record is selected	ON-POPULATE-DETAILS
Message handling	Events that display messages, such as inserting a record or reporting an error	ON-ERROR
Navigational	Actions that change the form focus, such as moving to a new block or new form record, or internal form actions, such as inserting a new blank record	WHEN-NEW-RECORD-INSTANCE
Query time	Data block form query processing involving retrieving records	POST-QUERY
Transactional	Database transaction events, such as inserting, updating, or deleting records	PRE-INSERT
Validation	Database transaction events that require form data validation, such as checking to make sure required primary key values have been entered	WHEN-VALIDATE-ITEM

Table 7-2   Form trigger categories

Most of the triggers you will create in this book are interface triggers. These triggers generally have the following format for the trigger name: WHEN-*object-action*. Interface triggers can fire in response to keyboard actions. For example, the KEY-F1 trigger fires when the user presses F1. There is a keyboard trigger named KEY-OTHERS that fires when the user presses any key other than ones for which specific keyboard triggers have been defined. The KEY-OTHERS trigger is used to disable function keys that the developer does not want to define in an application. Interface triggers can also involve mouse events, such as WHEN-MOUSE-CLICK.

Trigger Timing

Trigger timing is related to when a trigger fires—just before, during, or after its triggering event. PRE- triggers, which are triggers with PRE- as the first three characters of their name, fire just before an event successfully completes. For example, a PRE-FORM trigger fires just before a form appears in the Forms Runtime window, and a PRE-BLOCK trigger fires just before the user successfully navigates to a new form block. POST- triggers fire just after an event successfully completes. For example, a POST-QUERY trigger fires just after a query in a data block form retrieves records. In contrast, the ON-, WHEN-, and KEY- triggers fire in response to actions. For example, the ON-DELETE trigger fires in response to deleting a record in a data block form, and the WHEN-BUTTON-PRESSED trigger fires in response to a user clicking a button.

Trigger Scope

Trigger scope defines where an event must occur in order for the trigger to fire. The scope of a trigger includes the object to which the trigger is attached, as well as any objects within the trigger object. For example, when a trigger is attached to a block, the trigger's scope extends to the items within the block. If the triggering event occurs within any block item, then the block trigger will fire. For example, suppose you create a KEY-F1 trigger associated with a data block. The trigger will fire whenever the user presses the F1 key, as long as the form insertion point is in any item in the block. Similarly, if you create a KEY-F1 trigger associated with a form, the trigger will fire whenever the user presses the F1 key anywhere in the form.

Trigger Execution Hierarchy

Trigger execution hierarchy defines which trigger fires when an object within a form object contains the same trigger that the form object contains. For example, suppose a data block has a KEY-F1 trigger, and a text item within the data block also has a KEY-F1 trigger. By default, the trigger in the higher-level object, which is the block, overrides the trigger in the lower-level object, which is the text item. If you anticipate a conflict and want both triggers to execute, you can specify a custom execution hierarchy in the Property Palettes of both triggers and have one trigger execute immediately before or after the other.

Form Navigational Triggers

In a form, two types of navigation, or movement between form objects, occur: external navigation and internal navigation. **External navigation** occurs when the user causes the form focus to change by making a different form item active. A form item has the form focus when it is the item that is currently selected on the form. For example, a text item has the form focus when the insertion point is in the text item. A command button has the form focus when the user presses the Tab key and causes the button to be selected. External navigation occurs when, for example, the user clicks the insertion point in a different data block or clicks the Next block button .

> When a form item, such as a command button, has the form focus, it is high-lighted. If the user presses the Enter key when a button has the form focus, it is equivalent to clicking the button.

Internal navigation is performed by the internal form code that responds to external navigation operations or trigger commands. For example, when a user opens a data block form, a series of triggers fire that cause the form to open and a new blank record to appear.

Navigational triggers fire when internal navigation occurs. These triggers exist for different types of form items and fire at different times depending on internal and external navigation events. Table 7-3 summarizes the Form Builder navigational triggers.

Trigger Name	Fires	Example Usage	Scope
PRE-FORM	When form first appears in Forms Runtime	Initializing form values and parameters	Form
POST-FORM	Just before form closes	Displaying an exit message or returning memory allocated to form objects	Form
PRE-BLOCK	When a block is first entered	Allowing or disallowing access to the block	Block, Form
POST-BLOCK	When a block is exited	Validating current block values	Block, Form
PRE-TEXT-ITEM	When user navigates to a text item, but before he or she changes the text item value	Storing current text item value for future reference	Item, Block, Form
POST-TEXT-ITEM	When user leaves a text item	Calculating or changing text item values	Item, Block, Form
WHEN-NEW-FORM-INSTANCE	When a form is opened	Initializing form values and parameters	Form
WHEN-NEW-BLOCK-INSTANCE	When navigation moves to a new block	Setting values of block items	Block, Form
WHEN-NEW-RECORD-INSTANCE	When a new record is created	Inserting a sequence value into a new record	Block, Form
WHEN-NEW-ITEM-INSTANCE	When an item is entered	Calling message triggers	Item, Block, Form

Table 7-3   Form navigational triggers

When you run a form, a series of navigational triggers fire, starting with triggers associated with the highest-level object (the form module), and ending with triggers associated with the lowest-level objects (items). Figure 7-12 illustrates how the navigational triggers fire when a form starts. After the user starts the form, the PRE-FORM and PRE-BLOCK triggers fire, the form appears on the screen, but no data is visible. The next four triggers fire, and the form appears and becomes available for use.

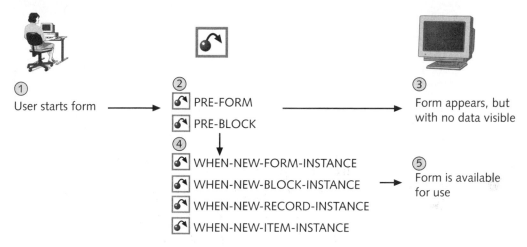

Figure 7-12   Navigational triggers that fire at form start-up

Figure 7-13 illustrates the triggers that fire when a user navigates among form items by clicking the form insertion point in a new text item or by clicking the Next Record button ▶. When a user places the insertion point in a text item, the WHEN-NEW-ITEM-INSTANCE trigger fires, and the insertion point appears in the text item. When the user clicks ▶, the WHEN-NEW-RECORD-INSTANCE and WHEN-NEW-ITEM-INSTANCE triggers fire, and the next record appears.

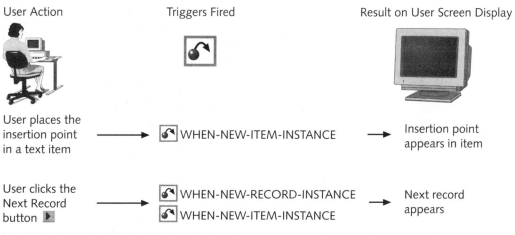

Figure 7-13   Navigational triggers that fire as a result of external navigation

Figure 7-14, illustrates that when a user closes a form, the POST-BLOCK and POST-FORM triggers fire, and the Forms Runtime window closes.

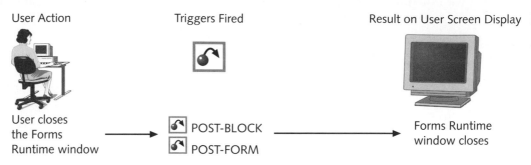

User Action   Triggers Fired   Result on User Screen Display

User closes
the Forms   →   POST-BLOCK ──────────────→   Forms Runtime
Runtime window   POST-FORM   window closes

Figure 7-14   Navigational triggers that fire as a result of closing a form

Now you will create a navigational trigger. For all of the forms you have created, the user has had to manually maximize the Forms Runtime window. You will create a PRE-FORM trigger associated with the form module that fires when the form first opens. The trigger will use the SET_WINDOW_PROPERTY command to automatically maximize the Forms Runtime window when the form first appears. To maximize a window, you use the following syntax: **SET_WINDOW_PROPERTY(*window_name*, WINDOW_STATE, MAXIMIZE);**. The Forms Runtime window is a multiple-document interface (MDI) window. Multiple forms can be active within the Forms Runtime MDI window at the same time. In Forms Runtime, it is referenced as FORMS_MDI_WINDOW.

> MDI is the Microsoft Windows window management system that allows programs to display an outer "parent" window (called the application window) and multiple inner windows.

To create the PRE-FORM trigger for maximizing the Forms Runtime MDI window:

1. Open the Object Navigator window in Ownership View, select the **Triggers node** that is directly under the RECEIVING_FORM module, right-click, point to **Smart Triggers**, and then click **PRE-FORM**. The PL/SQL Editor window opens.

2. Type the following code in the PL/SQL Editor to maximize the Forms Runtime window at runtime.

   ```
   SET_WINDOW_PROPERTY(FORMS_MDI_WINDOW, WINDOW_STATE,
   MAXIMIZE);
   ```

3. Compile the code, correct any syntax errors, close the PL/SQL Editor, and save the form.

4. Run the form to confirm that the Forms Runtime window maximizes automatically, and then close Forms Runtime.

5. Close the form in Form Builder.

DIRECTING EXTERNAL NAVIGATION WITHIN FORMS

In most Windows applications, the user can press the Tab key to navigate between text fields and other form items. Usually the navigation order is top to bottom, left to right. To set the tab order of items in a custom form, you need to place the items in the correct order under the Items node in the Object Navigator window. The text item in which the insertion point will appear at form start-up should be listed first, the item where the insertion point will go when the user subsequently presses the Tab key should be listed next, and so on. Now, you will check the navigational order of the items in a form named Receiving_NAV.fmb. You will open the form, save it using a different filename, and then run the form and navigate through the text items.

To open and run the form:

1. Open **Receiving_NAV.fmb** from the Chapter7 folder on your Data Disk, and save the file as **7AReceiving_NAV.fmb** in your Chapter7\TutorialSolutions folder.

2. Run the form. The form appears in the Forms Runtime window. Note the current position of the insertion point: It is in the Description text item.

3. Press **Tab**. The insertion point moves to the Size text item.

4. Press **Tab** again. The insertion point moves to the Date Received text item.

5. Continue to press **Tab** until the form focus is on the Exit button, and then press **Enter** to exit the form.

As you can see, the navigational order of the form items is not correct: As the user presses the Tab key, the insertion point should move down the column of text items on the left side of the form. When the user has moved through all of the text items, the form focus should move to the first command button (Update) and then move down the list of command buttons. Now, you will adjust the navigational order of the block items by moving the items into the correct order in the Object Navigator.

To adjust the tab order of the form items:

1. In the Object Navigator, click ➕ beside the Data Blocks node, click ➕ beside RECEIVING_BLOCK, and click ➕ beside the Items node to display the block items.

2. Select **SHIPMENT_ID_TEXT**, drag it so the mouse pointer is on the Items node, and drop it so it is the first item listed under the Items node in the RECEIVING_BLOCK.

3. Select the **EXIT_BUTTON**, and drag and drop it so it is the last item listed under Items in the RECEIVING_BLOCK.

4. Continue to move the form items until they are in the order shown in Figure 7-15, and then save the form.

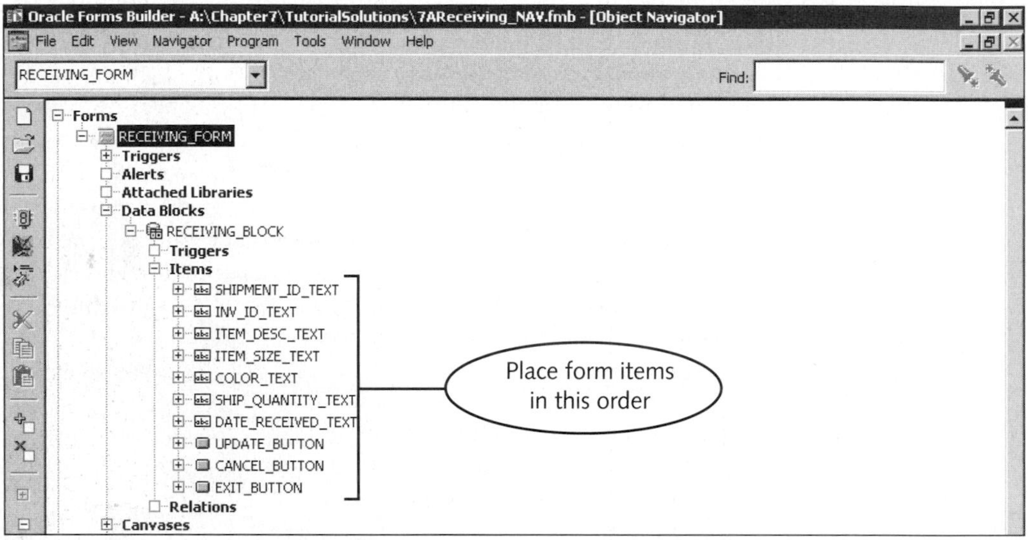

Figure 7-15   Specifying the external navigation order of form items

5. Run the form to check the navigation order. The order in which the items receive the form focus should be the same as the order in which the items are listed in Figure 7-15.

6. Close Forms Runtime.

Form Builder provides several built-in subprograms, which are usually called built-ins, that you can use to direct external form navigation. These built-ins can be used to programmatically place the insertion point or form focus on a specific item, move to a different block or record, and so forth. The navigational built-in subprograms are summarized in Table 7-4.

Note that in the GO_ITEM, GO_BLOCK, and GO_FORM built-ins, the target item is placed within single quotation marks. This target item is not verified when the trigger using the built-in is compiled, but is only verified at runtime. If the item does not exist or is specified incorrectly, a runtime error will occur. Navigational built-in subprograms cannot be used in navigational triggers, because navigational triggers fire in response to navigation events, and the navigational built-ins cause navigation events to occur. Using a navigational built-in in a navigational trigger will not cause a compile error, but an error will occur when the form runs and the navigational trigger fires.

Built-in Name	Description	Usage
GO_ITEM	Moves focus to a specific form item	`GO_ITEM('receiving_block.` `shipment_id_text');`
GO_RECORD	In a data block, moves focus to a specific record number	`GO_RECORD(1);`
GO_BLOCK	Moves focus to a specific block	`GO_BLOCK('receiving_block');`
GO_FORM	In a multiple-form application, moves focus to a specific form	`GO_FORM('receiving_form');`
FIRST_RECORD	In a data block form, moves focus to the first record in the current block	`FIRST_RECORD;`
LAST_RECORD	In a data block form, moves focus to the last record in the current block	`LAST_RECORD;`
NEXT_RECORD	In a data block form, moves focus to the next record	`NEXT_RECORD;`
PREVIOUS_RECORD	In a data block form, moves focus to the first item in the previous record	`PREVIOUS_RECORD;`
UP	In a data block form, moves focus to the current item in the previous record	`UP;`
DOWN	In a data block form, moves focus to the current item in the next record	`DOWN;`

Table 7-4 Navigational built-in subprograms

Now you will use the GO_ITEM built-in to automatically place the form insertion point in the SHIPMENT_ID_TEXT item after the user clicks the Cancel button.

To use the GO_ITEM built-in in the Cancel button trigger:

1. In the Object Navigator, click ✚ beside the Triggers node, click ✚ beside CANCEL_BUTTON, and double-click the **Trigger icon** 📝 to open the Cancel button trigger code. Add the following command as the last line of the trigger:

   ```
   GO_ITEM('receiving_block.shipment_id_text');
   ```

2. Compile the trigger, close the PL/SQL Editor, and save the form.

3. Run the form, open the LOV display, and select the first record. The values appear on the form.

4. Place the form insertion point in the Inventory ID text item.

5. Click **Cancel**. Note that the insertion point now appears in the Shipment ID text item, as a result of the GO_ITEM navigational built-in.

6. Click **Exit** to close the form, close the form in Form Builder, and close all open Oracle applications.

SUMMARY

- ❑ A custom form is processed using triggers that contain SQL commands that retrieve, insert, update, and delete data values.

- ❑ To create a custom form, you have to create the form manually, "paint" the form items on the canvas, and write the code that controls the form functions.

- ❑ Custom forms contain control blocks that are not connected to a particular database table.

- ❑ A text item that displays a database field must have the same data type as the corresponding data field, and its maximum length must be large enough to accommodate the maximum length of the data, plus any characters included in a format mask.

- ❑ You can use system variables to automatically retrieve and display the current system date in a form text item.

- ❑ To clear the text items in a form, you can use the CLEAR_FORM built-in procedure, which clears all of the form text items; or you can clear selected text items, while retaining the current values in others, by setting the value of the text items to a blank text string.

- ❑ Complex triggers should be divided into smaller program units that are easier to understand and debug.

- ❑ The Forms Debugger allows you to step through PL/SQL programs one line at a time to examine values of variables during execution.

- ❑ Breakpoints are places in PL/SQL programs where you can pause program execution, step through the code, and examine current variable values.

- ❑ The stack node in the Forms Debugger allows you to examine the values of local variables. The modules node allows you to examine the values of form text items.

- ❑ To debug errors in form triggers, first look up explanations for FRM- and ORA-messages. Then, run the form in the Forms Debugger and step through the code to determine the program line that is causing the error. Finally, run the form again in the Debugger, and just before the line containing the error executes, carefully examine the variable values that are used in the program line causing the error.

- ❑ Triggers can be categorized as supporting block processing, interface events, master-detail processing, message handling, navigational events, query processing, transaction processing, and validation.

❑ Trigger scope defines where an event must occur in order for a trigger to fire. The scope of a trigger includes the object to which the trigger is attached, as well as any objects within the trigger object.

❑ External navigation occurs when the user causes the form focus to change. Internal navigation is performed by internal form code within the form in response to external navigation operations.

❑ Navigational triggers fire when internal navigation occurs. Navigational triggers are associated with forms, data blocks, and form items, and they fire at different times, depending on internal and external navigation events.

❑ MDI (multiple-document interface) is the Microsoft Windows window management system in which applications display an outer "parent" window, called the application window, and multiple inner windows. You can automatically maximize the MDI runtime window using a PRE-FORM navigational trigger.

❑ To set the external navigation order of items in a form, you need to place the items in the correct order under the Items node in the Object Navigator window.

❑ Form Builder provides several built-in subprograms, or built-ins, that you can use to direct external form navigation. Two of the most commonly-used navigational built-ins are GO_ITEM and GO_BLOCK.

7

REVIEW QUESTIONS

1. What is the difference between a custom form and a data block form?

2. When do you need to create a custom form?

3. What is a control block?

4. When should you use the CLEAR_FORM built-in procedure, and when should you write a custom program unit to clear specific form text items?

5. What is a breakpoint? On what kinds of program lines can you set breakpoints? On what kinds of lines can you not set breakpoints?

6. Describe the recommended strategy for using the Forms Debugger to debug form triggers.

7. What is the difference between external navigation and internal navigation?

8. List the navigational triggers that fire in the order that they fire when a user opens a form.

9. List the navigational triggers that fire when a user closes a form.

10. What is the MDI window? How can you cause it to automatically maximize at runtime?

PROBLEM-SOLVING CASES

The following cases use the Clearwater Traders (see Figure 1-11), Northwoods University (see Figure 1-12), and Software Experts (see Figure 1-13) sample databases. Rename all form components using descriptive names, and format forms using the guidelines described in Chapter 6. For all forms, create an Exit command button to exit Forms Runtime, and create a PRE-FORM trigger to automatically maximize the Forms Runtime window at start-up. Make all form items have a top-down, left-to-right external navigation order. Save all form files in your Chapter7\CaseSolutions folder. Save all SQL commands in a file named 7AQueries.sql.

1. Create a custom form to insert new records into the SKILL table in the Software Experts database that has a user interface like the one shown in Figure 7-16. When the user clicks the Create button, a new skill ID appears. The user then enters the skill description. When the user clicks the Save button, the record is inserted into the SKILL table. Save the file as 7ACase1.fmb.

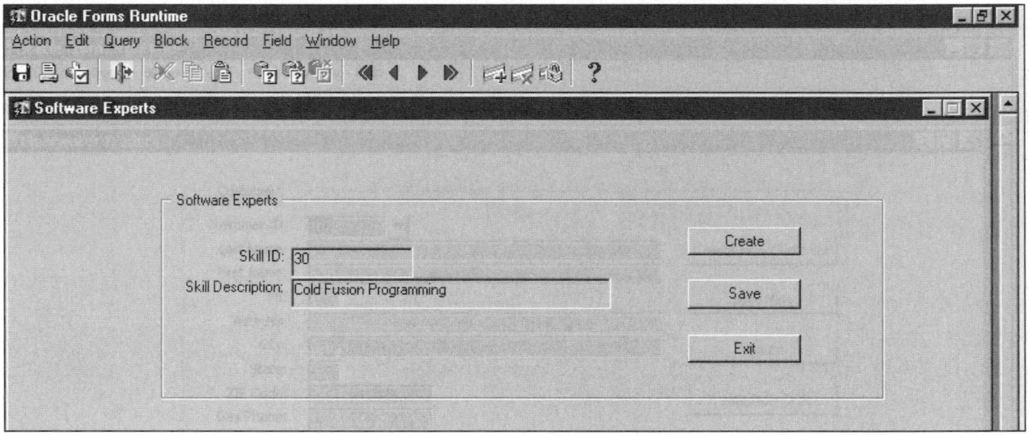

Figure 7-16   Navigational built-in subprograms

a. In SQL*Plus, create a new sequence named SKILL_ID_SEQUENCE that starts with 10 and has no maximum value, and use the sequence to automatically generate the skill ID values. (If you already created this sequence in a case in a previous chapter, use the existing sequence.) Then, create a trigger for the Create button that automatically displays the next sequence value in the Skill ID text item. (*Hint*: Use the command SELECT *sequence_name* INTO *:block_name.text_item* FROM DUAL; in the trigger.)

b. Create a trigger for the Save button that inserts the current values that appear in the Skill ID and Skill Description text items into the SKILL database table, commits the inserted record to the database, and then clears the form text items.

2. Create a custom form to insert new records into the TERM table in the Northwoods University database. Make command buttons with the labels Create, Save, and Exit, as shown in the form in Figure 7-17. When the user clicks the

Create button, a new term ID appears. The user then enters the term description and status. Represent the STATUS field using a check box. When the user clicks the Save button, the record is saved in the TERM table. Save the file as 7ACase2.fmb.

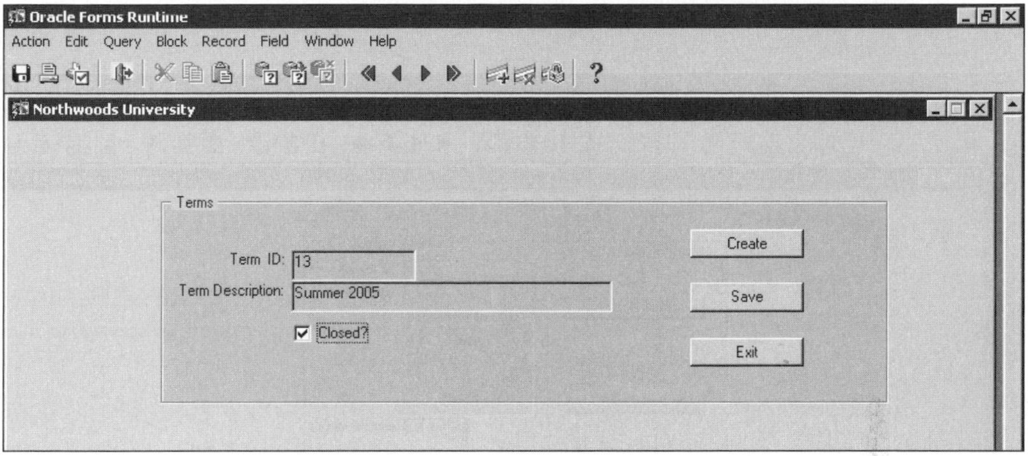

Figure 7-17

a. Draw the check box on the form canvas using the Check Box tool [x] on the Layout Editor tool palette, and then modify the check box properties as needed. Have the box appear unchecked when the form first appears and after the user saves a new record. (*Hint*: To clear the check box, set the check box item to the value that the check box has when it is cleared.)

b. In SQL*Plus, create a new sequence named TERM_ID_SEQUENCE that starts with 10 and has no maximum value, and use the sequence to automatically generate the term ID values. Then, create a trigger for the Create button that automatically displays the next sequence value in the Term ID text item. (*Hint*: Use the command SELECT *sequence_name* INTO :*block_name.text_item* FROM DUAL; in the trigger.)

c. Create a trigger for the Save button that inserts the current values that appear in the form items into the TERM table, commits the inserted record to the database, and then restores the form items to their initial startup state. (*Hint*: Use the name of the check box item for the STATUS value.)

3. Create a custom form to insert new records into the CUST_ORDER table in the Clearwater Traders database. Make command buttons with labels Create, Save, and Exit, as shown in Figure 7-18. When the user clicks the Create button, a new order ID appears. The user then enters the values for the CUST_ORDER fields. Represent the payment method field using a Combo box. When the user clicks the Save button, the record is saved in the CUST_ORDER table. Save the file as 7ACase3.fmb.

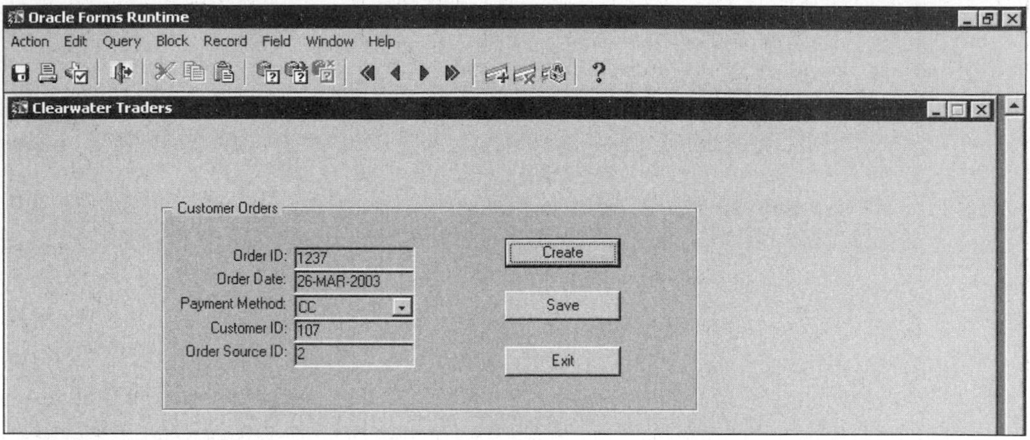

Figure 7-18

a. Make the order date field automatically display the current date when the form first opens.

b. Draw the Combo box on the form canvas using the list tool ▦ on the Layout Editor tool palette, and then change the properties as needed.

c. In SQL*Plus, reate a new sequence named ORDER_ID_SEQUENCE that starts with 2000 and has no maximum value, and use the sequence to automatically generate the order ID values. (If you already created this sequence in a case in a previous chapter, use the existing sequence.) Then, create a trigger for the Create button that automatically displays the next sequence value in the Order ID text item. (*Hint*: Use the command SELECT *sequence_name* INTO :*block_name.text_item* FROM DUAL; in the trigger.)

d. Create a trigger for the Save button that inserts the current text item values into the CUST_ORDER table, and then commits the inserted record to the database. (*Hint*: Use the name of the Combo box list item for the METH_PMT value.)

e. Create an LOV associated with the customer ID text item that displays the customer ID and customer first and last name for the customer ID field. Allow the user to press F9 to display the LOV.

f. Create an LOV associated with the order source ID text item that displays the order source ID and source description for the order source ID field. Allow the user to press F9 to display the LOV.

g. In the Save button trigger, create a program unit that clears all form text items except the order date after the record is committed to the database.

4. Create a custom form to update records in the INVENTORY table in the Clearwater Traders database that has a user interface like the one shown in Figure 7-19. When the user clicks the LOV command button next to the

Inventory ID text item, a list of values for all inventory items appears. The user can select a specific item, and its associated data values appear on the form. The user can modify the values as necessary, and then click the Update button to save the changes to the database. Save the file as 7ACase4.fmb.

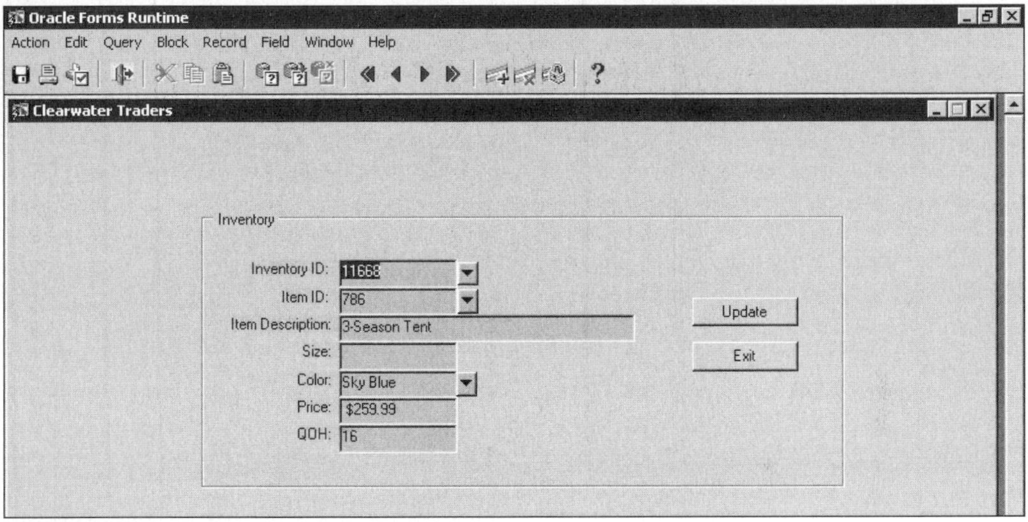

Figure 7-19

    a. Create an LOV associated with the Inventory ID text item that displays the inventory ID, item description, size, and color. The LOV should return the selected record's INV_ID, ITEM_ID, ITEM_DESC, ITEM_SIZE, COLOR, PRICE, and QOH to the form text items. (*Hint*: Your LOV query will have some fields in the SELECT statement that do not appear in the LOV display. To hide a column in the LOV display, set its display width equal to zero.)

    b. Create an LOV associated with the Item ID text item that displays the item ID and item description, and returns the selected values to the form text items.

    c. Create an LOV associated with the Color text item that displays the color values in the COLOR table and returns the selected value to the form text items.

5. Create a custom form to update records in the ENROLLMENT table in the Northwoods University database that has a user interface like the one shown in Figure 7-20. When the user clicks the LOV command button next to the Student ID text item, a list of values appears that shows student information. When the user clicks the LOV command button next to the Section ID text item, a list of values appears that shows information for all course sections in which the student is enrolled. The user can select a specific course section, and its associated data values appear on the form. The user can change the GRADE value and click the Update button, and the record is then updated in the database. Save the file as 7ACase5.fmb.

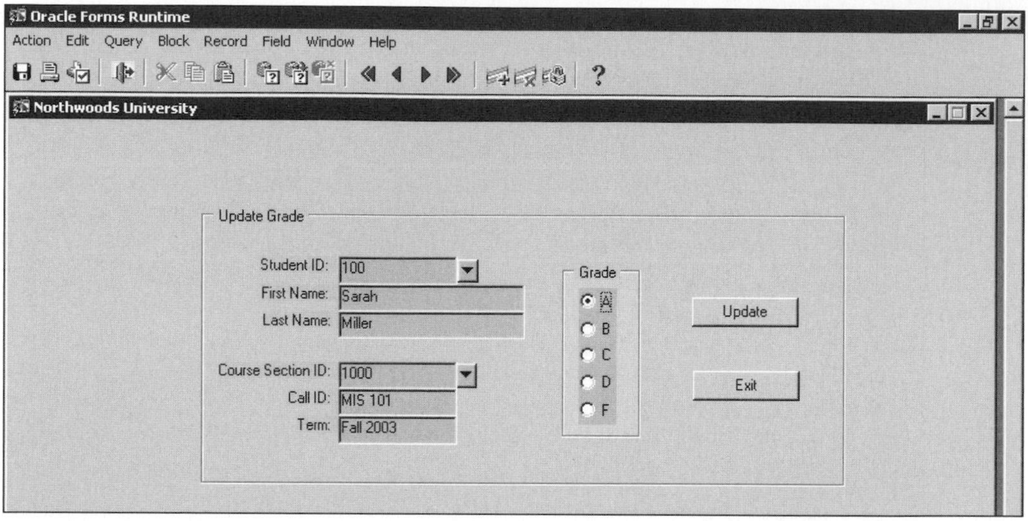

Figure 7-20

a. Draw the radio buttons corresponding to grade values using the Radio Button tool [⊙] on the Layout Editor tool palette, and then modify the properties of the buttons and the radio button group. Recall that the radio button group name references the value of the radio button that is currently selected.

b. Create an LOV associated with the Student ID text item that displays the student ID, last name, and first name. The LOV should return the selected record's S_ID, S_FIRST and S_LAST values to the form text items.

c. Create an LOV associated with the Course Section ID text item that displays the section ID, term description, and call ID for all courses in which the student has ever been enrolled. The LOV should return the selected C_SEC_ID, CALL_ID, TERM_DESC, and GRADE values to the form items. In the LOV query, specify the search condition so that S_ID is equal to the current value of the Student ID text item.

d. When the user clicks the Update button, update the ENROLLMENT table with the modified grade value for the selected student ID and course section ID, commit the change, and then clear the form text items. Use the radio button group name to represent the current GRADE value in the UPDATE command.

6. Create a custom form to insert, update, and delete records in the SHIPMENT table in the Clearwater Traders database that has a user interface like the one shown in Figure 7-21. When the user clicks the Create button, a new Shipment ID value appears. The user enters the expected date value and clicks the Save New button, and the new record is saved in the database. The user can click the LOV command button next to the Shipment ID text item to display a list of existing shipment records, which can then be modified, using the Update button, or deleted, using the Delete button. Save the file as 7ACase6.fmb.

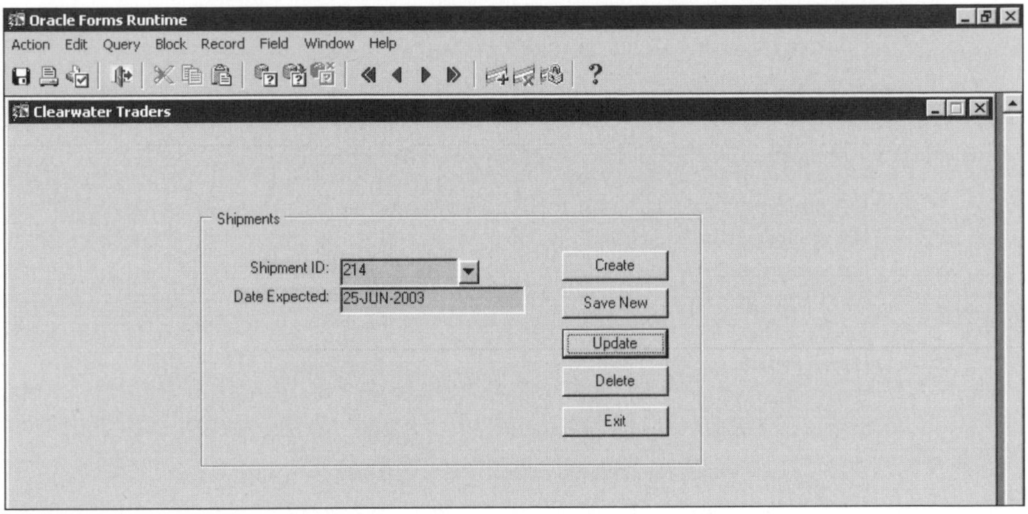

Figure 7-21

a. In SQL*Plus, create a new sequence named SHIPMENT_ID_SEQUENCE that starts with 300 and has no maximum value, and use the sequence to automatically generate the Shipment ID values. (If you already created this sequence in a case in a previous chapter, use the existing sequence.) Then, create a trigger for the Create button that clears the form text items, automatically displays the next sequence value in the Shipment ID text item, and automatically displays the system date in the Date Expected text item. (*Hint*: Use the command SELECT *sequence_name* INTO :*block_name.text_item* FROM DUAL; in the trigger.)

b. Create an LOV associated with the Shipment ID text item that displays the shipment ID and date expected for all shipments. The LOV should return the selected record's values to the form text items.

c. When the user clicks the Save New button, insert the form text item values into the database, commit the changes, and then clear the form text items.

d. When the user clicks the Update button, update the SHIPMENT table with the modified date expected value for the selected shipment ID, commit the change, and then clear the form text items.

e. When the user clicks the Delete button, delete the records in the SHIPMENT_LINE table where the current Shipment ID is referenced as a foreign key, delete the record in the SHIPMENT table, commit the change, and then clear the form text items.

f. Set the Initial Value property of the Date Expected text item so the current operating system date appears when the form first opens.

7. Create a custom form to insert and update records in the PROJECT_CONSULTANT table in the Software Experts database that has a user interface like the one shown in Figure 7–22. The user first clicks the LOV

command button next to the Project ID text item and selects a project. The user then clicks the LOV command button next to the Consultant ID text item, and the names of consultants who are not already assigned to the project appear. After selecting a consultant, the user enters the roll-on and roll-off dates and total hours, and clicks the Save New button, and the record is saved in the database. If the user clicks the Display Existing Consultants button, a LOV display shows the consultant IDs and first and last names of all consultants who are associated with the project ID that currently appears in the Project ID text item. The user can select a consultant and then modify the consultant's roll-on and roll-off dates and total hours, and click the Update button to save the changes. Save the file as 7ACase7.fmb.

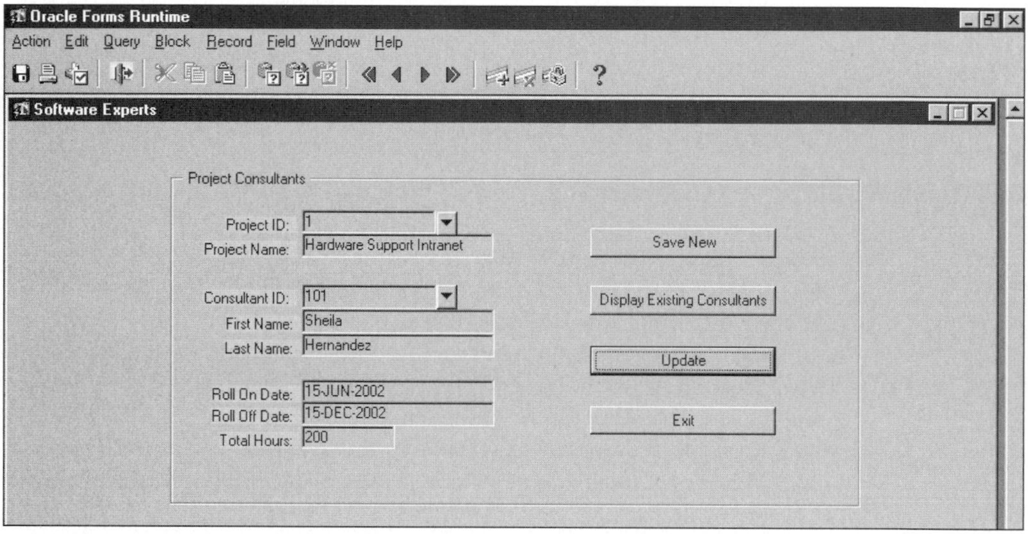

Figure 7-22

a. Create an LOV associated with the Project ID text item that displays the Project ID and project name for all projects in the database. The LOV should return the selected record's values to the corresponding form text items.

b. Create an LOV associated with the Consultant ID text item that displays the consultant ID and the first and last names of all consultants who are not currently working on the selected project. The LOV should return the selected values to the corresponding form text items. (*Hint*: Use a nested query that contains the NOT IN clause.)

c. Configure the Roll On Date and Roll Off Date text items so they display the current operating system date when the form first opens and after the form text items are cleared.

d. Create a trigger for the Save New button that inserts the form text item values into the PROJECT_CONSULTANT table, clears the form text items, and redisplays the system date in the date text items.

e. Create a trigger for the Display Project Consultants button that displays an LOV that retrieves the C_ID, C_LAST, C_FIRST, ROLL_ON_DATE, ROLL_OFF_DATE, and TOTAL_HOURS values for all consultants who are currently working on the project ID that appears in the Project ID text item. In the LOV display, show only the consultant ID, first name, and last name. (*Hint*: Your LOV query will have some fields in the SELECT statement that do not appear in the LOV display. To hide a column in the LOV display, set its display width equal to zero. Associate the LOV with any form text item that does not currently have an associated LOV, and write the commands in the button trigger to place the insertion point in this text item, and then open the LOV display.)

f. When the user clicks the Update button, update the roll-on date, roll-off date, and total hours fields in the PROJECT_CONSULTANT table using the current project ID and consultant ID values as search conditions.

8. **Debugging Problem**: A custom form saved in a file named 7ACase8_ERR.fmb in the Chapter7 folder on your Data Disk inserts and updates records in the FACULTY table in the Northwoods University database. When the user clicks the Create button, a form trigger retrieves a new faculty ID value from a sequence named F_ID_SEQUENCE and displays the sequence value in the Faculty ID text item. When the user clicks the Save New button, the new record is saved in the database. When the user clicks the Update button, the record for the current faculty ID is updated with the values that currently appear in the form items.

a. Save the form as 7ACase8_CORR.fmb.

b. Debug the form and correct all of the syntax and runtime errors. List the errors you find, and how you corrected them, in a text file named 7ACase8_CORR.txt.

9. **Debugging Problem:** A custom form that is saved in a file named 7ACase9_ERR.fmb in the Chapter7 folder on your Data Disk inserts, updates, and deletes records in the PROJECT table in the Software Experts database. When the user clicks the Create button, a new Project ID value appears. The user types a project name; selects a valid client ID, manager ID, and optional parent project ID; and clicks the Save New button, and the new record is saved in the database. The user can click the LOV command button next to the Project ID text item to display a list of existing project records, which can then be modified using the Update button.

a. Save the form as 7ACase9_CORR.fmb.

b. Debug the form and correct all of the syntax and runtime errors. List the errors you find and how you corrected them, in a text file named 7ACase9_CORR.txt.

◀ LESSON B ▶

Objectives

- ◆ Understand and control default system messages
- ◆ Create alerts and messages to provide system feedback
- ◆ Create applications that avoid user errors
- ◆ Learn how to trap common runtime errors
- ◆ Learn how to validate form data

DEFAULT SYSTEM MESSAGES

In a form, a **message** is the text that appears in the status line on the Forms Runtime window. By now, you are familiar with the default system messages with FRM- and ORA- prefixes that appear in the status line when you execute forms. These messages are useful for determining if a record is successfully inserted, updated, or deleted, as well as determining the nature of errors that occur while you are running a form. Oracle system messages are categorized by severity level and message type, as summarized in Table 7-5.

Message Severity Level	Description	Example
5	Informative message that describes what has happened; does not require user intervention	FRM-40400: Transaction complete: 1 records applied and saved.
10	Informative message that identifies a procedural mistake made by the user.	FRM-40201: Field is full. Can't insert character.
15	Informative message that identifies a mistake made by the user, such as entering an incorrect data value in a text item	FRM-50016: Legal characters are 0-9 - + E.
20	Condition that keeps a trigger from working correctly	FRM-40602: Cannot insert into or update data in a view.
25	Condition that causes the form to operate incorrectly	FRM-40919: Internal SQL statement execution error: %d.
>25	Extreme severity; condition must be corrected immediately. Messages of this level cannot be suppressed programmatically.	FRM-40024: Out of memory.

Table 7-5 System message severity levels

Level 5 error messages provide information about what is happening and usually do not require user action. Error messages that appear as a result of user errors have message severity levels of 10, 15, or 20 and require user intervention to correct the condition. Error messages caused by errors within a form trigger have levels of 25 or greater and require intervention by the form developer or DBA.

The error severity levels are assigned by Oracle, and the scale simply compares relative severity levels, with 5 being low severity, 15 somewhat higher, 20 somewhat higher still, and >25 the highest severity. Message severity levels do not correspond to error code numbers—the error code numbers are assigned sequentially by Oracle as new errors are documented.

Sometimes you might want to suppress the default system messages and replace them with your own messages. Alternatively, you might want to display a dialog box that poses a question that a user can respond to, such as whether to commit a change to the database or roll it back. Form Builder determines which messages to display in the message area using a system variable named `:SYSTEM.MESSAGE_LEVEL`. By default, all system messages appear. In a PRE-FORM trigger, you can set the level of this variable to a value of 25 to suppress all messages except the most severe (Level > 25) ones, which cannot be programmatically suppressed.

Now you will refresh your database tables, and open and rename a form that you can use to insert, update, and delete records in the PROJECT table in the Software Experts database. Then you will modify its PRE-FORM trigger to set the `:SYSTEM.MESSAGE_LEVEL` variable value to 25 so that only the most severe messages (severity level >25) appear. The form is stored in a file named Project.fmb in the Chapter7 folder on your Data Disk.

To refresh your database tables, open the form, rename the form, and modify the PRE-FORM trigger:

1. Start SQL*Plus, log onto the database, and run the **clearwater.sql**, **northwoods.sql**, and **softwareexp.sql** scripts in the Chapter7 folder on your Data Disk.

2. In SQL*Plus, type the following command to create a sequence to automatically generate project ID values:

   ```
   CREATE SEQUENCE project_id_sequence START WITH 10;.
   ```

3. Start Form Builder, open **Project.fmb** from the Chapter7 folder on your Data Disk, and save the file as **7BProject.fmb** in your Chapter7\TutorialSolutions folder.

4. If necessary, maximize the Object Navigator window. Click ➕ beside the Triggers node, and double-click the **Trigger icon** 🔊 next to the PRE-FORM trigger. The PL/SQL Editor opens with the current PRE-FORM trigger displayed.

5. Modify the PRE-FORM trigger by typing the following code as the last line of the trigger:

   ```
   :SYSTEM.MESSAGE_LEVEL := 25;
   ```

6. Compile the trigger, correct any syntax errors, close the PL/SQL Editor, and save the form.

Now you will run the form to see the effect of modifying the PRE-FORM trigger. You will change the project manager for project ID 1 to Sheila Hernandez and see if the Level 5 "FRM-40401: No changes to save" message appears.

To test the form:

1. Run the form, and log onto the database if necessary.

2. Click the **LOV command button** next to the Project ID text item, select the record for Project ID **1**, and then click **OK**. The data values for the project appear in the form text items.

 In order for the icons to appear on the LOV command buttons, you must copy the down3.ico file to your Chapter7\TutorialSolutions folder. You might need to change the Icon Filename property in the LOV command button to specify the complete folder path and filename of the .ico file.

3. Click the **LOV command button** next to the Manager ID text item, select the record for consultant **Sheila Hernandez**, and then click **Update**. Notice that the usual confirmation message ("FRM 40401: No changes to save.") does not appear.

4. Click the **LOV command button** next to the Project ID text item, select the record for Project ID **1** again, and then click **OK**. Notice that the updated value for project manager appears.

5. Click **Exit** to close Forms Runtime.

The form text items were cleared when you clicked the Update button, but a confirmation message did not appear in the message area. The update was successful, and the modified PRE-FORM trigger suppressed the confirmation message. However, the form needs to provide users with explicit system feedback. You can provide explicit feedback by replacing the system messages with custom messages or by displaying dialog boxes, called alerts, that enable users to respond in different ways.

CREATING ALERTS AND MESSAGES TO PROVIDE SYSTEM FEEDBACK

An important principle of graphic user interface design is to give users feedback as to what is happening and to make applications forgiving, so users can undo unintended operations. Right now, the project form does not provide any confirmation that an update operation was successful, nor does it provide an opportunity for canceling the update. To provide this additional functionality, you will create an **alert**, which is a dialog box that displays information to users and allows them to choose different ways to proceed. You will create an alert named UPDATE_ALERT that informs users that the PROJECT table was updated and gives them the option of continuing or canceling the operation. If a user chooses to continue, a custom message will appear in the message area confirming that the record was updated successfully. If the user chooses to cancel, a message confirming that the change has been canceled will appear.

You could make alerts to inform users that the operation was confirmed or canceled, but this would force them to click an OK button to acknowledge the alert. If you want to provide a short, informative message that does not require any user choices or actions, it is better to display a message than an alert. To display a message, you use the MESSAGE built-in subprogram, which has the following syntax: **MESSAGE('*text of message*');.** Now you will create the UPDATE_ALERT.

To create the alert:

1. Open the Object Navigator window in Ownership View.

2. Select the **Alerts** node under the PROJECT_FORM node.

3. Click the **Create button** 🔲 on the Object Navigator toolbar to create a new alert object.

4. Rename the alert **UPDATE_ALERT**.

Alert Properties

The alert properties you will modify are Title, Style, Button, and Message. The *Title* determines the title that appears in the alert window title bar. The *Style* determines the icon that appears on the alert; possible styles are Note, Caution, and Stop. Note alerts display an "i" for information, Caution alerts display an exclamation point (!), and Stop alerts display a red "X" or a red stoplight. Note that alerts convey information to the user, such as confirming that a record has been inserted. Caution alerts inform the user that he or she has to make a choice that cannot be undone and could lead to a potentially damaging situation, such as deleting a record. Stop alerts inform the user that he or she has instructed the system to perform an action that is not possible, such as trying to delete a record that is referenced as a foreign key in another table. Figure 7-23 shows an example of a Caution alert.

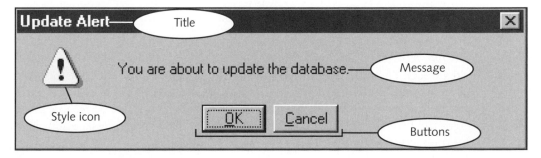

Figure 7-23   Caution alert

The *Button Label* property determines how many buttons will appear on the alert and the labels that will appear on the buttons. An alert can have a maximum of three buttons, so there are three Button Label properties named Button 1 Label, Button 2 Label, and Button 3 Label. If you delete the label for a given button, then that button will not appear on the alert. The *Message* property defines the text that appears in the alert.

Now you will change the properties of the UPDATE_ALERT so the alert appears like the one in Figure 7-23. You will specify the button labels for Button 1 and Button 2, and delete the label for Button 3.

To change the properties of the new alert:

1. Double-click the **Alert icon** ▌ beside UPDATE_ALERT to open its Property Palette, and then change its properties as follows:

Title	**Update Alert**
Message	**You are about to update the database.**
Alert Style	**Caution**
Button 1 Label	**OK**
Button 2 Label	**Cancel**
Button 3 Label	(deleted)

2. Close the Property Palette, and then save the form.

Displaying Alerts

To display an alert in a form, you use the SHOW_ALERT built-in function. In programming, a function always returns a value. The SHOW_ALERT function returns a numeric value. To display an alert during the execution of a trigger, you need to declare a numeric variable and then assign to this variable the value returned by the SHOW_ALERT function. This value corresponds to the alert button that the user clicks. If the user clicks the first button on the alert, this value is assigned to a variable named ALERT_BUTTON1. If the user clicks the second button on the alert, this value is assigned to ALERT_BUTTON2, and if the user clicks the third button, this value is assigned to ALERT_BUTTON3. You can specify different program actions (such as COMMIT or ROLLBACK) depending on which button the user clicks.

Now you will create a new program unit named DISPLAY_ALERT that will use the SHOW_ALERT function to display the UPDATE_ALERT, which has two buttons: OK and Cancel. The program unit will use an IF/THEN/ELSE structure to specify the correct action, depending on which alert button the user clicks. If the user clicks the OK button, the transaction is committed and a confirmation message appears. If the user clicks the Cancel button, the transaction is rolled back and a message appears stating that the transaction was rolled back.

To create the program unit:

1. In the Object Navigator window, select the **Program Units node**, and then click the **Create button** 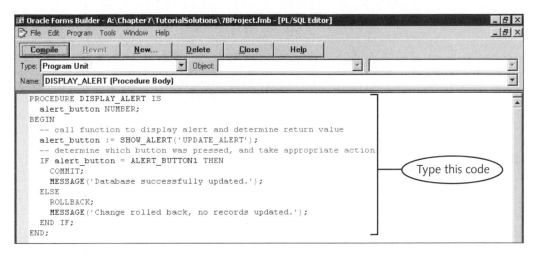 to create a new program unit.

2. Type **DISPLAY_ALERT** for the new program unit name, make sure that the Procedure option button is selected, and then click **OK**.

3. Type the code shown in Figure 7-24 in the PL/SQL Editor, compile the code, and correct any syntax errors if necessary. Do not close the PL/SQL Editor.

Figure 7-24   Code to display alert

Next, you need to modify the trigger for the Update button so that it calls the DISPLAY_ALERT program unit instead of committing the transaction to the database. You will navigate to this trigger directly in the PL/SQL Editor.

To modify the Update button trigger:

1. In the PL/SQL Editor, open the Type list to display the types of code objects, and select **Trigger**. By default, the form-level PRE-FORM trigger opens.

2. Open the Object list to select the form object, and select **PROJECT_BLOCK**. Since there are no block-level triggers in this form, no code appears in the PL/SQL Editor.

3. Open the (Data Block Level) list, scroll down in the list, and select **UPDATE_BUTTON**. The code for the Update button trigger appears.

4. Delete the program line that says COMMIT, and replace it with the following program line that calls the DISPLAY_ALERT program unit:

DISPLAY_ALERT;

5. Compile the trigger, correct any syntax errors, close the PL/SQL Editor, and save the form.

Now you will update a project record to confirm that the alert appears and that the correct messages appear depending on which alert button the user clicks.

To test the alert:

1. Run the form, click the **LOV command button** next to the Project ID text item, select the record for Project ID **1**, and then click **OK**. The data values for the project appear in the form text items.

2. Click the **LOV command button** next to the Manager ID text item, select the record for consultant **Brian Zhang**, click **OK**, and then click **Update**. The UPDATE_ALERT appears.

3. Click **OK**. The confirmation message "Database successfully updated" appears in the message area.

4. Click the **LOV command button** next to the Project ID text item, select the record for Project ID **1** again, and then click **OK**. The data values for the project appear in the form text items, showing the updated project manager values.

5. Click the **LOV command button** next to the Manager ID text item, select the record for consultant **Sarah Carlson**, click **OK**, and then click **Update**. The alert appears again. You decide not to update the record, so you will cancel your changes.

6. Click **Cancel**. The rollback confirmation message appears.

STRATEGIES FOR AVOIDING USER ERRORS

Your forms should include ways to keep users from making mistakes and ways to help them correct their mistakes. To implement strategies for avoiding user errors, you can modify form properties or add program code that prevents a user from performing an illegal action or detects an illegal action after it is performed and then provides suggestions to the user for correction.

Now you will intentionally perform an action that causes an error in the project form. You will change the value of the Client ID to a value that does not exist in the CLIENT table. Then you will click the Update button and try to update the database.

To try to update the Client ID to a value that does not exist:

1. Click the **LOV command button** next to the Project ID text item, select the record for Project ID **2**, and then click **OK**. The data values for the project appear in the form text items.

2. Select the current Client ID value, change the value to **15**, which is a value that does not exist in the CLIENT table, and then click **Update**. The message for the ORA-02291 error appears. If you looked up the explanation for

this error code in the Oracle Help application, you would find that the error was caused by attempting to insert a foreign key value that does not exist in the parent table.

3. Click **Exit** to close the Forms Runtime window.

You do not want to give users the opportunity to directly update or delete text items that contain primary and foreign key values. One strategy is to make these text items **nonnavigable**, which means the user cannot press the Tab key to place the insertion point in the text item. To make a text item nonnavigable, you need to set the item's Keyboard Navigable property to No, so the only way a user can place the insertion point in the item is by clicking the mouse pointer. Then you need to create a trigger that moves the insertion point, or form focus, to another form item whenever the user clicks the text item using the mouse.

Now you will make the Client ID text item nonnavigable and create a trigger associated with the WHEN-MOUSE-UP event to move the insertion point to a different form item. You could also use the WHEN-MOUSE-CLICK event, because users usually click the mouse button to place the form insertion point in a text item. However, you would have to create an identical trigger for the WHEN-MOUSE-DOUBLECLICK event, because some users might double-click the field to place the form insertion point in a text item. The WHEN-MOUSE-UP event occurs whenever any mouse button is clicked in the item and the mouse button is released. The WHEN-MOUSE-UP trigger will switch the form focus to the Client ID LOV command button whenever the user clicks the Client ID text item. This prompts the user to click the LOV command button and select the Client ID value from the LOV display. You will use the GO_ITEM built-in subprogram to move the form insertion point to a different form item.

To make the Client ID text item nonnavigable and create the trigger to move the form focus:

1. In the Object Navigator, double-click the **Text Item icon** beside CLIENT_ID_TEXT under the PROJECT_BLOCK node. The text item Property Palette opens.

2. Change the Keyboard Navigable property to **No**, and close the Property Palette.

3. Right-click **CLIENT_ID_TEXT**, and then click **PL/SQL Editor**. The list of event triggers for the text item appears.

4. Type **WHEN-MOUSE-UP** in the Find box, press **Enter**, and then press **OK**. The PL/SQL Editor opens.

5. Type `GO_ITEM('project_block.client_id_lov_button');`, compile the trigger, correct any syntax errors, and close the PL/SQL Editor.

6. Save the form, and then run the form. The insertion point initially appears in the Project ID text item.

7. Press **Tab** three times to try to move the form insertion point to the Client ID text item. Note that the Client ID text item is skipped, because it is now nonnavigable.

8. Try to place the form insertion point in the Client ID text item by clicking on the Client ID text item using the mouse pointer. The form focus moves to the Client ID LOV command button.

9. Click **Exit** to close the form.

Another way to help users avoid errors is by programmatically disabling and enabling command buttons. Currently, when a user clicks the Create button, a new Project ID value appears on the form. The user must enter values into the other form text items, and then click the Save New button, to save the new record. If the user enters new values and then clicks the Update button, an error will occur, because the trigger code is trying to update a record that has not yet been inserted. Therefore, when the user clicks the Create New button, the Update button should be disabled (grayed-out) and not be enabled until after the user has clicked the Save New button. To disable a command button, you open the command button's Property Palette and set the Enabled property to No.

You can also set item property values programmatically using the SET_ITEM_PROPERTY built-in, which has the following general syntax: **SET_ITEM_PROPERTY('*item name*'**, *property_name, property_value***);**.

 The property value parameter in this command usually is not the same as the values that appear in the item Property Palette. For a complete list of property values to use in the SET_ITEM_PROPERTY command, click Help on the Form Builder menu bar, click Form Builder Help Topics, and search for SET_ITEM_PROPERTY.

Now you will modify the trigger for the Create button so it programmatically disables the Update button. Then, you will modify the trigger for the Save New button so it programmatically reenables the Update button after the form items are cleared.

To change the button properties programmatically:

1. In the Object Navigator, right-click **CREATE_BUTTON**, and then click **PL/SQL Editor**. The button trigger code appears.

2. Place the insertion point at the beginning of the first line of code, press **Enter**, and then type the following code as the new first line of the trigger:

```
SET_ITEM_PROPERTY('UPDATE_BUTTON', ENABLED,
PROPERTY_FALSE);
```

3. Compile the trigger, and debug it if necessary. Do not close the PL/SQL Editor.

4. Open the Object list (where CREATE_BUTTON is currently selected), and select **SAVE_BUTTON**. The Save button trigger code appears.

5. To reenable the Update button, add the following code as the last line of the trigger:

```
SET_ITEM_PROPERTY('UPDATE_BUTTON', ENABLED,
PROPERTY_TRUE);
```

6. Compile the trigger, debug it if necessary, close the PL/SQL Editor, and save the form.

Now you will test the triggers to make sure your changes work correctly. You will click the Create New button and confirm that the Update button is disabled. Then, you will enter a new project record, save the record, and confirm that the Update button is reenabled.

To test the triggers:

1. Run the form, and click **Create**. A new value appears in the Project ID text item, and the label of the Update button is grayed-out, indicating the button is disabled.

2. Select **1** for the Client ID and **101** for the Manager ID values, and click **Save New**. The Update button is reenabled.

3. Click **Exit** to exit Forms Runtime.

4. Close the form in Form Builder.

TRAPPING COMMON RUNTIME ERRORS

There are common errors that occur in many forms, such as trying to delete a record that is referenced as a foreign key by other records or trying to insert a new record when fields with NOT NULL constraints have been left blank. It is a good practice to **trap** these errors, which means to intercept the default system error message and replace it with a custom error message. The custom error message gives more detailed information to the user about how to correct the error. To investigate how to trap and handle these errors, you will use a custom form created to retrieve, update, and delete records from the ITEM and INVENTORY tables in the Clearwater Traders database. First you will open this form and run it.

To open and run the form:

1. Open **Inventory.fmb** from the Chapter7 folder on your Data Disk, save it as **7BInventory.fmb** in your Chapter7\TutorialSolutions folder, and then run the form.

2. Click the **LOV command button** beside the Item ID text item, select Item ID **894** (Women's Hiking Shorts), and then click **OK**. The values appear in the form text items. You could edit the item data and click the Update button in the Items frame to edit the record, or you could click the Delete button in the Items frame to delete the record.

7

3. Click the **LOV command button** beside the Inventory ID text item. The inventory items associated with the selected Item ID appear in the LOV display. Select Inventory ID **11776** (size M, color Khaki), and then click **OK**. The selected record's data values appear in the text items in the Inventory Items frame.

4. Delete the current price, type **29.50** in the Price text item, and then click **Update** within the Inventory Items frame. The Updating Database alert appears in the messaging area. Click **OK**. A customized confirmation message ("Database change successfully committed") appears in the message area.

5. Click **Clear** in the Inventory Items frame to clear the inventory data.

6. Click **Clear** in the Items frame to clear the item data.

7. Click the **LOV command button** beside Inventory ID again. This time, a Stop alert appears stating that you must select an Item ID before you can select an Inventory ID. Click **OK**.

Now you will deliberately make some errors while updating and deleting records to view the error messages. First, you will try to delete an ITEM record that has associated INVENTORY records in the INVENTORY table.

To try to delete the ITEM record:

1. Click the **LOV command button** beside Item ID, select Item ID **786** (3-Season Tent), and then click **OK**. The selected values appear in the form text items.

2. Click **Delete** in the Items frame. The error message "FRM-40735: WHEN-BUTTON-PRESSED trigger raised unhandled exception ORA-02292" appears.

Recall that Forms Runtime has error-handling routines that intercept DBMS (ORA-) error messages and then generate the associated Forms (FRM-) error messages. Now you will look up FRM-40735 in the Form Builder Help system to view an explanation of the error.

To look up an FRM- error explanation:

1. Click the **Oracle Forms Builder** button on the taskbar to switch to Form Builder, click **Help** on the Form Builder menu bar, click **Form Builder Help Topics**, type **FRM-40735** as the search string, click **Display**, select **FRM-40735** in the Topics Found window, and then click **Display**. The explanation states that the trigger raised an "unhandled exception." Close the Help window.

The FRM- error explanation indicates that a database (ORA-) error occurred, and there is no associated FRM- error message for this error. Therefore, you need to use the ORA-error message code to determine the cause of the error. If you look up the ORA-02292 error code in the Oracle8i Help application, it explains that the record you tried to delete is referenced in other tables and cannot be deleted. This record is referenced in the INVENTORY table, and the user cannot delete it. It probably is not reasonable to expect users to look up ORA- error code messages to learn why they cannot complete a task, so you need to intercept the DBMS error message in your form, and replace it with a more understandable message.

Next, you will generate an FRM- error message. FRM- errors occur when the user violates constraints specified in the form. Usually, these are violations of properties set in a text item's Property Palette, such as not entering the correct data type or format mask. Now you will generate an FRM- error by entering an incorrect data type in a form field.

To generate an FRM- error:

1. Switch back to Forms Runtime, click the **LOV command button** beside the Inventory ID text item in the Inventory Items frame, select Inventory ID **11668**, and then click **OK**.

2. Delete the current QOH, and type the letter **a** for the new QOH. This should generate a data type error, since the QOH_TEXT form field is specified as a NUMBER data type.

3. Click **Update** in the Inventory Items frame. The Form error message "FRM-50016: Legal Characters are 0-9-+E." appears. Note that the form insertion point moves back to the QOH field and highlights the current value. Since this error occurs within the form, you cannot change the form focus to a different item until you correct the condition that is causing the error.

4. Change the QOH value back to **16**, and then click **Exit** to exit Forms Runtime.

You may need to scroll down on the form display to view the Exit button.

You can intercept both FRM- and ORA- error messages by creating a form-level trigger that corresponds to the ON-ERROR event. Whenever an ORA- or FRM- error occurs while a form is running, the ON-ERROR event fires a corresponding trigger. Form Builder has several built-in procedures that return information about error conditions that can be used in this trigger. Table 7-6 summarizes these built-in procedures.

Procedure Name	Data Returned
DBMS_ERROR_CODE	Error number of the most recent database (ORA) error, represented as a negative integer
DBMS_ERROR_TEXT	Message number and text of the most recent ORA error
ERROR_CODE	Error number of the most recent Forms Runtime (FRM) error, represented as a positive integer
ERROR_TEXT	Message number and text of the most recent FRM error
ERROR_TYPE	Three-character code indicating the type of the most recent error (FRM or ORA)
MESSAGE_CODE	Message number of the most recent error (FRM or ORA)
MESSAGE_TEXT	Message text of the most recent error (FRM or ORA)

Table 7-6 Form Builder built-in procedures for handling errors

If an FRM- error occurs, the corresponding FRM- error code is returned by the ERROR_CODE function. If an ORA- error occurs, the corresponding ORA-error code is returned by the DBMS_ERROR_CODE function. By testing to find the values of ERROR_CODE and DBMS_ERROR_CODE, you can anticipate common errors and display custom alerts to provide users with informative messages and alternatives. The trigger code tests for specific FRM- error codes. ORA- errors are considered to be FRM-40735 (unhandled form exceptions) errors, so when an ORA- error occurs, the form returns an FRM-40735 error code. The trigger code must then use a nested IF/THEN structure to identify the specific ORA- error code that occurred.

Next, you will create an ON-ERROR trigger that traps FRM- error 50016 (entering an illegal data type in a form field) and ORA- error 02292 (trying to delete a record that is referenced in another table) and displays customized alerts informing the user of the cause of the error. When an error occurs, the trigger first tests to see if the value of ERROR_CODE is 50016. If it is, an appropriate alert appears. If ERROR_CODE 40735 (unhandled exception) occurs, then the trigger tests for the value of DBMS_ERROR_CODE, which corresponds to an ORA- error code. If DBMS_ERROR_CODE is 02292, then a different alert appears. Note that FRM- error code values are represented as positive integers, while ORA- error code values are represented as negative integers.

To create the ON-ERROR trigger:

1. In Form Builder, open the Object Navigator window in Ownership View, if necessary, and then select the **Triggers node** directly below the form module to create a form-level trigger.

2. Click the **Create button** on the Object Navigator toolbar, and create a new form-level trigger that corresponds to the **ON-ERROR** event. The PL/SQL Editor opens.

3. Type the code shown in Figure 7-25 in the PL/SQL Editor.

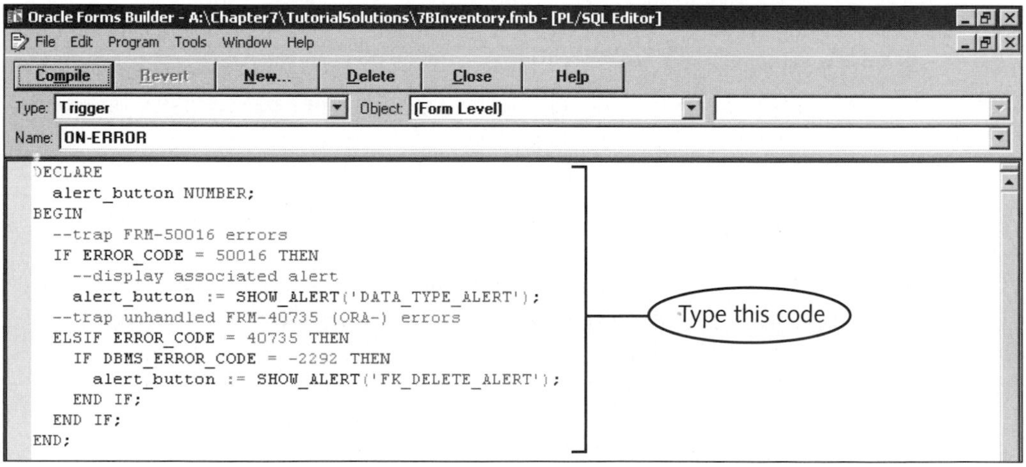

Figure 7-25   ON-ERROR trigger code

4. Compile the code, correct any syntax errors, and then close the PL/SQL Editor.

Next, you need to create new alerts named DATA_TYPE_ALERT and FK_DELETE_ALERT that will appear instead of the FRM-50016 and ORA-02292 error codes and messages.

To create the alerts:

1. Return to the Object Navigator window, select the **Alerts node** under the form module, and click the **Create button** 🔳 on the Object Navigator toolbar to create a new alert.

2. Change the name of the new alert to **DATA_TYPE_ALERT**.

3. Select the **Alerts node** again, click 🔳 again, and name the new alert **FK_DELETE_ALERT**.

4. Open the Property Palette of each new alert, change its properties as follows, and then close the Property Palette.

Alert	DATA_TYPE_ALERT	FK_DELETE_ ALERT
Title	**Data Type Error**	**Delete Error**
Message	**You have entered a value in the text item that is the wrong data type.**	**You cannot delete this record because it is referenced by other database tables.**
Alert Style	**Stop**	**Stop**
Button 1 Label	**OK**	**OK**
Button 2 Label	(deleted)	(deleted)
Button 3 Label	(deleted)	(deleted)

5. Save and then run the form.

6. Click the **LOV command button** beside Item ID, select Item ID **786** (3-Season Tent), click **OK**, and then click **Delete** in the Items frame. The Delete Error alert appears. Click **OK** to close the alert.

7. Click the **LOV command button** beside Inventory ID, select Inventory ID **11668**, and click **OK**. Change the QOH to the letter **a**, and click **Update** in the Inventory Items frame. The Data Type Error alert appears. Click **OK** to close the alert, change the QOH back to **16**, and then click **Exit**.

Next, you need to modify the ON-ERROR trigger to display an alert for unhandled exceptions, which are errors that you don't anticipate. Currently, if an error other than FRM-50016 or ORA-02292 occurs, the ORA- or FRM- error code appears in the message area. Now you will modify the ON-ERROR trigger to display an alert with a message that is the error description associated with an unhandled FRM- or ORA- code. To do this, you will first create an alert named UNHANDLED_ERROR_ALERT. Instead of specifying the alert's Message property in the Property Palette, the Message property will be set at runtime in the ON-ERROR trigger, and the alert's Message property will be left blank in the Property Palette.

To create the alert to display messages for unhandled errors:

1. In the Object Navigator window, create a new alert named **UNHANDLED_ERROR_ALERT**.

2. Open the Property Palette of the new alert, change its properties as follows, and then close the Property Palette.

Property	Value
Title	**Error**
Message	(leave blank)
Alert Style	**Stop**
Button 1 Label	**OK**
Button 2 Label	(deleted)
Button 3 Label	(deleted)

Recall from Table 7-6 that when an FRM- error occurs, the ERROR_CODE_TEXT function returns the text of the error message, and when an ORA- error occurs, the DBMS_ERROR_TEXT function returns the message text. Now you will modify the ON-ERROR trigger so that when an unhandled error occurs, the system will display the UNHANDLED_ERROR_ALERT showing the appropriate system error description. To show the correct system error description, you will use the SET_ALERT_PROPERTY built-in procedure, which can be used to set alert properties at runtime. It has the following syntax: **SET_ALERT_PROPERTY('alert_name', alert_property_name, property_value);**. The alert name must be placed in single quotation marks. In this trigger, the alert property name is ALERT_MESSAGE_TEXT, and the property value will be the system error description returned by ERROR_TEXT or DBMS_ERROR_TEXT.

To modify the ON-ERROR trigger to display a customized alert for unhandled exceptions:

1. Double-click the **Trigger icon** 🖅 beside the ON-ERROR trigger. The PL/SQL Editor opens.

2. Modify the ON-ERROR trigger as shown in Figure 7-26. Compile the code, correct any syntax errors, close the PL/SQL Editor, and save the form.

Now you will run the form and test to see what happens when an error that is not explicitly handled in the ON-ERROR trigger occurs. You will delete the Item Description value for Item 786 and then attempt to update the record. The Item Description field has a NOT NULL constraint in the ITEM table, so an ORA- error should occur.

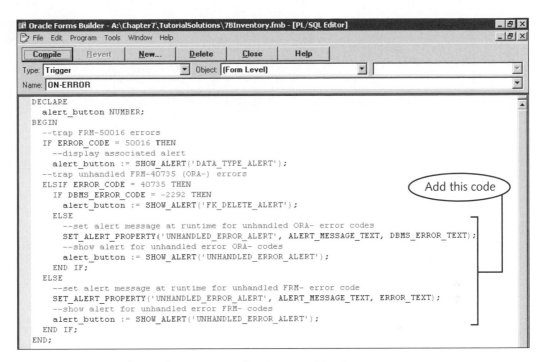

Figure 7-26   Handling other errors in the ON-ERROR trigger

To generate an unhandled ORA- error:

1. Run the form, click the **LOV command button** beside the Item ID text item, and select Item **786**.

2. Delete the Item Description value for Item 786.

3. Click **Update** to generate an error by not entering data in the NOT NULL field. The alert for the unhandled error appears, showing the dynamically generated Oracle error message associated with the ORA-01407 database error. Note that the error message describes the database user schema, table, and fieldname that cannot be NULL.

4. Click **OK** to close the alert, and then click **Exit** to close Forms Runtime.

5. Close the form in Form Builder

FORM VALIDATION

Validation is the process of ensuring that data meets specific preset requirements to avoid introducing incorrect values into the database. For example, values entered for date fields must contain characters that represent valid dates, and values entered in numerical fields must contain numbers. In a form, data validation is performed at the item level based on the input item properties. Table 7-7 summarizes the Form Builder validation properties for text items.

Item Property Node	Property	Allowable Values	Function
Data	Data Type	Char, Number, Date, Alpha, Integer, Date-time, Long, Rnumber, Jdate, Edate, Time	Ensures that entered values are of the specified data type
Data	Maximum Length	Integer	Defines the maximum number of characters the item will accept
Data	Fixed Length	Integer	Requires input to be exactly the specified length
Data	Required	Yes or No	Specifies whether or not the value can be NULL
Data	Lowest Allowed Value	Integer	For numerical fields, specifies the lowest value that will be accepted
Data	Highest Allowed Value	Integer	For numerical fields, specifies the highest value that will be accepted
Data	Format Mask	Legal format masks	Ensures that user input matches the specified format mask
Database	Insert Allowed	Yes or No	Specifies whether the item allows input at all
Database	Update Allowed	Yes or No	Specifies whether the item's value can be changed
Database	Update Only If NULL	Yes or No	Specifies that the item can be changed only if its previous value was NULL
List of Values	Validate From List	Yes or No	Specifies that the value entered by the user should be validated against the item's LOV

Table 7-7   Form Builder item validation properties

You can use all of these properties to validate values that are entered in both data block and custom forms. The Data Type property specifies that the value that the user enters in the item must be of a specific data type. So far, you have used this property in custom forms to match the text item property with the associated database field data type. You can use this property to further restrict input values. For example, you can specify the *Alpha* data type if the field should only allow nonnumerical characters, the *Datetime* data type if the value must include both a date and time component, or the *Time* data type if the value must include only a time component. Other choices specify the data value format: *Edate* specifies a European date format (DD/MM/YY); *Jdate* specifies a Julian calendar date that is stored in a NUMBER data field and is included for compatibility with prior Oracle versions; and *Rnumber* specifies a right-justified number.

The Validate From List property causes Forms Runtime to compare the value entered by the user with the first column of the LOV associated with the item. If the LOV does not contain a matching value, the LOV display opens, and the value entered by the user is used as a search condition. For example, if the user enters the character *M* in a field that contains a LOV with state abbreviations, all state abbreviations that begin with M appear in the LOV display.

Now you will open a form that is used to insert, update, and delete records in the STUDENT table in the Northwoods University database, and you will modify properties of some of the text items to provide additional validation. You will specify that the State and Class text items accept exactly two characters of character (Alpha) data. You will also specify that the F_ID text item must be validated using the LOV associated with the text item.

To open the form and modify its validation properties:

1. Open **Student.fmb** from the Chapter7 folder on your Data Disk, and save the file as **7BStudent.fmb** in your Chapter7\TutorialSolutions folder.

2. Click **Tools** on the menu bar, and then click **Layout Editor** to open the form in the Layout Editor.

3. Select the **State text item**, press and hold the **Shift** key, and then click the **Class text item** to select the text items as an object group. Click **Tools** on the menu bar, and then click **Property Palette** to open the intersection Property Palette.

4. Scroll down to the Data property node, and change the Data Type to **Alpha**. Confirm that the Maximum Length property value is **2**, change the Fixed Length property value to **Yes**, and then close the Property Palette.

5. Open the Property Palette for the Faculty ID text item, scroll down to the List of Values (LOV) property node, change the Validate from List property to **Yes**, and close the Property Palette.

6. Save the form.

Validation can occur each time the user moves the insertion point out of an item, record, or block, or each time he or she exits a form. The level at which validation occurs is called the **validation unit**, which represents the largest chunk of data that a user can enter before the entered values are validated. By default, the validation unit is at the item level, so each time the user moves the insertion point out of a specific item, the current value is validated based on the item's validation properties. The validation unit is set in the form Property Palette using the form's Validation Unit property. This property has possible values of Form, Data Block, Record, or Item. Now you will run the form, create a new record, and test the validation properties that you have just set.

7

To test the validation properties:

1. Run the form, and click **Create**. A new value appears in the Student ID text item.

If an error message appears, it is probably because the S_ID_SEQUENCE does not exist in your database schema. Click OK, create the sequence in SQL*Plus using the following command: CREATE SEQUENCE s_id_sequence START WITH 200;, and then run the form again.

2. Place the insertion point in the State text item, type **22**, and then press **Tab** to move the insertion point out of the item, and validate the item value. The message "FRM-50001: Acceptable characters are a-z, A-Z, and space" appears in the status line, indicating that numerical characters cannot be used in this text item.

3. Delete the current State value, type **W**, and then press **Tab**. The message "FRM-40203: Field must be entered completely" appears, indicating that the entered value must be two characters long.

4. Type **WI** for the State value.

5. Press **Tab** six times to place the insertion point in the Faculty ID text item, type **10** (which is not a legal F_ID value), and then click **Save New**. The Faculty LOV display opens. Select ID **1** (Kim Cox) and then click **OK**. The confirmation message ("No changes to save") appears, indicating that the record was successfully inserted.

6. Click **Exit** to close Forms Runtime.

You cannot use the text item validation properties to specify that only numeric values can be entered in a character field, or to specify that only certain characters can be entered. You can perform complex validation operations like these using validation triggers that are associated with the WHEN-VALIDATE-ITEM event. An item validation trigger is an item-level trigger that fires when the item is validated, as determined by the form validation unit. The trigger code tests the current item value to see if it satisfies the validation condition or conditions, displays a message if it does not, and then raises a built-in exception named FORM_TRIGGER_FAILURE, which automatically aborts the current trigger. Now you will create an item validation trigger for the Class text item on the Student form to confirm that the entered value is one of the legal class codes of FR, SO, JR, or SR. Then, you will test the trigger.

To create and test an item validation trigger:

1. In the Layout Editor, select the **Class text item**, right-click, point to **SmartTriggers**, and then click **WHEN-VALIDATE-ITEM**. The PL/SQL Editor opens.

2. Type the code for the validation trigger as shown in Figure 7-27, compile the trigger, correct any syntax errors, close the PL/SQL Editor, and save the form.

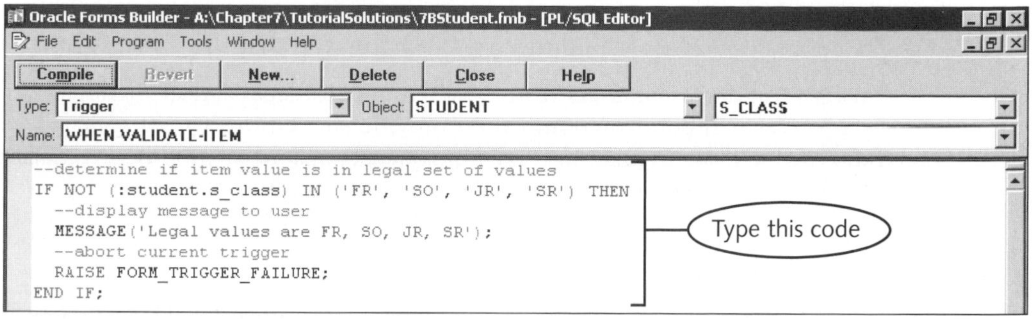

Figure 7-27   Item validation trigger

3. Run the form, and click **Create**. A new Student ID value appears.

4. Place the insertion point in the Class text item, type **AA**, and then click **Save New**. The message "Legal values are FR, SO, JR, SR" appears in the message area, indicating that the item validation trigger fired correctly.

5. Delete the current Class value, and then click **Exit** to exit Forms Runtime.

6. Close the form in Form Builder, and then exit Form Builder and all other open Oracle applications.

SUMMARY

◻ Default system messages are categorized by severity level and message type. Messages with severity levels of 5 provide information about what is happening, and usually do not require any user action. Messages with severity levels of 10, 15, 25, and >25 require user, developer, or DBA action.

◻ You can suppress all messages except those with severity levels greater than 25 by setting the :SYSTEM.MESSAGE_LEVEL variable to a higher value in a PRE-FORM trigger.

◻ An important principle of graphic user interface design is to provide feedback to users about form operations while they are working and to make applications "forgiving," so users can cancel unintended operations.

◻ Alerts are dialog boxes that provide information to users and allow them to choose from different options for proceeding.

◻ If you want to provide a short, informative message that does not involve any choices, display the message in the message area in the status line using the MESSAGE built-in. If you want to include more than one line of text in the message, give the user different action options, or alert the user to a serious error or problem, create an alert.

❏ To display an alert in a form, you use the SHOW_ALERT function, which returns a numeric value corresponding to the alert button that the user clicks.

❏ Forms should prevent users from performing an illegal action or detect an illegal action after it is performed and then provide suggestions to the user for correction. This can be done by making text items nonnavigable or by programmatically enabling and disabling command buttons at certain times.

❏ It is a good practice to trap common form errors, which intercepts the default system message and replaces it with a custom message.

❏ System runtime errors correspond to the form-level ON-ERROR event, so you can create a trigger associated with this event to trap system error messages and substitute your own message or alerts.

❏ ORA- errors represent errors that violate database rules, such as trying to delete a record that is referenced by another table. FRM- errors violate form properties, such as trying to enter an incorrect data type in a form text item.

❏ To display error messages for unhandled system messages, you can display the Oracle system error message description in an alert by assigning the system message as the alert message at runtime.

❏ Validation is the process of ensuring that data meets specific preset requirements to avoid introducing incorrect values into the database.

❏ Form validation can be performed using text item validation properties or by creating validation triggers. The form validation unit determines whether validation occurs at the item, record, block, or form level.

REVIEW QUESTIONS

1. What is the purpose of setting the :SYSTEM.MESSAGE_LEVEL variable to a different value?

2. Which default system messages cannot be suppressed?

3. What is the difference between an alert and a message?

4. What value does the SHOW_ALERT built-in function return?

5. What are two strategies for helping users avoid making errors?

6. What is the purpose of the ON-ERROR trigger?

7. What is the difference between an FRM- and an ORA- error?

8. List two ways of performing form validation.

9. Define form validation unit. What is the default validation unit?

10. How does the Validate from List item property work?

PROBLEM-SOLVING CASES

For the following cases, use the Clearwater Traders (see Figure 1–11), Northwoods University (see Figure 1–12), and Software Experts (see Figure 1–13) sample databases. The form files required for all cases are stored in the Chapter7 folder on your Data Disk. Save all solutions files in your Chapter7\CaseSolutions folder. Save all SQL commands in a file named 7BQueries.sql.

1. A custom form has been created to update records from the STUDENT table in the Northwoods University database. Open the file named 7BCase1.fmb, make the following modifications, and save the modified file as 7BCase1_DONE.fmb.

   a. Add a Caution alert that appears after the user clicks the Update button, but before the UPDATE command is committed. The alert should display the following message: "You are about to change the database." If the user clicks OK, commit the update and display a message stating "Record successfully changed." If the user clicks Cancel, do not commit the change, and display a message stating "Change rolled back." (*Hint*: Do not forget to modify the SYSTEM.MESSAGE_LEVEL value.)

   b. Prevent the user from changing the value of the Student ID text item by directing the form focus to the LOV command button associated with the text item.

   c. Modify the PIN text item validation properties to ensure that the data value is a four-digit integer.

2. A custom form has been created to insert new records into the FACULTY table in the Northwoods University database. When the user clicks the Create button, a new Faculty ID is retrieved from a sequence. The user enters the data values for the new faculty member, clicks the Save button, and the new record is saved in the database. Open the file named 7BCase2.fmb, make the following modifications to the form, and save the modified file as 7BCase2_DONE.fmb.

   a. If necessary, create a new sequence named F_ID_SEQUENCE, starting with 10 and with no maximum value. (If you already created this sequence in a previous case, use the existing sequence.)

   b. Create a Caution alert that appears when the user clicks the Save button but before the INSERT command is committed to the database. The alert message should be "Add the new record to the database?" If the user clicks the alert OK button, the trigger will save the record, clear the form, and show a message in the message area stating "Record successfully inserted." If the user clicks the alert Cancel button, do not commit the change, clear the form text items, and show a message in the message area stating "Changes not saved." (*Hint*: Do not forget to modify the value for :SYSTEM.MESSAGE_LEVEL.)

   c. Validate the LOC_ID value using the LOV associated with the text item.

   d. Configure the form so the Save button is disabled when the form first appears and is not enabled until the user clicks the Create button. After the user clicks the Save button, the Save button should again be disabled.

    e. Add an ON-ERROR trigger that displays a Note alert with an OK button to provide information on the following errors: ORA-00001 (trying to insert a record with a primary key value that already exists in the table), FRM-50016 (entering an incorrect data type in a form field), and ORA-01407 (trying to insert a NULL data value into a field with a NOT NULL constraint). Create customized alerts with informative messages for each of these errors. Configure the ON-ERROR trigger so it displays the associated Oracle system error description for all unhandled FRM- and ORA- errors.

    f. Create a validation trigger to confirm that the user entered a value of ASST, ASSO, FULL, or INST in the Rank text item. If the user did not enter an appropriate value, provide an informative message prompting him or her to do so.

3. Figure 7-28 shows a custom form that has been created to insert, update, and delete records in the CONSULTANT_SKILL table in the Software Experts database. The LOV command button allows the user to select existing consultant/skill combinations for update or delete, and the Select Consultant, Select Skill, and Save New buttons allow the user to select a consultant and a skill, and then insert a new consultant/skill combination. Open the file named 7BCase3.fmb, make the following modifications, and save the file as 7BCase3_DONE.fmb.

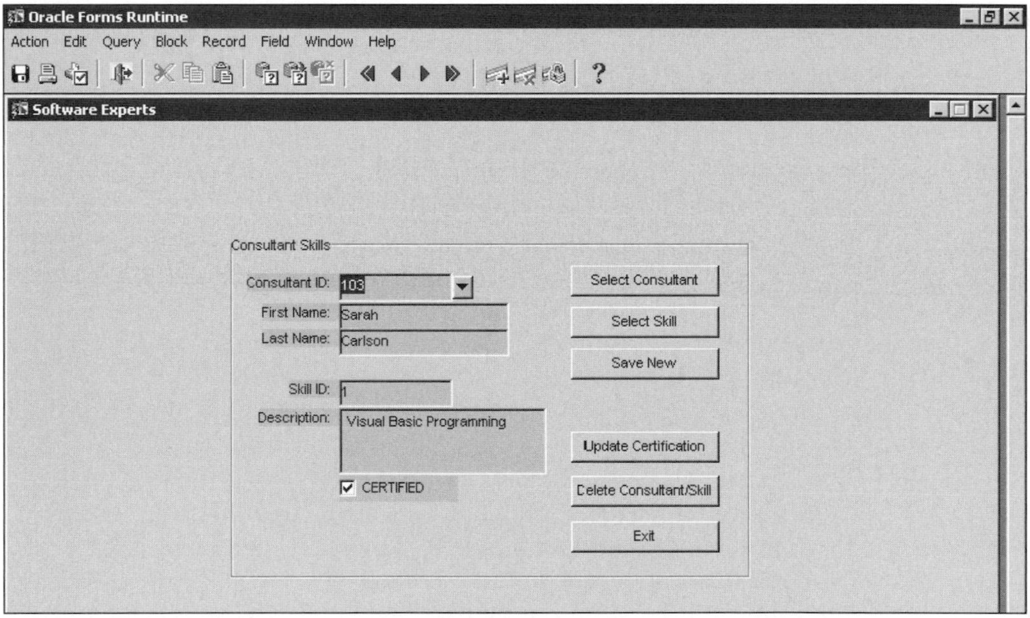

Figure 7-28

    a. Modify the form so that when the form first appears, the Select Consultant button, the LOV command button next to the Consultant ID text item, and the Exit button are enabled, and the other buttons are disabled.

b. After the user clicks Select Consultant and selects a consultant, programmatically enable the Select Skill button, which displays an LOV showing skills that the current consultant has not yet been associated with in the CONSULTANT_SKILL table.

c. After the user clicks Select Skill and selects a skill, programmatically enable the Save New button, so the user can save the new consultant/skill combination.

d. After the user clicks the Save New button and the new record is committed to the database, return the form buttons to their initial start-up state. (*Hint:* You cannot programmatically change the state of the current item, so move the form focus to a different item.)

e. After the user clicks the LOV command button beside the Consultant ID text item and selects an existing consultant/skill combination, disable the Select Consultant button, and enable the Update Certification and Delete Consultant/Skill buttons.

f. After the user either updates or deletes a consultant/skill combination and the change is committed to the database, return the form buttons to their initial start-up state.

g. Make all text items nonnavigable. Add triggers so the user cannot use the mouse pointer to place the insertion point in any text item.

4. A form has been created to insert new records into the EVALUATION table in the Software Experts database and is shown in Figure 7-29. When the user clicks the Create button, a new Evaluation ID is retrieved from a sequence. The user then selects a project using the LOV associated with the Project ID text item. When the user clicks the LOV command buttons to select the evaluator ID and evaluatee ID, records appear only for consultants associated with the selected project. When the user clicks the Save button, the new record is saved in the database. Open the file named 7BCase4.fmb, make the following modifications, and save the modified file as 7BCase4_DONE.fmb.

a. In SQL*Plus, create a new sequence named E_ID_SEQUENCE, starting with 200 and with no maximum value.

b. Create a Caution alert that appears when the user clicks the Save button but before the INSERT command is committed to the database. The alert message should be "Add the new record to the database?" If the user clicks the alert's OK button, the trigger will save the record, clear the form, and show a message in the message area stating "Record successfully inserted." If the user clicks the alert's Cancel button, do not commit the change, clear the form fields, and show a message in the message area stating "Changes not saved." (*Hint:* Do not forget to modify the value for :SYSTEM.MESSAGE_LEVEL.)

c. Validate the Project ID, Evaluator ID, and Evaluatee ID values using the LOVs associated with the text items.

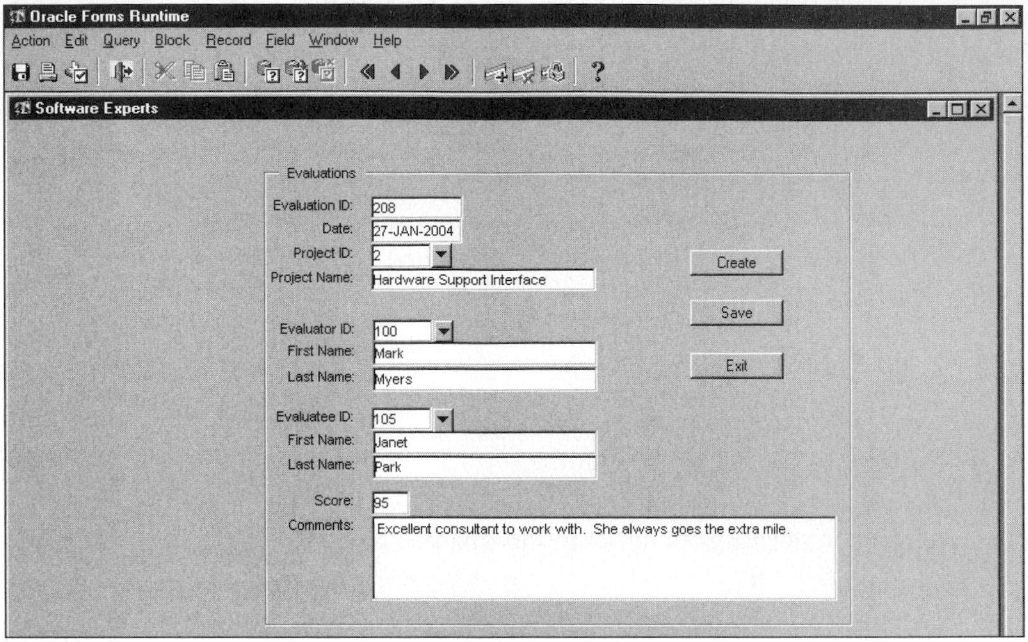

Figure 7-29

d. Configure the form so the Save button is disabled when the form first appears and is not enabled until the user clicks the Create button. After the user clicks the Save button, the Save button should again be disabled.

e. Configure the form so that when a user clicks the LOV command buttons beside Evaluator ID or Evaluatee ID, the LOV displays do not appear unless the user has previously selected a Project ID. If the user has not yet selected a project ID, display a message prompting him or her to do so.

f. Add an ON-ERROR trigger that displays a Note alert with an OK button to provide information on the following errors: FRM-50016 (entering an incorrect data type in a form text items) and ORA-01407 (trying to insert a NULL data value into a field with a NOT NULL constraint). Create customized alerts with informative messages for each of these errors. Configure the ON-ERROR trigger so it displays the associated Oracle system error description for all unhandled FRM- and ORA- errors.

g. Modify the properties of the Score text item to ensure that a value is entered, and that the value is greater than or equal to 0 and less than or equal to 100.

h. Create an error trap that determines, before a new record is inserted, if an evaluation has already been entered for the selected project by the selected evaluator for the selected evaluatee. If it has, display an alert that gives the user the option of replacing the previous evaluation or canceling the operation.
(*Hint*: Use an explicit cursor to try to retrieve the existing record.)

5. The custom form shown in Figure 7-30 has been created to insert new records, update existing records, and delete records from the COURSE_SECTION table of the Northwoods University database. When the user clicks the Create New Section ID button, a new Course Section ID appears. The user selects the data values for the course ID, Term ID, Faculty ID, and Location ID; enters values for the Sec. Number, Day, Time, and Maximum Enrollment; clicks the Save New Record button; and the new record is saved in the database. The user can click the LOV command button next to the Course Section ID text item to select one of the existing records, edit the data values, and then click the Update Existing Record button to update the record, or click the Delete Existing Record button to delete the record. Open the file named 7BCase5.fmb, make the following modifications, and save the modified file as 7BCase5_DONE.fmb.

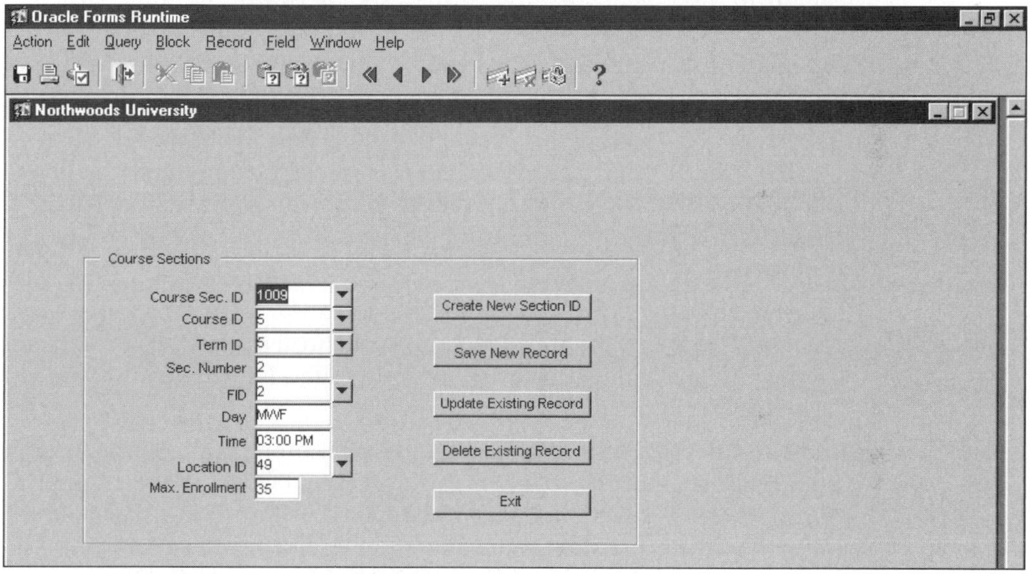

Figure 7-30

a. In SQL*Plus, create a new sequence named C_SEC_ID_SEQUENCE that starts with 2000 and has no maximum value.

b. When the form first appears, specify that the only buttons that are enabled are Create New Section ID, the Course Section ID LOV command button, and the Exit command button. After the user clicks Create New Section ID, programmatically enable the other LOV command buttons, and enable the Save New Record button. After the user clicks Save New Record, return the form buttons to their start-up state.

c. After the user clicks the Course Section ID LOV command button and selects an existing course section record for update or delete, enable the other LOV command buttons, and enable the Update Existing Record and Delete Existing Record command buttons. After the user has clicked the Update or Delete buttons, return the form buttons to their start-up state. (*Hint*: You cannot pro-grammatically change the state of the current item, so move the form focus to a different item.)

d. Add a Caution alert that appears when the user clicks the Save New Record button and before the INSERT command is committed. The alert should have the following message: "Add the new record to the database?" If the user clicks OK, commit the insert, and show a message stating "Record successfully inserted." If the user clicks Cancel, do not commit the change, and show a message stating "Changes not saved." (*Hint*: Do not forget to modify the value for :SYSTEM.MESSAGE_LEVEL.)

e. Add a Caution alert to the Update Existing Record button that appears before the UPDATE command is committed. The alert should display the message "Update the current record?" If the user clicks OK, commit the update and show a message stating "Record successfully updated." If the user clicks Cancel, do not commit the change, and show a message stating "Changes not saved."

f. Add a Caution alert to the Delete Existing Record button that appears before the DELETE command is committed. The alert should display the message "Delete the current record?" If the user clicks OK, commit the delete, clear the form, and show a message stating "Record successfully deleted." If the user clicks Cancel, do not commit the change, and show a message stating "Changes not saved."

g. Prevent the user from changing the Course ID, Term ID, Course Section ID, Faculty ID, or Location ID values except through the LOV command buttons by making the items nonnavigable and redirecting the form focus to the asso-ciated LOV command buttons.

h. Add an ON-ERROR trigger that displays a Note alert to provide additional information about the following errors: ORA-00001 (trying to insert a record with a primary key value that already exists in the table), ORA-02292 (trying to delete a record that is referenced in another table), and ORA-01407 (trying to insert NULL into a field with a NOT NULL constraint). Create customized alerts with informative messages for each of these errors. Display the associated Oracle system error description for all unhandled FRM- and ORA- errors.

6. The custom form shown in Figure 7-31 has been created to insert new records, update existing records, and delete records from the CUSTOMER, CUST_ORDER, and ORDER_LINE tables of the Clearwater Traders database. When the user clicks Create New Customer in the Customers frame, a new Customer ID is retrieved from a sequence. The user enters the data values for the new customer and clicks Save New Customer, and the new customer record is saved in the database. The user can click the LOV command button next to the Customer ID field to select an existing CUSTOMER record, edit the data fields, and then click Update Customer to update the customer record, or click Delete Customer to delete the customer record.

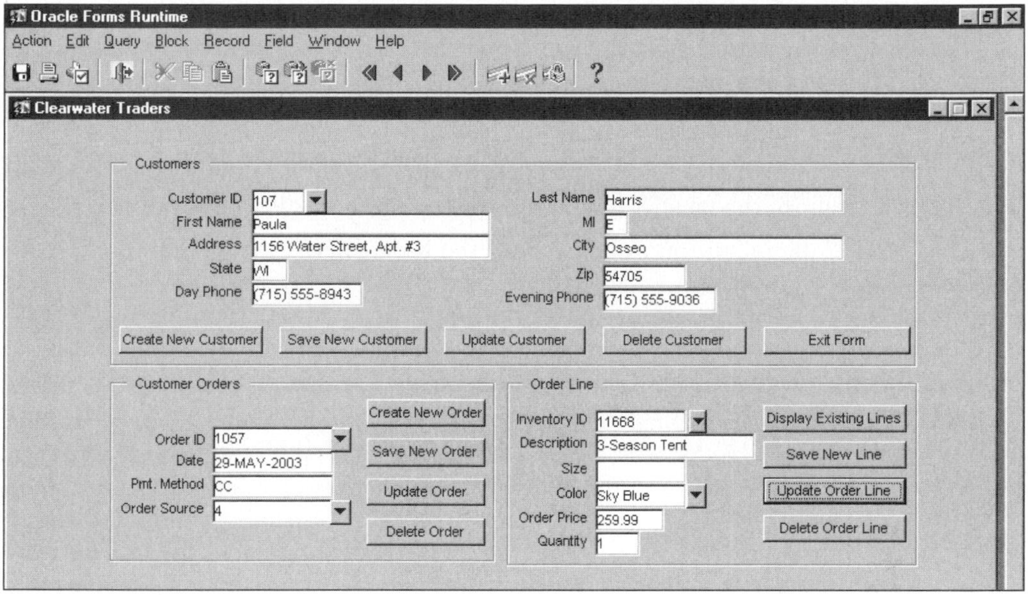

Figure 7-31

After the user creates a new customer record or selects an existing customer record, he or she can click Create New Order in the Customer Orders frame to display a new customer order ID, enter the order data in the data fields, and then click Save New Order to save the order. If the user wants to edit an existing order, he or she clicks the LOV command button next to the Order ID field to view the selected customer's orders. The user selects an order, edits the order information, and clicks Update Order to update the order, or Delete Order to delete the order.

After the user creates or selects a customer order, he or she can create new order lines for the order by clicking the LOV command button next to Inventory ID in the Order Line frame, selecting an Inventory ID and entering a quantity, and then clicking the Save New Line button. (The primary key of the ORDER_LINE table is the combination of ORDER_ID and INV_ID, so no new primary key needs to be generated.) If the user clicks the Display Existing Lines button, the order line information for the Customer Order ID that is currently selected is displayed. The user can select an existing order line and edit or delete it as needed. Open the file named 7BCase6.fmb, make the following modifications, and save the modified file as 7BCase6_DONE.fmb.

a. In SQL*Plus, create a new sequence named CUST_ID_SEQUENCE that starts with 200 and has no maximum value.

b. In SQL*Plus, create a new sequence named ORDER_ID_SEQUENCE that starts with 1100 and has no maximum value. (If you created this sequence previously, use the existing sequence.)

c. Add a Caution alert to notify the user that a record is about to be inserted. The alert should appear before the INSERT command is committed, and show the message "Add the new record to the database?" If the user clicks the OK button, commit the insert, and show a message stating "Record successfully inserted." If the user clicks the Cancel button, do not commit the change, and show a message stating "Changes not saved." Display the alert when the user clicks the Save New Customer, Save New Order, or Save New Line button.

d. Add a Caution alert to notify the user when a record is going to be updated. The alert should appear before the UPDATE command is committed and display the message "Update the current record?" If the user clicks OK, commit the update, and show a message stating "Record successfully updated." If the user clicks Cancel, do not commit the change, and show a message stating "Changes not saved." Display the alert when the user clicks the Update Customer, Update Order, or Update Order Line button.

e. Add a Caution alert to notify the user when a record is about to be deleted. The alert should appear before the DELETE command is committed and show the message "Delete the current record?" If the user clicks the OK button, commit the delete, and show a message stating "Record successfully deleted." If the user clicks the Cancel button, do not commit the change, and show a message stating "Changes not saved." Display the alert when the user clicks the Delete Customer, Delete Order, or Delete Order Line button. After the user deletes an order record, the text items in the Customer Orders and Order Line frames should be cleared, but the text items in the Customer frame should still appear. After the user deletes an order line record, the text items in the Order Line frame should be cleared, but the text items in the Customers and Customer Orders frames should still appear. (*Hint*: Program units named CLEAR_CUSTORDER_FIELDS and CLEAR_ORDERLINE_FIELDS are included in the file to clear these fields correctly.)

f. Do not allow the user to change the values of CUST_ID or ORDER_ID, except through the LOV command buttons, by redirecting the form focus to the LOV command buttons associated with the text items.

g. Validate the Order Source, Inventory, and Color values using their associated LOVs.

h. Create a Stop alert with only an OK button that appears if the user clicks the Create New Order button or the LOV command button beside Order ID in the Customer Orders frame when no value has been selected for customer ID. The alert should remind the user that he or she must select a customer before he or she can create or select a customer order.

i. Create a Stop alert with only an OK button that appears if the user clicks the Display Existing Lines button or the LOV command button beside Inventory ID in the Order Line frame when no value has been selected for Order ID. The alert should remind the user that he or she must select an order ID before he or she can create or select an order line.

j. Create a Stop alert with only an OK button that appears if the user clicks the LOV command button beside the Color field in the Order Line frame when no value has been selected for Inventory ID. The alert should remind the user that he or she must select an inventory item before he or she can select a color for that item.

k. Add an ON-ERROR trigger that displays a customized alert to handle the following errors: ORA-00001 (trying to insert a record with a primary key value that already exists in the table), ORA-02292 (trying to delete a record that is referenced in another table), and ORA-01407 (trying to insert NULL into a field with a NOT NULL constraint). Create customized alerts with informative messages for each of these errors. Display the associated Oracle system error description for all unhandled FRM- and ORA- errors.

l. Create a validation trigger to ensure that the value entered in the Payment Method text item is either CC or CHECK.

7

◀ LESSON C ▶

Objectives

♦ Link data blocks to control blocks
♦ Work with a form that has multiple canvases
♦ Create tabbed and stacked canvases

CREATING A RELATIONSHIP BETWEEN A DATA BLOCK AND A CONTROL BLOCK

Recall that a data block is directly associated with a database table or view and contains text items that correspond to the table fields. You can make data blocks quickly and easily using the Data Block and Layout Wizards, and the Forms Runtime application has built-in programs for viewing, inserting, updating, and deleting data in data block forms. However, data block forms do not have the flexibility that is sometimes needed for custom form applications, because data block forms are based on the structure of the database tables rather than on the organizational processes that they support. When you need to work with fields from multiple tables, you must create a control block, which is not associated with a particular database table, but instead has fields that can contain data values that can be referenced in SQL queries written by the form developer. Sometimes it is useful to combine data blocks and control blocks on the same canvas to take advantage of the strengths of both block types.

To learn about linking control blocks and data blocks, you will work with a form that creates customer orders and associated order line information. First, you will refresh your database tables. Then, you will create a sequence to automatically generate order ID values and open and run the form to see how it works.

To refresh your database tables, create the sequence, and open and examine the form:

1. Start SQL*Plus, log onto the database and run the **clearwater.sql**, **northwoods.sql** and **softwareexp.sql** scripts that are stored in the Chapter7 folder on your Data Disk.

2. In SQL*Plus, create a new sequence using the following command:

```
CREATE SEQUENCE order_id_sequence
START WITH 2000;
```

 If you already created this sequence in a previous lesson, use the existing sequence.

584

3. Start Form Builder, open **CustOrder.fmb** from the Chapter7 folder on your Data Disk, and save the form as **7CCustOrder.fmb** in your Chapter7\TutorialSolutions folder.

4. Run the form and connect to the database. The form appears, as shown in Figure 7–32.

Recall that to make the icon appear on the LOV command button, you must change the Icon Filename property in the LOV command button to specify the complete folder path and filename of the .ico file.

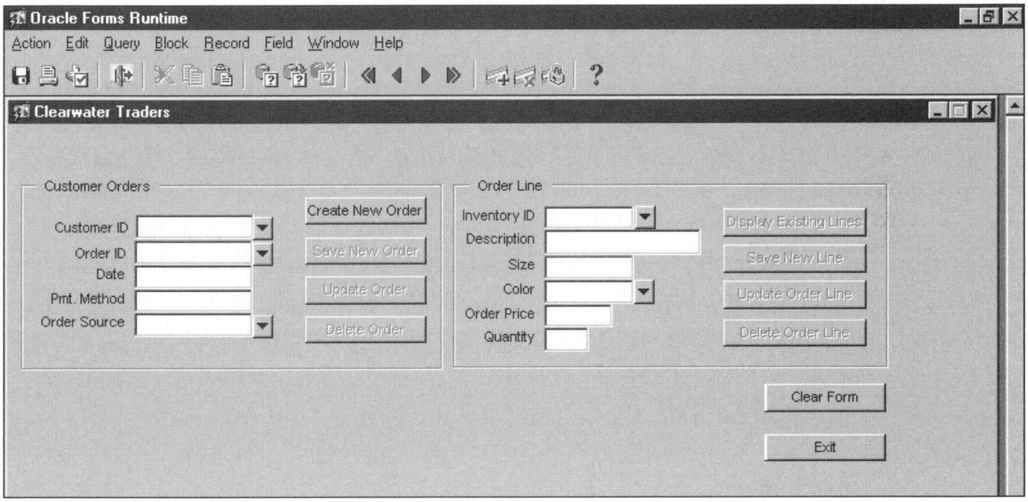

Figure 7-32   Customer order form

The form has two frames: Customer Orders, which corresponds to a control block named CUST_ORDER, and Order Line, which corresponds to a control block named ORDER_LINE. Currently, the Create New Order and LOV command button beside the Customer ID text item are enabled. You can create a new order. You can also select a customer, then select an existing customer order, and edit or delete the order or the corresponding order lines. To see how the form works, you will create a new order and add some order lines to the order.

To create the new order:

1. Click **Create New Order**. A new order ID value appears, and the current system date appears for the order date.

2. Click the **LOV command button** beside Customer ID, and select Customer ID **107** (Paula Harris) and then click **OK**.

3. Type **CC** for Payment Method, click the **LOV command button** beside the Order Source text item, select Order Source ID **4** (Outdoor 2003), click **OK**, and then click **Save New Order**. The confirmation message "FRM-40401: No Changes to Save" that appears in the status line indicates that the new order is saved in the database.

4. Click the **LOV command button** beside Inventory ID, and select ID **11669** (3-Season tent, color Light Grey) and then click **OK**. Type **1** for Quantity, and then click **Save New Line**. The confirmation message "FRM-40401: No Changes to Save" indicates the line was successfully saved.

 If you receive the error message "FRM-40735: WHEN-BUTTON-PRESSED trigger raised unhandled exception ORA-02291," you probably forgot to click the Save New Order button to save the order information. Click Save New Order, and then click Save New Line.

5. Click the **LOV command button** beside Inventory ID, and select ID **11775** (Women's Hiking Shorts, color Khaki, size S) and then click **OK**. Type **1** for Quantity, and then click **Save New Line**. The confirmation message "FRM-40401: No Changes to Save" indicates the line was successfully saved.

6. Click **Display Existing Lines**. The information about the order lines for the current order appears in the LOV display. Select the order line corresponding to inventory ID **11775**.

7. Change the quantity to **2**, and then click **Update Order Line**. The confirmation message indicates that the order line was updated successfully.

8. Click **Display Existing Lines**. The updated order line information appears. Click **Cancel**.

9 Click **Exit** to exit Forms Runtime.

This form would be easier to use if the order line information appeared directly on the form, rather than in a LOV display. To make this change, you will create a view-only order summary that appears at the bottom of the canvas and shows information about each ordered item, including the inventory ID, quantity ordered, description, size, color, and price. This information will appear in a data block with a tabular layout. Since this display involves information from the ITEM, INVENTORY, and ORDER_LINE database tables, you will have to create a database view in SQL*Plus as the data block source. The view must contain the ORDER_ID, because the data block will have a master-detail relationship with the ORDER_ID text item in the CUST_ORDER block so that it displays the order line information only for the current customer order, rather than for all order lines in the database. The view must also contain the inventory ID, price, order quantity, item description, color, and item size fields from the INVENTORY, ITEM, and ORDER_LINE tables. Now you will create the view and corresponding data block to display the order summary information.

To create the view and data block:

1. Switch to SQL*Plus, and type the following command to create the ORDER_SUMMARY_VIEW on which the data block will be based:

```
CREATE VIEW order_summary_view AS
SELECT order_id, inventory.inv_id, price,
order_quantity, item_desc, color, item_size
FROM order_line, inventory, item
WHERE order_line.inv_id = inventory.inv_id
AND inventory.item_id = item.item_id;
```

 This query is also stored in the order_summary_view.sql script file in the Chapter7 folder on your Data Disk. Instead of typing the command, you could run the script file in SQL*Plus by typing start a:\chapter7\order_summary_view.sql at the SQL prompt. (The path to the Chapter7 folder on your Data Disk might be different if you are not storing your Data Disk files on the A: drive.)

2. Switch back to Form Builder. Open the Object Navigator window if necessary, and select the **CUST_ORDER_FORM** module. Since you are going to create a new data block using the Data Block Wizard, you do not want to have any data block currently selected. If a data block is selected, the Wizard will open the selected data block in reentrant mode for editing, rather than creating a new data block.

3. Click **Tools** on the menu bar, click **Data Block Wizard**, and then click **Next** when the Data Block Wizard Welcome page appears. When the Type page appears, make sure the **Table or View** option is selected, and then click **Next**.

4. On the Table page, click **Browse**, check the **Views** check box, select **ORDER_SUMMARY_VIEW**, and then click **OK**.

5. Select all of the view data fields to be included in the block, and then click **Next**. The Master-Detail page appears.

When you display a control block and a data block on the same canvas, you cannot directly create a master-detail relationship between the data block and the control block. Instead, you must create the data block without a master-detail relationship and then later modify the detail data block's WHERE property on its Property Palette so that the matching key field on the detail data block is equal to the value of the master key field on the master control block. For now, you will omit the master-detail relationship and finish creating the data block.

To omit the master/detail relationship and finish the data block:

1. Click **Next** to pass the Master-Detail page, and then click **Finish**. The Layout Wizard Welcome page appears.

Next, you will create the block layout using the Layout Wizard. The block will show five records at a time and have a scroll bar. You will display all of the block items except the ORDER_ID. The ORDER_ID field was included in the data block to create the master-detail relationship, but it does not need to appear in the order summary, because the order summary shows order items only for the current customer order.

7

To create the block layout:

1. Click **Next** on the Layout Wizard Welcome page.

2. On the Canvas page, make sure that ORDER_CANVAS is the selected canvas, and then click **Next**.

3. On the Data Block page, select all of the block items for the layout except ORDER_ID. Drag and drop the items so they are in the order shown in Figure 7-33, and then click **Next**.

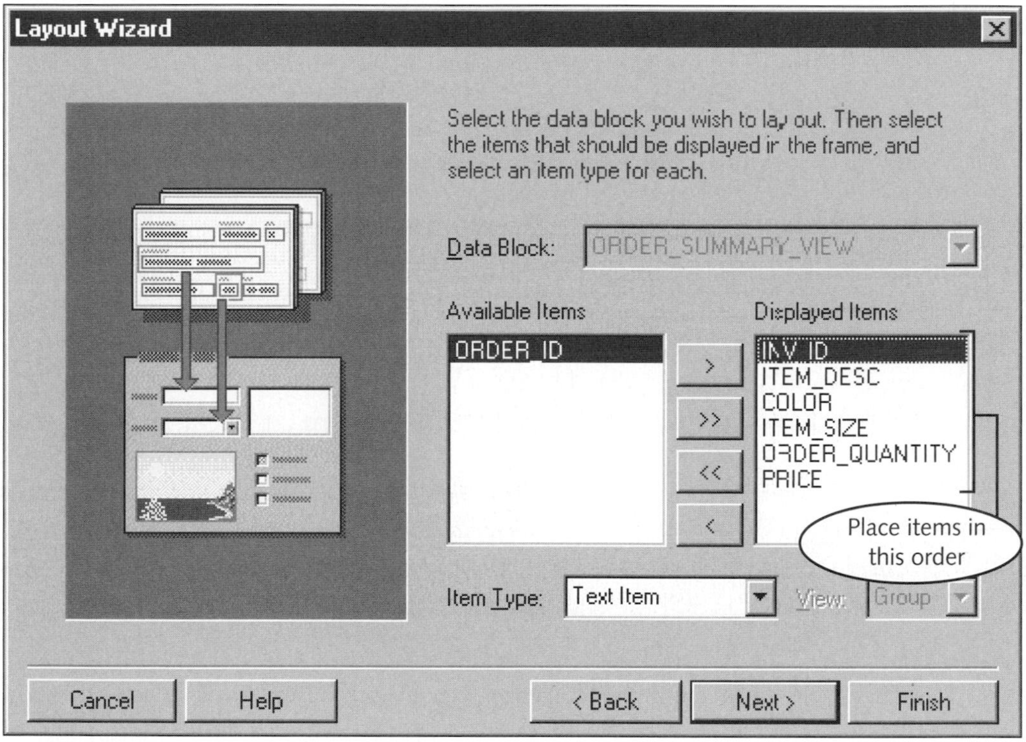

Figure 7-33 Specifying the layout items

4. On the Items page, modify the column prompts and widths as follows, and then click **Next**.

Name	Prompt	Width
INV_ID	Inv. ID	35
ITEM_DESC	Description	80
COLOR	Color	75
ITEM_SIZE	Size	35
ORDER_QUANTITY	Quantity	25
PRICE	Price	41

5. On the Style page, select the **Tabular option button**, and then click **Next**.

6. On the Rows page, type **Order Summary (View Only)** for the Frame Title. Type **5** for Records Displayed, and check the **Display Scrollbar check box**. Click **Next**, and then click **Finish**. The Order Summary appears on the canvas.

7. Move the Order Summary display so it is directly under the Customer Order frame, and then save the form.

Creating the Relationship Between the Control Block and the Data Block

Recall that when you create a relationship between a master control block and a detail data block, you do not specify a master-detail relationship in the Data Block Wizard when you create the detail data block. Instead, you modify the detail data block's WHERE property on its Property Palette so that the matching key field on the data block is equal to the value of the master key field on the control block. The general syntax of the WHERE property is *table_field_name = :control_block.text_item_name* In the ORDER_SUMMARY_VIEW detail block, you will set the order ID field equal to the value of the order ID text item in the CUST_ORDER master block.

Next, you will create the relationship between the CUST_ORDER master control block and the ORDER_SUMMARY_VIEW detail data block. You will open the Property Palette for the ORDER_SUMMARY_VIEW detail data block and specify that the value for ORDER_ID in the detail data block is equal to the value of ORDER_ID in the CUST_ORDER master control block.

To create the relationship between the data block and control block:

1. Open the Object Navigator window in Ownership View, and double-click the **Data Block icon** 🔡 beside ORDER_SUMMARY_VIEW to open the data block Property Palette.

2. Type the following statement for the WHERE Clause property:

```
order_id = :cust_order.order_id
```

3. Close the Property Palette, and save the form.

Displaying and Refreshing the Data Block Values

After you create a data block that is associated with a control block, you need to add commands to initially populate the data block display and periodically refresh the data block display when the data in the master block or the underlying view changes. To populate and refresh the display, you move the form focus to the data block and then refresh the display using the EXECUTE_QUERY built in, which automatically **flushes** the block, or makes its information consistent with the block's corresponding database data. In this form, you will need to refresh the Order Summary data block when the user creates a new order, selects an order ID, or updates or deletes an order line within the existing order. First, you will modify the trigger associated with the LOV command button beside Order ID so the Order Summary is refreshed when the user selects an order ID.

To add the code to refresh the Order Summary when the user selects a different order ID:

1. Click **Tools** on the menu bar, and then click **Layout Editor** to open the Layout Editor. Right-click the **LOV command button** beside Order ID, and click **PL/SQL Editor**. The PL/SQL Editor opens, displaying the current trigger code. The current code places the insertion point in the ORDER_ID text item, displays the LOV, and then enables the buttons for updating and deleting order and order line information.

2. Add the following code as the last two lines of the trigger to move the form focus to the data block, and then execute the query to refresh the data block display:

```
GO_BLOCK ('order_summary_view');
EXECUTE_QUERY;
```

3. Compile the trigger, correct any syntax errors, and then close the PL/SQL Editor.

4. Save the form.

Now, you will test the trigger to confirm that the order line information appears when the user selects an order and that the order line information is refreshed when the user selects a different order.

To test the trigger that refreshes the order line display:

1. Run the form, click the **LOV command button** beside Customer ID, and select Customer ID **107** (Paula Harris).

2. Click the **LOV command button** beside Order ID, and select the order with an order date for the current date, which is the order that you created earlier. The order summary information should appear in the data block, as shown in Figure 7-34. Now you will display information for a different order.

3. Click **Clear Form** to clear the form text items, click the **LOV command button** beside Customer ID, and select Customer ID **232** (Mitch Edwards).

4. Click the **LOV command button** beside Order ID, and select order ID **1058**. The order summary information appears in the data block.

5. Click **Exit** to close Forms Runtime.

6. Click **File** on the menu bar, and then click **Close** to close the form in Form Builder.

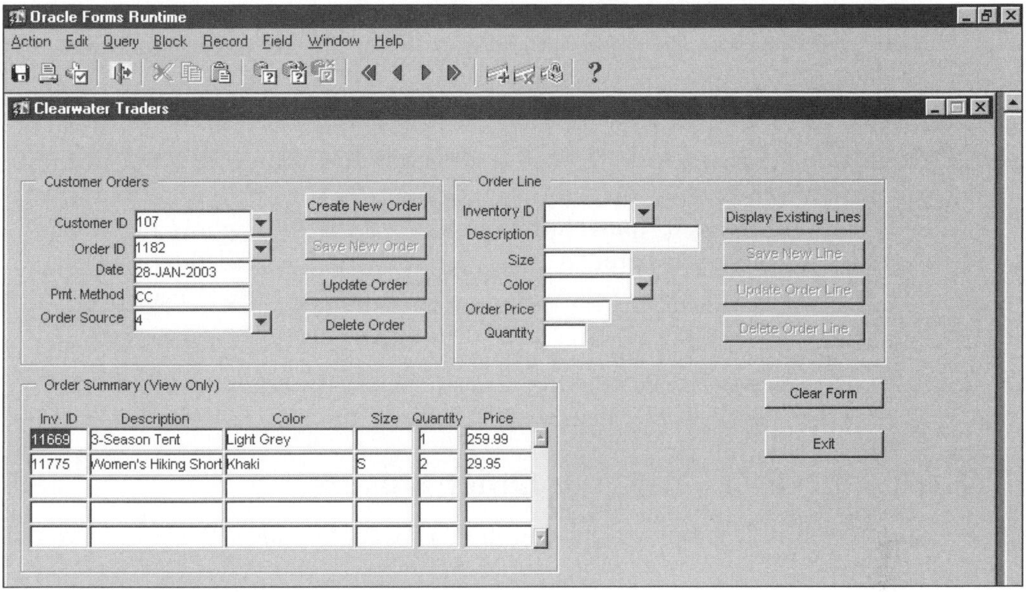

Figure 7-34   Displaying order summary information

WORKING WITH A FORM WITH MULTIPLE CANVASES

As database applications become more complex, it is a good practice to display information on different screens. A general rule is to not place more information on a single screen than a user can view without scrolling. Otherwise, the user can become disoriented about her or his position on the form and can suffer from information overload.

For example, when a Clearwater Traders customer calls the company to place a new order, the customer service representative processes the order by entering information in an on-screen form. The first form screen is used to enter customer order information. The customer service representative can access a second screen to enter a new customer record or to modify customer information, if necessary. The application then displays a third screen that is used to specify information about the individual order items.

Form Builder provides two approaches for creating applications with multiple screens. The **single-form approach** involves creating one form with multiple canvases. This approach allows data to easily be shared among the different canvases but makes it impossible for multiple programmers to program different canvases of the same application at the same time, since the .fmb file can only be opened and modified by one programmer at a time.

The **multiple-form approach** involves creating multiple forms with a different .fmb file for each canvas in the application. This approach works well when multiple programmers are collaborating to create a complex application. It also enables a form to be used in many different applications, thus avoiding redundant programming efforts. Data sharing among forms is more difficult in the multiple-form approach than in the single-form approach and will be discussed in Chapter 11.

In general, you use the single-form approach to create application components that are closely related and have to share many variable values. You use the multiple-form approach to integrate several form applications into a single database system. For example, you might create individual form applications to support the Clearwater Traders merchandise receiving process and sales order process, and then integrate them into a single system.

Now you will create an application to support the Clearwater Traders sales process that uses the single-form approach. You will learn how to use the multiple-form approach to create an integrated system in Chapter 11.

The Clearwater Customer Order Process

When you design a new form, you should always first identify the business processes that the form will support and the associated database tables. When the customer service representative at Clearwater Traders takes a customer order, he or she first must either create a record for a new customer or retrieve the CUST_ID for an existing customer. Then, the representative must enter the payment method and order source ID (catalog number or Web site) and create a new CUST_ORDER record. Next, the customer service representative must enter the inventory number and desired quantity for each order item and insert a corresponding record in the ORDER_LINE table.

Figure 7-35 illustrates the customer order process in a flowchart. The application starts on the CUSTOMER canvas. If the customer is an existing customer, an LOV retrieves the CUST_ID. The sales representative verifies the customer information, and updates the information in the CUSTOMER table, if necessary. If the customer is a new customer, the sales representative records the new customer information and inserts a new record into the CUSTOMER table. The process then moves to the CUST_ORDER canvas, where the method of payment and order source are entered, and the CUST_ORDER record is inserted.

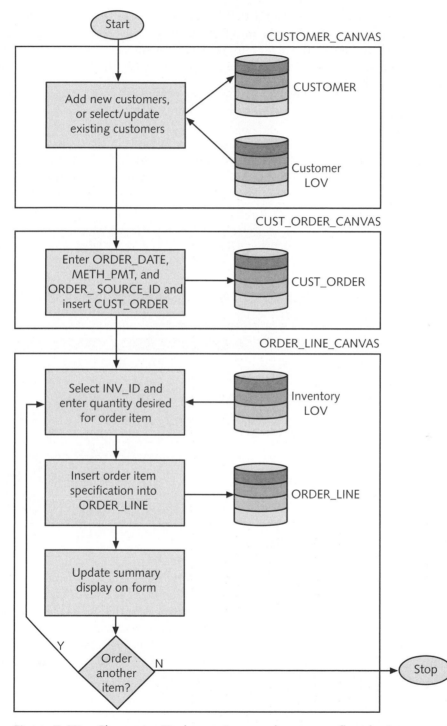

Figure 7-35   Clearwater Traders customer sales process flowchart

The process then moves to the ORDER_LINE canvas. The sales representative retrieves the first item's inventory ID using the inventory LOV. The representative enters the order quantity and then inserts the order item specification for the first item into the ORDER_LINE table. A summary of items ordered so far appears on the form. This process repeats until all items are ordered.

Interface Design

Figure 7-36 shows the CUSTOMER canvas layout. When the sales representative clicks Create New Customer, a new customer ID appears on the form. The sales representative can then enter the new customer data and click Save New Customer to insert a new customer record into the CUSTOMER table. Alternately, the sales representative can click the LOV command button next to the Customer ID text item to select from a list of existing customers, verify the information, and click Update Customer to update the selected record. The sales representative then clicks Create New Order to move to the CUST_ORDER canvas.

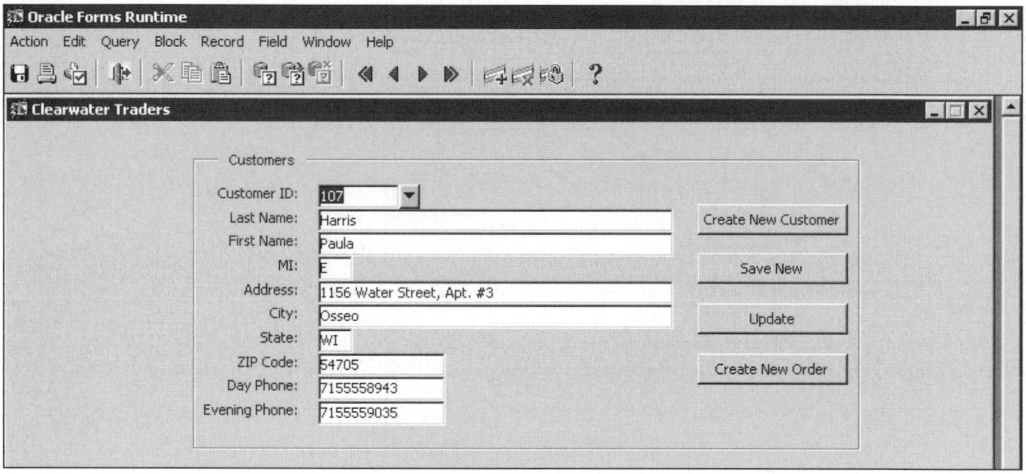

Figure 7-36   CUSTOMER canvas layout

The layout for the CUST_ORDER canvas is shown in Figure 7-37. When the canvas first appears, the values for the new order ID and the current system date automatically appear in the form text items. The sales representative uses the option buttons to select the payment method (check or credit card type) and selects the order source ID using an LOV that will show valid order sources from the ORDER_SOURCE table. After entering this information, the sales representative clicks Save Order to insert the new order record into the CUST_ORDER table. The Enter/Edit Items button (which is the second button in the button group, and is currently disabled) is then enabled. The sales representative clicks Enter/Edit Items to display the ORDER_LINE canvas, which is used to specify the order items. Alternately, the sales representative can click Edit Customer to redisplay the CUSTOMER canvas and change the customer information if necessary.

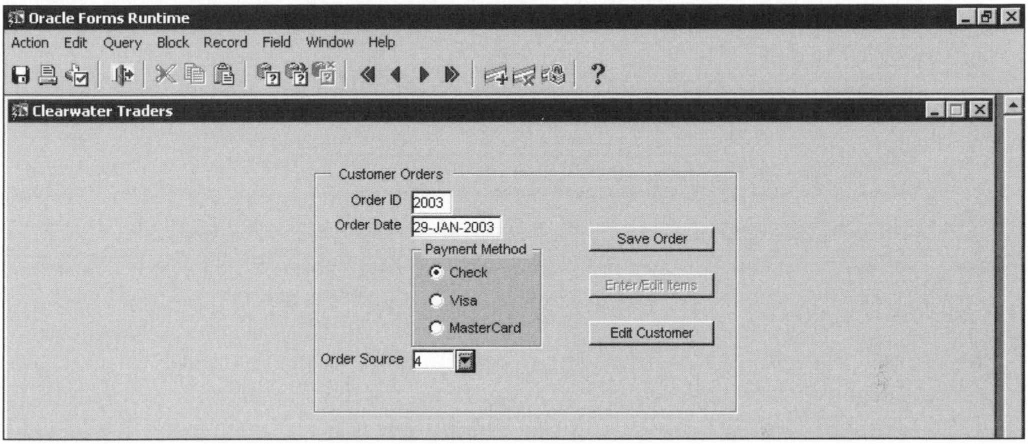

Figure 7-37   CUST_ORDER canvas layout

The ORDER_LINE canvas (see Figure 7-38) has text items that enable the sales representative to specify the data for a specific order line (inventory ID, description, size, color, price, and desired quantity). When the sales representative clicks Add Order Line, the order line information is inserted into the ORDER_LINE table. The sales representative can click Select Order Line to display an LOV that shows all of the order lines for the current order and allows the sales representative to select a specific order line and retrieve its data into the Order Items fields. The sales representative can then modify the order line data and update it using the Update Line button or delete it using the Delete Line button. When all order items are entered, the sales representative clicks Next Order to clear the form text items and redisplay the CUSTOMER canvas to begin entering information for a new order or Exit to exit the form.

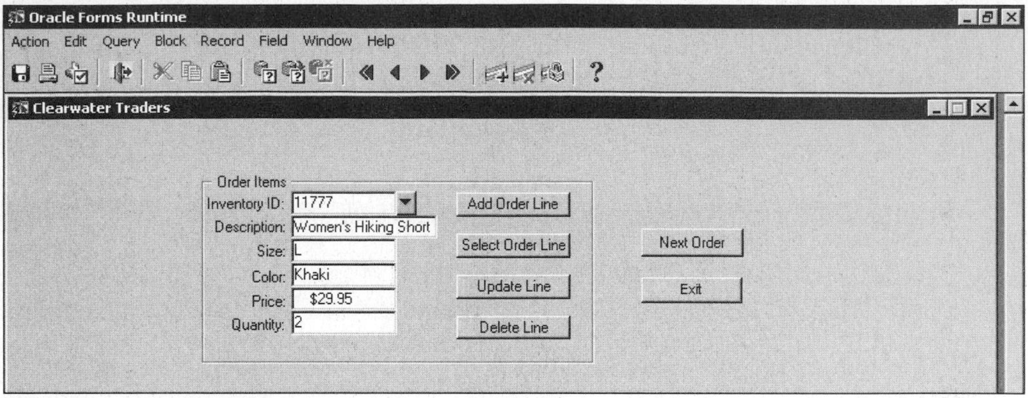

Figure 7-38   ORDER_LINE canvas layout

Working with Multiple Canvases and Data Blocks

Now you will open a form that is saved in a file named Sales.fmb in the Chapter7 folder on your Data Disk. This form contains the layouts for the CUST_ORDER and ORDER_LINE canvases. You will modify this form by adding the CUSTOMER canvas and then adding the trigger code to all of the form buttons so the canvases work together to make a finished application. Now, you will open the form file and examine its contents.

To open and examine the form:

1. Open **Sales.fmb** from the Chapter7 folder of your Data Disk and save it as **7CSales.fmb** in your Chapter 7\TutorialSolutions folder.

2. In the Object Navigator, click ➕ beside the Data Blocks node to examine the form data blocks. Note that there are currently two data blocks: CUST_ORDER and ORDER_LINE. When you are creating a form with multiple canvases, you should always place the items for each canvas in a separate data block to keep the block item lists smaller and more manageable.

> **Tip** You cannot place two items that have the same name in the same data block. For example, you might want to display a text item named CUST_ID on two different canvases. To do so, you must place the items that have the same name in separate data blocks.

3. Click ➕ beside the Canvases node, and then double-click the **Canvas icon** ▭ beside CUST_ORDER_CANVAS. The canvas appears in the Layout Editor.

4. Select the **ORDER_ID text item**, and note the values that appear in the Canvas and Block lists at the top of the Layout Editor window. The Canvas list displays CUST_ORDER_CANVAS, which is the name of the canvas that is currently displayed, and the Block list displays CUST_ORDER, which is the name of the block that is associated with the item that is currently selected.

5. Open the Canvas list, and select **ORDER_LINE_CANVAS**. The canvas displaying the order line items appears. Note that the Block list still displays CUST_ORDER. That is because you have not yet selected an item on the ORDER_LINE_CANVAS, and the item that was previously selected (ORDER_ID on the previous canvas) is in the CUST_ORDER block.

6. Select the **INV_ID text item**. Note that the block name changes to ORDER_LINE.

When you are working in forms with multiple blocks and multiple canvases, you must always be aware of what block the current item is in. Sometimes when you create a new item on a canvas, it is mistakenly placed in a different block than the rest of the canvas items because the current block in the Block list does not correspond to the items on the current canvas.

Converting a Data Block to a Control Block

First, you will create the CUSTOMER block and canvas. Instead of manually creating the canvas and all of the text items, you will create a data block and then convert it to a control block. To do this, you will create the data block in the usual way and then change some of the properties in the block and block items to convert the data block to a control block. First, you will create the CUSTOMER block using the Data Block and Layout Wizards.

To create the CUSTOMER data block and layout using the wizards:

1. In the Object Navigator, click the **SALES_FORM** module, click **Tools** on the menu bar, and then click **Data Block Wizard**. The Welcome page appears. Click **Next**.

2. Make sure that the Table or View option button is selected, and then click **Next**. On the Source page, click **Browse**, select the **CUSTOMER** database table, click **OK**, select all of the database items for the block, and then click **Next**. The Relationships page appears. Since this is a custom form and the data block will not have a master-detail relationship with any other data blocks in the form, click **Next** to skip creating a relationship, and then click **Finish**.

3. The Layout Wizard Welcome page appears. Click **Next**.

4. On the Canvas page, open the Canvas list, and note that both of the existing form canvases are listed. Since you want the customer items to appear on a different canvas, select **(New Canvas)**, and then click **Next**.

5. Select all of the data block items for the layout, and then click **Next**.

6. Change the prompts so they appear as shown in Figure 7-36, and then click **Next**.

7. Accept Form as the layout style, click **Next**, type **Customers** for the frame title, click **Next**, and then click **Finish**. The canvas layout appears in the Layout Editor.

8. Open the Object Navigator, and change the canvas name to **CUSTOMER_CANVAS**.

9. Save the form.

When you create a data block using the Data Block Wizard, the block Database Data Block property value is Yes. The Required property of the text item associated with the block table's primary key is also set to Yes, because every record must have a primary key value. Now, you will modify the block and layout so it can be used as a control block. You will change the block's Database Data Block property to No and the CUST_ID text item's Required property to No. (When a text item is required, text item validation does not allow the user to move the insertion point out of the text item until he or she enters a value.) Finally, you will change the Update Layout property of the layout frame to Manually, so you can add custom items to the layout. Recall that if you do not change the frame Update Layout property to Manually, the frame items will automatically be stacked on top of each other in the order they appear in the Object Navigator whenever you resize or move the frame.

To modify the block and layout so it can be used as a control block:

1. In the Object Navigator, double-click the **Data Block icon** 🖳 beside the CUSTOMER data block to open its Property Palette, scroll down to the Database property node, change the Database Data Block property to **No**, and then close the Property Palette.

2. Double-click the **Canvas icon** 🖼 beside CUSTOMER_CANVAS, select the **CUST_ID text item**, right-click, and then click **Property Palette**. Scroll down to the Data property node, change the Required property value to **No**, and then close the Property Palette.

3. In the Layout Editor, select the frame that surrounds the customer items, right-click, and then click **Property Palette**. Change the Update Layout property to **Manually**, and then close the Property Palette.

4. Save the form.

Next, you will create the LOV that retrieves the current customer values. You will also create the LOV command button that is beside the customer ID text item.

To create the LOV:

1. Click **Tools** on the menu bar, and then click **LOV Wizard**. Make sure the New Record Group based on a query option button is selected, and then click **Next**.

2. In the SQL Query Statement box, type **SELECT * FROM customer**, and then click **Next**. Select all of the record group columns for the LOV columns, and then click **Next**.

3. Modify the LOV column properties as follows, and then click **Next**.

Column	Title	Width	Return Value
CUST_ID	**Customer ID**	30	**CUSTOMER.CUST_ID**
LAST	**Last Name**	80	**CUSTOMER.LAST**
FIRST	**First Name**	80	**CUSTOMER.FIRST**
MI	(not displayed)	0	**CUSTOMER.MI**
ADDRESS	(not displayed)	0	**CUSTOMER.ADDRESS**
CITY	(not displayed)	0	**CUSTOMER.CITY**
STATE	(not displayed)	0	**CUSTOMER.STATE**
ZIP	(not displayed)	0	**CUSTOMER.ZIP**
D_PHONE	(not displayed)	0	**CUSTOMER.D_PHONE**
E_PHONE	(not displayed)	0	**CUSTOMER.E_PHONE**

4. Type **Customers** for the frame title, change the Width to **270** and the Height to **160**, and then click **Next**.

5. Accept the default properties on the Advanced page, and then click **Next**. Select **CUSTOMER.CUST_ID** for the assigned text item, click **Next**, and then click **Finish**.

6. Open the Object Navigator, and change the names of the new LOV and record group to **CUSTOMER_LOV**.

Now you will create the LOV command button. You must make sure that you place this button in the CUSTOMER data block. If you do not place the button in the CUSTOMER data block, the user cannot use the Tab key to navigate in the form. To specify the block in which an item is placed, you must select the desired block from the Block list at the top of the Layout Editor window.

To create the LOV command button:

1. In the Layout Editor, select **CUSTOMER** from the Block list at the top of the window.

2. Select the **Button tool** ▦ on the tool palette, and draw a new command button for the LOV command button. Position the button on the right edge of the CUST_ID text item, as shown in Figure 7-36, change the button properties in the Property Palette as follows, and then close the Property Palette:

Name	**CUSTOMER_LOV_BUTTON**
Label	(deleted)
Iconic	**Yes**
Icon Filename	**a:\chapter7\down3** (or the path to the Chapter7 folder on your Data Disk)
Width	**12**
Height	**14**

3. Create a WHEN-BUTTON-PRESSED trigger for the button that contains the following commands:

```
GO_ITEM('customer.cust_id');
LIST_VALUES;
```

4. In the Object Navigator, confirm that the CUSTOMER_LOV_BUTTON is listed under the CUSTOMER data block, and drag the button so it is directly below the CUST_ID item in the Items list so the form navigation order is correct.

5. Save the form.

To complete the CUSTOMER_CANVAS layout, you need to create the command buttons shown in Figure 7-36 for creating and saving a new customer record, updating an existing customer record, and creating a new order. You will resize the Customers frame to make it larger, draw the command buttons, and then modify the button properties. Again, you must be careful to ensure that the command buttons are placed in the CUSTOMER data block, or you will have problems with form navigation later.

To create the command buttons:

1. In the Layout Editor, select the **Customers frame**, and drag the right-center selection handle to the right edge of the canvas to make the frame larger, as shown in Figure 7-36.

2. Make sure that the CUSTOMER block is selected in the Block list, and then draw four command buttons on the right side of the text items, as shown in Figure 7-36.

3. Open the Property Palette for each button, modify the button properties as follows, and then close the Property Palette.

Name	Label	Width	Height
CREATE_CUSTOMER	**Create New Customer**	86	18
SAVE_CUSTOMER	**Save New Customer**	86	18
UPDATE_CUSTOMER	**Update Customer**	86	18
CREATE_ORDER	**Create New Order**	86	18

4. Select all of the buttons as an object group, click **Arrange** on the menu bar, click **Align Objects**, select the Horizontally **Align Left** and Vertically **Distribute** options buttons, and then click **OK**.

5. Click **Edit** on the menu bar, click **Select All**, and then reposition the frame and the enclosed text items so they are centered on the canvas, as shown in Figure 7-36.

6. Save the form.

Specifying the Block Navigation Order

Before you can run the form, you need to ensure that the Customer canvas will appear as the first canvas when the form opens. When you create a form with multiple canvases, the canvas that appears first is the canvas whose block items appear first under the Data Blocks node in the Object Navigator window. The order of the canvases in the Object Navigator does not matter, only the order of the data blocks. Now you will open the Object Navigator window, examine the order of the blocks, and if necessary, make the CUSTOMER block the first block that appears in the block list. Since CUST_ORDER will be the second block that is used in the application, you will make it the second block that appears in the block list. Then, you will run the form to ensure that the canvas containing the customer block items appears first.

To modify the order of the form blocks and test the form:

1. Open the Object Navigator window in Ownership View.

2. Select the **Data Blocks node**, and then click the **Collapse All button** 🔲 on the toolbar.

3. Click ➕ beside the Data Blocks node to examine the block order.

4. If CUSTOMER is not the first block in the list, select the **CUSTOMER** data block, and drag it toward the top edge of the screen display. The pointer changes to the move block pointer 🖑. When the pointer is on the Data Blocks node and the Data Blocks node appears selected, drop the CUSTOMER data block to make it the first block in the list.

5. If CUST_ORDER is not currently the second block, select it, drag it up until the mouse pointer is on the CUSTOMER block, and drop the CUST_ORDER block so it appears directly below the CUSTOMER block. Your form data block list should appear as shown in Figure 7-39.

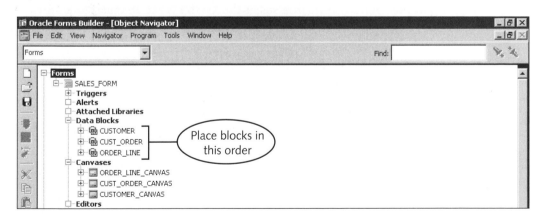

Figure 7-39   Specifying the form data block order

6. Save the form, run the form, and click the **LOV command button** beside the Customer ID text item. The LOV display opens.

7. Select Customer ID **107** (Paula Harris), and then click **OK**. The customer data values appear, as shown in Figure 7–36.

8. Close Forms Runtime.

Navigating in Different Form Blocks and Canvases

Recall that users can press the Tab key to navigate to different items in a form data block. You can set the **Navigation Style** property of a data block to specify that when the user has navigated to the last item in the current block, he or she can press the Tab key to move to the first item in the next data block. Items, such as text items, have a **Canvas** property, which specifies the name of the canvas on which they appear. If the next item is on a canvas that is different from the canvas that currently appears in the Forms Runtime window, the new canvas appears when the user presses the Tab key and navigates to the next text item.

The default value for the Block Navigation Style property is **Same Record**, which means that once the user navigates through all items in the block, item navigation cycles back to the first item in the same block and moves through the same block items again. Alternately, you can set this property to **Change Data Block**, to specify that after the user navigates to the last item in the current block, navigation moves to the first item in the next block that appears in the Object Navigator Data Blocks list.

Now you will change the Block Navigation Style property in the form data blocks to Change Data Block, so the user can press the Tab key to move to the next block. Then, you will run the form and navigate through all of the form blocks and their associated items.

To change the Block Navigation Style property and navigate through the form blocks:

1. In the Object Navigator, select the **CUSTOMER** data block, press and hold the **Shift** key, and select the **CUST_ORDER** and **ORDER_LINE** data blocks to select all three blocks as an object group.

2. Click **Tools** on the menu bar, and then click **Property Palette** to open the intersection Property Palette.

3. Under the Navigation property node, open the **Navigation Style** property list, select **Change Data Block**, and then close the Property Palette.

4. Save and then run the form. The canvas with the Customers frame appears. Press **Tab** to navigate through the canvas text items until the Create New Order button has the form focus.

If one or more of the form buttons is skipped, then you placed the skipped buttons in the wrong data block or the buttons are in the wrong order in the data block. Exit Forms Runtime, open the Object Navigator, look in the CUST_ORDER and ORDER_LINE data blocks for the missing buttons, and drag and drop them into the CUSTOMER data block.

5. Press **Tab** again. The canvas with the Customer Orders frame appears. Press **Tab** to navigate through the canvas items until the Edit Customer button has the form focus.

6. Press **Tab** again. The canvas with the Order Items frame appears. Press **Tab** to navigate through the canvas items until the Exit button has the form focus.

7. Press **Tab** again. The canvas with the Customers frame appears, indicating you have navigated through all of the form block items and have now cycled back to the first item in the first block.

8. Close Forms Runtime.

Sometimes, you want to prevent users from navigating to different canvases using the Tab key, because certain operations must be completed on one canvas before the user should move to the next canvas. For example, in this application, the user must click the Create New Order button on the CUSTOMER canvas to retrieve a new order ID value before the CUST_ORDER canvas appears. To disable Tab key navigation among different canvases, you set the block Navigation Style property to Same Record. Then, to move to the next canvas when the required operations are complete, you write commands to programmatically display the next canvas. To disable Tab key block navigation in this form, you will change the CUSTOMER Navigation Style property back to Same Record, so the user can only navigate within the CUSTOMER block items. Then, you will add a command to the WHEN-BUTTON-PRESSED trigger of the Create New Order button so that when the user clicks the Create New Order button, the CUST_ORDER_CANVAS appears.

To programmatically display a different canvas within a form, you can use the GO_ITEM built-in to switch the form focus to an item (such as a text item or button) on the new canvas. The canvas that contains the item that has the form's focus then appears. When the user clicks the Create New Order button on the CUSTOMER_CANVAS, the application should retrieve the next value from the sequence that generates new Order ID values, set the value of the ORDER_DATE text item on the CUST_ORDER_CANVAS equal to the current system date, and then display the CUST_ORDER_CANVAS. Now you will change the Navigation Style property for the CUSTOMER block and create the trigger for the Create New Order button.

To change the Navigation Style property and create the Create New Order button trigger:

1. Open the Property Palette for the CUSTOMER data block, change the Navigation Style to **Same Record**, and then close the Property Palette.

7

2. Create a WHEN-BUTTON-PRESSED trigger for the Create New Order button on the CUSTOMER_CANVAS using the PL/SQL commands shown in Figure 7-40. Compile the trigger, correct any syntax errors, and then close the PL/SQL Editor.

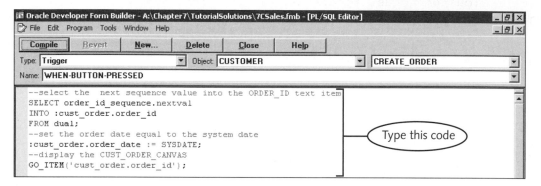

Figure 7-40   Code to navigate to a different canvas

 If a message appears stating that the ORDER_ID_SEQUENCE does not exist, create the sequence in SQL*PLUS using the following command:
CREATE SEQUENCE order_id_sequence
START WITH 2000;

3. Save the form.

Now you will test the trigger. You will run the form, select an existing customer on the CUSTOMER_CANVAS, and then click the Create New Order button to move the form focus to the CUST_ORDER_CANVAS.

To test the trigger:

1. Run the form. The customer canvas appears.

2. Click the **LOV command button** beside the Customer ID text item, select customer **232** (Mitch Edwards), and then click **OK**.

3. Click **Create New Order**. The customer orders canvas appears, with the new order ID displayed and the current system date displayed in the order date text item.

4. Close Forms Runtime. If a dialog box opens asking if you want to close the form, click **Yes**.

Referencing Items from Multiple Blocks

Next, you will create the trigger for the Save Order button on the CUST_ORDER_CANVAS. This trigger will insert a new record in the CUST_ORDER table. All of the values required for the INSERT command, except CUST_ID, are in the

CUST_ORDER block and appear on the CUST_ORDER_CANVAS. CUST_ID is in the CUSTOMER block and appears on the CUSTOMER_CANVAS. To reference CUST_ID in the Save Order button trigger, you preface the item name with the block name, just as you have done before in other triggers in custom forms, using the reference `:customer.cust_id`. After you have committed the CUST_ORDER record to the database, you must enable the Enter/Edit Items button so the user can navigate to the ORDER_LINE canvas and specify the order items.

To create the Save Order trigger:

1. Create a WHEN-BUTTON-PRESSED trigger for the Save Order button on the CUST_ORDER_CANVAS using the PL/SQL commands shown in Figure 7-41. Compile the trigger, correct any syntax errors, and then close the PL/SQL Editor.

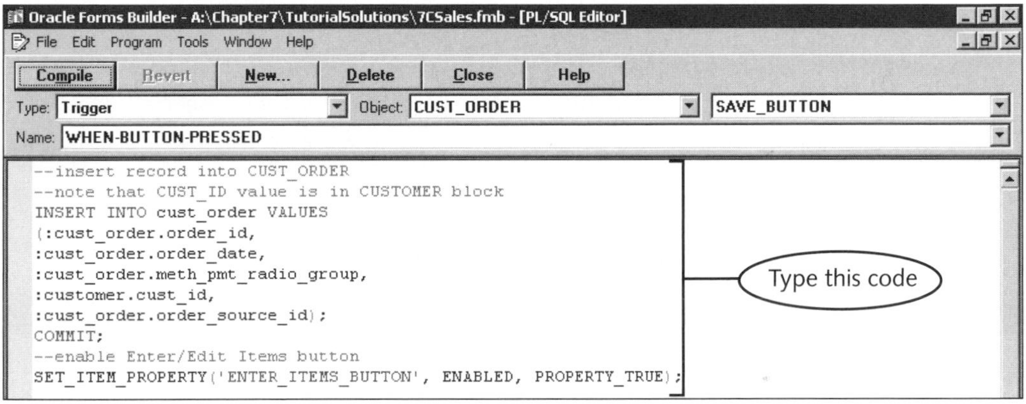

Figure 7-41   Code for Save Order button trigger

2. Save the form.

Next, you will create the trigger for the Enter/Edit Items button. This button needs to programmatically display the ORDER_LINE_CANVAS. To do this, you will use the GO_ITEM built-in and navigate to an item on the ORDER_LINE_CANVAS. You could navigate to any canvas item. You will navigate to the LOV command button beside the INV_ID text item, because that is logically the first item that the user would access on the form.

To create the Enter/Edit Items button trigger:

1. Create a WHEN-BUTTON-PRESSED trigger for the Enter/Edit Items button on the CUST_ORDER_CANVAS using the following PL/SQL command:

```
GO_ITEM('ORDER_LINE.INV_ID');
```

2. Compile the trigger, correct any syntax errors, close the PL/SQL Editor, and save the form.

Now you will test the form. You will create a new order for customer Mitch Edwards that is dated today, using a check as the payment method and order source 4 (Outdoor 2003). You will save the new order record and then click Enter/Edit Items to navigate to the ORDER_LINE_CANVAS.

To test the form:

1. Run the form. The customer canvas appears.

2. Click the **LOV command button** beside the customer ID text item, select customer **232** (Mitch Edwards), and then click **OK**.

3. Click **Create New Order**. The customer orders canvas opens, the new order ID appears, and the current system date appears for the order date.

4. Select the **Check option button**, click the **LOV command button** beside the order source ID text item, select order source **4** (Outdoor 2003), and then click **OK**.

5. Click **Save Order**. The confirmation message "FRM-40401: No Changes to Save" appears, which indicates that the record was successfully inserted, and the Enter/Edit Items button is now enabled.

6. Click **Enter/Edit Items**. The Order Items frame appears.

7. Click **Exit** to exit the form.

8. Close the form in Form Builder.

TAB AND STACKED CANVASES

So far, all of the forms you have created have used content canvases, which provide the basic background for a form window. Form Builder also supports **tab canvases**, which are multiple-page canvases that allow users to move between multiple canvas surfaces by clicking tabs at the top of the canvas, and **stacked canvases**, which are canvases that can be laid on top of content canvases. You can use a tab canvas to display a large number of related items in a modular way or to direct a user through a sequence of steps for performing a task. You can use a stacked canvas to make objects appear and disappear as needed, to create a toolbar, or to create scrolling subwindows within a single window. Now you will create a tab canvas to support the Clearwater Traders sales application and add a stacked canvas to one of the tab pages to display the current order items.

Creating a Tab Canvas

A tab canvas lies on top of a content canvas within the form window. The tab canvas contains two or more **tab pages**, which are objects that represent surfaces that display form items and have a tab label identifier at the top. Figure 7-42 shows the components of the tab canvas for the Clearwater Traders sales application that you will create. The application will have three tab pages that will be used to enter the customer information, order information, and order line information.

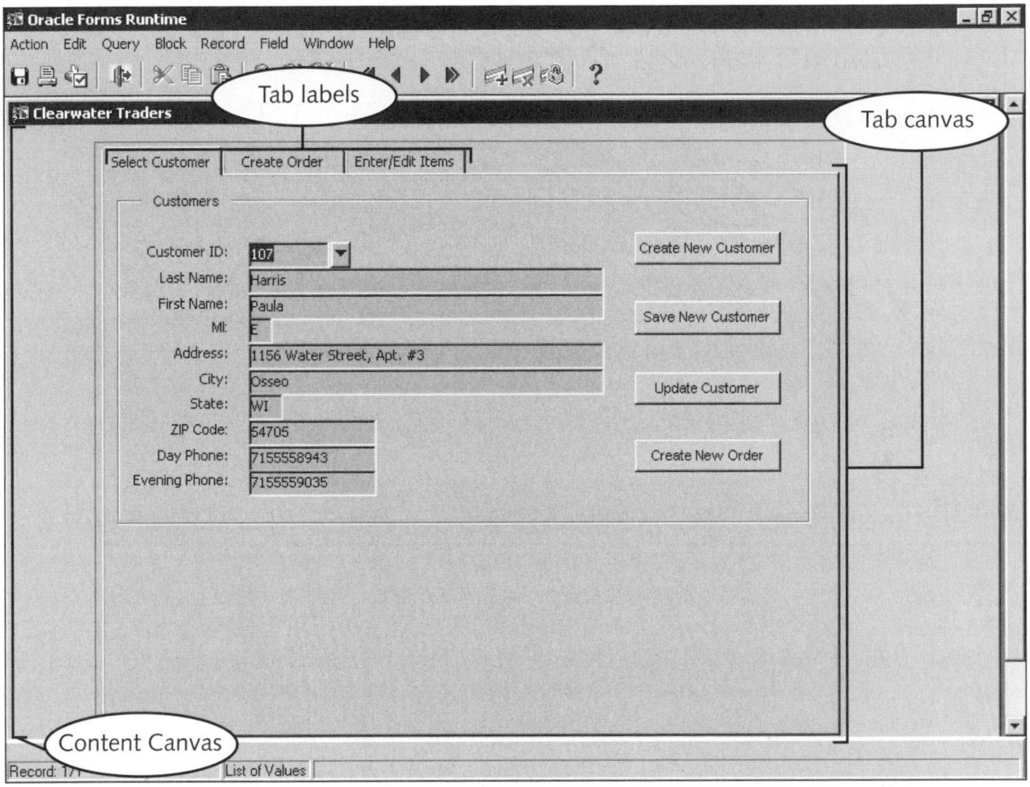

Figure 7-42   Tab canvas application components

To create this form, you will open a custom form that contains a blank content canvas. You will create a tab canvas on this content canvas and configure the tab pages. Then, you will open a second form that contains the form items for all three tab pages. Currently, the form items are on separate canvases. You will copy the items into your tab canvas form. First, you will open the custom form and create the tab canvas.

To open the form and create the tab canvas:

1. Open **Sales_TAB.fmb** from the Chapter7 folder on your Data Disk, and save the form file as **7CSales_TAB.fmb** in your Chapter7\TutorialSolutions folder.

2. In the Object Navigator, click **+** beside the Canvases node. Currently, the form contains one canvas named SALES_CANVAS. This is the content canvas on which you will create the tab canvas.

3. Double-click the **Canvas icon** 🖬 beside SALES_CANVAS to display the canvas in the Layout Editor. Currently, the canvas is blank.

4. Click the **Tab Canvas tool**  on the tool palette, and draw a large rectangle on the canvas to represent the tab canvas, as shown in Figure 7-43. Note that by default, a new tab canvas contains two pages.

Figure 7-43   New tab canvas with default pages

5. Open the Object Navigator, and note that the new tab canvas appears under the Canvases node. Change the name of the tab canvas to **SALES_TAB_CANVAS**.

6. If necessary, click ⊞ beside SALES_TAB_CANVAS. The tab canvas contains one object node, called Tab Pages, with two page objects below it.

Currently, there are two tab page objects. Since the tab canvas sales application has three pages, you need to create another tab page. Then, you will change the names and labels of the tab pages. The order of the tab pages in the Object Navigator specifies the order that the tab labels appear in the form layout. The first tab page label appears on the left edge of the tab canvas, the second tab page label is on the next tab to the right, and so forth.

To create a new tab page and change the tab page properties.

1. Select the **Tab Pages node**, and click the **Create button** on the toolbar to create a new tab page. The new page appears under the Tab Pages node.

2. Select the first Tab Page object, right-click, and then click **Property Palette** to open the first tab page Property Palette. Change the Name property to **CUSTOMER_PAGE**, the Label property to **Select Customer**, and then close the Property Palette.

> You cannot open a tab page Property Palette by double-clicking the Tab Page icon in the Object Navigator. If you double-click, the tab page appears in the Layout Editor.

3. Change the properties of the second and third tab pages as follows:

Tab Page	Name	Label
Second	**CUST_ORDER_PAGE**	**Create Order**
Third	**ORDER_LINE_PAGE**	**Enter/Edit Items**

4. Double-click the **Tab Page tool** beside CUSTOMER_PAGE. The tab page appears in the Layout Editor. Note that the content canvas that borders the tab canvas does not appear in this view.

5. Open the Object Navigator, and double-click the **Canvas icon** beside SALES_CANVAS, which is the content canvas that the tab canvas appears on. The content canvas appears with the tab canvas on it.

6. Save the form.

At this point, you could create items on the tab canvas pages using the Data Block Wizard or by creating a control block and drawing text items directly on the tab canvases. To speed up development of the application and to learn how to copy and paste objects from one form to another, you will open a form that contains all of the form objects that will appear on the tab canvas pages. This form will be called the existing form. You will copy the objects from the existing form, paste them into the new form, and then modify the object properties so they appear correctly in the new form. First, you will copy the block items, which includes the form text items, LOVs, and buttons, from the existing form. The existing form contains three data blocks: CUSTOMER, CUST_ORDER, and ORDER_LINE. You will copy all of the blocks and paste them into the new form.

To copy and paste the blocks:

1. With the SALES_TAB_FORM still open, click the **Open button** on the toolbar and open **Sales_DONE.fmb** from the Chapter7 folder on your Data Disk. This form contains all of the form items that are used in the Clearwater Traders sales application. The SALES_DONE form module appears in the Object Navigator.

2. Click ✚ beside the Data Blocks node under SALES_DONE. The existing form's blocks appear.

3. Click **CUSTOMER**, press and hold the **Shift** key, and click **CUST_ORDER** and **ORDER_LINE** to select all of the data blocks in the existing form.

4. Click the **Copy button** 📋 on the toolbar to copy the existing form's blocks.

5. Select the **Data Blocks** node in the SALES_TAB_FORM module, and then click the **Paste button** 📋 to paste the blocks into the new form.

6. Select **SALES_TAB_FORM** in the Object Navigator, and then click the **Save button** 💾 to save the form.

When multiple forms are open in Form Builder, the form that is currently selected is saved when you click the Save button 💾.

Recall that when you create a form with multiple canvases, the canvas that appears first when the form opens in the Forms Runtime window is the canvas whose block items appear first under the Data Blocks node in the Object Navigator window. This same principle is true for tab canvases: The tab page that appears on top when the form first appears in the Forms Runtime window is the page whose items are associated with the block listed first. Now, you will view the data blocks in the SALES_TAB_FORM to confirm that they are in the correct order so that the Customer page appears first.

To view the data blocks to determine if they are in the correct order:

1. In the Object Navigator, view the blocks under the Data Blocks node for the SALES_TAB_FORM, and confirm they are listed in this order:

CUSTOMER
CUST_ORDER
ORDER_LINE

2. If the blocks are not in the correct order, drag and drop them so they are listed correctly, and then save the form.

Recall that every block item has a Canvas property that specifies the canvas on which it appears. Currently, the Canvas values for the pasted block items list the name of the canvas that they originally appeared on in the existing form. Now you need to modify the Canvas property of each item in the new form, so the item appears on the correct canvas. When an item appears on a tab canvas, you must also specify the tab page on which the item appears, using the item's Tab Page property. All of the CUSTOMER block items will appear on the CUSTOMER_PAGE of the tab canvas, all of the CUST_ORDER block items will appear on the CUST_ORDER_PAGE, and all of the ORDER_LINE items will appear on the ORDER_LINE_PAGE. You will select all of the block items in each block as a group and then change their Canvas and Tab Page properties to the correct values on the intersection Property Palette.

To modify the Canvas and Tab Page properties of the block items:

1. Click ✚ beside the CUSTOMER block in the SALES_TAB_FORM, and then click ✚ beside the Items node to display the block items.

2. Select **CUST_ID**, which is the first block item, press and hold the **Shift** key, and then select **CREATE_ORDER**, which is the last block item. All of the block items appear selected.

3. Click **Tools** on the menu bar, and then click **Property Palette** to open the intersection Property Palette.

4. Scroll down to the Physical property node, open the Canvas list, and change the Canvas property to **SALES_TAB_CANVAS**. Open the Tab Page list, and change the Tab Page property to **CUSTOMER_PAGE**. Close the intersection Property Palette.

5. Repeat Steps 1 and 2 to select all items in the CUST_ORDER block, and then change the Canvas property for all of the block items to **SALES_TAB_CANVAS** and the Tab Page property to **CUST_ORDER_PAGE**. Close the intersection Property Palette.

6. Repeat Steps 1 and 2 to select all items in the ORDER_LINE block, and then change the Canvas property for all of the block items to **SALES_TAB_CANVAS**, and the Tab Page property to **ORDER_LINE_PAGE**. Close the intersection Property Palette.

7. View the SALES_CANVAS in the Layout Editor to confirm that the block items appear on the correct canvases. Reposition the items as necessary so they are centered on the canvases, and then save the form.

As you view the tab canvas pages, you might have noticed that the frames around the block items, as well as the frame around the radio group in the CUST_ORDER block, are missing. Recall that frames are not block items; rather, they are objects that are painted on the canvas. Next, you will copy the frames from the existing form and paste them onto the tab pages in the new form. You will copy and paste all of the graphic objects in the Object Navigator rather than directly onto the tab canvas pages in the Layout Editor, because often when you try to copy and paste objects directly onto tab canvas pages, the objects do not appear on the correct page.

To copy and paste the frames:

1. In the Object Navigator, scroll down to the SALES_DONE form module, click ✚ beside the Canvases node, click ✚ beside CUSTOMER_CANVAS to display its Graphics node, and then click ✚ beside Graphics. The canvas frame appears.

2. Select the frame, and then click the **Copy button** 📋 to copy the frame.

3. Scroll up to the SALES_TAB_FORM module, click ▬ beside the Data Blocks node to close the Data Blocks node, click ➕ beside the Canvases node to display the form canvases, select **CUSTOMER_PAGE** under the SALES_TAB_CANVAS, and then click the **Paste button** 📋 to paste the frame onto the page.

4. Double-click the **Tab Page icon** 📁 beside CUSTOMER_PAGE to display the page in the Layout Editor. Reposition and resize the frame as necessary.

5. Repeat Steps 1 through 4 to copy both of the frames on the CUST_ORDER_CANVAS in the existing form to the CUST_ORDER_PAGE in the new form.

6. Repeat Steps 1 through 4 to copy the frame on the ORDER_LINE_CANVAS in the existing form to the ORDER_LINE_PAGE in the new form.

7. Save the SALES_TAB_FORM.

The final objects that must be copied from the existing form to the new form are the LOVs and their associated record groups. Now you will copy these objects from the existing form, and paste them into the new form.

To copy the LOVs and record groups:

1. In the Object Navigator, scroll down to the SALES_DONE form module, click ➕ beside the LOVs node, select all of the form LOVs as an object group, and then click the **Copy button** 📋 to copy the LOVs.

2. Scroll up to the SALES_TAB_FORM module, select the **LOVs node**, and then click the **Paste button** 📋 to paste the LOVs. The pasted LOVs appear in the list.

3. Repeat Steps 1 and 2 to copy and paste all of the record groups from the existing form to the new form.

4. Select the **SALES_DONE form module** in the Object Navigator, click **File** on the menu bar, and then click **Close** to close the SALES_DONE form.

5. Save the SALES_TAB_FORM.

Now you will run the tab canvas form and confirm that it works correctly. You will select customer 232 (Mitch Edwards) and create an order dated today, paid by check, using order source 4 (Outdoor 2003).

To test the tab canvas form:

1. Run the form. The tab canvas appears, with the Select Customer tab page visible.

This form requires that sequences named CUST_ID_SEQUENCE and CUST_ORDER_SEQUENCE exist in your database schema. If you have not created these sequences in a previous exercise or case, a message will appear stating that the sequence object cannot be found. Use the following commands to create the required sequences: CREATE SEQUENCE cust_id_sequence START WITH 500; and CREATE SEQUENCE cust_order_sequence START WITH 2000;.

2. Click the **LOV command button** beside the Customer ID text item, select customer **232** (Mitch Edwards) and then click **OK**. The customer values appear in the canvas text items.

3. Click **Create New Order**. The Create Order tab appears with a new Order ID inserted. Select the **Check radio button**, click the **LOV command button** beside Order Source, select order source **4** (Outdoor 2003), and then click **OK**.

4. Click **Save Order**. The "FRM-40401: No Changes to Save" message appears, which confirms that the record was saved, and the **Enter/Edit Items** button is enabled.

5. Click **Enter/Edit Items**. The Enter/Edit Items tab page appears.

6. Click the **Create Order** tab to move back to the Create Order page.

7. Click the **Select Customer** tab to move back to the Select Customer page.

8. Close Forms Runtime.

It is useful to be able to easily navigate back to previous pages to view the information displayed there. Note, however, that users cannot be allowed to change the value for the Order ID once the CUST_ORDER record has been created. If they do, an error message will appear when the user tries to insert a record into the ORDER_LINE table, because the parent key value for the order ID will not be valid. Therefore, once the CUST_ORDER record has been inserted, the ORDER_ID text item should be disabled. You will do this next. You will open the WHEN-BUTTON-PRESSED trigger for the Save Order button on the CREATE_ORDER_PAGE and add the code to disable the ORDER_ID text item. Then, you will reenable the text item when the user clicks the Next Order button on the ORDER_LINE_PAGE.

To disable the ORDER_ID text item:

1. In the Object Navigator, click **Tools** on the menu bar, and then click **Layout Editor** to open the form in the Layout Editor. When the SALES_TAB_FORM: Canvases dialog box opens, select **SALES_CANVAS** from the canvas list, and then click **OK**.

2. If necessary, click the **Create Order tab** to display the CUST_ORDER_PAGE, select the **Save Order** button, right-click, and then click **PL/SQL Editor** to open the button trigger in the PL/SQL Editor.

3. Add the following command as the last command in the trigger. Then compile the trigger, correct any syntax errors, and close the PL/SQL Editor.

```
SET_ITEM_PROPERTY('ORDER_ID', ENABLED, PROPERTY_FALSE);
```

4. Click the **Enter/Edit Items tab** to display the ORDER_LINE_PAGE, select the **Next Order** button, right-click, and then click **PL/SQL Editor** to open the button trigger in the PL/SQL Editor.

5. Add the following command as the last command in the trigger, and then compile the trigger, correct any syntax errors, and close the PL/SQL Editor.

```
SET_ITEM_PROPERTY('ORDER_ID', ENABLED, PROPERTY_TRUE);
```

6. Save the form.

Now you will run the form and confirm that your change successfully disables the ORDER_ID text item until the user completes the order.

To test the form:

1. Run the form. The tab canvas appears with the Select Customer tab page visible.

2. Click the **LOV command button** beside the Customer ID text item, select customer **232** (Mitch Edwards) and click **OK**. The customer values appear in the canvas text items.

3. Click **Create New Order**. The Create Order tab appears with a new Order ID inserted. Select the **Check** radio button, click the **LOV command button** beside Order Source ID, select order source **4** (Outdoor 2003), and then click **OK**.

4. Click **Save Order**. The "FRM 40401: No Changes to Save" message indicates that the record was saved, the **Enter/Edit Items** button is enabled, and the order ID text item is disabled.

5. Close Forms Runtime.

6. Close the form in Form Builder.

Creating a Stacked Canvas

Recall that a stacked canvas lays on top of a content canvas and is used to display or hide canvas objects. You create a stacked canvas by painting it on an existing content canvas in the desired position. You can then configure the properties of the stacked canvas and create the form items that will appear on it. You use program commands to display the stacked canvas when it is needed and to hide the stacked canvas when it is not needed. Once you have created and configured the stacked canvas, you create triggers containing program commands to display and hide the canvas. Now, you will create a stacked canvas to optionally display the order detail information on the ORDER_LINE canvas of the Clearwater Traders sales application, as shown in Figure 7-44. When the ORDER_LINE canvas first appears, the order summary detail information does not appear. When the user clicks the Show Summary button, the stacked canvas containing the summary information appears. When the user clicks Hide Summary on the stacked canvas, the stacked canvas no longer appears on the screen display. This is similar to the canvas that you created in Figure 7-34, except that on that form, the order summary detail information always appeared. In this form, the order summary detail information will appear only when it is needed.

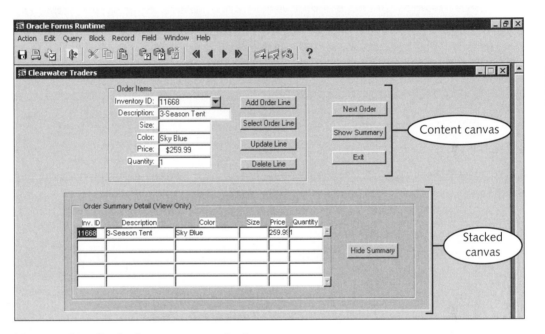

Figure 7-44   Stacked canvas on content canvas

First, you will open the Sales_DONE.fmb form from the Chapter7 folder on your Data Disk and create the stacked canvas. Then, you will modify its properties.

To open the form and create the stacked canvas:

1. In Form Builder, open **Sales_DONE.fmb** from the Chapter7 folder on your Data Disk, and save the form as **7CSales_STACK.fmb** in your Chapter7\TutorialSolutions folder.

2. In the Object Navigator, click **+** beside the Canvases node to display the form canvases. The form has three content canvases: ORDER_LINE_CANVAS, CUST_ORDER_CANVAS, and CUSTOMER_CANVAS.

3. Double-click the **Canvas icon** ▦ beside ORDER_LINE_CANVAS to display the canvas in the Layout Editor. This is the content canvas on which the stacked canvas will appear.

A stacked canvas is not associated with a particular content canvas, but appears on top of the content canvas that currently appears in the form window.

4. Click the **Stacked Canvas tool** 🗗 on the tool palette, and draw the stacked canvas surface on the content canvas surface, as shown in Figure 7-45. The edges of the canvas appear as selection handles.

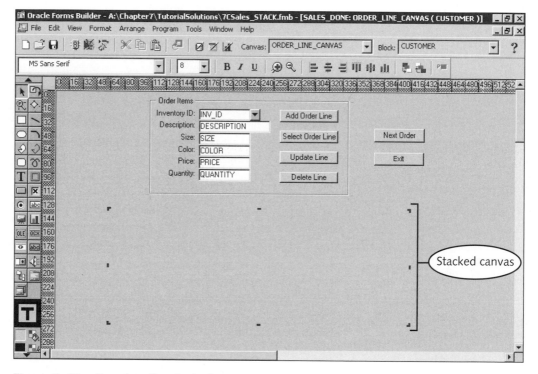

Figure 7-45   Drawing the stacked canvas

The area that you draw on the content canvas is called the stacked canvas **viewport**. This is the portion of the canvas that is visible when the stacked canvas appears, and it is independent of the stacked canvas width and height. You can specify the Viewport Width and Viewport Height properties in the canvas Property Palette or by adjusting the canvas size in the Layout Editor. You can also adjust the viewport X- and Y-position on the background of the stacked canvas either by modifying the Viewport X-Position and Viewport Y-Position values in the stacked canvas Property Palette or by moving the canvas in the Layout Editor. The viewport can be larger than the size of the stacked canvas. However, if any canvas items are placed off of the stacked canvas background, an error will occur.

Now, you will modify the properties of the new stacked canvas. You will change the canvas name and change the canvas's Visible property to No, so the canvas does not appear when the form first opens. To make the canvas easier to work with while you are developing the form, you will change the canvas background color to white. You will also modify the canvas width and height.

To modify the stacked canvas properties:

1. Open the Object Navigator, and examine the form canvases. Note that the icon beside the new stacked canvas is the **Stacked Canvas** icon 🖿, which differentiates it from the content canvases.

2. Change the name of the stacked canvas to **DETAIL_STACKED_CANVAS**.

3. Select the **DETAIL_STACKED_CANVAS**, right-click, and then click **Property Palette** to open the stacked canvas Property Palette.

4. Under the Physical property node, change the Visible property to **No** so the canvas does not appear when the form first opens, the Width to **580**, and the Height to **375**.

5. Under the Color property node, select **Background Color**, click the **More button** [...] to display the Fill Property Palette, and then select a white square for the background color.

6. Close the Property Palette, and then save the form.

Next, you need to create the form items that appear on the stacked canvas. The stacked canvas will display a data block based on the ORDER_SUMMARY_VIEW, which you created earlier in this chapter, that displays the order line detail information. You will create the data block and layout using the Data Block and Layout Wizards.

To create the data block and layout on the stacked canvas:

1. In the Object Navigator, select the **SALES_DONE** form module so the Data Block Wizard does not open in reentrant mode, click **Tools** on the menu bar, click **Data Block Wizard** and then click **Next**. Make sure the Table or View option button is selected, and then click **Next**.

2. Click **Browse**, clear the **Tables check box** and check the **Views check box**, select **ORDER_SUMMARY_VIEW**, and then click **OK**.

 If ORDER_SUMMARY_VIEW is not in your list of views, start SQL*Plus, and create the view by running the order_summary_view.sql script that is in the Chapter7 folder on your Data Disk.

3. Select all of the view fields for the data block, and click **Next**. Since you will not create a master-detail relationship for this block, click **Next** to skip the Relationships page, and then click **Finish**. The Layout Wizard Welcome page appears.

4. Click **Next** to skip the Layout Wizard Welcome page.

5. On the Canvas page, open the Canvas list, select **DETAIL_STACKED_CANVAS** to place the data block items on the stacked canvas, and then click **Next**.

6. Select all of the block items except ORDER_ID for the layout, and place the items in the following order in the Displayed Items list: INV_ID, ITEM_DESC, COLOR, ITEM_SIZE, PRICE, ORDER_QUANTITY. Click **Next**.

7. Modify the item prompts and column widths as follows, and then click **Next**.

Name	Prompt	Width
INV_ID	Inv. ID	35
ITEM_DESC	Description	80
COLOR	Color	75
ITEM_SIZE	Size	35
PRICE	Price	41
ORDER_QUANTITY	Quantity	25

8. Select the **Tabular option button**, click **Next**, type **Order Summary Detail (View Only)** for the frame title, type **5** for Records Displayed, check the **Display Scrollbar check box**, click **Next**, and then click **Finish**. The stacked canvas appears in the Layout Editor.

9. Open the Canvas list, and select **ORDER_LINE_CANVAS** to view how the stacked canvas will look on the content canvas. If necessary, adjust the position of the frame and block items so they are positioned as shown in Figure 7-44.

10. Save the form.

Next, you need to create the relationship between the control block and the data block. In this form, the order line records will be retrieved based on the ORDER_ID value that appears on the CUST_ORDER_CANVAS. This text item is in the CUST_ORDER block. Recall that to create a relationship between a control block and a data block, you open the data block Property Palette and modify its WHERE property to specify the relationship.

To create the relationship between the control block and the data block:

1. Open the Object Navigator, and double-click the **Data Block icon** 🖼 beside ORDER_SUMMARY_VIEW to open the data block Property Palette.

2. Under the Database property group, change the WHERE Clause property to `order_id = :cust_order.order_id`.

3. Close the Property Palette, and then save the form.

The final step is to create the Show Summary and Hide Summary buttons that display and hide the stacked canvas. To display a stacked canvas, you use the GO_BLOCK built-in to move the form focus to the data block whose items are associated with the stacked canvas. You then use the SHOW_VIEW built-in, which has the following syntax: **SHOW_VIEW('*stacked_canvas_name*');**. To hide a stacked canvas, you move the form focus to a data block that does not have form items on the stacked canvas, and then use the HIDE_VIEW built-in, which has the following syntax: **HIDE_VIEW('*stacked_canvas_name*');**. Now you will create the Show Summary and Hide Summary command buttons, and write the trigger code to display and hide the stacked canvas. First, you will create the Show Summary button. This button will be on the ORDER_LINE_CANVAS, and will be placed in the ORDER_LINE block. The button will move the form focus to the ORDER_SUMMARY_VIEW block, display the stacked canvas using the SHOW_VIEW built-in, and then issue the EXECUTE_QUERY command to refresh the data block contents.

To create the Show Summary button:

1. Double-click the **Canvas icon** 🖼 beside ORDER_LINE_CANVAS to display the canvas in the Layout Editor.

2. Select **ORDER_LINE** in the Block list, click the **Button tool** ⬜ on the tool palette, and draw a new command button that is the same size as the New Order and Exit buttons. Reposition the buttons so the new button is between the existing buttons, as shown in Figure 7-44.

3. Open the new button's Property Palette, change the Name to **SHOW_SUMMARY_BUTTON** and the Label to **Show Summary** and then close the Property Palette.

4. Create a WHEN-BUTTON-PRESSED trigger to show the stacked canvas using the following commands:

```
GO_BLOCK ('ORDER_SUMMARY_VIEW');
SHOW_VIEW ('DETAIL_STACKED_CANVAS');
EXECUTE_QUERY;
```

5. Compile the trigger, debug it if necessary, close the PL/SQL Editor, and then save the form.

Next, you will create the Hide Summary button. This button will be on the stacked canvas, so it is not visible unless the stacked canvas is visible. To place the button on the stacked canvas rather than on the content canvas you must select the stacked canvas, and then draw the button on the stacked canvas while it is selected. You will create the Hide Summary button in the ORDER_SUMMARY_VIEW block. The ORDER_SUMMARY_VIEW block displays five records at a time. When you create a new item in a block that uses a tabular style layout and shows multiple records, every block item displays once for each record. When you create the new button, it will appear five times, once for each record. To make the button only appear once, you must change the button's Number of Items Displayed property to 1.

To create the Hide Summary button:

1. Display the ORDER_LINE_CANVAS in the Layout Editor, select the **Order Summary Detail (View Only) frame**, and make it wider by dragging the right center selection handle to the right edge of the canvas so the button can be enclosed in the frame.

2. Select the stacked canvas so selection handles appear on its edges, and select **ORDER_SUMMARY_VIEW** in the Block List. This places the new button on the correct canvas and in the correct data block.

 If you do not place the button on the stacked canvas when you draw it, you can open its Property Palette later and change its Canvas property to the name of the stacked canvas. If you do not place the button in the correct block initially, you can move it to the correct block in the Object Navigator.

3. Click the **Button tool** 🔲 on the tool palette, and draw the Hide Summary button on the right side of the Order Summary Detail frame. Five buttons appear on the canvas.

4. Select any one of the new buttons, right-click, and then click **Property Palette**. Change the following property values, and then close the Property Palette.

Property	Value
Name	HIDE_SUMMARY_BUTTON
Label	Hide Summary
Number of Items Displayed	1

5. Create a WHEN-BUTTON-PRESSED trigger to show the stacked canvas using the following commands:

```
GO_BLOCK ('ORDER_LINE');
HIDE_VIEW ('DETAIL_STACKED_CANVAS');
```

6. Compile the trigger, debug it if necessary, and close the PL/SQL Editor.

7. Select the stacked canvas, right-click, and click **Property Palette**. Scroll down to the Color property node, select the **Background Color** property, click the **More button** 🔲, and select a gray square to change the stacked canvas fill color to the background color of the content canvas. Close the Property Palette

8. Save the form.

Now you will run the form to confirm that the stacked canvas appears and hides correctly. You will select customer 232 (Mitch Edwards) and create a new order dated today, paid by check, and using order source 4 (Outdoor 2003). Then, you will order one unit of item 11669 (3-Season Tent, color Light Grey), and click the Show Summary button to confirm that the stacked canvas displays the order summary information. Finally, you will click the Hide Summary button to hide the summary information.

To test the stacked canvas:

1. Run the form, click the **LOV command button** beside Customer ID, select customer **232**, and then click **Create New Order**.

2. Select the **Check option button**, click the **LOV command button** beside Order Source, select Order Source **4**, click **OK**, and click **Save Order**. The "FRM-40401: No Changes to Save" confirmation message appears, indicating that the new order was saved successfully. Click **Enter/Edit Items**.

3. Click the **LOV command button** beside Inventory ID, select item **11669**, click **OK**, type **1** in the Quantity text item, and then click **Add Order Line**. The "FRM-40401 confirmation message indicates that the new order was saved successfully.

4. Click **Show Summary**. The stacked canvas appears, showing the order detail information and the Hide Summary button.

5. Click **Hide Summary**. The stacked canvas no longer appears.

6. Close Forms Runtime.

7. Close the form in Form Builder, click **Yes** to save your changes, and close Form Builder and all other open Oracle applications.

A problem with using stacked canvases in applications with multiple canvases is that when the stacked canvas appears, it appears on all canvases, not just the one on which you intended it to appear. For example, in this application, if you do not hide the stacked canvas before you move to the Customer canvas to take the next order, the stacked canvas will still appear. You would need to programmatically hide the stacked canvas using the HIDE_VIEW built-in before displaying a different canvas.

7

Summary

❏ To create a relationship between a control block and a data block, create the data block without a master-detail relationship, and then modify the data block's WHERE property so the matching key field on the data block is equal to the value of the master key field on the control block.

❏ To programmatically update a data block in a form, you must first move the form focus to the block using the GO_BLOCK built-in, and then issue the EXECUTE_QUERY command, which automatically flushes the block and makes its information consistent with the block's corresponding database data.

❏ It is a good practice to display information on different display screens so as to not place more information on a single screen than a user can view without scrolling.

❏ To create an application with multiple canvases, you can use either a single-form approach or a multiple-form approach. With a single-form approach, you create a single form with multiple canvases. With a multiple-form approach, you create multiple forms with a single canvas and then integrate the forms into a single application.

❏ When you create a single form with multiple canvases, data can be easily shared among the different canvases. However, this approach makes it impossible for different programmers to program different screens of the same application at the same time, since a form's .fmb file can be modified by only one programmer at a time.

❏ The multiple-form approach is appropriate when several people are collaborating to create a complex application.

❏ When you create a single form with multiple canvases, the canvas that appears first is the canvas whose block items appear first in the Object Navigator Data Block node.

❏ The Navigation Style property of a data block determines if the user can press the Tab key to move among different data blocks in a form.

❏ The Canvas property of a form item determines the canvas on which the item appears.

❏ To programmatically display a different canvas in a multiple-canvas form, use the GO_ITEM built-in to move the insertion point to a text item on the canvas to be displayed.

❏ In a form with multiple blocks, every item within a data block must have a unique name so the combination of the block name and the item name uniquely identifies every form item.

❏ A tab canvas is a multiple-page canvas that allows users to move between multiple canvas surfaces by clicking tabs at the top of the canvas. A tab canvas is used to display a large number of related items in a modular way or to direct a user through a sequence of steps for performing a task.

❏ A stacked canvas is a canvas that can be laid on top of a content canvas and can be used to make objects appear and disappear as needed, to create a toolbar, or to create scrolling subwindows within a single window.

REVIEW QUESTIONS

1. How do you create a relationship between a control block and a data block?

2. When you have a data block that is related to a control block, how do you refresh the data in the data block when the key value in the control block changes?

3. When should you use the multiple-form approach for creating an application with multiple canvases?

4. In a form with multiple canvases, how do you move from one canvas to another?

5. In a form with multiple canvases, how do you share a data value among triggers attached to items on different canvases in the form?

6. In a form with multiple canvases, which canvas appears first when the form starts?

7. What are the different values for the Data Block Navigation Style property, and what do they specify?

8. What is the relationship between a tab canvas and a tab page?

9. How do you specify the order that the tab labels appear on the top of the tab canvas? How do you specify the tab that is selected when the form is first started?

10. When should you use a stacked canvas?

11. What is a viewport? How does the viewport width and height relate to the height and width of a stacked canvas?

PROBLEM-SOLVING CASES

All cases refer to the Clearwater Traders (see Figure 1-11), Northwoods University (see Figure 1-12), and Software Experts sample databases (see Figure 1-13). Run the clearwater.sql, northwoods.sql, and softwareexp.sql scripts in the Chapter7 folder on your Data Disk to refresh the tables in the sample databases. Rename all form components using descriptive names, and format forms using the guidelines described in Chapter 6. For all forms, create an Exit command button to exit Forms Runtime, and create a PRE-FORM trigger to automatically maximize the Forms Runtime window at start-up. Make all form items have a top-down, left-to-right external navigation order. The required files for all cases are stored in the Chapter7 folder on your Data Disk. Save the text for all SQL commands in a file named 7CQueries.sql in your Chapter7\CaseSolutions folder. Save all solutions in your Chapter7\CaseSolutions folder.

1. Modify the 7CCase1.fmb form so that when the user selects a project in the Software Experts database, the details about the consultants working on the project appear in a view-only Consultant Summary frame, as shown in Figure 7-46. Save the completed file as 7CCase1_DONE.fmb.

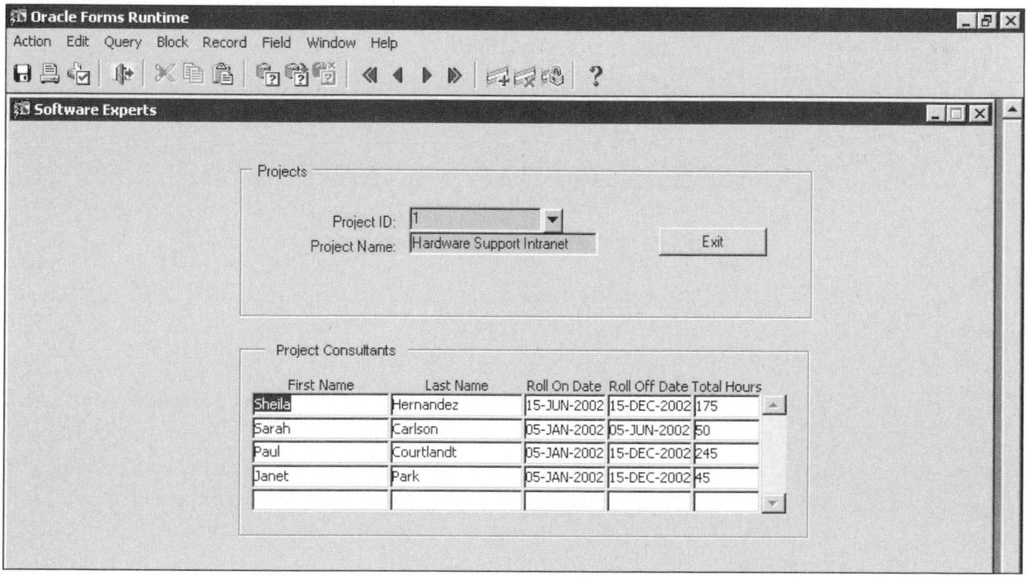

Figure 7-46

a. In SQL*Plus, create a view named PROJECT_CONSULTANTS_VIEW that contains the project ID, consultant first and last name, consultant roll-on date, consultant roll-off date, and consultant project hours.

b. Create a data block based on the view, and then create a relationship between the existing form control block and the new data block so that when the user selects a different project ID, the consultant details for the selected project appear in the data block. (*Hint*: Place the commands to refresh the data block display in the LOV command button trigger.)

2. Modify the 7CCase2.fmb form so that when users select a shipment in the Clearwater Traders database, they can optionally display or hide the shipment line details using a stacked canvas that displays the view-only Shipment Line Details frame, as shown in Figure 7-47. Save the completed file as 7CCase2_DONE.fmb.

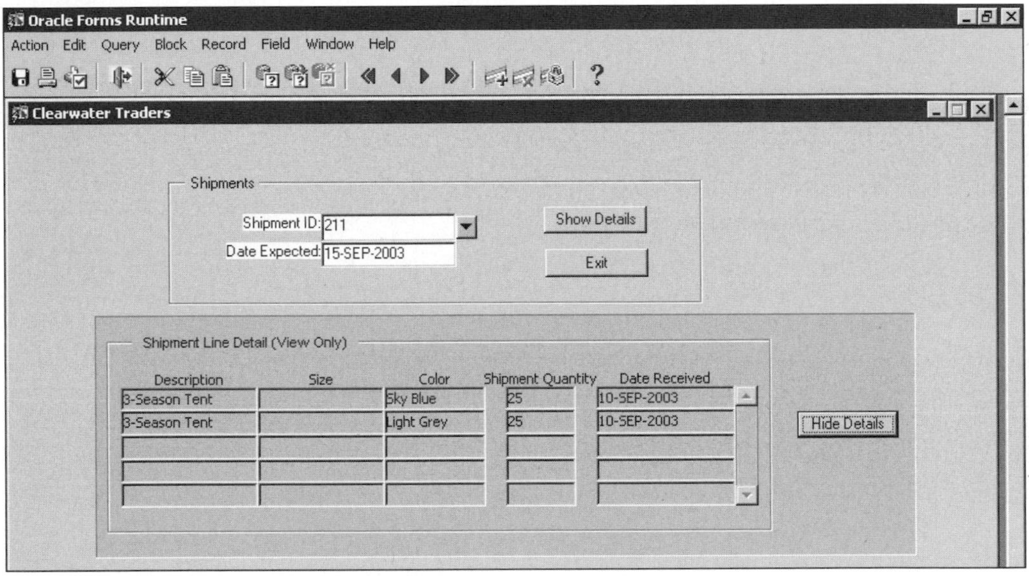

Figure 7-47

a. In SQL*Plus, create a view named SHIPMENT_LINE_VIEW that contains the shipment ID, item description, item size, color, shipment quantity, and date received.

b. Create a stacked canvas that appears below the shipment information. When the user clicks the Show Details button, the stacked canvas appears. When the user clicks the Hide Details button, which is on the stacked canvas, the stacked canvas is hidden. Configure the canvas so it does not appear when the form first starts. (*Hint*: When the data block on a stacked canvas is the current block, the stacked canvas always appears.)

c. Create a data block based on the view, and display the data block on the stacked canvas.

d. Create a relationship between the shipment ID in the existing form control block and the new data block so that when the user selects a different shipment ID, the shipment line details for the selected shipment appear in the data block. Be sure to configure the form so the data block display is refreshed if the user selects a new shipment when the shipment line display is visible. (*Hint*: Place the commands to refresh the data block display in the LOV command button trigger.)

3. Modify the 7CCase3.fmb form so that when the user selects a consultant in the Software Experts database, he or she can optionally display or hide the details about the consultant's skills or about the projects the consultant is currently working on, which appear on stacked canvases. Figure 7-48 shows the form display when the consultant's skills appear, and Figure 7-49 shows the form when the consultant's projects appear. Save the completed file as 7CCase3_DONE.fmb.

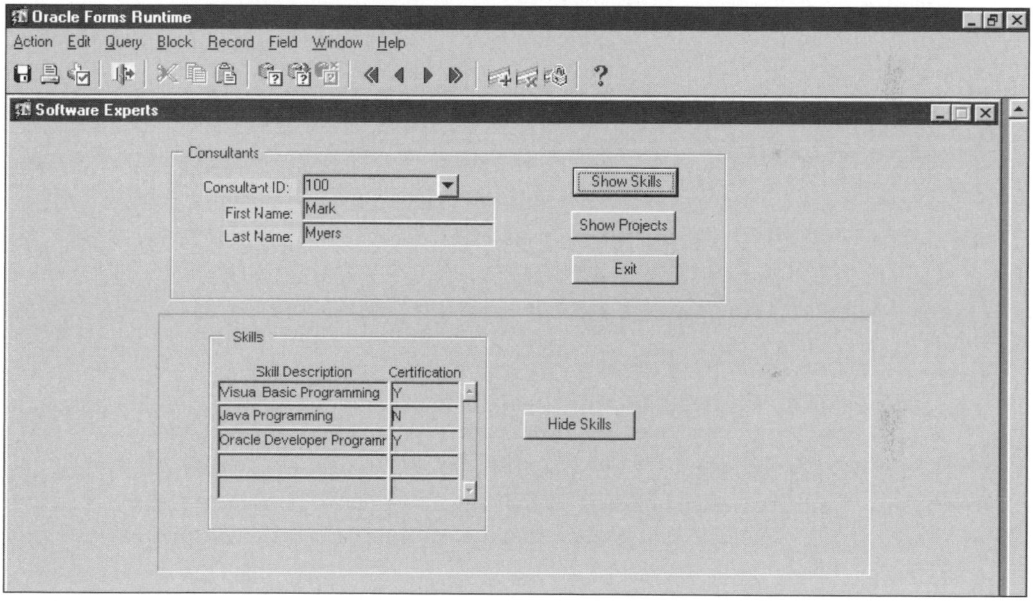

Figure 7-48

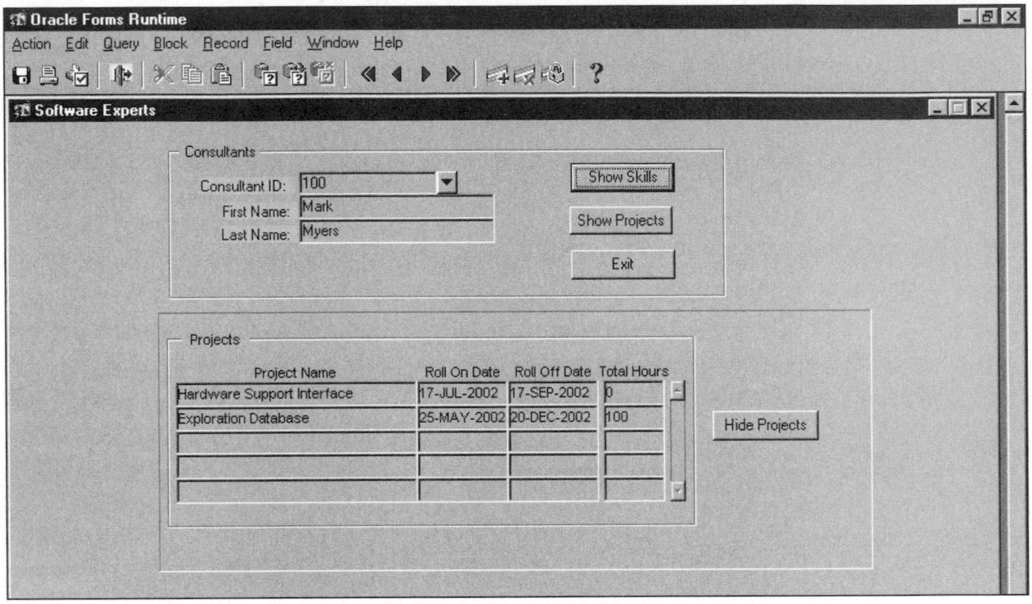

Figure 7-49

a. In SQL*Plus, create a view named SKILL_VIEW that contains the consultant ID, skill description, and certification status for every consultant skill. Create a second view named PROJECT_VIEW that contains the consultant ID, project name, roll-on date, roll-off date, and total hours for every project on which the consultant has worked.

b. Create two stacked canvases that appear below the consultant information on the current content canvas. One canvas will display the skill information, and the other will display the project information. When the user clicks the Show Skills button, the stacked canvas that displays the skill information appears, and when the user clicks Show Projects, the stacked canvas that displays the project information appears. Make the viewports the same size for both stacked canvases, and place the stacked canvases in the same position on the content canvas, so the canvases hide each other when one appears on top of the other.

c. Create a Hide button on both stacked canvases that hides the canvas on which it appears. Configure the canvases so they do not appear when the form first starts. (*Hint*: When the data block on a stacked canvas is the current block, the stacked canvas always appears.)

d. Create a separate data block for each view, and display each data block on the associated stacked canvas. Then create a relationship between the consultant ID in the existing form control block and the new data block so that when the user selects a different consultant ID, the skill or project details for the selected consultant appear in the associated data block. Be sure to configure the form so the data block display refreshes if the user selects a new consultant when the detail display is visible. (*Hint*: Place the commands to refresh the data block displays in the consultant ID LOV command button trigger.)

4. In this case, you will create a form that contains a tab canvas that can be used to create new projects in the Software Experts database. You will use tab pages to specify the project details, specify the skills required for the project, and select and assign project consultants. Configure all form blocks so the user cannot navigate from one block to another using the Tab key and must use the canvas tabs instead. Save the form file as 7CCase4_DONE.fmb.

a. Create a tab canvas that has three tab pages, as shown in Figure 7-50.

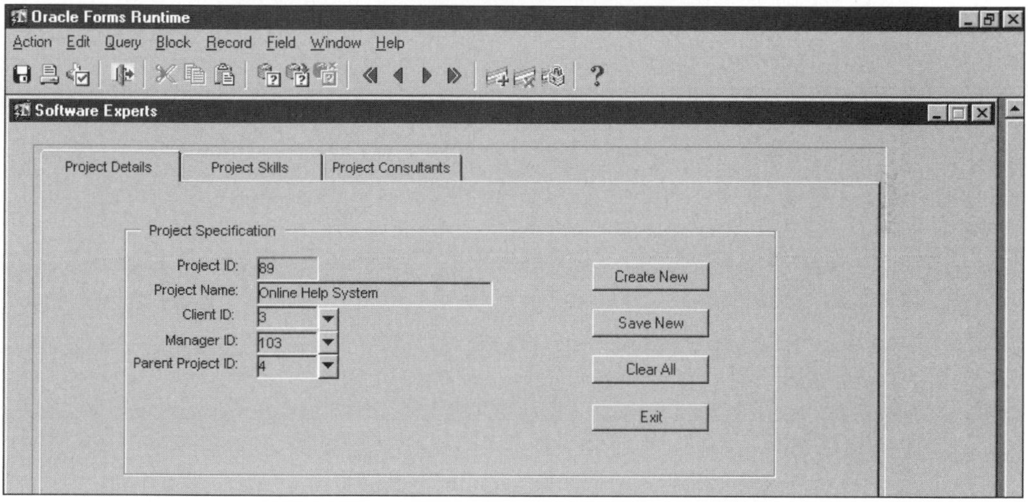

Figure 7-50

b. Create a control block named PROJECT to display the items on the Project tab page, as shown in Figure 7-50. When the user clicks the Create New button, a new project ID value appears in the Project ID text item. The project ID value is retrieved from a sequence named PROJECT_ID_SEQUENCE. (If you have not created this sequence previously, create it now, and start the sequence at 10.) Create LOVs to display information to enable the user to select the client ID, manager ID, and parent project ID from existing values in the CLIENT, CONSULTANT, and PROJECT tables. Write the commands to insert the new project record into the PROJECT table when the user clicks the Save New button. Once a new project is created, disable the Create New button, Save New button, and Project ID field so the user cannot change the value. Create a trigger so that when the user clicks the Clear All button, all form text items are cleared, and the disabled items in the PROJECT block are reenabled.

c. Create a control block named SKILL to display the items in the Add Skills frame on the Project Skills tab page shown in Figure 7-51. Create an LOV to display the skill ID and description of every skill, and return the value to the block text items. When the user clicks the Add Skill to Project button, insert the associated values into the PROJECT_SKILL table.

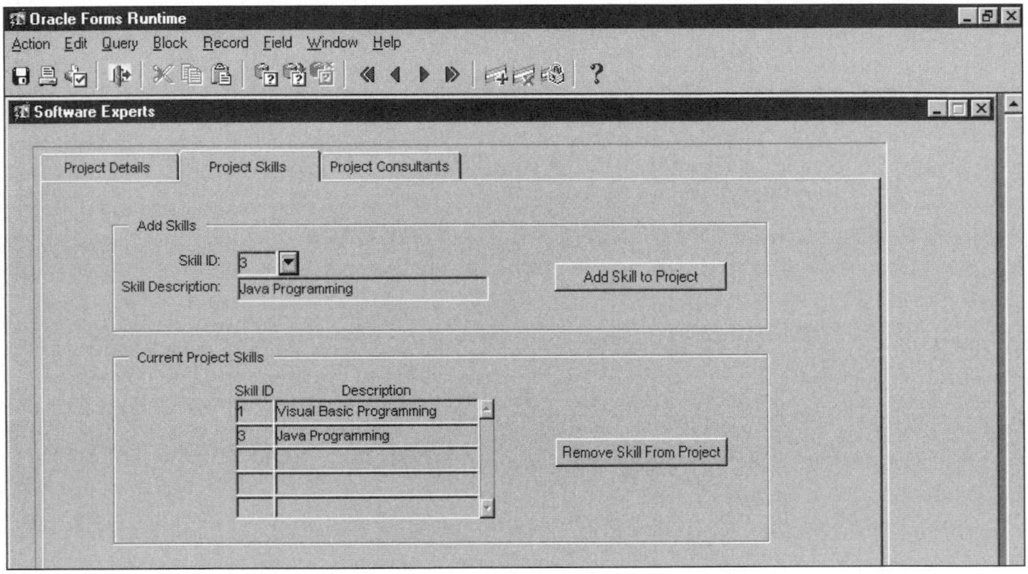

Figure 7-51

d. In SQL*Plus, create a view named PROJECT_SKILL_VIEW that contains the project ID, skill ID, and skill description for all skills associated with the current project. Create a data block based on the view, display the view on the Project Skills tab page, and create a relationship between the view and the control block that contains the P_ID text item on the Project Details tab page. When the user adds a new skill to the project, dynamically refresh the display in the Current Project Skills frame. Create a trigger for the Remove Skill From Project button that removes the skill that is selected in the view block and then dynamically refreshes the view block display.

e. Create a control block named CONSULTANT to display the items in the Add Consultants frame on the Project Consultants tab page, as shown in Figure 7-52. Create an LOV to display the consultant ID, first name, and last name of all consultants, and return the selected values to the block text items. Whenever the user selects a new consultant, automatically display the current system date for the roll-on and roll-off dates, and 0 for the total hours. When the user clicks the Add Consultant to Project button, insert the associated values into the PROJECT_CONSULTANT table.

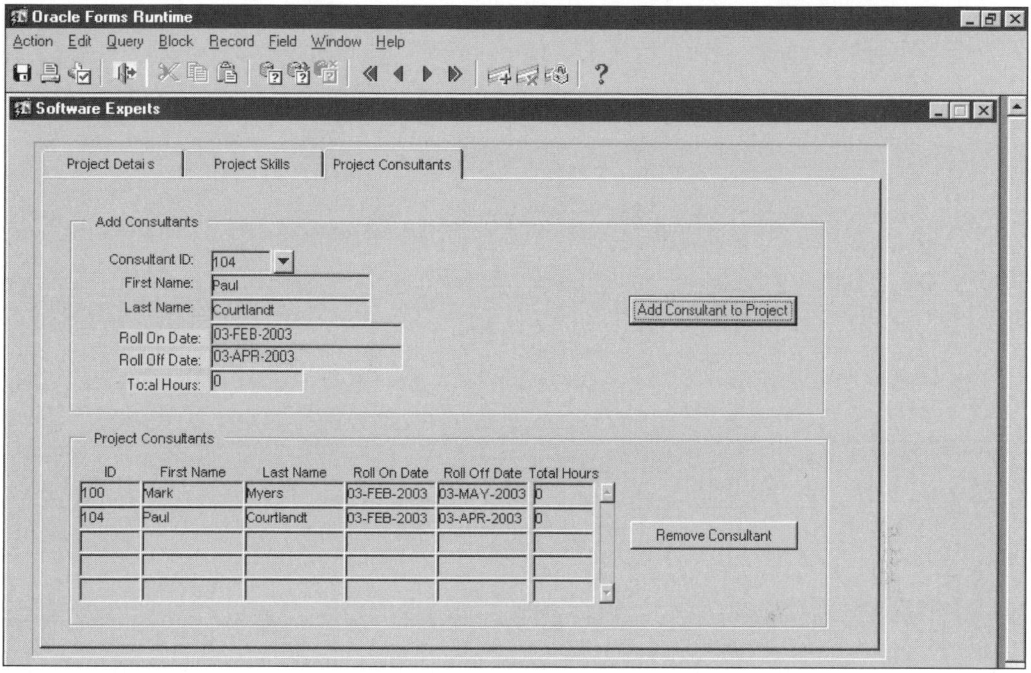

Figure 7-52

f. In SQL*Plus, create a view named PROJECT_CONSULTANT_VIEW that contains the project ID, consultant ID, consultant first and last names, roll-on date, roll-off date, and total hours for all consultants associated with the current project. Create a data block based on the view, display the view on the Project Consultants tab page, and create a relationship between the view and the control block that contains the P_ID text item on the Project Details tab page. When the user adds a new consultant to the project, dynamically refresh the view block display. Create a trigger for the Remove Consultant button that removes the consultant that is selected in the view block and then dynamically refreshes the view display.

5. In this case, you will create a form that contains a tab canvas that creates new course sections in the Northwoods University database. You can use tab pages to select the faculty member who teaches the course section, select the associated course, location, and term, and enter the other course section information. Save the form file as 7CCase5_DONE.fmb.

a. Create a tab canvas that has five tab pages, as shown in Figure 7-53.

b. Create a control block to display all of the fields in the COURSE table. Display the block layout on the Course tab, as shown in Figure 7-53. Create an LOV to display all COURSE records, and return the values for the selected record to the block items.

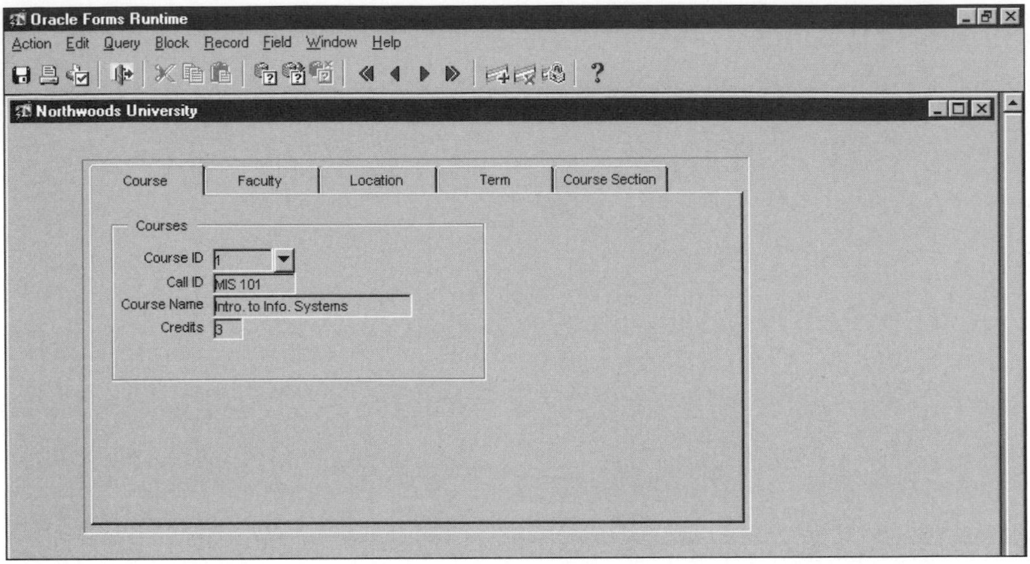

Figure 7-53

c. Create a control block to display all of the fields except the F_IMAGE field in the FACULTY table, and display the block layout on the Faculty tab using a style similar to the one shown on the Course tab. Create an LOV to display all FACULTY records, and return the values for the selected record to the block items.

d. Create a control block to display all of the fields in the LOCATION table, and display the block layout on the Location tab using a style similar to the one shown on the Course tab. Create an LOV to display all LOCATION records, and return the values for the selected record to the block items.

e. Create a control block to display all of the fields in the TERM table, and display the block layout on the Term tab using a style similar to the one shown on the Course tab. Create an LOV to display all TERM records, and return the values for the selected record to the block items.

f. Create a control block to display all of the fields in the COURSE_SECTION table, and display the block layout on the Course Section tab as shown in Figure 7-54. Create a Create New button that retrieves the next value from a sequence named C_SEC_ID_SEQUENCE and displays the value in the Course Section ID text item. (If you have not created this sequence previously, create it now, and start the sequence at 2000.) Create a Save New button that inserts the values that currently appear on the tab page into the COURSE_SECTION table.

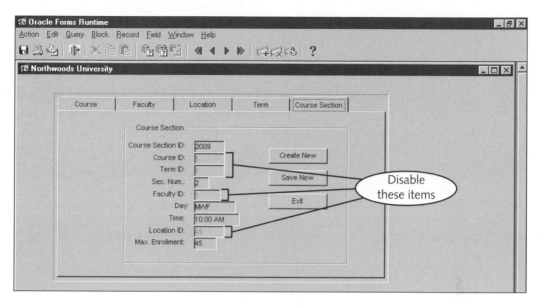

Figure 7-54

   g. Display the Course ID, Faculty ID, Location ID, and Term ID values that the user selected on the individual tab pages in the associated text items on the Course Section tab page. For example, if the user selects Course ID 1 on the Course page, then the value 1 will appear in the Course ID text item on the Course Section page. (*Hint*: Use an assignment statement in the LOV command button triggers on the individual tab pages.)

   h. Disable the foreign key fields on the Course Section page so the user cannot change the values directly, but can only select the values on the other tab pages.

6. In this case, you will create a form with multiple canvases that will allow a Northwoods University faculty member to log onto the database using a logon screen. If the faculty member logs on successfully, he or she is presented with a menu to access forms to view and update personal information, advisee information, and class list information.

   a. Create a new form named FACULTY_FORM, and save it as 7CCase6_DONE.fmb.

   b. Create a canvas and associated control block named LOGON_CANVAS and LOGON_BLOCK containing the items shown in Figure 7-55. When the user clicks the Log On button, the system should check to see if the entered values for F_ID and F_PIN are valid for any of the current FACULTY records and display a message stating "User not found." if the F_ID/F_PIN combination is not valid. If the logon information is correct, a canvas with the menu selections shown in Figure 7-56 appears. The user should have three chances to successfully enter his or her ID and PIN. After the third unsuccessful attempt, the form should exit. Configure the block so that the user cannot use the Tab key to move to the next block.

(*Hint*: To conceal the PIN data, change the text item's Conceal Data property to Yes.

Hint: Create an explicit cursor and use the %FOUND attribute to check if the faculty ID and PIN entered by the user are included in the FACULTY table.

Hint: To track the number of logon attempts, create a text item on the form, and set its Visible property to No.

Hint: To exit the form without having the system ask for user confirmation, use the following command: EXIT_FORM (NO_VALIDATE);.)

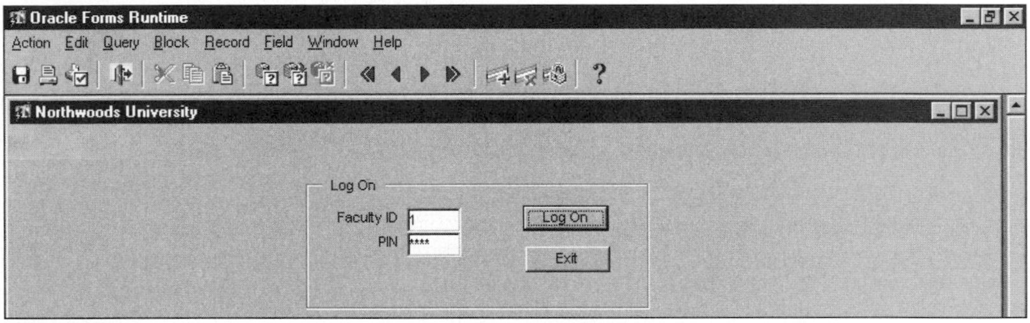

Figure 7-55

c. Create a Menu canvas named MENU_CANVAS and an associated control block containing the command buttons and layout items shown in Figure 7-56.

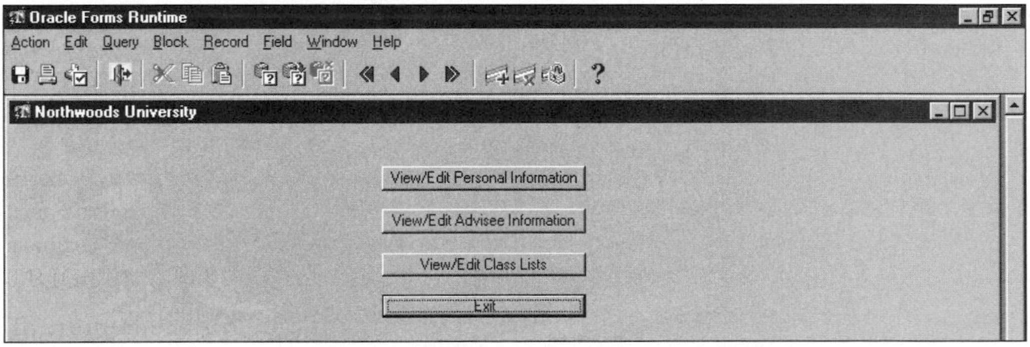

Figure 7-56

d. When the user clicks the View/Edit Personal Information button on the Menu canvas, the canvas shown in Figure 7-57 appears with the user's personal information displayed. Create a control block containing the fields shown. Create a trigger for the Update button to save updated information in the FACULTY table. Create an LOV to display all LOCATION building codes and room numbers, and another LOV to display all choices for F_RANK. When the user clicks the Return button, the Menu canvas appears again. (*Hint*: Create an implicit cursor in the View/Edit Personal Information button trigger to populate the canvas text items.)

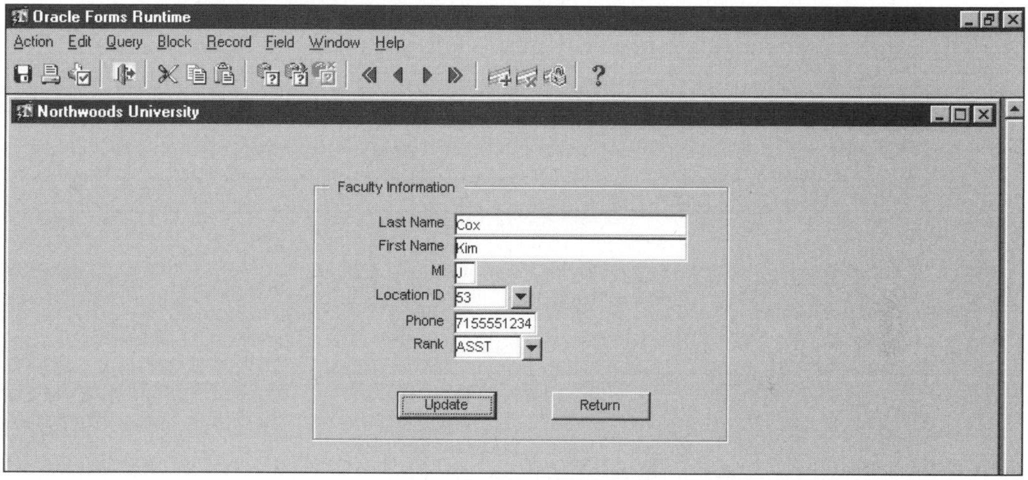

Figure 7-57

e. When the user clicks the View/Edit Advisee Information button on the Menu canvas, the canvas shown in Figure 7-58 appears, and the faculty member can view information about his or her advisees. Create a data block and associated layout that displays the student information for the current faculty member's advisees. When the user clicks the Return button, the Menu canvas should appear again.

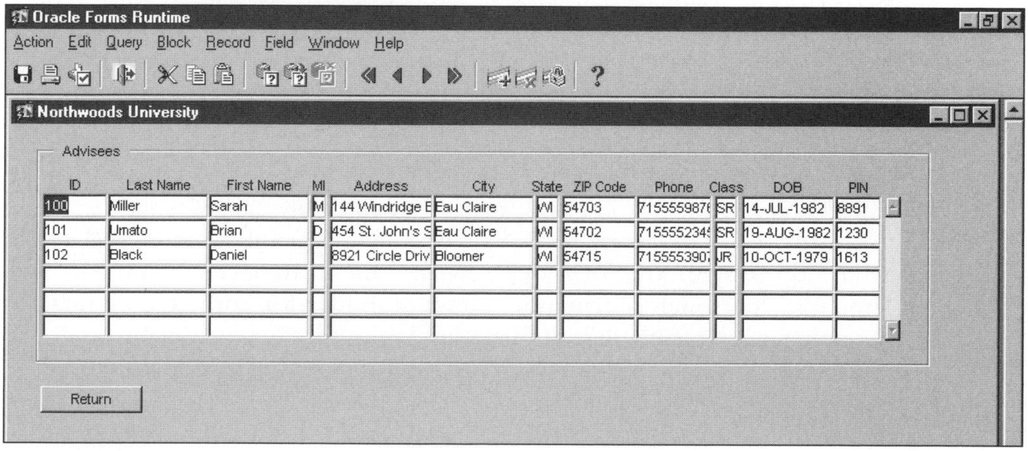

Figure 7-58

f. When the user clicks the View/Edit Class Lists button, the canvas shown in Figure 7-59 appears. The Course List frame lists information about all course sections ever taught by the current faculty member. Create a data block based on a database view named COURSE_VIEW that contains the faculty ID, course section ID, term description, course call ID, and section number. Do not display the faculty ID in the layout. Make sure that the data block retrieves course information only for the current faculty member. Create a second data block based on a database view named CLASS_LIST that contains the course section ID, student ID, student last name, student first name, student middle initial, student class, and course grade. Do not display the course section ID in the layout. Create a master-detail relationship between the course section ID displayed in the Course List data block and the course section ID in the Class List data block. When the user clicks the Return button, the Menu canvas appears again.

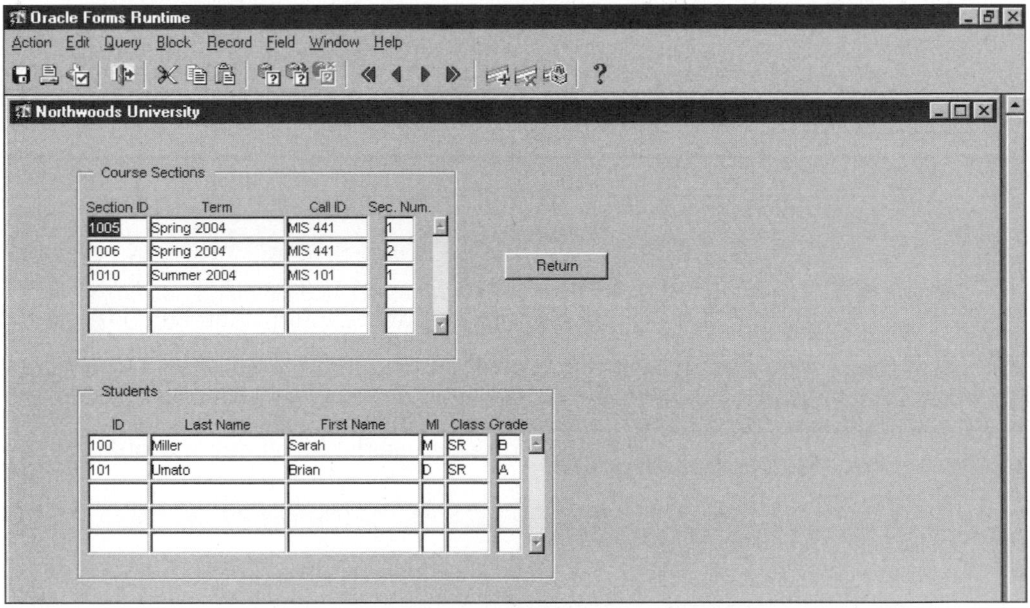

Figure 7-59

7. In this case, you will create a form with multiple canvases that will allow a Northwoods University student to view student information. When the student starts the system, the form opens with a menu to access forms for viewing and updating personal information, viewing grades for completed courses, and viewing current schedule information.

a. Create a new form named STUDENT_FORM, and save it as 7CCase7_DONE.fmb.

b. Create a canvas and associated control block named LOGON_CANVAS and LOGON_BLOCK containing the items shown in Figure 7-60. When a student clicks the Log On button, the system should check to see if the student ID and PIN values are valid for any of the current STUDENT records. If they are, a canvas with the menu selections shown in Figure 7-61 appears. The student should have three chances to successfully enter his or her ID and PIN. After the third unsuccessful attempt, the form should exit.

(*Hint:* To track the number of logon attempts, create a text item on the form, and set its Visible property to No.

Hint: To conceal the PIN data, change the text item's Conceal Data property to Yes.

Hint: Create an explicit cursor and use the %FOUND attribute to check if the student ID and PIN entered by the user are included in the STUDENT table.

Hint: To exit the form without having the system automatically ask for user confirmation, use the following command: `EXIT_FORM(NO_VALIDATE);`.)

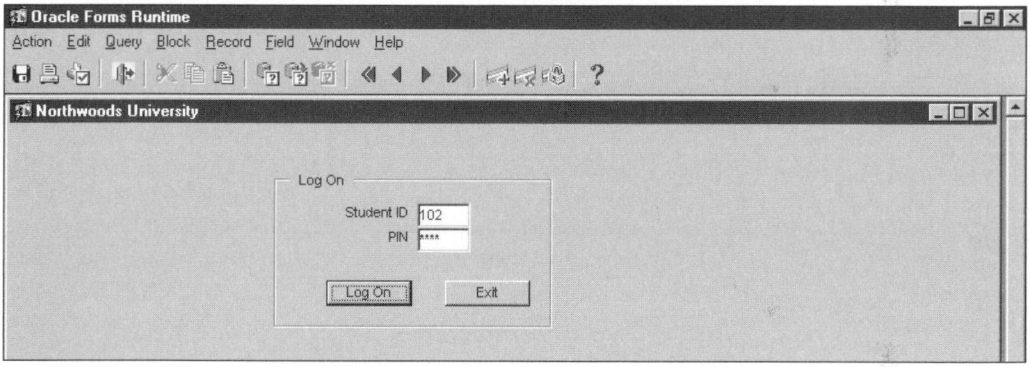

Figure 7-60

c. Create a canvas named MENU_CANVAS and an associated control block containing the command buttons and layout items shown in Figure 7-61.

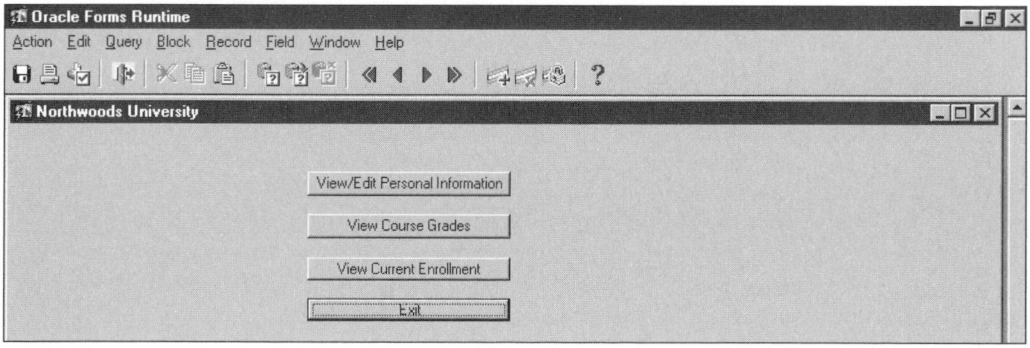

Figure 7-61

d. When the user clicks the View/Edit Personal Information button on the Menu canvas, the canvas shown in Figure 7-62 appears with the student's personal information displayed in the form text items. Create a data block that contains all of the STUDENT data fields and displays the fields as shown by making a relationship with the control block where the current S_ID value appears. When the user clicks the Return button, the Menu canvas appears again.

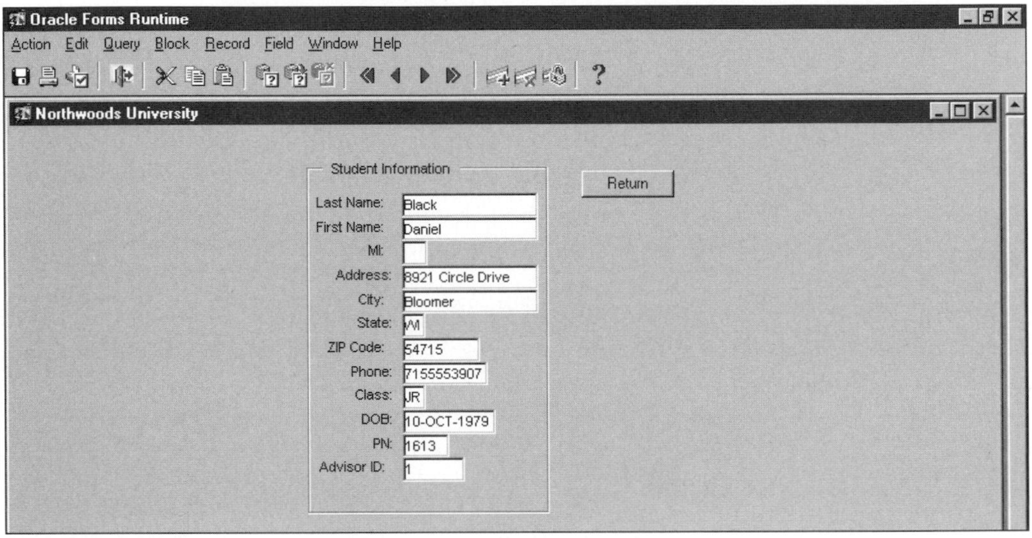

Figure 7-62

e. When the student clicks the View Course Grades button on the Menu canvas, the canvas shown in Figure 7-63 appears, allowing the student to view information for all past course grades. Create a data block based on a database view named GRADE_VIEW that contains the student ID, term ID, term description, course call ID, and grade. Do not retrieve records for course enrollment records that have not yet been assigned a grade. Sort the retrieved course grade information by term ID, and do not display grade information for any other students except the one who has logged onto the system. Do not display the student ID in the layout. When the user clicks the Return button, the Menu canvas appears again.

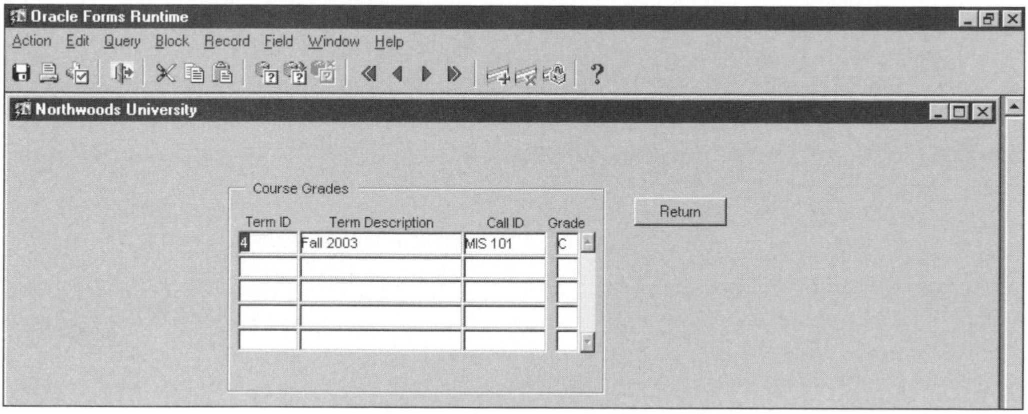

Figure 7-63

7

f. When the student clicks the View Current Enrollment button on the Menu canvas, the canvas shown in Figure 7-64 appears, which shows courses that the student is currently enrolled in but has not yet received a grade for. Create a data block based on a database view named CURRENT_ENROLLMENT_VIEW to display the Current Schedule frame fields. CURRENT_ENROLLMENT_VIEW should contain the student ID, call ID, section number, faculty last name, day, time, building code, and room number, and only include records for which a grade value has not yet been assigned. Do not display the student ID in the layout. This Current Schedule frame will display the schedule information for only the current student. When the user clicks the Return button, the Menu canvas appears again.

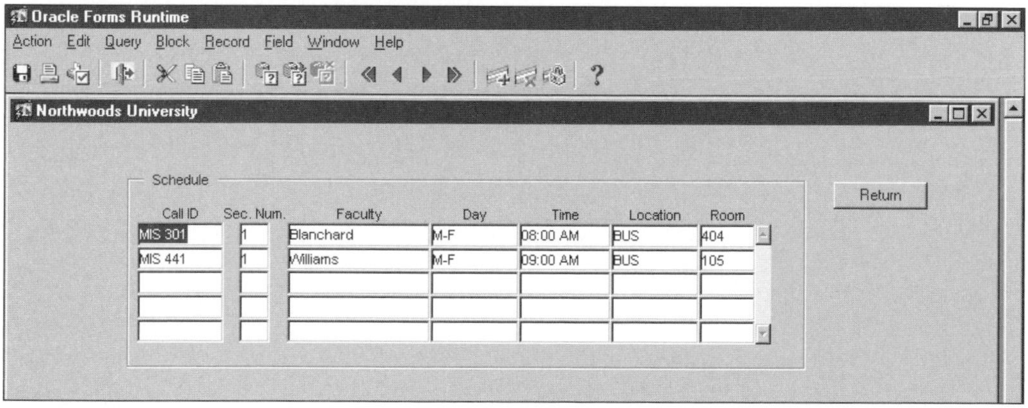

Figure 7-64

8

ADVANCED FORM BUILDER TOPICS

◀ LESSON A ▶

Objectives

- ♦ Become familiar with form non-input data items that display images and sounds
- ♦ Create boilerplate items and calculated form fields
- ♦ Create static and dynamic image items that display graphics stored in the file system or database
- ♦ Create forms that insert sound objects into the database and play sounds

In the previous two chapters, you learned the basics of how to create and use data block and custom forms. This chapter explores advanced topics for creating forms that display and manipulate different types of data in different ways. In Lesson A, you will learn how to create forms that contain items to display calculated data, images, and sounds. Lesson B investigates creating data blocks using alternate data block sources, such as views and PL/SQL procedures. It also addresses using forms to control transactions and configuring forms that retrieve large data sets. Lesson C explores how to control mouse actions and data block relationships, and how to effectively use record groups to enhance form functionality.

NON-INPUT FORM ITEMS

So far, all of the forms items you have created (text items, radio group, check boxes, and so forth) allow users to input and change data values. Form Builder also supports a number of items that display data but do not allow the user to change the displayed value. Table 8-1 summarizes the different types of non-input form items, along with the tool on the Layout Editor tool palette that you use to create the item.

 You can change an item's type by changing its Item Type property value in the item Property Palette.

Item Type	Description	Usage	Layout Editor tool
Boilerplate Text	Form text that appears directly on the form	Form titles and other nondata text	T Text tool
Boilerplate Object	Frames and shapes (rectangles, circles, lines, etc.)	Enhance form appearance	▢ Frame tool ▢ Rectangle tool ◯ Ellipse tool ◥ Line tool
Display Item	Displays text data in a text box, but does not allow input	Calculated fields; read-only data	abc Display Item tool
Image Item	Displays graphic images	Display graphic image data or file images to enhance form appearance	🖳 Image Item tool
Sound Item	Provides access to audio data	Play sound data retrieved from the file system or database	◥ Sound Item tool

Table 8-1   Non-input form items

In the following sections, you will learn how to create and use these non-input item types.

CREATING BOILERPLATE ITEMS AND DISPLAY ITEMS

Recall that boilerplate items are form items that do not display database data, but are used to enhance form functionality and appearance. You created boilerplate frames around radio groups in a previous chapter. In this section, you will create other types of boilerplate items. You will also create display items that display calculated values. A **display item** displays text data in a text box. This text cannot be changed by the user. Display items display text that might be changed in the program by a trigger or values that are calculated using retrieved data values.

To gain experience with creating boilerplate and display items, you will create the form shown in Figure 8-1, which retrieves data from the INVENTORY table in the Clearwater Traders database using a data block. The form has a display item that shows the value for each inventory item, calculated as the price times the quantity on hand. A second display item sums the values of all retrieved items. A boilerplate rectangle and text label describe and emphasize the display item that displays the total values.

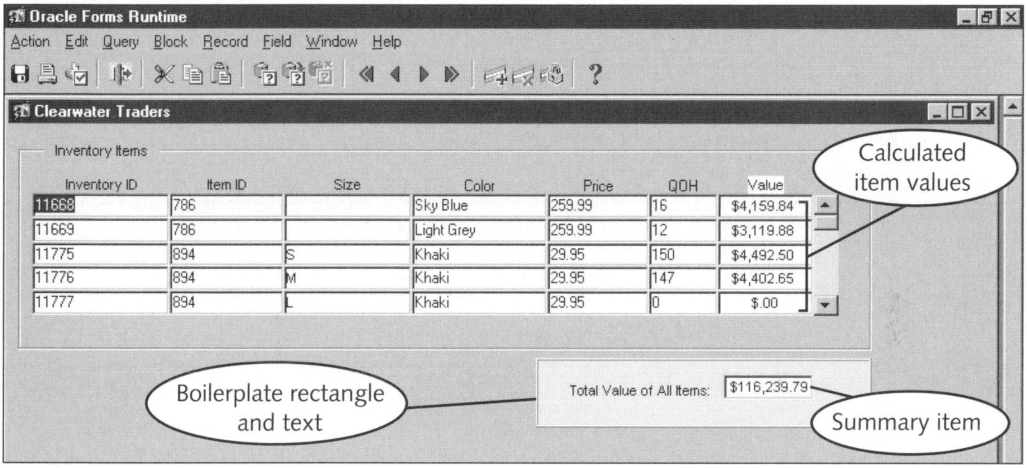

Figure 8-1   Form with boilerplate objects and calculated items

First, you will refresh your database tables. Then, you will open a form named Inventory.fmb in the Chapter8 folder on your Data Disk that contains the data block form for displaying all of the fields in the INVENTORY table. After you open the form and change the filename, you will create the display item that shows the calculated value for each inventory item. You will create this display item in the INVENTORY data block. Currently, the INVENTORY data block uses a tabular-style layout, with five records appearing on the canvas at one time. When you draw the new display item in the data block, five display items will appear, one for each record.

 The files on the Chapter8 Data Disk are stored on three floppy disks, so if you are using floppy disks for your Data Disks, you will need to insert the disk that has the required files. Your Chapter 8 TutorialSolutions files will have to be stored on two separate floppy disks. If you are using floppy disks for your TutorialSolutions disk, you will need to insert the second floppy disk when the first disk becomes filled.

To open the form and create the display item:

1. Start SQL*Plus, log onto the database, and run the **clearwater.sql**, **northwoods.sql**, and **softwareexp.sql** scripts in the Chapter8 folder on your Data Disk to refresh your database tables.

2. Start Form Builder, open **Inventory.fmb** from the Chapter8 folder on your Data Disk, and save the file as **8AInventory.fmb** in your Chapter8\TutorialSolutions folder. If necessary, maximize the Object Navigator window.

3. Open the Layout Editor, select the **Inventory Items frame**, and make it wider by dragging the center selection handle on the right edge of the frame toward the right edge of your screen display.

4. Select the scrollbar, and move it to the right edge of your screen display so there is an empty space wide enough for the Value column.

5. Open the Block list at the top of the Layout Editor window, and select **INVENTORY**, so the new display item will be in the INVENTORY block.

6. Select the **Display Item tool** [abc] on the tool palette, and draw a rectangle that is the same height as the other text items and as wide as the Value field in Figure 8-1. Since the display item is in the INVENTORY block and the block displays five records at a time, five individual display items appear, even though you drew only one display item.

Note that the Display Item tool [abc] shows a text box with a gray background, while the Text Item tool [abc] shows a text box with a white background.

7. Resize and reposition the display items so they look like the Value column display items in Figure 8-1.

8. Save the form.

Next, you will modify the properties of the new display item. You will change the name, data type, maximum length, format mask, and prompt.

To modify the display item properties:

1. Double-click any of the display items on the Layout Editor canvas to open the display item Property Palette. Modify the following properties. Do NOT close the Property Palette yet.

Property	Value
Name	**VALUE_DISPLAY_ITEM**
Data Type	**Number**
Maximum Length	**15**
Format Mask	**$999,999.99**
Prompt	**Value**
Prompt Justification	**Center**
Prompt Attachment Edge	**Top**

The next step is to specify how the display item calculates its value. To do this, you will change the **Calculation Mode** property, which can have one of three values: **None**, which indicates the item is not a calculated value; **Formula**, which indicates the item's value will be calculated by a PL/SQL formula; or **Summary**, which indicates the item's value will be calculated using a summary operation (such as SUM or AVG) on a single form item. This display item will contain inventory values calculated using a formula written in PL/SQL. In the code, you reference form items within the formula in the usual way, using the block name and item name, separated by a period and prefaced with a colon. The formula will show that the calculated value is the product of the PRICE and QOH text items.

A display item that displays a calculated value does not have to be in the same data block as the source items used in the calculation formula.

After you change the calculation properties of the display item, you will run the form, retrieve all of the records from the INVENTORY table, and view the calculated inventory values.

To specify the display item calculation properties and test the form:

1. In the VALUE_DISPLAY_ITEM Property Palette, scroll down to the Calculation property node, and change the Calculation Mode property to **Formula**.

2. Select the **Formula** property, click the **More button** , type `:INVENTORY.PRICE * :INVENTORY.QOH` as the formula, and then click **OK**.

3. Close the Property Palette, and save the form.

4. Run the form, connect to the database if necessary, click the **Enter Query button** ![], do not enter a search condition, and then click the **Execute Query button** ![] to retrieve all of the INVENTORY records. The records appear, along with the calculated values.

5. Close Forms Runtime.

Next, you will create the display item that shows the total of all of the item values. This item will use the Summary calculation mode. You must specify the name of the item being summarized, the block the item is in, and the summary function (such as SUM or AVG) that will be applied. A summary item can be in the same data block as the source item, or you can place it in a separate control block. You will place the summary item in a separate control block.

If you place the summary item in the same data block as the source item, and the data block uses the tabular display style and displays multiple records, you must change the display item's Number of Items Displayed property value to 1, or a summary item will appear for each record displayed.

To place the summary item in a separate control block, you must set the control block's Single Record property to Yes. This specifies that the block contains only one value and cannot contain more than one value. Now you will create a control block that contains the summary item and modify the control block properties.

To create the control block for the summary item:

1. Open the Object Navigator, select the **Data Blocks node**, and then click the **Create button** to create a new block. Select the **Build a new block manually option button**, and then click **OK**.

2. Open the new block's Property Palette, change the Name property to **SUMMARY_BLOCK**, the Database Data Block property to **No**, the Single Record property to **Yes**, and then close the Property Palette.

Next, you will create the summary display item. This time, you will create the display item in the Object Navigator by creating a new item object in the SUMMARY_BLOCK, and then modifying its properties. By default, when you create a new item in the Object Navigator, it is a text item. Therefore, you will need to change the Item Type property to Display Item. Also, you will need to specify that the new item appears on the INVENTORY_CANVAS—otherwise, the item will exist in the block but will not appear on the canvas. You will also modify the calculation properties and then run the form and test the summary item.

To create and test the summary display item:

1. In the Object Navigator, select the **Items node** under the SUMMARY_BLOCK, and click the **Create button** to create a new item. Open the new item's Property Palette, modify the item's properties as follows, and then close the Property Palette:

Property	Value
Name	**SUMMARY_ITEM**
Item Type	**Display Item**
Data Type	**Number**
Maximum Length	**15**
Format Mask	**$999,999.99**
Calculation Mode	**Summary**
Summary Function	**Sum**
Summarized Block	**INVENTORY**
Summarized Item	**VALUE_DISPLAY_ITEM**
Canvas	**INVENTORY_CANVAS**

2. Open the Layout Editor, and move the new summary item so it is below the calculated item values, as shown in Figure 8-1.

3. Save the form, and then run the form. Click the **Enter Query button**, type **786** in the first Item ID text item for the search condition, and then click the **Execute Query button**. The records and calculated values

appear, along with the summary item value, which should display $7279.72 as the total value for all inventory items associated with Item ID 786.

If the display item is not wide enough to show the entire value, close Forms Runtime, select any one of the display items, and make the display items wider. Then, repeat Steps 3 and 4.

4. Close Forms Runtime.

To finish the form, you need to create the boilerplate objects. By default, input items (such as text items and radio buttons) and non-input items (such as display items) always appear on top of boilerplate objects. You can place boilerplate objects, such as text and shapes, on top of other boilerplate objects, and they appear in the order they are created. For example, if you create the boilerplate text for the summary display item ("Total Value of All Items:") and then draw the boilerplate rectangle on top of the text, the text will be hidden. Next, you will draw and format the boilerplate rectangle around the summary item. Then, you will add the boilerplate text.

To create and format the boilerplate objects:

1. Select the **Rectangle tool** ☐ on the tool palette, and draw the boilerplate rectangle as shown on Figure 8-1.

2. Select the rectangle, click **Format** on the menu bar, point to **Bevel**, and click **Outset**.

3. With the rectangle still selected, select the **Fill Color tool** on the tool palette, and select a lighter shade of gray than the background color.

Always use subtle colors when creating form enhancements that involve colors. Intense, saturated colors look less professional than subtle shades and can cause eyestrain. Always ensure that there is sufficient contrast between foreground text and background colors.

4. Select the **Text tool** T on the tool palette, click in the rectangle at approximately the point where the boilerplate text starts in Figure 8-1, and type **Total Value of All Items:**. Reposition and resize the boilerplate items as necessary so your form looks like Figure 8-1.

5. Save the form, run the form, click the **Enter Query button**, do not enter a search condition and then click the **Execute Query button** to retrieve all of the INVENTORY records. The completed form should look like Figure 8-1.

6. Close Forms Runtime, and then close the form in Form Builder.

CREATING IMAGE ITEMS TO DISPLAY GRAPHIC IMAGES

In a form, you can use image items to display graphic images that enhance the appearance of the form or to retrieve images that are stored in the database or file system. There are two ways to display a graphic image on a form: as a **static imported image**, which incorporates the image data into the form design (.fmb) file and compiles it into the .fmx file; or as a **dynamic image**, which loads the image data into the form at runtime. Usually, you use static imported images to add graphic enhancements that stay the same regardless of the data that currently appears on the form. You use dynamic images to display images that are retrieved from the database or file system while the form is running, or to retrieve and display large static images that are not loaded every time you run the form. Dynamic images can also be used within data blocks to store image data associated with text data. For example, you might store and display a photograph of a person or a product along with the text data describing the person or product.

Static imported images make the .fmb and .fmx files larger, so you should always select small (less than 100 KB) images for graphic enhancements. When you use dynamic image items, keep in mind that they make the form load more slowly because the image must load each time it is retrieved, and the graphic file from which the image loads must be available either in the file system or the database when the user runs the form.

To learn how to create image items, you will create the form shown in Figure 8-2. This is a data block form associated with the ITEM table in the Clearwater Traders database. It contains a static imported image that shows the Clearwater Traders company logo and a dynamic image that displays a photograph of the current item.

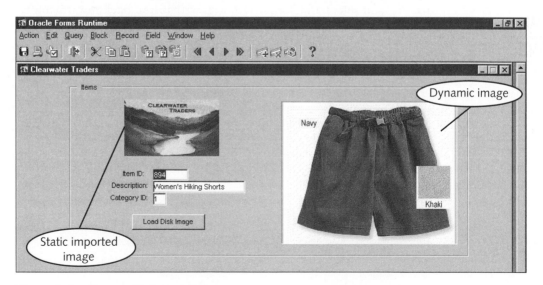

Figure 8-2   Form with image items

The image will be retrieved from a file in the file system, and then stored in the database. Users can click the Load Disk Image button to insert a new image or update the current image. When a user clicks the button, a file dialog box opens that allows him or her to specify the image file that will load into the image item. After selecting an image file from the file system and displaying it in the image item, the user can click the Save button 🖫 on the Forms Runtime toolbar to store the image in the database. When the user retrieves a record that already has a stored image, the image will automatically appear in the image item.

To create this form, you will first open a file named Item.fmb stored in the Chapter8 folder on your Data Disk. Currently, this form contains no data blocks. You will create a data block and layout for the ITEM table in the Clearwater Traders database. Recall from previous chapters that the ITEM table has a field named ITEM_IMAGE that is a BLOB (binary large binary object) data type, and that this field currently contains a locator for the BLOB field. (Recall that a locator is a structure that contains information about the BLOB type and points to the alternate memory location where the BLOB image data will eventually be stored.) Since the ITEM_IMAGE field has the BLOB data type, the Layout Wizard will automatically display the ITEM_IMAGE field in the form layout using an image item. Now you will open the form, save the file using a different name, and create the data block and layout.

To open the form and create the data block and layout:

1. In Form Builder, open **Item.fmb** from the Chapter8 folder on your Data Disk, and save the form as **8AItem.fmb** in your Chapter8\TutorialSolutions folder.

2. In the Object Navigator, select the **Data Blocks node**, click the **Create button** 🛬 to create a new data block, confirm that the Use the Data Block Wizard option button is selected, and then click **OK**.

3. On the Data Block Wizard Welcome page, click **Next**. Confirm that the Table or View option button is selected, click **Next**, click **Browse**, select the **ITEM** table as the data block source, and then click OK. When the ITEM fields appear in the Available Columns list, note that the data type icon for the ITEM_IMAGE is the LOB icon 🔲, which indicates that the column has one of the LOB data types.

4. Select all of the fields in the ITEM table to be included in the data block, click **Next**, and then click **Finish**.

5. Click **Next** on the Layout Wizard Welcome page. On the Canvas page, make sure ITEM_CANVAS is the selected display canvas, and click **Next**.

6. On the Data Block page, select all of the available items for display in the layout, and click **Next.**

7. On the Items page, change the prompts and widths as follows, and then click **Next**.

Name	Prompt	Width	Height
ITEM_ID	Item ID:	68	14
ITEM_DESC	Description	80	14
CATEGORY_ID	Category ID:	27	14
ITEM_IMAGE	(deleted)	186	165

8. On the Style page, be sure that the Form layout style is selected, and click **Next**. On the Rows page, type **Items** for the Frame Title and accept the other default values, click **Next**, and then click **Finish**. The layout appears in the Layout Editor.

9. Select the frame, open its Property Palette, change the Update Layout property to **Manually**, and then close the Property Palette.

10. Resize the frame, reposition the form items so your canvas looks like Figure 8-3, and then save the form.

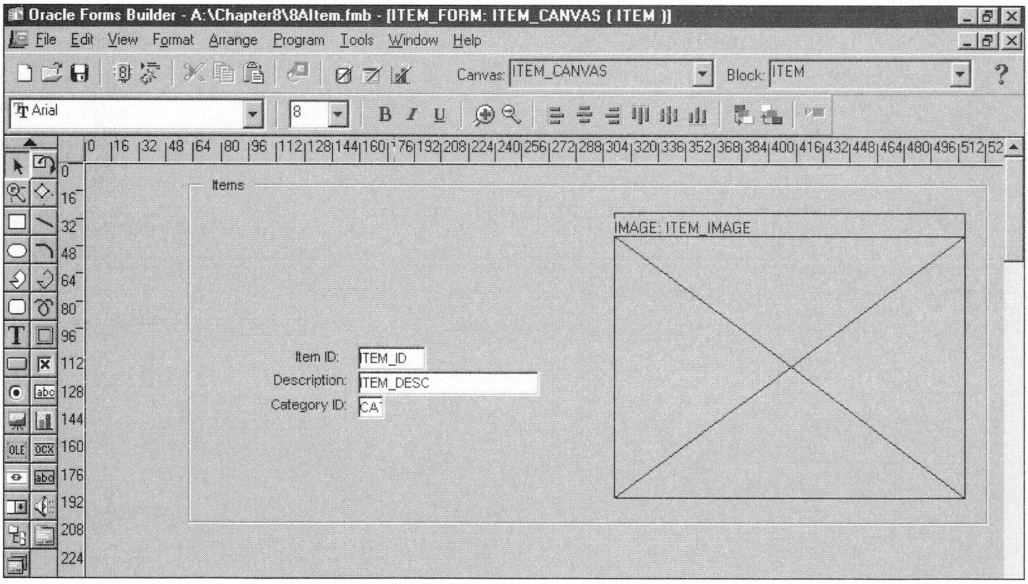

Figure 8-3   Positioning the form items

Next, you will import a graphic image of the Clearwater Traders logo. Since this is a fairly small graphics file (69 KB), and the same image will appear regardless of what item values also appear on the form, you will treat the image as a static imported image. To import a clip art image as a static imported image, you click File on the Form Builder menu bar, point to Import, and then click Image to open the Import Image dialog box, as shown in Figure 8-4. This dialog box enables you to import an image that is stored either in the file system or in the database by specifying the image file and file format. The **image format** property specifies the file format, which is determined by the file extension. Popular image types include bitmaps (.bmp), PC Paintbrush (.pcx), GIF (.gif), and TIFF (.tif) files. Bitmap files are usually uncompressed, while .pcx, .gif, and .tif files use different compression methods. Most graphics applications support one or more of these file types. If you select the "Any" option, the Form Builder application will automatically import the image using the correct image type.

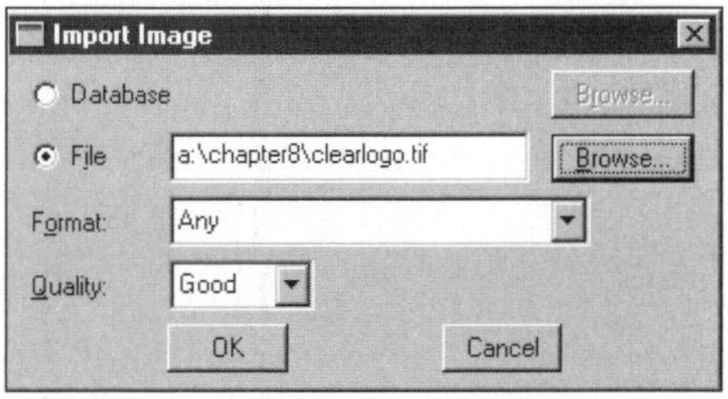

Figure 8-4   Import Image dialog box

The **image quality** property determines how the image is stored in terms of image resolution and number of colors. Possible image quality choices include Excellent, Very Good, Good, Fair, and Poor. If you import an image using the Excellent quality selection, Form Builder will save the image using the maximum number of colors and highest possible resolution. It will also use the maximum amount of file space and take the longest time to load. The default image quality is Good, which is satisfactory for most graphic images. You can experiment with the different quality levels to determine the minimum acceptable quality for each image.

Now you will add the Clearwater Traders logo to the form as a static imported image. The image is stored as clearlogo.tif in the Chapter8 folder on your Data Disk. You will import the image using the default (Good) image quality selection.

To add the logo as a static imported image:

1. Click **File** on the menu bar, point to **Import**, and then click **Image**. The Import Image dialog box opens, as shown in Figure 8-4.

2. Make sure that the File option button is selected, click **Browse** beside the File text box, navigate to the Chapter8 folder on your Data Disk, select **clearlogo.tif**, and then click **Open**. Make sure that Any is the selected image format, make sure Good is selected for the image quality, and then click **OK**. The image appears in the upper-left corner of the form.

3. In the Layout Editor, select the image, resize it if necessary, and drag it so that it is positioned in the frame as shown in Figure 8-2. Then save the form.

4. Run the form, and note that the imported image appears in the Forms Runtime window. Click the **Enter Query button** 🖳, type **786** in the Item ID text item for the search condition, and then click the **Execute Query button** 🖳. The data for the 3-Season Tent appears. Note that no image appears in the dynamic image item on the form, because no images have been loaded into the ITEM_IMAGE field yet.

5. Close Forms Runtime.

Now you will create the Load Disk Image button, which has an associated trigger that displays a file dialog box and allows the user to select an image file that is then loaded into the current record's ITEM_IMAGE field. When the user clicks the Save button 🖫 on the Forms Runtime toolbar, the data for the current image is saved in the record's ITEM_IMAGE field in the database.

The code in this trigger declares a variable that represents the complete specification of the image file, which includes the drive letter, the names of all folders and subfolders associated with where the file is stored, separated by back slashes (\), and the filename and extension. To enable the user to specify this information using a file dialog box, you will use the GET_FILE_NAME built-in function. This function enables the user to navigate in the file system and select the desired file, rather than having to type in the file specification. The function returns a text string that represents the complete specification for the selected file.

To load the selected file image into the form image item, you will use the READ_IMAGE_FILE built-in procedure, which has the following general syntax: **READ_IMAGE_FILE (*filename*, '*file_type*', '*item_name*');**. This command requires the following parameters:

- *filename* is the complete specification of the graphic image file. You can pass this value to the procedure as the filename returned by the GET_FILE_NAME function, or as a literal character string. When you pass the value as a literal character string, you must place the value in single quotation marks. For example, the complete specification for the shorts.jpg file stored in the Chapter8 folder on your Data Disk might be 'a:\chapter8\shorts.jpg'. Note that the specification is not case sensitive.

- *file_type* is the type of image file being used. You pass this value as a character string in single quotation marks. Legal values are the following file types: ANY, BMP, CALS, GIF, JFIF, JPG, PICT, RAS, TIFF, or TPIC. If ANY is specified, Forms Runtime will attempt to determine the file type by examining the image. If you use this value with a known file type, you should always specify the file type rather than using the ANY value, because making Forms Runtime determine the file type slows down performance.

The different image file types depend on the graphics art application used to create the file and how the image is compressed to make it occupy less space. Most popular graphics applications support one of these types of files, which are identified by the file extension.

- *item_name* is the name of the image item where the file will appear and is passed as a character string in single quotation marks using the format '*block_name.item_name*'.

Now you will create the Load Disk Image button and create the trigger to allow the user to select the image file. Then, you will run the form and load the image for Item ID 894 (Women's Hiking Shorts).

To create the button and button trigger, and load the image:

1. In the Layout Editor, draw the Load Disk Image button shown in Figure 8-2. Open the button's Property Palette, and change the button Name value to **LOAD_IMAGE_BUTTON** and the button Label value to **Load Disk Image**.

2. Create a new WHEN-BUTTON-PRESSED trigger for the button, and type the code shown in Figure 8-5 to display the file dialog box, retrieve the image filename, and then load the file data into the form item image. Compile the trigger, correct any syntax errors, close the PL/SQL Editor, and then save the form.

```
Oracle Forms Builder - A:\Chapter8\8AItem.fmb                    _ 8 X
File  Edit  Program  Tools  Window  Help
 Compile    Revert    New...    Delete    Close    Help
Type: Trigger              ▼  Object: ITEM        ▼  LOAD_IMAGE_BUTTON    ▼
Name: WHEN-BUTTON-PRESSED                                              ▼
 DECLARE
    image_filename VARCHAR2(255);
 BEGIN
    --allow user to specify image filename using a file dialog box
    image_filename := GET_FILE_NAME;             Type this
    --read the image file into the form item image   code
    READ_IMAGE_FILE(image_filename, 'ANY', 'ITEM.ITEM_IMAGE');
 END;
```

Figure 8-5 Code to retrieve file and load data into image item

3. Run the form, click the **Enter Query button** 🔲, type **894** for the Item ID text item search condition, and then click the **Execute Query button** 🔲. The data for Item ID 894 appears.

4. Click **Load Disk Image**. The Open dialog box opens. Navigate to the Chapter8 folder on your Data Disk, select **shorts.jpg**, and then click **Open**. The image appears in the form image item, as shown in Figure 8-2.

5. Click the **Save button** 🔲 on the Forms Runtime toolbar to save the image in the database, and then close Forms Runtime.

The displayed image should be about the same size as the form image item. However, different images can have different sizes. To enable all images to appear correctly in the form item image, you need to adjust the image item's Sizing Style property. This property can have one of two values: **Crop**, which means that images that are too large for the image item will be cropped, or cut off; or **Adjust**, which means that the image is scaled to fit correctly within the image item. Now, you will change the ITEM_IMAGE item's Sizing Style to Adjust, so the image's size is automatically adjusted. Then, you will run the form and insert the images for the rest of the items in the ITEM table.

To adjust the image item's Sizing Style and insert the additional images:

1. In the Layout Editor, select **IMAGE: ITEM_IMAGE**, right-click, and then click **Property Palette** to open the image's Property Palette. Under the Functional property node, change the Sizing Style to **Adjust**, and then close the Property Palette.

2. Save the form, and then run the form. Click the **Enter Query button** 🔳, do not enter a search condition, and then click the **Execute Query** button 🔳 to retrieve all of the table records. The data values for Item ID 894 appear, along with the image of the item that you inserted in the last set of steps.

3. Click the **Next Record** button ▶ to display the next record in the table. The values for Item ID 897 appear. The image data has not been loaded yet.

4. Click **Load Disk Image**, navigate to the Chapter8 folder on your Data Disk, select **fleece.jpg**, and then click **Open**. The image appears on the form.

5. Click the **Save button** 🔳 to save the image in the database, and then click ▶ to move to the next record, which is Item ID 995.

6. Click **Load Disk Image**, navigate to the Chapter8 folder on your Data Disk, select **sandals.jpg**, and then click **Open**. The image appears on the form. Click 🔳 to save the image in the database.

7. Load the image data for the remaining ITEM records using the filenames specified in Figure 1-11.

8. Close Forms Runtime.

9. Close the form in Form Builder.

CREATING SOUND ITEMS

You can store sound data in an Oracle database in fields that have the BLOB (binary large object) data type. You can store, retrieve, and play this sound data using a form **sound item**. The components of a sound item are shown in Figure 8-6.

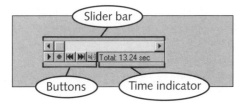

Figure 8-6   Sound item components

The **slider** moves from left to right on the screen display as the sound plays, and the user can use it like a scrollbar to move to different locations within the sound clip. The buttons allow the user to operate the sound. The **Play button** ▶ starts and stops the sound. The **Record button** ● allows users to record new sounds, provided they have the required hardware and software to do so. The **Rewind button** ◀◀ places the slider at the beginning of the sound clip, and the **Fast Forward button** ▶▶ quickly moves the slider toward the end of the sound clip. The **Volume Control button** 🔊 enables the user to adjust the sound volume. The **time indicator** indicates the duration of the sound clip, in seconds. You can optionally display or hide all of these components.

To gain experience working with sound data, you will use the FORM_BUILDER_TOPICS database table shown in Table 8-2, which contains audio tutorial data about different Form Builder topics.

TOPIC_ID	TOPIC_NAME	TOPIC_RECORDING
1	Object Navigator	Object_Navigator.WAV
2	Layout Editor	Layout_Editor.WAV
3	Property Palette	Property_Palette.WAV
4	PL/SQL Editor	PLSQL_Editor.WAV

Table 8-2   Database table with sound data

The table fields include TOPIC_ID, which is a surrogate primary key; TOPIC_NAME, which identifies the topic; and TOPIC_RECORDING, which is a BLOB data field that contains a sound clip that describes the topic. The names of the sound files that will be stored in the TOPIC_RECORDING field are shown. These files use the WAVE encoding format, which is identified by the .WAV file extension. Now, you will run the script to create the FORM_BUILDER_TOPICS table. The script creates the table and inserts the TOPIC_ID and TOPIC_NAME data values. It also inserts a locator for the sound BLOB, but it inserts no sound data.

To create the table:

1. In SQL*Plus, run the **formbuilder.sql** script stored in the Chapter8 folder on your Data Disk. The output indicates that the table was created and the four data rows were inserted.

If the message "ERROR at line 1: ORA-00942: table or view does not exist." appears, don't worry, because the script contains the DROP TABLE command to first drop the FORM_BUILDER_TOPICS table and then create the table. The DROP TABLE command is included in case you need to run the script multiple times or already have created a table named FORM_BUILDER_TOPICS.

2. Type **SELECT topic_id, topic_name FROM form_builder_topics;** at the SQL prompt to check the table structure and contents. The output should look like Table 8-2. The TOPIC_RECORDING field does not appear, because LOB data cannot be shown in SQL*PLUS.

Sound items are similar to dynamic image items: You create a data block form based on a table that has a LOB field that will store the binary sound data. Then, you insert or update the sound data from a file in the file system. Supported sound formats include AU (.au), AIFF (.aiff), AIFF-C (.aifc), or WAVE (.wav). (The different sound formats are based on compression techniques used to create the sound files.) Now, you will create the form shown in Figure 8-7 to display the data in the FORM_BUILDER_TOPICS table, insert the sound recordings into the database, and play the sound recordings.

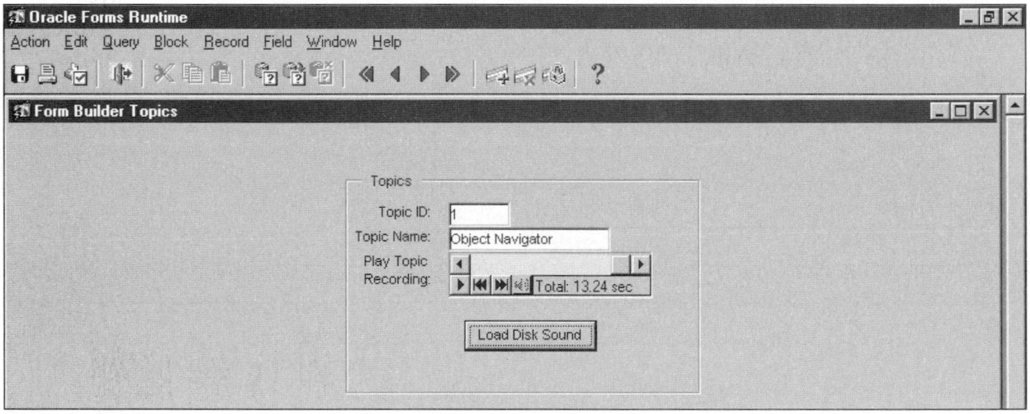

![Oracle Forms Runtime window showing the Form Builder Topics form with a Topics frame containing Topic ID field showing "1", Topic Name field showing "Object Navigator", a Play Topic Recording control showing "Total: 13.24 sec", and a "Load Disk Sound" button.]

Figure 8-7   Form to insert and play sound recordings

When the user clicks the Load Disk Sound button, a file dialog box will appear to allow the user to select the correct sound file. When the user saves the data block changes, the sound will be inserted into the database. To create this form, you will modify a form named Topics.fmb stored in the Chapter8 folder on your Data Disk. This form does not currently contain any data blocks. You will create the data block and layout, and specify that the TOPIC_RECORDING field will appear as a sound item.

To open the file and create the data block and layout:

1. In Form Builder, open **Topics.fmb** from the Chapter8 folder on your Data Disk, and save the file as **8ATopics.fmb** in your Chapter8\TutorialSolutions folder.

2. In the Object Navigator, select the **Data Blocks node**, click the **Create button** 🔳 to create a new data block, confirm that the Use the Data Block Wizard option button is selected, and then click **OK**.

3. On the Data Block Wizard Welcome page, click **Next**. Confirm that the Table or View option button is selected, click **Next**, click **Browse**, select the **FORM_BUILDER_TOPICS** table as the data block source, and click **OK**. Note that the data type icon for the TOPIC_RECORDING field is the LOB icon 🔳, which indicates that the column has an LOB data type.

4. Select all of the fields in the table to be included in the data block, click **Next**, and then click **Finish**.

5. On the Layout Wizard Welcome page, click **Next**, make sure TOPIC_CANVAS is the selected display canvas, and then click **Next**.

6. Select all of the available items for display in the layout. Then select **TOPIC_RECORDING** in the Displayed Items list, change the Item Type to **Sound**, and click **Next**.

 The Layout Wizard displays all LOB objects as image items by default.

7. Change the prompts and widths as follows, and then click **Next**.

Name	Prompt	Width	Height
TOPIC_ID	**Topic ID:**	36	14
TOPIC_NAME	**Topic Name:**	95	14
TOPIC_RECORDING	**Play Topic Recording:**	120	28

8. Be sure that the Form-style layout is selected, click **Next**, type **Topics** for the frame title, and accept the other default values. Click **Next**, and then click **Finish**. The layout appears in the Layout Editor.

9. Select the frame, open its Property Palette, change the Update Layout property to **Manually**, and then close the Property Palette.

10. Resize the frame and reposition the form items so your form items look like the ones in Figure 8-7, and then save the form.

Next, you will configure the sound item display. You can display or hide the components on the sound item display using the sound item's Property Palette. By default, the slider, time indicator, Play button ▶, and Volume Control button 🔊 appear. The Record button ⏺, Fast Forward button ⏩, and Rewind button ⏪ do not appear. You will change the default display by specifying that the Fast Forward button ⏩ and the Rewind button ⏪ should appear.

To configure the sound item display:

1. In the Layout Editor, select the sound item and open its Property Palette. Scroll down to the Physical property node, change the Show Fast Forward Button property to **Yes** and the Show Rewind Button property to **Yes**, close the Property Palette, and save the form. The additional buttons appear on the canvas.

Now you will create the Load Disk Sound button and create the trigger to allow the user to select and load the sound file. To load the sound file data into the sound item, you use the READ_SOUND_FILE built-in, which has the following syntax: **READ_SOUND_FILE (*filename*, '*file_type*', '*item_name*');** . The *filename* parameter must specify the complete path to the sound file, including the drive letter and folder path, and is placed in single quotation marks. The *file_type* parameter can be any of the supported sound file types (AU, AIFF, AIFF-C, or WAVE). The *file_type* value can also be ANY, which is the value that you should use if the user can select from different sound file types. The *file_type* value must be enclosed in single quotation marks. If the READ_SOUND_FILE command is used with a known file type, you should always specify the file type rather than using the ANY value, because making Forms Runtime determine the file type slows performance. The *item_name* parameter specifies the sound item block name and item name, and is also enclosed in single quotation marks. After you create the trigger, you will run the form and load the recording for Topic ID 1, which is Object Navigator.

To create the button trigger and load the sound:

1. In the Layout Editor, draw the Load Disk Sound button shown in Figure 8-7. Open the button's Property Palette, and change the button Name property value to **LOAD_SOUND_BUTTON** and the button Label property value to **Load Disk Sound**. Close the Property Palette.

2. Create a new WHEN-BUTTON-PRESSED trigger for the button, and type the following code, which displays the file dialog box, retrieves the sound filename, and then loads the file data into the form sound item.

```
DECLARE
  sound_filename VARCHAR2(255);
BEGIN
  sound_filename := GET_FILE_NAME;
READ_SOUND_FILE (sound_filename, 'ANY',
  'FORM_BUILDER_TOPICS.TOPIC_RECORDING');
END;
```

3. Compile the trigger, correct any syntax errors, close the PL/SQL Editor, and then save the form.

4. Run the form, click the **Enter Query button** 🔲, do not enter a search condition and then click the **Execute Query button** 🔲 to retrieve all of the records. The data for the first topic (Topic ID 1, Object Navigator) appears. Note that the sound item is disabled, which indicates that no sound data was retrieved for the record.

5. Click **Load Disk Sound**. The Open dialog box opens. Navigate to the Chapter8 folder on your Data Disk, select **Object_Navigator.WAV**, and click **Open**. The buttons on the sound item are enabled, indicating that a sound is available for playing.

6. Click the **Play button** ▶ on the sound item. The sound item plays.

7. Click the **Save button** 🖫 on the Forms Runtime toolbar to save the sound data in the database. The confirmation message appears, indicating that the record was saved.

8. Close Forms Runtime, close the form in Form Builder and close all open Oracle applications.

SUMMARY

❐ A display item displays text data in a text box. This cannot be directly changed by the user, but can be changed in the program by a trigger or values that are calculated using retrieved data values.

❐ You can create a calculated field by specifying a display item's Calculation Mode property to be Formula, which means that the item's value will be calculated by a PL/SQL formula, or Summary, which indicates that the item's value will be calculated using a summary operation.

❐ Input items always appear on top of boilerplate items. Boilerplate items can be placed on top of other boilerplate items and appear in the order they are created.

❐ Images can appear in forms as static imported images, for which the image data is incorporated into the form design (.fmb) file and compiled into the .fmx file, or as dynamic images, for which the image data is loaded into the form at runtime.

❐ Static imported images are used to add graphic enhancements that stay the same regardless of the data that currently appears on the form. Dynamic images are used to display images retrieved from the database or file system at runtime.

❐ Dynamic image data values are stored in BLOB (binary large binary object) data fields in the database.

❐ To load dynamic image data into a form from the file system, you use the READ_IMAGE_FILE built-in procedure.

❐ Sound data can be stored in a BLOB database field and inserted, retrieved, and played in a form using a sound item.

❐ A form sound item displays a slider, which can be used like a scrollbar to move to different positions within the sound clip, as well as buttons to play, record, rewind, and fast forward the sound clip, and adjust the sound volume.

❐ To load sound data into a form from a file in the file system, you use the READ_SOUND_FILE built-in procedure.

8

REVIEW QUESTIONS

1. What is a boilerplate object?

2. What is the difference between a text item and a display item? What is the difference in the appearance of the tools on the tool palette that are used to create each type of item?

3. Describe the two types of calculated fields that you can create using display items.

4. What is the difference between a static imported image and a dynamic image? When should each be used?

5. Why should static images be limited to files that are less than 100 KB in size?

6. What database data type is used to store sounds?

7. What sound item components can appear on a form? What components appear by default?

8. In the LOAD_IMAGE_FILE and READ_SOUND_FILE built-ins, what does the ANY parameter value specify? When should this value be avoided, and why?

PROBLEM-SOLVING CASES

The cases reference the Clearwater Traders (see Figure 1-11), Northwoods University (see Figure 1-12), and Software Experts (see Figure 1-13) sample databases. The data files needed for all cases are stored in the Chapter8 folder on your Data Disk. Store all solutions in your Chapter8\CaseSolutions folder. Save the text for all SQL commands in a file named 8AQueries.sql in your Chapter8\CaseSolutions folder.

1. The 8ACase1.fmb form currently contains a control block that displays consultant information from the Software Experts database and an associated data block that displays project information. When the user selects a consultant using the LOV command button in the Consultants frame, the associated information about the projects that the consultant is working on appears in the Consultant Projects frame. Modify the form to include the calculated fields shown in Figure 8-8 that display the total number of days that the consultant is scheduled to work on each project, the total number of project days the consultant has been assigned for all projects, and the total number of project hours that the consultant has completed so far. Add the boilerplate text and rectangle as shown to highlight the item totals. (*Hint*: Calculate the total number of days for each project as the roll-off date minus the roll-on date.) Save the completed form as 8ACase1_DONE.

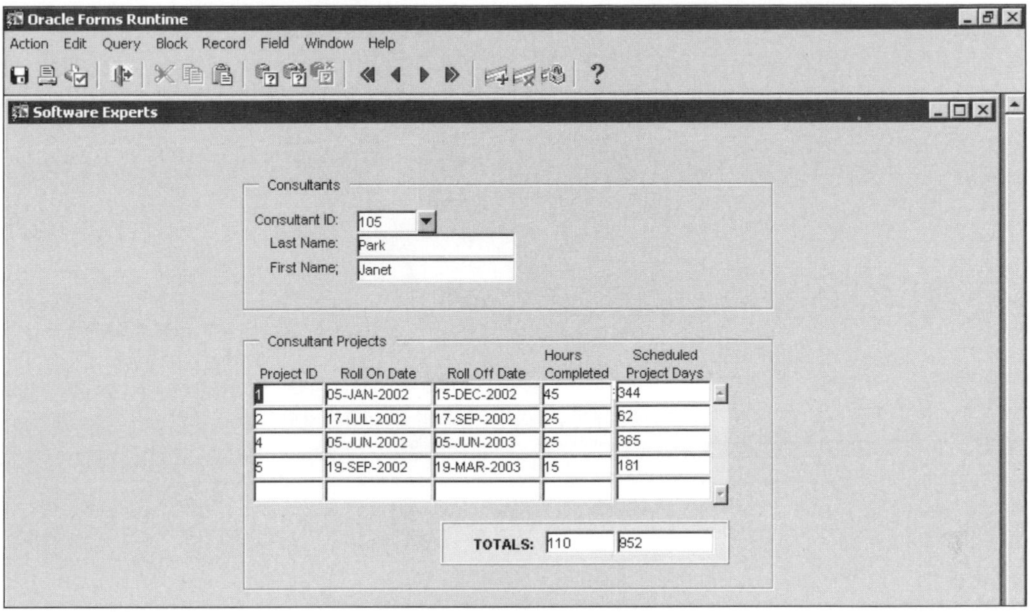

Figure 8-8

2. Modify the 8ACase2.fmb form so that when the user selects a Clearwater Traders customer, and then selects an order for that customer, the order line details appear in the view-only Order Summary frame, as shown in Figure 8-9. The extended total (quantity times price) appears for each order line, along with the order subtotal, sales tax, and final total. Save the completed form as 8ACase2_DONE.fmb.

   a. Create a view named ORDER_LINE_VIEW that contains the order ID, inventory ID, price, order quantity, item description, color, and size for every record in the ORDER_LINE table. Then, create a data block to display the view as shown in the Order Line Detail frame in Figure 8-9, and create a relationship between the form control block and the data block so that when the user selects an order, the order line values for the selected order appear.

   b. Create a display item to show the extended total (quantity ordered times price) for each order line.

   c. Create display items to calculate the order subtotal (sum of all extended totals), sales tax (6% of the order subtotal), and total order cost (subtotal plus sales tax). Add the boilerplate text items and the filled rectangle object to highlight the summary data as shown.

   d. Import the Clearwater Traders logo as a static imported image from the clearlogo.tif file in the Chapter8 folder on your Data Disk. Resize and reposition the image and the other form items as shown.

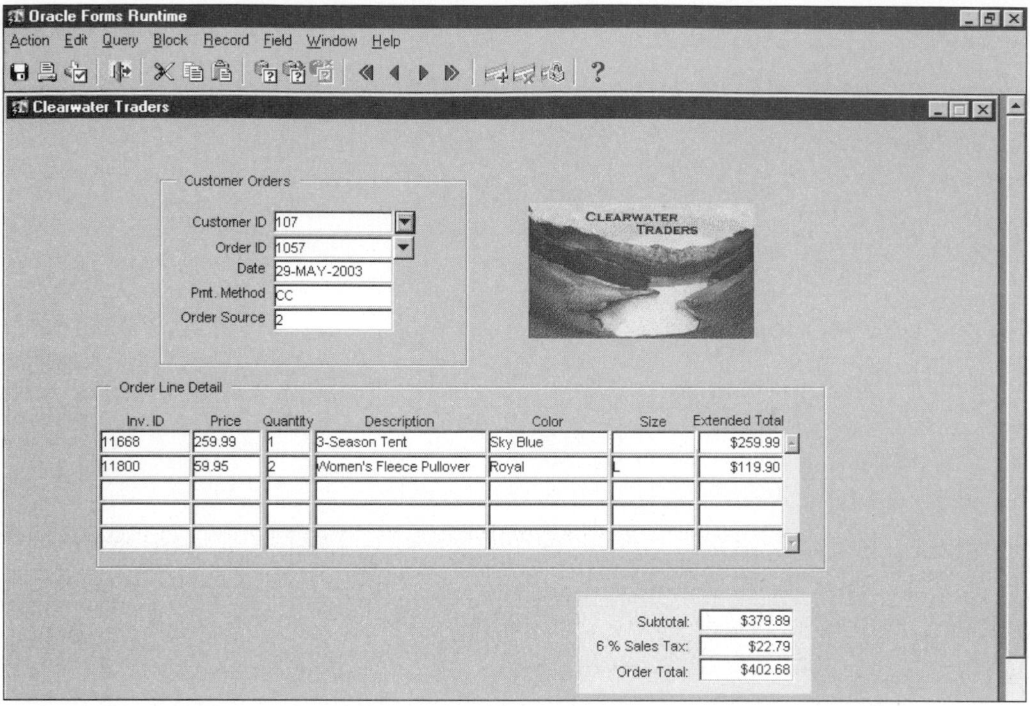

Figure 8-9

3. Modify the 8ACase3.fmb form so that when the user selects a student in the Northwoods University database, the details about the courses that the student has completed appear in the view-only Completed Courses frame, as shown in Figure 8-10. The student's total course credits and cumulative grade point average also appear as shown. Save the completed form as 8ACase3_DONE.fmb.

a. Create a view named COMPLETED_COURSE_VIEW that contains the student ID, term description, call ID, credits, and grade for every course for every student where the grade is not NULL. Then, create a data block to display the view as shown in Figure 8-10, and create a relationship between the form control block and the data block so that when the user selects a student, only courses for that student appear.

b. Create a display item to show the total number of course credits that the student has completed. Place the display item in the COMPLETED_COURSE_VIEW block. Add the boilerplate text items and the filled rectangle object to highlight the summary data as shown.

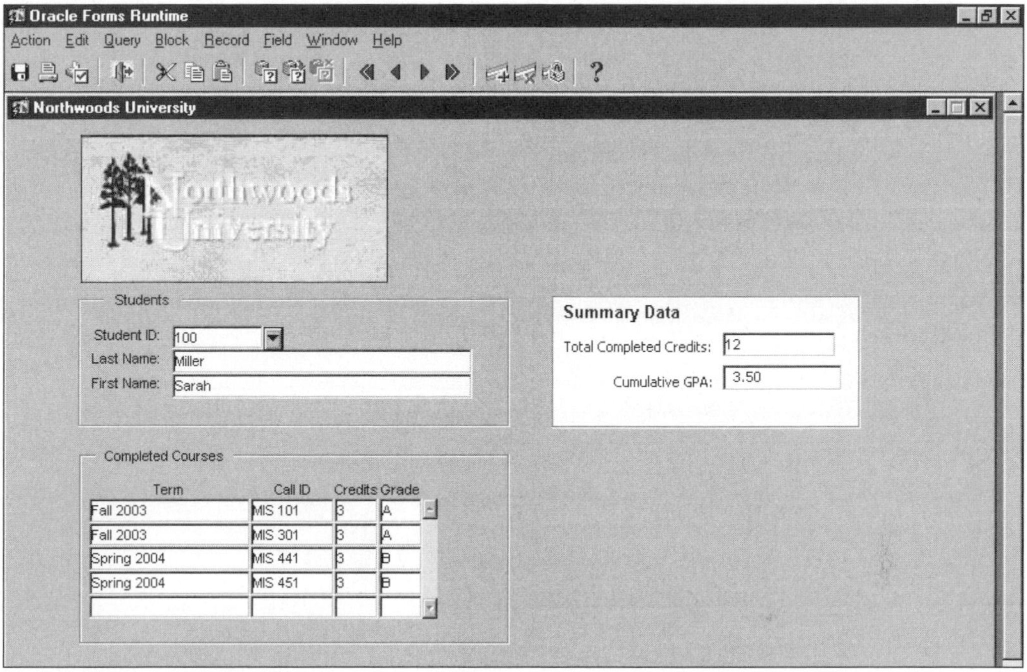

Figure 8-10

   c. Create a second display item in the COMPLETED_COURSE_VIEW block to display the student's cumulative grade point average. Specify that the GPA display item is a Formula calculated field, and use it to call a program unit function that receives the current student ID as an input parameter and returns the calculated grade point average. Recall that grade point average is calculated as follows:

$$\frac{\text{SUM(Course Credits * Course Grade Points)}}{\text{SUM(Course Credits)}}$$

Course grade points are awarded as follows:

Grade	Grade Points
A	4
B	3
C	2
D	1
F	0

Create an explicit cursor that retrieves the data for the selected student directly from the database.

   d. Import the Northwoods University logo as a static imported image from the nwlogo.jpg file that is in the Chapter8 folder on your Data Disk. Resize and reposition the image and the other form items as shown.

4. Create the form shown in Figure 8-11 that uses a data block form to load and display data and images from the FACULTY table in the Northwoods University database. To enhance the form's appearance, import the Northwoods University logo as a static imported image. The logo image is stored in the nwlogo.jpg file in the Chapter8 folder on your Data Disk. After you have created the form, load each faculty image into the FACULTY table using the image files stored in the Chapter8 folder on your Data Disk. The name of the JPEG image file for each faculty member is the same as the faculty member's last name.

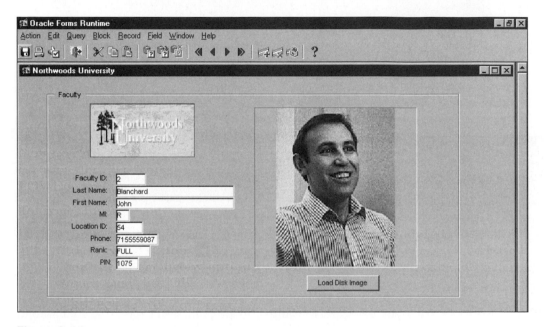

Figure 8-11

5. In this case, you will modify the COLOR table in the Clearwater Traders database so it contains a new field that contains a graphic image representing the associated color. Then, you will create the form shown in Figure 8-12 that loads and displays the names of the colors and their associated images. Save the completed form as 8ACase5_DONE.fmb.

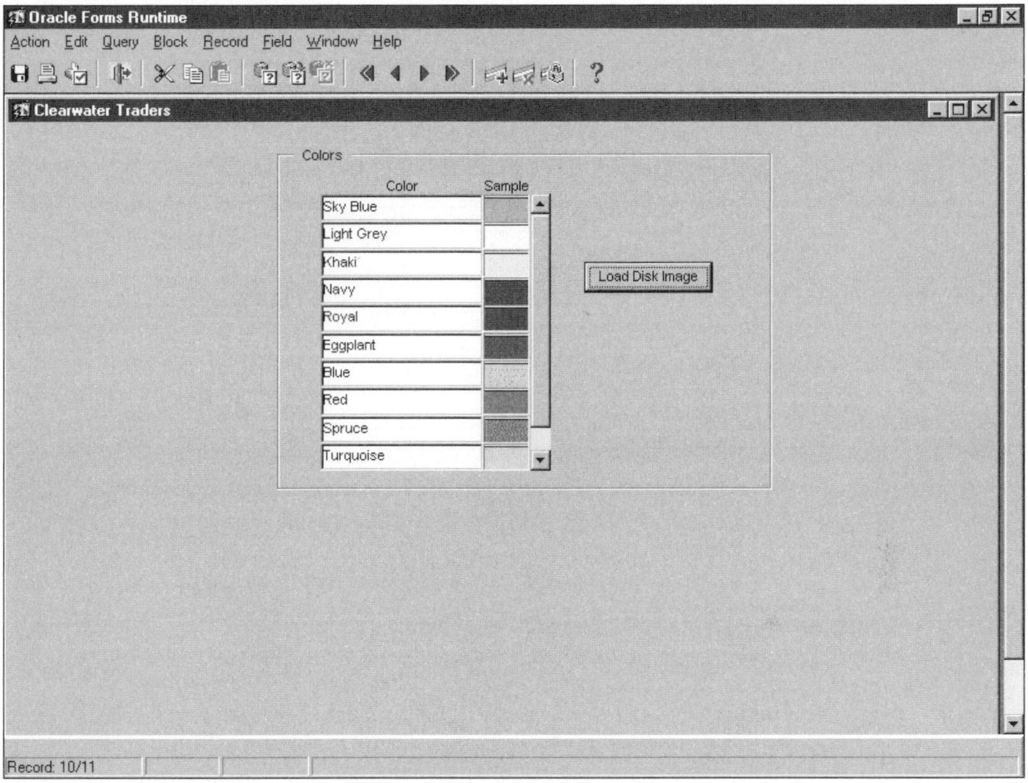

Figure 8-12

a. In SQL*Plus, modify the COLOR table by adding a new field named COLOR_IMAGE that has the BLOB data type.

b. Create the form shown in Figure 8-12 to load and display the color images. After you have created the form, load each color image into the COLOR table using the image files that are stored in the Chapter8 folder on your Data Disk. To load a specific color image, select the record associated with the color name in the tabular form, and then click the Load Disk Image button. The name of the TIFF image file on your Data Disk for each color is the same as the color name.

6. In this case, you will create the PROGRAM_SOUNDS table, which contains WAVE files representing sounds that are commonly used in Windows-based programs. Then, you will create the form shown in Figure 8-13 so users can load and preview the sounds.

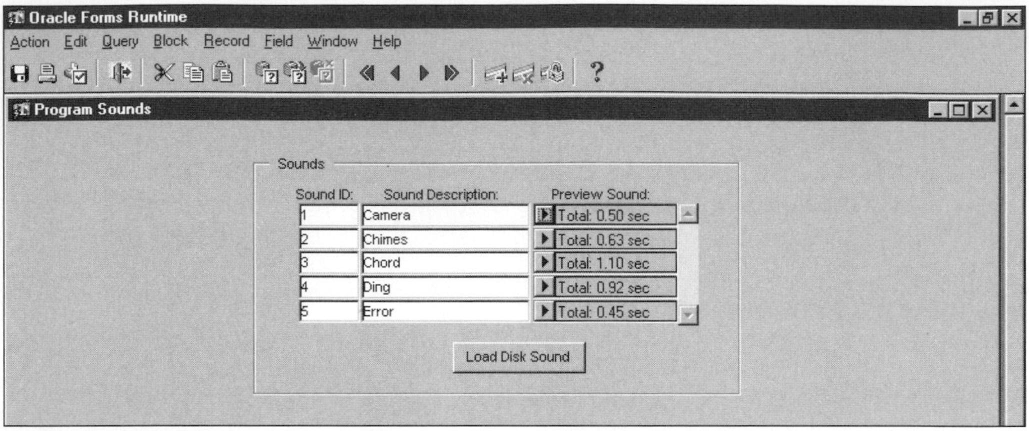

Figure 8-13

a. Create the PROGRAM_SOUNDS table in SQL*Plus. The table has the following columns and associated data type specifications:

Column Name	Data Declaration
SOUND_ID	NUMBER(6)
SOUND_DESC	VARCHAR2(30)
SOUND_DATA	BLOB

b. Create the form shown in Figure 8-13 to load the following data values into the table. Load the sound data from the specified files, which are stored in the Chapter8 folder on your Data Disk. Insert additional program sound file data records if you wish. Save the completed form as 8ACase6_DONE.fmb.

SOUND_ID	SOUND_DESC	Sound Data File
1	Camera	Camera.WAV
2	Chimes	Chimes.WAV
3	Chord	Chord.WAV
4	Ding	Ding.WAV
5	Error	Error.WAV

◄ LESSON B ►

Objectives

♦ Learn about alternate data block sources
♦ Program function keys in a form
♦ Understand how forms handle transaction processing
♦ Learn how to develop forms that retrieve and manipulate large data sets

DEFINING ALTERNATE DATA BLOCK SOURCES

So far, the data blocks you have created have been based on database tables or views. Table 8-3 summarizes all of the data sources on which you can base data blocks.

Block Data Source	DML Operations	Query?	Advantages	Disadvantages
Table	Yes	Yes	Easy to create	Can be slow when retrieving large data sets
View	Yes, but only as simple views created from a single table	Yes	Easy to create	View must be created; can be slow when retrieving large data sets
FROM clause query	No	Yes	Can display data from multiple tables without creating a view	View only
Stored Procedure (creates PL/SQL Table or REF cursor)	No	No	Can provide more complex displays and better performance; encapsulates logic in a subprogram	View only; requires procedure or cursor to be created beforehand
Transactional Triggers	Yes	Yes	Can be used to create a form that is based on a non-Oracle database source	Requires custom programs to be written to intercept all commands and replace them with ON- triggers

Table 8-3   Data block sources

The DML Operations column in Table 8–3 indicates whether the specified data block source can support DML (Data Manipulation Language) operations, such as INSERT, UPDATE, or DELETE. The Query? column indicates whether the block can retrieve data when the user clicks the Enter Query button 🔲, types an exact or inexact search condition in a text item, and then clicks the Execute Query button 🔲.

The first alternate data source is the **FROM clause query**, which enables you to create a data block that contains fields from multiple tables based on a SELECT command. This is similar to basing a data block on a view, except that you don't have to create a view. Sometimes database administrators are reluctant to let users or form developers create database views. The disadvantage of basing a data block on a FROM clause query is that DML operations are not supported. (Recall that this same disadvantage also exists for data blocks that are based on views that are created by joining multiple tables.) To create a data block that uses a FROM clause query as its data source, you create a data block manually, set the block's Query Data Source Type property to *FROM clause query*, and then change the block's Query Data Source Name property to the text of the SELECT command that retrieves the data block data. Finally, you manually create the form layout by creating text items corresponding to the fields listed in the query SELECT clause.

The second alternate data source is a **stored procedure** that creates a PL/SQL table or REF cursor that the data block is then based on. (Recall that a stored procedure is a PL/SQL procedure that is stored and processed on the database server.) You can use stored procedures to create PL/SQL tables, which are table data structures stored in the memory of the server rather than in the database. You can also use stored procedures to create a **REF cursor**, which is a PL/SQL data type that references a cursor. (Recall that a cursor is a pointer that references a memory location that contains one or more retrieved records). The advantage of using a stored procedure to create one of these alternate data block sources is that you can create more complex data block displays than are possible using tables or views. For example, you could create a procedure that generates a table that contains one column that displays the months of the current year, and a second column that displays the revenue generated by all customer orders placed during each month. All processing is performed on the database server, so performance is enhanced because only the final data values are transmitted to the client workstation. The drawbacks of this approach are that it requires creating complex PL/SQL programs, and the display can only be viewed, not updated. To create a data block based on a stored procedure, you create the stored procedure using PL/SQL, then use the Data Block Wizard to create the data block, and specify that the data block source is a stored procedure rather than a table or view.

The third alternate data block source is transactional triggers. A **transactional trigger** is a trigger that fires in place of a DML command (INSERT, UPDATE, DELETE, or SELECT) for a specific table.

Transactional triggers are similar to the INSTEAD-OF triggers that you created in Chapter 5, which fired when a user attempted to insert, update, or delete a record in a complex view.

In most circumstances, you would use transactional triggers as a data block source only if you were creating a data block form that is based on a non-Oracle database table. To create a data block based on transactional triggers, you create the transactional triggers in PL/SQL or Procedure Builder. Then, you create a data block manually and set the block Query Data Source Type property to Transactional Triggers. The triggers automatically fire instead of the DML command operations within the form. For example, when the user changes a value in a text item and then clicks the Save button 🔲, the transactional trigger corresponding to the SAVE command for the table fires.

To gain experience with creating a data block using an alternate data source, you will create the data block form shown in Figure 8-14 that is based on a FROM clause query. This form displays the order ID, order date, customer first and last name, and total order cost, which is calculated as SUM(PRICE * ORDER_QUANTITY), for all orders in the CUST_ORDER table in the Clearwater Traders database. The records appear in a tabular format, with five records appearing at one time. The form data comes from four separate database tables (ORDER, CUSTOMER, ORDER_LINE, and INVENTORY) and involves a calculated field.

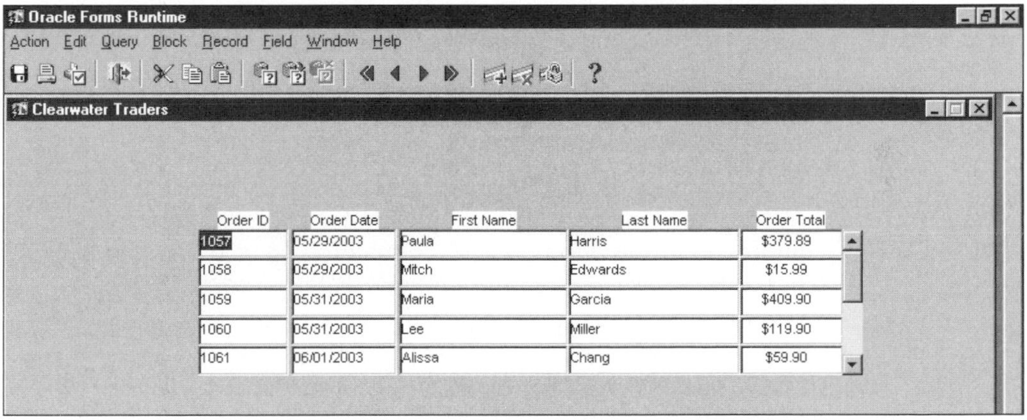

![Oracle Forms Runtime window showing Clearwater Traders form with order data in tabular format]

Order ID	Order Date	First Name	Last Name	Order Total
1057	05/29/2003	Paula	Harris	$379.89
1058	05/29/2003	Mitch	Edwards	$15.99
1059	05/31/2003	Maria	Garcia	$409.90
1060	05/31/2003	Lee	Miller	$119.90
1061	06/01/2003	Alissa	Chang	$59.90

Figure 8-14   Data block form based on FROM clause query

You could also display this data using a data block form that has master-detail relationships and a calculated text item, or using a data block that is based on a view.

First, you will refresh your database tables by running the scripts in the Chapter8 folder on your Data Disk. Then, you will open a form named CustOrder.fmb saved in the

Chapter8 folder on your Data Disk. Currently, this form does not contain any data blocks. You will save the form using a different filename, create a data block manually, then change the block properties and specify the FROM clause query. You will change the block's Query Data Source Name property to *FROM clause query* and the block's Query Data Source Name property to the text of the SELECT command that retrieves the records that will appear in the data block. You will also specify that the data block displays five records at one time.

To open the form and create the data block and FROM clause query:

1. Start SQL*Plus, log onto the database, and run the **northwoods.sql**, **clearwater.sql**, and **softwareexp.sql** scripts that are stored in the Chapter8 folder on your Data Disk.

2. Start Form Builder, open **CustOrder.fmb** from the Chapter8 folder on your Data Disk, and save the file as **8BCustOrder.fmb** in your Chapter8\TutorialSolutions folder.

3. In the Object Navigator, select the **Data Blocks node**, and click the **Create button** to create a new data block. Select the **Build a new data block manually** option button, and then click **OK**. A new data block appears in the Object Navigator.

4. Double-click the **Data Block icon** to open the new data block's Property Palette, and change the following property values:

Property	Value
Name	**CUST_ORDER_BLOCK**
Query Data Source Type	**FROM clause query**
Number of Records Displayed	**5**

5. Click the **Query Data Source Name** property, click the **More button**, and then type the following command to specify the data block columns and rows. Do not type a semicolon (;) at the end of the command. Click **OK**, and then close the Property Palette and save the form.

```
SELECT cust_order.order_id, order_date, first, last,
SUM(price * order_quantity) AS order_total
FROM cust_order, customer, order_line, inventory
WHERE cust_order.cust_id = customer.cust_id
AND cust_order.order_id = order_line.order_id
AND order_line.inv_id = inventory.inv_id
GROUP BY cust_order.order_id, order_date, first, last
```

 Recall that you must use the GROUP BY clause when some of the columns in the SELECT clause involve a group function and some do not. All columns not involved in the group function must be listed in the GROUP BY clause.

The next steps are to create the text items in the data block and to create the scrollbar. You will draw the items on the canvas and then modify the item properties. The name of each text item must be exactly the same as the corresponding field name in the data block query, and the data type and maximum length properties must be the same as the corresponding database field. In a data block form based on a FROM clause query, one of the text items must be specified as the block primary key. You will designate the ORDER_ID text item as the block primary key by changing its Primary Key property value to *Yes*. You will also give each text item a descriptive prompt and modify the Prompt Attachment Edge to Top, so the prompt appears above each text item group rather than on the left side of the first item. To display a scrollbar, you will change the data block's Display Scrollbar property to *Yes* and specify that the scrollbar should appear on the form canvas.

To create the block text items:

1. Open the Layout Editor, select the **Text Item tool** [abc], and draw the text item for the Order ID field, as shown in Figure 8-14. Five text items appear, because you set the block's Number of Records Displayed property to 5 in the previous set of steps.

2. Double-click any text item in the group to open its Property Palette, change the property values as follows, and then close the Property Palette.

Property	Value
Name	**ORDER_ID**
Data Type	**Number**
Maximum Length	**8**
Primary Key	**Yes**
Prompt	**Order ID**
Prompt Attachment Edge	**Top**
Prompt Alignment	**Center**

3. Create the four remaining text item groups shown in Figure 8-20, and modify the properties of the text items as follows. Note that the item name must always be exactly the same as the field name in the SELECT clause of the query in the Query Source Name property.

Property	Item 1	Item 2	Item 3	Item 4
Name	**ORDER_DATE**	**FIRST**	**LAST**	**ORDER_TOTAL**
Data Type	**Date**	**Char**	**Char**	**Number**
Maximum Length	**30**	**30**	**30**	**15**
Format Mask	**MM/DD/YYYY**			**$999,999.99**
Primary Key	**No**	**No**	**No**	**No**
Prompt	**Order Date**	**First Name**	**Last Name**	**Order Total**
Prompt Attachment Edge	**Top**	**Top**	**Top**	**Top**
Prompt Alignment	**Center**	**Center**	**Center**	**Center**

8

4. In the Object Navigator, open the CUST_ORDER_BLOCK Property Palette. Change the Show Scrollbar property to **Yes** change the Scrollbar Canvas property to **ORDER_CANVAS** to display the scrollbar on the form, and then close the Property Palette. View the form in the Layout Editor, and resize and reposition the scrollbar so your form looks like Figure 8-14.

5. Save the form.

Now, you test the form. First, you will retrieve all of the records. Then, you will retrieve the records for orders for which the order total is over $100. Finally, you will observe how the form behaves when you try to perform a DML operation by modifying a record.

To test the form:

1. Run the form and connect to the database if necessary. Click the **Enter Query button** , do not enter a search condition, and then click the **Execute Query button** to retrieve all of the form records. The form records appear as shown in Figure 8-14.

 If the message "ORACLE Error: Unable to retrieve record" appears, then your query has an error or you did not specify one of the form text items correctly. Run the query in SQL*Plus to make sure it is correct, and debug it if it has an error. Then, check all of the text items to make sure they have the same name as the columns in the SELECT clause and the same data type as the associated database columns. As a last resort, delete all of the text items except the first one, and try to run the form. If it runs correctly, add the next text item, and try to run the form again. Continue this process until all form text items appear.

2. To retrieve only the orders that have an order total over $100, click , type **>100** as the query search condition in the first Order Total text item, and then click . The records for Order ID values 1057, 1059, 1060, and 1062 appear.

3. To see what happens when you try to change a record value, select the Order Total value for Order ID 1057, delete the existing value, and try to type **$400.00** for the new value. The error message "FRM-40501: ORACLE error: unable to reserve record for update or delete" appears, and the original value appears in the text item. Recall that you cannot insert, update, or delete values using a data block based on a FROM clause query.

4. Close the Forms Runtime window, click **No** to discard your changes if necessary, and then close the form in Form Builder. (If you do not click No, you will repeatedly see the error message stating that you cannot update or delete the record.)

PROGRAMMING FORM FUNCTION KEYS

You can perform most form actions in Forms Runtime using menu selections, but many users like to use keyboard function keys (such as F1) or shortcut keys (such as Ctrl + E) to expedite repeated operations. In a form, these actions are controlled using **key triggers**, which execute when the user presses a particular key sequence, such as a function key, a shortcut key combination, or combination of a function key plus another key (such as Shift + F2). Table 8-4 lists several Forms Runtime actions associated with different key sequences, along with the action descriptions, the name of the key trigger event associated with the action, and the default key sequences used to perform the action. The Associated Form Command or Built-in column specifies the command that executes when the user presses the key sequence, and the Action Code specifies the numeric code that Oracle has assigned to the action.

Form Action Name	Description	Associated Key Trigger Event	Default Key Sequence	Associated Form Command or Built-in	Action Code
Accept	Saves changes made to current form values	Key-COMMIT	F10	COMMIT;	36
Clear Block	Clears all text items in current block	Key-CLRBLK	Shift + F5	CLEAR_BLOCK;	70
Clear Form	Clears all form text items	Key-CLRFRM	Shift + F7	CLEAR_FORM;	74
Clear Record	Clears all text items in current record	Key-CLRREC	Shift + F4	CLEAR_RECORD;	63
Delete Record	Deletes current record	Key-DELREC	Shift + F6	DELETE_RECORD;	63
Display Error	Displays error message description corresponding to most recent error	None	Shift + F1	DISPLAY_ERROR;	78
Down	Moves insertion point to current text item in next record	Key-DOWN	Down arrow	DOWN;	7
Edit	Opens editor associated with current text item	Key-EDIT	Ctrl + E	EDIT_TEXTITEM;	22
Enter Query	Places form in Enter Query mode	Key-ENTQRY	F7	ENTER_QUERY;	76
Execute Query	Places form in Execute Query mode	Key-EXEQRY	F8	EXECUTE_QUERY;	77

Table 8-4   Forms Runtime actions and associated key triggers

8

Form Action Name	Description	Associated Key Trigger Event	Default Key Sequence	Associated Form Command or Built-in	Action Code
Exit	Closes Forms Runtime window	Key-EXIT	Ctrl + Q	EXIT_FORM;	1002
Help	Opens Forms Runtime Help window	Key-HELP	F1	HELP;	1004
Insert Record	Inserts a new blank record in current block	Key-CREREC	F6	INSERT_RECORD;	65
List of Values	Opens LOV associated with current text item	Key-LISTVAL	F9	LIST_VALUES;	29
Next Block	Moves form focus to the next block	Key-NXTBLK	Ctrl + Page Down	NEXT_BLOCK;	71
Next Item	Moves insertion point to next text item	Key-NXT-ITEM	Tab	NEXT_ITEM;	1
Next Primary Key	Moves insertion point to next text item containing a primary key	Key-NXTKEY	Shift + F3	NEXT_KEY;	61
Next Record	Moves form focus to next record	Key-NXTREC	Shift + Down arrow	NEXT_RECORD;	67
Previous Block	Moves form focus to previous block	Key-PRVBLK	Ctrl + Page Up	PREVIOUS_BLOCK;	72
Previous Item	Moves insertion point to previous text item	Key-PRV-ITEM	Shift + Tab	PREVIOUS_ITEM;	2
Previous Record	Move form focus to previous record	Key-PRVREC	Shift + Up arrow	PREVIOUS_RECORD;	68
Print	Prints a screen capture of current form display	Key-PRINT	Shift + F8	PRINT;	79
Scroll Down	Scrolls display in tabular style layout so previously-hidden record at bottom of display appears	Key-SCRDOWN	Page Down	SCROLL_DOWN;	13
Scroll Up	Scrolls display in tabular style layout so previously hidden record at top of display appears	Key-SCRUP	Page Up	SCROLL_UP;	12
Up	Moves insertion point to currently-selected text item in next record		Up arrow	UP;	6

Table 8-4   Forms Runtime actions and associated key triggers (continued)

To display a complete list of form operations and their associated key sequences in Forms Runtime, click Help on the menu bar, and then click Keys, or press Ctrl + F1.

You have already used some of these default key sequences in your applications. For example, you pressed Tab to move to the next text item, Ctrl + e to open an item editor, and F9 to open an LOV display. Form Builder allows you to redefine existing key sequence operations or to create new key triggers for key sequences currently undefined.

To gain experience with key triggers, you will use a data block form named ProjectConsultants.fmb that is in the Chapter8 folder on your Data Disk. First, you will open the form, save it using a different filename, and examine some of its default key sequence operations. Then, you will use the form to redefine existing key sequences and define new key sequences.

To open the form and examine its default key sequence operations:

1. In Form Builder, open **ProjectConsultants.fmb** from the Chapter8 folder on your Data Disk, and save the file as **8BProjectConsultants.fmb** in your Chapter8\TutorialSolutions folder.

2. Run the form. The form appears in the Forms Runtime window, where it shows frames for displaying information about projects, project consultants, and details about the selected consultant.

3. Press **F7**, which places the form in Enter Query mode and is the equivalent of clicking the Enter Query button ⬚. The form status appears on the status bar.

4. Do not type a search condition, and then press **F8** to execute the query and retrieve all of the table records. This is the equivalent of clicking the Execute Query button ⬚. The data for Project ID 1 (Hardware Support Intranet) appears, along with its associated detail records.

5. Place the insertion point in the Project Name text item, change the project name to **Project Support Intranet**, and then press **F10** to save your changes. This is equivalent to clicking the Save button ⬚. The message "FRM-40400: Transaction Complete: 1 records applied and saved" appears in the status bar, confirming that the save was committed to the database.

6. Close Forms Runtime.

Redefining Existing Key Sequence Operations

To redefine the operation associated with an existing key sequence, you create a key trigger using the key trigger event currently associated with the existing key sequence. For example, to redefine the operation of the F10 key, you would create a trigger associated with its current key event, which Table 8-4 shows to be *Key-COMMIT*. You then write the trigger code to specify the alternate operation. Now you will redefine an existing key sequence operation.

Suppose that at Software Experts, the managers decide that all database applications will allow the user to press the F10 key to move to the next form text item. Table 8-4 shows that the F10 key is currently associated with the Key-COMMIT key trigger event. To redefine the F10 key, you must create a key trigger that is associated with this event. The trigger code must specify the command for the alternate operation, which in this case is NEXT_ITEM. (You can determine that this is the correct code since you know this operation is also accomplished by pressing the Tab key, and Table 8-4 shows that the NEXT_ITEM command is currently associated with the Tab key.)

You can specify key triggers at the form or block level. If you define a key trigger at the form level, the trigger fires when the user presses the associated key sequence anywhere in the form. If you define the key trigger at the block level, it fires only when the user presses the key sequence when the form focus is in the block where the trigger exists. Now, you will create a form-level key trigger associated with the Key-COMMIT event (which is associated with the F10 function key). The trigger code will specify that when the user presses F10, the NEXT_ITEM command executes, and the insertion point moves to the next text item. After you create the trigger, you will test it to make sure the operation of the F10 key is redefined.

To create a form-level key trigger to redefine an existing key sequence:

1. In Object Navigator, select the **Triggers node** under the PROJECT_ CONSULTANTS_FORM, and then click the **Create button** to create a new form-level trigger. The PROJECT_CONSULTANTS_FORM: Triggers dialog box opens.

2. Place the insertion point before the % in the Find text box, type **key-**, and then press **Enter** to display all of the key trigger events. Select **Key-COMMIT**, and then click **OK**. The PL/SQL Editor opens.

3. Type the command NEXT_ITEM; as the trigger code, compile the trigger, debug it if necessary, and then close the PL/SQL Editor.

4. Save the form, and then run it. Press **F7** to place the form in Enter Query mode, do not enter a search condition, and then press **F8** to execute the query. The form data appears, with the insertion point in the Project ID text item in the Projects frame.

5. Press **F10**. The insertion point moves to the Project Name text item, which is the next text item. Recall that before you redefined the trigger, the F10 key saved database changes. You have successfully redefined the function key.

6. Close Forms Runtime, and then close the form in Form Builder.

FORM TRANSACTION PROCESSING

Recall that a transaction is a series of DML commands (SELECT, INSERT, UPDATE, or DELETE) that constitute a logical unit of work. All commands in the transaction must succeed, or none of them can succeed. For example, consider the process of creating new customer orders in the Clearwater Traders database. When you insert a record for a new order, you must also insert a record for at least one new order line, because an order without order lines doesn't make sense. Inserting a new record into the CUST_ORDER table and at least one associated record into the ORDER_LINE table constitutes a transaction.

When you have entered all of the commands in a transaction, you can either commit that transaction to save all of the changes or roll back the transaction to discard all of the changes. In an Oracle database, a transaction has two phases: the posting phase and the committing phase. In the **posting** phase, the database server receives and acknowledges the DML command, places it on a list of uncommitted transactions, and makes the new record available for the current user to manipulate and reference as a foreign key. A posting phase happens when, for example, you insert a new record using SQL*Plus, but before you issue the COMMIT or ROLLBACK command. In the **committing** phase, the change is made permanent in the database and the result of the change is made available to other users, or the change is rolled back and discarded. The committing phase happens in SQL*Plus when you issue the COMMIT or ROLLBACK commands.

In a form, Form Builder uses transaction processing, which is also called **commit processing**, to make the data in the database consistent with the data that appears in the form. When you are working with a form, Form Builder's normal operations are:

1. Read records from the database.

2. Allow the user to make tentative insertions, updates, and deletions, which appear only in the form, while the database remains unchanged.

3. Post changes to the database.

4. Commit the posted changes so they become permanent changes in the database.

In a form, commit processing specifically refers to Steps 3 and 4, and these steps normally occur together. Form Builder performs commit processing when:

- The COMMIT_FORM built-in executes as a result of the user clicking 🖫, pressing F10, or executing any trigger that contains the COMMIT_FORM command.

- Items in the current data block have changed since the last commit processing cycle, and a trigger executes the CLEAR_BLOCK or CLEAR_FORM built-in.

- The user clicks Yes when the alert with the message "Do you want to commit the changes you have made?" appears.

8

Form Builder provides several **transaction triggers**, which are triggers used to control commit processing and record auditing information within forms. Table 8-5 lists transactional triggers that fire just before an event occurs; these triggers have a PRE- prefix.

Trigger Name	Description	Example Usage	Scope
PRE-COMMIT	Fires just before a COMMIT action and only fires if records have been inserted, updated, or deleted	Special data validation or record locking	Form
PRE-DELETE	Fires once for every record deleted	Determining if deleted records contain foreign key references	Block, Form
PRE-INSERT	Fires once before each record is inserted using either a post or commit operation	Creating auditing information, such as the user who inserted the record and the date it was inserted	Block, Form
PRE-QUERY	Fires once just before a query operation	Modifying or validating query criteria	Block, Form
PRE-SELECT	Fires during a query after the SELECT clause is constructed, but before the query is processed	Preparing a query string that is constructed as a text string and used in a non-Oracle database	Block, Form
PRE-UPDATE	Fires once before each updated record is posted or committed	Creating auditing information	Block, Form

Table 8-5   PRE- transaction triggers

The Scope column in Table 8-5 indicates the object level at which the trigger can be created and the associated event that causes it to fire. If a trigger is created at the form level, it fires regardless of which form block has the current form focus. A trigger created at the block level fires only if the event occurs when the form focus is in the current block.

Table 8-6 lists POST- transaction triggers, which fire immediately after an event occurs.

Recall that in Chapter 7, you learned about navigational triggers, which are triggers that fire automatically in response to user navigation operations. During form commit processing, both navigational triggers and transaction triggers fire. Figure 8-15 lists the order in which the navigational and transaction triggers fire when the user performs transaction processing operations on a form and shows the result that appears on the screen display after the triggers fire.

Trigger Name	Description	Example Usage	Scope
POST-DATABASE-COMMIT	Fires once after a commit action	Displaying confirmation messages	Form
POST-CHANGE	Fires when item values are changed	Validating the value of an item	Item, Block, Form
POST-FORMS-COMMIT	Fires when a change is posted, but before it is committed	Creating auditing information	Form
POST-INSERT	Fires once for each record successfully committed in an INSERT command	Creating auditing information	Block, Form
POST-QUERY	Fires once for every record retrieved in a SELECT command	Modifying or formatting data values before they are displayed	Block, Form
POST-SELECT	Fires once when a query is submitted to the database, but before the actual data values are retrieved and displayed	Counting the number of records to be retrieved before actually retrieving them	Block, Form
POST-UPDATE	Fires once for each row updated in either a post or commit operation	Creating auditing information	Block, Form

Table 8-6   POST- transaction triggers

When the user commits a query that involves a combination of INSERT, UPDATE, and DELETE operations, Oracle processes the DELETE operations first. This ordering might speed up subsequent UPDATE operations, since there is no point in updating a record that will ultimately be deleted. The form processes the UPDATE transactions next and the INSERT transactions last.

The transaction triggers fire based on the values of the form STATUS system variables. Form Builder maintains several system variables that track the internal state of the form and control how the form behaves. Recall that you reference a system variable with the syntax **:SYSTEM.*variable_name***. Table 8-7 summarizes the system STATUS variables. You can retrieve the values of these variables in transaction triggers and use them to determine the current commit processing status and direct further processing.

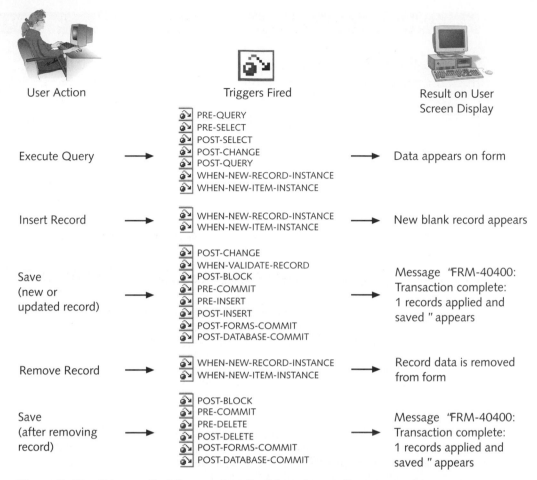

Figure 8-15   Triggers that fire as a result of form transaction processing

System Variable Name	Values
FORM_STATUS	NEW, if all form records are new CHANGED, if one or more form records have been changed QUERY, if all form records are unchanged since being retrieved
BLOCK_STATUS	NEW, if all block records are new CHANGED, if one or more block records have been changed QUERY, if all block records are unchanged since being retrieved
RECORD_STATUS	New, if current record is new CHANGED, if record values have been changed but not validated QUERY, if record was just retrieved from database INSERT, if record contains data that has not been validated or saved

Table 8-7   Form system STATUS variables

Now, you will create a transaction trigger to ensure that when a user inserts a record for a new order in the CUST_ORDER table in the Clearwater Traders database, the user has also specified a record for at least one new order line. You will use a form named CustOrderLine.fmb saved in the Chapter8 folder on your Data Disk. First, you will open the form and save it using a different filename. Then, you will run the form, retrieve the current order information for customer ID 107 (Paula Harris), and insert a new order that does not contain any associated order lines, to confirm that this operation is currently possible, although undesirable.

To open the form and insert a new order record with no order lines:

1. In Form Builder, open **CustOrderLine.fmb** from the Chapter8 folder on your Data Disk, and save the file as **8BCustOrderLine.fmb** in your Chapter8\TutorialSolutions folder.

2. Run the form. This form displays data from the CUSTOMER, CUST_ORDER, and ORDER_LINE tables, and it has two master-detail relationships: one between CUSTOMER and CUST_ORDER, and one between CUST_ORDER and ORDER_LINE.

3. Press **F7** to place the form in Enter Query mode, type **107** in the Customer ID field in the Customers frame, and then press **F8** to execute the query. The form displays the order records for customer Paula Harris.

4. To insert a new order, type the following values in the blank record below Order ID 1057 in the Customer Orders frame. Note that currently, this record does not have any associated values in the Order Line Detail frame.

Order ID	**2000**
Order Date	**06/15/2003**
Meth Pmt	**CC**
Order Source ID	**3**

5. Click the **Save button** 🔲 on the toolbar. The confirmation message "FRM-40400: Transaction complete: 1 records applied and saved" appears, indicating that currently you can insert a customer order record with no associated order line records.

6. Close Forms Runtime.

To ensure that the user enters at least one record in the Order Line Details frame before saving a CUST_ORDER record, you must create a trigger associated with an event that occurs when the user attempts to save a new record. These events are illustrated in Figure 8-15. You will use the PRE-INSERT trigger, because this trigger fires after all data validation is performed and just before Oracle attempts to insert the record. This trigger will be attached to the CUST_ORDER block, because it is only relevant when the user is inserting a new CUST_ORDER record. The trigger will test to see if the current ORDER_ID value in the form ORDER_LINE block is the same as the ORDER_ID value in the CUST_ORDER block. If the values are not the same, then an order line for the new CUST_ORDER record has not been specified. The form will display an alert to instruct the user that he or she must enter an order line for the new order. Then, the trigger will issue the RAISE FORM_TRIGGER_FAILURE command. This command causes the current trigger to fail, which immediately halts the current COMMIT process.

To create the PRE-INSERT transaction trigger:

1. In the Object Navigator, select the **Triggers node** under the CUST_ORDER block, and then click the **Create button** 🖽 to create a new block-level trigger. The CUSTOMER_ORDER_FORM: Triggers dialog box opens.

2. Place the insertion point before the % in the Find box, type **pre-insert** to select the trigger event, press **Enter**, and then click **OK**. The PL/SQL Editor opens.

3. Type the trigger commands shown in Figure 8-16, compile the code and debug it if necessary, and then exit the PL/SQL Editor, and save the form. (You will create the alert that the trigger calls in a later set of steps.)

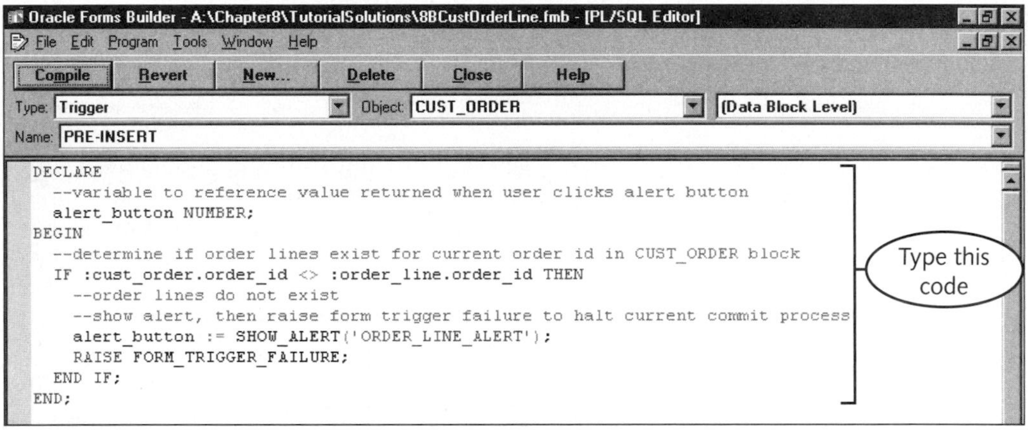

Figure 8-16 PRE-INSERT trigger

Next, you will create the alert that appears when the user has failed to specify an order line. You must use an alert rather than a message that appears on the status line, because a status line message appears only briefly and is overwritten when the FORM-TRIGGER-FAILURE exception is raised.

To create the alert:

1. In the Object Navigator, select the **Alerts node** under the CUSTOMER_ORDER_FORM, and then click the **Create button** 🖽 to create a new alert. The new alert appears under the Alerts node.

2. Double-click the **Alert icon** ▐▌ to open the alert Property Palette, and change the following property values:

Property	Value
Name	**ORDER_LINE_ALERT**
Title	**Order Line Error**
Alert Style	**Stop**
Message	**You must enter at least one order line before you can save a new order.**
Button 1 Label	**OK**
Button 2 Label	(deleted)

3. Close the Property Palette, and then save the form.

Now, you will test your PRE-INSERT trigger to make sure that it works correctly. You will run the form, retrieve the information for customer ID 107 (Paula Harris), and attempt to insert a new order record that does not have any associated order line information.

To test the PRE-INSERT trigger:

1. Run the form, press **F7** to place the form in Enter Query mode, type **107** in the Customer ID text item in the Customers frame for the search condition, and then press **F8** to execute the query. The form displays the records for customer Paula Harris.

2. Type the following values in the blank record below Order ID 2000 (which is the order you inserted earlier) in the Customer Orders frame.

Order ID	**2001**
Order Date	**06/20/2003**
Meth Pmt	**CC**
Order Source ID	**3**

3. Click the **Save button** 🖫 to commit the order information to the database. The Order Line Error alert appears. Click **OK** to close the alert. The values for the new customer order (Order ID 2001) still appear, but they are not committed to the database.

4. Close Forms Runtime. When the dialog box asking if you want to save your changes appears, click **Yes**. This attempts to commit the record to the database, so the Order Line Error alert appears again. Do not close Forms Runtime or delete the new customer order record.

The PRE-INSERT trigger fires whenever the user attempts to commit the new customer order record to the database. At this point, the user can exit Forms Runtime and not save the record or enter an order line for the record. Now, you will enter an order line for the new record.

To enter a new order line for the customer order:

1. In Forms Runtime, place the insertion point in the Order ID text item in the Order Line Details frame. The new order ID value (2001) appears.

2. Type **11668** for the Inventory ID and **1** for the Order Quantity, and then click the **Save button** 🖫. The confirmation message "FRM-40400: Transaction complete: 2 records applied and saved" appears, indicating that the new CUST_ORDER and ORDER_LINE records were both successfully inserted.

3. Close Forms Runtime, and then close the form in Form Builder.

USING FORMS WITH LARGE DATA SETS

So far, all of the forms you have created involve databases that contain only a few sample records. Production databases often contain thousands, or hundreds of thousands, of data records. Forms that retrieve large data sets process very slowly. For example, suppose that the Northwoods University database contains 200,000 records in the STUDENT table. Depending on current database activity, network speed, and other factors, a query that retrieves all of the records into a simple data block form based on the STUDENT table might take several minutes. This performance significantly degrades in a master-detail form that retrieves additional student information, such as course schedules or advisor information. Form developers need to estimate the maximum number of records that form queries might retrieve, and then design their applications to ensure that form performance remains satisfactory.

Some approaches for improving data retrieval performance in forms are:

- Create indexes on fields used for searching or joining tables. Indexes were described in Chapter 3.

- Encourage users to count **query hits**, which are the number of records a query will retrieve, before actually retrieving the records. The Oracle database can locate and mark a record to be retrieved much faster than it can actually retrieve the data values.

- Limit the number of records that queries retrieve by requiring users to enter one or more search conditions to filter the retrieval set.

- Configure form LOVs so they allow the user to filter data and so they handle large retrieval sets more efficiently.

- Use **array processing**, which enables records to be processed in groups, rather than one at a time, which improves performance.

- Base form retrievals on **asynchronous queries**, which retrieve and display part of the requested data, and then continue to retrieve the rest of the data in the background as needed.

To explore these approaches, you will open a data block form that displays data from the ENROLLMENT table in the Northwoods University database. The form is in the Chapter8 folder on your Data Disk in a file named Enrollment.fmb. You will open the file, rename it, run the form, examine its values, and determine the current number of query hits.

To open and examine the form:

1. In Form Builder, open **Enrollment.fmb** from the Chapter8 folder on your Data Disk, and save the form as **8BEnrollment.fmb** in your Chapter8\TutorialSolutions folder.

2. Run the form, press **F7** to place the form in Enter Query mode, do not enter a search condition, and then press **F8** to execute the query. The table records appear and show the student ID, course section ID, and grade for all courses.

3. Click **Query** on the menu bar, and then click **Count Hits** to display the total number of records that were retrieved. The message "FRM-40355: Query will retrieve 20 records" appears in the status bar.

4. Close Forms Runtime.

Limiting Retrievals by Counting Query Hits and Using Search Conditions

Data block forms retrieve records when you place the form in Enter Query mode, enter an optional search condition, and then execute the query. Sometimes, users unknowingly execute a query that retrieves a large number of records and are then "stuck" for several minutes (or even hours) waiting for the data to appear. To avoid this scenario, you can advise users to always count query hits prior to actually executing the query. Alternately, you can create a transaction trigger that requires users to enter a search condition for any query that can potentially retrieve a large number of records.

Suppose you are designing the ENROLLMENT form for the administrators at Northwoods University. You need to determine the maximum number of records that users might retrieve using this form in a production setting. Northwoods University enrolls 10,000 students per year, and every year, each student enrolls in an average of 10 courses. Therefore, 10 times 10,000, or 100,000, new records are inserted into the ENROLLMENT table each year. The administrators at Northwoods would like to have 40 years worth of enrollment data available for online retrievals, so you must plan to store 40 times 100,000, or 4,000,000 records, in the ENROLLMENT table. This analysis indicates that the ENROLLMENT form is a good candidate for a trigger that requires a search condition.

You will implement this trigger as a PRE-QUERY trigger that fires just before the records are retrieved. Now, you will create the trigger that requires users to specify a query search condition. This will be a block-level trigger that is associated with the ENROLLMENT block. The trigger will test to see if the user has entered a value in either the Student ID (S_ID) or course section ID (C_SEC_ID) form text items. If not, the trigger will display the alert, and then abandon the query by raising the FORM_TRIGGER_FAILURE exception. (Recall that this exception halts the processing of the current transaction. You will create the alert in a later set of steps.)

8

To create the transactional trigger to require the user to enter a search condition:

1. In the Object Navigator, click ➕ beside the Data Blocks node, click ➕ beside ENROLLMENT, select the **Triggers node**, and then click the **Create button** 🔲 to create a new trigger for the ENROLLMENT data block. The ENROLLMENT_FORM: Triggers dialog box opens.

2. Place the insertion point before % in the Find box, type **PRE-QUERY** to find the trigger event, press **Enter**, and then click **OK**. The PL/SQL Editor opens.

3. Type the trigger code shown in Figure 8-17, compile the trigger, debug it if necessary, and then close the PL/SQL Editor.

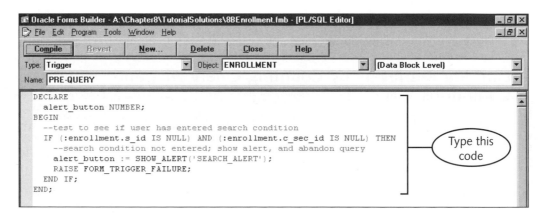

Figure 8-17   Trigger code requiring user to enter a search condition

Next, you will create the alert that appears when the number of query hits exceeds the threshold level. As before, you must use an alert rather than a message that appears on the status line, because a status line message appears only briefly and is overwritten by other system messages.

To create the alert:

1. In the Object Navigator, select the **Alerts node** under the ENROLLMENT_FORM, and then click the **Create button** 🔲 to create a new alert. The new alert appears.

2. Double-click the **Alert icon** 🔳 to open the alert Property Palette, and change the following property values:

Property	Value
Name	**SEARCH_ALERT**
Title	**Data Retrieval Error**
Alert Style	**Stop**

Message	**This query could potentially retrieve a very large number of records. Please enter a search condition.**
Button 1 Label	**OK**
Button 2 Label	(deleted)

3. Close the Property Palette, and then save the form.

Now you will run the form and try to retrieve all of the table records. The alert should appear, advising you to enter a search condition. You will then enter a search condition specifying to retrieve only the records for student ID 100 (Sarah Miller).

To test the form:

1. Run the form, press **F7** to place the form in Enter Query mode, do not enter a search condition, and then press **F8** to execute the query. The Data Retrieval Error alert appears. Click **OK**. The insertion point appears in the Student ID text item.

2. Type **100** in the Student ID text item, and then press **F8** to execute the query. The enrollment records for Student ID 100 appear. DO NOT close the Forms Runtime window yet.

Configuring LOVs to Handle Large Retrieval Sets

Users can also unwittingly have to endure performance delays or system malfunctions when they open an LOV display that retrieves a large data set. An LOV does not retrieve records asynchronously; it must retrieve all records before the LOV display appears. The maximum number of records that an LOV can retrieve is 32,767. If you attempt to open an LOV display that retrieves more than this number of records, the form application usually crashes.

The enrollment form that you have been working with has LOVs associated with both the Student ID and Course Section ID text items. Both of these LOVs have the potential to retrieve a lot of records. Now, you will open the LOV display associated with the Student ID text item and analyze its behavior.

To open the Student ID LOV display:

1. In Forms Runtime, press **F7** to place the form in Enter Query mode. Make sure that the insertion point is in the Student ID text item, and then press **F9** to open the text item's LOV. The Students LOV display opens, showing the names of all of the students in the STUDENT table.

2. Click **Cancel**, and then close Forms Runtime.

Since the STUDENT table contains only six records, the LOV display appears almost instantly. However, if the student table contained thousands or tens of thousands of records, a long delay would occur before the LOV display would appear and be ready for use. To avoid this delay, you can configure the LOV to minimize retrieval time and allow users to always specify a search condition before making retrievals. Now, you will open the LOV associated with the Student ID in reentrant mode using the LOV Wizard in Form Builder and examine the LOV properties that can be configured to help handle large data sets.

To open the LOV Wizard in reentrant mode:

1. In Form Builder, open the Object Navigator, click ➕ beside the LOVs node, select **S_ID_LOV**, click **Tools** on the menu bar, and then click **LOV Wizard**. The LOV Wizard opens in reentrant mode, and tabs appear on the Wizard pages.

2. Click **Next** several times, until the Advanced page appears, as shown in Figure 8-18. This page allows you to configure parameters to enable the LOV to handle large data sets. DO NOT close the LOV Wizard.

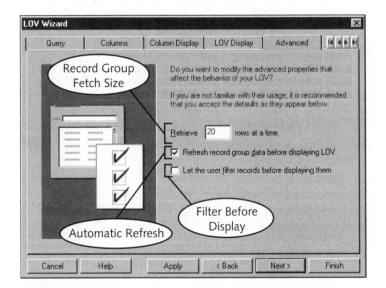

Figure 8-18   LOV Wizard Advanced page parameters

The **Record Group Fetch Size** property specifies the number of records fetched in each query processing cycle. Oracle query processing involves three phases: parse, execute, and fetch. During the **parse** phase, the DBMS checks the query syntax to make sure that the syntax is correct and that the user has adequate system and object privileges for the objects specified in the query. During the **execute** phase, the DBMS prepares to retrieve the requested records. During the **fetch** phase, the DBMS retrieves the records and returns them to the user. The user can specify the maximum number of records that the query can retrieve in each cycle. For queries that retrieve large data sets, you can improve performance by making the record group fetch size larger.

The **Automatic Refresh** property specifies whether the LOV display refreshes each time the user opens the LOV display. If the check box associated with this parameter is cleared, the LOV retrieves the records only once, when the user opens the LOV display the first time, and the values are saved in memory. When the user opens the LOV display subsequent times, the display shows the saved values from the first retrieval. This box should be checked for an LOV display in which the values do not change very often and that potentially retrieves a large data set.

The **Filter Before Display** parameter specifies whether or not records appear before the user filters the display using a search condition. If the check box associated with this parameter is checked, no records appear in the LOV display without first allowing the user to enter a search condition in the Find text box.

You can also specify the Record Group Fetch size on the Property Palette of the record group associated with the LOV. You can enable or disable the Automatic Refresh and Filter Before Display properties on the LOV Property Palette.

Now, you will modify the S_ID_LOV Filter Before Display Property value to give the user the option of filtering the data before the LOV display appears. (You will not modify the other properties, because you cannot see a change in performance unless you are retrieving a large data set.) An LOV can only perform data filtering on the first column in the LOV display. To specify the LOV search field, you must list the search field first in the LOV record group query and display the search field first in the LOV display. To display the search column first, place it first in the LOV Columns display on the LOV Column page in the LOV Wizard. You will modify the S_ID_LOV so the student last name (S_LAST) is the first column that the LOV query retrieves and is the first column that the LOV display shows, so users can search on this column.

To modify the S_ID_LOV:

1. In the LOV Wizard, check the **Let the user filter records before displaying them** check box to enable the Filter Before Display property.

You can also allow users to filter LOV records before displaying them by changing the Filter Before Display property to Yes in the LOV Property Palette.

2. To specify the student last name as the search column in the LOV, click the **Query tab**, and modify the query so it reads **SELECT s_last, s_first, s_id FROM student**.

3. To specify the S_LAST column as the first column that appears in the LOV column display, click the **Columns** tab, and drag S_ID to the bottom of the list, so the column order is as follows: S_LAST, S_FIRST, S_ID.

4. Click **Finish** to save your changes, and close the LOV Wizard.

5. Save the form.

Now you will run the form and open the LOV display. When the LOV display first opens, it will show the Find text box, which allows the user to enter a search condition before retrieving any LOV records. You will type the letter *M* as the first character in the search condition to retrieve only records for students whose last name begins with *M*.

To open the LOV display and enter a search condition:

1. Run the form, and press **F7** to place the form in Enter Query mode. Make sure that the insertion point is in the Student ID text item, and then press **F9** to open the LOV display. The Students LOV appears, with no records retrieved. Make sure that the insertion point is in the Find text box just before the %, and then type **M** to specify the search condition so that only records with M as the first character are retrieved. (Recall that % is the Oracle wildcard character representing a variable number of characters.)

2. Click **Find**. The records for the students whose last names begin with M (Sarah Miller and Amanda Mobley) appear. Select the record for **Sarah Miller**, and then click **OK**. Student ID 100 (which corresponds to Sarah) appears in the form.

3. Press **F8** to execute the query. The form displays the ENROLLMENT records for Student ID 100.

4. Close Forms Runtime.

Array Processing

By default, Form Builder applications process each record that is inserted, updated, deleted, or retrieved, one record at a time. To improve performance for large data sets, you can enable array processing, which processes groups of records as single units. The form enables array processing by creating an array data structure to hold the records that are being processed. An **array** is like a table, and it contains rows and columns. The records are loaded into the array, and the array is sent as a unit to the server for processing. Data blocks based on tables, views, and FROM clause queries support array processing for query operations.

Array processing is enabled differently for DML (INSERT, UPDATE, DELETE) and query (SELECT) operations. Array processing is available for SELECT operations for blocks based on tables, views, procedures, and subqueries. Array processing is not supported for blocks based on transactional triggers. Only data blocks based on tables support array processing for DML operations.

Enabling Array Processing for DML Operations

Whenever you insert, update, or delete multiple records in a data block form, all of the transaction triggers associated with the operation fire once for every record. For example, if you insert 100 new records into the ENROLLMENT form and then click the Save button 🖫, the save will probably take a long time because eight triggers fire for each new record (POST-CHANGE, WHEN-VALIDATE-RECORD, POST-BLOCK, PRE-COMMIT, PRE-INSERT, POST-INSERT, POST-FORMS-COMMIT, POST-DATABASE-COMMIT). Since you inserted 100 records before saving, a total of 800 triggers have to fire!

To enable DML array processing in a data block, you set the block DML Array Size property to a value larger than 1. By default, the DML Array Size property is 1, and each record is processed separately. When you enable DML array processing, all triggers that fire prior to the execution of the DML operation, which are called **PRE- triggers**, fire at the same time for every record being changed. The records are then processed in units that are the size of the array. All triggers that fire after the execution of DML operations, which are called **POST- triggers**, fire at the same time for every record being changed.

For example, suppose you set the DML Array Size property to 10 and then insert 100 new records. All of the PRE- triggers fire for all 100 records to be inserted. Then, an array of 10 records is inserted, a second array of 10 records is inserted, and so forth, until all 100 records are inserted. Finally, all 100 POST- triggers fire.

Array DML processing is automatically disabled if an ON-INSERT, ON-UPDATE, or ON-DELETE trigger is present, because these triggers execute in place of the default processing operations.

Now, you will enable DML array processing on the ENROLLMENT data block. You will open the ENROLLMENT data block Property Palette and change the DML Array Size property to 10.

To enable DML array processing:

1. In the Object Navigator, double-click the **Data Block icon** 🏫 beside the ENROLLMENT data block to open the data block Property Palette.

2. Scroll down to the Advanced Database property node, and change the DML Array Size property to **10**. Close the Property Palette, and save the form.

You will not test your changes, because you will not see a visible difference in performance unless you try to insert, update, or delete a large number of records.

Enabling Array Processing for Query Operations

Recall that when you learned how to improve LOV retrieval performance, you could change the Record Group Fetch Size property to fetch multiple records in each query processing cycle and improve performance. You can enable query array processing, which fetches multiple records in a query processing cycle, by setting the Query Array Size property to a value greater than 1. It is interesting to note that a Query Array Size of 1 provides the fastest initial response time, because Form Builder fetches and displays only 1 record at a time. However, setting the Query Array size to 10 fetches up to 10 records before displaying any records. This larger size actually reduces overall processing time by making fewer calls to the database for records. Query array processing is automatically disabled if a PRE-QUERY or POST-QUERY trigger is present, because these triggers execute in place of the default processing operations.

Before working with query array processing, you need to save the ENROLLMENT form using a different filename. You will delete the PRE-QUERY trigger you created earlier, because you cannot use query array processing with a block that has PRE-QUERY or POST-QUERY triggers. Then, you will change the Query Array Size to 10 to enable query array processing. You will not run the form, because you will not be able to see a change in the form performance unless you retrieve a large number of records.

To enable query array processing:

1. In Form Builder, save the 8BEnrollment.fmb file as **8BEnrollment_ARRAY.fmb** in your Chapter8\TutorialSolutions folder.

2. In the Object Navigator, click ➕ beside ENROLLMENT under the Data Blocks node to view the data block items if they are not displayed already. Then click ➕ beside the Triggers node to view the data block triggers if they do not already appear. Select the **PRE-QUERY** trigger, click the **Delete button** ✖ to delete the trigger, and then click **Yes** to confirm the deletion.

3. Double-click the **Data Block icon** 🖼 beside the ENROLLMENT block to open the block Property Palette. Under the Records property node, change the Query Array Size property to **10**, and then close the Property Palette and save the form.

4. Close the form in Form Builder, and close Form Builder and all other open Oracle applications.

Performing Asynchronous Queries in Forms

Recall that another approach for managing large data sets in form applications is to use asynchronous queries, which retrieve and display part of the requested data, and then retrieve the rest of the data as needed. This approach requires using advanced PL/SQL programming techniques that are beyond the scope of this book, but the basic approach is as follows: To implement a data block that uses an asynchronous query, you create a data block that is based on a stored procedure. The stored procedure creates multiple PL/SQL tables that are populated using a cursor that retrieves only a specific number of records at a time. Each table contains a different block of records. For example, for a query that might retrieve 10,000 records, you would create four tables, referenced as

Table1, Table2, Table3, and Table4, and you would plan to retrieve the records in blocks of 2,500 records. Table1 would contain the first 2,500 records, Table2 would contain the second 2,500 records, and so forth. Then, you would write a trigger to manually populate the form records using retrievals from each of the four tables.

SUMMARY

- Data blocks can be based on FROM clause queries, stored procedures, and transactional triggers, as well as on tables and views.

- Data blocks based on FROM clause queries can display data from multiple tables, and do not support DML operations.

- Data blocks based on stored procedures derive their data from PL/SQL tables or REF cursors that are created in the stored procedure. These data blocks can contain more complex values than data blocks created using other sources.

- Data blocks based on transactional triggers fire alternate triggers created in PL/SQL instead of executing the associated form triggers.

- You can perform many form actions using keyboard function keys or shortcut keys that are based on key triggers, which execute when the user presses a particular key sequence.

- To redefine the operation associated with an existing key sequence, you create a key trigger using the key trigger event that is currently associated with the existing key sequence.

- A form-level key trigger fires when the user presses the associated key sequence anywhere in the form. A block level key trigger fires only when the user presses the key sequence and the form focus is in the block where the trigger exists.

- A transaction has a posting phase, in which the DML command is received and acknowledged by the database server, placed on a list of uncommitted transactions, and made available for the current user to manipulate and reference as a foreign key.

- A transaction has a committing phase, in which the change is made permanent in the database and the result of the change is made available to other users, or the change is rolled back and discarded.

- Form Builder uses transaction processing, which is also called commit processing, to make the data in the database consistent with the data that appears in the form.

- Form Builder provides transaction triggers to control transaction processing and record auditing information within forms.

- System variables track the internal state of the form and control how the form behaves. Transaction triggers fire based on the values of the form STATUS system variables.

❑ When the user commits a query that involves a combination of INSERT, UPDATE, and DELETE operations, Oracle processes the DELETE operations first, the UPDATE operations second, and the INSERT operations last.

❑ Whenever a form processes a DML command, a specific sequence of navigational and transaction triggers fire.

❑ Form developers need to estimate the maximum number of records that form queries might retrieve and design applications to ensure that form performance remains satisfactory even when users retrieve large data sets.

❑ You can improve form performance by creating indexes on tables associated with forms, advising users to determine the number of query hits before actually executing a query, requiring users to enter search conditions to filter large retrieval sets, and performing asynchronous queries in forms.

❑ To avoid performance delays in LOV displays, you can specify a large fetch size, not automatically refresh the LOV display each time the user opens the LOV display, and require users to always filter LOV records before they display them.

❑ You can improve form performance by basing data blocks on stored procedures that use asynchronous queries to retrieve and display part of the requested data, and then continue to retrieve the rest of the data in the background as needed.

REVIEW QUESTIONS

1. When is it desirable to base a data block on a FROM clause rather than on a table or view?

2. What is a REF cursor? How can you use one in a form?

3. What is an advantage of creating a data block that is based on a stored procedure? What is a disadvantage?

4. How do you define a new action for an existing key sequence?

5. Describe the posting and committing phases in transaction processing.

6. List and describe five approaches for improving form performance with large data sets.

7. List and define the three phases of Oracle query processing.

8. How does increasing the number of records fetched by an LOV query improve form performance?

9. What is array processing, and how does it improve performance when you are working with large data sets?

10. What is an asynchronous query, and how do you implement one in a form?

PROBLEM-SOLVING CASES

The cases reference the Clearwater Traders (see Figure 1-11), Northwoods University (see Figure 1-12), and Software Experts (see Figure 1-13) sample databases. The data files needed for all cases are stored in the Chapter8 folder on your Data Disk. Store all solutions in your Chapter8\CaseSolutions folder.

1. Create the form shown in Figure 8-19 that contains a data block based on a FROM clause that displays the project ID, project name, consultant ID, consultant first and last name, roll-on and roll-off dates, and total project hours from the Software Experts database. Save the form in a file named 8BCase1.fmb.

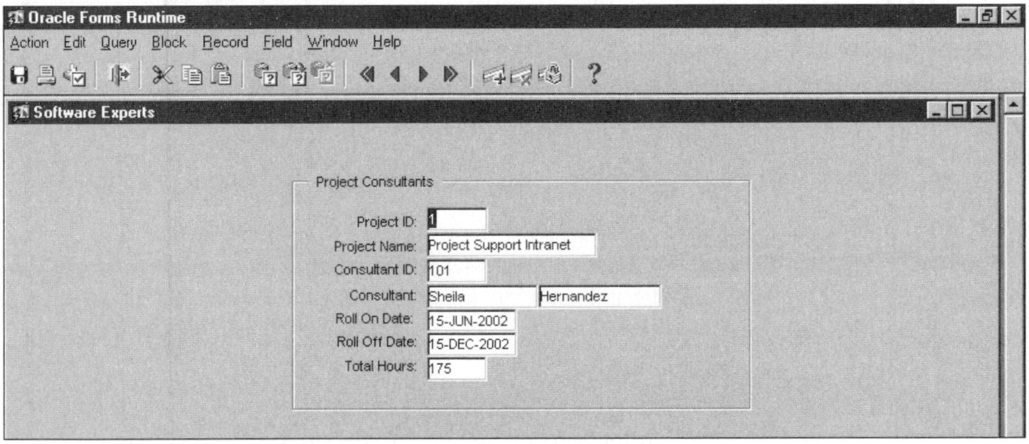

Figure 8-19

2. Create the form shown in Figure 8-20 that contains a data block based on a FROM clause that displays the course section ID, call ID, section number, term description, faculty last name, day, time, building code, room, and maximum enrollment from the Northwoods University database. Save the form in a file named 8BCase2.fmb.

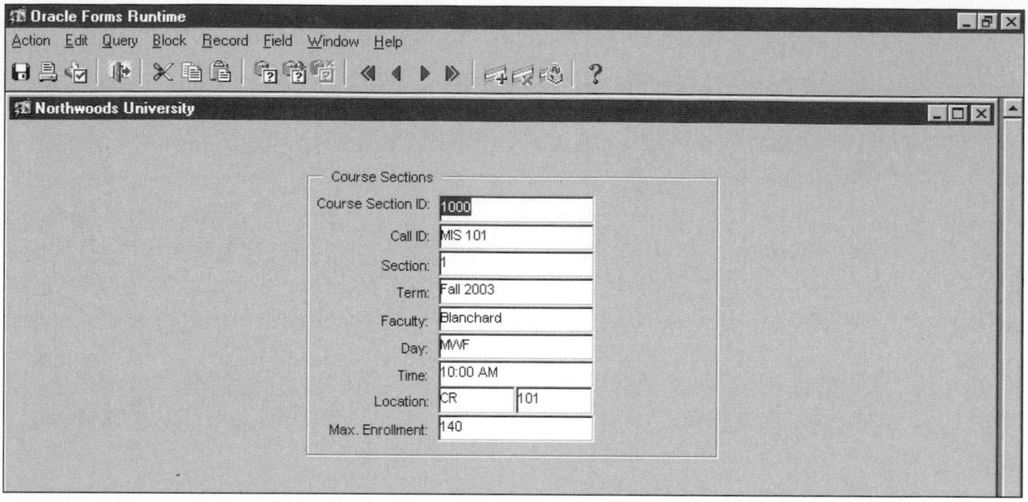

Figure 8-20

3. A form that displays records from the INVENTORY table in the Clearwater Traders database is stored in a file named 8BCase3.fmb in the Chapter8 folder on your Data Disk. Open the file in Form Builder, save it as 8BCase3_DONE.fmb, and then associate the following actions with the specified key sequences.

Key Sequence	Action
F10	Place the insertion point in the Item ID text item and then open the associated LOV display
Page Down	Move to the next text item
Page Up	Move to the previous text item

4. The custom form shown in Figure 8-21 allows users to create new records in the PROJECT table in the Software Experts database and to update existing projects. This form is stored in a file named 8BCase4.fmb in the Chapter8 folder on your Data Disk. When the user clicks Create, the form retrieves the next P_ID value from a sequence named PROJECT_ID_SEQUENCE and displays the value in the Project ID text item. (If you have not created the PROJECT_ID_SEQUENCE in a previous chapter, create it now, and specify that it starts with the value 10.) When the user clicks Save New, the form inserts the information in the form text items into the PROJECT table. When the user selects an existing project, modifies the project data, and then clicks Update, the form updates the selected project record.

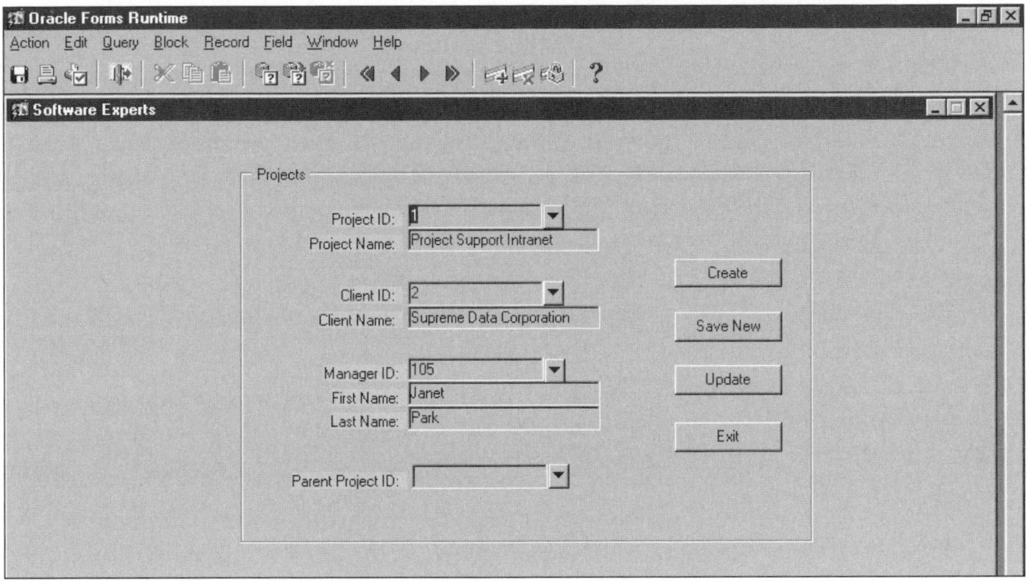

Figure 8-21

Open the file in Form Builder, save it as 8BCase4_DONE.fmb, and then associate the following actions with the specified key sequences.

Key Sequence	Action
F6	Create a new record (same as clicking the Create button)
F10	Save the current record (same as clicking the Save New button)
F7	Place the insertion point in the Project ID text item and open its associated LOV
F8	Update the current record (same as clicking the Update button)
F1	Displays the alert in Figure 8-22, which describes the new function key definitions

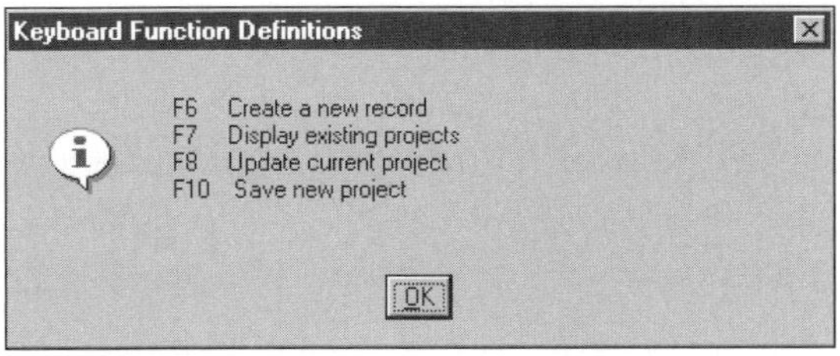

Figure 8-22

5. A data block form based on the COURSE_SECTION table in the Northwoods University database is saved in a file named 8BCase5.fmb in the Chapter8 folder on your Data Disk. The form includes LOVs that allow the user to select values for the Course ID, Term ID, Faculty ID, and Location ID fields by pressing F9 when the form insertion point is in the associated text item. In this case, you will create transaction triggers that check to make sure that the status of the selected term is OPEN and that the capacity of the selected location is adequate to handle the maximum course enrollment. Save the completed form as 8BCase5_DONE.fmb.

   a. Create a PRE-INSERT trigger that checks to make sure that the STATUS value associated with the current term is OPEN and that the CAPACITY value associated with the selected location is greater than or equal to the maximum enrollment value. If either condition is not met, abandon the insert operation, and display an alert explaining why the insert failed. (*Hint*: Use implicit cursors to retrieve the current values for term status and location capacity.)

   b. Create two separate STOP alerts to explain the insert failures. Each STOP alert should display an OK button and no other buttons. One alert should display the message "The location you have chosen does not have adequate capacity for the maximum enrollment. Change the maximum enrollment, or choose a different location." The second alert should display "You must choose a term for which the term status is "OPEN"."

   c. Create a PRE-UPDATE trigger that checks for the same conditions, does not update the record if either condition is not met, and displays the associated alert.

6. A data block form based on the ENROLLMENT table in the Northwoods University database is saved in a file named 8BCase6.fmb in the Chapter8 folder on your Data Disk. In this case, you will create a table named ENROLLMENT_AUDIT that tracks when a user updates a record in this table. This table will record the date the change was made, the user who made the change, the type of operation (INSERT, UPDATE, DELETE), and the values before and after the change was made. Then, you will create transaction triggers that record the audit information. Save the modified form as 8BCase6_DONE.fmb.

   a. In SQL*Plus, create a new table named ENROLLMENT_AUDIT that contains the following columns. Save the command text in a file named 8BCase6.sql

Field Name	Data Specification	Constraint
UPDATE_ID	NUMBER(10)	Primary Key
UPDATE_DATE	DATE	
UPDATE_USER	VARCHAR2(30)	
OPERATION_TYPE	VARCHAR2(10)	

   b. In SQL*Plus, create a sequence named UPDATE_ID_SEQUENCE that automatically generates primary key values for the ENROLLMENT_AUDIT table. Start the sequence with 1, and save the command to create the sequence in the 8BCase6.sql file.

c. Create a POST-INSERT trigger that inserts a record into the audit table when a new record is inserted into the ENROLLMENT table. (*Hint*: Use the keyword USER to reference the username of the current user.)

d. Create a POST-UPDATE trigger that inserts a record into the audit table whenever an ENROLLMENT record is updated.

e. Create a POST-DELETE trigger that inserts a record into the audit table whenever an ENROLLMENT record is deleted.

7. A data block form based on the COURSE_SECTION table in the Northwoods University database is saved in a file named 8BCase7.fmb in the Chapter8 folder on your Data Disk. The form includes LOVs that allow the user to select values for the Course ID, Term ID, Faculty ID, and Location ID fields by pressing F9 when the form insertion point is in the associated text item. The COURSE_SECTION table contains a record detailing course section data each time a course is offered, so this table could contain a large number of records. In this case, you will modify the form to manage large data sets. Save the modified form as 8BCase7_DONE.fmb.

a. Create a PRE-QUERY trigger to ensure that the user enters a search condition in the Course Section ID, Course ID, Term ID, Faculty ID, or Location ID text items. If a search condition exists in any one of these items, execute the query. If no search condition exists in any of these items, display a Stop alert instructing the user to enter a search condition in one of these text items.

b. Modify all of the form LOVs so they always allow the user to filter data and so the LOV display is refreshed only the first time the LOV opens.

c. Modify the LOVs so that the user can search on following columns in the associated LOVs:

LOV	Search Column
Course ID	Call ID
Term ID	Term description
Faculty ID	Faculty last name
Location ID	Building code

d. Modify the COURSE_SECTION block to enable DML array processing in units of 20 records.

8

◀ LESSON C ▶

Objectives

♦ Learn about mouse events
♦ Learn how to control the appearance of the mouse pointer
♦ Work with record groups
♦ Understand how to control data block relationships

FORM BUILDER MOUSE OPERATIONS

In applications with graphical user interfaces, users perform a variety of operations with the mouse pointer. Everything that a user can do with a mouse involves a **mouse event**, which is an action that the user performs with the mouse pointer that has an associated trigger. Table 8–8 describes the mouse events and lists the name of the associated trigger that fires when the user performs the mouse event.

Mouse Event	Mouse Trigger
User clicks mouse button and holds it down	WHEN-MOUSE-DOWN
User releases mouse button after clicking it	WHEN-MOUSE-UP
User clicks mouse button and releases it	WHEN-MOUSE-CLICK
User double-clicks mouse button by clicking and then releasing it two times rapidly	WHEN-MOUSE-DOUBLECLICK
User places mouse pointer onto a canvas item	WHEN-MOUSE-ENTER
User moves mouse pointer out of a canvas item	WHEN-MOUSE-LEAVE
User moves mouse pointer within a canvas item	WHEN-MOUSE-MOVE

Table 8-8   Form Builder mouse events and triggers

Some mouse triggers are described as **mouse-click triggers**, and they fire when the user presses a mouse button. The mouse-click triggers are WHEN-MOUSE-DOWN, WHEN-MOUSE-UP, WHEN-MOUSE-CLICK, and WHEN-MOUSE-DOUBLECLICK. Other mouse triggers are described as **mouse move triggers**, and fire when the user moves the mouse pointer across the screen display. The mouse move triggers include WHEN-MOUSE-ENTER, WHEN-MOUSE-LEAVE, and WHEN-MOUSE-MOVE. Mouse click triggers do not fire when the user moves the mouse pointer, and mouse move triggers do not fire when the user clicks the mouse.

Whenever a mouse event occurs, Form Builder updates the value of one or more mouse-related system variables and then executes the appropriate mouse trigger. The mouse-related system variables are summarized in Table 8-9. Some of the mouse system variables are updated in response to all mouse events, and some are updated only in response to mouse click events.

System Variable Name	Description	Example Values	Associated Mouse Events
MOUSE_BUTTON_PRESSED	Numerical text string corresponding to the mouse button that was clicked	Left mouse button = '1' Middle mouse button = '2' Right mouse button = '3'	Mouse click events
MOUSE_BUTTON_MODIFIERS	Name of key or keys that were pressed at the same time the mouse button was clicked, such as Shift, Alt, or Ctrl	Shift Ctrl + Alt Ctrl + A	Mouse click events
MOUSE_CANVAS	Name of canvas mouse is currently on	ITEM_CANVAS	All mouse events
MOUSE_ITEM	Name of item mouse is currently on	ITEM_IMAGE	All mouse events
MOUSE_RECORD	Number of the record containing the form insertion point	3	Mouse click events
MOUSE_X_POS	Mouse X-axis location on canvas or in item	437	All mouse events
MOUSE_Y_POS	Mouse Y-axis location on canvas or in item	132	All mouse events

Table 8-9   Mouse-related system variables

To learn how to create triggers based on mouse events and work with mouse system variables, you will use a data block form associated with the ITEM table in the Clearwater Traders database. First, you will start SQL*Plus and run the scripts to refresh your sample databases. Then, you will open the form, which is stored in a file named Item_MOUSE.fmb in the Chapter8 folder on your Data Disk. You will save this form using a different filename and then run the form and insert a new item.

To run the form and insert a new item:

1. Start SQL*Plus, log onto the database, and run the clearwater.sql, northwoods.sql, and softwareexp.sql scripts in the Chapter8 folder on your Data Disk.

2. Start Form Builder, open **Item_MOUSE.fmb** from the Chapter8 folder on your Data Disk, and save the file as **8CItem_MOUSE.fmb** in your Chapter8\TutorialSolutions folder.

3. Run the form, and connect to the database if necessary. The form appears with the insertion point in the Item ID text item.

4. Type **1000** for the Item ID, **Hockey Skates** for the Item Description, and **4** for the Category ID.

5. Click **Load Disk Image**, navigate to the Chapter8 folder on your Data Disk, select **skates.jpg**, and click **Open**. The image appears on the form.

6. Click the **Save button** 🖫. The confirmation message appears, confirming that the record was inserted into the database.

7. Close Forms Runtime.

Creating Mouse Triggers

Mouse triggers can be associated with a form, block, or item. If a mouse trigger is associated with a form, the trigger fires anytime the user performs the associated mouse action anywhere in the form. If the mouse trigger is associated with a block, the trigger fires only when the user performs the mouse action while the form focus is in an item in that block. If the mouse trigger is associated with a specific item, the trigger fires only when the user performs the action while the form focus is on that item. In the following sections, you will create a mouse click trigger and a mouse move trigger.

Creating a Mouse Click Trigger

Often in forms with images, users can double-click the image to view a larger version of the image. Now, you will create a trigger that fires when the user double-clicks the mouse button when the mouse pointer is on a specific form item.

You will modify the ITEM form by adding a new canvas that displays an image item with a larger version of the item image. In this form, the current canvas will be called the main canvas, and the new canvas will be called the detail canvas. When the user double-clicks the image on the main canvas, the detail canvas appears. First, you will create the detail canvas and modify its properties. Then, you will create a large image item on the detail canvas to display the item image and a command button that the user will click to return to the main canvas.

To create the detail canvas and canvas items:

1. In the Object Navigator, select the **Canvases node**, and then click the **Create button** 🔲 to create a new canvas. The new canvas appears.

2. Right-click the new canvas, and then click **Property Palette** to open the canvas Property Palette. Change the following canvas property values, and then close the Property Palette.

Property	Value
Name	**IMAGE_DETAIL_CANVAS**
Width	**580**
Height	**375**

3. In the Object Navigator, double-click the **Canvas icon** ![icon] to display the canvas in the Layout Editor. Select the **Image Item tool** ![icon] on the tool palette, and draw the image item as shown in Figure 8-23.

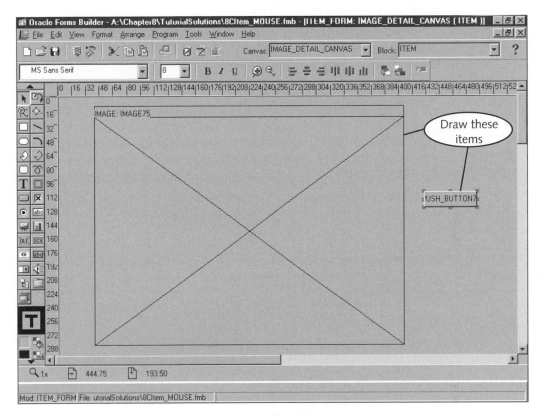

Figure 8-23   Creating items on the IMAGE_DETAIL_CANVAS

4. Select the **Button tool** ![icon] on the tool palette, and draw the button shown in Figure 8-23.

5. Save the form.

Next, you will modify the properties of the new image item. You will change the image item's name to DETAIL_IMAGE and its Sizing Style property to Adjust, so it automatically enlarges the image to the size of the image item. You will also modify its Synchronize With property. This property allows you to synchronize a form item so it displays the same data value as another form item. Since this new image item displays the same data value as the IMAGE_ITEM on the main canvas, the new image item can retrieve its image directly from the item on the main canvas, rather than retrieving it from the database.

To modify the image item properties:

1. In the Layout Editor, select the new image item, right-click, and then click **Property Palette**. Change the image item properties as follows:

Property	Value
Name	**DETAIL_IMAGE**
Sizing Style	**Adjust**
Synchronize with Item	**ITEM_IMAGE**

2. Close the Property Palette, and save the form.

Now you will modify the properties of the command button on the detail canvas. You will change the button name to RETURN_BUTTON and the button label to Return. Then, you will create a WHEN-BUTTON-PRESSED trigger for the button, so when the user clicks the button, the main form canvas appears. Recall that to display a different form canvas, you navigate to an item on the different canvas using the GO_ITEM built-in. You will redisplay the main canvas by navigating to its ITEM_ID text item.

To modify the command button on the detail canvas:

1. In the Layout Editor, right-click the command button on the detail canvas and click **Property Palette**. Change the Name value to **RETURN_BUTTON**, change the Label value to **Return**, and then close the Property Palette.

2. In the Layout Editor, right-click the command button, point to **SmartTriggers**, and then click **WHEN-BUTTON-PRESSED** to create a new trigger.

3. Type the following command in the PL/SQL Editor: **GO_ITEM('item.item_id');**. Compile the trigger, debug it if necessary, and then close the PL/SQL Editor and save the form.

To complete the changes, you need to create a mouse trigger that fires when the user double-clicks the image on the main canvas and displays the detail canvas. To do this, you will create a WHEN-MOUSE-DOUBLECLICK trigger associated with the IMAGE_ITEM on the main canvas. This trigger will contain the code to navigate to the DETAIL_IMAGE on the detail canvas, using the GO_ITEM built-in.

To create the mouse trigger:

1. In the Layout Editor, open the **Canvas list** at the top of the window, and select **ITEM_CANVAS**. The main form canvas appears.

2. Select the **IMAGE: ITEM_IMAGE**, right-click, and then click **PL/SQL Editor**. The ITEM_FORM: Triggers dialog box opens and lists the trigger events associated with the item.

3. To find the mouse event, click the insertion point in the Find text box just before the %, type **WHEN-MOUSE-DOUBLECLICK**, press **Enter**, and then click **OK**. The PL/SQL Editor opens for the trigger event.

4. Type **GO_ITEM('item.detail_image');** in the PL/SQL Editor source code pane. Compile the trigger, debug it if necessary, and then close the PL/SQL Editor and save the form.

You will run the form and test your changes. You will retrieve the record for the hockey skates that you inserted earlier. Then, you will double-click the item image and view the detail canvas, which shows a larger version of the item image.

To run the form and test your changes:

1. Run the form. The form opens and a new blank record appears. Click the **Remove Record** button 🖼 to remove the blank record.

2. Click the **Enter Query** button 🖼 to place the form in Enter Query mode, type **1000** in the Item ID text item to specify the search condition, and then click the **Execute Query** button 🖼 to retrieve the record. The item's data values and image appear.

3. Place the mouse pointer anywhere on the hockey skates image, and then double-click. The detail canvas appears, showing the larger version of the image.

4. Click **Return** to redisplay the main canvas, and then close Forms Runtime.

Creating a Mouse Move Trigger

Like the mouse click triggers, you can create the mouse movement triggers at the form, block, or item level. Now, you will create triggers associated with the WHEN-MOUSE-ENTER and WHEN-MOUSE-LEAVE events. When the user places the mouse pointer on the item image in the main canvas, the message "Double click to view enlarged image" will appear in the form status line. When the user moves the mouse pointer off the image, the message will no longer appear. Recall that to display a message, you use the MESSAGE built-in, followed by the text that is displayed. To clear a message, you use the CLEAR_MESSAGE built-in.

To create the mouse movement triggers.

1. In the Object Navigator, select the **Triggers node** under ITEM_IMAGE, and then click the **Create button** 🖼 to create a new trigger associated with the item image. The ITEM_FORM: Triggers dialog box opens. Select the **WHEN-MOUSE-ENTER** event, and then click **OK**.

2. In the PL/SQL Editor, type the following trigger code: **MESSAGE('Double-click to view enlarged image');**. Compile the trigger, and debug it if necessary. DO NOT close the PL/SQL Editor.

3. To create another trigger associated with the ITEM_IMAGE directly in the PL/SQL Editor, click **New** on the button bar, select the **WHEN-MOUSE-LEAVE** event, and then click **OK**. The PL/SQL Editor window opens for the new event trigger.

8

4. Type the following code for the WHEN-MOUSE-LEAVE trigger: **CLEAR_MESSAGE;**. Compile the trigger, debug it if necessary, close the PL/SQL Editor, and then save the form.

Now, you will run the form and place the mouse pointer on the item image in the main canvas. The message prompting you to double-click to view the enlarged item should appear. Then, you will move the mouse pointer off the image to confirm that the message is cleared. Since the triggers are associated with the IMAGE_ITEM, the triggers fire whenever the user performs the trigger actions, regardless of whether an image is currently displayed in the form.

To test the form:

1. Run the form. A new blank record appears.

2. Move the mouse pointer onto the item image. The message prompting you to double-click to view the enlarged image appears in the status line.

3. Move the mouse pointer off the item image. The message clears.

4. Close Forms Runtime.

Customizing Forms with Mouse System Variables in Mouse Triggers

You can use the mouse system variables in conjunction with the mouse triggers to customize form functionality and appearance. For example, you can create different triggers that fire depending on whether the user clicks the right or left mouse button. In addition, you can dynamically change the mouse behavior depending on the current canvas displayed or the X- or Y-position of the mouse pointer.

Suppose that you would like to customize the Clearwater Traders item form so that when the user clicks the right mouse button anywhere in the form, the form displays the next record. To do this, you will create a form-level WHEN-MOUSE-CLICK trigger. By default, this trigger responds to left-button mouse clicks. To make the trigger move to the next record only when the user clicks the right mouse button, you must create an IF/THEN conditional structure that tests the current value of the MOUSE_BUTTON_PRESSED system variable. Recall that this variable's value is '1' for the left mouse button, '2' for the middle mouse button, and '3' for the right mouse button. If the user clicks the right mouse button, this value will be '3', and the trigger will execute the NEXT_RECORD built-in. After you create the trigger, you will run the form, retrieve all of the ITEM records, and then test the trigger.

To create and test the trigger to respond to right-button mouse clicks:

1. Select the **Triggers node** under ITEM_FORM, and then click the **Create button** to create a form-level trigger. Select the **WHEN-MOUSE-CLICK** event, and then click **OK**. The PL/SQL Editor opens.

2. Type the code shown in Figure 8-24, compile the code, and debug it if necessary. Then close the PL/SQL Editor.

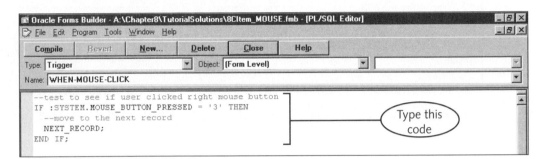

Figure 8-24   Mouse trigger using mouse system variable

3. Save the form, and then run it. Click the **Enter Query button** 🔲 to place the form in Enter Query mode, do not enter a search condition, and then click the **Execute Query button** 🔲 to retrieve all of the table records. The first record (Item ID 894, Women's Hiking Shorts) appears.

The item image will not appear because you ran the clearwater.sql script to refresh the Clearwater Traders database tables at the beginning of this lesson. The images that you loaded in Lesson A are no longer stored in the database.

4. Right-click anywhere on the form. The next record (Item ID 897, Women's Fleece Pullover) appears.

5. Continue to right-click anywhere on the form until you display all of the table records. Then close Forms Runtime.

Controlling the Mouse Pointer Appearance

Sometimes, you need to change the appearance of the mouse pointer to indicate that the application is busy or to indicate that the user can enter text in a specific location. To change the appearance of the mouse pointer in a form, you use the SET_APPLICATION_PROPERTY built-in, which has the following syntax: **SET_APPLICATION_PROPERTY (*property_name*, '*property_value*');**. To change the appearance of the pointer, you use CURSOR_STYLE for the *property_name* parameter, and place the *property_value* parameter in single quotation marks. Table 8-10 summarizes the different CURSOR_STYLE property values available in Form Builder. It also shows the default Windows pointer images associated with each style.

CURSOR_STYLE Property Value	Default Windows Pointer
DEFAULT	
BUSY	
CROSSHAIR	+
HELP	
INSERTION	I

Table 8-10   Form Builder CURSOR_STYLE values and associated Windows default pointers

Your pointer images might be different if custom pointer images have been installed on your workstation.

Now, you will change the pointer from the DEFAULT style to the HELP style when the user moves the mouse pointer into the ITEM_IMAGE on the main form. To set the mouse pointer style to HELP, you will add the SET_APPLICATION_PROPERTY command to the WHEN-MOUSE-ENTER trigger. Then, to change the pointer back to the DEFAULT style when the user moves the pointer out of the ITEM_IMAGE, you will modify the WHEN-MOUSE-LEAVE trigger and set the mouse pointer style back to DEFAULT. Then, you will run the form to test your changes.

To change the mouse pointer style:

1. In the Object Navigator, double-click the **Trigger icon** beside the WHEN-MOUSE-ENTER trigger under the ITEM_IMAGE node. The trigger code appears in the PL/SQL Editor.

2. Add the following line of code to the trigger to change the mouse pointer style to HELP: **SET_APPLICATION_PROPERTY(CURSOR_STYLE, 'HELP');**.

3. Compile the trigger, and correct any syntax errors. DO NOT close the PL/SQL Editor.

4. To open the WHEN-MOUSE-LEAVE trigger, open the **Name** list, which shows the names of all triggers associated with the current item. Select **WHEN-MOUSE-LEAVE**. The trigger code appears.

5. Add the following line of code to the trigger to change the mouse pointer back to the DEFAULT style: **SET_APPLICATION_PROPERTY (CURSOR_STYLE, 'DEFAULT');**. Compile the trigger, correct any syntax errors, close the PL/SQL Editor, and save the form.

6. Run the form, and then move the mouse pointer onto the ITEM_IMAGE. The pointer changes to ⓘ?. Move the mouse pointer out of the ITEM_IMAGE. The pointer changes back to ⓘ.

7. Close Forms Runtime, and then close the form in Form Builder.

FORM RECORD GROUPS

A record group is a form object that represents data in a tabular format. There are three types of record groups:

- **Static**, in which you specify the contents when you design the form. The values cannot be changed while the form is running.

- **Query**, in which you base the record group contents on a SQL SELECT command that executes while the form is running. You have already used query record groups to populate LOVs.

- **Nonquery**, in which you insert the contents using program commands that execute while the form is running.

You can use a record group to define the data that appears in an LOV or to pass records from one form application to another. In the following sections, you will learn how to create and work with the different types of record groups.

Static Record Groups

In a static record group, you specify the record group contents at design time and the user cannot change the contents while the form is running. You should use static record groups to create LOVs in which the display values do not change.

Now, you will open a form with a data block based on the CONSULTANT_SKILL table in the Software Experts database. This table contains three columns: C_ID, SKILL_ID, and CERTIFICATION. The CERTIFICATION column always contains either Y (indicating the consultant is certified in the skill), or N (indicating the consultant is not certified in the skill). You will create a static record group that contains one column to represent the certification status, with two values, Y and N. Then, you will create an LOV based on this static record group.

To open the form and create the static record group:

1. Open **ConsultantSkill.fmb** from the Chapter8 folder on your Data Disk, and save the file as **8CConsultantSkill.fmb** in your Chapter8\TutorialSolutions folder.

2. In the Object Navigator, select the **Record Groups node**, and then click the **Create button** 🔲 to create the new record group. The CONSULTANT_SKILL_FORM: New Record Group dialog box opens.

3. Select the **Static Values option button** to specify that this is a static record group, and then click **OK**. The CONSULTANT_SKILL_FORM: Column Specification dialog box opens.

In this dialog box, you first specify the name of each record group column in the Column Names list. Then, you select each column name individually and specify the column's data type and maximum length. Finally, you enter the static data values for each column in the Column Values list. The Data Type list and Length text box represent the column currently selected in the Column List, and the Column Values list displays the values for the column currently selected in the Column List. Now, you will create the CERTIFICATION_STATUS column and specify its data type and values.

To create the record group column and values:

1. Place the insertion point in the first item in the Column Names list, delete the current value, and type **CERTIFICATION_STATUS**.

2. Make sure that the Data Type is Character, and change the Length value to **1**.

3. Place the insertion point in the first item in the Column Values list, and type **Y**.

4. Place the insertion point in the second item in the Column Values list, and type **N**. Your completed record group column specification should look like Figure 8-25.

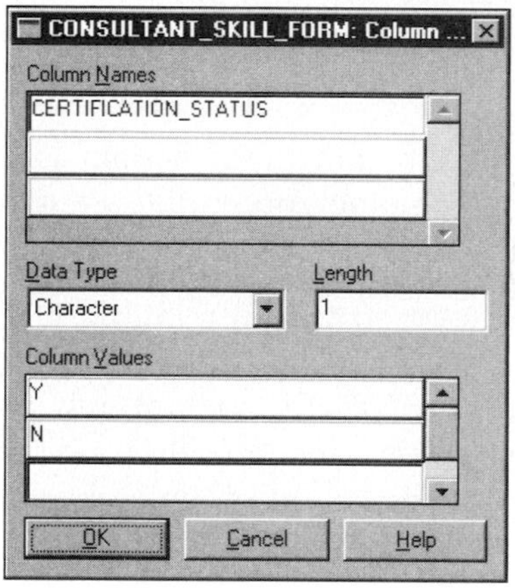

Figure 8-25   Static record group column specification

5. Click **OK** to save the column specification. The record group appears in the Object Navigator.

6. Save the form.

Next, you will give the record group a descriptive name. Then, you will create an LOV based on the static record group.

To rename the record group and create the LOV:

1. In the Object Navigator, rename the new record group **CERTIFICATION_RECORD_GROUP**.

2. Select the **LOVs node**, click **Tools** on the menu bar, and then click **LOV Wizard**. The LOV Wizard LOV Source page appears.

3. On the LOV Source page, select the **Existing Record Group option button**, and confirm that CERTIFICATION_RECORD_GROUP is selected in the record group list. Click **Next**.

4. On the Column Selection page, select **CERTIFICATION_STATUS** in the Record Group Columns list, and then click the **Select button** [>] to select it as the LOV column. Click **Next**.

5. On the Column Display page, change the column title to **Status** and the Width to **40**. Click in the Return value space, select **Look up return item**, select **CONSULTANT_SKILL.CERTIFICATION** as the return value for the LOV, click **OK**, and then click **Next**.

6. On the LOV Display page, type **Certification Status** for the window title, change the Width to **135**, and click **Next**.

7. On the Advanced Options page, click **Next**.

8. On the Items page, select **CONSULTANT_SKILL.CERTIFICATION** in the Return Items list, click the **Select button** [>] to select it as the assigned item, click **Next**, and then click **Finish**. The LOV appears in the Object Navigator.

9. Change the LOV name to **CERTIFICATION_LOV**, and then save the form.

Now, you will run the form and open the LOV display. You will enter a new CONSULTANT_SKILL record and open the LOV display by placing the insertion point in the Certification text item and pressing F9.

To run the form and open the LOV display:

1. Run the form. The form opens, with the insertion point in the Consultant ID text item.

2. Type **100** for the consultant ID and **9** for the Skill ID.

3. Place the insertion point in the Certification text item, and press **F9**. The LOV display opens, showing the static values from the record group. Make sure that Y is selected, and then click **OK**. The value Y appears in the Certification text item.

4. Close Forms Runtime, and do not save your changes.

8

Query Record Groups

Query record groups are based on a SQL query. A query record group can be created manually by the form developer or programmatically by commands that execute while the form is running. Either way, the actual record group values are retrieved while the form is running. In the following sections, you will learn how to create query record groups both manually and programmatically.

Creating Query Record Groups Manually

To create a new query record group manually, you create a new record group in the Object Navigator and then specify a SQL query as the record group source. When you use the LOV Wizard to create a new LOV, the Wizard gives you the option of automatically creating a new query record group that will populate the LOV. You used this option to create the LOVs you have worked with in previous chapters. You can also base an LOV on an existing query record group. This approach is useful if you are creating several LOVs that display values from the same record group.

Now, you will manually create a query record group to retrieve the consultant ID and last name for every consultant in the CONSULTANT table. Then, you will create an LOV based on this record group.

To create a query record group manually:

1. In the Object Navigator, select the **Record Groups node**, and then click the **Create button** 🔲 to create a new record group. Make sure the **Based on the query below option button** is selected, then type the following query, and click **OK**.

   ```
   SELECT c_id, c_first, c_last
   FROM consultant
   ```

2. In the Object Navigator, change the record group name to **CONSULTANT_RECORD_GROUP**, and then save the form.

Now, you will use the LOV Wizard to create an LOV based on this query record group.

To create the LOV based on the query record group:

1. In the Object Navigator, select the **LOVs node**, click **Tools** on the menu bar, and then click **LOV Wizard**. The LOV Wizard LOV Source page appears.

2. On the Source page, click the **Existing Record Group** option button, select **CONSULTANT_RECORD_GROUP** in the record group list, and then click **Next**. The Record Group page appears.

The Record Group page appears only when you are creating an LOV based on an existing query record group or when you edit an LOV using the LOV Wizard in reentrant mode. This page gives you the option of creating a new record group specifically for this LOV or using the existing record group. If you modify the LOV in the LOV Wizard,

Form Builder automatically modifies the record group on which the LOV is based. If you are using this record group to retrieve records for other LOVs, and the record group is modified, then the other LOVs will not work correctly. Therefore, you should always use the option of creating a new record group when a record group is associated with more than one LOV. Since you are using this record group only for this LOV, you will select the option to allow the Wizard to modify the record group. Now, you will complete the LOV specification.

To complete the LOV specification:

1. Make sure the **Modify existing Record Group option button** is selected, and click **Next**. The Query page appears. Click **Next**.

2. On the Column Selection page, click the **Select All button** `>>` to select all of the record group columns as LOV columns, and then click **Next**.

3. On the Column Display page, change the column specifications as follows, and then click **Next**.

Column	Title	Width	Return Item
C_ID	**ID**	30	**CONSULTANT_SKILL.C_ID**
C_FIRST	**First Name**	60	
C_LAST	**Last Name**	60	

4. On the LOV Display page, type **Consultants** for the window title, change the Height to **160**, and click **Next**.

5. On the Advanced Options page, click **Next**.

6. On the Items page, select **CONSULTANT_SKILL.C_ID** in the Return Items list, click the **Select button** `>` to select it as the assigned item, click **Next**, and then click **Finish**. The LOV appears in the Object Navigator.

7. Change the LOV name to **CONSULTANT_LOV**, and then save the form.

Now, you will run the form and open the LOV display that shows consultant information. You will place the insertion point in the Consultant ID text item and press F9.

To run the form and open the LOV display:

1. Run the form. The form opens, with the insertion point in the Consultant ID text item.

2. Press **F9**. The LOV display opens, showing the values from the query record group. Select ID **100** (consultant Mark Myers), and then click **OK**. The value 100 appears in the Consultant ID text item.

3. Close Forms Runtime, and click **Yes** to confirm closing.

4. Close the form in Form Builder.

To make forms more flexible and useful, you can allow the user to dynamically change the name of the record group associated with an LOV. Consider the form shown in Figure 8-26, which is used to assign consultants to projects at Software Experts.

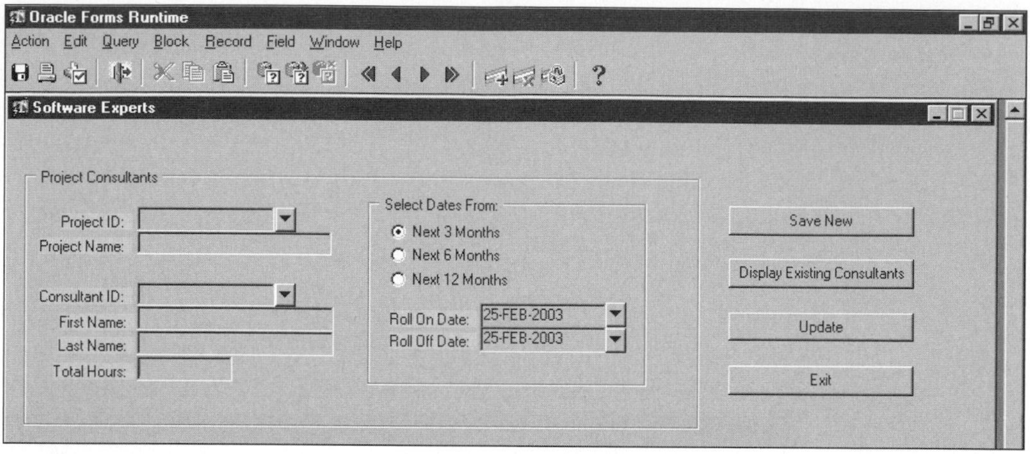

Figure 8-26   Form to allow user to specify LOV display values

To specify the roll-on and roll-off dates for new projects, it would be useful if users could select from a list of future dates using an LOV display that opens when a user clicks the LOV command buttons by the Roll On Date and Roll Off Date text items. Furthermore, it would be helpful if users could control the number of future date values that appear by selecting one of the option buttons in the frame. If the user selects the Next 3 Months option button, the LOV display shows a list of dates representing every day during the next three months. If the user selects the Next 6 Months option button, the LOV display shows a list of dates representing every day during the next six months. If the user selects the Next 12 Months option button, the LOV display shows a list of dates representing every day during the next year.

To learn how to dynamically change the record group associated with an LOV, you will create the LOV and record groups associated with the Roll On Date text item. First, you need to create a table that provides the date values for the Roll On Date text item. To do this, you will run a script named date_table.sql stored in the Chapter8 folder on your Data Disk. This script creates a table that has two columns: DATE_ID, and DATE_VALUE, as shown in Table 8-11.

DATE_ID	DATE_VALUE
1	25-FEB-2003 (or your current system date)
2	26-FEB-2003 (or your current system date plus one day)
3	27-FEB-2003
...	
365	24-FEB-2004 (or your current system date plus 364 days)

Table 8-11   Table to provide date values for LOV

The DATE_ID column is the primary key of the table and provides a unique numerical identifier for each date. The DATE_VALUE column contains the actual date values. The script contains a PL/SQL program that automatically inserts the DATE_ID and DATE_VALUE records for the next 365 days into the table. When the script runs, the first date will be your current system date. The next date will be the day after your current system date. After the script inserts all of the records, the table will contain a date for each day during the next year.

To run the script to create the table that stores date values:

1. Switch to SQL*Plus, and type the following command to run the script: **START a:\chapter8\date_table.sql;**. The confirmation message "PL/SQL procedure successfully completed" appears to indicate the script ran successfully. (The path to your Chapter8 Data Disk files might be different.)

 Do not worry if the message "ERROR at line 1: ORA-00942: table or view does not exist" appears. The script contains a command to automatically drop the table before creating it. If this is the first time you have run the script, this message will appear.

2. To view the table contents, type the following command at the SQL prompt: **SELECT * from date_table;**. The table appears as shown in Table 8-11, starting at your current system date and containing dates for the next year.

Now, you will modify the project consultants form so that users can configure the LOV associated with the Roll On Date text item while the form is running. First, you will open the form, which is stored in a file named ProjectConsultants_LOV.fmb in the Chapter8 folder on your Data Disk, and save it using a different filename. Then, you will create three record groups. The first record group will retrieve dates from the DATE_TABLE for the next three months; the second record group will retrieve dates for the next 6 months; and the third record group will retrieve dates for the next 12 months. To do this, you will use the ADD_MONTHS SQL function in the record group query search condition, and specify that the retrieved dates are less than (that is, they occur before) the current date plus 3, 6, or 12 months.

To open the form and create the record groups:

1. In Form Builder, open **ProjectConsultants_LOV.fmb**, which is in the Chapter8 folder on your Data Disk, and save the form as **8CProjectConsultants_LOVa.fmb** in your Chapter8\TutorialSolutions folder.

2. In the Object Navigator, select the **Record Groups node**, and click the **Create button** to create a new record group. Type the following query in the New Record Group dialog box to specify that the record group

retrieves the DATE_VALUE column from the DATE_TABLE for the next three months, and then click **OK**.

```
SELECT date_value
FROM date_table
WHERE date_value < ADD_MONTHS(SYSDATE,3)
```

3. Change the record group name to **NEXT3_RECORD_GROUP**.

4. Create a second record group using the following query, and then change the record group name to **NEXT6_RECORD_GROUP**.

```
SELECT date_value
FROM date_table
WHERE date_value < ADD_MONTHS(SYSDATE,6)
```

5. Create a third record group using the following query, and then change the record group name to **NEXT12_RECORD_GROUP**.

```
SELECT date_value
FROM date_table
WHERE date_value < ADD_MONTHS(SYSDATE,12)
```

6. Save the form.

Next, you will create the LOV associated with the Roll On Date text item. To specify the LOV display properties, you will initially base the LOV on the NEXT3_RECORD_GROUP. However, the record group that is actually associated with the LOV will be specified while the form is running.

To create the LOV for the Roll On Date text item:

1. In the Object Navigator, select the **LOVs node**, click **Tools** on the menu bar, and then click **LOV Wizard**. The LOV Wizard opens.

2. On the Source page, select the **Existing Record Group option button**, select **NEXT3_RECORD_GROUP**, and then click **Next**.

3. On the Record Group page, make sure the Modify existing Record Group option button is selected, and then click **Next**. When the Query page appears, click **Next** again.

4. On the Column Selection page, click **DATE_VALUE** in the Record Group Columns list, and then click the **Select button** ⟩ to select it as the LOV column. Click **Next**.

5. On the Column Display page, change the column title to **Date**. Click in the Return value space, click **Look up return item**, select **PROJECT_BLOCK.ROLL_ON_TEXT** as the return value for the LOV, click **OK**, and then click **Next**.

6. On the LOV Display page, type **Roll On Date** for the window title, change the Height to **200**, and click **Next**.

7. On the Advanced Options page, click **Next**.

8. On the Items page, select **PROJECT_BLOCK.ROLL_ON_TEXT** in the Return Items list, click the **Select button** [>] to select it as the assigned item, click **Next**, and then click **Finish**. The LOV appears in the Object Navigator.

9. Change the LOV name to **ROLL_ON_LOV**, and then save the form.

To complete the form, you need to write the trigger for the LOV command button beside the Roll On Date text item so that when the user clicks the LOV command button, the trigger associates the correct record group with the LOV. Recall that the user specifies the record group using the form radio buttons. The form radio group is named DATES_RADIO_GROUP and contains three radio buttons with the following properties:

Radio Button Name	Label	Value
NEXT3_RADIO_BUTTON	Next 3 Months	NEXT3
NEXT6_RADIO_BUTTON	Next 6 Months	NEXT6
NEXT12_RADIO_BUTTON	Next 12 Months	NEXT12

Recall that the value of the radio group is the value of the selected radio button. For example, when the radio button labeled "Next 3 Months" is selected, the DATES_RADIO_GROUP has the value NEXT3. To determine which record group to associate with the LOV, you will need to create an IF/THEN/ELSIF conditional structure that tests for the current value of DATES_RADIO_GROUP. If the radio group value is NEXT3, then the NEXT3_RECORD_GROUP (which retrieves dates for the next 3 months) must be associated with the LOV. If the radio group value is NEXT6, then the NEXT6_RECORD_GROUP (which retrieves dates for the next 6 months) must be associated with the LOV. If the radio group value is NEXT12, then the NEXT12_RECORD_GROUP (which retrieves dates for the next 12 months) must be associated with the LOV.

After the trigger code determines which record group to use, it must issue the command to associate the record group with the LOV. To do this, you use the SET_LOV_PROPERTY built-in, which has the following syntax:

```
SET_LOV_PROPERTY ('LOV_name', property_name, property_value);
```

This SET_LOV_PROPERTY command has the following parameters:

- *LOV_name*, which represents the name of the LOV to which the record group is being assigned. In this case, the value will be 'ROLL_ON_LOV'. This value must be placed in single quotation marks.

- *property_name*, which is the name of the LOV property being set. To change the name of the LOV record group, you use GROUP_NAME for the property name.

8

- *property_value*, which represents the name of the new record group, which might be 'NEXT3_RECORD_GROUP', 'NEXT6_RECORD _GROUP', or 'NEXT12_RECORD_GROUP'. This value must be placed in single quotation marks.

 When you replace the record group associated with an LOV with a new record group, the new record group must have the same number of columns as the existing record group, and the columns must have the same data types and names.

Now, you will modify the trigger associated with the LOV command button beside the Roll On Date text item. Before the trigger executes the LIST_VALUES command, which opens the LOV display, the trigger will determine which record group should appear and then modify the LOV RECORD_GROUP property.

To modify the LOV command button trigger to dynamically specify the LOV record group:

1. Open the Layout Editor, right-click the **LOV command button** beside the Roll On Date text item, and then click **PL/SQL Editor**. The button's WHEN-BUTTON-PRESSED trigger code appears.

2. Add the code shown in Figure 8-27. Compile the code, debug it if necessary, close the PL/SQL Editor, and save the form.

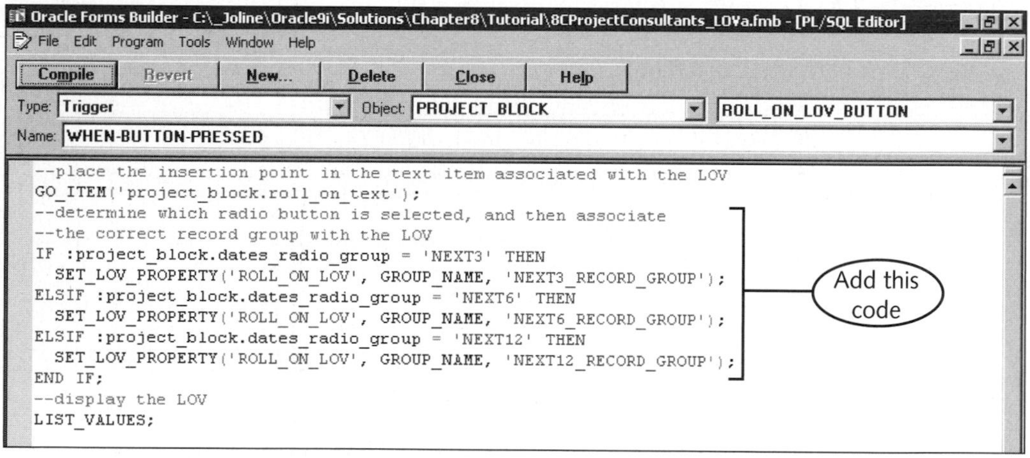

```
Oracle Forms Builder - C:\_Joline\Oracle9i\Solutions\Chapter8\Tutorial\8CProjectConsultants_LOVa.fmb - [PL/SQL Editor]
File   Edit   Program   Tools   Window   Help

  Compile      Revert      New...      Delete      Close      Help

Type: Trigger                              ▼   Object: PROJECT_BLOCK          ▼   ROLL_ON_LOV_BUTTON       ▼
Name: WHEN-BUTTON-PRESSED                                                                                  ▼

--place the insertion point in the text item associated with the LOV
GO_ITEM('project_block.roll_on_text');
--determine which radio button is selected, and then associate
--the correct record group with the LOV
IF :project_block.dates_radio_group = 'NEXT3' THEN
  SET_LOV_PROPERTY('ROLL_ON_LOV', GROUP_NAME, 'NEXT3_RECORD_GROUP');
ELSIF :project_block.dates_radio_group = 'NEXT6' THEN
  SET_LOV_PROPERTY('ROLL_ON_LOV', GROUP_NAME, 'NEXT6_RECORD_GROUP');
ELSIF :project_block.dates_radio_group = 'NEXT12' THEN
  SET_LOV_PROPERTY('ROLL_ON_LOV', GROUP_NAME, 'NEXT12_RECORD_GROUP');
END IF;
--display the LOV
LIST_VALUES;
```

Add this code

Figure 8-27 Code to dynamically associate correct record group with LOV

Now, you will run the form and confirm that the correct values appear in the LOV display.

To test the LOV display:

1. Run the form. The form appears with the first radio button (Next 3 Months) selected.

2. Click the **LOV command button** beside the Roll On Date text item. Scroll down the LOV display to confirm that dates appear for the next three months. Click **Cancel**.

3. Select the **Next 6 Months** radio button, click the **LOV command button** beside Roll On Date, confirm that dates appear for the next six months, and then click **Cancel**.

4. Select the **Next 12 Months** radio button, click the **LOV command button** beside Roll On Date, confirm that dates display for the next 12 months, and then click **Cancel**.

5. Click **Exit** to close Forms Runtime.

Creating a Query Record Group Programatically

In the previous example, you created a separate record group corresponding to each radio button selection. An alternate method is to create a query record group using program commands, in which the SQL query can be structured based on user input. This way, you can use a single record group to handle all three options. Now, you will replicate the form you just created, in which the user can specify the number of dates that the project roll-on date LOV displays. Instead of creating three different record groups, you will create a single record group. You will then write program commands for a SQL query for this record group that specifies the number of dates to display based on the radio button the user selects.

To create a query record group programmatically, you use the CREATE_GROUP_FROM_QUERY built-in function, which has the following syntax:

```
group_id := CREATE_GROUP_FROM_QUERY('record_group_name',
'SQL_query_text', record_group_scope,
number_of_fetch_records);.
```

The function parameters include:

- *group_ID*, which is the value returned by the function that uniquely identifies the record group within the form. You must declare a variable to reference this value in the DECLARE section of the PL/SQL program block. This variable has the RECORDGROUP data type, which is a special Form Builder data type used to identify record groups that are generated programmatically.

- *record_group_name*, which is the name of the record group and must be enclosed in single quotation marks. This name cannot be the same as any existing record groups in the form and must follow the Oracle naming standard.

- *SQL_query_text*, which is the text of the SQL command that specifies the record group values. You enclose the query text in single quotation marks and omit the ending semicolon.

8

- *record_group_scope*, which is an optional parameter that can have two values: FORM_SCOPE and GLOBAL_SCOPE. FORM_SCOPE specifies that the record group is visible only inside the current form, and GLOBAL_SCOPE specifies that the record group is visible to all forms currently running. The default value is FORM_SCOPE.

- *number_of_fetch_records*, which is an optional parameter that specifies how many records are fetched in each query fetch cycle. The default value is 20.

After you create a record group in a program, you must populate the record group by executing the query using the POPULATE_GROUP function. This function has the following syntax:

```
return_value := POPULATE_GROUP('record_group_name');.
```

The function parameters include:

- *return_value*, which is a declared variable of the NUMBER data type that is returned by the function to indicate whether the function successfully executed. If the return value is 0, the record group was successfully populated. If the function was not successful, the function returns the value of the corresponding error code.

- *record_group_name*, which is the name of the record group, enclosed in single quotation marks.

Suppose a user clicks the LOV command button to create the record group. The CREATE_GROUP_FROM_QUERY command executes and creates the record group. Then, suppose the user clicks the LOV command button again. The CREATE_GROUP_FROM_QUERY command executes again, and tries to create another record group with the same name as the first record group. This generates an error. When you create a record group programmatically using the CREATE_GROUP_FROM_QUERY built-in, you must first delete the group before you can create it again in the same program. To delete a record group, you use the DELETE_GROUP procedure, which has the following syntax:

```
DELETE_GROUP('record_group_name');.
```

When you use the programmatic approach to create a record group associated with an LOV, you must first create the LOV using the LOV Wizard and base the LOV values on a SQL query that has the same number of columns with the same data types in the SELECT statement as the record group created using the CREATE_GROUP_FROM_QUERY command. The LOV will then be populated at runtime using the query specified in your program commands.

Now, you will modify the project consultants form so that is uses programmatically created record groups to populate the LOV for the roll-on dates. Since the LOV already exists and has a single column that holds DATE values, you do not need to create a new LOV. You will modify code in the roll-on date LOV command button so that the

IF/THEN/ELSIF control structure that determines which option button the user selected now specifies the text of the SQL query that retrieves the correct number of dates. Then, you will add code to programmatically create and populate the record group, associate the LOV with the record group, and display the LOV. After the user closes the LOV display, the trigger will delete the record group, so the record group can be created again using a different search condition.

To modify the form to use a programmatically generated record group:

1. Save the 8CProjectConsultants_LOVa.fmb form as **8CProjectConsultants_LOVb.fmb** in your Chapter8\TutorialSolutions folder.

2. In the Layout Editor, right-click the **LOV command button** beside the Roll On Date text item, and then click **PL/SQL Editor** to open the trigger code. Modify the code as shown in Figure 8-28.

```
DECLARE
  date_group_id RECORDGROUP;
  query_text VARCHAR2(500);
  query_result NUMBER;
BEGIN
  GO_ITEM('project_block.roll_on_text');
  --determine which radio button is selected, and then specify the correct query
  IF :project_block.dates_radio_group = 'NEXT3' THEN
    query_text := 'SELECT date_value FROM date_table
                   WHERE date_value < ADD_MONTHS(SYSDATE, 3)';
  ELSIF :project_block.dates_radio_group = 'NEXT6' THEN
    query_text := 'SELECT date_value FROM date_table
                   WHERE date_value < ADD_MONTHS(SYSDATE, 6)';
  ELSIF :project_block.dates_radio_group = 'NEXT12' THEN
    query_text := 'SELECT date_value FROM date_table
                   WHERE date_value < ADD_MONTHS(SYSDATE, 12)';
  END IF;
  --create and populate the record group
  date_group_id := CREATE_GROUP_FROM_QUERY ('DATE_RECORD_GROUP', query_text);
  query_result := POPULATE_GROUP('DATE_RECORD_GROUP');
  SET_LOV_PROPERTY('ROLL_ON_LOV', GROUP_NAME, 'DATE_RECORD_GROUP');
  --display the LOV
  LIST_VALUES;
  DELETE_GROUP('DATE_RECORD_GROUP');
END;
```

Figure 8-28   Code to create and populate a query record group programmatically

3. Compile the trigger, correct any syntax errors, then close the PL/SQL Editor, and save the form.

Now, you will run the form and confirm that the correct values appear in the LOV display.

To test the LOV display:

1. Run the form. The form appears, with the first radio button (Next 3 Months) selected.

2. Click the **LOV command button** beside the Roll On Date text item. Scroll down the LOV display to confirm that dates appear for every day during the next three months. Click **Cancel**.

3. Select the **Next 6 Months option button**, click the **LOV command button** beside Roll On Date, confirm that dates appear for the next six months, and then click **Cancel**.

4. Select the **Next 12 Months option button**, click the **LOV command button** beside Roll On Date, confirm that dates display for the next 12 months, and then click **Cancel**.

5. Select **Exit** to close Forms Runtime.

Nonquery Record Groups

A **nonquery record group** is a record group that contains values that cannot be directly retrieved using a SQL query. You create nonquery record groups using program commands, and you populate them by executing commands that specify the individual values of each field in each record. For example, you could create a record group that contains calculated values, such as the total amount of all customer orders placed during each month of the current year. An advantage of using nonquery record groups to display date values is that you do not need to create a table like the DATE_TABLE. Instead, you can programmatically generate future dates and then add them to the nonquery record group.

Creating a nonquery record group involves four steps:

1. Create the record group

2. Define the record group columns

3. Add new blank rows to the record group

4. Specify the values of each field in each row

The following sections describe the commands for each step.

Creating a Nonquery Record Group

To create a nonquery record group, you use the CREATE_GROUP built-in function, which has the following syntax:

```
group_id := CREATE_GROUP('record_group_name',
record_group_scope, number_of_fetch_records);
```

The CREATE_GROUP function parameters include:

- *group_id*, which is a variable that identifies the record group and is declared using the RECORDGROUP data type, which is a special Form Builder data type used to identify record groups that are created programmatically.

- *record_group_name*, which is the name of the record group, enclosed in single quotation marks. This name cannot be the same as any existing record groups and must follow the Oracle naming standard.

- *record_group_scope*, which is an optional parameter that can have the values FORM_SCOPE and GLOBAL_SCOPE. FORM_SCOPE specifies that the record group is visible only within the current form, and GLOBAL_SCOPE specifies that the record group is visible to all forms. The default value is FORM_SCOPE.

- *number_of_fetch_records*, which is an optional parameter that specifies how many records are fetched in each query fetch cycle. The default value is 20.

Adding Columns to a Nonquery Record Group

After you have created a nonquery record group, you must add each column to the record group using a separate call to the ADD_GROUP_COLUMN built-in function, which has the following syntax:

```
column_id := ADD_GROUP_COLUMN ('record_group_name',
'column_name', column_data_type_specification,
column_width);
```

The ADD_GROUP_COLUMN function parameters are:

- *column_id*, which is a variable that identifies the column. This variable is declared using the GROUPCOLUMN data type, which is a special Form Builder data type used to identify nonquery record group columns generated programmatically.

- *record_group_name*, which is the name of the record group to which the column is being added. You enclose the record group name in single quotation marks. If the record group does not exist, an error occurs.

- *column_name*, which is the name of the column being added. Record group column names must follow the Oracle naming standard, and each column name must be unique within a record group. If you are creating a column that will appear in an existing LOV display, the column name must be the same as the column in the original record group on which the LOV is based.

- *column_data_type_specification*, which specifies the type of data that the column will store. Legal values are CHAR_COLUMN, which specifies that the column contains VARCHAR2 data values; DATE_COLUMN, which specifies that the column contains DATE values; and NUMBER_COLUMN, which specifies that the column contains NUMBER values.

8

■ *column_width*, which is only required for the CHAR_COLUMN data type specification and specifies the maximum column width.

Now, you will modify the project consultants form so it uses a nonquery record group to display the future dates. You will save the form using a different filename, and then modify the code in the roll-on date LOV command button so it creates the nonquery record group and defines its columns. The record group will have only one column, which will be named DATE_VALUE and will contain DATE data values. You will use the same nonquery record group for all three date options (3 months, 6 months, and 12 months of dates). You will modify the IF/THEN/ELSIF control structure that determines which option button the user selected. Depending on the selection, the trigger will call a program unit that contains the code to populate the record group. For now, you will only implement a program unit named ADD_3_MONTHS, which inserts three months of future dates. You can implement the other options as an end-of-chapter case.

To modify the form to create a nonquery record group:

1. In Form Builder, save the 8CProjectConsultants_LOVb.fmb file as **8CProjectConsultants_LOVc.fmb** in your Chapter8\TutorialSolutions folder.

2. In the Layout Editor, right-click the **LOV command button** beside the Roll On Date text item, and then click **PL/SQL Editor** to open the editor and view the trigger code. Modify the code as shown in Figure 8-29. Note that the commands to call the program units to populate the record group for the 6 and 12 month selections are commented out, because you will not implement them now.

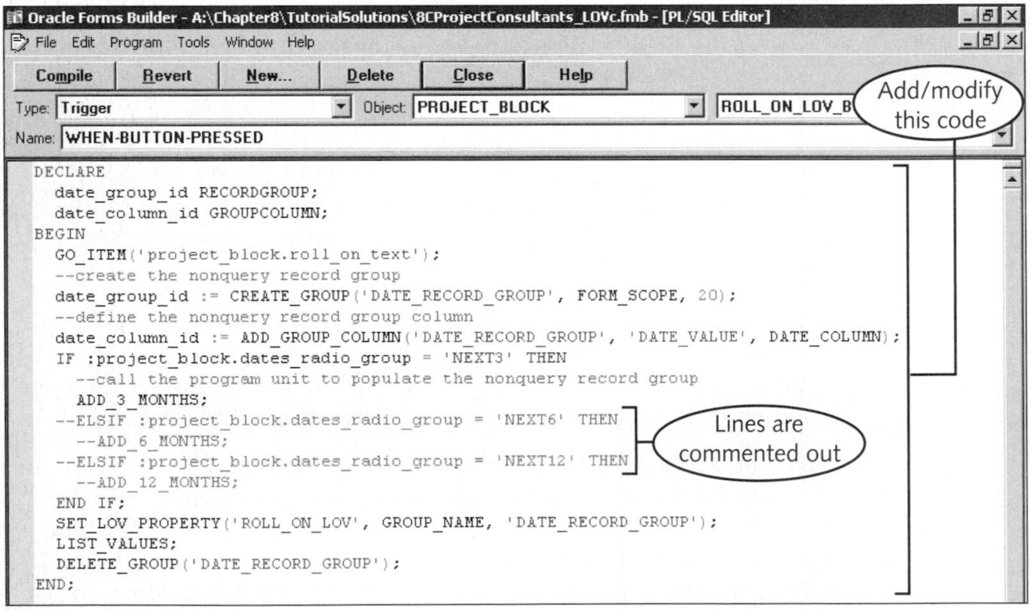

Figure 8-29   Code to create nonquery record group and define its column

3. DO NOT compile the trigger—it will not compile successfully until you create the ADD_3_MONTHS program unit. Close the PL/SQL Editor window, and then save the form.

Adding Blank Rows to a Nonquery Record Group

After you have created a nonquery record group and defined its columns, the next step is to add a new blank row. To do this, you use the ADD_GROUP_ROW built-in procedure, which has the following syntax:

```
ADD_GROUP_ROW ('record_group_name', row_number);
```

The ADD_GROUP_ROW procedure has the following parameters:

- *record_group_name,* which specifies the name of the record group to which you want to add the new row and is enclosed in single quotation marks.

- *row_number,* which specifies an integer that is the position in the record group where the new row is to be inserted. You can insert a new row anywhere in the record group, and all rows below the inserted row are automatically renumbered. To add a row to the end of a group, specify END_OF_GROUP as the *row_number* value.

Adding Data Values to a Nonquery Record Group

The final step in creating a nonquery record group is to add the data values to the new blank rows. To do this, you use the SET_GROUP_CHAR_CELL, SET_GROUP_DATE_CELL, and SET_GROUP_NUMBER_CELL built-in procedures. Each procedure specifies the data type of the value it adds. The syntax for these procedures is:

```
SET_GROUP_datatype_CELL ('record_group_name.column_name',
row_number, value);
```

The parameters for the procedures are:

- *datatype,* which, in the procedure name, indicates the type of data being inserted and can have the values CHAR, DATE, or NUMBER

- *record_group_name.column_name,* which specifies the name of the record group and the column name, separated by a period, in which the value will be inserted. This parameter is enclosed in single quotation marks. An example value would be 'DATE_RECORD_GROUP.DATE_VALUE'.

- *row_number,* which specifies the integer number of the record group row in which the value is being inserted.

- *value,* which specifies the value to be inserted. The value's data type must correspond with the procedure data type: SET_GROUP_CHAR_CELL requires a character value, SET_GROUP_DATE_CELL requires a date value, and SET_GROUP_NUMBER_CELL requires a number value.

Now, you will create a program unit named ADD_3_MONTHS that adds future date values for the next three months to the DATE_RECORD_GROUP that you created using the code in Figure 8-29. The code in the program unit declares the following variables: *counter*, which counts the number of dates inserted into the record group; *todays_date*, which retrieves the current system date; and *next_date*, which represents the current date being inserted into the record group and is calculated as the sum of *todays_date* and *counter*. In the record group, Row 1 contains the current date, Row 2 contains the current date plus 1, Row 3 contains the current date plus 2, and so forth. The trigger uses an EXIT WHEN loop that exits when the trigger inserts the date that is three months from the current system date.

To add the future date values to the record group:

1. In the Object Navigator, select the **Program Units node**, and then click the **Create button** to create a new program unit. The New Program Unit dialog box opens.

2. Type **ADD_3_MONTHS** for the program unit name, be sure that the Procedure option button is selected, and then click **OK**. The PL/SQL Editor opens for the new program unit.

3. Type the program unit code shown in Figure 8-30. Compile the code, correct any syntax errors, and then close the PL/SQL Editor, and save the form.

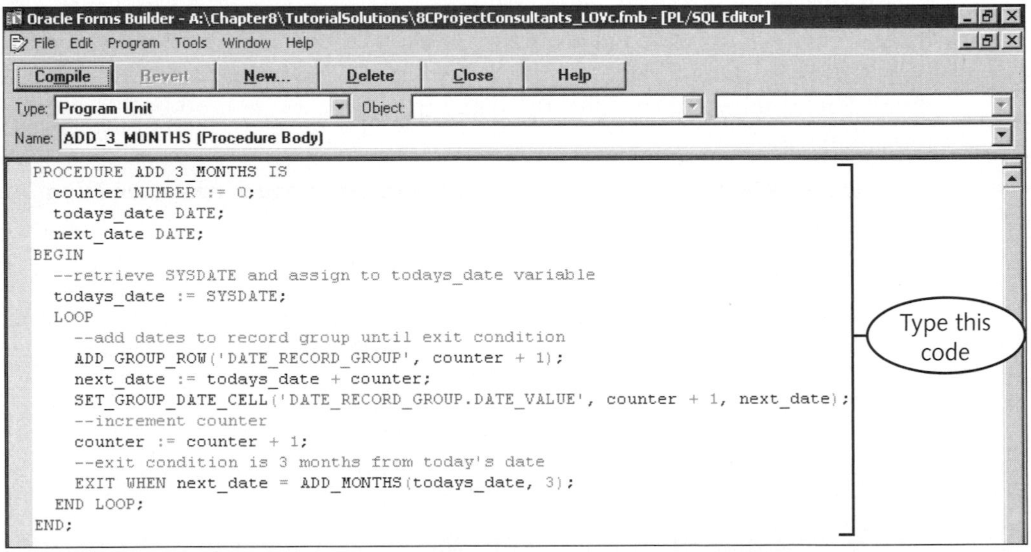

```
PROCEDURE ADD_3_MONTHS IS
  counter NUMBER := 0;
  todays_date DATE;
  next_date DATE;
BEGIN
  --retrieve SYSDATE and assign to todays_date variable
  todays_date := SYSDATE;
  LOOP
    --add dates to record group until exit condition
    ADD_GROUP_ROW('DATE_RECORD_GROUP', counter + 1);
    next_date := todays_date + counter;
    SET_GROUP_DATE_CELL('DATE_RECORD_GROUP.DATE_VALUE', counter + 1, next_date);
    --increment counter
    counter := counter + 1;
    --exit condition is 3 months from today's date
    EXIT WHEN next_date = ADD_MONTHS(todays_date, 3);
  END LOOP;
END;
```

Type this code

Figure 8-30   Code to add blank rows and values to nonquery record group

Now, you will run the form and confirm that the correct values appear in the LOV display.

To test the LOV display:

1. Run the form. The form appears with the first radio button (Next 3 Months) selected.

2. Click the **LOV command button** beside the Roll On Date text item. Scroll down the LOV display to confirm that dates appear for the next three months. Click **Cancel**. You cannot test the other option button LOV displays, because you have not yet implemented the program units that define the nonquery record groups for these options.

3. Click **Exit** to close Forms Runtime, and if necessary click **No** if you are asked if you want to save your changes. Then close the form in Form Builder.

CONTROLLING DATA BLOCK RELATIONSHIPS

In Chapter 6, you learned how to create data block forms with master-detail relationships using the Data Block Wizard. Recall that in a master-detail relationship, a master database record can have multiple related detail records that are defined through foreign key relationships. For example, a record in the CUSTOMER table in the Clearwater Traders database could have multiple related customer order records in the CUST_ORDER table. The CUST_ID field is the primary key in the CUSTOMER table and is a foreign key in the CUST_ORDER table. In a master-detail form, the master block and the detail block are **coordinated**, which means that the master record and the detail record are associated by a relationship. When a user performs a **coordination-causing event**, which is any operation that causes the current record in the master block to change, the detail block values also change. For example, when the user retrieves a specific record in the CUSTOMER master data block, the information about the selected customer's orders appears in the CUST_ORDER detail data block. If the user selects a different customer, the customer order information in the detail block changes to show the order information for the new customer.

In this section, you will become familiar with the form objects and triggers that define master-detail relationships, and you will learn how to modify these relationships. To do this, you will open a form named CustomerOrder.fmb stored in the Chapter8 folder on your Data Disk. This form contains a master block associated with the CUSTOMER table and a detail block associated with the CUST_ORDER table. You will save the file using a different filename and then run the form to review how a master-detail form works.

To open and run the master-detail form:

1. In Form Builder, open **CustomerOrder.fmb** in the Chapter8 folder on your Data Disk, and save the file as **8CCustomerOrder.fmb** in your Chapter8\TutorialSolutions folder.

2. Run the form. The form appears in the Forms Runtime window, with the Customer master block items in the Customers frame, and the Customer Orders detail block items in the Customer Orders frame.

3. Click the **Enter Query button** 🔢 to place the form in Enter Query mode, do not enter a search condition, and then click the **Execute Query button** 🔢 to execute the query. The data values for the first CUSTOMER record (Customer ID 107, Paula Harris) appear, as shown in Figure 8-31. Note that the detail frame shows the associated CUST_ORDER records for customer ID 107.

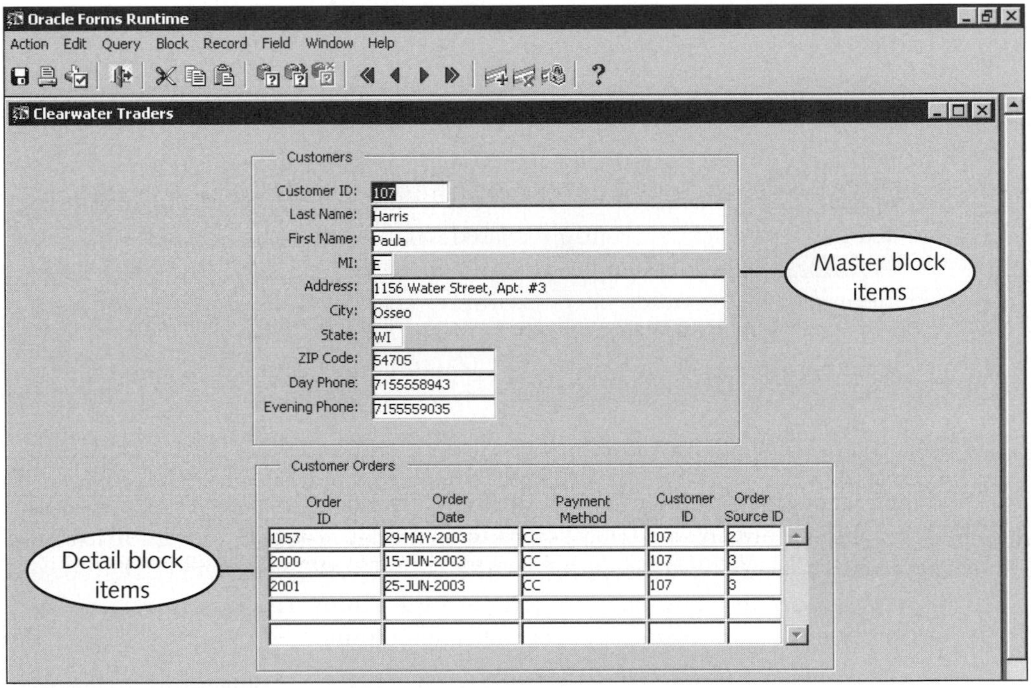

Figure 8-31   CUSTOMER-CUST_ORDER master-detail form

4. Click the **Next Record button** ▶. The data values for the next CUSTOMER record (Customer ID 232, Mitch Edwards) appear. Note that the detail block values now show Mitch's customer order record.

5. Close Forms Runtime.

The master–detail relationship in this form was created using the Data Block Wizard. Recall from Chapter 6 that when you create a master–detail relationship between two blocks using the Data Block Wizard, Form Builder automatically creates a relation object in the master block. Form Builder also automatically creates several triggers and PL/SQL procedures that enforce coordination between the form master and detail blocks. In the following sections, you will learn more about the relation object and about relation-handling triggers and program units. You will also learn about the system variables that control master-detail processing and how to use commands to modify properties of a master-detail relationship while a form is running.

Data Block Relation Properties

When you create a new master-detail relationship using the Data Block Wizard, Form Builder creates a relation object in the master block that is named using the following syntax: *<master_block_name>_<detail_block_name>*. In the master-detail form you just ran, the master block is named CUSTOMER, and the detail block is named CUST_ORDER, so the relation name is CUSTOMER_CUST_ORDER. The relation specifies properties about the relationship, such as the name of the detail block, the SQL join condition that specifies the relationship between the master block and the detail block, and how the form handles deletions of master block records. Now, you will open the Property Palette of the CUSTOMER_CUST_ORDER relation and examine its properties.

To examine the relation properties:

1. In the Object Navigator, click ➕ beside the Data Blocks node, click ➕ beside the CUSTOMER data block, and then click ➕ beside the Relations node. The CUSTOMER_CUST_ORDER relation appears.

2. Double-click the **Relations icon** ⬛ beside CUSTOMER_CUST_ORDER to open the relation Property Palette.

Relation properties include:

- **Relation Type**, which can be either **Join**, which specifies that the relationship is based on joining two key fields in the block, or **Ref**, which specifies that the relationship was created using a REF pointer. A REF pointer is an item in an Oracle8 object table that creates a relationship between two tables by specifying the location of the related field. In this relationship, the Relation type is Join.

- **Detail Data Block**, which specifies the name of the detail block in the relationship. In this relationship, the Detail Data Block is CUST_ORDER.

- **Join Condition**, which specifies the names of the master block item and detail block item on which the blocks are joined. The Join Condition property uses the syntax *detail_block_name.join_item_name = master_block_name.join_item_name*. In this relationship, the Join Condition is CUST_ORDER.CUST_ID = CUSTOMER.CUST_ID.

- **Delete Record Behavior**, which specifies how deleting a record in the master block affects records in the detail block. Possible values are **Non Isolated**, which prevents deleting a master record when associated detail records exist in the database; **Isolated**, which allows deleting a master record when associated detail records exist; and **Cascading**, which performs a **cascading delete**, in

8

which the master record is deleted and all of the associated detail records in the detail block's base table are also deleted. In this relationship, the current value is Non Isolated, which is the default value.

- **Prevent Masterless Operations**, which specifies whether users are allowed to query or insert records in a detail block when there is no master record in the master block. When this value is set to **Yes**, Forms Runtime displays an error message stating that the form cannot insert or query detail records without a parent (master) record present. When this value is set to **No**, detail blocks can be used independently of master blocks when no master record is selected. The default value is No, which is also the current value for this relation.

- **Deferred** and **Automatic Query**, which work together to determine whether the detail block records automatically change when the user selects a new master record, or if the user has to explicitly navigate to the detail block and issue the EXECUTE_QUERY command to refresh the detail records. When Deferred is set to **No**, which is the default value, Form Builder fetches the detail records immediately, regardless of the value of Automatic Query. When Deferred is set to **Yes** and Automatic Query is set to **Yes**, Form Builder defers fetching the detail records until the user navigates to the detail block. At that point, Form Builder automatically fetches the new records. When Deferred is set to **Yes** and Automatic Query is set to **No**, Form Builder does not automatically fetch the detail records, and the user must navigate to the detail block and explicitly execute a query using the EXECUTE_QUERY built-in.

Now, you will run the form again and examine how the default Delete Record Behavior and Prevent Masterless Operations relation properties behave. You will attempt to delete a master record that has associated detail records. You will also try to query the detail block when no master record exists.

To examine the default relation properties:

1. Close the Property Palette, and run the form. Click the **Enter Query button** 🔳 to place the form in Enter Query mode, do not enter a search condition, and then click the **Execute Query button** 🔳 to execute the query. The data values for the first CUSTOMER record (Customer ID 107, Paula Harris) appear.

2. Make sure that the insertion point is in the Customer ID text item, and click the **Remove Record button** 🔳 to delete the master record. The message "Cannot delete master record when matching detail records exist" appears briefly.

3. Click the **Save button** 🔳 to try to commit the change. The message "FRM-40401: No changes to save" appears, indicating that the record was not deleted.

4. Click **Block** on the menu bar, and then click **Clear** to clear the master block.

5. To perform a masterless operation, place the insertion point in the Order ID text item, which is in the detail block. Click the **Enter Query button** 🔲 to place the form in Enter Query mode, do not enter a search condition, and then click the **Execute Query button** 🔲 to execute the query. All of the records in the CUST_ORDER form appear, indicating that you successfully performed a masterless operation.

6. Close Forms Runtime.

Under most circumstances, you should accept the default relationship properties. In almost all situations, you should leave the Delete Record Behavior property set to Non Isolated. It is dangerous to allow users to delete master records or perform cascade deletes, because these settings can allow users to delete records that they do not intend to delete. One possible reason to set the Prevent Masterless Operations property to Yes is to keep users from trying to insert detail records when no corresponding master record exists. For example, if a user tries to insert a new customer order record for a customer ID that does not exist in the database, an error will occur. One possible reason for deferring retrieval of detail records or forcing the user to explicitly issue the EXECUTE_QUERY command is to improve form performance if the detail records constitute a large data set.

Relation Triggers and Program Units

When you create a master–detail relation, Form Builder automatically generates several triggers and program units to manage the relation. Table 8-12 summarizes the relation-handling triggers.

Trigger Name	Scope	Purpose	Delete Record Behavior Value(s)
ON-CLEAR-DETAILS	Form	Clears detail block records	Non Isolated, Isolated, Cascading
ON-POPULATE-DETAILS	Master Block	Coordinates values in master and detail blocks	Non Isolated, Isolated, Cascading
ON-CHECK-DELETE-MASTER	Master Block	Prohibits deleting master record when detail records exist	Non Isolated
PRE-DELETE	Master Block	Executes cascading deletes	Cascading

Table 8-12 Relation-handling triggers

The ON-CLEAR-DETAILS and ON-POPULATE-DETAILS triggers are always present in a form with a master–detail relationship. Which specific trigger is created for controlling the deletion of master records depends on the value of the relation's Delete Record Behavior property. When the Delete Record Behavior property value is set to Non Isolated, an ON-CHECK-DELETE-MASTER trigger is created, which fires when the user attempts to delete the master record. The trigger checks for the existence of detail records, and if they exist, the deletion is aborted. When the Delete Record Behavior property is set to Cascading, the PRE-DELETE trigger is created, which executes the cascading delete. When the Delete Record Behavior property is set to Isolated, neither of these triggers is present, and master records are deleted without trigger intervention.

Table 8-13 summarizes the relation-handling program units. These program units are called by the relation-handling triggers or by one another and handle the details of the master-detail processing.

Program Unit Name	Purpose	Called By
CLEAR_ALL_MASTER_DETAILS	Clears detail records	ON-CLEAR-DETAILS trigger
QUERY_MASTER_DETAILS	Fetches detail records	ON-POPULATE-DETAILS trigger
CHECK_PACKAGE_FAILURE	Determines if trigger successfully executes and displays corresponding error message if it does not	ON-POPULATE-DETAILS trigger, CLEAR_ALL_MASTER_DETAILS and QUERY_MASTER_DETAILS program units

Table 8-13 Relation-handling program units

Now, you will examine the relation-handling triggers in the customer order form. Then, you will change the Delete Record Behavior property and observe how the triggers change.

To examine the form's relation-handling triggers:

1. In the Object Navigator, click **+** beside the Triggers node under CUSTOMER_ORDER_FORM. The form-level triggers appear, including the ON-CLEAR-DETAILS relation-handling trigger, as shown in Figure 8-32.

2. Click **+** beside the Data Blocks node, and click **+** beside CUSTOMER, which is the master block. Then click **+** beside the Triggers node to examine the master block's triggers. The ON-POPULATE-DETAILS and ON-CHECK-DELETE-MASTER triggers appear. Recall that the relation's Delete Record Behavior property is set to Non Isolated, which means that the ON-CHECK-DELETE-MASTER trigger fires when the user attempts to delete the master record.

3. Click **+** beside Program Units to display the relation-handling program units. The three relation-handling program units appear.

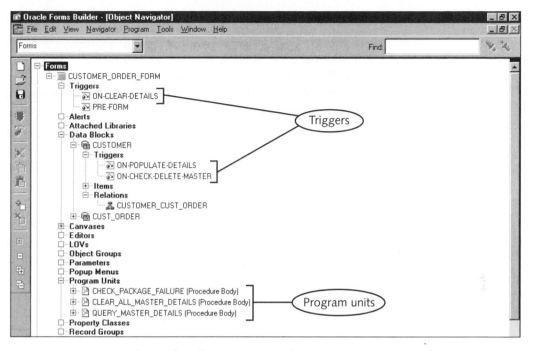

Figure 8-32   Form relation-handling triggers and program units

Now, you will change the relation's Delete Record Behavior and observe how the form's relation-handling triggers change. First, you will change the Delete Record Behavior value to Isolated. This means that the user can freely delete master records. When you do this, the CHECK-DELETE-MASTER trigger will be removed from the form. Then, you will change the relation's Delete Record Behavior property to Cascading. When you do this, Form Builder creates a PRE-DELETE trigger for the form.

To change the Delete Record Behavior property and observe how the triggers change:

1. In the Object Navigator, double-click the **Relation icon** ⚏ beside CUSTOMER_CUST_ORDER to open the relation Property Palette, change the Delete Record Behavior property to **Isolated**, and then close the Property Palette.

2. If necessary, click ➕ beside the Triggers node under the CUSTOMER master block, and observe that the ON-CHECK-DELETE-MASTER trigger no longer appears.

3. Double-click ⚏ beside CUSTOMER_CUST_ORDER to open the relation Property Palette again, change the Delete Record Behavior property to **Cascading**, and then close the Property Palette.

4. Click ➕ beside the Triggers node under the CUSTOMER master block, and observe that a PRE-DELETE trigger has been created, which handles cascading deletes.

5. Double-click ⬚ beside CUSTOMER_CUST_ORDER to open the relation Property Palette again, change the Delete Record Behavior property back to **Non Isolated**, and then close the Property Palette.

6. Click ➕ beside the Triggers node under the CUSTOMER master block, and observe that the PRE-DELETE trigger has been deleted and the ON-CHECK-DELETE-MASTER trigger has been created again.

Many applications require complex master-detail relations that involve more than two blocks. To create such relations, you create multiple individual relations. When you add a new master-detail relationship to a form that already has an existing master-detail relationship, the new relationship uses the existing triggers and program units and does not create new ones, as long as the correct trigger is present to handle the new relationship's Delete Record Behavior property.

System Variables Involved in Relation Processing

The relation-handling triggers and program units use three system variables to track the current status of master-detail processing. These variables are:

- :SYSTEM.BLOCK_STATUS, which reports the status of the records in the current data block. The status value can be **NEW**, which means that the block contains new records that must be committed or rolled back; **CHANGED**, which means that at least one of the record values has been changed and must be either committed or rolled back; or **QUERY**, which means that the block contains records that have been retrieved from the database, but not yet changed. The :SYSTEM.BLOCK_STATUS value is used in the CLEAR_ALL_MASTER_DETAILS program unit to determine if the user should be prompted to save modified records.

- :SYSTEM.COORDINATION_OPERATION, which identifies which coordination event fired the ON-CLEAR-DETAILS trigger. This variable is used in the CLEAR_ALL_MASTER_DETAILS program unit and helps coordinate master-detail processing.

- :SYSTEM.MASTER_BLOCK, which returns the name of the current master block. This variable is used in the CLEAR_ALL_MASTER_DETAILS program unit and helps coordinate master-detail processing.

These three variables are used in the relation-handling triggers and program units, and you will probably not directly use them in the code that you write.

Changing Relation Properties Dynamically

You can allow the user to specify the retrieval properties of a relation while a form is running. For example, to improve form performance, a user might prefer not to automatically retrieve detail records every time the master record changes. Recall that you can configure a relation's Deferred and Automatic Query properties to determine whether the detail block records automatically change when the user selects a new master record. The default value of the Deferred property is No, so Form Builder fetches detail records immediately. You can use program commands to change the values of the Deferred and Automatic Query properties so the user can control the detail block's fetch behavior. To do this, you use the SET_RELATION_PROPERTY built-in, which has the following syntax:

```
SET_RELATION_PROPERTY('relation_name', property, value);
```

This command uses the following parameters:

- *relation_name*, which references the name of the relation in which the property value is to be changed. The relation name is enclosed in single quotation marks.

- *property*, which references the name of the property to be changed. This value is DEFERRED_COORDINATION for the Deferred property and AUTOQUERY for the Autoquery property. To find the names of other relation properties used in this command, search in the Form Builder online Help system for the SET_RELATION_PROPERTY built-in.

- *value,* which is the new value for the property. For the Deferred and Autoquery properties, this value can be either PROPERTY_TRUE, which corresponds to Yes in the Property Palette, or PROPERTY_FALSE, which corresponds to No in the Property Palette. To find the values of other relation properties used in this command, search in the Form Builder online Help system for the SET_RELATION_PROPERTY built-in.

Now, you will modify the customer order form so that it has two radio buttons, as shown in Figure 8-33. These radio buttons will allow the user to control the master-detail relationship retrieval behavior.

8

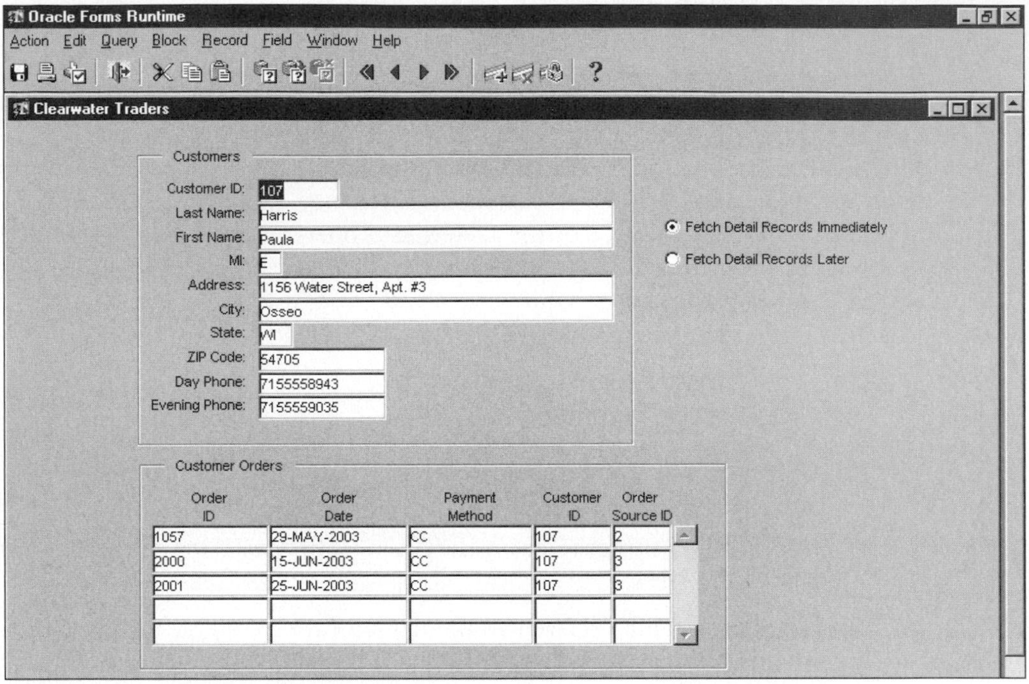

Figure 8-33   Form with radio buttons to allow user to control retrieval actions

When the user selects the Fetch Detail Records Immediately option button, the detail records update whenever the master record changes. When the user selects the Fetch Detail Records Later option button, the detail records do not update until the user navigates to the detail block.

First, you will create a control block that will contain the radio buttons. Then, you will create the radio buttons and associated radio group, and configure their properties.

To create the control block and radio buttons:

1. In the Object Navigator, select the **Data Blocks node**, and then click the **Create button** 🔲 to create a new block. Select the **Build a new data block manually option button**, and then click **OK**.

2. Double-click the **Data Block icon** 🔳 beside the new data block to open its Property Palette, and change the Name property value to **QUERY_CONTROL_BLOCK** and the Database Data Block property value to **No.** Then close the Property Palette.

3. Open the Layout Editor, and confirm that QUERY_CONTROL_BLOCK is selected in the Block list. Select the **Radio button tool** 🔘 on the tool palette, and draw the Fetch Detail Records Immediately radio button on the canvas, as shown in Figure 8-33.

4. Double-click the radio button to open its Property Palette, change its properties as follows, and then close the Property Palette:

Property   **Value**
Name   **IMMEDIATE_RADIO_BUTTON**
Label   **Fetch Detail Records Immediately**
Radio Button Value   **IMMEDIATE**

5. If necessary, make the radio button larger so its label is completely displayed, and change its fill color to the same color as the canvas.

6. Select the new radio button, click the **Copy button** 📋 on the toolbar, and then click the **Paste button** 📋. The Radio Groups dialog box opens. Accept the default radio group name, and then click **OK**. The pasted radio button appears on top of the existing button.

7. Drag the new radio button so it is below the existing button, as shown in Figure 8-33.

8. Double-click the new radio button to open its Property Palette, change its properties as follows, and then close the Property Palette.

Property   **Value**
Name   **LATER_RADIO_BUTTON**
Label   **Fetch Detail Records Later**
Radio Button Value   **LATER**

9. Open the Object Navigator, select the new radio group under the QUERY_CONTROL_BLOCK right-click, and then click **Property Palette**. The radio group Property Palette opens. Change the Name property value to **QUERY_RADIO_GROUP**, change the Initial Value property to **IMMEDIATE**, close the Property Palette, and save the form.

Now, you need to create the trigger that changes the CUSTOMER_CUST_ORDER relation's Deferred and Automatic Query properties based on the selected radio button. Recall that when the relation's Deferred property is set to No, Form Builder fetches the detail records immediately. When the relation's Deferred property is set to Yes and the Automatic Query property is set to Yes, Forms Runtime waits to fetch the detail records until the user navigates to the detail block. You will create a trigger associated with the radio group's WHEN-RADIO-CHANGED event that fires whenever the user selects a different radio button. This trigger will evaluate the current value of the radio group. If the value of the radio group is IMMEDIATE, then the first radio button is selected, and the relation's Deferred property will be set to No using the SET_RELATION_PROPERTY built-in. If the value of the radio group is LATER, then the second radio button is selected, and the relation's Deferred property and Automatic Query property will be set to Yes. After the relation property is set, the trigger will execute the GO_BLOCK built-in to place the insertion point back in the CUSTOMER block so the user can execute the next query.

To create the trigger to dynamically change the relation properties:

1. In the Object Navigator, select the **Triggers node** under the QUERY_RADIO_GROUP, and then click the **Create button** to create a new trigger. The CUSTOMER_ORDER_FORM: Triggers dialog box opens.

2. Type **WHEN-RADIO-CHANGED** in the Find text box, press **Enter**, and then click **OK**. The PL/SQL Editor opens.

3. Type the code shown in Figure 8-34, compile the trigger, and correct any syntax errors.

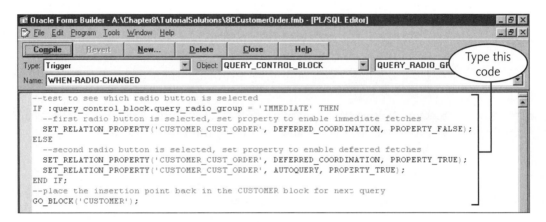

Figure 8-34   Code to dynamically control the relation retrieval properties

4. Close the PL/SQL Editor, and then save the form.

Now, you will run the form and confirm that you can change the relation retrieval properties using the radio buttons. You will retrieve all of the form records using the default retrieval action, which populates the detail block as soon as the master block value changes. Then, you will select the option button that delays detail block fetches until you navigate to the detail block, and confirm that the fetch is delayed when you select a new master record.

To test the trigger to change the relation properties:

1. Run the form, place the insertion point in the Customer ID text item in the Customers frame, click the **Enter Query button** to place the form in Enter Query mode, do not enter a search condition, and then click the **Execute Query button** to execute the query. The data values for the first CUSTOMER record (Customer ID 107, Paula Harris) appear, along with her associated order information.

2. Click the **Next Record button** ▶ to retrieve the next master record. The values for Customer ID 232 (Mitch Edwards) appear, along with his associated order information.

3. Select the **Fetch Detail Records Later option button** to delay retrieving the detail records, and then click ▶ to retrieve the next master record. The data values for customer 133 (Maria Garcia) appear, but note that her order records do not appear.

4. Place the insertion point in the Order ID text item in the Customer Orders frame. After a moment, the detail records appear for the current master record.

5. Place the insertion point in the Customer ID text item, and then click ▶ to retrieve the next master record. Again, the order records do not appear. Place the insertion point in the Order ID text item to retrieve the order records.

6. Click the **Fetch Detail Records Immediately option button** to switch back to fetching the detail records immediately.

7. Place the insertion point in the Customer ID text item, and then click ▶ again to retrieve the next master record. The records for Customer ID 179 (Alissa Chang) display, and her order values appear immediately.

8. Close Forms Runtime, and click **No** if you are asked if you want to save your changes. Close the form in Form Builder, and close Form Builder and all other open Oracle applications.

8

Summary

❑ A mouse event is a mouse pointer action that has an associated trigger.

❑ Mouse trigger types include mouse click triggers, which fire when the user presses a mouse button, and mouse move triggers, which fire when the user moves the mouse pointer across the screen display.

❑ Whenever a mouse event occurs, Form Builder updates the value of one or more mouse-related system variables.

❑ You can use mouse system variables in conjunction with mouse triggers to customize a form's functionality and appearance.

❑ You can use the SET_APPLICATION_PROPERTY built-in procedure to customize the appearance of the mouse pointer in applications. The pointer options include DEFAULT ⬉, BUSY ⧗, CROSSHAIR +, HELP ⬉?, and INSERTION ⊺.

❑ A record group is a form object that represents data in a tabular format. You can use record groups to define the data that appears in an LOV or to pass records from one form application to another.

❑ There are three different types of record groups: static, query, and nonquery.

❑ You specify the contents of a static record group when you create the record group, and the values cannot be changed while the form is running.

❑ The contents of a query record group are based on a SQL SELECT command that executes while the form is running. You can create a query record group manually at design time or programmatically using commands that execute while the form is running.

❑ You create a nonquery record group programmatically using program commands that execute while the form is running and specify the individual column values for each record group row.

❑ You can use the SET_LOV_PROPERTY built-in to dynamically change the record group associated with an LOV.

❑ You use the CREATE_GROUP_FROM_QUERY built-in to programmatically create a query record group while a form is running.

❑ You can use the POPULATE_GROUP built-in to programmatically populate a query record group that is created using the CREATE_GROUP_FROM_QUERY built-in.

❑ You use the DELETE_GROUP built-in to programmatically delete a record group.

❑ To create a nonquery record group, you must create the record group, define the record group columns, add new blank rows to the record group, and specify the values of each field in each row.

❑ You use the CREATE_GROUP built-in to create a nonquery record group and the ADD_GROUP_COLUMN built-in to add individual columns to the record group.

❑ You use the ADD_GROUP_ROW built-in to add new blank rows to a nonquery record group.

❑ You use the SET_GROUP_CHAR_CELL, SET_GROUP_DATE_CELL, and SET_GROUP_NUMBER_CELL built-ins to add new character, date, or number values to fields in a nonquery record group. You must use the built-in with the name that corresponds to the data type of the value you are adding.

❑ In a master–detail form, the master block and the detail block are coordinated, which means that the master record and the detail record are associated by a relationship.

❑ A coordination-causing event is any operation that changes the current record in the master block, which then causes the detail block to change.

❑ When you create a master–detail relationship between two blocks using the Data Block Wizard, Form Builder automatically creates a relation object in the master block and several triggers and PL/SQL program units that enforce coordination between the master and detail blocks.

❑ A relation object specifies properties about the relationship, such as the name of the detail block, the SQL join condition that specifies the relationship between the master block and the detail block, and how deletions of master block records are handled.

❐ A relation's Delete Record Behavior property specifies how deleting a record in the master block affects records in the detail block. Non Isolated deletions prevent deleting a master record when associated detail records exist in the database. Isolated deletions allow deleting a master record when associated detail records exist. Cascading deletions perform a cascading delete, in which all detail records are deleted along with the master record.

❐ A relation's Prevent Masterless Operations property specifies whether or not users can query or insert records in a detail block when there is no master record in the master block.

❐ A relation's Deferred and Automatic Query properties work together to determine whether the detail block records automatically change when the user selects a new master record.

❐ The ON-CLEAR-DETAILS and ON-POPULATE-DETAILS triggers are always present in a form with a master-detail relationship.

❐ When the Delete Record Behavior property value is set to Non Isolated, an ON-CHECK-DELETE-MASTER trigger is created, which fires when the user attempts to delete the master record.

❐ When the Delete Record Behavior property is set to Cascading, a PRE-DELETE trigger is created, which executes a cascading delete.

❐ The relation-handling triggers and program units use the :SYSTEM.BLOCK_STATUS, :SYSTEM.COORDINATION_OPERATION, and :SYSTEM.MASTER_BLOCK system variables to track the current status of master-detail processing.

❐ You can use the SET_RELATION_PROPERTY built-in to dynamically change the properties of a relation while a form is running.

REVIEW QUESTIONS

1. What is a mouse event?
2. List and describe the two different types of mouse triggers.
3. What are the differences in the behavior of a form-level WHEN–MOUSE–DOUBLECLICK trigger and a block-level WHEN–MOUSE–DOUBLECLICK trigger?
4. Define the three different types of record groups.
5. When should you use a static record group to populate an LOV?
6. Describe the two ways to create a query record group.
7. When do you need to delete a record group using the DELETE_GROUP built-in?
8. When do you need to create a nonquery record group?

9. How do you specify a relation that handles detail block updates by requiring the user to navigate to the detail form and then explicitly execute an EXECUTE_QUERY statement?

10. When does a form have an ON-CHECK-DELETE-MASTER trigger? When does it have a PRE-DELETE trigger?

PROBLEM-SOLVING CASES

The cases reference the Clearwater Traders (see Figure 1-11), Northwoods University (see Figure 1-12), and Software Experts (see Figure 1-13) sample databases. The data files needed for all cases are stored in the Chapter8 folder on your Data Disk. Store all solutions in your Chapter8\CaseSolutions folder.

1. In the tutorial, you created a form based on the PROJECT table in the Software Experts database that creates a nonquery record group to display future project roll-on date values for 3 months, 6months, and 12 months. The incomplete form code is saved in a file named 8CCase1.fmb. Implement the program units to create the 6 month and 12 month date record groups. Then modify the trigger code for the LOV command button associated with the Roll On Date text item (see Figure 8-29) so it calls the new program units. Save the completed form as 8CCase1_DONE.fmb.

2. Figure 8-35 shows a data block form stored in a file named 8CCase2.fmb that is associated with the LOCATION table in the Northwoods University database.

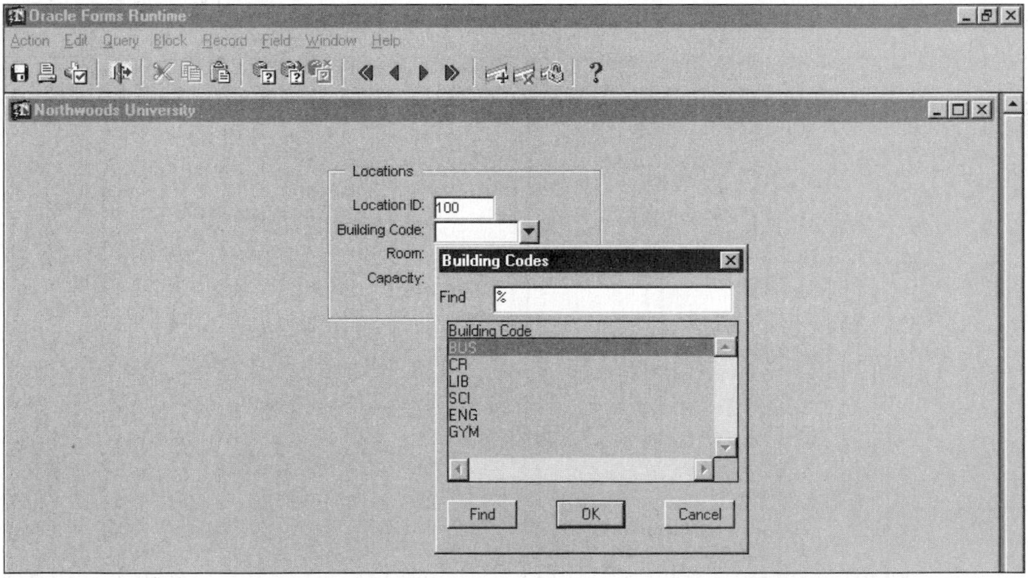

Figure 8-35

Create a static record group named BLDG_CODE_RECORD_GROUP that provides values for the BLDG_CODE field, and populate the record group using the following values: BUS, CR, LIB, SCI, ENG, GYM. Then, create an LOV named BLDG_CODE_LOV that is based on the static record group, and associate the LOV with the Building Code text item on the form. Add the code to display the LOV to the LOV command button trigger associated with the Building Code text item. Save the completed form as 8CCase2_DONE.fmb.

3. Figure 8-36 shows a data block form stored in a file named 8CCase3.fmb that is based on the INVENTORY table in the Clearwater Traders database. In this case, you will implement the check box shown on the form. When the user clears the check box and then opens the LOV display, the LOV associated with the Color text item displays values from the COLOR lookup table. When the user checks the check box and then opens the LOV display, the LOV retrieves the existing COLOR values from the INVENTORY table. Also modify the form so that when the user clicks the right mouse button when the insertion point is in the Color text item, the LOV associated with the Color text item appears. Save the completed form as 8CCase3_DONE.

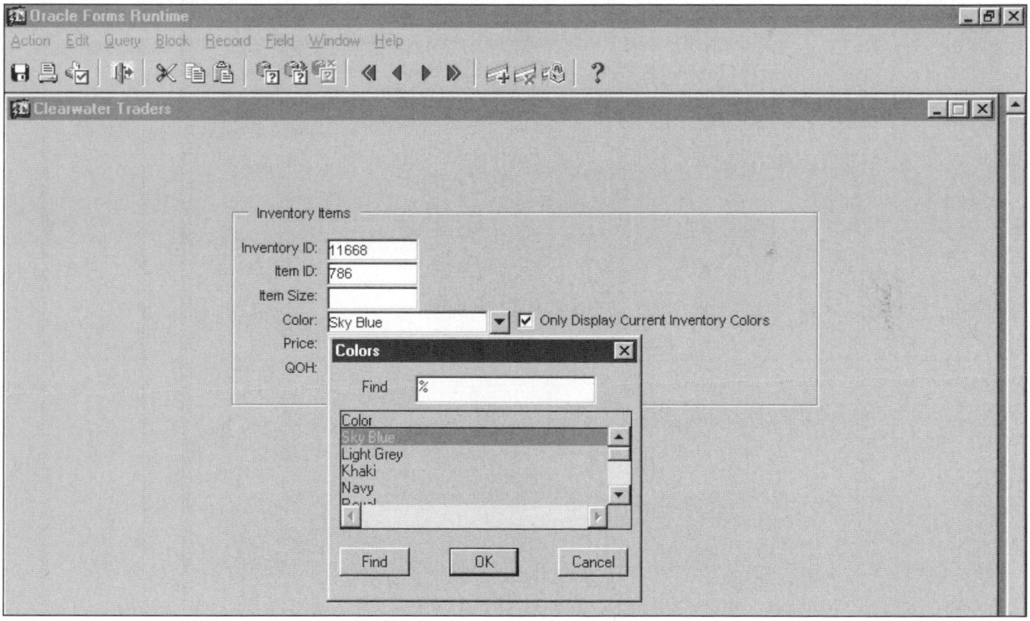

Figure 8-36

a. Create the check box as shown on the form.

b. Create a query record group object named INVENTORY_COLORS_RECORD_GROUP that is based on a SQL query that retrieves all COLOR values from the INVENTORY table. Suppress duplicate color values.

c. Create a second query record group object named COLOR_RECORD_GROUP based on a SQL query that retrieves all COLOR values from the COLOR table.

d. Create an LOV based on either record group that displays the COLOR values. Associate the LOV with the Color text item, and create an LOV command button to open the LOV display.

e. Add the commands to the LOV command button trigger to dynamically change the record group associated with the LOV, based on whether the check box is checked or cleared.

f. Create a WHEN-MOUSE-CLICK trigger so that when the user right-clicks the mouse when the insertion point is in the Color text item, the LOV associated with the COLOR text item appears. Be sure to configure the LOV display based on whether the form check box is checked or cleared.

4. Figure 8-37 shows a data block form based on the COURSE_SECTION table in the Northwoods University database. This form is stored in a file named 8CCase4.fmb. In this case, you will create an LOV that retrieves values from the TERM table based on the radio button that the user selects. If the user selects the Show Open Terms radio button, only TERM records for which the STATUS value is 'OPEN' appear in the LOV display. If the user selects the Show All Terms radio button, all TERM records appear in the LOV display.

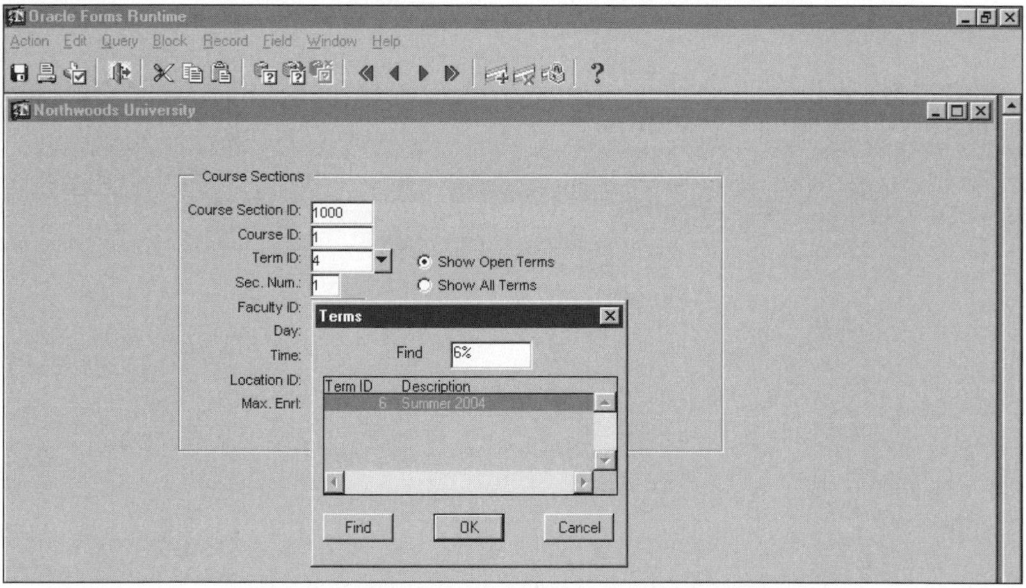

Figure 8-37

a. Create the radio buttons shown on the form.

b. Use the LOV Wizard to create a new LOV named TERM_LOV that retrieves the TERM_ID and TERM_DESC columns from the TERM table. For now, retrieve all of the TERM table records.

c. Modify the trigger of the LOV command button beside the Term ID text item so it creates and populates a query record group named TERM_RECORD_GROUP. Specify the query text so that the correct records are retrieved and appear in the TERM_LOV based on the selected radio button.

5. A data block form associated with the EVALUATION table of the Software Experts database is stored in a file named 8CCase5.fmb. In this case, you will create an LOV associated with the SCORE field. The LOV will display possible score values ranging from 0 to 100, incrementing by 5. You will create a nonquery record group to specify the LOV data values. The nonquery record group will display values of 0, 5, 10, 15, and so forth, to a maximum value of 100. Save the completed file as 8CCase5_DONE.fmb.

a. Create an LOV named SCORE_LOV that is associated with the Score text item on the form and displays the current values in the EVALUATION table's SCORE field for now. (You will change the LOV's source record group later.)

b. Add commands to the form's PRE-FORM trigger to create a nonquery record group named SCORE_RECORD_GROUP. Create a NUMBER column in the record group, and populate the record group using a loop that inserts score values beginning with 0, ending with 100, and incrementing by 5. The score values will be 0, 5, 10, 15,..., 95, 100. Include the command to associate the nonquery record group with the SCORE_LOV. Be sure to name the record group column SCORE, so the existing LOV will display the new record group values.

c. Create an LOV command button beside the Score text item on the form. Add commands to the button's trigger so it moves the insertion point into the Score text item, and displays the SCORE_LOV.

6. A master-detail form that displays records from the CLIENT table in the Software Experts database in the master block and associated PROJECT records in the detail block, is stored in a file named 8CCase6.fmb. Modify the master-detail relationship so that the form does not allow masterless operations. Then, change the pointer properties so the BUSY mouse pointer appears when the user places the mouse pointer anywhere in the detail block when no master record is currently displayed. Save the modified form as 8CCase6_DONE.fmb.

7. Figure 8-38 shows a master-detail form that displays FACULTY records from the Northwoods University database in the master block and related STUDENT records in the detail block. In this case, you will add the check box that lets the user specify how detail records are fetched. Save the modified form as 8CCase7_DONE.fmb.

a. Create the check box as shown, and configure the check box so it is checked when the form first opens. (*Hint:* You will need to place the check box in a control block.)

b. Configure the form so that when it first opens, when the user retrieves master block records, the form immediately fetches detail records.

c. Create a trigger that executes when the user checks or clears the check box. Add code to set the form relation properties according to the user preference regarding how the form fetches detail records. If the check box is checked, fetch detail records immediately. If the check box is cleared, do not fetch detail records until the user places the form insertion point in the detail block.

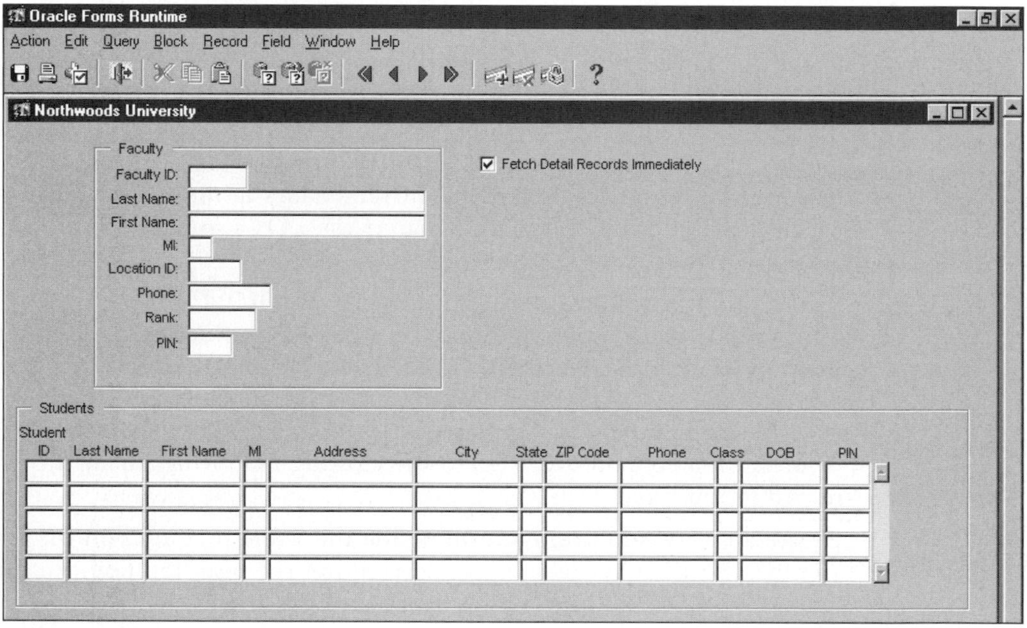

Figure 8-38

USING REPORT BUILDER

◀ LESSON A ▶

You have learned how to create forms that allow users to insert, update, delete, and view data in a variety of ways. In this chapter, you will learn how to use Report Builder, which is the Oracle Developer utility for creating reports. A **report** is a summary view of database data that can easily be printed on paper or sent to a file for electronic distribution. Reports can retrieve database data using SQL queries, perform mathematical or summary calculations on the retrieved data, and format the output to look like invoices, form letters, or other business documents. Reports are usually not interactive like forms: The report developer specifies the report contents at design time. Users can customize report contents in limited ways at runtime by entering search condition parameters directly into the report or by using a form to select search conditions and then running the report from the form.

INTRODUCTION TO REPORT BUILDER

A report is a static view of database data at a specific point in time that users can view or print. In Report Builder, you can format report data using a tabular display or make reports look like business documents, such as form letters or invoices. Reports can also display data to show database master-detail relationships. Report Builder supports the following specific report layout styles:

- **Tabular**, which presents data in a table format with columns and rows. An example of a tabular report appears in Figure 9-1, which shows a partial listing of the inventory ID, item description, size, color, price, and quantity on hand for the inventory items in the Clearwater Traders database.

Clearwater Traders Inventory

Report run on: March 3, 2003 9:35 AM

Inv. ID	Description	Item Size	Color	Price	QOH
11668	3-Season Tent		Sky Blue	259.99	16
11669	3-Season Tent		Light Grey	259.99	12
11775	Women's Hiking Shorts	S	Khaki	29.95	150
11776	Women's Hiking Shorts	M	Khaki	29.95	147
11777	Women's Hiking Shorts	L	Khaki	29.95	0
11778	Women's Hiking Shorts	S	Navy	29.95	139
11779	Women's Hiking Shorts	M	Navy	29.95	137
11780	Women's Hiking Shorts	L	Navy	29.95	115
11795	Women's Fleece Pullover	S	Eggplant	59.95	135
11796	Women's Fleece Pullover	M	Eggplant	59.95	168
11797	Women's Fleece Pullover	L	Eggplant	59.95	187
11798	Women's Fleece Pullover	S	Royal	59.95	0
11799	Women's Fleece Pullover	M	Royal	59.95	124
11800	Women's Fleece Pullover	L	Royal	59.95	112

Figure 9-1   Tabular report

- **Form-like**, which resembles a Form Builder form-style layout. This report style displays one record per page and shows data values to the right of field labels.

- **Mailing label**, which prints mailing labels in multiple columns on each page.

- **Form letter**, which includes the recipient's name and address (which are stored in the database), as well as other database values embedded in the text of a letter.

- **Group left** and **group above**, which display master-detail relationships. In a master-detail relationship, one master record might have several associated detail records through a foreign key relationship. The relationship between the ITEM and INVENTORY tables in the Clearwater Traders database is an example of a master-detail relationship: Each ITEM record can have many associated INVENTORY records. An example of a group left report for this relationship is shown in Figure 9-2.

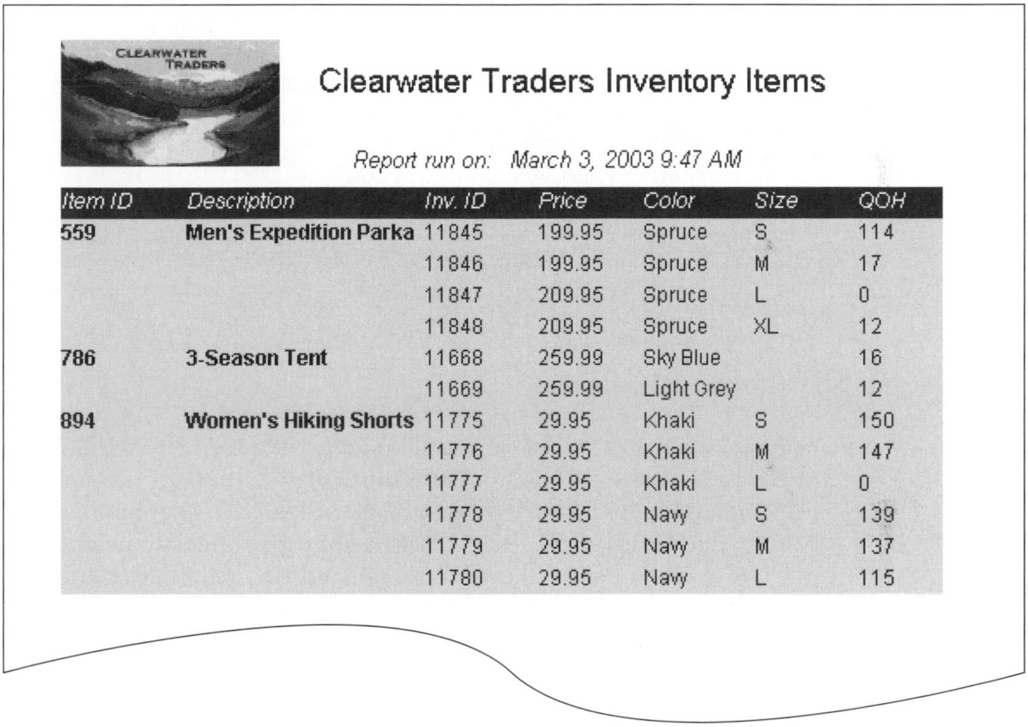

Clearwater Traders Inventory Items

Report run on: March 3, 2003 9:47 AM

Item ID	Description	Inv. ID	Price	Color	Size	QOH
559	Men's Expedition Parka	11845	199.95	Spruce	S	114
		11846	199.95	Spruce	M	17
		11847	209.95	Spruce	L	0
		11848	209.95	Spruce	XL	12
786	3-Season Tent	11668	259.99	Sky Blue		16
		11669	259.99	Light Grey		12
894	Women's Hiking Shorts	11775	29.95	Khaki	S	150
		11776	29.95	Khaki	M	147
		11777	29.95	Khaki	L	0
		11778	29.95	Navy	S	139
		11779	29.95	Navy	M	137
		11780	29.95	Navy	L	115

Figure 9-2 Group left report

Each item ID (the master item) and item description appears on the left side of the report, and a detailed listing of the corresponding INVENTORY records appears to the right of the master item. The same data is shown in a group above report in Figure 9-3, in which each master item appears above the detail lines.

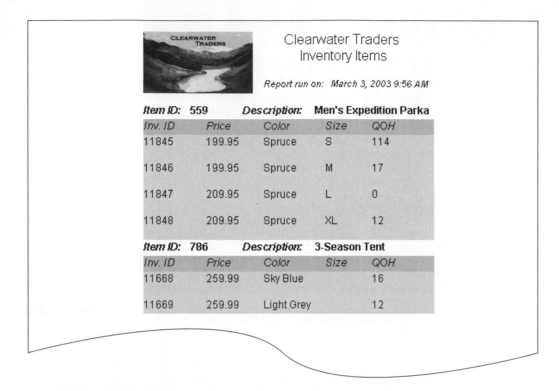

Figure 9-3   Group above report

- **Matrix**, which displays field headings across the top and down the left side of the page. A matrix layout displays data at the intersection point of two data values. Figure 9-4 shows an example of a matrix report in which Northwoods University student names appear in the row headings, course call IDs appear in the column headings, and the specific student's grade for the associated course appears at the intersection.

- **Matrix with group**, which displays a detail matrix for a master-detail relationship. Figure 9-5 shows an example of this kind of report, in which the master record is the student name, and a detail matrix shows course call IDs in the rows, term descriptions in the columns, and the student's grade at the intersection.

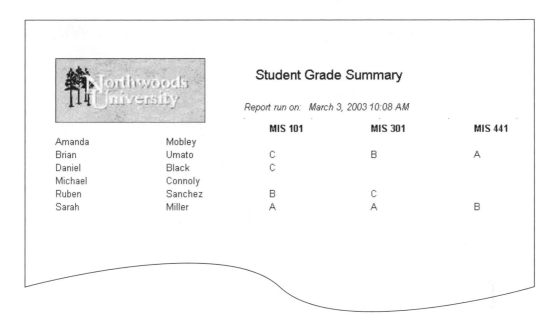

Figure 9-4   Matrix report

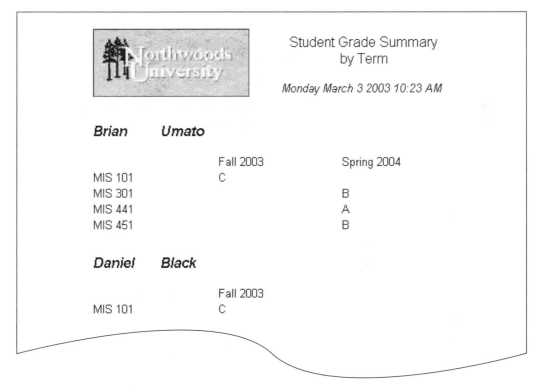

Figure 9-5   Matrix with group report

CREATING A SINGLE-TABLE REPORT USING THE REPORT WIZARD

To create a report, you must specify the source of the data that will appear in the report and then define the report layout style. In this book, you will use SQL queries for all report data sources. Now you will create a tabular report that uses a SQL query to retrieve each student's last name, first name, middle initial, address, city, state, ZIP code, phone number, class, date of birth, and advisor's last name from the Northwoods University database. To automate the report creation process, you will use the Report Wizard, which leads you through a series of pages to specify the query and layout. To be sure that you are starting with a complete database, you will first run the north-woods.sql script. Then you will start Report Builder.

To run the script and start Report Builder:

1. Start SQL*Plus and log onto the database.

2. Run the **northwoods.sql** script from the Chapter9 folder on your Data Disk to refresh your database tables.

3. Exit SQL*Plus.

4. Click **Start** on the taskbar, point to **Programs**, point to **Oracle Reports 6i**, and then click **Report Builder**. The Welcome to Report Builder dialog box opens.

If you are storing your Data Disk files on floppy disks, you will need to use two floppy disks for this chapter. When your first disk becomes full, save your solutions files to the second disk.

You also can start Report Builder by starting Windows Explorer, changing to the *Developer_Home*\BIN folder, and then double-clicking the rwbld60.exe file.

This dialog box enables you to build a new report using the Report Wizard, build a new report manually, or open an existing report. Next, you will use the Report Wizard to specify the report style and enter the SQL query to return the fields from the STU-DENT and FACULTY tables that will appear in the report. You can enter SQL queries using one of three methods. You can type a SQL query, import a query from a script file, or build a SQL query using Query Builder.

Query Builder is a high-level tool for creating SQL queries using a graphical user interface.

To start the Report Wizard, specify the report style, and enter the SQL query:

1. Make sure the Use the Report Wizard option button is selected, and then click **OK**.

2. When the Welcome to the Report Wizard page appears, click **Next**.

3. The Report Wizard Style page appears. This page allows you to specify the report title and style. Type **Northwoods University Students** in the Title box, make sure that the Tabular report style option button is selected, and then click **Next**.

4. The Type page appears. This page allows you to specify whether the report data will be based on a normal SQL query or an Express query. Make sure that the SQL statement option button is selected, and then click **Next**.

 An Express query is created by an Express server, which provides online analytical processing tools for forecasting, creating what-if scenarios, and performing financial modeling. You will not use Express queries in this book.

5. The Data page appears. This page allows you to type a SQL query, import a query from a script file, or build a SQL query using Query Builder. Type the following command to retrieve the desired data fields, and then click **Next**. Note that you do not need to type the semicolon at the end of the command.

```
SELECT s_last, s_first, s_mi, s_add, s_city, s_state,
s_zip, s_phone, s_class, s_dob, f_last
FROM student, faculty
WHERE student.f_id = faculty.f_id
```

6. Since you have not yet connected to the database, the Connect dialog box opens. Log onto the database in the usual way, and then click **Connect**. If your SELECT command is correct, the Fields page showing the fields the query returns appears. If the Fields page does not appear, debug your command until it works correctly.

In the Available Fields list, the Report Wizard Fields page shows the data fields returned by the SELECT command that are available to display in the report. Note that the icons to the left of the field names indicate the field data types. The next steps in building the report are selecting the fields that will appear in the report and specifying the field labels.

To select the report fields and specify the labels:

1. Click the **Select All button** ⟫ to select all query fields for the report. The Available Fields list clears, and all of the query fields now appear in the Displayed Fields list. Click **Next**.

2. The Totals page appears to allow you to specify one or more fields for which you might want to calculate a total. None of the fields in this report require totals, so don't select any fields. Click **Next**.

3. The Labels page appears, which allows you to specify the report labels and field widths. Modify the field labels and widths as follows:

Labels	Width
Last	10
First	10
MI	1
Address	7
City	7
State	2
ZIP Code	9
Phone	10
Class	2
DOB	9
Advisor	10

Since these labels will appear as column headings in a table, the label names do not end with a colon. Click **Next**. The Templates page appears.

The final step in creating the report is to select a **report template**, which defines characteristics of the report appearance, such as fonts, graphics, and color highlights. You can use one of the predefined templates, select a template that you have created, or opt to format the report manually by not selecting a template. You will use one of the predefined templates.

To use a predefined template:

1. Make sure the Corporate1 predefined template is selected, click **Next**, and then click **Finish**. The report appears in the Report Editor Live Previewer window.

2. If necessary, maximize the Report Builder window, and maximize the Live Previewer window. Your screen display should look like Figure 9-6.

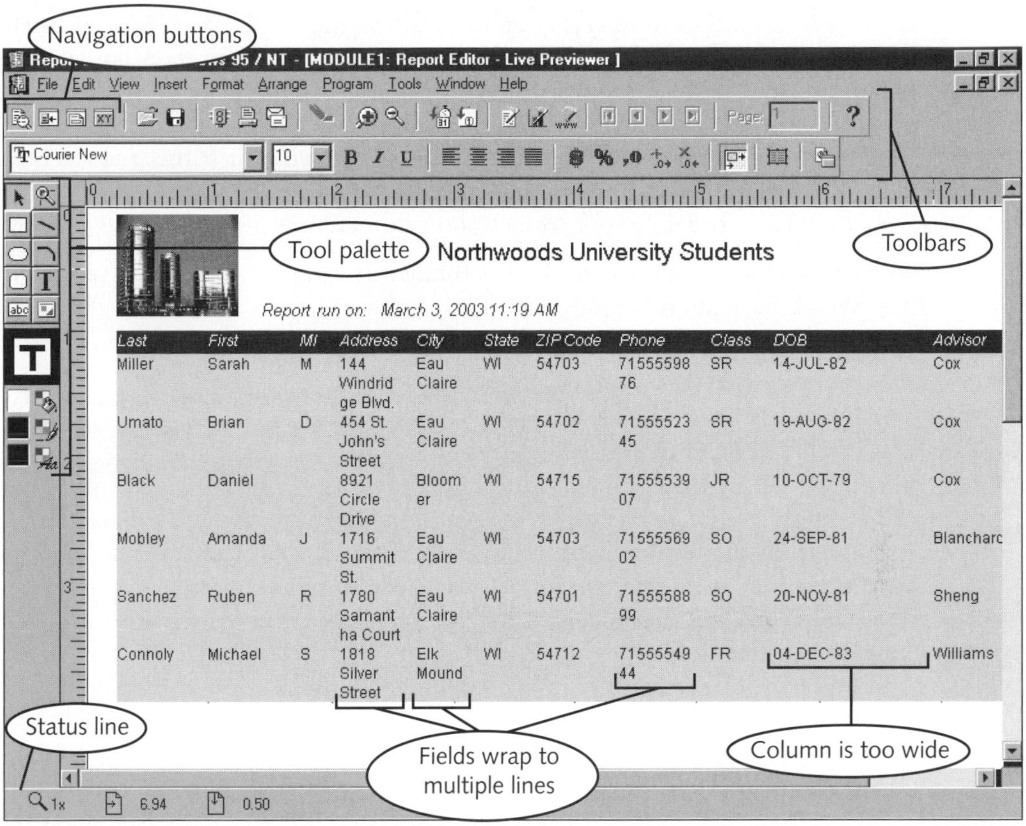

Figure 9-6   Report display in Live Previewer

> You might need to click View on the menu bar, and then click Status Bar or Tool Palette if some of the Live Previewer window objects do not appear.

The Live Previewer window shows how the report will look when it prints or appears on the user's screen display. It also provides an environment for refining the report's appearance. This window has toolbars for working in the Report Builder environment, a tool palette for altering the report's appearance, and a status line for showing the current zoom status and pointer position.

The navigation buttons on the Live Previewer toolbar allow you to move to different windows within the Report Builder environment. The navigation buttons include the Live Previewer button, the Data Model button, the Layout Model button, and the Parameter Form button. You use the **Live Previewer button** 🔲 to open the Live Previewer. The **Data Model button** 🔲 allows you to open the Data Model view, which you can use to modify the report data. You use the **Layout Model button** 🖻 to view the report in the Layout Model view, which displays the report components as symbols and shows the relationships among the report components. The **Parameter Form button** xy allows you to move to the report parameter form, which allows you to view and customize the parameter form, which is a form on which the user selects parameter values to customize the form appearance and functionality at runtime. You will learn more about the Data Model, Layout Model, and parameter forms later in the chapter.

Now you will save your report. Report Builder design files are saved as **report definition files**, which have an .rdf extension.

To save the report:

1. Click the **Save button** 🖫 on the toolbar, and save the report as **9AStudent.rdf** in your Chapter9\TutorialSolutions folder.

Next, you will edit the report in the Live Previewer window by changing the position of the report title. The margin is the area on the page beyond where the report data retrieved from the database appears. It can contain boilerplate text and graphics, such as a company logo, the date the report was created, or the current report page number. Along with the logo, the student report top margin contains the report title (Northwoods University Students) and the date label (Report run on: *todays_date*). You will change the justification of the title so it is left-justified, and move it so that it is aligned with the data label.

To modify the report top margin items:

1. In the Live Previewer, select the **Northwoods University Students** report title label. Currently, the field is center-justified in its text box. Click the **Start justify button** ▤ on the toolbar to left-justify the label.

2. Drag the Northwoods University Students report title label so that its left edge is about 0.25 inches from the right edge of the logo.

 You can make fine adjustments to an object's position by selecting the object and then moving it using the arrow keys on the keyboard.

Next, you will adjust the data field widths. Figure 9-6 shows that the Address, City, and Phone columns wrap to multiple lines. The report would be easier to read if these columns were wider. You will make the City and Phone fields wide enough so that all of the data values appear on a single line. Since the Address data values contain several characters, you will make this field wider, but it will still wrap to multiple lines. In addition, the DOB column is wider than it needs to be. You will adjust the column widths by changing them in the Report Wizard. Like the other Developer wizards, the Report Wizard is reentrant, which means that after you have completed specifying a report, you can go back into the Wizard and edit the report properties. Now you will modify the field column widths using the Report Wizard.

You should modify reports using the Report Wizard in reentrant mode with caution. If you modify a report using the Report Wizard in reentrant mode, you will lose custom formatting, such as resizing columns widths.

To modify the field column widths using the Report Wizard:

1. Click **Tools** on the menu bar, and then click **Report Wizard**. The Report Wizard opens in reentrant mode, with different tabs representing each of the Report Wizard pages. These tabs allow you to go directly to the page that contains the feature you want to edit.

2. Click the **Labels** tab to view the page for specifying field labels and widths. Change the following field widths:

   Address   **15**
   City   **10**
   Phone   **12**
   DOB   **7**

3. Click **Finish** to save your changes and close the Report Wizard. The report appears in the Live Previewer window with the column widths modified.

Clicking Apply in the Report Wizard saves your changes without closing the Report Wizard.

You can also modify column widths directly in the Live Previewer. Suppose you decide to make the DOB column narrower still. You can select the column in the Live Previewer and drag the column selection handles so the column is the desired width. You can also change the position of columns in the Live Previewer.

To modify the column widths and positions in the Live Previewer:

1. In the Live Previewer, select any of the DOB data values to select the column as a group. Selection handles appear around all of the DOB values.

2. Select the center selection handle on the right side of any of the DOB values, and then drag the mouse pointer toward the left edge of the screen display to make the column narrower.

If a field value appears as all asterisks (*), it means that you have made the column too narrow to display the data values in that field. Make the column wider so all of the data values appear.

3. Notice that since you made the DOB column narrower, the Advisor values need to be moved toward the left side of the report. Select any Advisor data value so selection handles appear around all of the column data values, and then drag the column toward the left edge of the screen display so the values are beside the DOB values. Your formatted report should look like Figure 9-7.

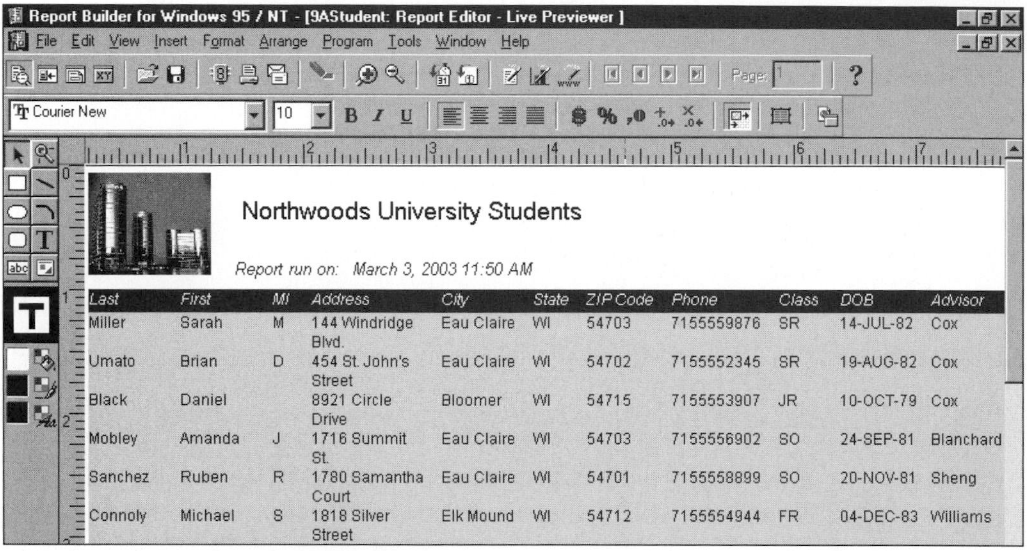

Figure 9-7   Formatted student report

4. Click the **Save button** to save your changes.

You can easily print reports to create hard copy output. Now you will print the report and then close the report file.

To print the report and close the file:

1. Click the **Print button** on the toolbar. The Print dialog window for your printer opens. Adjust the printer properties if necessary, and then click **OK**. Your report should print on your printer.

2. Click **File** on the menu bar, and then click **Close**. If necessary, click Yes to save your changes. The Report Builder Object Navigator window appears.

After you close your report, the Report Builder Object Navigator window appears. The following section discusses the components of this window and how you can use it as you work with Report Builder.

THE REPORT BUILDER OBJECT NAVIGATOR WINDOW

The Report Builder Object Navigator window is shown in Figure 9-8. Like the Object Navigator window in Form Builder, you use it to access different components in the Report Builder environment, as well as the components of an individual report.

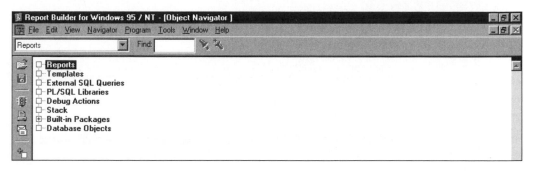

Figure 9-8   Report Builder Object Navigator window

The top-level node is **Reports**. Currently, no reports are open in Report Builder, so there is no ➕ next to Reports. Other environment objects include Templates, External SQL Queries, PL/SQL Libraries, Debug Actions, Stack, Built-in Packages, and Database Objects. **Templates** specify report formatting, and **External SQL Queries** allow you to use queries created in other environments. **PL/SQL Libraries** are collections of related PL/SQL functions or procedures, and **Debug Actions** are actions you can enable or disable while using the Forms Debugger. The **Stack** node shows current values of local variables during a debugging session. **Built-in Packages** are code libraries provided by Oracle to simplify common tasks, and **Database Objects** enable you to access all the objects in the database, such as users, tables, sequences, and triggers.

CREATING A MASTER-DETAIL REPORT

Recall that you can use Report Builder to create reports that show master-detail data relationships in which one record has many associated detail records through foreign key relationships. Most reports that retrieve data from multiple tables involve master-detail relationships. Now you will create a report that involves master-detail relationships. The report, which is shown in Figure 9-9, will list Northwoods University term IDs and term descriptions, and the call IDs and course names of all courses taught during each term. The report also lists detailed information (section number, instructor, day, time, and location) about each course.

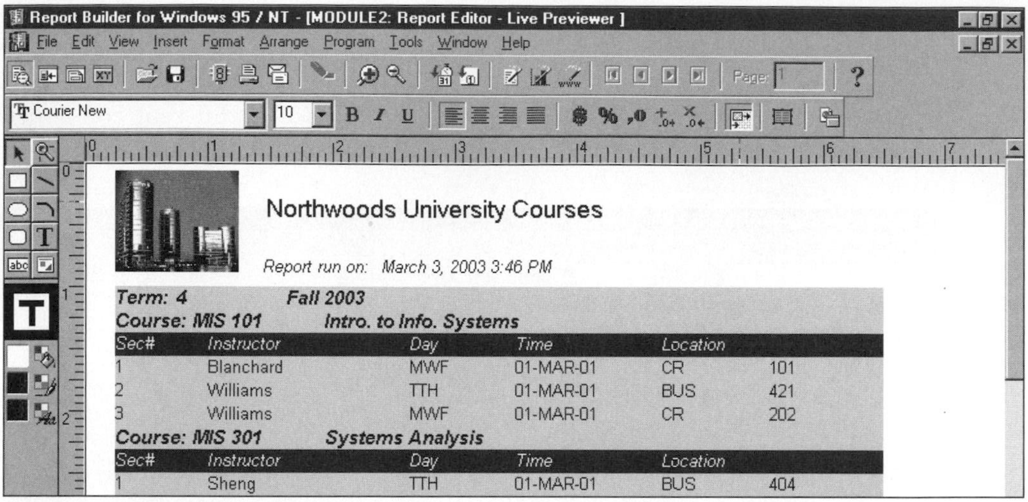

Figure 9-9   Master-detail report

This report involves two master–detail relationships. A term can have multiple courses, so in this relationship, the term data is the master record, and the course data is the detail record. Each course can have multiple course sections, so in this relationship, the course data is the master record, and the course section data is the detail record.

To create a new report in the Object Navigator, you select the Reports node and click the Create button. You are given the option of creating the report manually or creating the report using the Report Wizard. Under most circumstances, you use the Report Wizard to create a new report and then modify the report manually if necessary. Now you will create the master-detail report.

To create the master-detail report:

1. In the Object Navigator, select the **Reports node**, and then click the **Create button** to create a new report. The New Report dialog box opens.

2. Make sure the Use the Report Wizard option button is selected, and then click **OK**.

3. When the Welcome page appears, click **Next**. The Style page appears.

Recall that there are two main report styles for master-detail reports: group left, in which the master records appear on the left side of the report, and the detail records appear in columns to the right of the master records (see Figure 9-2); and group above, in which the detail records appear below the master records (see Figure 9-3). For this report about courses at Northwoods University, you will use the group above style. Now, you will specify the report title and style.

To specify the title and style:

1. Type **Northwoods University Courses** in the Title text box.

2. Select the **Group Above option button**, and then click **Next**. The Type page appears.

3. You will use a SQL command to retrieve the report data, so make sure the SQL statement option button is selected, and then click **Next**. The Data page appears.

The report's SQL query must return all master and detail records that appear in the report. Therefore, the SQL query will include the term, course, and course section data. Next, you will specify the report's SQL query.

To specify the report's SQL query:

1. Type the following SQL command, and then click **Next**. Since this query returns data with master-detail relationships (one term might have multiple courses, and one course might have multiple course sections), the Groups page appears.

```
SELECT term.term_id, term_desc, call_id, course_name,
sec_num, f_last, day, time, bldg_code, room
FROM term, course, course_section, faculty, location
WHERE term.term_id = course_section.term_id
AND course.course_id = course_section.course_id
AND course_section.f_id = faculty.f_id
AND course_section.loc_id = location.loc_id
```

The Groups page specifies how the master-detail values appear in the report. Data in a master-detail report has multiple levels. This report shows term information in the top level, information about courses offered during each term in the next level, and information about individual course sections corresponding to each offered course in the most detailed level. Each data level is called a **group**. The top-level master group is called Level 1, the next level is called Level 2, the next level is called Level 3, and so forth.

The first field you select from the Available Fields list and move into the Group Fields list is automatically put in Level 1, and a Level 1 heading appears above the selected field's name in the Group Fields list. To add additional fields to the Level 1 group, you must select one of the fields currently in the Level 1 group, select the new field from the Available Fields list, and then add the new field to the Group Fields list. To create a Level 2 heading, you select the Level 1 heading in the Group Fields list, then add a field from the Available Fields list that will be in Level 2. The Level 2 heading will appear, with the selected field under it. If you place all of the fields from the Available Fields list into the Group Fields list, an error occurs. So, you must leave the most detailed data fields in the Available Fields list. First, you will specify the Level 1 report group.

To specify the Level 1 report group:

1. Make sure that term_id is selected in the Available Fields list, and then click the **Select button** > . The Level 1 heading appears in the Group Fields list with term_id below it.

2. Now you will add the term_desc field to the Level 1 group. To add another field to Level 1, you must first select an existing Level 1 field in the Group Fields list, so if necessary, select term_id in the Group Fields list.

3. Select **term_desc** in the Available Fields list, and then click > . The term_desc field appears under term_id in the Level 1 group definition, as shown in Figure 9-10.

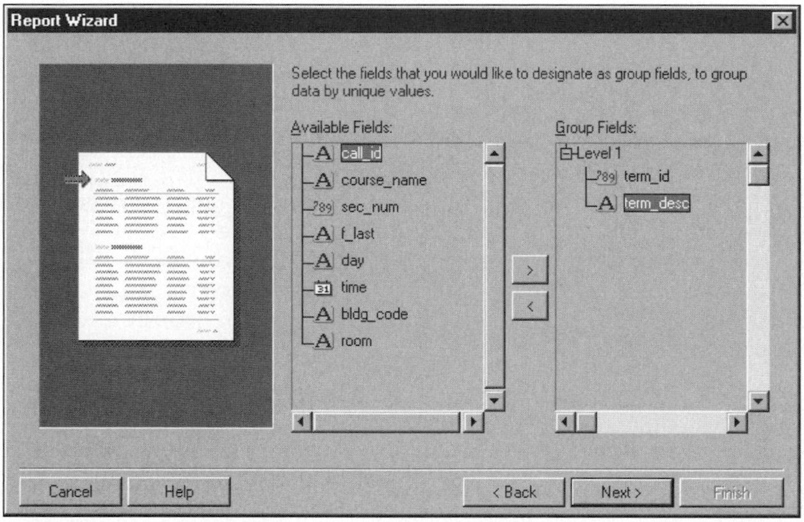

Figure 9-10   Selecting the Level 1 group fields

If term_desc appears under a Level 2 heading, select it, click the Deselect button < to remove it from the Group Fields list, and then repeat Steps 2 and 3.

Now you will create the Level 2 group. To create a new group, you must select its parent group in the Group Fields list, select the first field of the new data group from the Available Fields list, and then click > . For the Level 2 group in this report, you will select Level 1 in the Group Fields list and then select the call ID field in the Available Fields list.

To create the Level 2 group:

1. Select **Level 1** in the Group Fields list, make sure that call_id is selected in the Available Fields list, and then click the **Select button** > . The Level 2 heading appears in the Group Fields list, with call_id under it.

2. To add the course_name field to the Level 2 group, make sure that call_id is selected in the Group Fields list, select **course_name** in the Available Fields list, and then click [>]. The completed Groups page looks like Figure 9-11.

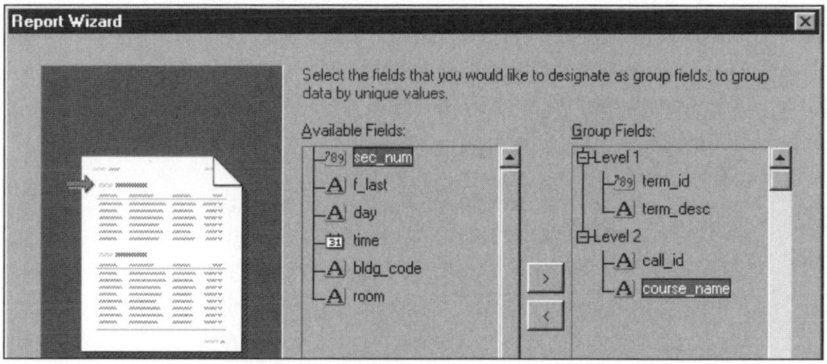

Figure 9-11   Completed Groups page

Recall that the most detailed data fields must remain in the Available Fields list. Since all of the remaining fields in the Available Fields list reference a specific course section, they are in the Level 3 group, and the group specifications are complete.

The order in which the fields appear in the Group Fields list and Available Fields list is the same order that the fields will appear on the report. If you want to change the order of the fields, you can select a field that is higher in the list and drag it down to a lower position. To practice doing this, you will now change the order of the Level 1 group fields.

To change the order of the Level 1 group fields:

1. Select **term_id** in the Group Fields list, drag the mouse pointer so that the tip of the mouse pointer is positioned below term_desc, and then release the mouse button. The term_id field now appears below term_desc.

2. Return the fields to their original order by selecting **term_desc** and then dragging the mouse pointer so the tip of the mouse pointer is below term_id. The fields should now appear in the order shown in Figure 9-11, with term_id first and term_desc second in the Level 1 group list.

3. Click **Next**. The Fields page appears.

To finish the report, you must specify the display fields, totals fields, field labels, and template. Then, you will modify the report's appearance in the Live Previewer by repositioning the margin labels.

To finish the report:

1. On the Fields page, click the **Select All button** [>>] so that all of the report fields appear in the Displayed Fields list. Click **Next**. The Totals page appears.

2. Since none of the data fields will be totaled, do not select any fields, and then click **Next**. The Labels page appears.

3. Modify the field labels and widths as follows:

Field and Totals	Labels	Width
term_id	**Term:**	4
term_desc	(deleted)	10
call_id	**Course:**	5
course_name	(deleted)	10
sec_num	**Sec#**	4
f_last	**Instructor**	10
day	**Day**	5
time	**Time**	7
bldg_code	**Location**	5
room	(deleted)	6

Note that the labels for the Level 1 and Level 2 data fields have a colon after the label, because they will appear horizontally on the form (see Figure 9-9). The data labels in the most detailed (Level 3) data group do not have a colon, because the data appear in a tabular format.

4. Click **Next**. The Templates page appears.

5. Make sure the Corporate1 predefined template is selected, click **Next**, and then click **Finish**. The report appears in the Live Previewer window.

6. If necessary, reposition the report title so that your report looks like Figure 9-9. Note the relative positions of the group fields, with the Level 2 fields grouped under the Level 1 fields and the Level 3 fields grouped under each Level 2 field.

 The Time data field currently displays date values because it does not use a format mask that displays the time components of the dates. You will learn how to change the format mask of a report field later in this chapter.

7. Save the report as **9ACourses.rdf** in your Chapter9\TutorialSolutions folder, and then close the report in Report Builder.

REPORT TEMPLATES

When you are creating many reports that need to have a similar appearance in terms of fonts, graphics image, and background color, it is useful to create a custom template to specify the report appearance. Along with giving all of your reports a similar appearance, using a custom template keeps you from having to perform the same formatting tasks over and over again. In the following sections, you will learn how to create a custom template and how to apply custom templates to reports.

Creating a Custom Template

When you apply a template to a report, objects in the report (such as the report title, date and time the report was generated, and page numbers) always appear in the same locations in the report and have the same label fonts and colors. Boilerplate objects, such as graphics, are also applied to the report through a template. Template definitions are stored in a template definition file that has a .tdf extension. Now you are going to create a template for Northwoods University reports by modifying the existing Corporate1 template. The modified template will use a different logo, display the page numbers in a different position, and use different background colors. First, you will open the corporate1.tdf file and save the file using a different name.

To open and save the corporate1.tdf file:

1. In the Object Navigator, click the **Open button** 🖼 on the toolbar, navigate to the Chapter9 folder on your Data Disk, and open **corporate1.tdf**. CORPORATE1 appears under the Templates node in the Object Navigator.

2. Save the template file as **northwoods.tdf** in your Chapter9\TutorialSolutions folder. The template name changes to NORTHWOODS.

In the Object Navigator, you can see that a template has five components: Data Model, Layout Model, Report Triggers, Program Units, and Attached Libraries. In this book, the main template component you will work with is the Layout Model, which defines template objects such as text, graphics, the date on which the report was run, and the report page numbering.

Recall that the Live Previewer window in Report Builder shows a report on the screen the same way it will appear when it prints and is similar to the Forms Runtime window. In contrast, the Report Builder **Layout Model** is similar to the Layout Editor in Form Builder and represents objects symbolically to highlight their types and relationships. You cannot view or edit a template in the Live Previewer window, because the template always runs with a report and cannot run as a stand-alone object. You must use the Layout Model to view and edit a report template. Now you will open the template in the Layout Model Template Editor, which is an environment within the Layout Model that you can use for editing templates.

To open the template in the Layout Model Template Editor:

1. Double-click the **Layout Model icon** 📄 under the NORTHWOODS template in the Object Navigator. The report template appears in the Layout Model Template Editor, as shown in Figure 9-12.

9

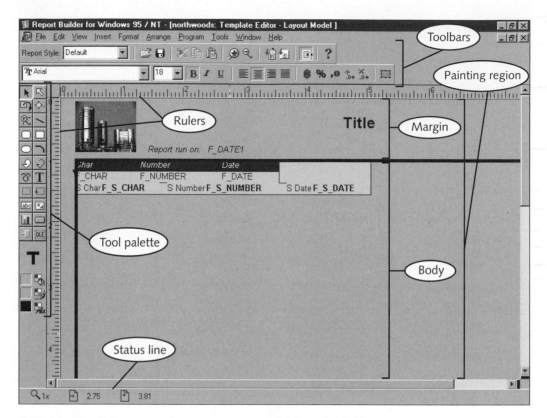

Figure 9-12 Report template in Layout Model Template Editor

The Template Editor, like the Layout Editor in Form Builder, has a tool palette, toolbars, painting region, and status line. The painting region has rulers to help you position report components. There are two areas within the painting region in a report template: the **margin**, where the report title, date the report was generated, and page number appear; and the **body**, which contains the report data. When you are editing a template in the Template Editor, you can edit the objects in only one area at a time. The Template Editor toolbar has a **Margin button** [icon], also called the Edit Margin button, that enables you to toggle between the margins and the body for editing. Currently, the Margin button is pressed (see Figure 9-12), and the report margins are visible and available for editing.

Now you will edit the objects in the template's top margin. You will move the report title and date labels so they appear near the right side of the logo. You will delete the current logo and replace it with the Northwoods University logo. The Northwoods University logo will appear as a static graphic image, so you will import it into the template the same way you imported static art images into forms in Form Builder.

To edit the top margin objects:

1. Select the **Report run on** label in the top margin, press **Shift**, then select the **F_DATE1** field while keeping the Shift key pressed to select the date label and date field as an object group. Drag the label and date field toward the right edge of the report so the left edge of the word *Report* aligns about 2.5 inches from the left edge of the report, as shown in Figure 9-13.

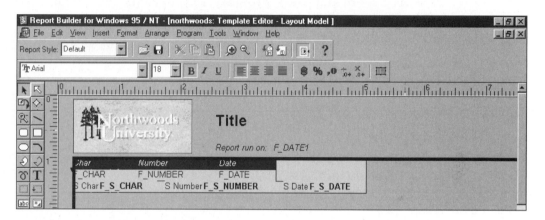

Figure 9-13   Editing the template's top margin

2. Click the current logo to select it, and then press **Delete** to delete the current logo.

3. Click **File** on the menu bar, point to **Import**, then click **Image**. Click **Browse**, navigate to the Chapter9 folder on your Data Disk, select the **nwlogo.jpg** file, click **Open**, and then click **OK**. The Northwoods University logo appears on the template.

4. Resize the logo so it fits in the top margin area, as shown in Figure 9-13.

5. Select the **Title** text object, and then click the **Start justify button** ▣ so the title is left-justified. Drag the Title text object so it starts about 0.25 inches from the right edge of the logo, as shown in Figure 9-13, and so its left edge aligns with the Report run on: label. Resize the Title object if necessary so it fits on the painting region.

6. Click the **Save button** 🖫 to save the template file.

Now you will modify the field that displays the date the report was generated so that the time component of the date does not appear. To modify the date field, you must change the format mask of the date data field in the field Property Palette. In Report Builder, you can specify format masks for fields that display NUMBER and DATE data, but you cannot specify format masks for fields that display CHAR and VARCHAR2 data. To change the format mask of the date field, you will open the field Property Palette and then enter the desired format mask. The format mask property allows you to choose from a list of predefined format masks for a particular data field or to enter a customized format mask.

To change the date field's format mask:

1. Select **F_DATE1** in the top margin, right-click, then click **Property Palette**. The field Property Palette opens.

2. Select the **Format Mask property**, then open the property list. A list of possible format masks for the date field appear. You can select from the predefined format masks or enter a custom format mask. You will edit the current format mask to create a custom format mask, so don't select a new format mask from the list.

3. Close the list, and then edit the current format mask by deleting its time portion, so it appears as **fmMonth DD, RRRR**.

4. Close the Property Palette, and save the template file.

On the Layout Model toolbar, the **Insert Date and Time button** 🔳 enables you to insert the current date and time in the report, and the **Insert Page Number button** 🔳 enables you to insert the current page number. These buttons open dialog boxes that specify the format of the date or page number and its placement position. These buttons are available only when the Margin button 🔳 is pressed, because date/time values and page numbers can only appear in the report margin. Currently, the report displays the page number in the center of the bottom margin. Next, you will delete the page number from the template's bottom margin and create a new page number specification that displays the current page number, as well as the total number of report pages, on the right edge of the top margin.

To delete the current page number and create a new page number:

1. Scroll to the bottom of the Layout Model window to display the bottom margin. Select **Page &<PhysicalPageNumber>**, which specifies the format and position of the current page number, and then press **Delete** to delete the current page number.

2. Click the **Insert Page Number button** 🔳 on the toolbar. The Insert Page Number dialog box opens.

3. Select **Top-Right** from the placement list to specify the page number placement position.

4. Click the **Page Number and Total Pages option button**, and then click **OK**. The page number appears on the right side of the top margin.

5. If necessary, select the new page number field and change the font to 10 point Arial.

6. Save the template file.

Next, you will modify the colors in the template body. You will change the background color of the column headings from dark blue to dark green. To make this change, you must click the Margin button 🔳 to toggle to the template body and make it available for editing. Then you will select the column headings and change their fill color to dark green.

To change the column heading fill color:

1. Click the **Margin button** 🔳 to deselect it and make the template body available for editing. The Template Editor window changes so that only the body is visible and the toolbar buttons that can be used only while editing the template margins no longer appear.

2. Move the mouse pointer so that it is just before the word *Number* in the column headings, and then click to select the column heading background. Selection handles appear around the columns as a group, as shown in Figure 9-14, and the fill color should appear as dark blue on the tool palette.

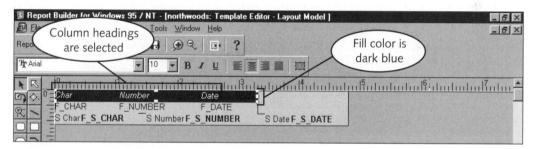

Figure 9-14   Selecting the column heading background

If selection handles appear around one of the column headings and the selected fill color on the tool palette is not dark blue, then you selected an individual column heading rather than the background area behind all of the headings. Try to click in the area between the columns to select the column heading background.

3. Click the **Fill Color tool** 🪣 on the tool palette, and change the fill color to dark green. The column heading background color changes to the shade you selected.

4. Click anywhere on the Layout Model painting region to deselect the column heading background, and then save the template file.

A report template has two types of attributes: default and override. **Default attributes** define the default visual attributes for all report styles, which can include object placement, font types and sizes, background colors, and so forth. **Override attributes** define attributes for individual report styles, such as tabular reports or form letters. When you apply a template to a report, all default attributes are applied to the report. Override attributes are applied to the report only if it is the report style for which the override attribute is defined.

To define override attributes, you use the Report Style list, which appears in the Layout Model window on the top-left edge of the toolbar. This list specifies the current report style. To be able to customize the template for different report styles, you must modify the template body for each style in the list. So far, you have modified the template for the Default, or Tabular, report style. Whenever you create or modify a template, you must modify the template for every report style to which the template will be applied. Next, you will select a different report style (the group above style) and examine how the template looks. Then you will modify the template so that the column heading background is dark green for the group above style as well.

To modify the template for the group above report style:

1. Open the Report Style list to display the different report styles, and select **Group Above**. Note that for this report style, the column heading background color has not been changed and is still dark blue.

2. Select the column heading background, and change the fill color to dark green.

3. Save the template file, click **File** on the menu bar, and then click **Close** to close the template file.

Applying Custom Templates to Reports

You can select a predefined template to specify report formatting when you create a new report, or you can apply a template to an existing report to change its formatting. If you apply a template to a report and then later apply another template, the original template formatting will be deleted or replaced by the formatting in the most recent template. You can apply custom templates to reports by specifying the path to the template file or by registering the template in Report Builder. The following sections discuss both approaches.

Applying Templates Using the Template File Specification

One way to apply a custom template to a report is to specify the template file on the Report Wizard Templates page. To do this, you select the Template file specification option, and then enter the full folder path and filename of the template file, including the drive letter. This approach provides a quick and easy way to apply custom templates to forms. The disadvantage to this approach is that the template file must be available at the specified location whenever the report file is opened in Report Builder.

Now you will open the 9ACourses.rdf report that you created earlier in the lesson, start the Report Wizard in reentrant mode, and apply the northwoods.tdf template file to the report using the template file specification.

To apply the template to the 9ACourses.rdf report using the template file specification:

1. Click the **Open button** 🖻 on the toolbar, and open the **9ACourses.rdf** report from your Chapter9\TutorialSolutions folder. The report appears in the Object Navigator.

2. Click **Tools** on the menu bar, and then click **Report Wizard** to open the Report Wizard in reentrant mode so you can modify the report template.

3. Click the **Template tab** to open the Template page. If necessary, click Next until the Template tab appears.

4. Select the **Template file option button**, click **Browse**, select the **northwoods.tdf** file from your Chapter9\TutorialSolutions folder, and then click **Open**. Click **Finish** to save your changes and close the Report Wizard. The report appears in the Live Previewer using the formatting specified in the new template.

Note that the report now displays the Northwoods logo and has page numbers at the top of the page. (You might need to scroll to the right edge of the window to see the page numbering.) However, the report displays two report titles: one from the template and one from the original report. You need to delete the title from the original report and show only the title from the template. You also need to adjust the position and justification of the page number.

To delete the original report title and adjust the page number position:

1. In the Live Previewer, select the **Northwoods University Courses** report title that is partially covered by the logo, which is the original report title, and then press **Delete** to delete it.

2. Select the page number, and then click the **Start justify button** to make it left-justified. Adjust the size and position of the page number so the page number appears as shown in Figure 9-15.

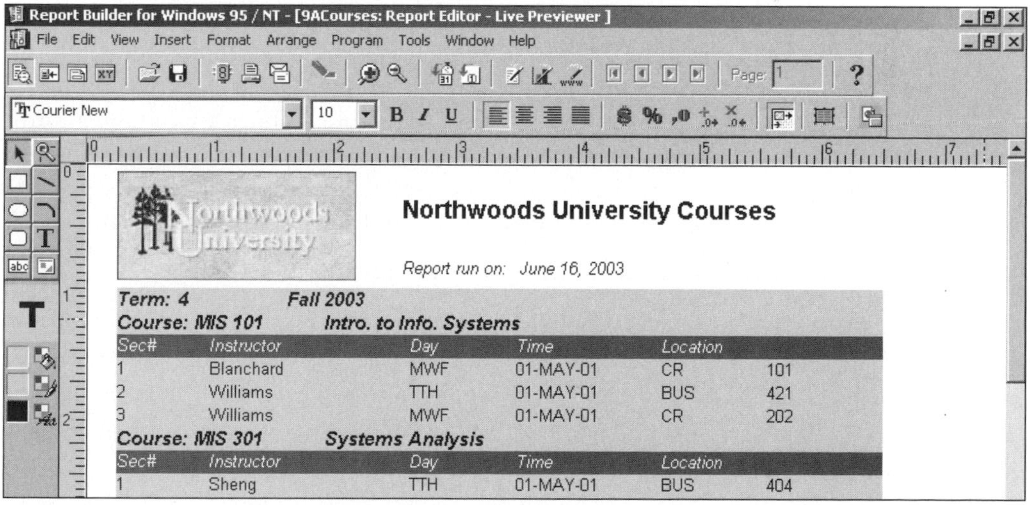

Figure 9-15   Report with formatting from template

3. Save the report, and then close the report file.

Registering Custom Templates in Report Builder

Another way to make custom templates available to reports is to **register** the custom template, so it appears in the Predefined Templates list on the Templates page in the Report Wizard. The advantage of registering templates is that you don't have to specify the path to the template file, and the file does not always have to be available at the specified location when the report opens. To register a custom template, you need to modify the Developer global preferences file. The **global preferences file** is a text file that specifies configuration information about all of the Developer utilities, including Form Builder and Report Builder. This information includes user preferences, such as whether the Wizard Welcome pages appear or which format masks appear in the Property Palette format mask lists.

In Report Builder, you can register a custom template file by modifying the **cagprefs.ora** global preferences file, which is stored in the *Developer_Home* folder on your workstation. Every time Report Builder opens, the application reads the global preferences file and configures the environment based on the file contents. All users who run Report Builder on your workstation share this file. Therefore, you will first make a backup copy of the existing cagprefs.ora file and place it in the Chapter9 folder on your Data Disk. When you finish modifying and using the file, you will replace the modified cagprefs.ora file with the original file.

To make a backup copy of the cagprefs.ora file:

1. Start Windows Explorer, navigate to the *Developer_Home* folder, and copy **cagprefs.ora** to the Chapter9 folder on your Data Disk.

To modify the global preferences file to display a custom template name in the Predefined Templates list in the Report Wizard, you need to specify the **template description**, which is the description that appears in the Predefined Templates list. You also need to specify the name of the associated template .tdf file. You have to specify a separate template description and filename for each report style (tabular, group above, group left, and so forth). For now, you will specify a custom template only for the tabular report style. Now, you will open the global preferences file and examine the commands for specifying a template description and corresponding filename.

To examine the commands in the global preferences file:

1. Start Notepad or an alternate text editor, click **File** on the menu bar, click **Open**, navigate to the *Developer_Home* folder, select **All files** in the Files of type list, and then open **cagprefs.ora**. The file opens.

2. Scroll down until you see the `Reports.Tabular_Template_Desc =` command, as shown in Figure 9-16.

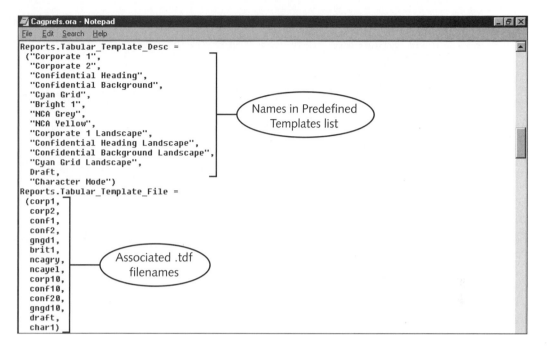

Figure 9-16 Global preferences file commands to register templates

The `Reports.Tabular_Template_Desc` = command specifies the descriptions that appear in the Predefined Templates list for tabular-style reports. You will add the description "Northwoods" to the description list. The `Reports.Tabular_Template_File` = command specifies the names of the corresponding template filenames. The first description in the description list corresponds with the first filename, the second description corresponds with the second filename, and so forth. You will add the filename "northwoods" to specify that the name of the template file is *northwoods.tdf*. Note that you do not specify the .tdf extension in the filename.

To add the commands to the global preferences file to register the Northwoods custom template:

1. In Notepad, add the commands to the cagprefs.ora file shown in Figure 9-17. The first entry specifies the description that appears in the Predefined Templates list, and the second entry specifies the corresponding template filename.

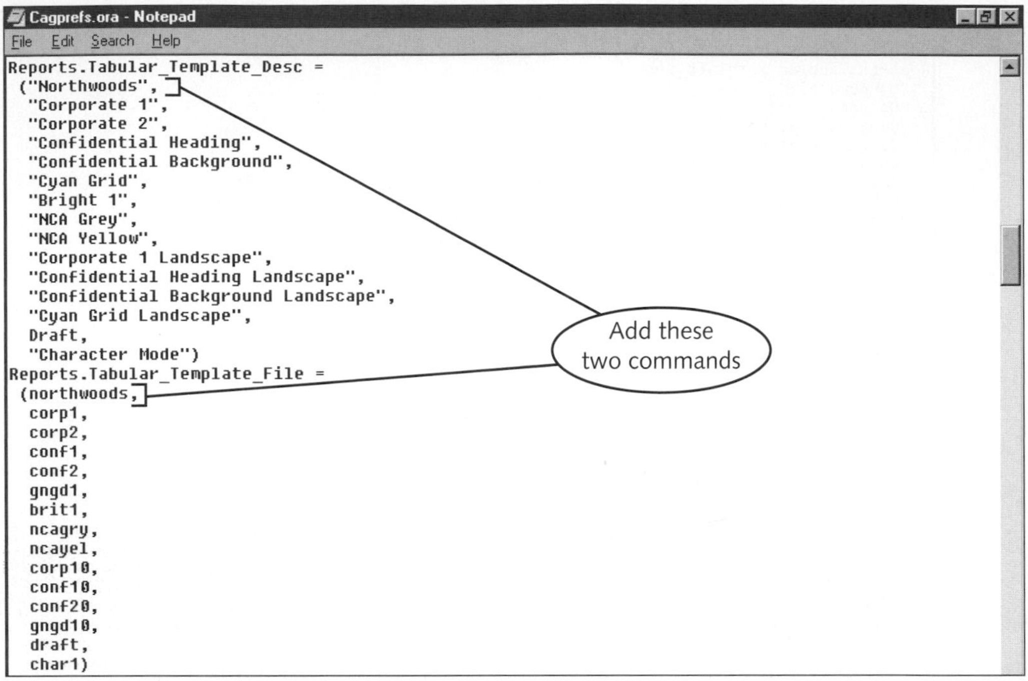

Figure 9-17   Modifying the global preferences file

2. Save the file, and then exit Notepad.

The last step in registering the custom template is to place a copy of the template .tdf file in the default Report Builder templates folder, which is *Developer_Home*\Report60\ Admin\Template\Us. Now, you will copy the northwoods.tdf file from your Chapter9\TutorialSolutions folder to the default templates folder.

To copy the template file to the default templates folder:

1. In Windows Explorer, copy **northwoods.tdf** from your Chapter9\TutorialSolutions folder to the ***Developer_Home*\report60\admin\template\us** folder.

Now, you will test to see if you successfully registered the Northwoods template in Report Builder. First, you must exit Report Builder and start it again to enable the changes in the global preferences file. Then, you will open the 9AStudent.rdf report file that you created earlier in the lesson, start the Report Wizard in reentrant mode, and select the Northwoods template from the Predefined Template list.

To test the template registration:

1. Exit Report Builder, and then start Report Builder again. When the Welcome to Report Builder window appears, select the **Open an existing report option button**, and then click **OK**.

2. Navigate to your **Chapter9\TutorialSolutions** folder, and open **9AStudent.rdf**.

3. Click **Tools** on the menu bar, and then click **Report Wizard** to open the Report Wizard in reentrant mode.

4. Click the **Template tab** to display the Templates page. The Northwoods template should appear in the Predefined Templates list. Select **Northwoods**, and then click **Finish**.

 To create a thumbnail image of a custom template that appears on the Template tab when you select a template from the Predefined template list, make a screenshot of a small sample area of a report that uses the template, and save the image using the .bmp file type. Name the image *template_filename*x.bmp, where x represents the report style: substitute *t* for tabular, *a* for group above, *f* for form-like, *g* for matrix with group, *l* (lowercase L) for group left, *m* for mailing label, *r* for form letter, or *x* for matrix. Store the file in the default template file directory. The image will appear on the Templates tab when you select the template from the Predefined template list.

9

5. Log onto the database. The Students report appears, with the Northwoods template formatting applied. Delete the extra report title, and adjust the page number position if necessary.

6. Save the report, close the report in Report Builder, and then close Report Builder and all other open Oracle applications.

Recall that you need to restore the backup copy of the cgaprefs.ora file to the *Developer_Home* folder on your workstation. You should also delete your northwoods.tdf template file from the default template folder so that it won't be used by other students.

To restore the backup global preferences file and delete the template file:

1. In Windows Explorer, copy **cagprefs.ora** from the Chapter9 folder on your Data Disk to the *Developer_Home* folder on your workstation hard drive to restore the original global preferences file.

2. Navigate to the *Developer_Home***\Report60\Admin\Template\Us** folder, delete **northwoods.tdf**, and then close Windows Explorer.

SUMMARY

❑ Reports retrieve database data using SQL queries, perform mathematical or summary calculations on the retrieved data, and format the output to look like invoices, form letters, or other business documents.

❑ A tabular report presents data in a table format with columns and rows.

❑ Group left and group above reports display master-detail relationships, whereby one master record might have several associated detail records through a foreign key relationship. In a group left report, each master item is listed on the left side of the report, and the multiple detail items appear to the right of the master item. In a group above report, each master item appears above the multiple detail lines.

❑ To create a report, you create a SQL command to specify the data that will appear in the report, and then define the report layout style.

❑ A report template defines the report appearance in terms of fonts, graphics, and fill colors in selected report areas. A report template also defines how the date the report is generated and how the report page numbers appear on the report.

❑ The Report Builder Live Previewer window shows how the report will look when it is printed and provides an environment for refining the report's appearance.

❑ Report Builder design files are saved as report definition files with an .rdf extension.

❑ The Report Wizard is reentrant, which means that after you have completed specifying a report, you can reopen the Wizard and edit the report's properties if necessary.

❑ When you create a report with a master-detail relationship, the report's SQL command must return all master and detail records that appear on the report.

❑ When you create a master-detail report, you use the Report Wizard Groups page to specify which fields are in the master group and which are in the detail group. A report can have multiple master-detail relationships.

❑ A report template has a body and a margin region, and you can only edit one region at a time.

❑ Template formatting definitions are stored in template definition files, which have a .tdf extension.

❑ The Report Builder Layout Model view represents objects symbolically to highlight their types and relationships.

❑ In Report Builder, you can specify format masks for fields that display number and data values, but you cannot specify format masks for fields that display character values.

❑ In a report template, default attributes define the default visual attributes for all report styles, and override attributes define attributes for individual report styles, such as tabular reports or form letters.

❑ To be able to apply a predefined template to different report styles, you must specify the template properties for each report style in the template body.

❑ To apply custom templates to reports using the Report Wizard Templates page, you can specify the template's complete file specification, or you can register the custom template so it appears in the Predefined Templates list.

❑ To register a template file, you must modify the Developer global preferences file (cagprefs.ora) so it contains both the template description and filename. You also must place the template file in the default templates folder.

REVIEW QUESTIONS

1. What is the difference between a form and a report?

2. Select the appropriate Report Builder report style (tabular, form-like, mailing label, form letter, group left, group above, matrix, matrix with group) for:

   a. A report that is a letter informing Northwoods University students of their grades for the past term and their current grade point average.

   b. A list of Northwoods University student names, with associated course call IDs and grades listed below each student name.

   c. A report that shows Clearwater Traders order source names on the column headings, inventory IDs on the row headings, and the number of each inventory item ordered from a specific order source at the intersection of the column and row.

   d. A list of courses offered during the Summer 2002 term at Northwoods University.

3. What is the difference between .rdf and .tdf files?

4. State the template region (body or margin) where you would place the following items:

   a. data values

   b. page numbers

   c. report title

   d. column headings

   e. date the report was run

5. What is the difference between the Live Previewer window and the Layout Model window? Why can't you view a template file in the Live Previewer window?

6. Which data types can you specify format masks for in Report Builder?

7. What do the levels on the Groups page in a master-detail report represent?

8. Describe the two ways that you can make a custom template available to reports, and discuss the advantages and disadvantages of each approach.

PROBLEM-SOLVING CASES

Cases reference the Clearwater Traders (see Figure 1-11), Northwoods University (see Figure 1-12), and Software Experts (see Figure 1-13) sample databases. Run the clearwater.sql and softwareexp.sql scripts stored in the Chapter9 folder on your Data Disk to refresh the sample databases. All required files are stored in the Chapter9 folder on your Data Disk. Save all files in your Chapter9\CaseSolutions folder.

1. In this case, you will first create a predefined template for Clearwater Traders. Then, you will create a report showing a listing of Clearwater Traders customers, as shown in Figure 9-18, and apply the template to the report.

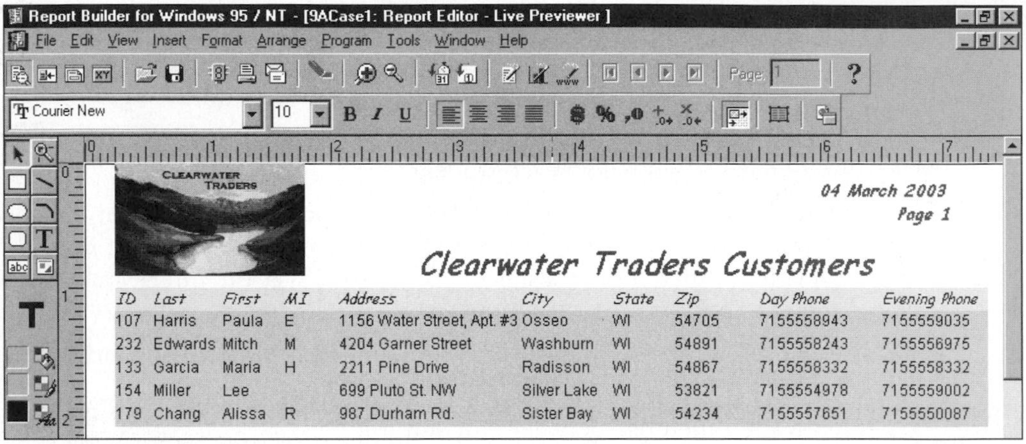

Figure 9-18

a. Open the casual1.tdf template file, and save it as clearwater.tdf.

b. Replace the current logo with the Clearwater Traders logo (clearlogo.tif).

c. Change the title position and justification so that the title is centered at the top of the page. Change the title font color to dark blue.

d. Insert the current date at the top-right corner of the report. Right-justify the date so its right edge aligns with the right edge of the report body. Format the date as DD MONTH YYYY. Use a 10-point italic Comic Sans MS font, and make sure the font color is dark blue. (If this font is not available on your system, substitute a different font.)

e. Insert the current page number in the top-right corner of each report page, just under the date, using the format "Page *current_page_number*." Do not display the total number of pages. Right-justify the page number so its right edge aligns with the right edge of the report body. Format the page number using a 10-point italic Comic Sans MS font. Make the page number font color dark blue.

f. For the tabular, group left, and group above report styles, change the background of the column headings to a light pink, and the text color to black. (*Hint*: To change the text color, you will have to select each column individually.)

g. Create a new report named 9ACase1.rdf that displays all of the fields from the Clearwater Traders CUSTOMER table.

h. Change the report title to "Clearwater Traders Customers," use descriptive column headings, and apply the clearwater.tdf template file by specifying the path to the template file from your Chapter9\TutorialSolutions folder. Modify the report formatting as necessary so the finished report looks like Figure 9-18.

2. Create an incoming shipment report for Clearwater Traders that lists item description, inventory ID, item size, and color values, and then lists the shipment ID and date expected for all incoming shipments that have not yet been received for that item. Format the report as shown in Figure 9-19. Apply the clearwater.tdf template you created in Case 1. (If you did not create the template, apply the Corporate1 predefined template, delete the current logo, and replace it with the Clearwater Traders logo, which is stored in the clearlogo.tif file.) Save the report as 9ACase2.rdf.

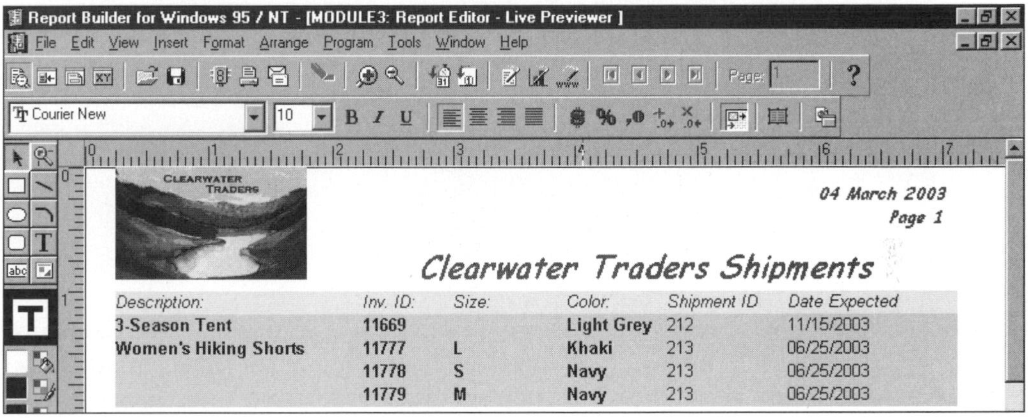

Figure 9-19

3. Create a report that summarizes the customer orders for Clearwater Traders. The report should list the customer ID and last and first name, then the order ID and order date for each order placed by that customer. For each order, list the inventory ID, item description, color, size, price, and quantity ordered. Change the report title to "Customer Orders," and use the group above report style. Apply the clearwater.tdf template file you created in Case 1. (If you did not create the template, apply the Corporate1 predefined template, delete the current logo, and replace it with the Clearwater Traders logo stored in the clearlogo.tif file.) Format the report as shown in Figure 9-20, and save the report as 9ACase3.rdf.

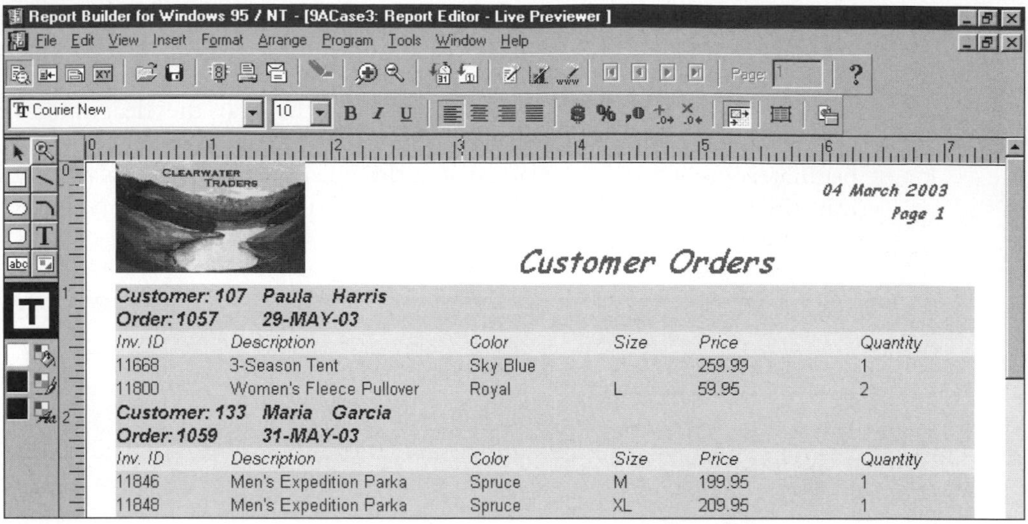

Figure 9-20

4. Create a report that makes mailing labels for the customers in the Clearwater Traders database, as shown in Figure 9-21. Format the labels using a 10-point Arial font. Do not apply a template to the report. Save the report as 9ACase4.rdf.

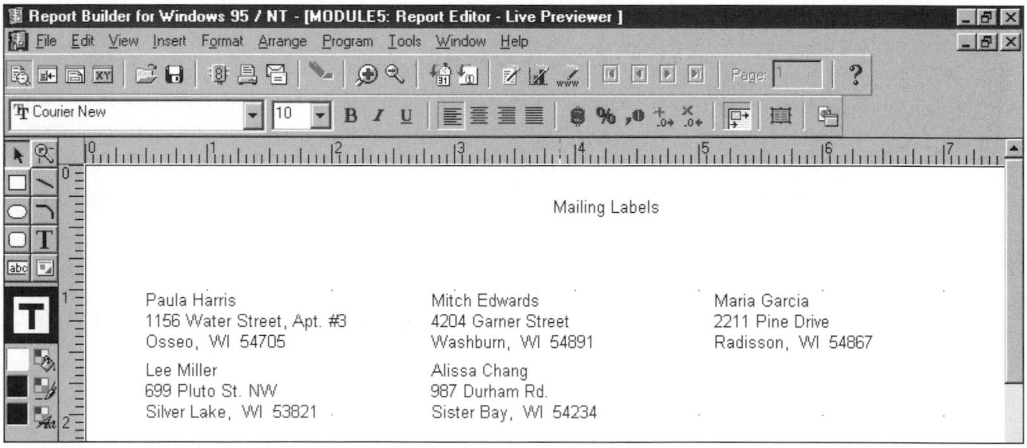

Figure 9-21

5. In this case, you will first create a predefined template for Software Experts. Then, you will create a report showing a listing of Software Experts consultants, as shown in Figure 9-22, and apply the template to the report.

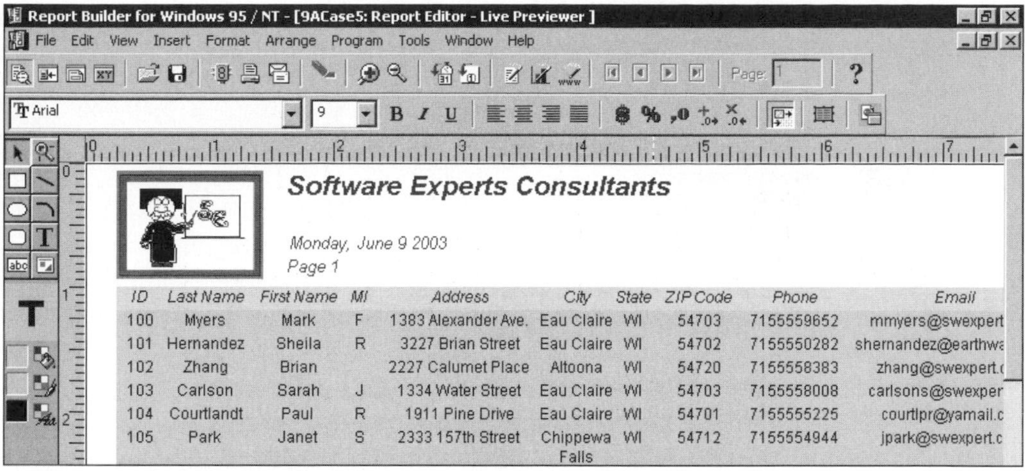

Figure 9-22

    a. Open the corporate1.tdf template file, and save it as softwareexp.tdf.

    b. Replace the current logo with the Software Experts logo, which is stored in a file named swelogo.tif.

    c. Adjust the template title so it is left-justified and appears to the right of the logo. Change the font to 16-point italic Arial, and make the font color dark blue.

    d. Format the date using the format mask shown, and make the date font color dark blue.

    e. Delete the current page number specification, and create a page number specification that displays the current page number and appears on the top margin, just below the date. Format the page number using a 10-point italic Arial font. Make sure the font color is dark blue.

    f. For the tabular, group left, group above, and matrix report styles, change the background of the column headings to light blue and the text color to black. (*Hint:* To change the text color, you will have to select each column individually.)

    g. Create a new report named 9ACase5.rdf that displays all of the fields from the Software Experts CONSULTANT table.

    h. Change the report title to "Software Experts Consultants," use descriptive column headings, and apply the softwareexp.tdf template file by specifying the path to the file from your Chapter9\CaseSolutions folder. Modify the report formatting as necessary to make the report look like Figure 9-22.

6. Create a report that lists Software Experts client name values, along with the project information for the client's projects, as shown in Figure 9-23. Apply the softwareexp.tdf template you created in Case 5. (If you did not create the template, apply the Corporate1 predefined template, delete the current logo, and replace it with the Software Experts logo, which is stored in the swelogo.tif file.) Save the report as 9ACase6.rdf.

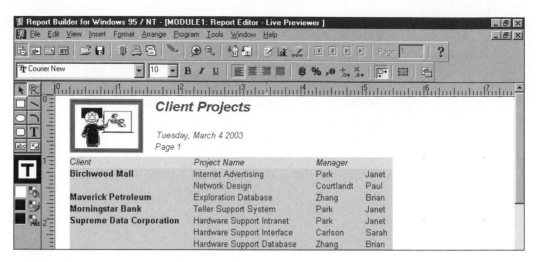

Figure 9-23

7. Create a report that summarizes Software Experts consultant skills and certifications, as shown in Figure 9-24. Use the matrix report style, and apply the softwareexp.tdf template you created in Case 5. (If you did not create the template, apply the Corporate1 predefined template, delete the current logo, and replace it with the Software Experts logo, which is stored in the swelogo.tif file.) Save the report as 9ACase7.rdf.

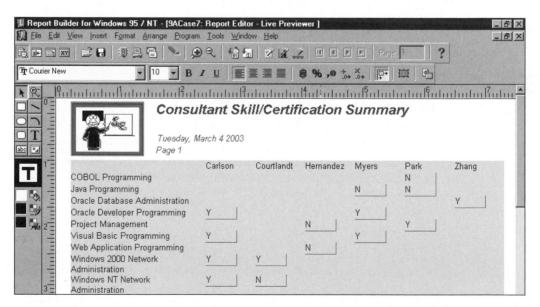

Figure 9-24

◀ LESSON B ▶

Objectives

♦ Understand the different components of a report
♦ Modify report components directly
♦ Modify the format of master-detail reports
♦ Create a user parameter to allow the user to customize report data
♦ Create multipanel reports that span multiple horizontal pages
♦ Learn how to display image, sound, and video data in a report

REPORT COMPONENTS

In Lesson A, you learned how to create reports quickly and easily using the Report Wizard. Sometimes, however, you need to customize the report output that the Report Wizard generates. For example, you might want to modify a report so there is blank space between master records in a master-detail report, or you might want to change the data that appears in the report. You can modify some report specifications, such as the query that retrieves the report data, using the Report Wizard in reentrant mode. The problem with using the Wizard in reentrant mode is that when you close the Wizard and apply your changes, the report is reformatted. All custom formatting changes that you made before opening the Wizard in reentrant mode (such as resizing column widths) are lost.

To customize report output and change report specifications without using the Report Wizard in reentrant mode, you need to become familiar with the report components and learn how to directly modify these components. In this lesson, you will learn about the following report components: the Data Model, which specifies the data that the report displays; the Layout Model, which displays the report components as symbolic objects; and the report frames, which group the report objects. To explore these report components, you will work with a tabular report created using the Report Wizard that shows the location ID, building code, room, and capacity values from the LOCATION table in the Northwoods University database. First, you will start Report Builder and open the report file.

To start Report Builder and open the file:

1. Start Report Builder. Click the **Open an existing report option button**, and then click **OK**.

2. Open **Locations.rdf** from the Chapter9 folder on your Data Disk. The Object Navigator window opens. If necessary, maximize the Object Navigator window. Save the file as **9BLocations.rdf** in your Chapter9\TutorialSolutions folder.

3. In the Object Navigator, double-click the **Live Previewer icon** 🖳, and connect to the database in the usual way. The report appears in the Live Previewer window, as shown in Figure 9-25.

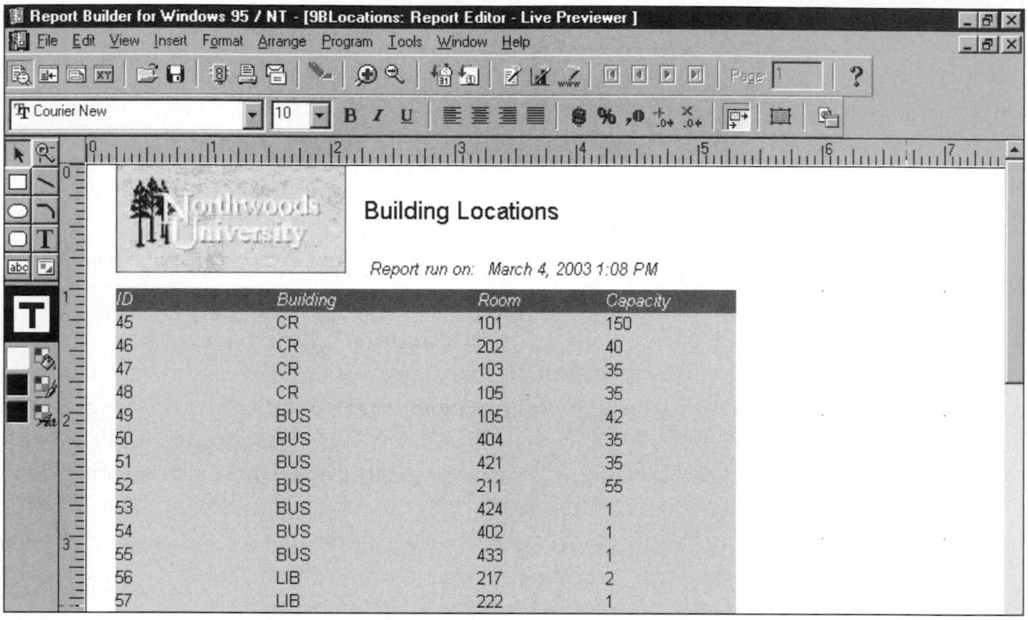

Figure 9-25   Building locations report

When you examine the report, note that the report title and date heading are in the top margin. The body contains a row with the column headings and multiple rows showing data from each record in the LOCATION table. The information in the report body comes from the report's Data Model, which the next section describes.

The Report Data Model

When you create a report using the Report Wizard, you specify a SQL query to retrieve the data records that appear in the report. In Chapter 8, you learned about form record groups, which are form objects that represent data in a tabular format. Reports also contain record groups. Report Builder organizes the returned data fields into one or more **report record groups**, which are sets of records that represent the data fields retrieved by the query. The report's **data model** consists of the SQL query that retrieves the report data and its associated record groups.

Now you will open the data model for the building locations report and examine its structure. You will do this using the **Data Model view**, which is the Report Builder environment for viewing and editing report data model components.

To examine the report data model:

1. Click the **Data Model button** on the Live Previewer toolbar. The Data Model view opens, as shown in Figure 9-26. If necessary, maximize the inner Data Model view window.

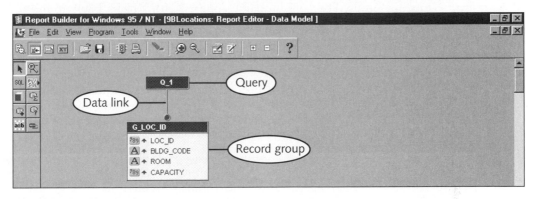

Figure 9-26   Report data model

> You can also open the Data Model view by clicking View on the menu bar and then clicking Data Model, or by double-clicking the report's Data Model icon in the Object Navigator.

The Data Model view has a tool palette for creating queries and report data fields. It also has a toolbar for navigating to other windows within Report Builder and for opening and saving files and running and printing reports. Presently, there are two objects in the report's Data Model: **Q_1**, which represents the SQL query that was typed in the Report Wizard when the report was created, and **G_LOC_ID**, which represents the record group created by the query. The line between Q_1 and G_LOC_ID is called a **data link** and shows that the Q_1 query determines the data that appears in the G_LOC_ID record group.

The Q_1 query retrieves the LOC_ID, BLDG_CODE, ROOM, and CAPACITY fields, and the G_LOC_ID record group contains these fields in the data model, as shown in Figure 9-26. The general format for the name of a record group is **G_*first_query_column***. Since LOC_ID is the first column retrieved in the query, the record group name is G_LOC_ID.

> The terms *column* and *field* are used interchangeably in Report Builder.

The icon in front of each column name indicates the column data type, and the break order arrow beside the icon indicates the column break order, which controls the order in which the column data values appear. Now you will open the Property Palette for the LOC_ID column in the G_LOC_ID record group and view its properties.

To open the Property Palette for the LOC_ID record group column:

1. In the Data Model view, select the **LOC_ID** column in the G_LOCID record group. The column name appears with a black background to show that it is selected.

2. Right-click, and then click **Property Palette**. The LOC_ID column's Property Palette opens.

The Name property (LOC_ID) is the same as the corresponding database field. The Column Type property describes the type of data that the column displays. Table 9-1 summarizes the different record group column types.

Column Type	Description
Database – Scalar	Discrete data value retrieved from a database table
Summary	Data value calculated by applying a summary function (such as SUM, AVG, or COUNT) to other report columns
Formula	Data value calculated by applying a user-defined formula to values in other report data columns

Table 9-1   Report column types

The LOC_ID Column Type is Database-Scalar, which indicates that this column displays a discrete value from a database table. The Datatype and Width properties describe the data type and maximum width of the data column, respectively. The Value if Null property allows the developer to substitute a different value for the data field if the retrieved data value is NULL.

You can use the report data model to modify the report query or to modify the record group structure. The following sections describe how to modify the report query, modify the order that data values appear in the report, and create a group filter to filter the data that the report query retrieves.

Modifying the Report Query

Now you will use the Data Model to modify the report query. You will open the query in the Data Model view and modify it by adding an ORDER BY clause to order the records by BLDG_CODE.

To modify the query in the Data Model:

1. Close the Property Palette, and then double-click **Q_1** in the Data Model view. The SQL Query Statement dialog box opens, showing the report's SQL query.

2. Modify the query so it appears as follows, and then click **OK**. The Data Model window appears again.

```
SELECT * from location
ORDER BY bldg_code
```

3. Click the **Live Previewer button** 🔎 on the toolbar to view the result of your change. The data records are now sorted by building codes.

 You could also change the order in which the records appear by opening the Report Wizard in reentrant mode and modifying the query on the Data page.

4. Click the **Data Model button** 🔳 to reopen the Data Model view, and then save the report.

Using Break Groups to Change the Data Order

In the previous exercise, you added the ORDER BY clause to the report SQL command to specify the order of the report data values. You can also control the order in which data values appear in a report by creating a break group. A **break group** is a user-defined record group that you create by moving, or breaking, one of the existing columns out of an existing record group. You can then sort the report data values using the break group column as the sort key. Suppose you want to sort the records in the location report by room capacity, in descending order. You would create a break group on the CAPACITY column and specify the column's Break Order property as Descending. Now, you will create a break group to sort the data based on room capacity.

To sort the report data by creating a break group:

1. In the Data Model view, select the **CAPACITY** column, and drag it upward so it is on the data link between Q_1 and G_LOC_ID. Drop the CAPACITY column. A new break group named G_CAPACITY appears in the Data Model.

2. Select the **G_CAPACITY** break group, and move it toward the right edge of the screen. Move the Q_1 query and G_LOC_ID record group as necessary so the Data Model objects appear as shown in Figure 9-27.

9

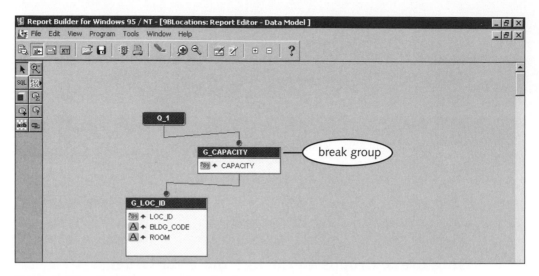

Figure 9-27   Creating a new break group

Note that the break order arrow beside the CAPACITY column points upward. When you create a user-defined break group, you can change the Break Order property value in the break group column's Property Palette to modify the data sort order. If a column is *not* in a user-defined break group, then the ORDER BY clause in the record group query controls the order in which the column data values appear, and changing the column's Break Order property value will have no effect on the order in which the column values appear in the report. Now you will change the Break Order property value of the CAPACITY column to Descending. Then you will view the report in the Live Previewer and examine the sorted data values.

To change the CAPACITY column Break Order property:

1. In the Data Model view, select the **CAPACITY column**, right-click, and then click **Property Palette** to open the column Property Palette.

2. Change the Break Order property value to **Descending**, select another property to save the changed value, and then close the Property Palette. Note that the break order arrow beside the CAPACITY column now points downward.

 In Report Builder, after you change a property value in the Property Palette, you often have to select another property to save the change.

3. Save the report, and then click the **Live Previewer button** 🖳 to view the report in the Live Previewer. The report data values now appear sorted by descending capacity.

4. Click the **Data Model button** on the toolbar to reopen the Data Model view.

To delete a user-defined break group, you must move the break group column back into the original record group. Then, you delete the break group. Now, you will delete the G_CAPACITY break group.

To delete the break group:

1. In the Data Model, select the **CAPACITY column** in the G_CAPACITY break group, drag it downward, and drop it back into the G_LOC_ID record group. CAPACITY now appears in the G_LOC_ID record group.

2. Select **G_CAPACITY**, and press **Delete**. The G_CAPACITY record group no longer appears in the Data Model.

3. Save the report.

Using a Group Filter to Control Report Data

Sometimes, you need to limit the data that appears in a report. For example, if a report might retrieve a very large data set, you might want to limit the number of retrieved records so the user is not forced to wait a long time for the query to complete. To do this, you can create a **group filter**, which is a structure that uses some criteria to limit the number of records that a report query retrieves. To create a group filter, you assign a value to the Filter Type property in the Property Palette of the associated report record group. You can set the record group Filter Type property value to *First*, which specifies that the report displays a specific number of records from the beginning of the retrieved data set; *Last*, which specifies that the report displays a specific number of records from the end of the retrieved data set; or *PL/SQL*, which allows you to write a PL/SQL function that evaluates each individual record to determine whether or not it appears in the report.

> Group filters defined using PL/SQL functions are very inefficient. When you need to filter data based on retrieved values, you should perform the filtering using a SQL search condition in the report query.

Now you will create a group filter to display the first ten Northwoods University locations that the data set retrieves. You will set the G_LOC_ID record group Filter Type property to First and specify to display the first 10 records. Then, you will view the report in the Live Previewer to see the result of the group filter.

To create a group filter:

1. In the Data Model Editor, click the border of the **G_LOC_ID** record group to select the record group. Selection handles appear on the borders of the record group.

2. Right-click, and then click **Property Palette** to open the record group Property Palette.

3. Select the **Filter Type** property, open the property list, and select **First**.

4. Change the Number of Records property value to **10**, select another property to save the change, and then close the Property Palette.

5. Click the **Live Previewer button** 📄 on the toolbar to open the Live Previewer. The report now displays only the first ten records from the LOCATION table.

6. Save the report, and then click the **Data Model button** 🔲 on the toolbar to reopen the Data Model view

The Report Layout Model

Often after creating a report using the Report Wizard, you need to manually reposition report items to improve the report appearance. For example, you might need to move related fields (such as first name, middle initial, and last name) so they appear beside one another on the report. Master-detail reports often are more attractive and easier to understand if there is blank space between the data for each new master record value. To reposition report items, you need to understand the underlying report structure and how the report layout relates to the data model.

When you examine the building locations report in the Live Previewer (see Figure 9-25), note that it has rows that show the data from each record in the LOCATION table. These data rows are called **repeating rows**, since each row shows the same data fields (LOC_ID, BLDG_CODE, ROOM, CAPACITY) but with different data values. Now you will view the report layout in the Layout Model view, which is also simply called the Layout Model. The Layout Model shows the report layout items as objects, rather than as actual data values. You can use the Layout Model to analyze and modify the report structure.

To view the report in the Layout Model:

1. Click the **Layout Model button** 🔲 on the toolbar. The report objects appear in the Layout Model, as shown in Figure 9-28.

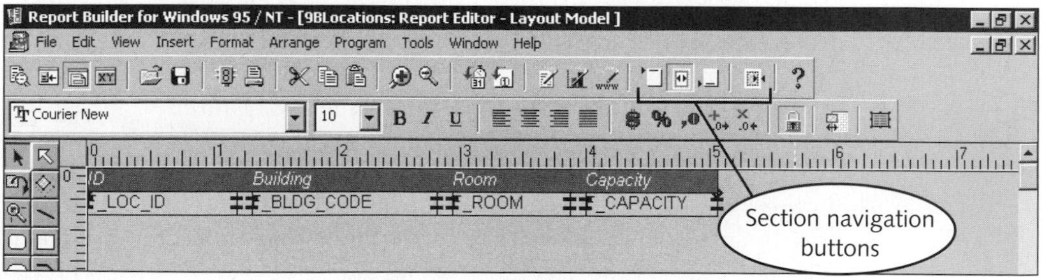

Figure 9-28   Building locations report in Layout Model

The Report Layout Model is similar to the Template Editor Layout Model and to the Form Builder Layout Editor: it has a tool palette, toolbar, painting region, and status line. In addition, the top toolbar has four section navigation buttons that allow you to move to different sections of the report.

A report has four sections: header, body, margins, and trailer. The **header** is an optional page (or pages) that appears at the beginning of the report and can contain text, graphics, data, and computations to introduce the report. The report **body** usually has multiple pages and contains the report data and computations. The report body has **margins** on each page, where you can put text, such as titles and page numbers. In the two reports you have created so far, the company logo, report title, date, and date label were in the report margin. The report **trailer**, like the header, is an optional page (or pages) that appears at the end of the report. The trailer can include summary data, or for a report with hundreds of pages, it could contain text to mark the end of the report.

The first button in the navigation button group is the **Header Section button** 🔲, which enables you to activate the report header and make it available for editing. The second button is the **Main Section button** 🔲, which enables you to activate the report body. The third button is the **Trailer Section button** 🔲, which allows you to activate the report trailer. The fourth button is the **Edit Margin button** 🔲, which allows you to toggle the margins portion of the active report section on or off. (You used 🔲 earlier to toggle between the body and the margins of the report template.) When you click 🔲, you can view and edit the margins of the report section (header, body, or trailer) that are currently active.

Now you will examine the individual components of the report body. In the Layout Model, you can see that there are boxes around both the repeating fields and the column headings. Actually, there are several boxes, or frames, on the report.

Report Frames

Frames are containers for grouping related report items so that you can set specific properties for a group of items, rather than having to set the property for each item. For example, all of the report column headings might be placed in a frame so that you can apply the same background color to all of the headings. In the Layout Model, individual frames are not always visible, because they are on top of each other. Figure 9-29 shows a schematic representation of the frames in the building locations report.

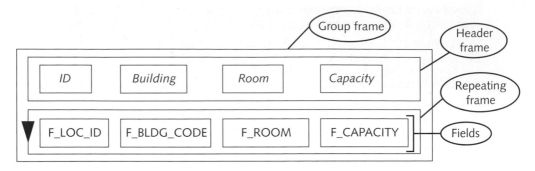

Figure 9-29   Building location report frames

Each record group in a report has a corresponding **group frame**, which encloses a repeating frame and an optional header frame. A **repeating frame** encloses repeating data rows and is designated with a down-pointing arrow on its left side. Each repeating row is comprised of multiple data fields, which are report items that display data values. When you create a report using the Report Wizard, a field is created for each column in the report record group, and each field is given the default name **F_*column_name***. Figure 9-29 shows that the record group data fields (F_LOC_ID, F_BLDG_CODE, F_ROOM, and F_CAPACITY) are enclosed in a repeating frame. Repeating frames have **variable sizing**, which means that they can shrink or stretch vertically depending on how many data records appear in the report. For example, if the Building Locations report query retrieves 13 records, then the repeating frame stretches so that all 13 records appear.

A **header frame** contains all of the column headings for a record group. In Figure 9-29, the column headings (ID, Building, Room, and Capacity) are enclosed in a header frame. Header frames are created only when report values appear in a tabular format.

If an object within a frame is moved outside of its frame, an error message appears in Report Builder when you view the report in the Live Previewer. For example, if you move the F_LOC_ID field outside of its repeating frame, an error message will appear. Furthermore, objects in a frame must be totally enclosed by their surrounding frames, or an error message will appear when you view the report in the Live Previewer. For example, in Figure 9-29, the header frame and repeating frame must be completely enclosed by the group frame. If one of these frames is outside of the group frame or if its borders overlap the borders of the group frame, an error message will appear.

Frame names are derived from the name of their associated record groups. Table 9-2 shows the general format used by the Report Wizard for naming report frames, as well as the names of the frames in the building locations report. (Recall that in the building locations report, the record group name is G_LOC_ID.)

Frame Type	Default Name Format	Frame Name in building location report
Group	M_record_group_name_GRPFR	M_G_LOC_ID_GRPFR
Header	M_record_group_name_HDR	M_G_LOC_ID_HDR
Repeating	R_record_group_name	R_G_LOC_ID

Table 9-2   Report default frame names

To become familiar with working with frames, you will select the different frames in the building locations report. It is difficult to select individual frames in a report, because they are on top of each other. The best way to select a specific report frame in the Layout Model is to select an item that is in the frame and then select the item's parent frame by clicking the **Select Parent Frame button** 🔲 on the Layout Model toolbar. A **parent frame** is the frame that directly encloses an object. In Figure 9-29, the header frame directly encloses the individual column headings, so the header frame is the parent frame of each column heading. Similarly, the group frame directly encloses the header frame, so the group frame is the header frame's parent frame. First, you will select the report header frame.

To select the header frame:

1. In the Layout Model, click the **Edit Margin button** 🔲, if necessary, to toggle the report body margins off and make the report body available for editing. The margins no longer appear, and the body becomes the active report section.

2. Click the **ID** column heading so that selection handles appear around its edges, and then click the **Select Parent Frame button** 🔲 on the toolbar to select the ID column heading's parent frame, which is the header frame. Selection handles appear around all of the column headings.

3. Click **Tools** on the menu bar, and then click **Property Palette**. The Property Palette opens, confirming that the header frame is selected and that the frame name property is M_G_LOC_ID_HDR.

 You cannot open a frame Property Palette by right-clicking and then clicking Property Palette, because right-clicking deselects the frame.

4. Close the Property Palette.

Recall that the group frame is the parent frame of the header frame. Since the header frame is currently selected, clicking 🔲 again will select the frame that directly encloses the header frame, which is the group frame. Next, you will select the group frame and examine its Property Palette.

To select the group frame and open its Property Palette:

1. In the Layout Model, with the M_G_LOC_ID_HDR header frame currently selected, click the **Select Parent Frame button** 🔲 to select the header frame's parent frame, which is the group frame. Selection handles appear around all of the report objects.

2. Click **Tools** on the menu bar, and then click **Property Palette**. The Property Palette for the group frame opens. The frame's name is M_G_LOC_ID_GRPFR.

3. Close the Property Palette.

Finally, you will select the report's repeating frame and open its Property Palette. To do this, you will select one of the report data fields, and then click 🔲 to select its parent frame, which is the repeating frame.

To select the report's repeating frame:

1. In the Layout Model, select **F_LOC_ID**. Selection handles appear around the data field.

2. Click the **Select Parent Frame button** 🔲 to select the field's parent frame, which is the repeating frame. Selection handles appear around all of the data fields.

3. Click **Tools** on the menu bar, and then click **Property Palette**. The Property Palette for the repeating frame opens. The repeating frame's name is R_G_LOC_ID. The Source property is G_LOC_ID, which is the record group that is the source of the repeating frame's data. Do not close the Property Palette.

You can change frame properties on the frame Property Palette. For example, you might want to change the frame background color, specify that the frame items will print on a new page, or specify special printing instructions, such as leaving a set amount of blank space between rows in a repeating frame. Now you will modify some of the properties of the R_G_LOC_ID repeating frame. You will modify the repeating frame so that only five records appear per page. Then you will increase the vertical spacing between each record so that 0.25 inches of blank space appears between each data row.

To modify the repeating frame properties:

1. On the R_G_LOC_ID repeating frame Property Palette, change the Maximum Records per Page property to **5**.

2. Change the Vert. Space Between Frames value to **.25**. Select another property to save the change, and then close the Property Palette.

3. Click the **Live Previewer button** 🔲 to view the report in the Live Previewer. The first five records appear on the first report page, with 0.25 inches of blank space between each row, as shown in Figure 9-30.

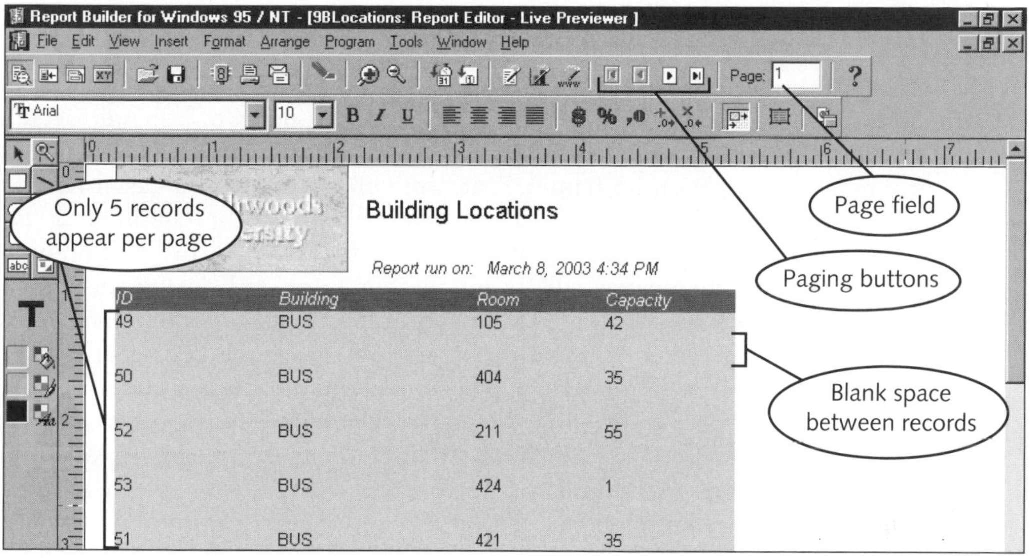

Figure 9-30   Building locations report with modified repeating frame properties

If red dashed lines appear in the Live Previewer, it means you currently have a frame selected in the Layout Model. Open the Layout Model, deselect the frame, and then open the Live Previewer again.

Since you modified the report to display five records per page, your report now has multiple pages. When a report has more than one page, the Live Previewer paging buttons are enabled (see Figure 9-30). The **Next Page button** ▶ allows you to step through the report pages, one page at a time. When you are on any report page except the first page, the **Previous Page button** ◀ is enabled, which allows you to step through the report pages in reverse order. The **Last Page button** ▶| displays the last page of the report, and the **First Page button** |◀ displays the first page of the report. The **Page field** displays the page number of the current report page. To jump to a specific report page number, you can type the page number in the page field and then press Enter. Now you will step through the report pages using the report paging buttons and Page field. Then, you will close the report and save your changes.

To step through the report pages, and then close the report:

1. Click the **Next Page button** ▶ on the toolbar. The next five records appear, and the Page field displays page 2.

2. Click the **First Page button** |◀ to navigate back to the first page of the report.

3. Type **2** in the Page field, and then press **Enter**. The second report page appears.

4. Click **File** on the menu bar, click **Close** to close the report, and then click **Yes** to save your changes.

COMPONENTS OF A MASTER-DETAIL REPORT

Now that you understand how to view and modify the components of a single-table report, you need to learn how to modify the components of a master-detail report to make it more readable. Reports with master-detail relationships have multiple record groups and multiple group frames. First, you will open a master-detail report that provides class lists for all of the classes offered during the Summer 2004 term at Northwoods University. The report output includes the term description and detail records showing the course call ID for all courses offered during the term. For each course call ID, detail records show information about the section number, instructor first and last name, day, time, building code, and room for each course offered during the term. For each section number, detail records show the student ID, last name, and first name of each student enrolled in the course. Therefore, the report has a total of three master-detail relationships: a term has multiple courses; a course has multiple course sections; and a course section has multiple students. First, you will open the report file, save it using a different filename, and then view the report in the Live Previewer.

To open, save, and view the master-detail report:

1. In the Object Navigator, click the **Open button** ![icon], navigate to the Chapter9 folder on your Data Disk, select **ClassLists.rdf**, and then click **Open**. The file opens in the Object Navigator.

2. Click **File** on the menu bar, click **Save As**, and save the report as **9BClassLists.rdf** in your Chapter9\TutorialSolutions folder.

3. Double-click the **Live Previewer icon** ![icon] to display the report in the Live Previewer. The report opens in the Live Previewer, as shown in Figure 9-31.

Figure 9-31 highlights some of the problems with the report's current format. Since the class list needs to be distributed to individual instructors, the information for each course should be printed on a separate page. The report would be easier to understand if the course data fields were stacked on top of each other vertically, rather than spread horizontally across the page. Fields that should appear adjacent to each other, such as the instructor first and last name, should be anchored together instead of being spaced to accommodate the largest possible field width. Also, the report would look better if there were blank lines between the course section and student information, and if the student fields were indented on the page. To make these changes, you will need to reposition the data fields, which requires modifying the frames generated by the Report Wizard. You will also need to modify some properties of the report frames. First, you will examine the report's master-detail data model.

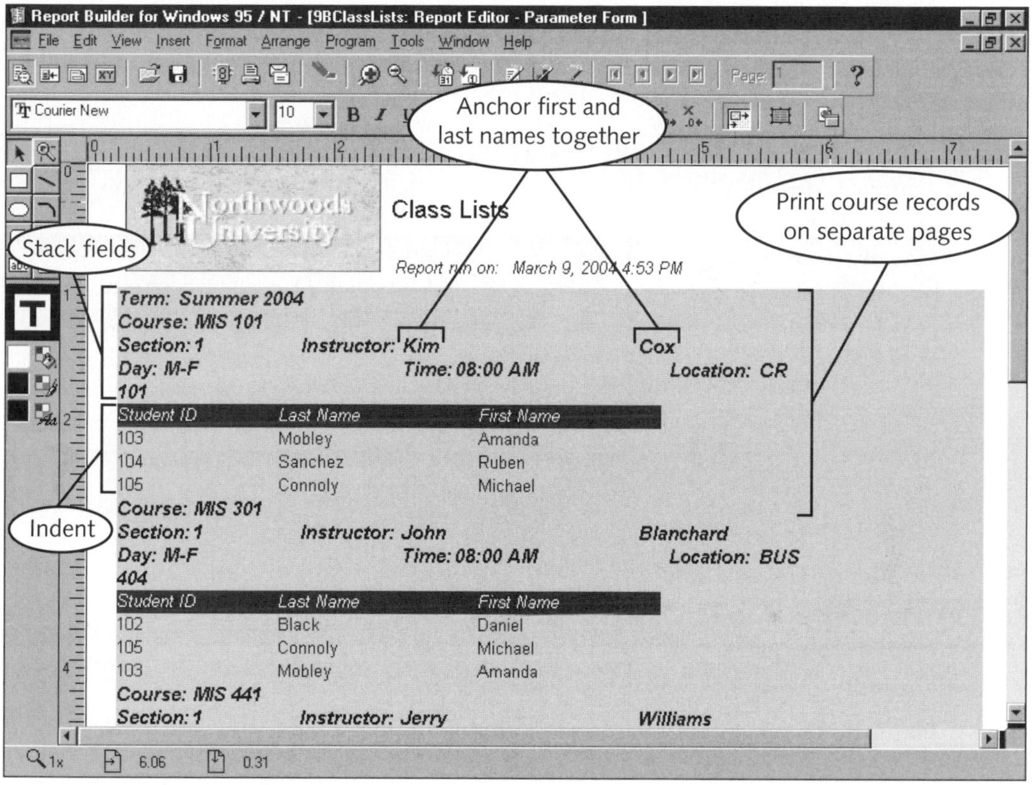

Figure 9-31   Class lists report

Master-Detail Data Model

The data model in a master–detail report is more complex than the data model for a single-table report, because it contains multiple record groups. For each master–detail relationship, the master records are in one record group, and the detail records are in a separate record group. Now you will examine the report data model.

To examine the data model:

1. Click the **Data Model button** ![icon] on the toolbar to open the Data Model view. The report record groups appear.

The report contains four separate record groups. Some of the record groups might currently appear off-screen. First, you will move the record groups so that they are all visible on the screen display. To move a record group, you select it and then drag it to the desired position.

To move the record groups so that they are all visible:

1. Select **Q_1**, and move it to the top-left corner of the Data Model painting region, as shown in Figure 9-32.

2. Select **G_term_desc**, and move it so that it is to the right and slightly below Q_1, as shown in Figure 9-32.

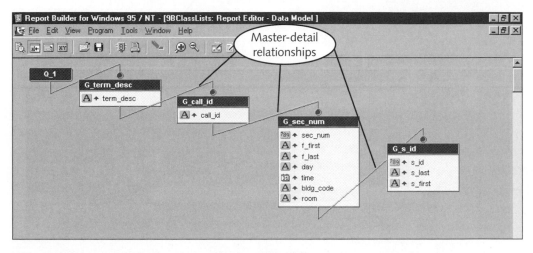

Figure 9-32   Repositioning the report record groups

3. Move the other record groups so that they are positioned as shown in Figure 9-32.

 If a scroll bar appears on the bottom or right side of one of the record groups, it means that the record group is not large enough to display all of its columns. Select the record group by clicking it, and then drag the right-center handle horizontally to the right or drag the bottom center handle vertically to the bottom so that all fields appear and the scroll bar no longer appears.

Note that there are four record groups: G_term_desc, G_call_id, G_sec_num, and G_s_id. Report data fields are grouped according to master-detail relationships, and each data link between two record groups represents a master-detail relationship. The highest-level (Level 1) group contains the term description, which is Summer 2004. Since the term has multiple course call ID values, the second (Level 2) record group contains the course call IDs. One course call ID might have several section numbers, so all of the data fields that are unique for a given course section (SEC_NUM, F_FIRST, F_LAST, DAY, TIME, BLDG_CODE, ROOM) appear in the Level 3 G_sec_num record group. Each section number might have several students, so the student fields appear in the Level 4 (most detailed) G_s_id record group.

Master-Detail Report Frames

Recall that in a report, each record group has an associated group frame. The class lists report contains four separate record groups, so it has four associated group frames that have the same hierarchical relationship as the record groups in the data model. The group frame associated with the G_s_id record group is enclosed by the group frame associated with the G_sec_num record group. The group frame associated with G_sec_num is enclosed by the group frame associated with G_call_id, and so forth. The group frame associated with the G_term_desc record group encloses all of the other group frames.

To make the report formatting modifications specified in Figure 9–31, you will have to reposition the data fields within the frames. You will also need to modify properties of individual frames. Now you will view the report frames in the Layout Model so you can select individual frames and modify their properties.

To view the report in the Layout Model:

1. Click the **Layout Model button** 🖻 on the toolbar. The master-detail report layout appears in the Layout Model, as shown in Figure 9–33. If the report body margins appear, click the **Edit Margin button** 🖼 to toggle the margins off.

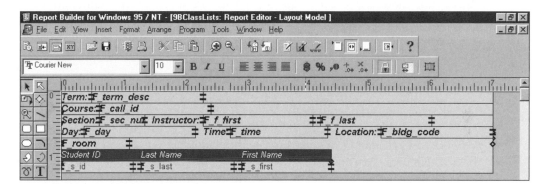

Figure 9-33   Master-detail report layout

Currently, it is difficult to interpret the report frame relationships in the Layout Model, because many of the frames are stacked on top of each other. You will need to modify the sizes of some of the frames so you can move items around in the frames and create space between different record group items. You will begin by examining a series of schematic diagrams that explain the frame structure of the report. Then, you will examine the individual report frames in the Layout Model.

First, you will examine the layout of the G_s_id record group's student fields. Recall that when record group fields appear in a tabular form, a header frame encloses a record group's column headings, a repeating frame encloses the record group's data fields, and a group frame encloses the record group header frame and repeating frame. Figure 9–34 shows a schematic diagram of the frames that were created for the G_s_id record group. These frames exist in the Layout Model in Report Builder, but they are hard to see because the header frame and repeating frame are stacked directly on top of each other, and the group frame is directly on top of these two frames.

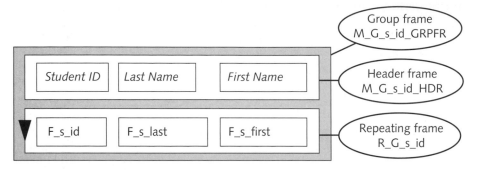

Figure 9-34   G_s_id record group frames

Figure 9–34 shows that the data fields that display the student data (F_s_id, F_s_last, F_s_first) are enclosed by a repeating frame named R_G_s_id. The column headings associated with these data fields (Student ID, Last Name, and First Name) are enclosed by a header frame named M_G_s_id_HDR. Both the repeating frame and header frame are enclosed by a group frame named M_G_s_id_GRPFR, which is shaded in the figure.

In a report with multiple record group levels, frames for more detailed record groups are nested inside frames for less detailed record groups. Recall that in this report, the Level 1 record group (G_term_desc) shows information for a term description. The Level 2 record group (G_call_id) shows information for a particular course call ID. The Level 3 record group (G_sec_num) shows information for a specific course section associated with that term and course call ID. The Level 4 record group (G_s_id) lists the students that are in a specific course section. The Level 1 record group's group frame contains the Level 2 record group's group frame, and the Level 2 record group's group frame contains the Level 3 record group's group frame. Figure 9-35 shows a schematic representation of these frames. To contrast the relationships between repeating frames and group frames, the repeating frames are filled in white, and the group frames are shaded.

Now you will examine the frames in the Layout Model so that you become familiar with the frame relationships and gain more experience selecting specific frames. Recall that these frame relationships are hard to see because many of the frames appear directly on top of each other. Recall that the best way to select a frame is to select an item inside the frame, and then click ▣. Since sometimes a frame and its parent frame are directly on top of each other, there might be no visual change on your screen when you click ▣. To keep track of which frame is currently selected, you can open the frame Property Palette immediately after you select the frame and check the frame name.

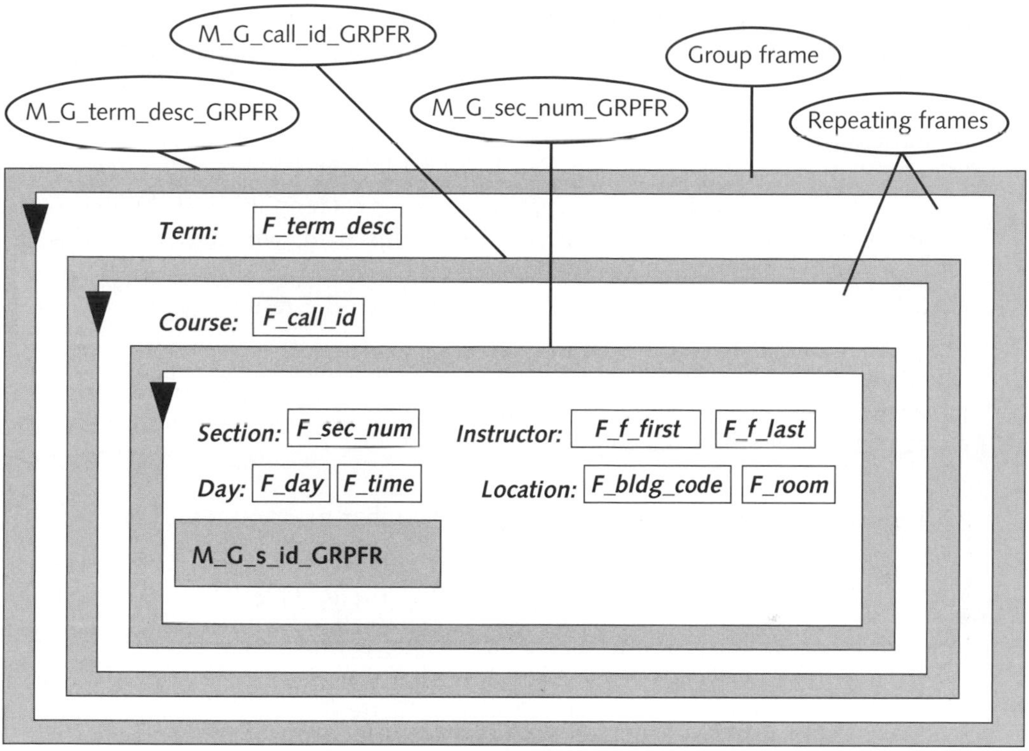

Figure 9-35   Master-detail report group frame relationships

To examine the G_term_desc record group frames:

1. In the Layout Model, select the **F_term_desc** data field, and then click the **Select Parent Frame button** 🔲 to select its parent frame, which is the G_term_desc repeating frame (R_G_term_desc). Since this repeating frame surrounds all of the report items, selection handles appear around all of the report items.

2. Click 🔲 to select the repeating frame's parent frame, which is the G_term_desc group frame (M_G_term_desc_GRPFR, as shown in Figure 9-35). There is no visible change in the selection handles that appear in the Layout Model, because the group frame is directly on top of the repeating frame.

To select a specific repeating frame, you can use this process of selecting an item in the frame, and then clicking 🔲 to select its parent frame. To select a group frame, you can select an item in the frame, click 🔲 to select its repeating frame, and then click 🔲 again to select the repeating frame's parent frame, which is the group frame. Now that you understand the report structure, you will format the report by selecting specific report frames and modifying their properties.

Printing Report Records on Separate Pages

The first change you will make to the report format is to make the data for each course print on a separate page. To create a page break between sets of repeating records, you must open the Property Palette for the repeating frame that contains the records that you want on each page and change its Maximum Records per Page Property to 1. For this report, you want to print separate pages for each course call ID. To do this, you will select the repeating frame which encloses this field, which is R_G_call_id, open its Property Palette, and change its Maximum Records per Page property to 1. Then you will view the report in the Live Previewer to see the result.

 Sometimes you have to experiment to determine which repeating frame's Maximum Records per Page property must be changed to make report page breaks appear as desired.

To make the data for each course section number print on a separate page:

1. In the Layout Model, select the **F_call_id** data field, and then click the **Select Parent Frame button** ▦ to select the R_G_call_id repeating frame.

2. Open the frame Property Palette, and confirm that you have selected R_G_call_id. Change the Maximum Records per Page property to **1**, select another property in the Property Palette to save the change, and then close the Property Palette.

3. Click the **Live Previewer button** ▧ to display the report in the Live Previewer. The report output appears as shown in Figure 9-36, with the course data and class list for MIS 101 Section 1 on the first report page.

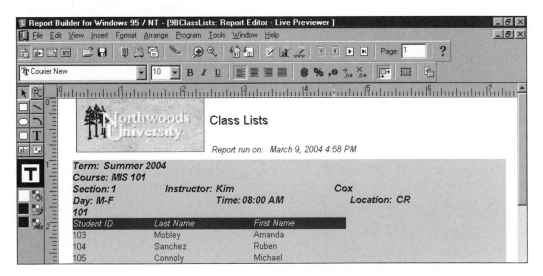

Figure 9-36   Report with course data on a separate pages

4. Click the **Next Page button** ▶ to view the next class list. The class list for MIS 301 Section 1 appears.

5. Click ▶ again. The class list for MIS 441 Section 1 appears.

6. Click the **First Page button** ◀. The first page of the report appears again.

7. Click the **Save button** 🖫 on the toolbar to save the report.

Repositioning the Report Objects

The next formatting task is to reposition the report objects to improve the report appearance. You want to move the course section-specific fields and labels (section number, instructor, day, time, and location) so that they stack on top of each other. You also want to leave some blank space between the course section fields and the student fields, and indent the student fields. When you reposition report objects, you must be careful not to move objects outside of their enclosing frames. If you move an object outside of its enclosing frame, the report layout no longer is consistent with the report data model, and an error message appears when you try to view the report in the Live Previewer.

Report Builder has a property called **confine mode**, which determines whether you can move objects outside of their enclosing frames. When confine mode is enabled, you cannot move an object outside of its enclosing frame. When confine mode is disabled, you can freely drag objects anywhere on the Layout Model painting region. You can enable confine mode by clicking the **Confine Mode button** 🔒 on the Layout Model toolbar so that it is selected. You can disable confine mode by clicking 🔒 again. It is safest to always leave confine mode enabled, because it is difficult to locate an object that has been moved outside of its enclosing frame and place it back into the correct frame.

Another way to enable confine mode is to right-click anywhere in the Layout Model, and then click Confine Mode on the pop-up menu so that Confine Mode is checked. To disable confine mode, right-click anywhere in the Layout Model, and then click Confine Mode on the pop-up menu so that Confine Mode is no longer checked.

Another important Report Builder property is flex mode. When **flex mode** is enabled, an enclosing frame automatically becomes larger when you move an enclosed object beyond the enclosing frame's boundary. To enable flex mode, you click the **Flex Mode button** 🔲 on the Layout Model toolbar. (You can determine if flex mode is enabled or disabled by right-clicking on the Layout Model painting region and seeing if Flex Mode is checked on the pop-up menu.) To disable flex mode, you click 🔲 so the button is not pressed. Flex mode overrides confine mode: When flex mode is enabled, frames are automatically resized regardless of whether confine mode is enabled or disabled.

A problem with enabling flex mode is that when you move a report field, flex mode automatically resizes all of the surrounding frames. Sometimes this causes the outermost frames to extend beyond the boundaries of the report body, which generates an error when you view the report in the Live Previewer. Flex mode works well when you need to make a frame longer vertically, because there is usually sufficient extra space on the length of the report page to resize all of the surrounding frames. However, when you need to make a frame wider, it is best to leave flex mode off and resize the frame manually, so you can ensure that report frames do not extend beyond the report body boundaries. Now you will enable flex mode, and then you will move a report column lower on the report to make its enclosing frames longer.

To use flex mode to make report frames longer:

1. In the Layout Model, click the **Flex Mode button** 🔲 on the Layout Model toolbar so that it appears pressed. If it is already pressed, leave it pressed.

2. Right-click in the blank area of the Layout Model painting region below the report objects, and confirm that Flex Mode has a check mark before its name on the menu, which indicates that it is enabled. If Flex Mode already has a check mark, do not click Flex Mode again or you will disable it.

3. Click the **F_s_id** field to select it, press **Shift**, click the **F_s_last** field while keeping the Shift key pressed, and then click the **F_s_first** field to select all three columns as a group. Drag the columns down until their bottom edges are 1.5 inches from the top of the report, as shown in Figure 9-37. Notice that as you drag the columns down, their parent frame (R_G_s_id) automatically resizes to accommodate the new column locations.

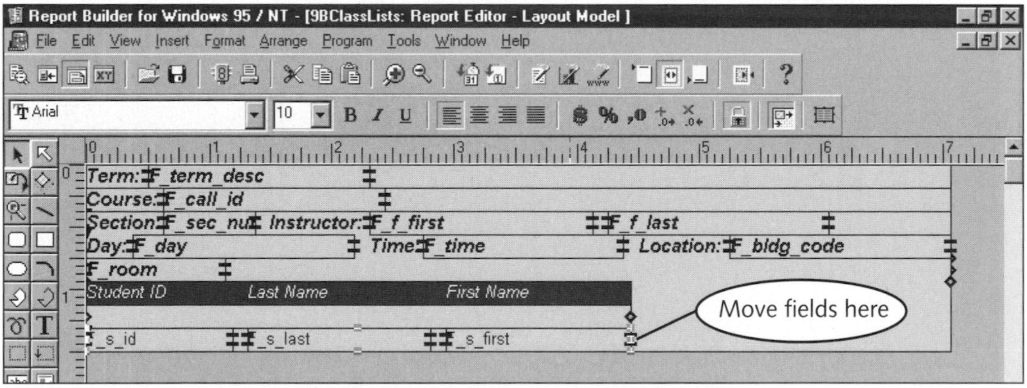

Figure 9-37   Using flex mode to make a report frame longer

4. Click the **Live Previewer button** 🔲 to view the report in the Live Previewer, and note the new positions of the student data fields. There is now about 0.5 inch of white space between each student record as a result of vertically enlarging the repeating frame and moving the student columns lower in the frame.

5. Click **File** on the menu bar, and then click **Revert** to revert the report back to its state the last time you saved it. Click **Yes** to undo your changes. The Object Navigator window opens.

6. Double-click the **Layout Model icon** 📄 to reopen the Layout Model.

Now you will examine the effects of using flex mode to make report frames wider. You will move a report column to the right edge of the report, which causes the enclosing frames to extend horizontally. Recall that this sometimes causes the report frames to extend beyond the report body boundary, as you will see next.

To examine the effects of using flex mode to make report frames wider:

1. Click the **Flex Mode button** 🔲 to enable flex mode.

2. Click **F_s_first** to select it, and then drag it to the right edge of the report so that the right edge of F_s_first is 5½ inches from the left edge of the report, as shown in Figure 9-38. Note that the frames enclosing F_s_first resize as you drag the data column to the right edge of the report.

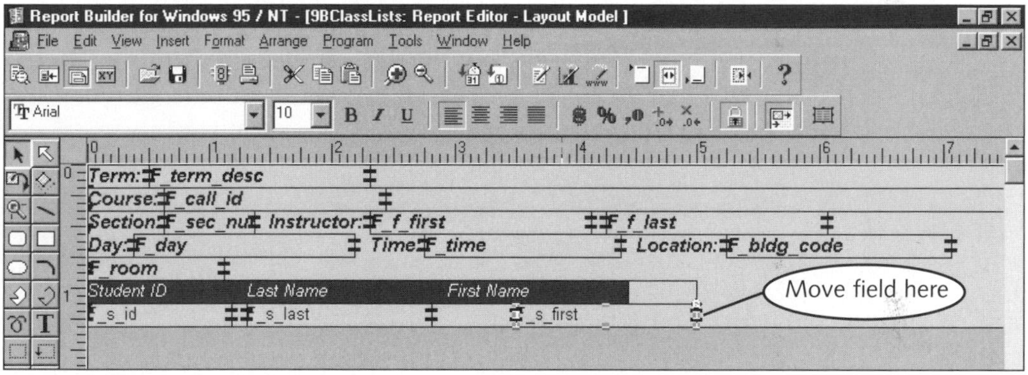

Figure 9-38   Moving F_s_first to the right edge of the report

3. View the report in the Live Previewer. The error message "REP-1212: Object 'Body' is not fully enclosed by its enclosing object room." appears. This error message indicates that one of the report objects extends beyond the edges of the report body.

4. Click **OK**, and then open the Layout Model and scroll to the right edge of the report. You will see that some of the outer report frames now extend beyond the report body boundaries.

5. Click **File** on the menu bar, click **Revert**, and then click **Yes** to undo your changes. The report appears in the Object Navigator.

6. Double-click the **Layout Model icon** 📄 to reopen the Layout Model.

Now that you understand the intricacies and peculiarities of confine mode and flex mode, you are ready to reposition the report objects. First, you will use flex mode to make the R_G_sec_num repeating frame and its surrounding frames longer, so that there is room to stack the course section fields vertically.

To use Flex Mode to resize the R_G_sec_num repeating frame and its surrounding frames:

1. In the Layout Model, click the **Flex Mode button** 🖳 to enable flex mode.

2. Click **F_sec_num** to select it, and then drag it to the lower edge of the report so that the bottom edge of the outermost frame is about 2½ inches from the top of the report, as shown in Figure 9-39.

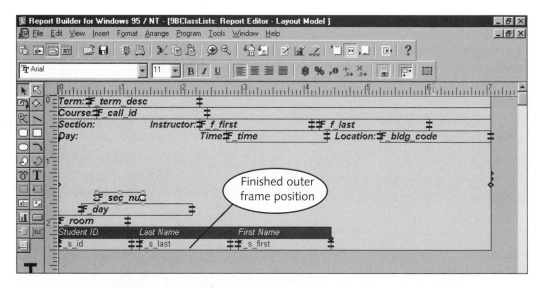

Figure 9-39   Resizing the report frames

3. Click the **Live Previewer button** 🔖 to view the report in the Live Previewer and confirm that all objects are properly within their enclosing frames. Then save the report.

Now you will reposition the report fields and labels. You will indent the student fields and vertically stack the course section fields and labels. First, you will select the student group frame and move it to the right so that the student fields are indented. There is enough room in the frames that enclose the student group frame, so you will do this with Flex Mode disabled. (If you moved the group frame to the right with Flex Mode enabled, it would enlarge the surrounding frames.) Then you will reposition the course fields and labels.

To indent the student group frame and reposition the course fields and labels:

1. In the Layout Model, make sure that flex mode is disabled. Select **F_s_id**, and then click the **Select Parent Frame button** 🔲, which selects the student repeating frame. Click 🔲 again, which selects the student group frame.

2. Click **Tools** on the menu bar, click **Property Palette**, and confirm that the student group frame (M_G_s_id_GRPFR) is selected. (If M_G_s_id_GRPFR is not selected, repeat steps 1 and 2.) Close the Property Palette.

3. Press the **right arrow key** to move the group frame so that its left edge is indented about 0.25 inches from the left edge of the report body, as shown in Figure 9–40.

If you try to use the mouse to drag a frame, it will be deselected. Always use the arrow keys to move frames in the painting region.

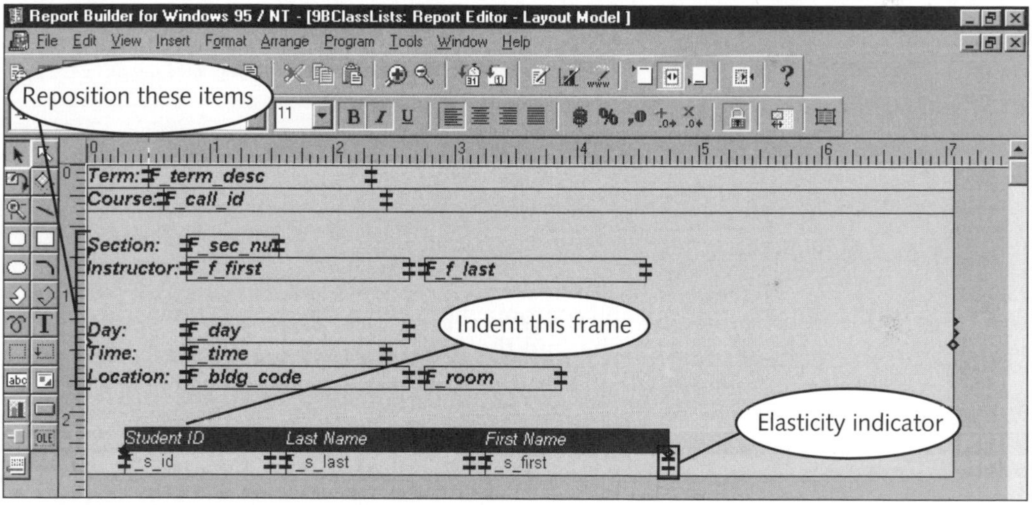

Figure 9-40   Repositioning the report fields and labels

4. Reposition the report fields and labels as shown in Figure 9–40.

5. Run the report to make sure it runs correctly.

6. Save the report, and then click the **Layout Model button** 🔲 to view the report in the Layout Model.

Adjusting the Spacing Between Report Columns

Another formatting task is to adjust the spacing between report fields so that data values that should appear next to each other (such as the instructor's first and last name) are not separated by blank space. You have probably noticed that some of the report fields have horizontal tick marks on their right and left edges, as indicated in Figure 9-40. These marks reflect the field's **elasticity**, which determines whether a field's size is fixed on the printed report, or whether the field can expand or contract automatically, depending on the height and width of the retrieved data value. Figure 9-41 shows the different elasticity options and associated markings on report field horizontal and vertical borders.

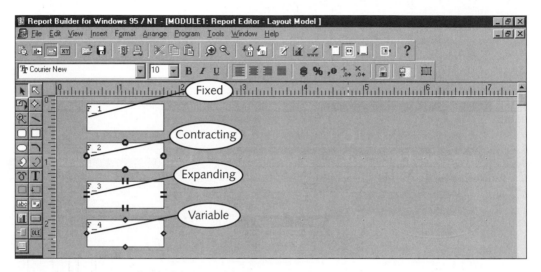

Figure 9-41   Report field elasticity indicators

The markings on the top and bottom edges specify the **horizontal elasticity property**, or how wide the field can be, and the markings on the right and left edges specify the **vertical elasticity**, or how tall the field can be. **Fixed elasticity** means that the column will have the exact size shown in the layout, and extra text will be truncated. **Contracting elasticity** means that the column will contract if the data value is smaller than the layout size, but wider data will be truncated. **Expanding elasticity** means that the field will expand automatically to accommodate wider data values, but narrower data values will still occupy the entire space shown in the layout, and extra blank space will appear. **Variable elasticity** means that the report field will contract or expand as needed to fit the data value. A field can have **mixed elasticity**, which means that the vertical and horizontal elasticity can be different. By default, data fields created by the Report Wizard have horizontally expanding and vertically fixed elasticity.

To make adjacent report fields, such as the instructor first and last name, appear directly next to each other on the report regardless of the width of individual data values, you must change the horizontal elasticity of the fields to *variable*. Now you will change the horizontal elasticity for the F_f_first, F_f_last, F_bldg_code, and F_room fields to variable. You will select the fields as a group and then change the elasticity property using an intersection Property Palette.

To adjust the field elasticity:

1. In the Layout Model, click **F_f_first**, press **Shift**, and then click **F_f_last**, **F_bldg_code**, and **F_room** to select the columns as an object group.

2. Click **Tools** on the menu bar, and then click **Property Palette** to open the intersection Property Palette. Change the Horizontal Elasticity property to **Variable**, and then click anywhere on the Property Palette to save the change.

3. Close the Property Palette and click anywhere on the blank region of the Layout Model to deselect the fields. Note that a diamond symbol now appears on the top and bottom edges of all the selected data fields to indicate that the horizontal elasticity is variable.

4. Click the **Live Previewer button** 🖹 to view the report in the Live Previewer. The Instructor and Location fields appear next to each other.

5. Save the report.

Separating Tabular Data with Variable-Length Lines

Sometimes when report data appears in a tabular format, you might want to add grid-lines between columns to separate data values and make report data appear in a grid. For example, you might want to enclose the Student ID, Last Name, and First Name columns in a grid. To do this, you draw boilerplate lines on the report in the Layout Model painting region. You draw a vertical line on the left edge of each data field and a vertical line on the right edge of the right-most data field. The vertical lines must be the same height as the repeating frame that contains the data values that you want to separate. You draw a single horizontal line under all of the data fields. To make the vertical lines stretch based on the number of retrieved records, you open the Property Palettes for the vertical lines and set the Line Stretch with Frame property to the name of the repeating frame that determines the line height.

Now, you will draw vertical boilerplate lines that appear on the left edges of the F_s_id and F_s_last fields and on the left and right edges of the F_s_first field. You will open an intersection Property Palette for the lines and change the Line Stretch with Frame property to the name of the repeating frame that encloses the fields, which is R_G_s_id (see Figure 9-34). You will also draw a horizontal line on the bottom edge of the R_G_s_id frame for the grid's bottom border.

To draw the lines to format the tabular data in a grid:

1. Click the **Layout Model button** 🖻 to open the Layout Model.

2. Click the **Zoom In button** 🔍 on the toolbar to magnify the Layout Model objects.

3. Click the **Line tool** ◥ on the tool palette, and then draw the four vertical lines shown in Figure 9-42. (You will need to scroll to the right in the magnified view to draw the line on the right edge of the F_s_first field.)

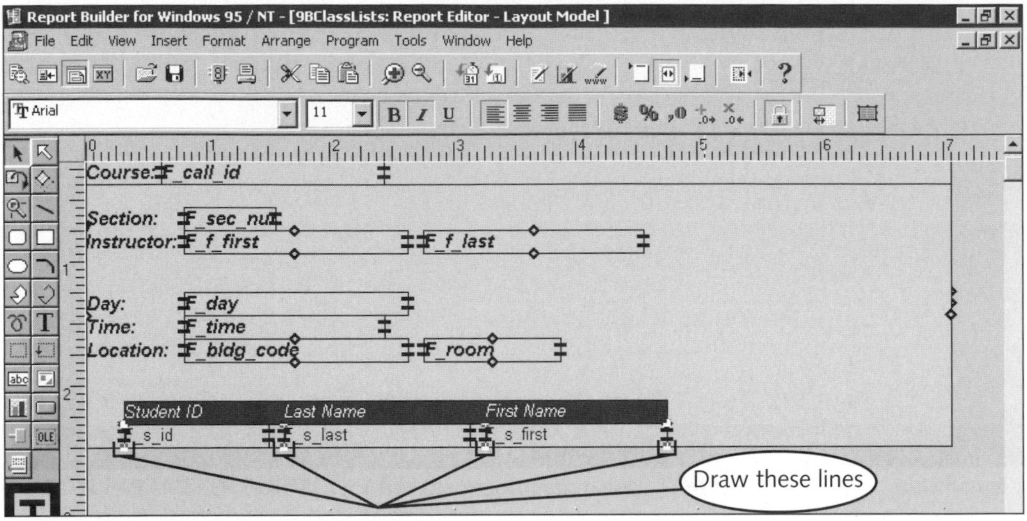

Figure 9-42   Drawing the vertical lines

4. Click **View** on the menu bar, click **Normal Size** to return the Layout Model to normal magnification, and then draw a horizontal line on the lower edge of the R_G_s_id frame.

5. Select the vertical line on the left edge of the F_s_id field, press and hold the **Shift** key, and then select the rest of the vertical lines as an object group. Your screen display should look like Figure 9-42.

6. Click **Tools** on the menu bar, click **Property Palette** to open the intersection Property Palette, change the Line Stretch with Frame property to **R_G_s_id**, select another property to save the change, and then close the Property Palette.

> If the Line Stretch with Frame property does not appear on the intersection Property Palette, then you selected an object other than a line. If you have trouble selecting the lines as an object group, select each line individually, and perform Step 6 separately for each line.

7. Click the **Live Previewer button** 🔲 to view the lines in the Live Previewer. The tabular data appears in a grid, as shown in Figure 9-43.

8. Save the report.

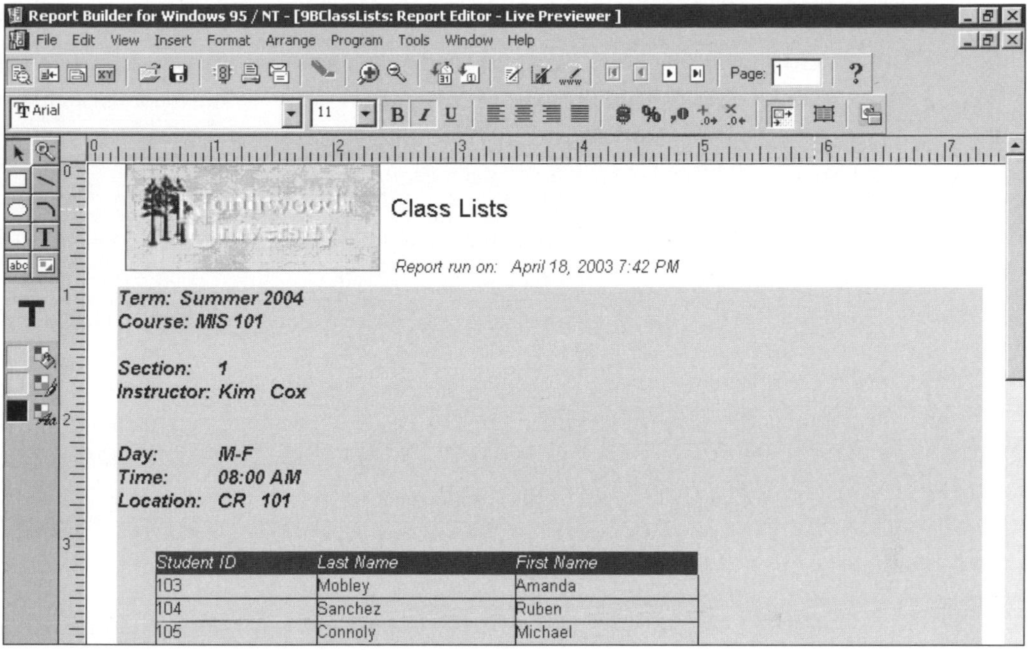

Figure 9-43   Report with tabular data in grid

CREATING AND USING REPORT PARAMETERS

Reports have two types of parameters: system parameters and user parameters. **System parameters** specify properties that control how the report appears in the user display and how the report application environment behaves. Examples of system parameters include the currency symbol that appears and whether the print dialog box opens when the user prints the report. **User parameters** allow the user to select values that specify the report data content. For example, a user could open a list of values and select a specific term ID as a user parameter, and the report would display the class lists for that term only.

System Parameters

Table 9-3 summarizes the report system parameters and describes what the parameters specify.

Parameter Name	Specifies
BACKGROUND	Whether the report should run in the foreground or the background.
COPIES	Number of report copies that are made when the report is printed.
CURRENCY	Symbol for the currency indicator, such as "$".
DECIMAL	Symbol for the decimal indicator, such as ".".
DESFORMAT	Definition of the output device format. You might use this parameter when sending a report to a file to specify whether to create PDF or HTML output.
DESNAME	Name of the output device, such as the filename, printer name, or email userid.
DESTYPE	Type of device to which to send the report output, such as screen, file, email, or printer.
MODE	Whether the report should run in character or bitmap mode.
ORIENTATION	Report print direction (landscape or portrait).
PRINTJOB	Whether the Print Job dialog box should appear before the report is run.
THOUSANDS	Symbol for the thousands indicator, such as ",".

Table 9-3   Report system parameters

You can modify report system parameter values by opening the Property Palette of the target system parameter in the Object Navigator. To open a system parameter Property Palette, you open the report Object Navigator, click **+** beside the Data Model node, click **+** beside the System Parameters node, and then
open the Property Palette for the system parameter you wish to modify. In most cases, you accept the default system parameter values.

User Parameters

User parameters allow the user to customize the data that the report displays. Currently, the class lists report shows class lists for only the Summer 2004 term. The report would be more flexible if users could select from a list of term descriptions, and the report would display the class lists for the selected term. To implement a report that allows users to specify the data that appears, you will create a user parameter. In a report, a user parameter is a value that the user selects from a parameter list. A **parameter list** shows possible values that can be used in a report search condition. After the user selects the value, it is inserted in the report query as a search condition. The report then retrieves data based on the selected search condition value.

To create a user parameter, you create a user parameter object in the report Data Model. Then, you open the user parameter Property Palette and specify the SQL query that retrieves the values that appear in the parameter list. You should always use number fields rather than character fields for user parameters, because sometimes Oracle applications add blank spaces to pad out character fields, making it difficult to get an exact match between the parameter value in the search condition and the data value in the database column. Therefore, you will use TERM_ID for the actual parameter and display the associated term description (TERM_DESC) along with the TERM_ID in the parameter list.

Now you will create a user parameter associated with the TERM_ID field. After you create the user parameter object, you will change the object name. This name appears on the parameter form on which the user selects the parameter value, so it should describe the contents of the parameter list. Although the name appears in the Object Navigator in all capital letters, it appears in the parameter list heading using mixed-case letters, and Report Builder replaces the underscores with blank spaces. You will name the user parameter TERM_DESCRIPTION in the Object Navigator, and it will appear in the parameter list heading as Term Description. You will also change the user parameter data type to NUMBER, since the parameter data type must match the data type of the TERM_ID field in the database.

To create the user parameter:

1. Click **Window** on the menu bar, and then click **Object Navigator**. The report Object Navigator opens.

2. Click **+** beside the Data Model node, and then select the **User Parameters node**.

3. Click the **Create button** on the Object Navigator toolbar to create a new user parameter. A new parameter appears.

4. Click the new parameter, and then click it again, if necessary, so that the background of the parameter name turns blue. Change the name of the new user parameter to **TERM_DESCRIPTION**.

5. Double-click the **User Parameter icon** beside TERM_DESCRIPTION to open the user parameter Property Palette.

6. Make sure the Data type value is **Number**, and change the Width property to **5** to match the Width property of the TERM_ID field in the TERM table.

The next step is to specify the values that appear in the parameter list. To do this, you select the user parameter List of Values property, which causes the More button to appear. The More button allows you to open a dialog box in which you can define the list of values from which the user can choose the report input parameter value. You can enter a static list of predetermined input values, or you can enter a SQL query that returns a list of values. You will enter a SQL query that retrieves term IDs and term descriptions. The first field returned by the query must be the user parameter field (TERM_ID) that will be used as a search condition in the report's data model query. You will select the option of not displaying the TERM_ID in the user parameter list of values, so the selection list that the user sees will display only the second field returned by the query, which is the term description (TERM_DESC). Now you will specify the user parameter list of values.

To specify the user parameter list of values:

1. Click the space next to the List of Values property, and then click the **More button** [...]. The Parameter List of Values dialog box opens.

2. Select the **SELECT Statement option button**.

3. Type the command shown in Figure 9-44 in the SQL Query Statement text box.

4. Clear the **Restrict List to Predetermined Values check box**, because you want the list to change dynamically as new terms are added to the TERM table.

5. Check the **Hide First Column check box** to hide the term ID in the list of values. Your completed Parameter List of Values dialog box should look like Figure 9-44.

6. Click **OK**, close the Property Palette, and then save the report.

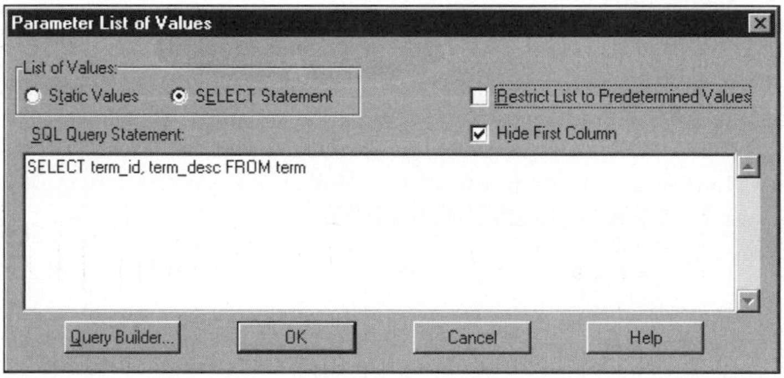

Figure 9-44   User parameter List of Values specification

Now you need to modify the report query to retrieve only the records for the term selected by the user and assigned to the user parameter. Currently, the query clause that contains the search condition to determine which term's records are retrieved is AND term_desc = 'Summer 2004'. You will change the search column to TERM_ID, and you will set the search value to reference to the user parameter. To reference a user parameter in a query, you preface the parameter name with a colon, using the following syntax: **:parameter_name**. The TERM_DESCRIPTION user parameter will be referenced as **:TERM_DESCRIPTION**.

To modify the report query:

1. In the Object Navigator, double-click the **Data Model icon** to open the Data Model, and then double-click **Q_1** to view the data model SELECT statement.

2. Modify the SELECT statement by replacing the line `AND term_desc = 'Summer 2004'` with the following line:

`AND term.term_id = :TERM_DESCRIPTION`

3. Click **OK** to save the changes and close dialog box, and then save the report.

Now you will view the report. When a report has a user parameter, the report parameter form appears before the report appears. The **parameter form** allows the user to specify values for system and user parameters. For a user parameter with a parameter list, the user can directly enter a value for the user parameter, or open the parameter list and select a value. To ensure that the parameter form appears, you should view the report in the Runtime Previewer window. The **Runtime Previewer** window duplicates how the report will appear to the user by showing the parameter form first, and then displaying the report in the Live Previewer. Now you will run the report in the Runtime Previewer, select the Spring 2004 term from the parameter list, and then view the report.

To run the report using the Runtime Previewer, select a parameter value, and view the report:

1. Click **View** on the menu bar, and then click **Runtime Preview** to run the report using the Runtime Previewer. The report parameter form opens and displays a text box for entering the Term Description parameter value and a list for selecting a value from the parameter list, as shown in Figure 9-45.

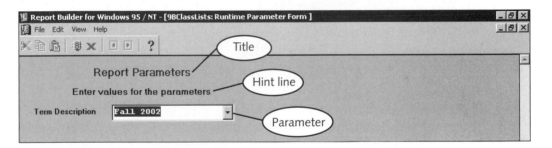

Figure 9-45 Report parameter form

The parameter form displays a title and hint line and a list of one or more parameter text boxes with associated prompts. Since you specified a list of values for the term description parameter, the parameter's text box appears as a list. The parameter form also has a toolbar that allows you to cut, copy, and paste parameter values. To run the report and display the values in the Live Previewer, you select a parameter value, and then click the **Run Report button** on the toolbar. To close the parameter form without running the report, you click the **Cancel Run button**. Now, you will select the Summer 2004 term from the parameter list and run the report.

To select the parameter value and run the report:

1. Open the parameter list, select the **Spring 2004** term, and then click the **Run Report button** to display the report. The class lists for the Spring 2004 term appear.

No report data values will appear for the Fall 2002, Spring 2003, and Summer 2003 terms, because there are no course section records for these terms in the database.

Using the Parameter Form Builder to Modify Report Parameters

You can use the Parameter Form Builder utility to customize the appearance of the report parameter form. This utility allows you to specify the parameter form title, and the text that appears on the hint line and on the status line when the parameter form opens. You can also specify to display one or more system parameters on the parameter form and allow the user to modify the system parameter values at runtime. For example, you might want to allow the user to select the DESTYPE, which specifies whether the report is routed to the screen, to a printer, and so forth. Now you will open the Parameter Form Builder and examine its properties.

To open the Parameter Form Builder and examine its properties:

1. Click the **Close Previewer** button ✕ to close the Runtime Previewer.

2. Click **Tools** on the menu bar, and then click **Parameter Form Builder**. The Parameter Form Builder window opens, as shown in Figure 9-46.

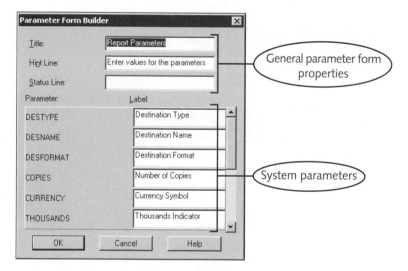

Figure 9-46   Parameter Form Builder window

The Parameter Form Builder allows you to specify the general properties for the report parameter form. The parameter form title is the top text line that appears on the parameter form. The hint line is the second line of text that appears on the parameter form, and the status line is the third line of text that appears on the parameter form. The Parameter Form Builder also lists all of the system parameters that you can allow the user to modify at runtime. To specify to display a system parameter at runtime, you select the system parameter on the

Parameter Form Builder window so the parameter name appears on a black background. To change the name of the prompt that appears beside the text box for the system parameter, you type the new prompt in the text box in the Parameter Form Builder window. You can also use the Parameter Form Builder to change the prompt that appears beside user parameters and to hide user parameters on the parameter form if you wish to disable them.

Now, you will customize the class list report parameter form by changing the parameter form title, hint line, and status line. You will also select the DESTYPE system parameter value and allow the user to specify the report destination (screen, printer, and so on) at runtime. Then, you will run the form and view the parameter form changes.

To customize the class list report parameter form:

1. In the Parameter Form Builder window, change the Title to **Northwoods University**.

2. Change the Hint Line to **Class List Report Parameters**.

3. Change the Status Line to **Enter parameter values**.

4. Select **DESTYPE** in the Parameter list so its background changes to black.

5. Scroll down to the bottom of the Parameter list, and note that the Term Description user parameter that you previously created is selected for display.

6. Click **OK** to save your changes, and close the Parameter Form Builder window. The parameter form appears in the Report Builder parameter form view.

7. Save the report.

The Report Builder **Parameter Form view** displays the layout of customized parameter forms that you create in the Parameter Form Builder. You can use this view to edit the appearance of the objects that the Parameter Form Builder generates and to add additional boilerplate text, images, or shapes to customize the parameter form. You can open the Parameter Form view by clicking the **Parameter Form button** 🔲 on the navigational toolbar or by clicking View on the menu bar and then clicking Parameter Form.

Now, you will customize the parameter form in the Parameter Form view. You will move the existing parameter form objects so they appear lower on the parameter form and then import the Northwoods logo onto the parameter form.

To customize the parameter form:

1. In the Parameter Form view, click **Edit** on the menu bar, and then click **Select All**. All of the parameter form objects appear selected. Move the objects downward on the form to the positions shown in Figure 9-47.

2. To import the Northwoods logo, click **File** on the menu bar, point to **Import**, and then click **Image**. Click **Browse**, navigate to the Chapter9 folder on your Data Disk, select **nwlogo.jpg**, click **Open**, and then click **OK**. The logo appears on the parameter form. Modify the image size and appearance so your parameter form looks like Figure 9-47, and then save the report.

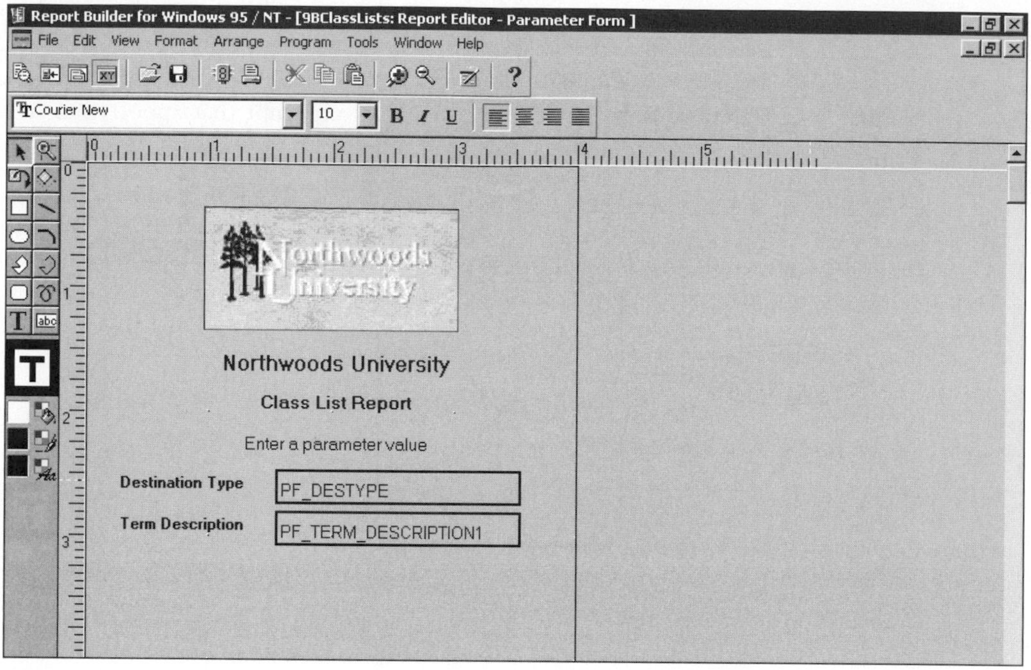

Figure 9-47 Customized parameter form in Parameter Form view

Now, you will run the form and view the customized parameter form. You will specify to send the report output to the screen and select Summer 2004 for the term description parameter value.

To run the form and view the customized parameter form:

1. Click **View** on the menu bar, and then click **Runtime Preview**. The report parameter form appears.

2. Select **Screen** for the Destination Name, select **Summer 2004** for the Term Description, and then click the **Run Report button** 📇 to run the report. The report appears in the Runtime Previewer.

3. Click the Close Previewer button ✕ to close the Runtime Previewer, and then close the report in Report Builder. If necessary, click **Yes** to save the changes to the report.

CREATING MULTIPANEL REPORTS

Sometimes when you create reports, the column data is too wide to fit across a single page. To accommodate the data, you might print it on multiple pages and then attach the pages together. For example, suppose you create a report that displays all of the columns in the STUDENT table in the Northwoods University database. The table has 12 columns, and most standard-sized printer paper can display only about 7 or 8 columns on a report page. You would display the first 6 columns on one page, the final 6 columns on the next page, and then attach the pages together, as shown in Figure 9-48. The report shown in Figure 9-48 is called a **multipanel report**, which is a report in which the columns span multiple horizontal pages. Each report page that displays horizontal data is called a **panel**.

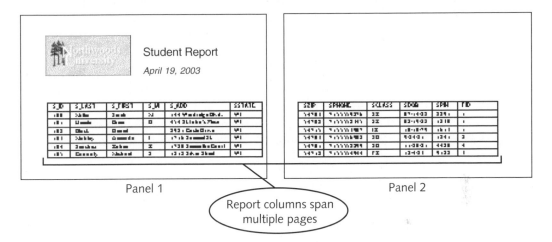

Figure 9-48   Multipanel report

In this section, you will create a multipanel report that displays the Northwoods University student data. First, you will open an existing student report in Report Builder and save the file using a different filename. Currently, the report is a single panel report, and the data appears on a single page because the column widths are narrow and the data values wrap to multiple lines. After you open the report, you will try to make the Address column wider and see what happens.

To open the report, save it using a different filename, and try to make the Address column wider:

1. In Report Builder, open **Student_PANEL.rdf** from the Chaper9 folder on your Data Disk, and save the file as **9BStudent_PANEL.rdf** in your Chapter9\TutorialSolutions folder.

2. Double-click the **Live Previewer icon** 📖 to view the report in the Live Previewer. Currently, the report column widths are very narrow, and the data values in several columns wrap to multiple lines.

3. Click any data field in the Address column to select the Address column, select one of the selection handles on the right edge of the column, and drag the mouse pointer toward the right edge of the screen to make the column wider. The mouse pointer changes to the Unavailable mouse pointer ⊘, indicating that you cannot make the column any wider because report data fills all of the available report width.

Internally, report values are represented using both physical and logical pages. A **physical page** corresponds to your printer's page size, such as 8.5 inches wide and 11 inches long. A **logical page** is a value that is a multiple of the physical page width and height. For example, a logical page can be twice as wide and twice as long as a physical page. For the report shown in Figure 9-48, the logical page needs to be wider than 8.5 inches to show all of the report columns. You determine a report's logical page size based on the width and length of the data that will appear in the report. Usually, the physical and logical page sizes are the same. For multipanel reports the logical page size is larger than the physical page size.

To create a multipanel report, you use the Layout Model Property Palette to change the logical page size by specifying that the report is comprised of multiple horizontal and/or vertical panels. Recall that the Layout Model has three sections: header, main, and trailer. Each section has a separate Property Palette on which you can specify section properties, including the number of horizontal and vertical panels. Now you will open the Property Palette for the main report section, and examine its properties.

To open the main section Property Palette:

1. Click **Window** on the menu bar, and then click **Object Navigator** to open the Object Navigator.

2. If necessary, click ➕ beside the Layout Model node. Nodes for the Header Section, Main Section, and Trailer Section appear.

3. Select the **Main Section node**, right-click, and then click **Property Palette**. The Property Palette for the Main Section opens.

In the Main Section Property Palette, the Section property node lists the following properties:

- **Width** and **Height**, which specify the physical page size in the report measurement unit. The default report measurement unit is inches.

- **Horizontal** and **Vertical Panels per Page**, which specify the number of physical pages that comprise a logical page. Currently, the values are both 1, because by default the physical and logical page size are the same size. You can change these values to any whole number between 1 and 50.

- **Orientation**, which specifies whether the report appears in a portrait or landscape orientation.

- **Distribution**, which specifies different report destinations from which the user can choose, such as printer or screen.

Now, you will change the Horizontal Panels per Page value to 2 to specify that the report logical page is two physical pages wide. Then, you will open the Live Previewer and take advantage of the larger logical page size by making some of the report columns wider.

To change the Horizontal Panels per Page property and widen the report columns:

1. In the Main Section Property Palette, change the Horizontal Panels per Page property value to **2**, select another property to save the change, and then close the Property Palette.

2. In the Object Navigator, double-click the **Live Previewer icon** ![icon] to open the report in the Live Previewer.

3. Click any data field in the Address column to select the column, select one of the selection handles on the right edge of the column, and drag the mouse pointer toward the right edge of the screen to make the column wider. Adjust the column width so all of the data values appear on a single line. Note that you were successful this time, because the report's logical page is now wider.

4. Scroll to the right edge of the report, and note that a heavy line indicates the position of the physical page break. Some of the report data values now appear on the right side of the physical page break. This shows the data that will appear on the second report panel.

5. Adjust the column widths for the City and Phone data columns so the data values appear on a single line.

6. Adjust the column width for the Class data column so the physical page break is between the Class and DOB columns and does not split any data values, as shown in Figure 9-49.

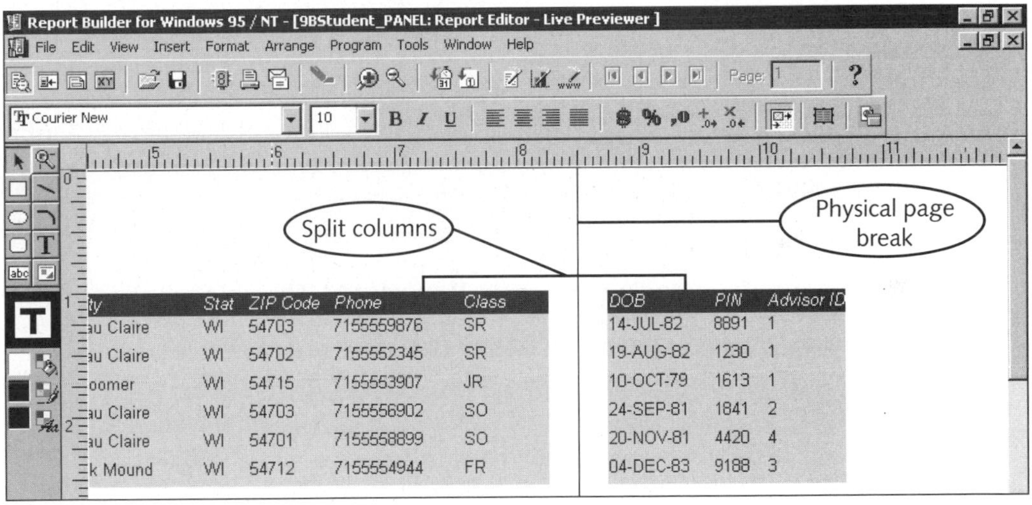

Figure 9-49   Splitting columns between physical pages

7. Save the report, and then close the report in Report Builder.

When you print a multipanel report, the report's **Panel Print Order** property determines the order in which the report panels print. This property can have a value of **Across/Down**, which specifies that the report panels print horizontally across all of the horizontal logical pages first, and then down to the next physical page, and across all of the second page's horizontal logical pages, and so forth. The Panel Print Order property can also be Down/Across, which specifies that the report panels print down through all of the report panels in the first column, and then print down the second column of panels, and so forth.

When you create a tabular report that contains multiple pages, the column headings repeat at the top of each new physical page. If you specify that the vertical logical page is made up of multiple physical pages, then the report column headings do not repeat for each new logical page.

DISPLAYING MULTIMEDIA DATA IN REPORTS

You can display images, sound, and video clips in reports. You can display image data that is stored in the database directly in a report layout. You can also display image, sound, and video data by creating a command button that references the source data files in the file system on your workstation. The following sections describe how to display image data in report data fields and how to create a command button to reference files that contain sound and video data.

Displaying Image Data

In Chapter 8, you learned how to store and retrieve image data in BLOB database columns. You can retrieve and display image data directly in a report by using the Report Wizard to create a report based on a table with a BLOB column that contains image data. Now, you will open a form that allows you to insert item images into the ITEM table in the Clearwater Traders database and insert the images for each item into the ITEM table in your database schema. Then, you will use the Report Wizard to create a report based on the ITEM table and view the image data in the report. First, you will open the form and insert the ITEM image data.

To insert the ITEM image data:

1. Start Form Builder, and open **Item_IMAGE.fmb** from the Chapter9 folder on your Data Disk. Run the form, and log onto the database when you are prompted to do so.

 If your workstation does not have enough memory to run Form Builder and Report Builder at the same time, close Report Builder. Complete Steps 1 through 6, then close Form Builder, and restart Report Builder.

2. Click the **Enter Query button** , do not enter a search condition, and then click the **Execute Query button** to retrieve all of the table records. The record for item ID 894 (Women's Hiking Shorts) appears.

3. Click the **Load Disk Image button**, navigate to the Chapter9 folder on your Data Disk, select **shorts.jpg**, and click **Open**. The image appears on the form.

4. Click the **Save button** to save the image in the database, and then click the **Next Record button** to view the next record.

5. Repeat Steps 3 and 4 for the rest of the ITEM records, using the following image files for each record:

Item ID	Image Filename
897	fleece.jpg
995	sandals.jpg
559	parka.jpg
786	tents.jpg

6. Close the form in Forms Runtime, and then close Form Builder. If necessary, click No to discard any form changes.

Now, you will use the Report Wizard to create a tabular report based on the ITEM table in the Clearwater Traders database.

To create a report based on the ITEM table:

1. In the Object Navigator, select the **Reports node**, and then click the **Create button** 🔲 to create a new report. The New Report dialog box opens. Make sure the Use the Report Wizard option button is selected, and then click **OK**. On the Report Wizard Welcome page, click **Next**.

2. On the Style page, type **Items** for the report title, make sure the Tabular option button is selected, and then click **Next**.

3. On the Query page, make sure the SQL statement option button is selected, and then click **Next**.

4. On the Data page, type **SELECT * FROM item** for the report query, and then click **Next**.

5. On the Fields page, click the **Select All button** 🔲 to select all of the table fields for the report. Note that the ITEM_IMAGE field icon appears as the LOB icon 🔲, which indicates that the field has the BLOB data type. Click **Next**.

6. You will not display any totals for any of the data fields, so do not select any fields on the Totals page, and then click **Next**.

7. On the Labels page, accept the default column labels, and then click **Next**.

8. On the Template page, make sure that the Corporate1 predefined template is selected, click **Next**, and then click **Finish**. The report appears in the Live Previewer.

9. Save the report as **9BItem_IMAGE.rdf** in your Chapter9\TutorialSolutions folder.

9

Currently, the data in the Item Image column appears as a data field that displays the characters "MM." When you create a report using the Report Wizard, all data fields appear as text. When a column contains multimedia data, the data appears as the characters "MM," to represent multimedia data. To display the actual images, you must modify the field by opening its Property Palette and changing its File Format property to *Image*. Now, you will open the F_ITEM_IMAGE field Property Palette and change its File Format property to Image. Then, you will view the report in the Live Previewer and change the ITEM_IMAGE field width and height to format the image data.

To change the ITEM_IMAGE field's File Format and modify its width and height:

1. In the Live Previewer, select any of the MM data fields, right-click, and then click **Property Palette**. The F_ITEM_IMAGE Property Palette opens.

2. Under the Column property node, change the File Format property value to **Image**. The Report Progress dialog box opens while Report Builder reformats the report layout. When the formatting completes, close the Property Palette. The report appears in the Live Previewer and displays the image data.

3. If necessary, select any image so selection handles appear around all of the images. Select the right center selection handle of the Women's Hiking Shorts image, and drag it toward the right edge of the screen display, so your report looks like Figure 9-50. The right edge of the report should be at about 7 inches, and the bottom edge of the Women's Hiking Shorts image should be at about 3.25 inches.

Displaying Sound and Video Data

You cannot display sound and video data directly in a report data field. To display sound and video clips, you must create a command button on the report layout and then associate the command button with the file that contains the sound or video clip. The command button can appear on the first report page, the last report page, or all report pages. The command button cannot be associated with a specific data record. Now, you will open the report in the Layout Model view and create a command button that displays a video clip. You will place the command button in the upper margin of the report, beside the report title.

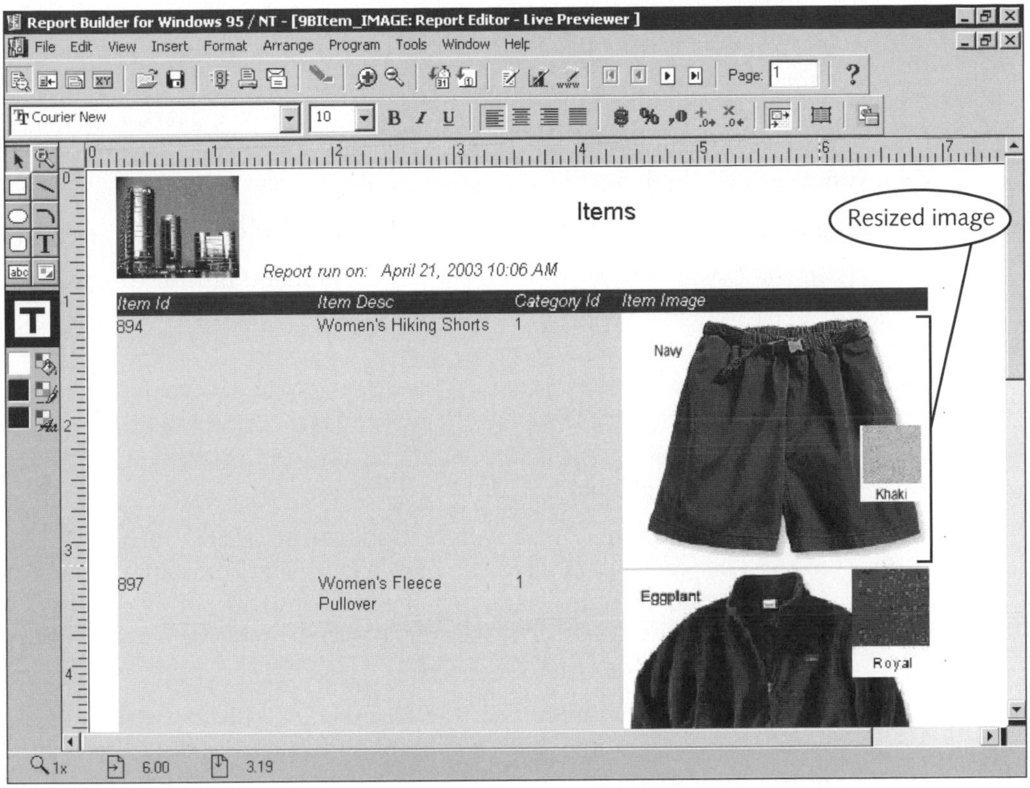

Figure 9-50   Resizing the image data fields

To create a command button that displays a video clip:

1. In the Live Previewer, click the **Layout Model button** 🖻 to open the Layout Model. The report appears in the Layout Model.

2. Click the **Edit Margin button** 🖩 to make the report margin available for editing. The report margin appears in the Layout Model.

3. Click the **Button tool** ▢ on the tool palette, and then draw a button in the top right corner of the margin.

Now, you will modify the command button properties. In Report Builder, command buttons can display text labels or icons. You will configure the button to display a text label and change the command button Label property. To specify the data that the button displays, you must specify values for the following properties:

- **Multimedia File Type**, which specifies whether the button displays image, sound, or video data and can have values of *Image*, *Sound*, or *Video*.

- **Multimedia File**, which specifies the location and filename of the multimedia file in the user's file system.

You will open the button Property Palette and change the button name and label property values. You will also change its Multimedia File Type and Multimedia File property values to specify that when the user clicks the button, a video clip that is stored in a file named globe.avi in the Chapter9 folder on your Data Disk appears. Then, you will click the button and play the video clip. To test a command button that displays multimedia data, you must run the report in the Runtime Previewer.

To configure the command button and display the video clip:

1. In the Layout Model, double-click the command button to open its Property Palette, and change the following properties.

Property	Value
Name	**VIDEO_BUTTON**
Label Type	**Text**
Text	**Play Video**
Multimedia File Type	**Video**

2. Select the **Multimedia File** property, click the **More button** [...], navigate to the Chapter9 folder on your Data Disk, select **globe.avi**, and click **Open**. Click another property to save the change, and close the Property Palette. If necessary, adjust the button size so all of the label text appears, and then save the report.

3. Click **View** on the menu bar, and then click **Runtime Preview** to run the report in the Runtime Previewer. The button appears as a three-dimensional object on the report.

 If the button appears as a two-dimensional object and nothing happens when you click it, then you are viewing the report in the Live Previewer. Repeat Step 3 to view the report in the Runtime Previewer.

4. Click **Play Video**. The Video Player window opens and shows an image of a globe.

5. Click **Play** to play the video image, and then close the Video Player window.

6. Click the **Close button** [X] to close the Runtime Previewer.

CREATING EXPLICIT ANCHORS TO LINK REPORT OBJECTS

Sometimes, report items that should appear beside each other become separated. This often happens with image items. The text data for the record might appear on one report page, and the image might appear on a different page. Now, you will view the records in the report that contains the image data for the Clearwater Traders ITEM table and examine a record in which the text and item data have become separated across different pages.

To view the separated data:

1. Click the **Live Previewer button** 🔍, and scroll down the first page of the report until you see the text data for item ID 559 (Men's Expedition Parka). Note that the image for the parka does not appear by the text data.

2. Click the **Next Page button** ▶ on the toolbar, and scroll to the top of the next page. Note that the parka image appears at the top of the next page, but no text data appears beside the image.

To keep text item and image data from being separated across report pages, you can create an **explicit anchor**, which is a report layout object that explicitly links report items that should always appear together. To create an explicit anchor, you designate one of the objects in the linking pair as the master object and the other object as the detail object. Then, you open the report in the Layout Model and select the **Anchor tool** 🔲 on the tool palette. To create the anchor, you use 🔲 to create a link from the master object to the detail object. You click the mouse pointer on the bottom-right corner of the master object, drag the mouse pointer to the bottom-left corner of the detail object, and then double-click the mouse pointer on the bottom-left corner of the detail object. This anchors the detail object to the master object.

Now, you will create a series of explicit anchors between the fields in the ITEM record to link all of the fields on the same page. You will create the anchors to link all of the fields from left to right. The first anchor will link the F_ITEM_ID field to the F_ITEM_DESC field, the second anchor will link the F_ITEM_DESC field to the F_CATEGORY_ID field, and the third anchor will link the F_CATEGORY_ID field to the F_ITEM_IMAGE field. First, you will open the report in the Layout Model and magnify the report so you can see the report field edges better. Then, you will create the anchors.

To open the Layout Model, magnify the report objects, and create the anchors:

1. In the Live Previewer, click the **Layout Model button** 🔳 to open the Layout Model.

2. Click the **Margin button** 🔳 to disable margin editing and make the main report section available for editing.

3. Click the **Zoom In button** 🔍 on the toolbar to magnify the Layout Model. Scroll to the top-left corner of the screen to view the report objects.

4. Click the **Anchor tool** 🔲 on the tool palette. The mouse pointer changes to the Precision pointer ╌┼╌ as you move it across the painting region.

5. Click the lower-right corner of F_ITEM_ID, drag the mouse pointer to the lower-left corner of F_ITEM_DESC, and double-click. The explicit anchor appears between the two fields, as shown in Figure 9-51.

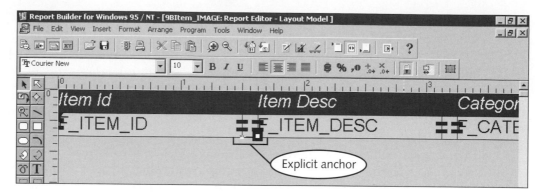

Figure 9-51   Creating an explicit anchor

6. Repeat Step 5 two times to create explicit anchors between F_ITEM_DESC and F_CATEGORY_ID, and between F_CATEGORY_ID and F_ITEM_IMAGE. (You may need to scroll to the right edge of your screen display to create the anchor between F_CATEGORY_ID and F_ITEM_IMAGE.)

7. Save the report.

Now, you will view the report in the Live Previewer and confirm that the image appears on the same page as the other fields.

To view the report in the Live Previewer and confirm that the record fields display on the same page as the image:

1. Click the **Live Previewer button** to view the report in the Live Previewer and click the **First Page button** to move to the first page of the report. Note that the data for each record is now aligned with the bottom edge of the item image.

2. Scroll to the bottom of the first report page. Note that the data values for Item ID 559 no longer appear on the first page.

3. Click the **Next Page** button. The data values for Item ID 559 now appear beside the image.

4. Close the report in Report Builder, and then close Report Builder and all other open Oracle applications.

SUMMARY

❐ When you create a report, Report Builder organizes the data fields returned by the SQL query into one or more record groups. Record groups are sets of records with the same field headings, but different data values.

❐ You can use the report Data Model to modify the report query, and to create user-defined break groups that you can use to change the order in which data values appear in the report.

❐ When a report has master-detail relationships, the fields for the master records and the fields for the detail records are placed in separate record groups.

❐ You can create a break group to change the order of the report data without modifying the report query.

❐ You can create a group filter to limit the number of records that a report query retrieves. The group filter can limit the retrievals based on a specific number of records to retrieve or based on a PL/SQL function that evaluates the individual data values in each record.

❐ Frames are containers for grouping report record group data items and column headings.

❐ Reports can have header frames, repeating frames, and group frames. Header frames contain all the column headings for a record group and are created only when report values appear in a tabular format. Repeating frames enclose the data fields that display the actual data values. Group frames group associated header and repeating frames together.

❐ An object's parent frame is the frame that directly encloses the object. To select a frame, you select one of its enclosed objects, then click the Select Parent Frame button ▣.

❐ Frame names are derived from the record groups they enclose. The general format for the name of a group frame is M_*record_group_name*_GRPFR. The general format for the name of a header frame is M_G_*record_group_name*_HDR. The general format for the name of a repeating frame name is R_G_*record_group_name*.

❐ In a report with master-detail relationships, frames for higher-level (more detailed) record groups are nested inside frames for lower-level (less detailed) record groups.

❐ When confine mode is enabled, you cannot move an object out of its enclosing frame.

❐ When flex mode is enabled, an enclosing frame becomes larger automatically when you move an enclosed object beyond the enclosing frame's boundary.

❐ Field elasticity determines whether a field's size is fixed on the printed report, or whether it can expand or contract automatically, depending on the size of the retrieved data value.

❐ To make data values appear directly next to each other on a report regardless of the width of the individual data fields, you must change the fields' column horizontal elasticity property to variable.

❐ Reports have system parameters, which specify properties concerning how the report display looks and how the environment works, and user parameters, which specify a search condition that controls the data that appears in the report.

❐ The report parameter form allows the user to enter values for system and user parameters. You can use the Parameter Form Builder utility to customize the appearance of a report's parameter form.

❐ A multipanel report is a report in which the report columns span multiple horizontal report pages.

❐ A physical page corresponds to the printer page size, and a logical page corresponds to a size needed to accommodate the width of the data that will appear in the report. In a multipanel report, a logical page is comprised of multiple physical pages.

❐ You can display image data directly in report data fields. You can display image, sound, and video data in reports using a command button that references a file that contains the sound or video clip.

❐ An explicit anchor is a report layout object that explicitly links report items that should always appear together.

REVIEW QUESTIONS

1. List and define the three different types of data that can appear in a report field.

2. List the three ways you can use a group filter to limit the data that appears in a report.

3. List the names of the record groups created when you use the following queries to create master-detail reports using the Report Wizard in Report Builder. (Note: some queries might have more than one record group.)

   a. SELECT inv_id, item_size, color, qoh
      FROM item, inventory
      WHERE item.item_id = inventory.item_id;

   b. SELECT c_last, c_first, skill_description, certification
      FROM consultant, skill, consultant_skill
      WHERE consultant.c_id = consultant_skill.c_id
      AND consultant_skill.skill_id = skill.skill_id;

   c. SELECT cust_id, last, first, address, city, state, zip, order_date,
      cust_order.order_id, inventory.inv_id, item_desc, quantity
      FROM customer, cust_order, order_line, item, inventory
      WHERE customer.cust_id = cust_order.cust_id
      AND cust_order.order_id = order_line.order_id
      AND order_line.inv_id = inventory.inv_id
      AND inventory.item_id = item.item_id;

4. List the names of each repeating frame created by the queries in Question 3a.

5. List the names of each group frame created by the queries in Question 3a.

6. What is the relationship between the record groups that appear in a report Data Model and the group, header, and repeating frames that appear in the Layout Model?

7. What is the difference between confine mode and flex mode?

8. When flex mode is enabled, does confine mode also have to be enabled?

9. What is the difference between a system parameter and a user parameter?

10. How do you create a relationship between a report's user parameter and the report's SQL query?

11. How do you determine what data type and width to assign to a user parameter?

12. How do you format tabular report data in a grid?

13. How can you display sound and video clips in a report?

14. What is an explicit anchor? How do you create an explicit anchor?

PROBLEM-SOLVING CASES

Cases reference the Clearwater Traders (see Figure 1-11), Northwoods University (see Figure 1-12), and Software Experts (see Figure 1-13) sample databases. All required files are stored in the Chapter9 folder on your Data Disk. Save all solution files in your Chapter9\CaseSolutions folder.

1. In this case, you will create the report shown in Figure 9-52 that lists each item category in the Clearwater Traders database and the corresponding values, item descriptions and a detailed listing of the corresponding inventory, size, color, current price, and quantity on hand.

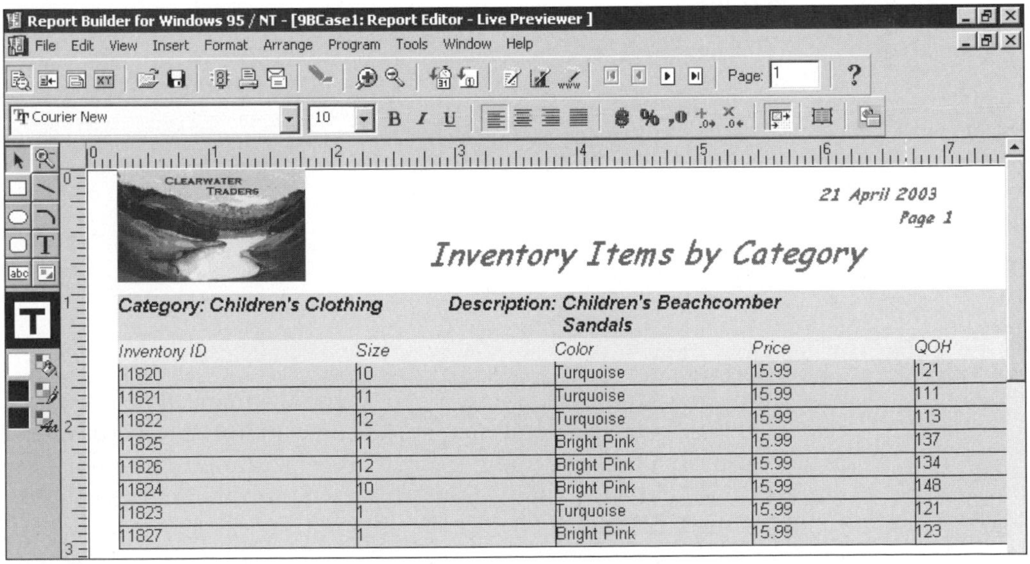

Figure 9-52

a. Use the Report Wizard to create a report that lists item categories and descriptions, and then lists individual inventory IDs, sizes, colors, prices, and quantities on hand, as shown in Figure 9-52. Apply the clearwater.tdf custom template or another template to the report, and save the report as 9BCase1.rdf.

b. Create a schematic diagram of the report objects and frames that is similar to the class lists report diagrams in Figures 9-34 and 9-35. Label the data fields, repeating frames, header frames, and group frames using the default names assigned by the Report Wizard.

c. Move the frames enclosing the inventory records down so that there are 0.25 inches of blank space between the Category and Description fields and the inventory column headings.

d. Format the report so that each category appears on a different report page. The first page of your completed report should look like Figure 9-52. Place the tabular data in a grid as shown.

2. In this case, you will create the report shown in Figure 9-53 that lists each consultant in the Software Experts database and the consultant's corresponding skills and certification status.

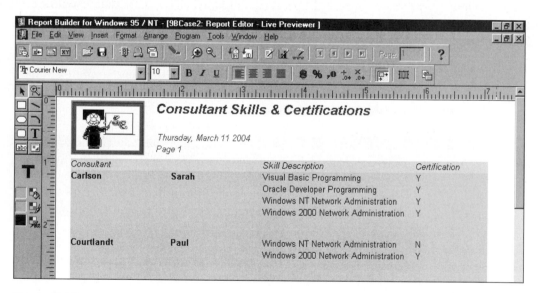

Figure 9-53

a. Use the Report Wizard to create a report that lists consultant first and last names, and then lists the consultant's skill descriptions and certification status, as shown in Figure 9-53. Apply the softwareexp.tdf custom template that you created in Lesson A, or apply another template to the report, and save the report as 9BCase2.rdf.

b. Create a schematic diagram of the report objects and frames that is similar to the ones in Figures 9–34 and 9–35. Label the data fields, repeating frames, header frames, and group frames using the default names assigned by the Report Wizard.

c. Modify the repeating frame that encloses the consultant records so that there are .25 inches of blank space between the records for each consultant.

3. In this case, you will create a report that lists each building code and room number in the Northwoods University database and shows the room usage according to course sections during each term. The report will have a user parameter that allows the user to select a specific building and view records for only that building.

a. Use the Report Wizard to create a report that displays building codes and room numbers followed by term descriptions, and then displays the course call IDs, section numbers, days, and times that the room is used during the term, as shown in Figure 9–54. Apply the northwoods.tdf template or another template to the report, and save the report as 9BCase3.rdf.

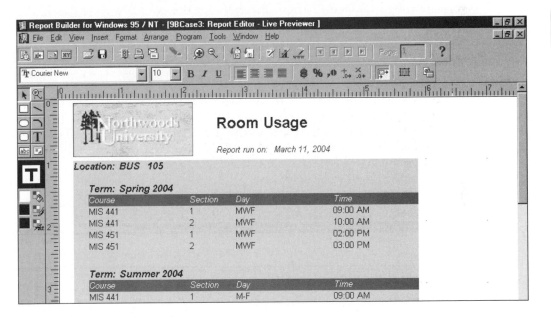

Figure 9-54

b. Create a user parameter so that the user can select the name of a specific building, and view only the output for that selection. Use the value of the LOC_ID primary key for the user parameter and query search condition, but display only the BLDG_CODE and ROOM values in the parameter list.

c. Change the elasticity of the building code and room fields so that they appear directly next to each other.

d. Modify the other components of the report so that the output looks like Figure 9-54.

e. Format the parameter form so it looks like Figure 9-55.

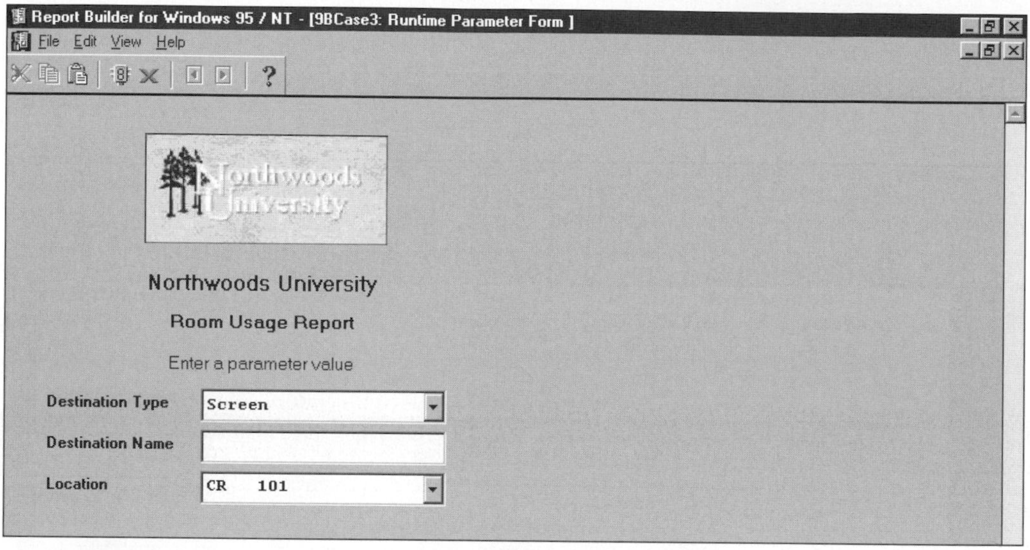

Figure 9-55

4. In this case, you will create a report that lists the name of each client in the Software Experts database, shows the client projects, and shows information about each consultant working on each project. The report will have a user parameter that allows the user to select a specific client and view records for only that client.

a. Use the Report Wizard to create a report that displays client names, followed by project names, and then displays the first and last name, roll-on date, roll-off date, and total hours for every consultant working on the project, as shown in Figure 9-56. Apply the softwareexp.tdf template or another template to the report, and save the report as 9BCase4.rdf. Format the report as shown in Figure 9-56.

b. Create a user parameter so that the user can select a specific client name, and view only the output for that client. Use the value of the CLIENT_ID primary key for the user parameter and query search condition, but only display the client names in the parameter list.

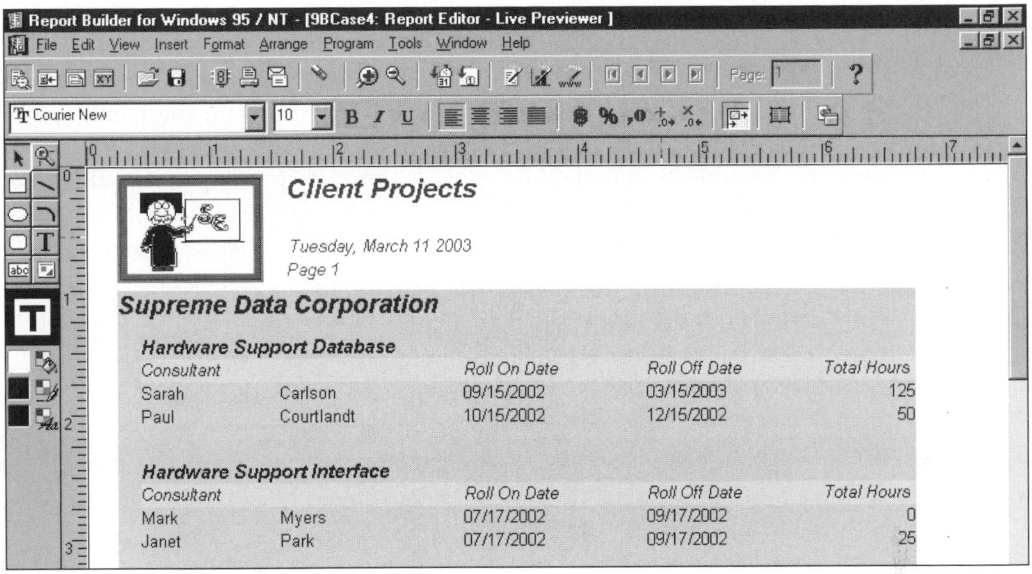

Figure 9-56

5. Create the report shown in Figure 9-57 that provides a schedule of courses offered at Northwoods University during a selected term. The report should list the term, then show each course call ID, name, and credits, and then show the section ID, section number, instructor, day, time, building code, and room for each course section.

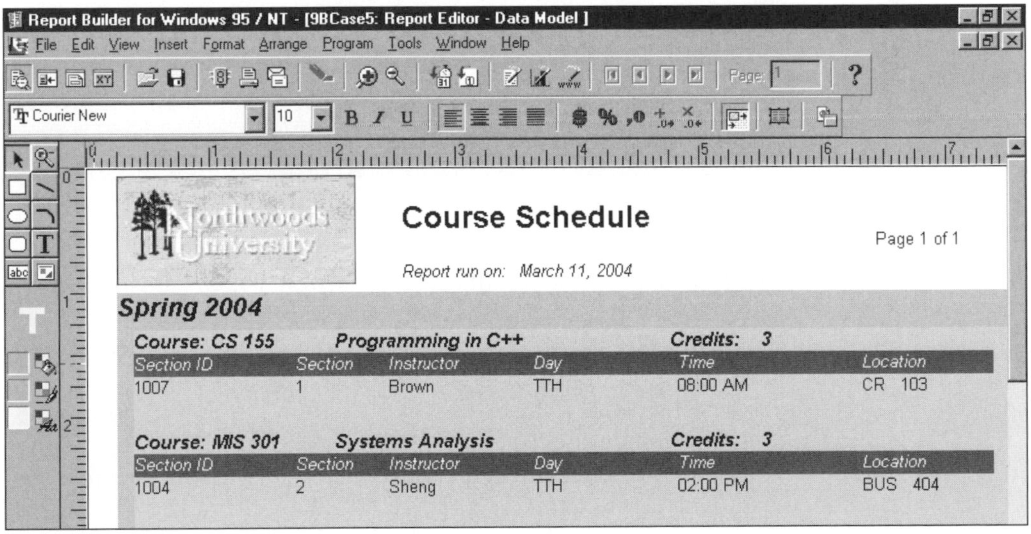

Figure 9-57

The output for each term should start on a new page, and there should be about 0.25 inches of blank space between the data for each course. The location columns (building code and room) should appear directly beside each other. Create a user parameter that lists the term description in the parameter list and passes the associated TERM_ID value to the report as a search condition. Apply the northwoods.tdf template or another template to the report, and save the report as 9BCase5.rdf.

6. Create a grade report for students at Northwoods University that lists the student's complete name and address, and then shows the term description, course call ID, course name, and grade. Only include courses for which a grade has been assigned. Create the report using the Report Wizard, and format the report so that the output looks like Figure 9-58.

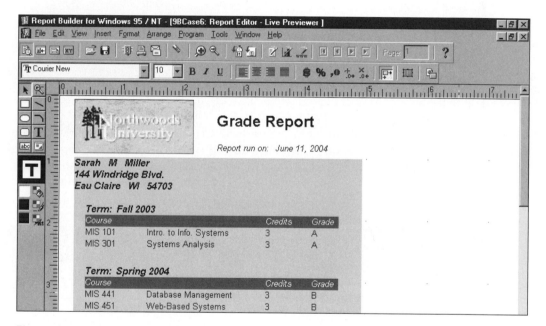

Figure 9-58

The output for each student should start on a new page, and there should be about 0.25 inches of blank space between the records for each term. The student first name, middle initial, and last name fields should appear directly beside each other. Similarly, the student's city, state, and ZIP code should appear directly beside each other. Create a user parameter to allow the report to be run for only one student at a time. Use S_ID as the search condition, and display only the student last and first names in the parameter list of values. In the parameter list, order the student names alphabetically by last name using the ORDER BY clause in the parameter list query. Save the report in a file named 9BCase6.rdf.

7. In this case, you will create the multipanel report shown in Figure 9-59 that displays data about Northwoods University courses. The report will display the course call ID, day, time, building code, room, instructor first and last name, and an image of the instructor.

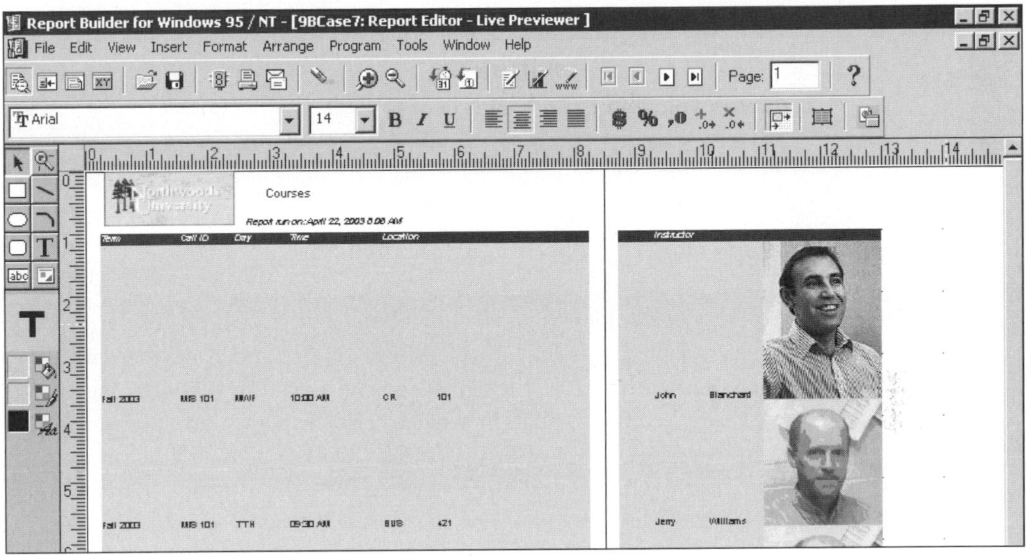

Figure 9-59

a. Start Form Builder, and open the Faculty_IMAGE.fmb form file. Run the form, and load the image data for each faculty member. The image filename for each faculty member is the same as the faculty member's last name.

b. Use the Report Wizard to create the report. Use a tabular report style, and format the report as shown in Figure 9-59. Format the Time column using a format mask that shows time values. Specify that the report has two panels. Size the columns so the page break places the instructor first and last name and image on the second (right-most) panel and the rest of the data on the first (left-most) panel. Save the report in a file named 9BCase7.rdf.

c. Create explicit anchors to ensure that the instructor image is always adjacent to the other course data fields.

◀ LESSON C ▶

Objectives
- ◆ Understand different report file types
- ◆ Create calculated columns
- ◆ Understand report triggers
- ◆ Manually create queries and data links
- ◆ Create a link file object to display the contents of a file in a report
- ◆ Run a report from a form

REPORT FILE TYPES

When you use Report Builder to create a database report, you create a **report definition**, which specifies the data values that appear in the report and how the data appears on the user's screen or when the report prints. Report Builder allows you to save the report definition using a variety of file formats. Users can also display report output in different formats. For example, a user can save a report as a word processor document or display the report as a Web page. The following sections explore the options for creating and saving report definitions and outputs.

Report Definition File Types

So far, you have saved report definitions using Report Builder report definition files that have an .rdf extension. The report definition file specifies what data the report contains and how the report output appears. Report Builder provides alternate ways to save report definitions. This is useful if you want to create reports that can appear in applications other than Report Builder, if you want to display a report on a non-Windows-based computer, or if you want to change the report definitions outside of the Report Builder environment. Table 9-4 summarizes the different report definition file types.

File Type	Extension	Description
Report definition	.rdf	Binary file containing report definitions that can be displayed and modified in Report Builder.
Report definition text	.rex	Text file containing report definitions
Report definition rich text format	.rtf	Rich text file containing report definitions

Table 9-4   Report definition file types

When you create a new report in Report Builder, you save it using the .rdf filetype that can be modified in Report Builder. You can convert .rdf files to text files, which have an .rex extension, and to rich text files, which have an .rtf extension.

The rich text file (.rtf) format is a text file format that has enhanced formatting capabilities that are recognized by many applications.

To convert a report definition file to a different format, you must use the Report Builder Convert utility, which specifies the source filename, the destination filename, and the conversion format. To convert a report definition file to a different format, you do not need to open the report in Report Builder—you can open the Convert utility, specify the .rdf file to be converted as the source file, and specify the name of the new file as the destination file. Now you will convert a report .rdf file to a text report definition file. For the source file, you will use an .rdf file named Student.rdf that is stored in the Chapter9 folder on your Data Disk.

To create a text report definition file:

1. Start Report Builder. When the Welcome to Report Builder window opens, click **Cancel**. The Object Navigator window opens.

2. In the Object Navigator, click **File** on the menu bar, point to **Administration**, and then click **Convert**. The Convert dialog box opens.

3. Make sure that the Source Type is Report Binary File (RDF). To select the source file, click **Browse**, navigate to the Chapter9 folder on your Data Disk, select **Student.rdf**, and then click **Open**. The path to the Student.rdf report definition file appears in the Source text box on the Convert dialog box.

4. Open the Destination Type list, and select **Report ASCII File (REX)**.

5. To specify the new report definition filename, place the insertion point in the Destination text box, click **Browse**, navigate to your Chapter9\TutorialSolutions folder, type **9CStudent.rex** in the File Name box, and then click **Open**. The destination filename appears in the Destination text box.

6. Click **OK** to perform the conversion, and then click **OK** to acknowledge the Conversion Successful message.

7. Click **Cancel** to close the Convert dialog box.

This .rex file contains text definitions that define the report display. You can use these definitions to display the report on different Oracle database platforms, such as UNIX or mainframe operating systems. Savvy report developers can also modify the report definitions directly using a text editor. Now you will start Notepad, and open the 9CStudent.rex report definition file to examine its contents.

9

To open and view the text report definition file:

1. Start Notepad, click **File** on the menu bar, click **Open**, navigate to your Chapter9\TutorialSolutions folder, open the Files of type list and select **All Files (*.*)**, select **9CStudent.rex**, and then click **Open**.

2. Since this is a very large text file, an Error dialog box might open, prompting you to open WordPad instead, which is a Windows-based text editor that can handle large text files. If this happens, click Yes. The text report definition file opens. Scroll down the file, and examine the report definitions.

3. Close Wordpad.

This file contains text definitions that specify all of the elements of the report, including column, color, and font specifications.

Report Output File Types

So far, you have viewed report output in the Report Builder Live Previewer and the Runtime Previewer. Users shouldn't open and view reports in Report Builder, because they might accidentally change the report definitions and cause the report to malfunction. Once you finish developing a report, you can create a report executable file, which appears in the Reports Runtime window. Reports Runtime is the Oracle application that allows users to display and manage database reports. A report executable file has an .rep extension. To create an executable .rep report file, you generate a compiled version of the report in Report Builder. Now you will generate a report executable file for the student report.

To generate a report .rep file:

1. In the Object Navigator, click the **Open button** 🖼, navigate to the Chapter9 folder on your Data Disk, and open **Student.rdf**.

2. Click **File** on the menu bar, point to **Administration**, and then click **Compile Report**. The Compile Report dialog box opens.

3. Navigate to your Chapter9\TutorialSolutions folder, change the filename to **9CStudent.rep**, and then click **Save**.

Now you will view the compiled report in Reports Runtime by starting the Reports Runtime application, and then opening the report .rep file.

To view the compiled report in Reports Runtime:

1. Click **Start** on the taskbar, point to **Programs**, point to **Oracle Reports 6i**, and then click **Reports Runtime**. The Reports Runtime window opens. Maximize the window.

2. Click **File** on the menu bar, and then click **Run**. The Open dialog box opens. Navigate to your Chapter9\TutorialSolutions folder, select **9CStudents.rep**, and then click **Open**. The Connect dialog box opens.

3. Connect to the database in the usual way. The report appears in the Reports Runtime window.

 If your system has an association between the .rep filetype and the Reports Runtime application, which is stored in a file named *Developer_Home*\BIN\ rwrun60.exe, you can display reports in Reports Runtime by double-clicking the report .rep file in Windows Explorer.

In Reports Runtime, users can save the report output to a file using a variety of file types that can appear in different applications or appear as Web pages. Now, you will save the student report output in a rich text file format. Then, you will view the contents in Microsoft Word, or another word processor that supports rich text format (.rtf) files.

To save the report output as a rich text file, and view the output in a word processor:

1. In Reports Runtime, click **File** on the menu bar, point to **Generate to File**, and then click **RTF**. The Save dialog box opens.

2. Navigate to your Chapter9\TutorialSolutions folder, make sure the filename is 9CStudent.rtf, and then click **Save**.

3. Start Microsoft Word or another word processor, click **Open** on the menu bar, navigate to your Chapter9\TutorialSolutions folder, select **9CStudent.rtf**, and then click **Open**. The report opens in the word processor. You might need to adjust your page setup or fonts so all of the report fields appear on the page.

4. Close the word processor, and close the Reports Runtime window.

5. In Report Builder, click **File** on the menu bar, and then click **Close** to close the student report.

CREATING REPORT COLUMNS TO DISPLAY CALCULATED VALUES

Sometimes you need to display calculated values on reports. For example, every Northwoods University student receives a student grade report each term that summarizes his or her total credits and term and cumulative grade point averages. To display calculated values in a report, you can create formula columns, which display values that are calculated using PL/SQL functions that use report data fields as input parameters; summary columns, which display values that summarize data that appears in repeating frames; and placeholder columns, which are similar to formula columns, and display data values that have a specific data type and can be defined using a PL/SQL report trigger. (You will learn about report triggers later in this lesson.)

To learn about the different report column types, you will work with a report that displays transcript information for students at Northwoods University shown in Figure 9-60.

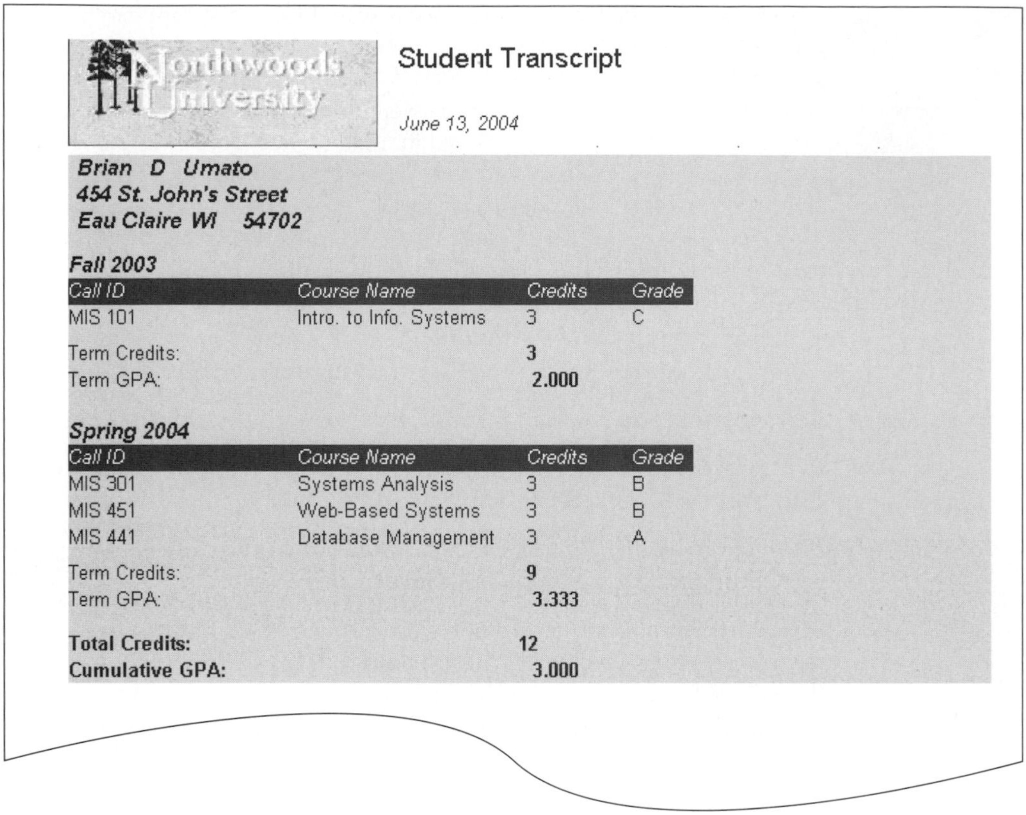

Figure 9-60   Student transcript report

The student's name and address appear at the top of the transcript report, followed by the term description and a list of each course, the course credits, and the grade received for the course. Course credits are summed for each term, and the student's grade point average (GPA) is calculated for each term. Total credits for all terms and the student's cumulative GPA also appear. Now you will open the student transcript report and save it using a different filename.

To open and save the transcript report:

1. In Report Builder, open **Transcript.rdf** from the Chapter9 folder on your Data Disk, and save the file as **9CTranscript.rdf** in your Chapter9\TutorialSolutions folder.

2. In the Object Navigator, double-click the **Live Previewer icon** 📚 to view the report in the Live Previewer. Connect to the database, if necessary.

Currently, the transcript report displays the student name and address, the term description, and the course and course grade data. In the following sections, you will add calculated columns that calculate and display the student credit points, which is used to calculate student GPA, and to sum the student credits and credit points.

Recall that in a report, the terms column and field are used interchangeably. Summary, placeholder, or formula columns might display only a single value and not appear in a columnar format, but are still called columns.

Note that the final report formatting has not yet been done: The student's first and last names are not adjacent to each other, and there are no blank lines between the student address and the term and course information. You do not want to perform the final formatting tasks yet, because whenever you make a modification using the Report Wizard, all custom formatting—such as changing the size and spacing of fields in frames or adjusting the elasticity property of data fields—is lost.

 Whenever you create a report that contains calculated report columns, always make sure that all of the report values appear correctly before you perform the final formatting.

Before you create the calculated report columns, you will open the report data model and become familiar with the report record groups. You need to understand the structure of the report data, because you will be working with the report record groups and frames to make the required modifications.

To view the data model:

1. In the Live Previewer, click the **Data Model button** 📧 on the toolbar. The Data Model view opens.

2. Rearrange the data model objects so that they are all visible on the screen, as shown in Figure 9-61.

3. Save the report.

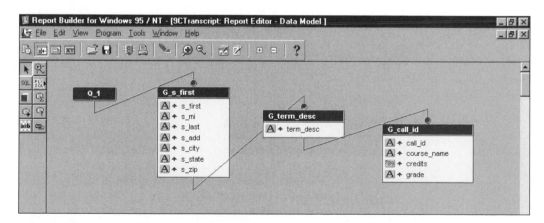

Figure 9-61   Transcript report data model

The data model shows that the report has three record groups. The Level 1 group, G_s_first, contains the student information. The Level 2 group, G_term_desc, shows the term description, and the Level 3 group, G_call_id, displays the course and grade information. In the following sections, you will create columns to display the calculated report data values.

Formula Columns

A **formula column** displays a value that is returned by a user-defined function that performs mathematical computations on report data values. This function is written in PL/SQL, using syntax similar to what you used in earlier chapters. For example, you might want to calculate and display a Northwoods University student's age by subtracting the student's date of birth from the current system date, or display the value of inventory items at Clearwater Traders by multiplying the item price times the quantity on hand.

In the transcript report, you will use a formula column to calculate a student's grade point average for each term. Grade point average is calculated as follows:

SUM(Course Credits * Course Grade Points)
SUM(Course Credits)

Course grade points are awarded as follows: A = 4 grade points, B = 3, C = 2, D = 1, F = 0. First, you will create a formula column to calculate the total course credit points for each term, which is equal to course credits * (multiplied by) course grade points. For example, if a student received an A in a 3-credit course, the credit points would be 3 (the course credits) * 4 (the grade points for an A), or 12. Since the course credit points are calculated using the CREDITS and GRADE data fields, this formula will be placed in the G_call_id record group.

Creating and displaying a formula column is a three-step process:

1. Open the report data model, create the new formula column in the record group that contains the data values used in the calculations, and modify the formula column in the new column's Property Palette. Formula column names are usually prefaced with the identifier *CF*. For example, a formula column that calculates the student term GPA might be named *CF_term_gpa*.

2. Write the formula function using PL/SQL.

3. Open the Layout Model, and create a field on the report layout to display the formula value.

The following sections describe these steps.

Creating a Formula Column in the Data Model

To create a formula column in the data model, you determine the record group that the formula column will be associated with, and then create the formula column using the **Formula Column tool** 🔲 on the Data Model tool palette. A formula column must be placed in the same record group as the values that are used in the formula, because the formula column calculates a value for each record in the record group. You will now create a formula column in the transcript report data model to calculate the student credit points for each course. You will create the formula column in the G_call_id record group, since this record group contains the CREDITS and GRADE fields that will be used in the formula. After you create the formula column, you will open the formula column Property Palette and change the name of the new formula column.

To create the formula column to calculate course credit points:

1. Click the **Formula Column tool** on the Data Model tool palette. The pointer changes to the Precision pointer ━┿━ when you move it across the painting region.

   If the tool palette does not appear in your Data Model window, click View on the menu bar, and then click Tool Palette.

2. Click the mouse pointer on the **G_call_id** record group. A new formula column named CF_1 appears in the record group. Resize the record group if necessary so that all of the data columns are visible.

3. Select **CF_1**, right-click, and then click **Property Palette**. Change the new column's Name property to **CF_credit_points**. Do not close the Property Palette.

Writing the Formula Column Function

Every formula column has an associated PL/SQL function. The function can reference columns in the report's record groups, including data fields, summary columns, and other formula columns. To reference a column in the PL/SQL formula column function, you preface the column name (as it appears in the data model) with a colon. For example, you would reference the CREDITS column as **:credits**. You would reference the CF_credit_points column as **:CF_credit_points**.

   You can only reference columns that are in the same record group as the formula column or that are in higher record groups.

Recall that a function always returns a single value. To create a formula column function, you must declare a variable in the function DECLARE section, perform the required calculations, assign the value to be returned to the declared variable, and then use the RETURN command to instruct the function to return this value. Now you will write the PL/SQL function to calculate the course credit points for a single record in the G_call_id record group. You will create a summary column that sums the course credit points for all of the records later.

To write the PL/SQL function to calculate the course credit points for a record:

1. In the CF_credit_points Property Palette, select the **PL/SQL Formula** property so that a button appears, and then click the button. The PL/SQL Editor opens.

2. Type the code shown in Figure 9-62 in the PL/SQL Editor to return the credit points for a student's course section record.

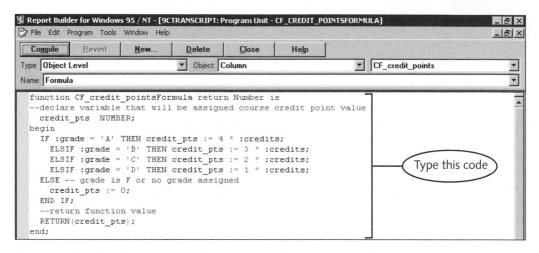

Figure 9-62   Function to calculate course credit points

3. Compile the code, correct any syntax errors, close the PL/SQL Editor, and then close the Property Palette.

4. Save the report.

Creating a Field to Display the Formula Column

Currently, the formula column exists in the data model but does not appear anywhere on the report. To display a formula column on a report, you must use the **Field tool** on the Layout Model tool palette to draw a new report data field in the Layout Model. You must place the field in the same repeating frame as the source values for the formula. Then, you must set the new field's Source property to the name of the formula column.

The data values used in the credit points formula are GRADE and CREDITS, which are in the G_call_id record group. Therefore, the field that displays the formula column must be in the R_G_call_id repeating frame. Now you will open the report Layout Model, make the R_G_call_id repeating frame wider, draw a new field on the painting region, and then assign the CF_credit_points formula column as its data source. First, you will make the R_G_call_id repeating frame wider.

Recall from the previous lesson that when you made a report frame larger, you enabled flex mode, and then moved an item inside the frame to automatically enlarge the surrounding frames. Now you will open the Layout Model, enable flex mode and make the R_G_call_id frame wider.

To make the R_G_call_id frame wider:

1. Click the **Layout Model button** to open the Layout Model.

2. Confirm that the Confine Mode button is pressed and that the Flex Mode button is pressed. If necessary, adjust the buttons so both confine mode and flex mode are enabled.

3. Select **F_grade**, and then click the **Select Parent Frame button** to select the R_G_call_id repeating frame.

4. Click the center selection handle on the right edge of the R_G_call_id frame, and drag it toward the right edge of the screen display so the repeating frame is about 5 inches wide, as shown in Figure 9-63.

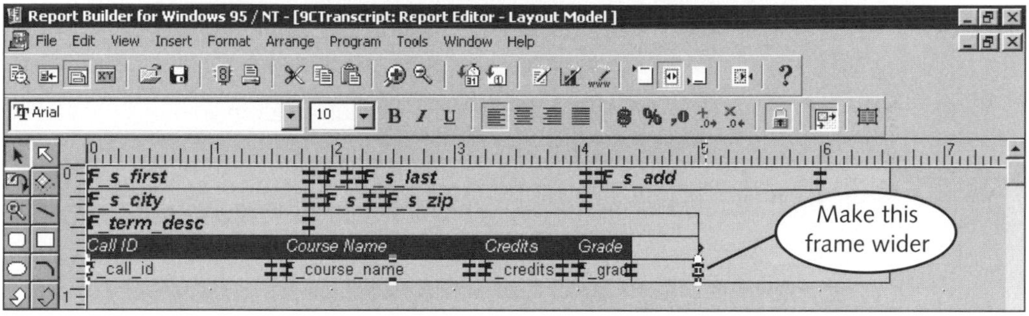

Figure 9-63   Widening the R_G_call_id repeating frame

If an error appears when you try to make the repeating frame 5 inches wide, make the repeating frame as wide as possible, but do not make is so wide as to generate the error.

5. Click the **Live Previewer button** to view the report in the Live Previewer. The report should appear with no error messages and show the enlarged frame boundaries.

If an error message appears, click File on the menu bar, then click Revert. Click Yes to undo your changes, then repeat Steps 1 through 5.

6. Save the report.

Now you will draw the field in the R_G_call_id repeating frame that will display the value of the credit points for each course section. Then, you will modify the field properties. You will change the field name and assign its data source to be the CF_credit_points formula column.

To draw the field and modify its properties:

1. Click the **Layout Model button** 🔲 to open the Layout Model, select the **Field tool** [abc] on the tool palette, and draw a new field inside the R_G_call_id frame, as shown in Figure 9-64. Make sure that the field is totally enclosed by the frame.

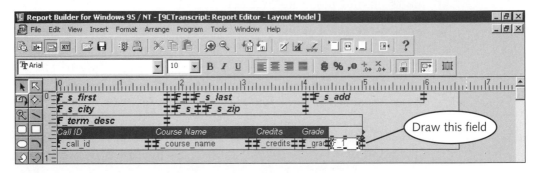

Figure 9-64   Drawing the field to display the formula column

2. Right-click the new field, and then click **Property Palette** to open its Property Palette. Change the Name property to **F_credit_points**.

3. Click the **Source** property, then scroll down the list and select **CF_credit_points** as the field data source. Select another property to save the change, and then close the Property Palette.

4. Click the **Live Previewer button** 🔍 to view the report in the Live Previewer. Brian Umato's transcript on page 1 of the report should look like Figure 9-65 and display the credit points formula column values correctly.

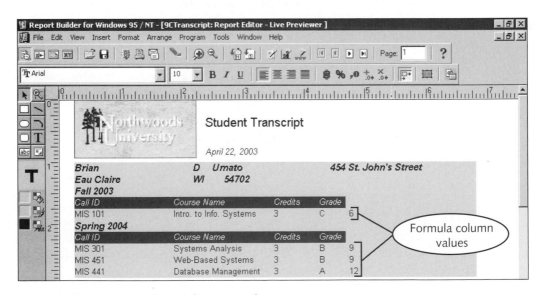

Figure 9-65   Viewing the credit point values

If a border appears around the formula column values, select the formula column field in the Live Previewer, select the Line Color tool 🖍 on the tool palette, and then click No Line. If the field has a different background color than the report, select formula column field, select the Fill Color tool 🖍, and then click No Fill. If the field has a different font name or size than the other report fields, select the formula column field, open the font list at the top of the Live Previewer window, and select Arial. Open the font size list, and select 10.

If an error message appears stating that a display field references a data column at a frequency below its group, this means that you placed the field in the wrong frame. For example, you cannot place a column that displays each course's credit points (which are calculated using data values from individual course section records) in the repeating frame for the term, because there are multiple course section records in a term.

5. Save the report.

Summary Columns

A **summary column** is a calculated column that returns summary values, such as the sum or average of a series of data fields in a repeating frame. You can create summary columns for any report data field using the Totals page in the Report Wizard. You can also create summary columns manually. The following sections describe how to create summary columns using both approaches.

Creating a Summary Column Using the Report Wizard

To create a summary column using the Report Wizard, you select the field to be summarized and the summary function, such as sum or average, on the Report Wizard Totals page. When you create a summary column using the Report Wizard, the Wizard automatically modifies the report data model, and creates fields to display the summary column on the report layout. Now you will use the Report Wizard to create a summary column to sum the total credits per term and the total credits for each student. Before you do this, you will open the Report Wizard in reentrant mode and examine the Totals page.

To open the Wizard in reentrant mode and examine the Totals page:

1. In the Data Model, click **Tools** on the menu bar, and then click **Report Wizard** to open the Report Wizard in reentrant mode. The Report Wizard opens, with tabs for each page across the top of the window.

2. Click the **Totals tab**. The Totals page appears.

The Totals page displays all of the report fields in the Available Fields column. To create a report summary column, you select the field to be summarized from the Available Fields list, and then click the button that describes the summary function you would like to apply to the selected field: Sum, Average, Count, Minimum, Maximum, or %

Total. When you create a summary column using the Report Wizard, the Wizard creates a summary column in each group level above the record group that contains the source field and a field that summarizes the data for the overall report. In this report, the summary column source field is CREDITS, which is in the G_call_id (Level 3) record group. The Report Wizard will create a summary column that sums the credits at the term level (Level 2), which corresponds to the G_term_desc record group, and at the student level, which corresponds to the G_s_first (Level 1) record group. It will also create a summary column that summarizes the credits for all students in the report. Now you will create the summary column that sums the student credits. Then, you will view the summary fields.

To create and view the CREDITS summary column:

1. On the Report Wizard Totals page, scroll down in the Available Fields list, and select **credits**.

2. Click the **Sum >** button. Sum(credits) appears in the Totals list, indicating that a summary column to sum the CREDITS field has been created.

3. Click **Finish** to close the Report Wizard and save the change. In the Live Previewer, note the total credits appear for each term and for each student. If necessary, delete the Corporate1 logo, and reposition the report margin items.

Note that the formula column values no longer appear. When you enter the Report Wizard in reentrant mode, recall that all custom formatting objects are lost.

4. Click the **Last Page button** ▶ to view the last report page, and note the field that displays the total credits for the entire report, which has the value 36.

5. Save the report.

The general format for the names of the new summary fields in each record group is **F_Sum*fieldname*Per*record_group***. The *fieldname* is the name of the field being summed, and *record_group* is the name of the record group corresponding to the summary level. In this report, the summary column that sums the credits for each term is named **F_SumcreditsPerterm_desc**. It appears once per term, which explains the logic behind the name "Perterm_desc." The summary column that sums the credits for each student is named **F_SumcreditsPers_first** and appears once per student on the report. The name of the summary column for the entire report is **F_SumcreditsPerReport** and appears only once at the end of the report. Now you will examine the report data model and view the new summary fields.

To view the summary fields in the report data model:

1. Click the **Data Model button** ▦ to open the Data Model view. If some of the names of the new summary columns are not visible, select the record group, and drag the selection handles to make the record group longer or wider.

2. Resize and reposition the data model items, as shown in Figure 9-66.

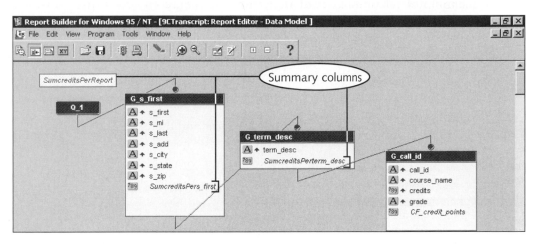

Figure 9-66   Data model with summary columns

The data model displays the "per report" summary column above the report query Q_1. The "per s_first" summary column is in the G_s_first record group, and the "per term_desc" summary column is in the G_term_desc record group. The G_call_id record group does not have a summary column because this record group contains the CREDITS field that is being summarized.

Recall that record groups have associated repeating frames, which enclose the record group data columns, and header frames, which enclose the record group column headings. A group frame encloses the record group repeating frame and header frame. When you create summary columns using the Report Wizard, the Wizard creates a new kind of frame, called a footer frame, to enclose the summary fields. A record group **footer frame** encloses the summary field that displays the summary value for the data in the record group. The general format for the name of a footer frame is **M_*record_group*_FTR**.

In this report, the call ID record group footer frame encloses the G_term_desc summary column (SumcreditsPerterm_desc), which sums the CREDITS column values for all courses for each student for each term. The term description record group footer frame encloses the G_s_first summary column (SumcreditsPers_first), which sums the CREDITS values for each student for all terms. The G_s_first record group footer frame encloses the SumcreditsPerReport summary column, which sums the CREDITS for all students in the report.

Viewing Summary Columns in the Object Navigator

The report frame structure is probably starting to seem a little complex. Viewing the report objects in the Object Navigator will help you understand the relationships among the report frames and the summary columns. The Object Navigator shows the frame hierarchical relationships and indicates how frames enclose other frames and report objects. You can view the report frames and other report objects and see their relationships in the Layout Model node in the Object Navigator.

To examine the summary columns and footer frames in the Object Navigator:

1. Click **Window** on the menu bar, and then click **Object Navigator** to open the Object Navigator.

2. Select the **9CTRANSCRIPT node**, and then click the **Collapse All button** to collapse all of the report objects.

3. Click + next to 9CTRANSCRIPT, click + next to the Layout Model node, select the **Main Section node**, then click the **Expand All button** to expand all of the report body components. The report body components and their hierarchical relationships appear, as shown in Figure 9-67.

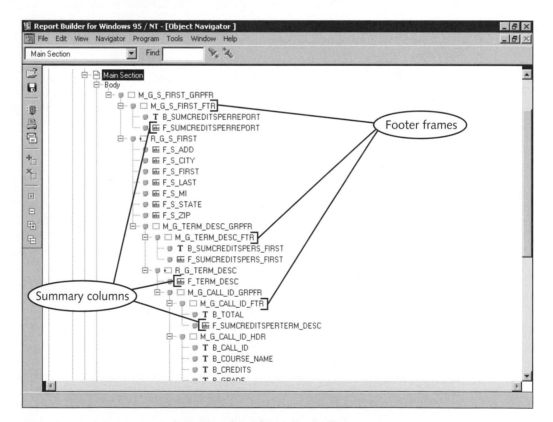

Figure 9-67   Summary columns in the Object Navigator

In the Report Builder Object Navigator, you can identify different object types by the icon that appears beside the object's name. [...] denotes a group frame or footer frame, [...] denotes a repeating frame, **T** denotes a boilerplate label (such as a column heading), and ⊡ denotes a report field that displays data. Also, note the convention for naming report items: Display field names (such as F_S_ADD) are prefaced with the letter *F* (for field), and boilerplate object names (such as the text label B_SUMCREDITSPERREPORT) are prefaced with the letter *B* (for boilerplate). Figure 9-67 shows that the M_G_s_first footer frame contains the "per report" summary column (F_SUMCREDITSPERREPORT) and its associated label. This footer frame is within the outermost group frame on the report, M_G_S_FIRST_GRPFR. The report-level summary column sums the credits for all students shown on the report, and it appears only once per report.

The M_G_term_desc_FTR footer frame contains the "per student" summary column (F_SUMCREDITSPERS_FIRST) and its associated label. This summary column summarizes the credits for all terms for a particular student, and it appears once for every student. The M_G_call_id_FTR footer frame contains the "per term" summary column (F_SUMCREDITSPERTERM_DESC) and its associated label. This summary column summarizes the credits for all courses for a particular term, and it appears once for every student for every term.

Since the "per report" summary column, which sums the credits for all students, is not needed in this report, you will hide it by opening its Property Palette in the Object Navigator, and setting its Visible property to No. (If you delete the "per report" summary column, it will be created again the next time you open the Report Wizard in reentrant mode, so changing its Visible property is a more permanent solution.) Then, you will delete its label. The label will not be generated again when you open the Report Wizard in reentrant mode.

To hide the per report summary column and delete its label:

1. In the Object Navigator, select **F_SUMCREDITSPERREPORT**, right-click, and then click **Property Palette** to open its Property Palette.

2. Set the Visible property to **No**, select another property to save the change, and then close the Property Palette.

3. Select the **B_SUMCREDITSPERREPORT** boilerplate label item, which is the label associated with the per report summary column, click the **Delete button** ⊡ to delete the label, and then click **Yes** to confirm the deletion.

4. Select the **9Ctranscript node**, and then click the **Run button** ⊡ to view the report in the Live Previewer. If necessary, click the **Last Page button** ⊡ to navigate to the last report page, and confirm that the per report total column and label (which displayed 36 credits) no longer appear.

5. Save the report.

Creating a Summary Column Manually

When you use the Report Wizard to create a summary column, the Report Wizard automatically generates summary fields and footer frames for all report record groups above the summary column source field. You might not need all of these summary fields and footer frames, and they make your report more complex. The Report Wizard also eliminates all of your custom formatting and deletes custom objects, such as the field you created to display the course credit points. An alternative to creating summary columns using the Report Wizard is to create a summary column manually. Creating a summary column manually allows you to control the summary column placement and frequency, and enables you to retain other custom objects. To create a summary column manually, you first create the new summary column in the report data model. The summary column must be in the record group that is one level higher than the values it sums. After you create the summary column, you must modify its properties to specify its name, the report field that it summarizes, and the summary function (SUM, AVG, and so forth). Usually, the names of summary columns that you create manually are prefaced with CS, followed by the name of the report field being summed. For example, the name of a summary column that sums term credits would be **CS_credits**. To display the summary column value on the report, you draw a field on the report layout and set its data source as the name of the summary column.

Now you will manually create a summary column in the transcript report that sums the student credit points for each term. Since the sum of the credit points per term will appear once per term, the summary column will be placed in the G_term_desc record group. Summary columns have a *Reset At* property, which specifies the record group level at which the summary column is reset to zero and begins summing again. You want to calculate the sum of credit points for each term, so you will set the summary column's Reset At property to **G_term_desc**. Now you will create the summary column in the data model and modify its properties.

You do not need to recreate the field to display the credit points per course that you lost by opening the Report Wizard in reentrant mode, because this value will not appear on the finished report. Even though the value no longer appears on the report, the formula column still calculates the course credit point values correctly.

To create the summary column in the data model:

1. Click the **Data Model button** ⊞ to open the Data Model, and then select the **Summary Column tool** ⬚ on the tool palette. The pointer changes to the Precision pointer + as you move it across the painting region.

2. Click the mouse pointer at the bottom of the G_term_desc record group. A new summary column named CS_1 appears in the G_term_desc record group. If necessary, select the new summary column and drag it down so that it is the last field in the record group.

3. Select **CS_1**, right-click, and then click **Property Palette** to open the column Property Palette. Change the name of the new summary column to **CS_CF_credit_points**.

4. Select the **Source** property, and select **CF_credit_points**.

5. Since you want the summary column to calculate the total credit points for every term, open the Reset At list, and select **G_term_desc**.

6. Select another property to save the changes, and then close the Property Palette and save the report.

Drawing a Field to Display the Summary Column Value

Next, you will draw a field on the report layout to display the sum of the credit points per term. You will draw the field so that it appears in the M_G_call_id_FTR frame, directly beside the summary column that sums the credits per term. First, you will make the footer frame wider to make room for the credit points summary field.

To widen the footer frame:

1. Click the **Layout Model button** 🖻 to open the Layout Model, and confirm that confine mode and flex mode are enabled.

2. Select the **F_SumcreditsPerterm_desc** summary column, which is directly below the F_credits field, and then click the **Select Parent Frame button** 🔲 to select the call ID footer frame (M_G_call_id_FTR). Open its Property Palette to confirm you have selected the correct frame, and then close the Property Palette.

3. Click the mouse pointer on the center selection handle on the right edge of the frame, and make the frame wider by dragging its right edge about 0.75 inches to the right, so that the frame is about 5 inches wide (see Figure 9-68).

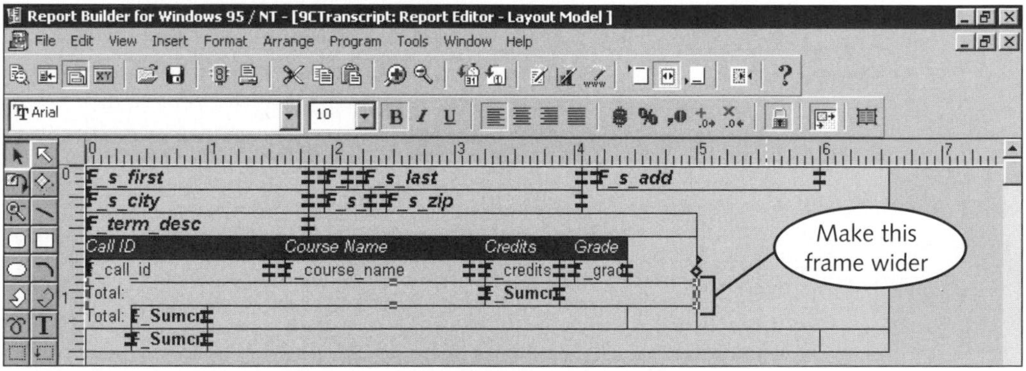

Figure 9-68   Widening the call ID footer frame

4. Click the **Live Previewer button** 📷 to view the report in the Live Previewer, and confirm that the frame is still enclosed by its parent frame and no errors appear.

5. Save the report.

Next, you will draw the field to display the credit points summary column. You will modify the field properties by changing its name and source.

To draw and configure the field to display the summary value:

1. Click the **Layout Model button** 🖼 to open the Layout Model, and click the **Field tool** abc on the tool palette. Draw a new field beside F_SumcreditsPerterm_desc. Figure 9-69 shows the position of the new field.

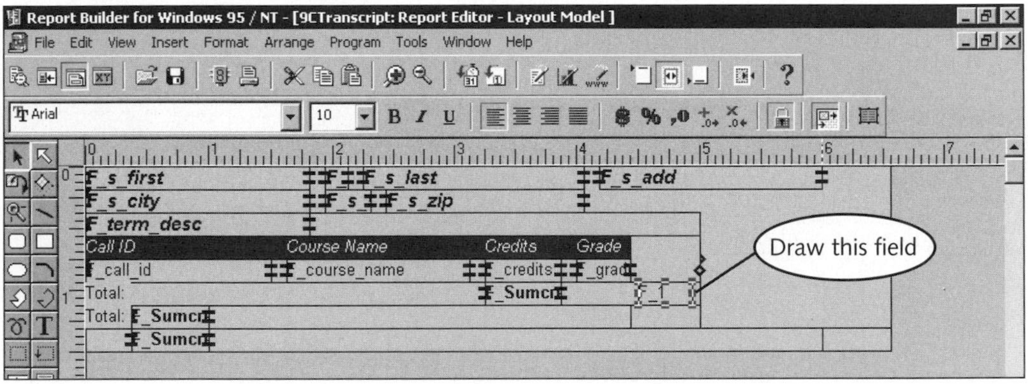

Figure 9-69   Field to display summed credit points per term

2. Open the new field Property Palette, and change its name to **F_CS_CF_credit_points**.

3. Change the field Source property to **CS_CF_credit_points**, click another property to save the change, and then close the Property Palette.

4. Click the **Live Previewer button** 📷 to view the report in the Live Previewer and confirm that the new summary column sums the credit point values correctly. The summary column should appear once per term. If necessary, click the **First Page button** 🔘, and examine the total credit points for Brian Umato. Brian's values should be 6 for the Fall 2003 term and 30 for the Spring 2004 term, as shown in Figure 9-70. If necessary, adjust the field's outline, fill color, and font specifications to match the rest of the report fields.

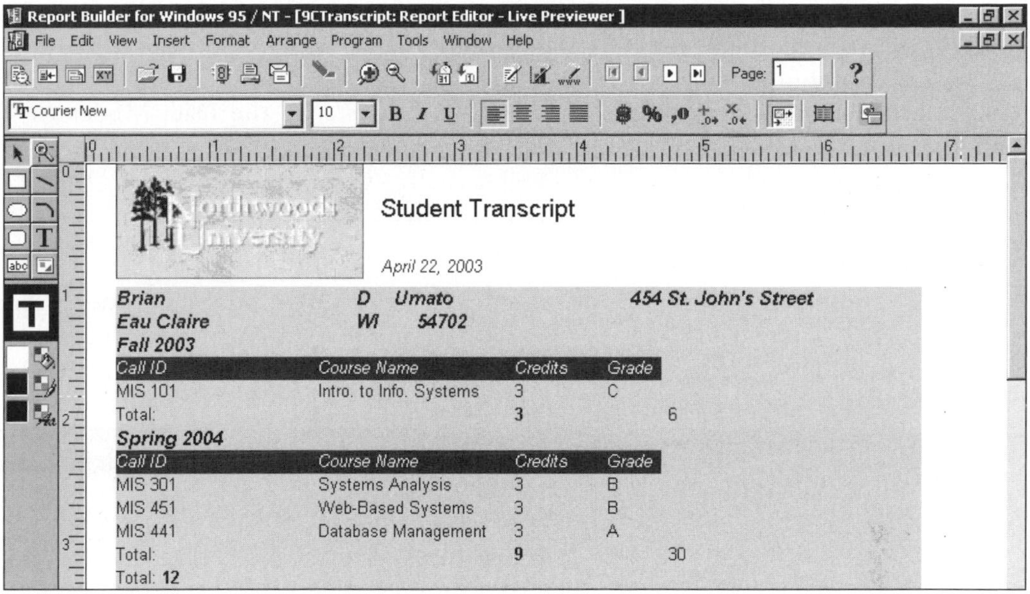

Report Builder for Windows 95 / NT - [9CTranscript: Report Editor - Live Previewer]

File Edit View Insert Format Arrange Program Tools Window Help

Courier New   10   B *I* U

Student Transcript

April 22, 2003

| Brian | | D | Umato | | 454 St. John's Street |
| Eau Claire | | WI | 54702 | | |

Fall 2003

Call ID	Course Name	Credits	Grade
MIS 101	Intro. to Info. Systems	3	C
Total:		3	6

Spring 2004

Call ID	Course Name	Credits	Grade
MIS 301	Systems Analysis	3	B
MIS 451	Web-Based Systems	3	B
MIS 441	Database Management	3	A
Total:		9	30
Total: 12			

Figure 9-70   Viewing the summed credit points per term

5. Save the report, and then close the report in Report Builder.

In the student transcript report, the student credit points per course and the sum of the student credit points per term are intermediate totals that do not appear on the final report. When you are creating a report with calculated columns that represent intermediate values that do not appear on the final report, it is a good practice to display the intermediate values during development to confirm that the values are correct. For example, in Figure 9-70, you can view the summed credit points per term and confirm that the summed field values are correct. To suppress this intermediate value in the final report display, you can delete the field on the Layout Model, since the calculated column still exists in the report data model. Or, you can change the display field's Visible property value to No in the field Property Palette.

Finishing the report GPA calculations, and formatting the report will be left as an end-of-chapter case.

Placeholder Columns

Placeholder columns are report columns similar to formula columns and can display values calculated using either a PL/SQL function or a report trigger. (The next section describes report triggers.) You can associate a placeholder column with a specific record group, or you can place it at the report level, if it represents a value for the entire report.

To create a placeholder column, you open the report data model, select the **Placeholder Column tool** 🔲, and click the record group in which you want to create the placeholder column. If you want to create a placeholder column that is not associated with any specific record group, you click 🔲 anywhere on the Data Model view painting region to create a report-level placeholder column.

REPORT TRIGGERS

A **report trigger** is a trigger that fires and executes PL/SQL code as a result of a specific report event. Table 9-5 describes the report trigger events, when the triggers fire, and how the triggers are used.

Trigger	Triggering Event	Example Use
AFTER PARAMETER FORM	After the report parameter form opens	Access, validate, and change parameter values
AFTER REPORT	After user exits the Runtime Previewer, or after the user sends output to a file, printer, or e-mail account	Delete temporary tables created by the report
BEFORE PARAMETER FORM	Just before parameter form opens	Validate input parameters
BEFORE REPORT	After report queries are parsed but before data is retrieved	Error handling
BETWEEN PAGES	Before report pages after Page 1 are formatted	Customized page formatting

Table 9-5   Report triggers

To create a report trigger, you click ➕ beside the Report Triggers node in the Object Navigator, select the trigger, right-click, click PL/SQL Editor, and then enter the trigger code.

CREATING QUERIES AND DATA LINKS MANUALLY

Earlier in the chapter, you created reports in which the report data model contained master-detail relationships. To do this, you retrieved all of the report data in a single query, and then used the Report Wizard Groups page to specify the master-detail relationships and create the associated master-detail record groups and data links. Another way to create reports that display master-detail relationships is to manually create a query that retrieves the master records, manually create a query that retrieves the detail records, and then manually create a data link between the two queries. To gain experience creating queries and data links manually, you will use a report that currently displays records from the ITEM table in the Clearwater Traders database. First, you will open the report, save the report file using a different filename, and then view the report.

To open, save, and view the report:

1. In Report Builder, open **Items.rdf** from the Chapter9 folder on your Data Disk, and save the file as **9CItems.rdf** in your Chapter9\TutorialSolutions folder.

2. Double-click the **Live Previewer icon** to view the report. Currently, the report displays the data for each ITEM record on a separate report page, which includes the text data for the item and the item image.

> If the message "Unable to display image data" appears when you try to view the report, the item image data is not available in the ITEM table in your database schema. To load the image data, run the Item_IMAGE.fmb form in Form Builder, and load the image data from the associated image files. All required files are in the Chapter9 folder on your Data Disk.

To create a query manually, you select the **SQL Query tool** on the Data Model tool palette, click in the painting region, and then specify the SQL command associated with the query. You can then draw fields on the report layout to display the query columns and create explicit data links between queries in the data model to manually create master-detail relationships. In a data link, the master side of the relationship is called the parent, and the detail side of the relationship is called the child. You can create the following types of data links:

- **Query to Query**, in which the relationship is defined at the query level, and the primary key of the parent record group is a foreign key in the child record group.

- **Group to Group**, which is defined at the record group level, and one of the columns in the parent record group, which is not necessarily the record group's primary key, has a foreign key relationship with one of the fields in the child record group. You would create this type of link when one of the queries in a master-detail relationship has additional master-detail relationships, and the data link involves only part of the query.

- **Column to Column**, which is defined at the column level, and has two identical columns that do not necessarily have a foreign key relationship.

> Sometimes deciding what type of data link to use is a challenging process. Try different data link types until you get the result you want.

To create a data link, you select the **Data Link tool** on the Data Model tool palette and then draw a link from the parent object to the child object. To create a Query to Query link, you draw the link from the parent query to the child query. To create a

Group to Group link, you draw the link from the parent record group title bar to the child record group title bar. To create a Column to Column link, you draw the link from the parent column to the child column.

Suppose you want to modify the item report to display the category description of the current item. You could modify the query in the Report Wizard so it retrieves the CATEGORY_DESC column from the CATEGORY table, but that would reformat the report layout. Instead, you will create a new query in the report data model that retrieves all of the records from the CATEGORY table and then create a link between the CATEGORY_ID field in the new query to the CATEGORY_ID field in the existing query. Since CATEGORY_ID is the primary key in the CATEGORY table and a foreign key in the ITEM table, you will create a Query to Query link. The query that retrieves the CATEGORY records will be the parent query, and the query that retrieves the ITEM records will be the child query.

To manually create the query and data link:

1. Click the **Data Model button** [≣+] to open the Data Model view, click the **SQL Query tool** [sql] on the tool palette, and click the painting region on the left side of the existing Q_1 query. The SQL Query Statement dialog box opens.

2. Type **SELECT * FROM category** in the SQL Query Statement box, and then click **OK**. A new query named Q_2 appears in the painting region. Note that the record group name is G_CATEGORY_ID1.

 When you add a new column to a data model that has the same name as an existing column, a number is appended to the column name. This ensures that every column in a report Data Model has a unique name.

3. Reposition the queries so they appear as shown in Figure 9-71.

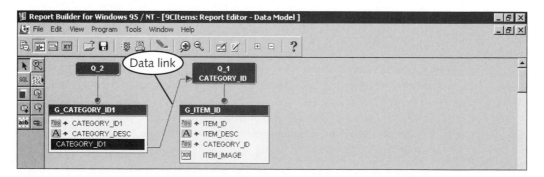

Figure 9-71 Creating a query and data link manually

4. To create the data link, select the **Data Link tool** on the Data Model tool palette, select **Q_2**, which is the parent query, and draw a link from Q_2 to Q_1. The data link appears, as shown in Figure 9-71, and shows that CATEGORY_ID1 in query Q_2 is linked to CATEGORY_ID in query Q_1.

5. Save the report.

When you create data links using the Data Link tool , you always draw the link from the parent object to the child object.

To display a new data model column in a report, you draw a new field in the report layout model and then set the field Source property to the name of the new data model column. Now you will draw a new field in the report Layout Model and specify the field Source property value to be the CATEGORY_DESC column in the report data model.

To display the new data column in the report:

1. Click the **Layout Model button** to open the Layout Model.

2. In the Layout Model, select the **Field tool** , and draw a new field on the right side of the F_CATEGORY field and just below the F_ITEM_DESC field. The new field should be about 1 inch wide and the same height as the other report fields. If necessary, adjust the field fill color and font, and remove the outline around the field.

3. Double-click the new field to open its Property Palette, change its Name value to **F_CATEGORY_DESC**, and its Source value to **CATEGORY_DESC**. Select another property to save the change, close the Property Palette, and then save the form.

4. Click the **Live Previewer button** to view the report. The category description value that was retrieved using the manually created query appears as shown in Figure 9-72. If necessary, resize the field in the Layout Model so all of the category description text appears.

9

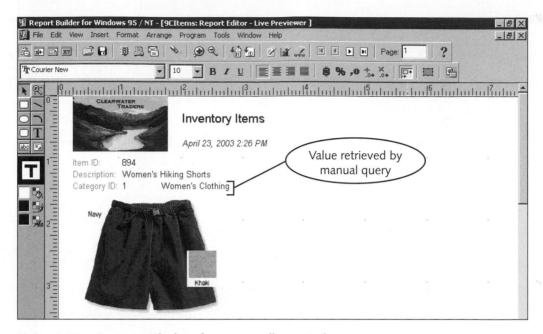

Figure 9-72   Report with data from manually created query

In the previous example, the manual query retrieves only a single value, which is the category description. You can also display values from a manual query that retrieves multiple values. To do this, you must create the query that retrieves the multiple values, manually draw a repeating frame on the report, and then set the repeating frame Source property as the manual query record group name. Suppose you want to modify the item report so the INVENTORY records associated with each item appear on the right side of the report. You can create a query to retrieve the INVENTORY records and then create a repeating frame and the associated data fields to display the inventory values. First, you will create the query to retrieve the INVENTORY records and create the data link. This time, you will create a Column to Column data link. The ITEM_ID column in the G_ITEM_ID record group will be the parent object, and the ITEM_ID1 column in the G_INV_ID group will be the child object.

To create the query and data link for the INVENTORY records:

1. Click the **Data Model button** 🖳 to open the Data Model.

2. To create the query, select the **SQL Query tool** 🔳 on the tool palette, click the painting region on the right of Q_1, type the following SQL command, and then click **OK**. A new record group named G_INV_ID appears.

   SELECT * FROM inventory

3. To create the Column to Column link, select the **Data link tool** 🔳 in the tool palette, click the ITEM_ID column in G_ITEM_ID, and drag the mouse

pointer to draw a line joining the ITEM_ID column in G_ITEM_ID to the ITEM_ID1 column in G_INV_ID. The data link appears in the report data model, as shown in Figure 9-73.

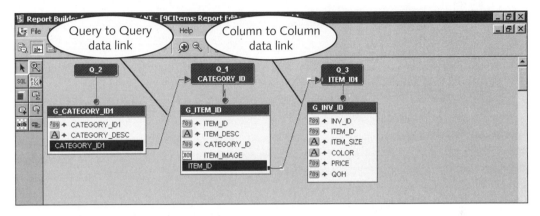

Figure 9-73   Creating a Column to Column data link

4. Save the report.

Now, you need to display the inventory columns on the report. To do this, you will open the Layout Model and draw a repeating frame, which will automatically size or shrink depending on the number of inventory records retrieved. You will set the repeating frame Source property to be the name of the record group that provides its values, which is G_INV_ID. Then, you will draw fields in the repeating frame to represent each of the G_INV_ID record group columns. You will also modify each field's Source property to be the name of the associated record group column.

To draw the repeating frame and fields to display the inventory columns:

1. Click the **Layout Model button** 🔲 to open the report layout model.

2. Select the **Text tool** T on the Layout Model tool palette, click the painting region in the area on the right side of F_ITEM_ID, and type **Inventory Detail** to create a label for the repeating frame, as shown in Figure 9-74.

3. Select the **Repeating Frame tool** 🔲 on the tool palette, and draw the repeating frame shown in Figure 9-74.

4. Select the repeating frame, right-click, click **Property Palette** to open the frame Property Palette, change the Source property value to **G_INV_ID**, click another property to save the change and then close the Property Palette.

5. Select the **Field tool** 🔲 on the tool palette, and draw five fields inside the repeating frame, as shown in Figure 9-74. These fields will display the inventory ID, size, color, price, and quantity on hand.

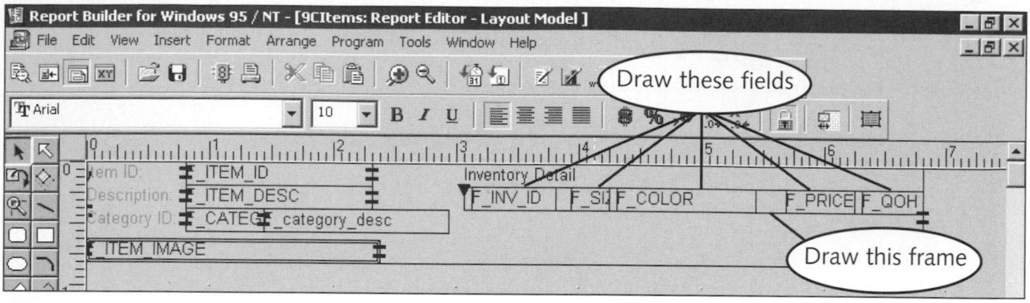

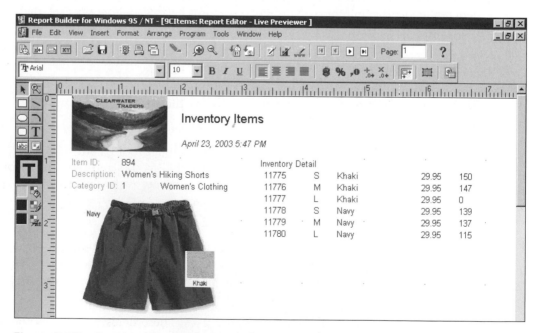

Figure 9-74   Drawing the label, repeating frame, and fields

6. Open the Property Palette for each of the fields, change the properties as follows, and then close the Property Palette.

	Field 1	**Field 2**	**Field 3**	**Field 4**	**Field 5**
Name	**F_INV_ID**	**F_SIZE**	**F_COLOR**	**F_PRICE**	**F_QOH**
Source	**INV_ID**	**ITEM_SIZE**	**COLOR**	**PRICE**	**QOH**

7. Save the report, and then click the **Live Previewer button** 🔲 to view the report in the Live Previewer. The report should appear as shown in Figure 9-75. If necessary, adjust the frame and field outline style and line fill color, and modify the font of the fields to match the rest of the report fields.

Figure 9-75   Report with detail records from manual query

USING LINK FILES TO DISPLAY FILE DATA

A **link file** is a report layout object that displays the contents of a file in a report. You would create a link file if your report has a related file in which the file contents change, and you always want to show the contents of the most current file. For example, suppose that Clearwater Traders makes the item report that you have been using available to customers on the company Web site. The marketing department would like to display an image file on the report that describes a current marketing promotion. You could create a link file on the report and then display the promotion information in the link file. When the marketing promotion changes, you could replace the file on the Web server with a different file and not have to modify the report code.

To create a link file object, you select the **Link File tool** 🔲 on the Layout Model tool palette, and draw the link file object on the painting region. Then, you can specify the following properties:

- **Source File Format**, which specifies the type of file that appears in the link file object. Supported file formats include text, image, CGM (computer graphics metafile), Oracle Drawing Format, and Image URL.

- **Source Filename**, which specifies the folder path and filename of the link file.

- **Print Object On**, which specifies the report page or pages on which the object appears. Values can be *All Pages*, *All But First Page*, *All But Last Page*, *Default*, *First Page*, and *Last Page*.

Now, you will add a link file to the item report layout. You will specify the file format as *Image*, and the source filename as a file named *promotion.tif* that is stored in the Chapter9 folder on your Data Disk. You will specify to display the file on all report pages.

To add a link file to the inventory items report:

1. Click the **Layout Model button** 🗐 to open the Layout Model.

2. Select the **Link File tool** 🔲 on the tool palette, and draw the link file object, as shown in Figure 9-76.

3. Select the link file object, click the **Line Color tool** 🖉 on the tool palette, and select a gray square that is a different shade than the report background color. (If you select No Line, you will not be able to see the link file object on the painting region.)

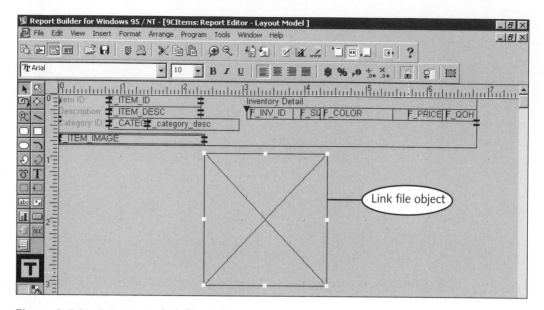

Figure 9-76 Drawing a link file object

4. Right-click the link file object, and then click **Property Palette** to open the link file object Property Palette. Modify the following property values:

Property	Value
Name	**B_PROMOTION_LINK_FILE**
Source File Format	**Image**
Print Object On	**All Pages**

5. To specify the link filename, select the **Source Filename** property, click the **More button** [...], navigate to the Chapter9 folder on your Data Disk, select **promotion.tif**, click **Open**, and then close the Property Palette.

6. Save the report, and click the **Live Previewer button** to view the report in the Live Previewer. The link file appears on the report display. (You might need to scroll down the report page to view all of the file image.)

7. Close the report in Report Builder.

RUNNING A REPORT FROM A FORM

In finished applications, reports are often called from Oracle Forms Runtime applications so that the user can select a parameter value in a form, and then display or print the related report information in Reports Runtime. Figure 9-77 shows a form in which a user selects a specific term and student. When the user clicks the Run Transcript button, the form opens the transcript report and passes the selected student ID and term

ID values as parameter values. When the report receives these values, it displays the transcript values for the selected term and student.

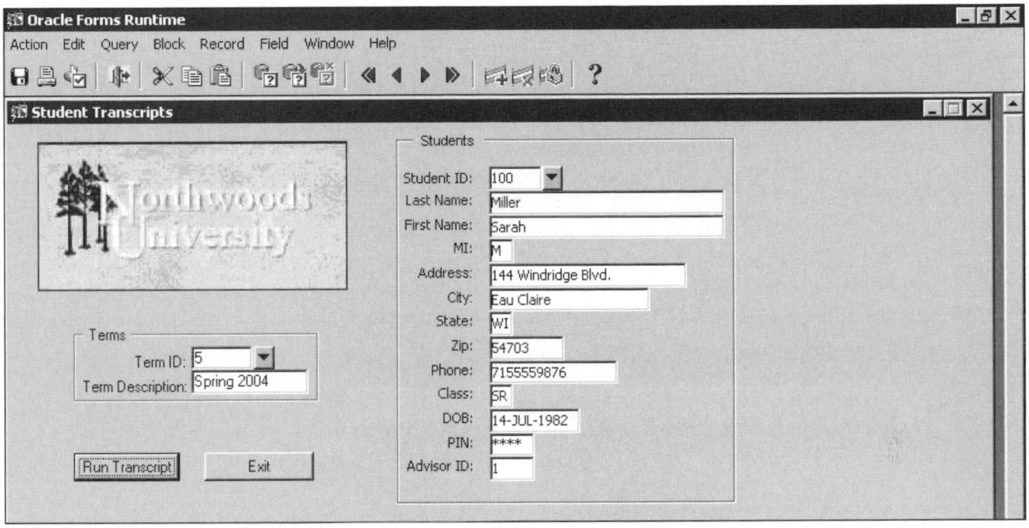

Figure 9-77   Form to call transcript report

To implement the form/report combination, you must modify the report so that it will accept the parameters passed from the form. You must add commands to the form to create a parameter list that contains the selected values for student ID and term ID. Then you must call the report, and pass the parameter list to it. Finally, you must destroy the parameter list. The following sections describe these steps.

Modifying the Report to Accept Parameters from a Form

To pass one or more variable values from a form to a report, you must create bind parameters in the report query. Recall that bind parameters are parameters whose values are determined at runtime. The values are then assigned, or bound, to variables within an application. Bind parameters are just like the report user parameters you created earlier, except that the values are passed to the report rather than selected from the report parameter form.

To create a report bind parameter, you replace one or more expressions in the report SQL query with a variable value preceded by a colon, and Report Builder automatically creates the parameter. Now you will open a copy of the transcript file report that contains all of the modifications performed so far and modify the report query. You will add search conditions for S_ID and TERM_ID, and then set these values equal to bind parameters whose values will be assigned when the form calls the report and passes the selected values for term ID and student ID. Often, parameter variables are prefaced with the letter *P* to indicate that they are parameters. You will name the bind parameters P_S_ID and P_TERM_ID.

You can create a bind parameter by explicitly creating a user parameter object in the Object Navigator and then referencing the parameter in the report query. You can also create a bind parameter by modifying the report query and replacing a search condition value with a parameter name.

To open the transcript report and create the bind parameters in the report query:

1. In Report Builder, open **TranscriptForm.rdf** from the Chapter9 folder on your Data Disk, and save the file as **9CTranscriptForm.rdf** in your Chapter9\TutorialSolutions folder.

2. In the Object Navigator, double-click the **Data Model icon** ⊞+ to open the report data model, and then double-click **Q_1** to open the SQL query statement. Modify the query as shown in Figure 9-78 to create the bind parameters, and then click **OK**.

SQL Query Statement   `_ □ ×`

| Query Builder... | Import SQL Query... | Connect... |

SQL Query Statement:

```
SELECT s_first, s_mi, s_last, s_add, s_city, s_state, s_zip, term_desc, call_id,
course_name, credits, grade
FROM student, course, course_section, enrollment, term
WHERE student.s_id = enrollment.s_id
AND enrollment.c_sec_id = course_section.c_sec_id
AND course_section.term_id = term.term_id
AND course_section.course_id = course.course_id
AND grade IS NOT NULL
AND term.term_id = :P_TERM_ID
AND student.s_id = :P_S_ID
```

Add these commands

| OK | Cancel | Help |

Figure 9-78   Adding bind parameters to report query

---

3. The informational message "Note: The query 'Q_1' has created the bind parameter(s) 'P_TERM_ID', 'P_S_ID'." appears to confirm that the bind parameters have been created. Click **OK**.

4. Click **Window** on the menu bar, and then click **Object Navigator** to open the Object Navigator. Click ➕ beside the User Parameters node under the Data Model node. The bind parameters P_S_ID and P_TERM_ID appear.

5. Save the report.

Creating a Form Parameter List

Next, you will open the form file in Form Builder and create the parameter list that will pass the term ID and student ID from the form to the report. In a form, a parameter list is a list of variables that is used to pass data values from a Form Builder application to another application. Parameter lists are created in triggers or program units using PL/SQL. Whenever you create a new object in Form Builder, Form Builder assigns a unique **object ID** to the object. Form Builder then uses this ID whenever it references the object in subsequent operations. To create a parameter list, you first must declare a variable with the data type PARAMLIST to reference the list's object ID, using the following syntax:

```
DECLARE
    list_ID PARAMLIST;
```

The *list_id* value can be any legal Oracle variable name.

The next step is to create the list. This is done in the body of the PL/SQL procedure, using the following syntax:

```
BEGIN
    list_ID := CREATE_PARAMETER_LIST('list_string');
```

The *list_id* is the parameter list variable you declared before. The *list_string* value can be any legal Oracle variable name, and is enclosed in single quotation marks.

After creating the parameter list, you use the ADD_PARAMETER built-in procedure to add the parameter data values to the list. You can use a single parameter list to pass multiple parameters. Each parameter is added to the list individually, using a separate call to the ADD_PARAMETER procedure. Parameters passed using a parameter list can only be text (character) data or record groups. If you want to pass a date or number field, you must first convert it to a character using the TO_CHAR data conversion function.

To use the ADD_PARAMETER procedure to add a parameter to a parameter list, you use the following syntax:

```
ADD_PARAMETER(list_ID, 'key', PARAMTYPE, value);
```

This command uses the following items:

- *list_ID* is the name of the list ID variable for the parameter list that you defined when you created the parameter list;

- *key* is the name of the bind parameter as it was defined in the report that is receiving the parameter. *Key* is passed as a character string, so it must be enclosed in single quotation marks. For the transcript report, the parameter keys are P_TERM_ID and P_S_ID, because these are the names of the bind parameters that were used in the report query (see Figure 9-78).

- *PARAMTYPE* can have a value of either TEXT_PARAMETER (for parameters that reference character data values) or DATA_PARAMETER (for parameters that are record groups).

- *value* represents the actual parameter character string or record group name. In the form that will call the transcript report, *value* will reference the values for term ID and student ID that appear on the form. Since both of these values have a NUMBER data type, they must be converted to character strings before they are added to the parameter list.

9

When you pass a parameter list from a form to a report, the report parameter form appears before the report appears and shows the parameter values that were passed in the parameter list. To suppress showing the parameter form, you must add an additional parameter to the parameter list to instruct the report to suppress showing the parameter list. The syntax for adding this additional parameter is:

```
ADD_PARAMETER( _list_id, 'PARAMFORM',
TEXT_PARAMETER, 'NO');
```

Now you will start Form Builder, open the form shown in Figure 9-77, and create a trigger for the Run Transcript button. In the DECLARE section, the trigger will declare the parameter list variable and the character variables that will reference the TERM_ID and S_ID values when they are converted to character strings. In the trigger body, the code will initialize the parameter list, add the TERM_ID and S_ID form values to the list, and add the parameter to suppress the parameter form. You will not compile the trigger yet, because it is not complete.

To open the form and initialize the parameter list:

1. Start Form Builder, open the **StudentReport.fmb** file from the Chapter9 folder on your Data Disk, and save the form as **9CStudentReport.fmb** in your Chapter9\TutorialSolutions folder.

If your computer does not have enough memory to run Form Builder and Report Builder at the same time, you will need to close Report Builder.

2. Click **Tools** on the menu bar, and then click **Layout Editor** to open the Layout Editor. Select the **Run Transcript button**, right-click, point to **SmartTriggers**, and then click **WHEN-BUTTON-PRESSED** to create a new trigger for the Run Transcript button.

3. Type the code shown in Figure 9-79 to declare and initialize the parameter list. DO NOT compile the trigger yet.

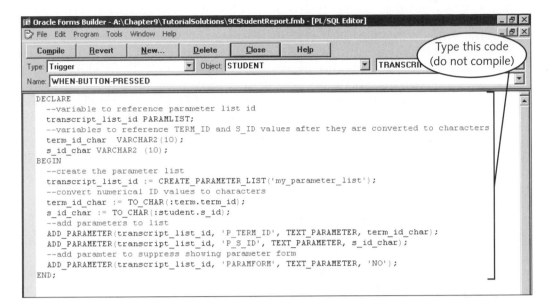

Figure 9-79   Code to create form parameter list

Calling a Report from a Form and Passing a Parameter List

To call a report from a form, you use the RUN_PRODUCT built-in procedure. RUN_PRODUCT starts the Reports Runtime application, specifies the name of the report definition file that will appear, and passes the parameter list that contains data values that customize the report output. The general syntax for the RUN_PRODUCT procedure is:

```
RUN_PRODUCT(product, 'document', communication_mode,
  execution_mode, location, list_id, display);
```

This procedure uses the following parameters:

- *product* specifies which Oracle Developer application you want to run. RUN_PRODUCT can be used to run either Report Builder or Graphics Builder applications, so its value will be either REPORTS or GRAPHICS.

- *document* specifies the complete folder path and filename of the report .rdf file as a character string enclosed in single quotation marks. The filename and folder path cannot contain any blank spaces. If the report filename is 9CTranscript.rdf and is stored in your Chapter9\TutorialSolutions folder on the A: drive, the document specification would be 'a:\chapter9\tutorialsolutions\9ctranscript.rdf'.

- *communication_mode* can be either SYNCHRONOUS or ASYNCHRONOUS. SYNCHRONOUS specifies that program execution control returns to the form only after the called product (Reports or Graphics) is closed. ASYNCHRONOUS specifies that control returns to the form while the product is still running, and users can multitask between the form and report applications.

- *execution_mode* can be either BATCH or RUNTIME. RUNTIME specifies that the called product's runtime environment (such as Reports Runtime) runs, while BATCH specifies that the called product generates an executable file that appears directly on the form. Use RUNTIME when running a Report Builder application, so that control is transferred to the Reports Runtime environment. Use BATCH when running a Graphics application when a graphic appears directly in a form.

- *location* can be either FILESYSTEM or DB (database) and specifies where the target document is stored. In the exercises in this book, all documents will be stored in the file system on the user's workstation.

- *list_id* is the ID of the parameter list that contains the data values that are passed to the called product. For the transcript form, this variable is `transcript_list_id` (see Figure 9-79). If there is no parameter list, this value must be passed as NULL.

- *display* is used only when you are running a Graphics chart. (Graphics charts will be discussed in Chapter 10.) This property specifies the name of the form image that will display the chart on the form. The DISPLAY value must always be passed as NULL when you are calling a report.

Now you will add the RUN_PRODUCT command to the Run Transcript form button to run the report and pass the parameter list.

To add the RUN_PRODUCT command to the button trigger:

1. In the PL/SQL Editor in Form Builder, add the following command as the last line of the Run Transcript button trigger code, just after the third ADD_PARAMETER command, and just before the END command. You can enter this command on multiple lines, as long as you do not break the line within a character string that is enclosed in single quotation marks.

```
RUN_PRODUCT (REPORTS,
'a:\chapter9\tutorialsolutions\9ctranscriptform.rdf',
SYNCHRONOUS, RUNTIME, FILESYSTEM, transcript_list_id,
NULL);
```

2. Compile the trigger, and correct any syntax errors. Do not close the PL/SQL Editor.

3. Click **File** on the menu bar, and then click **Save** to save the form.

Destroying Parameter Lists

After a trigger passes a parameter list to a report, the trigger must destroy the parameter list, or the list will remain in the server's memory. The problem with a parameter list remaining in the server's memory is that you cannot create a new parameter list with the same name as an existing parameter list. Suppose the user selects a specific term ID and student ID, clicks the Run Transcript button, views the report, and then closes the report. The form appears again. Now, suppose the user selects a different term ID and student ID, and then tries to run the transcript report again using the new values. An error will appear, because the form cannot create the parameter list again, since the list has the same name as the parameter list that was created the first time the user clicked the Run Transcript button. When you destroy the parameter list, the database server deletes the reference to the list name, releases the memory space used by the list, and makes that memory space available for other program structures.

To destroy a parameter list, you use the DESTROY_PARAMETER_LIST built-in, which has the following syntax: `DESTROY_PARAMETER_LIST(list_ID);`. You will add the command to destroy the parameter list as the last command in the button trigger.

To add the command to destroy the parameter list:

1. In the PL/SQL Editor in Form Builder, add the following command as the last line of the Run Transcript button trigger code, just after the RUN_PRODUCT command, and just before the END command:

   `DESTROY_PARAMETER_LIST(transcript_list_id);`

2. Compile the trigger, correct any syntax errors, close the PL/SQL Editor, and then save the form.

Now you will test the form/report combination. You will run the form and select the Spring 2004 term and student Sarah Miller. When you click the Run Transcript button, the transcript report for the selected term and student should appear.

To test the form/report combination:

1. In Form Builder, click the **Run Form Client/Server button** 🔲 to run the form. Connect to the database if necessary.

2. Click the **LOV command button** beside Term ID, and select **Spring 2004**, and then click **OK**.

3. Click the **LOV command button** beside Student ID, select student ID **100** (Sarah Miller), and then click **OK**.

4. Click **Run Transcript**. The report appears for the values for the selected student and term.

 If the report does not appear, check to make sure that the report is named 9CTranscriptForm.rdf, and that the drive and path specification are correct. If you have saved the file under a different name or using a different path, then change the document string in the RUN_PRODUCT command accordingly.

5. Click the **Close Previewer button** ☒. The Student Transcripts form appears again. Click **Exit**.

6. Close the form in Form Builder, and close Form Builder and all other open Oracle applications.

SUMMARY

- ❑ A report definition file is used by developers to define report contents and how they appear. Users use a report output file to view report data.

- ❑ You can save report definitions in Report Builder .rdf files, as well as in text and rich text formats.

- ❑ You can create compiled report files that have an .rep file extension and appear in Reports Runtime. In Reports Runtime, users can save report output files in a variety of different formats.

- ❑ You can create formula, summary, and placeholder columns to display calculated values that are based on report data values.

- ❑ Formula columns display a value that is returned by a PL/SQL function that performs mathematical computations on report data.

- ❑ To create a formula column, you create the column in the data model, write the formula function in PL/SQL, and then create a field on the report layout to display the calculated value.

- ❑ You must place formula columns in the record group that contains the columns that the formula references.

- ❑ To reference report columns in a PL/SQL formula column function, you preface the column name, as it appears in the data model, with a colon.

- ❑ To display a formula column on a report, you must create a new report data field that is in the same repeating frame as the source values for the formula and that uses the formula column as its data source.

- ❑ A summary column summarizes data fields using group functions, such as SUM, AVG, or COUNT. Summary columns can also be used to find the minimum or maximum value in a data series or to find the number of records retrieved by a query.

- ❑ When you use the Report Wizard to create a summary column, it creates a summary field that displays the summary value in each group level above the summarized field's record group in the report. It also creates a summary field that displays the summary value for the entire report.

- ❑ The Report Wizard places a new summary column in a footer frame that is in the same group frame as the one that encloses the repeating records being summarized.

- ❑ You can view a report in the Object Navigator to examine how frames enclose other frames and report objects.

- ❑ To create a summary column manually, you create the summary column in the data model, draw a field on the report layout to display the summary column value, and then set the new field's source property as the summary column name.

❏ A placeholder column is similar to a formula column and can display the result of a report trigger or a PL/SQL function.

❏ A report trigger is a trigger that fires and executes PL/SQL code as a result of a specific report event.

❏ You can manually create queries and data links in the Data Model view and then create fields to display the values that the queries retrieve. Data links that you create manually can be joined using Query to Query, Group to Group, or Column to Column links.

❏ You can create link file objects in the Layout Model view to display the contents of a file on a report. Link file objects are useful for displaying file contents when the file contents often change.

❏ Users can run reports as stand-alone executable programs or from Form Builder applications. To run a report from a form, you must modify the report so that it will accept parameters passed from the form. You must add commands to the form to create a parameter list that contains the parameters that will be passed to the report, call the report, and pass the parameter list to it.

❏ Parameters passed from a form to a report can be only character data or record group data. If you want to pass a date or number field, you must first convert it to a character using the TO_CHAR data conversion function.

❏ A single parameter list can pass multiple parameters. Each parameter value must be added to the parameter list using a separate ADD_PARAMETER command.

❏ To call a report from a form, you use the RUN_PRODUCT procedure, which specifies the path to the report file, the name of the parameter list, and other values.

❏ After you create a parameter list, you must destroy the list before you try to create it again, or an error will occur.

REVIEW QUESTIONS

1. What is a report definition file? What are the different types of report definition files that you can create in Report Builder?

2. What is the Reports Runtime application used for?

3. When you create a summary column using the Report Wizard, in which record groups does the wizard create new summary fields?

4. List the steps for creating a formula column.

5. In which record group in the data model do you place a new formula column?

6. How do you reference a report column in a formula column function?

7. What is a placeholder column?

8. List the three types of data links, and describe how each is created.

9. What is a link file object?

10. Which report triggers validate Parameter Form values?

11. Which of the following data field values from the CUST_ORDER table in the Clearwater Traders database could be passed directly in a parameter list without using any data type conversion functions?

   a. ORDER_ID

   b. ORDER_DATE

   c. METH_PMT

   d. ORDERSOURCE_ID

12. When is it a good idea to manually create a summary column?

13. List the steps for modifying an existing report and form so you can call the report from the form and pass form values to the report to control its output.

PROBLEM-SOLVING CASES

The following cases use the Clearwater Traders (see Figure 1-11), Northwoods University (see Figure 1-12), and Software Experts (see Figure 1-13) sample databases. All required files are in the Chapter9 folder on your Data Disk. Store all solutions in your Chapter9\CaseSolutions folder.

1. In this case, you will open the incomplete transcript report created in the tutorial and complete the report so it looks like the student transcript report shown in Figure 9-60.

   a. Open the 9CCase1.rdf file, and save it as 9CCase1_DONE.rdf.

   b. Calculate and display the student's term GPA. (*Hint:* Create a formula column that divides the sum of the term credit points by the sum of the term course credits. Reference the summary fields in the formula just like any other record group field.)

   c. Calculate and display the student's cumulative GPA. (*Hint:* Sum all course credit points and divide the result by the sum of all course credits.)

   d. Format the report so it looks like Figure 9-60.

2. Create a report that displays building codes at Northwoods University and then shows all associated location IDs, rooms, and capacities for each building code, ordered by room numbers. Create a summary column that sums the capacity of all of the rooms in a specific building and also shows the capacity of all rooms in all buildings. Format the report as shown in Figure 9-80, and apply the northwoods.tdf template file you created earlier in the chapter, or apply a different template. Save the file as 9CCase2.rdf. Create a text version of the report definition file named 9CCase2.rex, and create a compiled version of the report output file named 9CCase2.rep.

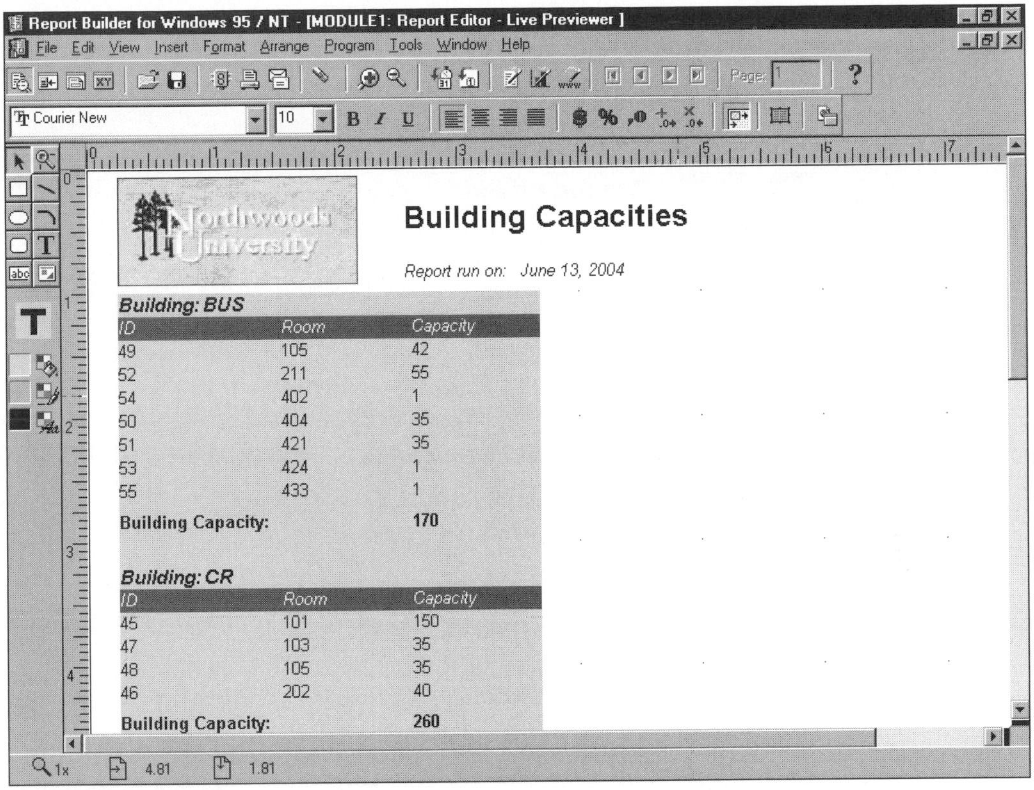

Figure 9-80

3. Create a report that displays information about Software Experts consultants and their associated skill certifications. Create a summary column that displays the total number of skills that each consultant is certified in. (*Hint*: Create a formula column that displays the number 1 in a placeholder column if a consultant is certified in a specific skill, and the number 0 if the consultant is not certified in that skill. Then sum the values in the placeholder columns.) Format the report as shown in Figure 9-81, and apply the softwareexp.tdf template that you created earlier in the chapter, or apply a different template to the report. Save the report as 9CCase3.rdf. Create a text version of the report definition file named 9CCase3.rex, and create a compiled version of the report output file named 9CCase3.rep.

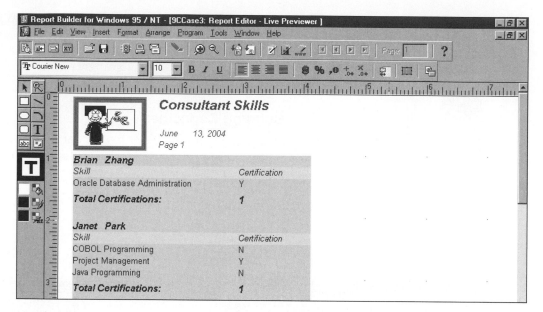

Figure 9-81

4. Create the report shown in Figure 9–82 that displays information about students within different class ranks at Northwoods University.

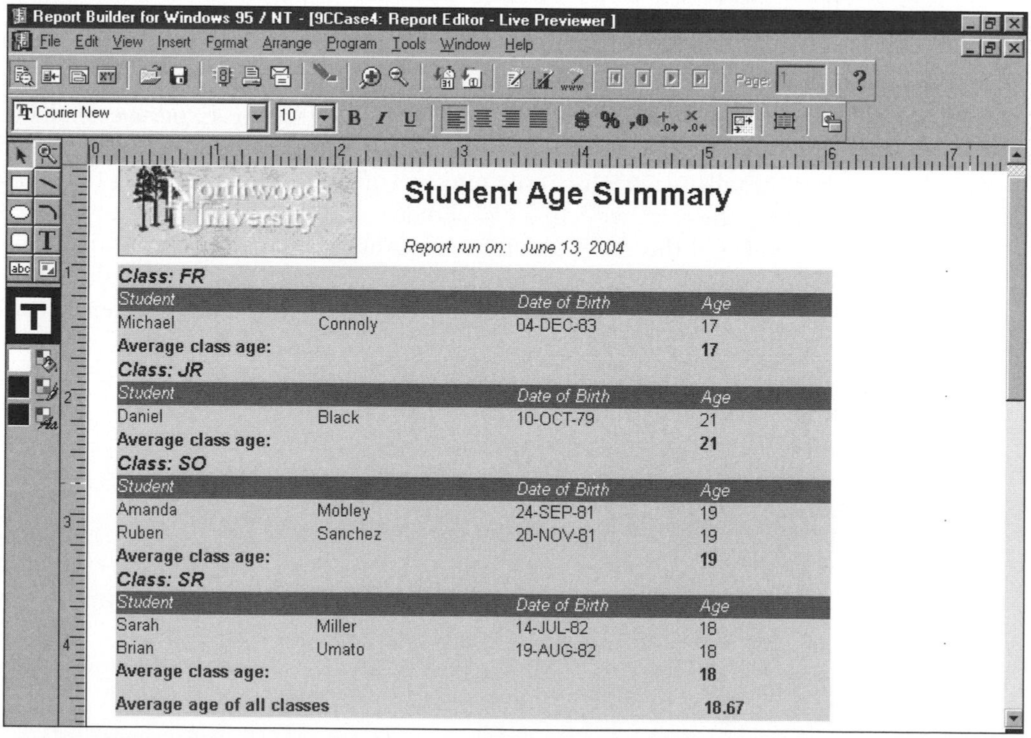

Figure 9-82

The report calculates the age of each student, based on the student date of birth. The report also displays the average student age for all students in each class and for all students in all classes in a summary column. Format the report as shown in Figure 9-82, and apply the northwoods.tdf template file you created earlier in the chapter, or apply a different template. Save the report as 9CCase4.rdf.

5. Create a report that displays customer order information for Clearwater Traders, as shown in Figure 9-83. The top section of the report should display the customer name and address, as well as the order date, order ID, and payment method. The detail section should show the item ID, item description, size, color, order price, and quantity, and calculate the extended total. The summary section at the bottom of the report should show a subtotal that sums all of the extended totals, calculates sales tax and shipping and handling, and sums the subtotal, tax, and shipping and handling to create a final order total. Format the report as shown in Figure 9-83, and apply the clearwater.tdf template file you created earlier in the chapter, or apply a different template. Save the report as 9CCase5.rdf.

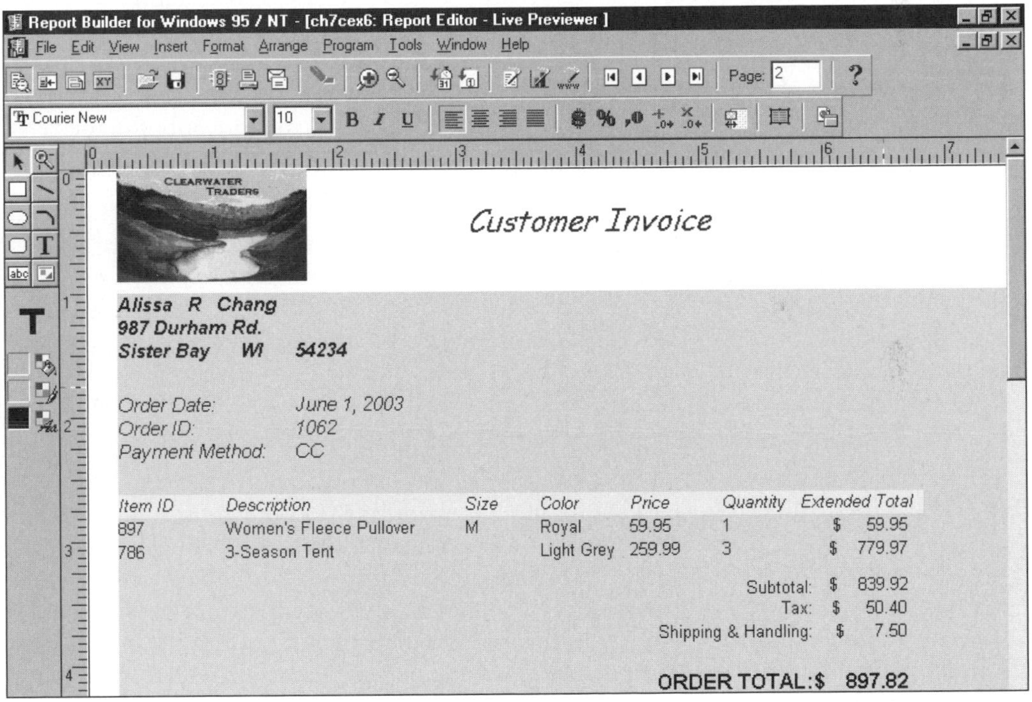

Figure 9-83

a. Create the report using the Report Wizard to display the required data fields. Each customer order should appear on a separate report page.

b. Create a formula column to calculate the extended total, which is the order price times quantity.

c. Create a summary column to calculate the subtotal, which is the sum of all of the extended totals.

d. Create a formula column that calculates the tax as 6% of the subtotal for Wisconsin residents and as 0 for residents of all other states.

e. Create a formula column to calculate shipping and handling as follows: If the order subtotal is less than $25, then the shipping cost is $3.50. If the subtotal is $25 or more, but less than $75, then the shipping cost is $5. If the subtotal is $75 or more, then the shipping cost is $7.50.

f. Create a column to calculate the order total, which is the sum of the subtotal, tax, and shipping and handling.

6. In this case, you will modify a form that has an LOV that allows users to select from a list of Clearwater Traders item ID and description values. After the user selects a specific item ID and clicks a button labeled "Show Inventory," the form calls a report and passes the selected item ID value as a parameter. The report receives the parameter and shows inventory information only for the selected item.

a. Open the 9CCase6.rdf report file, and modify the report so it receives the item ID value as a bind parameter. Save the modified report file as 9CCase6_DONE.rdf.

b. Open the form in the file named 9CCase6.fmb, and modify the form so that when the user clicks the Show Inventory button, the form calls the report and passes the selected item ID value as a parameter. The resulting report displays only inventory information for the selected item ID. Suppress displaying the parameter form. Save the completed form file as 9CCase6_DONE.fmb.

7. In this case, you will modify the form shown in Figure 9-84 that allows the user to select either a project or a consultant from the Software Experts database, and then run a report showing the details about the project consultants. If the user selects a project, the report shows consultant details about only the selected project. If the user selects a consultant, the report shows details about only that consultant's projects. You will create two reports: one will accept a project ID as a bind para-meter, and the other will accept a consultant ID as a bind parameter. Then, you will add the form triggers to call the correct reports, based on whether the user selects a project ID or a consultant ID.

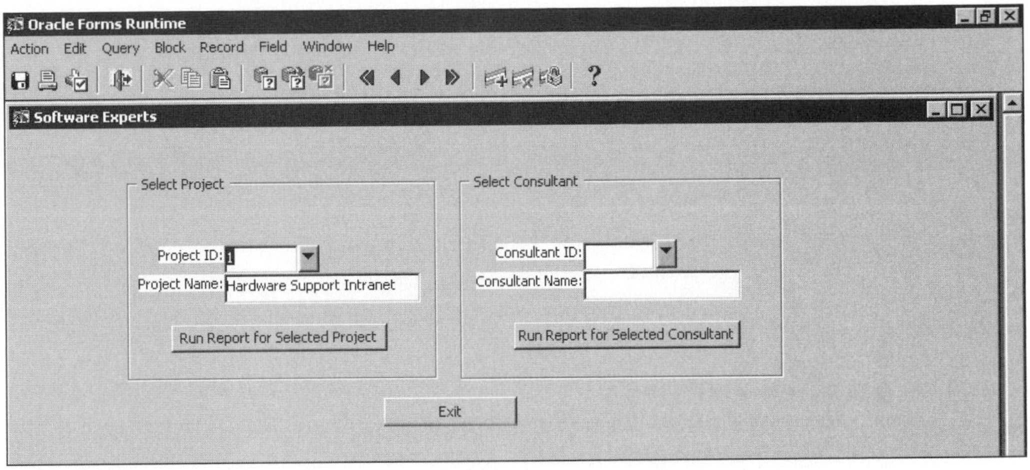

Figure 9-84

a. Open the 9CCase7.rdf report file. Save the report as 9CCase7_PROJECT.rdf, and then modify it so it accepts a project ID bind parameter.

b. Open the 9CCase7.rdf report file again, save the report as 9CCase7_CONSULTANT.rdf, and then modify it so it accepts a consultant ID bind parameter.

c. Open the 9CCase7.fmb form file, save it as 9CCase7_DONE.fmb, and then create a trigger for the Run Report for Selected Project button that calls the 9CCase7_PROJECT.rdf report file, and passes the selected project ID value to it. When the user selects a specific project (such as project ID 1, Hardware Support Intranet), the report appears as shown in Figure 9-85. Suppress displaying the parameter form.

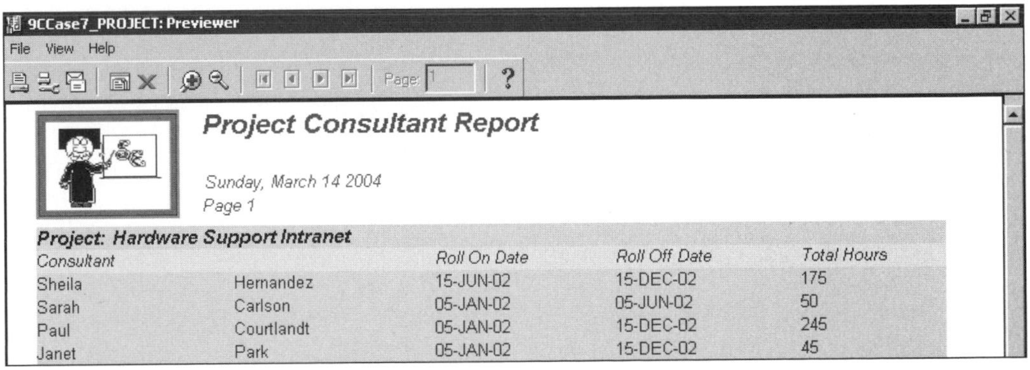

Figure 9-85

d. Create a trigger for the Run Report for Selected Consultant button that calls the 9CCase7_CONSULTANT.rdf report file, and passes the selected consultant ID value to it. When the user selects a specific consultant (such as consultant ID 104, Paul Courtlandt), the report appears as shown in Figure 9-86. Suppress displaying the parameter form.

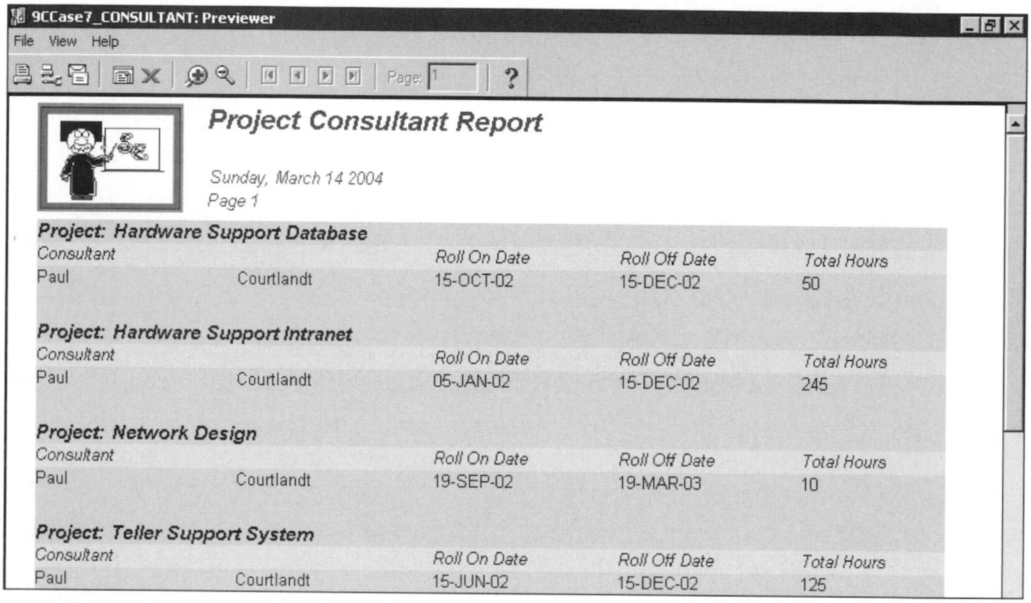

Figure 9-86

8. In this case, you will modify a report stored in a file named 9CCase8.rdf that currently shows information from the STUDENT table in the Northwoods University database. You will manually create a query that retrieves the F_FIRST and F_LAST columns from the FACULTY table and a second query that retrieves the call ID, course name, and grade value for every course that the student has completed. Your completed report will look like Figure 9-87.

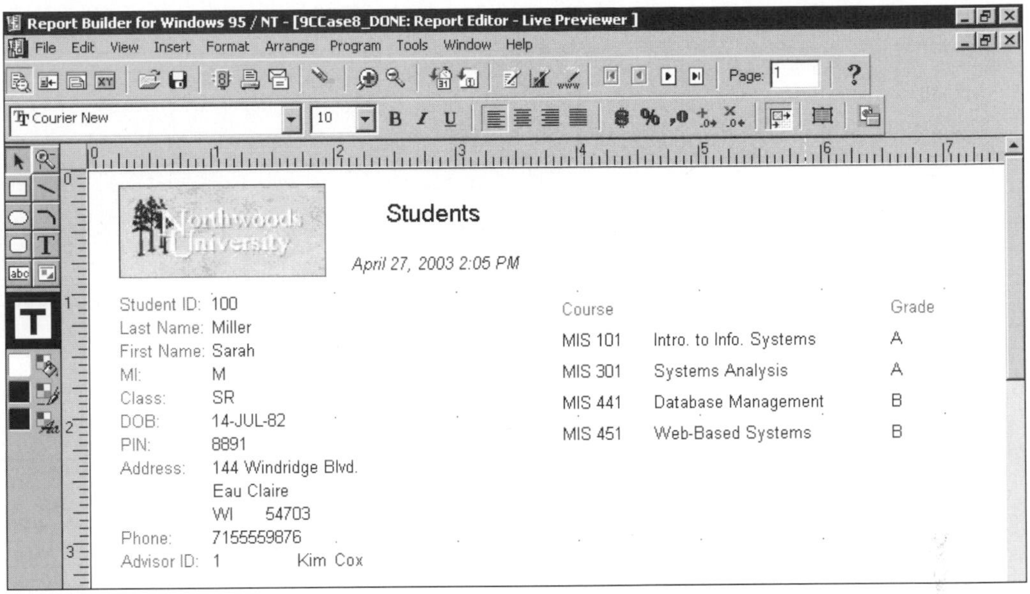

Figure 9-87

a. Open 9CCase8.rdf, and save it as 9CCase8_DONE.rdf.

b. Manually create a new query to retrieve the F_LAST and F_FIRST columns from the FACULTY table, and then create a Query to Query link between the new query and the existing query. Then, create display fields on the report layout to display the first and last name of the student's faculty advisor. Be sure to format the fields using variable elasticity so the first and last names appear directly beside each other on the report.

c. Manually create a new query to retrieve the S_ID, CALL_ID, COURSE_NAME, and GRADE values for all enrollment records for which a grade has been assigned. Then, create a Column to Column link between the S_ID column in the existing query and the S_ID1 column in the new query.

d. Manually create a repeating frame on the report to display the student grade information.

e. Manually create fields inside the repeating frame to display the call ID, course name, and grade for each course that the student has completed.

f. Add boilerplate labels so the report format appears as shown in Figure 9-87.

10

USING GRAPHICS BUILDER

Objectives

♦ Use Graphics Builder to create charts to display database data visually

♦ Create a form that passes user inputs to a chart

♦ Create a form that displays a chart

♦ Create a report that displays a chart to illustrate report data

It is often easier to understand and interpret numerical data presented in a visual format, such as a bar, pie, or line chart, than the same data presented in a textual or numerical format. For example, you might want to show Clearwater Traders sales trends for the past year by using a line chart. Or, you might want to show the percentage of sales from different Clearwater Traders order sources by using a pie chart. You can use Oracle's Graphics Builder utility to create **charts**, which are visual displays of database data. These charts can run as standalone applications, or they can be called from forms. A chart can also appear directly on a form or report. In this chapter, you will learn how to create a variety of charts and how to integrate charts with forms and reports.

Graphics Builder lets users retrieve database data and display values visually using a variety of chart types. Before you create a chart, it is important to consider the type of data you want to display and the purpose of the chart. Table 10-1 lists the different chart types that you can create using Graphics Builder, explains what each chart illustrates, and describes an example of an application.

Chart Type		Description	Example
Column		Shows discrete values using vertical columns	Display current quantities on hand for all Clearwater Traders inventory items
Bar		Shows discrete values using horizontal bars	Same as column chart
Gantt		Shows task or project scheduling or progress information	Display Software Experts consultant roll-on and roll-off dates for various projects
High-Low		Shows multiple Y-axis values for a single X-axis point	Display Northwoods University student highest, lowest, and average GPAs
Line		Shows data values as points connected by lines to show trends	Display Clearwater Traders total order revenues for each item category for each month of the past year
Mixed		Combines a column and line chart	Display Clearwater Traders total order revenues for each item category for each month of the past year as columns, with lines connecting the tops of the columns to show trends
Pie		Shows how individual data values contribute to an overall total amount	Illustrate the percentage of sales that each Clearwater Traders item category (Women's Clothing, Men's Clothing, and so forth) contributes to total sales revenue
Scatter		Plots two sets of data points to identify potential trends	Plot Northwoods University student ages on the X-axis versus the corresponding student's GPA on the Y-axis, to look for trends
Table		Shows textual data in a tabular format	Show a textual listing of all Northwoods University course call IDs, descriptions, days, times, and locations, that is similar to the information that appears in a form or report
Double Y		Shows a line chart with two different sets of Y-axis data for the same X-axis data	Show months on the X-axis, and show Clearwater Traders revenue dollars and units sold on the Y-axis, to illustrate how revenue dollars and units sold have varied over the same time period

Table 10-1   Graphics Builder chart types and uses

To illustrate the concepts and skills involved in making charts with Graphics Builder, this chapter describes how to create pie and bar charts, which are two of the most common chart types. Once you are familiar with the basic steps for creating a chart, you can easily create the other chart types, according to the needs of your application.

CREATING A PIE CHART

The managers at Clearwater Traders need an application that displays the portion of total order revenue that comes from specific order sources and from the Clearwater Traders Web site. A pie chart is appropriate for this application because it compares individual components, which are the proportions of revenue from each order source, to an overall total, which is the total order revenue. Figure 10-1 shows a design sketch for the order revenue source pie chart. Suppose Clearwater Traders had $1,000 in total revenue, and one-third came from the Spring 2003 catalog, one-third came from the Summer 2003 catalog, and one-third came from the Web site. Each of these three sources contributes one-third to the total revenue, so each represents one-third of the pie. The chart labels show the source name, the value that each contributes to the total revenue, and the percentage that each contributes to the total revenue.

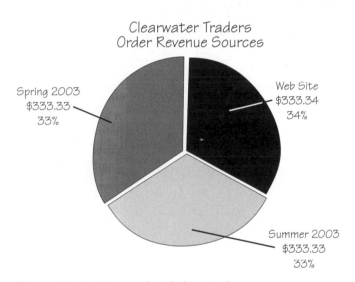

Figure 10-1   Design sketch for pie chart

The following sections describe how to create this pie chart. First, you will start SQL*Plus, and run the scripts to refresh your database tables. Then, you will start Graphics Builder.

To start SQL*Plus, and refresh your database tables:

1. Start SQL*Plus, and log onto the database.

2. Run the **clearwater.sql**, **northwoods.sql**, and **softwareexp.sql** scripts that are in the Chapter10 folder on your Data Disk, and then close SQL*Plus.

Starting Graphics Builder and Creating the Chart

The first step in creating the pie chart is to start Graphics Builder and create the chart object. Just as with Form Builder and Report Builder, after you start Graphics Builder, you must connect to the database.

To start Graphics Builder and connect to the database:

1. Click **Start** on the taskbar, point to **Programs**, point to **Oracle Forms & Reports 6i**, and then click **Graphics Builder**. The Graphics Builder Object Navigator opens. Maximize the Object Navigator window, so your screen display looks like Figure 10-2.

2. Click **File** on the menu bar, click **Connect**, and then connect to the database in the usual way.

 Another way to start Graphics Builder is to open Windows Explorer, change to the *Developer_Home*\BIN folder, and then double-click Gobld60.exe.

 If the Graphics Builder [Graphics: Disp1: Layout Editor] window opens instead of the Object Navigator window when you start Graphics Builder, click Window on the menu bar, click Object Navigator, and then maximize the Object Navigator window. Do not close the Layout Editor window.

Figure 10-2   Graphics Builder Object Navigator window

The Graphics Builder Object Navigator window looks a lot like the Object Navigator windows in Form Builder and Report Builder. A top-level Graphics Builder chart object is called a **display**. When you start Graphics Builder, it automatically creates a new chart display named Displ1. The Object Navigator shows the new chart, as well as the external Oracle resources (PL/SQL libraries, built-in packages, debug actions, stack, and database objects) that you can access within Graphics Builder. First you will save the chart. When you save the chart, the display object name automatically changes to the saved filename.

To save the chart:

1. In the Object Navigator, select **Disp1** if necessary.

2. Click the **Save button** on the Object Navigator toolbar. The Save dialog box opens.

3. If necessary, click the File System option button, and then click **OK**. The Save As dialog box opens.

4. Navigate to your Chapter10\TutorialSolutions folder, type **10OrderSource** in the File name text box, and then click **Save**. The display name in the Object Navigator window changes to the filename.

To specify the chart properties, you will use the Chart Genie in the Layout Editor, which automates the chart creation process. The Chart Genie is similar to the wizards you used in Form Builder and Report Builder. You can only start the Chart Genie from the Layout Editor window. Now, you will open the Layout Editor and start the Chart Genie.

To start the Chart Genie:

1. In the Object Navigator, click **Window** on the menu bar, click **Graphics: 10OrderSource.ogd:Layout Editor** to open the Layout Editor, and then maximize the Layout Editor window, if necessary. The Graphics Layout Editor is similar to the Form Builder Layout Editor.

2. Click **Chart** on the menu bar, and then click **Create Chart**. The Chart Genie - New Query dialog box opens.

 You can also start the Chart Genie by selecting the Chart tool on the Layout Editor tool palette, and then clicking the mouse pointer on the Layout Editor canvas.

Creating a new chart is a two-step process: First you create the query to retrieve the data that appears in the chart, and then you define the chart properties. The Chart Genie guides you through this process.

Creating the Chart Query

When you specify the query to retrieve the chart data, you must first visualize what the chart will look like. Then, you can determine which data fields the query needs to retrieve. A chart query must retrieve the numerical data values that appear in the pie slices of a pie chart, or on the X- and Y-axes of a column, bar, or line chart. The query must also retrieve the values that appear as labels on the pie slices, or on the X- and Y-axes of the other chart types.

For a pie chart, Graphics Builder will automatically calculate the proportion that each source contributes to the total.

Since you are working on a pie chart, the query must retrieve the data fields that will appear as the pie slice labels, which are the order source descriptions (SOURCE_DESC, from the Clearwater Traders ORDER_SOURCE table in Figure 1-11). To display the different proportions of total order revenue from the different order sources, the query must also retrieve the total revenue amount for each individual order source. The total revenue amount for an individual order source is calculated as PRICE, from the INVENTORY table, multiplied by ORDER_QUANTITY for each record in the ORDER_LINE table. You will use the GROUP BY clause to group the output for each order by the order source description. Since the total revenue column is a calculated value, the SQL command must create an alias for this column. Recall that an alias is an alternate name for a column retrieved in a SELECT command.

When you use a calculated value in a Graphics chart query, you should create an alias for the column with calculated values, or some Graphics Builder functions will not work correctly.

To define your chart query, you can directly enter a SQL command, import a SQL query from a text file, or import the data from a variety of other source file types, such as Microsoft Excel. Now you will define the query. First, you will define the query name, which specifies how Graphics Builder references the query in the Object Navigator. Then, you will enter the SQL command to retrieve the data values for the chart.

To define the chart query:

1. In the Chart Genie-New Query dialog box, place the insertion point in the Name text box, delete the current value, and type **ordersource_query** for the query name.

2. Confirm that SQL Statement is selected in the Type list. Since you will enter a SQL command to retrieve the data for the pie chart, the File text box is disabled.

3. Place the insertion point in the SQL Statement text box, and then type the command shown in Figure 10-3.

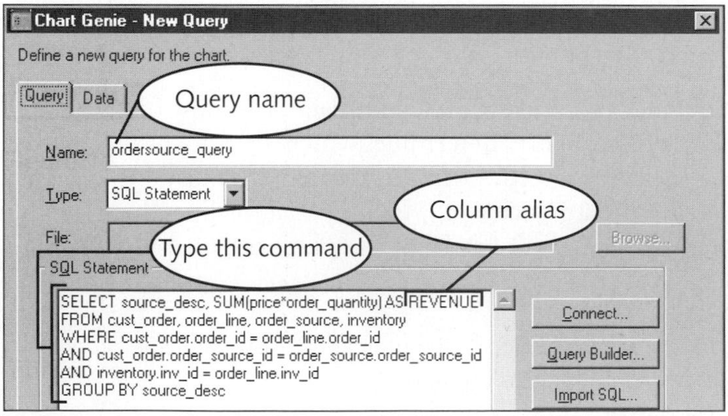

Figure 10-3   Defining the chart query

4. Click **Execute** to execute the SQL statement. The Data tab of the Chart Genie – New Query dialog box displays the query's output, as shown in Figure 10–4.

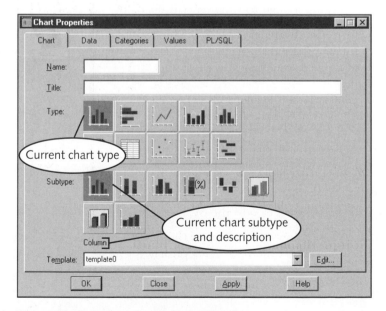

Figure 10-4   Chart query data tab

If a DBMS (ORA-) error message appears when you try to execute a query, or if your query does not return the data values shown, then you probably made an error in your query. Select the query text, press Ctrl + C to copy it, paste the query text into a text editor, and then debug it using SQL*Plus. When the command retrieves the data values shown in Figure 10-4, copy the corrected query text into the SQL Statement text box on the Query tab, and execute the query again.

5. Click **OK** to accept the data and open the Chart Properties dialog box, as shown in Figure 10–5.

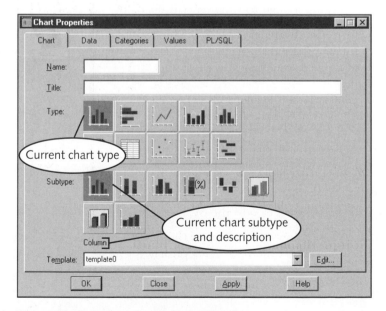

Figure 10-5   Chart Properties dialog box

Defining the Chart Properties

The Chart Properties dialog box lets you configure the chart's appearance. You use the Chart tab of the Chart Properties dialog box to specify the chart name, title, type, subtype, and template, which allow you to apply a predefined style to the chart. The **chart name** defines how the chart is referenced in the Graphics Builder Object Navigator. The chart name should be different from the query name, because you might have several charts that use the same query to define the data that they display. The **chart title** appears at the top of the chart. The **chart type** defines the type of chart you want to create—pie, column, line, and so forth. The **subtype** defines the different display and shading options for the selected chart type. The default chart type is column, which is currently selected. The default chart subtype is also column. The name of the current subtype appears above the template list box. The **template list box** lets you select a predefined template for the chart, which is useful if you want to create many charts using the same chart properties.

Now you will define the order source revenue chart properties. You will specify the chart name and title. You will also define the chart to be a pie chart, and select the pie with depth subtype so the pie chart appears three-dimensional.

To name the chart and change its properties:

1. Place the insertion point in the Name text box, and type **ordersource_chart** for the chart name.

2. Place the insertion point in the Title text box, and type **Order Revenue Sources** for the chart title.

3. Select the **Pie chart** type (the first chart type in the second row). Note that the Subtype options change to show variations of the pie chart.

4. Select the **Pie with depth** subtype (the third subtype from the left). Do NOT click OK, because this will close the Chart Properties dialog box, and you still need to specify additional chart properties.

 If you accidentally click OK, right-click on the gray chart background within the selection handles that define the chart's perimeter, and then click Properties to redisplay the Chart Properties dialog box.

Next, you will define how the query data appears on the chart. Besides the Chart tab, the Chart Properties dialog box has the following tabs to define chart properties:

- **Data**, which enables you to edit the SQL query

- **Categories**, which allows you to select the data values that will appear as the chart labels from a list of the data fields returned by the SQL query

- **Values**, which allows you to specify the query column that will be used to determine the pie slice proportions for the chart

- **PL/SQL**, which is used to create a PL/SQL trigger that executes when the user performs a specific action, such as clicking the mouse button

Now you will specify the rest of the chart properties. You will first examine the contents of the Data tab. Then, you will use the Categories tab to specify the source description (SOURCE_DESC) column values for the chart labels, and the Values tab to specify the REVENUE values for the pie slice proportions. You will not use the PL/SQL tab, because it is used for advanced chart features, which will not be covered in this book.

To specify the rest of the chart properties:

1. Click the **Data tab**. The name of the chart query (ordersource_query) appears in the Query text box. The Edit button enables you to edit the SQL query. The Filter Function list enables you to select a predefined PL/SQL function that filters the query data values so that only certain rows are returned. The Mapping option buttons specify how the data values are mapped for a Gantt or High-Low chart type. The Data Range option buttons allow you to specify that only a set number of records are returned. This is useful for speeding up the performance of queries that might return hundreds or thousands of records.

2. Click the **Categories tab** to select the chart labels from a list of the data fields returned by the SQL query. The Chart Genie selected the SOURCE_DESC data field for the Chart Categories data source, since this was the only character data field returned by the query. This is the data field that you want to use for the pie slice labels (see Figure 10-1), so you will not change the Chart Categories selection.

3. Click the **Values tab** to specify the query column that will determine the pie slice proportions for the chart. Since REVENUE is the only numerical column in the query SELECT statement, the Chart Genie selected it as the default choice for the chart values.

4. Click **OK** to close the Chart Properties dialog box and display the chart in the Layout Editor, as shown in Figure 10-6.

10

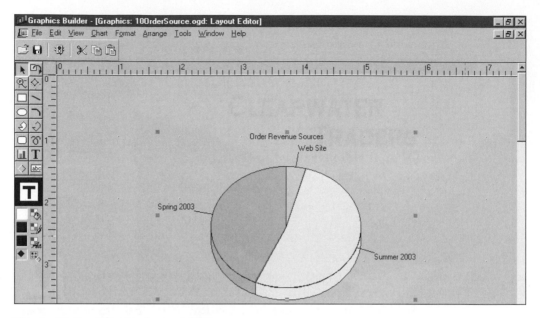

Figure 10-6 Chart in Layout Editor

> 5. Click the **Save button** 🔲 on the toolbar to save the chart.

Editing and Formatting the Chart Display

After you create a chart using the Chart Genie, you usually have to fine-tune the formatting of the chart display. Currently, each slice of the pie chart shows the approximate proportion of total revenue for each of the different sources, but the chart would be more informative if it also listed the actual dollar figures and the percent that each category contributes to overall revenue. The chart would be more attractive if the title were larger and positioned higher on the display. Now, you will modify the chart labels and format the chart's appearance.

Modifying Pie Chart Labels

You can reopen the Chart Properties dialog box to modify the data that appears on chart labels. First you will reopen the Chart Properties dialog box, and edit the display to show the actual revenue values along with the category values on the pie slice labels.

To show revenue values on the pie slice labels:

> 1. If necessary, select the pie chart by clicking outside of the pie, but within the area of the pie slice labels. Selection handles appear around a square perimeter of the chart, as shown in Figure 10-6, to indicate that the chart is selected.

If selection handles appear around the individual pie slices as well as the chart, you selected the chart's individual components rather than the entire chart. Click outside of the pie, but within the area of the pie slice labels to select the entire chart.

2. Within the selected area of the chart, right-click anywhere on the gray background, and then click **Properties** on the menu to reopen the Chart Properties dialog box.

You also can double-click anywhere on the background area around the chart to reopen the Chart Properties dialog box.

3. Click the **Categories tab**, click **REVENUE** in the Query Columns list box, and then click **Insert** to copy REVENUE into the Chart Categories list. Click **OK** to close the Chart Properties dialog box and save your changes. The chart now shows revenue amounts along with the data source labels.

Next, you will use the Frame Properties dialog box to add the proportion percentages to the labels. The **Frame Properties** dialog box allows you to format the chart frame, which is the chart area delimited by selection handles when you select the chart. The Frame Properties dialog box has two tabs. The **Frame tab** allows you to fine-tune the appearance of column charts and specify whether a legend appears on the column charts. The frame tab is not applicable to pie charts. The **Pie Frame tab** allows you to specify how Graphics Builder calculates and displays the pie slice proportions. The Pie Frame tab also allows you to specify whether category values, data values, or percentage values appear on pie slice labels.

Sometimes while you are modifying the format of a chart, you specify a change, but the change does not appear when you view the chart in the Layout Editor. When a specified change does not appear, you need to refresh the chart display. Now, you will open the Frame Properties dialog box, and specify to display percentage labels on the pie slices. Then, you will refresh the chart display.

To display percentage values on pie slice labels using the Frame Properties dialog box:

1. If necessary, select the pie chart so selection handles appear around the chart perimeter, right-click, and then click **Frame** to open the Frame Properties dialog box.

2. Select the **Pie Frame tab**, check the **Show Percent Values check box**, and then click **OK**.

3. If the percentage values for each pie slice proportion do not appear below the category and value labels, click **Chart** on the menu bar, and then click **Update Chart** to refresh the chart display. The chart appears as shown in Figure 10-7.

You can also right-click anywhere on the chart display, and then click Update to refresh the display.

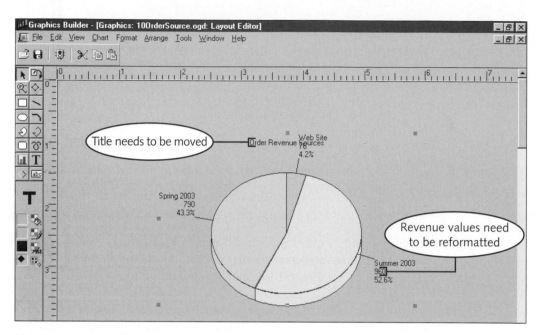

Figure 10-7   Pie chart showing values and percentages on labels

Formatting the Chart Display

Now you need to perform the final chart formatting. Figure 10-7 highlights the final formatting tasks. Currently, the revenue values appear as whole numbers rather than currency. You need to apply a format mask to these values so they appear with dollar signs and two decimal places. You also need to move the chart title so it appears higher on the chart, and is not overwritten by the Web Site pie slice label. You will also change the fonts of the title and labels to make them more prominent.

First, you will modify the revenue value format. You can apply format masks to values that have the NUMBER or DATE data types. Now, you will apply a currency format mask to the revenue amounts.

To format the revenue amounts as currency:

1. If necessary, select the chart, and then click any one of the revenue values to select them as a group. Selection handles should appear around each revenue value.

2. With the revenue values selected, click **Format** on the menu bar, click **Number** to open the Number Format dialog box, and click **$999,990.99**. Click **OK** to apply the selected format mask to the revenue values.

If the format mask you want to use does not appear in the list of available format masks, click Format on the menu bar, and then click Number. Type the format mask you want to use in the Format text box, click Add, and then click OK.

3. If necessary, click **Chart** on the menu bar, and then click **Update Chart** to refresh the chart display. The revenue values should now appear as currency.

Next, you will reposition the chart on the canvas, move the chart title, and format the chart labels to make them more prominent.

To reposition the chart, resize the chart, and format the labels:

1. Select the chart so that selection handles appear around the entire chart area. Drag the chart to the top-left corner of the Layout Editor, drag the lower-right selection handle, and use the rulers to resize the chart so it is about 4.8 inches wide and 3.5 inches tall. The icons in the bottom-left corner of the Layout Editor window show the exact dimensions of the chart as you resize it. If the proportions of the chart change, adjust the chart height and width as necessary to make the pie appear round.

If the rulers do not appear at the top and left edges of the Layout Editor window, click View on the menu bar, and then click Rulers to display the rulers.

10

2. Select the entire chart, and then reposition the chart so the upper-left corner of the chart is 0.5 inches from the top and 0.5 inches from the left edge of the Layout Editor painting region.

3. Select the chart title so selection handles appear around it, and drag the chart title to the top of the display so the title is about .3 inches from the top of the painting region.

4. With the chart title still selected, click **Format** on the menu bar, point to **Font**, and click **Font**. Change the font to 10-point Arial bold, and then click **OK**. (If your system does not have Arial, substitute another sans-serif font.)

5. Click any one of the order source labels to select them as a group, press **Shift**, and then click any one of the revenue figures to select them as a group. Both the category labels and the data values should appear selected.

6. Click **Format** on the menu bar, point to **Font**, and click **Font**. Change the font to 8-point Arial bold, and then click **OK**. The formatted pie chart should appear as shown in Figure 10-8.

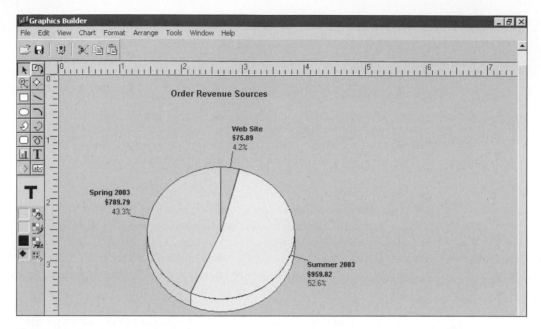

Figure 10-8 Formatted pie chart

> 7. Save the chart.

Chart File Types

Graphics Builder allows you to save charts both as Oracle Graphics Design files (with an .ogd extension) and as Oracle Graphics Runtime files (with an .ogr extension). Users can run .ogr files in the Graphics Runtime application. Users can also open a chart by double-clicking the .ogr file in Windows Explorer if .ogr files are a registered file type on the user's workstation. In the Graphics Runtime window, users can view and print the chart, and export the chart image to other image file types.

When you click the Run button ⟨🔧⟩ on the Oracle Graphics toolbar, the chart appears in the Graphics Debugger window, which enables you to preview charts during development and see how they will appear to users in the Graphics Runtime environment. Now you will display the order source revenue pie chart in the Graphics Debugger window.

To preview the chart in the Graphics Debugger window:

> 1. Click the **Run button** ⟨🔧⟩ on the Layout Editor toolbar. The chart appears in the Graphics Debugger window.

If the Layout Editor toolbar is not visible, scroll to the top of the Layout Editor window, and maximize the window.

You can also display the chart in the Graphics Debugger window by clicking File on the menu bar, and then clicking Run.

2. Close the Graphics Debugger window by clicking **File** on the menu bar, and then clicking **Close**.

 You can also close the Graphics Debugger window by clicking ☒ on the inner window in the Graphics Debugger window. If you click ☒ on the outer window, you will close Graphics Builder.

Now you will generate an Oracle Graphics Runtime (.ogr) file. Then, you will start Graphics Runtime, and open the chart.

To create and view an Oracle Graphics Runtime file:

1. Click **File** on the menu bar, point to **Administration**, point to **Generate**, and then click **File System**. The Save As dialog box opens.

2. If necessary, navigate to your Chapter10\TutorialSolutions folder, confirm that the filename is 10OrderSource.ogr, and then click **Save** to save the .ogr file.

3. Click **Start** on the taskbar, point to **Programs**, point to **Oracle Forms & Reports 6i**, and then click **Graphics Runtime**. The Graphics Runtime window opens.

4. Click **File** on the menu bar, point to **Open**, and then click **File System**. The Open dialog box appears. Navigate to your Chapter10\TutorialSolutions folder, select **10OrderSource.ogr**, and then click **Open**. The Connect dialog box opens.

5. Connect to the database in the usual way. The chart appears in the Graphics Runtime window.

6. Click **File** on the menu bar, and then click **Exit** to close Graphics Runtime.

7. In Graphics Builder, click **File** on the menu bar, and then click **Close** to close the chart.

CREATING A COLUMN CHART

To further analyze the revenues generated by different order sources, the managers at Clearwater Traders would like to view a detailed analysis of the revenue generated by each item (Women's Hiking Shorts, Women's Fleece Pullover, and so forth) sold through the Clearwater Traders Web site. To enable them to do this, you will create a revenue detail column chart that displays the revenue value for each item as a vertical column. Figure 10-9 shows a design sketch for the revenue detail column chart. In the chart, each item appears as a separate column on the X-axis, and total revenue generated determines each column's height on the Y-axis.

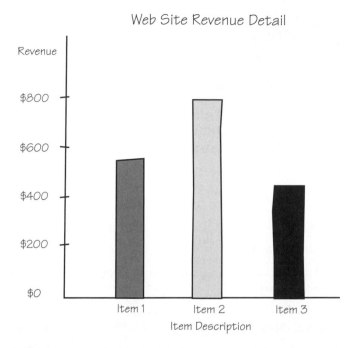

Web Site Revenue Detail

Figure 10-9   Design sketch for revenue detail column chart

First, you will create a new chart display object in the Object Navigator and save the chart design file. Then, you will define the chart query and specify the chart properties.

To create and save a new chart:

1. Maximize the Object Navigator if necessary. Then, select the **Displays node**, and click the **Create button**  to create a new display object. The new chart display appears in the Layout Editor.

2. Click the **Save button** . Make sure that the File System option button is selected, click **OK**, navigate to your Chapter10\TutorialSolutions folder if necessary, and save the new display as **10RevenueDetail.ogd**.

Defining the Chart Query

For this chart, you will display the item descriptions on the X-axis, and show the total revenue generated by each item as the column value on the Y-axis. Therefore, the query will need to retrieve item descriptions from the ITEM_DESC column in the INVENTORY table. The query will also need to retrieve total revenues for each item, which is calculated as the sum of the PRICE from the INVENTORY table times ORDER_QUANTITY from the ORDER_LINE table. This value must be grouped by each item description, and the query will only retrieve values for orders placed using the Web site (ORDER_SOURCE_ID 6) as the order source. Now you will start the Chart Genie and define the chart query.

To start the Chart Genie and define the chart query:

1. Click **Chart** on the menu bar, and then click **Create Chart**. The Chart Genie – New Query dialog box opens. Type **revdetail_query** for the query name in the Name text box.

2. Make sure that SQL Statement is selected in the Type list, place the insertion point in the SQL Statement text box, and then type the following command:

```
SELECT item_desc, SUM(price * order_quantity) AS REVENUE
FROM item, inventory, order_line, cust_order
WHERE item.item_id = inventory.item_id
AND inventory.inv_id = order_line.inv_id
AND cust_order.order_id = order_line.order_id
AND order_source_id = 6
GROUP BY item_desc
```

3. Click **Execute** to execute the SQL statement and display the Data tab. The data values should appear as follows:

ITEM_DESC	REVENUE
Children's Beachcomber Sandals	15.99
Women's Hiking Shorts	59.9

 If your values are different, double-check to make sure you typed the SQL command correctly.

4. Click **OK** to accept the data and to close the Query Properties dialog box. The Chart Properties dialog box opens.

Specifying the Chart Properties

Next you will specify the chart properties. You will specify that this will be a three-dimensional column chart, that the item description values will appear as the X-axis labels, and that the revenue values will determine the column values.

To specify the chart properties:

1. In the Chart Properties dialog box, type **revdetail_chart** in the Name text box as the chart name and **Web Site Revenue Detail** in the Title text box as the chart title.

2. Confirm that the Column chart type (the first chart in the first type row) is selected, and select the **Column with depth** subtype (the first chart in the second subtype row).

3. Select the **Categories tab**. Recall that the Categories tab specifies the labels that appear on the X-axis of the column chart. ITEM_DESC is already selected as the chart category for each bar, which agrees with the X-axis labels on the chart design sketch in Figure 10-9.

10

4. Click the **Values tab**. REVENUE is listed as the source for the Y-axis columns, which is also correct. Click **OK** to accept the data, close the dialog box, and create the column chart.

5. Reposition the chart on the painting region, as shown in Figure 10-10, and then save the chart.

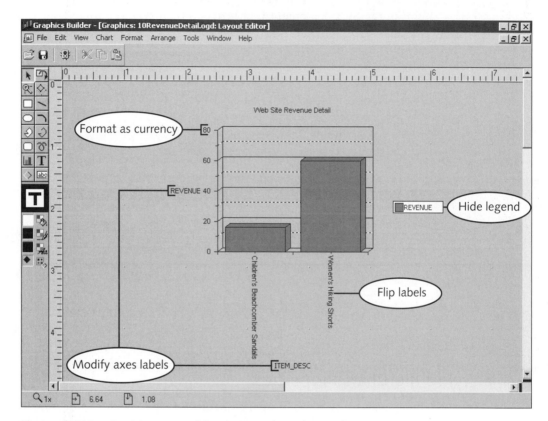

Figure 10-10 Positioning and formatting the column chart

Formatting the Column Chart Layout

Figure 10-10 highlights some of the problems with the present chart format. Currently the axis labels are the same as the query column titles—ITEM_DESC and REVENUE. You will create custom axis labels so the labels match the design sketch in Figure 10-9. The X-axis **tick labels**, which are the labels on the tick marks for each column, should be flipped so they are upright and aligned with the right edge of the screen display. You need to add a currency format mask for the Y-axis revenue values, and change the font size and style for the chart title and axis labels. In addition, since there is only one data series, you will hide the chart legend.

A column chart has different formatting requirements than a pie chart because it has an X- and Y-axis. To modify the axis labels and tick labels, you will open the **Axis Properties** dialog box. To format an axis, you select either the X- or Y-axis, and then specify the axis display properties, such as the axis label, direction, tick mark style, tick label rotation, and number of tick marks per interval. A chart axis that displays variables that have one and only one value, such as an item description, is called a **discrete axis**. An axis that displays numerical or date values that increase as you move along the axis, such as the revenue amount, is called a **continuous axis**. For a continuous axis, you can specify the axis minimum and maximum values and the step size. You can also specify whether a continuous axis displays values using a linear or logarithmic scale. Currently, the revenue values appear using a linear scale, the minimum value is 0, the maximum value is 80, and the step size is 20. Usually, you let Graphics Builder automatically determine the axis minimum and maximum values and the step size for a continuous axis, because it needs to adjust these properties to accommodate the data that the chart retrieves from the database.

Now, you will open the Axis Properties dialog box, change the axis labels to the values shown in Figure 10-9, confirm that the Y-axis minimum and maximum values and step size are automatically specified, and modify the X-axis labels so the tick labels are displayed upright and parallel to the right edge of the screen display.

To format the chart axes:

1. Select the column chart, right-click, and then click **Axes**. The Axis Properties dialog box opens. You will modify the Y-axis properties first.

2. Open the Axis list, and select **Y1**. Note that the tabs are Axis, which is currently selected, and Continuous Axis. Also note that the Data Type list displays the current axis data type value as Continuous.

The Y-axis is labeled Y1 because you could have two Y-axes in a Double-Y chart.

3. Place the insertion point in the Custom Label text box, and type **Revenue** for the Y-axis custom label.

4. Select the **Continuous Axis tab**, and confirm that the auto check box is checked for the Minimum, Maximum, and Step Size text boxes.

5. Click **Apply** to save your changes, but leave the Axis Properties dialog box open. You might have to scroll to the bottom of the screen to see the Apply button.

When you click Apply, the changes are saved, but the dialog box remains open. When you click OK, the changes are saved and the dialog box closes.

6. Click the **Axis tab**, open the Axis list, and select **X**. Note that the tabs are now Axis, which is currently selected, and Discrete Axis, which indicates that the X-axis displays discrete data values.

7. Place the insertion point in the Custom Label text box, and type **Item Description** for the axis label.

8. Select the first **Tick Label Rotation option button** so the tick labels are displayed upright and parallel to the right edge of the page or screen display. Click **OK** to save your changes and close the Axis Properties dialog box. The axes labels are now formatted correctly.

9. Save the chart.

Next you will modify the format mask of the Y-axis tick labels. You will do this directly within the Layout Editor.

To modify the format mask and chart label fonts:

1. In the Layout Editor, click any one of the Y-axis revenue value labels to select the labels as a group. Click **Format** on the menu bar, click **Number**, click **$999,990** in the Number Format dialog box, and then click **OK**. The labels now appear as currency values that are rounded to the nearest dollar amount.

2. Save the chart.

A column chart can display multiple data columns for each discrete value. The data for each separate column is called a **data series**. For example, you could show multiple columns for each item description, with each column showing the revenue for different order sources. By default, column charts display a **legend** that identifies the current data in the columns. Since there is only one data series (REVENUE) in this chart, you will hide the legend. To hide the legend, you use the Frame properties dialog box. (You used this dialog box earlier to display percentage values on the pie chart.) It allows you to customize the appearance of aspects of the chart frame for column charts, such as the bar shadow depth and legend display.

To hide the chart legend:

1. Select the column chart, right-click, and then click **Frame** to open the Frame Properties dialog box.

2. Clear the **Show Legend check box** to hide the legend, and then click **OK**. The legend no longer appears in the Layout Editor.

Currently, the columns are the same color on their front, side, and top surfaces. To give two-dimensional objects a three-dimensional appearance, you can create the illusion of light shining on one surface with the other surfaces in shadow. You can enhance the appearance of the three-dimensional columns if you choose a light shade for the front surface and a slightly darker shade of the same color for the side and top surfaces. Now you will modify the column colors to enhance their appearance.

To enhance the appearance of the column bars:

1. Select the front bar surface on one of the column bars in the column chart. Selection handles appear around the front edges of all column bars. Select the **Fill Color tool** on the tool palette, and then change the fill color to light blue.

2. Select the side surface of any column bar. Selection handles appear around the side surfaces of all column bars. Select , and then change the fill color to a slightly darker blue shade than the shade you used on the front surface of the bars.

> To achieve the best three-dimensional effect, choose colors that are directly beside each other on the Fill Color palette.

3. Select the top surface of any column bar, click ⬛, and then change the fill color to the same darker blue shade you used in Step 2.

> You can select the side and top surfaces of the column bars as a group, and then change the fill color of both surfaces in a single step.

4. Save the chart, and then click the **Run button** 🔳 to view your chart in the Graphics Debugger window. Your formatted chart should appear as shown in Figure 10-11.

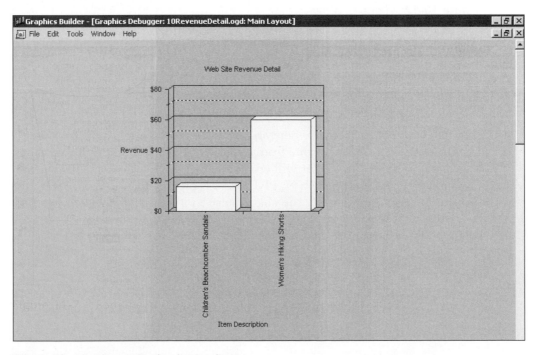

Figure 10-11   Formatted column chart

> 5. Close the Graphics Debugger window, and then close the chart in Graphics Builder.

CREATING A FORM THAT CALLS A CHART

Oracle Graphics Builder charts can be called from a Form Builder application, or they can appear directly on a form. In this section, you will use the form shown in Figure 10-12 that allows a user to select an item ID from the Clearwater Traders database using an LOV.

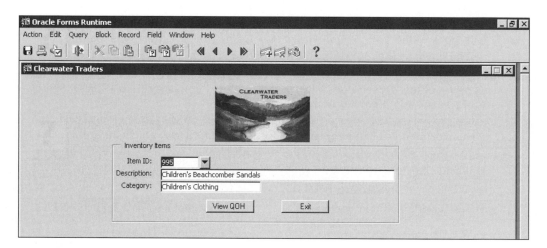

Figure 10-12   Form to display chart

When the user clicks the LOV command button beside the Item ID text item, an LOV display presents a list of the database inventory items. When the user selects a specific item—in this case, Children's Beachcomber Sandals—and then clicks the View QOH button, the chart shown in Figure 10-13 appears, which shows the inventory quantity on hand for each inventory item corresponding to the selected item.

Recall from Chapter 9 that when you called a report from a form, you used a parameter list to pass a parameter value from a form to the report. In the form that displays the chart, the form button trigger will create a parameter list that contains the selected item ID and then execute the RUN_PRODUCT procedure to open a chart instead of a report. First, you will modify the chart query so it uses a parameter instead of a specific search condition value for item ID. Then you will modify the form so it creates a parameter list containing the selected item ID, and then calls the chart.

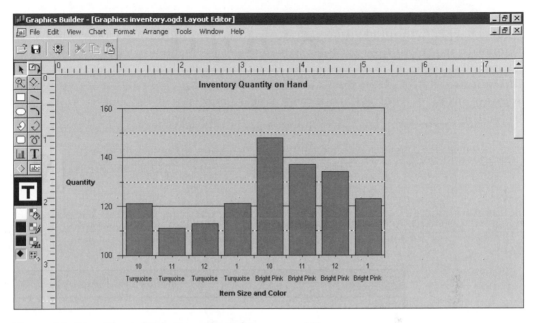

Figure 10-13   Chart displaying selected item quantities

Creating a Chart Bind Parameter

The query in the current chart display is coded so that the search condition is for Item ID 995 (Children's Beachcomber Sandals). To pass the item ID variable values from the form to the chart, you must create a bind parameter in the chart, just as you created a bind parameter in the report when you called a report from a form in Chapter 9. To create a bind parameter in a chart, you create the parameter in the Object Navigator window, and then replace the search condition in the chart's SQL query with the parameter name, preceded by a colon. First you will open the chart display in Graphics Builder, and create a parameter.

To open the chart display and create a parameter:

1. In Graphics Builder, open the **Inventory.ogd** file from the Chapter10 folder on your Data Disk, and save it as **10Inventory.ogd** in your Chapter10\TutorialSolutions folder.

2. Click **Window** on the menu bar, and then click **Object Navigator** to open the Object Navigator window.

3. Select the **Parameters node** under 10Inventory.ogd, and then click the **Create button** ⬚. (If necessary, maximize the Object Navigator window so the toolbar is visible.) The Parameters dialog box opens.

The Parameters dialog box allows you to specify the parameter name, data type, and initial value. It currently displays a default parameter name, such as PARAM1 (although your default parameter name might be different). Since the parameter will be assigned to the value of ITEM_ID in the ITEM table, you will change the parameter name to P_ITEM_ID and its data type to NUMBER. Its initial value can be any valid ITEM_ID data value. You will set it to inventory ID 995 (Children's Beachcomber Sandals). After you create the parameter, you will modify the chart query so that the chart displays the data for the item ID value that is assigned to the parameter. Now you will specify the parameter properties and modify the chart query.

To specify the parameter properties and modify the query:

1. Place the insertion point in the Name text box, delete the current default parameter name, and type **P_ITEM_ID** for the parameter name.

2. Open the Type list, and select **Number**.

3. Place the insertion point in the Initial Value text box, delete the current value, and type **995**. Your completed parameter specification should look like Figure 10-14. Click **OK**. The new parameter appears in the Object Navigator window.

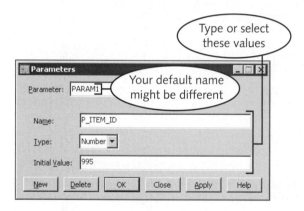

Figure 10-14   Chart parameter specification

4. In the Object Navigator, click ➕ beside the Queries node, if necessary, to display the chart queries, and then double-click the **Queries icon** 🔲 beside qoh_query to open the Query Properties dialog box.

5. If necessary, click the **Query tab**, modify the query search condition as follows, and then click **OK**.

```
SELECT item_size, color, qoh
FROM inventory, item
WHERE inventory.item_id = item.item_id
AND item.item_id = :P_ITEM_ID
```

6. Save the chart.

Modifying the Form

Next, you will open the form file and create the trigger for the View QOH button. The View QOH button trigger will create the parameter list that will pass the selected item ID to the chart and execute the RUN_PRODUCT command to display the chart. When the form executes the RUN_PRODUCT command, it starts Graphics Runtime and displays the chart in the Graphics Runtime window. First, you will open the form, and create a new trigger.

To open the form and create the View QOH button trigger:

1. Start Form Builder, open **Item.fmb** from the Chapter10 folder on your Data Disk, save the file as **10Item.fmb** in your Chapter10\TutorialSolutions folder, and connect to the database.

 If your computer does not have enough memory to run Form Builder and Graphics Builder at the same time, close Graphics Builder and then reopen it when necessary.

2. In Form Builder, open the Layout Editor, select the **View QOH button**, right-click, point to **SmartTriggers**, and then click **WHEN-BUTTON-PRESSED** to create a trigger for the View QOH button. The PL/SQL Editor opens.

In the trigger code, you will create the parameter list the same way you created the parameter list that was passed to a report in Chapter 9, Lesson C. (If necessary, review the materials on creating parameter lists in Chapter 9.) After you create the parameter list, you will use the RUN_PRODUCT procedure to call the chart. In this trigger, you will use the RUN_PRODUCT procedure a little differently, because you are calling a chart instead of a report. Recall that the general syntax of the RUN_PRODUCT procedure is as follows:

```
RUN_PRODUCT(product, 'document', communication_mode,
    execution_mode, location, list_ID, display);
```

When you are calling a chart, this procedure requires the following items:

- *product* specifies which Oracle Developer application you want to run. Since you are calling a Graphics Builder application, its value will be GRAPHICS.

- *document* specifies the complete path and filename of the chart .ogd file as a character string enclosed in single quotation marks. The filename and folder path cannot contain any blank spaces. If the chart filename were 10Inventory.ogd, and was stored in your Chapter10\TutorialSolutions folder on the A: drive, the document specification would be 'a:\chapter10\tutorialsolutions\10inventory.ogd'.

- *communication_mode* can be either SYNCHRONOUS or ASYNCHRONOUS. SYNCHRONOUS specifies that program execution control returns to the form only after the called product (Reports or Graphics) is

10

closed. ASYNCHRONOUS specifies that control returns to the form while the product is still running, and users can multitask between the form and report applications. You will set this property value to SYNCHRONOUS, so the user must exit the chart before continuing work in the form.

- *execution_mode* can be either BATCH or RUNTIME. RUNTIME specifies that the called product's runtime environment (such as Graphics Runtime) runs, while BATCH specifies that the called product generates an executable file that appears directly on the form. You will use RUNTIME in this application, so that control is transferred to the Graphics Runtime environment.

- *location* can be either FILESYSTEM or DB (database), and specifies where the document to be run is stored. In the exercises in this book, all documents will be stored in the file system on the user's workstation.

- *list_ID* is the object ID of the parameter list that contains the data values that are passed to the called product. For the inventory chart, this variable is `item_list_id`. If there is no parameter list, this parameter value must be passed as NULL.

- *display* specifies the name of the form image object that will contain the chart, and display the chart directly on the form. Since you are displaying the chart in the Graphics Runtime window, the DISPLAY value will be NULL.

Now, you will enter the code for the View QOH button trigger. The trigger will create a parameter list that contains the current value of the item ID. Then, the trigger will call the chart, using the RUN_PRODUCT built-in. Finally, the trigger will destroy the parameter list. (Recall that after a trigger passes a parameter list to a report, the trigger must destroy the list, or the list will remain in the server memory. The problem with a parameter list remaining in server memory is that you cannot create a new parameter list with the same name as an existing parameter list while the form is running. You must destroy the list and then create a new list each time the user selects a different item and displays its associated chart.)

To enter the View QOH button trigger code:

1. Type the code shown in Figure 10-15 to create the parameter list, run the chart, and destroy the parameter list. Modify the path to the .ogd file, if necessary, to reflect the location of your Chapter10\TutorialSolutions folder.

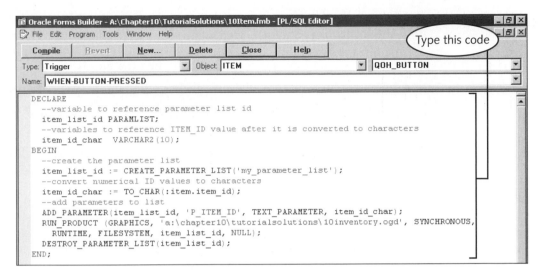

```
DECLARE
  --variable to reference parameter list id
  item_list_id PARAMLIST;
  --variables to reference ITEM_ID value after it is converted to characters
  item_id_char  VARCHAR2(10);
BEGIN
  --create the parameter list
  item_list_id := CREATE_PARAMETER_LIST('my_parameter_list');
  --convert numerical ID values to characters
  item_id_char := TO_CHAR(:item.item_id);
  --add parameters to list
  ADD_PARAMETER(item_list_id, 'P_ITEM_ID', TEXT_PARAMETER, item_id_char);
  RUN_PRODUCT (GRAPHICS, 'a:\chapter10\tutorialsolutions\10inventory.ogd', SYNCHRONOUS,
    RUNTIME, FILESYSTEM, item_list_id, NULL);
  DESTROY_PARAMETER_LIST(item_list_id);
END;
```

Figure 10-15   View QOH button trigger code

2. Compile the code, and correct any syntax errors. Then close the PL/SQL Editor, and save the form.

Now, you will run the form and confirm that it successfully displays the chart and passes the selected item ID. You will select item ID 559 (Men's Expedition Parka). Then, you will click the View QOH button to view the chart, which displays the quantities on hand for the inventory of the selected item.

To run the form and display the chart:

1. Run the form, click the **LOV command button** next to item ID, select item **559** (Men's Expedition Parka), and then click **OK**.

2. Click **View QOH**. The inventory quantity on hand for each size of the selected item appears in the Graphics Runtime window. (This may take a few moments to load and display, and you might need to maximize the Graphics Runtime window.)

 If the chart does not appear, check to make sure the chart .ogd file is named 10Inventory.ogd, that it is stored in your Chapter10\TutorialSolutions folder, and that the path specified in the code in Figure 10-15 is correct.

 If the Graphics Runtime window opens but your chart is not visible, scroll to the bottom of the Graphics Runtime window, and then maximize the inner window that contains the chart.

3. In the Graphics Runtime window, click **File** on the menu bar, and then click **Exit** to close the Graphics Runtime window.

4. In the Forms Runtime window, click the **LOV command button** beside Item ID again, select Item ID **786** (3-Season Tent), click OK, and then click **View QOH**. The Graphics Runtime window opens again, this time displaying the chart for the tent inventory items. Close the Graphics Runtime window.

5. In the Forms Runtime window, click **Exit** to exit the form.

6. In Form Builder, click **File** on the menu bar, and then click **Close** to close the form file in Form Builder.

7. Switch to Graphics Builder, click **File** on the menu bar, and then click **Close** to close the inventory chart.

CREATING A FORM WITH AN EMBEDDED CHART

Another way to display charts with forms is to display the chart directly on the form. Now you will work with an inventory form that allows the user to select an item from the Clearwater Traders' ITEM table using an LOV. The corresponding inventory quantities on hand appear in a column chart directly on the form, as shown in Figure 10-16.

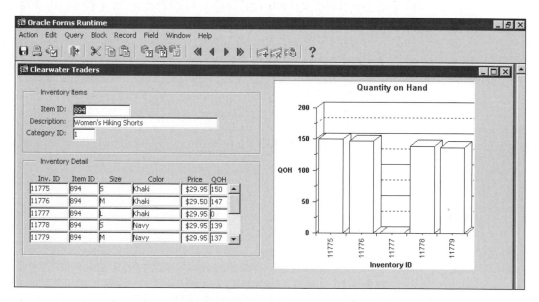

Figure 10-16   Chart that appears directly on form

To create a chart that appears directly on a form, the chart data must appear on the form in a data block. The form shown in Figure 10-16 is a master-detail form. The Inventory Items frame items are in the master block, and the Inventory Detail frame items are in the detail block. The chart derives its data from the detail block. When the data in the detail block changes, the chart also changes. When a chart appears directly on a form, a Graphics Builder application named Oracle Graphics Batch starts when the form runs.

This application dynamically updates the chart when the data values in the detail block change. After you exit the form, Oracle Graphics Batch continues running and must be closed manually. Now you will open the inventory master-detail form, save it using a different filename, and then run the form to become familiar with its operation.

To open, save, and run the form:

1. In Form Builder, open **Inventory.fmb** from the Chapter10 folder on your Data Disk, and save the file as **10Inventory.fmb** in your Chapter10\TutorialSolutions folder.

2. Run the form, click the **Enter Query button** 🔲, do not enter a search condition, and then click the **Execute Query button** 🔲. Item ID 894 (Women's Hiking Shorts) appears in the Inventory Items frame, and its associated inventory information appears in the Inventory Detail frame.

3. Click the **Next Record button** ▶. Data for the next item ID (897, Women's Fleece Pullover) appears, along with its detailed inventory information.

4. Continue to scroll through the records until there are no more records, and then close the Forms Runtime window.

You will use the Chart Wizard in Form Builder to create the chart item on the form. The Chart Wizard has the following pages to lead you through the chart creation process:

- **Type page**, which allows you to specify the chart title, type (bar, column, line, and so forth), and subtype options

- **Block page**, where you specify the name of the form data block that is the source of the chart data

- **Category page**, where you specify the form item that provides the data for the X-axis

- **Value page**, where you specify the form item that provides the data for the Y-axis

- **File Name page**, where you specify the file that stores the chart .ogd file. After you create the chart using the Chart Wizard, you can modify the chart properties directly in Graphics Builder by opening this file in Graphics Builder.

Like the other Oracle wizards, the Chart Wizard is reentrant, so you can use it to modify an existing chart created with the Chart Wizard. However, be aware that if you use the Chart Wizard in reentrant mode, the Chart Wizard regenerates the chart .ogd file, and you will lose any custom formatting that you have added to the chart. Now, you will use the Chart Wizard to create the chart.

To use the Chart Wizard to create the chart:

1. In Form Builder, click **Tools** on the menu bar, and then click **Layout Editor** to open the Layout Editor.

10

2. Make sure no items are selected on the layout, and then click the **Chart Wizard button** on the toolbar. The Chart Wizard Welcome page opens. Click **Next**. The Type page opens.

 You can also start the Chart Wizard by clicking Tools on the menu bar, and then clicking Chart Wizard. Alternately, you can select the Chart tool on the tool palette, click the mouse pointer on the canvas, make sure the Use the Chart Wizard option button is selected, and then click OK.

As soon as you start the Chart Wizard, the Oracle Graphics Batch application opens, and is visible on the taskbar.

3. On the Type page, type **Quantity on Hand** in the Title box, be sure that Column is the selected chart type, and select **Depth** as the subtype. Click **Next**. The Block page appears.

4. The Block page shows that the form has two blocks, ITEM and INVEN-TORY. The chart data will come from the INVENTORY block, so the chart will be part of this block. Select **INVENTORY** as the block name, and then click **Next**. The Category page appears.

5. The Category page shows all of the items in the INVENTORY block in the Available Fields list. Inventory ID numbers will appear on the X-axis, so select **INV_ID** from the Available Fields list, click the **Select button** so that INV_ID appears in the Category Axis list, and then click **Next**. The Value page appears.

When you use the Chart Wizard to create a chart, you can only display one Category value for column labels. You can select multiple categories, but only the first selection will appear on the chart.

6. Quantity on hand (QOH) values will appear on the Y-axis, so select **QOH** from the Available Fields list, click so that QOH appears in the Value Axis list, and then click **Next**. The File Name page appears.

7. Click **Save As**, navigate to your Chapter10\TutorialSolutions folder, type **10FormChart.ogd** in the File name text box, and click **Save**. The filename appears on the File Name page. If you are saving your solutions on the A: drive on your workstation, then the filename should appear as a:\chapter10\tutorialsolutions\10FormChart.OGD.

 If you accept the default chart filename, the chart will be placed in the default chart directory and will be difficult to locate later for editing.

8. Click **Finish**. The new chart appears on the form.

9. If necessary, select the chart so that selection handles appear around its edges. Resize and reposition the chart so your form looks like Figure 10-17.

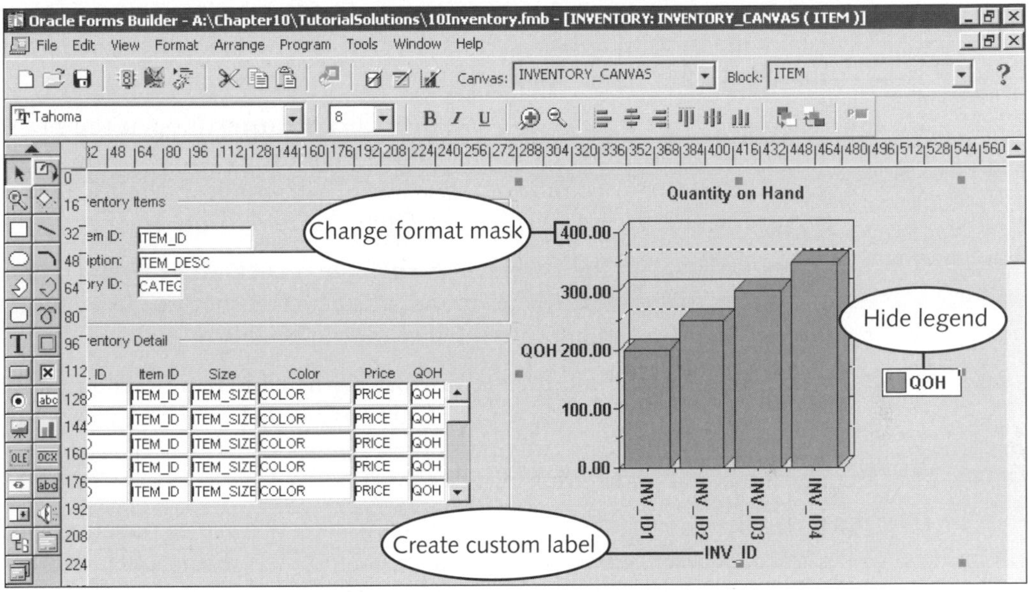

Figure 10-17 Chart created in Form Builder using Chart Wizard

> If a white box labeled CHART_ITEM appears rather than the template that looks like a column chart, an error occurred with the Chart Wizard. Exit Form Builder, open the form again, and repeat Steps 1 through 9. If the white box appears again, ask your instructor or technical support person for help.

10. Save the form.

Next, you need to fine-tune the chart's formatting, as noted on Figure 10-17. You need to hide the chart legend, create a custom label for the X-axis, and change the format mask on the Y-axis tick labels. You can modify the chart title, type, subtype, and data values that the chart displays in Form Builder by using the Chart Wizard in reentrant mode. However, you cannot modify chart formatting using the Chart Wizard. To modify chart formatting, you have to open the chart file in Graphics Builder and make the formatting modifications there. Recall that if you open the Chart Wizard in reentrant mode and apply changes after you have changed the chart formatting in Graphics Builder, you will lose all of your custom formatting. Therefore, it is a good practice to create the chart using the Chart Wizard, but to make *all* formatting modifications in Graphics Builder.

To modify the chart properties in Graphics Builder:

1. Switch to Graphics Builder, or if necessary, close Form Builder, open Graphics Builder, and open the **10FormChart.ogd** chart file from your Chapter10\TutorialSolutions folder. Maximize the Layout Editor, if necessary. Since the chart is based on a form data block rather than a query, the X-axis labels appear as template INV_ID values. The actual values will appear when the chart runs.

2. Select the chart, right-click, and then click **Frame** to open the Frame Properties dialog box. Clear the **Show Legend** check box, and then click **OK**. If necessary, maximize the Layout Editor window, click **Chart** on the menu bar, and then click **Update Chart** to refresh the layout.

3. If necessary, select the chart, right-click, then click **Axes** to open the Axis Properties dialog box. Make sure X is selected in the Axis list, and type **Inventory ID** as the Custom Label for the X-axis.

4. Select the first **Tick Label Rotation option button**, so the tick mark labels are upright and parallel with the right edge of the display, and then click **OK** to close the Axis Properties dialog box. If necessary, maximize the Layout Editor and refresh the display.

5. In the Layout Editor, click one of the X-axis tick labels to select the labels as a group, click **Format** on the menu bar, point to **Font**, click **Font**, and change the font to 8-point Arial Regular.

6. Select the Y-axis tick labels, and change the font to 8-point Arial Regular.

7. Change the font of the X- and Y-axis labels to 8-point Arial Bold.

8. Select the front surface of the columns so selection handles appear around the front surfaces of all of the columns, select the **Fill Color tool** 🪣 on the tool palette, and change the fill color of the front of the columns to pale yellow.

9. Change the fill color of the sides and tops of the columns to a slightly darker yellow shade. (You might need to try several different yellow shades before you find one that looks correct.)

10. Select the Y-axis tick labels, click **Format** on the menu bar, click **Number**, select **999,990,** and then click **OK** to change the QOH format mask so that the QOH labels do not display decimal values. Update the chart display if necessary.

11. Select the X-axis tick labels, click **Format** on the menu bar, click **Number**, type **999999** for the format mask, and then click **OK** to change the INV_ID format mask so that the ID numbers appear without commas or decimal points. This change will not appear until you run the chart in Form Builder and display actual INV_ID values.

12. Save the chart. Your formatted chart should look like Figure 10-18.

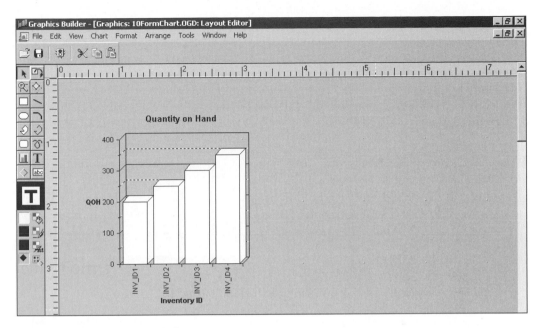

Figure 10-18 Formatted chart

Now, you will switch back to Form Builder and run the form to view the chart and determine if any other modifications are necessary. When you view the chart in the Form Builder Layout Editor, the chart image will not show the formatting changes you made in Graphics Builder. This happens because when you create a chart using the Chart Wizard, the Chart Wizard creates a chart template to show the chart as a symbol, so you can adjust its size and position. This chart template is only updated from the chart .ogd file when you close the form in Form Builder, and then reopen the form file. Now, you will close the form, then reopen the 10Inventory.fmb form file, and view the modified chart template. Then, you will run the form.

To close the form, open it again, and run the form:

1. In Form Builder, click **File** on the menu bar, and then click **Close** to close the form. If necessary, close Graphics Builder and open Form Builder.

2. Click the **Open button** , navigate to your Chapter10\TutorialSolutions folder, and open **10Inventory.fmb**.

3. Click **Tools** on the menu bar, and then click **Layout Editor** to open the Layout Editor. The modified chart template appears on the canvas.

 If the chart template is not modified, right-click the chart object, click Property Palette, and confirm that the chart is stored as 10FormChart.ogd in your Chapter10\TutorialSolutions folder. Then, switch to Graphics Builder, and confirm that the chart you are editing is 10FormChart.ogd in your Chapter10\TutorialSolutions folder. Make sure that you save the chart in Graphics Builder after you make the formatting modifications.

4. Click the **Run Form Client/Server button** 🔲 to run the form, and if necessary, connect to the database in the usual way. The form appears in the Forms Runtime window. Currently, no data appears on the form, and the chart is blank.

 If an error message appears stating that some of the form items are not on the canvas, click OK, and then open the Layout Editor and resize and reposition the chart so it is entirely on the canvas. Then run the form again.

5. Click the **Enter Query button** 🔲, do not enter a search condition, and then click the **Execute Query button** 🔲. The data appears for item 894 (Women's Hiking Shorts), and the associated inventory records appear in the Inventory Detail frame and on the chart, as shown in Figure 10-19.

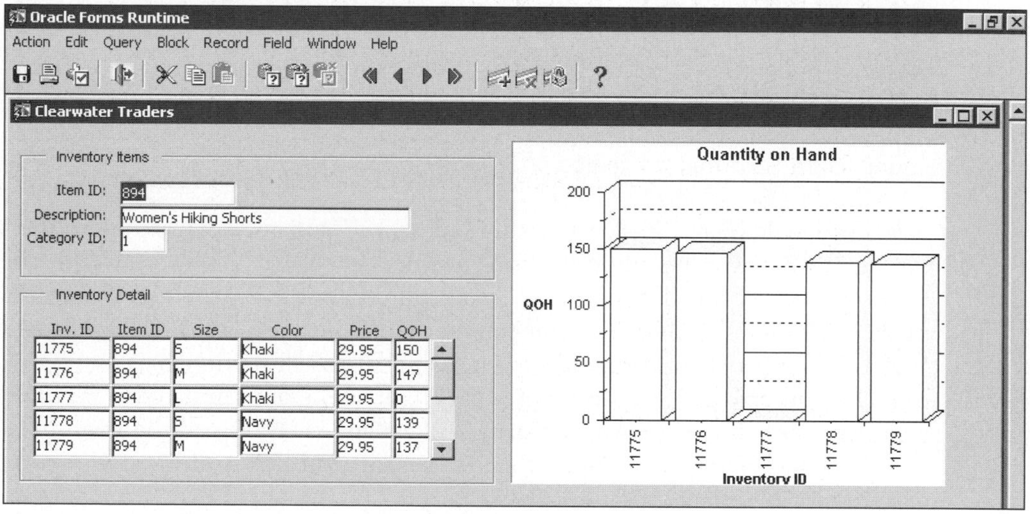

Figure 10-19   Formatted chart display on form

6. Scroll through the inventory item records to verify that the chart is updated correctly.

7. Close Forms Runtime.

8. Close Form Builder, and click **Yes** to save your changes.

9. Click the **Oracle Graphics Batch** application button on the taskbar, and then close the Oracle Graphics Batch application.

10. Switch to Graphics Builder, if necessary, and close the 10FormChart.ogd chart.

CREATING A REPORT WITH AN EMBEDDED CHART

Sometimes you need to display data visually on a report. For example, suppose the managers at Clearwater Traders need an inventory report that shows each item ID and description, and lists all of the associated inventory numbers, sizes, colors, current prices, and quantities on hand. They would like to have the report display a chart that shows the associated quantity on hand for each inventory item, so they can visually see when an item is out of stock or when the stock is running low. The report design sketch is shown in Figure 10-20. The report data values appear in a master-detail layout on the left side of the page, with the item information as the master record and the inventory information as repeating detail records.

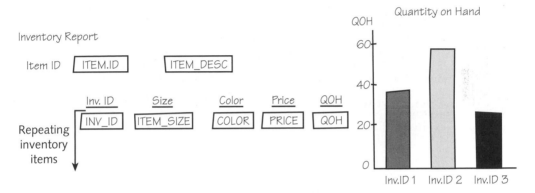

Figure 10-20   Design sketch for report with chart display

A report has been created that displays all of the data values, and you need to add the quantity on hand chart. Now, you will start Report Builder, open the report, save the report with a different filename, and view the report in the Live Previewer.

To start Report Builder and view the report:

1. Start Report Builder, open the **Inventory.rdf** report file from the Chapter10 folder on your Data Disk, and save the report as **10Inventory.rdf** in your Chapter10\TutorialSolutions folder. The report opens in the Object Navigator. If necessary, maximize the Object Navigator window.

2. Click **File** on the menu bar, click **Connect**, and then connect to the database as usual.

3. Double-click the **Live Previewer icon** to view the report in the Live Previewer.

You will create the chart using the Report Builder Chart Wizard. All values that appear on the chart come from report data fields. The chart shows the inventory ID on the X-axis, and the columns indicate the quantity on hand for each inventory item. The Report Builder Chart Wizard pages, which are similar to the pages in Form Builder Chart Wizard, include:

- **Type page**, which allows you to specify the chart title, type (bar, column, line, and so forth), and subtype options.

- **Category page**, on which you specify the report record group field that provides the data for the X-axis.

- **Value page**, on which you specify the report record group field that provides the data for the Y-axis;

- **Breaks page**, which allows you to specify the frequency with which the chart will appear in the report. A chart can appear once at the beginning of the report, once at the end, or once within any of the report's repeating frames. For this report, you want the chart to appear once for each item ID value.

  - **File Name page**, on which you specify the file that stores the chart .ogd file. After you create the chart using the Chart Wizard, you can modify the chart properties directly by opening this file in Graphics Builder.

Now, you will start the Chart Wizard, and create the chart. In Report Builder, you can only start the Chart Wizard from the Live Previewer window or Layout Model window.

To use the Chart Wizard to create the chart in the report:

1. In the Live Previewer, click the **Chart Wizard button** on the toolbar. The Chart Wizard Welcome page opens.

2. Click **Next**. The Type page appears.

3. Type **Quantity on Hand** in the Title text box, and confirm that Column is the selected chart type and Plain is the subtype. Click **Next**. The Category page appears.

4. The Category page lists all report data fields in the Available Fields list. Since inventory ID numbers will appear on the X-axis, select **inv_id** from the Available Fields list, click the **Select button** so that **inv_id** appears in the Category Axis list, and then click **Next**. The Value page appears.

5. The Value page lists all report numerical fields in the Available Fields list. Since quantity on hand values will appear on the Y-axis, select **qoh** from the Available Fields list, click so that **qoh** appears in the Value Axis list, and then click **Next**. The Break page appears.

6. The Break page lists the options for the chart frequency. Click **once per Item ID:,item_desc (R_G_item_id)** in the Chart list box to specify that the chart will appear once for every item ID, and then click **Next**. The File Name page appears.

7. Click **Save As**, navigate to your Chapter10\TutorialSolutions folder, change the filename to **10ReportChart.ogd** in the File name box, and click **Save**. The complete path to the chart file appears on the File Name page. If you are saving your solutions on the A: drive, the path should appear as a:\Chapter10\TutorialSolutions\10reportchart.ogd.

8. Click **Finish**. The new chart appears on the report below the inventory records.

9. Save the report.

Next, you will move the chart so that it appears to the right of the inventory records. Since the chart is inside the R_G_item repeating frame, you will need to enable flex mode so that the repeating frame automatically resizes when you drag the chart to the right edge of the report. You will do this next.

To move the chart:

1. If necessary, click the Flex Mode button 🖳 on the toolbar to enable flex mode.

2. If necessary, click the chart to select it, drag the chart toward the right edge of the report, and then toward the top edge of the report, so that it appears on the right side of the inventory records, as shown in Figure 10-21.

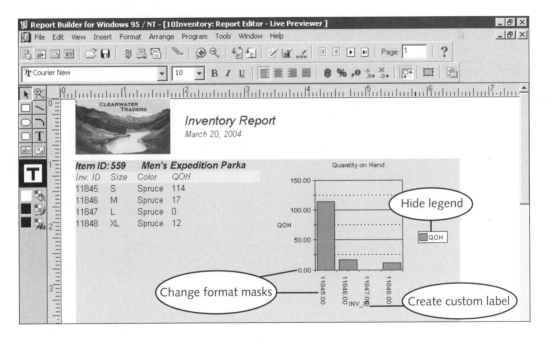

Figure 10-21   Repositioning the report chart

3. Save the report.

Figure 10-21 shows that you need to modify the chart to hide the legend, create a custom label for the X-axis, and change the format masks for the tick labels on both axes so that the numbers do not display decimal points. Now, you will switch to Graphics Builder, open the report chart file, and modify the chart properties.

To modify the report chart properties in Graphics Builder:

1. Switch to Graphics Builder, and open the **10ReportChart.ogd** chart file from your Chapter10\TutorialSolutions folder. If necessary, maximize the Layout Editor window. Like the chart on the form, the chart appears as a template rather than with actual data values, because the actual data values come from the report.

 If your computer does not have enough memory to run Report Builder and Graphics Builder at the same time, close Graphics Builder and then reopen it as necessary.

2. Select the chart, right-click, then click **Frame** to open the Frame Properties dialog box. Clear the **Show Legend check box**, and then click **OK**. If necessary, update the chart display and maximize the Layout Editor window.

3. Select the chart if necessary, right-click, and then click **Axes** to open the Axis Properties dialog box. Type **Inventory ID** as the custom label for the X-axis, and then click **OK** to close the Axis Properties dialog box. If necessary, maximize the Layout Editor window.

4. In the Layout Editor, select the X-axis tick-labels, click **Format** on the menu bar, and then click **Number**. The Number Format dialog box opens. Type **999999** in the Number text box, and then click **OK** to apply the new format mask to the chart labels.

5. In the Layout Editor, select the Y-axis tick-labels, and change the format mask to **999,990**.

6. Select the column bars, and change the fill color to dark blue.

7. Save the chart.

Now, you will view the report in Report Builder to check the formatting changes. You will need to reload the report in the Live Previewer to refresh the chart display.

To view the report and check the formatting:

1. Switch to Report Builder, click **View** on the menu bar, and then click **Refresh Data**. The formatted chart appears on the report, as shown in Figure 10-22.

2. Save the report, and then exit Report Builder and all other open Oracle applications.

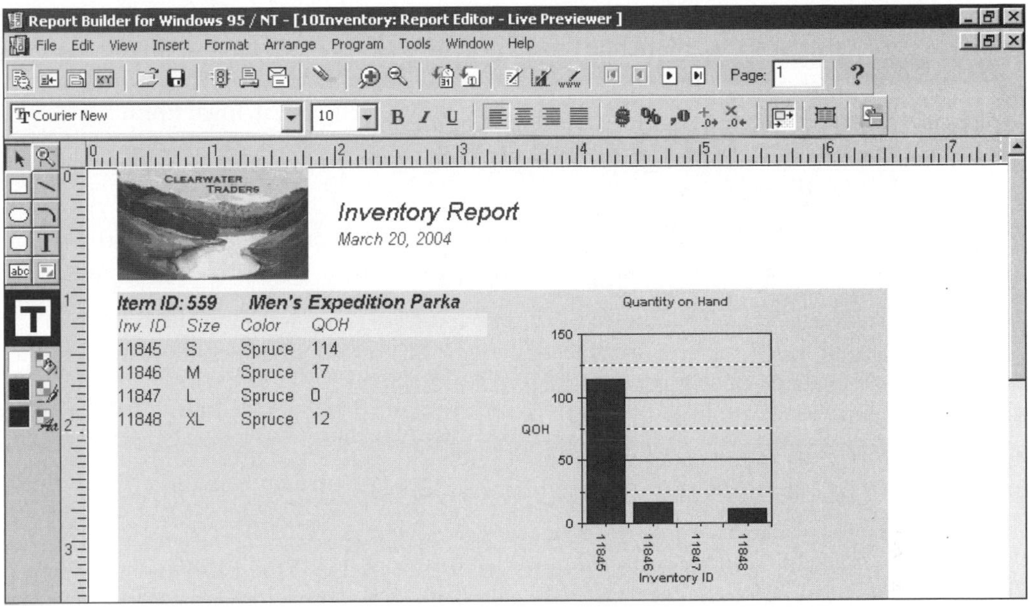

Figure 10-22   Formatted report chart

SUMMARY

❐ Data is easier to understand and interpret when it appears in graphical formats using bar, pie, and line charts rather than in textual or tabular formats.

❐ In Graphics Builder, a top-level chart object is called a display.

❐ To create a new chart, create the query to define the chart, and then define the chart display properties.

❐ For pie charts, the different pie slice labels will be the values of one query data column, and the corresponding percentage of the whole pie will be the values in the second query column.

❐ For column charts, the X-axis item labels will be the values of one query data column, and their Y-axis values will be the values in another query column.

❐ In a chart query, when you use calculated column values based on SQL arithmetic or aggregate functions, you should assign the calculated column an alias using the SQL AS clause.

❐ When defining chart properties, categories refer to the chart labels, and values specify the numerical values that determine the data that appears on the chart.

❐ When you are formatting a chart, the Frame Properties dialog box allows you to format the frame in which the chart appears. The Axis dialog box allows you to format the properties of the chart X- and Y-axes.

❏ A discrete chart axis displays discrete values, which have one and only one value. A continuous axis has numerical or date values that increase as you move along the axis. On a continuous axis, you can specify the axis minimum value, maximum value, step size, and whether the axis shows a linear or logarithmic scale.

❏ In Graphics Builder, you can apply format masks to values that have a DATE or NUMBER data type.

❏ Graphics Builder allows you to save charts both as design files (with an .ogd extension) and as runtime files (with an .ogr extension).

❏ To pass a data value from a form to a chart, you must modify the chart so that its query uses a bind parameter for a search condition instead of a specific search condition value.

❏ You can use the RUN_PRODUCT procedure to pass a parameter list from a form to a Graphics Builder chart, and then display the chart.

❏ You can use the Chart Wizard to create charts that appear directly on forms and reports.

❏ When you use the Chart Wizard to create a chart on a form, the chart data comes from a data block on the form. When you use the Chart Wizard to create a chart on a report, the chart data comes from report fields.

❏ To change the appearance of a chart created using the Chart Wizard, open the chart file in the Graphics Builder and modify its properties.

❏ You can specify that a chart will appear in a report at the beginning of the report, at the end of the report, or once in every repeating frame in the report.

REVIEW QUESTIONS

1. Specify the chart type that best illustrates the following information from the Clearwater Traders (Figure 1-11), Northwoods University (Figure 1-12), and Software Experts (Figure 1-13) databases:

    a. How the total enrollment for each term varies across terms at Northwoods University

    b. The total dollar amount of all orders that each Clearwater Traders customer has placed

    c. The proportion of Software Experts consultants who are certified in each skill

2. What is a chart type? What is a chart subtype?

3. When you create a pie chart, what values does the query need to retrieve?

4. When you create a column chart, what values does the query need to retrieve?

5. Describe the difference between a discrete axis and a continuous axis.

6. How do you hide the legend on a column chart?

7. When do you need to use a column alias in a chart query?

8. Describe two ways to integrate charts with form applications.

9. When you create a chart in a form or report using the Chart Wizard, why do you need to format the chart in Graphics Builder?

10. When you create a chart on a form using the Chart Wizard, what is the chart data source?

PROBLEM-SOLVING CASES

All required files are stored in the Chapter10 folder on your Data Disk. Save all files in your Chapter10\CaseSolutions folder. The cases reference the sample data in the Clearwater Traders database (see Figure 1-11), Northwoods University database (see Figure 1-12), and Software Experts database (see Figure 1-13).

1. Create a pie chart that shows each building code in the LOCATION table of the Northwoods University database as a pie slice, with the total capacity of the building's rooms as the proportion of the pie. Format the chart as shown in Figure 10-23, and save the chart as 10Case1.ogd.

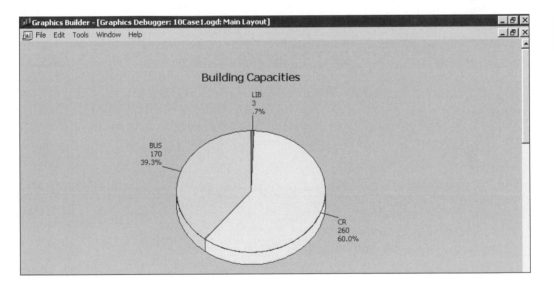

Figure 10-23

2. Create a pie chart that shows the proportion of consultants currently working on different projects in the Software Experts database. Format the chart as shown in Figure 10-24, and save the chart as 10Case2.ogd.

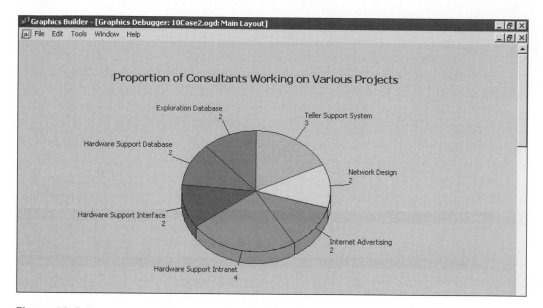

Figure 10-24

3. Create a column chart that shows the faculty ranks (ASSO, FULL, INST, ASST) from the Northwoods University FACULTY table on the X-axis, and shows how many faculty members are in each rank as column values on the Y-axis. Format the chart as shown in Figure 10-25. (*Hint*: Use the SQL COUNT function, and modify the Y-axis on the Continuous Axis tab on the Axis Properties dialog box so that the Minimum Value is 0 and the Step size is 1.) Modify the column colors to achieve a three-dimensional appearance. Save the chart as 10Case3.ogd.

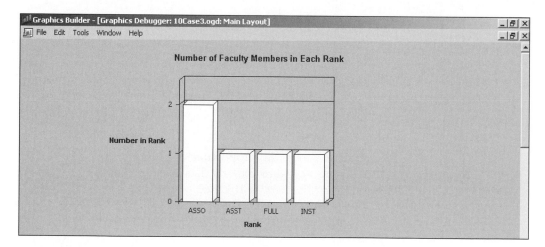

Figure 10-25

4. Create a line chart to show how total enrollments in all courses have varied over all terms at Northwoods University. (*Hint*: Sum enrollment values for each course section.) Show the term description on the X-axis and the enrollment values on the Y-axis. Format the chart as shown in Figure 10-26, and save the chart as 10Case4.ogd.

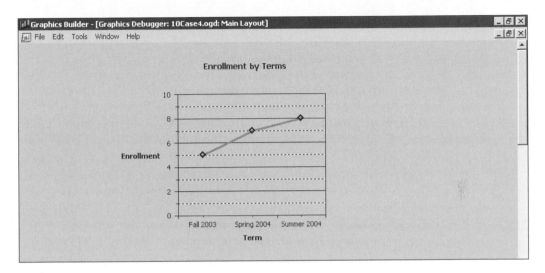

Figure 10-26

5. In this case, you will modify a form that allows the user to select a Northwoods University term. The form then calls a chart that compares each course's maximum enrollment with its location capacity for all courses offered during the selected term.

a. Create the column chart shown in Figure 10-27 that shows course call ID values on the X-axis, and two series of data on the Y-axis that represent the maximum enrollment and maximum location capacity for each course. Initially, create the chart query to retrieve data for TERM_ID 6 (Summer 2004). Save the chart as 10Case5.ogd.

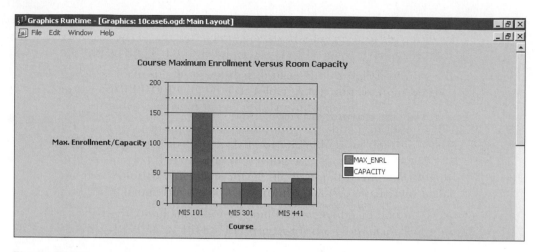

Figure 10-27

b. Create a bind parameter named P_TERM_ID for the chart search condition, and modify the query so the search condition uses the TERM_ID field and searches for the P_TERM_ID parameter value.

c. Open the form named 10Case5.fmb. This form has a control block that displays term IDs and term descriptions for Northwoods University, as shown in Figure 10-28. Save the form as 10Case5_DONE.fmb. (No chart data will appear for term ID values 1, 2, or 3, because these terms do not have any associated course section data.)

d. Create a trigger for the Show Enrollments/Capacities command button that creates a parameter list containing the TERM_ID value for the term that the user selects, and then calls the chart. The chart should display the maximum enrollments and capacities for the selected term.

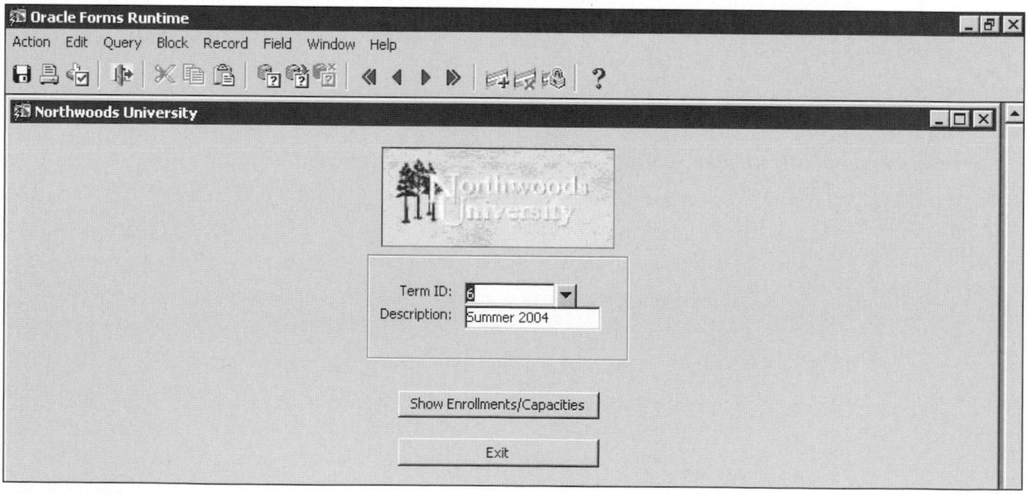

Figure 10-28

6. In this case, you will modify a form that allows the user to select a Software Experts consultant. The form then calls a chart that displays a Gantt chart that shows the consultant's schedule for all project assignments.

a. Create the Gantt chart shown in Figure 10-29 that shows roll-on and roll-off dates. Initially, create the chart query to retrieve data for consultant ID 103 (Sarah Carlson). Save the chart as 10Case6.ogd.

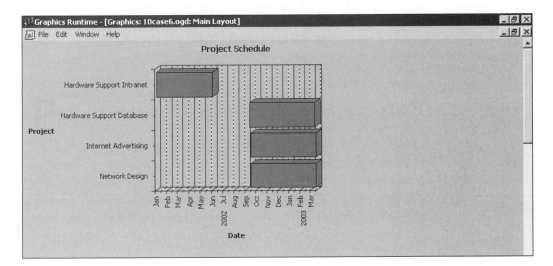

Figure 10-29

b. Create a bind parameter named P_C_ID for the chart search condition, and modify the query so the consultant ID search condition uses the parameter.

c. Open the form named 10Case6.fmb. This form has a control block that displays consultant IDs and first and last names, as shown in Figure 10-30. Save the form as 10Case6_DONE.fmb.

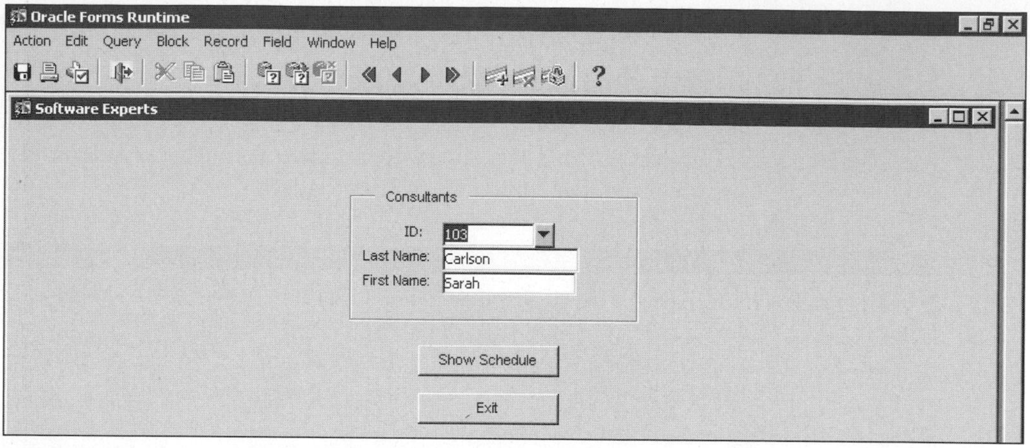

Figure 10-30

d. Create a trigger for the Show Schedule command button that creates a parameter list containing the consultant ID value for the consultant that the user selects, and then calls the chart.

7. In this case, you will add a chart to a master-detail data block form that displays consultant information from the Software Experts database in the master block and consultant project information in the detail block.

a. Open the 10Case7.fmb form, and save it as 10Case7_DONE.fmb.

b. Use the Chart Wizard to create a column chart on the form that displays project ID values on the X-axis and total project hours on the Y-axis. Use the PROJECT_CONSULTANT data block as the chart data source. Save the chart as 10Case7.ogd.

c. Use Graphics Builder to format the chart as shown in Figure 10-31.

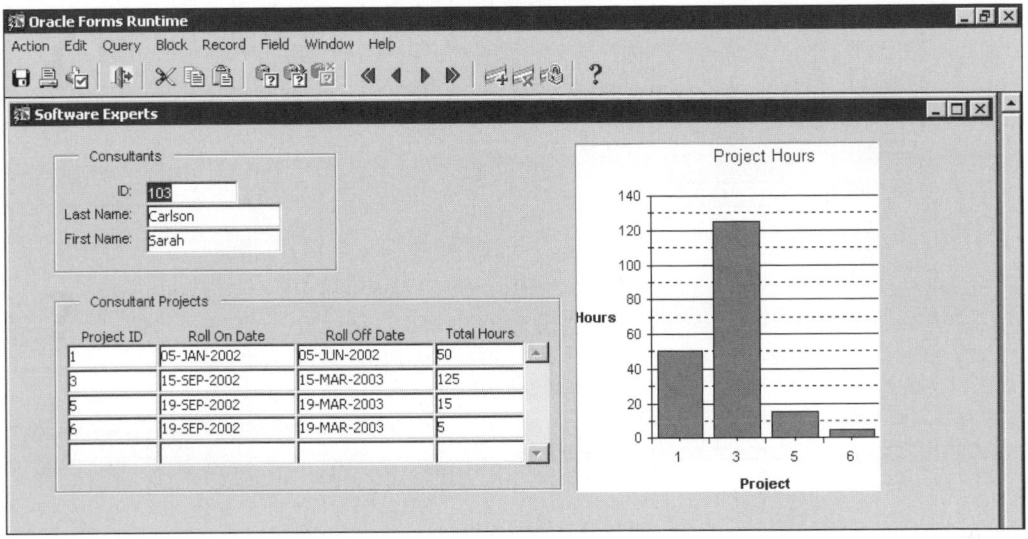

Figure 10-31

8. In this case, you will add a chart to a master-detail data block form that displays course information from the Northwoods University database in the master block and course section information in the detail block.

a. Open the 10Case8.fmb form, and save it as 10Case8_DONE.fmb.

b. Use the Chart Wizard to create a column chart on the form that displays course section ID values on the X-axis and maximum enrollments on the Y-axis. Use the COURSE_SECTION data block as the chart data source. Save the chart as 10Case8.ogd.

c. Use Graphics Builder to format the chart as shown in Figure 10-32.

10

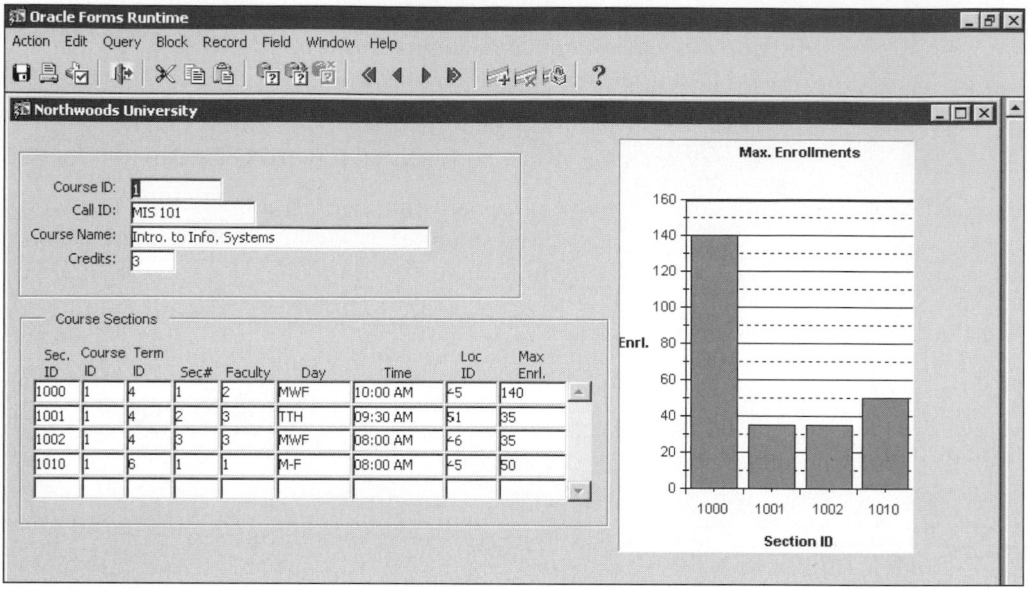

Figure 10-32

9. In this case, you will modify a report that lists all building codes and the associated room numbers and capacities for each location at Northwoods University.

a. Open the 10Case9.rdf report file, and save it as 10Case9_DONE.rdf.

b. Use the Chart Wizard to create a chart on the report that is saved in the file named 10Case9.ogd. The chart shows each building's room numbers on the X-axis and the associated capacities on the Y-axis. Use Graphics Builder to format the chart as shown in Figure 10-33.

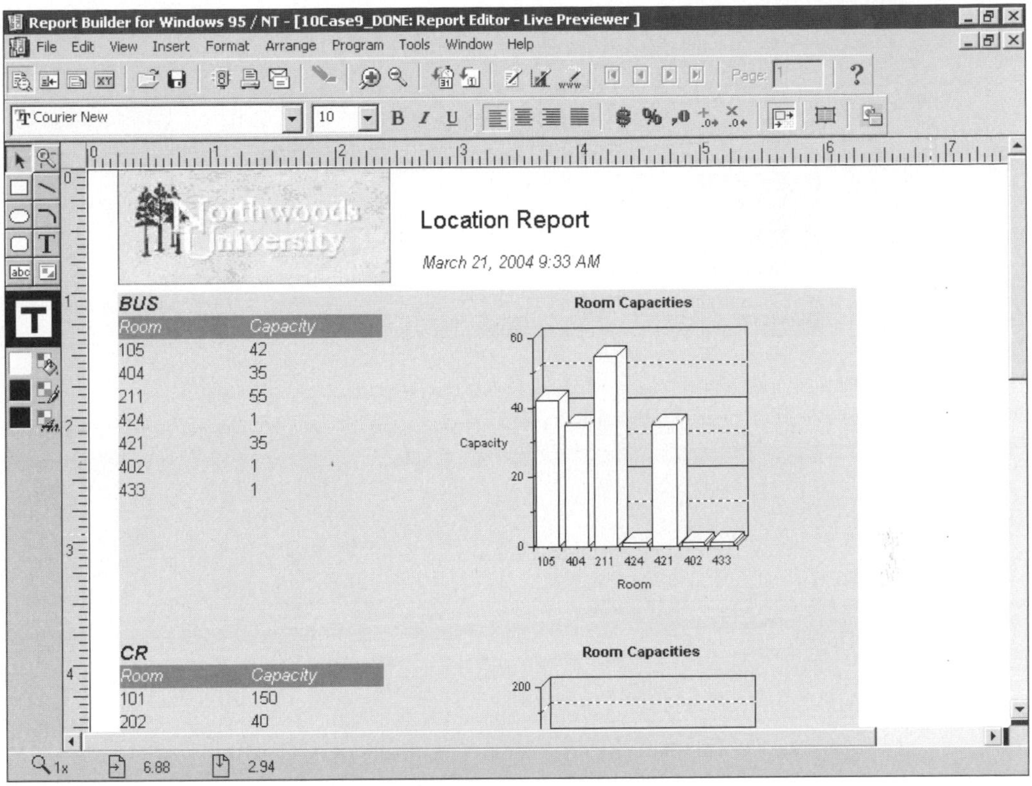

Figure 10-33

10. In this case, you will modify a report that lists Software Experts projects and the associated project evaluation information. (The sample database contains evaluation records for only the Hardware Support Intranet project.)

a. Open the 10Case10.rdf report file, and save it as 10Case10_DONE.rdf.

b. Use the Chart Wizard to create a chart on the report that is saved in the file named 10Case10.ogd. The chart shows each evaluatee ID on the X-axis and the associated evaluation score on the Y-axis. Use Graphics Builder to format the chart as shown in Figure 10-34. Modify the Y-axis so the maximum value on the scale is 100.

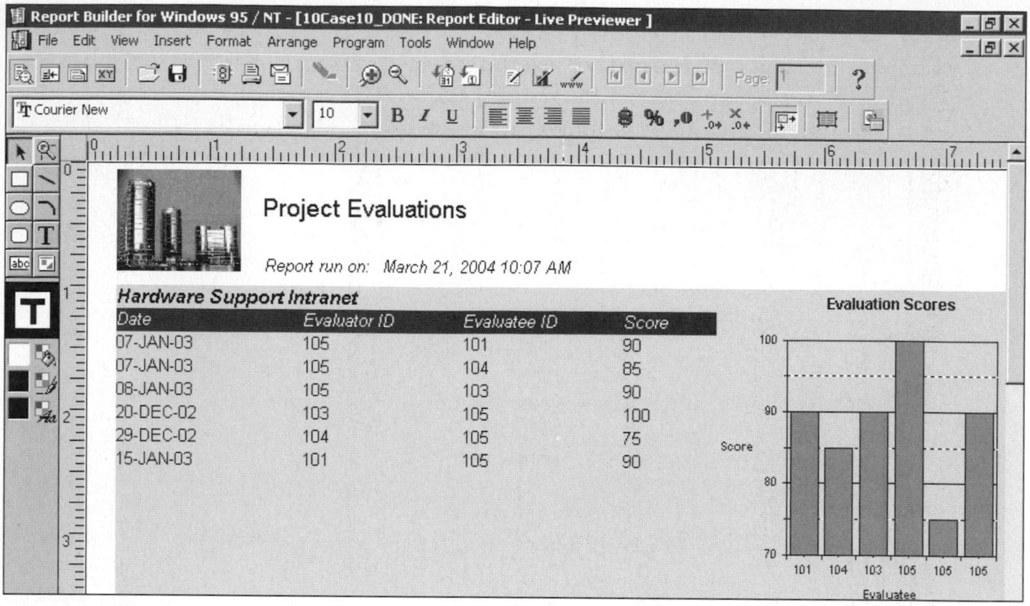
Report Builder for Windows 95 / NT - [10Case10_DONE: Report Editor - Live Previewer]
File Edit View Insert Format Arrange Program Tools Window Help
Courier New
10
Project Evaluations
Report run on: March 21, 2004 10:07 AM
Hardware Support Intranet
Date Evaluator ID Evaluatee ID Score
07-JAN-03 105 101 90
07-JAN-03 105 104 85
08-JAN-03 105 103 90
20-DEC-02 103 105 100
29-DEC-02 104 105 75
15-JAN-03 101 105 90
Evaluation Scores
Score
Evaluatee

Figure 10-34

11

CREATING AN INTEGRATED DATABASE APPLICATION

◀ LESSON A ▶

Objectives

- ♦ Understand the steps for developing a database application
- ♦ Design a database application interface
- ♦ Create a global path variable to specify the location of form, report, and graphics files
- ♦ Use timers in a Form Builder application to create a splash screen
- ♦ Create form templates and visual attribute groups to ensure consistency across application modules
- ♦ Understand how to open, navigate among, and close form modules in a multiple form application

You have learned that when you develop a new database system, you begin by identifying the business processes that the database system will support and the data items required to support these processes. Then you create the database tables and load the data. Finally, you develop the forms, reports, and charts to manage the data and support the business processes. In this chapter, you will learn how to combine the individual form, report, and chart components into a single integrated system. To learn how to build an integrated system, you will create a single Form Builder form module that serves as the entry point for the individual components that compose the system.

In Lesson A, you will learn how to use features in Form Builder to develop an integrated database application made of multiple form modules. You will learn about the elements of an integrated database application, and learn how to create them in Form Builder. You will also learn the commands for opening, navigating to, and closing forms in a multiple-form application. In Lesson B, you will learn how to use Project Builder, which is the Oracle utility for managing and deploying integrated database applications. In Lesson C, you will learn how to create pull-down menus, pop-up menus, and menu toolbars to integrate your application. You will also learn how to control menus programmatically, and enforce application security using menus.

DESIGNING AN INTEGRATED DATABASE APPLICATION

When you create a new database system, you create a variety of individual forms, reports, and charts and integrate them into a single application that allows users to access all of these items from a single entry point. Developing database applications usually involves the following phases:

- *Design*, which involves creating the specifications for the application components, based on user needs, input from project managers and development team members, and user requests for enhancements or bug fixes

- *Module development*, which involves creating the individual modules, including forms, reports, and charts

- *Module integration*, which involves integrating the individual modules into a single application

- *Testing*, which occurs in two stages: unit testing and system testing. **Unit testing** involves testing the individual modules, such as individual forms or reports. **System testing** evaluates the integration of the modules.

- *Deployment*, which involves packaging the integrated modules in an installable format that can be delivered to customers

For large database applications, you usually repeat these phases iteratively and do not necessarily follow the sequence exactly as listed. For example, during the module integration phase, developers might determine that a new module is needed, which would require moving back to the module development phase.

In this chapter, the tutorial exercises illustrate how to create an integrated database application for Clearwater Traders. For this application, the main business processes that the system will support are merchandise receiving and customer sales. The system must also have data block forms for managing the data in the individual database tables, and reports and charts for helping managers direct operations. (Recall that the details of the design and development process for the individual application modules have been addressed in previous chapters.)

When you create a complex database application, it is a good practice to use separate form files rather than combine all of the project forms into a single .fmb file with many different canvases. With separate form files, the project is broken into modules that are independent of one another. Each module can be developed, tested, and debugged independently, and then can be integrated into the overall application. Smaller modules are easier to work with, and make it easier for project teams with multiple team members to split up the development effort, since different team members can simultaneously work on different form files. Smaller modules also create smaller .fmx files, which occupy less space on the client workstation's hard drive and in the client workstation's main memory. This is an important consideration when users have older workstations in which the main memory and hard drive capacities are close to the minimum amounts required for running Oracle applications.

 Tip When you are designing individual form modules, keep in mind that data blocks with master-detail relationships must be in the same form and cannot be split between multiple form modules.

After the individual modules are complete, you should place them in a single folder that all developers can access. Table 11-1 summarizes all of the form, report, and chart files that will be used in the Clearwater Traders database application. All of these files, along with the form design (.fmb) files, are included in the CWPROJECT subfolder in the Chapter11 folder on your Data Disk.

11

Design File	Description
\CWPROJECT\Sales.fmb	Custom form that records information about new customer orders in the CUST_ORDER, ORDER_LINE, and INVENTORY tables
\CWPROJECT\Receiving.fmb	Custom form that records information about new inventory item shipments in the INVENTORY, SHIPMENT, and SHIPMENT_LINE tables
\CWPROJECT\Customer.fmb	Data block form that displays and edits CUSTOMER data records
\CWPROJECT\Item.fmb	Data block form that displays and edits ITEM data records
\CWPROJECT\Inventory.fmb	Data block form that displays and edits INVENTORY data records
\CWPROJECT\Inventory.rdf	Report that displays INVENTORY information
\CWPROJECT\OrderSource.ogd	Pie chart that shows the amount of revenue generated from different order sources

Table 11-1   Module files in Clearwater Traders database application

To integrate individual Oracle form, report, and chart modules into a single application, you create a single form module, called the main form, from which users can access all of the individual application components. The main form can have a **splash screen**, which is a front-end entry screen that introduces the application. Then, the application displays the main form screen display. The main form screen display often has a **switchboard**, which consists of command buttons that enable users to quickly and easily access the most commonly used forms, reports, or charts. All application components should also be accessible through pull-down menus for users who prefer to use the keyboard rather than the mouse pointer. Pull-down menus also provide access to features that are used less often than the ones on the switchboard and to features that have multiple levels of choices. For example, you could have a pull-down menu titled Reports. The next-level menu would list different reports that the user could access. Also, you can use different pull-down menus to restrict users from accessing specific forms. For example, you might not want all users to be able to insert, update, and delete data using certain forms.

Figure 11-1 shows the design for the main form screen display for the Clearwater Traders database system. The top section of the design shows the pull-down menu structure. The left panel on the bottom section of the design shows the system switchboard, and the right panel on the bottom section is reserved for displaying a graphic image to enhance the application's appearance.

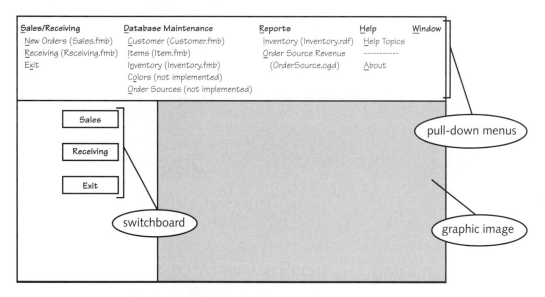

Figure 11-1   Main form screen design

To make applications easier to maintain and debug, they should always have a single entry and exit point. For the Clearwater Traders system, the main form is this point. Users will navigate to all application forms from the main form, and they will exit the application from this form. Users can access the form modules that support the two main business processes—processing new sales orders and receiving incoming shipments—on the switchboard using the Sales and Receiving command buttons. Users can close the application by using the switchboard Exit button. The first pull-down menu item in Figure 11-1 is Sales/Receiving. This menu contains selections for processing new orders, receiving incoming shipments, and exiting the form. Note that the name of the form file that will be called by the menu selection appears beside the selection name. The Database Maintenance menu provides access to the basic forms that let users insert, update, and delete records in several tables in the database. Note that forms for two of the selections, Colors and Order Sources, have not been implemented yet. You will learn how to develop application placeholder elements called **stubs**, which are programs or messages that handle undeveloped system features, in Lesson C of this chapter. The Reports menu provides access to the Inventory report and the Order Source Revenue chart. The final two menus on the Clearwater Traders database application menu are Help and Window. The Help menu has two second-level selections: Contents, which provides access to the Help search engine, and About, which gives details about the application. The Window menu selection allows users to move between windows in a multiple-window application.

Most Windows applications have Help as the last menu choice, but Form Builder automatically places Window as the last pull-down menu selection.

Now, you will start Form Builder, and create the main form module, shown in Figure 11-1, that integrates the rest of the application modules. This form will be saved in a file named Main.fmb that you will store in a folder named CWPROJECT_DONE in your Chapter11\TutorialSolutions folder.

The application that you will create in this chapter requires integrating many different files. To improve application performance, if you have been storing your Data Disk files on a floppy disk, it is a good idea to copy the Chapter11 folder and its contents from your floppy disk to a hard drive and save your files there. The tutorial exercises will illustrate that the tutorial solutions are stored in a folder named Chapter11\TutorialSolutions\CWPROJECT_DONE located in a folder named OracleSolutions on the C: drive.

To start Form Builder and create the main form module:

1. Start Windows Explorer, create a folder named **CWPROJECT_DONE** in your Chapter11\TutorialSolutions folder, copy the contents of the CWPROJECT folder on your Data Disk to this folder, and then exit Windows Explorer.

2. Start Form Builder. When the Welcome to Form Builder window opens, select the **Build a new form manually option button**, and then click **OK**. The Object Navigator window opens. Connect to the database.

3. Change the form module name to **MAIN**.

4. To create the trigger to automatically maximize the form window when the form runs, select the **Triggers node** under the form module, right-click, point to **SmartTriggers**, and click **PRE-FORM**. Type the following code in the PL/SQL Editor, compile the trigger, correct any syntax errors, and then close the PL/SQL Editor.

```
SET_WINDOW_PROPERTY (FORMS_MDI_WINDOW, WINDOW_STATE,
MAXIMIZE);
```

5. To configure the template canvas, select the **Canvases node**, and then click the **Create button** 🔲 to create a new canvas. Change the canvas name to **MAIN_CANVAS**, right-click the **Canvas icon** 🔲, and then click **Property Palette**. Under the Physical property node, change the Width to **580**, the Height to **375**, and then close the Property Palette.

6. To configure the template window, click ➕ beside the Windows node, and change the form window name to **MAIN_WINDOW**. Double-click the **Window icon** 🔲 to open the MAIN_WINDOW Property Palette, change the following property values, and then close the Property Palette.

Title	Clearwater Traders
Resize Allowed	No
Maximized Allowed	No
Width	580
Height	375

7. Save the form as **Main.fmb** in your Chapter11\TutorialSolutions\ CWPROJECT_DONE folder.

Now you will create a control block that will contain the form command buttons. Then, you will create the buttons and import the graphic image.

To create the command buttons and import the graphic image:

1. In the Object Navigator, select the **Data Blocks node**, and click the **Create button** 🔲 to create a new block. Select the **Build a new data block manually option button**, and then click **OK**. The new block appears in the form's Data Block list. Change the block name to **MAIN_BLOCK**.

2. Double-click the **Canvas icon** 🔲 beside MAIN_CANVAS to open the MAIN_CANVAS in the Layout Editor.

3. To create the Sales button, confirm that the MAIN_BLOCK is selected in the Block list box in the Layout Editor, select the **Button tool** 🔲 on the tool palette, and draw a button on the canvas in the approximate position of the Sales button in Figure 11-1.

4. Right-click the new button, click **Property Palette**, change the following properties, and then close the Property Palette.

Property	Value
Name	**SALES_BUTTON**
Label	**Sales**
Width	**63**
Height	**16**

5. Select the button, click the **Copy button** on the toolbar, and then click the **Paste button** two times to make the next two command buttons. Move the pasted buttons so they are in the approximate positions of the Receiving and Exit buttons in Figure 11-1.

When you paste a button, the pasted button is placed directly on top of the copied button.

6. Open the Property Palettes for the two new buttons, modify their properties as follows, and then close the Property Palettes.

Button	Button Name	Label
First	**RECEIVING_BUTTON**	**Receiving**
Second	**EXIT_BUTTON**	**Exit**

11

7. To import the graphic image that appears on the right side of the form, click **File** on the menu bar, point to **Import**, and then click **Image**. Click **Browse**, navigate to your Chapter11\TutorialSolutions\CWPROJECT_DONE folder, select **clearimage.gif**, click **Open**, and then click **OK**.

8. Scroll to the right and bottom edge of the canvas, and reposition the image as needed so that the image is completely on the canvas. Your form should look like Figure 11-2.

If any part of the image is off the canvas surface, an error will occur when you run the form.

9. Save the form.

10. Run the form, note if the item positions and sizes need to be adjusted, close Forms Runtime and then adjust the sizes and positions in Form Builder, if necessary.

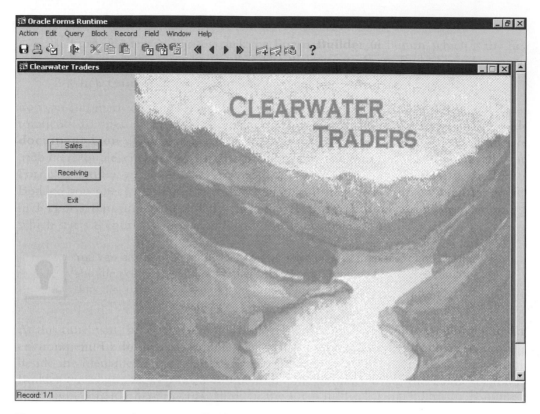

Figure 11-2   Main form screen display

CREATING A GLOBAL PATH VARIABLE

A **global variable** is a form variable that is created in the code of one form, and can then be referenced in any other open form. A global variable remains in memory during the entire duration of the user's database session, and is available to other forms, even when the form that created the global variable closes. A global variable is a character variable, with a maximum length of 255 characters. Because global variables consume system resources, as a general rule, you should not use more than five global variables in an integrated database application.

A form global variable is like a public variable in a package. You can also use public package variables to share data values among different forms in a database application.

When you create an integrated database application, the main form application needs to call other form, report, and chart application files using their complete folder path specifications. The individual modules may also need to reference files that contain data objects, such as graphic images or sounds. It is useful to create a **global path variable**, which is a global variable that references a text string that specifies the complete path to the drive and folder where you store all the application files. The global path variable enables you to place the path information in a single location. If you move the application files to a new location, you only have to change the specification in the global path variable, rather than changing the path statements in many different locations.

You reference a global variable using the syntax `:GLOBAL.variable_name`. The *variable_name* must follow the Oracle naming standard. To create a new global variable, you simply assign a value to it. You do not need to formally declare a global variable. If you want to create a global variable that has a NUMBER or DATE data type, you must assign the data value as a character, and then convert the value to the desired data type using the PL/SQL TO_DATE or TO_NUMBER conversion function. If you try to reference a global variable value that has not yet been assigned a value, an error will occur. To assign a value to a global variable, you use the following syntax:

```
:GLOBAL.variable_name := variable_value;
```

Suppose that the Clearwater Traders database application files are stored in the \OracleSolutions\Chapter11\TutorialSolutions\CWPROJECT_DONE folder on the C: drive on your workstation. To assign that folder location to a global variable named PROJECT_PATH, you would use the following command:

```
:GLOBAL.project_path :=
'c:\oraclesolutions\chapter11\tutorialsolutions\
cwproject_done\';
```

Note that the global project path specification includes a trailing back slash (\) in the file path.

To specify the complete path to a file that is stored in the application project folder, you concatenate the path variable to the filename. For example, if you want to reference the customer.fmx file that is in the \Chapter11\TutorialSolutions\CWPROJECT_DONE subfolder, you would specify the complete folder path as follows: `:GLOBAL.project_path || 'customer.fmx'`. Using the value for :GLOBAL.project_path assigned in the previous example, this command specifies the complete path to the file as `'c:\oraclesolutions\chapter11\tutorialsolutions\cwproject_done\customer.fmx'`.

Now you will modify the PRE-FORM trigger in the main application form. The trigger will assign the global project path to a text string that specifies the drive letter and folder path where you store your project application files. You must specify the drive letter where your folder is stored, along with the complete path specification, including all subfolders.

11

The global project path specification cannot contain any blank spaces in the folder names, and you cannot break the text string that specifies the folder path across multiple lines of code.

To create the global project path in the PRE-FORM trigger:

1. Click **Window** on the menu bar, and then click **Object Navigator** to open the Object Navigator.

2. Double-click the **Trigger icon** beside PRE-FORM trigger under the form module. The PRE-FORM trigger opens in the PL/SQL Editor. If necessary, maximize the PL/SQL Editor window.

3. Add the following command as the last command in the trigger. Compile the trigger, close the PL/SQL Editor, and save the form.

```
:GLOBAL.project_path := 'c:\oraclesolutions\chapter11\
tutorialsolutions\cwproject_done\';
```

Be sure to change the path specification as necessary to reflect the exact location of the folder that contains your tutorial solutions. Be sure to place the path specification on a single code line.

CREATING A SPLASH SCREEN

Recall that a splash screen is the first image that appears when you run an application. It introduces the application and usually identifies the system author(s) and copyright information. To implement the splash screen for the Clearwater Traders database application, two separate windows will exist in the main application form. When the user starts the program, the splash screen window shown in Figure 11-3 will appear for five seconds. Then the application focus will switch to the second window, and the main form screen will appear. To implement the splash screen, you will modify the PRE-FORM trigger to load the splash screen image and then set a timer. The splash screen window will appear first, and when the timer expires, a form-level trigger named WHEN-TIMER-EXPIRES will execute. This trigger sets the application's focus to an item in the main form screen, which hides the splash screen window and causes the main form screen to appear.

The first step in developing the splash screen is to create the window, canvas, and block for the splash screen.

To create the splash screen window, canvas, and block:

1. In the Object Navigator, select the **Windows node**, and then click the **Create button** to create a new window in the form. Change the name of the new window to **SPLASH_WINDOW**.

2. To create the new canvas, select the **Canvases node**, and then click . Change the name of the new canvas to **SPLASH_CANVAS**.

3. To create the new block, select the **Data Blocks node**, and then click . Select the **Build a new data block manually option button**, click **OK**, and then change the name of the new block to **SPLASH_BLOCK.**

4. Save the form.

Figure 11-3   Splash screen

To create the splash screen, you must first configure the splash screen window and create the splash screen canvas items. Then you must add the commands that load the splash image and create the timer to the main form's PRE-FORM trigger. Finally, you must create the trigger that fires when the timer expires. This trigger hides the splash window, and places the application focus on an item on the main window. The following sections address each of these steps.

Configuring the Splash Screen Window

In the Windows operating system, applications appear in a window that has a title bar and buttons that allow the user to minimize, maximize, and close the window. This window can run in a normal state, in which windows for multiple applications are visible on the screen display, or in a maximized state, in which the window for a single application fills the screen display. So far, the windows that you have made in Form Builder applications have been **document windows**, which appear inside the Forms Runtime MDI window. If the user resizes the Forms Runtime MDI window so that it is smaller than the document window inside it, the document window is clipped and scroll bars appear. Document windows are **non-modal** windows, which are windows that can appear side by side on the screen display and allow the user to multitask between different forms.

A window in a Form Builder application can also be a dialog window. **Dialog windows** are **modal windows**, which are application windows that demand the user's full and undivided attention. When a modal window opens, the user can interact only with the modal window until he or she closes the modal window. An LOV display is an example of a modal window.

You will configure the splash screen window as a dialog window. You will modify the window's Hide on Exit property so that the window no longer appears on the screen when the application focus switches to the main application window. Splash screen windows typically do not fill the entire screen display, so you will also change the window size to 320 by 200 pixels.

To configure the splash screen window:

1. Double-click the **Window icon** ▢ beside the SPLASH_WINDOW to open its Property Palette, and type **Clearwater Traders** as the Title property.

2. Scroll to the Window Style property, click the space next to the property title, and select **Dialog** from the list.

3. Change the Hide on Exit property value to **Yes**.

4. Change the Move Allowed, Resize Allowed, and Maximize Allowed property values to **No**. Change the Width property to **320**, and the Height property to **200**. Do not close the Property Palette.

Next, you will change the X and Y Position property values so that the window appears centered on the screen display. To calculate the X and Y positions to center the window, use the following formulas: X position = [(screen resolution width) - 320]/4; Y position = [(screen resolution height) - 200]/4. For example, if your screen resolution is 800 by 600, your X position would be (800 − 320)/4, or 120, and Y position would be (600 - 200)/4, or 100.

To center the splash window on the screen display:

1. In the SPLASH_WINDOW Property Palette, change the X Position property value to **120**, and the Y Position property value to **100**, or substitute the correct values for your screen resolution.

2. Close the SPLASH_WINDOW Property Palette, and save the form.

Since you have resized the SPLASH_WINDOW to 320 pixels wide by 200 pixels high, you must also resize the SPLASH_CANVAS to the same size. In addition, you must confirm that the canvas Window property is SPLASH_WINDOW, which specifies that the canvas appears in the SPLASH_WINDOW. You will make these changes next.

To configure the SPLASH_CANVAS:

1. Right-click the **Canvas icon** ▦ beside SPLASH_CANVAS, click **Property Palette**, and change the Width to **320**, and the Height to **200**.

2. Confirm that the Window property value is SPLASH_WINDOW. If it is not, select the Window property, open the list, and select **SPLASH_WINDOW**.

3. Close the Property Palette, and save the form.

Creating the Splash Screen Canvas Items

The canvas that appears in a splash screen window usually displays boilerplate objects, and can include text, objects such as shapes and lines, and graphic images. The splash screen in the Clearwater Traders database application will display a graphic image and a display item describing the application's copyright information.

Creating the Splash Screen Image

There are two approaches you can use to display graphic art on a form. The first approach is to create the image as a boilerplate image, as you did when you imported the Clearwater Traders and Northwoods University logos into forms in previous chapters. This approach stores the image inside the .fmb file when you design the form. However, this approach will not work for a splash screen. The splash screen canvas will display only boilerplate objects, and Forms Runtime will not display a canvas that only contains boilerplate objects and no block items. To display a graphic image on a splash screen, you must use a second approach, which is to create a dynamic image item on the splash screen canvas. The dynamic image item is a block item rather than a boilerplate item, so Forms Runtime will display the canvas. Then, you write a trigger that retrieves the image from the database or file system while the form is running, and loads the image into the image item.

Now you will create an image item on the SPLASH_CANVAS. Later, you will create a trigger to dynamically load the file image into the form image item. First, you will create the image item on the canvas and adjust its properties. You will change the image Sizing Style property value to Adjust, so that the item automatically adjusts the image size to fit correctly. You will also modify the image item size so that it is the same size as the canvas, and change its X and Y positions so that it is in the top-left corner of the canvas.

To create an image item for the SPLASH_CANVAS and modify its properties:

1. Double-click the **Canvas icon** ▣ beside SPLASH_CANVAS to view the canvas in the Layout Editor. If necessary, select SPLASH_BLOCK from the block list.

2. Select the **Image Item tool** ▣ on the tool palette, and draw an image item that is about the same size as the canvas and just inside the canvas borders.

3. Right-click the new image item, then click **Property Palette** to open its Property Palette. Change the Name property value to **SPLASH_IMAGE**.

4. To automatically adjust the graphic size so it is the same size as the image item on the canvas, click the space next to the Sizing Style property, and select **Adjust** from the list.

5. To modify the position for the SPLASH_IMAGE so that it is positioned in the upper-left corner of the canvas, change the X Position and Y Position properties to 0, if necessary.

6. To make the image item the same size as the canvas, change the Width property to **320**, and the Height property to **200**.

7. Close the Property Palette, and save the form.

Creating the Splash Screen Display Item

Recall from Chapter 8 that a display item displays text data that cannot be changed by the user. Now you will create a display item on the graphic image on the splash canvas. You will draw the display item on the canvas, and change its properties so that it displays the name of the system, the system developer, and the copyright date. You will modify the display item name, fill color, and Maximum Length property. The Maximum Length property specifies the maximum length of the text string that can appear in the display item. You will also specify that the display item text appears in an 8-point Arial Regular font.

To create the splash screen display item:

1. In the Layout Editor, select the **Display Item tool** [abc] and draw a display item in the lower-right corner of the canvas, as shown in Figure 11-3.

 Be sure to use the Display Item tool [abc]. If you use the Text Item tool [abc], an error will occur when you run the application.

2. Select the new display item, click the **Fill Color tool** [▨] on the tool palette, and change the display item's fill color to white.

3. Open the display item Property Palette, and change the following property values:

Property	Value
Name	**SPLASH_ITEM**
Maximum Length	**500**
Font Name	**Arial**
Font Size	**8**

4. Close the Property Palette, and save the form.

Displaying the Splash Screen

To display the splash screen, you will modify the form's PRE-FORM trigger to load the graphic image file into the splash canvas image item before the form appears, assign a text string to appear in the SPLASH_ITEM display item, and create a timer that determines how long the splash screen appears on the screen. Then, you will create a trigger that fires when the timer expires. This trigger will hide the splash screen window. In this section, you will modify the form's PRE-FORM trigger to load the graphic image file into the splash canvas image item, and assign the value to the text string that appears in the display item.

To load an image file into an image item, you use the READ_IMAGE_FILE built-in procedure, which has the following syntax:

```
READ_IMAGE_FILE('filename', 'file_type', 'item_name');
```

This procedure requires the following parameters:

- *filename* is the complete path and filename of the graphic art image file. If the filename is passed as a literal value, such as 'c:\myfilename.tif', the value must be enclosed in single quotation marks. The filename can also be passed as a variable value that references a character string that specifies the path and filename. In this application, the image filename is cwsplash.tif, and it will be stored in the folder specified by the :GLOBAL.project_path variable. Therefore, its value will be **:GLOBAL.project_path || 'cwsplash.tif'**.

- *file_type* is the type of image file being used, and is passed as a character string enclosed in single quotation marks. Legal values are the following file types: .BMP, .PCX, .PICT, .GIF, .CALS, .PCD, and .TIF. For the TIF file used in the previous example, the parameter would be passed as 'TIFF'.

 The different image file types reflect the graphics art applications used to create the files and how the images are compressed to make them take up less file space. Bitmap (.BMP) files are usually uncompressed, while .PCX, .GIF, and .TIF files use different compression methods. .PICT and .PCD files are made by specific graphics applications, and .CALS files are used to compress black and white images for fax transmissions. Most popular graphics applications support one of these types of files.

- *item_name* is the name of the form image item in which the file will appear, and is passed as a character string in single quotation marks using the format '*block_name.item_name*'. For the image item you just created, the parameter value would be **'splash_block.splash_image'**.

You also need to specify the text that appears in the display item on the splash canvas. In Chapter 8, you created display items to display calculated values on forms. On the splash canvas, you will display a text string in the display item. To specify the text that appears in a text item, you assign the text string to the display item, using the following syntax:

:block_name.display_item_name := 'display text';

Note that in the display item in Figure 11-3, the text appears on three separate lines. To programmatically insert hard returns into the text string, you need to use the CHR built-in function, which converts ASCII numerical codes to characters that can be inserted into a text string. In the ASCII numerical coding scheme, the value 13 represents a carriage return, which moves the insertion point to the beginning of the current line. The value 10 represents a line feed, which moves the insertion point to the line below the current line. If you concatenate the string value CHR(13) || CHR(10) within a text string, the function places the text that follows the codes at the beginning of the next line.

Now you will modify the PRE-FORM trigger to load the image into the image item on the splash canvas, and specify the text that appears in the display item. You will modify the splash display item text to display your name as the system developer and to show the current year for the copyright year.

To modify the PRE-FORM trigger:

1. Click **Window** on the menu bar, and then click **Object Navigator** to switch to the Object Navigator. If necessary, click **+** beside the Triggers node, and then double-click the **Trigger icon** beside PRE-FORM to open the PRE-FORM trigger in the PL/SQL Editor. Modify the PRE-FORM trigger using the code shown in Figure 11-4.

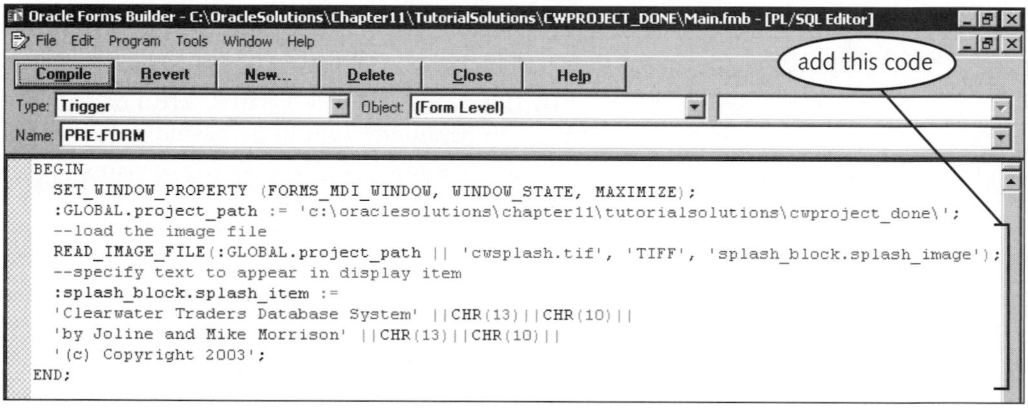

Figure 11-4   Code to configure the splash screen

2. Compile the trigger, and correct any syntax errors. Do not close the PL/SQL Editor.

Creating and Using Form Timers

To control how long the splash image appears and to hide the splash screen image, you will use a form timer. A **form timer** is a form construct that you create programmatically. You use it to control time-based events, such as displaying objects for a set time interval or creating animated objects. When you create a timer, you specify the time interval that will elapse before the timer expires. When the timer expires, a form-level trigger named WHEN-TIMER-EXPIRED fires. This trigger is not associated with any particular timer, so you cannot create and use multiple timers that expire at different times because all timers cause the same trigger to fire. The following sections discuss how to create, modify, and delete timers, and how to create a WHEN-TIMER-EXPIRED trigger.

Creating a Timer

To create a timer, you use the CREATE_TIMER function, which has the following syntax:

```
timer_id := CREATE_TIMER('timer_name', milliseconds,
iteration_specification);
```

This function requires the following parameters:

- *timer_id* is a previously declared variable of data type TIMER.
- *timer_name* can be any legal Oracle variable name, and is enclosed in single quotation marks. An example of a timer name is 'splash_timer'.
- *milliseconds* is a numeric value that specifies the time duration, in milliseconds, until the timer expires. When the timer expires, it calls a form-level trigger named WHEN-TIMER-EXPIRED, which you will create later in this lesson. To create a timer that expires in five seconds, you would specify the *milliseconds* value as 5000.
- *iteration_specification* specifies if the timer should be reset immediately after it expires. Valid values are **REPEAT**, meaning it should be reset immediately and start counting down again, and **NO_REPEAT**, meaning it should stay expired. You can use the REPEAT option to create animated graphics that appear repeatedly. Since the splash screen will appear only once, you will use the NO_REPEAT value.

Now you will modify the PRE-FORM trigger and create and set the timer. You will modify the splash display text to show your name as the system developer and to show the current year for the copyright year.

To modify the PRE-FORM trigger:

1. In the PL/SQL Editor, modify the PRE-FORM trigger using the code shown in Figure 11-5 to declare, create, and set the timer.

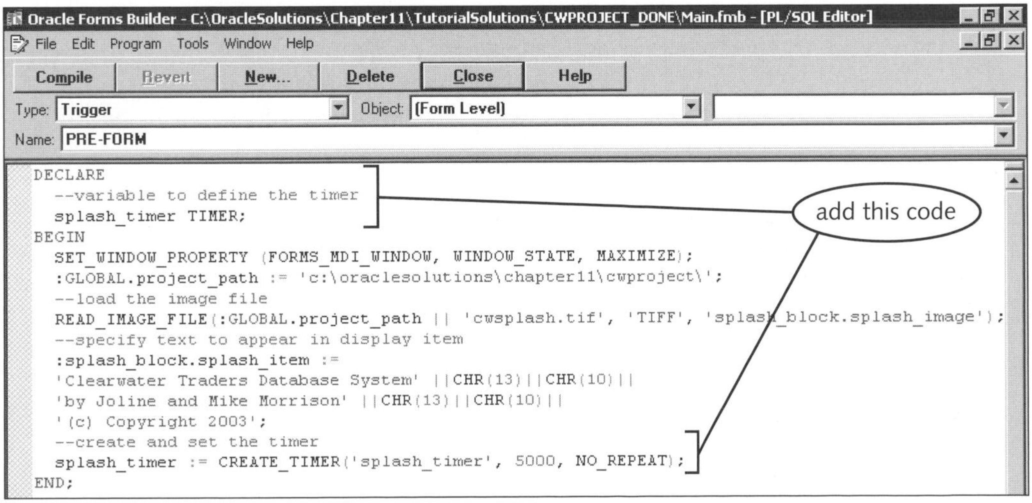

Figure 11-5   Code to declare, create, and set the timer

2. Compile the trigger, correct any syntax errors, and then close the PL/SQL Editor.

3. Save the form.

Modifying and Deleting Timers

Sometimes, you might want to change the properties of a timer while a form is running. For example, you could allow the user to reconfigure the timer in the Clearwater Traders database application so the splash screen appears for a longer duration or does not appear at all. To allow users to change timer properties while a form is running, you use the SET_TIMER built-in procedure, which has the following syntax:

```
SET_TIMER('timer_name',
milliseconds, iteration_specification);
```

The procedure parameters are the same as the parameters used in the CREATE_TIMER function.

Sometimes you need to delete an existing timer. For example, you might want to create a timer that displays a message a specific number of times and then stops. To delete an existing timer, you use the DELETE_TIMER built-in, which has the following syntax:

```
DELETE_TIMER('timer_name');
```

Creating the WHEN-TIMER-EXPIRED Trigger

When a form timer expires, the form-level WHEN-TIMER-EXPIRED trigger fires. If no WHEN-TIMER-EXPIRED trigger exists, nothing happens. If this trigger exists, then the code in the trigger executes. When the timer that you created in the PRE-FORM trigger expires, you want the application to display the main form window. To show a different window in a multiple-window application, you add commands to show the window using the SHOW_WINDOW command and then move the form insertion point to an item on that window using the GO_ITEM command. Now you will create the WHEN-TIMER-EXPIRED trigger, add commands to show the MAIN_WINDOW, and use the GO_ITEM procedure to switch the application focus to the SALES_BUTTON.

To create the WHEN-TIMER-EXPIRED trigger:

1. In the Object Navigator, select the **Triggers node**, and then click the **Create button** to create a new form-level trigger. The Main: Triggers dialog box opens.

2. Scroll down the list, select **WHEN-TIMER-EXPIRED**, and then click **OK**. The PL/SQL Editor opens.

3. Type the code shown in Figure 11-6 to create the trigger. Compile the trigger, correct any syntax errors, and close the PL/SQL Editor.

4. Save the form.

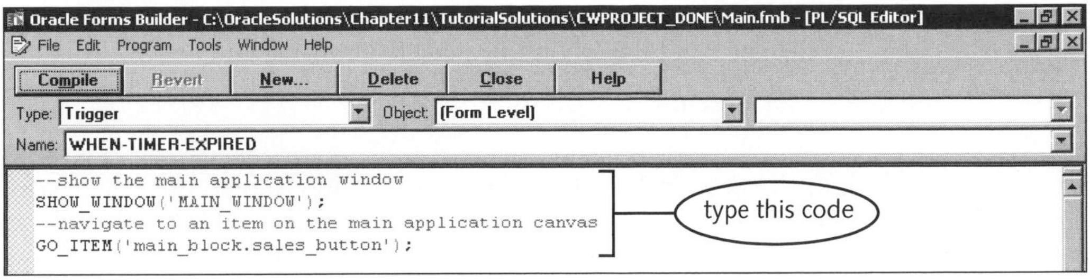

```
Oracle Forms Builder - C:\OracleSolutions\Chapter11\TutorialSolutions\CWPROJECT_DONE\Main.fmb - [PL/SQL Editor]
File  Edit  Program  Tools  Window  Help
[Compile]  [Revert]  [New...]  [Delete]  [Close]  [Help]
Type: Trigger                       ▼  Object: (Form Level)                ▼                                ▼
Name: WHEN-TIMER-EXPIRED                                                                                   ▼

--show the main application window
SHOW_WINDOW('MAIN_WINDOW');                           type this code
--navigate to an item on the main application canvas
GO_ITEM('main_block.sales_button');
```

Figure 11-6   WHEN-TIMER-EXPIRED trigger code

When a Form Builder application opens, Forms Runtime displays the window that contains the item that is first in the form's navigational sequence. Recall that Form Builder determines the form navigation sequence by the order in which the blocks and items are listed in the Object Navigator. Therefore, SPLASH_BLOCK must be the first block listed. SPLASH_ITEM and SPLASH_IMAGE must be in the SPLASH_BLOCK, and the other form items must be in the MAIN_BLOCK. Next you will double-check the navigational sequence to make sure that the SPLASH_BLOCK is the first block listed in the Object Navigator and confirm that all of the form objects are in the correct blocks.

To check the navigational sequence and form objects:

1. If necessary, open the Object Navigator window in Ownership View.

2. Confirm that the navigational sequence looks like the one shown in Figure 11-7 and that the form objects are in the correct blocks, as shown. If your Object Navigator window does not look like Figure 11-7, drag and drop the objects into the correct positions.

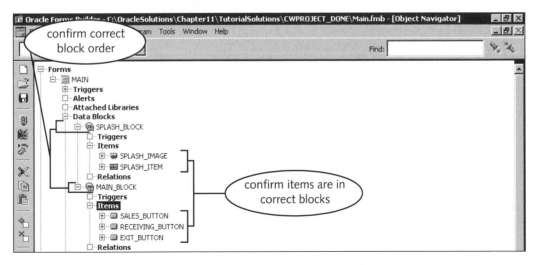

Figure 11-7   Form navigation sequence and object relationships

3. Save the form.

Now you will test the splash screen. You will run the form, confirm that the splash window appears for five seconds, and confirm that the main window then appears.

To test the splash screen:

1. Run the form. The splash screen should appear for five seconds, as shown in Figure 11-3, and then the main form window should open. The splash window should appear centered in the screen display, and the display item should appear as shown. If necessary, close Forms Runtime, and adjust the size and position of the display item.

If the error "FRM-30041: Position of item places it off of canvas" appears, then either your splash image or splash item is not entirely on the SPLASH_CANVAS surface. Open the SPLASH_CANVAS in the Layout Editor, and adjust the size and position of the image item or display item so it is on the canvas.

If the SPLASH_WINDOW opens, but no image appears, double-check to make sure that the cwsplash.tif image file is in the Chapter11\TutorialSolutions\ CWPROJECT_DONE folder, and that you correctly specified the global project path location in the PRE-FORM trigger. An error message will appear in the status line showing the location from which the application is trying to load the image file.

2. Close Forms Runtime

ENSURING A CONSISTENT APPEARANCE ACROSS FORM MODULES

A large database application can include hundreds of different form modules that might be created by many different form developers. It is important that all of the forms have a consistent "look and feel," both to make the forms appear as a polished and integrated application and to reduce user training time and frustration. For example, an application will frustrate users if in some forms, they must press the F9 key to open an LOV display, while in other forms, they must press F7 to open an LOV display. Similarly, the application will confuse users if some forms display an Exit button for returning to the main application menu, while other forms display a Return button. Two ways to standardize the appearance of multiple forms in an application are to use template forms and to use visual attribute groups. The following sections describe how to create template forms and visual attribute groups.

Template Forms

A **template form** is a generic form that includes standard form objects, such as graphics, toolbars, program units, and standard window configurations that will appear on every form in an application. You store the template form in a location that is accessible to all developers. When you create a new form in an application, you open and modify the template form, rather than creating a new form each time. Having a template form speeds up the form development process, because you and your fellow developers do not have to make the same modifications, such as resizing the window or importing the corporate logo, each time you create a new form. Having a template form also ensures a consistent form appearance, since all developers build from the same template.

 You already learned how to create a report template in Chapter 9, which creates a standard appearance for application reports.

Now you will create a template form for the Clearwater Traders database. The specifications for the Clearwater template form are shown on Figure 11-8.

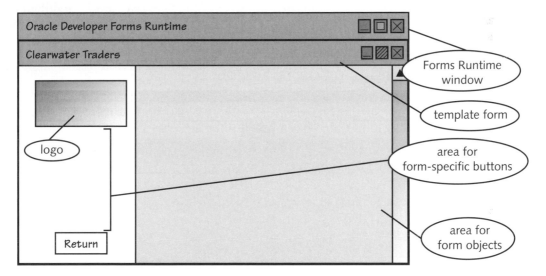

Figure 11-8 Clearwater template form specifications

The background color of the left one-fourth of the canvas will be a lighter color than the rest of the canvas. The Clearwater Traders logo will be in the top-left corner, and command buttons for the specific application will be placed under the logo. Under the form-specific buttons, the template will have a Return button that will exit the current form and return the user to the main form. The form window and canvas are configured according to the design specifications outlined in earlier chapters: The window fills the outer Forms Runtime window and can be minimized, but cannot be maximized or resized. The window title is "Clearwater Traders," and the form will automatically be maximized when it opens. The area on the right side of the form will display form objects, such as text items and command buttons, that are specific for each individual form. Now you will create a new form that will be the Clearwater template form.

To create the template form:

1. In the Object Navigator, click ⊟ beside MAIN to collapse the MAIN form objects.

2. Select the top-level **Forms node**, and then click the **Create button** 🔳 to create a new form. Change the form module name to **CLEARWATER_TEMPLATE**.

3. To create the trigger to automatically maximize the form window when the form runs, select the **Triggers node** under the form module, right-click, point to **SmartTriggers**, and click **PRE-FORM**. Type the following code in the PL/SQL Editor, compile the trigger, correct any syntax errors, and then close the PL/SQL Editor.

   ```
   SET_WINDOW_PROPERTY (FORMS_MDI_WINDOW, WINDOW_STATE,
   MAXIMIZE);
   ```

4. To configure the template canvas, select the **Canvases node**, and then click 🔳 to create a new canvas. Change the canvas name to **TEMPLATE_CANVAS**. Then right-click the **Canvas icon** 🖼, and click **Property Palette**. Under the Physical property node, change the Width to **580**, change the Height to **375**, and then close the Property Palette.

5. To configure the template window, click ➕ beside the **Windows node**, and change the form window name to **TEMPLATE_WINDOW**. Double-click the **Window icon** ⬜ to open the Property Palette, change the following property values, and then close the Property Palette.

Title	**Clearwater Traders**
Resize Allowed	No
Maximized Allowed	No
Width	580
Height	375

6. Save the form as **CLEARWATER_TEMPLATE.fmb** in your Chapter11\TutorialSolutions\CWPROJECT_DONE folder.

Next, you will add the template canvas objects shown in Figure 11-8. You will open the form canvas in the Layout Editor, and create a boilerplate rectangle that will cover the left one-fourth of the canvas. This rectangle will be a lighter shade than the default canvas color. Then, you will import the Clearwater Traders logo and create the Return button. You will create a trigger for the Return button that will exit the form.

To add the template canvas objects:

1. In the Object Navigator, double-click the **Canvas icon** 🖼 beside TEMPLATE_CANVAS to open the form canvas in the Layout Editor.

2. To create the boilerplate rectangle, select the **Rectangle tool** on the tool palette, and draw the rectangle as shown in Figure 11-9. The rectangle should start at the top-left corner of the canvas, and should extend the entire length of the canvas. The rectangle's bottom edge should be even with the bottom edge of the canvas.

 Do not place the rectangle so it extends beyond the canvas surface, or an error will occur when you run the form.

Figure 11-9   Creating the template form objects

3. Select the rectangle if necessary, select the **Fill Color tool** on the tool palette, and change the rectangle fill color to a lighter gray shade than the form background color. The filled rectangle should appear slightly lighter than the rest of the canvas.

 Do not use bright or saturated colors on screen displays, because they can cause eyestrain.

4. Make sure that the rectangle is still selected, select the **Line Color tool** 🖫 on the tool palette, and then click **No Line**.

5. To import the Clearwater Traders logo, click **File** on the menu bar, point to **Import**, and then click **Image**. Click **Browse**, navigate to your Chapter11\TutorialSolutions\CWPROJECT_DONE folder, select **clearlogo.tif**, click **Open**, and then click **OK**. The logo appears on the canvas. Resize and reposition the logo as shown in Figure 11-9.

6. To create the Return button, select the **Button tool** 🔲 on the tool palette, and draw the button as shown in Figure 11-9. Double-click the button to open its Property Palette, change the button properties as follows, and then close the Property Palette.

Name	**RETURN_BUTTON**
Label	**Return**
Width	**90**
Height	**16**

7. To create the button trigger, select the button, right-click, point to **SmartTriggers**, and then click **WHEN-BUTTON-PRESSED**. Type the following trigger code, compile the trigger, debug it if necessary, and then close the PL/SQL Editor.

    EXIT_FORM;

8. Click **Window** on the menu bar, and then click **Object Navigator** to open the Object Navigator. Note that when you created the Return button, Form Builder automatically created a new control block that contains the button. Change the block name to **TEMPLATE_BLOCK**.

9. Save the form.

Visual Attribute Groups

The form template ensures a uniform overall appearance for application forms, but developers can still specify different properties for individual block items. For example, one developer might use a 10-point Arial font for text items or command button labels, while another might use an 8-point font. To ensure a standard appearance for block items, you can create a **visual attribute group**, which is a form object that defines object properties, such as text item colors, font sizes, and font styles. Visual attribute groups can be assigned to form windows, canvases, and items. You can create a visual attribute group in the form template, and then apply the visual attribute group to the associated form items.

Now you will create a visual attribute group in the Clearwater template form that specifies the font and color attributes for form text items in Clearwater Traders forms. Then, you will create a data block form based on the Clearwater template, apply the visual attribute group to the form text items, and examine how the visual attribute group defines the appearance of the form text items. First, you will create the custom visual attribute group that will specify the standard properties for form text items.

To create the visual attribute group for form text items:

1. In the Object Navigator window, select the **Visual Attributes node**, and then click the **Create button** ⬚. A new visual attribute object appears.

2. Double-click the **Visual Attribute icon** ⬚ beside the new visual attribute object to open its Property Palette.

In the Visual Attribute Property Palette under the General property node, the *Visual Attribute Type* property defines the type of attributes to which the defined attribute properties apply. The possible values are *Common*, which means that the defined properties apply to all attributes in the object; *Prompt*, which means that the defined properties apply only to prompts; and *Title*, which means that the defined properties apply only to titles. For example, for a visual attribute that corresponds to a text item, you could create a Common type visual attribute group to define visual attribute properties for both prompt and data values. You could also create a Prompt type visual attribute group to define only the prompt properties.

The Visual Attributes property node includes a Character Mode Logical Attribute property and a Black and White property. These properties specify the appearance of visual attributes that you define for forms that appear in character-only environments, such as terminals that connect to mainframe computers. The properties in the Color and Font property node specify the font foreground (text) color and background color, and the font size and style. Now you will specify the properties for the visual attribute group that will define text items in Clearwater Traders forms.

To specify the properties of the text item visual attribute group:

1. In the visual attribute group object Property Palette, change the Name value to **TEXT_ITEM_VISUAL_ATTRIBUTES**.

2. Make sure that the Visual Attribute Type property value is Common.

3. Select the **Foreground Color** property, click [...], and select any other color. Click [...] again, and then select the black square that is the first square in the first row. Make sure that the Foreground Color property value is "black." If it is not, click [...] again, and select the black square again.

4. Select the **Background Color** property, click [...], and select any other color. Click [...] again, and then select the white square that is the second square in the first row. Confirm that the Background Color property value is "white."

5. Select the **Font Name** property, click **More**, select Arial, and click **OK**. Select the **Font Weight** property and type **8**.

6. Close the Property Palette, and save the form.

11

Now you will create a new form, and use the Clearwater template as the basis for the new form. To do this, you will save the template file using a different filename. Since you will use this form in the Clearwater Traders integrated database application, you will save the form in your Chapter11\TutorialSolutions\CWPROJECT_DONE folder. You will use the Data Block and Layout Wizards to create a new data block and layout based on the INVENTORY table in the Clearwater Traders database. Then, you will apply the text item visual attribute group to the form text items.

To create a new form using the template:

1. In the Object Navigator, click **File** on the menu bar, click **Save As**, and save the form as **11AInventory.fmb** in your Chapter11\TutorialSolutions\ CWPROJECT_DONE folder. Change the form module name to **INVENTORY_FORM**.

2. To create the new data block, select the **Data Blocks node**, click the **Create button** 🔳, make sure the **Use the Data Block Wizard option button** is selected, and then click **OK**. When the Block Wizard Welcome page appears, click **Next**.

3. When the Type page appears, make sure that the Table or View option button is selected, and then click **Next**. On the Source page, click **Browse**, connect to the database if necessary, select the **INVENTORY** database table, and click **OK**. Click the **Select All button** 🔲 to select all of the table fields for the data block, do not check the Enforce data integrity check box, and then click **Next**.

4. When the Master-Detail page appears, click **Next**, because you do not want to create a master-detail relationship. When the Finish page appears, make sure that the Create the data block, then call the Layout Wizard option button is selected, and then click **Finish**.

5. When the Layout Wizard Welcome page appears, click **Next**. Accept the default values on the Canvas page, and then click **Next**.

6. On the Data Block page, click 🔲 to select all of the data block fields to appear in the layout, and then click **Next**. On the Items page, accept the default prompt values, and then click **Next**.

7. On the Style page, make sure that the Form option button is selected, and then click **Next**. On the Rows page, type **Inventory** for the frame title, leave the Records Displayed value as 1, click **Next**, and then click **Finish**.

8. In the Layout Editor, the new data block objects are not visible because they appear under the existing form template objects. Scroll to the bottom of the window, select the **Inventory frame**, and drag it to the top of the canvas. Resize the frame, and format the text item labels so your form looks like Figure 11-10.

To change the text item prompts so they are on a single line, select the prompt, click the prompt again to make the text available for editing, and delete the hard returns that wrap the text to a new line. To make the text items appear in a single column, make the frame narrower and longer.

9. Since there will be no other command buttons on the form, move the Return button toward the top of the canvas, so it is positioned directly under the logo, as shown in Figure 11-10.

10. Save the form.

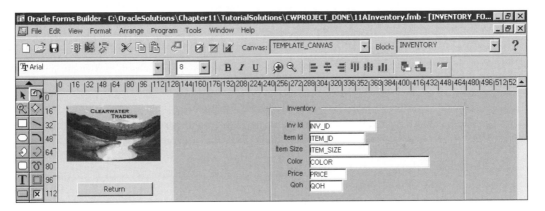

Figure 11-10 Inventory form created using form template

Next, you will apply the visual attribute group to the form text items. You will select all of the text items as an object group, and open the intersection Property Palette. Then, you will change the Visual Attribute Group property value to the name of the custom visual attribute group that you created earlier.

To apply the custom visual attribute to the form text items:

1. In the Layout Editor, select the **INV_ID** text item, press and hold the **Shift key**, and then select **ITEM_ID**, **ITEM_SIZE**, **COLOR**, **PRICE**, and **QOH** so all of the text items are selected as an object group. You do not need to select the item prompts, because their properties are defined with their associated text items.

2. Click **Tools** on the menu bar, and then click **Property Palette** to open the intersection Property Palette for the text item object group.

3. Scroll down to the Visual Attributes property node, select the **Visual Attribute Group** property, open the list, and select **TEXT_ITEM_VISUAL_ATTRIBUTES**. Note that you can assign various visual attribute group names to specify the visual attributes of the overall item, as well as the item prompts and how the item appears in a character-only environment.

4. Close the Property Palette, and save the form. Note that the form text items now appear as black text on a white background, in an 8-point Arial Regular font, as specified in the custom visual attribute group.

5. Run the form, click the **Enter Query button** 📷, do not enter a search condition, and then click the **Execute Query button** 📷. Step through the table records to view the visual attributes, and then close Forms Runtime.

6. Click **Window** on the menu bar, and then click **Object Navigator** to open the Object Navigator. Select the **INVENTORY_FORM** module in the Object Navigator, click **File** on the menu bar, and then click **Close** to close the form.

OPENING, CLOSING, AND NAVIGATING TO FORMS IN MULTIPLE-FORM APPLICATIONS

To complete the main form in the Clearwater Traders database application, you need to write the triggers for the Sales and Receiving buttons on the application switchboard. Form Builder provides a number of built-in procedures that enable you to call one form from another, change the application focus to a specific form, and close application forms. The following sections describe these procedures. After you are familiar with these procedures, you will create triggers that contain commands to open the sales and receiving form modules from the main application form and then close the form.

Opening a Form From Another Form

Table 11-2 summarizes the built-in procedures that allow you to open a form from another form. The form that is calling the second form is referred to as the parent form, and the form that is called is referred to as the child form.

Procedure Name	Description
CALL_FORM	Opens a child form and immediately switches the application focus to the child form; has options for hiding the parent form, displaying the menu from the parent form, and passing a parameter list from the parent form
OPEN_FORM	Opens a child form, with the option of not immediately changing the application focus to the child form; has an option for creating a new database session for the child form or using the parent form's database session
NEW_FORM	Opens a child form and exits the parent form

Table 11-2   Built-in procedures to open a child form from a parent form

The CALL_FORM procedure has the following syntax:

```
CALL_FORM('form_specification', display, switch_menu,
query_mode, parameter_list_id);
```

This procedure has the following parameters. All of the parameters are optional except for *form_specification*.

- *form_specification* is either the name of the child form, as specified in the Object Navigator, enclosed in single quotes, or the full path and filename, including the drive letter, to the child form's .fmx file, enclosed in single quotes. To specify the SALES form, you would use the expression `'SALES'`. To specify the sales.fmx file that is stored in the file location specified in the global project path, you would use the expression `:GLOBAL.project_path || 'sales.fmx'`. The .fmx file extension is optional.

- *display* specifies whether the parent form is hidden or not hidden by the child form. Valid values are HIDE and NO_HIDE. You would specify the *display* parameter value as NO_HIDE when you want the two forms to appear side by side on the screen display. The default value is HIDE.

- *switch_menu* specifies whether the child form's standard pull-down menus are inherited from the parent form, or if the child form will have different menus. Valid values are NO_REPLACE (pull-down menus are inherited from the calling form) and DO_REPLACE (different pull-down menus are used). The default value is NO_REPLACE. You will learn how to replace pull-down menus in Lesson C of this chapter. For now, the calling form will use the standard Forms Runtime menu selections, and you will use the NO_REPLACE value.

- *query_mode* specifies whether the child form will run in normal mode, in which the user can insert, update, or delete values, or in query mode, in which the user can only view data. Valid values are NO_QUERY_MODE, which runs the child form in normal mode, and QUERY_MODE, which runs the child form in query mode. The default value is NO_QUERY_MODE.

- *parameter_list_id* specifies the identifier for an optional parameter list that is passed from the parent form to the child form.

You can also use the OPEN_FORM procedure to open a new form. This procedure is preferred by many developers, because it allows the user to multitask between parent and child forms in the application. The OPEN_FORM procedure has the following syntax:

```
OPEN_FORM('form_specification', activate_mode,
session_mode, parameter_list_id);
```

This procedure has the following parameters. Like the RUN_FORM procedure, all of the parameters are optional except for *form_specification*.

- *form_specification* is the same parameter used in the CALL_FORM procedure.

- *activate_mode* specifies whether the application focus switches to the child form or is retained by the parent form. Values can be ACTIVATE, which switches the application focus to the child form, and NO_ACTIVATE, which retains the application focus in the parent form. The default value is ACTIVATE.

11

- *session_mode* specifies whether the child form uses the same database session as the parent form, or whether the child form starts a new database session. Values can be NO_SESSION, which specifies that the child form shares the parent form's session, and SESSION, which specifies that the child form starts a new session. The default value is NO_SESSION. When a child form is opened in the same session as the parent form, all uncommitted values are committed in both forms when a COMMIT command is executed. A benefit of opening a child form in a separate session is that whenever a COMMIT command is issued, transactions are only committed in the form in which the COMMIT command executes. This allows the developer to control how the forms commit transactions.

- *parameter_list_id* specifies the identifier of an optional parameter list that is passed from the parent form to the child form.

The final built-in procedure for opening new forms is NEW_FORM. The NEW_FORM procedure opens a child form, and immediately exits the parent form. This procedure has the following syntax:

```
NEW_FORM ('form_specification', rollback_mode, query_mode,
parameter_list_id);
```

The *form_specification*, *query_mode*, and *parameter_list_id* parameters are the same as the ones used in the CALL_FORM built-in. The *rollback_mode* parameter specifies whether uncommitted records in the parent form are automatically committed or rolled back. Values can be TO_SAVEPOINT, which specifies that all uncommitted transactions in the parent form are rolled back to the last savepoint; NO_ROLLBACK, which specifies that any uncommitted transactions in the parent form are not rolled back, and any locks obtained by the parent form are not released; and FULL_ROLLBACK, which specifies that all uncommitted transactions that were made anytime in the parent form runtime session are rolled back. The default value is TO_SAVEPOINT.

Navigating Among Forms

Table 11-3 summarizes the procedures that allow you to navigate among current open forms.

Procedure Name	Description
GO_FORM	Moves the application focus to the specified form
NEXT_FORM	Moves the application focus to the next form
PREVIOUS_FORM	Moves the application focus to the previous form

Table 11-3   Procedures to navigate among open forms

The GO_FORM procedure places the application focus on the specified form. This procedure has the following syntax:

```
GO_FORM ('form_identifier');
```

When different forms are open in a multiple-form application, each form is given a unique identifier, called the form_ID, which is a sequence number based on the order in which the form was opened. The *form_identifier* parameter can be either the form_ID or the name of the form, as specified in the Object Navigator. If the form name is used, it must be enclosed in single quotation marks. If the specified form is not currently open, an error occurs. To retrieve the current form_ID value, you use the FIND_FORM built-in, which has the syntax `FIND_FORM('form_name');`. The *form_name* parameter is the form module name as it appears in the Object Navigator.

When multiple forms are opened, the assigned form_ID values are sequential, based on the order in which forms are opened. The first form has the lowest form_ID value, the next form that opens has the next form_ID value, and so forth. The NEXT_FORM procedure navigates to the form that has the next form_ID value, and the PREVIOUS_FORM procedure navigates to the form that has the previous form_ID value.

Closing Forms

Table 11-4 summarizes the procedures that allow you to close forms.

Procedure Name	Description
CLOSE_FORM	Closes the specified form, which might not be the current form
EXIT_FORM	Closes the current form, and provides options for committing or rolling back the uncommitted data

Table 11-4   Procedures to close forms

The CLOSE_FORM procedure closes a specified form, and has the following syntax:

```
CLOSE_FORM ('form_identifier');
```

The *form_identifier* parameter is the form_ID or name of the form module, as specified in the Object Navigator. If the form name is used, it must be enclosed in single quotation marks.

The EXIT_FORM closes the current form, and has the following syntax:

```
EXIT_FORM (commit_mode, rollback_mode);
```

This procedure has the following optional parameters:

- *commit_mode*, which specifies how uncommitted form data is handled. Valid values are ASK_COMMIT, which causes the form to ask the user to save uncommitted changes; DO_COMMIT, which automatically commits unsaved data; and NO_COMMIT, which automatically discards uncommitted changes. The default value is ASK_COMMIT.

11

- *rollback_mode*, which specifies if uncommitted records in the parent form are automatically committed or rolled back. Values can be TO_SAVEPOINT, which specifies that all uncommitted transactions in the parent form are rolled back to the last savepoint; NO_ROLLBACK, which specifies that any uncommitted transactions in the parent form are not rolled back, and any locks obtained by the parent form are not released; and FULL_ROLLBACK, which specifies that all uncommitted transactions that were made anytime in the parent form runtime session are rolled back. The default value is TO_SAVEPOINT.

Creating Triggers for the Main Form Switchboard Buttons

Now, you will create the triggers for the buttons on the application switchboard. Since it is a good practice for every application to have a single entry and exit point, you will use the CALL_FORM procedure to call the forms named SALES and RECEIVING, while keeping the main form open. You will accept the default values for the optional parameters in the CALL_FORM command. You will also create a trigger for the Exit button, to exit the application.

To create the button triggers:

1. In the Object Navigator, click ➕ beside MAIN to expand the main form objects. If necessary, click ➕ beside the Canvases node, and then double-click the **Canvas icon** 🖼 beside MAIN_CANVAS to open the canvas in the Layout Editor.

2. Select the **Sales button**, right-click, point to **SmartTriggers**, and then click **WHEN-BUTTON-PRESSED** to create a new button trigger. Type the following code in the PL/SQL Editor: **CALL_FORM(:GLOBAL.project_path || 'sales');**.

3. Compile the trigger, correct any syntax errors, and then close the PL/SQL Editor.

4. Select the **Receiving button**, right-click, point to **SmartTriggers**, and then click **WHEN-BUTTON-PRESSED** to create a new button trigger. Type the following code in the PL/SQL Editor: **CALL_FORM(:GLOBAL.project_path || 'receiving');**.

5. Compile the trigger, correct any syntax errors, and then close the PL/SQL Editor.

6. Select the **Exit button**, right-click, point to **SmartTriggers**, and then click **WHEN-BUTTON-PRESSED** to create a new button trigger. Type the following code in the PL/SQL Editor: **EXIT_FORM;**.

7. Compile the trigger, correct any syntax errors, and then close the PL/SQL Editor.

8. Save the form.

Currently, your project folder contains only the Form Builder design (.fmb) files for the sales and receiving form. You will need to compile these files to generate the executable (.fmx) files that are called by the switchboard button triggers. You will do this next.

To create the sales and receiving form executable files:

1. In Form Builder, click the **Open button** on the toolbar, navigate to your Chapter11\TutorialSolutions\CWPROJECT_DONE folder if necessary, select **Receiving.fmb**, and then click **Open**. The RECEIVING form opens in the Object Navigator.

2. Make sure that the RECEIVING form module is selected, click **File** on the menu bar, point to **Administration**, and then click **Compile File**. After a few moments, the "Module built successfully" message appears on the status line.

3. Make sure that the RECEIVING form module is selected, click **File** on the menu bar, and click **Close** to close the receiving form. Click **Yes** if you are asked if you want to save your changes.

4. Repeat Steps 1, 2, and 3 for the Sales.fmb file to open, compile, and close the sales form.

Now you will run the main form and confirm that the switchboard button triggers work correctly. When you click the Sales button, the sales form (sales.fmx) should open. When you click the Receiving button, the receiving form (receiving.fmx) form should open. When you click the Exit button, the application should close.

To test the switchboard buttons:

1. Run the form. When the main form appears, click **Sales**. The sales form (sales.fmx) should open.

> If the sales form does not open, check the PRE-FORM trigger to be sure that you correctly entered the path for the global path variable to the location where your project files are stored. Also make sure you entered the path to the .fmx file correctly in the Sales button trigger, and confirm that the sales.fmx file is in the project folder.

2. Click **Return** to exit the sales form and return to the main form application.

3. Click **Receiving** to open the receiving form (receiving.fmx). The Merchandise Receiving form should appear. Click **Return** to exit the receiving form and return to the main form application.

4. In the main form application, click **Exit** to close the application.

WHEN-WINDOW-CLOSED Triggers

Currently, when the user clicks the Close button ☒ on the title bar of any of the inner application form windows (rather than on the Forms Runtime window title bar), nothing happens. To allow users to close individual forms by clicking ☒ on the form window, you need to create a WHEN-WINDOW-CLOSED trigger. This is a form-level trigger that fires whenever the user clicks ☒ on any form window. If you do not want users to be able to close an application by clicking ☒, then you should set the window's Close Allowed property to No on the window's Property Palette, which disables ☒.

Now, you will open the sales form (sales.fmb) in Form Builder, and create a WHEN-WINDOW-CLOSED trigger for the form window. This trigger will contain the code `EXIT_FORM;`, so that when the user clicks ☒, the form closes. Then, you will recompile the sales form to create a new .fmx file. Finally, you will run the main form application, call the sales form, and close the sales form by clicking ☒.

To create the WHEN-WINDOW-CLOSED trigger:

1. Click the **Open button** 🗁, navigate to your CWPROJECT_DONE folder, select **Sales.fmb**, and click **Open**. The SALES form appears in the Object Navigator.

2. Select the **Triggers node** under the SALES form module, and click the **Create button** 🔩 to create a new form trigger. Scroll down the trigger list, select **WHEN-WINDOW-CLOSED**, and click **OK**. The PL/SQL Editor opens.

3. Type **EXIT_FORM;** in the source code pane, compile the trigger, correct any syntax errors, and then close the PL/SQL Editor window.

4. To compile the sales form, confirm that the SALES form module is selected in the Object Navigator, click **File** on the menu bar, point to **Administration**, and then click **Compile File**.

5. To test the modified sales form, run the main application by selecting **MAIN** in the Object Navigator, and then clicking the **Run Form Client/Server button** 📇. The main application form opens. Click **Sales** to open the sales form.

6. When the sales form opens, click the **Close button** ☒ on the inner form window. You might need to scroll up to the top of outer window so the title bar of the inner window is visible. The sales form closes, and the main form application appears again.

7. Click **Exit** to close the main form application.

8. Close the forms in Form Builder, and close Form Builder and all other open Oracle applications.

SUMMARY

- Developing a database application requires the following phases: design, module development, module integration, testing, and deployment.

- Unit testing involves testing individual modules, and system testing evaluates the integration of the modules.

- When you create a complex database application, it is a good practice to separate the system into individual form files that can be developed, tested, and debugged independently, and then integrated into the overall application.

- To integrate individual Oracle form, report, and chart modules into a single application, you create a main form module from which you access all of the individual application components. This main form module has a switchboard to access the most commonly used components, and menus for accessing all system components.

- When you create an integrated Oracle database application, you should place all of the form, report, and chart files in the same project folder.

- Stubs are programs or messages that handle application access to other programs that have not yet been implemented.

- A global variable is a form variable that you can create in one form, and can then reference in any other open form. A global variable remains in memory and is available to other forms even when the form that created the global variable closes.

- When you create an integrated database application, it is a good idea to create a global path variable to represent the folder path where all of the individual form, report, and chart files are stored. Then, if you change the location of the application files, you only have to modify the path specification in the global path variable.

- A splash screen is the first image that appears when you run an application. It introduces the application and usually identifies the system author(s) and copyright information.

- To implement the splash screen, you create two separate form windows. The splash screen appears in a dialog window, which is a modal window. The main application window is a document window, which is a non-modal window.

- A timer is a form construct that is created using program code. A timer waits a set period of time before it expires. When a timer expires, the WHEN-TIMER-EXPIRED trigger fires. If no WHEN-TIMER-EXPIRED trigger exists, nothing happens. If the trigger does exist, then the trigger code executes.

- To create a splash screen, you load the splash image and display item text and set a timer in the form's PRE-FORM trigger. Then, you create a WHEN-TIMER-EXPIRED trigger that displays the main application window after a preset time interval.

11

❐ When a Form Builder application runs, the Forms Runtime application displays the window that contains the item that is first in the form's navigation sequence.

❐ A template form contains standard form objects, such as graphics, toolbars, program units, and standard window configurations, that will appear on every form in an application. After you create a template form, you use it as the starting point when you create all other application forms.

❐ A visual attribute group is a form object that defines object properties, such as text item colors, font sizes, and font styles. Visual attribute groups can be assigned to form windows, canvases, and items.

❐ You can use the CALL_FORM procedure to open a child form and switch the application focus to the child form. You can use the OPEN_FORM procedure to open a child form and not immediately change the application focus to the child form. You can use the NEW_FORM procedure to open a child form and immediately exit the parent form.

❐ You can use the GO_FORM procedure to move the application focus to a specific form. You can use the NEXT_FORM procedure to move the application focus to the next form in the form sequence and the PREVIOUS_FORM procedure to move the focus to the previous form in the form sequence.

❐ You can use the CLOSE_FORM procedure to close a specific form and the EXIT_FORM procedure to close the current form.

❐ To allow users to close a form by clicking the Close button ✖ on the form window, you need to create a form-level WHEN-WINDOW-CLOSED trigger that fires whenever the user clicks ✖ on any form window.

REVIEW QUESTIONS

1. When you make a complex database application, why should you create multiple-form modules rather than one large form module with multiple canvases?

2. In an integrated database application, which applications should be accessed using the switchboard, and which applications should be accessed using pull-down menus?

3. Why should you create a global project path variable in an integrated database application?

4. What is the difference between the dialog and document window styles?

5. What is a visual attribute group?

6. You are creating an integrated Oracle database application, and have stored all of the form, report, and graphics files in a folder named \Oracle\MyProject that is on the D: drive of your workstation.

   a. Write the assignment statement for the `:GLOBAL.project_path` variable.

   b. Write the command that you will use to open a form file named Inventory.fmx that is stored in the D:\Oracle\MyProject folder, using the global project path variable. Specify the parameters so that the calling form will be hidden and the new form's menus will be replaced.

   c. You decide to move all of the project files to the F:\DatabaseProjects\Project1 folder on your network server. What code do you need to change in your integrated database application?

7. Which window opens first when you start a form module that has multiple windows?

8. Write the command to open a child form named Customer.fmx. You do not want to immediately activate the child form, and you want to open a new database session for the child form.

9. Write the command to close a form named Customer.fmx. This form does not currently have the application focus.

10. What event causes the WHEN-WINDOW-CLOSED trigger to fire?

PROBLEM-SOLVING CASES

The cases refer to the Northwoods University (see Figure 1-12) and Software Experts (see Figure 1-13) sample databases. Place all solution files in the specified folder in your Chapter11\CaseSolutions folder.

1. In this case, you will create a template form for the Northwoods University database. Then, you will use the template form to create a form that displays data from the STUDENT table.

   a. Create a folder named NWPROJECT_DONE in your Chapter11\CaseSolutions folder, and copy all of the files from the Chapter11\NWPROJECT folder on your Data Disk to this folder.

   b. Create a form named NORTHWOODS_TEMPLATE, and save the form in a file named NORTHWOODS_TEMPLATE.fmb in the new folder.

   c. Create a PRE-FORM trigger to automatically maximize the form window.

   d. Create a window named TEMPLATE_WINDOW, a canvas named TEMPLATE_CANVAS, and a block named TEMPLATE_BLOCK. Configure the window and canvas properties using the form appearance guidelines described in Chapter 6. Change the window title to "Northwoods University."

e. Create a boilerplate rectangle on the right edge of the canvas, as shown in Figure 11–11. Fill the rectangle with a pale yellow color.

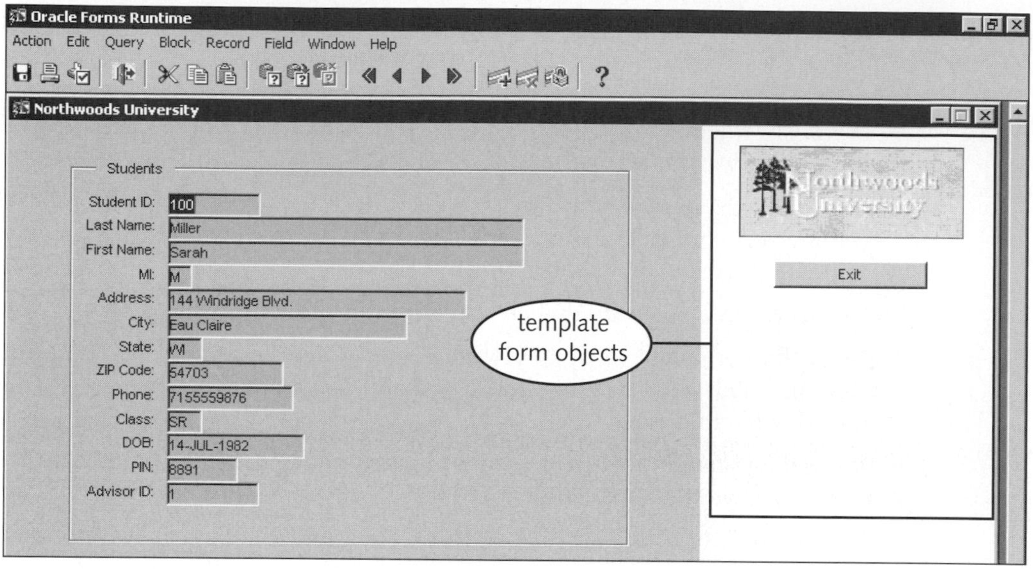

Figure 11-11

f. Import the Northwoods University logo (nwlogo.jpg) from your NWPROJECT_DONE folder, and place it on the template canvas as shown.

g. Create the Exit button shown on the canvas, and create a WHEN-BUTTON-PRESSED trigger that exits the current form.

h. Create a form-level WHEN-WINDOW-CLOSED trigger so the form is exited when the user clicks the Close button ☒ on the form window.

i. Create a visual attribute group named TEXT_ATTRIBUTES that formats text items using an 8-point Arial Demilight font.

j. Save the template form. Then create a new form named STUDENT_FORM that is based on the template form. Save the new form as Student.fmb in your NWPROJECT_DONE folder. Create a data block and layout based on the STUDENT table, as shown in Figure 11–11. Apply the TEXT_ATTRIBUTES visual attribute group to all form text items.

2. In this case, you will create a template form for the Software Experts database. Then, you will use the template form to create the consultant form shown in Figure 11–12.

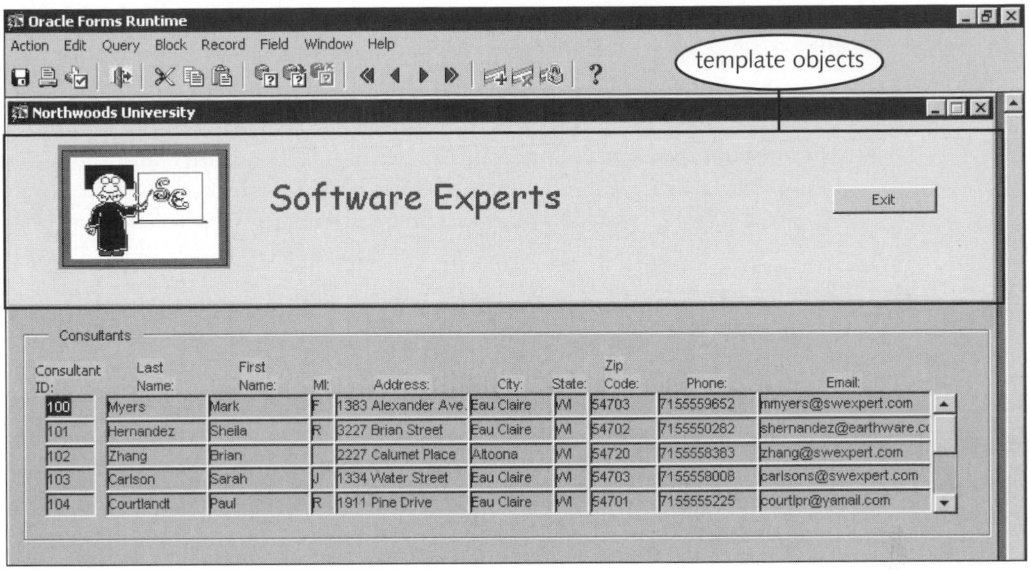

Figure 11-12

a. Create a folder named SWEPROJECT_DONE in your Chapter11\CaseSolutions folder, and copy all of the files from the Chapter11\SWEPROJECT folder into the new folder.

b. Create a form named SOFTWARE_EXPERTS_TEMPLATE, and save the template form in a file named SOFTWARE_EXPERTS_TEMPLATE.fmb in the new folder.

c. Create a PRE-FORM trigger to automatically maximize the form window.

d. Create a window named TEMPLATE_WINDOW, a canvas named TEMPLATE_CANVAS, and a block named TEMPLATE_BLOCK. Configure the window and canvas properties using the form appearance guidelines described in Chapter 6. Change the window title to "Software Experts."

e. Create a boilerplate rectangle on the top edge of the canvas, as shown in Figure 11-12. Fill the rectangle with a pale blue color.

f. Import the Software Experts logo (swelogo.tif) from the SWEPROJECT_DONE folder, and place it on the template canvas as shown in Figure 11-12. Create the boilerplate "Software Experts" text. Format the text using a 20-point Comic Sans font, and change the text color to blue. (If your system does not have this font, substitute a different font.)

g. Create the Exit button as shown, and create a WHEN-BUTTON-PRESSED trigger for the button that exits the current form.

h. Create a form-level WHEN-WINDOW-CLOSED trigger so the form is exited when the user clicks the Close button ⊠ on any form window.

  i. Create a visual attribute group named TEXT_ATTRIBUTES that formats text items using a dark blue foreground color and 8-point Arial Demilight font.

  j. Save the template form, and then create a new form named CONSULTANT_FORM that is based on the template form. Save the new form as Consultant.fmb in your SWEPROJECT_DONE folder, and then create a data block and layout based on the CONSULTANT table, as shown in Figure 11-12. Apply the TEXT_ATTRIBUTES visual attribute group to all form text items.

3. In this case, you will create an integrated database application for the Northwoods University database that allows users to access student and faculty information using the application switchboard shown in Figure 11-13.

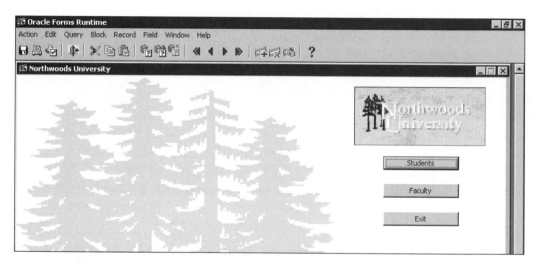

Figure 11-13

  a. Create a folder named NWPROJECT_DONE in your Chapter11\CaseSolutions folder, and copy all of the files from the Chapter11\NWPROJECT folder on your Data Disk into the new folder. (If you already made this folder in Case 1, use the existing folder.)

  b. Create a new form named Main.fmb that will be the main application form. Save the form in the NWPROJECT_DONE folder.

  c. Create a PRE-FORM trigger that maximizes the window and initializes a global project path variable to specify the location of the project files in the NWPROJECT_DONE folder.

  d. Create a form splash screen that displays the nwsplash.tif image that is stored in the NWPROJECT_DONE folder. Add a display item on the image item that provides copyright name and date information.

  e. Format the main form application canvas as shown in Figure 11-13. Use the trees3.gif image as the background image on the canvas, and place the

Northwoods logo (nwlogo.jpg) in the top-left corner. (Both images are stored in the NWPROJECT_DONE folder.) Change the canvas background color to a pale ivory shade.

f. Create the Students, Faculty, and Exit switchboard buttons as shown. Create a WHEN-BUTTON-PRESSED trigger for the Students button so that when the user clicks the button, a custom form that is stored in a file named custom_student.fmx opens that allows students to log onto the system and access their information. Also create a WHEN-BUTTON-PRESSED trigger for the Faculty button so that when the user clicks the button, a custom form named custom_faculty.fmx opens and allows faculty members to log onto the system and access their information. Create a WHEN-BUTTON-PRESSED trigger for the Exit button so that when the user clicks the button, the main application form closes. The Custom_Faculty.fmx form requires a view named CLASS_LIST, which displays the C_SEC_ID, S_ID, S_LAST, S_FIRST, S_MI, S_CLASS, and GRADE fields. Create this view in SQL*Plus. (You will need to create the required .fmx files by compiling the associated .fmb files that are stored in the NWPROJECT_DONE folder.)

4. In this case, you will create an integrated database application for the Software Experts database that allows users to access and modify consultant and project information using the application switchboard shown in Figure 11-14.

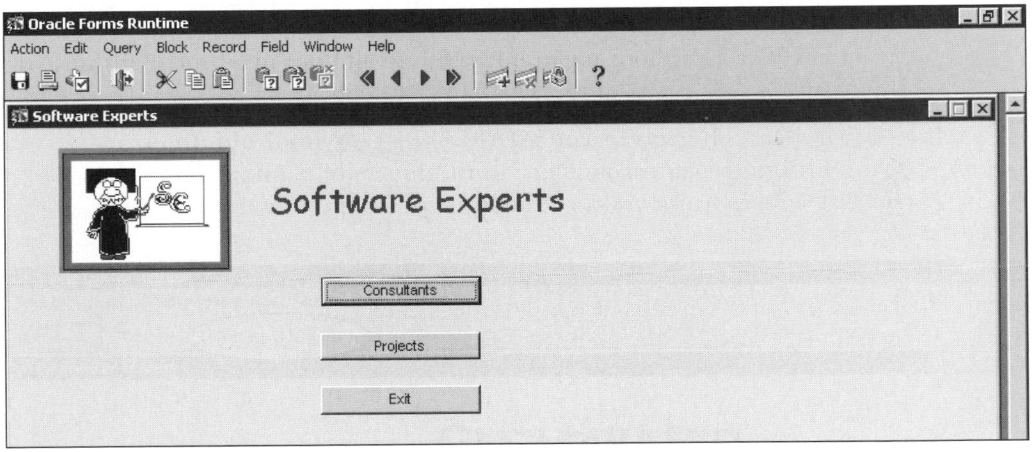

Figure 11-14

a. Create a folder named SWEPROJECT_DONE in your Chapter11\CaseSolutions folder, and copy all of the files from the Chapter11\SWEPROJECT folder on your Data Disk into the new folder. (If you already made this folder in Case 2, use the existing folder.)

b. Create a new form named Main.fmb that will be the main application form. Save the form in the SWEPROJECT_DONE folder.

c. Create a PRE-FORM trigger that maximizes the window and initializes a global project path variable to specify the location of the project files in the SWEPROJECT_DONE folder.

d. Create a form splash screen that displays the swesplash.tif image that is stored in the SWEPROJECT_DONE folder. Add a display item on the image item that provides copyright name and date information.

e. Format the main form application canvas as shown in Figure 11-14. Place the swelogo.tif image in the top-left corner, and create the "Software Experts" boilerplate text. Format the text using a 20-point Comic Sans font and change the text color to blue. (The image file is stored in the SWEPROJECT_DONE folder.) Change the canvas background color to a pale blue shade.

f. Create the Consultants, Projects, and Exit switchboard buttons as shown. Create a WHEN-BUTTON-PRESSED trigger for the Consultants button so that when the user clicks the button, a custom form named custom_consultants.fmx opens that displays consultant information. Create a WHEN-BUTTON-PRESSED trigger for the Projects button so that when the user clicks the button, a custom form named custom_projects.fmx opens that displays project information. (You will need to create the required .fmx files by compiling the associated .fmb files that are stored in the SWEPROJECT_DONE folder.) Create a WHEN-BUTTON-PRESSED trigger for the Exit button so that when the user clicks the button, the main application form closes.

◀ LESSON B ▶

Objectives

- ♦ Use Project Builder to create an integrated database application project
- ♦ Become familiar with the Project Builder environment
- ♦ Learn how to manage project components
- ♦ Learn how to deliver an Oracle project
- ♦ Customize the Project Builder environment

Project Builder is an Oracle utility that helps database developers organize and manage the individual forms, reports, and graphic files that compose an integrated database application. It provides a central point from which multiple software developers can view and launch application files in the different Developer utilities, and then recompile modified files so that the integrated application works correctly.

CREATING PROJECTS IN PROJECT BUILDER

To create a new project in Project Builder, you first define the project parameters, such as the project title and the folder location where the project files are stored. Then, you select the individual project files. To create a new project and define the project parameters, you can use the Project Wizard, which guides the project creation process using the following pages:

- *Project filename*, which specifies the name of the project registry file. When you create a new project, Project Builder creates a **project registry file**, which is a file with a .upd extension. This file stores the project parameters, such as the project name, along with a series of pointers that specify the locations of the individual project files.

- *Project definition*, which specifies the project title and the project directory (the drive letter and folder path where the project files are stored). When you have created multiple projects, this page also allows you to specify whether a project is a standalone project, or whether it is a subproject of an existing project.

- *Default database connection*, which specifies the project database connection information (username, password, and connect string), so a developer can access the individual project components without having to log onto the database each time.

- *User information*, which specifies the project author and contains comments about the project, such as the date on which it was last modified.

- *Finish*, which provides the option of just creating the project registry file, or opening a dialog box that allows you to immediately add files to the project.

Now you will start Project Builder, and use the Project Wizard to create a project for the Clearwater Traders database application. First, you will specify the project parameters. You will enter the project title and the name of the project registry file, which will be ClearwaterProject.upd. You will store the project registry file and the other project files in the CWPROJECT_DONE folder that you created in Lesson A in this chapter. You will also specify the database connection information.

To start Project Builder and create the project:

1. Click **Start** on the Windows taskbar, point to **Programs**, point to **Forms & Reports 6i**, and then click **Project Builder**. The Welcome to Oracle Developer dialog box opens.

You can also start Project Builder by starting Windows Explorer, changing to the *Developer_Home*\BIN folder, and then double-clicking the PJ60.exe file.

If a dialog box opens stating that new launcher options are available and asking if you would like to install them, click OK.

2. Make sure the Use the Project Wizard option button is selected, and then click **OK**. The Welcome to the Project Wizard page appears. Click **Next**. The Project filename page appears.

If you have created a project before, then the Project definition page appears instead of the Project filename page. Make sure the Create a standalone project option button is selected on the Project definition page, and then click Next to display the Project filename page.

3. On the Project filename page, click **Browse**, and then navigate to the CWPROJECT_DONE folder in your Chapter11\TutorialSolutions folder.

If you did not create the CWPROJECT_DONE folder in Lesson A of this chapter, start Windows Explorer, create a folder named CWPROJECT_DONE in your Chapter11\TutorialSolutions folder, and copy all of the files from the Chapter11\CWPROJECT folder on your Data Disk into the new folder.

The tutorial exercises will illustrate that the tutorial solutions are stored in the C:\OracleSolutions\Chapter11\TutorialSolutions\CWPROJECT_DONE folder.

11

4. To specify the project registry filename, type **clearwaterproject** in the File name text box, and then click **Save** to save the project registry file. The Project Wizard will automatically add the .upd extension to the filename when you save the file. The complete path to the project registry file appears in the Project Registry Filename text box. Click **Next**.

5. The Project definition page appears. Currently, the project title is the same as the project filename. Delete the current project title, and type **Clearwater Traders Database System** for the project title. By default, the Project Directory text box specifies that the project files will be stored in the same folder as the project registry file. Since you will store the project files and the project registry file in the same folder, accept the default project directory, and then click **Next**.

6. The Default database connection page appears. This page allows you to connect to the database. You can select a predefined (existing) connection or create a new connection. Make sure that <Current connection> is selected in the Pre-defined connection list, and then type your usual connection information to connect to the database. This will apply your current connection information to the project. Click **Next**.

7. The User Information page appears. Type your name in the Author text box. Type **Project created on** *today's_date* in the Comments box, and substitute the current date for *today's_date*. Click **Next**. The Finish page appears.

8. Make sure the Select files to add to the project option button is selected, and then click **Finish**. The Add Files to Project dialog box opens.

Now you will select the project files. The project files will include the following .fmb, .rdf, and .ogd files that are shown in the main form screen design in Figure 11-1 and summarized in Table 11-1: Sales.fmb, Receiving.fmb, Customer.fmb, Item.fmb, Inventory.fmb, Inventory.rdf, and OrderSource.ogd.

You will add only the Form Builder development source code (.fmb) files to the project. You do not need to add the associated form executable (.fmx) files, because Project Builder knows that every .fmb file must have an associated .fmx file, and it automatically includes the associated .fmx file for every .fmb file. Along with the individual application files, you will also include the main form module that you created in the previous lesson that contains the splash screen and system switchboard. Now you will add the files to the project:

To add the files to the project:

1. In the Add Files to Project dialog box, navigate to your CWPROJECT_DONE folder. Select the **Customer.fmb** file, press and hold the **Ctrl key**, and then select the following files: **Inventory.fmb**, **Inventory.rdf**, **Item.fmb**, **Main.fmb**, **OrderSource.ogd**, **Receiving.fmb**, and **Sales.fmb**. Note that the project registry file (clearwaterproject.upd) appears in the window, which confirms that the project registry file was created.

2. Click **Open**. The Project Builder – Project View window opens. If necessary, maximize the Project View window. If necessary, click ✚ beside Projects, and then click ✚ beside Clearwater Traders Database System. Your screen display should look like Figure 11-15.

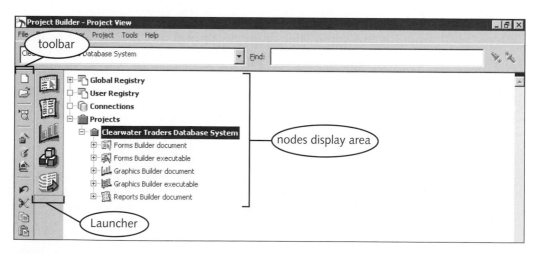

Figure 11-15 Project View window

THE PROJECT BUILDER ENVIRONMENT

In Project Builder, you can access and edit projects and project components from the Project View window, which is also called the Project Navigator. The following sections describe the Project Navigator, and explore different project components.

The Project Navigator Window

The Project Navigator has three main components: the toolbar, which provides access to common Project Builder commands; the **Launcher**, which is a secondary toolbar on the right side of the regular toolbar that provides access to other Developer utilities; and the **nodes display area**, which shows Project Builder components and provides a hierarchical view of the components in individual projects.

The Project Navigator displays top-level nodes for the following components:

- *Global Registry*, which specifies the different types of files that you can include in a project. You can customize the Global Registry to include almost any file type, including Oracle applications (such as Form Builder and Report Builder), and external applications, such as Excel spreadsheets and HTML files.

- *User Registry*, which stores configuration information to allow individual users to configure the Project Builder environment. Each user registry inherits information about the types of applications that can be included in a project from the Global Registry. Users can define additional file types in their individual user registries. A user registry can also store environment and preference settings.

- *Connections*, which specify usernames, passwords, and database connect strings. Connections can be assigned across projects and to different projects so that developers do not have to log onto the database each time they start a project application.

- *Projects*, which displays nodes and corresponding information about individual projects

Viewing Project Properties

In the Project Navigator, you can view and modify properties for an overall project or for individual files within a project. The following sections describe how to view project and project file information.

Viewing Project Information

To view and modify the parameters of a project, you can open the project Property Palette. Now, you will examine the Property Palette of the Clearwater Traders Database System project you just created.

11

To open the Clearwater Traders Database System project Property Palette:

1. Select **Clearwater Traders Database System**, right-click, and then click **Property Palette**. The project Property Palette opens, as shown in Figure 11-16. If necessary, resize and reposition the window as shown, so all property values are visible.

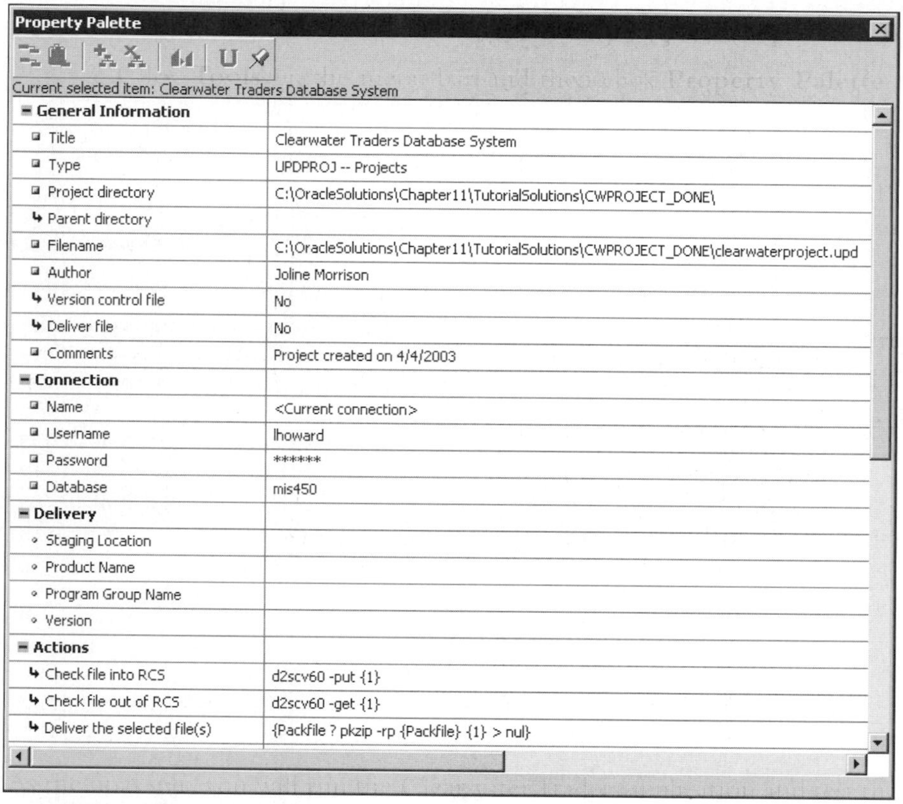

Property Palette	
Current selected item: Clearwater Traders Database System	
= General Information	
▫ Title	Clearwater Traders Database System
▫ Type	UPDPROJ -- Projects
▫ Project directory	C:\OracleSolutions\Chapter11\TutorialSolutions\CWPROJECT_DONE\
↳ Parent directory	
▫ Filename	C:\OracleSolutions\Chapter11\TutorialSolutions\CWPROJECT_DONE\clearwaterproject.upd
▫ Author	Joline Morrison
↳ Version control file	No
↳ Deliver file	No
▫ Comments	Project created on 4/4/2003
= Connection	
▫ Name	<Current connection>
▫ Username	lhoward
▫ Password	******
▫ Database	mis450
= Delivery	
∘ Staging Location	
∘ Product Name	
∘ Program Group Name	
∘ Version	
= Actions	
↳ Check file into RCS	d2scv60 -put {1}
↳ Check file out of RCS	d2scv60 -get {1}
↳ Deliver the selected file(s)	{Packfile ? pkzip -rp {Packfile} {1} > nul}

Figure 11-16   Project Property Palette

The project Property Palette shows the project properties that you just specified. The General Information property node shows the project title, project directory, registry filename, and project author. The Connection property node shows the current project connection information, which you specified in the Project Wizard. Currently, the project connection value is <Current connection>, which means that the project assumed the properties of the current database connection. You will learn how to assign a specific database connection to a project later. The Delivery property node shows information about delivering the project, which you will learn about later in this lesson. The Actions node allows developers to check out specific project components.

Viewing Project File and Dependency Information

Next you will examine the file objects in the Clearwater Traders Database System project. Project file objects have **dependencies**, which are conditional relationships that show that specific files depend on the presence of other files. Executable files, called **targets**, depend on their associated source code files, which are called **inputs**. For example, a Forms Runtime executable (.fmx) file is a target that depends on its associated Form Builder source code (.fmb) file, which is also called a Form Builder document file. Project Builder automatically detects these dependencies and deduces that any target files in your project depend on their associated input files. If an input file changes, it must be recompiled to create a new target file. When you add files to a project, you only need to add the input files. Project Builder will automatically display the associated target files, because their presence in the project is implied by the existence of the input files. Project Builder always assumes that all project input and target files are stored in the project default directory.

You can view the project files in the Project Navigator in two different views. The **Project View** shows project items organized by item types, such as Form Builder document files, Form Builder executable files, Graphics Builder document files, and so forth. Project view is useful for tracking the different types of files in a project. The **Dependency View** shows files in the Project Navigator according to their dependencies. Project nodes are at the highest point in the hierarchy. Each project's target nodes are shown below the project, and each target node's input components are shown beneath the target node. In both views, input filenames are always shown in boldface type.

Now you will view the current project files and their dependencies. By default, the Project View window opens in Project View, so you will first examine the Clearwater Traders Database System's project files in the Project Navigator window in Project View.

To examine the project files and dependencies in Project View:

1. Click the **Close button** ☒ on the Property Palette window to close the project Property Palette.

2. Confirm that the Project Navigator is in Project View by clicking **Navigator** on the menu bar. Project View should be selected. If Project View is not selected, select it.

3. Click ✚ beside Forms Builder document. A listing of the Form Builder input (.fmb) files appears.

4. Click ✚ beside Forms Builder executable. A listing of the Form Builder target (.fmx) files appears.

5. Click ✚ beside Graphics Builder document. The Graphics Builder input (.ogd) file appears.

6. Click ✚ beside Graphics Builder executable. The Graphics Builder executable (.ogr) file appears.

11

7. Click ➕ beside Reports Builder document. The Report Builder input (.rdf) file appears. The screen display of the project components in Project View with all nodes expanded should look like Figure 11-17.

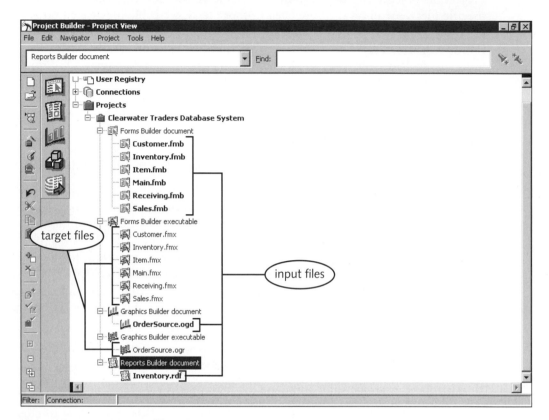

Figure 11-17 Project files in Project View

Recall that you did not explicitly add the .fmx files to the project. Project Builder automatically deduced that since the input files exist in the project, the target files also exist in the project. Note that the input files appear in boldface type, and the target files appear in regular type.

Next, you will view the project components in Dependency View. Dependency View shows the hierarchical relationships among target files and input files.

To view the project components in Dependency View:

1. In the Project Navigator, click **Navigator** on the menu bar, and then click **Dependency View**. Only the target .fmx, .rdf, and .ogr files appear under the top-level project node.

2. Click ➕ beside Sales.fmx. Its associated input file (Sales.fmb) appears below it, as shown in Figure 11-18. This indicates that the Sales.fmx target file is dependent on the Sales.fmb input file.

3. Expand the rest of the target nodes to view the dependency relationships between the other project target files and their associated input files. The screen display of the project components in Dependency View with all nodes expanded should look like Figure 11-18. (Your nodes may appear in a different order.)

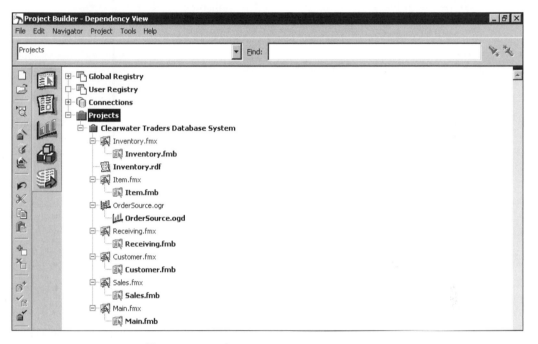

Figure 11-18   Project files in Dependency View

Each project file has a Property Palette that specifies two types of file properties: values that are specific to the file (such as the filename, the date the file was created, the file size, and the file type) and values that the file inherits from the project by virtue of being in the project (such as the directory, author, and database connection information). Now you will open the Property Palette of the Sales.fmx project file and examine its properties.

To open the Sales.fmx Property Palette:

1. In the Project Navigator, select the **Sales.fmx node**, right-click, and then click **Property Palette**. The Property Palette opens. If necessary, resize the window so all of the properties and associated values are visible, as shown in Figure 11-19.

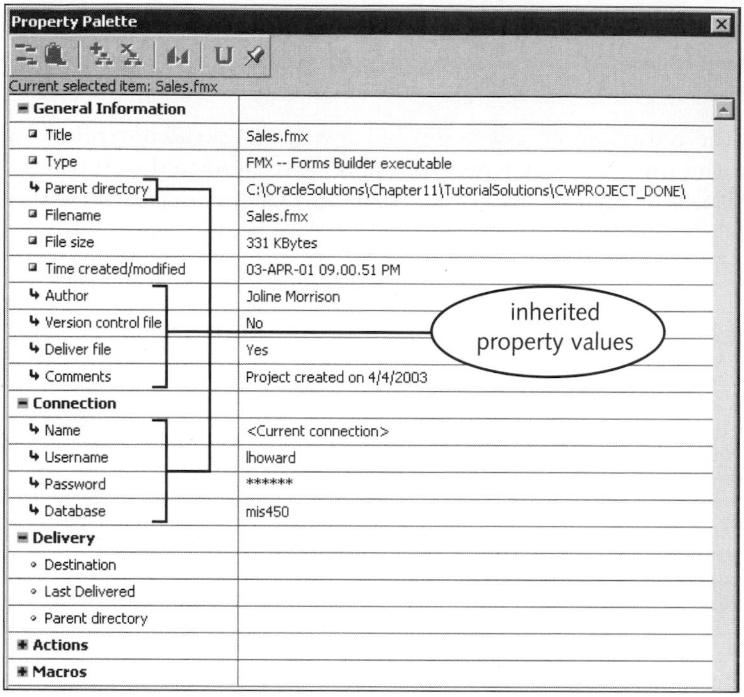

Figure 11-19   Project file Property Palette

Recall that when you change a property value in the Property Palette, the property node marker is colored green. In the Sales.fmx project file Property Palette, properties that are derived from the specific project file, such as the filename, file size, and time created, have a green node marker. Properties that are derived from the project specifications have an **inheritance arrow** ⤷ as the property node marker. Note that properties that you specified when you created the project, such as the project directory, author, and connection information, have ⤷ beside the property name. This indicates that their value was inherited from the project.

USING PROJECT BUILDER TO MANAGE PROJECT COMPONENTS

Project Builder provides a central point from which you can manage project components. You can add and delete project files, modify project files, create and assign predefined database connections to projects, create subprojects, and rebuild project files. The following sections describe how Project Builder supports these project management tasks and how to open and close project files in Project Builder.

Adding and Deleting Project Files

When you created the Clearwater Traders Database System project, you added the project files using the Project Wizard. After you create a project, you often need to add new files and remove existing files. To add a project file, you click the **Add Files To Project button** 🖹⁺ on the toolbar. To remove a file from a project, you select the file in the Project Navigator, and then click the **Delete button** 🗙 on the toolbar.

Recall that when you add an input file to a project, the target file automatically appears. Likewise, when you add a target file to a project, its input file automatically appears. When you remove files, the same thing happens. When you remove a target file from a project, its input file is automatically removed, and when you remove an input file from a project, its target file is automatically removed. Adding and removing files works this way because input and target files have an **implicit dependency**. The compiled code in the target file depends on the source code in the input file, and vice versa. Whenever you remove a file, the file that has an implicit dependency on that file is removed also.

When you manually add a file to a project, you can create an explicit dependency between two input files. An **explicit dependency** indicates that a child input file depends on a parent input file because the parent file calls the child file. Recall that you can call a child form from a parent form using the CALL_FORM or OPEN_FORM procedures, and that you can call a report or chart from a form using the RUN_PRODUCT procedure. The child form has an explicit dependency on the parent form. For example, suppose you create a form file that calls a report. You would add the form input file to the project, and then add the report to the form to show that the report is called by the form and has an explicit dependency on the form. This way, if you decided to remove the form from the project, the called report is automatically removed as well.

You can use 🖹⁺ to add a file to a project, or to add a file to a file within a project and create an explicit dependency. 🖹⁺ is only enabled when a specific project is selected in the Project Navigator or when a specific project file is selected. After you add a new file to a project, it is not necessary to save the project. The project registry file is automatically saved every time you make a change to the project.

Now, you will add a new form, named ItemReportForm.fmb, to the Clearwater Traders project. This form allows the user to select an item ID and description from an LOV display, and then run a report named ItemReport.rdf that displays inventory information for the selected item. You will add the form to the project, and then add the report to the form, which creates an explicit dependency between the form and the report.

To add the form and report to the project:

1. Click the **Close button** 🗙 on the Property Palette window title bar to close the Sales.fmx file Property Palette.

2. In the Project Navigator in Dependency View, select **Clearwater Traders Database System** in the Project Navigator window to select the project. If you do not select the project before adding files, the new files will be added to whatever item is currently selected.

To make sure the Project Navigator is in Dependency View, click Navigator on the menu bar, and make sure Dependency View is selected.

3. Click the **Add Files To Project button** on the toolbar. The Add Files to Project dialog box opens.

Another way to add files to a project is to click Project on the menu bar and then click Add Files To Project.

4. If necessary, navigate to your CWPROJECT_DONE folder, select **ItemReportForm.fmb**, and then click **Open**. The ItemReportForm.fmx file appears in the list of project target files.

5. Click ⊞ beside ItemReportForm.fmx. The ItemReportForm.fmb input file appears, showing the implicit dependency between the ItemReportForm.fmx file and the ItemReportForm.fmb file (see Figure 11-20).

6. To create the explicit dependency between the ItemReportForm.fmb file and the ItemReport.rdf file, select **ItemReportForm.fmb**, click, select **ItemReport.rdf**, and then click **Open**.

7. Click ⊞ beside ItemReportForm.fmb. The ItemReport.rdf file appears, showing the explicit dependency between the ItemReportForm.fmb and ItemReport.rdf input files (see Figure 11-20).

Since there is an explicit dependency between the ItemReportForm.fmb file and the ItemReport.rdf file, if you remove the ItemReportForm.fmb file from the project, the ItemReport.rdf report file will also be removed. Now, you will remove the ItemReportForm.fmb file from the project. When you do this, the file that has the explicit dependency, which is the report file, will automatically be removed. The form's target .fmx file will also be removed.

To remove the file and its associated dependencies from the project:

1. If necessary, select ItemReportForm.fmb, and then click the **Delete button** on the toolbar. The ItemReportForm.fmb, ItemReportForm.fmx, and ItemReport.rdf files are all removed from the project. If necessary, click **Yes** to confirm the deletion.

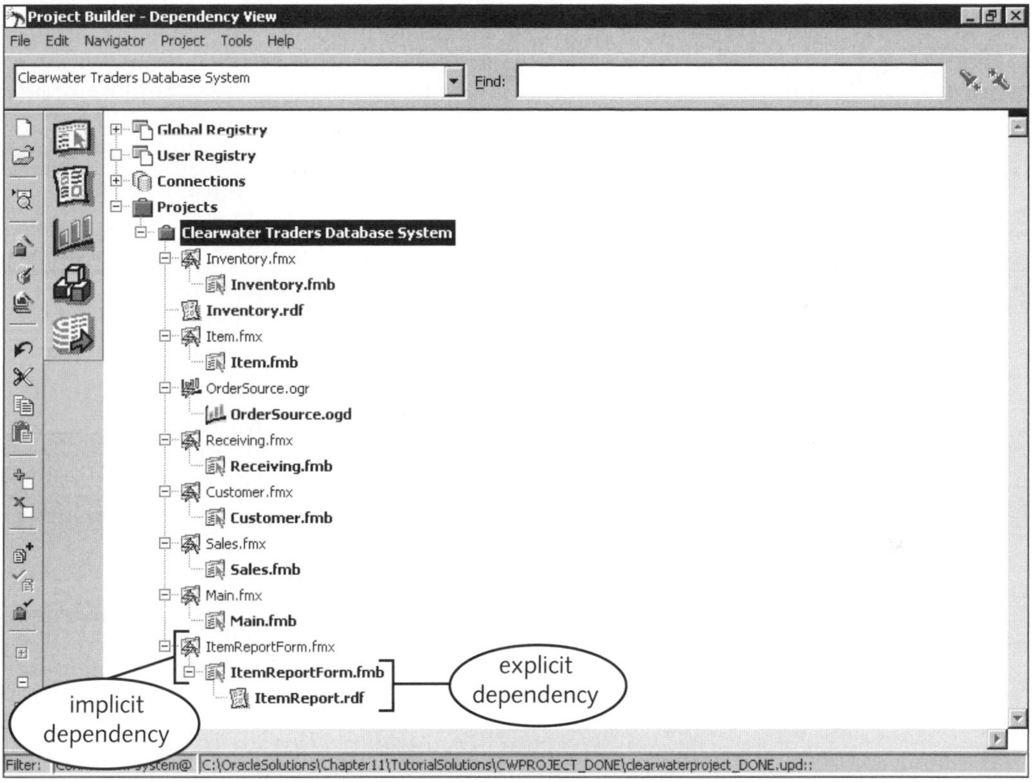

Project Builder - Dependency View

File   Edit   Navigator   Project   Tools   Help

Clearwater Traders Database System   Find:

- Global Registry
- User Registry
- Connections
- Projects
  - **Clearwater Traders Database System**
    - Inventory.fmx
      - **Inventory.fmb**
    - **Inventory.rdf**
    - Item.fmx
      - **Item.fmb**
    - OrderSource.ogr
      - **OrderSource.ogd**
    - Receiving.fmx
      - **Receiving.fmb**
    - Customer.fmx
      - **Customer.fmb**
    - Sales.fmx
      - **Sales.fmb**
    - Main.fmx
      - **Main.fmb**
    - ItemReportForm.fmx
      - **ItemReportForm.fmb**
        - **ItemReport.rdf**

implicit dependency

explicit dependency

Filter: ...ystem@   C:\OracleSolutions\Chapter11\TutorialSolutions\CWPROJECT_DONE\clearwaterproject_DONE.upd::

Figure 11-20   Creating an explicit dependency

Opening and Modifying Project Components

One of the main benefits of Project Builder is that it provides a central point from which you can start all of the Oracle development environments (Form Builder, Report Builder, Graphics Builder, Procedure Builder, and Query Builder). It also allows you to open specific project application files in their associated development and runtime environments to view and modify the project components.

 Query Builder is a visual environment for creating SQL queries based on Oracle database tables.

To start a specific development environment with no project file loaded, you click the button associated with the environment on the Launcher toolbar. When you place the mouse pointer on the Launcher toolbar buttons, a ToolTip appears that describes the associated development environment. Now you will start Form Builder using the Launcher toolbar.

11

To start Form Builder using the Launcher toolbar:

1. In the Project Navigator, click the **Form Builder 6i** button, which is the first button on the Launcher toolbar. The Welcome to Form Builder window opens.

2. Click **Cancel**, and then close Form Builder.

You can automatically open specific project files in their development and runtime environments. To open an .fmb file in Form Builder, you double-click the **Forms Builder document icon** which automatically starts Form Builder and loads the associated .fmb file. Double-clicking the **Forms Builder executable icon** automatically starts Forms Runtime and loads the associated .fmx file. To open a report file in Report Builder, you select the report file, right-click, and then click Edit. To open report files in Reports Runtime, you double-click the **Report Builder document icon**, which starts Reports Runtime and loads the associated .rdf file.

 You can also start Form Builder files by selecting the .fmb file, right-clicking, and then clicking Edit. You can also start Forms Runtime files by selecting the .fmx file, right-clicking, and then clicking Run.

At this time, you can open Graphics Builder documents only in the Graphics Runtime environment. To do this, you double-click the **Graphics Builder document icon** beside the filename. If you want to open a Graphics Builder document in the Graphics Builder environment, you need to open Graphics Builder using the Launcher toolbar and then explicitly load the file.

When you open a specific project file from the Project Navigator, the database connection information for the file that is specified in the file Property Palette automatically creates a database connection in the associated application. For example, when you start a Forms Runtime .fmx file, you do not need to log onto the database first. The connection is automatically created when the application starts.

Now, you will open the Main.fmx form file in Forms Runtime. Then, you will open the Inventory.rdf report file in Report Builder.

To open project files in their associated environments:

1. In Project Builder, click **Navigator** on the menu bar, and then click **Project View** to open the Project Navigator in Project View.

2. Click ➕ beside the Forms Builder executable node, and then double-click the **Forms Builder executable icon** beside the Main.fmx file node. The Clearwater Traders database application opens in Forms Runtime. Note that you were not prompted to connect to the database, because Project Builder automatically created the database connection based on the project database connection information. Click **Exit** to close the form.

If an error message stating "Unable to read file" appears, double-check to make sure that the Main.fmx file exists in your CWPROJECT_DONE folder. If the file is not present, open Main.fmb in Form Builder, and run the form to create a new .fmx file.

3. If necessary, click the ➕ beside the Reports Builder document node to expand the node. Select **Inventory.rdf**, right-click, and then click **Edit**. Report Builder opens, with the INVENTORY report loaded. Close Report Builder.

Creating a Predefined Database Connection

In Project Builder, you can define a **project database connection** in the Project Navigator that specifies a database username, password, and connect string. A project database connection is useful when you want to assign the same connection information to different projects or when you want to change the database connection information for all the objects in a project. Now you will create a project database connection.

To create a project database connection:

1. Select the top-level **Connections node**, and then click the **Create button** 📑 on the toolbar. The New Connection dialog box opens.

 Another way to create a new connection is to select the Connections node, right-click, and then click New Connection.

11

2. Delete the current title, and type **Clearwater Connection** in the title text box.

3. Type your usual username, password, and database connect string, and then click **OK** to save the connection. The new connection appears in the Project Navigator window.

 When you create a connection in Project Builder, the password is saved in the project registry file. This file could be exported to another file format and someone could view your password, so it is a good idea to always keep project files in a secure location.

After you have created a project database connection, you can apply it to a project. To do this, you drag and drop the **Connection icon** 📇 onto the project in the Project Navigator window. Now you will apply the Clearwater Connection to the Clearwater Traders Database System project.

To apply the project database connection to the project:

1. Select **Clearwater Connection**, and then drag the mouse pointer down on the screen display. The pointer changes to ⬚. When the Clearwater Traders Database System project appears selected, release the mouse button to drop the connection onto the project.

2. Select the project, right-click, and then click **Property Palette** to open the project Property Palette. The project Connection property value should now be Clearwater Connection. All project files will inherit this connection information.

3. Close the project Property Palette.

Creating Subprojects

To make large projects easier to manage, they are often divided into subprojects. After you create your first project in Project Builder, every subsequent time you use the Project Wizard to create a new project, the Subprojects page appears. The **Subprojects page** allows you to specify whether the new project will be a subproject of the existing project. Creating subprojects reduces the number of files that are in an individual project. When you create a new subproject, Project Builder does not create a new project registry file. Rather, it adds an entry to the parent project's registry file that contains the subproject information.

Now you will use the Project Wizard to create a subproject of the Clearwater Traders Database System. You will not add any files to the subproject. Like other Oracle wizards, the Project Wizard is reentrant, so you can use it to edit the properties of an existing project. When you want to create a new project, you must be sure that an existing project is not selected, so that the Project Wizard will not open in reentrant mode.

To create a new subproject using the Project Wizard:

1. Click **Projects** in the Project Navigator window to ensure that no objects in the existing project are selected and the Project Wizard does not open in reentrant mode.

2. Click **Tools** on the menu bar, and then click **Project Wizard**. The Welcome to the Project Wizard page appears. Click **Next**.

3. Select the **Create a sub-project under an existing project option button**, and then click **Next**.

4. Make sure that the Clearwater Traders Database System is selected as the Parent Project, and then click **Next**.

5. Type **Clearwater Subproject** in the Title text box, click **Browse**, navigate, if necessary, to your CWPROJECT_DONE folder, and then click **OK**. The folder path appears in the Project Directory text box. Note that you are not prompted to enter the project registry filename, because the subproject uses the parent project's registry file. Click **Next**.

6. Open the Pre-defined Connection list, and select **Clearwater Connection**. The connection information appears. Click **Next**.

7. Type your name in the Author text box, and then click **Next**.

8. Since you will not add any files to the project, click the **Create an empty project option button**, and then click **Finish**. The subproject appears as a new object under the Clearwater Traders Database System project in the Project Navigator window.

Building Input Files

When you create a project in Project Builder, you can open individual form input files and edit them in the Form Builder environment. After you modify an .fmb file in Form Builder, the corresponding modified .fmx file is not built, or recompiled, unless you explicitly run the form compile the .fmb file in Form Builder. To avoid having to explicitly build project .fmx files in Form Builder after you modify their input files, you can build .fmb files in the Project Navigator.

There are three options for building input files in Project Builder: **Build Selection**, which builds a selected target file; **Build Incremental**, which builds all target files whose associated input files have been changed but have not yet been rebuilt; and **Build All**, which builds all of the project's input files. To build a selection, you select the target file in the Project Navigator window and click the **Build Selection button** 🗒️ on the toolbar. For example, to rebuild Sales.fmb, you select Sales.fmx in the Project Navigator window and click 🗒️. To incrementally build the project files, you select the project in the Project Navigator window, and then click the **Build Incremental button** 🗒️ on the toolbar. To build all of the project files, you select the project in the Project Navigator window, click Project on the menu bar, and then click Build All. Now, you will rebuild the Main.fmx target file in Project Builder. Then, you will rebuild all of the project files.

To rebuild the Main.fmx target file, and then build all of the project target files:

1. In the Project Navigator window, if necessary click ➕ beside the Forms Builder executable node, select **Main.fmx**, and then click the **Build Selection button** 🗒️ to build the selected file. The Build dialog box opens, and then the Forms Compiler window opens. When the Build dialog box displays the message "1 item built.", click **OK**.

 Another way to build a selected file is to select the target file, right-click, and then click Build Selection. Alternately, you can select the target file, click Project on the menu bar, and then click Build Selection.

2. To rebuild all of the project files, select the **Clearwater Traders Database System project node**, click **Project** on the menu bar, and then click **Build All**. You will see various forms flash on your screen display as Project Builder builds the files. When the Build dialog box displays the message "7 items built.", click **OK**.

Another way to rebuild all of the project files is to select the project, right-click, and then click Build All.

Opening and Closing Projects in Project Builder

Recall that when you add project files to or delete project files from a project, you do not need to explicitly save the project registry file, because all changes to the project structure are automatically saved when you perform them. When you exit Project Builder, your projects will automatically appear in the Project Navigator window the next time you start Project Builder, as long as the project registry files have not been deleted or moved. If you do not want a project to appear in the Project Navigator, you select the project, click File on the menu bar, and then click Close Project. To add a project to the Project Navigator display, you click File on the menu bar, click Open Project, and open the project registry (.upd) file.

Now you will close the Clearwater Traders Database System project in Project Builder. Then, you will open it again.

To close and then reopen the Clearwater Traders Database System project:

1. Select any object in the Clearwater Traders Database System project, click **File** on the menu bar, and then click **Close Project** to close the project. The project no longer appears in the Project Navigator.

To remove a project from the Project Navigator, you can also select the project, and then click the Delete button or press the Delete key.

2. Click **File** on the menu bar, click **Open Project**, navigate if necessary to your CWPROJECT_DONE folder, select **clearwaterproject.upd**, and then click **Open**. The project appears again in the Project Navigator.

DELIVERING A PROJECT

Delivering a project is the process of preparing a completed application for distribution and deployment to users. In Oracle, project delivery involves two steps:

1. Preparing the files for delivery, which involves placing the files in a staging area. A **staging area** is a file system folder where the files are placed for eventual copying to a distribution media, such as a CD-ROM.

2. Generating scripts so the project can be installed using Oracle Installer. **Oracle Installer** is a utility that is used to install all Oracle products. Usually, Oracle Installer is distributed on CDs that are used to distribute

Oracle products. The installation scripts might specify to package and install the Forms or Reports Runtime environments with the project, so users can install both the project files and the runtime environments in a single installation operation.

You can use the Project Builder Delivery Wizard to prepare the project files for delivery and place the files in the staging area. The Delivery Wizard has the following pages:

- Project setup page, which specifies the project name and whether you want to deliver all project files or only the files that have changed since the last delivery. For a new project delivery, you will deliver all project files. If you are distributing a **patch**, which is an incremental update that corrects errors or omissions in existing files, you deliver only changed files.

- Staging area page, which specifies the staging area folder. The staging location can be in the local file system, in a remote location using FTP, or in a custom location specified in a .sql script. By default, the staging area is the project directory.

- Files page, which specifies the project files to be delivered.

- Subdirectories page, which specifies the folder where the product will be delivered and the text that will appear on the user's Start menu when he or she installs the product. The menu text can use mixed case letters, and include blank spaces. The default folder is %PROD_HOME%, which specifies that all of the project folders are installed in the product's home directory. The user specifies the product's home directory when he or she installs the product using the Oracle Installer.

- Finish page, which summarizes the delivery information.

After you have used the Delivery Wizard to prepare a project for delivery, each time you start the Delivery Wizard with that project selected, the wizard automatically opens in reentrant mode, so you do not need to specify the delivery information again.

Now, you will use the Delivery Wizard to prepare the files in the Clearwater Traders Database System project for delivery. First, you will create a folder in your CWPROJECT_DONE folder named CWSTAGING, which will be the project staging location. Then, you will start the Delivery Wizard and prepare and move the files.

To use the Delivery Wizard to prepare the project for delivery:

1. Start Windows Explorer, navigate to your CWPROJECT_DONE folder, and create a subfolder named **CWSTAGING**. Do not close Windows Explorer.

2. In Project Builder, click **Tools** on the menu bar, and then click **Delivery Wizard**. The Delivery Wizard Welcome page appears. Click **Next**.

3. On the Project setup page, make sure that the Clearwater Traders Database System project is selected in the project list. Since this is a new project delivery, select the **Deliver all files option button**, and then click **Next**.

If a message appears stating that one or more of the selections do not have attached files, it means that some of the project files need to be rebuilt. Click Cancel to exit the Delivery Wizard, rebuild all of the project files, and then start the Delivery Wizard again.

4. On the Staging area page, make sure that the Deliver to a local staging area option button is selected. To specify the staging area location, click **Browse**, navigate to the **CWSTAGING** folder in your CWPROJECT_DONE folder, and click **OK**. The staging area location appears on the Location page. Click **Next**.

5. On the Files page, note that by default, all of the project target files appear in the Files to Deliver list. Since this is a new delivery rather than a patch, accept the default file selection, and then click **Next**.

6. On the Subdirectories page, note that the default subdirectory for each project file is %PROD_HOME%. Since all of the system files will be installed in the project's home folder on the user's workstation, you will not change the Subdirectories specification.

7. Since the only entry on the user's Start menu will be to the main form module (Main.fmx), type **CT Database System** in the Start menu label space beside Main.fmx. Leave the Start menu label entries for the other files blank, and then click **Next**.

8. The Finish page appears, which confirms the location of the files, the number of files, and the file sizes. Click **Finish** to prepare the project files and copy them to the staging area. A status bar showing the percentage complete appears briefly.

9. Switch back to Explorer, open the CWSTAGING folder, and note that the project .fmx files now appear in the staging area.

10. Close Explorer.

CUSTOMIZING THE PROJECT BUILDER ENVIRONMENT

You can customize the Project Builder environment to meet your needs and preferences. To customize the Project Builder environment, you open the Preferences dialog box by clicking Tools on the menu bar, and then clicking Preferences. The Preferences dialog box has the following tab pages:

- *Display*, which allows you to specify the toolbars that appear in the Project Navigator and to suppress certain pages in wizards or features in dialog boxes.

- *Projects*, which allows you to enter default project information that automatically appears when you create a new project, such as the author name, comments, and default connection.

- *Navigator*, which allows you to specify how the Project Navigator displays project data. Specifically, you can specify whether or not to show file types in Project View and whether or not to automatically show files that are included in the project as a result of implicit dependencies.

- *Launcher*, which allows you to configure the appearance of the Launcher toolbar by specifying the buttons that appear and where the toolbar is docked.

Each tab page has an OK button, which saves the changes and closes the Preferences dialog box, and a Cancel button, which closes the Preferences dialog box without saving the changes. Now you will open the Preferences dialog box and configure the Project Builder environment setup. You will specify to hide the Welcome page on all of the Project Builder wizards and enter default project author and connection information. You will also specify to hide file types in Project View, to change the appearance of the Launcher toolbar so the Query Builder button does not appear, and to dock the Launcher toolbar on the top edge of the Project Navigator window.

To configure the Project Builder environment:

1. Click **Tools** on the menu bar, and then click **Preferences**. The Preferences dialog box opens.

2. Make sure that the Display tab is selected, clear the **Show initial "Welcome" dialog check box**, and clear the check boxes that specify to show the "Welcome" screen in the Project Wizard, Module Type Wizard, and Delivery Wizard.

3. Select the **Projects tab**. Type your name in the Author text box, open the Connection list, and select **Clearwater Connection**. Do not click OK, because this will close the Preferences dialog box, and you want to specify additional configuration changes.

4. Select the **Navigator tab**, and clear the **Show implied items check box**.

5. Select the **Launcher tab**. In the Entries list, select the **Query Builder 6i node**, click **Remove**, and then click **Yes** to confirm removing the node.

6. In the Layout frame, select the **Docked at Top option button**, and then click **OK**. The Preferences dialog box closes, and the reconfigured Project Navigator appears as shown in Figure 11-21.

11

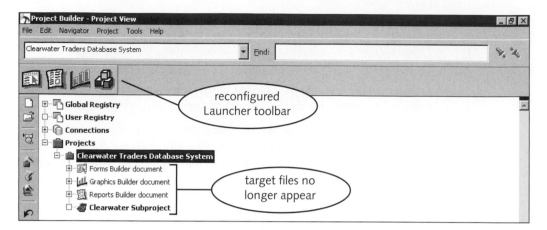

Figure 11-21   Reconfigured Project Navigator enviroment

In the Project Navigator window, note that the target file types (Forms Builder executable and Graphics Builder executable) no longer appear in the nodes display area. These file types are included because of implicit dependencies, and they no longer appear because you cleared the Show implied items check box on the Navigator preferences tab page. When you do not show implicit dependencies, you cannot rebuild .fmx files in Project Builder, because to rebuild files, you must select the target (.fmx) files. Also, note that the Query Builder button no longer appears on the Launcher toolbar, and the toolbar is docked at the top of the window, instead of on the left side of the window. Now, you will reconfigure the Project Builder environment to restore the implied items so you can recompile files in Project Builder as needed.

To restore the implied items in the Project Builder configuration:

1. Click **Tools** on the menu bar, and then click **Preferences** to reopen the Preferences dialog box.

2. Select the **Navigator tab**, check the **Show implied items check box**, and then click **OK**. The Forms Builder executable and Graphics Builder executable nodes appear again in the Project Navigator window.

3. Close Project Builder and all other open Oracle applications.

SUMMARY

❏ Project Builder provides a central point from which developers can view and edit project components in the Developer utilities, and easily rebuild executable files.

❏ When you create a new project, Project Builder creates a project registry file, which is a file with a .upd extension that contains information about the project, along with a series of pointers specifying the locations of individual project applications.

❐ When you add files to a project, you need to add only the source code (input) files. The executable (target) files will automatically appear in the project in Project Builder.

❐ The Project Navigator has a toolbar to access application commands, a Launcher toolbar that provides access to other Developer applications, and a nodes display area, which shows project components.

❐ Project files have dependencies, which show that certain files depend on the presence of other files. Executable files are called targets, and design files are called inputs. Target files depend on their associated input files. If an input file changes, it must be rebuilt to create a new target file.

❐ You can view project files in the Project Navigator in Project View, which lists files based on file types, and in Dependency View, which lists files based on file dependencies.

❐ A project file has specific properties, such as the file size and type, as well as properties that it inherits from the project, such as the project author and connection information.

❐ Project files can have implicit dependencies, which define the relationship between target and input files, and explicit dependencies, which define the relationship between a child file that is called by a parent file using form navigation commands or the RUN_PRODUCT command.

❐ When you remove a file from a project, all files that depend on the file are automatically removed.

❐ You can use the Project Navigator Launcher toolbar to start Developer applications with no specific file opened, or you can click the icon beside a specific project file to start its associated Developer application with the file loaded.

❐ A project database connection specifies a username, password, and database connect string that can be assigned to different projects, so developers don't have to log onto the database each time they open a new project file.

❐ Projects can have subprojects, which helps to reduce the number of files in a single project. A subproject does not have a unique project registry file, but is defined as an entry in the registry file of the parent project.

❐ Project Builder project registry files are automatically saved every time you change the project structure. When you exit Project Builder, your projects will automatically appear in the Project Navigator window the next time you start it, as long as the project registry files have not been deleted or moved.

❐ When you rebuild input files in Project Builder, you can choose Build Selection, which builds a selected target file; Build Incremental, which rebuilds all target files whose associated input files have been changed, but not yet rebuilt; or Build All, which rebuilds all of the project's input files.

11

❐ Delivering a project is the process of preparing a completed application for distribution and deployment to users. To deliver an Oracle project, you must prepare the project files, move them to the staging area, and then generate scripts that guide the installation process.

❐ The Project Builder Delivery Wizard automatically prepares completed project files for delivery and moves them to the staging area.

❐ You can customize the Project Builder environment by changing properties such as the buttons that appear on the Launcher toolbar, the file types that appear in the nodes display area, and whether or not implied file dependencies appear.

REVIEW QUESTIONS

1. What does the project registry file contain?
2. Describe the difference between an implicit dependency and an explicit dependency.
3. What is the purpose of creating a database connection in Project Builder?
4. What is the purpose of creating subprojects?
5. _____ files are dependent on _____ files.
   a. .fmx, .fmb
   b. .ogr, .ogd
   c. .mmx, .mmb
   d. all of the above
6. What is the difference between Project View and Dependency View in the Project Navigator?
7. Describe the circumstances in which you would use the Build All, Build Incremental, and Build Selection options.
8. What is the staging area?
9. How do you customize the Project Builder environment?

PROBLEM-SOLVING CASES

The cases refer to the Northwoods University (see Figure 1-12) and Software Experts (see Figure 1-13) sample databases. Place all solution files in the specified folder in your Chapter11\CaseSolutions folder.

To complete Case 1, you must have completed Case 3 in Lesson A of this chapter.

1. In this case, you will create a project in Project Builder for the Northwoods University student and faculty database system. Table 11-5 summarizes the individual module files.

Design File	Description
\NWPROJECT_DONE\main.fmb	Application switchboard to access the student and faculty custom forms (see Figure 11-13)
\NWPROJECT_DONE\custom_student.fmb	Custom form that allows students to log onto the system and access their information
\NWPROJECT_DONE\custom_faculty.fmb	Custom form that allows faculty members to log onto the system and access their information
\NWPROJECT_DONE\location.fmb	Data block form that displays and edits LOCATION data records
\NWPROJECT_DONE\course.fmb	Data block form that displays and edits COURSE data records
\NWPROJECT_DONE\class_lists.rdf	Report that displays class lists for each faculty member for a selected term
\NWPROJECT_DONE\transcripts.rdf	Report that displays transcript information for each student
\NWPROJECT_DONE\enrollments.ogd	Line chart that displays enrollment trends over time

Table 11-5   Module files in Northwoods University database application

11

a. Use Project Builder to create a new project titled "Northwoods University Student/Faculty System." Name the project registry file nwproject.upd, and store the registry file in the NWPROJECT_DONE folder that you created in Case 1 in Lesson A of this chapter. Specify that the NWPROJECT_DONE folder is the default project directory. Identify yourself as the project author, and specify the date the project was created in the Comments text box.

b. Add the input files shown in Table 11-5 to the project. These files are stored in the NWPROJECT_DONE folder in your Chapter11\CaseSolutions folder.

c. Create a project database connection named Northwoods University Connection that uses your username, password, and connect string, and apply the connection to the project.

d. Create a subproject named Course Section Items, and specify that the subproject uses the same project directory as its parent project. Apply the Northwoods University Connection to the subproject. Identify yourself as the project author, and specify the date the project was created in the Comments text box. Do not add any files to the subproject.

e. Use Project Builder to rebuild all of the project files.

f. Use Windows Explorer to create a subfolder named NWSTAGING in your NWPROJECT_DONE folder in your Chapter11\CaseSolutions folder, and then use the Delivery Wizard to prepare the project files for delivery. Specify to deliver all project files, and do not specify any Start menu labels.

 To complete Case 2, you must have completed Case 4 in Lesson A of this chapter.

2. In this case, you will create a project in Project Builder for the Software Experts database system. Table 11-6 summarizes the individual module files.

Design File	Description
\SWEPROJECT_DONE\main.fmb	Application switchboard to access the system custom forms (see Figure 11-14)
\SWEPROJECT_DONE\custom_consultants.fmb	Custom form that displays consultant information
\SWEPROJECT_DONE\custom_projects.fmb	Custom form that displays project information
\SWEPROJECT_DONE\client.fmb	Data block form that displays and edits CLIENT data records
\SWEPROJECT_DONE\skill.fmb	Data block form that displays and edits SKILL data records
\SWEPROJECT_DONE\client_projects.rdf	Report that displays information about clients and their associated projects
\SWEPROJECT_DONE\consultant_skills.rdf	Report that displays information about consultant skills
\SWEPROJECT_DONE\schedule_form.fmb	Form that allows the user to select a consultant, and then call a chart that displays the consultant's schedule on a Gantt chart
\SWEPROJECT_DONE\schedule_chart.ogd	Chart that displays each consultant's schedule on a Gantt chart; called by schedule_form.fmb

Table 11-6   Module files in Software Experts database application

a. Use Project Builder to create a new project titled "Software Experts System." Name the project registry file sweproject.upd, and store the registry file in the SWEPROJECT_DONE folder in the Chapter11\CaseSolutions folder that you created in Case 2 in Lesson A of this chapter. Specify that the SWEPROJECT_DONE folder is the default project directory. Identify yourself as the project author, and specify the date the project was created in the Comments text box.

b. Add the input files shown in Table 11-6 to the project, except for schedule_chart.ogd. These files are in your SWEPROJECT_DONE folder.

c. View the project in Dependency View, and add the schedule_chart.ogd file to the schedule_form.fmb file to create an explicit dependency. Your project files should look like Figure 11-22.

d. Create a project database connection named Software Experts Connection that uses your username, password, and connect string, and apply the connection to the project.

e. Create a subproject named New Projects, and specify that the subproject uses the same project directory as its parent project. Apply the Software Experts Connection to the subproject. Identify yourself as the project author, and specify the date the project was created in the Comments text box. Do not add any files to the subproject.

f. Use Project Builder to rebuild all of the project files.

g. Use Windows Explorer to create a subfolder named SWESTAGING in your SWEPROJECT_DONE folder, and then use the Delivery Wizard to prepare the project files for delivery. Specify to deliver all project files, and do not specify any Start menu labels.

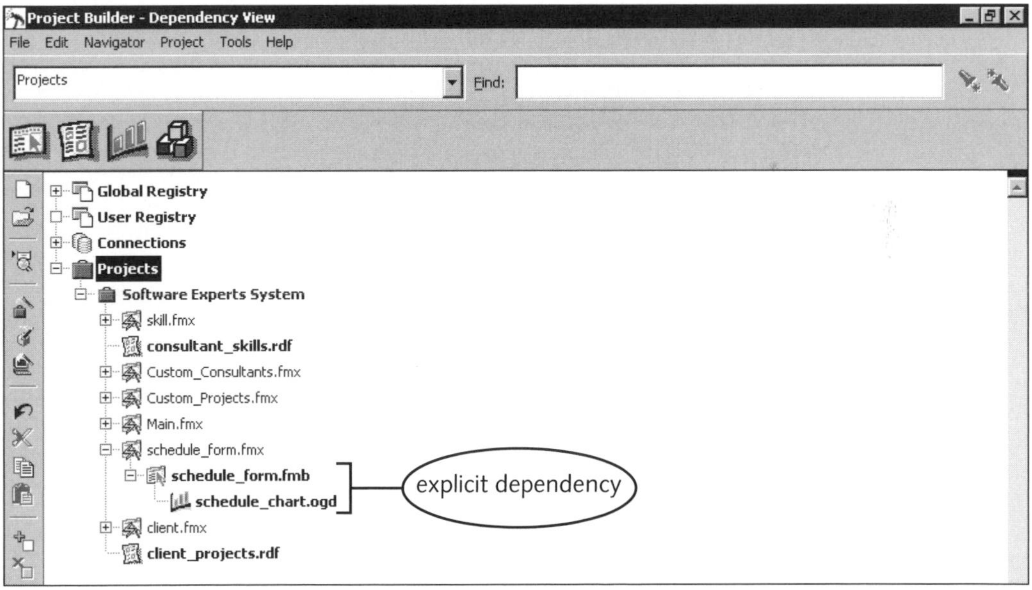

Figure 11-22

◀ LESSON C ▶

Objectives

- ♦ Understand form menu components
- ♦ Create and display custom pull-down menus on forms
- ♦ Create a menu toolbar
- ♦ Create context-sensitive pop-up menus
- ♦ Learn how to control menu items programatically
- ♦ Learn how to use menus to enforce form security

A Windows application usually has a menu bar that contains **pull-down menus** that allow the user to click a top-level choice and display related selections. Applications can also have **menu toolbars**, which are toolbars that display iconic buttons that provide the same functions as pull-down menu selections. Many applications have **pop-up menus**, which appear when the user right-clicks the mouse pointer. Pop-up menus are context-sensitive, and different menu selections appear, based on the current screen selection. In this lesson, you will learn how to create these menu components in your database applications. You will also learn how to use menus to enforce application security.

CREATING CUSTOM PULL-DOWN MENUS

Currently, whenever you display a form in Forms Runtime, the default Forms Runtime pull-down menu appears on the menu bar. This menu has the following selections: Action, Edit, Block, Field, Record, Query, Window, and Help. These menu selections are appropriate for data block forms, but often you need to create custom menus for custom form applications.

To replace the default Forms Runtime pull-down menu choices with custom pull-down menu choices, you create a menu module. A **menu module** is a separate module, which you create in Form Builder, that is independent of any specific form. It can contain individual menus and objects such as program units and parameters. A menu module is saved in your workstation's file system as a design file with an .mmb extension and as an executable file with an .mmx extension. You attach the executable (.mmx) menu file to a form module in the form module Property Palette, and the custom menu selections appear when you open the form in Forms Runtime.

A menu module contains one or more menu items. A **menu item** is a set of menu selections that appears horizontally on the menu bar. Each menu module has a Main Menu property, which specifies the name of the menu item that appears when the menu is attached to a form. Since a menu module can have multiple menu items, you can store several different menu items in a single menu module, and then programmatically specify the menu item that appears when the menu is attached to a form. This reduces the number of menu modules that you have to create.

11

The design sketch for the Clearwater Traders database application in Figure 11-1 shows the design for the application's custom pull-down menu selections. In the following sections, you will learn how to create this menu. First you will start Project Builder, where the Clearwater Traders Database System project that you created in the previous lesson will appear. Then, you will start Form Builder from the Project Navigator window by using the Launcher toolbar. You will create a new menu module, and a menu item for the Clearwater Traders database application. Since the menu module is a separate module, distinct from any form in the application, you do not need to open or create a form while you are working with the menu module.

To start Project Builder, and then start Form Builder and create the menu module and menu item:

1. Start Project Builder. The Clearwater Traders Database System project appears in the Project Navigator. If the project does not appear, click **File** on the menu bar, click **Open Project**, navigate to your Chapter11\CWPROJECT_DONE folder, and open **clearwaterproject.upd**.

 To complete the exercises in this lesson, you must have previously completed the exercises in Lesson B to create the project in Project Builder.

2. Click the **Form Builder 6i button**, which is the first button on the Launcher toolbar. The Welcome to Form Builder window appears. Click **Cancel**. The Object Navigator window opens, and a new form named MODULE1 appears. If necessary, maximize the Object Navigator window.

3. If necessary, click MODULE1 to select it, and then click the **Delete button** [×] to delete the form module.

4. Select the top-level **Menus node**, and then click the **Create button** [+] to create a new menu module. Rename the new menu module **CW_MENU**.

5. Select the **Menus node** under CW_MENU, and then click [+] to create a new menu item in the menu module. Rename the new menu item **MAIN_MENU**. Your Object Navigator window should look like Figure 11-23.

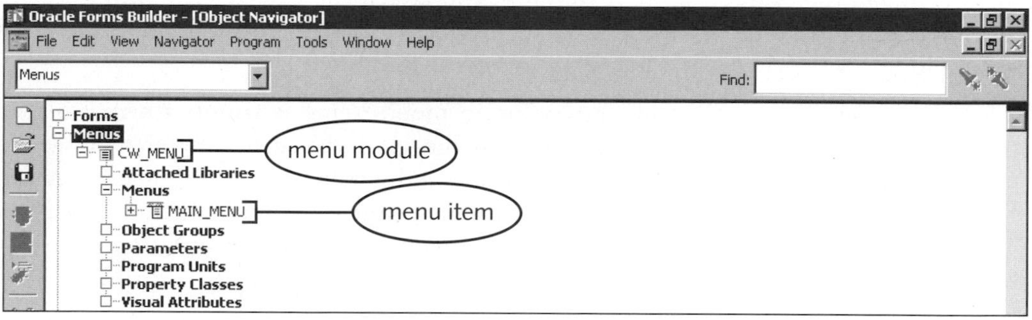

Figure 11-23   Creating a new menu module and menu item

6. Click **File** on the menu bar, click **Save**, and save the menu module as **CW_MENU.mmb** in the CWPROJECT_DONE folder in your Chapter11\TutorialSolutions folder.

The next step is to specify the pull-down menu selections. To do this, you use the Form Builder Menu Editor.

The Form Builder Menu Editor

The Form Builder Menu Editor allows you to visually specify the menu bar structure and to define the underlying triggers that fire when a user selects a menu choice. To open the Menu Editor, you double-click the **Menu module icon** [≣] or double-click the **Menu item icon** [≣]. Figure 11-24 shows the components of a menu.

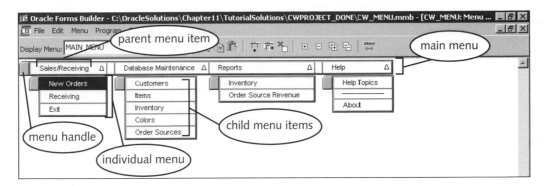

Figure 11-24   Menu components

The **main menu** defines the **parent menu items**, which are the menu selections that appear horizontally on the menu bar. Each parent menu item has an **individual menu** that contains its **child menu items**, which are the choices that appear when the user clicks the parent menu item. When the user clicks a child menu item, a menu code trigger runs. A **menu code trigger** is a PL/SQL program that is associated with a child menu item and contains commands to perform actions such as calling another form, calling a report, or clearing the form. Both the main menu and each individual menu have a **menu handle**, which allows you to detach the menu and move it to a different location in the Menu Editor. In a form menu, child menu items can have associated individual menus. For example, you could click a top-level parent menu item, point to a second-level child menu item to view its menu items, and then click one of the third-level child menu items. Menu items can be nested to as many levels as necessary. It is a good design practice to nest menu items no deeper than three levels.

You can use the Menu Editor to create the menu items by creating the parent menu items across the menu bar and the child menu items below their associated parents. When the Menu Editor first opens, a default parent menu item appears in the top-left corner. You create the new parent menu items from left to right, using the default parent menu item as the starting point. To create the first new parent menu item, you select the default parent menu item and click the **Create Right button** on the Menu Editor toolbar to create the new parent menu item that appears on the right side of the default parent menu item. You can continue to click to create parent menu items from left to right across the top of the menu bar. To change the label of a menu item, you select the item, delete the default label text, and then type the desired label text. To create a new child menu item, you select the associated parent menu item, and then click the **Create Down button** on the toolbar to create the first child menu item. To create the next child menu item, you select the child menu item that you just created, and then click . To delete a menu item, you select the item, and then click the **Delete button** .

Now you will open the Menu Editor and create the parent and child menu items shown in the design sketch in Figure 11-1. (These menu items also appear in the Menu Editor window in Figure 11-24.)

Note that the Window selection, which is the rightmost parent menu item in the design sketch in Figure 11-1, is not shown in Figure 11-24, and you should not create it. In the Form Builder Menu Editor, you do not explicitly create the Window menu selection. It is automatically added at runtime and is always the rightmost parent menu item.

To open the Menu Editor and create the parent and child menu items:

1. Double-click the **Menu module icon** ▤ beside CW_MENU. The Menu Editor opens. The default parent menu item named <New_Item> appears.

To open the Menu Editor, you can also select a menu module or menu item, click Tools on the menu bar, and then click Menu Editor.

2. Click **<New Item>** so the insertion point appears in the menu item label, and the background appears blue. Delete the current label (<New Item>, including the angle brackets), and type **Sales/Receiving** to specify the first parent menu item label.

If you press Delete when the menu item has a black background, you will delete the entire menu item rather than just the label text.

3. Click the **Create Right button** 🖪 on the Menu Editor toolbar to create a new parent menu item to the right of the Sales/Receiving menu item.

4. Delete the current item label, and type **Database Maintenance** to rename the new parent menu item.

5. Select the **Sales/Receiving** parent menu item so its background turns black, and then click the **Create Down button** 🖩 on the Menu Editor toolbar to create a child menu item under the Sales/Receiving menu item.

6. Delete the current menu item label, and type **New Orders** to change the child menu item label.

7. Click 🖩 again to create the second child menu item, and change the new child menu item's label to **Receiving.**

8. Click 🖩 again to create another child menu item, and change the new child menu item label to **Exit**.

9. Select the **Database Maintenance** parent menu item, and then create its child menu items, as shown in Figure 11-24.

10. Repeat Steps 3 through 9 to create all of the parent and child menu items shown in Figure 11-24. Under the Help parent menu item, just create the Help Topics and About child menu items. You will create the divider line between the Help Topics and About child menu items later.

To edit a menu label, select the menu item, and then click the label again to place the insertion point in the label.

11. Click **File** on the menu bar, then click **Save** to save the menu module.

Menu Item Properties

Each menu item has an associated Property Palette in which you can specify properties to customize the item appearance and functionality. Table 11-7 summarizes some of the properties of menu items.

Property Node	Property Name	Description
Functional	Enabled	Specifies whether the menu item is enabled or disabled when the menu first opens
Functional	Menu Item Type	Determines how the menu item appears in the menu and whether the item can have associated menu code. Allowable values are Plain, Check, Radio, Separator, and Magic.
Functional	Visible in Menu	Specifies whether the menu item is visible or hidden
Functional	Visible in Horizontal Menu Bar Visible in Vertical Menu Bar	Specifies whether a button corresponding to a menu item appears on a horizontal or vertical menu toolbar
Menu Security	Item Roles	Allows you to implement menu security based on database roles
Menu Security	Display Without Privilege	Allows you to specify if the menu item appears to users who do not have the required privileges to use the item

Table 11-7   Menu item properties

Menu toolbars and menu security will be discussed in detail in a later section.

The Menu Editor supports the following different menu item types:

- Plain, which displays a text label and has an associated action trigger that fires when the user selects the menu item.

- Check, which specifies a property that can be enabled or disabled. For example, you could have a check menu selection labeled "Toolbar" that specifies whether a toolbar appears or is hidden.

- Radio, which specifies a selection in a group of menu selections that behave like radio buttons, in which only one selection can be activated at a time. For example, you could create a radio menu selection that specifies whether the switchboard buttons appear as command buttons or as iconic buttons.

- Separator, which specifies that the menu selection appears as a separator bar.

- Magic, which allows you to specify that the menu selection is one of the following predefined *magic types*: Cut, Copy, Paste, Clear, Undo, About, Help, Quit, or Window. The Cut, Copy, Paste, Clear, Window, and Quit magic menu items have built-in functionality supplied by Form Builder, while the other magic menu item types can have commands automatically associated with them.

By default, all menu items are Plain item types unless you specify otherwise. To change a menu item's item type property, you create the menu item, open its Property Palette, and modify the Menu Item Type property. Now you will create the separator bar in the Help submenu that separates Help Topics and About in Figure 11-24. You will create a new child menu item between the Help and About menu items. Then you will open the new item Property Palette and change its Menu Item Type property to Separator. You will also change the Exit menu item to a magic item type. The Exit item will be a Quit magic type, so the form closes when the user selects Exit.

To create the separator bar menu item and modify the Exit menu item type:

1. Select the **Help Topics** child menu item, and then click the **Create Down button** to insert a new child menu item under Help Topics.

2. Select the new menu item, right-click, and then click **Property Palette** to open the new menu item Property Palette.

> You can also open the Property Palette of a menu item by double-clicking the menu item or by selecting the menu item, clicking Tools on the menu bar, and then clicking Property Palette.

3. Change the name of the menu item to **SEPARATOR_BAR**.

4. Open the **Menu Item Type** property list, select **Separator**, and close the Property Palette.

5. Double-click the **Exit** menu item to open its Property Palette, change its Menu Item Type value to **Magic**, and its Magic Item value to **Quit**. Close the Exit Property Palette.

6. Save the menu module.

Closing and Reopening Menu Module Files and Individual Menu Items

Recall that menu modules are stored separately from form applications. Therefore, to open a menu module, you need to retrieve it from the file system independently of the form in which it will appear. For example, suppose you decide to take a break and close your menu module file and work on it later. In the following steps, you will close the menu module and then reopen it.

To close the menu module file then reopen it:

1. Click **File** on the menu bar, then click **Close** to close the CW_MENU menu module. The Object Navigator window appears, and the CW_MENU menu module object no longer appears under the Menus object.

2. Click the **Open button** on the Object Navigator toolbar, click the **Files of type** list arrow, select **Menus (*.mmb)**, navigate if necessary to your CWPROJECT_DONE folder, select **CW_MENU.mmb** from the file list, and then click **Open** to open the file.

3. To open the Menu Editor, double-click the **Menu module icon** beside CW_MENU.

In the Menu Editor, you can open individual parent menu items to show their associated child menu items, and close individual parent menu items to hide their associated child menu items. When a parent menu item has child menu items, and the child menu items are not currently open, a **down-pointing tab** appears on the parent menu item. When a menu item has child menu items, and the child menu items are open, an **up-pointing tab** appears on the parent menu item. When a menu item does not have child items, no tab appears on the menu item. Now, you will open and close individual parent menu items.

To open and close parent menu items:

1. Notice that each parent menu item with underlying child selections has an up-pointing tab beside the label name if the child menu items are visible, and a down-pointing tab beside the label name if they are not. If necessary, click on the Sales/Receiving menu item to display its child menu items. Note that the menu item's tab is now an up-pointing tab, indicating that its child items now appear. Also, note that the child items do not have tabs beside them, which indicates that they do not have child items.

2. Click on the Sales/Receiving menu item tab to hide the item's child items.

11

Creating Access Keys

Most pull-down menu selections have an underlined letter in the selection label. This letter is called the menu item's **access key**, and allows the user to open or select the menu item by using the keyboard instead of the mouse pointer. In Forms Runtime, users can press the Alt key plus the access key to open parent menu items. Once a parent menu item is opened, users can select a child menu item simply by pressing the child item's access key, without pressing the Alt key.

The first letter of each menu item label is the default access key. If two menu selections have the same first letter, then the access key will only work for the first selection. For example, in the Database Maintenance parent menu item in Figure 11-24, the Item and Inventory child menu items both start with the letter *I*. If the user opens the Database Maintenance parent menu item, and then presses I, the Items menu item will be selected every time. Therefore, it is sometimes necessary to change the default access key, because it already has been used or because another key seems more intuitive. For example, the access key for Exit might be *X* rather than *E* because *X* sounds more like Exit. To override the default access key choice, you type an ampersand (&) before the desired access key letter in the menu label. The ampersand will not appear on the menu when the menu appears in Forms Runtime.

If you want an ampersand to appear in a menu choice, type the ampersand twice. For example, to create the menu label Research & Development, you would type the label as Research && Development. Now, you will add the access key definitions shown in the menu design sketch in Figure 11-1. You will also change the label for the Sales/Receiving menu item to Sales & Receiving, by entering the ampersand two times.

To change the menu access keys and add the ampersand to the Sales/Receiving label:

1. Open the **Sales/Receiving** menu item so the Exit selection appears.

2. Select the **Exit** child item by clicking it, click it again to open the label for editing, and type **&** before the x so the label now reads E&xit.

3. Open the Database Maintenance menu item so its child menu items appear, and change the label for the Inventory selection to **I&nventory**, and the label for the Colors selection to **C&olors**.

4. Select the Sales/Receiving menu item, click it again to open the label for editing, and change the label to **Sales && Receiving** so an ampersand appears in the menu label when the menu appears in Forms Runtime. Your modified menu labels should look like the labels in Figure 11-25.

5. Save the menu module.

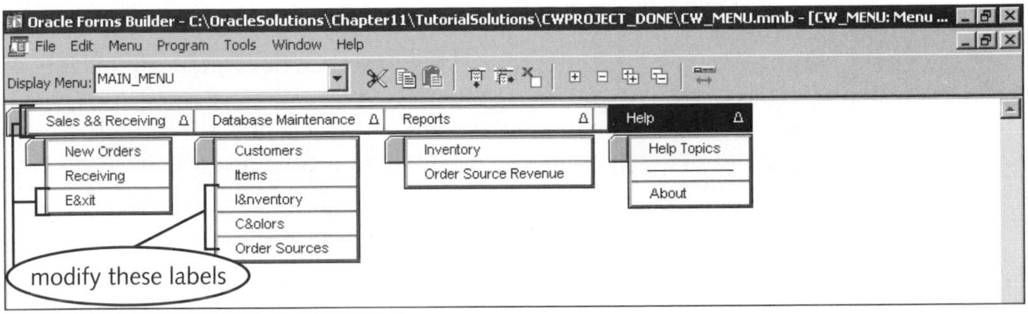

Figure 11-25 Modifying the menu labels

Creating Menu Code Triggers

Recall that child menu items have associated menu code triggers, which are PL/SQL programs that contain commands that execute when the user selects the child menu item. Menu code triggers can contain the following types of PL/SQL commands:

- Built-in commands that do not reference specific form items, such as CLEAR_FORM or LIST_VALUES. Menu code can also contain keyboard triggers, which are summarized in Table 8-4.

- Module navigation commands, such as the CALL_FORM and OPEN_FORM commands to navigate to other forms, and the RUN_PRODUCT command to open reports and charts.

Menu code commands can reference global variable values, such as the global project path that specifies the location of the application files. Menu code triggers cannot contain commands that directly reference specific form items, such as GO_ITEM(:inventory_block.inv_id). This is because the menu is not associated with a specific form, so the form item is not visible to the menu module, and a compile error will occur when you compile the menu code trigger.

Every child menu item must have an associated menu code trigger, or an error will occur when you compile the menu module. Now you will implement the menu code trigger for the New Orders selection under the Sales & Receiving parent menu item. To create a menu code trigger for a child menu item, you select the menu item, right-click, and then click PL/SQL Editor. This opens the PL/SQL Editor, in which you specify the trigger code. Since the New Orders menu selection calls the Sales.fmx form, which is stored in your CWPROJECT_DONE folder, you will use the CALL_FORM built-in to call the form, and use the global project path variable to specify the form file location.

11

To create the New Orders item menu code trigger:

1. In the Menu Editor, right-click the **New Orders** menu item, and then click **PL/SQL Editor** on the menu. The system automatically creates a menu code trigger for the menu item, and you can enter the code that will execute when the user selects the menu item. Maximize the PL/SQL Editor window if necessary.

2. Type the following code in the source code pane, compile the trigger, debug it if necessary, and close the PL/SQL Editor.

```
CALL_FORM(:GLOBAL.project_path || 'sales.fmx');
```

3. Save the menu module.

Next, you will create the menu code trigger for the Receiving child menu item under the Sales & Receiving parent menu item, and the triggers for the Customers, Items, and Inventory selections under the Database Maintenance parent menu item.

To create the menu code triggers:

1. Right-click the **Receiving** menu item, and then click **PL/SQL Editor** on the menu. Type the following code in the Source code pane, compile the trigger, and close the PL/SQL Editor.

```
CALL_FORM(:GLOBAL.project_path || 'receiving.fmx');
```

2. Create menu code triggers with the associated code for the following child menu items:

Child Menu Item	Code
Customers	`CALL_FORM(:GLOBAL.project_path || 'customer.fmx');`
Items	`CALL_FORM(:GLOBAL.project_path || 'items.fmx');`
Inventory	`CALL_FORM(:GLOBAL.project_path || 'inventory.fmx');`

3. Save the menu module.

You also need to create the menu code triggers for the Inventory and Order Source Revenue child menu items, which are found under the Reports parent menu item. These triggers will use the RUN_PRODUCT built-in procedure, which you learned about in Chapters 9 and 10, to call reports and charts from a form. Since you will not pass a parameter list to either the report or chart, you pass a blank text string, which is defined by two single quotation marks, as the parameter list ID value.

To create the Report child menu item triggers:

1. In the Menu Editor, open the Reports parent menu item if necessary, right-click the **Inventory** menu item, and then click **PL/SQL Editor** to open the PL/SQL Editor and create the trigger.

2. Type the following command in the source code pane. Note that the next to last parameter value is a blank text string, defined by two single quotation marks.

```
RUN_PRODUCT(REPORTS, :GLOBAL.project_path ||
'inventory.rdf', ASYNCHRONOUS, RUNTIME,
FILESYSTEM, '', NULL);
```

3. Compile the trigger and correct any syntax errors. Do NOT close the PL/SQL Editor.

4. In the PL/SQL Editor, open the Name list, and select **ORDER_SOURCE_REVENUE** to open the PL/SQL Editor for the Order Source Revenue child menu item.

5. Type the following command in the source code pane:

```
RUN_PRODUCT(GRAPHICS, :GLOBAL.project_path ||
'ordersource.ogd', ASYNCHRONOUS, RUNTIME,
FILESYSTEM,'',NULL);
```

6. Compile the trigger, correct any syntax errors, and then close the PL/SQL Editor.

7. Save the menu module.

To complete the menu code triggers, you need to create the stubs for the child menu selections that do not yet have menu code. These items are the Color and Order Sources child menu items under the Database Maintenance parent menu item, and the Help Topics and About child menu items under the Help parent menu item. To implement the stubs, you will display a message on the status line to inform the user that the feature has not yet been implemented.

To create the menu code stubs for the unimplemented menu items:

1. Create a menu code trigger for the Colors menu item using the following code: **MESSAGE('Feature not implemented yet.');**

2. Compile the trigger and correct any syntax errors, then close the PL/SQL Editor.

3. Repeat Steps 1 and 2 for the Order Sources child menu items under the Database Maintenance parent menu item, and for the Help Topics and About child menu items under the Help parent menu item.

4. Save the menu module.

Creating an Executable Menu File

When you attach a pull-down menu module to a form, you specify the full path to the menu module's compiled .mmx file in the form Property Palette. Before you can attach a pull-down menu module to a form, you must compile the menu design (.mmb) file into an executable (.mmx) menu file. In addition, every time you modify the menu module, you must recompile the .mmb file into a new .mmx file. Otherwise, your form will use your old .mmx file and will not show the most recent changes. It is best to compile your menu module in Form Builder the first time you compile it, so you can correct errors immediately. When you make incremental changes later on, you can recompile the menu module directly in Project Builder. Now you will compile the menu module and generate the executable menu file.

To generate the executable menu module file:

1. In the Menu Editor, click **File** on the menu bar, point to **Administration**, and then click **Compile File**. If you have not logged onto the database yet, you will be prompted to do so. The message "Module built successfully" on the status line indicates that the executable file was created successfully.

Another way to compile a menu module is to select one of the menu items in the Object Navigator, click File on the menu bar, point to Administration, and then click Compile File. If you select a form item rather than a menu item, this command will compile the current form rather than the menu module.

If a menu compile error message like the one shown in Figure 11-26 appears, it indicates that you forgot to create a menu code trigger for one of the child menu items. The "No PL/SQL source code in menu item NEW_ORDERS" message indicates that the New Orders menu item is missing its menu code. Recall that every child menu item must have a menu code trigger. If you encounter a similar error, click OK, add the necessary trigger code under the menu selection indicated in the error message and then recompile the menu file.

```
Compilation Errors                                                    ⊠
┌─────────────────────────────────────────────────────────────────┐▲
│  - Creating menu module                                         │
│C:\OracleSolutions\Chapter11\TutorialSolutions\CWPROJECT_DONE\CW_M│
│ENU.mmx.                                                         │
│  - Inserting menu MAIN_MENU.                                    │
│  - Inserting menu SALESRECEIVING_MENU.                          │
│FRM-31651: No PL/SQL source code in menu item NEW_ORDERS.        │
│                                                                 │▼
└─────────────────────────────────────────────────────────────────┘
                              [ Search ]  [ OK ]  [ Cancel ]
```

Figure 11-26   Menu compile error

2. After you have successfully compiled your menu module, close Form Builder. Click **Yes** if you are asked if you want to save your changes.

Adding the Menu Module to the Project

To attach a custom pull-down menu module to a form, you open the form's Property Palette and specify the full path to the menu module's compiled .mmx file as the form's Menu Module property value. Now you need to add the menu module files to the project, attach the menu module to the form in the main form module's Property Palette, and test the menus. First you will add the menu module design (.mmb) and executable (.mmx) files to the project in Project Builder.

To add the menu module files to the project:

1. If necessary, click Project Builder on the taskbar to open Project Builder.

2. If necessary, click Clearwater Traders Database System to select the project, and then click the **Add Files To Project button** 📄. The Add Files to Project dialog box opens.

3. Click **CW_MENU.mmb** to add the menu module design file to the project, and then click **Open**. Note that both the Forms Builder menu document and Forms Builder menu executable file types appear in the Project Navigator, because the target (.mmx) file is automatically added to the project when the design (.mmb) file is added.

4. Click ➕ beside Forms Builder menu document, and click ➕ beside Forms Builder menu executable to view the menu module files.

11

Displaying a Custom Menu on a Form

Next, you will modify the main form module file so that it displays the custom menu. To do this, you will open the MAIN.fmb document file, and change the form's Menu Module property. The default value for the form Menu Module property is DEFAULT&SMARTBAR. The DEFAULT specification indicates that the form displays the default Forms Runtime pull-down menus. The SMARTBAR specification indicates that the Forms Runtime window displays the Forms Runtime toolbar that you use for working with data block forms. Since the main module form is not a data block form, you will not display the smartbar. Now you will modify the form's Property Palette to specify the full folder path to the cw_menu.mmx file. The property value must include the full path specification to the menu .mmx file, including the drive letter.

You cannot use the :GLOBAL.project_path variable in the form Property Palette. You must specify the complete path to the menu file.

To display the custom menu on the main form module:

1. In the Project Navigator, if necessary, click ➕ beside the Forms Builder document node to display the project's .fmb files. Double-click the **Forms Builder Document file icon** 📄 beside MAIN.fmb to open the main form

module in Form Builder. Form Builder opens, and the MAIN form module appears in the Object Navigator.

2. In the Form Builder Object Navigator, double-click the **Form icon** next to MAIN to open the form Property Palette. Confirm that the Menu Source property value is File.

3. Scroll down to the Menu Module property, and change the value to the full path and filename of your newly generated menu module. If your CWPROJECT_DONE folder is stored on the C drive in the OracleSolutions\Chapter11\TutorialSolutions folder path, the full path would be C:\OracleSolutions\Chapter11\TutorialSolutions\CWPROJECT_DONE\CW_MENU.mmx.

 You cannot specify the menu module location using the :GLOBAL.project_path variable, so if you change the location of the project files, you will need to change the menu module path in the form Property Palette.

4. Close the Property Palette, save the form, and then close Form Builder.

Now you must rebuild the main module .fmb file. To rebuild the file, you will use the Build Selection option in Project Builder. Whenever you make a change to a single project file, you should build only the changed file, because it is time-consuming to rebuild files that do not need to be rebuilt. Recall that to build a single file, you select the target (.fmx) file in the Project Navigator window, and then click the **Build Selection button** on the Project Navigator toolbar. After you rebuild the file, you will run the form to confirm that the custom menus appear.

To rebuild the main module form file and test the menus:

1. If necessary, click Project Builder on the taskbar. In the Project Navigator, if necessary, click **+** beside the Forms Builder executable node, and then select **MAIN.fmx**.

2. Click the **Build Selection button** on the toolbar to build the selected file. When the "1 item built." message appears, click **OK**.

3. Double-click the **Forms Builder executable icon** beside MAIN.fmx under the Forms Builder executable node to run the form. Your custom pull-down menus should appear on the main application menu bar, as shown in Figure 11-27. Note that the Forms Runtime toolbar no longer appears.

Figure 11-27   Form with custom menus

 If the error message "FRM-10221: Cannot read file *path_to_mmx_file*" appears, it is because you did not specify the filename or path of the menu executable file correctly in the form Menu Module property or because the menu module .mmx file is not in the specified folder. Open the main module form, correct the path specification, save the form, rebuild the form, and then run the form again.

4. Test all of your pull-down menu choices to confirm that they call the correct form, report, or chart, and that the "Not implemented yet" message appears for unimplemented selections. Note that all of the called forms display the custom menu selections also.

 The data block forms (Customers, Items, and Inventory) will not work correctly because they require the default Forms Runtime menu module and the smartbar. You will learn how to restore the default menus to these forms later in the lesson.

5. Test all of the pull-down menu choices using the access keys to confirm that the access keys work correctly. Recall that to open a top-level menu selection, you press Alt plus the access key, and that to execute a child-level menu selection, you simply press the access key.

6. Click **Exit** to close Forms Runtime.

Specifying Alternate Pull-down Menus in Called Forms

In the Clearwater Traders application, the Menu Module property of all of the child forms (Sales.fmb, Receiving.fmb, Customer.fmb, Item.fmb, and Inventory.fmb) has the default Menu Module value, which is DEFAULT&SMARTBAR. However, every time you call a child form from the main form module, the child form displays the custom menu module that appears in the parent form. To enable a child form to display its own menu module, rather than its parent's menu module, you need to modify the CALL_FORM command that calls the child form. Recall that the syntax for the CALL_FORM command is as follows:

```
CALL_FORM ('form_specification', display, switch_menu,
    query_mode, parameter_list_id);
```

Recall that the *display* parameter specifies whether or not the parent form is hidden when the child form appears. Valid values are HIDE (the parent form is not visible) and NO_HIDE (the parent form remains visible). The *switch_menu* parameter specifies whether the child form's standard pull-down menus are inherited from the parent form, or if the child form displays its own menu module, as specified in the child form's Menu Module. Valid values are NO_REPLACE (pull-down menus are inherited from the calling form) and DO_REPLACE (different pull-down menus are used). Now, you will modify the CALL_FORM command in the menu code for the Customers child menu item, which is under the Database Menu parent item selection. You will add the HIDE parameter to specify that the parent form is not visible, and you will add the DO_REPLACE parameter to the command to specify that the child form will display its own menus. You will accept the default values for the *query_mode* and *parameter_list_id* parameters by omitting their values. Since you are not accepting the default value for *switch_menu*, you must explicitly include the value for *display*, because *display* appears before *switch_menu* in the parameter list for the CALL_FORM command.

To edit the menu code for the Customers menu item:

1. In the Project Navigator, double-click the **Menu item icon** beside CW_MENU.mmb to open the menu module in Form Builder. Form Builder opens, and the CW_MENU module appears in the Object Navigator.

2. In the Form Builder Object Navigator, double-click the **Menu Module icon** beside CW_MENU. The Menu Editor opens. If necessary, maximize the Menu Editor window.

3. If necessary, open the Database Maintenance parent menu item, right-click the **Customers** child menu item, and then click **PL/SQL Editor** to open the menu item's action trigger.

4. Modify the menu code command as follows:

```
CALL_FORM(:GLOBAL.project_path || 'customer.fmx', HIDE,
    DO_REPLACE);
```

5. Compile the trigger, correct any syntax errors, and close the PL/SQL Editor.

6. Save the menu module, and then exit Form Builder.

Now you will rebuild the menu module in Project Builder, run the main form module from Project Builder, and then select the Customers child menu item to confirm that the default data block menus appear in the Customer form.

To rebuild the menu module, run the application, and test the modified menu module:

1. Switch to Project Builder. In the Project Navigator window, select **CW_MENU.mmx** under Forms Builder menu executable node, and then click the **Build Selection button** to rebuild the menu module. Click **OK** when the "1 item built." message appears.

> If a message stating "No items were built" appears, open the CW_MENU.mmb file in Form Builder, click File on the menu bar, point to Administration, and click Compile File to compile the menu and generate the new CW_MENU.mmx file in Form Builder. Then close Form Builder.

2. Double-click the **Forms Builder executable icon** beside MAIN.fmx to run the application. The main form module opens.

3. Click **Database Maintenance** on the menu bar, and then click **Customers** to open the customer form. The default Forms Runtime pull-down menu and toolbar appear on the form, confirming that the child form now displays its own menus rather than the menus of the parent form.

> If the customer form still displays the custom menu selections, make sure that you modified the CALL_FORM command in the Customers menu code to use the DO_REPLACE option. Also make sure that you rebuilt the modified menu module's .mmx file.

4. Click **Return** to exit the Customer form and display the main application form.

5. Click **Exit** to close the form.

11

CREATING A MENU TOOLBAR

A form menu toolbar displays iconic buttons that appear horizontally under the pull-down menus or vertically on the left edge of the window. Each button on a menu toolbar is associated with a specific child menu item. When the user clicks a button, it is the equivalent of selecting the child menu item on the pull-down menu.

To create a menu toolbar, you open the menu module that contains the menu items that will correspond to the menu toolbar selections. To create a toolbar selection, you open the Property Palette of the selected menu item. To display the item in a horizontal toolbar, you change the menu item's Visible in Horizontal Menu Toolbar property to Yes. To display the item in a vertical toolbar, you change the item's Visible in Vertical Menu Toolbar property to Yes. Then, you specify the complete path specification to the icon (.ico) file that will represent the menu selection on the toolbar. The buttons appear on the toolbar in the same order in which their commands appear on the menu, from the top down first, and then from left to right. When the user moves the mouse pointer onto a toolbar button in Forms Runtime, a ToolTip appears that shows the text of the menu label associated with the menu selection.

Now you will create a horizontal toolbar in the main application form in the Clearwater Traders database system. You will create menu toolbar selections associated with the New Orders and Receiving child menu items under the Sales & Receiving parent menu item, and with the Inventory child menu item under the Reports parent menu item. Since the buttons will appear in a top-down, left-to-right order, the buttons will appear in the following order: New Orders, Receiving, and Inventory.

To create the horizontal toolbar selections, you will open the Property Palettes for the New Orders, Receiving, and Inventory (report) menu items. You will change the Visible in Horizontal Menu property values for all three items to Yes. You will also specify the filenames of the icon files that will appear on the toolbar. The New Orders menu item will be represented by an icon that is stored in a file named sales.ico. The Receiving menu item will be represented by an icon that is stored in a file named receiving.ico, and the Inventory report item will be represented by an icon that is saved in a file named inventory.ico. All three icon files are stored in your CWPROJECT_DONE folder. First, you will create the horizontal toolbar button for the New Orders menu item.

To create the horizontal toolbar button for the New Orders item:

1. In the Project Navigator, double-click the **Menu item icon** 🗒 beside CW_MENU.mmb to open the menu module in Form Builder. Form Builder opens, and the CW_MENU module appears in the Object Navigator. If necessary, maximize the Object Navigator window.

2. Double-click the **Menu module icon** 🗐 beside CW_MENU to open the Menu Editor.

3. If necessary, open the Sales & Receiving parent menu item, and then double-click the **New Orders** child menu item to open the child menu item's Property Palette.

4. Change the Visible in Horizontal Menu Toolbar property value to **Yes**.

5. In the Icon Filename property, type the complete folder path specification to your CWPROJECT_DONE folder, followed by the icon filename, which is **sales**. You do not need to include the .ico file extension. If your CWPROJECT_DONE folder is saved in the C:\OracleSolutions\Chapter11\TutorialSolutions folder, then your complete filename specification will be c:\oraclesolutions\chapter11\tutorialsolutions\cwproject_done\sales.

6. Close the Property Palette, and save the menu module.

Now, you will create the horizontal toolbar buttons for the Receiving and Inventory menu item 3. The associated icon files are named receiving.ico and inventory.ico, and are stored in your CWPROJECT_DONE folder.

To create the toolbar buttons for the Receiving and Inventory menu items:

1. In the Menu Editor, open the Property Palette for the Receiving child menu item, change the Visible in Horizontal Menu Toolbar property value to **Yes**, and change the Icon Filename property to the complete folder path specification to your CWPROJECT_DONE folder, followed by the icon filename, which is **receiving**. You do not need to include the .ico file extension.

2. Close the Property Palette.

3. If necessary, open the Reports parent menu item, and then open the Property Palette for the Inventory child menu item. Change the Visible in Horizontal Menu Toolbar property value to **Yes**, and change the Icon Filename property to the complete folder path specification to your CWPROJECT_DONE folder, followed by the icon filename, which is **report**.

4. Close the Property Palette, and save the menu module.

5. Click **File** on the menu bar, point to **Administration**, and then click **Compile File** to rebuild the menu module .mmx file.

6. Close the Menu Editor, and then close Form Builder.

Now, you will run the main form module in Forms Runtime, and confirm that the menu toolbar appears. Then, you will click the toolbar buttons to confirm that they work correctly.

To run the main form module and test the menu toolbar:

1. If necessary, switch to Project Builder, and double-click the **Forms Builder executable icon** beside MAIN.fmx to open the main form module in Forms Runtime. The application switchboard opens, and the menu toolbar appears below the pull-down menus, as shown in Figure 11-28.

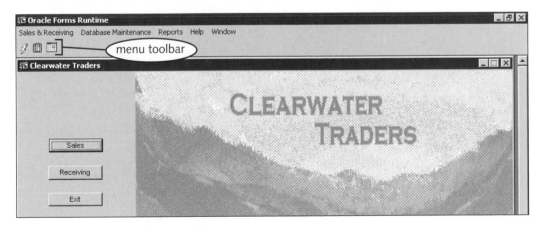

Figure 11-28   Viewing the menu toolbar

2. Move the mouse pointer onto the first button on the toolbar. Notice that the ToolTip "New Orders" appears.

3. Click the **New Orders button** . The sales form opens. Click **Return** to close the form and return to the main application form.

4. Move the mouse pointer onto the second button, and note that the ToolTip "Receiving" appears. Click the **Receiving button** . The receiving form opens. Click **Return** to close the form and return to the main application form.

5. Move the mouse pointer onto the third button, and note that the ToolTip "Inventory" appears. Click the **Inventory button** . The inventory report opens. Close the Reports Runtime window to return to the main application form.

6. Click **Exit** to close the application.

CREATING POP-UP MENUS

Recall that pop-up menus are context-sensitive menus that appear when the user right-clicks a specific screen display item. In Form Builder, pop-up menus are defined as top-level form objects.

> A pop-up menu is associated with a specific form and can appear only in the form in which it is defined. In contrast, a pull-down menu module is an independent object that is not associated with a specific form and can be used in any form.

To create a pop-up menu, you create a new *Popup menu* object in the Object Navigator. Then, you open the object in the Menu Editor, define the menu items, change the menu item labels, and create the menu code triggers. To assign the pop-up menu to a form item, you open the Property Palette of the item to which you want to attach the menu item, and change the item's Popup Menu property value to the name of the associated pop-up menu object.

Now, you will create a pop-up menu in the receiving form in the Clearwater Traders database application. Currently, the receiving form allows the user to select unreceived shipment ID values using an LOV that appears when the user places the insertion point in the Shipment ID text item and presses F9. You will create a pop-up menu that displays the menu selection "List Values" when the user right-clicks the Shipment ID text item. If the user clicks List Values, the LOV display for the Shipment ID LOV will open. First, you will open the receiving form (Receiving.fmb) in Form Builder. Then, you will create the pop-up menu object, change its name, and specify its menu item label and trigger code.

To create and configure the pop-up menu:

1. If necessary, switch to Project Builder, and click ➕ beside the Forms Builder document node. Double-click the **Forms Builder document icon** 📄 beside Receiving.fmb to open the receiving form in Form Builder. Form Builder opens, and the RECEIVING form appears in the Object Navigator.

2. Select the top-level **Popup Menus node**, and then click the **Create button** 🔲 to create a new pop-up menu. The new menu appears in the node list. Change the menu item name to **SHIPMENT_ID_MENU**.

3. Double-click the **Menu item icon** 📇 beside SHIPMENT_ID_MENU to open the menu in the Menu Editor. The Menu Editor opens, and the default <New Item> menu item appears.

4. Change the <New Item> label to **List Values**.

5. Select the **List Values** menu item, right-click, and then click **PL/SQL Editor** to open the PL/SQL Editor. Type the following command in the Source code pane:

   LIST_VALUES;

6. Compile the trigger, correct any syntax errors, and close the PL/SQL Editor.

7. Close the Menu Editor, and save the form.

Now, you need to attach the pop-up menu to the SHIPMENT_ID_TEXT form item. You will open the Property Palette of the SHIPMENT_ID_TEXT text item, and change its Popup Menu property to the name of the pop-up menu, which is SHIPMENT_ID_MENU.

To attach the pop-up menu to the form item:

1. In the Object Navigator, click ➕ beside the Data Blocks node, click ➕ beside RECEIVING_BLOCK, and click ➕ beside the Items node. The block items appear.

2. Double-click the **Text Item icon** 🔠 beside SHIPMENT_ID_TEXT to open its Property Palette.

3. In the Functional property node, open the Popup Menu property list, and select **SHIPMENT_ID_MENU** to attach the pop-up menu to the text item.

4. Close the Property Palette, and save the form.

Now you are ready to test the form and confirm that the pop-up menu works correctly. Since the pop-up menu exists only in the receiving form, you will run the form as a standalone form in Form Builder, rather than running it within the integrated application from Project Builder.

11

To test the pop-up menu:

1. In Form Builder, run the form. The form opens in the Forms Runtime window.

2. Right-click the **Shipment ID text item**. The pop-up menu appears.

3. Click **List Values**. The LOV display for the Shipment ID LOV opens.

4. Click **Cancel**, and then click **Return** to close the form.

5. Close Form Builder.

IMPLEMENTING MENU SECURITY

Database security is enforced by roles and privileges assigned to database objects. You can also enforce application security through menus. For example, you might specify that only DBAs can interact with the Clearwater Traders database application through the pull-down menus, and force all other users to interact with the application only through the switchboard items. You can also implement security for individual menu items. For example, you might allow DBAs to use all menu selections, and allow other users to use all menu selections except Database Maintenance.

Implementing security in a menu module involves the following steps:

1. Enable the security function in the menu module.

2. Specify the database roles that can access the menu module in the menu module Property Palette.

3. Specify the database roles that can access individual menu items in the menu item Property Palettes.

To enable or disable security on the overall menu module, you set the menu module Use Security property to Yes or No. You can set this property to No to disable menu access to all application users. You use the menu module Module Roles property to list the specific database roles that will be granted access to the menu module. (Recall from Chapter 2 that a role is a database object that is associated with a set of system privileges and that a role can be assigned to multiple users.) Then, you grant access to individual menu items by specifying the role name in the menu item's Item Role property.

Now, you will modify the menu security in the CW_MENU.mmb menu module. You will specify that the DBA role can access all of the menu items, and that users with the CT_SALES role can access all menu items except the Database Maintenance menu item.

If you are connecting to a remote database and you do not have the DBA role, your database administrator needs to create a role named CT_SALES and assign the role to your user account. If you are connecting to a remote database and have the DBA role, then you will not see any changes after you implement the menu security.

If you are using Personal Oracle, you already have the DBA role, so you will not be able to see any differences when you change the menu security properties. Therefore, Personal Oracle users will need to change the database connection information and use a different database account that does not have the DBA role. If you are a Personal Oracle user, you need to perform the following steps to configure your project and your database to implement menu security:

1. Right-click the Clearwater Traders Database System object in Project Builder, and then click Property Palette. Change the Username to scott, and the Password to tiger. (This is a default Oracle user account that does not have the DBA role.) Close the Property Palette.

2. Create a role named CT_SALES, and assign it to the scott/tiger account. To do this, start SQL*Plus, and log onto the database as username *system*, password *manager*. (This account has the DBA role.) Type CREATE ROLE ct_sales; at the SQL prompt.

3. In SQL*Plus, type GRANT ct_sales TO scott; at the SQL prompt to grant the new role to user scott.

4. In SQL*Plus, run an initialization script named frm60sec.sql that is stored in the Chapter11 folder on your Data Disk. Don't worry if one or more error messages appear when you run the script, because this is normal.

5. In SQL*Plus, type GRANT SELECT ON frm50_enabled_roles TO PUBLIC;

6. Exit SQL*Plus, start SQL*Plus again, log onto the database using the scott/tiger account, and run the clearwater.sql script that is in the Chapter11 folder on your Data Disk to create the Clearwater Traders sample database tables.

7. Exit SQL*Plus

First, you will open the CW_MENU.mmb menu module in Form Builder. You will then open the menu module's Property Palette, enable menu security, and add the DBA and CT_SALES roles to the Module Roles property to specify that users with these roles will be able to use selected menu items within the menu module. You will associate specific roles with specific menu items later.

To open the menu module, enable menu security, and add the module roles:

1. In Project Builder, double-click the **Menu item icon** 📋 beside CW_MENU.mmb to start Form Builder and open the menu module. Form Builder opens, and CW_MENU appears in the Object Navigator.

2. Right-click the **CW_MENU** module, and click **Property Palette** to open the menu module's Property Palette.

3. Change the Use Security property value to **Yes**.

11

4. Select the **Module Roles property**, and then click **More**. The CW_MENU: Menu Module Roles dialog box opens.

5. Place the insertion point in the first item in the Roles list, and type **DBA**. Place the insertion point in the second item in the Roles list, and type **CT_SALES**.

6. Click **OK** to save your changes and close the Menu Module Roles dialog box.

7. Close the Property Palette, and save the menu module.

Next, you will specify the roles that can use specific menu items. To do this, you open the Property Palette for the item to which you want to assign the role, and select the role or roles in the menu item's Item Role property list. You must explicitly assign at least one role to all parent and child menu items. If a menu item does not have any assigned roles, the item will not be available to any users.

To speed up the role assignment process, you can use an intersection Property Palette. You can assign roles to multiple parent menu items using an intersection Property Palette, and you can assign roles to multiple child menu items using an intersection Property Palette.

 You cannot use an intersection Property Palette that combines parent and child menu items to assign roles to both parent and child menu items at the same time.

First, you will assign the roles to the parent menu items. You will specify that the Sales & Receiving, Reports, and Help parent menu items can be accessed by both the DBA and CT_SALES roles, and you will specify that the Database Maintenance menu item can be accessed only by the DBA role.

To assign roles to the parent menu items:

1. In the Object Navigator, double-click the **Menu module icon** ▤ beside CW_MENU to open the menu module in the Menu Editor.

2. In the Menu Editor, select **Sales && Receiving**, press and hold the **Ctrl** key, select **Reports**, and select **Help**. The three menu items appear selected as an object group.

3. Click **Tools** on the menu bar, and then click **Property Palette** to open the intersection Property Palette.

4. Select the **Items Roles property**, and click **More**. A dialog box stating "Multiple objects selected. Edited value will be applied to all of them" appears. Click **OK**. The CW_MENU: Menu Item Roles dialog box opens, as shown in Figure 11-29.

Figure 11-29   Menu Item Roles dialog box

The Menu Item Roles dialog box lists the roles that have been assigned to the menu module and are available to be assigned to the menu item. When a role is assigned to a menu item, the role appears selected, and its background color is dark blue. Currently, no roles are assigned to the menu item. To assign a role, you select the role in the Menu Item Roles list. You can select multiple roles in the list by selecting the first role, pressing and holding the Ctrl key, and then selecting the additional roles. Now you will assign both the DBA and CT_SALES roles to the Sales & Receiving, Reports, and Help menu items.

To assign the roles to the menu items:

1. In the Menu Item Roles dialog box, select **DBA**, press and hold the **Ctrl key**, and then select **CT_SALES**. Both roles appear selected.

2. Click **OK** to save the selections and close the Menu Item Roles dialog box.

3. Close the intersection Property Palette, and save the menu module.

Now, you will assign the rest of the menu item roles. You will open the Database Maintenance parent menu item, and assign to it the DBA role. Then, you will open an intersection Property Palette for all of the child menu items under the Sales & Receiving, Reports, and Help parent menu items, and assign both the DBA and CT_SALES roles. Finally, you will open an intersection Property Palette for all of the child menus under the Database Maintenance parent menu item and assign the DBA role.

To assign the rest of the menu item roles:

1. In the Menu Editor, open the Database Maintenance menu item Property Palette, click **Item Roles**, click **More**, make sure that DBA is selected, and click **OK**. Close the Property Palette.

2. Select the **New Orders** menu item, press and hold the **Ctrl key**, and then select the following menu items: **Receiving, Exit, Inventory** (under Reports), **Order Source Revenue, Help Topics, (Separator Bar), About**.

3. Click **Tools** on the menu bar, and then click **Property Palette** to open the intersection Property Palette. Select the **Item Roles property**, click **More**, and then click **OK** to acknowledge that multiple items are selected.

4. In the Menu Item Roles dialog box, select **DBA**, press and hold the **Ctrl** key, select **CT_SALES**, click **OK** to assign the roles to the menu items, and then close the Property Palette.

5. In the Menu Editor, select **Customers** under the Database Maintenance parent menu item, press and hold the **Shift key**, and then select **Order Sources**. All of the Database Maintenance child menu items appear selected.

6. Click **Tools** on the menu bar, and then click **Property Palette** to open the intersection Property Palette. Select the **Item Roles property**, click **More**, and click **OK** to acknowledge that multiple items are selected.

7. In the Menu Item Roles dialog box, select **DBA**, click **OK** to assign the role, and then close the Property Palette.

8. Save the menu module.

9. Click **File** on the menu bar, point to **Administration**, and then click **Compile File** to rebuild the CW_MENU.mmx executable file.

10. Close Form Builder.

As the final step, you will run the Clearwater Traders application and test the menu security. Since you are not logged onto the database as a DBA, the Database Maintenance menu item should appear disabled.

To run the application and test the menu security:

1. In Project Builder, double-click the **Forms Builder executable icon** ▧ beside MAIN.fmx to open the main form module in Forms Runtime. The application switchboard opens, and the menu appears as shown in Figure 11-30, with the Database Maintenance menu disabled.

Figure 11-30   Form menu with menu security enforced

2. Click **Exit** to exit the form.

The Database Maintenance menu selection appears visible but disabled to unauthorized users. You can also configure the menu module so that users cannot even see the menu selections that they are not authorized to use by changing the menu item's Display Without Privilege property to No. Then, if a user does not have the privilege to access a menu item that is represented by a menu toolbar button, then the button does not appear on the user's menu toolbar.

CONTROLLING MENU ITEMS PROGRAMATICALLY

Form Builder provides built-in programs that you can use to access and modify menu properties while a form is running. Table 11-8 summarizes the Form Builder built-in programs for menus.

Name	Description
MENU_PARAMETER	Displays the current menu's parameters in the Enter Parameter Values dialog box
FIND_MENU_ITEM	Returns the object ID for the specified menu item
GET_MENU_ITEM_PROPERTY	Retrieves the current value of a menu item, such as the label, or whether the item is enabled or visible
REPLACE_MENU	Replaces the current menu module in all application windows with the specified menu module
SET_MENU_ITEM_PROPERTY	Dynamically modifies the property of a menu item

Table 11-8   Form Builder menu built-in programs

11

You can also control menu items by using **substitution parameters**, which are parameters that menu code triggers can reference to retrieve values about the current environment. Table 11-9 summarizes the menu substitution parameters.

Substitution Parameter	Value
AD	Folder where current menu's .mmx file is stored
LN	Current user's language preference
PW	Current user's password
SO	Menu item that is currently selected
TT	Current user's terminal type
UN	Current user's username

Table 11-9   Menu substitution parameters

You can use the menu built-in programs along with the substitution parameters to dynamically change the menus to match the needs of the current user. For example, you could create menu modules with item labels in different languages, use the LN substitution parameter to determine the current user's language preference, and then use the REPLACE_MENU built-in program to retrieve the correct menu for that user.

Now, you will use the SET_MENU_ITEM_PROPERTY built-in program to dynamically change a menu property at runtime. Suppose that a user opens the sales form for the Clearwater Traders database application, and then selects the Sales & Receiving parent menu item. One of the child menu items is New Orders, which allows the user to open the sales form again. If the user opens the sales form again, an error occurs, and the entire application closes abruptly. To avoid this error, you will disable the New Orders menu item when the user opens the sales form.

To dynamically disable the New Orders menu item, you will use the SET_MENU_ITEM_PROPERTY built-in procedure. This procedure has the following syntax:

```
SET_MENU_ITEM_PROPERTY(menu_specification, property_name,
value);
```

This command uses the following parameters:

- *menu_specification*, which can be either the menu ID or the menu item name, specified using the following format: '*parent_menu_name.menu_item_name*'. When the menu ID is used, the value does not need to be enclosed in single quotation marks. When the menu item name is used, the specification must be enclosed in single quotation marks. You can determine the names of individual menu items by consulting the menu structure in the Object Navigator. For example, to reference the New Orders menu item under the Sales & Receiving parent menu item, you would use `'salesreceiving_menu.new_orders'` as the menu specification.

- *property_name* is the name of the property to be modified. This value might be different from the property name in the Property Palette, so you should open the Form Builder online Help system to determine the exact syntax for referencing the property that you want to modify.

- *value* is the desired property value.

Now, you will modify the menu property in the PRE-FORM trigger in the sales form. You will add the SET_MENU_ITEM_PROPERTY command to the trigger, and use the command to set the New Orders menu item's Enabled property to False. You will also modify the form's Return button to reenable the menu item when the user exits the form.

To modify the sales form to disable and then reenable the New Orders menu item:

1. In Project Builder, double-click the **Forms Builder document icon** 📝 beside Sales.fmb to open the sales form in Form Builder. If necessary, maximize the Object Navigator window.

2. In the Object Navigator, click ➕ beside the Triggers node, and then double-click the **Trigger icon** 🔧 beside PRE-FORM to open the PRE-FORM trigger in the PL/SQL Editor.

3. In the PL/SQL Editor, add the following code as the last command in the trigger:

   ```
   SET_MENU_ITEM_PROPERTY('salesreceiving_menu.new_orders',
   ENABLED, PROPERTY_FALSE);
   ```

4. Compile the trigger, correct any syntax errors, and then close the PL/SQL Editor.

5. Click **Tools** on the menu bar, and then click **Layout Editor** to open the form in the Layout Editor. Select **TEMPLATE_CANVAS** from the Canvases list, and then click **OK**. The form opens in the Layout Editor.

6. Select the **Return button**, right-click, and then click **PL/SQL Editor** to display the button's trigger code in the PL/SQL Editor. Add the following command as the first line of the trigger:

   ```
   SET_MENU_ITEM_PROPERTY('salesreceiving_menu.new_orders',
   ENABLED, PROPERTY_TRUE);
   ```

7. Compile the trigger, correct any syntax errors, and then close the PL/SQL Editor.

8. Save the form, and then close Form Builder.

Now, you will rebuild the sales form in Project Builder. Then, you will run the application, open the sales form by clicking the Sales button on the application switchboard, and confirm that the New Orders menu item is now disabled.

11

To rebuild the main form, and then run the application and confirm that the menu item is disabled:

1. In Project Builder, select **Sales.fmx** under the Forms Builder executable node, and then click the **Build Selection button** to rebuild the file. When the "1 item built" message appears, click **OK**.

> **?** **Help** If a message appears stating that you do not have an object named ORDER_ID_SEQUENCE, start SQL*Plus, log onto the database, and create the sequence using the following command: `CREATE SEQUENCE order_id_sequence START WITH 1100;`. Then, open Sales.fmb in Form Builder, recompile the file, and close Form Builder.

2. Double-click the **Forms Builder executable icon** beside MAIN.fmx to run the application. The application switchboard appears.

3. Click **Sales**. The sales form opens.

4. Click **Sales & Receiving** on the menu bar, and note that the New Orders selection is now disabled.

5. Click **Return** on the sales form to return to the application switchboard.

6. Click **Sales & Receiving** on the menu bar, and note that the New Orders selection is enabled again.

7. Click **Exit** to exit Forms Runtime, and then close Project Builder and all other open Oracle applications.

SUMMARY

- Windows applications usually have pull-down menus, which appear on the menu bar, and pop-up menus, which are associated with a specific form item and appear when the user right-clicks the item with the mouse pointer.

- To define custom pull-down menus in a form, you create a menu module, which is independent of any specific form and is stored in the file system as a file with an .mmb extension.

- A menu module contains one or more menu items, which are sets of menu selections that appear on the menu bar. A menu item contains parent menu items, which are the menu selections that appear on the menu bar, and child menu items, which are the choices that appear when the user clicks a parent menu item.

- A menu item has a Property Palette on which you can specify how the menu item appears and functions, whether the menu item appears on a menu toolbar, and the security properties of the menu item.

- Menu items have access keys, which allow the user to open or select the menu item using the keyboard instead of the mouse pointer. The first letter of each menu item label is the default access key. In Forms Runtime, users can press the Alt key plus the access key to open parent menu items. Users can simply press the access key to select child menu items.

❐ Every child menu item must have an associated menu code trigger, which is a PL/SQL program that runs when the user selects the item. Menu code triggers can contain built-in commands that do not reference specific form items and module navigation commands. Menu code triggers can reference global variable values.

❐ To attach a pull-down menu module to a form, you specify the full path to the menu module's compiled .mmx file as the Menu Module property value in the form Property Palette.

❐ Whenever you modify a menu module design (.mmb) file, you must rebuild it so the modifications will be reflected in the menu module executable (.mmx) file.

❐ Form Builder automatically places Window as the last pull-down menu selection. In a menu module, the Window menu selection is added automatically at runtime, so you don't need to create it.

❐ To display a different menu module in a called form, you specify the new menu module filename in the form's Menu Module property, and then call the form using the DO_REPLACE option in the CALL_FORM command.

❐ A menu toolbar displays iconic buttons and appears horizontally under the pull-down menus or vertically on the left edge of the window. Each button on a menu toolbar is associated with a specific pull-down menu child item. When the user clicks a button, it is the equivalent of selecting the child menu item on the pull-down menu.

❐ In Form Builder, pop-up menus are defined as top-level form objects and can be assigned to specific form items.

❐ You can enforce application security through menus by implementing security for an overall menu module or by implementing different security properties for individual menu items. You enforce menu security by assigning specific database roles to menu modules and menu items.

❐ Form Builder provides several built-in procedures that allow you to modify menu properties while a form is running. You can use menu substitution parameters to determine the values of form environment properties, such as the current user's username and password, to help configure menus to match user needs.

11

REVIEW QUESTIONS

1. What is the difference between how you implement a pull-down menu and how you implement a pop-up menu in Form Builder?
2. Define the terms menu module, menu item, and individual menu.

3. What is a parent menu item? What is a child menu item?

4. What is the default access key for pull-down menu selections? How do you override the default access key?

5. Why must you rebuild your menu module every time you modify it?

6. When do you need to use the DO_REPLACE parameter in the CALL_FORM command?

7. What is a menu code trigger? What kinds of commands can a menu code trigger contain, and what kinds of commands can't it contain?

8. How do you specify the menu module that appears on a form?

9. What is a menu toolbar? How can you change the order in which the buttons appear on a menu toolbar?

10. How many pop-up menus do you need to create to display the same identical pop-up menu when the user right-clicks four different text items on a canvas?

11. What property enables or disables menu security in a menu module, and where do you set the property value?

12. What is a substitution parameter?

PROBLEM-SOLVING CASES

The cases refer to the Northwoods University (see Figure 1-12) and Software Experts (see Figure 1-13) sample databases. Place all solution files in the specified folder in your Chapter11\CaseSolutions folder.

To complete Case 1, you must have completed Case 1 in Lesson B of this chapter.

1. In this case, you will create a new menu module that displays the pull-down menu selections for the Northwoods University Student/Faculty Database System, as shown in Figure 11-31.

a. Create a new menu module named NW_MENU, and save the module as NW_MENU.mmb in your NWPROJECT_DONE folder. Create a menu item named MAIN_MENU, and then create the parent and child menu items shown in Figure 11-31. Be sure to specify the menu access keys as shown.

b. Create menu code triggers that enable the child menu items to call the associated target files. Use the :GLOBAL.project_path variable in the CALL_FORM commands. Create stubs that display the message "Not implemented yet." for the Terms, Course Sections, Contents, and About child menu items.

c. Configure the Exit child menu item as a magic menu item that automatically exits the form.

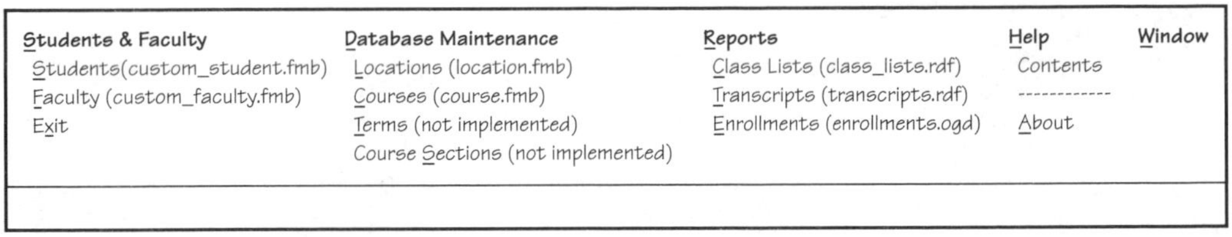

Students & Faculty	Database Maintenance	Reports	Help	Window
Students(custom_student.fmb)	Locations (location.fmb)	Class Lists (class_lists.rdf)	Contents	
Faculty (custom_faculty.fmb)	Courses (course.fmb)	Transcripts (transcripts.rdf)	------------	
Exit	Terms (not implemented)	Enrollments (enrollments.ogd)	About	
	Course Sections (not implemented)			

Figure 11-31

d. In Project Builder, add the menu module to the Northwoods University Student/Faculty Database project, and modify the Main.fmb form file so the menu module appears on the application switchboard.

e. Configure the CALL_FORM commands in the menu and switchboard triggers so that all of the child forms display their own menu modules rather than the menus on the parent form.

f. Create a vertical menu toolbar that displays iconic buttons for the Students, Faculty, Class Lists, and Transcripts pull-down menu selections. Use the icon images that are stored in the following .ico files, which are in your NWPROJECT_DONE folder.

Menu Item	Icon Filename
Students	students.ico
Faculty	faculty.ico
Class Lists	list.ico
Transcripts	transcript.ico

2. Modify the custom_faculty.fmb form file that is in the NWPROJECTS folder in the Chapter11 folder on your Data Disk so that when the user right-clicks any form canvas except the Logon canvas, a pop-up menu named FORM_MENU appears that shows the menu canvas command button selections (View/Edit Personal Information, View/Edit Advisee Information, View/Edit Class Lists). (Do not show the Exit selection on the pop-up menu.) Implement menu code triggers so that the pop-up menu selections work correctly. (*Hint:* Copy the code for the command button triggers into the menu code triggers.) If you have completed Case 1 in Lesson B of this chapter, store the modified file as custom_faculty.fmb in your NWPROJECTS_DONE folder. If you have not completed Case 1 in Lesson B, save the modified file as 11CCase2.fmb in your Chapter11\CaseSolutions folder.

 To complete Case 3, you must have completed Case 2 in Lesson B of this chapter.

3. In this case, you will create a new menu module that displays the pull-down menu selections for the Software Experts System shown in Figure 11-32.

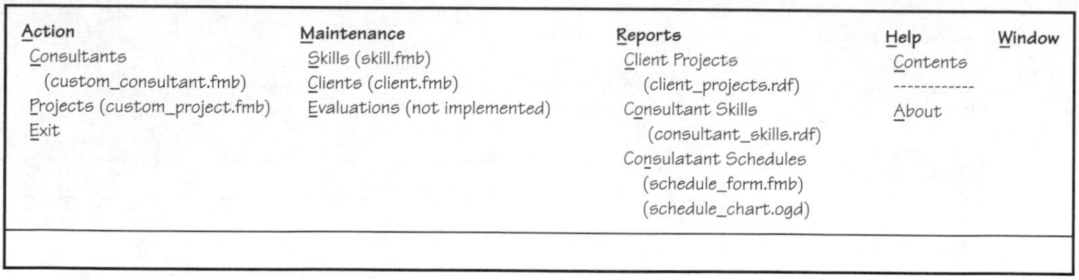

Action	Maintenance	Reports	Help	Window
Consultants	Skills (skill.fmb)	Client Projects	Contents	
(custom_consultant.fmb)	Clients (client.fmb)	(client_projects.rdf)	------------	
Projects (custom_project.fmb)	Evaluations (not implemented)	Consultant Skills	About	
Exit		(consultant_skills.rdf)		
		Consulatant Schedules		
		(schedule_form.fmb)		
		(schedule_chart.ogd)		

Figure 11-32

a. Create a new menu module named SWE_MENU, and save the module as SWE_MENU.mmb in your SWEPROJECT_DONE folder. Create a menu item named MAIN_MENU, and then create the parent and child menu items shown in Figure 11-32. Be sure to specify the menu access keys as shown.

b. Create menu code triggers that enable the child menu items to call the associated target files. Use the :GLOBAL.project_path variable in the CALL_FORM commands. Create stubs that display the message "Not implemented yet." for the Evaluations , Contents, and About child menu item.

c. Configure the Exit child menu item as a magic menu item that automatically exits the form.

d. In Project Builder, add the menu module to the Software Experts System project, and modify the Main.fmb form file so the menu module appears on the application switchboard.

e. Configure the CALL_FORM commands in the menu and switchboard triggers so that all of the child forms display their own menu modules rather than the menus on the parent form.

f. Create a horizontal menu toolbar that displays iconic buttons for the Consultants, Projects, and Consultant Schedules pull-down menu selections. Use the icon images that are stored in the following .ico files, which are in your SWEPROJECT_DONE folder.

Menu Item	Icon Filename
Consultants	consultants.ico
Projects	projects.ico
Consultant Schedules	schedule.ico

4. Modify the schedule_form.fmb form file that is in the SWEPROJECTS folder in the Chapter11 folder on your Data Disk so that when the user right-clicks the form canvas, or any of the form text items, a pop-up menu named LOV_MENU appears that shows the menu item "List Values." Create a menu code trigger that opens the form LOV display if the user clicks "List Values" on the pop-up menu. If you have completed Case 2 in Lesson B in this chapter, store the modified file as schedule_form.fmb in your SWEPROJECTS_DONE folder. If you have not completed Case 2 in Lesson B, save the modified file as 11CCase4.fmb in your Chapter11\CaseSolutions folder.

CHAPTER 12

FORM BUILDER OBJECTS AND FLEXIBLE CODE

◀ LESSON A ▶

Objectives

- ♦ Review object-oriented principles
- ♦ Learn how to use property classes to share object properties
- ♦ Store reusable objects in object groups
- ♦ Create object libraries to store reusable objects
- ♦ Use PL/SQL libraries in forms to share PL/SQL program units

As your form applications become large and complex, you need to use Form Builder features that allow you to produce sophisticated and consistent forms in a minimum amount of time. In this chapter, Lesson A describes how to create form objects that you can reuse in many different forms. Specifically, you will learn how to create and use property classes, object groups, object libraries, and PL/SQL libraries that contain many of the basic building blocks that make up a form. Lesson B describes how to write flexible code, which is code that contains commands that makes the code easier to reuse, maintain, and expand.

As you develop integrated database applications, you perform similar repetitive tasks. For example, every time a form requires an LOV command button, you must create the same kind of object, which is a button. The button has the same properties in terms of being an iconic button of a certain size, and the trigger code is very similar, regardless of the form application. Similarly, whenever you create a form button trigger that creates a parameter list that is passed to a report or chart, the trigger code is essentially the same, with only a few minor modifications to tailor it for each particular application. To speed up application development and promote application consistency, Form Builder supports ways to enable you to reuse form objects and code. All of these approaches involve object-oriented principles.

OBJECT ORIENTED PRINCIPLES

In object-oriented terminology, an **object** is an abstract representation of something in the real world. An example of an object that you have already used in Oracle is a library, which is a collection of programs, and is similar to its brick-and-mortar counterpart that contains collections of books. In the object-oriented world, a **class** defines the structure and properties of similar objects. In Form Builder, the Object Navigator provides a list of different classes of form objects: data blocks, canvases, windows, and so forth. A **class instance** is an object that belongs to a class. For example, each actual data block, canvas, or window within a form is a class instance of the associated object class.

An important aspect of object orientation is **inheritance**, which specifies that when you create a new class instance, the object's class determines the new object's structure and properties, and the object has the same structure as all other objects in its class. For example, when you create a new canvas, the canvas inherits the properties of the canvas class. Each canvas has the same list of properties, such as Name and Canvas Type. Recall that the Canvas Type property specifies whether the canvas is a content canvas, a stacked canvas, or a tab canvas. No other Form Builder object has this property. Different canvas instances have different values for this property, but every canvas has the same property list.

In an object-oriented system, a class can have **subclasses**, which have similar properties that are the same as those of their parent class, but also have special properties called **specializations**, that make them unique. An example of a class with subclasses is the Item class of form objects. Items can be text items, check boxes, radio groups, and so forth. Each different item type is a subclass of the item class. Each item type has similar properties, such as a data type, maximum length, and default value. Each item type also has specific properties that make it unique. For example, a radio group has an Initial Value property that specifies which radio button is selected when the form first opens. A check box has a Value when Checked property that specifies the data value when the box is checked, and a Value when Unchecked property that specifies the data value when the box is cleared. A text item does not have these specific properties, but it has other properties that make it unique, such as the List of Values property, which allows you to attach an LOV to the text item.

Form Builder supports four approaches to enable developers to reuse form properties, objects, and code:

- Inheriting object properties from property classes
- Storing reusable objects in object groups
- Storing reusable objects in object libraries
- Storing reusable PL/SQL program units in PL/SQL libraries

The following sections explore each of these approaches.

INHERITING OBJECT PROPERTIES FROM PROPERTY CLASSES

A **property class** is a top-level Form Builder object that represents an object class that specifies a collection of properties and associated values. When you create a property class instance in Form Builder, you specify the name of the property class, its properties, and their associated values. When you associate the property class with a form object, the object inherits the properties of the property class and becomes a subclass of the property class. Associating an object with a property class is called **subclassing** the object.

Property classes are similar to visual attribute groups, which you learned about in Chapter 11. In some ways, property classes are more powerful than visual attribute groups: Property classes enable you to specify default values for any property of any type of object, while visual attribute groups allow you to specify default values only for colors, font sizes, and font styles of items, canvases, and windows. However, visual attribute groups are more flexible than property classes, because visual attribute group values can be modified while the form is running. In contrast, property class values can be modified only at design time.

Suppose that the Clearwater Traders database application development team decides that all form text items that contain currency values will be formatted using the format mask $99,999.99, and that the currency value will be right justified within the text item. They decide to create a property class in the form template to define these properties, and instruct developers to always subclass text items that contain currency data values based on the new property class. Now you will create this property class and use it to subclass form objects.

Creating Property Classes

To format currency data values in the Clearwater Traders database application, you will create a property class named CURRENCY_PROPERTY_CLASS in the Clearwater template form that you created in Chapter 11. This property class will have the following properties and associated values:

Property	Value
Item Type	Text Item
Data Type	Number
Maximum Length	12 (which is wide enough to accommodate the format mask)
Justification	Right
Format Mask	$99,999.99

There are two ways to create a property class:

- Create a form object with the desired subclass properties, and then create a property class based on the object

- Manually create a new property class object in the Object Navigator, and explicitly specify its properties and their values

The easiest way to create a new property class is to use the first approach, and base the class on an existing form object. To create the CURRENCY_PROPERTY_CLASS, you will first create a text item in the form that has the desired properties.

To create a text item to display currency values:

1. Start Form Builder, open the **ClearwaterTemplate.fmb** file from the Chapter12 folder on your Data Disk, and save the file as **12AClearwaterTemplate.fmb** in your Chapter12\TutorialSolutions folder. Connect to the database.

2. Under the CLEARWATER_TEMPLATE form module, click ➕ beside the Data Blocks node, click ➕ beside TEMPLATE_BLOCK, select the **Items node**, and then click the **Create button** 📑. A new text item appears under the Items node. Change the text item name to **CURRENCY_ITEM**.

3. Double-click the **Text Item icon** 🔡 to open the text item Property Palette, and modify the following properties. Do not close the Property Palette.

Property	Value
Justification	**Right**
Data Type	**Number**
Maximum Length	**12**
Format Mask	**$99,999.99**

Now, you will create a new property class based on the properties of the new text item. This property class will contain the property values that you want all currency value text items to inherit. To create the property class, you will select the desired properties in the text item Property Palette, and then click the **Property Class button** 📑 on the Property Palette toolbar. This creates a new form property class that has the desired properties and associated values.

To select the item properties and create the new property class:

1. In the text item Property Palette, select the **Item Type property**, press and hold the **Ctrl key**, and then select the following properties while keeping the Ctrl key pressed: **Justification**, **Data Type**, **Maximum Length**, **Format Mask**.

 A property appears selected in the Property Palette when its upper border becomes darker and wider.

2. Click the **Property Class button** 🔲 on the Property Palette toolbar. The message "Creating property class PROPERTY_CLASS2" appears. (Your property class number may be different.) Click **OK** to acknowledge the message that the new property class was created, and then close the Property Palette.

3. Scroll down in the Object Navigator window and click ➕ beside the Property Classes node. The new property class appears. Change the property class name to **CURRENCY_PROPERTY_CLASS**.

4. Double-click the **Property class icon** 🔲 beside CURRENCY_PROPERTY_ CLASS to open its Property Palette. Note that the property class has the selected properties and values. Do not close the Property Palette.

You can click the Add Property button 🔲 on the property class Property Palette toolbar to add an additional property to the property class and specify its value. You can select an existing property and click the Delete Property button 🔲 to delete a property.

To manually create a property class that is not based on an existing item, you create a new property class object in the Object Navigator, and then click 🔲 to add each property. Now you will delete the property class you just created, along with the text item on which it was based. Then, you will recreate the property class manually.

To delete the property class and text item, and recreate the property class manually:

1. Close the property class Property Palette.

2. To delete the existing property class and text item, select **CURRENCY_PROPERTY_CLASS**, press and hold the **Ctrl key**, select the **CURRENCY_ITEM** text item under the TEMPLATE_BLOCK, and then click the **Delete button** 🔲 to delete the items. Click **Yes** to confirm the deletions. The objects no longer appear in the Object Navigator.

3. To manually create the new property class, select the **Property Classes node** under the CLEARWATER_TEMPLATE form module, and then click the **Create button** 🔲 to create a new property class. Change the name of the new property class to **CURRENCY_PROPERTY_CLASS**.

4. Double-click the **Property class icon** 🔲 beside CURRENCY_PROPERTY_CLASS to open its Property Palette. Note that the new property class does not have any properties except the General node properties, which are the properties that are shared by all property classes.

12

Next, you will define the property class's properties by selecting each desired property from a list of all available properties. This causes the desired property to appear on the property class Property Palette. Then, you specify the value for the property.

To define the property class properties and values:

1. To add a property to the property class, click the **Add Property button** on the toolbar. The Properties dialog box opens.

2. First, you will select the Item Type property. Place the insertion point before the % in the Find text box, type **Item Type** as the property search condition name, and then press **Enter**. The Item Type property appears in the list. Click **OK**. The Item Type property appears on the Property Palette.

3. Make sure that Text Item is the value of the Item Type property, and then select another property to save the value.

4. To add the next property to the property class, click , place the insertion point before the % in the Find text box, type **Justification** as the property search condition name, press **Enter**, and then click **OK**. The Justification property appears on the Property Palette.

5. Select **Right** for the Justification property value, and then select another property to save the value.

6. Add the following properties and associated values to the property class. Your completed property class Property Palette should look like Figure 12-1. Do not close the Property Palette.

Property	Value
Data Type	**Number**
Maximum Length	**12**
Format Mask	**$99,999.99**

If you add a property to the property class by mistake, select the undesired property, and then click the Delete Property button on the toolbar to remove the property.

If you find that you cannot change the value of a property, close the Property Palette, open it again, and then change the property value.

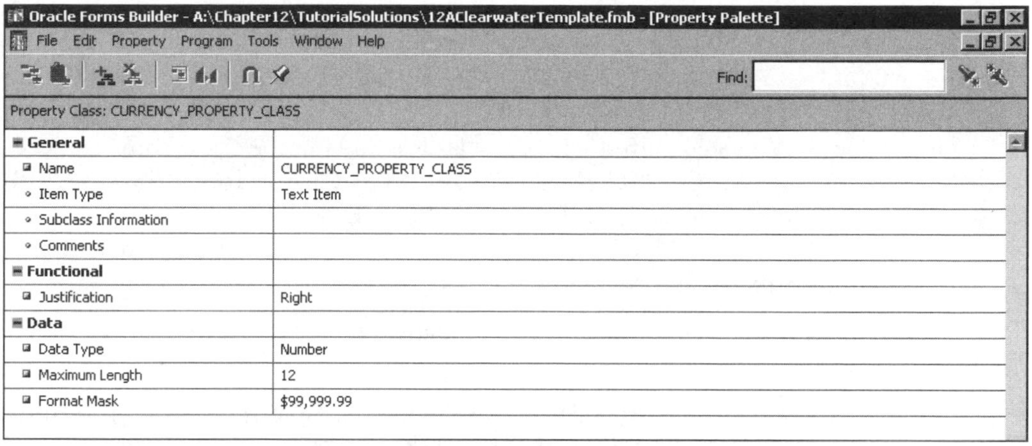

Oracle Forms Builder - A:\Chapter12\TutorialSolutions\12AClearwaterTemplate.fmb - [Property Palette]   _ |8| X|

File   Edit   Property   Program   Tools   Window   Help   _ |8| X|

Find:

Property Class: CURRENCY_PROPERTY_CLASS

General	
Name	CURRENCY_PROPERTY_CLASS
Item Type	Text Item
Subclass Information	
Comments	
Functional	
Justification	Right
Data	
Data Type	Number
Maximum Length	12
Format Mask	$99,999.99

Figure 12-1   CURRENCY_PROPERTY_CLASS Property Palette

Using Property Classes to Subclass Form Objects

Recall that a form object becomes subclassed when you apply a property class to the object and cause the object to inherit some of its properties from the property class. To subclass a form object, you open the Property Palette of the form object, select the Subclass Information property, open the Subclass Information dialog box, and select the class from which the object will be subclassed.

Notice that in the property class Property Palette on your screen display, the nodes beside some of the properties are green. A green node indicator shows that you changed the property's default value. When you apply a property class to a form object, the subclassed form object inherits the property values with green nodes. Note that the Item Type node is not green. The item type of an item cannot be specified through subclassing.

When you change a default property value, you cannot restore the default node indicator to the property, even when you set the property value back to its default value.

When you select a changed property with a green node on the property class Property Palette, the **Inherit button** on the toolbar is enabled, which indicates that the property is available to be inherited by subclassed objects. If you want to disable a property in a property class so that it will not be inherited by subclassed objects, you select the property in the property class Property Palette, and click, so the property node is no longer green. You should do this with care, however, because you cannot re-enable inheritance except by deleting the property in the property class and then adding it again.

To subclass an object, you change its Subclass Information property to the name of the desired property class. An object can be subclassed using any property class in any form, as long as the form that contains the property class is currently open in Form Builder.

12

Since you cannot always count on another form being open, it is a good practice to sub-class objects using a property class that is in the same form as the subclassed object. Now, you will open a form named Inventory.fmb that is stored in the Chapter12 folder on your Data Disk, and save the file using a different filename. This form is a data block form that is associated with the INVENTORY table in the Clearwater Traders database. You will copy the CURRENCY_PROPERTY_CLASS to the inventory form, and subclass the text item associated with the PRICE database column in this form using the CURRENCY_PROPERTY_CLASS. Before you subclass the text item, you will examine its Subclass Information dialog box.

To open the form, copy the property class, and examine the text item's Subclass Information dialog box:

1. Close the Property Palette, and then, open **Inventory.fmb** from the Chapter12 folder on your Data Disk. Save the file as **12AInventory.fmb** in your Chapter12\TutorialSolutions folder.

2. In the Object Navigator, select the **CURRENCY_PROPERTY_CLASS** in the CLEARWATER_TEMPLATE form module, and click the **Copy button** to copy the property class.

3. Select the **Property Classes** node in the INVENTORY_FORM, and then click the **Paste button** . The CURRENCY_PROPERTY_CLASS appears in the INVENTORY_FORM.

4. Select the **CLEARWATER_TEMPLATE** form module, click **File** on the menu bar, and then click **Close** to close the template form in Form Builder. If necessary, click **Yes** if you are prompted to save your changes.

5. In the INVENTORY_FORM, click ⊞ beside the Data Blocks node, click ⊞ beside the INVENTORY data block, click ⊞ beside the Items node and double-click the **Text Item icon** beside the PRICE text item to open its Property Palette.

6. Select the **Subclass Information property**, and then click the **More button** . The Subclass Information dialog box opens, as shown in Figure 12-2.

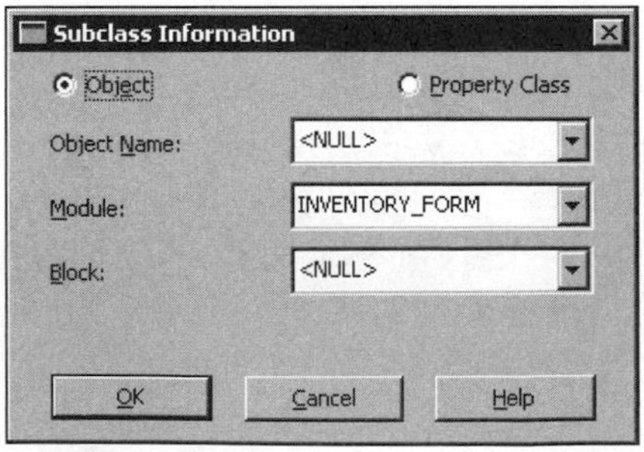

Figure 12-2 Subclass Information dialog box

Form objects can be subclassed using either other form objects or property classes. The Object and Property Class option buttons on the Subclass Information dialog box allow you to specify whether the object will inherit information from another object or from a property class. If the object will inherit its properties from another object, the form that contains the object must be open in Form Builder, and you must specify the source object name, form module name, and block name. If the object will inherit its properties from a property class, you must specify the form module that contains the source property class and the property class name. Now, you will subclass the PRICE text item using the CURRENCY_PROPERTY_CLASS.

To subclass the PRICE text item:

1. In the Subclass Information dialog box, select the **Property Class option button**.

2. Confirm that the current Module value is INVENTORY_FORM, which is the name of the current form. If your Module value is different, open the Module list, and select **INVENTORY_FORM**.

3. Open the Property Class Name list, select **CURRENCY_PROPERTY_CLASS**, and then click **OK** to close the Subclass Information dialog box.

4. Scroll down in the Property Palette to the Data property node, and note the appearance of the properties that inherited values from the property class. Do not close the Property Palette.

When a property inherits a value from another source, its property node appears as an **Inheritance Arrow** ⬇, instead of as a normal or changed (green) node. You can manually override an inherited property value by entering a new value. When you do this, the property node of the overridden property appears as a **Broken Inheritance Arrow** ⬇. Now, you will override an inherited property by changing the Maximum Length property value to 13. Then, you will close the Property Palette, and save the form.

To override an inherited property value:

1. In the PRICE text item Property Palette, change the Maximum Length property value to **13**, and then select another property to save the change. Note that the property node changes to a Broken Inheritance Arrow ⬇.

 If you cannot change the value, close the Property Palette, open it again, and repeat Step 1.

2. Close the Property Palette, and save the form.

Note that in the Object Navigator, the text item icon for the PRICE text item now appears as a Subclassed Text Item icon 🔧. This indicates that the text item currently inherits properties from another object or property class. As a final step, you will run the

12

form and retrieve all of the inventory records. You will examine the PRICE text item values to confirm that they successfully inherited the property class's property values.

To run the form, retrieve the records, and examine the PRICE text item values:

1. Run the form, click the **Enter Query button** 📷, do not enter a search condition, and then click the **Execute Query button** 📷 to retrieve all of the table records. The first inventory record (ID 11668) appears in the form. The price text item should appear as $259.99, and should be right-justified in the text item box.

If the price text item does not appear right-justified, it is because the form text item display is wider than the maximum specified length of the text item. To fix this problem, close the form in Forms Runtime, open the PRICE text item Property Palette, scroll down to the Physical property node, make the Width value smaller, and then run the form again.

2. Click the **Next Record button** ▶ to view the next record and observe the price text item, and then click **Return** to close Forms Runtime. Do not close the INVENTORY_FORM.

STORING REUSABLE OBJECTS IN OBJECT GROUPS

Property classes are useful for allowing form objects to inherit specific properties, but sometimes you need to make entire objects available for sharing. For example, it would be useful to create an LOV with specific properties and make it available to all developers. As a result, each developer would not have to independently specify the LOV query, prompts, and display properties. In addition, the LOV appearance would be uniform in all applications.

To enable developers to share objects, you can create an **object group**, which is a top-level Form Builder object type that is a container for a group of similar objects. For example, the Clearwater Traders development team might create an object group that contains commonly used LOVs, a second object group that contains commonly used data blocks, and a third object group that contains commonly used alerts. Developers can then create subclassed objects that are based on the objects in the object groups. In the following sections, you will learn how to create and use object groups.

Creating an Object Group

Form object groups should be placed in a central object source form that is accessible to all developers. You should not place objects in the template form, because the template form will be used to build new applications, and you do not want to have a copy of every form object in the template. Rather, developers will reference the required objects from the object source form.

Now you will create a new object source form named CLEARWATER_OBJECTS that will contain reusable objects that will be used in Clearwater Traders database applications. After you create this object source form, you will create a new object group in the form. To create an object group, you select the Object Groups node in the Object Navigator, and create a new object group. The new object group will contain LOVs used in Clearwater Traders applications, so you will name the new object group LOV_OBJECT_GROUP. First, you will create the object source form and object group.

To create an object source form to store the objects and an object group:

1. In the Object Navigator, select the top-level **Forms node**, and click the **Create button** 🔳 to create a new form module. A new form module appears. Change the form name to **CLEARWATER_OBJECTS**, and save the form as **CLEARWATER_OBJECTS.fmb** in your Chapter12\TutorialSolutions folder.

2. Select the new form's **Object Groups node**, and then click the **Create button** 🔳. A new object group appears. Rename the object group **LOV_OBJECT_GROUP**.

3. Click **+** beside LOV_OBJECT_GROUP. The Object Group Children node appears.

To add objects to the object group, you create the desired objects in the object source form, and then place the objects in the object group. When you place an object in an object group, it must be an object in the object source form. You cannot add an object from a different form to a form's object group. Placing the object in the object group only creates a reference to the object, and does not create an actual physical copy of the object. The objects in the Object Group Children node provide an interface that makes the objects available to other forms.

An object group can only contain the top-level Form Builder objects that appear in the Object Navigator Ownership view: Triggers, Alerts, Attached Libraries, and so forth. Individual objects within a data block cannot be placed in an object group. For example, you cannot place a command button directly in an object group. A work-around for this limitation is to add the command button to a data block, and then place the data block in an object group.

Now you will create an LOV in the CLEARWATER_OBJECTS form that displays all of the item ID values and associated descriptions from the Clearwater Traders ITEM table. Then, you will place this LOV in the LOV_OBJECT_GROUP. First, you will use the LOV Wizard to create the LOV. You will not specify any return values or a return item for the LOV, because these items are specific to the form in which the LOV is used. Other developers will specify these values when they use the LOV object group in a form.

To create the LOV to retrieve the item values:

1. Select the **CLEARWATER_OBJECTS form module node** to ensure that the LOV is created in the object source form. Then, click **Tools** on the

menu bar, and click **LOV Wizard**. The Source page appears. Accept the default values, and then click **Next**.

2. On the SQL Query page, type the following query, and then click **Next**.

```
SELECT item_id, item_desc
FROM item
```

3. On the Column Selection page, click the **Select All button** ⟩⟩ to select all of the query fields for display in the LOV, and then click **Next**.

4. On the Column Display page, change the column properties as follows, and then click **Next**.

Column	Title	Width	Return value
ITEM_ID	Item ID	30	(not specified)
ITEM_DESC	Description	80	(not specified)

5. On the LOV Display page, type **Items** in the Title text box, change the Width to **180**, the Height to **180**, and then click **Next**.

6. On the Advanced Options page, accept the default values, and click **Next**.

7. On the Items page, there are no values in the Return Items list because you did not specify any return values for the LOV display items. Other developers will specify the return item when they use the LOV object group in a specific form. Click **Next**, and then click **Finish**.

8. If necessary, open the Object Navigator, and rename the new LOV **ITEM_LOV**, and the new record group **ITEM_LOV**.

9. Save the form.

Now, you will add the new LOV and record group to the object group. To add an object to an object group, you select the object, drag it to the object group's Object Group Children node, and drop it onto the node. The object then appears as one of the object group's child nodes.

To add the LOV and record group to the object group:

1. In the Object Navigator, select the LOV named **ITEM_LOV**, and drag it to the Object Group Children node under LOV_OBJECT_GROUP.

2. Release the mouse button to drop the LOV onto the node. The ITEM_LOV appears under the Object Group Children node.

3. Select the record group named **ITEM_LOV**, and move the record group into the LOV_OBJECT_GROUP. Both the LOV and record group associated with the LOV appear under the LOV_OBJECT_GROUP's Object Group Children node, as shown in Figure 12-3. The order of the objects in your object group might be different.

4. Save the form.

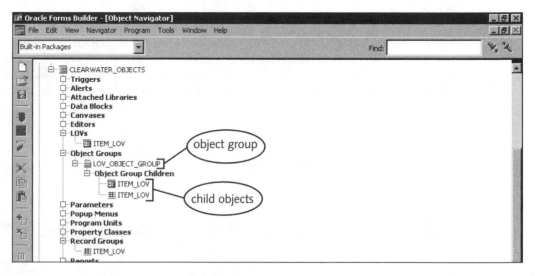

Figure 12-3   Object group and child objects

When you delete an object from the source object form, the object is automatically removed from the object group. For example, if you delete the ITEM_LOV that is under the LOVs node in the form, the ITEM_LOV that is under the LOV_OBJECT_GROUP will also be removed. If you delete an object in an object group, the object is not removed from the form module. If you delete the ITEM_LOV that is under the LOV_OBJECT_GROUP, the ITEM_LOV that is under the LOVs node will still remain in the form.

Using Object Groups

To reference an object in an object group in another form, called the target form, you select the object group in the source form, drag it to the Object Groups node in the target form, and drop it there. A dialog box opens asking if you want to copy or subclass the object group. The difference between copying and subclassing an object group involves how Form Builder handles changes to the original object group. If you copy an object group to a target form, and then change a property in an object in the original object group, the change does not propagate to the copy. If you subclass the object group in a target form, changes to the original object group propagate to all subclassed object groups in all target forms. To keep all form objects consistent, it is a good practice to always subclass object groups rather than copying them.

Now you will subclass the LOV_OBJECT_GROUP in the INVENTORY_FORM. You will select the object group in the CLEARWATER_OBJECTS form, drag it onto the Object Groups node in the INVENTORY_FORM, and drop it there. When the dialog box asks if you want to copy or subclass the object group, you will select the subclass option.

12

To subclass the object group in the target form:

1. In the Object Navigator, select the **LOV_OBJECT_GROUP** in the CLEARWATER_OBJECTS form, and drag it onto the Object Groups node in the INVENTORY_FORM. (Currently, the INVENTORY_FORM does not have any Object Groups.) The mouse pointer changes to the Object Group mouse pointer ⬚.

2. Drop the object group onto the Object Groups node. The Forms alert appears, asking if you want to copy or subclass the object group. Click **Subclass**. The subclassed object group appears in the INVENTORY_FORM, as shown in Figure 12-4. The object group's child objects (the ITEM_LOV record group and LOV) also appear as subclassed objects in the target form.

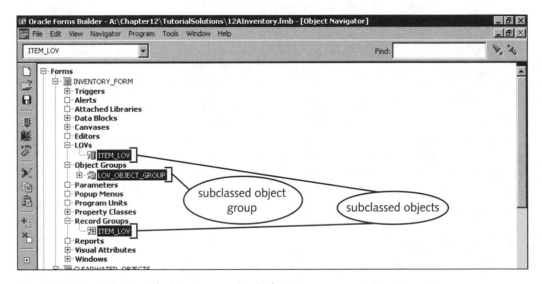

Figure 12-4   Subclassed object group and objects

Note that when you create a subclassed object group, the object group appears in the Object Navigator using the **Subclassed Object Group icon** 🖥. The subclassed objects appear with an inheritance arrow on their object icons.

When you subclass an object in a target form, you usually need to modify properties of the subclassed object so it works correctly in the target form. As you work with sub-classed objects, you will learn by experience what properties need to be modified. For example, when you subclass an LOV, you need to modify the return values and return item of the LOV so it is assigned to a text item on the new form. Now you will open the INVENTORY_FORM's ITEM_LOV in the LOV Wizard in reentrant mode, and modify its properties by assigning its return value and return item. Then, you will test the form to confirm that the LOV works correctly.

To modify the LOV and test the form:

1. In the Object Navigator, select the **ITEM_LOV** subclass under the LOVs node in the INVENTORY_FORM, click **Tools** on the menu bar, and then click **LOV Wizard** to open the wizard in reentrant mode.

2. Click **Next** several times to scroll through the LOV Wizard pages until the Column Display page appears. Place the insertion point in the Return value column for the Item ID column, click **Look up return item**, select **INVENTORY.ITEM_ID**, and then click **OK**.

3. Click **Next** several times to scroll through the LOV Wizard pages until the Items page appears. Select **INVENTORY.ITEM_ID** in the Return Items list, click the **Select button** to move the value into the Assigned Items list, and then click **Finish**.

4. Save the form, and then run the form.

5. Click the **Enter Query button**, do not enter a search condition, and then click the **Execute Query button** to retrieve all of the table records.

6. Place the insertion point in the Item Id text item, and press **F9** to open the LOV display. When the LOV display opens, make sure that item ID 894 (Women's Hiking Shorts) is selected, and then click **OK**. The selected value appears in the form's Item Id text item.

7. Click **Return** to close the form, and click **No** to discard your changes.

Recall that an object group contains only references to objects, not actual copies of the objects. When you use subclassed objects in a target form in Form Builder, the source form for the object class must be open in Form Builder. If you close the source form in Form Builder, then the subclassed objects no longer appear in the target form and are not available when you run the form. Now, you will close the CLEARWATER_OBJECTS form in Form Builder, which will cause the subclassed objects to no longer be available to the target form. Then, you will examine what happens to the target form when you try to run it, and then close and reopen it.

To close the object source form in Form Builder and see the effect on the subclassed objects in the target form:

1. In the Object Navigator, select the **CLEARWATER_OBJECTS** form module, click **File** on the menu bar, and then click **Close**. (If necessary, click Yes if you are asked if you want to save your changes to the form.) The FRM-18108 "Failed to load the following objects" error message appears, indicating that the object group source form is no longer available. Click **OK**.

2. Save the INVENTORY_FORM as **12AInventory_ERR.fmb** in your Chapter12\TutorialSolutions folder, and then run the INVENTORY_FORM. A Compilation Errors dialog box opens, which reports the FRM-30048 ("Unable to find record group LOV ITEM_LOV") and FRM-30085: ("Unable to adjust form for output") errors. Click **OK**. These errors occur because you closed the object source form.

12

3. Close the INVENTORY_FORM, and then reopen the 12AInventory_ERR.fmb form file. The FRM-10108 "Failed to load the following objects" error message appears. Click **OK**. The form loads, but the subclassed LOV and record group objects no longer appear in the Object Navigator.

4. Click ⊞ beside the Object Groups node, click ⊞ beside the LOV_OBJECT_GROUP, and then click ⊞ beside the Object Group Children node. The two ITEM_LOV child objects appear, but their icons appear as the Undefined Objects icon 🔳, indicating they are undefined object types.

5. Close the INVENTORY_FORM in Form Builder.

When you lose a reference to a subclassed object, the only way to restore the reference is to open the source object form, move the object group into the target form again, and modify the subclassed objects as necessary.

When you create a Forms Runtime (.fmx) file from a form that uses subclassed objects, the objects are still available, even if the source form is not open in Form Builder or Forms Runtime. Now, you will start Forms Runtime and open the 12AInventory.fmx runtime file that you created when you ran the ITEM_FORM while the CLEARWATER_OBJECTS form was open in Form Builder.

To open the .fmx file with the subclassed objects in Forms Runtime:

1. Click **Start** on the taskbar, point to **Programs**, point to **Oracle Forms 6i**, and then click **Forms Runtime**. The Forms Runtime Options window opens.

2. Click **Browse**, navigate to your Chapter12\TutorialSolutions folder, select **12AInventory.fmx**, and click **Open**.

3. Type your username in the Userid text box, your password in the Password text box, your usual database connection string in the Database text box, and then click **OK**. The form opens in Forms Runtime.

4. Click the **Enter Query button** 🔳, do not enter a search condition, and then click the **Execute Query button** 🔳 to retrieve all of the table records.

5. Make sure that the insertion point is in the Item Id text item, and then press **F9**. The LOV display opens, confirming that the Forms Runtime file still successfully references the subclassed objects. Click **Cancel** to close the LOV display, and then click **Return** to close Forms Runtime.

STORING REUSABLE OBJECTS IN OBJECT LIBRARIES

Another approach for creating reusable form objects is to store the objects in an object library. An **object library** is a top-level Form Builder object that is independent of any specific form module and contains form objects that can be referenced in target forms.

You save an object library in a file with an .olb (Object Library) extension, and place it in a location where it is accessible to all form developers. When developers use an object library rather than an object class, they can develop new form applications without having to open a source object form in Form Builder and risk losing the referenced objects if they inadvertently close the source form. Another advantage of object libraries is that object libraries can contain item-level objects, such as command buttons or text items, rather than only top-level objects. The following sections describe how to create and use an object library.

Creating an Object Library

To create an object library, you select the Form Builder Object Library node in the Object Navigator, and then create a new library object. Since the Object Library node is at the same level as the Forms node in the Object Navigator, the object library is independent of any specific form, and can be referenced by multiple forms. An object library is saved as an independent file with an .olb extension in the workstation file system. After you create a new object library, or open an existing object library, the library automatically opens every time you start Form Builder.

To learn about object libraries, you will use the ITEM_FORM, which is a data block form associated with the Clearwater Traders ITEM table. Currently, the form allows the user to select an item ID from an LOV display by placing the insertion point in the ITEM_ID text item and pressing F9. You will use this form to create an LOV command button, and then you will store the LOV command button in an object library. Now, you will open the ITEM_FORM, which is saved in a file named Item.fmb in the Chapter12 folder on your Data Disk, and save the form using a different filename. Then, you will create and save a new object library named CLEARWATER_LIBRARY.

To open the ITEM_FORM, save the form using a different filename, and then create and save a new object library:

1. In Form Builder, no forms should currently be open. Click the **Open button** ![open icon], navigate to the Chapter12 folder on your Data Disk, and open **Item.fmb**.

2. Save the file as **12AItem_OL.fmb** in your Chapter12\TutorialSolutions folder.

3. In the Object Navigator, select the **Object Libraries node**, and then click the **Create button** ![create icon] to create a new object library. The new object library appears. Change the name of the object library to **CLEARWATER_LIBRARY**.

4. In the Object Navigator, select the new library, and then click the **Save button** ![save icon] on the toolbar. The Save As dialog box opens. If necessary, navigate to your Chapter12\TutorialSolutions folder, make sure the filename is CLEARWATER_LIBRARY.olb, and then click **Save** to save the library.

Currently, the object library contains an empty Library Tabs node. A **library tab** is an object that represents a page on a tab canvas that contains one or more objects. You create a separate library tab for each object type that you store in the object library. For example, you might create a Buttons tab, an Alerts tab, and a Blocks tab. To create a new tab, you create a new library tab object in the Object Navigator, and then change the tab object name. You change the tab label by opening the library tab Property Palette and changing the Label property to the desired label text. Now, you will create two library tab objects, one to represent button objects and the other to represent alert objects. Then, you will modify the tab labels.

To create the library tabs and modify the tab labels:

1. In the Object Navigator, select the **Library Tabs node**, which is currently empty, and click the **Create button** 🔳 two times to create two new library tabs. The new library tabs appear in the Object Navigator.

2. In the Object Navigator, change the name of the first tab to **BUTTON_TAB**, and change the name of the second tab to **ALERT_TAB**.

3. Right-click the **Library Tabs icon** 🔳 beside BUTTON_TAB, and then click **Property Palette** to open the tab Property Palette. Change the Label property value to **Buttons**, and then close the Property Palette.

4. Right-click 🔳 beside ALERT_TAB, click **Property Palette**, change the Label property value to **Alerts**, and then close the Property Palette.

5. Select **CLEARWATER_LIBRARY** in the Object Navigator, and then click the **Save button** 🔳 to save the library. The library specification appears as shown in Figure 12-5.

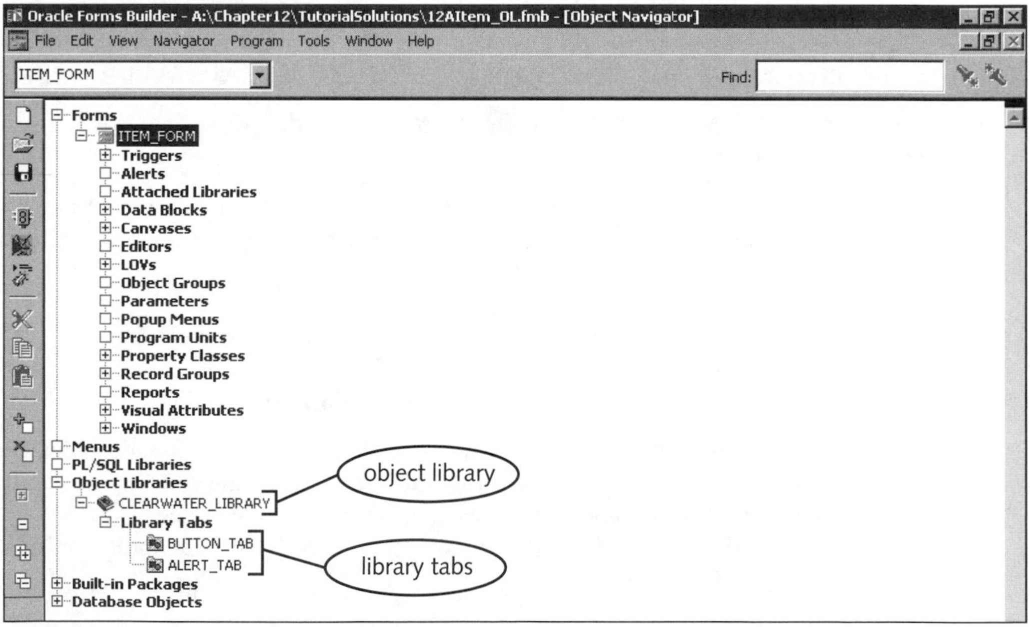

Figure 12-5   Object library specification

To open the Object Library window and view the first tab page that is listed in the Object Navigator, you double-click the **object library icon** in the Object Navigator. To open the Object Library window with a specific tab page selected, you double-click the Library Tabs icon beside the desired tab page. Now, you will double-click to open the Object Library window and view the library tab pages, with the first tab page, which is the Buttons page, selected.

To open the Object Library window and view the library tab pages:

1. Double-click the **object library icon** beside the CLEARWATER_LIBRARY to open the Object Library window. The library tab pages appear as shown in Figure 12-6, with the Buttons page selected.

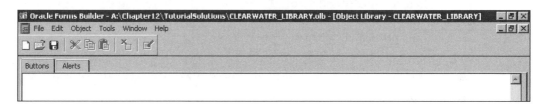

Figure 12-6   Object library window and tab pages

 If your tab labels do not appear as shown in Figure 12-6, click ☒ on the inner window to close the window that displays the tab pages, right-click the Library Tabs Icon 🔳 beside the tab page in which the label is not correct, click Property Palette, and change the Label property value to the correct tab label.

2. Click the **Close button** ☒ on the inner Object Library window to close the window and display the Object Navigator window again.

 If a dialog box opens asking if you want to save your changes to the library, you clicked ☒ on the outer Form Builder window. Click Cancel, and then click ☒ to close the inner window.

The Object Library window has a toolbar with buttons for manipulating the library and for adding and configuring library objects. The **New button** 🗋 creates a new library tab page. The **Open button** 📂 opens a new library, and the **Save button** 🖫 saves the current library. The remaining toolbar buttons are currently disabled, because there are no objects in the library yet. The remaining buttons are used to cut, copy, and paste library objects. The **Remove Object button** 🗙 can be used to delete an object from the library, and the **Edit Comment button** 📝 can be used to create or edit comments to document a library object.

Adding Objects to an Object Library

To add an object to an object library, you create the object in any form, and then drag and drop the object onto the target object library tab page where you want to save the object. Now, you will create an LOV command button in the ITEM_FORM, and then store the button as an object on the Buttons page in the object library. First, you will create the LOV command button.

To create the LOV command button:

1. In the Object Navigator, select the **ITEM_FORM**, click **Tools** on the menu bar, and then click **Layout Editor** to open the Layout Editor.

2. Select the **Button tool** 🔲 on the tool palette, and draw an LOV command button on the right edge of the ITEM_ID text item. If necessary, resize and reposition the button.

3. Double-click the new button to open its Property Palette, change the following property values, and then close the Property Palette.

Property	Value
Name	**LOV_COMMAND_BUTTON**
Label	(deleted)
Iconic	**Yes**
Icon Filename	**a:\chapter12\down3 (or the path to the Chapter12 folder on you Data Disk)**
Width	**12**
Height	**14**

4. Right-click the new button, point to **SmartTriggers**, and then click **WHEN-BUTTON-PRESSED**. The PL/SQL Editor opens.

5. Type the following code in the PL/SQL Editor, compile the trigger, and debug it if necessary. Then close the PL/SQL Editor, and save the form.

   ```
   GO_ITEM('item.item_id');
   LIST_VALUES;
   ```

Next, you will add the button to the object library by dragging the button onto the library's BUTTON_TAB page. You must drop the object onto the actual tab page. You cannot drop the object onto the library tab page in the Object Navigator. To drop the object onto the actual tab page, you must position the library tab page and the Object Navigator window that displays the form object side by side on your screen display. First, you will open the Object Navigator. Then, you will open the Object Library window with the Buttons page selected. Finally, you will click the Restore button 🔳 on the Object Library window, so that it appears beside the Object Navigator window.

To add the LOV command button to the object library:

1. In the Layout Editor, click **Window** on the menu bar, and then click **Object Navigator** to open the Object Navigator. If necessary, maximize the Object Navigator.

2. Scroll down to the object library, and double-click the **Library Tabs icon** 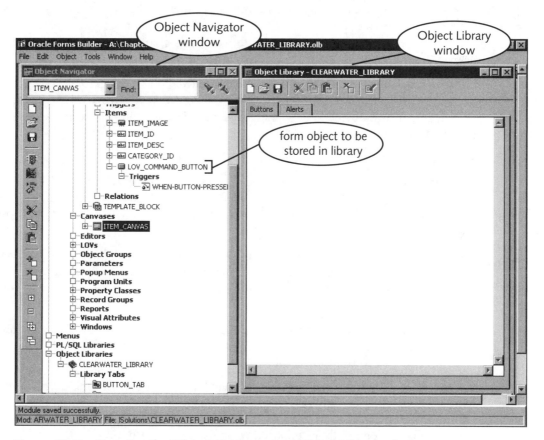 beside the BUTTON_TAB library tab page to open the Object Library window and display the Buttons tab page.

3. Click the **Restore button** on the inner tab page window. The Object Library window and the Object Navigator window should appear side by side on your screen display, as shown in Figure 12-7. If necessary, resize and reposition the windows so your screen display looks like Figure 12-7.

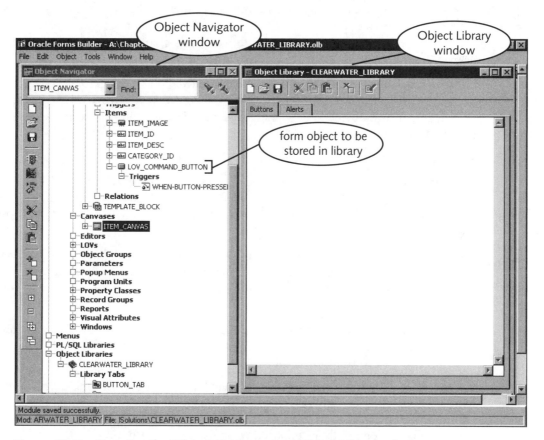

Figure 12-7   Opening the Object Navigator and Object Library windows

4. Select the **LOV_COMMAND_BUTTON** in the Object Navigator window, and drag it to the Buttons tab page. The pointer changes to the Object Move pointer.

5. Drop the object onto the Buttons tab page. The object icon and name appear on the library tab page.

6. Close the Object Library window, and maximize the Object Navigator window.

When you place an object in an object library, all of the object's properties and sub-objects (such as triggers associated with a button) are included with the object in the object library. You cannot edit the object directly in the object library. To edit the object, you must copy the object into a form, modify it, drag the modified object back into the library, and instruct the library to replace the existing object with the modified object.

Using Object Library Objects in Forms

To use an object library object in a target form, you select the object in the Object Library window, and then drag it into the target form. Like objects in object classes, you have the option to copy or subclass the object. As before, it is a good practice to always subclass objects rather than copying them, so that changes to the original objects are propagated to the subclasses.

Now, you will subclass the LOV_COMMAND_BUTTON object in the ITEM_FORM. First, you will delete the existing LOV_COMMAND_BUTTON that you created as the object source. Then, you will place the button from the object library into the form and subclass it.

To subclass an object library object in a target form:

1. In the Object Navigator, select the **LOV_COMMAND_BUTTON** item under the ITEM data block in the ITEM_FORM, and then click the **Delete button** 🔲 to remove the button from the form. Click **Yes** to confirm the deletion.

2. Scroll down in the Object Navigator, double-click the **Library Tabs icon** 🔲 beside BUTTON_TAB to open the Object Library window, and click the **Restore button** 🔲 to open the Object Library window and the Object Navigator window side by side.

3. Select the **LOV_COMMAND_BUTTON** object in the Object Library window, and drag it to the ITEM block in the ITEM_FORM. The pointer changes to the Object Move pointer 🔲.

4. Drop the object in the block. When the alert opens, asking if you want to copy the object or subclass it, click **Subclass** to subclass the object. The LOV_COMMAND_BUTTON appears in the ITEM block as a subclassed block item, with an inheritance arrow on its icon.

5. Click ➕ beside LOV_COMMAND_BUTTON, and then click ➕ beside the Triggers node. The button trigger appears with an inheritance arrow on its icon.

6. Close the Object Library window, and maximize the Object Navigator window.

Now you will run the form, and confirm that the subclassed LOV command button works correctly. You will retrieve all of the table records, and then click the LOV command button.

To test the subclassed LOV command button:

1. Run the form, click the **Enter Query button** 🔲, do not enter a search condition, and then click the **Execute Query button** 🔲. The first ITEM record appears.

2. Confirm that the insertion point is in the Item ID text item, and then click the **LOV command button**. The LOV display opens.

3. Click **Cancel**, and then click **Return** to close the form.

The LOV command button works correctly in this form, because the GO_ITEM command in the LOV command button trigger references the text item associated with the form, which is the ITEM_ID in the ITEM block. And, the LOV command button's Canvas property was already specified to display the button on the form's canvas. If you subclass this button in other forms, you will have to modify the trigger command so it references the correct form text item. You will also have to modify the button's Canvas property, so that the button appears on the correct form canvas. As with other object types, experience will dictate the modifications you will have to perform to make objects work correctly in different forms.

Creating and Using SmartClasses in Object Libraries

You can use object libraries to define **SmartClasses**, which are object library objects that are defined as standard object types for an application. A **standard object type** is an object that is always subclassed when an application needs that type of object. For example, the managers at Clearwater Traders could specify that the LOV_COMMAND_BUTTON object in the CLEARWATER_LIBRARY is the standard object type, and that when an application needs an LOV command button, this object must always be used.

To use a SmartClass object, you create a new object in the target form, select the object, right-click, and then point to SmartClasses. A list of all SmartClass objects that are of the same object type as the new object appears. You can then select one of the SmartClass objects, and the new object is automatically subclassed from the selected SmartClass object. Using SmartClasses speeds up development, because you do not need to open the Object Library window, select the library object, and drag it to the form.

To gain experience with using SmartClasses, you will now designate the LOV_COMMAND_BUTTON as a SmartClass object. Then, you will create a new command button in a form, and subclass it using a SmartClass.

To designate the LOV_COMMAND_BUTTON as a SmartClass object:

1. In the Object Navigator, double-click the **Library Tabs icon** 🔲 beside BUTTON_TAB to open the Buttons tab page in the Object Library window.

2. Select the **LOV_COMMAND_BUTTON**, click **Object** on the menu bar, and then click **SmartClass**. A green check mark appears to the left of the object, indicating it is now a SmartClass.

 You can tell when an Object Library object is selected because the upper border of the object becomes thicker and darker.

3. Close the Object Library window, and save the library.

To use a SmartClass, you must create an object in a target form that is the same object type as the SmartClass object. Now, you will delete the existing LOV command button in the ITEM_FORM. Then, you will create a new button, select the new button, right-click, point to SmartClasses, and select the SmartClass object to create a new LOV command button.

To use the SmartClass object:

1. In the Object Navigator, select the **ITEM_FORM**, click **Tools** on the menu bar, and then click **Layout Editor** to open the Layout Editor.

2. Select the existing **LOV command button**, and press **Delete** to delete the button.

3. Select the **Button tool** 🔲 on the tool palette, and draw a new button of any size anywhere on the canvas.

4. Right-click the new button, point to **SmartClasses**, and then click **LOV_COMMAND_BUTTON**. The new button becomes an LOV command button. If necessary, move the button so that it is positioned beside the ITEM_ID text item.

5. Save the form, and then close the form in Form Builder.

6. Select the **CLEARWATER_LIBRARY**, click **File**, and then click **Close** to close the library. If necessary, click **Yes** to confirm saving changes to the library.

 If you do not explicitly close the library, it will load every time you start Form Builder.

STORING REUSABLE PL/SQL CODE IN PL/SQL LIBRARIES

Often developers need to share PL/SQL program code. For example, the developers at Clearwater Traders might want to store and share the program code to calculate the sales tax or shipping charges that are added to customer orders. They could create a button that has a trigger with this code, and store the button in an object class or object library, but since the code might be used by triggers other than command buttons triggers, it makes more sense to use a PL/SQL library stored program unit.

Recall that a PL/SQL library is a collection of PL/SQL named program units, such as functions, procedures, and packages. In Form Builder, you can create and use PL/SQL libraries that are stored in files where all developers can access them. You can create a new library, and then add program units to the library, or attach existing PL/SQL libraries to the library. Since library program units can be used in many different forms, they always receive parameter inputs from the calling form and return output parameters to the calling form.

In this section, you will create a PL/SQL library named CLEARWATER_PLSQL_LIBRARY. You will create a program unit in the library that contains the code to receive a customer order ID as an input parameter, and returns the amount of sales tax that should be charged for the order as an output parameter. Then, you will use the library program unit in a form.

Creating a PL/SQL Library in Form Builder

To create a new PL/SQL library in Form Builder, you create the library in the Object Navigator. Like object libraries, PL/SQL libraries are top level objects that are not associated with a specific form. The code in the library can be used in any form. You save the library in the file system in a library file with a .pll extension. Now, you will create a new PL/SQL library in Form Builder, and save the library.

To create and save a new PL/SQL library:

1. In the Object Navigator, select the top-level **PL/SQL Libraries node**, and then click the **Create button** to create a new library. The new library appears in the Object Navigator.

2. Make sure that the new library is selected, and then click the **Save button**. If necessary, navigate to your Chapter12\TutorialSolutions folder, and save the library as **CLEARWATER_PLSQL_LIBRARY.pll**. Note that the library name in the Object Navigator changes to the library filename. Your library should look like the library shown in Figure 12-8.

You cannot change a PL/SQL library name directly in the Object Navigator. You can only change a PL/SQL library name by changing the library filename.

Figure 12-8   Creating a PL/SQL library in Form Builder

Figure 12-8 shows that a Form Builder PL/SQL library has two child nodes: Program Units and Attached Libraries. The Program Units node enables you to create a new program unit and specify the PL/SQL code directly in Form Builder. The Attached Libraries node enables you to attach an existing PL/SQL library .pll file to the library. Now you will create a new program unit that calculates the amount of sales tax that will be charged for a Clearwater Traders customer order. The calling program will pass the customer order ID value to the program unit as an input parameter and receive a single output value, which is the sales tax amount. Since the calling program receives a single output value, the program unit will be structured as a function.

> Recall that a function is a program unit that returns a single value to the calling program. Chapter 5 describes how to create user-defined PL/SQL functions.

After you create the new program unit, you will copy the program unit code from a text file named CalcTax.txt that is stored in the Chapter12 folder on your Data Disk. This code determines the customer's state of residence, based on the order ID value. If the customer is a Wisconsin resident (STATE = 'WI'), then the customer is charged 6% sales tax on the total order amount. If the customer is not a Wisconsin resident, then he or she will not be charged sales tax. The code retrieves the total order amount from the database by summing the price times quantity ordered from all order lines, and then calculates and returns the sales tax amount.

To create the PL/SQL library program unit:

1. Select the **Program Units node** under the CLEARWATER_LIBRARY, and then click the **Create button** . The New Program Unit dialog box opens.

2. Type **CALC_TAX** in the Name text box, select the **Function option button**, and then click **OK**. The PL/SQL Editor opens, and displays the function template.

3. You will be copying all of the function code from another file, so highlight all of the template code, and press **Delete** to delete the current function template code.

Do not click Delete on the PL/SQL button bar, or you will delete the program unit. If you accidentally delete the program unit, repeat Steps 1 and 2 to create the program unit again.

4. Start Notepad, and open **CalcTax.txt** from the Chapter12 folder on your Data Disk. Click **Edit** on the menu bar, and then click **Select All** to select all of the file text. Click **Edit** on the menu bar again, and then click **Copy**.

5. Close Notepad, and switch back to Form Builder. If necessary, place the insertion point in the PL/SQL Editor, click **Edit** on the menu bar, and then click **Paste**. The copied code appears in the source code pane.

6. Compile the code, correct any syntax errors, and close the PL/SQL Editor.

7. Select **CLEARWATER_PLSQL_LIBRARY** in the Object Navigator, and click the **Save button** 🖫 to save the library.

Attaching a PL/SQL Library to a Form

To use a PL/SQL library in a form, you attach the library to the form. You can then call program units in the library, just as you call form program units that are created in the form and appear under the form's Program Units node. When you attach a library to a form, the form stores only a reference to the program unit code, and does not store the actual code. As a result, the code in library program units can only be modified within the library and cannot be modified within the form. Since the library code is stored in a central location, it can easily be modified and kept consistent across all applications. When you attach a library to a form, Form Builder loads library program units into your workstation's memory only when they are called in the form. This reduces the amount of memory required by the user's workstation to run the form.

When you attach a library to a form, an alert appears asking if you want to store path information to the library. When you store a library in the default Forms library folder, which is specified in the system Registry, Form Builder looks for the library file in this folder whenever a form that uses the library requests a library program unit. If the library is stored in another folder, then you need to include the path information so the form can locate the library in the file system at runtime.

As a general practice, you should always store PL/SQL libraries in the default PL/SQL library folder, so the library is available to all form applications. If you are working in a computer laboratory where many students use the same workstation, you should store the library in a different folder and include the path information, so other students can create and save their own libraries.

In Forms 6i, the default PL/SQL library folder is *Developer_Home*\TOOLS\OPEN60.

12

Now you will open a form file named Sales.fmb that is in the Chapter12 folder on your Data Disk and save the file using a different filename. The Sales.fmb form is a custom form that Clearwater Traders salespeople use to take customer orders. To take an order, the salesperson enters the customer information or selects the customer from the existing records in the CUSTOMER table. Then, the salesperson determines the method of payment and the order source for the order. Next, the salesperson selects the order line items and quantities. Finally, the salesperson determines the final order amount, including sales tax. Now you will open the form, save it using a different filename, and run the form to become familiar with its use.

To open, save, and run the sales form:

1. In Form Builder, open **Sales.fmb** from the Chapter12 folder on your Data Disk, and save the file as **12ASales.fmb** in your Chapter12\TutorialSolutions folder.

2. Run the form. The form appears with four tab pages: Select Customer, Create Order, Enter/Edit Items, and Finish Order. First, you will select a customer.

 This form uses a sequence named ORDER_ID_SEQUENCE. If you did not create this sequence in a previous chapter, an error will appear when you try to run the form. If this happens, start SQL*Plus, and create the sequence using the following command: `CREATE SEQUENCE order_id_sequence START WITH 2000;`. Then, run the form again.

3. On the Select Customer page, click the **LOV command button**, select customer ID **107** (Paula Harris), and then click **OK**. The customer data appears on the page. Click **Create New Order** to create a new order.

4. The Create Order tab page appears with a new order ID value in the Order ID text item. To specify the details about the customer order, select the **Check option button**, click the **LOV command button** beside the Order Source text box, select order source **6** (Web Site), and then click **OK**. Click **Save Order** to save the order information. The "FRM-40401: No changes to save" message appears, confirming that the order was successfully saved.

5. Click **Enter/Edit Items** to specify the order line items. The Enter/Edit Items tab page appears.

6. Click the **LOV command button** to select the first order item, select Inv. ID **11668** (3-Season Tent, color Sky Blue), and then click **OK**. Place the insertion point in the Quantity text box, type **1** for the order quantity, and then click **Add Order Line**. The "No changes to save" confirmation message appears, confirming that the order line was inserted into the database, and the page text items are cleared.

7. To add the next order item, click the **LOV command button** again to select the next item, select Inv. ID **11775** (Women's Hiking Shorts, color Khaki, size S), and then click **OK**. Place the insertion point in the Quantity text box, type **2**, and then click **Add Order Line**. The "No changes to save"

confirmation message appears, confirming that the order line was inserted into the database, and the page text items are cleared.

8. Click **Finish Order**. The order lines appear in the Order Line Detail frame, and the order subtotal, which is the sum of the order quantities times the item prices, appears in the Order Subtotal text box. Currently, the sales tax and total order amount do not appear, because the code to calculate these amounts has not been added to the form yet.

9. Close Forms Runtime.

To finish the form, you need to attach the CLEARWATER_PLSQL_LIBRARY to the form, and call the CALC_TAX function to calculate and display the order sales tax. To attach the library to the form, you select the form's Attached Libraries node in the Object Navigator, and then specify the name of the library file to be attached. When you attach the library, an alert will appear asking if you want to store folder path information for the library. Since you might be working in a computer laboratory, you will include the path information. You will need to have the library file available at this location whenever you run a form that uses the library. Now, you will attach the CLEARWATER_PLSQL_LIBRARY to the sales form.

To attach the PL/SQL library to the form:

1. In the Object Navigator, select the **Attached Libraries node** under the SALES_TAB_FORM, and click the **Create button** to attach a new library. The Attach Libraries dialog box opens.

2. Make sure that the File System option button is selected, and click **Browse**. If necessary, navigate to your Chapter12\TutorialSolutions folder, select **CLEARWATER_PLSQL_LIBRARY.pll**, click **Open**, and then click **Attach**. The alert appears, and the alert message states that the library name contains a non-portable path specification.

3. Since you want to include the library path information, click **No**, which specifies that the path information will not be removed. The library appears under the form's Attached Libraries node.

4. Click ➕ beside CLEARWATER_PLSQL_LIBRARY under the Attached Libraries node in the SALES_TAB_FORM. The CALC_TAX (Function Body) program unit appears.

5. Click ➕ beside CALC_TAX (Function Body). The function specification appears as shown in Figure 12-9. The function specification shows that the function receives an input variable of data type NUMBER and returns a NUMBER value.

12

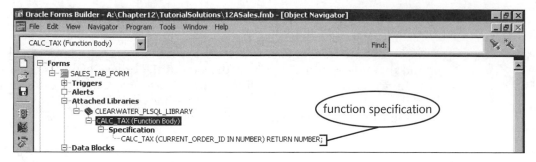

Figure 12-9   Attached library function specification

6. Save the form.

Note that when you attach a library to a form, the form stores only a reference to the program unit code, and does not store the actual code. Since the library code is stored in a central location, it can be easily modified and kept consistent across all applications. For example, if the sales tax rate changes, you just need to modify the tax rate in the library program unit, and the new tax rate will automatically propagate to all applications that use the library program unit.

Calling a Library Program Unit from a Form Trigger

To use the CALC_TAX program unit in the form, users will call the function from the trigger associated with the Finish Order button on the Enter/Edit Items tab page. When you click the Finish Order button, the Finish Order tab page appears and displays the order line details, the order subtotal sales tax, and total amounts. Now, you will modify this trigger to call the CALC_TAX function, pass the order ID value to the function as an input parameter, and assign its return value to the TAX_AMT text item on the Finish Order tab page. The order ID value appears in the ORDER_ID text item, which is in the form's CUST_ORDER block, so you will pass the input parameter using the reference **:cust_order.order_id**. The TAX_AMT text is in the ORDER_TOTALS form block, so you will assign the function return value to **:order_totals.tax_amt**. Now, you will modify the form trigger to call the library function.

To modify the form trigger to call the library function:

1. In the Object Navigator, click ➕ beside Canvases, and then double-click the **Tab Canvas icon** 📄 beside SALES_TAB_CANVAS to open the form tab canvas in the Layout Editor.

2. Select the **Enter/Edit Items tab**, select the **Finish Order button**, right-click, and then click **PL/SQL Editor**. The PL/SQL Editor opens, showing the current button trigger code.

3. Add the following code as the last line in the trigger:

   ```
   :order_totals.tax_amt := CALC_TAX(:cust_order.order_id);
   ```

4. Compile the trigger, correct any syntax errors, then close the PL/SQL Editor, and save the form.

Now you will run the form and confirm that the attached library code works correctly. You will create another new order for customer Paula Harris, select two order line items, and note the value for the order sales tax.

To run the form and test the attached library code:

1. Run the form. On the Select Customer page, click the **LOV command button**, select customer ID **107** (Paula Harris), and then click **OK**. The customer data appears on the page. Click **Create New Order** to create a new order.

2. To specify the order details, select the **Check option button**, click the **LOV command button** beside the Order Source text box, select order source **6** (Web Site), click **OK**, and then click **Save Order** to save the order information. Click **Enter/Edit Items** to specify the order line items.

3. On the Enter/Edit Items page, click the **LOV command button** to select the first order item, select Inv. ID **11668** (3-Season Tent, color Sky Blue), and click **OK**. Place the insertion point in the Quantity text box, type **1** for the order quantity, and then click **Add Order Line** to insert the order line into the database. The "No changes to save" confirmation message appears to confirm that the order line was inserted.

4. To add the next order item, click the **LOV command button** again to select the next item, select Inv. ID **11775** (Women's Hiking Shorts, color Khaki, size S), and click **OK**. Place the insertion point in the Quantity text box, type **2**, and then click **Add Order Line**. The "No changes to save" confirmation message appears again to confirm that the order line was inserted.

5. Click **Finish Order**. The order lines appear in the Order Line Detail frame, and the order subtotal, tax amount, and order total appear on the form, as shown in Figure 12-10. The order tax amount was successfully calculated using the library function.

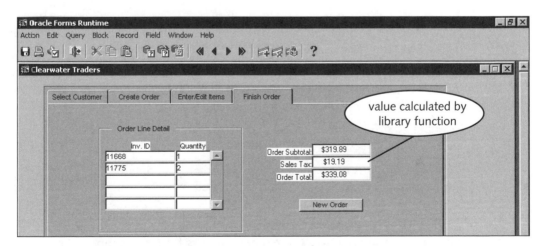

Figure 12-10   Order totals display

6. Close Forms Runtime, close the form in Form Builder, and save your changes if necessary.

7. Select **CLEARWATER_PLSQL_LIBRARY**, and click the **Delete button** to close the library in Form Builder. Close Form Builder and all other open Oracle applications.

SUMMARY

❑ An object is an abstract representation of something in the real world. A class defines the structure and properties of similar objects, and a class instance is an object in a class.

❑ An object inherits its structure from its class. Every object in a class has the same structure as all other objects in the class.

❑ A subclass has properties that are similar to the properties of its parent class, but also has additional properties, called specializations, that make it unique.

❑ Form Builder provides property classes, object groups, object libraries, and PL/SQL libraries to enable developers to create form objects that can be reused in different forms.

❑ A property class is a top-level form object that specifies properties and associated values that can be inherited by other form objects.

❑ Subclassing an object means that the object inherits properties from another object or from a property class. Changes to the original object propagate to all subclassed objects.

❑ You can base a property class on an existing object, or you can manually create the property class, and then define its properties and their associated values.

❑ To subclass a form object using a property class, you open the object Property Palette, select the Subclass Information property, open the Subclass Information dialog box, and specify the class from which the object will be subclassed.

❑ When you apply a property class to a form object, the properties that the object inherits appear in the object Property Palette with an inheritance arrow by the property node.

❑ To subclass an object from a property class, the form in which the property class exists must be open in Form Builder.

❑ An object group is a top-level form object that is a container for reusable objects. Developers can create subclassed objects based on objects in the object group.

❑ Form object groups should be placed in a central object source form that is accessible to all developers. Developers will use the object source form as they create applications and reference the required objects from the source form.

❐ You create an object group in the Object Navigator, and then add objects to the object group by dragging them onto the object group's Object Group Children node and dropping them there. Objects can only be added from the same form that contains the object group.

❐ Placing an object in an object group creates a reference to the object, and does not create an actual physical copy of the object.

❐ Objects in an object group can only be top-level form objects.

❐ To reference an object in an object group in a target form, you select the object group in the source form, drag it onto the Object Groups node in the target form, and drop it there.

❐ When you subclass an object, changes made to the object propagate to all sub-classed objects in target forms.

❐ If you close an object group's source form in Form Builder, subclassed objects no longer appear in target forms in Form Builder, although the subclassed objects still exist in compiled form .fmx files.

❐ An object library is a top-level Form Builder object that is independent of any specific form module and contains form objects that can be referenced in target forms. Object libraries are saved in files with .olb extensions.

❐ You create an object library in Form Builder. An object library has tab pages, and each tab page contains a different object type.

❐ To add an object to an object library, you create the object in any form, then drag the object onto the object library tab page where you want to save the object, and drop it there.

❐ To use an object library object in a target form, you select the object in the Object Library window, and then drag it into the target form. You can either subclass or copy the object.

❐ SmartClasses are object library objects that are defined as standard object types for an application. You can create an item, and then easily subclass it by using the SmartClasses pop-up menu selection.

❐ You can create PL/SQL libraries in Form Builder that contain named program units with code that can be used by many applications.

❐ You create a PL/SQL library as a top-level Form Builder object that is saved in a file with a .pll extension. You can add PL/SQL program units to the library, or attach other PL/SQL libraries to the library.

❐ To use a PL/SQL library in a form, you attach the library to the form, and then write code in form triggers to call program units in the library.

❐ When you attach a PL/SQL library to a form, the form stores only a reference to the library, and does not store the actual program unit code. The form loads the code when the library program unit is called by a form trigger.

12

REVIEW QUESTIONS

1. What is the relationship between an object and a class?

2. What is the relationship between a class and a subclass?

3. When should you create a property class, and when should you use an object group?

4. How can you tell when a form item property is inherited from a property class?

5. List three ways that an object library is different from an object group.

6. What is the difference between subclassing an object and creating a copy of an object?

7. What is a SmartClass, and what is the advantage of creating one?

8. Indicate whether you should use a property class, object group, object library, or PL/SQL library to make the following reusable form objects. (If you can use multiple approaches to create some of the items, list all possible approaches.)

    a. A window that you configure by changing its properties in its Property Palette.

    b. A radio group that contains multiple radio buttons.

    c. A command button with associated trigger code.

    d. A procedure that calls a report from a form and passes a parameter list.

    e. A data block based on a view that appears in multiple forms.

PROBLEM-SOLVING CASES

Cases refer to the Clearwater Traders (see Figure 1-11), Northwoods University (see Figure 1-12), and Software Experts (see Figure 1-13) sample databases. All required data files are stored in the Chapter12 folder on your Data Disk. Store all solution files in your Chapter12\CaseSolutions folder.

1. In this case, you will open a template form for the Northwoods University database system that is stored in a file named 12ACase1.fmb. You will create property classes to format text items that contain telephone number and date values. Then, you will use the template to create a data block form based on the STUDENT table and apply the property classes to the S_PHONE and S_DOB form text items.

    a. Open 12ACase1.fmb, and save it as 12ACase1_DONE.fmb.

    b. Create a property class named PHONE_PROPERTY_CLASS that specifies the properties for a text item that has the Character data type, Maximum Length 16, and a format mask such that student Sarah Miller's telephone number appears as (715) 555-9876.

    c. Create a property class named DATE_PROPERTY_CLASS that specifies the properties for a text item that has the Date data type and a format mask such that student Sarah Miller's birth date appears as 07/14/1982.

d. Use the Data Block Wizard to create a data block in the template form based on the STUDENT table. Display all of the table fields, create descriptive item prompts, use a form-style layout, and change the frame title to "Students."

e. Apply the PHONE_PROPERTY_CLASS to the S_PHONE text item and the DATE_PROPERTY_CLASS to the S_DOB text item.

2. In this case, you will open a template form for the Software Experts database system that is stored in a file named 12ACase2.fmb. You will create property classes to format all text items using a specific font color, name, and weight, and to format date items using a specific format mask. Then, you will use the template to create a data block form based on the PROJECT_CONSULTANT table.

a. Open 12ACase2.fmb, and save it as 12ACase2_DONE.fmb.

b. Create a property class named TEXT_ITEM_PROPERTY_CLASS that specifies the properties for a text item that has an 8-point Comic Sans MS font using the Demilight font weight. Specify the font color as dark blue. (If your system does not have the Comic Sans MS font, substitute a different font.)

c. Create a property class named DATE_PROPERTY_CLASS that is a subclass of the TEXT_ITEM_PROPERTY_CLASS. (*Hint*: Create a new property class, and then change its Subclass Information property to TEXT_ITEM_PROPERTY_CLASS.) Specify that the property class has the Date data type, and a format mask value so that the roll-on date for P_ID 1 and C_ID 101 appears as 15 June , 2002. (The month will appear padded to eight total characters.)

d. Use the Data Block Wizard to create a data block in the template form based on the PROJECT_CONSULTANT table. Display all of the table fields, create descriptive item prompts, use a form-style layout, and change the frame title to "Project Consultants."

e. Apply the DATE_PROPERTY_CLASS to the ROLL_ON_DATE and ROLL_OFF_DATE text items, and apply the TEXT_ITEM_PROPERTY_CLASS to the remaining data block text items.

3. In this case, you will open a template form for the Northwoods University database system that is stored in a file named 12ACase3.fmb. You will create an object group that contains a formatted canvas object and a second object group that contains a formatted window object. Then, you will subclass the two object groups in a new form, and create a data block in the new form that is based on the COURSE table and that displays data using the subclassed canvas and window.

a. Open 12ACase3.fmb, and save it as 12ACase3_DONE.fmb.

b. Create a new object group named CANVAS_OBJECT_GROUP. Create a new form canvas named LOGO_CANVAS, import the Northwoods University logo (nwlogo.jpg) into the canvas, and position the logo in the top-left corner of the canvas. Then, place the LOGO_CANVAS in the new object group.

12

   c. Create a second object group named WINDOW_OBJECT_GROUP. Create a new form window named NORTHWOODS_WINDOW, and specify the window properties using the window formatting guidelines described in Chapter 6. Change the window title to "Northwoods University."

   d. Create a new form named COURSE_FORM, and save the form as 12ACase3_COURSE.fmb.

   e. Subclass the CANVAS_OBJECT_GROUP and WINDOW_OBJECT_GROUP in the COURSE_FORM.

   f. Use the Data Block Wizard to create a data block based on the COURSE table. Display all of the table fields, create descriptive item prompts, use a tabular-style layout, display five records at a time and show a scrollbar, and change the frame title to "Courses."

   g. Display the data block on the subclassed canvas, and display the canvas in the subclassed window. (*Hint*: To display a canvas in a specific window, specify the window name in the canvas Window property.)

4. In this case, you will open a template form for the Software Experts database system that is stored in a file named 12ACase4.fmb. You will create an object group that contains data blocks corresponding to the CONSULTANT and PROJECT_CONSULTANT tables. Then, you will subclass the object group in a new form, and modify the form so that the subclassed data blocks have a master-detail relationship.

   a. Open 12ACase4.fmb, and save it as 12ACase4_DONE.fmb.

   b. Create a new object group named BLOCK_OBJECT_GROUP.

   c. Use the Data Block Wizard to create a data block based on the CONSULTANT table. Include all of the table fields in the data block. Do not create a layout for the data block.

   d. Use the Data Block Wizard to create a second data block based on the PROJECT_CONSULTANTS table. Do not create a master-detail relationship between the PROJECT and PROJECT_CONSULTANTS data blocks. Do not create a layout for the data block.

   e. Add both data blocks to the BLOCK_OBJECT_GROUP.

   f. Create a new form named PROJECT_CONSULTANTS_FORM, and save the form as 12ACase4_PROJ_CONSULTANTS.fmb. This new form will be the target form.

   g. Subclass the BLOCK_OBJECT_GROUP in the target form.

   h. Create a new canvas in the target form named PROJ_CONSULTANTS_CANVAS.

   i. Use the Data Block Wizard in reentrant mode to create a master-detail relationship between the blocks in the target form, in which the CONSULTANT block is the master block and the PROJECT_CONSULTANTS block is the detail block. (*Hint*: Recall that you always create a master-detail relationship in the detail block.)

j. Use the Layout Wizard to create a layout for the CONSULTANT data block in the target form. Specify descriptive item prompts, use a form-style layout, and change the frame title to "Consultants."

k. Use the Layout Wizard to create a layout for the PROJECT_CONSULTANT data block in the target form. Specify descriptive item prompts, use a tabular-style layout, display five records at one time and show a scrollbar, and change the frame title to "Consultant Projects."

5. In this case, you will create an object library that contains tab pages for alerts and radio groups that are used in Northwoods University database application forms. Then, you will subclass the objects into a custom form that is used to update and delete records in the STUDENT table.

a. Create a new object library named NORTHWOODS_OBJECTS. Create tab pages named ALERT_TAB and RADIO_GROUP_TAB. Label the tabs as "Alerts" and "Radio Groups." Save the library as 12ACase5.olb.

b. Open the COURSE_SECTION_FORM that is stored in the 12ACase5_ALERTS.fmb form file, add all of the form alerts to the Alerts tab page, and then close the COURSE_SECTION_FORM.

c. Open the forms listed below, and add the form radio groups to the Radio Groups tab page.

Form Name	Form Filename	Radio Group Name
STUDENTS	12ACase5_RADIO1.fmb	S_CLASS
FACULTY	12ACase5_RADIO2.fmb	F_RANK

d. Open the STUDENT_FORM form that is stored in the 12ACase5_STUDENT.fmb file. Save the file as 12ACase5_STUDENT_DONE.fmb. Currently, the form does not work correctly, because it does not have the alerts that are referenced in the trigger code in the Update and Delete buttons.

e. Subclass the required alerts from the object library so the Update and Delete form buttons work correctly. (You will need to add a new student record before you can test the Delete button, because all of the current students are referenced as foreign key values in other records.)

f. Subclass the S_CLASS radio group from the object library, and modify the form so that the S_CLASS value in the STUDENT_FORM appears as a radio group rather than as a text item. (You will need to delete the existing S_CLASS text item before you subclass the radio group.)

6. In this case, you will create a PL/SQL library that contains a program unit to calculate the sales tax on Clearwater Traders customer orders, and a second program unit to calculate the shipping and handling on customer orders. Then, you will modify the sales form you worked with in the tutorial so that it calls the program units to calculate the order sales tax and shipping and handling. (You might

12

already have created and used the library program unit to calculate the sales tax in the tutorial, but you will repeat it again in this case.)

a. Create a new PL/SQL library named 12ACASE6 that is saved in a file named 12ACase6.pll.

b. Create a new function in the library named CALC_TAX, which calculates the tax for a customer order. Replace all of the function template code with the code that is stored in the CalcTax.txt file.

c. Create a second new function in the library, named CALC_SH, which calculates the order shipping and handling based on the order subtotal. Replace all of the function template code with the code that is stored in the CalcSH.txt file.

d. Open the sales form, which is stored in a file named 12ACase6.fmb, and save the file as 12ACase6_DONE.fmb.

e. Attach the PL/SQL library to the sales form. Do not remove the path information.

f. Modify the sales form so that when the user clicks the Finish Order button on the Enter/Edit Items canvas, the button trigger calls the PL/SQL library program unit functions and displays the sales tax and shipping and handing values on the Finish Order tab canvas.

7. In this case, you will create a PL/SQL library that contains a program unit to calculate the cumulative grade point average for a specific Northwoods University student, and a second program unit to calculate the grade point average for a student for a specific term. Then, you will attach the library to a form that allows the user to select a specific student and a specific term, and then display the student's term and cumulative grade point averages.

a. Create a new PL/SQL library named 12ACASE7 that is saved in a file named 12ACase7.pll.

b. Create a new function in the library named CALC_CUM_GPA, which calculates the student's cumulative GPA. The function receives the student ID as an input parameter and returns the cumulative GPA value. The complete function code is provided in a file named CalcCumGPA.txt.

c. Create a second new function in the library named CALC_TERM_GPA, which calculates the student's GPA for a specific term. The function receives the student ID and term ID as input parameters, and returns the term GPA. The complete function code is provided in a file named CalcTermGPA.txt.

d. Open the form that allows the user to select the student and the term, which is stored in a file named 12ACase7.fmb. Save the file as 12ACase7_DONE.fmb.

e. Attach the PL/SQL library to the form. Do not remove the path information.

f. Modify the form so that when the user clicks the Display GPA Values button on the canvas, the button trigger calls the PL/SQL library program unit functions and displays the term and cumulative GPA values in the associated form text items.

◀ LESSON B ▶

Objectives

♦ Use the built-in calendar class to create a form calendar window
♦ Understand flexible code
♦ Learn how to use system variables to create flexible code
♦ Use built-in subprograms to create flexible code
♦ Understand how to use indirect referencing to write flexible code

In the previous lesson, you learned how to create reusable objects that reduce the amount of time it takes to create complex applications and that make forms consistent within an integrated application. In this lesson, you will learn how to use the Form Builder calendar class, which is a built-in object class that Oracle provides to help developers create a pop-up calendar window to enable users to search for and select dates. This lesson also explores **flexible code,** which contains commands that are dynamically specified at runtime, depending on the current properties and state of the form.

USING THE FORM CALENDAR CLASS TO CREATE A CALENDAR WINDOW

Oracle Forms 6i provides a built-in object class and PL/SQL library that enables you to create a calendar window to allow users to select date values from a visual representation of a calendar. When the user selects a date on the calendar, the selected date value is returned to a form text item. Figure 12-11 shows the form calendar window.

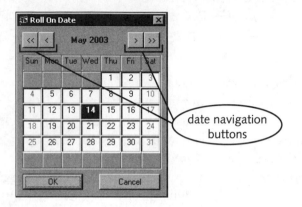

Figure 12-11   Form calendar window

The calendar window has a configurable window title bar that you can customize to fit your application. In Figure 12-11, the calendar is used to select project roll-on dates for the PROJECT_CONSULTANT table in the Software Experts database, so the window title is "Roll On Date." The calendar window contains date navigation buttons that allow the user to move to different calendar pages and dates within a page. The **Previous Year button** « and **Next Year button** » allow the user to move to the calendar page for the previous or next year, respectively. The **Previous Month button** < and **Next Month button** > allow the user to move to the previous or next month, respectively, on the current calendar page. The OK button closes the calendar window and returns the selected value to the associated form text item, and the Cancel button closes the calendar window but does not return the selected value.

Creating a calendar class object involves the following steps:

- Open the STNDRD20.OLB object library, and copy or subclass the Calendar object group. This object group contains the data blocks, canvases, and windows that display the calendar objects.

- Attach the CALENDAR.PLL PL/SQL library to the form. This library contains a package named DATE_LOV, which contains variables and procedures that display the calendar.

- Create a trigger associated with the text item to which the selected date value is returned. This trigger calls the GET_DATE procedure in the DATE_LOV package, which displays the calendar object. The GET_DATE

procedure calls other procedures in the DATE_LOV package that control the
calendar display and return the selected value to the form text item.

 If you are using a client/server database, the standard Form Builder object and
PL/SQL libraries can be installed from the Oracle Forms and Reports 6i Demos
CD. If you are using the Personal Oracle database, or do not have access to
this CD, you can find these standard library files in the Chapter12 folder on
your Data Disk.

Now you will add a calendar window to a data block form associated with the
PROJECT_CONSULTANT table in the Software Experts database. First you will open
the form, and save it using a different filename. Then, you will open the
STNDRD20.OLB object library, and subclass the CALENDAR object group in the
form. Recall that when you subclass an object group in a form, you must display the
Object Library window beside the form Object Navigator window, drag the object
group, and drop it on the Object Groups node in the form.

To open and save the form, and open and subclass the object library:

1. Start Form Builder, open **ProjectConsultants.fmb** from the Chapter12
   folder on your Data Disk, and save the file as **12BProjectConsultants.fmb**
   in your Chapter12\TutorialSolutions folder. Connect to the database.

2. Click the **Open button** 🖬 on the toolbar, navigate to the Chapter12
   folder on your Data Disk, select **STNDRD20.OLB**, and click **Open**. The
   STANDARDS object library appears in the Object Navigator.

3. Click ➕ beside STANDARDS, and then click ➕ beside Library Tabs to
   display the library tab pages.

4. Double-click the **Library Tabs icon** 🖾 beside COMPONENTS to open
   the Object Library window with the COMPONENTS tab page displayed.

5. Click the **Restore button** 🗗 on the Object Library window to display the
   library tab page beside the Object Navigator window. If necessary, adjust the
   window sizes and positions so that the objects in both windows are visible.

6. In the Object Library window, select the **CALENDAR** object group, drag it
   to the Object Navigator window, and drop it on the Object Groups node in
   the PROJECT_CONSULTANT form. The Form alert opens. Click
   Subclass to subclass the object in the form. Several new subclassed form
   objects appear in the form.

7. Close the Object Library window, maximize the Object Navigator window,
   and then save the form.

The next step is to attach the CALENDAR.PLL PL/SQL library to the form. Recall that
to attach a PL/SQL library to a form, you select the Attached Libraries node in the Object
Navigator, and then attach the library. Since you might be working on a workstation that

12

you share with other students, you will retain the path information for the library, which is stored in the Chapter12 folder on your Data Disk.

To attach the library to the form:

1. Select the **Attached Libraries node** in the Object Navigator, and then click the **Create button** to attach a PL/SQL library to the form.

2. In the Attach Library dialog box, click **Browse**, and navigate to the **Chapter12** folder on your Data Disk. Select **CALENDAR.PLL**, click **Open**, and then click **Attach**. Click **No** when the alert opens asking if you want to remove the library path information. The CALENDAR library appears under the form Attached Libraries node.

3. Save the form.

The CALENDAR PL/SQL library contains a package named DATE_LOV. The DATE_LOV package contains several public variables and procedures to create, display, and manipulate the calendar window. To display the calendar window in the PROJECT_CONSULTANT form, you will need to call a procedure named GET_DATE, which has the following syntax:

```
DATE_LOV.GET_DATE(initial_date, 'return_item',
window_x_position, window_y_position, 'window_title',
'OK_button_label', 'Cancel_button_label', highlight_
weekend_days, autoconfirm_selection, autoskip_after_
selection);
```

This procedure uses the following parameters:

- *initial_date* specifies a date that defines the month that appears when the calendar window first opens. Usually, this value is specified using the SYSDATE system variable.

- *return_item* specifies the form text item to which the selected date is returned. This value uses the format ***block_name.item_name***, and must be enclosed in single quotation marks. For the roll-on date text item in the PROJECT_CONSULTANT form, this value will be `'project_consultant.roll_on_date'`.

- *window_x_position* and *window_y_position* specify the X- and Y-position of the calendar window on the form. These values are specified as integers and usually reflect the position of the top right corner of the text item associated with the calendar.

- *window_title* specifies the title that appears on the calendar window, and is a text string enclosed in single quotation marks. For the calendar window associated with the roll-on date text item, this value will be `'Roll On Date'`.

- *OK_button_label* and *Cancel_button_label* specify the labels that appear on the OK and Cancel buttons in the calendar class window.

These values are specified as text strings that are enclosed in single quotation marks. For the calendar window that you will create in this application, you will use `'OK'` and `'Cancel'`.

- *highlight_weekend_days* specifies whether dates that fall on weekends are highlighted using a different font color. Valid values are TRUE or FALSE.

- *autoconfirm_selection* specifies how the calendar window behaves when the user selects a date. When the user selects a date and the autoconfirm_selection value is TRUE, the user must click OK to close the window and return the selected date to the form. When this value is FALSE, and the user selects a date, the window automatically closes and the selected date appears in the form text item.

- *autoskip_after_selection* specifies whether the form insertion point automatically moves to the next block item when the calendar window closes. Valid values can be TRUE or FALSE.

Now, you will create a **Show Calendar button** ▦ that appears on the right side of the roll-on date text item on the form. The button will display an icon that is stored in a file named calendar.ico in the Chapter12 folder on your Data Disk. The button trigger will call the GET_DATE procedure in the DATE_LOV package, and pass the required parameters to configure the calendar window and return the selected date to the roll-on date text item.

To create the button to display the calendar window:

1. In the Object Navigator, double-click the **Canvas icon** 🖼 beside TEMPLATE_CANVAS to open the main form canvas in the Layout Editor.

2. Place the tip of the mouse pointer on the top-right corner of the Roll On Date text item, note the X- and Y-position values on the status line, and write the values down. You will use these values to position the calendar window on the form.

3. Click the **Button tool** ▢ on the tool palette, and draw button on the right side of the Roll On Date text item. Do not worry about the button size, because you will adjust it in the button Property Palette.

4. Double-click the new button to open its Property Palette, change the following property values, and then close the Property Palette.

Property	Value
Name	**CALENDAR_BUTTON**
Label	(deleted)
Iconic	**Yes**
Icon Filename	**a:\chapter12\calendar.ico** (substitute the path to the Chapter12 folder on you Data Disk if necessary)
Width	**15**
Height	**15**
Tooltip	**Show Calendar**

12

5. In the Layout Editor, select the button if necessary, right-click, point to **SmartTriggers**, and then click **WHEN-BUTTON-PRESSED** to create a new button trigger. Type the code shown in Figure 12-12 to configure and open the calendar window. Use the X- and Y-position values that you noted in Step 2 for the window X position and Y position values. Compile the code, correct any syntax errors, and then close the PL/SQL Editor.

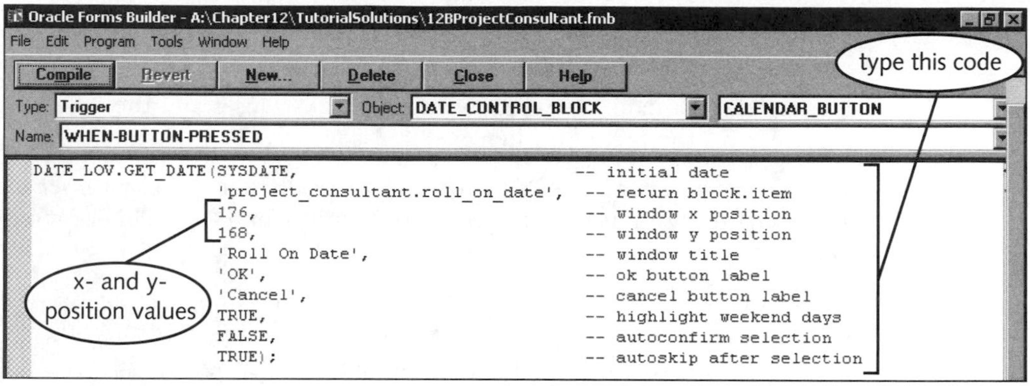

Figure 12-12   Code to configure and open the calendar window

6. Save the form.

Recall that when a form contains multiple blocks, the block that appears first in the Object Navigator Data Blocks list is the block whose items appear first when the form first opens. When you add a calendar window to a form, you must always ensure that the blocks that contain the calendar object do not appear first in the Data Blocks list. If they appear first, then the calendar window will open first, rather than the form window, and the form will not work correctly. Now, you will modify the block navigation order in the form so that the PROJECT_CONSULTANT block appears first in the form's Data Blocks list. Then you will run the form, insert a new record into the PROJECT_CONSULTANT table, and select the roll-on date value using the calendar window.

To modify the block navigation order and run the form:

1. Click **Window** on the menu bar, and then click **Object Navigator** to open the Object Navigator window.

2. Select the **PROJECT_CONSULTANT node** under the Data Blocks node, drag it upward, and drop it onto the Data Blocks node so that PROJECT_CONSULTANT appears as the first data block in the Data Blocks list. The order of the other data blocks in the form does not matter, so long as the block that displays the main form items appears first.

3. Save the form, and then run the form. The form appears in the Forms Runtime window.

4. Type **1** in the Project ID text item, press **Tab**, and type **102** in the Consultant ID text item.

5. Click the **Show Calendar button** 🖩. The calendar window opens. Select a date that is one week from today, and then click **OK**. The selected date appears in the Roll On Date text item.

6. Click **Exit** to close the form, and then click **No** to discard your changes.

7. Close the form in Form Builder, and, if necessary, click **Yes** to save your changes.

8. Select the **STANDARDS** object library, and then click the **Delete button** 🗙 to remove the object library from Form Builder.

FLEXIBLE CODE

Flexible code allows you to use objects in different modules without modifying the object properties or code. You can create flexible code by using system variables, which return values based on current system properties, such as the current date and time. You can also create flexible code by using built-in form subprograms that dynamically retrieve and set form object properties, and retrieve and display values from other forms. The following sections explore these approaches for creating flexible code.

Using System Variables to Create Flexible Code

You have already used form system variables to retrieve the current system data and time, set the system error message level, and determine which mouse button the user clicks. You can find a summary of the system status variables in Table 8-7 and a summary of the system mouse status variables in Table 8-9. Table 12-1 lists additional form system variables, which have not been presented in conjunction with other topics, and the value that each variable returns.

System Variable	Return Value
:SYSTEM.COORDINATION_OPERATION	Type of event that is maintaining coordination between master-detail blocks
:SYSTEM.CURRENT_BLOCK	Name of current block, if current navigational unit is a block, record, or item; NULL if current navigation unit is a form
:SYSTEM.CURRENT_DATETIME	Current operating system date and time, in the default DD-MON-YYYY HH24:MI:SS format
:SYSTEM.CURRENT_FORM	Name of current form

Table 12-1   Additional form system variables

System Variable	Return Value
:SYSTEM.CURRENT_ITEM	Name of current item, in the format *item_name*, if current navigation unit is an item; NULL if current navigation unit is a form, block, or record. Is included for compatibility with older versions of Oracle. New code should use :SYSTEM.CURSOR_ITEM or :SYSTEM.TRIGGER_ITEM.
:SYSTEM.CURRENT_VALUE	Value of the current item, if current navigation unit is an item; NULL if current navigation unit is a form, block, or record
:SYSTEM.CURSOR_BLOCK	Name of block that contains the form insertion point, if current navigation unit is a block, record, or item; NULL if current navigation unit is a form
:SYSTEM.CURSOR_ITEM	Name of item that contains the insertion point, in the format *block_name.item_name*, if current navigation unit is a an item; NULL if current navigation unit is a form, block, or record
:SYSTEM.CURSOR_RECORD	Number representing the physical position of the record that contains the form insertion point
:SYSTEM.CURSOR_VALUE	Value of the item that contains the insertion point, if current navigation unit is a an item; NULL if current navigation unit is a form, block, or record
:SYSTEM.EVENT_WINDOW	Name of the last window that fired a window event trigger
:SYSTEM.LAST_QUERY	Text of most recent SELECT query used to populate a block
:SYSTEM.LAST_RECORD	TRUE, if current record is last record in block; FALSE if current record is not the last record
:SYSTEM.MASTER_BLOCK	Name of master data block involved when an ON-CLEAR-DETAILS trigger fires
:SYSTEM.MODE	NORMAL during normal processing, QUERY when a query is being processed, and ENTER-QUERY when a form is in Enter Query mode
:SYSTEM.TAB_NEW_PAGE	Name of destination tab page in tab page navigation
:SYSTEM.TAB_PREVIOUS_PAGE	Name of source page in tab page navigation
:SYSTEM.TRIGGER_ITEM	Name of current item when current trigger fired
:SYSTEM.TRIGGER_RECORD	Number of current record when current trigger fired

Table 12-1   Additional form system variables (continued)

To gain experience with using other system variables, you will open a form named CustomerOrder.fmb that is in the Chapter12 folder on your Data Disk. This is a master-detail report, which shows Clearwater Traders CUSTOMER records in the master block and associated CUST_ORDER records in the detail block. You will modify the form as shown in Figure 12-13, so that it contains text items that display system variable values showing form status information. Specifically, the text items will display the current block, current item, current item value, and current form mode.

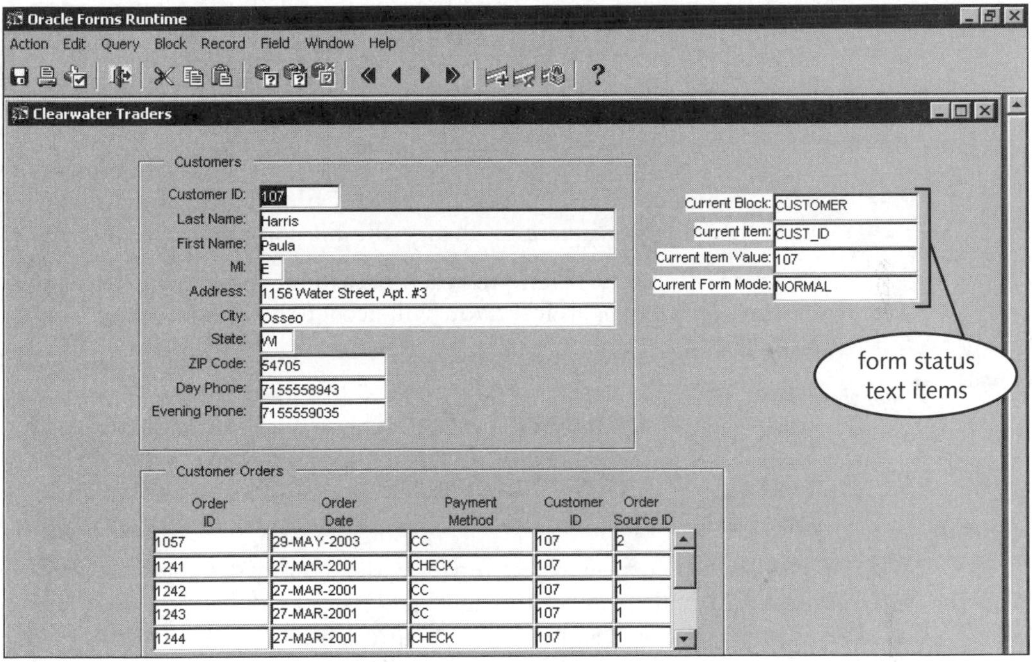

Figure 12-13   Form with text items showing system variable values

> This form is designed to allow you to gain experience with using these system variables. You probably would not display the system variables directly on the form in an actual application.

First, you will open the file, and save it using a different filename. Then, you will add a new control block to the form and the new text items to the control block.

To open and rename the form, and create the block and text items:

1. In Form Builder, open **CustomerOrder.fmb** from the Chapter12 folder on your Data Disk, and save the form as **12BCustomerOrder.fmb** in your Chapter12\TutorialSolutions folder.

2. In the Object Navigator, select the **Data Blocks node**, click the **Create button** 🔲, select the **Build a new data block manually option button**, and then click **OK**. The new data block appears in the Object Navigator. Change the block name to **STATUS_CONTROL_BLOCK**.

3. In the Object Navigator, select the **STATUS_CONTROL_BLOCK**, drag it downward, and drop it on the CUST_ORDER data block so STATUS_CONTROL_BLOCK is the last block in the Data Blocks list.

4. Click ➕ beside Canvases, and then double-click the **Canvas icon** 🔳 beside CUSTOMER_ORDER_CANVAS to open the form in the Layout Editor.

5. If necessary, open the Block list, and select **STATUS_CONTROL_BLOCK**.

6. Select the **Text Item tool** 🔲 on the tool palette, and draw the four form status text items as shown in Figure 12-13. If necessary, select the new text items as an object group, select the **Fill Color tool** 🔳 on the tool palette, and select a white square to change the fill color of the text items to white.

7. Open the Property Palette for each text item, and modify its Name and Prompt property as follows. You will accept the default values for all other properties.

Item	Name	Prompt
Item 1	**CURRENT_BLOCK**	**Current Block:**
Item 2	**CURRENT_ITEM**	**Current Item:**
Item 3	**CURRENT_VALUE**	**Current Item Value:**
Item 4	**MODE_TEXT**	**Current Form Mode:**

Next, you will create a trigger that retrieves the system variable value associated with each text item. You will create a form-level key trigger to retrieve these values. (You cannot use a form object, such as a command button, because when the user clicks the button, the button gets the form focus and will determine the current block, item, and value.) You will use the KEY-HELP trigger, which fires when the user presses F1. This trigger will retrieve the system variable for each form status text item, and assign the retrieved value to the text item.

To create the key trigger to display the form status values:

1. Click **Window** on the menu bar, and then click **Object Navigator** to open the Object Navigator window.

2. In the Object Navigator, select the **Triggers node** under CUSTOMER_ORDER_FORM, and click the **Create button** 🔲 to create a new trigger. The CUSTOMER_ORDER_FORM: Triggers dialog box opens. Place the form insertion point in front of % in the Find text box, type **key-help**, press **Enter**, and then click **OK**. The PL/SQL Editor opens.

3. Type the following code in the source code pane to retrieve the system variable values, and display them in the form text items:

```
:status_control_block.current_block :=
:SYSTEM.CURRENT_BLOCK;
:status_control_block.current_item :=
:SYSTEM.CURRENT_ITEM;
:status_control_block.current_value :=
:SYSTEM.CURRENT_VALUE;
:status_control_block.mode_text := :SYSTEM.MODE;
```

4. Compile the trigger, correct any syntax errors, then close the PL/SQL Editor and save the form.

Now you will run the form, and view the form status text items. To update the text items, you will press F1.

To run the form and view the form status text items:

1. Run the form. When the Forms Runtime window opens, press **F1**. The form status text items show that the current block is CUSTOMER, the current item is CUST_ID, and the form mode is NORMAL.

2. Click the **Enter Query button** , do not type a search condition, and then press **F1** again to update the form status text items. The form mode value changes to ENTER-QUERY.

If you place the form insertion point in one of the STATUS_CONTROL_BLOCK text items, an error message will appear if you click , because STATUS_CONTROL_BLOCK is not a data block.

12

3. Click the **Execute Query button** . The first form record (customer ID 107, Paula Harris) appears. Press **F1** again to update the form status text items. The current item value appears as 107, and the form mode returns to NORMAL.

4. Close Forms Runtime.

5. Close the form in Form Builder. If necessary, click **Yes** if you are asked if you want to save your changes.

To create flexible code that can be placed in reusable objects, these system variables are often used in conjunction with form built-in subprograms, which the next section describes.

Using Built-in Subprograms to Create Flexible Code

Form Builder provides several built-in subprograms that allow you to dynamically retrieve and set form object properties. The built-in functions that allow you to retrieve object properties are sometimes called the **GET_ built-ins**. Some of the most commonly used GET_ built-ins include:

- GET_APPLICATION_PROPERTY
- GET_BLOCK_PROPERTY
- GET_CANVAS_PROPERTY
- GET_FORM_PROPERTY
- GET_ITEM_PROPERTY
- GET_LIST_ELEMENT_LABEL
- GET_LIST_ELEMENT_VALUE
- GET_LOV_PROPERTY
- GET_MENU_ITEM_PROPERTY
- GET_RADIO_BUTTON_PROPERTY
- GET_RECORD_PROPERTY
- GET_RELATION_PROPERTY
- GET_TAB_PAGE_PROPERTY
- GET_VIEW_PROPERTY
- GET_WINDOW_PROPERTY

The GET_APPLICATION_PROPERTY built-in function returns information about the current Form Builder application, which can consist of multiple integrated forms. Returned values can include the name of the current form, the user connect string, and so forth. The general syntax for the GET_APPLICATION_PROPERTY built-in is:

```
GET_APPLICATION_PROPERTY(property_name);
```

You can retrieve a complete listing of the *property_name* values that you can pass as input parameters to this built-in in the Form Builder online help system.

The other GET_ built-in functions return property values for a specific form, block, canvas, and so forth, and have the following general syntax:

```
return_value :=
GET_object_PROPERTY(object_ID | 'object_name',
property_name);
```

These built-ins require the following input parameters:

- *return_value*, which is a previously declared variable that matches the data type of the value to be retrieved. If the retrieved value is a number, such as the height of a window, then the variable must be declared as a number. If the retrieved value is text, such as the window title, the variable must be declared as a character.

- *object*, which describes the object type, such as BLOCK or CANVAS.

- *object_ID or object_name*, which identifies the object for which you want to retrieve the property value. *Object_ID* references a Form Builder **object ID**, which is an internal identifier that Form Builder creates every time you create a new object. *Object_name* references the value of the object's Name property on its Property Palette, and must be enclosed in single quotation marks. Form Builder internally references all form objects using the object ID. When you use the object ID to reference an object in a GET_ built-in, form processing is faster than when you use the object name. When you use the object name, Form Builder must use the object name to look up the object ID.

- *property_name*, which is the name of the property. The exact format of the property name depends on the object type and the property name.

For detailed information on the syntax to use for the *property_name* parameter for different object properties, click Help on the Form Builder menu bar, click Form Builder Help Topics, and then type the name of the built-in as the search string.

You can retrieve the object ID of a form object by using the FIND_ built-in functions, which include:

- FIND_ALERT
- FIND_BLOCK
- FIND_CANVAS
- FIND_EDITOR
- FIND_FORM
- FIND_ITEM
- FIND_LOV
- FIND_MENU_ITEM
- FIND_TAB_PAGE
- FIND_TIMER
- FIND_TREE_NODE
- FIND_WINDOW

The general syntax for the FIND_ built-ins is

```
return_value := FIND_object('object_name');
```

In this syntax, *return_value* is a previously declared variable that has the same data type as the object in the FIND_ command. For example, if you are retrieving the object ID of an alert, you declare the *return_value* variable using the ALERT data type. If you are retrieving the object ID of a block, you use the BLOCK data type. *Object_name* is the object name, as specified in the object's Property Palette. The *object_name* parameter must be enclosed in single quotation marks.

Form Builder also provides several built-in subprograms that you can use to dynamically set form object properties while a form is running. These built-ins are called the **SET_ built-ins**. The most commonly used SET_ built-ins include:

- SET_ALERT_BUTTON_PROPERTY
- SET_ALERT_PROPERTY
- SET_APPLICATION_PROPERTY
- SET_BLOCK_PROPERTY
- SET_CANVAS_PROPERTY
- SET_FORM_PROPERTY
- SET_ITEM_PROPERTY
- SET_LOV_COLUMN_PROPERTY
- SET_LOV_PROPERTY
- SET_MENU_ITEM_PROPERTY
- SET_RADIO_BUTTON_PROPERTY
- SET_RECORD_PROPERTY
- SET_RELATION_PROPERTY
- SET_TAB_PAGE_PROPERTY
- SET_VIEW_PROPERTY
- SET_WINDOW_PROPERTY

The SET_ built-ins have the following general syntax:

```
SET_object_PROPERTY(object_ID | 'object_name',
property_name, property_value);
```

As with the GET_ built-ins that retrieve form object information, the SET_ built-ins require input parameters of the *object_ID* or *object_name* for which you want to set the property. The *property_name* is the target property, and the *property_value* is the new property value. If the property value is a character value, then it must be enclosed in single quotation marks. Recall that you have already used the SET_WINDOW_PROPERTY built-in to automatically maximize the Forms MDI window when Forms Runtime starts, and the SET_ITEM_PROPERTY to enable and disable form items.

To gain experience using these built-ins, you will open a custom form to retrieve and update information from the STUDENT table in the Northwoods University database.

First you will open the form, save the form file using a different filename, and then examine the form canvas.

To open and save the form, and examine the form canvas:

1. In Form Builder, open **Student.fmb** from the Chapter12 folder on your Data Disk, and save the file as **12BStudent.fmb** in your Chapter12\TutorialSolutions folder.

2. Click **Tools** on the menu bar, and then click **Layout Editor**. The form opens in the Layout Editor, as shown in Figure 12-14.

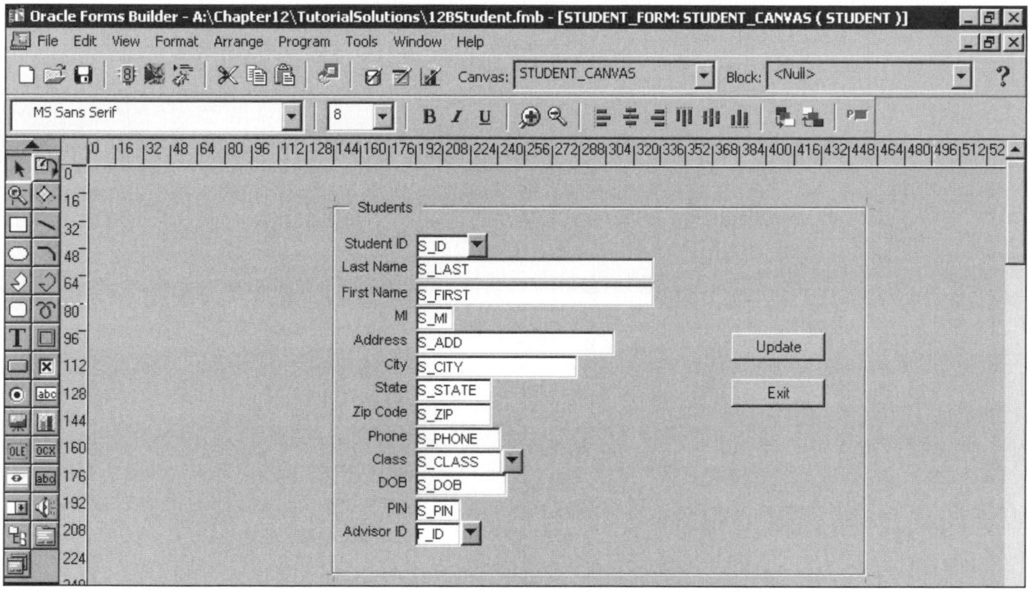

Figure 12-14   Student form layout

Currently, the form has an LOV command button to allow the user to select an existing student ID. The user can modify the student's data values by typing different values, by selecting an alternate class value using the LOV command button beside the S_CLASS text item, or by selecting an alternate advisor ID using the LOV command button beside the F_ID text item. The user can then click the Update button to save the changes. When the user clicks Update, an alert appears stating that the user is about to update the database. If the user clicks the OK alert button, the form saves the changes. If the user clicks the Cancel alert button, the form discards the changes.

To gain experience using a GET_ built-in, you will use the GET_FORM_PROPERTY built-in to retrieve the current form's filename. To gain experience using a FIND_ built-in, you will use the FIND_ALERT function to retrieve the object ID of the alert, which has the object name UPDATE_ALERT. You will then dynamically display the filename value in the alert message, using the SET_ALERT_PROPERTY built-in. You will call the SET_ALERT_PROPERTY built-in using the retrieved alert object ID, rather than the alert object name. You will add these commands to the Update button trigger, which updates the student record and then displays the alert. Then, you will run the form, retrieve and update a student record, and view the dynamically generated alert message.

To add the commands to the Update button trigger:

1. In the Layout Editor, select the **Update button**, right-click, and then click **PL/SQL Editor** to display the button trigger code in the PL/SQL Editor.

2. Modify the trigger code as shown in Figure 12-15 to declare the variables that are returned by the GET_ and FIND_ functions, to declare the variable that represents the alert message, and to include the GET_, FIND_, and SET_ built-ins.

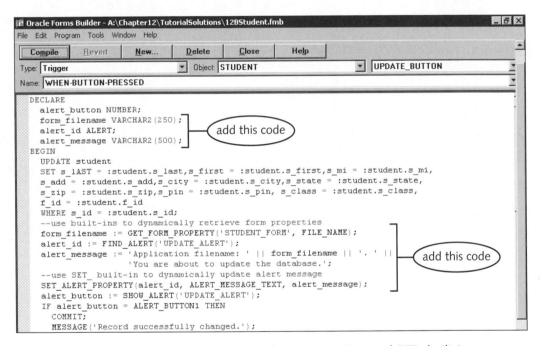

Figure 12-15   Modified trigger code using the GET_, FIND_, and SET_ built-ins

3. Compile the trigger, correct any syntax errors, and then close the PL/SQL Editor and save the form.

4. Run the form, click the **LOV command button** beside the Student ID text item, select student ID **100** (Sarah Miller), and then click **OK**. Sarah's data values appear in the form.

5. Click the **LOV command button** beside the Advisor ID text item, select ID **2** (John Blanchard), and then click **OK**. Sarah's Advisor ID value changes to 2.

6. Click **Update**. The Updating Database alert opens, as shown in Figure 12-16. Note that the alert displays the application filename, which was retrieved using the GET_FORM_PROPERTY built-in.

Figure 12-16   Alert with dynamically generated message

7. Click **Cancel** to close the alert, and then click **Exit** to close the form.

8. Close the form in Form Builder, and if necessary, click **Yes** to save your changes.

Using Indirect Referencing to Write Flexible Code

Form code uses **direct referencing** when it references form objects using the standard *:block_name.object_name* format. You can only use direct referencing when an object exists in the current form. For example, suppose you are working with a data block form associated with the FACULTY table in the Clearwater Traders database, and you want to assign the value that is stored in the LOC_ID text item to a variable named CURR_LOCATION. You would use the following assignment statement:

```
curr_location := :faculty.loc_id;
```

If you tried to use this assignment statement in a PL/SQL code block in a menu module associated with the form, a compile error message appear, stating that the :FACULTY.LOC_ID object does not exist in the current module. (Recall that menu modules are independent of the form modules in which they appear.)

Indirect referencing allows you to set and retrieve form object values without explicitly specifying their block and item names at design time. The actual values are established while the form is running. This allows you to reference form items in menu modules. To use indirect referencing in forms, you use the COPY and NAME_IN built-in procedures, which the following sections describe.

12

Using the COPY Built-in Procedure

To indirectly set the value of a form item, you use the COPY command, which has the following syntax:

```
COPY(value, 'destination');
```

In this syntax, the *value* parameter is the desired value, and the *destination* parameter specifies the item to which the value is copied. The destination can be a form text item or a global variable. When the destination is a form text item, it is specified using the format '*block_name.item_name*'. Since the destination is enclosed in single quotation marks, you can successfully compile code that contains a reference to an object that does not exist in the current module, because the compiler does not verify that objects enclosed in single quotation marks exist in the current module. For example, to use indirect referencing to set the value of LOC_ID in the FACULTY block to the value 43, you would use the following command:

```
COPY(43, 'faculty.loc_id');
```

Now, you will open a custom form that creates and saves new records in the CUST_ORDER table in the Clearwater Traders database. You will create a menu module for this form that has selections that duplicate the form command buttons, so users can control all form actions by using either the menu or the command buttons. Since the menu module is separate from the form, you must use indirect referencing within the menu code to reference form items. First, you will open the form and save it using a different filename. Then, you will open the form in the Layout Editor to examine its command buttons.

To open and rename the form, and to view the form in the Layout Editor:

1. Open **CustomerOrder_MENU.fmb** from the Chapter12 folder on your Data Disk, and save the form as **12BCustomerOrder_MENU.fmb** in your Chapter12\TutorialSolutions folder.

2. Click **Tools** on the menu bar, and then click **Layout Editor** to view the form in the Layout Editor. The form appears as shown in Figure 12-17.

Currently, the form has three command buttons: Create, which retrieves the next value from a sequence named ORDER_ID_SEQUENCE and displays the value in the Order ID text item; Save, which inserts the new order information into the database; and Exit, which closes the form. Now you will create a new menu module that contains a menu item that has a parent menu item selection labeled "Action", and child menu selections for each of the form command buttons (Create, Save, and Exit).

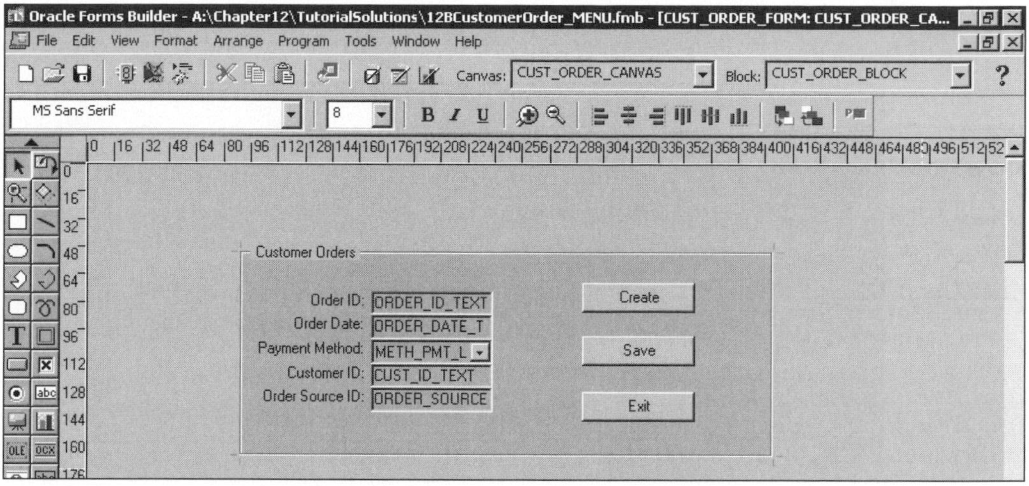

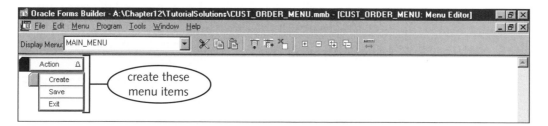

Figure 12-17   Customer order form

To create the menu module, menu item, and menu selections:

1. In Form Builder, click **Window** on the menu bar, and then click **Object Navigator** to open the Object Navigator.

2. Select the top-level **Menus node**, and then click the **Create button** to create a new menu module. The new menu module appears. Change the module name to **CUST_ORDER_MENU**, and save the menu module as **CUST_ORDER_MENU.mmb** in your Chapter12\TutorialSolutions folder.

3. Select the **Menus node** under CUST_ORDER_MENU, and then click to create a new menu item. Change the menu item name to **MAIN_MENU**.

4. Double-click the **Menu Item icon** beside MAIN_MENU to open the Menu Editor.

5. In the Menu Editor, create the parent and child menu item selections shown in Figure 12-18, and then save the menu module.

Figure 12-18   Creating the menu item selections

Recall that to change a menu item label, you select the menu item, and then click it again so the insertion point appears in the current label. To create a child menu item, click the Create Down button [icon].

Next, you will create the menu code for the child menu item selections. For the Create menu item, you will retrieve the next value from the ORDER_ID_SEQUENCE, and then use the COPY built-in procedure to copy the retrieved sequence value into the form text item, which is referenced as :CUST_ORDER.ORDER_ID_TEXT. You will create a stub for the Save menu selection, and create the code to exit the form for the Exit menu selection.

To create the menu code:

1. In the Menu Editor, select the **Create** menu selection, right-click, and then click **PL/SQL Editor**. The PL/SQL Editor opens.

2. Type the code shown in Figure 12-19, compile the trigger, and correct any syntax errors. Do not close the PL/SQL Editor.

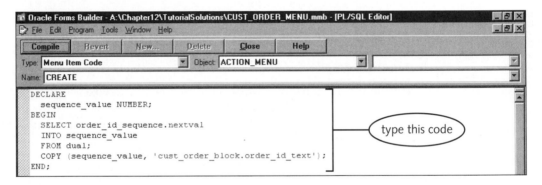

Figure 12-19   Menu code with indirect referencing

3. Open the Name list, and select **Save** to open the source code pane for the Save menu item. Type the following code to create the menu stub: `MESSAGE('Not implemented yet.');`.

4. Compile the code, and correct any syntax errors. Do not close the PL/SQL Editor.

5. Open the Name list, and select **Exit** to open the source code pane for the Exit menu item. Type the following code to exit the form: `EXIT_FORM;`.

6. Compile the code, correct any syntax errors, close the PL/SQL Editor, and then save the menu module.

7. To compile the menu module, click **File** on the menu bar, point to **Administration**, and then click **Compile File**. The message "Module built successfully" appears on the status line.

To test the menu module, you will attach the menu module to the form by specifying the complete path to the module in the form's Menu Module property. Then, you will run the form, and test the Create menu item.

To attach the menu module to the form and test the Create menu item:

1. Click **Window** on the menu bar, and then click **Object Navigator** to open the Object Navigator.

2. Double-click the **Form icon** beside the CUST_ORDER_FORM node to open the form Property Palette. Change the Menu Module property to **a:\chapter12\tutorialsolutions\cust_order_menu.mmx**, and then close the Property Palette. (If necessary, change the path specification to reflect the location of your Chapter12\TutorialSolutions folder.)

3. Save the form, and then run the form. The form opens, and the custom menu selections appear.

4. Click **Action** on the menu bar, and then click **Create**. The next sequence value appears in the Order ID text item.

> **?**
> **Help**
>
> An error will appear if you have not yet created the ORDER_ID_SEQUENCE that retrieves the next order ID value. To create the sequence, start SQL*Plus, and type the following command: CREATE SEQUENCE order_id_sequence START WITH 1300;. Then, repeat Step 4.

5. Click **Action** on the menu bar, and then click **Exit** to close the form in Forms Runtime.

Using the NAME_IN Built-in Procedure

To indirectly retrieve the current value of a form item, you use the NAME_IN function. This function allows you to indirectly retrieve "the name that is in" a form object. The NAME_IN function has the following syntax:

```
return_value := NAME_IN(source);
```

In this syntax, the *return_value* parameter is a previously declared variable that has the VARCHAR2 data type. The *source* parameter is the item whose value you want to determine. The *source* can be a variable or a form text item. When the *source* is a form text item, it is specified using the format '*block_name.item_name*', and is enclosed in single quotation marks. As stated before, when the *source* is enclosed in single quotation marks, the compiler does not verify that the target exists in the current module.

To use indirect referencing to assign the value of the BLDG_CODE text item in the FACULTY block to a variable named **CURRENT_BLDG_CODE**, you would use the following command:

```
current_bldg_code := NAME_IN('faculty.loc_id');
```

12

Since the NAME_IN function always returns a character string, you might need to convert the character string to a NUMBER or DATE data type if the actual value is a number or date. For example, recall that the LOC_ID database column has the NUMBER data type. To use indirect referencing to assign the value of LOC_ID in the FACULTY block to a variable named CURR_LOCATION that is declared using the NUMBER data type, you would use the following command:

```
curr_location := TO_NUMBER(NAME_IN('faculty.loc_id'));
```

Similarly, to use indirect referencing to assign the value of S_DOB in the STUDENT block to a variable named CURR_S_DOB that is declared using the DATE data type, you would use the following command:

```
curr_s_dob :=
TO_DATE(NAME_IN('faculty.loc_id'), 'DD-MON-YY');
```

This command assumes that the date in the S_DOB text item appears using the default DD-MON-YY format mask. If the date appears using an alternate format mask, you would need to change the format mask in the code.

Now, you will use the NAME_IN function to implement the Save menu selection in the customer order form. To insert the values that appear on the form, you will need to declare variables in the menu code to represent each of the form text items. Then, you will retrieve the values in the form text items by using the NAME_IN function. You will convert the retrieved values for the order ID, customer ID, and order source ID text items to NUMBER data types, and you will convert the retrieved value for the order date to a DATE data type.

To implement the Save menu selection by using the NAME_IN function:

1. Click **Window** on the menu bar, and then click **CUST_ORDER_MENU: Menu Editor** to open the Menu Editor.

2. Select the **Save** menu item, right-click, and then click **PL/SQL Editor** to open the PL/SQL Editor. Delete the current command, and type the commands shown in Figure 12-20 to specify the menu code for the selection.

3. Compile the code, correct any syntax errors, and then close the PL/SQL Editor.

4. Save the menu module, and then compile the menu module.

Now, you will run the form, create a new order, enter data values for the order, and then use the menu to save the order in the database.

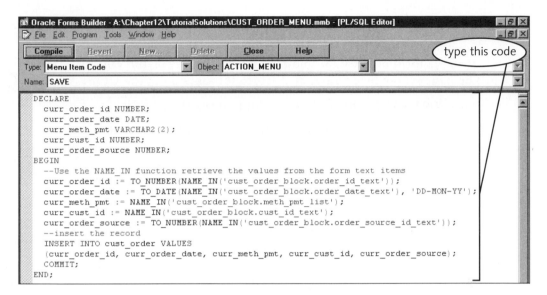

Figure 12-20   Menu code for Save menu selection

To run the form and test the Save menu code:

1. Click **Window** on the menu bar, and then click **Object Navigator** to open the Object Navigator.

2. Select the **CUST_ORDER_FORM**, and then run the form. The form opens in the Forms Runtime window.

3. Click **Action** on the menu bar, and then click **Create** to retrieve the next order ID value from the sequence.

4. Open the Payment Method list, and select **CC**.

5. Place the insertion point in the Customer ID text item, press **F9** to open the LOV, select customer ID **107** (Paula Harris), and click **OK**.

6. Place the insertion point in the Order Source ID text item, press **F9** to open the LOV, select order source ID **1** (Winter 2002), and click **OK**.

7. Click **Action** on the menu bar, and then click **Save**. The confirmation message "FRM-40401: No changes to save" appears, confirming that the record was successfully inserted.

8. Click **Exit** to close the form in Forms Runtime.

12

Nesting the NAME_IN Command in the COPY Command

You can nest the NAME_IN function in the COPY command when you need to retrieve a form object value and then immediately copy the value back to a form item. The basic syntax for nesting the NAME_IN function in the COPY command is:

```
COPY(NAME_IN(source), destination);
```

In this command, the *source* parameter is the data source of the NAME_IN function, and the *destination* parameter is the location to which the source value is copied. For example, to retrieve the value of the LOC_ID text item in the FACULTY block and copy it to the LOC_ID text item in the LOCATION block, you would use the following command:

```
COPY(NAME_IN('faculty.loc_id'), 'location.loc_id');
```

Now you will modify the customer form by adding a display item to the form that displays a message that confirms when the form successfully saves an order, as shown in Figure 12-21. Recall that a display item displays text that the user cannot change.

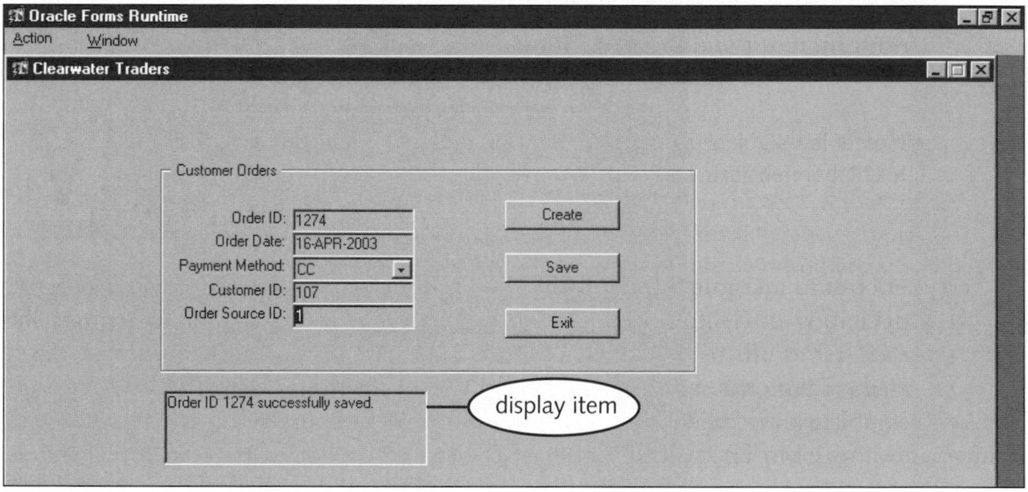

Figure 12-21 Display item confirming inserted order ID

You will use the COPY command in the Save menu selection to specify the text that appears in the display item. The display item text shows the order ID for the order that was just inserted, so the command will nest the NAME_IN function within the COPY command. Now you will create the display item and modify the menu code to display the confirmation message in the text item.

To create the display item and modify the menu code:

1. Click **Tools** on the menu bar, and then click **Layout Editor** to open the form in the Layout Editor.

2. Select the **Display Item tool** 🔤 on the tool palette, and draw the display item as shown in Figure 12-21.

3. Double-click the new display item to open its Property Palette, change the following property values, close the Property Palette, and then save the form.

Property	Value
Name	**ORDER_SUMMARY_DISPLAY**
Maximum Length	**300**

4. Click **Window** on the menu bar, and then click **CUST_ORDER_MENU: Menu Editor** to open the Menu Editor.

5. Select the **Save** menu selection, right-click, and then click **PL/SQL Editor** to open the PL/SQL Editor.

6. Add the following command as the last line of the menu code (after the COMMIT; command but before the END; command). Then compile the trigger, correct any syntax errors, and close the PL/SQL Editor window.

```
COPY('Order ID ' ||
NAME_IN('cust_order_block.order_id_text') ||
' successfully saved.', 'cust_order_block.order_
summary_display');
```

7. Save the menu module, and then compile the menu module.

Now, you will run the form, and add a new customer order record using the Save pull-down menu item. When you click Save, the confirmation message should appear in the form display item.

To run the form and test the display item:

1. Click **Window** on the menu bar, and then click **Object Navigator** to open the Object Navigator.

2. Select the **CUST_ORDER_FORM**, and then run the form. The form opens in the Forms Runtime window.

3. Click **Action** on the menu bar, and then click **Create** to retrieve the next order ID value from the sequence.

4. Open the Payment Method list, and select **CC**.

5. Place the insertion point in the Customer ID text item, press **F9** to open the LOV, select customer ID **107** (Paula Harris), and click **OK**.

6. Place the insertion point in the Order Source ID text item, press **F9** to open the LOV, select order source ID **1** (Winter 2002), and click **OK**.

7. Click **Action** on the menu bar, and then click **Save**. The confirmation message "FRM-40401: No changes to save" appears, confirming that the record was successfully inserted. The confirmation message should also appear in the display item.

12

8. Click **Exit** to close the form in Forms Runtime.

9. Close the form and menu in Form Builder, and then close all other open Oracle applications.

SUMMARY

❑ The Form Builder calendar class enables you to create a calendar window that allows users to select a date from a visual calendar display and return the selected date to a form text item.

❑ To create a calendar class object, you must copy or subclass the Calendar object group from the STNDRD20.OLB object library, attach the calendar.pll PL/SQL library, and then create a trigger to call the GET_DATE procedure in the DATE_LOV package.

❑ Flexible code is code in which values within the code are dynamically specified at runtime. Flexible code allows you to use objects in different modules without modifying the object's properties or code and to reference objects that are contained in separate modules.

❑ You can use system variables to retrieve form values that describe a variety of form properties.

❑ You can use the GET_ built-ins to retrieve the values of object properties while a form is running.

❑ When you use the GET_ built-ins, you can reference the target object using the object ID or the object name.

❑ The object ID is an internal identifier that Form Builder creates whenever you create a new object and uses whenever you reference a form object. Whenever possible, you should use the object ID in the GET_ built-ins to speed up form processing.

❑ You can use the FIND_ built-ins to retrieve the object ID of an object, based on the object name.

❑ You can use the SET_ built-ins to dynamically set properties of Form Builder objects.

❑ When you reference a form object by specifying the object's block name and item name, you are using direct referencing.

❑ Indirect referencing allows you to set and retrieve form object values without explicitly specifying their block and item names at design time, which allows you to manipulate a form's objects in other modules.

❑ To copy a value into a form object using indirect referencing, you use the COPY command.

❑ To retrieve the value of a form object using indirect referencing, you use the NAME_IN command.

❑ The NAME_IN command always returns a VARCHAR2 value, so you must convert its output using the TO_NUMBER and TO_DATE conversion functions when you retrieve number or date values.

❑ You can nest the NAME_IN function in the COPY command to retrieve a value, and then immediately copy the value to a form item.

REVIEW QUESTIONS

1. What is a calendar class used for?

2. What external libraries are required to create a calendar class?

3. What is flexible code? List three ways for implementing flexible code in Form Builder.

4. What is the difference between the SET_ built-ins and the GET_ built-ins?

5. What is the purpose of the FIND_ built-ins?

6. What is an object ID, and when is it created?

7. What is the advantage of using the object ID instead of an object name in the SET_ and GET_ built-ins?

8. What is the difference between direct and indirect referencing?

9. What is the purpose of the NAME_IN function? What is the purpose of the COPY procedure?

10. What data type does the NAME_IN function return?

11. What single command would you use in the menu code trigger in a menu module to read the value that appears in a text item in Form A and then display it on Form B? Assume that neither form is the currently active form.

PROBLEM-SOLVING CASES

Cases refer to the Clearwater Traders (see Figure 1-11) and Software Experts (Figure 1-13) sample databases. All required data files are stored in the Chapter12 folder on your Data Disk. Store all solution files in your Chapter12\CaseSolutions folder.

1. In this case, you will modify a custom form saved in a file named 12BCase1.fmb that is used to process incoming shipments in the Clearwater Traders database. Create a calendar class in the form so that the user can select the date the shipment was received from a calendar window. Create an iconic button beside the Date Received text item to display the calendar window using the calendar.ico file for the icon image. Configure the calendar window so that when the user selects a date, the calendar window immediately closes and returns the selected date. Save the completed form as 12BCase1_DONE.fmb.

12

2. In this case, you will modify the Software Experts database system template form so that the current date appears in a form display item, as shown in Figure 12-22.

Figure 12-22

a. Open the 12BCase2.fmb form file, and save the file as 12BCase2_DONE.fmb.

b. Create a display item named DATE_DISPLAY_ITEM on the template form.

c. Add the code to the PRE-FORM trigger to display the current system date in the display item by retrieving the value by using a system variable. Format the date value as shown in Figure 12-22. (*Hint*: You will need to convert the retrieved value to the DATE data type.)

3. In this case, you will determine the number of records that will be retrieved by an LOV query that displays information from the INVENTORY table in the Clearwater Traders database. Then, you will use the SET_LOV_PROPERTY built-in to dynamically change the Height property of the LOV based on the number of retrieved records. (*Hint*: You will need to retrieve a record before you can open the LOV.)

a. Open the 12BCase3.fmb file, and save the file as 12BCase3_DONE.

b. Add commands to the LOV command button beside the Inventory ID text item to count the number of records in the INVENTORY table

c. Use the SET_LOV_PROPERTY built-in to change the LOV Height property based on the number of retrieved records. As a general rule, each record requires about 13.5 points of display height. Use Form Builder online Help to determine the required parameter values for the SET_LOV_PROPERTY built-in. Use the GET_LOV_PROPERTY value to determine the current LOV display Width value. (Do not worry about adjusting the LOV display position.)

4. You have probably noticed that sometimes when you open a form, the iconic arrow images on the form LOV command buttons do not appear. This happens because the path and filename specification for the LOV icon file is not correct in the LOV command button's Icon Filename property. In this case, you will modify the form shown in Figure 12-23 so the PRE–FORM trigger uses the SET_ITEM_PROPERTY command to dynamically set the Icon Filename property for the LOV command buttons based on a global project path variable.

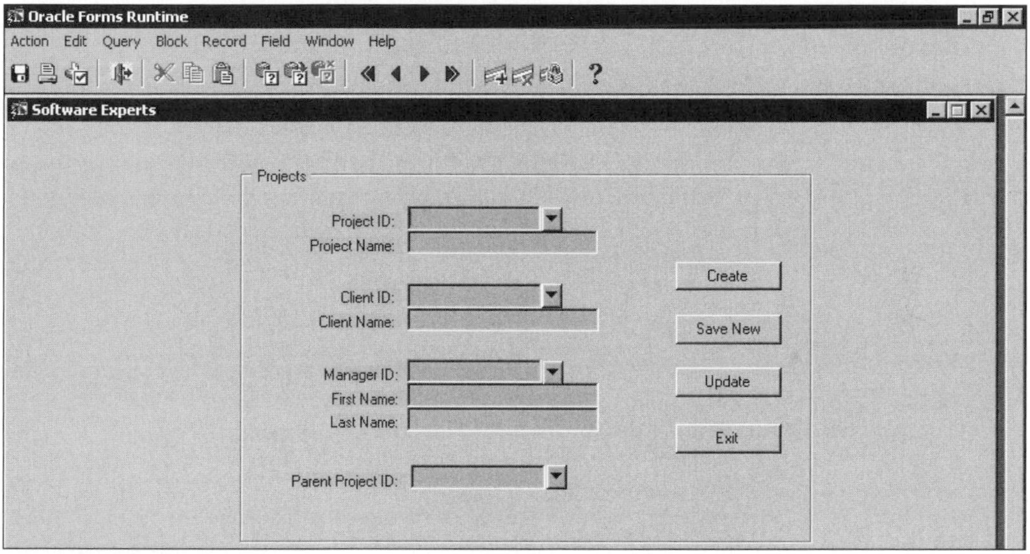

Figure 12-23

a. Open the 12BCase4.fmb file, and save it as 12BCase4_DONE.fmb.

b. In the PRE–FORM trigger, initialize the global project path variable to specify the path location for the Chapter12 folder on your Data Disk. (In practice, the global project path would be set in an application switchboard and then passed to the form when the switchboard form calls the target form.)

c. In the PRE–FORM trigger, use the SET_ITEM_PROPERTY built-in to dynamically set the Icon Filename property for the LOV command buttons.

5. Figure 12-24 shows the layout of a custom form to insert, update, or delete records in the CONSULTANT_SKILL table in the Software Experts database. In this case, you will create a menu module that will appear with this form and provide menu access for all of the commands that are currently only accessible through the form command buttons. You will write the commands in the menu module using indirect referencing.

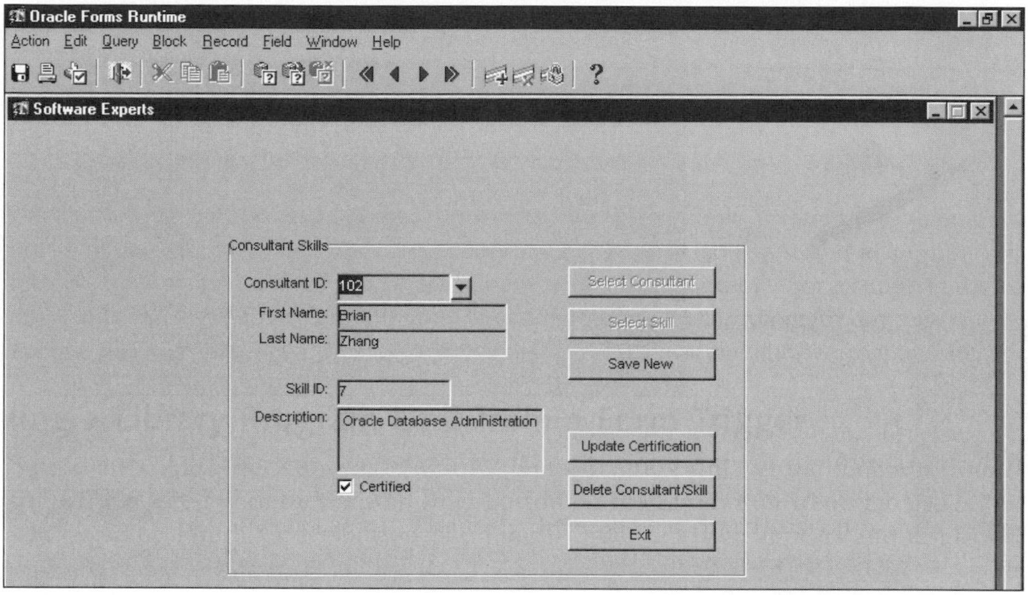

Figure 12-24

a. Open 12BCase5.fmb, and save the file as 12BCase5_DONE.fmb.

b. Create a menu module named CONSULTANT_SKILL_MENU that contains a menu item named MAIN_MENU. In the menu item, create an Action parent item, and the following child menu selections: Select Consultant, Select Skill, Save New Consultant/Skill, Update Certification, Delete Consultant/Skill, and Exit. Save the menu file as 12BCase5_MENU.mmb.

c. Change the menu selections so they are enabled or disabled at form startup, based on the startup state of their associated command buttons.

d. Write menu code for each of the child menu items, based on the code in the associated command button triggers. Replace direct references with indirect references whenever necessary.

e. Use the SET_MENU_ITEM_PROPERTY built-in to enable or disable menu items, based on whether their corresponding command buttons are enabled or disabled when the form is in specific states. You will need to add commands to enable the associated menu selections when the user clicks the LOV command button beside the Consultant ID text item and when the user clicks the form command buttons.

Index